Environmental Literacy in Science and Society

From Knowledge to Decisions

In an era where humans affect virtually all of the Earth's processes, questions arise about whether we have sufficient knowledge of human–environment interactions. How can we sustain the Earth's ecosystems to prevent collapses and what roles should practitioners and scientists play in this process? These are the issues central to the concept of environmental literacy.

This unique book provides a comprehensive review and analysis of environmental literacy within the context of environmental science and sustainable development. Approaching the topic from multiple perspectives, it explores the development of human understanding of the environment and human–environment interactions in the fields of biology, psychology, sociology, economics, and industrial ecology.

The discussion emphasizes the importance of knowledge integration and transdisciplinary processes as key strategies for understanding complex human–environment systems (HES). In addition, the author defines the HES framework as a template for investigating and transforming sustainably coupled HES in the 21st century.

Roland W. Scholz chairs the Natural and Social Science Interface in the Department of Environmental Sciences at the ETH (Swiss Federal Institute of Technology), Zurich. A mathematician, psychologist and decision theorist by training, he is particularly interested in environmental systems analysis, human–environment interactions, environmental decisions, and risk assessment. He has led numerous large-scale transdisciplinary processes to foster sustainable transitions of urban and regional systems.

Environmental Literacy in Science and Society

From Knowledge to Decisions

Roland W. Scholz
ETH Zurich
Institute for Environmental Decisions
Chair of Natural and Social Science Interface

Some chapters are coauthored by Claudia R. Binder, Fridolin Brand, Justus Gallati, Daniel J. Lang, Quang Bao Le, Roman Seidl, Timo Smieszek and Michael Stauffacher

CAMBRIDGE UNIVERSITY PRESS
Cambridge, New York, Melbourne, Madrid, Cape Town,
Singapore, São Paulo, Delhi, Tokyo, Mexico City

Cambridge University Press
The Edinburgh Building, Cambridge CB2 8RU, UK

Published in the United States of America by Cambridge University Press, New York

www.cambridge.org
Information on this title: www.cambridge.org/9780521192712

© R. W. Scholz 2011

This publication is in copyright. Subject to statutory exception
and to the provisions of relevant collective licensing agreements,
no reproduction of any part may take place without the written
permission of Cambridge University Press.

First published 2011
Reprinted 2011

Printed in the United Kingdom at the University Press, Cambridge

A catalog record for this publication is available from the British Library

Library of Congress Cataloging in Publication data
Scholz, Roland W.
  Environmental literacy in science and society : from knowledge to decisions / Roland W. Scholz ;
  some chapters are coauthored by Claudia R. Binder ... [et al.].
    p. cm.
  Includes bibliographical references and index.
  ISBN 978-0-521-19271-2 (hardback) – ISBN 978-0-521-18333-8 (paperback)
  1. Environmental education. 2. Environmental sciences. I. Gallati, Justus. II. Title.
  GE70.S35 2011
  304.2–dc22   2011011288

ISBN 978-0-521-19271-2 Hardback
ISBN 978-0-521-18333-8 Paperback

Cambridge University Press has no responsibility for the persistence or
accuracy of URLs for external or third-party internet websites referred to in
this publication, and does not guarantee that any content on such websites is,
or will remain, accurate or appropriate.

To Maya

Contents

List of boxes ix
Acknowledgments xiii
Praise for the book xv
Preamble xvii
Overview: roadmap to environmental literacy xviii
Legend of the roadmap xxii

Part I Invention of the environment: origins, transdisciplinarity, and theory of science perspectives

1 What knowledge about what environment? 3
2 From environmental literacy to transdisciplinarity 15
3 Basic epistemological assumptions 29

Part II History of mind of biological knowledge

4 Emerging knowledge on morphology, ecology, and evolution 45
5 From molecular structures to ecosystems 94

Part III Contributions of psychology

6 Psychological approaches to human–environment interactions 137
7 Drivers of individual behavior 190

Part IV Contributions of sociology

8 Traditional sociological approaches to human–environment interactions 215
9 Modern sociological approaches to human–environment interactions 231

Part V Contributions of economics

10 Origins of economic thinking and the environment 257
11 Contemporary economic theories dealing with the environment 281

Part VI Contributions of industrial ecology

12 The emergence of industrial ecology 307
13 Industrial agents and global biogeochemical dynamics 320

Part VII Beyond disciplines and sciences

14 Integrated systems modeling of complex human–environment systems 341
Roland W. Scholz, Justus Gallati, Quang Bao Le, and Roman Seidl
15 Transdisciplinarity for environmental literacy 373

Part VIII A framework for investigating human–environment systems (HES)

16 **The HES Postulates**   407

17 **The HES framework**   453
Roland W. Scholz, Claudia R. Binder, and Daniel J. Lang

18 **Applying the HES framework**   463
Roland W. Scholz, Claudia R. Binder, Daniel J. Lang, Timo Smieszek, and Michael Stauffacher

19 **Comparing the HES framework with alternative approaches**   509
Roland W. Scholz and Fridolin Brand

Part IX Perspectives for environmental literacy

20 **New horizons: environmental and sustainability sciences**   525

Glossary   537
References   551
Index   626

Boxes

Part I Invention of the environment: origins, transdisciplinarity, and theory of science perspectives 1

1.1 The ability to read the environment: **Polynesian navigation** 4

1.2 Correcting negative human impacts: **Ascension's spring** 5

1.3 Ignoring environmental collateral damage: **British forests** 7

1.4 What type of rationale underlies human decisions? **Models of man and rationality** 9

1.5 The double environment: **Ossianic dreams** 11

2.1 Origins of a sustainable relationship with the environment: **mining and brewing** 17

2.2 The literacy of environmental system vulnerability: **Einstein's honeybees** 18

2.3 Conflicting incentives on the macro and the micro level: **population growth** 20

2.4 Does unintended ecological suicide cause societal collapse? **Easter's canoes and Greenland's fish** 21

2.5 Knowledge integration as a means of environmental literacy: **from disciplines to cultures** 23

2.6 An architecture of knowledge for environmental literacy: **from case empathy to propositional logic** 25

3.1 What is part of me and what is part of the environment? **My pig heart valve** 30

3.2 No human–environment complementarity: **Maya constructions of the world** 35

Part II History of mind of biological knowledge 43

4.1 The difficulty of discriminating between humans and animals in archaic thinking: **totemism** 49

4.2 At the crossroads between naturalist and magic impacts: **ancient Egypt's medical knowledge** 56

4.3 Mayans' physical and symbolic treatments of diseases: "me'**winik" or gallbladder cancer?** 57

4.4 How many types – and where do they come from? **Semen and genders** 65

4.5 Emerging geology meets biology: **catastrophes, infinite time, and evolution** 71

4.6 The dilemma of empiricism, religious dogma, and theory: **Linné's mule plants** 76

4.7 Driving species to adaptation: **Lamarckism** 81

4.8 Frauds, sleights of hand or simplifications in embryonic similarities? **Evo-devo and abortion** 85

4.9 Too stupid to adapt: **dodos and kakapos** 88

4.10 Climate variability: **Dansgaard–Oeschger events** 89

4.11 Perceived environments: **tick cybernetics** 92

5.1 A continuum of sexes in plants and animals? **American Holly and Hollywood** 98

5.2 From basic via applied research and public experimentation: **Pasteur's Quadrant** 102

5.3	The central dogma of molecular biology: **the protein as a sink?** 105	6.13	Correct perceptions but incorrect judgments: **group power** 183
5.4	Why should a mouse gene work in a fly? **Eys and Seys** 107	6.14	Signs of environmental compliance: **buttons** 184
5.5	Eating intelligence: **memory transfer?** 111	6.15	Resource dilemmas need social solutions: **Mongolian grassland** 185
5.6	The immune system: **the unsung hero** 118	7.1	Why don't we have coherent preferences? **Arrow's cyclic triads** 194
5.7	Understanding fundamentals of ecosystems: **the ten principles of New Ecology** 124	7.2	Not following expected value: **the Allais paradox** 195
5.8	Biodiversity and extinction: **the rise and fall of species** 127	7.3	Emotions and cognition: **anxiety-reducing snails** 199
5.9	Sociobiological and eco-ethological adaptation: **Tibet fraternal polyandry** 131	7.4	Flashbulb memory, risk perception, and fatal crashes: **the impacts of September 11** 201

Part III Contributions of psychology 135

6.1	Do aesthetics follow a natural law? **The golden section** 140	7.5	What drives cognitive performance? **Beyond savantism** 203
6.2	The chemical and symbolic notion of odor: **you stink** 142	7.6	Do values affect perception: **"le waldsterben"** 208
6.3	The wonder of crisp flowers: **the probabilistic nystagmus** 147	7.7	Fuel efficiency in car purchase: **Hummer's symbolic motives** 210

Part IV Contributions of sociology 213

6.4	Anticipating environmental information: **networking "Ferrari cones"** 149	8.1	The myth of permanent population growth: **pestilence and Old World diseases** 221
6.5	Behavior is a function of the person and the environment: **field theory** 151	8.2	People, not experts, have brought the toxification of nature and people to the forefront: **what role should experts take?** 224
6.6	Coherent patterns of environmental setting and behavior: **is this a church, Mommy?** 156		
6.7	How do we order the world? **Gestalts, prototypes, and misgestalts** 159	8.3	Resource scarcity drives societal and ecological decline: **West Africa's emptied sea** 228
6.8	Stages of environmental literacy: **the invisible man** 161	9.1	Environmental injustice: **landfills, Chernobyl, and the north–south problem** 237
6.9	An intriguing complex concept: **time** 167		
6.10	What means one million years? **Nuclear waste** 168	9.2	The environment as an input–output system: **New York's Fresh Kills waste disposal** 239
6.11	When and why does environmental noise harm? **Trains and planes** 174		
6.12	Group decision schemes matter: the Bay of Pigs fiasco 182	9.3	Humanization of the environment: **the home alone fish** 247

Part V Contributions of economics 255

10.1 Mercantilism's collateral damage: **changed landscapes** 261

10.2 Energy in economics: **the curse of coal** 264

10.3 Incorporating transportation costs in rents: **Von Thünen circles** 270

10.4 Economics as a kind of chess game: **Nash equilibria** 275

10.5 Internalization of negative external effects: **Pigovian taxes** 277

10.6 Constructing new agents for improving environmental quality: **Pareto optimality arguments** 278

11.1 No more endless plains: **the end of cowboy economics** 282

11.2 Mineral reserves become larger if prices rise: **Earth's mineral resources economic cycle** 287

11.3 Cost–benefit analysis: **a Faustian bargain?** 293

11.4 The second law of thermodynamics and economic processes: **Ayres vs. Daly** 295

11.5 Cost, not physical availability, matters: **unlimited solar energy?** 298

Part VI Contributions of industrial ecology 305

12.1 Cycling drivers for recycling: **GDR's SERO** 311

13.1 A critical element of sustainable agriculture: **phosphorus** 322

13.2 Trading and auctioning pollution: **acid rain** 331

Part VII Beyond disciplines and sciences 339

14.1 Climate impacts of the agrarian society: **pest impacts?** 343

14.2 Societal complexity, unsustainable resources flows, and indebtedness: **the western Roman Empire** 345

14.3 Dynamic patterns in coupled HES: **collective irrigation management** 354

14.4 Emerging systems: **slime mold** 362

14.5 The Prisoner's Dilemma: **strategies and the emergence of cooperation** 365

14.6 Social dilemmas: **contributing to a land reclamation system or not?** 370

14.7 Interactive household and landscape agents: **Vietnam deforestation** 371

15.1 A transdisciplinary process needs methods: **sustainable future of industry** 383

15.2 From parsons to Napoleon: **variants of mediation** 387

15.3 Science, not only a matter of universities: **building the Tower of Babel** 396

15.4 Academic–industrial transformation of the world food production: **Haber–Bosch** 398

Part VIII A framework for investigating human–environment systems 405

16.1 The human animal is building environments: **small huts and Las Vegas** 409

16.2 Level hierarchies in biological systems: **controlled genes?** 417

16.3 An impact of technological and societal changes: **cosmopolitical institutions?** 422

16.4 Parallel societies: **Sinti, Roma, and Amish people** 423

16.5 Interfering interests between nations, banks, and individuals: **green housing** 428

16.6 The origins of feedback loops: **Ktesibios' water clock** 435

16.7 Multiple feedback loops in complex systems: **Forrester's world model** 438

16.8 Rebound effects in introducing ecoefficient mobility systems: **Swissmetro** 440

16.9	Emerging environmental awareness: **nuclear eyewatch** 441	18.1	Different views on biomass energy potential: **biofuel** 488
16.10	Secondary feedback loop management in agriculture: **crop-field rotation** 442	18.2	The technological view on bioethanol production: **T Fords** 489
16.11	Loops in technical systems: **Three Mile Island** 447	18.3	A neutral CO_2 balance means not renewable: **bioethanol boomerangs** 492
16.12	Climate changes in an uncertain world: **sea level rise** 449	18.4	Endocannibalism and inbreeding: **kuru, BSE, and vCJD** 497
16.13	Understanding the environmental impacts of disasters: **the Dust Bowl phenomenon** 450	**Part IX**	**Perspectives for environmental literacy** 523
		20.1	The HES framework in brief: **a short manual** 530

Acknowledgments

Accidents can happen, often when one least expects them. On April 14, 2006, Dana, an escaped Doberman pinscher, knocked me off my bicycle when I was training for the forthcoming racing season. When viewing my fight to recover from brain injury, it was René Schwarzenbach, the Dean of the Department of Environmental Sciences, ETH, who seized the chance. Noticing that I was not yet prepared to struggle through the daily research, teaching and transdisciplinary project obligations of my institute, he suggested: "Now you have more time than before. Stop writing papers and write a comprehensive book on what you have elaborated in the last two decades." I took the challenge. Thus, one could say that this book is a product of an accident.

"You," in this place, is definitely not only "I, myself." As it is with staged races in cycling, which I experienced as a late competitive (hobby) racing cyclist, you can only finish if you have an excellent team. A staged race asks you to cope with a multitude of exigencies in a wide range of profiles, including flat and mountain stages. Thus you must have a strong team of specialists and helpers for all situations, especially if the race has 20 stages. Sometimes, you even need people to push you uphill!

However, preparing for a race is as important as running it. The 20 chapters that this book offers for the reader are the result of 29 "reader circle" exercises. I consider these to be a kind of pre-race exercise, in which the territory for each of the 20 chapters, or "stages," of the book was thoroughly scouted.

The very idea of writing this book emerged from the insight that environmental and sustainability sciences needed a "theory and resource book" that supports coupled human–environment and transdisciplinary research. Just before the bicycle accident, a blueprint of this book took shape through my work with Claudia R. Binder, Daniel J. Lang, and Michael Stauffacher and other former senior researchers from my team.

Let us go back to the cycle race metaphor. Race teams need coaches, proper training and workout partners, and a dedicated technical staff.

I consider the reviewers of different track sections to be my coaches and the members of the "reader circle" to be my training partners.

In the biology part, which tendered some previously unknown terrain and steep heliclines, I needed some more coaches. Here Peter Edwards, Theo Koller, Jukka Jokkela, Bo Samuelson, Beda Stadler, and Josef Zeyer drilled me. Patricia Holm, Walter Schaffner, and Gottfried Schatz helped me in difficult parts of the route. In psychology, Michael Siegrist and Paul Vlek, and in sociology, Andreas Diekmann, Matthias Gross, Gerhard Lenski, and Klaus Seeland were my coaches. Economics again was a demanding section, and I want to thank Catharina Bening, Stefanie Engel, Bernard Lehmann, Markus Ohndorf, and John Tilton. In the new terrains of industrial ecology, Volker Hoffmann and Reid Lifset, and in the modeling section, Andreas Ernst and Wander Jager took the coaching job. My special thanks go for Tim McDaniels, Cliff Davidson, and Anton J. M. Schoot Uiterkamp, and particularly Charles Vlek, who meticulously challenged me in most of the stages.

The team mates in the "reader circle" were Fridolin Brand, Thomas Flüeler, Peter de Haan, Justus Gallati, Bastien Girod, Fadri Gottschalk, Berit Junker, Thomas Köllner, Daniel Lang, Quang Bao Le, Marco Morosini, Siegmar Otto, Roman Seidl, Timo Smieszek, and Michael Stauffacher, who ran many or all legs of the race with me. As one can see, the last stages of writing the books' contents, involved breaking new ground, and I was grateful to find among these team mates several experienced external coaches and a team of inspiring "co-racers."

I was also fortunate that several PhD and masters students dedicated some of their time and focus to helping me find the straightest and surest track for my message. Here, this team of "junior cyclists" includes

xiii

Acknowledgments

Laura de Baan, Mónica Berger González, Yann Blumer, Julia Brändle, Matthias Dhum, Rainer Gabriel, Martin Hitziger, Grégoire Meylan, Corinne Moser, Matthias Näf, Anja Peters, Alexander Scheidegger, Andy Spörri, Anna Stamp, Saša Parađ, Evelina Trutnevyte, Andrea Ulrich, Timo von Wirth, Stefan Zemp, and, in particular, Pius Krütli, who provided most significant support in the transdisciplinarity stage.

The technical staff included mechanics, masseurs, and a procurement team. Here Sandro Bösch and Rebecca Cors took on special roles. Sandro meticulously tracked mountains of book material, which were permanently upgraded to a whole text, and designed the large numbers of figures. Rebecca was the kneader of the text. She worked through all parts of the book as a sparring partner, checking for cohesion and coherence. She cooperated with Stefanie Keller, Erin Day, and Devon Wemyss, who not only checked the English but helped to improve the understandability of this multidiscipline book.

As racers need many bottles of liquid to drink each day, a writer of a book needs many resource books each week. Here my thanks go to Ursula Müller and her team, who acquired many, sometimes difficult to find, books, and to Robert Bügl, Silvia Cavelti, Marco Huber, and Andrea Ziegler, who ensured that all the bottles put into the reference list were properly prepared. Cyclists of today need sophisticated electronic equipment for keeping records and communicating with other team mates, before and during the race. We were fortunate to have team mate Andarge Aragai, who provided just the right software and hardware to keep our work going around the clock.

Special thanks go to my personal assistant Maria Rey. One of her main tasks before the bicycle accident was to coordinate my ambitious cycling training schedule with my full research and teaching agenda. While this book introduced new and unexpected challenges here, she managed it with an agility and reliability that makes her an invaluable part of the team, and we learned that managing a multistage book writer is even more difficult than managing the scheduling of an ambitious hobby cyclist.

Dominic Lewis, from Cambridge University Press, may be seen as a "tour-de-book-chef." The team was surprised as to how easy it was to communicate with him and how patiently and clearly he responded to special requests. It was most impressive to see how Dominic's technical (editorial) staff – Sarah Beanland, Abigail Jones, and Megan Waddington – allowed for an absolutely frictionless and satisfactory concourse.

Naturally, of utmost importance if you undertake such a big project, is your home base. Writing such a book changes one's life and entails an intense two-shift work week, which, in this case, endured over several years. Here my wife Maya – to whom this book is dedicated – has been the ultimate backup. She has been a counterpart in many vivid discussions about the book's essential messages, and a wonderful helper in the long course of recovering after the bicycle accident.

What began as an opportunity soon developed into an obligation. Those who share the experience with me and know what it means to write a challenging book or to participate in a staged race would most certainly agree that these trials often become an obsession.

While I have many to thank for contributing to the writing of the book who are not mentioned above, the responsibility for its content, however, remains my own.

International praise for Environmental Literacy in Science and Society

"Roland Scholz has written a visionary book that for the first time comprehensively approaches modern sustainability challenges by recognizing the critical role of integrated human, natural, and built domains in the complex systems that characterize the Anthropocene. It is an important step forward in our ability to understand, and respond ethically and rationally to the demands of environment, technology, and society in a context of complexity that is increasingly beyond traditional disciplinary and policy approaches for linking theory and practice."
Braden Allenby, Lincoln Professor of Engineering and Ethics and Professor of Law, Department of Civil and Environmental Engineering, Ira A. Fulton School of Engineering, Arizona State University, USA

"Society has yet to make the 'great transition' toward sustainability. We still increasingly appropriate the world's non-renewable resources, fail to safeguard ecosystem services on which civilization depends, and elect irresponsible government leaders. What is the solution? In this brilliant work, Roland Scholz addresses these issues head-on in a remarkably open and honest exploration of human-environment systems. Scholz argues that we need new knowledge and new science to tackle these challenges: the 'environment' must be redefined as a co-evolving system coupled to human systems. Furthermore, he demonstrates that interdisciplinary research is not enough – we need transdisciplinary research to integrate our scientific knowledge in a way that results in sustainable decision making. The book is critically important in providing a roadmap to begin the transition to a sustainable world; the reader experiences an unforgettable journey toward ecological literacy, achieving a sufficient understanding of human-environment interactions to manage the earth's biogeochemical cycling in a sustainable way. With over 7 billion people on the planet, it is a journey we have no choice but to take. This is a must-read for anyone who relies on planetary resources and ecosystem services."
Cliff Davidson, Thomas C. and Colleen L. Wilmot Professor of Engineering and Director of the Center for Sustainable Engineering, Center for Energy and Environmental Systems and Department of Civil and Environmental Engineering, Syracuse University, USA

"Developing adequate solutions for human-environmental problems requires both substantive expertise and a deeply interdisciplinary perspective. Anyone who doubts this assertion need spend but a few minutes reading almost any part of Roland W. Scholz' monumental work on Environmental Literacy to have their doubts erased. In addition to thoughtful theoretical discussion they will find case after case of detailed worked-out examples that illustrate both the complexity, and the exciting intellectual challenges, that face students and professionals working to create a better and more sustainable world."
M. Granger Morgan, Lord Chair Professor in Engineering; Professor and Department Head, Engineering and Public Policy; Professor, Electrical and Computer Engineering and Heinz College, Carnegie Mellon University, USA

"Half a century ago Rachel Carson published her book *Silent Spring*. It marked not only the beginning of the modern environmental movement but it also laid the foundation for the strongly interdisciplinary field of environmental sciences. The field emerged from a range of contributing natural and social sciences disciplines. Over the years it was shaped and codified in innumerable papers and books. Following the emergence of the concept of sustainable development it also further developed into specific disciplines like sustainability science. Still one of the grand challenges of any field of science remained. No one had yet succeeded in creating an overarching synthesis of the field. That is until now. In his monumental magnum opus *Environmental Literacy* Roland Scholz not only presents

an outstanding in-depth analysis and splendid review of the field but he goes far beyond it. He also presents a strategic framework to address the many challenges we 21st-century humans are facing in our interactions with the environment. Moreover he convincingly shows the preconditions for using a framework for effective and feasible strategic decision making and action. It requires a good integrated knowledge of disciplines like biology, psychology, sociology, economics and industrial ecology as well as a genuine understanding of the transdisciplinary processes that characterize human–environment systems. In conclusion, Scholz's book is both a sparkling sourcebook and an advanced textbook for sustainability science. It is also the first successful attempt to produce a convincing theory of coupled human-environment systems. And finally, it presents a strategic framework for environmental decision making and action based on that theory."
Ton Schoot Uiterkamp, Professor of Environmental Sciences, Center for Energy and Environmental Studies, University of Groningen, The Netherlands

"Collective environmental and social problems constitute the dark side of increasing wealth for growing human populations. Roland Scholz strongly pleads for broad, multi- and interdisciplinary thinking about human–environment interactions. In the author's view, human and environmental systems cannot be separated. Rather, their interaction should be the central topic of our visions, methodologies and strategies. For natural scientists and technologists this requires a basic familiarity with how human individuals and societies function. For behavioral and social scientists it demands a solid appreciation of specific environmental problem domains. By consequence, actual policy-making should rest upon integrative teamwork. Scholz's book provides for an inspiring boost to our own environmental literacy, what it is and how it historically developed. It's a fruitful basis for extensive student courses. And it may well serve as a reference book for scientists, policy-makers and other key actors who want to improve and reflect on sustainable transitions."
Charles Vlek, Professor of Environmental Psychology and Decision Research, Department of Behavioural and Social Sciences. University of Groningen, The Netherlands

Preamble

> **Key questions**
>
> This book addresses three major questions:
>
> Q1 Who *invented the* (concept of) *environment*, why, when and in what manner?
>
> Q2 What rationales do we find in different human systems and environmental systems, and how do they interact?
>
> Q3 Do we need a "*disciplined* (i.e. discipline-grounded) *interdisciplinarity*" in *transdisciplinary* (i.e. theory–practice-based) *processes* to cope successfully with the challenging *environmental problems* of the twenty-first century?

Figures

There are a few figures that represent essentials of this book and which are referred to throughout the text. These figures are marked with an asterisk and can be found separately on the foldout page at the end of the book.

Overview: roadmap to environmental literacy

Natural and social environments are constantly adapting to changing demands from human systems. This particularly holds true as we see increasing impacts on the natural environment and resources from human systems. A key question is whether societies and their subunits have sufficient knowledge about the structure, dynamics, limits, and potential of human–environmental systems to function and evolve in a sustainable manner. And what role can science take to help in this venture? We deal with these fundamental problems under the heading of environmental literacy. "Environmental literacy" means the capacity to perceive, appropriately interpret, and value the specific state, dynamics, and potential of the environmental system, as well as to take appropriate action to maintain, restore, or improve these states. This book elaborates what knowledge and capabilities should be available in science and society to develop suitable strategies for coping with critical interactions of human systems with environmental systems. This should ultimately help to avoid the unintended and unpleasant environmental rebound effects of human action, and allow us to cope with conflicting interests which may hamper sustainable transitions.

Given these societal and scientific challenges, this book is a source book for those interested in the following questions related to environmental literacy:

- Why and when was the concept of the environment developed?
- Do we have to redefine the environment when facing that most processes in the material–biophysical layer of the Earth are affected by human action?
- How do various scientific disciplines deal with the interrelationship between human systems and the environment?
- How can we distinguish the material–biophysical environment, which includes the built environment, from the social-epistemic environment, which is historically and culturally shaped?
- When, why, and in what context are human systems concerned about the state, dynamics, potential, and negative impacts of environmental states and resources?
- What drives human systems (such as individuals, groups, companies, and societies) to exploit, protect, or sustainably cope with the environment?
- How can conflicts and dilemmas between individual and societal environmental behavior (i.e. interferences between the micro and the macro levels of human systems) be explained?
- What constitutes sound environmental literacy, sustainability learning, and sustainable behavior?

Preparing for map reading

Part I of this book consists of three chapters that illuminate why environmental literacy is of interest and what it is. We define different types of human and environmental systems and explain how these systems relate and interact. We further introduce some tools (i.e. epistemological assumptions) that are helpful to better understand what the reader will encounter in the different stages of the journey.

From the origins to the future of environmental literacy

To read our answers to the questions above, the reader can continue in Parts II–VI on a journey from the history to the future of environmental literacy in science and society. We introduce epistemological assumptions as prerequisites for coping better with the challenges of examining environmental literacy.

After this initial descent into the origins of environmental literacy, we wander through a handful of scientific disciplines. These stages of our journey will not always be the most convenient ones. Depending on the disciplinary background of the reader, he or she will sometimes have to row upstream. We start with biology, where we learn about the origins of human knowledge about organismic environments and about how societies can successfully conserve and develop this knowledge. We also explore those biological principles that are of special interest for mastering sustainable development. We make an excursion to the frontline of research on microstructures, such as the cell, and the immune system. Here we can discover how important the environment is for these systems and how they process environmental information. A comparative view of large-scale biological systems reveals that they are strongly affected by human systems and miss the essential self-regulating, homeostatic properties of microstructures.

We then look to psychology to gain an understanding of the biophysical, social, and cognitive foundations of human perception, decisions, and behavior. The sections on psychology also provide insight into the drivers of individuals and small groups when interacting with the material and social environment.

In the next stage, sociology, we focus on theories that consider the natural environment and technology as significant factors of societal development. Just as with psychology, we meet approaches that provide insights into the drivers of societies in human–environment interaction and that explain why and when environmental issues raise concern. However, we will learn that we can find a material–biophysical layer in many but not all sociological approaches. In some they are hidden, in some they do not exist.

The last stage in looking at social sciences is economics. Following our own curiosity we explore the roles that material, biophysical, land and other resources play from the view of classical and neoclassical economics, and the types of goods that are dealt with in this discipline. In new terrains of economics, we will see that economists make highly controversial assumptions about how one should deal with natural resources and the material–biophysical environment. Some seem to be reckless, whereas others seem overly troubled about their future markets and companies. We also learn that many ideas from other sciences have invaded the new subfields of economics that deal with the environment.

Next, we look at industrial ecology, a small but steadily growing discipline investigating how companies, industrial branches, and trade can reinvent themselves to reduce environmental impacts resulting from production, business, and services. Here we encounter some engineering methods that allow the assessment of manmade environmental impacts. But we also look at some special sites, such as eco-industry parks, from which we can learn to better cope with material flow. What is more, we see how industrial ecology offers broad, long-term perspectives and strategies for the future of environmental literacy. This offers a global view that highlights the fundamental changes that the landscape of human–environment interaction has undergone in the past and also the new structures and components that this landscape might exhibit in the future.

Our journey could have taken another route with different stops. Besides biology, we could also have looked, for instance, at geology, and, instead of industrial ecology, at civil and environmental engineering. Naturally, other disciplines, such as geography or anthropology, could deserve their own chapter. However, we think that the selected disciplines allow the demonstration of why and when various types of contributions to environmental literacy emerged in academic disciplines.

The reader might ask why we are not visiting some new, exotic and more exciting domains, such as environmental or sustainability sciences. These two disciplines are the current home base of the author, and scientists from these disciplines are the most likely to use our roadmap to promote environmental literacy. As key question 3 of the Preamble indicates, we explore each discipline both for its unique knowledge and for the value it can bring to investigations that involve more than one discipline. We call this perspective, common to environmental and sustainability sciences work, "disciplined interdisciplinarity."

Provisions for traveling beyond the boundaries of disciplines and sciences

The first part of the journey highlights how many issues from different disciplines are indispensable to cultivate environmental literacy. But knowledge from lone disciplines only takes us partway toward answers to today's environmental problems. In Part VII, this book introduces three new perspectives that are required to take us further.

The first is that today's environmental problems cannot be managed without incorporating analysis of human systems that affect many processes on all levels, from molecular up to global biogeochemical processes of the material–biophysical environment. This asks for redefinition of the environment and the destination of the journey. We will discuss a new goal for the second stage of the journey, which includes an anthropocenic redefinition of the environment.

Second, environmental literacy requires an integrated view of knowledge about the environment and human–environment interactions. This requires new techniques of making and processing pictures. We will see how "integrated modeling" becomes a vital means of extending environmental literacy and allows a deeper understanding of what we have seen. This will facilitate in taking an *interdisciplinary* view.

Third, we will notice that, in the first stages, disciplinary scientists face limitations in acquiring all the information about the environment. Thus, we, along with these scientists, have to leave our travel route and step into real-world cases to gain valuable additional information from directly talking, interacting, collaborating with, and getting first-hand information from the people and human actors who are directly experiencing, benefiting from, and interacting with the environment. This provides a completely new perspective, which we call *transdisciplinarity*. We will see that the people incorporated in transdisciplinary processes benefit from and appreciate these mutual learning processes.

Extending environmental literacy: the human–environment systems Postulates and the HES framework

After this intermediate stop and reorientation, we explore new territory and the character of the journey changes. It will slightly resemble an excursion into more complex domains. Instead of the environment, human–environment interactions become the object and objective of the journey. It becomes difficult to keep track of our place, to figure out where to go first, and to know what we need to move toward these new destinations. Thus we offer Part VIII as a guidebook that outlines seven Postulates, or assumptions, for investigating human–environment systems (HES). This travel guide consists of seven Postulates, the HES Postulates. The HES Postulates depict the constituents of the HES world, how HES behave, how they can be classified, and of what they are and are not aware. We also examine the drivers of human systems and the conflicts that may exist between individual agents or subunits and superordinate systems. The latter can be communities or societies.

An important point of the guidebook is that we have to conceive HES as coupled, inextricably intertwined, systems. However, the guidebook also explains that we must first have a thorough look at the environment, in particular at the material–biophysical environment if we want to understand what a specific HES looks like, how it evolved and what future development might take place. The HES are explained by a Postulate that describes different types of feedback loops that may be at work in HES.

We will see that the HES Postulates draw on what we have learned when looking at the scientific disciplines in the first part of the journey. Our knowledge of social sciences, for example, will help us to understand the interests and values that underlie the rationale of human systems when interacting with the environment.

Having become familiar with the individual Postulates, we know how the Postulates relate to each other and how they can work together during an investigation of an environmental issue. We put forth this HES framework as a template for transdisciplinary collaboration.

Four cases for demonstrating HES literacy

Equipped with the HES framework, we are prepared to take a closer look at the challenges and threats of human–environment interactions. In Part VIII, we make four excursions, each of them demonstrating the improved environmental literacy gained by working with the HES Postulates.

Trip 1 looks at epidemic and pandemic threats. Using the HES Postulates we learn that the outbreak of pandemics is shaped by more than the mechanisms of viruses and bacteria. The type and severity of pandemics, and the unexpected rebound effects that may result from various pandemic management approaches, ask for an examination of the behavioral, contact, and mobility matrices of human systems. A look at micro and macro structures and their interactions is needed here.

Trip 2 involves an excursion to Switzerland. We scrutinize how transdisciplinary processes involving scientists and key agents, as well as people from the region, are helping government and industry to better adapt to market and environmental constraints. We learn that transdisciplinary processes are useful to identify robust orientations to find sustainable solutions.

Trip 3 looks at how HES manage basic supply services. The jaunt takes us to Sweden, where we encounter unexpected limits to biofuel resources. Here we see how proper identification of secondary feedback loops is a key feature of sustainability learning.

Lastly, trip 4 provides profound insights into the difficulties that human systems and societies face when making trade-offs and protecting natural resources in the anthropocenically formed environment. The case we look at is the dilemma of establishing a recycling management system for some minerals (i.e. phosphorus) when simultaneously aspiring to eliminate and dispose of material matter (e.g. carcass meal and bones), which includes high concentrations of the said mineral but also dangerous pathogens.

Upon our return from these trips, we check whether the guidebook and the HES framework have served as an effective compass for environmental literacy as we have defined it above. To do this we compare the HES framework with alternative travel guides.

The vision on future environmental literacy

The journey closes in Part IX by identifying key components that may promote environmental literacy ("Sustain–abilities", see Chapter 20.4.1). These are the new fields of environmental and sustainability sciences that cope with inextricably coupled HES. The HES framework, based on disciplined interdisciplinarity, allows a thorough investigation and understanding of complex environmental problems. Transdisciplinarity includes processes which use knowledge from theory and practice to generate socially robust solutions for sustainable development.

How to access the chapter overviews

After the novelist-like overview, the reader can best gain access to the content of the book by reading the sections "What to find in this part" on the front pages of all nine parts and of the 20 chapters. A closer look at the roadmap and its extended legend on pages xxii–xxiii may also help to understand the structure, main subject matters and the storyline of this sourcebook.

Who should use this book?

The overview suggests that this book might be of interest to those who are curious about how the environment interacts with human systems and how human systems, from the individual through groups, organizations, companies, communities and society to the whole human species, can become capable of adequately adjusting and adapting to the continuously changing and increasingly anthropocenically shaped environment.

These primary readers are researchers from the emerging fields of environmental and sustainability sciences who are interested in human-environment interactions or systems. Clearly, it is also relevant to anthropologists, human ecologists, geographers, and environmental planners, or people working in the hyphenized fields of sciences, such as environmental–psychology, -sociology, or -economics. Readers will also no doubt include people from the natural sciences, including ecology and those working with the climate and atmosphere. The book will also be of interest to those working in environmental chemistry and those in different branches of engineering sciences, such as industrial ecology, may learn from the comprehensive, integrative, coupled system perspective which looks at the constraints, feedback loops, and regulatory mechanisms of HES.

Environmental literacy is not only seen from an academic learning perspective but is rather focused here on what we call *societal didactics*. Thus, it should contribute to societal learning about how to cope with environmental challenges and provide access to the rationale of human–environment interactions. Further, this vision and the practice of transdisciplinarity were motivators for writing the book. However, as expressed in the Preamble, establishing a thorough, discipline-grounded interdisciplinary knowledge about HES, which favors transdisciplinary processes that deal with the current and future environmental challenges, is the very vision and mission of this book.

Legend of the roadmap

The concept map shows how the ideas in this book relate. The figures in the concept map come from informational boxes that are sprinkled throughout the book and tell stories from around the world, both historic and contemporary, that illustrate our message.

 Our starting point is the question, "Who invented the environment?" Chapter 1 describes how humans' awareness of their impacts on the environment developed and when and why the concept of the environment was invented. Chapter 2 provides a first definition of environmental literacy and introduces the value that transdisciplinarity can bring to how humans address environmental issues. Chapter 3 introduces the concept of environment based on an organismic, cell-based definition of the human individual and the complementarity of material–biophysical and social-epistemic levels of human and environmental systems. Here we describe the basic ontological and epistemological assumptions that underlie our world view and thesis.

 To discuss the issue, "What disciplines can and can't tell us," we review contributions to environmental literacy from five academic disciplines – biology, sociology, psychology, economics, and industrial ecology – in Parts II through VI. Taking two chapters to cover each discipline, we review the history of mind for each, examining which aspects of human–environment interactions were of interest during different time periods. We also review key theories from each discipline and prospective future perspectives that can inform environmental literacy.

 In Part VII, Chapters 14 and 15 describe the pivotal, integrating function of "managing interfaces to become literate." We examine how knowledge integration and transdisciplinarity help us to decrease the complexity of environmental issues, which warrant an "Anthropocenic redefinition of the environment in a coupled human–environment setting."

 We put forth seven Postulates, P1 to P7, to organize the complexity of today's environmental issues and related research.

 The HES framework is a methodological schema for employing the Postulates in an integrated manner when investigating environmental issues.

 To give readers a feeling for the HES framework in action, Chapter 18 presents four case studies.

 Chapter 19 compares the HES framework with alternative approaches and shows what added value it can provide.

 The last chapter presents "Perspectives for future research in human–environment systems," and links the coupled systems and the transdisciplinary perspective (see bottom left of the concept map and key question 3).

Legend of the roadmap

**CONCEPT MAP
for Environmental Literacy in Science and Society: From Knowledge to Decisions**

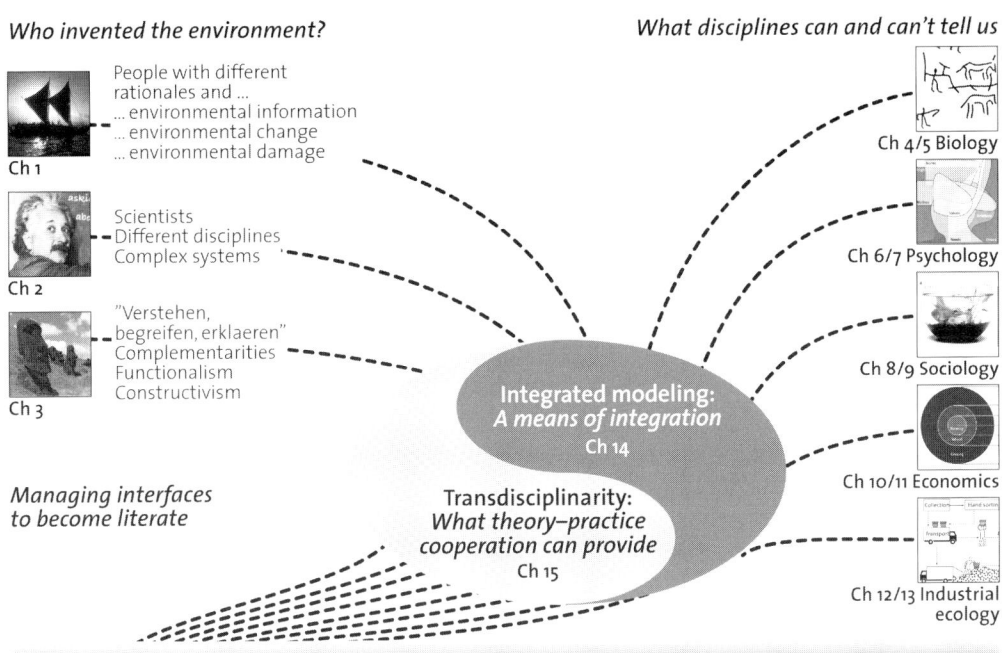

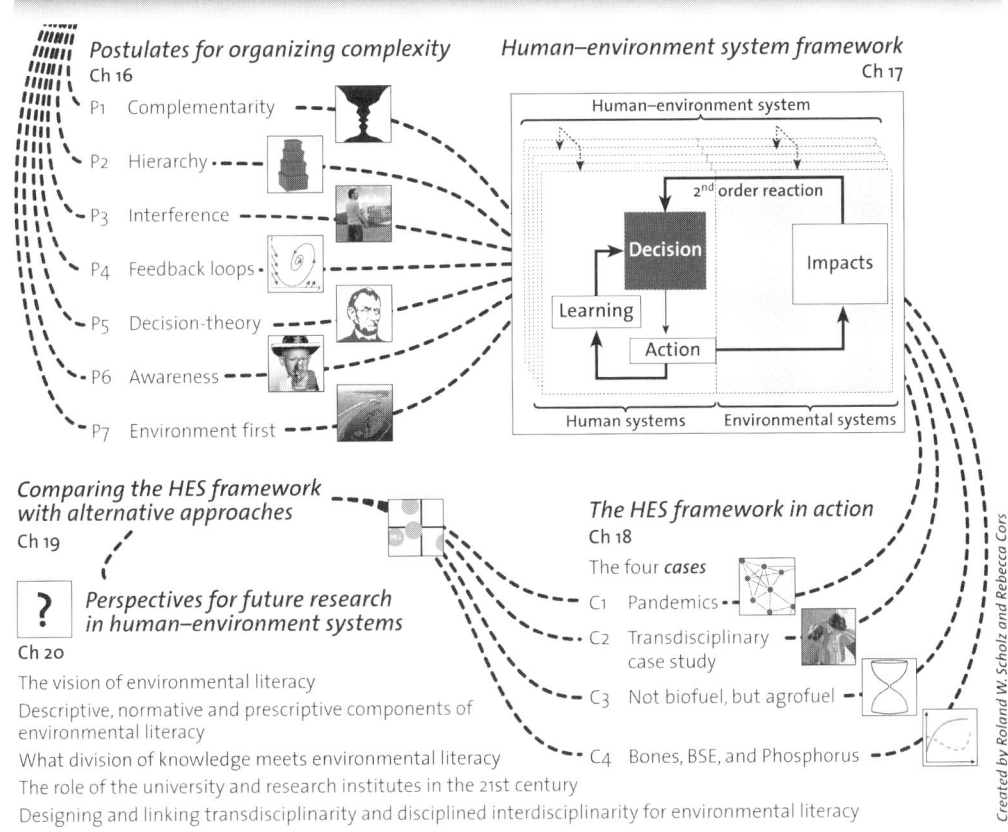

Created by Roland W. Scholz and Rebecca Cors

xxiii

# Part 1

Invention of the environment: origins, transdisciplinarity, and theory of science perspectives

1   What knowledge about what environment?   3
2   From environmental literacy to transdisciplinarity   15
3   Basic epistemological assumptions   29

Key questions (see key questions 1–3 of the Preamble)

What aspects of the key questions are answered in this part?

Q1   How do human systems *represent, store* and *retrieve* information about their environment?
Q2   How can we identify the drivers of human behavior?
Q3   What manmade environmental impacts can/should we change? What rebound effects can emerge from intervening? How can society generate appropriate knowledge and strategies for sustaining the Earth's environments?

What is found in this part

Chapter 1 introduces how environmental awareness emerges, and provides a first tight definition of environmental literacy as a necessary aptitude for human systems to sustain and to adapt to environmental change. We discuss from a historical point of view how the concept of environment was invented and reflect on the role of science in assuring that societies have the proper knowledge to master environmental challenges.

Chapter 2 takes a more differentiated view on what environmental literacy comprises. We show why thorough natural science knowledge is necessary but also why it is not sufficient as what becomes an environmental problem is based on human needs, interests, and values. We reveal that dealing with environmental problems often causes unwanted feedback loops and requires coping with trade-offs and conflicting interests of human systems. Coping with or integrating different interests and perspectives is one of the five types of knowledge integration included in environmental literacy. Another type is the use of different kinds of knowledge. Here we distinguish the experience-based understanding of practitioners, who often work with intuitive coping strategies for environmental challenges, and science-based knowledge, which is a specific type of analytical knowledge. This leads us to transdisciplinarity, which is a specific form of learning and cooperation between different parts of society and academia to meet the complex challenges of society.

Chapter 3 expands the basic ontological and epistemological assumptions underlying this book, which views environmental literacy as a dynamic learning process and offers the human–environment system (HES) framework for examining how humans interact with their environment. The framework is based on a specific definition of human systems that refers to the activities of human individuals which can be assigned to this system. This approach, together with a cell-based definition of human individuals, allows for a cohesive definition of human systems, environmental systems and of the human–environment complementarity.

Part I Invention of the environment

Chapter 1

What knowledge about what environment?

1.1 How did environmental awareness emerge?   3
1.2 What drives environmental awareness?   6
1.3 Who invented the environment?   10
1.4 Environmental literacy: from average knowledge to sciences and human systems' capabilities   13

Chapter overview

Humans' concerns about and interactions with the environment, from navigating the seas to managing a farm or operating a nuclear power plant, have emerged from the core need to survive by using environmental information and resources. This chapter explores the emerging awareness of the environment as a prelude to understanding human interactions with the environment. Central to this discussion is environmental literacy – the human aptitude for appropriately reading and using environmental information to use environmental resources properly and to adapt effectively to environmental dynamics. This chapter reveals what knowledge about the environment is relevant, what environmental literacy means, and introduces the drivers and rationales underlying environmental awareness.

1.1 How did environmental awareness emerge?

A few decades ago environmental issues surfaced, for the first time, as important public concerns. In this book we elaborate on how environmental awareness arose in science and society, and how this awareness shifted from simply perceiving the environment to interacting with and influencing it in a world with roughly seven billion people. Long before the time of Christ, there was already considerable environmental impact from large cities in Mesopotamia and elsewhere, such as Ur, Nineveh or Carthage, with populations up to 700 000. Most theories on the decline of the Maya and other ancient societies include ecological hypotheses such as environmental disasters, climate change or overpopulation. For example, suspicions that the spread of diseases such as plague and cholera were caused by human activities have been societal concerns since at least the early medieval era (Watts, 1999). Thus, it might be surprising that the term "environment," while quite colloquial, elementary and easy to define, is a rather recent concept, being less than 400 years old.

In this book we look at how human systems at the scale of the individual or greater cope with the environment. We use the somewhat unusual term "human systems" as a general denomination of individuals, groups, and human-made organizations such as companies, societies, or supranational systems. We can use different scientific disciplines, such as psychology, business science or sociology, thus broadening their use beyond only one level of human systems.

Whether humans can successfully cope with the environment depends on the environmental setting and on what they know and how they behave. The specific biophysical maritime environment of indigenous Polynesians, for instance, challenged them to develop geographically specific *environmental knowledge*, that of fishing societies (like navigation and boat-building; see Box 1.1). Amazonian Indians, on the other hand, had to develop the ability to read wildlife and flora to secure subsistence. To be successful in today's world, where competition and high-volume fishing are draining and, in some cases, damaging the very aquatic resources that sustain them, fishermen need to use modern technologies and techniques to be competitive. Such influences from technology on our environment can be seen in examples as vivid as how some

1 What knowledge about what environment?

> **Box 1.1** The ability to read the environment: Polynesian navigation
>
> Pacific islanders did not consider the sea a barrier but rather a highway. For navigational purposes, Samoans developed sophisticated multilayered knowledge. Star courses of many island destinations were committed to memory. By interpreting the heavens as a dome, they were able to find a bearing by reading how stars moved along paths over certain islands. On foggy days, long surface waves known as swells were used to navigate. These swells, sometimes formed by stable wind systems thousands of miles away, were considered a regular wave signal existing in the midst of other waves and chops.
>
> For long-distance voyages it is supposed that Polynesians followed the seasonal paths of birds (see Figure 1.1). Therefore, when choosing a route, information was taken from environmental systems such as waves, winds, stars, the Sun, the Moon, birds, and the water itself (Gehmacher, 1973; Gladwin, 1970; Richey, 1974).
>
> The Polynesians could neither read nor write and thus had to rely on their sense of sight, touch, smell, or hearing for their information. They did, however, name the stars and communicate their knowledge in songs and tales. The changing relationship between human and environmental systems can be seen by taking a wave as an example. The Polynesians had to use their eyes alone to read waves and used this information for navigation, whereas some modern surfers combine satellite remote control technology and modeling techniques to find the right monster waves (Stormsurf, 2007). 
>
> **Figure 1.1** (a) Ancient map of the southern night sky (Flamstéed, 1776). (b) Polynesian sailing canoe (photo by Sam Low/Polynesian Voyaging Society). (c) The physics of waves.

of today's Surui Indians use Google Earth to monitor their fields.

Thus the *material, biophysical, and technological* aspects of an environmental setting, which comprises the *material environment*, matters. The material world consists of an abiotic layer, including atoms and their parts, water, crystals, minerals, naturally occurring carbon-based (organic) compounds and material, polymers and manmade organic compounds, as well as other objects such as cosmic matter. The built environment and machines that humans have constructed to cope with the environment are part of the abiotic layer. The biotic layer includes organisms such as bacteria, plants, and animals (Cotterill, 2008). Knowledge about the material environment is a major component of environmental literacy and is important for human systems to survive.

Knowledge about how the material environment functions and how it can serve human needs is an important issue of environmental literacy. Another issue is knowledge about feedback loops between human actions and the biotic and abiotic environment resulting from environmental actions. The environment can fire back. Rebound effects related to human health, for instance, can be traced back 450 years. The risk of developing lung cancer from radon radiation by working in underground mines was documented by Georgius Agricola (1494–1555) in the sixteenth century (Agricola, 1565).

The need to understand humans' relationship with the environment attained another level with the beginning of the industrial age and the foundation of forestry and agricultural sciences. Concerns grew about the ways in which human action changed the environment: soil erosion, flooding, (regional) climate change, etc. (Boussingault, 1845; Marsh, 1864/2003). For instance, Goudie (2006) noted that the French engineer Boussingault (1802–87) posed questions about manmade climate change back in 1845 (see Box 1.2). The age of mining, industrial and agricultural engineering required an understanding of the ecosystem functions of woods, lakes, and rivers in evaluating the

1.1 How did environmental awareness emerge?

> **Box 1.2** Correcting negative human impacts: Ascension's spring
>
> Forestry and agricultural sciences were the first sciences that investigated the critical effects of large-scale human impact on the environment. Therefore, questions concerning human-induced climate change had already arisen in the first half of the nineteenth century: "A question of great importance and that frequently agitated at this time is, as to whether the agricultural labors of man are influential in modifying the climate of a country or not" (Boussingault, 1845, p. 673).
>
> The mining engineer Boussingault's original interest was in successful production and not in the environment. This is seen in the subtitle of his book, which reads: "Rural economy, in its relation with chemistry, physics, and meteorology; an application of the principles of chemistry and physiology to the details of practical farming." Thus he developed a specific environmental literacy for growing cash crops based on physiology, manure science, cultivation methods, and meteorological considerations. His questions about climate change were based on some personal hydrological observations he made between 1826 and 1830 on a small South Atlantic island:
>
> Ascension is a small island of 91 km^2 discovered in 1501 halfway between Africa and South America. In the nineteenth century it became a stopping point for ships. "In the Island of Ascension there was an excellent spring situated at the foot of a mountain originally covered with wood; the spring became scanty and dried up after the trees which covered the mountain had been felled. The loss of the spring was rightly ascribed to the cutting down of the timber. The mountain was therefore planted anew. A few years afterwards the spring reappeared by degrees, and by and by followed with its former abundance" (Boussingault, 1845, p. 685).
>
> The impact of deforestation was already a major topic of the new geography of the early nineteenth century (Boussingault, 1845; Marsh, 1864/2003, see Figure 1.2). Boussingault, for instance, argued that changes in vegetation on many islands such as Ascension, were a consequence of the erstwhile expansion of the Spanish empire, the freeing of slaves and land use changes such as the end of industry (i.e. plantations; Marsh, 1864/2003). Therefore, it is not only the number of people that matter but also the way they use the land.

Figure 1.2 Deforestation caused by agriculture expansion was considered to have a major impact on local and regional climate change in the early nineteenth century. (a) The Atlantic forest in Brazil by Spix and von Martius (1823–1831). (b) Jungle burned for agriculture in southern Mexico (photo by Jami Dwyer).

impact on the human species. This, in turn, challenged the major natural scientists of that century, prompting, for example, physicists to start measuring and comparing temperature dynamics in wooded and cleared areas. The Nobel Laureate Becquerel (1820–91) himself stated:

> ... forests act as frigorific cause ...; they shelter the ground against solar irrigation ... (Becquerel, 1853, quoted according to Marsh, 1864/2003, p. 140)
>
> They produce a cutaneous transpiration by the leaves. (Becquerel, 1853, p. vi)

Thus scientists became deeply concerned about the "momentous consequences" of "human action in the physical conditions of the globe we inhabit" and pointed out the "dangers of imprudence and the necessity of caution in all operations" (Marsh, 1864/2003, p. 3).

Following the current discourse about global warming, it is most interesting to see that there was a similar dispute 150 years ago concerning whether science can reliably predict climate change and on how climate change should be evaluated:

1 What knowledge about what environment?

In my opinion we have not yet a positive proof that the forest has, in itself, any real influence on the climate of a great country, or in particular locality. By closely examining the effects of clearing off the woods, we should perhaps find that, far from being evil, it is an advantage; but these questions are so complicated when they are examined in a climatological point of view that the solution of them is very difficult, not to say impossible. (Gay-Lussac, quoted according to Marsh, 1864/2003, p. 140)

Thus, the problem of assessing and evaluating human impact on global change is both a historic and current issue. Today, we can find exactly the same statements as elaborated by Gay-Lussac about the extraordinary difficulty in providing valid prognoses on impact on climate change resulting from land cover change. The essential difference is that we are dealing with the issue on a global level. The difficulty of reliably assessing how the extremely complex climate system reacts to land cover change is clearly reflected in a comparison of seven leading climate models:

> The imposed LCC [land cover change] led to statistically significant decreases in the northern hemisphere summer latent heat flux in three models, and increases in three models. (Pitman *et al.*, 2009, p. 1)

Even though the effects from the greenhouse model seem to be much less unsure, we start to understand the difficulty in acquiring a valid understanding of the environment. This will be acknowledged in the HES framework presented in Part VIII by the Environment-first Postulate P7, whose message is that a thorough understanding of the environment has to precede human action that affects and changes the natural environment.

Another critical issue is that human systems can be arranged in the material environment in completely different ways. What is considered environmentally harmful or desirable differs between individuals, groups, companies, and societies. In some cultures, for instance, cows are sacred, and in others beef is a favorite food. The priority given to the impact of climate change differs between political parties and nations. What humans judge as a critical impact, harm or benefit, and at what point environmental interventions are initiated, depends on human interests, values, and knowledge.

Clearly, *epistemics*, the knowledge available to humans, is an important component of environmental literacy. As humans' knowledge develops and is handed down to subsequent generations, it becomes embedded as social and cultural knowledge. The knowledge that constitutes environmental literacy is also in people's minds and not only in the books, computer files, or other media that represent and store signs. Thus the socio-epistemic and cultural aspects of the human system are of crucial interest if we want to understand what creates environmental literacy. A challenge, discussed in this book, is how the knowledge and values of different parts of society can be efficiently linked to scientific knowledge.

1.1.1 Key messages

- Different geographical and biophysical environments, as well as the continuously changing world and its technological equipment, ask for qualitatively different or new types of knowledge to cope with environmental challenges
- The severe impacts, including climate change, of human activities on the biosphere by land use changes were recognized in the middle of the nineteenth century.

1.2 What drives environmental awareness?

This book introduces a conceptual framework to understand the *drivers of environmental awareness* and to provide insights into the conflicting drivers that promote or prevent *sustainable behavior*. By sustainable behavior we mean those human activities that do not endanger global dynamics, resource availability, and the resilience of ecosystems in a way that can cause problems to the self-sustaining of the current population or of future generations (Laws *et al.*, 2004). This, of course, includes the stability of human systems, which can be imperiled by socially unstable or unjust settings. We are interested in how drivers of sustainable behavior can be conceptualized, investigated, and understood. However, such an analysis on human–environment interaction includes basic assumptions about the *model of man* (see Box 1.3) that underlies the self-sustaining kind of analysis. Thus the framework presented in this book draws from the views and conceptions we have of human systems. A key issue in this context is examining the rationality behind how human systems evolve.

While we can postulate that human action occurs in response to certain goals, we recognize that humans and human systems are complex and evolve in response to many driving factors. Human actions and underlying rationales depend on situational constraints and the capability of the human system (Scholz, 1987).

1.2 What drives environmental awareness?

Box 1.3 What type of rationale underlies human decisions? Models of man and rationality

When looking at the history of psychology, different concepts of man and rationality dominated leading scientific genres. Shulman and Carey (1984) stressed that changes of the idea or "model of man" (Simon, 1957) have been shaped by critical stages in history, particularly the twentieth century western wars. In particular, overcoming societal crises and wartimes has been the rationale behind many decisions and actions of science and society (see Figure 1.4).

After seeing the annihilation of human life during World War I, Sigmund Freud (1856–1939) put forth the death instinct theory, which describes humans as irrational beings. The theory asserts that human actions result from a perpetual conflict between a destructive "death instinct" (thanatos) and a sexually oriented "life instinct" (eros).

After World War II, there was a paradigm shift to viewing humans as rational beings. Full rationality would mean "to act consistently in its own interest – or in response to the consistency of its environment and to develop through education and learning, capacities for reason as it is mature" (Shulman & Carey, 1984, p. 501). This conception culminated in the view that human inference followed rules of mathematical models such as probability theory (Peterson & Beach, 1967).

Currently, the dominant view of humans is as bounded rational beings, with regard to judgment and decision-making. This comes from behavioral economics and other fields of behavioral sciences (Gigerenzer & Selten, 2001; Kahneman, 2003; Scholz, 1991, 1983). The principles of bounded rationality (Simon, 1982) suggest that individuals make active use of cognitive strategies and previous knowledge to deal with their memory and information processing limitations as well as their restricted operative and heuristic repertoire. There are "two camps on bounded rationality" (Jungermann, 1986). One stresses the biases and fallacies and takes a skeptical view on heuristics. The other is an optimistic view; working with a few, smart, simple, domain-specific, fast, and frugal heuristics is efficient and ecologically rational (Todd & Gigerenzer, 2007).

Another recent view, which describes humans as collectively rational beings, asserts that human ability and intelligence may only be practiced and investigated in the context of social interaction. Rather than examine human rationales via the human psyche, collective rationality and collective agency (Bandura, 2001) are dealt with on a societal level (Fischhoff *et al.*, 1993; Simon, 1956). This approach involves investigating relations between specific intentions and rationality on the level of the individual *and* the society rather than on the level of rational choice theories that are based on individualistic rationales, or collective rationality from the mere side of individuals (Mahon, 2001). Collective rationality and collective reasoning refer to the benefits of cooperative reasoning, which is based on a pool of commonly accepted reasons or criticized arguments.

It is clear that the rationale of an individual rather follows bounded rationality than full rationality. Which form of collective rationality is at work seems to depend on the situation and the societal rules, cultural norms and reward systems (see Box 6.15).

Figure 1.3 (a) World War I: Australian infantry wearing small box respirators at Ypres, Belgium in 1917 (photo by Frank Hurley). (b) J. Presper Eckert and John Mauchly with the "Electronic Numerical Integrator and Computer" ENIAC in 1947 (image courtesy of Computer History Museum).

On the level of the individual, for instance, decisions can result from multilayered, unconscious and intuitive processes that, more often than not, are based on multiple sets of goals that can even be at odds with one another.

The conceptual HES framework presented in Part VIII of this book emerges from game, decision, and system theoretical perspectives. It provides a specific language and a theoretical structure for describing human behavior from the perspective of a number of disciplines. This fits as the book contributes to what has been called "a second environmental science."

> [This] … field is the study of the feedbacks between humanity and the environment – the ways individuals, organizations, and governments act on the basis of experienced or anticipated environmental change to manage and preserve environmental values. (Stern, 1993, p. 1897)

1.2.1 Defining environmental literacy

One purpose of this book is to promote the understanding and the development of environmental literacy. By environmental literacy we mean the ability to appropriately read and to utilize environmental information, to anticipate rebound effects, and to adapt according to information about environmental resources and systems and their dynamics. Environmental literacy requires more than a fundamental understanding of the systems of the built and technical environments. It also requires a profound understanding of the potential and the limits of human systems to cope with essential settings of the biotic and the abiotic environment. Moreover, environmental literacy goes beyond understanding the impact of humans on the environment (Goudie, 2006) or assessing the effects of environmental hazards (Paustenbach, 1989). The focus is on the interaction of human and environmental systems, how humans learn from feedback and can avoid rebound effects, and what information they react to or ignore. In this manner environmental literacy is linked to learning, and so the question of how this literacy can be transmitted to future generations receives special attention. The breadth of environmental literacy, therefore, does not exclusively refer to human-induced problems. It also consists of the aptitude to adapt to natural hazards and to necessarily or reasonably cope with the changes or the potential of environmental systems.

Human systems – an individual, a company, a state agency or a nation – have to notice, discover, explore, investigate, represent, and, finally, to adapt appropriately to environmental systems. For many situations, environmental literacy is crucial to survival. As we can see from anthropology, environmental literacy has been an essential prerequisite to sustain life for presumably all indigenous people, as illustrated with the example of the Polynesian's ability to read and to *understand* a broad set of environmental information to navigate (see Box 1.1). Another example is the ability to *cope* with scarcity of resources. Today, there are mounting global concerns about the depletion of food supplies, energy, and minerals (see Chapter 11).

Environmental literacy is also essential to *assess* the human impact on natural systems such as climate, soil, water or ecosystems in a meaningful way and to *identify* rebound effects. Unintended environmental impacts and feedback inherent to human–environment systems (HES) can be seen through the example of global deforestation. Deforestation has resulted from settlement infrastructure and product development, and is also a form of collateral damage resulting from societal commerce and conflict (see Box 1.4; Schmithüsen, 2008).

1.2.2 Becoming aware of the anthropocene

This book explains that the (natural) environment cannot be considered or investigated independently of human activities. This is expressed by the term "anthropocene," which was suggested by Earth scientists to explicate that the human species has become a major geological factor (Crutzen, 2002a). This book elaborates that we need an anthropocenic redefinition of the environment. This will be done by conceptualizing HES as inextricably related and complementary systems (see Chapters 3.2, 14.1, and 16.2).

Coping with many environmental problems often calls for investigation as to which human behaviors and human systems are responsible for which environmental impact (Gardner & Stern, 1996). If we consider the western world's current concern about future energy supply, some scientists focus on individual consumerist development patterns as the lynchpin of energy depletion, climate change, loss of biodiversity, and other unwanted environmental impacts (Goldblatt, 2005). Others regard cultural norms or lifestyles of societal groups (Lutzenhiser, 1992), including sprawl settlement, mobility, or imprudent use of fossil energy for heating as the key issues. Those who deal with impacts of industrial products sometimes claim that

Box 1.4 Ignoring environmental collateral damage: British forests

The last Ice Age, which ended about 10 000 BC, covered two-thirds of the European continent with ice. Afterward, forests spread and, before agriculture began, closed forest covered up to 48.3 million km² and open woodland about 15.2 million km² worldwide. Since then, anthropogenic activity has reduced forest cover, and today closed forests have been reduced by about 7–8 million km² and open woodland by more than 2 million km² (Williams, 2000). The largest losses have come from agricultural activity and urban residential development, along with demand for timber and firewood.

McNeill (2007) elaborated that, historically, warfare played an important role in deforestation. Forests supplied the earliest weapons, such as clubs, spears, slings, bows, and arrows. Deforestation continued into the Bronze Age, as places like Cyprus, a major copper supplier in the Ancient Mediterranean, used timber to fuel and fire copper smelters. In addition, most agricultural societies built defensive fortifications from wood, and coastal societies protected the timber supplies they needed for shipbuilding. For instance, the seafaring Phoenicians defended Lebanon to maintain their access to high-quality ship timber (McNeill, 2007). Given that an "eighteenth-century-ship required four thousand mature oaks … or about twenty hectares of northern European forest" (McNeill, 2007, p. 8; see Figure 1.3), it is not surprising that forest ownership resulted from economic and imperial interests. Spain was an early global timber trader, importing wood from Italy, the Balkans, and Brazil. In addition, about 35% of its naval shipbuilding took place in Cuba, which became its most important colony. Timber remained a significant source of war material into the twentieth century as well (Gardner & Stern, 1996, 2002). Between 1916 and 1918, Britain felled half of its productive forests to meet the needs of the war.

There are other examples of unintentional, or "collateral," damage resulting from wartime activities. For instance, in 1741, some 10 000 men were assigned the task of clearing West African woods in preparation for a full-scale battle. As a tactic against guerilla resistance, forest burning has been a common practice by many groups since the time of Roman legions. During the Vietnam War, 23% of old forest was cleared (all data taken from McNeill, 2007).

All this occurred in spite of the (theoretical) knowledge that deforestation could lead to decreases in soil productivity and changed water balances ending with desertification and societal collapse on a large scale, as formulated by Marsh (1864/2003). Today, we have evidence that deforestation along the southern coast of West Africa induced desertification along the border with the Sahara (Zheng & Eltahir, 1997). Also, Amazon deforestation is a major component of climate change (Shukla et al., 1990), although there is uncertainty about the exact impacts on regional and global scales (Pitman et al., 2009).

Figure 1.4 (a) Shipbuilding in 1917 at Lockhart Shipyard in Nova Scotia (Deal, 2006). (b) In the naval battle of Salamis between the Greek city-states and Persia in 480 BC, more than 1300 triremes were involved (painting by unknown artist).

also responsible business companies or technology developers may become important agents of sustainable development (Robert et al., 2002; Schmidheiny, 1992). In principle, we elaborate that successfully coping with environmental impacts and feedbacks requires adequate mitigation and adaptation strategies on different levels of human systems (see Figure 14.1*).

Many researchers in the field of sustainable development look at highly aggregated systems and investigate uncontrolled world population and the related economic dynamics of the human species as origins for resource depletion (Ehrlich & Ehrlich, 1970; Ehrlich & Kennedy, 2005; Meadows et al., 2004; Robert et al., 2002). Interestingly, it seems that the issues of regulating world population and successfully negotiating compulsory mitigation and adaptation strategies among the nation state-governed world system has become taboo or at least a background issue. As elaborated in the sociology section in Chapter 8 and in the Hierarchy Postulate P2 in Chapter 16, supranational institutions can and should take a major role for a global development agenda (Beck, 2000; Beck & Sznaider, 2006; Bernauer, 1995; Tallberg, 2002). There are other challenges of environmental awareness and environmental literacy, such as being able to identify the most sustainable environmental protection interventions in situations where the trade-offs between different interventions are not clear. Another fundamental challenge is having the capability to recognize changes that signal irreversible shifts in ecosystems, such as noticing the disappearance of lynx from the Harz Mountains in Germany, which led to an overpopulation of deer. These are discussed in the following chapters.

1.2.3 Key messages

- Environmental literacy denotes "the ability to appropriately read and use environmental information and to anticipate rebound effects or to adapt according to information about environmental systems and their dynamics"
- There are different types of rationality which can be supposed to underlie human behavior
- Becoming aware of the biophysical environment and utilizing it is a basic prerequisite for human existence
- The impacts of human systems on environmental systems have attained a magnitude that suggests humans have become a geological factor. This calls for an anthropocenic redefinition of the environment from an inextricably coupled, complementary systems perspective
- Sustainable behavior requires appropriately reading of the limits, unwanted and long-term dynamics, and rebound effects emerging from the material, biophysical and technological environment, as well as from the social environment. Justice can be seen as an important trait of stabilizing human systems.

1.3 Who invented the environment?

Earth scientists suppose that the Big Bang occurred about 13–14 billion years ago (Spergel et al., 2003) and the emergence of life occurred about 3.8 billion years ago (Nisbet & Sleep, 2001). Paleontologists estimate that the first mammals appeared 65 million years ago (Rose, 2006), and anthropologists estimate that the emergence of *Homo sapiens* was between 195 000 and 50 000 years ago (McDougall et al., 2005). The founding of cities dates back about 5000 years (Benevolo, 1980).

In comparison, the term "environment" is a relatively recent concept that appeared less than 400 years ago. It has a most intricate history, with multiple, different meanings. Some may find this surprising, particularly against the background of the twentieth century in which the term environment has been ubiquitously used and thought of as having a historically constant state (Müller, 2001). Etymologically, "the state of being environed" in the sense of "nature, conditions in which a person or thing lives" (Harper, 2001) can be traced back to 1603 and was first recorded by the Scottish historian and freethinker Thomas Carlyle, who translated a description of Ossianic landscapes:

> In such an element with such an environment of circumstances[1], with studies and tastes of this sort; harassed by unsatisfied desires … (Carlyle, 1827; quoted according to Spitzer, 1942, p. 204)

The quote is typical for the time of *Sturm und Drang*, "storm and stress," a movement in German literature and music that took place from the late 1760s to the early 1780s, which dealt with extremes in emotions and feelings. Given the context of this phrase and the erotic shape of Ossianic landscapes used by Goethe (see Box 1.5), it was clear that the term environment (and landscape)

---

[1] German: "… bei solcher Umgebung bei Liebhabereien und Studien …" The translation "… in such an environment of desires …" seems to be a more appropriate translation.

was used to describe what might be called a (imagined) social environment (Slater, 1952). The focus here is on the complex of surrounding conditions, social and cultural conditions, the milieu, and influences affecting an individual. Thus, the terms "social milieu," "ambience," and "environment" have been used interchangeably in language (Spitzer, 1942, see Figure 1.5). Similarly, Carlyle (1831) employed the term "the environment" in a phrase to describe Bayreuth: "The whole habitation and environment looked overtrim and gay" (Carlyle, 1831, quoted according to Spitzer, 1942, p. 205).

1.3.1 From environment to total environment

The term environment had a "vague and intangible reference at the beginning" in the context of "biologico–sociological" discussions (Spitzer, 1942, p. 176). These discussions among biologists, psychologists and sociologists scrutinized the way in which the technical term of science "biologico–sociological" evolved into the notion of the term environment. Spitzer was not overly fond of Carlyle's translation of Goethe (1749–1832) and noted that the term environment took on new meaning during scientific discourse of pre-evolutionary topics. For instance, Jean-Baptiste Lamarck, a zoologist who invented the term "biology" by subsuming morphology, physiology and psychology (Ballauff, 1971), proposed that evolution has its origin in the loss or development of characteristics depending on the intensity of use of these organs stipulated by the milieu and circumstances. Lamarck's idea that an individual is adapting (see Box 4.7) was later replaced by Darwin's theory of species adaptation through the interplay of variation and extinction (see Chapter 4). Historically, in this context, the term environment

> took a new power ... in 1855, when Herbert Spencer, who coined the phrase of survival of the fittest, published his *Principles of Psychology* ... and his pre-Darwinian theory of evolution. We can see "environment" shifting in Part 3 of the book ... Spencer takes us from [a] single organism

---

Box 1.5 The double environment: Ossianic dreams

The first use in English of the term "environment" was by the Scottish poet and freethinker Thomas Carlyle in 1827. He created the term when translating desires and passions as they had been presented in Ossianic landscapes, which exuded a character of romantic idealism (see Figure 1.5).

From a semantic perspective, the term environment has several predecessors, such as circumstance, setting, ambient, milieu, milieu ambient, "tout ensemble," etc. (see Figure 1.6). When dealing with the etymology of many languages (e.g. Italian, English, French, German), the thorough historic analysis of Spitzer (1942) reveals that all of these terms were related to a physical, material, and geometrical (locus) meaning as well as to a social, immaterial notion. For instance, the term "milieu" (middle of a place, "lieu" being French for place) referred to the Latin *medius locus* (i.e. middle) and has been used by the mathematician Pascal to argue for the golden section. But all these terms, just as the later term environment, also had social, moral, and epistemic notions not substantiated by the ideal of a material–biophysical environmental rationale.

Figure 1.5 The dream of Ossian (painted 1813 by Ingres).

1 What knowledge about what environment?

Box 1.5 (cont.)

1225	*Circumstance:* "conditions and accompanying an event"; "surrounding conditions"; meaning a "persons's surrounding environment is from 1340"
1340	*Circumstance:* meaning "a person's surrounding environment"
1375	*Setting:* "fact or action of being set or setting"
1423	*Surround:* "to flood, overflow", "to shut in from all sides"; the first record of surroundings in the sense of environment is from 1861; **surroundings** 1877 from Fr., "middle, medium, mean"
1596	*Ambient:* "surrounding, encircling", "encircling, lying around"
1603	*Environment:* "state of being environed"
1827	*Environment* first recorded by **Carlyle** when translating **Goethe's** phrase on dynamic spiritual and inspiring settings
1855	*Environment:* "as the whole corresponding to the environment" **(Spencer)**
1866	*Ecology (Haeckel):* "as total science of the relationship of the organism with the outer world"
1921	*Ecology* differentiates **surrounding world** (German Umgebung) and **memorized world** and an effect world (i.e what has effect on an organism (von **Uexküll**)
1921	*Human/urban ecology (Park and Burgess; i.e, the sociological track):* Cities, built environments as well as nature for animals (accommodation needed)

t

Figure 1.6 First steps in the development of the concept of the "environment."

adapting, or "corresponding" to its own particular environment on to "life" as a whole corresponding to "the environment." Chapter 15 begins by saying that "all vital phenomena are directly or indirectly in correspondence with phenomena in the environment. (Owen, 2007, p. A15)

This latter concept is close to what we today call the material or biophysical environment. By material environment we refer to the complex of life-supporting physical, chemical, biotic, and technological factors that act upon human systems. Other labels that have historically been used for the material environment are "substance ambiente," "extra-organic milieu exterieur," "biological milieu," and "milieu ambient" (Spitzer, 1942).

1.3.2 From ecology to full ecology

Clearly, the exact notions of these concepts depend on the philosophical position, the disciplinary perspective, and the functions of the human system that are investigated. In more recent times, variants of human ecology introduced concepts such as nature (Boyden, 2001), resources (Ehrlich & Ehrlich, 1970) or ecosystems (Udo de Haes & Klijn, 1994). This book takes a full ecology perspective and not a natural or partial ecology view. Natural ecology integrates the biological, chemical, and physical aspects of the study of various plant and animal populations and communities in their environment. Human activities are traditionally kept separate. A full ecology approach considers human beings not only as a component but as the driving factor of the dynamics of every human use system (Bonnes & Bonaiuto, 2002). Full ecology acknowledges that today almost no biotic ecosystem can be investigated without integrating human activities (Yorque *et al.*, 2002). Within the full ecology approach, we study how the rationales behind different human systems interact with their environment. It is interesting to see that even biological ecology follows this view when acknowledging that dynamics of ecosystem patterns, structures, and processes cannot be understood without considering the human systems dynamics and rationales (Ellis & Ramankutty, 2008; Machlis *et al.*, 1997; Pickett & Cadenasso, 2008; Pickett *et al.*, 2001). It is worth noting that historic and anthropological studies show that world dynamics are characterized by a shift from nature-dominated to human-dominated impacts on ecology (Messerli *et al.*, 2000).

1.3.3 Key messages

- The term "environment" is less than 400 years old. The English term of environment resulted from Carlyle's translation error of a text by Goethe. The concept of "total environment" elaborated in Spencer's book *Principles of Psychology* (Spencer, 1855) refers to the concept of material–biophysical environment in this book

- A full ecology approach considers human beings not only as a component but as a key driving factor of the dynamics of every HES
- We have to reflect that the human species has become a major driver in world ecology.

1.4 Environmental literacy: from average knowledge to sciences and human systems' capabilities

The English term "illiteracy" dates back to 1660, appearing earlier than the term "literacy," which was itself first recorded in 1883 (Harper, 2001). The terms environment and literacy were first coupled in June 1968 by the US educational scientist Charles E. Roth (1968). Roth's paper dealt with "environmental illiterates" who were polluting the environment. Again, the negative antonym "environmental illiteracy" preceded the positive "environmental literacy" (Disinger & Roth, 2003; Roth, 1992).

We have provided a first definition of environmental literacy above. In general, environmental literacy is linked to the capacity to perceive, appropriately interpret, and value the specific state, dynamics, and potential of the environmental systems as well as to take appropriate action to maintain, restore, or improve these states (Golley, 1998; Roth, 1992). Thus, environmental literacy is based on the human systems' understanding of natural systems, the perception and valuation of human impacts on the environment, and the establishment of appropriate human–environment interactions.

In the frame of sustainable development, environmental literacy has also become an important educational issue in junior high and high schools (National Science Teachers Association (NSTA), 2007) and at university level (Rowe, 2002). Here environmental literacy is often primarily related to human health and ecosystem health. It is defined as:

> [Environmental literacy provides] … a basic understanding of the concepts and knowledge of the issues and information relevant to the health and sustainability of the environment as well as environmental issues related to human health. (Wolfe, 2001, p. 302)

Interestingly, educational researchers have noted that like other abstract nouns – such as "freedom," "justice," and "equality" – "literacy" refers to a value of human systems of a higher level, that is states and nations, for instance, not just individuals (Michaels & O'Connor, 1990). Some educational scientists took the average citizen as a representative of a nation's environmental literacy. This perspective is the subject of the book *Environmental Literacy in America* (Coyle, 2005), which summarizes 10 years of Roper research surveys. The Roper research surveys were formerly known as Times Mirror Magazines National Environmental Forum. The survey includes a cross-section of people in the USA. It focuses on the public understanding and the public encyclopedic knowledge of the origins of pollution and energy, the impact of human activities, and the functions of ecosystems such as wetlands. The Roper survey documents an irritating illiteracy about factual environmental knowledge:

> The NEETF/Roper (1988) found … that 45 million Americans think the ocean is a source of drinking water (Coyle, 2005, p. 20).
>
> Only 27% of Americans know that most of electricity (some 60% of all electricity) is produced by burning coal and other flammable materials. … But some 40% of people think that hydroelectric power is America's top source of energy (in reality it accounts for about 10% of the total). (Coyle, 2005, p. 27)

This book takes a different perspective. We conceive environmental literacy of society as the capability of an assembly of subunits of a society to jointly generate appropriate knowledge, proper strategies, and reasonable adaptations to changing environmental conditions and adequately anticipate unwanted effects of human actions on the environment. To achieve this capability, societies usually create textures of institutions, organizations, social subgroups, activities, and regulatory mechanisms to sustain themselves. A society consists of an interacting communality of politically autonomous people who are socially, socioculturally and economically related to and dependent on the material environment (Lenski *et al.*, 1991). Thus, environmental literacy must be seen as a fundamental aptitude of society and all its subsystems.

1.4.1 Environmental literacy and key components of society

We take specific views on the relation between science and society. From a macro-sociological perspective, societies have different principal components. These are the economic system, the political and legal system, the social and cultural order system, and the scientific and educational system (see Figure 1.7*). This perspective has been adopted by the sociologist Robert K. Merton in his paper "Science, technology and society in seventeenth century England" (Merton, 1938). The

Society

- Economic system
- Policy and legal system
- Social and cultural order system
- Scientific and educational system

Figure 1.7* Principal components of society.

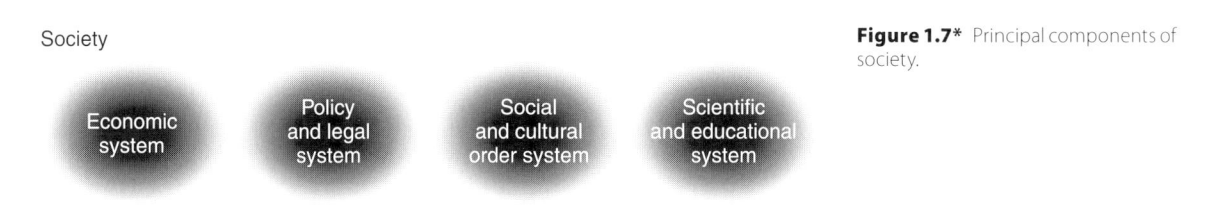

view that science is a subsystem of society is also taken when considering society as a specific hierarchy level of human or social systems. We also consider the science system complementary with other parts of society, as we deem science to be extraordinarily important for society's environmental literacy.

The relation and interaction between knowledge developed in science and in society is discussed further in the sections of this book that focus on transdisciplinarity (see Chapter 15). We particularly elaborate on what contributions and types of knowledge are developed in different scientific disciplines, how this knowledge is utilized, and how it interacts with knowledge and epistemics from business practice or other fields of society.

Given our definition of environmental literacy, sciences are of interest from at least two perspectives. On the one hand, sciences are the key pillars that equip society with abstract knowledge about how the environment works. On the other hand, sciences can also contribute to explaining how human and environmental systems interact and how environmental literacy emerges. We will see that we can identify some scientific theories which can explain why certain individuals, politicians, companies, or societies react to certain environmental issues or changes. However, we also meet approaches, in particular in sociology and economics, in which the material environment does not play any significant role. An important contribution of this book is that it screens some salient sciences for dealing with the social, economic, and cognitive–motivational side of human systems. This provides a deeper understanding of the drivers, perceptions, and behavior of human systems in different environmental settings.

Thus, for instance, the biology part also deals with the ethnobiologists' question of how, and in what ways, human societies perceive and use nature (the biophysical environment). From an environmental literacy view, biological knowledge is a key segment of the "appropriation of nature" (Ingold, 1986). Biological knowledge is also needed to answer the question whether, and how, to double the food supply in a world of limited natural resources (Tillman et al., 2002).

A key question dealt with in Parts III–V is: what drivers are inherent in human individuals, groups, societies, or economic agents? We consider industrial ecology as an important branch of engineering science that analyzes how companies and industries become aware of the material–biophysical environment. Industry and agriculture makes humans into a major biogeochemical agent while facing the challenge of reading and understanding global biogeochemical cycles (which are the grand nutrient cycles such as carbon, nitrogen, phosphorus, and sulfur; Rauch & Pacyna, 2009; Vitousek et al., 1997) or how human societies digest the *Dining at the Periodic Table* (Johnson et al., 2007).

To develop environmental literacy, Chapters 16 and 17 present a decision theoretical framework – the HES framework that allows us to utilize the presented disciplinary knowledge to analyze how environmentally literate human systems arrange human–environment interactions.

1.4.2 Key messages

- Environmental literacy goes beyond scholarly learning of an individual or the knowledge of the average citizen. Environmental literacy can be seen as the capability of human systems to appropriately read, utilize, and adapt to environmental information, resources, and system dynamics. It must be seen as a fundamental aptitude of society and all its subsystems
- There are four major components of society: the economic, policy and legal, social and cultural, and the scientific and educational systems
- Scientific systems serve multiple functions for environmental literacy. Sciences serve as key pillars that equip society with abstract knowledge about how the environment works. However, sciences can also contribute by explaining how human and environmental systems interact, how environmental awareness emerges, and which drivers and learning or decision processes underlie actions of different human systems.

Part I Invention of the environment

Chapter 2

From environmental literacy to transdisciplinarity

2.1 What does environmental literacy comprise? 15
2.2 Relating factual science with systemic knowledge 17
2.3 Conflicts between values and interests on environmental resources 19
2.4 Understanding environmental feedback loops in human–environment interactions 21
2.5 Knowledge integration as a prerequisite for environmental literacy 22
2.6 Transdisciplinarity and science–society cooperation in the frame of sustainable development 26

Chapter overview

This chapter defines environmental literacy and how it is related to information acquisition, becoming aware of environmental impacts, identifying possible changed actions (mitigation), anticipating rebound effects from changed actions, and becoming capable of forming a sustainable coupling of human and environmental systems.

We identify what type of epistemics (experiencing, understanding, conceptualizing, and explaining) and types of knowledge integration are at work in environmental literacy. Transdisciplinarity is briefly defined and seen as a central means of societal capacity building for sustainable transitions.

2.1 What does environmental literacy comprise?

Environmental literacy in science and society unfolds when human systems incorporate environmental aspects appropriately in their decisions and actions. Based on the definition of environmental literacy as the ability to appropriately read, utilize and to anticipate rebound effects or to adapt to environmental resources, system dynamics and information, we examine five key consecutive levels of questions for developing environmental literacy.

1. **What do/can we perceive? What is the problem in what system?** Human systems must be aware and be able to perceive, sample or acquire information from the material–biophysical environment (see Box 1.1). To benefit from experiences and use the information acquired during these experiences, human systems need to represent, store, and retrieve this information so that it is useful for successfully coping with future situations. However, perception and environmental awareness also includes not only a cognitive but a motivational component. As we show, perceiving, representing, and storing, and also the type of motivational incentives, vary between different human systems. These aspects, the competence and the willingness to perceive, define and frame an environmental problem.

2. **How do we understand human impacts on the environment and environmental dynamics?** In this step, developing an appropriate understanding of the human impacts on the environment and of the environmental feedbacks is required (see Box 1.2). Here, a critical issue is the appropriate understanding of the benefits (i.e. the utility or values) that are at work in environmental discourse. Depending on the knowledge and interests behind human actions (for instance, warfare interests), the environmental issues resulting from these actions may be completely ignored. This requires identification of the drivers of human

systems, in particular those which neglect environmental awareness.

3. **What manmade environmental impacts can/should we change?** Demands and values related to the environment emerge in certain historic, situational, and cultural settings. An historic example is forest protection (see Box 1.3; Schmithüsen, 2008). The emergence of forest protection (that is, the cradle of the idea of sustainability) was motivated by a severe shortage of wood in eighteenth-century Central Europe owing to high demands from housing, brewing, mining, and smelting (see Box 2.1). As we discuss in the third section of this chapter, human demands on environmental resources are often a matter of conflict between different human systems.

4. **What rebound effects can emerge from intervening?** Learning what we can have the day after tomorrow, if we take a certain amount today, is vital to sustainable human–environment relationships. We deal with this aspect as feedback loop learning (see Figure 16.12*) and consider this aptitude as an essential part of sustainability learning. As intervening is usually related to limited resources and benefits for certain stakeholders, diagnosing and coping with the different types of conflicts that may result is part of planning interventions.

5. **How can society generate the proper knowledge and strategies for sustaining untroubled human development?** A critical question is how human systems, in particular society, have to be organized to sufficiently grasp the mechanisms of changes of the material environment, the feedback loops between human and environmental systems (see Boxes 16.9 and 16.13), and the necessary mitigations and adaptations to be performed? Coping with these issues includes making this knowledge accessible to decision-makers and establishing interdisciplinarity and transdisciplinarity.

2.1.1 Environmental awareness

According to the functionalist perspective taken in this book (variants of functionalism are presented in Chapter 3), decisions or actions of human systems such as individuals, companies, institutions, nations or societies are made in relation to goals and benefits of human systems. In its simplest form, we have a direct causal link: if humans are hungry, they must know where to find food and how to get it. This also holds true on a regional, national or global level. At the same time, human action affects and changes the environment, and humans as a species tend rapidly to overtax any available natural resource, which can become a critical aspect of sustainable development (Haberl et al., 2002; Imhoff et al., 2004). Environmental literacy must therefore include appropriately assessing, anticipating, and coping with the effects of decisions or actions concerning the environment. We call this aspect environmental awareness. As elaborated in the Awareness Postulate P6 in Chapter 16, we can differentiate between a complete ignorance of environmental impacts, awareness of the impacts and changes resulting from action, and the awareness of the feedbacks to human systems. Environmental awareness is a key aptitude which, if it is fully developed, includes adequately anticipating environmental change and feedback resulting from human action. This holds true for both the material and the social environment.

However, environmental literacy also requires a motivational component (see Chapter 16.7). Human systems are most likely to adapt successfully to their environment if they are exposed to the right incentives and to the right environmental information.

2.1.2 Key messages

- Environmental literacy is a composite of appropriately: (1) perceiving, recording, monitoring, storing, and retrieving information about relevant environmental aspects; (2) identifying and anticipating causes for environmental impacts, dynamics, changes, and feedbacks; (3) identifying the actions of human systems which might become objects of change; (4) identifying rebound effects and trade-offs that might result from interventions; (5) efficiently organizing knowledge for coping with the forthcoming key environmental problems
- Environmental literacy is based on environmental awareness, which includes a motivational and a cognitive component. Fully developed environmental awareness requires adequately anticipating environmental change and feedbacks resulting from human behavior.

2.2 Relating factual science with systemic knowledge

Box 2.1 Origins of a sustainable relationship with the environment: mining and brewing

One should consider that specific functions of the environment, which became evident after critical human–environment interaction, were discussed long before Spencer's definition of environment. This is illustrated by the emergence of the term *sustainability*. This concept is credited to von Carlowitz (1645–1714), a Saxon "mining captain" in the silver mining business. He was concerned with "how to accomplish conservation and grow wood such that a continuous, steady and sustainable use can become possible" (von Carlowitz, 1713, p. 115–16, translated by Scholz). As the subtitle of his book *Sylvicultura Oeconomica* reveals (see Figure 2.1), he provided guidelines for overcoming the dramatic wood shortage for "housing, brewing, mining, and smelting". Thus, rules for preventing the overuse of natural resources and for efficiently utilizing the forest ecosystem's supply of wood for business purposes are far from new. Since the eighteenth century, von Carlowitz' idea of sustained yield forestry has become a basic principle of forestry science worldwide.

Von Carlowitz' forestry reference book *Sylvicultura Oeconomica* is based on a docile Lutheran rationale, basically a social reference system. Von Carlowitz refers to the nescience of people, who are only ignorant "renters of the world" and who behave "such as children of tender age who do not know where bread, food, clothes or livelihood comes from." Man is facing unwanted consequences if we do not "study the course of nature" (von Carlowitz, 1713, p. 399, translated by Scholz). He argues that common man would deplete young trees because he supposes that he does not benefit from them. Such behavior indicates that man follows the idea of wasteful commerce. As men tend to overuse, von Carlowitz concludes that "God's imposition will be that we will perform the hard and tedious reforestation" (ibid.).

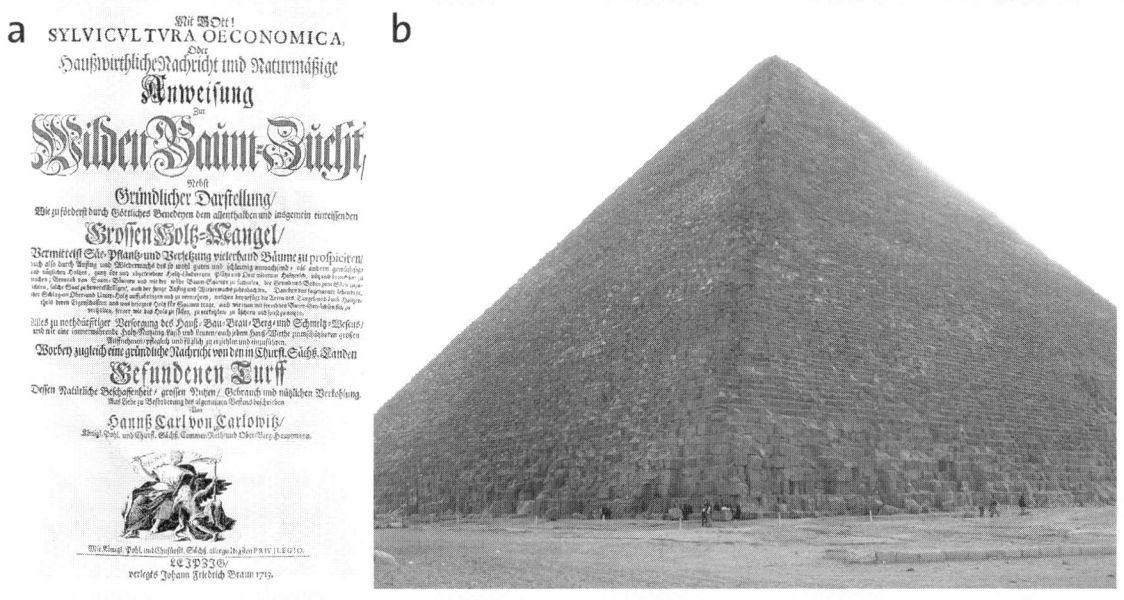

Figure 2.1 (a) Title page of *Sylvicultura Oeconomica* and (b) the Great Pyramid of Giza. Mining and brewing have been industries that ask for a steady flow of wood to sustain themselves. In 1640, a German brewery had an average timber consumption of 20–30 ha forest per year, which is about 4–6 times the base of the Great Pyramid of Giza (photo by D. Csörföly).

2.2 Relating factual science with systemic knowledge

Concerns about human domination and the rapid changes of global and local patterns of biodiversity or natural diversity have become a matter of scientific and public dispute for the past three decades (Kim & Byrne, 2006). Let us consider two quotes that require some material–biophysical and natural system environmental literacy. The first is attributed to Albert Einstein and the second to Edward O. Wilson, a pioneer in research on biodiversity.

> **Einstein**: "If the bee disappeared off the surface of the globe then man would only have four years of life left. No more bees, no more pollination, no more plants, no more animals, no more man."

Wilson: "If we would have only half of the species of today, the world would not die. It would be a less varied and interesting place to be."

A critical question is whether these statements are correct. The reader presumably notices that both quotes are overstated, sensationalist and, among themselves, discordant. In addition, both quotes can be used as flagship arguments for or against environmental conservation in political discourses. This is of interest, as sensitivity about the ecosystem has increased dramatically in recent decades (Yearley, 1991) and therefore biodiversity has become a top, albeit contested, research topic in biology and environmental conservation policy. Factually, Einstein's quote became a strategic argument in painting grim scenarios in the 2007 debate over the mysterious decline of honeybees. However, two basic scientific facts reveal that Einstein's statement is a fake which must have been constructed (see Box 2.2).

In both examples concerning biodiversity, environmental literacy demands both factual and systemic knowledge. Is it possible that the honeybee is the key species in the world's food chain? Historical environmental science literacy leads us to refer to the fact that there were no honeybees in the USA before 1450 (Smith, 1991). Biology tells us that ants, wasps, beetles, and hummingbirds, as well as various other insects and animals, pollinate flowers, and that many plants are pollinated by the wind. Thus, we might argue that we can contest the apocalyptic bee decline vision with two simple facts. Box 2.2, however, shows that the

---

Box 2.2 The literacy of environmental system vulnerability: Einstein's honeybees

"Bees vanish; scientists race for reasons" (Barrionuevo, 2007). "Are GM [genetically modified] crops killing bees?" (Latsch, 2007). "Are mobile phones wiping out our bees? Scientists claim radiation from handsets are to blame for mysterious 'colony collapse' of bees" (Lean & Shawcross, 2007). In 2007, newspapers and magazines such as *The New York Times*, *Der Spiegel*, and *The Independent* reported on colony collapse disorder in the USA, Germany, Switzerland, Spain, Portugal, and Greece (Lean & Shawcross, 2007).

Is it possible that the honeybee is the key species in the world's food chain? Assessing whether there is any evidence to support the assertions from these headlines requires a review of both systemic and factual knowledge of environmental literacy: (1) making a precise assessment of the extent of the losses; (2) explaining the causes; (3) describing the consequences and appropriate counteractions; (4) anticipating rebounds; and (5) adapting to the new situation.

(1) In April 2007, "more than a quarter of the country's 2.4 million bee colonies have been lost" in the USA (Lean & Shawcross, 2007). Others reported: "The West Coast is thought to have lost 60% of its commercial bee population, with 70% missing on the East Coast" (Barrionuevo, 2007). Although the reliability of these statements remains in question, one could obviously observe a colony collapse disorder (Vanengelsdorp et al., 2008).

(2) The reported potential causes included: the presence of parasitic mites, viral diseases, multiple, possibly unknown microorganisms, new types of pesticides (from the neonicotinoids group, for example), and mobile phone radiation that prevented bees from finding their way back to their hives. However, the media that described how radiation was actually causing environmental problems lacked, when published (Harst et al., 2006), grounding in scientific evidence and was based on a small pilot study on radiation conducted at an Educational Informatics unit of a small German university. Although this pilot study attracted major media coverage, it did not report any significant results, lacked contextual information (e.g. weather conditions), and was published in an online journal beyond the outer fringes of academic publications. Further, biophysical literacy shows that there were no honeybees in the USA before 1450 (Smith, 1991) and that many insects, hummingbirds, and other animals pollinate flowers.

(3) When reflecting on whether severe interventions are necessary when facing a honeybee decline, the fact that many plants are pollinated by other organisms

Figure 2.2 Albert Einstein, neither an entomologist nor a biologist, offered some environmentally literate insights about how humans depend upon nature. Media attention, perhaps inaccurate and sensationalist, focused on a statement about how humans would only continue to exist for 4 years after bees vanish.

or wind may alter first impressions. However, one should also take into account that US agriculture has fundamentally changed local ecosystems. Thus, when beekeepers were experiencing huge losses, the US National Academy of Agriculture questioned whether America was relying too heavily on one type of pollinator (Barrionuevo, 2007). Thus, proper counteractions require analysis of both the specific causes related to bee decline, an analysis of the structure of agroindustry, which possibly supported a honeybee or humble-bee decline (Goulson *et al.*, 2008), and the analysis of the vulnerability of world food production in the case of large-scale colony collapse.

Clearly (4) and (5), and appropriate counteractions and adaptations require a clear diagnosis, and thus we cannot fully report on this aspect for this example. However, it is evident that such situations need a new type of systemic research of human–environment systems (HES) combining discipline-based and system knowledge.

The origin of the Einstein quote "If the bee disappeared off the surface of the globe then man would only have four years of life left" is a mystery and has never been authenticated (Figure 2.2). The English version can be traced back to a 1994 Associated Press article of Paul Ames who refers to a "pamphlet of the National Union for French Apiculture" (Ames, 1994, January 24). Thus, presumably, Einstein did not sagely foresee the environmental crisis that was suspected in early 2007.

phenomenon of honeybee colony disorder may require a second look. Many agricultural crops worldwide are pollinated by specialized bee breeding. Thus, doubtlessly the issue deserves a look from a systemic perspective when assessing the economic and food supply vulnerabilities that might be linked to industrialized honeybee-based agricultural production. Evidently, this problem can serve as an example for the interrelationships of human shaped large-scale ecosystems and the dependency of human societies on a resilient functioning of these systems. Chapters 16 and 17 provide the HES framework, which facilitates adequately coping with the complexity and the systemic nature of this problem.

Still, the authentic quote by E. O. Wilson (personal communication, June 13, 2007) is tricky. It has been estimated that 99.9% of all species that have ever existed on Earth have disappeared (Raup, 1991). Paleontology shows that Earth went through extinction periods in which 90% of all species were wiped out. According to recent knowledge, 20 000–30 000 species become extinct annually (Wilson, 1992b). Given global estimates of 13–15 million species (Wright & Nebel, 2005) and using current mathematical models that integrate how speciation, extinction, dispersal, and ecological drift interact (Hubbell, 2001; McGill, 2003; Volkov *et al.*, 2003), one could expect that we would have half the number of today's species by the year 2100 (Wilson, 2002). Life on Earth may go on, but the organismic environment may become less interesting.

Clearly, the questions linked with the Einstein and Wilson statements are not exclusively of academic interest. Environmental literacy does not only involve scientific recording, monitoring, modeling, testing, and appropriately assessing and anticipating the bioecological potential of systems (Scholz & Tietje, 2002). Given the goal of maintaining vital resources and a stable and resilient world, science knowledge must be adequately incorporated in the strategies of practitioners, businessmen, politicians, etc. This leads us to transdisciplinarity.

2.2.1 Key messages

- Factual disciplinary knowledge about cause–impact relationships such as species decline caused by human impacts is a necessary prerequisite for an adequate appraisal of (human-caused) environmental change. Environmental literacy requires integrating this knowledge with systemic knowledge about human–environment interactions to assess its relevance.

2.3 Conflicts between values and interests on environmental resources

The above Wilson quote also includes a remark about the human interest, which is part of what we call the socio-epistemic dimension of human systems and which are based on human values. Thus environmental literacy should not only involve awareness of humans' impact on the environment but also, to some extent, awareness of the myriad of human interests and values that often conflict with one another. This is within the workings of the societal processes related to human–environment interactions. Human systems such as individuals, companies, and nations compete for limited resources. Theoretically, we can distinguish conflicts of human systems *within* hierarchy levels (e.g. between companies) from conflicts *between* hierarchy

2 From environmental literacy to transdisciplinarity

Box 2.3 Conflicting incentives on the macro and the micro level: population growth

In 1968, the biologist Garrett Hardin (1968) provided a game theoretical view of Thomas Robert Malthus' (1766–1834) population theory (Malthus, 1798). Malthus thought that population would outstrip food supply because of the geometric (exponential) growth of the population (e.g. if every couple had four children or more) but only arithmetic (linear) growth of food production. Malthus assumed that only war, epidemics (see Box 8.1), famine or moral restraint, such as late marriage or fraternal polygamism (see Figure 5.20) could hold back a population growth catastrophe. There are only a few recent examples of successfully managing the unhampered population growth in developing countries. However, research revealed, for instance, that traditional Chinese society had regulative demographic behavior since 1979, diverging from the Malthusian model (Lee & Feng, 1999). This is considered successful population policy for nearly one-quarter of the world's population, with some rebound effects, such as the sex-biased abortions resulting in 46% more male second child births in some rural areas (Zhu *et al.*, 2009).

Hardin considered the population problem a common dilemma for which no technical solution exists. Referring to Bentham's utilitarianism (see Chapter 3, Preferences and utilities), he postulated that each individual is self-absorbed and behaves according to his or her own self-interest. In the case of population growth, Hardin observed that it is rewarding for parents to have many children who can care for them in old age, especially in today's developing countries where there is a high infant mortality rate as well as an absence of a social security system. This, however, produces a tragedy of the commons, because this self-interest results in population growth and resource depletion. Hardin concludes: "… we need to reexamine our individual freedoms to see which ones are defensible" (Hardin, 1968, p. 1244).

Each individual is locked into a system to expand the individual family unit, thereby ruining a world that has limited resources. We should note that Thomas Malthus recognized the limits of procreating. This today is called "maximum sustainable yield," that is, the highest possible rate of use that the system can match with its own knowledge, technology, and resources (Wright & Nebel, 2005).

Hardin discussed a series of regulatory mechanisms. He argued that the possible repercussions of overbreeding (the parents' inability to care adequately for their large number of children) are unlikely to be comprehended. He is very skeptical because to him it "is a mistake to think that we can control the breeding of mankind in the long run by an appeal to conscience" (Hardin, 1968, p. 1246). Thus he targets appropriate societal regulations (see Figure 2.3) and concludes that one should deny the Universal Declaration of Human Rights which describes the family as the natural and fundamental unit of society. This has been similarly expressed by the system theorist Jay Forrester when hinting at the negative feedback loops that are necessary to stabilize population growth: "To try to raise quality of life without intentionally creating compensating pressures to prevent a rise in population density will be self-defeating" (Forrester, 1971a, p. 118).

Figure 2.3 Post 1980s Chinese population policy posters promoted the one-child family. (a) The poster promotes marrying when more mature and has an individualist western style, whereas (b) promotes following the basic national policy.

levels (e.g. between a society's and an individual car driver's view on speed limits). The tragedy of the commons (Hardin, 1968; Ostrom, 1990; see Box 2.3) can be seen as one of the biggest obstacles preventing sustainable development. There are many striking examples from history, such as the difficulty in coping with population dynamics, pollution control, common goods use (see Box 16.8), or the conflicting interests in settlement strategies (see Box 16.5).

Thus, environmental literacy is also knowledge of the interest and drivers of different human systems, and competency in knowing the conflict dynamics that result from interactions between the macro and the micro levels. Here game and decision theory (Axelrod, 1984; Osborne, 2004; Scholz, 1983; von Neumann & Morgenstern, 1944) and psychology provide insights on how malignant conflicts arise and how to deal with such dilemmas. Thus, knowledge on how to find the right policy process, for instance, is part of environmental literacy.

2.3.1 Key messages

- Environmental problems cause conflicts among human systems on the same level. This kind of "commons dilemma" conflict between human systems occurs on the micro and macro level. Thus environmental literacy requires competence to cope properly with these problems stemming from different individual, company, and society interests.

2.4 Understanding environmental feedback loops in human–environment interactions

A challenging issue in coping with the environment is that feedbacks are to be expected which differ in points of time, space, and scale. Some rewards and penalties appear immediately, while others manifest much later and at distant places. This requires insight into the rationale and dynamics of the environment and interpretation of what we call secondary or higher ordered feedback loops (Argyris, 1982; Scholz & Binder, 2004). In Part VIII of this book, we show that coping with secondary feedback loops is an important feature of understanding sustainable development and thus, what we call, sustainability learning. We show that decisions disregarding diachronic, secondary feedback loops are a major source of environmental degradation (see Box 2.4). Thus dealing with secondary feedback loops or rebound effects is essential, as no one should implement decisions and actions that destroy the ground upon which we stand. This is dealt with in detail in the Feedback Loops Postulate P4 in Chapter 16.

2.4.1 Key messages

- The anticipation of higher ordered feedback loops is fundamental for sustainability learning.

Box 2.4 Does unintended ecological suicide cause societal collapse? Easter's canoes and Greenland's fish

Jared Diamond, known for the Pulitzer Prize-winning *Guns, Germs, and Steel* (Diamond, 1994), is not a conventional scientist. He earned degrees in physiology and biophysics and simultaneously developed careers in evolution, ornithology, and ecology, and went on to become a professor of physiology and geography at UCLA. His book *Collapse* (Diamond, 2005) about how societies choose to fail or succeed, received dozens of scientific reviews from a wide range of disciplines.

The hypothesis of this book is that societies are endangered because of the inability of cultural decision-making and because of the limits of the physical environment. Diamond carefully points out that environmental damage is neither sufficient for, nor a necessary prerequisite of, societal collapse. The environment is included in the majority of the key factors of societal threat, which are: (1) environmental damage, including overpopulation and resource depletion; (2) climate change (see Box 4.10); (3) hostile neighbors; (4) absence of friendly trade partners; and (5) society's response to environmental problems.

He demonstrates his thesis with 12 historical and contemporary case studies, which he considers as natural experiments. The studies examined societies of varying complexity, including those of the Anasazi, Australia, China, Greenland Norse, Haiti, Montana, and Rwanda. Societal collapse is broadly conceived as "a drastic decrease in human population size and/or political/economic/social complexity, over a considerable area for an extended time" (Diamond, 2005, p. 3).

The Polynesian (see Box 1.1) founders of Easter Island trapped themselves after increasing in population and completely depleting their timber supply (see Boxes 1.2, 1.3 and 2.1, see also Figure 2.4) to build the large canoes necessary for harvesting deepwater food resources or exchanging goods with other countries. Another example is the failure

Box 2.4 (cont.)

of the Norse to adapt to the environmental shifts induced by the Little Ice Age in 1500 AD. The settlers, with their Scandinavian tradition and regard of cows as status symbols, maintained their animal husbandry practices instead of turning to a fishing or marine society, which could have mitigated the environmental stress.

Finally, Diamond shows how other contemporary countries succeeded or failed. He takes the example of Rwanda, which is a very densely populated country that has suffered a decline in land acreage per farm household of 0.89 acres in 1988 to 0.72 acres in 1993. According to Diamond, the genocide atrocity in 1994 was not a result of the elite wanting more wealth and power but the conflict was rather a result of people coveting their neighbors' bean patches, cows, and shoes.

Diamond's work is highly appreciated. None the less, some critiques argue, for instance, that institutions might "manage and prevent overuse" (Wiener, 2005, p. 885), that the work results from the choice of idiosyncratic cases (Rothman, 2005) with "marginal environments" (Greer, 2005, p. 441) that embody no truth about the "modern open economics integrated into world trade and politics" (Wiener, 2005, p. 888), or that he did not explicate uncertainties in his causations. We do not think that these critiques can really do much to reduce Diamond's contribution, although extrapolating from the case studies asks for critical reflection. He succeeded in linking cultural and political conditions to environmental conditions, which is an aspect often dismissed by many sociologists and historians who analyze the collapse of societies (see, for example, Tainter, 1988).

Figure 2.4 (a) Monolithic human figures (Mo'ai) carved from rock on Easter Island between the years 1250 and 1500 (photo by A. Urbina). (b) Scenery from Ravnefjeldet, Nanortalik, the southernmost part of Greenland (photo by J. B. Nielsen).

2.5 Knowledge integration as a prerequisite for environmental literacy

When coping with complex, environmental, real-world problems, different types of knowledge integration are required. This section discusses five important types of knowledge integration: (1) interdisciplinarity, or integration of different sciences; (2) integration of different systems; (3) integration of different types of knowledge or epistemics; (4) integration of different interests; and (5) integration or the interrelating of different cultures. These different types of knowledge integration are important for practicing environmental literacy on real-world problems in society when using science to study how human systems cope with environmental problems (see Box 2.5).

2.5.1 Interdisciplinarity and integrated modeling

Different sciences, such as environmental physics, chemistry, and ecology, create valuable knowledge

2.5 Knowledge integration as a prerequisite for environmental literacy

Box 2.5 Knowledge integration as a means of environmental literacy: from disciplines to cultures

Integration means the generation of something new, which goes beyond the summation of its parts. Integration and differentiation are basic complementary concepts in any developmental system-related research.

The most comprehensively investigated and clearly defined complementarity of epistemics (or modes of thought) is the complementarity between intuitive and analytic thinking, as operationalized by Scholz (1987; see Table 7.4; see also Gigerenzer & Murray, 1987; Hammond et al., 1983; and Hogarth, 2001). It is important to acknowledge that neither analytic nor intuitive thinking is better than the other. However, in complex settings, intuition, gut feeling or the intuition of lay persons can be better than analytic reasoning, which often leads one astray (Gigerenzer, 2007; Scholz, 1987).

Besides the five types of knowledge integration mentioned (see Figure 2.5 caption), language could be considered as a sixth important aspect. Language relates to the integration of modes of thought. Basil Bernstein showed that working class linguistic expressions describe concrete, experience-based issues in a narrative way (Bernstein, 1966), integrating emotional and subject matter issues while remaining descriptive. On the contrary, the elaborated code (such as formal language) is syntactically highly structured, context-free, and separated from extraverbal channels.

Methods that also allow for integrating quantitative and qualitative knowledge are offered in the volume *Embedded Case Study Methods* (Scholz & Tietje, 2002).

Figure 2.5 (a) The intersection of α, β, γ sciences (interdisciplinarity) by relating humanities, natural sciences, and social sciences; (b) modes of thought, represented by the left brain and right brain icon (e.g. intuition and analysis as different modes of thought); (c) the functions of different systems (e.g. water, air, and soil); (d) different perspectives or interests (i.e. by mediation at a round or oval table); and (e) cultures (here represented by different religions; clockwise from the top: Jewish Star of David, Hindu Aumkar, Buddhist wheel of Dharma, Islamic star and crescent, Christian cross), representing important types of knowledge integration.

about the environment and human–environment interaction. However, even simple questions, such as the modeling of rain or hail, demand interdisciplinarity (e.g. linking physics and chemistry). While acknowledging that interdisciplinarity is essential and at the core of environmental literacy and environmental sciences, this book contends that disciplinary knowledge is extremely valuable for modeling and understanding the rationale of different human and environmental systems and their interaction. We will see in Chapter 14 that integrated modeling is a helpful tool for linking systems and variables from different disciplines, and for anticipating developments and tipping points.

For instance, environmental psychology explains how individuals perceive, react, and act on the environment, whereas cultural sociology explains how society copes with resources or environmental conditions. Understanding complex environmental settings or environmental behavior in certain cultures ultimately requires integrating the knowledge from these disciplines and relating them to each other. If we define interdisciplinarity as the integration or even fusion of concepts and methods from different disciplines, environmental literacy on a scientific level strongly entails interdisciplinarity.

2.5.2 Integrating systems

The investigation of weather, soil pollution, or climate change, just to take three examples, asks for an integral view of dynamics or changes in the surface soil (e.g. deforestation), water (e.g. changes in ocean currents), and geological and cosmic radiation (e.g. large volcanic eruptions or cycles of solar radiation). Thus, environmental literacy necessitates an integrated view of the environment. Naturally, the same holds true for human systems. One aim of this book is to show how an integrated or holistic view and integration of systems is possible while simultaneously acknowledging the contributions of single disciplines.

Taking an integral, holistic view is often a prerequisite for proper understanding of complex human systems or the material–biophysical system. To answer why a client is ill, a society does not function well, an ecosystem approaches a critical state, or that the world biogeochemical cycles show critical flows requires an integrated, holistic view. An important way of supporting this is integrated modeling (see Chapter 14).

2.5.3 Relating different epistemics

In epistemology and in the philosophy of sciences, there are various suggestions on how to differentiate between the nature and grounds of knowledge. If we consider the level of an individual, one can, for instance, distinguish between an intuitive and an analytic mode of thought that is taken by an individual when coping with an environmental problem or an environmental decision (e.g. when buying a new car). These modes of thought differ in many respects; for instance, whether the processing of information is conscious, purely intellectual, independent of moods, sequential, or goal oriented, just to mention a few (a detailed description of these modes is provided in Table 7.4; Scholz, 1987, pp. 170–84; Scholz & Tietje, 2002, p. 233).

If we look at an epistemic level, one may distinguish between *verstehen* (understanding), *begreifen* (conceptualizing), and *erklären* (explaining). Scholz *et al.* (1997; see Box 2.6) introduced this differentiation to cope with complex environmental problems. The essence of this differentiation is that *verstehen* is characterized by empathy, intuition, and holistic comprehension. In the field of environment this can be attained by experience with complex, concrete systems. *Begreifen* is linked to a conceptual system, whereas *erklären* is linked to abstract systems and propositional logic; that is, by statements and inferences that follow the rules of formal systems. Clearly, this differentiation is linked to the concepts of rationality in Box 1.4, as full rationality is often reduced to the level of *erklären*. These epistemics follow different rationales. There is no general ranking of the strength and validity of the different epistemics. Whether environmental literacy is driven by one or the other or by a combination of these forms depends on the problem.

However, one should note that appropriate conceptualization of real-world problems without understanding is almost unthinkable, and *erfahren* (experiencing) precedes understanding (Kolb, 1984; Wellek, 1953).

Box 2.6 introduces the architecture of knowledge: distinguishing between *erfahren, verstehen, begreifen,* and *erklären*. We want to stress here that the highest level of epistemics, that is, expertise on complex, ill-defined problems, is built both on experience with the real world (real-world experience) and a competent mastery of analytic rules (analytic competence and experience). This would suggest that the proper combination of all levels of epistemics is targeted. This has also been suggested by the pioneers of artificial intelligence, who stress that real expertise goes beyond pure rule-based analytic knowledge and is built on extracting heuristics and working rules from experiential knowledge. Understanding based on expertise allows for reliable identification of the relevant aspects of a problem or system (Dreyfus & Dreyfus, 1986, 2005).

2.5.4 Integrating different interests

Different people, groups, organizations, nations, etc. have different perspectives, values, and preferences about what is critical in the environment or what should be changed. Thus, different humans have different interests that affect what aspects of the environment they look at and how they process the information they acquire. In this way, interests define how humans perceive the world, dictating how they deconstruct and reconstruct their environment. Depending upon their interests they will, for example, employ different languages and disciplines to interpret the world (Werner & Scholz, 2002).

2.5.5 Interrelating different cultures

The environment and various parts of the environment play different roles in different cultures. Customary beliefs, priorities, social forms, and material traits differ, for instance, between religious and social groups. For instance, westerners and East Asians perceive and think about the environment in different ways. Elaborate psychological experiments have shown that East Asians are prone to focus more broadly on the context; that is, the field in which central objects are located. In contrast, westerners are inclined to focus on the central object and to analyze the attributes and rules connected to it. There is experimental evidence that Asians are more likely to notice relationships and changes in a grouping and interplay of subjects (Nisbett & Masuda, 2003).

Originally, the meaning of integration was to create something new from different components. Thus, we should ask whether the term integration should also be used with respect to cultures or whether cultures

2.5 Knowledge integration as a prerequisite for environmental literacy

Box 2.6 An architecture of knowledge for environmental literacy: from case empathy to propositional logic

There are different types of epistemics at work when dealing with complex environmental problems or sustainable transitions of regions, cities, or organizations. This is illustrated by the architecture of knowledge that has been developed in the frame of transdisciplinary case studies (Scholz et al., 1997, p. 27; Scholz & Tietje, 2002, p. 30). These case studies (the top part of Figure 2.6 is an icon of the 15 000 working and residential places in districts of Zurich) include processes taking several months or years of mutual learning between university and case agents. The studies include transdisciplinary processes in which researchers are interested in the case as they want to investigate processes and prerequisites of sustainable development, and collaborate with case agents who are interested in gaining access to options and transition processes towards sustainability (Scholz et al., 2006; see Chapter 15).

At the top of Figure 2.6 is a guiding question that emerges from the priorities, concerns, and interests of both the case agents and scientists. This (value-based) guiding question should be consented to by the case agents and the scientists who cooperate with these practitioners in a transdisciplinary process. The guiding question is based on *verstehen*, a qualitative, experience-based process of understanding the case. As Radtke (1998) and others emphasized, there is no empathy, understanding, and conceptualizing without *erfahren*. Thus, *erfahren* precedes a reasonable definition of the guiding question.

Based on understanding, one can determine key aspects, subsystems, or layers of the case that have to be considered, analyzed, and changed for a sustainable transition. This is the conceptual level of *begreifen* (Wellek, 1953). On this level, the case is faceted or embedded in a conceptual grid which meets the relevant aspects of transition. Then, a qualitative system analysis is needed, which optimally incorporates the system knowledge of both the case agents and the scientists.

Various "embedded case study methods" (Scholz & Tietje, 2002) have been developed to serve different types of knowledge integration. Finally, the bottom level consists of *erklären* – that is, causation and verification of causal relationships that are of significance in establishing transitions on the different levels. An important epistemic on this level is propositional logic (Lambek & Scott, 1986) or methods of verification in a positivist sense (Popper, 1935/2005). A challenge of the "embedded case studies" for sustainable transitions is combining the different levels (see Box 2.5).

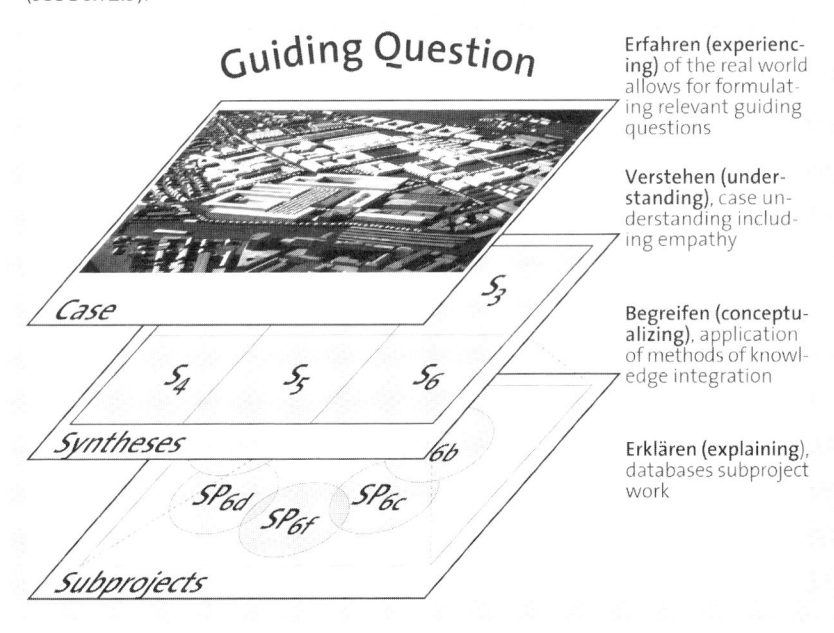

Figure 2.6 The architecture of knowledge presents three levels of epistemics applied in case studies on regional or organizational transitions. A guiding question is derived from a holistic *verstehen*. The medium level must link the qualitative top level of *verstehen* with the analytical level of *erklären*, which often includes quantitative, data-based reasoning (taken from Scholz & Tietje, 2002).

Erfahren (experiencing) of the real world allows for formulating relevant guiding questions

Verstehen (understanding), case understanding including empathy

Begreifen (conceptualizing), application of methods of knowledge integration

Erklären (explaining), databases subproject work

should be set in relation to each other. This has been aptly commented upon by the sociologist Ulrich Beck:

> We must avoid any glorification of the multi-ethnic society. It is less like a melting pot and more like a salad bowl in which cultural identities coexist with and against one another in a colorful way (Beck, 1999, p. 55).

The interrelating of different cultures is particularly important to achieve socially robust knowledge requiring consensus among different stakeholder groups and members of the scientific community. Why certain things happen in the world and why certain means are considered effective or acceptable may differ fundamentally between cultures and religions.

We should note that the same holds true for scientific disciplines. Different sciences use different reference systems; this may hold for the different types of constructing and validating knowledge in the humanities, natural, and social sciences (i.e. α, β, γ sciences). Within these fields, and presumably in each discipline, we find different cultures characterized by different conceptions of what is true and is considered as truth, valid, a theory, and science itself (Pickering, 1992). To express it in other words, different disciplines and different conceptions of sciences and cultures are based on different world views (cosmologies), basic assumptions, epistemics and reference systems.

2.5.6 Key messages

- Environmental literacy requires different types of knowledge integration, such as integrating disciplines, different (environmental) systems, modes of thought or epistemics, different perspectives and interests, and reconstructions of the environment from different cultures
- The understanding of complex, environmental real-world problems is based on experiential learning
- Profound environmental literacy should integrate experience-based (intuitive) understanding, (expertise-based) conceptualization, and (analytic) explaining.

2.6 Transdisciplinarity and science–society cooperation in the frame of sustainable development

In the past 20 years, a specific form of knowledge integration developed in the frame of sustainable development (United Nations (UN), 1987). Sustainable development can be defined as an ongoing inquiry for system limit management in the frame of intergenerational and intragenerational justice (Laws *et al.*, 2004). Environmental issues and the mastery of environmental problems are essential for sustainability. Clearly, sustainability is a normative issue and is highly value-laden. This also holds true for the proposed general definition, in particular for the aspects of what (security) standards are demanded for limit management and justice.

Thus, if environmental scientists want to cope with sustainability issues, they have to make reference to certain values taken from certain perspectives and stakeholders. Making reference to values and goals of different stakeholders is a main issue of transdisciplinarity, which is a new type of theory–practice cooperation:

> We consider transdisciplinarity as a type of scientific activity that supplements the traditional disciplinary and interdisciplinary scientific activities by incorporating processes … knowledge and goals of stakeholders … [in sciences and] organizes processes of mutual learning between science and society. (Scholz *et al.*, 2000, pp. 478–9)

There are different definitions of transdisciplinarity, which are discussed in detail in Chapter 15. Here we refer to the Zurich 2000 definition, which we consider to be significant for the future development of environmental literacy in science and society.

> Transdisciplinarity is a new form of learning and problem solving involving cooperation among different parts of society and academia in order to meet complex challenges of society. Transdisciplinarity research starts from tangible, real-world problems. … Through mutual learning, the knowledge of all participants is enhanced, including local knowledge, scientific knowledge, and the knowledge of concerned industries, businesses and non-governmental organizations (NGOs). (Häberli *et al.*, 2001, p. 7)

The primary goal of transdisciplinarity is to acquire socially robust knowledge (a comprehensive definition is given in Chapter 15; Nowotny *et al.*, 2001). Transdisciplinarity is related to Mode 2 Science (Gibbons *et al.*, 1994), a context-driven approach that focuses on real-world problems, and represents a new way of structuring knowledge (Thompson Klein, 2004). Transdisciplinarity is thus directly related to environmental literacy. In our definition, a solution is called socially robust if it emerges from knowledge integration, attracts consensus, meets state-of-the-art knowledge, and acknowledges the uncertainties inherent in, and constraints of, generating and using knowledge.

Transdisciplinarity, as we understand it, is different from applied sciences and consultancy (Funtowicz & Ravetz, 1993) as it includes joint problem definition and project leadership by members from theory and practice (see Chapter 15). Mutual learning, that is processes of joint problem definition, representation, and transformation, is considered a basic principle of transdisciplinarity (Scholz, 2000). Environmental risks and coping with low-probability and high-impact technological risks are key topics for transdisciplinarity and played a significant role in its emergence, as well as in the cooperation between science and society for the past 30 years. For instance, systemic uncertainties and values held by individuals and other human systems are key aspects of sustainability research. This contradicts the traditional science system.

> Now that the policy issues of risk and the environment present the most urgent problems for science, uncertainty and quality are moving in from the periphery, one might say the shadows, of scientific methodology, to become the central integrating concepts. (Funtowicz & Ravetz, 1993, p. 742)

Transdisciplinarity is also essentially different from (but usually includes) interdisciplinarity. Interdisciplinarity is a scientific activity that includes a fusion of concepts or methods from different disciplines, which is required when answering questions, solving problems or addressing topics that are too broad or complex to be dealt with adequately by a single discipline or profession (Klein & Newell, 1997).

Transdisciplinarity has been applied to complex environmental problems as well as questions of sustainability and has led to new forms of relationships between science and society. Two issues have emerged: the epistemo logical status of outcomes from transdisciplinary processes (Zierhofer & Burger, 2007) and the "adequacy of familiar knowledge-producing institutions" (Gibbons *et al.*, 1994, p. 1).

Critical questions have arisen regarding the epistemological status of statements, findings, and conclusions about the environment, its dynamics, resilience, quality, etc. These are generated in a transdisciplinary discourse or are derived from transdisciplinary research (based on a transdisciplinary process) and are discussed further in Chapter 15. What validity do such statements have? What criteria of scientific validity can or should be applied? Besides validity, other societal validation criteria also arise:

> In such participatory science, the goal is no longer truth per se, but responsible public decision making based upon understanding of complex situations where many key uncertainties remain to be resolved. (Gibbons *et al.*, 1994, p. 148)

Thus, the content of this book is not only of scientific interest. Although it is a book of theory, it also includes a pragmatic perspective. Transdisciplinary processes can also have an impact on changing the real world. We can consider it as a method for organizing sustainability learning in the abstract and the concrete.

An important institutional issue is linked to the changing role of universities and their internal structure, including the role of disciplines and funding.

> The modern university has become a hybrid institution, with multiple and sometimes incommensurable missions. (Scott, 2007, p. 214)

This is a critical issue both with respect to the role and function that universities will and should play in society. This includes defining what environmental problems are dealt with, in what manner, for whose interest, and under what goals. The university, as a cradle of science, has undergone major transformations. Universities are the "second oldest institution with a continuous history in the Western world" after the Roman Catholic Church (Rosenthal & Wittrock, 1993, p. 1; see Box 15.3). The first medieval university in Europe, Bologna (1088), was an "alma mater studiorum" (Latin for "fostering mother of studies," ibid.). Its purpose was to master and communicate existing knowledge. Dissertations consisted of learning a text and defending it in public. The role of the university was the preservation and transmission of knowledge.

This changed in the first academic revolution of the late nineteenth century when research became an accepted academic task (Etzkowitz, 2001). One new task taken on by universities was Wilhelm von Humboldt's conception of *Allgemeinbildung* (general science education). Here the focus was on understanding the general principles of electricity, engines, chemical processes, etc. Another primary goal of the institution was "general education of technically competent managers who both understand and can oversee the practical development and the long-term effects of the processes of industrial society" (Scholz *et al.*, 2004, p. 26).

Researchers have postulated an ongoing, second academic revolution for the past two decades. Some speculate this revolution will lead to a "triple helix" (Leydesdorff & Etzkowitz, 1996) of intersections between industry, government, and academia. This combination would result in a "third mission" of direct contributions to industry, which would make participation in the progress of economic development into a core value (Gibbons, 1999). This is based on observations that in some countries, such as the Netherlands, "professors who are researchers of high caliber in natural sciences and engineering are also consultants for industry" (Gibbons *et al.*, 1994, p. 141). The role that environmental knowledge and literacy will play and the ways in which the public or even future generations will be considered are still open issues. A critical question to be dealt with in this book is whether, as with the emergence of new roles for academia during the Industrial Age, twenty-first-century society is challenging universities to acquire and to transmit knowledge on critical problems of HES such as resource depletion, land use, water availability, biodiversity, climate change, etc. This issue is discussed in Chapter 15.

Whereas the institutional issue is related to transdisciplinarity, from a science perspective, the inner organizational issues are linked to interdisciplinarity. The question is what kind of knowledge the different disciplines, as well as interdisciplinary research, have produced, and should develop, conserve, and transmit. As we will elaborate, there have been a huge number of disciplines and interdisciplinary fields, from cybernetics to system theory, which have developed exceedingly valuable knowledge about HES. We postulate that profound knowledge about human–environment research demands an understanding of how human systems cope with the environment. As previously stated, the interface between the micro and the macro level is important in understanding environmental behavior.

2.6.1 Key messages

- Transdisciplinarity can be considered as a method for organizing sustainability learning and attaining socially robust knowledge
- Coping with twenty-first-century environmental problems requires efficient use of scientific knowledge. Thus environmental literacy requires transdisciplinarity, which includes mutual learning and joint problem-solving between science and society (see Chapter 15)
- The future role and institutionalization of universities becomes a decisive factor for developing environmental literacy and for coping with challenging large-scale problems, such as resource depletion, land use, water availability, biodiversity, climate change, etc.

Part I Invention of the environment

Chapter 3 Basic epistemological assumptions

3.1 Overview of epistemological assumptions 29
3.2 The human–environment complementarity 31
3.3 The material–social complementarity 33
3.4 The realist stance 36
3.5 Functionalism 37
3.6 Functional social constructivism in the frame of uncertainties 38
3.7 The assumption of human system hierarchy 38
3.8 Defining relations between environments and hierarchy levels 39

Chapter overview

The function of Chapter 3 is twofold. First, we introduce the epistemological assumptions underlying our exploration of the different disciplines that allow us to study environmental literacy. We assume that the social-epistemic and material–biophysical aspects of human systems and their environments can be viewed from a realist, objective perspective. At the same time, we assume that the realist view is compatible with a functional constructivist perspective.

Second, the chapter lays important groundwork that enables the reader to take in the generic framework for conducting research on human–environment systems (HES). Two complementarities are fundamental components of the blueprint that this book offers for coping with HES. First, the HES complementarity is based on the assumption that human and environmental systems are disjointed but inextricably coupled systems that cannot be well comprehended when considered singularly or disconnectedly. Second, the material–social complementarity, also called the material–biophysical versus social–epistemic or body versus mind complementarity, describes our assumptions about different types of processes within human systems. These complementarities are the building blocks of the HES framework put forward in Chapters 16–19.

3.1 Overview of epistemological assumptions

As holds true for many concepts such as uncertainty, nature, justice or love, the term "environment" is associated with multiple notions. In the following section we clarify what types of environment we are dealing with, whether they are considered real or constructed, and what functions these environments serve for which human systems. The reader will see that formulating a crisp definition of the concept of environment is by no means trivial.

Our discussion of the relationship between human and environmental systems is based on a number of assumptions. The distinction between human systems and the environment is a challenge. We show that the definition of the environment of specific human systems depends on how those human systems are defined. The definition of human systems in this book emerges from the activities of human cells. We conceive the human individual as the cells and the totality of (inter-)activities of those living cells which emerged from the fertilized ovum. This definition and its boundaries are illustrated in Box 3.1. Based on this definition of an individual we define superordinate human systems, such as a group or company, as all activities of the members of the superordinate system that can be conceived of as part of the group activity or company operation.

Box 3.1[1] What is part of me and what is part of the environment? My pig heart valve

The boundary between the human and the environment can be fuzzy. This also holds true for the proposed biological, cell-based definition of human systems. When breathing in oxygen or taking a big bite from a sandwich, there are states of phase transition where one cannot definitely determine whether a molecule is part of the human or the environment.

Although contradicting intuitive everyday knowledge, biologically, the gut flora (intestinal microbiota) are separate biological organisms living in symbiosis with the human organism to maintain health (Sears, 2005). Somewhat similarly, a cardiac pacemaker, a hip joint endoprosthesis, and a mechanical heart valve are all technical facilities. Like glasses, they are part of the environment, yet they are essential to maintain life.

What about heart valves? Some heart valves are produced biologically, whereas others are not. Heart valves produced by tissue engineering can be made autologously from one's own tissue or homologously from other living persons or animal tissue (Bloomfield *et al.*, 1991; Cohn *et al.*, 1989; Goldstein *et al.*, 2000; Schmidt & Hoerstrup, 2006, see Figure 3.1). Heart valves can also be produced from metal and plastics, much like pacemakers.

Deciding whether a pig heart valve is considered to be part of the environment or part of the human within which it resides also depends on how the porcine vessels are connected with the human vessels. Pig heart valves are not connected to any human vessels. Cartilage in general obtains its nutrients via diffusion, which is certainly the case in transplanted pig heart valves. Thus a pig heart valve can be seen, much like a pacemaker, to be a technical device in the environment.

But what about a transplanted organ that comes from another human or from an animal? If one takes a heart from another person (excluding a monozygous relation), it has a different genetic pattern. Thus, biologically it should not be seen as a genuine part of the person. Consider a thought experiment in which half of the cells or organs in one person are implanted from another person shows that there are situations in which it is impossible to decide. People who have a transplanted kidney know that they have another person's organ in them (and the host epithelium grows in the implanted organ). However, psychologically (i.e. from the socio-epistemic perspective), it is considered part of oneself.

Alternatively, consider the relationship between an embryo and its environment, the mother's womb. The embryo shares only half of its genetic material with its mother, yet it is not rejected from the mother because of the unique properties of the placenta. The embryo receives its oxygen via the mother's blood, but – from a certain developmental stage onward – has its own circulatory system. The circulatory systems of mother and fetus come into contact with one another occasionally to provide a route for immunization of the embryo. Again, there are fuzzy boundaries between fetus and mother.

A final example is that of conjoined, or Siamese, twins. There are different types of conjoined twins, ranging from those who can be easily separated to those where surgery results in the death of one or both. In the case of the famous entertainers Lori and George Schappell, separation is not feasible (see Figure 3.1). Although both share about 30% of their brain with their twin, each has developed an individual identity and has her own private space. For instance, George was baptized but also supported her sister's decision not to be baptized.

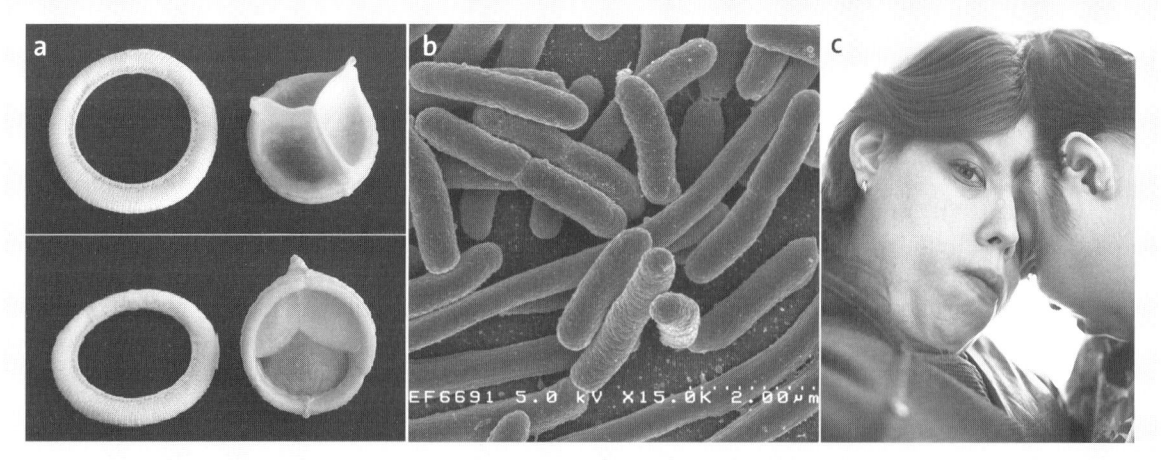

Figure 3.1 Borders between human and environmental systems can be fuzzy. Examples include: (a) a pig heart valve (a replaceable model of a cardiac biological valve prosthesis); (b) the gut flora (scanning electron micrograph of intestinal microbiota *Escherichia coli*; photo by Rocky Mountain Laboratories, NIAID, NIH); or (c) conjoined twins (Lori and George Schappell, © 2011 Stephen Barker All Rights Reserved).

---

[1] This box has been written together with Stefan Zemp.

Table 3.1 Concepts put forth in this book to describe the material and social dimensions of HES.

Basic labeling	Broader labeling	On the level of the individual	Examples
Material dimension	Material–biophysical (also: material–biophysical–technological)	Body	Cells, perceptual organs, and the central nervous system; object, such as a building or a tree, in the environment, materialized technological inventions, such as the steam engine
Social dimension	Social–epistemic (also socio-epistemic or social–cultural–epistemic)	Mind	Human interpretation of an object, mental "sign" for a building or a tree, knowledge necessary to build a steam engine

This chapter introduces the basic epistemological assumptions underlying our study of environmental literacy. Here, epistemology is essential, as it tells us why and how well we "know" what we know. First we examine assumptions in the two complementarity relationships mentioned above. The material–social complementarity is a well known and thoroughly discussed issue in the history of sciences, where it is more frequently referred to as the body–mind complementarity. We prefer to call it the material–social complementarity, as we not only deal with the level of the individual but consider a hierarchy of human systems. Table 3.1 provides examples of the material and social dimensions of HES. We sometimes call the material dimension the material–biophysical, to capture the technological knowledge, processes, and products of HES. Similarly, we sometimes call the social dimension the social–epistemic dimension to capture the why and how of what humans "know." We also put forth the body–mind complementarity, which becomes important when describing lower order levels of this complementarity.

Most of the epistemological assumptions are at work in the following chapters, particularly Chapter 16, which presents seven Postulates that should be considered when launching, conducting or reflecting on environmental research. Also note that the material–social or body–mind complementarity is further discussed in the Biosemiotics section of Chapter 4.12 where the cell is introduced as an entity which "reads the signs of nature."

3.2 The human–environment complementarity

A key assumption in this book is that human and environmental systems are considered to be complementary, coupled systems. The assumption of this complementary relationship is epistemologically relevant and fundamental to environmental literacy. This has been stated in the theoretical ideas of perceptual psychologist Egon Brunswik (1903–55). Brunswik investigated how capable people were of estimating the size of objects presented at different distances and in different situations. Chapter 6.3 offers a detailed description of his approach, called the theory of probabilistic functionalism. Brunswik strongly stressed the necessity of differentiating between the rationale of the internal human, the proximal processes, and the features, context, etc. of a distant body (Hammond, 2001). He argued:

> Both organism and environment will have to be seen as systems. Each with properties of its own ... Each has surface and depth, or overt and covert regions. It follows that, as much as psychology must be concerned with the texture of the organism or of its nervous properties and investigate them in depth, it must also be concerned with the texture of the environment. (Brunswik, 1957, p. 5)

Gibson (1979), another perceptual psychologist, went further and emphasized that no single individual, group, or species could exist without being functionally adapted to the environment surrounding it. He concludes that the individual is prepared to use the environmental information and setting properly.

> For an individual organism, the environment normally comprises three components: the non-living or abiotic world, the world of other species, and the world of conspecies. (Ingold, 1986, p. 3)

We also have to acknowledge that each organism and species, from the protozoon to the human, affects the environment. Today biologists, for instance, estimate that three-quarters of the terrestrial biosphere is altered and formed by the human species (Ellis & Ramankutty, 2008). In another example, Earth scientists consider human activity to be a major geological factor and talk about the anthropocene, the recent era in which human influence constitutes a human-made environment (Crutzen, 2002).

3.2.1 Operationalizing complementarity

Logic offers the most reduced, abstracted form of defining complementarity. Two sets, where one contains all the elements that are not an element of the other, define a complementarity. Thus the positive integers $\{1,2,3,...\}$ can be divided into two complementary sets, such as the even and the odd numbers. Here the elements of the one set are not part of the complementary set. This definition can also be applied to the human–environment complementarity. If a human system H consists of a finite set of atoms and the universe is denoted as U, then the complement of H consists of all atoms of U not belonging to H.

However, the boundary between real human and environmental systems is not that clearly or saliently defined, as the logical definition suggests. For instance, while intestinal flora reside in the human body in a symbiotic, mutualistic relationship with the human organism, they are considered to be separate organisms. Thus they are part of the material–biophysical environment of the individual. Cardiac pacemakers or prostheses, which are part of the technological environment, are examples of abiotic objects from the environment that lie within the human body that are meant to work in harmony with human body systems.

To understand the fuzziness of the boundaries of human and environmental systems more fully, we can turn to the logical, mathematical, theory definition of sets as either extensional or intensional. In an extensional definition, all elements are named or numbered. If we define a subset, it is clear that all names that cannot be found in it belong to the complementary set. In the intensional definition, sets are defined by rules. Here the challenge is to find a rule that allows definition of all parts of the universe and delineates which elements are part of the human system. We think that the biophysical, cell-based definition of a human individual is a proper intensional definition that also allows for defining human–environment complementarities at higher scales.

3.2.2 The material–biophysical dimension of human and environmental systems

We define H_m as the material–biophysical dimension of the human system, and E_m as the material–biophysical-technological dimension of the environment. As proposed, H_m consists of all cells (and the atoms constituting these cells) included in the human body, which emerge from the ovule and the sperm constituting an individual human being and their (inter-)activities. The environment consists of all parts of the universe not belonging to the human body. Here we can look for more constricted definitions such as the immediate (Miller, 1978) or the relevant environment when looking, for example, at those parts of the human universe affecting human life.

We also consider the machines and the built environment as part of the material environment. We call this the materialized technological dimension of the environment. Here the adjective "materialized" indicates that, in the literature, technology sometimes means technological knowledge and sometimes the processes related to generating machines and the built environment. If not explicitly noted or included in quotes, we understand technology to be the materialized dimension and include it under the label "material."

3.2.3 Fuzzy boundaries

As Box 3.1 illustrates, this biologically based definition of the human–environment complementarity has some fuzzy boundaries at the level of the individual. The same holds true if we consider superordinate human systems such as a group or a company (see Chapter 16.3). We define the human system of a company, for instance, by all company members' physical and mental activities (which we denote as the owners' or employees' cells and their interactions) that are assigned as company activities.

3.2.4 Complementarity: a modern construct

The human–environment complementarity assumption is epistemologically essential. We suspect that, historically, it emerged late in human development (see Figure 1.6). Whether or not such a complementarity is accepted depends on the type of world view one takes and, in particular, whether and what role one assigns to supernatural phenomena or forces. In the occidental world a monotheist entity was supposed to have created and to have governed the world. In the European Middle Ages humans thought of Earth as a transit station and nature as an embedded environment pervaded by demons, the counter-forces operating in the world. Thus, the religion–environment relationship presents a special but presumably important aspect of the human–environment interaction (Boersema et al., 2008). Supernatural forces and religion have been, and are still, inherent in the epistemic approach of many scientists as:

> ... first ... historically, modern science is an extrapolation of natural theology and second, that modern technology is at least partly to be explained as an occidental, voluntarist realization of the Christian dogma of man's transcendence of, and rightful mastery over, nature. (White, 1967, p. 1206)

An early, mundane, dualist view comes from the Italian scholar and poet Francesco Petrarca (1304–74), who described the climb up of Mont Ventoux. He stated that his soul is able to be pleased by a purely profane, areligious, aesthetic consideration. The German sociologist Max Weber (1864–1920) considered religion to be a major determinant in forming the structure of societies. But in his famous speech "Sciences as a profession" (German: Wissenschaft als Beruf; Weber, 1919/1984), Weber stated that science has the potential to examine and to explain every phenomenon of our material, biophysical environment, without referring to mystic, irregular, or supernatural systems. Here he introduced the concept of the "disenchantment of the world by science" (German: Die Entzauberung der Welt).

The modern, dualistic view on the human–environment relationship is not shared by all scientists (Popper & Eccles, 1977), nor is it by all cultures. The Maya culture can be seen as an example (see Box 3.2). Some scientists consider the human–environment complementarity to be a major component of the dominant, human-centered western paradigm. They contrast it to world views with ecocentric focus such as Native American, Taoism, Buddhism, deep ecology or eco-theologic approaches (Berkes, 2008; Devall & Sessions, 1999; Gardner & Stern, 2002).

3.3 The material–social complementarity

The material–social or the body–mind complementarity is of a different nature as the human–environment complementarity. We introduce this type of complementarity because the understanding of our environment necessitates going beyond the material. Environmental literacy must include an understanding of the rationale, drivers, intentions, epistemics, and so on, of human systems acting in the environment. This includes the development of symbols and language from at least 40 000 years ago (Takács-Sánta, 2004), which can be considered to be a major transition of environmental literacy.

We should note that the relationship between the body and the mind can be perceived from three distinct perspectives: from the body to the mind, from the mind to the body and from a complementarity view which does not include directionality. The body–mind perspective is aligned with the perspective that any human activity emerges from cell activities. From this perspective, we are interested in how the body or the material–biophysical side of an individual affects the mind, mentality, mental processes, or the socio-epistemic level, which is discussed further in Chapter 5. The mind–body relationship focuses on how ideas can affect the body (Shanon, 2008).

To better define the material–social complementarity, and to understand how this complementarity basically differs from the human–environment complementarity, we look at the wave–corpuscular dichotomy. The wave–corpuscular complementarity (which has also been sometimes called dualism) tries to explain, amongst other things, the nature of light. The physicist Niels Bohr stressed that a complete knowledge of phenomena of atomic components required a description of both wave and particle properties (Bohr, 1931). Both models, the wave and the corpuscular–particle model, represent different attributes of the same entity. Both intuitive (when thinking about water waves and billiard balls) and theoretical (when dealing with energetic and spatial dimensions) approaches deal with different concepts and can explain different properties of the atom. The quantum mechanical equation of motion could be expressed in terms of either a wave ontology or a particle ontology. If the electron is interpreted as a particle, certain phenomena of motion, momentum and the position of the electron at a given time cannot be explained. However, this can be done by modeling the electron as a wave.

> Both approaches turned out to be mathematically equivalent and led to the same results, convincing Heisenberg that the wave and particle representations of the electron were simply different ways of describing the same thing. (Camilleri, 2006, p. 299)

This view has also been called the "theory of complementarity" (Pauli, 1994, p. 7). The wave and particle theories could not be integrated (see Chapter 16.2) and Heisenberg's ontological assumption was that both theories belong to one and the same entity (Camilleri, 2007). However, because they cannot be thought of simultaneously, they are mutually exclusive. Thus we deal with one set of physical objects with complementary theories – that is, complementary epistemics. Only one epistemic can be used at a time. This is similar to the figure–ground relationship in psychology, where only one representation of an entity (i.e. a picture) can be cognized at one point of time (see Figure 16.2), though we are looking at one and the same entity.

In principle the figure–ground complementarity (see Figure 16.2) and the Chinese yin and yang relationship are of the wave–corpuscular type of complementarity as they cannot be conceived as physically separate entities. The philosophy of Taoism acknowledges that all elements of the universe are subject to the same physical laws. However, each object in the world has a yang – a causative, active, creative principle – and a complementary yin principle, representing the resultant, passive, destructive dynamics.

We should note that the material human–environment complementarity differs from the wave–corpuscular theory because the first deals with entities that are thought to be physically different, whereas the latter deals with one and the same entity. The body–mind complementarity can be seen as the complementarity of the material–biophysical versus the social–epistemic dimension of the human systems on the level of the individual. We contend that the same complementarity holds true for the environment. We find a material–social or body–mind complementarity in human individuals and other living systems (Miller, 1978). As we will see in Chapter 5.3–5.4, this distinction underlies the differentiation between biochemical and information-based views on the cell, which leads to the interpretation of the immune system as a cognitive system.

3.3.1 A non-Cartesian view

Historically, the body–mind complementarity has been dealt with through various interpretations and philosophical approaches. Besides the nature of the two complementary units, these approaches differ with respect to assumptions and in how these systems interact. Most modern approaches can be ascribed to the controversial interpretations provided by Homer (c. eighth century BC), Plato (428/427–348/347 BC), and Aristotle (384–322 BC). Homer differentiated between animate and inanimate bodies; Plato suggested a tripartition of the soul into reason, the eternal spirit, and appetite; and Aristotle postulated entelechy, which refers to a hypothetical, form-giving cause or agency directing the material existence, while skipping a double, separate existence (Specht, 1995).

Dualism differs from various types of monism, which assume that there are no fundamental differences between the systems and the sets of rules, or laws, that underlie their material–biophysical and the social–epistemic aspects. René Descartes (1596–1650) held that there is a material and a non-material substance, that is, a res cogitans (thinking thing). Thus he adopted Homer's position. Descartes suggested that the interconnection between body and mind was through the pineal gland and that each substance affects the other. Thus changes of the material–biophysical nature can effect change in the social–epistemic realm and vice versa. In contrast, the material–social complementarity view underlying this book is not a (pure) Cartesian one, as we do not assume two substances and an interfacing organ. Instead, it is like the wave–corpuscular complementarity and we provide a material and a non-material information-based conceptualization of the (one and the same) human system.

The study of environmental literacy deals with different human systems: the individual, group, organization, etc. We assume a material–social complementarity for all these systems. The material–biophysical dimension is called H_m. It consists of all cells belonging to the body of the people belonging to a human system (e.g. for an organization, H_m may be the employees and owners of a company working for the company). The nature of the social-epistemic dimension differs between human systems and is called H_s.

Similar terminology for the dimensions of human–environment systems also appear in the work of British anthropologist Mary Douglas (1970). Douglas describes how humans assign physical-based symbols to physical experiences and bodily functions, such as heartbeat, blood pressure, sleeping periods, dietary or other physiological processes (Scheper-Hughes & Lock, 1987). As humans interact with the (material and social) environment, the physical-based symbols can acquire social meaning. The newly emerged social symbols are shared and defended by members of societies. For example, the need by the human body to sleep has led to the development of the weekend as a feature of societal organization.

At the level of the individual's socioepistemic level H_s, the mind, we have mental states (symbols, beliefs, representations) which have meaning and function. One challenging question is the concept of consciousness and what knowledge and concepts we can attribute to the self, the body, and the mind, and also how these are related to, and controlled by, each other. One

3.3 The material–social complementarity

> **Box 3.2[1]** No human–environment complementarity: Maya constructions of the world
>
> There are a number of western world view frameworks that describe complementarities between human and environmental systems (see Chapter 16.2). Some world views, such as those of the Maya, do not utilize a human–environment complementarity. In the following, we refer predominantly to the Chiapas highland Maya in Mexico as described by Nigh (2002) and the Kaqchikel Maya of Guatemala.
>
> As in most Mesoamerican cultures, an individual Mayan considers himself or herself to be part of a complex universal network or matrix representing the entire universe (Reichel-Dolmatoff, 1996). In this way, Maya integrate the natural and the social world. Each individual is embedded so that everything in the universe resonates within the individual and each human activity affects the universe. So far everything could also be presented in terms of complementarity if we assume a reciprocal balance between the micro-cosmos of the individual and the macro-cosmos. One difference to the western view is that the Maya believe in a myriad of spirits and deities. In our terminology, supernatural forces rule the interplay between the entities of the universe. In the view of the Maya, there is a divine matrix governing the harmony of the world. This is an extension of the western assumption that laws of nature rule the biophysical world. Mayan individuals are embedded in nature, which comprises both the body and the mind, including the spirits. An essential aspect of caring for natural resources is the collective or individual rites for the spiritual identities that complement all material entities and for the spirits that exert authority on certain places. Interestingly, there is no term in "Mayan language that corresponds directly to our notion of 'nature' or the environment. If you ask a native Chiapas Mayan speaker to translate the concept of 'nature,' he will usually respond with a word such as *banamil* … [which] is usually translated as 'world,' 'land,' or 'earth' but also as vegetation" (Nigh, 2002, pp. 455–6). This is often the case in many premodern societies.
>
> To illustrate the inextricable principles of the universe in Mayan cosmology, we present the conception of the notion of reciprocal balance between components of the universe as an exemplar. As mentioned, the Maya do not dichotomize between material–organismic and supernatural causes. However, the soul of an individual has two principal components (see Figure 3.2): that is, the *ch'udel* and the *chanul*. The *ch'udel* is the part of the soul essential for life. It is the energy center residing at the "heart." The core is sometimes conceived of as a "heart bird," that is, a chicken or a dove (Nigh, 2002). However, the term heart is in some respects imprecise as notions of soul, center, essence, or anima are involved in the Mayan understanding of heart (Fischer, 1999). The *ch'udel* is the identity, including memory, feeling, emotions, etc. The *ch'udel* can leave the body, for instance during sleep, and delayed return may cause disorder and disease. According to common, ordinary belief, a duplicate *ch'udel* is living in a mountain under the control of ancestor spirits. Simplified, each person is linked to a companion animal life force, which is the *chanul* or *nahual*. The *chanul* is an animal spirit companion living in a cave of a sacred mountain governed by the highest-ranking official of ancient spirits. This, usually wild, animal is allowed to graze during the day.
>
> Admittedly rough, this simplified description touches on just a few aspects of Mayan cosmology. It illustrates, however, that societies can construct conceptions of humans that differ from the view of the human individual as the ensemble of all cells emerging from the fertilized ovum and the interactivity of these cells, as it is presented in Chapter 3.

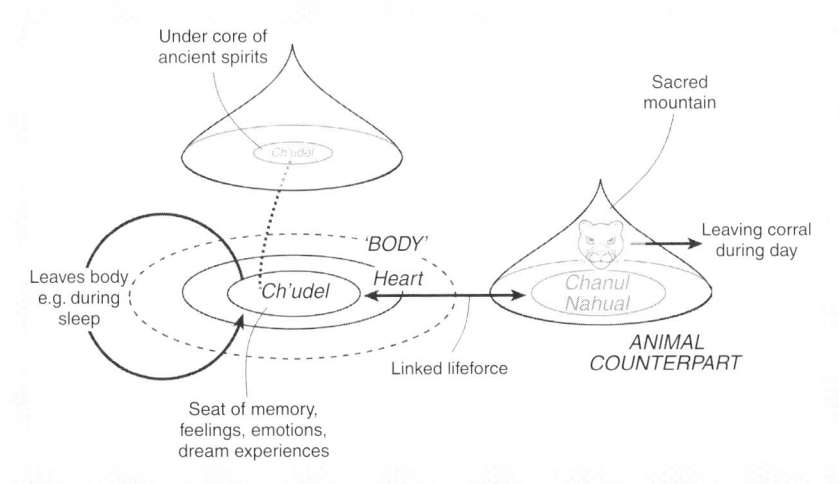

Figure 3.2 The Mayan non-complementary world view, which includes the relationship of the body with ancient spirits.

---

[1] This box has been written together with Mónica Berger González.

metaphor offered by Jerry A. Fodor (1968) is that the relationship between the brain and the mind is similar to the relationship between the hardware and the software of a computer.

> To use an analogy, the "psychosocial software" is not reducible to the "biological hardware." Each is governed by its own set of principles in their own rights. (Bandura, 2001, p. 19)

We consider the material–social complementarity to be a powerful concept for analyzing the emergence of environmental literacy. At the same time, we clearly acknowledge that the body–mind relationship must be seen as an inextricable unity (as it is with the human–environment complementarity). Although we do not delve into the issue here, the directionality of the complementarity is of importance. The mind–body complementarity becomes evident when you remember intense dreams, from which you wake with a quickened heartbeat.

On the level of the society, the material–social complementarity includes the questions: what knowledge about the environment is acquired and represented in which individuals, and how is it transmitted to future generations? Here, we should note that material–social complementarity clarifies externally stored (physical) data found in cave paintings, books, electronic media, and so on, or other physical entities which are produced by society and considered to be part of the material environment. The book you are reading is part of the environment. The information and interpretation you extract from the physical signs and signals are part of a socio-epistemic process that can serve to build up environmental literacy.

On the societal level our interest is also in the question: what information from the environment can the human species perceive through its natural senses or with the help of technology. Phylogenetically, this is linked to the development and evolution of the senses and the brain (including its size, structure, and evolution). The latter is the material–biophysical side of the human system H_m. The social side H_s refers to the epistemics of a society, including the interpretation (i.e. semiotics) of the information and the signs gathered from the environment. Besides the development of language, the knowledge about the natural environment (e.g. plants and animals and their usage), technological knowledge (about agriculture or information technologies), and the development of science knowledge can be subsumed in the socioepistemic side. In this context the role and type of supernatural forces are of interest, and involve different constructs ranging from animism, spiritualism, polytheism, monotheism, and variants. Each is linked to different conceptions of how supernatural forces govern the soul, spirit, psyche, or mind of a person or society. Some religions consider supernatural forces to be immaterial whereas others consider them as physical, and in some religions we find both.

The socio-epistemic dimension, which includes the cultural one, is an important layer of the principal components of society (see Figure 1.7*). The cultural aspect includes beliefs, language, history, religion, and what is considered as fashionable clothing or sustainable lifestyle. The social dimension of a society includes the type of economy, policy, and institutional rules, whereas science plays an important role in the epistemic and educational layers of society.

Finally, we should note that some scientists consider the human–environment complementarity to be a major component of the dominant, human-centered western paradigm and is in contrast with world views which have an ecocentric focus, such as Native American, Taoism, Buddhism, deep ecology or eco-theologic approaches (Berkes, 2008; Devall & Sessions, 1999; Gardner & Stern, 2002).

3.4 The realist stance

Clearly, the answers to these questions about HES depend on the ontological stance taken. This book takes a "realist" view in which we assume a mind-independent existence of reality. Thus, we assume that material entities, such as water or the Moon, exist independently of what we think or what conception we develop about these issues. But this does not only refer to the material world. An important aspect of a realist view is that it asserts "the autonomy of social entities from the conceptions we have of them" (de Landa, 2005, p. 1). This assumption is important from the point of view of the theory of science, as it is grounded in the idea that social theories may be objectively wrong. Only if we postulate a realist stance, we can provide judgments about which of two constructions is more adequate be answered from a non-subjective perspective.

A realist ontology of the social environment was first suggested by Durkheim (Durkheim, 1895/1982; see Chapter 8). Similarly, de Landa (2005) postulates "objective processes of assembly: a wide range of social entities, from persons to nation-states" considered through "very specific historical processes, processes in which language plays an important but not a constitutive role" (de Landa, 2005, p. 3).

The subtlety of assuming a complementarity between a social–epistemic and a material–biophysical environment from a realist perspective can be illustrated by the text you are reading at this moment. Without a doubt, the text is part of your material environment. It is represented on a sheet of paper, or a screen, or perhaps via some other media. If you are reading it on paper, there must be appropriate light-transmitting beams from the letters so that your eyes can perceive the text. Clearly, decoding is not only a physical matter. Your eyes and your brain have learned to decode single configurations of printed ink as letters or words. You have also learned to attach meaning to these signals. Language allows for the coding of objects, concepts, feelings, stories, and other issues, including their characteristics. Scriptural representations of these issues and the rules required to decode and attribute meaning to them are typical examples of social reality. This is reflected in the distinction between a material–biophysical and a social–epistemic environment from a realist perspective.

3.5 Functionalism

Functionalism was inspired by the naturalists Darwin, Galton, and Spencer, who all considered survival-targeted adaptation to environmental constraints to be a key element of organismic action. In its broadest sense, functionalism emphasizes the practical utility and functional relations between the entities of a system.

Functionalism, in the context of psychology, supposes that the desire to become literate about the environment does not depend on one's internal constitution but rather on a mental state that reflects the constraints and opportunities of the environment. Such an approach indicates that showing interest in the environment is beneficial. For instance, the father of probabilistic functionalism, Egon Brunswik (1952, 1955a, 1955b), postulated that mental activity, consciousness, and behavior were to be evaluated in terms of how they serve the organism in adapting to its environment (Chaplin, 2000; Dewey, 1896). A property of functional learning and adaptation is the use of environmental information to adapt behavior appropriately to the constraints and options of the corresponding environment (Figueredo *et al*., 2006). Ideas and cognitive activities are seen to exist primarily as tools or instruments to solve problems encountered in the environment. For instance, the ecopsychology approach developed by Barker (Barker, 1968; see Chapter 6.7) states that behavior, and what we do regarding said behavior, are predetermined by the situation or the form of the objects with which we are dealing.

In the North American interpretation of functionality, intention did not play a role. The European accent of functionalism, on the contrary, stressed the intentionality of things that "are remote from them in time and space" (Goldstein & Wright, 2001, p. 250). To cope adequately with the environment, individuals and other human systems develop an image or notion (German: Vorstellung), an evaluative feeling (German: Gefühl), and a judgment or decision (German: Urteil; Brentano, 1956).

Sociologists and anthropologists see the goal of survival as a primary driver at the societal level. According to the functionalistic sociologist Gerhard Lenski, societal change is driven by curiosity, imagination, and creativity. People are motivated to interact efficiently with their environment and to develop new technologies because they want to improve the conditions of their life (Lenski *et al*., 1991; Sanderson, 2007).

Famous functionalists were the British biologist and social philosopher Herbert Spencer (1820–1903), John Dewey (1859–1952), and the French sociologist Durkheim (1858–1917), who stressed that society is a forum for functional social ideas and more than the sum of its parts (Durkheim, 1895/1982). A structuralist functionalism proposes that society consists of parts, each of which have their own functions and work together to promote and guarantee the reproduction of society. According to the sociologist Talcott Parsons (1951), the adaptation of social structures and the building of functional subsystems are essential. Thus, institutions are special societal organizations that guarantee the reproduction of society. An evolutionary biologist's view on functional constructivism has been provided by Ernst Mayr, who reflects on 3000 million years of development and adaptation:

> The functional biologist is vitally concerned with the operation and interaction of structural elements, from molecules up to organs, and whole individuals. (Mayr, 1982, p. 69)

We should note that functionalism is not merely a matter of expedience. As anthropologists state, from a functionalist view it is also necessary to go beyond the practical, material–matter-based logic and to include social–cultural logic (Sahlins, 1976). Thus functionalism is related to and is shaped by a sociocultural dimension:

> The usefulness of an item or activity depends … on how it is understood and valued from a socially and historically contingent system of symbols. (Hutson & Stanton, 2007, p. 123)

3.6 Functional social constructivism in the frame of uncertainties

As functionalism is an answer to the question of *why* humans become environmentally literate, constructivism is an answer to *how* they do it. Constructivism has roots in philosophy, psychology, sociology, and education, and stresses learning and the genesis of knowledge. Basically, constructivism states that all of our knowledge about the environment is constructed and is not a simple representation of realist stances. Because the construction of knowledge does not start from *tabula rasa*, knowledge (previous experience), encounters with the environment, as well as what can be learned from the social environment, determines what people or human systems can construct. This not only holds true for children but also for societies and sciences (Knorr-Cetina, 1981; Piaget, 1970; Sismondo, 1993). Although constructivism in its most radical form criticizes realism (Berger & Luckmann, 1966; von Glasersfeld, 1996), it does not necessarily stand in contradiction to all forms of reality (Stauffacher *et al.*, 2006). As is the case with many approaches of weak constructivism, we think that humans construct their reality from a social, subjective, and functional perspective in a social setting according to their capabilities to access reality. Constructivists such as Berger and Luckmann acknowledge that:

> … as much as we are the producers of society, we are the products of the social and cultural contexts, in which we were raised and live. (Klüver, 2002, p. 28)

Thus the epistemological subjectivism can be combined with an ontological realism (see above) in a functionalistic frame. As we discuss in detail in the section on probabilistic functionalism (Brunswik, 1952; see Chapter 6.3), neither the process of construction nor any information acquisition from the material world are free of uncertainties. Thus, the coping of uncertainty in data and epistemics is essential.

3.7 The assumption of human system hierarchy

The hierarchy concept is a fundamental principle used in many sciences. Natural scientists speak of atoms, molecules, and crystals. The pathologist Rudolf Virchow suggested that life sciences must distinguish between cells, tissues, organisms, and social levels to understand diseases (Virchow, 1958). Biologists suggested that it makes sense to introduce the hierarchy concept to model biosystems through simplified models, as complex adaptive systems typically become organized hierarchically through non-linear interactions between their components (Levin, 1999). Others have modified the term and suggested using the term "panarchy" to indicate that there is a "destabilizing, creatively destructive" component in the interaction of hierarchical systems in nature (Holling *et al.*, 2002, p. 74). Landscape ecologists have divided the planet or biosphere into oceans, continents, and local ecosystems (Forman, 1995; Golley, 1998; Leser, 1997; Pumain, 2006). System theorists have suggested ordering the levels of energy sources in food chains (Odum, 1994). Finally, James G. Miller suggests that for the levels of human systems one should distinguish between the human individual, groups (e.g. families, teams, cohorts, tribes), organizations (of individuals or groups), communities, societies, and supranational systems (Miller, 1978; Miller & Miller, 1992).

We would like to note here that we have two reasons for our reference to the hierarchy assumption. The first reason is that hierarchical organizations are useful for reducing complexity. We discuss this issue in Chapter 16.3 (see the Hierarchy Postulate P2) and distinguish between different types of hierarchy and make primary reference to the concept of level hierarchy, which is important for many biotic and social systems. The second reason is that we have different disciplines and branches of social sciences that provide theoretical frameworks purely for certain levels. Introducing a hierarchy assumption for social systems simply makes the theoretical frameworks that are developed in different disciplines and branches of social sciences more accessible, in particular since their level of focus differs. For example, environmental psychology focuses on the individual, environmental management theorizes the level of a company and business organization, and environmental sociology and macroeconomics predominantly focus on the level of society or market nations, regions or the entire world.

3.8 Relations among environments and hierarchy levels

Table 3.2 Basic complementarities in human and environmental systems research: the human–environment complementarity is presented in the columns and the material–social complementarity in the rows.

	Human systems (*H*)	**Environmental systems** (*E*)
Material–biophysical dimension (*m*)	H_m	E_m
Social–epistemic dimension (*s*)	H_s	E_s

Key question 3 refers to different rationales in different human systems. An interesting example here is the relation of an individual to group processes in decision-making and the role of group decision schemes (Davis, 1973; see Chapter 6.21). The idea of decision schemes is that groups have a voting culture that is part of the rationale of a group. If there is dissent between group members on the preferred alternatives for a decision, the final decision depends not only on the distribution of the preferences and the power of the individual group members but also on the group's decision rule. This rule, which must be evident to all of the group members, determines which alternative is selected. Thus, groups with identical distributions of preferences on alternatives can end up with completely different solutions. At its most extreme, it is not the rules of the individual that matter; rather, it is the social rules of the immediate environment that matter. This can theoretically, lead to group preferences that are shared by none of the group members. An example regarding drinking alcohol or not (Feldmann, 1984) shows how group norms play an important role in environmental behavior. Thus, when defining the socio-epistemic environment of a human system, we have to include the rules or the rationale of higher systems, such as the groups to which an individual belongs.

3.8 Defining relations between environments and hierarchy levels

In this fashion, the framework of this book is based on two complementarities: (1) the complementarity between human and environmental systems (i.e. between *H* and *E*); and (2) the complementarity between a material–biophysical environment, abbreviated as material environment (E_m), and a social–epistemic environment, abbreviated as social environment (E_s). Both environments are considered from a realist perspective (see Table 3.2). We sometimes further distinguish between what is regarded as an environment and the "rest of the world." What is not counted in the environment includes all that is judged irrelevant for the human system under the timespan considered for both the social and the material worlds. The environment thus represents that which matters or is of interest from a certain perspective.

In Figure 3.3* we distinguish two types of arrows that are not marked with numbers. They represent different types of HES relations. The arrow (↔) describes the transmission of physical entities between the biophysical (organismic) dimension of human systems and the material–biophysical environment. The transmission can be realized by light rays, acoustic or blast waves, molecules, gravity forces or other physical entities. The second arrow (⇔) does not represent a physical transmission but the switch in the complementarity perspective from a material–biophysical to an immaterial social–epistemic dimension of human systems.

We illustrate the whole chain $H_s \Leftrightarrow H_m \leftrightarrow E_m \Leftrightarrow E_s$ by a simple communication on the level of the individual (see Figure 3.3*). Imagine, a person has an idea (e.g. we should go to a restaurant now) which can be represented by language. Both the idea and language are considered as part of H_s. By means of his larynx and mouth muscles (which are part of his body, H_m) the words representing the idea can be transformed into acoustic waves that are part of the material environment E_m. These waves might be perceived by the acoustic organs of another person who is part of H_m. If this person can decode the acoustic signals and interpret them, they become part of the socio-epistemic environment E_m. Similar chains can be constructed for groups, companies or societies if we refer to the definition of human systems provided in this book.

We have already argued that what human systems perceive about the environment, what they consider to be relevant, or what they factually do, does not solely depend on the material–biophysical environment. Group norms, organizational constraints or the societal role an individual has (or is supposed to have) can affect the individual. These norms can be explicit or implicit. Implicit means that none of the people involved or concerned by the norms, constraints, or roles may be consciously aware of them. Thus, the interest an individual or an other human system shows in certain environmental issues, such as in climate preservation, does not exclusively emerge from the interests or capability of the human systems. Rather, we propose that both the environment (i.e. both E_m and E_s) and the impacts of higher level human systems play a significant role.

In Figure 3.3*, we present the interaction of three layers (individual, group, and society) that can be at work if, for instance, a decision is made at the level of the individual. The solid arrow ❶ represents a subset–superset relationship, which, in the case of individual-to-group relationship, represents an inclusion hierarchy. Note that this would not necessarily hold true if we considered the relationship between a group and a society, as the members of a group may belong to different societies (e.g. nation states). In this case we would speak about level hierarchy (see Chapter 16.3).

According to the realist stance, higher level human systems (e.g. H_m^*) can directly or indirectly affect lower level systems by affecting their environment (e.g. E_m). The dotted arrow ❷ indicates that a group H_m^* without the individual H_m can become part of the environment of this individual H_m in the decision process. Likewise, the downward arrow ❸ indicates that the society level without the members of groups can be part of the environment of the group E_m^* and is thus affecting the group H_m^*. Naturally, there is usually also a direct arrow from the society H_m^{**} to the environment of the individual E_m; this is not drawn in Figure 3.3*.

The solid line ❹ between E_m^* and E_m again represents a subset–superset relationship if we assume that the environment of the group is the union of the environments of all its members.

The arrows ❺ and ❻ represent the body–mind complementarity. In this book, this complementarity is only related to the human members of the environment. The environment of an individual E_m includes people as material beings. These humans are consciously or unconsciously producing mental entities such as goals, intentions, thoughts, images, meanings of a symbolic kind, norms, rules, etc. Usually, the social environment of an individual (E_s) includes different social environments on various levels. Thus, for instance, the social environment of a specific individual is formed by the social system of other individuals (i.e. $H_{s,1}, H_{s,2}, \ldots$), of groups (i.e. $H_{s,1}^*, H_{s,2}^*, \ldots$), and of one or more societies (i.e. $H_{s,1}^{**}, H_{s,2}^{**}, \ldots$). All of these social environments affect the actions of the specific individual as part of E_m. Clearly, material environments E_m encompass the geographical setting, including radiation, air, water, and material resources, which affect the wellbeing of human systems.

Naturally, an individual can be a member of one or more groups, organizations, or countries. Let us consider a female, animal welfare activist working as a procurement agent for a grocery store chain. Through career advancement, she becomes the head of the department responsible for meat purchases. Here we meet what we call an interfering regulatory mechanism, or different rationales of environmental behavior. Personal level behavior is guided by personal norms and values (e.g. protect environment, unity with nature; see Figure 7.8). Given the extraordinarily low profit margins in the food trade, a manager in such a position interested in keeping her job would want to find the most lucrative purchases. Receiving an extraordinarily profitable offer from a somewhat questionable hog feeding company certainly causes interference between her personal environmental values and norms and the contractual agreement about objectives as a procurement manager. This is represented by the arrow ❹, assuming that the second level would be the level of the organization instead of the group, as companies are considered to be organizations (see Chapter 16.3, Hierarchy Postulate P2). The business goals of a company (H_s^*), which are represented by its management, may cause conflict with the internal values of the female manager (H_s). Here we are facing a typical conflict between what others (the social environment) want and what the self wants to do.

As we have seen, the material and the social environments at a certain level are affected by higher levels. In practice, or in the practice of research, it is mostly the impact of both the environment and of higher level systems that are defined to understand, model or predict certain effects of higher level systems or the interactions between lower and higher level systems. In Chapter 18 we present some cases, such as the Swedish government's decision to become an oil-free society by 2020 (Case 3). We will see that the interference between goals on the macro level (e.g. governments representing society) and micro level (e.g. companies and consumers) can present a major obstacle to sustainable development.

Thus the social–epistemic environment, including cultural norms, is a composite of inputs that come from all relevant higher level systems. From a realist stance, the social environment comprises the assembly of all social facts from higher levels and of systems from the same level.

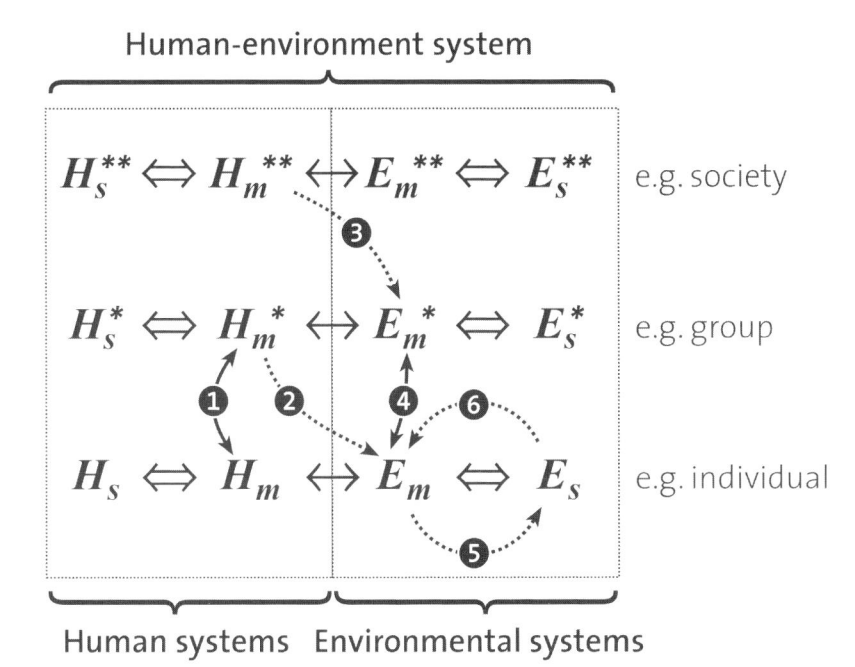

3.8 Relations among environments and hierarchy levels

Figure 3.3* Relations between human and environmental systems at different hierarchical levels. For each human system (H, H^*, etc.) we assume a complementarity between the social–epistemic (e.g. H_s) and material–biophysical layer (H_m). Actions or transmission of physically represented information build the interface ($\leftrightarrow$) with the material environment E_m. E_m includes inputs from higher level systems and interacts with the social environment. Environmental dynamics thus have to be understood as the interplay between different levels of social systems (which have different rationales) and material–environmental systems. Thus, the social environment is an assemblage of entities from social environments of the same and of higher levels. According to the concept of level hierarchy, there is upward and downward causation as represented by the vertical arrows. The causation dynamics represented by arrows ❶ to ❻ are explained in the text.

Figure 3.3* stresses the hierarchy aspect from top to bottom. Naturally, there is also interaction between human systems at the same level that do not necessarily impact higher levels. For instance, a dyadic bargaining between two individuals can be modeled if one bargainer is represented as H_m and the other as E_m. Whether to represent the impact on higher level human systems (e.g. bargaining conventions) depends on the situation and the question at hand.

The proposed definition of material environment coincides with the common, everyday definition of the environment if we define the top level as the human species. The environment then becomes the collective of physical, chemical, and biotic factors such as climate, soil, and living things that act upon an organism. It includes the natural environment, in addition to what is conceived as the built and technical environment. This definition coincides with Spencer's definition of the total environment of the human species (see Chapter 1.3).

Key messages

- Any research on HES and dynamics is related to a set of basic definitions and epistemological assumptions which should be explicated. This book is based on two complementary assumptions; it builds on a functionalist conception of human systems, an assumption of a hierarchy of human systems, and it takes a realist, functional constructivist perspective
- The human individual is defined by all living cells and their (inter-)activities that emerge from the fertilized ovum. The definition of human systems above the individual is based on the activities of the individuals that can be assigned to this system. The environment of human systems is a subset within the complementary set to the human system, referred to as the universe
- In addition to the human–environment complementarity, the material–social complementarity is a second essential ontological complementarity put forward in this book. In simple terms, this complementarity distinguishes between a material–biophysical dimension including the built environment and materialized technologies (i.e. the "hardware"), and a social–epistemic dimension including the cultural rule system (i.e. the "software") of human systems. On the level of the individual, this material–social complementarity has been called the body–mind complementarity
- Both the social–epistemic and the material–biophysical environment can be conceived from

a realist, objectivist perspective. The realist view implies that only the material–biophysical side of human systems can interact with their material–biophysical–technological environment
- Social constructivism (i.e. a subjectivist, constructivist epistemic) and a realist conception (i.e. an objectivist ontology) of both the social–epistemic and the material–biophysical environment are compatible
- This book is based on the propositions that human systems on different hierarchy levels follow different rationales to environmental systems, and that the behavior of human systems on each hierarchical level is related to different drivers. According to the functionalist perspective, the environmental literacy of human systems at and above the level of the individual is shaped by values and interests.

Part II
History of mind of biological knowledge

4 Emerging knowledge on morphology, ecology, and evolution 45
5 From molecular structures to ecosystems 94

Key questions (see key questions 1–3 of the Preamble)

Contributions from biology to environmental literacy

Q1 From both biophysical and epistemic (biosemiotic) dimensions, how do organisms conceptualize and interpret the signs of nature? How did knowledge about biosystems evolve?

Q2 What dynamics underlie biosystems and ecosystems? In what ways does socioecological research link the rationales of material–biophysical processes in the environment with the social–epistemic processes in human systems?

Q3 As ecosystems, unlike individual organisms, do not have built-in homeostatic responses to disturbances, what knowledge will enable researchers and practitioners to manage a balance for maintaining ecosystem functions in limits for supporting human life?

What is found in this part

Part II deals with the role of biological knowledge in environmental literacy. Known as "the science of life," biology offers methodologies and methods that are fundamental for investigating human–environment systems (HES). Because one of the research principles of the HES framework (see Chapters 6 and 17) is that an analysis of HES should begin with a thorough analysis of the environment itself, we begin our discussion of five relevant disciplines with biology.

Chapter 4 introduces the emergence of the field of biology and begins with a discussion of the knowledge that prehistoric societies had about their environment. As biological and medical knowledge coevolved, we also describe how knowledge about medical treatment emerged in ancient societies. Two key sections of Chapter 4 deal with classical, traditional biology. We discuss how knowledge about the form and structure of organisms (morphology), their emergence (evolution), and their pattern of interaction with the natural environment (ecology) has developed. Environmental literacy was propelled forward when humans developed a global view on the biotic environment, as we describe in a section about Alexander von Humboldt environments. It should be noted that biology does not merely deal with the biophysical aspects but also with the epistemic dimension that is biosemiotics, which grapples with the questions of what animals can perceive and what environmental information means for them.

Chapter 5 deals with contemporary biology. We first follow humans' progressing knowledge about cells, the basic building blocks of life. We see how knowledge and theory-building about the cell, its components and structure depend on the progress of magnification technology. The sections on genetics and immune defense introduce how cells respond to and interact with the environment. This is reflected when conceiving the immune system as a cognitive system.

The nature vs. nurture discussion is looked at from a natural science perspective. Quite remarkably, we find new theories which – similar to Jakob Johann von Uexküll's biosemiotics – assert that cell and cell systems are able "to read the signs of nature."

We then take a broad, "macro" view and follow how biologists conceptualize ecosystems and how they incorporate human species. We begin by discussing traditional biological ecosystem research and examine how humans have addressed their need to conceptualize human–environment interactions.

A fundamental difference between the micro and the macro level is the different role of regulatory systems. On the micro level of organisms or organs, internal physiological processes such as blood pressure in the human body are regulated and strive towards a homeostasis. This is not guaranteed in large ecosystems, in particular if they are affected or perturbed by human systems. Here, regulatory functions must be constructed with the help of the human mind, that is the socio-epistemic side of human systems (see Figure 3.3*).

Part II History of mind of biological knowledge

Chapter 4
Emerging knowledge on morphology, ecology, and evolution

- 4.1 Foundations of understanding life 45
- 4.2 The Herculean task of prehistoric research 46
- 4.3 Prehistoric biology and ethnobiology 50
- 4.4 Antiquity's conceptions of the biophysical dimension of life 59
- 4.5 European "Dark Ages": loss of environmental literacy 66
- 4.6 Rebirth of the natural science of organisms 67
- 4.7 Scientific revolution and the methodological foundations of modern biology 69
- 4.8 Classification 73
- 4.9 Biological exploration: Alexander von Humboldt environments 78
- 4.10 Evolution 80
- 4.11 Ecology 84
- 4.12 Biosemiotics: the first *Umwelt* research 90

Chapter overview

Chapter 4 starts with a section on environmental literacy in prehistoric times, but going beyond the simple analysis of ancient Greek or Egyptian knowledge common to many books on the history of biology. We take a developmental theory perspective to reflect on the types and levels of understanding that emerged in prehistoric times and which might still be around in different current understandings of life.

We deal in some detail with the fundamental changes of constructing cosmology, starting at the rise of science in ancient Greece, a period considered as the cradle of modern sciences, and follow through to the beginning of medieval times (200–1200). The rebirth of inquiry in Europe after the medieval period occurred following the establishment of universities in the twelfth century. The sixteenth century saw the emergence of modern botany and, soon after, of zoology including the mastery of anatomy. The scientific revolution of the seventeenth and eighteenth centuries introduced inventions like the microscope, which allowed people to see microbes, developing embryos, and capillaries. The discipline of biology developed in the nineteenth century in three streams of knowledge: theories of evolution, the naturalist conception, and the understanding of how human societies adapt to environmental settings.

Finally, this chapter deals with the emergence of ecology, which Ernst Haeckel (1834–1919) defined at the end of the nineteenth century as "the comprehensive science of the relationship of the organism to the environment" (Haeckel, 1866, p. 286; see Greene, 1961, p. 3) and also biosemiotics, the cognitive side of animal perception, which can be conceived as a biological conception, that is the body–mind complementarity on the side of the animal.

4.1 Foundations of understanding life

Biology deals with "the origin, development, structure, functions, and distribution of living matter as represented by plants and animals and to the generally recurrent phenomena of life, growth, and reproduction" (Merriam-Webster's, 2002). Thus biology can be seen as a major contribution to what Karl Popper (1902–94) called cosmology, that is, the attempt to provide explanations for how the world works and is organized:

> I, however, believe that there is at least one scientific problem in which all thinking men are interested. It is the problem of cosmology: *The problem of understanding the world – including ourselves, and our knowledge, as part of the world*. All science is cosmology, I believe, and for me the interest of philosophy as well as of science lies solely in the contributions which they have made to it. (Popper, 1959, p. 15)

Besides cosmology, *cosmogony* plays an important role in this chapter. Whereas cosmology provides a description of what happens in the natural order of the universe, cosmogony explains how the world came into being. This chapter reveals that basic assumptions about the origins of the universe and life (cosmogony) and about what are the basic processes of nature (cosmology) and how these assumptions strongly affected the building of biological theories. Thus, this chapter makes reference to these fundamental aspects, including the role of supernatural forces.

4.2 The Herculean task of prehistoric research

Investigating how people have interacted with plants and animals for subsistence requires combining ethnobiological, archeological, anthropological, and historical knowledge. Such a reflection about environmental literacy in prehistoric times is a challenge, as there are – by definition – no written documents and only sparse data from archeology. Whether data about today's indigenous people provides insights into prehistory is questionable, as all of today's tribes have presumably had direct or indirect contact with the modern world.

From a functionalist perspective, ancient biological knowledge about fauna and flora for nutrition, hunting, healing, safety and other purposes can be seen as a prerequisite of continued and improved existence of societies (tribes). People must have known, for example, which plants to eat and which to avoid, and improving such knowledge provided evolutionary advantages. This can be reconstructed from recent prehistoric times, for instance the agrarian revolution about 12 000 years ago.

4.2.1 Origins

The Stone Age saw the emergence of the first stone tools about 2.5 million years ago. Indications of the presence of *Homo sapiens* in Africa go back 160 000 years (White et al., 2003). *Homo neanderthalensis* went extinct about 30 000 years ago and *Homo floresiensis*, characterized by a small body and brain, went extinct about 12 000 years ago (Brumm et al., 2006; Morwood et al., 2005). Prehistory ended with the emergence of cuneiform and hieroglyph writing systems in the fourth millennium BC. As prehistoric time spans more than 99% of human development, we take a closer look at the emergence of environmental literacy in this period.

From an epistemological view, reconstructing environmental knowledge of prehistoric people is challenging (Schiffer, 1976). We have to deal with sparse data and a wide range of disciplines such as archeology, ethnology and earth sciences, as well with as different time scales, all making it difficult to arrive at unambiguous, substantiated scientific theories. We have to interpret limited data from the past and combine it with evidence from a few surviving contemporary indigenous peoples to infer what environmental knowledge might have been in the far past (see Figure 4.1).

Walking along these boundaries when integrating information about physiological, anatomical, and behavioral data from plants, animals, humans, climate, etc. has been nicely elaborated by Harris (2006). He stresses the difficulties in interpreting archeological evidence but also the possibility of constructing consistent inferences, for instance, when discussing the work of Mannino and Thomas (2002, 2008) to examine exploitation of mollusks within Mediterranean settlements 10 000 years ago:

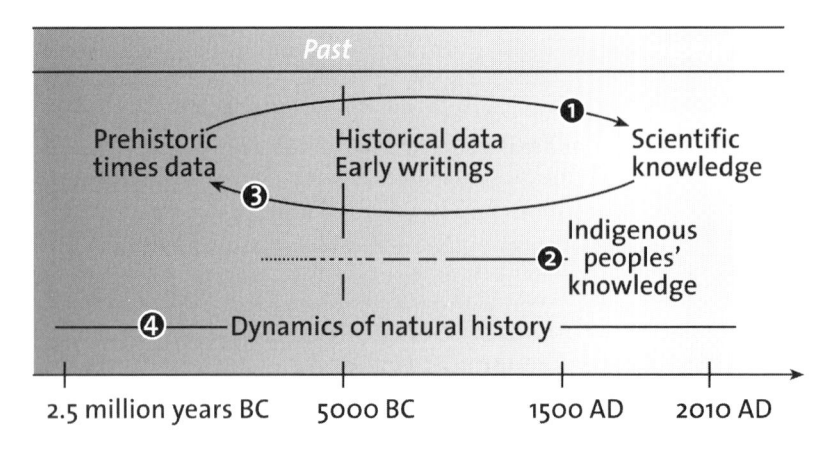

Figure 4.1 The interdisciplinary challenge to reconstruct the environmental literacy of prehistoric societies: (1) archeologists provide data; (2) cultural anthropologists supply linguistic and behavioral evidence; (3) anthropologists contribute interpretations; and (4) natural historians give insight into the dynamics of the constraints of the biotic and abiotic environments.

... data from ecological studies of present-day mollusk populations on the north-west coast of Sicily, combined with analyses of shells from Mesolithic and early Neolithic occupation layers in nearby caves, indicate that shellfish were an important seasonal food for the inhabitants of the caves 5000–10 000 years ago. (Harris, 2006, p. 64)

To reconstruct environmental knowledge of prehistoric people, we draw from five main pools of data: archeological data, cultural anthropology data, natural history records, ancient DNA analysis, and an interpretation of archeological and anthropological data through an epistemological lens.

First, we can look to archeological data, which includes cave paintings (see Box 4.1), hunting tools, and types of housing (caves, different types of buildings) that can be interpreted from a cultural anthropological perspective. Here, we look at *data from the past* to infer *past* knowledge. The radiometric data, for instance the ^{14}C method (Libby, 1961), and other modern dating methods such as the accelerator mass spectrometer method developed in the 1980s, were considered a quantum leap in archaeologists' ability to determine the precise age of objects (see Roberts, 1993). Because ideas, the main ingredients of socio-epistemic knowledge, do not fossilize, we depend upon present concepts to interpret what people might have known, thought, found inspiration in or communicated about thousands of years ago.

The paper "Eloquent bones of Abu Hureyra" (Molleson, 1994) can serve as an example of how we can infer past biological settings and interaction patterns. Molleson examined skeletons from 162 individuals of Hureyra dating from at least 7000 BC. Many of these skeletons showed malformations. Linking these findings with other evidence such as plant remains, he concluded that these malformations must be due to intense agricultural work. However, alternative hypotheses have also been put forth:

> For a time, we actually entertained the idea that the people of Abu Hureyra had engaged in some sport or athletics, but crippled ballerinas seemed unlikely to have appeared during the Neolithic. (Molleson, 1994, p. 71)

If we have "enough" data and evidence, we may dare to speculate about what mastery of nature (i.e. domestication of plants and animals) prehistoric people had. Animal bones found, for instance, in underground cellars of Hureyra can be seen as indicators of domestication and sedentariness. Unambiguous archeological excavations of large settlements suggest that there must have been knowledge of agriculture, in particular cereal domestication. Researchers can speculate from this that people must have been confronted with dung and fertilizing problems in the beginning of the agricultural age (Frossard *et al.*, 2009; Newman, 1997) if they were not settled in riverbanks, which were annually flooded (Löther, 2004).

Second, we look at data derived by cultural anthropological methods, in particular comparative linguistics (Bernard, 2006). We can study early Egyptian writings or the languages of indigenous people. Here the data are complete. For the study of prehistoric environmental literacy we look to writings from the *recent past* or on *present* language terms to infer about the *far past*. However, the study of language does not allow for an unbiased view of ancient times, considering that dynamic changes such as human migration come into play.

An illustrative example of how rapidly culture and knowledge adapts with a changing biological environment can be taken from anthropological research. Anthropologist Reichel-Dolmatoff (1996) provided a thorough analysis about how Kogi, a Colombian tribe of Indians, explained the history of creation in the midst of the twentieth century. According to reports from elders and shamans, honeycombs (in Kobi language: *hélyi*) were seen as a constitutive element of the lowest underworld. However, the honeybee was not present in the Americas before the fifteenth century (see Box 2.2). The much-studied European and African honeybee has been a very successful invasive species in America (Smith, 1991). The Native American Indians regarded the bee as a foreign insect which they called "white man's fly" (Pellet, 1938, p. 2). It becomes clear that there were no bees in North America when the Bible was translated into a Native American language. At this time there was no word in the Native American languages for wax or honey (Eliot, 1663).

Third, we should not forget about the data from natural history research. The environment changed profoundly in paleohistory (the past 2.5 million years). There is some evidence, for example, that rye had been already cultivated by 9000 BC in Abu Hureyra (see Figure 4.2). Such data provide coherent stories and also provide arguments about how human knowledge on the environment changed. For example, climatological data caused some anthropologists to assert that a drier climate during a certain period in Abu Hureyra caused the area's wild cereals to become scarce. From a functionalist interpretation, this would be considered a threat, causing people in that area to seek new knowledge to secure a food supply. This hypothesis is

4 Emerging knowledge on biology and evolution

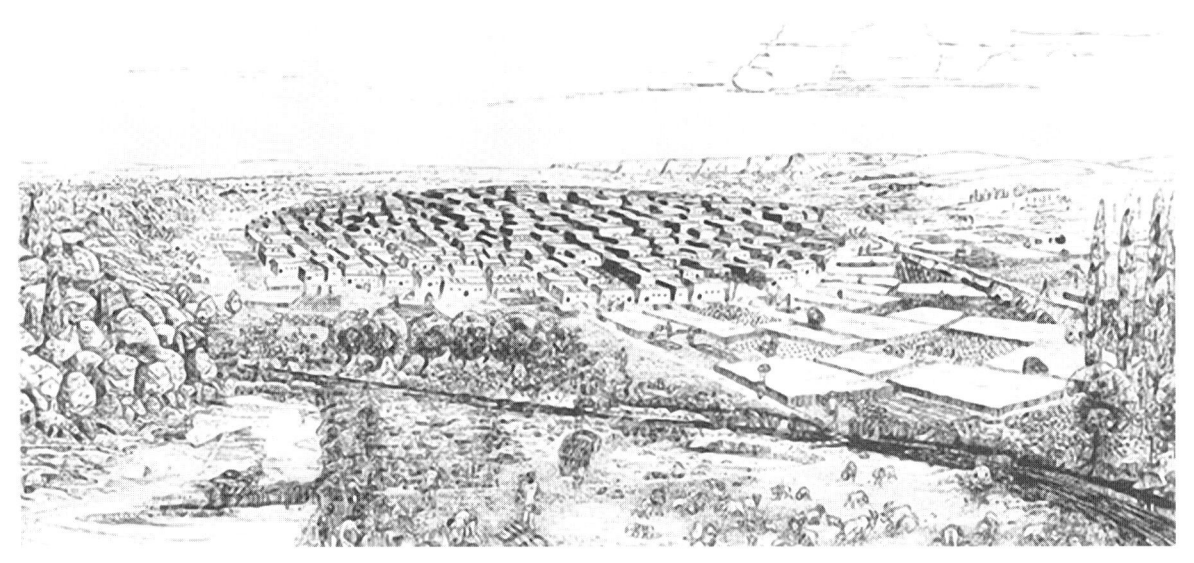

Figure 4.2 Archeological findings of the Near East Abu Hureyra (here a reconstruction referring to 7200 BC) suggest reconstructions that provide hypothetical insight into the environmental literacy of our ancestors, such as the domestication of plants and animals (Janick, 2002).

supported by agricultural scientists (with present data) showing that domestication could be achieved in two or three decades (Moore *et al.*, 2000).

New methods are improving data collection by archeological and natural history researchers. Besides the radiocarbon and DNA methods, there are novel techniques such as parenchyma, phytolith, and starch grain analysis. These methods focus on the analysis of fragmentary organic remains (plant assemblages), and can be of assistance in identifying species and their usage (Harris, 2006). Other important tools are pollen analysis which, for instance, was used to prove the deforestation of Easter Island palms (Roberts, 1993), paleomagnetic dating and other geological dating techniques. Nevertheless, as philosophers of science stress, reconstructing the past means striving to achieve some measure of authentic representation through historical narratives and pictures such as Figure 4.2 that tell stories "like it has been" (Norman, 1991).

Fourth, analysis of ancient DNA makes it possible to study the genetic relationships among extinct organisms. This allows, in principle, an insight into the impacts of hunting or agriculture and thus – implicitly – about how former societies interacted with fauna and flora.

Fifth, archeological and anthropological data have to be interpreted with respect to the cultural–epistemic settings (see Box 16.1). Here, in addition to the geographic setting (e.g. tropical forest vs. savannah), we also view data by considering prehistoric peoples' sets of beliefs and modes of thought or consciousness. With respect to the set of beliefs, hunters and gatherers such as the Penan, Malaysia, can provide insight into the type of knowledge available in indigenous people. The Penan, a tropical nomadic society, just recently became sedentary.

> [The] forest is alive … and plants are sacred, possessed by souls. It is the sophistication of their interpretation of biological relationships that is astounding. Not only do they recognize such conceptually complex phenomena as pollination and dispersal, they understand and accurately predict animal behavior. They anticipate flowering and fruiting cycles of all edible forest plants, know the preferred foods of most forest animals, and may even explain where any animal prefers to pass the night (Davis *et al.*, 1995, p. 35).

With a language of forest terms, where plants, humans, and animals are of the same nature, there is no reason to take a complementary view between human and environment systems. For Penans, plants "are born of the same earth that gave birth to people" (Davis, 1990, p. 99). The Penan are exceptional naturalists and do not differentiate the supernatural from reality. Thus, the supernatural, magical side of biotic systems is essential.

The Penan take on stewardship of certain important plants such as the sago palm (*Eugeissona utilis*), which is their carbohydrate staple. Plants are exploited in a manner that maintains their long-term availability. For example, fruit trees are marked with a wood marker or a machete cut, to signify effective ownership and to

4.2 The Herculean task of prehistoric research

Box 4.1 The difficulty of discriminating between humans and animals in archaic thinking: totemism

A critical question is, what kind of information about ancient people's perception of and relation to the natural environment is provided by the hundreds of cave paintings documenting animal species? In fact, paintings in more than 300 known Paleolithic caves include anthropozoomorphic figures (Jahn, 1990). "The Sorcerer of *les Trois Frères*" (Figure 4.3a) is a typical part-human, part-animal figure. It dates from 14 000 BC and is painted 5 m above the floor. The picture strongly deviates from earlier animal paintings which go back to 33 000 BC and mammoth ivory sculptures from southwestern Germany (Conard, 2003), which were much more realistic. The Sorcerer is a hybrid including antlers, owl eyes, lion or bear forepaws, a short foxtail and human feet. We can only speculate about what these ambiguous figures represent, including: (1) actual creatures that lived at this time; (2) fanciful beings invented by early humans; (3) hallucinations; (4) shamans in animal costume; (5) supersensory images of the shaman priest seen through the still-visionary consciousness of primal people; or (6) spirit beings (James, 2004). Given that antlered shamans are a common figure in Paleolithic paintings, the visualizations of shamans, as included in alternatives (4) to (6), seem likely.

One of the most narrative pictures is called the "Bird Hunted Man" (Figure 4.3b). There are many interpretations of it, ranging from a hunting accident to a symbolic presentation of a shaman's spirit flight. The latter refers to the interpretation that the disemboweled bison (one can see its hanging intestine) is a sacrifice to God. The shaman with erect penis is depicted entering a trance. The "bird staff is the magic symbol of the trance ascent, ... the vehicle upon which the shaman rises to the Upper World" (James, 2001, p. 228). Psychologists stress the symbolic and emotional contents (Klix, 1980) of such figures.

Early cave paintings have a powerful graphic quality and represent a strong relationship both to the animal world and to magic. The relation to the animal world had not only been a utilitarian, materialist, and food-based one, but also one that recognized animals as something "thinkable" (Lévi-Strauss, 1963). Archaic thinking on the biotic environment lacks the distinction of biological species and segments of society. This is called *totemism*, the belief in a kinship or mystical relationship between human individuals or tribes with animals (or plants), which provides "a code enabling man to express isomorphic properties between nature and culture" (Lévi-Strauss, 1963, p. 2), The latter is reflected in old folktales, where animals talk to each other and nature is filled with gods, demons, and spirits. Culture and nature are seen as isomorphic.

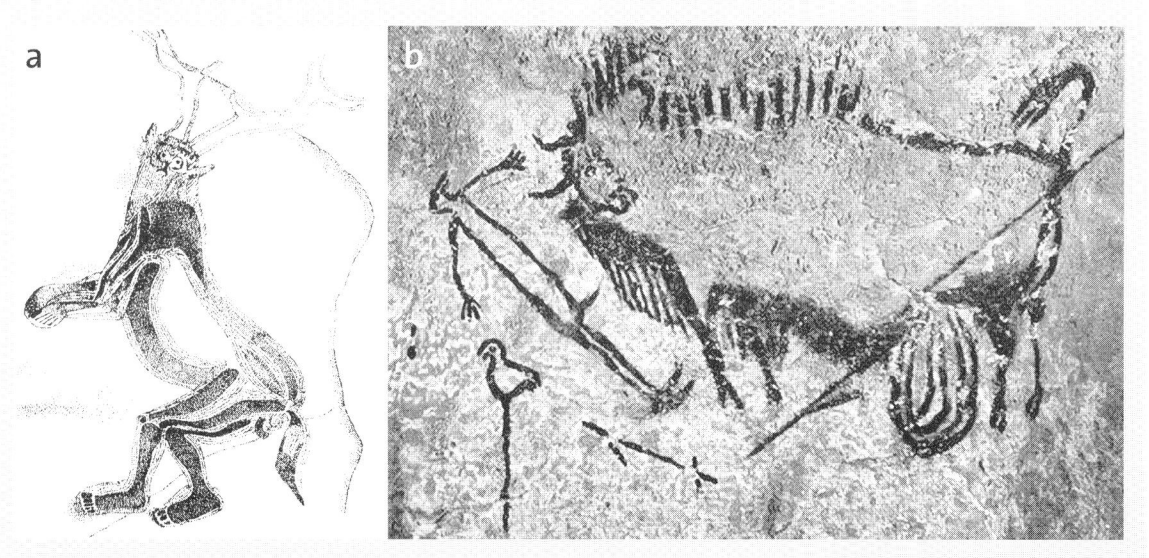

Figure 4.3 (a) "The Sorcerer of *les Trois Frères*" is an anthropozoomorphic figure that may depict shamans in ritual attire or spirit beings guiding shamans. (b) In the prehistoric cave painting "Bird Hunted Man" at Lascaux in France, the dead man besides a European bison has a bird-like head.

4 Emerging knowledge on biology and evolution

Figure 4.4 Letters and language are a carrier of biological knowledge (Frutiger, 1998, pp.112, 114, 122, reprinted by permission of Random House Group Ltd.)

Protohistoric "narrative," c. 10 000 BC

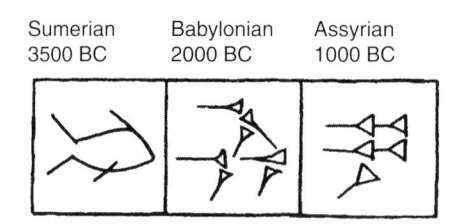

Sumerian 3500 BC | Babylonian 2000 BC | Assyrian 1000 BC

Fish

Development of Chinese characters

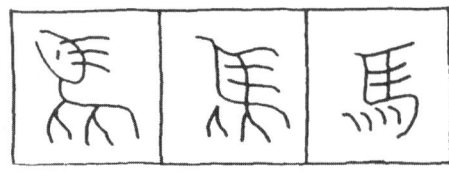

Horse (archaic) | Horse (modern)

Development of Latin characters

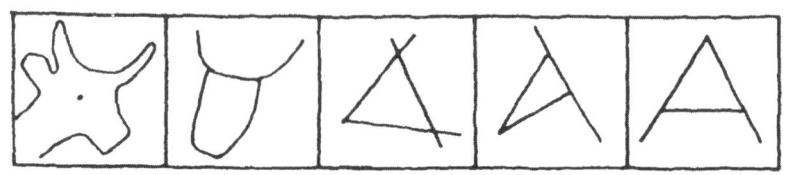

From the hieroglyph to the modern phonetic character A

serve as a public statement that the tree is to be preserved for harvesting at a future time (Brosius, 1997).

4.2.2 Key messages

The scientific reconstruction of biological knowledge held by prehistoric societies is an interdisciplinary, multi-methodological venture. It draws on new methods that relate cosmologies of the past with present natural and social science methods, worldview, and cosmology. This is a highly constructivist activity based on many explicit and implicit assumptions.

4.3 Prehistoric biology and ethnobiology

In this section we discuss how biological knowledge first emerged in human societies and describe five fundamental aspects of human development that allow researchers to gain insight into the environmental literacy of ancient civilizations. We also discuss ethnobiology, or how indigenous people organized and communicated about their environment. Through data from ancient Egypt, Ukraine, and Germany, we see that the need for medical treatment was a strong motivating force for early humans to learn about the human body. Data from ethnobiology and anthropology researchers show that early medical knowledge was, in some cases, basic and spiritual and, in other cases, exhibited some sophistication with anatomy, pathology and herbalism. Finally, we briefly review how functionalism and evolution drove ancient societies to develop more sophisticated biological knowledge.

4.3.1 Emerging biological knowledge in societies

Ambitious efforts to gain insight into the emergence of environmental literacy in prehistoric times (see Figures 4.3 and 4.4) show five aspects of development of early humans and their societies which seem to be most salient. In this section we discuss how these five aspects – brain development, patterns of subsistence, symbolism through language and paintings, agrarian knowledge, and religion – defined important foundations and constraints for generating knowledge about the organismic environment.

(1) Brain architecture: here we consider the biological component of human intelligence and evolution,

and we assume that the human brain developed in the past 150 000 years. Anthropologists concurrently state that there was a shift to a modern, "fully symbolic *sapiens*" 40 000–50 000 years ago (Henshilwood & Marean, 2003; Takács-Sánta, 2004). The size of the human brain did not increase in the past 150 000 years but rather declined in some areas due to meat in the diet (Ruff *et al.*, 1997). Though there are no physical data providing affirmative evidence, the expansion of analytic, higher-order thinking carried out in more elaborated brain structures seems likely. The dominating theory is that the human species spread from Africa 30 000–35 000 years ago to the rest of the Old World. It is disputed how homogeneous the brain is (Henshilwood & Marean, 2003; Lahr, 1994) in the human species. It is a paradox that there is relatively limited human genetic variation but extraordinarily high cultural differences. The human species today has a significantly smaller genetic variation worldwide than chimpanzee populations that are separated by a few kilometers (Jobling *et al.*, 2004; Mithen, 2006).

(2) The predating/foraging to hunting/gathering transition: the expansion of humans from Central Africa to other geographic areas of the world, along with 120 000 years of cyclical changes in climate zones (Holmes, 2008), meant that indigenous people had to develop geographically specific foraging and survival techniques, or ecological niches (Hardesty, 1977). There are differences in the way nature is appropriated as we can learn from the distinction between hunting/gathering and predating/foraging. According to cultural anthropologists, the hunter–gatherer sought socially acquired knowledge about biotic systems:

> The gatherer–hunter is a self-conscious agent, confronting a world of plants and animals from which he or she selects those that will furnish suitable raw materials for consumption. The forager–predator, to the contrary, merely responds to the presence of environmental objects. (Ingold, 1986, p. 3)

Whereas predating–foraging activities are driven by fixed, genetically programmed behavioral rules, the knowledge required for hunting–gathering is invented, modified, communicated, and transferred to others. Groups of 10 to 50 people developed physical (mimical and gestural) and phonetic means of expression. There are different evolutionary theories about the differentiation of abilities and knowledge such as the Savannah Theory (Orians, 1980), which postulates that humans developed specific skills and characteristics to help them survive in their regional environment.

Archeological studies focus on the development of tools, the development of sedentariness during the interglacial period between 60 000 and 120 000 years ago, and the social organization including rituals and the emerging of thinking (Klix, 1980).

Some cultural anthropologists stress that the switch from predating–foraging and hunting–gathering goes beyond pure extraction of biotic resources from the environment, because it is linked to an appropriative action of making something one's own property. This seems to be saliently evident for pastoral herders, who mastered biological behavior and even intervened in biological programming by selective breeding and killing.

Ancient biological knowledge about fauna and flora for nutrition, hunting, healing, safety and other reasons served continued and improved existence of tribes (societies) and can be considered from a functional ecosystem service and subsistence strategy perspective. Provision- or subsistence-related knowledge has been necessary for gathering wild plants and small land fauna, trapping and fowling, etc. Whereas the predator–forager simply responds to environmental objects, the hunter–gatherer constructs and transmits knowledge to the following generations. Ingold (1986) stresses that hunting and gathering is based on intentional action including a social organization and flexible, learned, practical operations for storing, aggregating and processing (e.g. cooking) the items. The activities of the hunting and gatherer society included "gathering of wild plants and small land fauna … trapping and fowling … shellfishing and the pursuit of large aquatic animals" (Murdock, 1967, p. 154), which means that action is guided by what people need to do rather than what to eat. The latter would feature the predator. As the knowledge has to be transferred from generation to generation without written documents, we meet a kind of collective memory about fauna and flora, which can be called ecological knowledge as a specific type of environmental literacy.

There are many theories stating that the geological setting required different cognitive skills and biological knowledge (Henshilwood & Marean, 2003). The spatial productivity of an ecosystem affected the size of hunting and gathering communities, cultural traits and the knowledge necessary to sustain communities (Ellen, 1982). Given their situated knowledge, one must acknowledge that non-literate, ancient people knew much about nature.

(3) Symbolism: the development of language and paintings, dating back at least 33 000 years (see Box

4.1) was a symbolic means of communication that went beyond gesture and animal-like communication. The agrarian revolution and a gradual development which goes beyond pure experiential knowledge, including abstracted models about life, are unthinkable without language (see Figure 4.4). We call this the capacity for symbolic behavior, which shaped the modern human species about 50 000 years ago, the socio-epistemic dimension H_s (see Figure 3.3*). As we will see and discuss below in the Ethnobiology section, some researchers consider certain features of language such as grammar as culturally invariant.

(4) Agrarian knowledge: the agricultural revolution, the domestication of plants and animals, started in different unconnected regions of the world remarkably synchronously about 12 000 BC, with horticulture. This revolution dramatically changed the relationship between humans and the environment. Knowledge, language, and culture had to be adapted. Division of labor can be seen as an important aspect for increasing the complexity of the social environment. And the new dominant type of reproduction called for an altered cultural superstructure of mixed farming systems and combining of agricultural and pastoral production. Biological knowledge was required for many respects, including separating and preparing toxic pulses and gaining insight into biological fertilizers and the energy cycle involved in mixed farming systems (Mithen, 2006).

(5) Religion: "Religion is a human quasi-universal" (Cohen et al., 2004, p. 735) and a basic component of the environmental literacy of prehistoric civilizations. The purpose of this section is to show how religion affected early humans' construction of nature and the structure of organismic beings. Dealing with ancient religions brings us to the present–past obstacle of anthropological research (see Figure 4.1). Given that prehistoric data is fragmented, we are compelled to make humanities-like inferences to construct plausible, meaningful "stories" consistent with the available data. For example, according to the phylogenetic perspective taken by anthropologists, we assume that animals and early stages of *Homo erectus* are too primitive to have been capable of religious ideas and practices, if we conceive religion as:

> the firm conviction on the believer's part … of the actual existence of supernatural, supersensory order of being, and of the actual or potential interplay, through a network of sacred symbols, of that order of being with the world on which normal life is lived. (Sharpe, 1983, p. 49)

To understand the role of supernatural systems in the emergence of biological knowledge and the meaning that early humans gave to biological systems, we introduce five stages of development. These stages provide a rough order and identify main types of religious beliefs without acknowledging specific variants or hybrid forms.

The first stage is one without sophisticated cosmological thinking and religious ideas and does not involve supernatural forces, at least for Paleolithic or pre-Paleolithic times (Watkins, 2001). At that time subsistence was relatively basic. Anthropologists assume that people had no access to complex interactions with nature and that surviving when hunting dangerous "prime adult animals … [such] as the Cape buffalo and the bushpig" (Henshilwood & Marean, 2003, p. 629) was a main challenge.

The second stage occurred during the Upper Paleolithic, when *Homo sapiens* were competing with *Homo neanderthalensis* and when language developed. This time is known for development of the "slower, serial, explicit and language-bound reasoning system" (Coolidge & Wynn, 2005, p. 22) thought to be based on genetic mutations affecting the prefrontal cortex which appeared 60 000 to 130 000 years ago (Coolidge & Wynn, 2005; Mithen, 1996, 2007). At that time the first anthropomorphic menhirs, cult statues, and cave paintings can be found. Anthropologists express what happened about 50 000 years ago in the following terms:

> Somehow things change around the beginning of the Neolithic, whether we are looking at the epi-paleolithic–Neolithic transition in south-west Asia or the Mesolithic–Neolithic transition in Europe. We may feel with Jaques Cauvin (2000) that the most important changes at the time concern *The Birth of Gods* and are "psycho-cultural", that is psychological changes relating to the operation of culture. (Watkins, 2001, p. 3)

Anthropologists think of this period as the time when mythic culture emerged. From the genetic epistemological perspective of Piaget (1970), totem cave paintings suggest how humans found the first signs or words to communicate via symbolic representations of the texture of the world (see Box 4.1). For simplicity, we call this second stage "nature religion," a time when magical, supernatural forces and the potency of the spirits formed the texture of thinking (Davis et al., 1995, p. 15) and when people placed animal parts into human graves. Mithen (2007) described this time as the origin of the body and mind, when the emergence of intellect allowed humans to read the behavior of others (including that of life-threatening prey such as the Cape buffalo) and to develop social intelligence and a reflexive conscience.

The third stage began around 6500 BC, at the beginning of the Neolithic. A recent shift to living in villages and using horticulture technologies also brought a focus on nature religion and ancient animistic traditions. "As the cult of the seed overthrew the hunt, so the priest displaced the shaman" (Davis et al., 1995, p. 15). A hierarchical religious system emerged, as did many monuments indicating different gods such as the gods of bird, bear, harvest, and fire. People thought that gods and goddesses were at work in the underworld, earth, and sky (Gimbutas, 1982; Lewis-Williams & Pearce, 2005). Shamans and ritual specialists were still around, but clues from Egyptian culture (see Box 4.2) show that the division of labor was also incorporated into the religious and healing systems, seen in the documented use of plants and animals for healing and religious purposes.

A fourth stage can be seen exemplified in the concept of monotheistic religions such as the Judeo-Christian system. In particular, Christianity had been conceived

> … the most anthropocentric religion the world has seen. … Christianity is in absolute contrast to ancient paganism and Asia's religions (except, perhaps Zoroastrianism), not only established a dualism of man and nature but also insisted that it is God's will that man exploit nature for his proper end.
>
> Man named all the animals, thus established his dominance over them. God planned all of this explicitly for man's benefit and rule … The spirits *in natural* objects, which formerly protected nature from man, evaporated. (White, 1967, p. 1205)

Lynn White's *Science* paper considers modern science and "man's transcendence of and … mastery over nature" (White, 1967, p. 1206) as an offspring of Christianity. Thus science can be viewed as a fifth stage of belief system, yet without transcendent or supernatural forces. The appropriation of nature including the exploitation of resources is the subject of science-based technological development.

4.3.2 Ethnobiology

Ethnobiology stands at the interface between biology, culture, and sociality (Ellen, 2006). It investigates how indigenous people classify and reason about the organic world. Recently, ethnobiology has turned to dealing with the ecological consequences and to "the rights of indigenous people to control their traditional knowledge" (Hunn, 2007, p. 1). With respect to prehistoric knowledge, ethnobiology is facing a present to past methodological challenge. Today indigenous peoples' knowledge is looked at from a modern perspective as a bridge to the past. Indigenous peoples' language is and has been a major source for ethnobiological research. This approach is often based on a nativist conception in line with Chomsky (1975) and Pinker's (2000) assumption of a "language instinct," the idea that humans are born with an innate capacity for learning language.

> Human minds appear to be pre-tuned for acquiring and processing information about animals and plants such that virtually no active matching is required. (Mithen, 2006, p. 53)

Such intuitive biology, and the markedly similar way that early people who were located in different parts of the world cognized different species, is documented by Brent Berlin's (1992) book *Ethnobiological Classification*. He showed that primordial people from different geographic regions utilized the same, usually, five-tiered taxonomic hierarchy for classifying plants and animals, segregating dogs from cats, trees from grass, etc. And the ranking is overly coherent with modern Linnéan classification (Sprockhoff, 1880). Berlin rejects a rough, direct functional perspective and instead postulates an equilibration-like curiosity when arguing:

> Human beings are drawn by some kind of innate curiosity to those groupings of plants and animals that represent the most distinctive chunks of biological reality. (Berlin, 1992, p. 290)

Berlin calls this phenomenon "folk biology." A special and interesting feature of folk biology is that it ranks species and that the indigenous people have common names for all birds. It is interesting that the classifications seem to be independent of the utility and the symbolic meaning of the plants and animals. The tribe of the Fore of New Guinea identified 136 out of the 137 species which were classified according to ornithologists such as Jared Diamond (1966; Berlin, 1992; Mayr, 1963).

Berlin identifies three types of common pre-tuning of the human mind in the formation of language (Berlin, 2006). The first is onomatopoeia, which is that language began as imitations of natural sounds, such as "moo" and "meow" (also called "bow-wow" theory) or emotive cries such as "oh!" for surprise or "ouch!" for pain (also called "pooh-pooh" theory, Müller, 1888). The second is metaphorical description, and the third, sound symbolism or phonosemantics. Sound symbolism refers to the Gestalt psychology (see Box 6.7) and "proposes that even the tiniest sounds comprising

4 Emerging knowledge on biology and evolution

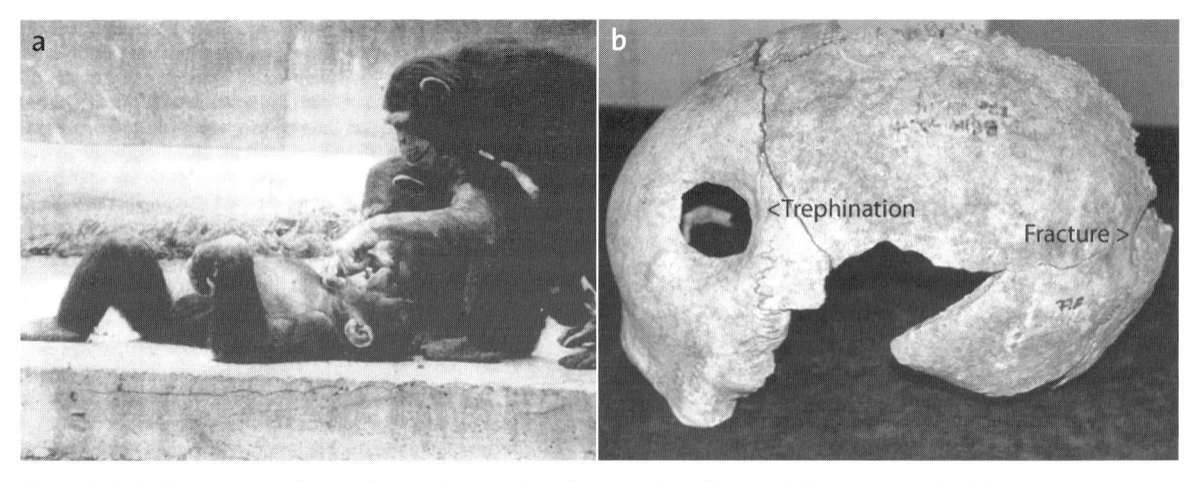

Figure 4.5 (a) Chimps (from McGrew, 1992) and (b) people of the Stone Age (from Piek et al., 2008) intervene in body function to cure diseases.

a word may suggest the qualities of the object which that word represents" (Abel & Glinert, 2008, p. 1836). Although there is some experimental evidence, various methodological questions arise.

The ethnobiological research about the Linnéan grid of category formation, classification and knowledge transmission has been investigated from the perspective of the evolution of human reasoning. The idea is that cross-cultural similarities in ethnobiological classification are a legacy for an "evolved predisposition", that is a strong nature component in the cognitive evolution of humans. Here we see special components of environmental literacy: The question of how "natural history intelligence might have evolved" (Ellen, 2006, p. 7), whether we can expect to find "universal cognitive systems for engaging with the natural world" (Mithen, 2006, p. 46), and what role the cultural process can take. At this point we should remind ourselves that understanding (German: *verstehen*, see Box 2.6) of life requires not only knowledge of biology, but also that knowledge from experienced sociocultural domains and observed biological domains are linked.

4.3.3 Medical ethnobiology and anthropology

Medical anthropology investigates why people get sick. Medical ethnobiology deals with understanding patterns of treatment (Waldstein & Adams, 2006). We assume that even in prehistoric times, the need for medical treatment motivated human systems to: (1) seek insight into the biology of the human body (How does the body function? Why is it out of balance?); (2) identify environmental biological causes of diseases (Why am I ill?); and (3) find effective treatments of the human organism by biophysical remedies (What can I do to cure my disease?).

With few data and no written records to guide our study of these goals (1) through (3), we rely heavily on data from the present and the near past, which are indirectly related to prehistoric times. One approach to understanding knowledge and fundamental patterns of human behavior is the ethology of human and other primates (Eibl-Eibesfeldt, 1989; Lorenz, 1981; Tinbergen, 1951). Figure 4.5 highlights an interesting behavioral pattern of chimpanzees and illustrates how disease is an object of social action and meaning without having speech (i.e. a phonetic, spoken language). One chimpanzee, Belle, has a disease. She can communicate the location to another, Bandit, who seems to know, or has experience with the treatment, whereas a third is watching the therapy. The therapy is supported by a technological device. This is a small stick, which might increase the efficacy of extraction. If non-human primates are aware of diseases, look for therapies and make highly cognitive investments, such as a chimpanzee's invention of a technology, we can conclude that at least finding effective treatments for physical disorders of the human body, i.e. (3), would have been a primary interest of prehistoric societies.

A different type of indirect evidence about how prehistoric people might have conceptualized diseases is provided by current anthropology on indigenous people (see Figure 4.7). Data show that early medical knowledge was, in some cases, basic and spiritual and, in other cases, exhibited some sophistication with anatomy, pathology and herbalism. Some researchers claim that Mayan healers such as the "Tzotzil have only vague and elementary knowledge of the human body …" (Holland & Tharp, 1964, p. 44). This view is supported by other anthropologists' statements, which claim that in certain cultures physiological aspects do not play an important role.

> Traditional healers, it would seem, have a vested interest in ignoring or downplaying physiological symptoms when they profit from dealing with personalistic causes of illness such witchcraft and soul losses. (Waldstein & Adams, 2006, p. 103)

This has been already pinpointed by Fabrega and Silver (1973, p. 211), who state "that Mayan traditional healers do not have specialized knowledge about the body" (Waldstein & Adams, 2006, p. 103). But, as described below, other researchers have found evidence of the efficacy of ancient, traditional healers and shamanist therapy.

Some direct data about Stone and Bronze Age people's interventions on the human body are provided by various skull findings that show trephinations. According to ^{14}C-dating, skull findings from Ukraine date back to 7300–6220 BC (Lillie, 1998), skulls found at the French–German border date back to 5100–4900 BC (Alt et al., 1997), and skulls found in Bölkendorf close to Berlin, Germany are from about 1940 BC (Piek et al., 2008). Given that the Bölkendorf skull had a fracture, we can speculate that the trephination was for medical reasons. Other evidence from regrowth of bones shows that many of the patients survived trephination, suggesting successful medical treatment. Other more speculative guesses about the motivation for these early trephinations include, "… to prevent complications, e.g. a suspected haematoma under the local bruise of the skin" (Piek et al., 2008) or to release evil spirits (see Box 4.1).

Archeological and historical research on Egyptian medicine (Nunn, 1996) provided excellent insights into past people's biological knowledge. John F. Nunn's fundamental research provides a comprehensive picture of ancient medicine, covering anatomy, physiology and pathology and the "biology knowledge based" remedies these people apply. From the papyri dating as far back as 1820 BC as well as more ancient stone engravings, we can infer that the Egyptians developed a very differentiated knowledge about the anatomy of the human body and its pathologies. There was a very wide range of plant- (in particular vegetable-), mineral-, and animal-based substances that were systematically applied and targeted to "eradicate the cause of a disease rather than merely relieving symptoms" (Nunn, 1996, p. 137). For instance, papyrus Ebers dating to 1500 BC documents a large set of plants just for disorders of the urinary system. There are about 40 pharmacological properties which could be attributed to the extracts, and the plants that could be identified based on the hieroglyphic. Thus, from today's pharmacological perspective, there seems to be a biophysical rationale for the application of many remedies (see Box 4.2).

Naturally not all remedies meet the standards of present medicine. Some appear strange and dubious. And some might have caused troubles for the ancillary staff, particularly the pharmacists who may have been responsible for producing them. Thus the papyrus Kahun (Kahun 21, 1820 BC) suggests gathering excrement from crocodiles, and the Ebers bile of tortoise (Ebers 347) or "the excrement of flies which is on the wall" (Ebers, 782, quoted according to Nunn, 1996, p. 138). Here, if we assume that the therapies have been successful, the symbolic function of the remedies appears to be more important than the biochemical one. Nevertheless, 160 plant products were used, but only 20 of them can be unambiguously identified (due to various reasons including difficulties of translating the hieroglyphic language). But some of them have definitely proven therapeutic efficiency.

At the same time, there is clear evidence, including hoodoo images drawn on Papyrus Ebers, that ancient Egyptians made great use of invoking deities in the curing of disease. Thus, when considering patterns of treatment by prehistoric people, two realms can be distinguished. First, we can identify a personalistic–psychic–symbolic approach focusing supernatural, magic, witchcraft-like, illness-causing spirits-based causes for disease and therapeutic success. Second, there is a naturalist–biophysical approach focusing on medical plants, animal substances, and minerals.

The personalistic–psychic–symbolic approach focuses on antagonistic causes such as malignant spirits. Key actors in this approach are shamans and folk healers. These approaches do not differentiate between the living body and the soul. From a contemporary perspective, they are targeting the mind (or what we have called the socio-epistemic side of the human system), if we consider the belief in supernatural forces as a (functional–social constructivist; see Chapter 3) mental entity or construct.

4.3.4 Meaning, minds, and placebo

Although double-blind studies are missing, evidence suggests that shamans and folk healers can sometimes

4 Emerging knowledge on biology and evolution

Box 4.2 At the crossroads between naturalist and magic impacts: ancient Egypt's medical knowledge

There are many medical papyri starting from 1820 BC including the famous 110 page papyrus (1500 BC), which Georg Ebers (1837–1898) purchased in 1862 in Luxor for the Leipzig museum in Germany. They include detailed descriptions of the treatment of skin, belly, stomach, anus, heart, teeth, and gynecological diseases, tumors and swellings due to various causes (such as snake bites), and pregnancy, but also depict hoodoos and many deities (see Figure 4.6c).

Doctors worked in a multilevel hierarchy that included administrators, overseers, inspectors of doctors, chiefs of doctors, and the doctors themselves. It is disputed whether doctors were trained in an institution such "per ankh" or "house of life," whether these served as a clinic, medical school, university, library, seminar and temple or whether they served just as scriptorium where texts, including medical documentations, were written (Nunn, 1996, p. 131).

Early papyri provided detailed descriptions of human anatomy and pathology. The important role of anatomical knowledge is reflected in the instance that the anatomical knowledge of mammals and diseases was shaping hieroglyphs, Egyptians' written language (Figure 4.6a). The papyri present an impressive and differentiated description of diseases and remedies.

Nunn surveyed 16 herbal remedies (including acacia leaves and roots of pomegranates), five minerals (including salt and desert oil), four honey remedies and vehicles such as milk, beer (of different types), and wine which had been given for the therapy of roundworms (*Ascaris*) and tapeworm (*Taenia*), which had been common diseases (Nunn, 1996, p. 72). From a modern perspective, the effect of many of the aforementioned medicines may seem limited but, in some cases, not unreasonable. In sum, we can conclude that Egypt ancient medicine laid the groundwork for modern natural science-based medicine.

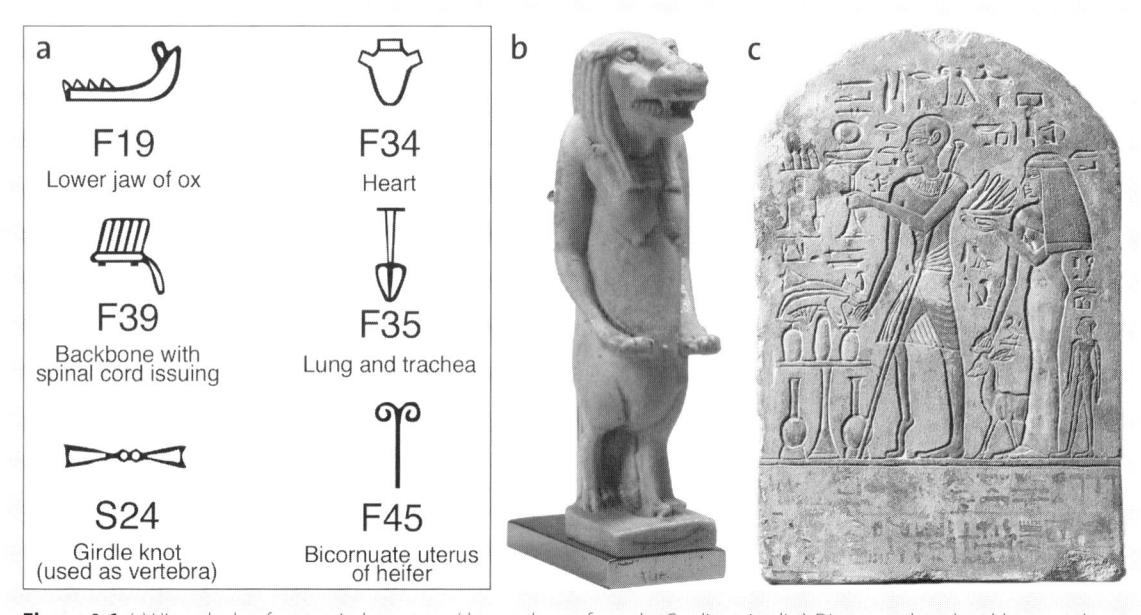

Figure 4.6 (a) Hieroglyphs of anatomical structures (the numbers refer to the Gardiner sign list). Diverse gods and goddesses such as the protective goddess Taweret (b, bpk The Metropolitan Museum of Art) and graphs that depicted diseases, malformations (such as club foot, talipes equinus), and therapies could be found in papyri, steles or cippi (c, © Ny Carlsberg Glyptotek).

be successful. This statement holds true, at least in part, due to the placebo effect (Bausell, 2007; Benedetti *et al.*, 2005; Moerman & Jonas, 2002). In addition to the design problems of doing research, there are multiple discussions and conceptual classifications regarding the shamans' use of placebo treatment. These new explanations go beyond the traditional expectation theory (the patient believes that the treatment will help) to focus on the psychosocial context, different neurobiological mechanisms in different contexts. Thus the term "contextual healing" (Kaptchuk, 2002; Miller & Kaptchuk, 2008) is introduced to reflect how the results of some studies showed that, overall, the effect of drugs to help patients heal is smaller than physicians thought.

Modern clinical approaches study "how the context of beliefs and values shape brain processes related

4.3 Prehistoric biology and ethnobiology

Box 4.3[1] Mayans' physical and symbolic treatments of diseases: "me'winik" or gallbladder cancer?

By examining the Maya people's conception of diseases, we can better understand non-western views of the human being. Some insight has been gained in the ethnomedical "me'winik" syndrome, which is sometimes linked to gallbladder cancer. As there is evidence that Amerindians, and in particular the Tzeltal and Tzotzil Maya populations of Highland Chiapas, have a higher rate of this cancer (Berlin *et al.*, 1993; Weiss *et al.*, 1984; Wiggins *et al.*, 2008), we exemplarily deal with cancer as an extraordinarily severe disorder.

Mayan cultural logic is based on the belief that continued human existence is predicated on the maintenance of a cyclic cosmic balance that refers to both an individual's and earthly conditions. The key issue is the harmony between the physical and the metaphysical world, which includes a harmony with the spirits and authorities of the world (see Box 3.2).

According to Mayan culture, Maya must treat each being and all natural resources with respect, and illnesses are thought to be a consequence of deviant behavior. Besides sociocultural deviations such as "disrespect towards elder or ... refusal to meet obligations of reciprocity with neighbors and relatives, rejection of Mayan culture and the adoption of characteristics considered *ladino* or foreign", disrespect shown to "the 'owners' of natural resources" (Nigh, 2002, p. 458), polluting water or cutting down a tree without asking the authorities' permission are also considered deviant behavior.

The Mayas think that cancer, as a severe disease, can – for instance – result from a stern unbalance or from causing pain to other aspects of creation. Thus, cancer can be considered as the cosmological payback to a local disorder linked to a person. As a part of nature, the law of action and reaction among the Mayans stipulates that he or she is basically causing his or her own distress. The person can only receive back from nature what he or she is giving to it, consciously or not, positively or negatively.

Mayan medicine is strongly shaped by medico-religious healing procedures, targeting efficient access to the life force (*ch'udel*, see Figure 4.7). Mayan people think that they can get access to the natural spirit world by ceremonies and specific rites involving the emotional, mental, and spiritual dimension. Thus, in the case of cancer, the symbolic treatment is by far the most important measure.

But also herbs play a role. There are some controversies about Mayan knowledge about herbs and the effectiveness of herbs. Nevertheless, more than 25% of the ingredients in today's cancer medicines were either discovered in rain forests or are analogues or derivatives of such compounds (Borchardt, 2004). However, Nigh warns against a simplified belief in the healing power of plants: "Herbal fetishism conceals the social relations and cultural context of Maya healing" (Nigh, 2002, p. 460). Blood pulsing is another component. Tedlock (1992) documents that Mayan healers have a sophisticated access to the pulse. They consider blood as a speaking system and healers are able to identify at least 28 different pulse types (Balick *et al.*, 2008). Finally, the hot–cold principle has been used as a good case where an external factor (e.g. cold drinks or hot foods) is mixing with an internal, symbolic one to balance disorders. Here gastrointestinal problems are a good example, which are mostly easily treatable.

Figure 4.7 "Me'winik" is an ethnoepidemiological phenomenon (Berlin *et al.*, 1993) of the gallbladder often related to cancer. Mayan medical treatment ranged from massage through herbal treatment to spiritual ceremonies (reproduced from Berlin *et al.*, 1993, p. 673).

[1] This box has been written together with Mónica Berger González.

to perception and emotion and, ultimately, mental and physical health" (Benedetti *et al.*, 2005, p. 10390). Similarly, an interesting approach which is compatible with the proposed mind–body (i.e. social–epistemic vs. material–biophysical layer of human systems, see Figure 3.3*) complementarity has been provided by the meaning concept. This concept proposes that the mind interprets meaning to information which activates brain functions (visible on functional magnetic resonance imaging) that activate psychosomatic processes supporting healing. Successful signs of meaning conveyed by present medicine are the "physician's costume (the white coat with stethoscope hanging out of the pocket) … and prognosis" such as "You will be fine, Mr. Smith" (Moerman & Jonas, 2002, pp. 473, 485).

There are many common metaphorical, symbolic cross-cultural phenomena. One is the humoral or hot–cold dichotomy, also found in Hippocratian Greek medicine. Here the dominant feature is that hot illnesses have to be healed by cold remedies and vice versa. Garlic, coffee, and alcohol are "hot" whereas beer, lime fruit, and cow's milk are perceived as "cold" (Foster, 1988).

The naturalist–biophysical approach of ethnomedicine should be looked at with some reservation. There are dozens of ethnomedical compendia about all parts of the world (Etkin, 1986). While few documented studies show what effects are prevalent (Ankli et al., 2002), Box 4.3 on the Mayans' conception of diseases shows that ethnomedicine includes physical, biological, and symbolic components.

4.3.5 Drivers of non-literate societies' biological knowledge

From an environmental literacy perspective, we wonder why ancient societies developed biological knowledge. One possible answer draws on a utilitarian perspective, which meets the functionalist and evolutionary perspective, and states that those who had better knowledge and a more realistic cosmology had advantages.

> In order to be effective agents in the natural world, animals require the guidance of a "world model", an internal representation of what the world is like and how it works. This model enables them to predict in advance the characteristics of "recognizable" objects, to anticipate the likely course of events in the environment, and to plan their behavior accordingly. The role of classification in this context is to help organize sensory experience and to introduce an essential economy into the description of the world. An effective classification system will reduce the "thought load" on the animal, expedite new learning, and allow rapid and efficient extrapolation from one set of circumstances to another (Humphrey, 1984, pp. 126–7).

In a biological context, for example, classification offers a reduction of complexity and a more economic coping with environmental resources as it is postulated by "ecological rationality" (see Box 1.4).

> People must be able to recognize, categorize, and identify examples of one species, group similar species together, differentiate them from others, and be capable of communicating this knowledge to others. (Greene, 1961, quoted by Berlin, 1992, p. 5)

Berlin's approach promotes the position that people are genetically predisposed to develop a useful classification of plants. But he also strongly argues that this performance includes a curiosity-based intellectualist component and is not solely a "utilitarianist's" venture. According to Berlin, the motivation is a cognitive, unconscious appreciation of nature's biological affinities among groups of plants and animals. The ethnobiologist Berlin also empirically tested his hypothesis against the alternative view that the classification of plants (including genus, species, and taxa of life form and intermediate rank) is based on a utilitarianist–economic or a utilitarianist–cultural–symbolic view in different hunter and gatherer societies of South America. This view contrasts positions which show that certain tribes classify plants that cannot be used for nutrition and maize to the same level as the existence or wealth of a society depends on it according to utilitarianist (Leontjew, 1964) or mythical, non-utilitarianist reasons (Levy-Brühl, 1921). We can see that a different nature–nurture controversy also underlies views on the emergence of biological knowledge (see Chapter 6.1).

4.3.6 Key messages

- Understanding prehistoric biological knowledge seeks to integrate fragmented data about the past material–biophysical environment with current, recent, or historic interpretations of data about the social–epistemic layer of human systems
- The representation of gods, demons, and spirits as anthropozoomorphic figures indicates that fauna–flora and human beings were not seen as a duality by prehistoric people, but rather as a kind of sameness which is governed by supernatural forces
- The main developmental components of *Homo sapiens*' biological knowledge in prehistoric times were the: (1) evolution of the human brain; (2) shift from hunting/gathering to predating/foraging patterns of subsistence; (3) development of language; (4) technology-driven agrarian transition; and (5) religious thought about how supernatural agents govern life
- Reconstructing religions identifies five stages of belief system on how supernatural agency affects biotic and living systems ranging through: (1) simple non-symbolic representations of supernatural agents; (2) totemism and animism, postulating a general life-controlling agency; (3) polytheism with

different gods, shamans, priests, etc.; (4) the prehistoric monotheistic stage; to (5) a science-based belief system without supranatural systems
- Ethnobiology documents some empirical evidence about a common "cognitive pretuning" of indigenous people, suggesting an experiential biological classification instinct resulting in a Linnéan grid
- From a functionalist perspective, curiosity- and utility-driven development of literacy about living systems was a prerequisite for the development of technologies and transitions of societies (e.g. predating/foraging to hunting/gathering to agrarian).

4.4 Antiquity's conceptions of the biophysical dimension of life

This section examines how people of earliest recorded history understood the biophysical world. We explore early cosmology and how, for example, ancient societies in China, Greece, and Egypt organized the elements, and also discuss emergence of the first research and early signs of a shift to secular inquiry and cosmogony. We review ideas from early theorists of biophysical and medical processes, including Hippocrates' four humors of medicine and Aristotle's early ideas for classifying animals. Finally, we consider, as the legacies of Greek thinking, how humans understood life around the year 150 BC.

4.4.1 Cosmology, cosmogony, and consciousness

In many parts of the world, the middle of the first millennium before Christ was shaped by the ideas of big thinkers, philosophers and prophets who offered fundamental ideas about how the world and life is organized. Thales (*c.* 625–547 BC), Buddha (*c.* 563–483 BC), Confucius (551–479 BC), and perhaps Zoroaster (between 1000 and 500 BC) all lived in the same period. Different questions dominated in the largest world dynasties.

Egypt and Mesopotamia had complex agricultural systems. Water, earth, and air were considered as the basic elements of the material world. Mystery and gods played a dominant role in constructing models to explain why, for instance, the floods of the Nile were predictable whereas those of the Tigris and Euphrates were not. The yin–yang complementarity was seen as an essential of Chinese thinking. Chinese philosophy supplemented water, earth and air with metal and wood in a circle of generation and destruction (see Magner, 2002, p. 5). In the cycle of generation, water produces wood, wood produces fire, fire creates earth (ash), earth is the source of metal, and metals – when heated – flow like water. In destruction, water puts out fire, fire melts metal, metal (axes) cut wood, wood(en) plows turn the earth, and earth(ern dams) stop water. Indian thought, and particularly the Hindu concept of *Ayurveda*, considered the universe to be of immense size and antiquity undergoing a continuous development and decay. Compared to Egyptian natural knowledge, Chinese and Hindu knowledge and medicine have been not highly developed. The primary elements of the Greek world model were water, earth, air, and fire which were considered as eternal principles of the cosmic cycle which generates everything (see Figure 4.8).

We exemplarily deal with the ancient Greek concept of organism, biology, and medicine. As we will see, medicine has been of interest. One can state that the first material, biophysical, and environmental impacts on organisms were discussed in the context of Hippocrates' medicine. From its beginning about 800 BC, Greek natural sciences have been mostly a top-down metaphysical theory of ideas.

> Hippocrates, for example, when trying to determine the cause of a disease, did not look for the divine influence but attributed it to natural causes such as climate and nutrition (Mayr, 1982, p. 25).

We present some concepts and processes for an understanding of the organism. Greek philosophy emerged from mathematics and physics (Pythagoras of Samos, *c.* 570–495 BC), to be supplemented by ethics (Socrates, *c.* 469–399 BC), and the medical, life and natural sciences, which were shaped in part by Hippocrates of Kos (*c.* 460–370 BC, Ballauff, 1954). Rather than systematically dealing with the different epochs, we introduce the year and dates which each philosopher or "scientist" is from.

Between 1000 and 200 BC, Greek territories were organized into city-states (Polisies). Smallholder agriculture (without irrigation systems), wine, and, later, olive farming were typical features of Polisies. Human subsistence was dependent on trade and imported goods and hunting. The resulting radical declines in game population and deforestation from mining and shipbuilding activities (see Box 1.3) were not a concern of these early societies and they did not become the focus of systematic investigation.

The Greeks used their technological sophistication with iron weapons to conquer rich Bronze Age cultures

4 Emerging knowledge on biology and evolution

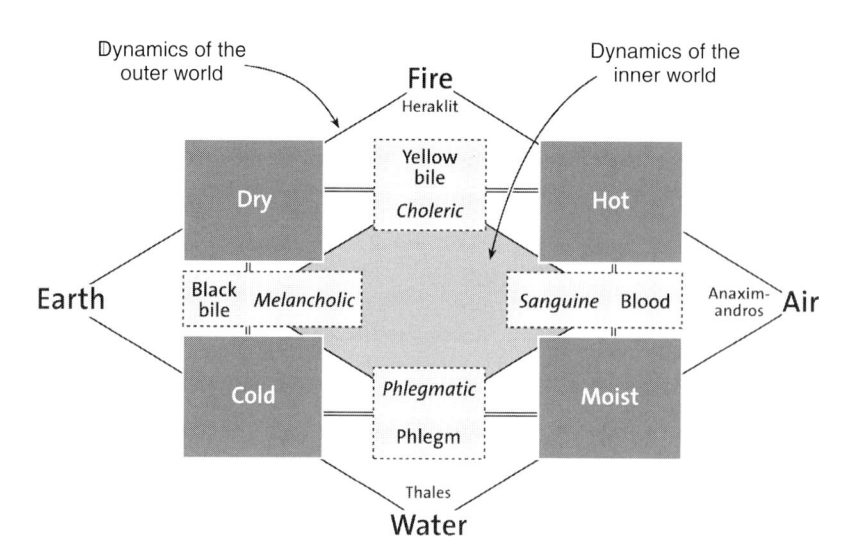

Figure 4.8 Basic elements of the world (earth, water, air, fire), of the person (black bile, phlegm, blood, yellow bile) and the humors of the world regime (dry, moist, hot, cold). Adapted from Bäumer, 1991.

and gain wealth and power (see Figure 8.3*). The Greek "Dark Ages" (1100–800 BC), a time of poverty and migration, are known through Homer's (about 850 BC) epic poems. The Greek city-state was not merely dependent on agriculture, but also conducted trade and shipping. Philosophers and their schools attained a special status. Some of them were related to the royal family, some got special funding for their research, and some of them founded schools such as Aristotle's Lyceum (334 BC). This increase in the production of ideas was facilitated by the increasing number of scrolls, and later in books. The Bibliotheca Alexandrina at the end of the Greek empire can be considered a research institution. By 200 BC, it housed about 500 000 scrolls and can be seen as a genuine scientific institution.

Philosophers considered different elements as dominant and ontologically existent. Thales of Milet chose water and considered god to be the mind that formed things from water. Anaximandros of Milet (610–547 BC) held that air is the most basic element and is the source of all that exists. Heraklit of Ephesos (535–475 BC) as well as Pythagoras of Samos believed that all things originated from primordial fire forming the center of the cosmos. Later, Empedocles of Acragas (494–434 BC) introduced the four elements of water, earth, air, and fire, and initially two forces – love and hate – that were thought to affect world dynamics.

Many Greek philosophers were involved in establishing secular inquiry into the natural world, starting with science and philosophy based on everyday experience. In some respects, the Greek polity was characterized by an emerging division of labor between priests and philosophers, though philosophers such as Pythagoras considered the soul immortal and religion and life as inseparable things, which both could be expressed by (divine) numbers as part of a new type of abstracted integer mysticism. The idea of reincarnation was also inherent in the Pythagorean approach, including the migration of souls to animals and some plants which were considered sacred (Bernal, 1970; Nordenskiöld, 1928). Some of the philosophers, such as Anaxagoras of Clazomenae (c. 499–428 BC), who proposed a heaven without gods were sentenced to death for impiety and godlessness, that is, of neglecting the official gods. Thus he got banished from Athens. Also Socrates, who saw knowledge as good and ignorance as evil, had to take care.

The disenchantment of the world when turning from Homer's time to pre-Socratic times can be highlighted by two examples. Homer thought that the god Poseidon caused earthquakes. Conversely, Thales of Milet argued that "just as a ship at sea is rocked by the waves, perturbations in the waters beneath the earth rock the surface" (Magner, 2002, p. 11). In another example, Xenophanes of Colophon (570–470 BC) noticed that man creates gods in his own image. Thus, Thracians' gods had light blue eyes and red hair, whereas Ethiopian gods are black and snub-nosed. Xenophanes proposed an alternative for wisdom-seeking through learning, an idea expressed later by Aristotle, that the highest faculty of man is reason. Here we see one of the earliest abstracted, rational argument-based sciences.

Around the same time, Alcmaeon of Croton (*c.* 500–450 BC) carried out the first known dissection and vivisection of animals in an effort to learn more about nature. He was the first to identify the brain as the seat of understanding and to distinguish understanding from perception. Alcmaeon thought that the sensory organs were connected to the brain by channels. He may have discovered the channels connecting the eyes to the brain (i.e. the optic nerve) by excising the eyeball of an animal. Perhaps based on this, he may have been the first to distinguish between perception and understanding. As the latter was considered a unique feature of man, he provided a first modern definition on the difference between human and animal.

The Greeks also provided many ideas about cosmogony and questions about where humans come from. Anaximandros, for instance, considered *aperion* a kind of primordial stuff where the key elements of the world texture (earth, water, air, fire) came from and are supposed to return. He proposed that fire acted on mud to produce dry land and mist whereas the Sun warmed up mud. This led to a dynamic process and brought forth the first animals, such as fish where water was predominant. But then the fish came onto land and, through adaptation, animals living on land. Man developed in these animals until he was nurtured to master his own life on earth. Xenophanes supplemented this model by referring to fossils as proof that earth had once been covered with mud (Magner, 2002; Nordenskiöld, 1928). As we can see, this theory combines Genesis and evolution. It might be interesting to note that Plato (of Athens, 428–348 BC) later suggested the counter idea of devolution, which states that animals degenerated from men. This degeneration occurred when those who did not use (or believe) in philosophy were consequently transformed into four-legged beasts (Magner, 2002).

An important issue of biological thinking is whether the world is seen as static or, as Heraclitus of Ephesos (*c.* 540–480 BC) has expressed by the phrase *panta rhei*, as in a state of flux or change. Heraclitus also supposed that man, including his soul, was subject to the same laws as nature.

4.4.2 Biophysical and medical processes

An interesting idea that Anaxagoras of Clazomenae (*c.* 500–428 BC) had about metabolism is found in the "everything" principle. One of his questions was, "For how, can hair come from what is not hair, and flesh from what is not flesh?" (quoted in Stanford Encyclopedia of Philosophy, 2009). He started from the idea that things emerge through dissociation. For example, human beings pre-exist in their food, which includes the basic ingredients. These earliest documented ideas about a biophysical process shared good company with Hippocrates' four humors of medicine, Democritus' (460–370 BC) atomic theories, Aristotle's notions about animals and their organs and about matter and form, and Theophrast's (372–287 BC) plant taxonomy and optimization techniques, all described below.

Hippocrates of Kos, without any doubt, was one of the most important Greek scientists to introduce ideas for a comprehensive approach to medicine. Two perspectives have been important. One is that he looked for ways to employ nature in curing the patient. The other is that he introduced a humoral medicine and pathology based on four humors (also called temperaments), which he thought governed the state of wellbeing. The first is yellow bile, symbolizing fire for the total world regime. Hippocrates saw yellow bile as medically related to the bitter fluid stored in the gallbladder which oscillates between hot and dry, regulating a choleric body and psyche. The second humor is black bile related to the dark, earth-like material sometimes found in vomit. Here the related humors are dry and cold. Third, phlegm related to mucus between cold and moist flaming activity. And fourth, blood, was an essential part of a positive, sanguine, vivid life (see Figure 4.8).

Hippocrates also proposed that the brain and not the heart was the center of the rationale, and his scholars provided a description of the anatomy of the brain that differentiated between cerebrum, cerebellum, meninges, etc. Scientifically, his greatest medical contribution is the distinction between anatomy, physiology and pharmacology.

Diogenes of Sinope (*c.* 412–323 BC) identified ramifications of the nerves in man and mammals. He also believed that the embryo developed in the uterus by the warmth of the mother and the semen of the father. This view was shared by Hippocrates, presumably based on dissection, which was done on bodies, mammals and presumably damned people. Hippocrates can be seen also as the first who related the ideas-based philosophy to observation-based empirics. He also considered medical herbs that were applied in folk medicine and prepared by rhizotoms (root collectors) and pharmacopols (apothecaries).

Democritus lived in the same period as Hippocrates and is known as one of the last major ancient Greek natural philosophers. Democritus saw all things as an arrangement of atoms. Atoms were compact, invisible, unchangeable, and impenetrable objects and too small to impinge directly on the senses. Democritus also introduced the *void* (space) between atoms, an idea with which many philosophers of the time disagreed.

Known for his contributions to physics, logic, metaphysics, ethics, and politics, Aristotle (*c.* 384–322 BC) also made significant contributions to science, with one-third of his writings focusing on biological topics (Bäumer, 1991). His five volumes on *History of Animals, Parts of Animals, and Generation of Animals* dealt with Nature, the reproduction of animals, and the relation among organisms. Aristotle aspired to provide a comprehensive overview of all animals and their organs (Balme, 1991; Bäumer, 1991).

Central for Aristotle was the distinction between matter and form, where form referred to real things whereas matter was the potentiality. This held true for both natural and manmade things. For example, signs written on paper with a pencil are the matter and the meaning of the text is the form. Also, the "seed is a potentiality, out of which the germinating plant develops reality" (Nordenskiöld, 1928, p. 36). Likewise the egg is potentiality, the bird the form. By introducing the idea that development starts with formless matter, this concept was a precursor to the idea of evolution. Aristotle regarded nature as an entelechial system, that is as a self-organizing teleological system with the potency to become an efficient form; thus nature was considered perfect. This idea that the morphology of an organ is solely affected by its function influenced biology for a long time (Harig & Kolletsch, 2004).

About 520 species were mentioned in Aristotle's writings and he introduced a rough classification system, using the concepts *genos* and *eidos*. *Eidos* is the individual animal form, for instance species (cow, dog, etc.) whereas *genos* is something broader, such as the family, but was not well defined. The vegetative soul of plants, for instance, was related to the *genos*.

Theophrast (371–287 BC) continued Aristotle's biological work and wrote two large manuscripts on *Enquiry into Plants* that elaborated a taxonomy of plants and are considered to be the cradle of botany. Theophrast also worked on the optimizing of growth of plants, optimal seed and harvest times, and procedures for preparing and applying manure as fertilizer. He thus provided important knowledge for improving agriculture. Theophrast realized the environmental impact when stating that "each semen is matching to the nature of the country" and thus noticed how the environment forms the specific organism. Contrary to his teacher Aristotle, Theophrast was an agnostic naturalist who "denied the existence of a dominant intelligence outside the universe" (Nordenskiöld, 1928, p. 45). He rather imagined that there are forces, such as the souls of humans, that govern operations of beings. A somewhat more radical view was taken by Strato of Lampsacus (*c.* 335–269 BC), the third director of the Lyceum, who considered man as an improved kind of animal.

4.4.3 Greek legacies in the understanding of life

The last period of the Greek world of thinking (300 BC–50 AD) was characterized by a decline of the city-states and an increasing incorporation of ideas from Egypt and Asian cultures, such as reincarnation. The center of science moved from Athens to Alexandria, the new Greek capital of Egypt, which was governed by the Ptolemies (305–30 BC). Alexandria's library housed more than 500 000 rolls and a Mouseion, a temple of goddesses of song and wisdom, and functioned as a research institute. Ideas about medical knowledge continued to evolve and the idea of textbook design was developed at that time (Magner, 2002).

Herophilus of Alexandria (325–255 BC) developed deep insight into anatomy, including understanding of the connections between the brain, spinal cord, and nerves based on 600 dissections. The Ptolemies allowed dissections on condemned criminals, something that was forbidden in the second century by Christians as a pagan action, in an attempt to devalue scientific activities.

Pharmacology, the application of empirically validated medicine, became the leading strategy promoted by Herophilus. There had been no fundamental new ideas in the time between 100 BC and 100 AD. It is worth noting that the Roman Titus Lucretius Carus (98–55 BC) offered a critique of the humoral theory based on Democritus' atomic model and applied it to the human body. He countered with his own humoral theory and also offered early ideas about genetics when assuming that there are hidden atoms, which can be latent for a long time but become evident in the offspring.

An important step in the development of medicine was the work of Galen of Pergamum (130–200) in

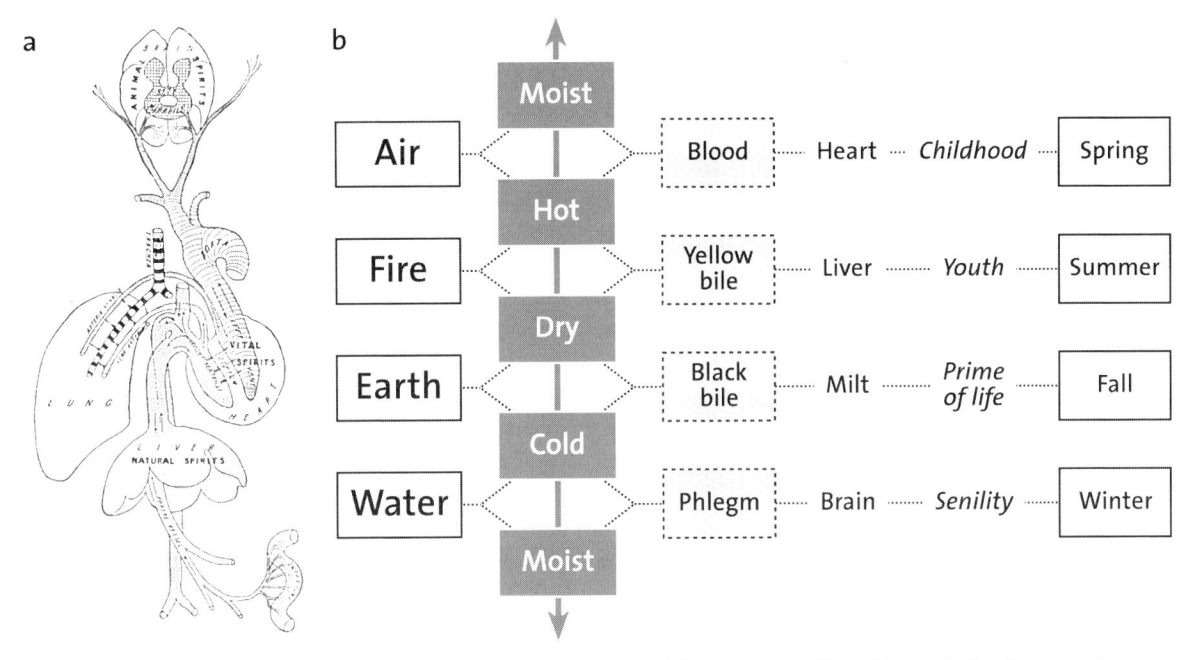

4.4 Antiquity's conceptions of the biophysical world

Figure 4.9 Galen's view (a) of anatomy and physiology (from Singer 1921, p. 183 by permission of Oxford University Press) relates to (b) organs, age, and season. Note the linkages to the elements of Aristotle and the humors of Hippocrates.

Rome (see Figure 4.9). Galen composed 600 treatises, which were heavily based on the Egyptian and Greek heritage of knowledge. Galen's work addressed four domains.

First, he succeeded in combining Hippocrates' physiological humoral pathology (for example, yellow bile is associated with choleric personality; see Figure 4.8) with Aristotle's four elements (for example, fire is associated with yellow bile; see Figure 4.9). Further, each element is linked to an organ, a phase of life, and a season. We call such a combination – earth, black bile, milt, prime of life, fall – a tuple. Each tuple represented human moods and conditions, such as irritableness, sleeplessness, and despondency (related to cold and dry).

He proposed a three-level theory in which the blood is generated in the liver, considered a natural spirit, and allocated to the peripheral flesh. The heart, the source of vital spirits, refines blood (which is not sent to the periphery) and allocates it to the lungs and other organs including the brain, which was called the psychic "pneuma" or the animal spirits governing the nervous system, and was needed for sensation and moving (see Figure 4.9). The role of the term "pneuma" is nicely illustrated in the following quote:

> The basic principle of life, in the Galenic physiology, is a spirit, *anima* or *pneuma*, drawn from the general world-soul in the act of respiration. It enters the body through the *rough artery* (ἀρτηρία τρᾱχεῖᾱ, *arteria aspera* of mediaeval notation), the organ known to our nomenclature as the trachea. From this trachea the pneuma passes to the lung and then, through the *vein-like* artery (ἀρτηρία φλεβώδης, *arteria venalis* of mediaeval writers, the pulmonary vein of our nomenclature), to the left ventricle. Here it will be best to leave it for a moment and trace the vascular system along a different route (Singer, 1959, p. 183).

Galen's framework dominated practical medicinein the Occident until the eighteenth century. "For Galen, the body was the instrument of the soul and proof of the existence and the wisdom of God" (Magner, 2002, p. 57). Environmental impacts appeared by the seasons, but have been of minor significance.

Second, drawing on Egyptian tradition, Galen tested the effects of simple drugs composed of minerals, plants, and animal extracts. In this task, the four elemental qualities (described above) served as heuristics or schema to classify the effectiveness and intensity of their impact (Harig, 1974).

Third, Galen refined knowledge about anatomy and physiology. As the dissection of human bodies was forbidden for legal reasons, Galen dissected apes, pigs, and other mammals.

Fourth, Galen was a surgeon who conducted operations on eyes and other organs; these surgeries were not tried again for almost 2000 years. His talent certainly became widely known when he worked as a successful physician to the gladiators.

In light of the achievements of Galen, the following centuries and the beginning of the European Medieval period saw a decline in the development of scientific thinking in the western world. For instance, drug recipes of the fourth and fifth century document immature, superstitious thinking but also the impact of the humoral theory (Harig & Kolletsch, 2004). In addition, blood-letting and extracting matter and liquids such as extracts from mummies, grease, bones, feces, urine, menstrual blood, hairs, nails, spittle, sanies, blood or bodies and cadavers were still practices in the *Dreckapotheke* ("Smud-apothecary") of Franz Paullini (1643–1712; see Lux, 2005; Scheible, 1847).

A similar decline and fall of ancient sciences can be observed until the late Medieval period with respect to knowledge about agriculture, which was an important sector in the Roman Empire. The Latin language volumes of the leading researchers of that period such as *De Agriculture* (Marcus Porcius Cato Censorius, 234–149 BC), *Res rustica* (Marcus Terentius Varro, 116–27 BC), *Georgica* ("Cultivation"; by Publius Vergilius Maro, 70–19 BC), *Cepurion* ("Horticulture"; Gaius Iulius Hyginus, 60 BC–10 AD), or the twelve books *De re rustica* (Lucius Iunius Collumella, first century AD) included biological issues but focused more on economic aspects of agriculture including administration and slavery (Harig & Kolletsch, 2004). The 37 books of *Natural History* by the encyclopedist Gaius Plinius II (23–79) can be taken as example. Magner considered his writings

> … rather as a random walk through observations about the heavens and the earth, geography and ethnography, natural history, stories of animals, fish, birds, and insects. (Magner, 2002, p. 50)

4.4.4 Key messages

- Antique Greek philosophy is characterized by the switch from mythological to logical reasoning. Gods were around and constructed for different purposes such as Asclepius for medicine and healing. At the same time, many purely secular theories of healing were developed
- Experimental biological/medical research started with vivisection, noticing the qualitative similarity of man and animals, identifying nerves, the brain as center of the mind, and providing classifications of fauna and flora
- The ancient Greek philosophers looked to different substances or material matter substances (e.g. atoms) to explain how the world functioned. Many such systems had a quadruple material matter structure: water, air, soil, and fire for the world; blood, phlegm, yellow and black bile for the individual; and sap, fibers, veins, and flesh for plants. These can all take different qualities such as hot, cold, dry, and moist. The equilibrium view was seen as a stable, desired state
- There were no elaborated models of human–environment interactions in ancient Greek philosophy. With few exceptions, such as Aristotle's reflections on seasonal impacts, the individual organism was widely seen as independent of the environment. Impacts of agriculture and deforestation did not arouse much interest.

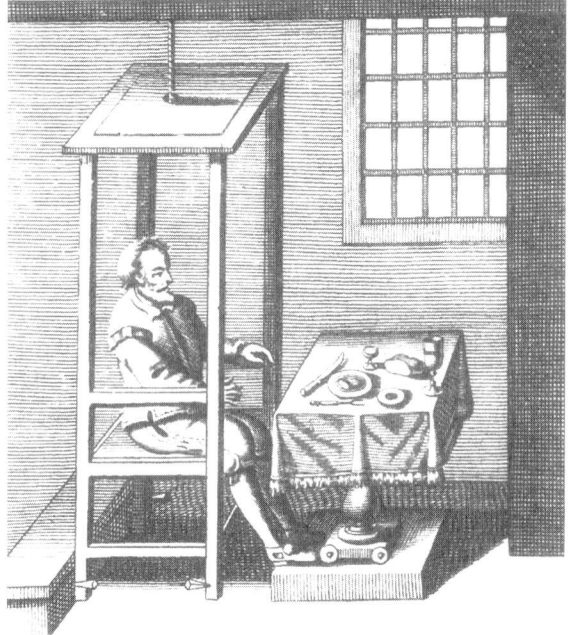

Figure 4.10 Santorio Sanctorius (1561–1636) in his weighing chair (Sanctorius, 1624/1711). Sanctorius was an Italian physician and the founder of quantitative measurement in medicine. He was the first to use a thermometer to measure body temperature.

4.4 Antiquity's conceptions of the biophysical world

Box 4.4[1] **How many types – and where do they come from? Semen and genders**

The mysteries of conception and birth have long been a topic of inquiry. In ancient times, pregnancy and intercourse were only stochastically linked. Thus, it seemed reasonable that other, supernatural forces had to be involved. A common magical explanation might have been that ancestral spirits in the form of living germs got into the maternal body.

There have been different one- and two-semen theories suggested by Greek philosophers. Anaxagoras supposed that sperm came from man and that the egg from a woman provides only the location for growth ("one-sperm theory"). Most theories are two-semen, gamete theories. But a critical question is where each gamete comes from. The enkephalo-myelogene theory of Kroton (about 500 BC) supposed that semen comes from the brain and that there are veins passing the ears and ending in the testicles. Eunuchs were explained by blood-letting from veins behind the ears during childhood, which interrupted the transport of semen to the testicles (Harig & Kolletsch, 2004).

Based on the four-element theory (see Fig. 4.9), the Pythagorean Empedocles (c. 490–430 BC) suggested a thermal model. He believed that men are of the warmer temperament and came into being in a southern climate whereas the cold-blooded women were created in more northerly climates. During reproduction, the embryo receives some parts of the body from the father's and the mother's seed. During genesis, dense male and thin female semen are fighting. The infertile mule is an example where the female gametes had won (Nordenskiöld, 1928).

Democritus' concept of atoms led to the first theory of pangenesis. Semen is supposed to be formed out of semen from all organs and representing a miniature copy of the organism (such as each gamete is a miniature copy of the universe; Harig & Kolletsch, 2004, p. 53). By subdivision and merging a mixture of father and mother, a *pangen* (i.e. a heredity-controlling set of particles) is built (Jahn, 1990, p. 63).

Aristotle differentiated between animal reproduction through sexual means (e.g. mammals), asexual means in the form of bud-formation (by hermaphroditic plants) or self-generation (e.g. by shellfishes), and spontaneous generation done by many insects. Semen, representing form, was only seen on the male side, whereas matter (the "cold" part) and potentiality was offered by the female side (Jahn, 1990, p. 70). This idea was still around with Thomas Aquinas (1225–1274), who believed that the male semen contributed the essential material. Physically he believed that the fetus was formed by a union of menstrual blood and sperm, a theory that had been slightly modified from Galen (Bullough, 2001, p. 207).

Finally, Theophrast noticed, by comparing plant and animal fertilization, that pollination of figs and dates by fig wasps resembles the act of fish mating (Nordenskiöld, 1928, p. 196).

But it was a long time till literacy about reproduction developed. An early natural science view by William Harvey (1578–1657) stated that "an egg is the common primordium of all animals" (Bullough, 2001, p. 208). With the advent of the microscope, Anton voan Leeuwenhoek (1632–1723) identified spermatozoa (Figure 4.11b), which were first called *seminal animalcules* and – according to the dominant preformation theory – given the image of a miniature human. However, early scientists were challenged to explain why there were so many more sperms than needed, since the

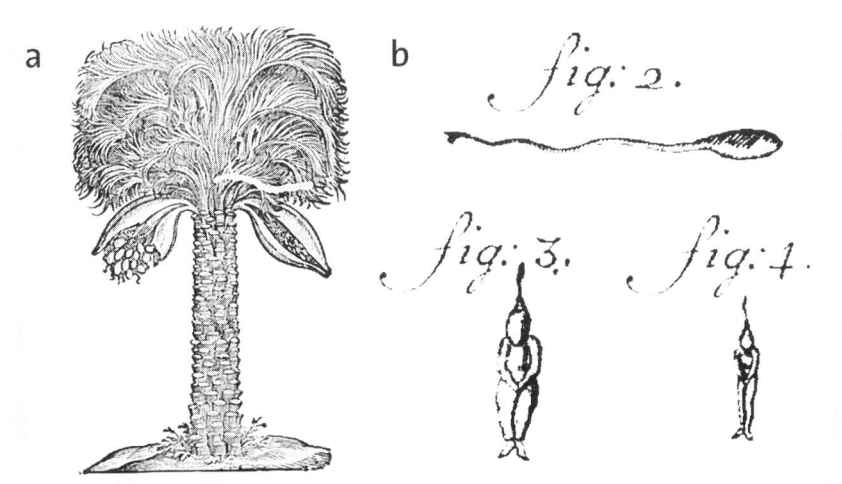

Figure 4.11 (a) An illustration of the date palm (*Phoenix dactylifera*), a dioecious plant species (Alpini, 1592). (b) Human spermatozoa painted according to microscopic observations by Jan Ham (1677).

---

[1] This box has been written together with Robert Bügl.

Box 4.4 (cont.)

notion that natural procreation involves a "Russian roulette" strategy was not compatible with Christian dogma. There was also no progress with empirical work during this time to describe, for example, embryonic development. Finally, in 1827, scientists identified the mammalian egg but only in 1875 did scientists discover that sperms invade eggs. Thus it was not until the beginning of the twentieth century that a coherent natural science model was developed about the mechanism of human reproduction (Bullough, 2001).

4.5 European "Dark Ages": loss of environmental literacy

The growth of environmental literacy was interrupted during the European "Dark Ages," a period that saw a shift from slavery to feudalism and which concentrated power and wealth with nobility, the Christian church, and the military. This section describes impediments to the development of knowledge during this period and how, nonetheless, scientific knowledge passed to subsequent generations and the first universities were founded.

Some historians point out that culture and ancient science were moribund between 300 and almost 1450:

> The productivity of Italian land diminished as a result of deforestation, erosion, and the neglect of irrigation canals; numerous farms and field were abandoned. Mining enterprises failed because the richer veins of minerals had been exhausted. Observers in Italy noted significant declines in population and in the live birth rate. (Magner, 2002, p. 64)

The Medieval period (approximately 500–1500) was characterized by a feudal exploitation of the vast majority (at least 95%) of people living in villages, by the Catholic doctrine, and by an agricultural system that used three-crop rotation as the only means to prevent impoverishment of the soil. Political leaders and the most important cultural institution, the church, had little or no curiosity about the natural, biophysical environment nor about the impacts and indirect, collateral damage caused by activities such as deforestation (see Box 1.3). Ideas about cosmology and cosmogony, about the relation between the human and the natural environment, and about biology were strongly shaped by Christian religion (see above; White, 1967).

As early centers of science, Greece and Egypt saw improved standards for society and agriculture. The neglect of science and education in the west during the subsequent period, between the fall of the Roman Empire and the fall of Constantinople (1453), was significant. During these Dark Ages in Europe, agricultural, empirical-inductive trials, for example, were not active. However, during the Golden Age of Arabic-Islamic governance in the Mediterranean belt between 632 and 750, a school system was introduced (Nabielek, 2004). Later, by translating the Greek heritage from Arabic to Latin, primarily in the Islamic period of Spain (711–1492), valuable knowledge was re-imported. Since the foundation of the European universities from the twelfth century onwards, biological knowledge was taught in the *studium generale*, to all medical faculties.

Historical analyses reveal how the dominance of clerical dogma hindered the development of biological knowledge in the Medieval period. A few systematic agricultural research projects, which never dealt with husbandry, were conducted at botanical gardens in monasteries and Arabic-Islamic governed parts of the Mediterranean (Nabielek, 2004). As the examples below illustrate, scientists of that time lacked appropriate technologies and methods for developing environmental literacy.

Since 1215, Franciscan and Dominican monks transferred the Greek onto the Medieval frame, where God came first and where plants and animals were made to serve humans, or to show him how weak (s)he is and how strong God is. Albertus Magnus of Cologne (1206–1280) translated, modified and extended Aristotle's theory, including heredity, where he provided detailed descriptions about anatomy and practice (Bäumer, 1991). Albertus Magnus also conducted empirical work with chicken eggs and embryos. Yet without a system for documenting the day-by-day changes, no systematic view could be developed. This was done almost three centuries later by Volcher Coiter (1534–1576; Singer, 1959), who provided new views of adaptation processes in embryology. Also Thomas Aquinas' (1225–1274) input is limited. Ballauff (1954) pointed that there had been a special conception of the body–mind complementarity in Thomas' writings. The body was seen as a technical work-piece of the soul, the relation has been such as between a *motor* and a *motum* (i.e. vehicle).

Historical analyses reveal the decline of biological knowledge in the Medieval period. Curiosity in biology and cosmology was obviously ruled by clerical dogma (Nabielek, 2004). Environmental literacy at that time lacked appropriate technologies and methods for taking a close view. Some impacts from other regions and world religions in the Medieval period, primarily linked to the emerging world trade from the third century, encouraged systematic science. However, the respect and protection of nature, an idea inherent to old world religions such as Hinduism and Buddhism, ideas that partly prohibited tilling land (Jahn, 1990), did not reach Europe.

4.5.1 Key messages

- The growth of environmental literacy was interrupted in the Medieval Age. The switch from slavery to feudalism, and the maintenance of power and wealth for the nobility based on the dogma of clerical and military leaders, impeded the development of knowledge.

4.6 Rebirth of the natural science of organisms

The movement to a secular, reasoned, non-mystic view of understanding organismic dynamics was initiated during the early Renaissance in the fifteenth and sixteenth centuries. To illuminate the progress attained in this period, we present some exemplary aftermaths.

Nicolaus Copernicus (1473–1543) provided a groundbreaking contribution to environmental literacy through his six volumes on *De revolutionibus orbium coelestium* ("About the revolutions of the heavenly spheres," 1543) launching the Copernican revolution. An important backdrop for his work was the ongoing discussion over the mismatch of the factual (average interval between vernal equinoxes) and the observed length of the year (the way in which it was defined by the calendar). A misunderstanding of a decree from Julius Caesar (100–44 BC) in 45 BC to improve the old Egypt–Alexandrian calendar and create the Julian calendar resulted in an incorrect recording of leap years. This led to a perceived mismatch of 10 days (with respect to the vernal equinox) and misalignment with the expected seasons, which had more practical implications (Hagedorn, 1994; Kuhn, 1985).

Giordano Bruno (1548–1600) promoted a fundamentally new conception of the Christian God when he considered that universal cause and effect has a unity, which is God. The "Holy Ghost is the soul of the world" (Turner, 1908), as the internally operating system. As Jahn (1990) pointed out, thus both Copernicus and Bruno introduced the idea of the equality of creatures to Christian thought, as the Holy Spirit is equally at work in all living subjects. Other fundamental changes in thinking were generated by the invention of the mathematical variable as a construct of calculation modeling by Franciscus Vieta (1540–1603) and by the Protestant Reformation (1517).

The spread of higher knowledge and thus the development of environmental literacy in society occurred through important innovations of technologies, such as Gutenberg's (*c.* 1398–1468) movable type printing technology. This printing press allowed the production of illustrated books, including texts on herbs and animals. Thus, we can see a shift from a scribal culture to a print one, paving the road for the Enlightenment (Magner, 2002).

In the fifteenth century, art employed science and promoted, in particular, the development of anatomy. Leonardo da Vinci (1452–1519) and Albrecht Dürer (1471–1528) were more than exceptional painters; they were also highly sophisticated mathematicians. Da Vinci stated that he could see nothing that could not be described by mathematics (Botnariuc & Jahn, 1985) and wrote a book on *Botany for Painters*. Dürer wrote four books on human proportions in 1528 (Jahn, 1990). Da Vinci was definitely an anatomist and botanist, and some of his paintings would not have been possible without experimentation and serial dissections of all kinds of animals. How close the leading minds of that time have been related can be taken from the fact that da Vinci was also involved in founding the heliocentric world view. He continued some of the work of Nuremberg astronomer Regiomontanus (1436–76), which provided basic data and ideas for the revolutionary contribution of Nicolaus Copernicus (Botnariuc & Jahn, 1985). Da Vinci's ingenuity also became apparent when he interpreted fossils as indicators of a world in continuous growth and decline, including fundamental changes at the Earth's surface.

The science of botany was launched via botanical atlases and aesthetic botanical gardens, which became places of documentation, demonstration, and learning throughout Europe. Systematic classification emerged and, just as an example, Jaques Daléchamp's (1523–88) *Historia generalis plantarum* recorded 3000 plants in 2700 paintings (Jahn, 2004b). With no cameras or televisions for depicting the environment, the role that

4 Emerging knowledge on biology and evolution

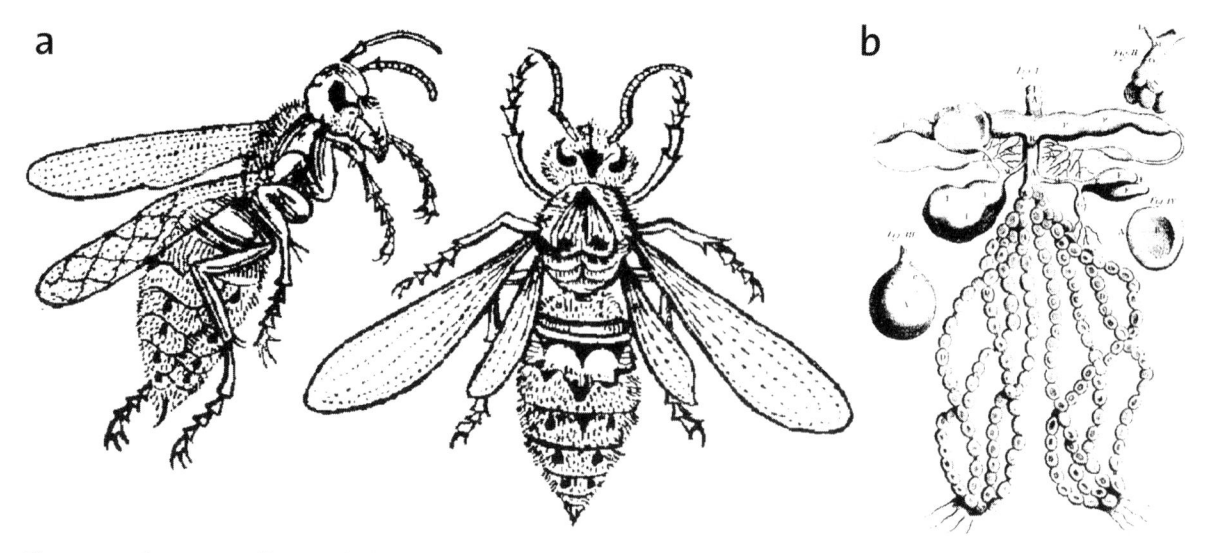

Figure 4.12 Environmental literacy is highly dependent on the technology employed for representing environmental information. A new dimension of knowledge was provided by naturalistic facsimile drawings of (a) hornets (Thomas Moufet, 1588; see Singer, 1959) and (b) the excretion organs of the bumble-bee, as seen by microscope (Malpighi, 1669; see Jahn, 2004b).

accurate and facsimile drawings played for opening windows to the natural environment should not be underestimated. Figure 4.12 shows how technology affects representation.

Paracelsus, also known as Theophrastus Bombastus of Hohenheim (1493–1541), built a theory of healing at the interface of Christian mysticism, the experience of nature, and *alchemic spagyrics* (i.e. the art of separating, dividing and associating fusing herbal and metallic tinctures). Diseases were thought to be caused by unfavorable cosmic constellations, toxins, a body's constitution, spirits, or God's hand. The Hippocratic four humors were part of Paracelsus' thinking (see Figure 4.8), but he based his own theory on the interaction of a triad of chemical elements including sulfur (for combustion), mercury (for liquefaction), and salt (for crystalization). While alchemist experimentation at this time assumed that disease must be a defect in body chemistry, we also see new, systematic treatment of the human body via (material) chemicals. In his world model, Paracelsus distinguished between humans (i.e. the microcosm) and the universe (i.e. the macrocosm) and introduced, at least in some domains of his work, a material, physical perspective on these systems. However, we should note that many of Paracelsus' alchemist treatments were similar to the smud-apothecary's. Paracelsus was a typical "mystical, half-experimenting, half-brooding" man (Nordenskiöld, 1928, p. 132), a personality described as "an errant star in the firmament of sixteenth-century aspirations" (Week, 1997, p. i) and an agglomerating spiritualist with Reformist elements and tendencies, who almost caused a crisis in the early Renaissance. The Renaissance was not only the age of the rebirth of science but also an age of occultism, mysticism, and superstition.

In the time of the early Renaissance, zoology and human anatomy were flourishing. The work of the Swiss physician Konrad Gessner (1516–1565), who succumbed to the Plague, can serve as an example. His book *Historiae animalium* (1565), thought to be "a starting point of modern zoology" (Singer, 1959, p. 96), included information on body structure, habits, diseases, and place of origin (Jahn, 1990; Mägdefrau, 1992). The *Kräuterbuch* ("Herbal book") of Otto Brunfels (1488–1534) had been referred to for centuries (Mägdefrau, 1992), though the work included primarily illustrations, with little text and a composition of old knowledge. Brunfels considered plants as heavenly bodies, refused classification, and provided a descriptive plant iconography (Bäumer, 1991). His illustrator, a scholar of Dürer, painted even withered leaves over-naturalistically.

Clearly, many of the books on plants were herbals that served medical purposes. But a general botany also emerged alongside a specialized, applied functional one. The technique of using detailed naturalistic drawings allowed differentiation between a specific medical

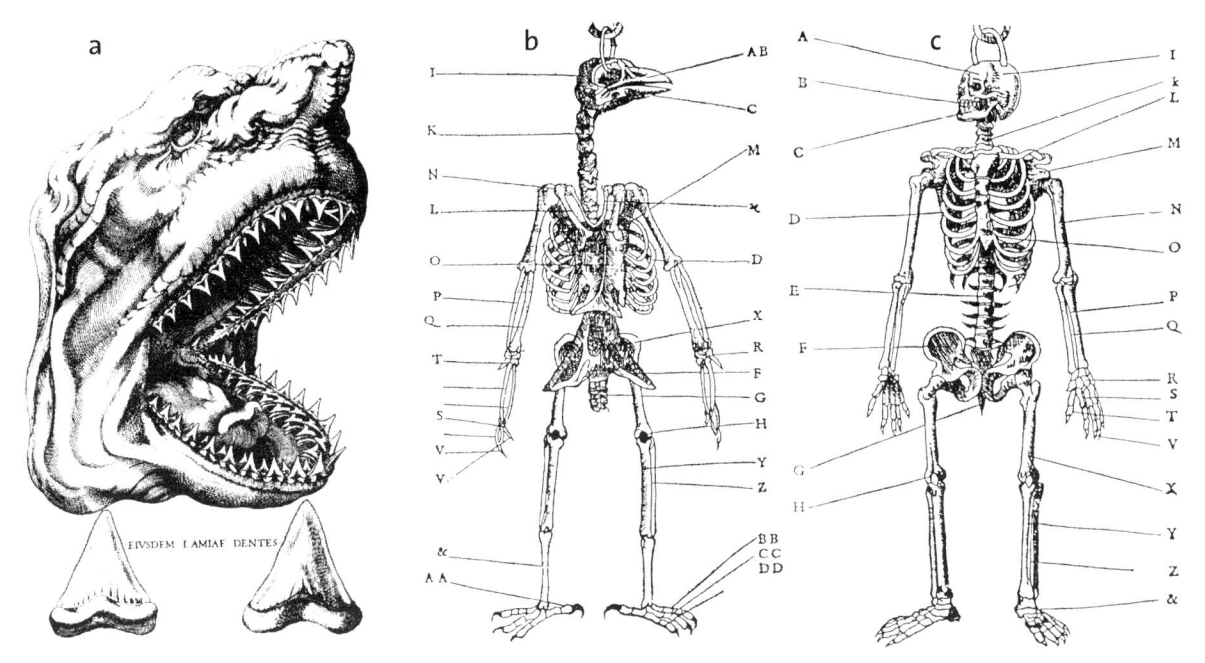

Figure 4.13 (a) A naturalistic illustration by Nicolaus Stenonis (1638–86) of the white shark *Carcharodon carcharias* (from Stenonis, 1669) and (b,c) comparative drawings by Pierre Belon (1517–64) (adapted from Belon, 1555; direction of human head rotated) seem to include implicit messages of nature (i.e. the threat of Figure 4.13a) or the extraordinary similarity between bird and human anatomy (Figures 4.13b & c).

(functional) botany and a general (comparative) botany. The latter defined leaves, blossoms, stem, etc. or (on a different level) petals, pistils, and stamen for different genera and species. Between 1523 and 1623, the number of species in botany books increased from 240 to 6000 (Mägdefrau, 1992).

It is interesting to see that during this period illustrations of organisms were not completely free of imagination, as conveyed by Figure 4.13; from today's perspective, the comparative illustrations also seem to include implicit messages, though evolutionary theory had not been developed.

4.6.1 Key messages

- The early Renaissance was characterized by a few fundamental changes in the world view, such as the Copernican revolution for explaining cause–effect relations, for instance of medical therapy. Materialist arguments emerged, although they included magic, spiritualist, or alchemist components
- Humans started to differentiate and to paint plants and flora naturalistically. Comparative studies of plants set the tone for a functional, medical botany.

4.7 Scientific revolution and the methodological foundations of modern biology

This period was characterized by a break from the straitjacket of the geocentric world view shaped by Christianity. From a theoretical perspective, the study of life gained new foundations through the mathematical–experimental method provided by Galileo Galilei (1564–1642), who also put forth a mechanistic conception of biology. René Descartes (1596–1650) introduced – among other ideas – a physiological mind–body dualism (see Chapters 3.3 and 6.2). He strongly pushed a mechanistic world view, which he applied to the operation of plants, animals, and human bodies, sensation, and behavior. His books *La description de corps humain de l'animal* (1647) or *De la formation de l'animal* (1648) assert that the same laws of motor activity (motion) and extension that are at work in organisms are also at work in inorganic materials, suggesting a physical *mechanomorphic* view of the universe. This Cartesian method was a deductive one, referring to Aristotle's principles. The mechanistic, corpuscular theory of Descartes, which did not allow for vacuums, was countered by Newton's (1642–1726)

Philosophiae Naturalis Principia Mathematica (1687) and his invisible gravitation law along with his theory of classical mechanics with the elementary categories of space, time, and motion. Along with Isaac Newton, Gottfried Wilhelm Leibniz (1646–1716) is seen as the founder of infinitesimal calculus.

Leibniz's metaphysics is centered on the claim that ultimate reality is composed of "individual substance, later monads, mind-like, immortal, containing all and reflecting" substances (Garber, 1985, p. 27). The monads are living, immaterial atoms (in the sense of smallest units), and exist everywhere in a kind of hierarchical order.

> The human soul has such a monad, which is consciousness; the life of monads consists of lower monads, but percipient; the monads of plants live, but are not percipient; the monads of inanimate nature are in a different state, as in a dreamless sleep, the human body is composed of these monads. (Nordenskiöld, 1928, p. 128)

Leibniz suggests the idea of a living goal-oriented and driving (i.e. entelechial) but immaterial agent on the micro level beyond the smallest or beyond the basic elements of humans, animals, and plants.

Francis Bacon (1561–1626), uncomfortable with the mathematical, deductive methods, proposed his own inductive approach for reasoning. He also criticized the gap between book-based scholarship and medical experience, and made inferences based on sensation, experiences, and data. Bacon's view contrasted with Aristotle's view and was ill-regarded by religious academies. He aspired to "establish the power and dominion of human race over the universe" (Bacon, quoted according to Whitney, 1989, p. 381) and abandoned the alliance of mechanistic science with magic action, as it has been drafted in the first Faust book in 1587 in Germany. Bacon, who stated that any effort to mix science with theology would not come to a good end, was characterized as a revolutionary pagan (Albritton, 1980). Not too long after Bacon lived, the Royal Society of London, founded in 1660, was the first research institution to split from the clerical frame explicitly.

The seventeenth and eighteenth centuries were a time of scientific giants who paved the way for mathematics and physics to become the leading sciences of the nineteenth and most of the twentieth century. They developed the demystified, rational methodological foundations of experimental research. The issues at hand were the development of a philosophy of nature and an understanding of what mechanisms and organs were at work in functioning organisms, along with their appearance and what drives their activities. From a biological perspective, the understanding of the organism and its organs offered through anatomy and physiology dominated. Environmental aspects were addressed when the functional perspective in the upcoming industrial society became of interest.

In the early decades of the seventeenth century, Friedrich Johannes Kepler (1571–1630) invented the telescope and compound microscopes (see Box 4.4). These inventions opened new scales of the environment for observation. One could follow the emergence of worms or the microstructure of insects (see Figure 4.12). Under the microscope, William Harvey (1578–1657) identified tiny networks of vessels that connected veins and arteries called capillaries (from Latin "hairlike"). Thus, the circulation system was visually detected, for instance, by watching the blood circulating in the tail of a tadpole, which led to theoretical verification. Marcello Malpighi (1628–94) identified blood cells as fatty globules and brain cells as adenoids. Additionally, the first cells or *cellulae* of sliced bottle cork were seen in 1665 by Robert Hooke (1635–1703), and the term *tissue* was coined for plant fibers. The development of insect grubs could be followed, although this was interpreted from a preformationist view, that is the belief that things grew without metamorphosis (Jahn, 2004b).

The last paragraph of Box 4.4 describes the detection of spermatozoa, which were called protozoa, taken from the Greek words meaning "first animals." We should note that *animalculism*, which is the theory that sperm contains all information or even all organs of the embryo, was opposed by those supporting *ovism*. Ovism postulated that the egg contains the whole of the embryo and that the sperm only awakens the growth of the offspring. Both theories reflect preformism and do not include ideas of epigenesis or morphogenesis, which include gradual diversification and differentiation of an initially undifferentiated entity.

The seventeenth century was a time of increasingly systematic experimentation which debunked theories that had been developed to explain what could not be seen by the naked eye. For example, Francseso Redi (1626–97) showed that maggots only emerge from meat on which flies could land and not from meat in flasks sealed with fine gauze (but still exposed to the air). From this, he inferred that Aristotle's assumption of spontaneous generation (see Box 4.4) was not valid for lower animals but that they leave eggs on the meat instead. Moreover, John Turberwill Needham's

4.7 Scientific revolution and new methodologies

Box 4.5[1] Emerging geology meets biology: catastrophes, infinite time, and evolution

In the seventeenth century, early geologists began to investigate the origins of the Earth and started to emancipate science from biblical doctrine, documented in the Old Testament, about how God created the Earth in 4004 BC (Halliday, 1997; Toulmin, 1975).

Reconciling the fossil record with, for example, the Great Flood described in the Bible meant that new ideas about the Earth and oceans would emerge. An approach called uniformitarianism postulated that changes to the Earth's surface follow processes that can be described by natural laws. Sedimentation and erosion, which have been at work in the past as they are at work today, were seen as examples. Hence, uniformitarians assumed that the Earth, as we perceive it today, has developed by a very slow process over an almost infinite period of time (Albritton, 1980; Weissert & Stössel, 2009). James Hutton (1726–97), for example, whose book *Theory of the Earth* (Hutton, 1788) is considered a seminal foundation of modern geology by many scholars, assumed that "geologic time is virtually infinite in duration" (Albritton, 1980, p. 89). The main message of uniformitarianism is that gradual, infinitely small changes over an infinite duration of time can result in dramatic changes.

A different approach is called catastrophism. Catastrophists assume that convulsive releases of energy caused revolutionary changes in the surface of the Earth as well as in species (Albritton, 1980). Georges Cuvier (1769–1832), for example, stated that the continents have been flooded repeatedly. In his view, this flooding did not happen in a slow and gradual way, but happened suddenly with a dramatic, catastrophic impact. He based his theory on observations of sequences of different fossils in sediment layers around Paris, from which he concluded that animals periodically suddenly went extinct and then new species appeared (Albritton, 1980).

Charles Lyell (1797–1875), a later uniformitarian, strongly criticized catastrophism. He claimed that uniformitarianism better explained the development of the Earth because it relies on real-world evidence, whereas catastrophism has to presume a supernatural power that elicited the catastrophies. In addition, catastrophism can be seen in line with the Bible, whereas uniformitarianism takes a different approach and detaches geology from theology (Anderson, 2007). Nevertheless, uniformitarianism was not applied to the development of species. Even when uniformitarians detected extinction and development of species in the biostratigraphy (sequences of different fossils in sediment

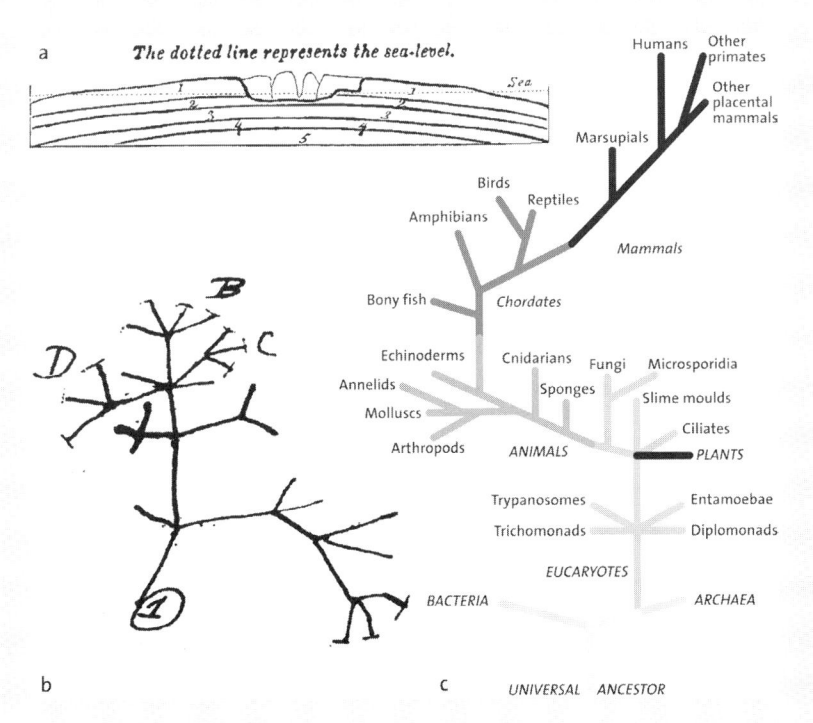

Figure 4.14 (a) A classic example of Lyell's gradualism – the "denudation of the Weald" (from Lyell, 1830–1833, p. 294/295). (b) Darwin's first diagram of an evolutionary tree of 1837. (c) A modern genealogical tree based on sequencing of 16S-rRNA (adapted from Cotterill, 2008, p. 421).

[1] This box has been written together with Corinne Moser.

> **Box 4.5** (cont.)
>
> layers), they predominantly believed in Creationism and hence, that these animals were created, went extinct and developed through acts of God (Albritton, 1980).
>
> The assumption of uniformitarians influenced the young Charles Darwin (1809–1882). On his Beagle expedition (Darwin, 2007/1845), Darwin read *Principles of Geology* by Charles Lyell (1830–1833), one of the primary proponents of uniformitarianism. On this expedition, Darwin recognized the uniformitarian principle in landscapes he visited. It is assumed that uniformitarianism inspired Charles Darwin inasmuch as he applied the principle not only to landscapes but also to the development of species. He "borrowed" the idea from Lyell, who stated that great changes happen in nature by small changes over extensive periods of time. The end of such processes may be spectacular: "… a new species or genus in one case, a great canyon in the other" (Albritton, 1980, p. 172).
>
> The stratigraphic thought is strong support of the theory of evolution. Most of the species which emerge from variation go extinct by evolution, as saliently presented by Darwin's first diagram of an evolutionary tree (see Figure 4.14b), which he drew right after the journey on the Beagle. All life emerges from one origin (see ❶ in Figure 4.14b), species ramify in different branches, and some may become extinct. Thus, the extinct species can be found below the former sediments of flooded area.

(1713–81) experiments also showed that boiled (sterilized) broth does not generate swarms of microorganisms, whereas broth that has not been boiled does.

4.7.1 Geology meets biology

Geological researchers from this period laid important groundwork for Darwin's biological theory of evolution (see Box 4.5). However, as the Age of Enlightenment began, they were challenged to align research results with Christian narratives of cosmogony. Nicolaus Stenonis (1638–86), for example, concluded from his stratigraphic data, which were mainly inspired by his investigations of sea animal fossils, that Tuscany must have been flooded twice in geological history. Of course, Stenonis had to match his calculations with biblical history, as the Vicar General of Florence asked for a peer review. Thus, he dated the second flood at the time of the Great Flood and concluded that the age of the Earth, that is the date of Creation, must have been about 4000 BC (Albritton, 1980). This was perhaps the first time that people in Europe took an analytical look at prehistoric times.

Not until the end of the eighteenth century did botany become its own discipline and zoology in the nineteenth century. This is surprisingly late, given that both disciplines had early beginnings through the works of Theophrastus and Aristotle (Botnariuc & Jahn, 1985). Systematic botany was actually established by Carolus Linnaeus (Carl Nilsson Linnæus, 1707–78, see Box 4.6), whose name has many slight modifications, for example the contemporary Carl von Linné.

From a history of science perspective, the eighteenth century is also a perfect example of realizing the limits of environmental literacy with respect to biochemistry. For example, Sanctorius of Padua (1561–1636), as described in Figure 4.10, conducted some well-known research on insensible perspiration. He spent years performing painstaking experimental procedures measuring himself before and after eating, drinking, sleeping, or exercising, in order to test the belief held by scientists at that time that weight loss was due to imperceptible vapors. His research had strong impacts on quantitative empirical methods, such as the use of thermometers and pulse clocks in medicine, and also laid the foundations for the idea of metabolism (Magner, 2002). Sanctorius was in a similar situation to the German physicist Ernst Stahl (1660–1734), who noticed that the weight of iron increased when it rusted (Bäumer, 1996). Here he first postulated that a substance called *phlogiston* would exist, which was released from metals during rusting and had negative weight. Thus, biological thinking took a role in detecting the mechanisms of air and gases.

4.7.2 Key messages

- The Cartesian deductive, mathematical method and the Baconian inductive method provided new tools for investigating and understanding the environment
- Invention of the microscope permitted insights into new scales of the environment and allowed users to observe the circulatory system and its

capillary structure, to identify insect eggs, as well as sperm and other cells, and to view microorganisms, all of which are necessary for understanding life processes
- The evolving experimental method of the seventeenth and eighteenth centuries led to an additional mindset supplementing the dominant Christian belief system. But theories and data had to be adapted to the Christian cosmology
- Experimentation, observation, and classification laid the groundwork for establishment of the fields of botany and zoology. However, explanations about basic physiological processes lacked as yet undeveloped biophysical knowledge.

4.8 Classification

Humans were driven to develop a classification system to facilitate their study of human–environment systems (see Figure 3.3*) and to give order to the diversity of beings, while allowing a focus on a system's parts, structure, or genesis. In this section about humanity's struggle to find order in the kingdoms, we look at how evolution became a theory, at the static species concept (fixity of species), at Linné's natural method of classification, and at some early pre-Linnéan approaches to classifying flora and fauna.

4.8.1 The struggle to find order in the kingdoms

Carl Linné was believed to have been obsessed with order. "He classified everything around him, be it alive or inanimate: Plants and animals, but also minerals, diseases, and colleague scientists" (Schmitt, 2008, p. 14). Thus, he was a data-driven researcher, and it is overly interesting to see that data (e.g. from plant studies) that do not fit the conventional approach of classification even urged Linné to introduce the chance concept into systematics (see Box 4.6). This was most remarkable but, considering the history of mind, it can be explained because the decade around 1660 was the birthtime of probabilistic thinking (Hacking, 1975, p. 11).

The classification of plants and animals was one of the most important developments in the field of biology. Until the Darwinian period in the nineteenth century, scientists subscribed to a static concept of species called the fixity of species, which meant that all species remained unchanged throughout the history of the Earth. The (written) foundations of classification were provided by Theophrastus (see Chapter 4.4). And as discussed in Chapter 4.3, botanists (Greene, 1909/1983) or ethnobiologists state that what "Theophrastus describes is strikingly reminiscent of what is found in the ethnobotanical systems of traditional societies today" (Berlin, 1992, p. 57).

Until the nineteenth century, biological classification did not consider the dynamic, evolutionary quality of biological relationships but was based on two operations: dichotomous and hierarchical subtype–supertype relationships. This system was proposed by Aristotle, who grouped animals according to whether or not they had blood, or were bloodless, or were four-legged. Aristotle also suggested a distinction between *eidos* and *genos* (see above), which later took on the meanings of species and family of a living being, respectively. However, their meaning in this context stems from a phenomenological level based on Aristotle's keen eye for affinity and differences.

If we rely on the historian's viewpoint of biology, the classification of fauna and flora and, in particular, of species, has been quite an erratic venture. Nordenskiöld (1928, all quotations in this paragraph from pp. 190–201) has provided an instructive description of different attempts in the pre-Linnéan period. Brunfels, for example, based plant classification on a "judgment according to its power" which represents "bearing witness of God's omnipotence". Leonhart von Fuchs (1501–66) started with characteristics and functions such as: (1) form; (2) habitat; (3) season (when it should be collected); (4) temperament; and (5) powers, but ended with a list following the Greek alphabet (Ballauff, 1954). Andrea Cesalpino (1519–1603), being obviously the only one who showed strong interest in fruits, saw that plants "feed, grow, and produce offspring" but do not move, and concluded that they have smaller organs than animals. Caspar Bauhin (1560–1624) is considered to have originated natural plant classification looking at "natural affinity" of external form. Joachim Jungius (1587–1657) looked at thorns, color, odor, taste, medical impact, sprout period, number of buds, and fruits, but failed to find a classification (Bäumer, 1996).

Finally, Joseph Pitton de Tournefort (1656–1708) made substantial progress by using a model which focused on plants as organic bodies that possess roots, seeds, mostly stalks, leaves, flowers, and fruits. Specifically, in Tournefort's approach all flowering plants are classified according to peculiarities of the corolla; some of his categorization is relevant even today. He used a dichotomous classification and was the first to employ the term *genus* and other more general, higher levels.

Pre-Linnéan classification obviously neglected two topics, one being the sexuality of plants (see Box 4.6 and 5.1). However, the ancient Egyptians and, presumably, the Greeks knew about dioecious plants, which were rediscovered in 1694 by Rudolph Jacob Camerarius (1665–1721) through his experimental work of crossing two male flowers, resulting in no fruit. The other neglected topic was the classification of animals. Animal classification was strongly impacted by John Ray (1627–1705), who in 1693 provided a logical classification of animals with respect to teeth, hooves, and toes (Mayr, 1982). Next, we exemplarily study how Linné succeeded in developing a classification of species and genus, which is referred to in thousands of papers such as: Sex pheromone of *Ectinus aterrimus* (Linné, 1761, see Tolasch, 2008).

4.8.2 Linné's "natural method" of classification

In the context of studying environmental literacy at the interface of science and society, Linné's work is of exceptional interest. It reveals how the traditional biblical conception of cosmogony affected his own theoretical underpinnings. But, as Linné's "natural method" did not ignore the reality of the material, biophysical environment ($E_m^{Linné}$, the reader is asked to follow the nomenclature of Figure 3.3*), we can show how his theory of classification and his world model ($E_m^{Linné}$) developed. This is elaborated in Box 4.6 and strongly refers to the hierarchy of human systems and the complementarities introduced in Figure 3.3*.

Linné based his work on a package of normative postulates in which plants were created by God. Thus, in his practice as a systematist, he was looking at a chain of discontinuity. At the beginning of his career, Linné rejected looking at any crossing of species and rejected the idea that any new species may emerge from the existing ones. But he did notice that plants are individuals and that some of them resemble each other more than others. For instance, offspring resemble their forebears. Thus, when looking at order, he did consider inter- and intra-specific variation. Linné also rejected spontaneous generation and stated that everything developed from eggs. His model of the organism included the concept that life consists of the circulation of fluids (transported in a kind of medulla, the essential part of a species) and targets preservation of the individual and the kind (Larson, 1967, 1968, 1994).

It is intriguing how Linné substantiated modern biological nomenclature and succeeded in providing a "natural method" representing all "natural affinities fundamental to botany" (Larson, 1967, p. 312) that endured for 300 years. What were the main ideas that enabled and allowed Linné to develop such a robust classification system without referring to the dynamic concept of speciation?

According to Cesalpino, Linné viewed reproductive function as an essential part of life. When viewing similarities between plant seeds and animal eggs, he considered the fruit as a crucial element of reproduction. But, he also went one step further and looked at the conditions necessary for reproduction and asked, "What are the generative organs in plants for producing seeds, that is the embryonic part of plants?"

As there is no fruit without a flower, he referred to Tournefort and considered the flowers as antecedents of fruits. Linné investigated the calyx, petal, stamen, apex, pistil, and fruit (see Figure 4.15b) and inferred that (see Larson, 1967): (1) most stems have stamens and pistils; (2) some plants have two kinds of flowers; and (3) there are a few plants with apices without pistils being sterile, and others with pistils without apices being fertile. Thus, he had already detected the reproductive mechanism of plants by 1735 (Linné, 1735/1788; see Box 4.6), though some differentiation was necessary because of differing visibility of the organs. In some plants the (female) pistils and/or the (male) stamens were not immediately perceivable with the eye. The original classification started from quantitative reasoning, the number of pistils. Plants from Lapland and those he saw during a visit to Holland were brought from Africa and India, and showed that these organs of generation were obviously general. A differentiated view was required for distinguishing between male, female and hermaphrodite flowers and their distribution. In this manner, 24 classes were created, the last of which consisted of Felices, Musci, Algae and Fungi. We should note that hermaphrodites are a biologically common, ubiquitous phenomenon also (rarely) appearing in the human species, as "both the Talmud and the Tosefta, the Jewish books of law, list extensive regulations for people of mixed sex" (Fausto-Sterling, 1993, p. 23).

Unfortunately, the sexual approach to classification led to contradictions with some plants such as species of the genus *Valerina*, which showed an unsystematic distribution of the number of stamens, whereas other plants showed this variation in the number of ovaries,

4.8 Classification

a

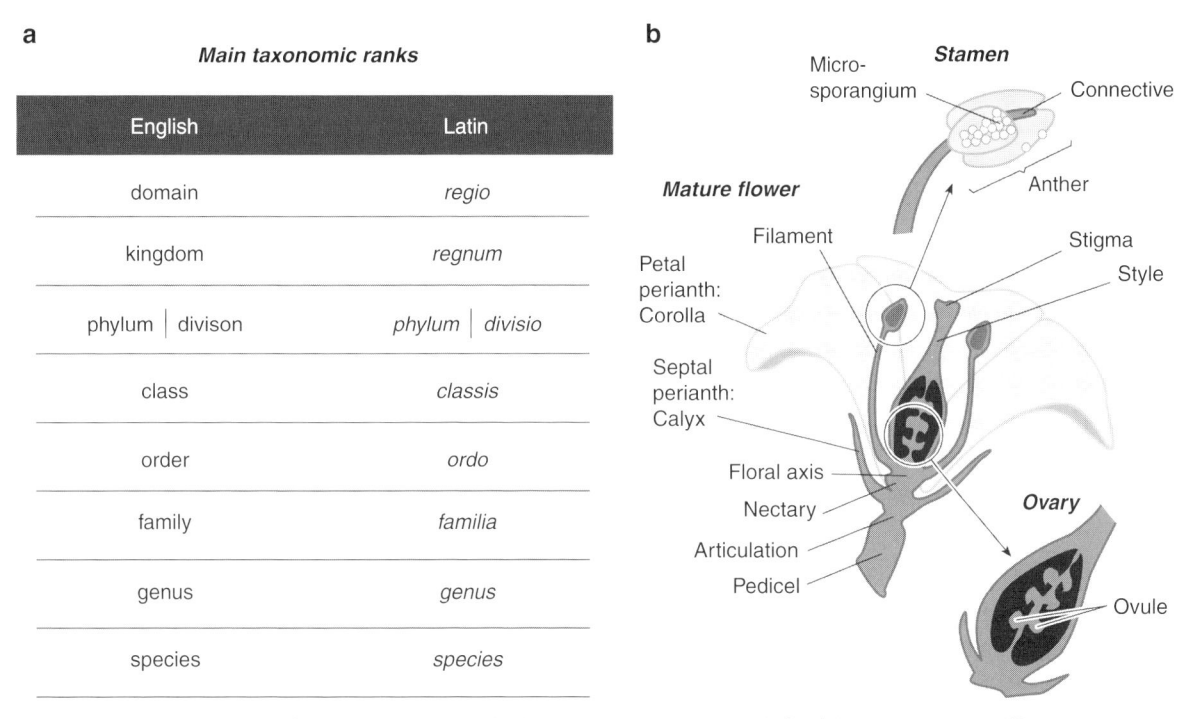

Main taxonomic ranks

English	Latin
domain	*regio*
kingdom	*regnum*
phylum | divison	*phylum* | *divisio*
class	*classis*
order	*ordo*
family	*familia*
genus	*genus*
species	*species*

Figure 4.15 The construction of a consistent hierarchical order or taxonomy (a) was a goal of early botany. The organs of flowers, including sexual organs (b), dominated in Linné's approach.

the female reproductive organs. Though the concepts of habitus (Linné, 1751, p. 101), the general outward appearance, and constitutional type held some ambiguity for Linné, he established characteristics for genera based on number, shape, proportion, and situation. In his book *Philosophica botanica* Linné classified plant groups and orders according to intuitive assignments of "habitus" (Linné, 1751, p. 101), providing 65 "unnamed and uncharacterized orders under which genera were unevenly disposed" (Larson, 1967, p. 313). This approach was criticized as unverifiable and unscientific during his lifetime, and the fructification–flower–stamen–pistil approach, which he considered to be most essential, did not work above the level of genus. Other approaches, such as referring to cotyledons, a part of the seed of a plant developed by the embryo (which have no sexual part), do not provide phytographic exactness but rather inconsistencies and unsystematic variations, an issue evidently inherent in all living entities. This also holds true for using the principle of symmetry for classification, which proved to be strongly applicable though showing many exceptions. At the end of his career, Linné obviously became aware of the heterogeneity of nature, perhaps when facing the "new" tropical species imported at that time. Thus he might have noticed the limits of his natural method. A thorough, deterministic, dichotomous, subtype–supertype classification based on similarity and divergence seemed unfeasible. In his autobiography Linné wrote:

> In this Linnæus himself set his masterpiece. Many have tried to refine it, but all to no avail. He who can give the key to this has found *methodum naturalem*, but [that] shouldn't happen before *quadratura circuli* is hit upon … (Linné, quoted according to Larson, 1967, p. 319)

The objective of Linné to construct a consistent hierarchical arrangement in nomenclature remained unattained, at least with respect to classes and orders (Bremekamp, 1953). From a practical view, his major contribution was the construction of a binomial nomenclature that was accepted by biologists. Stearn denoted the binomial nomenclature "a byproduct, almost an accident, of his task of providing definitions and means of identifying genera and species" (Stearn, 1959, p. 7).

By the introduction of his binomial system of nomenclature, Linnaeus gave plants and animals an essentially Latin nomenclature like vernacular nomenclature in style but linked to published and hence relatively stable and verifiable, scientific concepts and thus suitable for international use. This was his most important contribution to biology. (Stearn, 1959, p. 10)

Linné's classification would not have been possible without thorough theory-building, which was shaped by the following components. First, there was the Christian, literal, biblical code of belief about the creation of plants and animals through God's craft (discussed in Box 4.6). This is an example of a (societal) world model, which underlies any person's and any scientist's thinking. Second, Linné had keen eyes for perceiving similarity and affinity, and a genial capacity to intuitively recognize and organize environmental data and patterns. Third, he could utilize, modify or develop scientific models that already existed. We consider all three components as a prerequisite for environmental literacy.

Today, a phylogenetic (cladistic) systematics is established from DNA analysis based on the work of Hennig (1966). Systematists of today increasingly reject the Linnéan order and reveal the inconsistencies of those newer attempts to define categories above the level of species. One approach of DNA-based taxonomy-building only considers sister species, the most

> **Box 4.6** The dilemma of empiricism, religious dogma, and theory: Linné's mule plants
>
> Swedish Protestant society welcomed Carl Linné in 1707. Born the son of a clergyman, he was shaped by a "strong element of naïve religious faith" (Larson, 1968, p. 291). A strong religious education was a major component of his socio-epistemic environment. In 1743, Linné presented an essay with the following arguments (Linné, 1743, see Larson, 1968, p. 292): (1) God created a single pair for each organism, which can generate more than two offspring or more than one offspring in the case of hermaphrodites. Thought cannot go further back beyond creation; (2) to be efficient and follow the principle of least effort, God produced just one pair or hermaphrodite. The observed growth of species can thus be traced back to these beings; and (3) the Bible documents that Adam gave names to all animals. It should follow that all animals, plants, and insects must have already existed in Paradise. However, Linné concluded, if the world had looked as it did today, all plants would have been spread around the globe, and Adam would have had no chance to name them. This problem is solved if one supposes that the Earth was not covered by oceans as it is today, which is supported by the fact that many fossil shells indicate that it must have been flooded. Thus, it seems reasonable that the Earth was originally a small island large enough to contain all plants and animals.
>
> Thus, Linné inferred that Paradise must have been an equatorial island, so that Adam could name all creatures. The island would have been crowned with a very high mountain and proper soils for each plant to allow the varied habitat required for all animals and plants. According to the terminology of Chapter 3 (see Figure 3.3*), this can be represented as a strong impact of religious dogma affecting Linné's world model. This is represented by ❶ in Figure 4.16a.
>
> According to his belief in the biblical story of creation, Linné started with a discontinuous model for plants and animals. In 1742 he unexpectedly saw a five-spurred toadflax (*Linaria*), which he first believed to be a monster plant that he called Peloria (Gustafsson, 1979). Based on sightings of other hybrids, and thus subsequently overestimating the number of hybrid plants, Linné coined the term *mule plants* or *mule species*: "… I find hybrid plants are more common than hybrid animals and rather many in number. I believe I have [been allowed] to open the door to one of nature's extensive chambers, although it is not opened without creaking" (Linné, 1751/1907, p. 140). Here we can see that Linné acknowledged the factual material–biophysical environment in an empiricist manner (see ❷, Figure 4.16a). From a classification perspective, Linné concluded that many species belong to one genus. He followed the "from the simple to the complex" principle, but "he abandons neither the notion of creation by fiat nor faith in the constancy of that creation" (Larson, 1968, p. 297).
>
> Four years later, in 1755, Linné drew on the analogy to mules by asking whether hybrids, like mules, were infertile. In 1760, he suggested that new species are brought forward through hybridization (Larson, 1968). However, given that his natural science view had been shaped by scientific peers and ideals, he sought a theoretical explanation for hybridization. Referring to his "scientific peer group," Linné introduced the medulla–cortex theory. Applying a variant of Aristotle's *genos–eidos* complementarity, he supposed that fructification is part of the medulla, which was also supposed to exist in the plant. In hybrids, the medulla descends from the mother, whereas plant vegetation is part of the cortex, descending from the father. For the less essential cortical substance, he assumed variability and identified 64 kinds of cortexes. Plants are "issuing from the mother in fructification and belong to her group, although they resemble the father in outward appearance" (Larson, 1968, p. 297; see Figure 4.16b). Here we can suppose that the epistemic of his peer group, H_s^{peers}, strongly affected Linné's thinking, for instance by the texts E_m^{peers} he could read (see ❸, Figure 4.16a).

4.8 Classification

Now, Linné's challenge consisted of making his scientific model compatible with his faith, that is the Protestant Swedish society's interpretation of the Bible (see ❶ in Figure 4.16a). Fortunately for him, the medulla–cortex theory could be interpreted in such a way as to ensure that things came together. In *Genera Plantarum* (Linné, 1764; see Bremekamp, 1953, p. 243), he described how religious dogma, biophysical data, and theory can be brought together to provide a hierarchical order. Quite surprisingly, Linné became a probabilist to accomplish this and introduced chance as a property of nature. He argued:

1. ... the thrice exalted Creator ... covered the medullary substance ... with the various kinds of cortex ... and in thus way as many individuals were formed as there are now Natural Orders ...
2. The generic prototypes ... were mixed with each other by the Almighty and there are now so many genera in the Orders as in this way new plants were formed ...
3. The generic prototypes were mixed with each other by Nature, and in this way in every Genus ... Species were formed ...
4. The Species ... were mixed with each other by Chance and in this way the Varieties arose ...

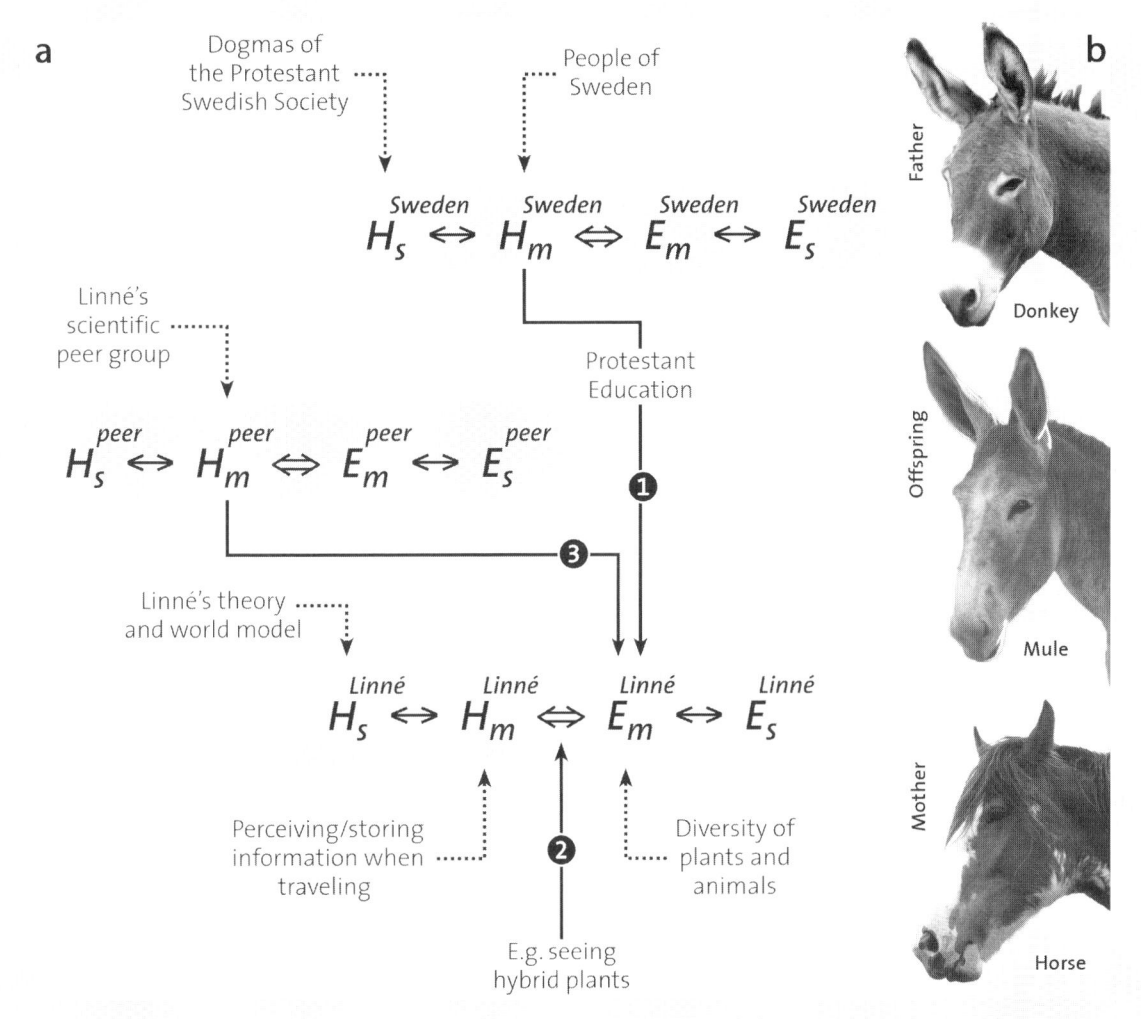

Figure 4.16 (a) From Linné's perspective the Protestant Swedish society's dogma H_s^{Sweden} was conveyed to him via visual and phonetic signals H_s^{Sweden}, for instance, the church bells he could hear which became part of his material environment $E_m^{Linné}$ and which represented his personal social environment $E_s^{Linné}$ (see also Figure 3.3*). (b) A mule and its parents illustrate the idea inherent in the medulla–cortex hypothesis, where, outwardly, the offspring of hybrids resemble the father more than the mother.

closely related populations in a given gene pool (Ax, 1996, 2003).

We should note that Linné was also working as political advisor to the king of Sweden and was pushing the idea of *oeconomi botanici*. He was convinced that Sweden's flora provided an abundance of food (Linné, 1947/1952). He considered the severe starving at that time not as God's will but as ignorance of the people, who did not know to select between healthy and dangerous plants (Koerner, 1999).

When studying Linné it is interesting to see how he acknowledges empiricism, and how he adjusted his taxonomy and its genesis with these observed data. But, like (almost) all other scientists of that time, he strove to make the results of theory consistent with the biblical world model. When Linné introduced chance as an important concept in his theory of classification, he became a probabilist (see Box 4.6). This can be seen as a first step towards a non-static understanding of the biophysical environment and as acknowledging dynamics inherent in the natural environment which are not under the control of supernatural forces.

4.8.3 Key messages

- Like many of his peers, Linné used an approach to classifying species that was strongly influenced by models of creation. In his early work he viewed species as static
- Classification has been pursued not only from an applied perspective, for example for serving medicine, but has obviously also been an issue of curiosity for understanding nature
- Linné founded his work on the idea of regeneration (life cycle) and sexual organs. He generated a viable binomial classification at the level of genus and family but not for higher classes
- In his later life, Linné noticed and acknowledged that chance has a creative role when crossing plants from different genera. Linné became a probabilist. This has to be considered as a major step towards a theory of evolution.

4.9 Biological exploration: Alexander von Humboldt environments

The turn of the nineteenth century was a time of major societal transitions. The industrial revolution, with inventions like the steam engine, effected a transition from manual labor-based to machine-based manufacturing, and instigated progress in chemical, physical, and technological knowledge. New industrial products and the knowledge revolution led to movements like the French Revolution, where common people demanded more rights and power, and called for the separation of church and state. This era also saw the earliest expeditions to describe the distribution of life over space and time. Explorer James Cook, naturalist Alexander von Humboldt and others paved the way for fields of study such as biogeography and oceanography, and for Darwin's expeditions and seminal work in evolution theory. Besides the technological progress promoting biochemical and physiological research, two aspects are essential.

One is that new industrial production activated worldwide exploration of resources, colonization, and tremendous development of trade and seafaring. In the slipstream of these developments, botanists and zoologists got the chance to explore new settings of fauna and flora. This period also saw the first significant introduction of exotic plants into European countries. *Rhododendron*, "now considered an alien junk weed" (De Almeida, 2004, p. 120), was brought from Asia to Britain in 1763 and became a symbol of British gardens.

> Carl Linnaeus ... sought to grow bananas, coffee and other tropical fruits in frigid Sweden, and lusted for coconuts that would sprout in Nordic air – as he declared: "Should coconuts chance to come into my hands ... it would be as if fried Birds of Paradise flew into my throat when I opened my mouth". (De Almeida, 2004, p. 120)

Cotton, rubber, and cocoa plantations have been a major driver for colonization.

The other aspect is the French Revolution (1789–99), which introduced a shift away from feudal and church privilege towards more rights and power for common people. The separation of state and church was introduced and the church lost land, properties, and the authority to claim the tithe (i.e. to levy taxes on crops). The idea of the creation of man and the world by God, which had underlain the theory-building of Linné and others, was questioned, laying the groundwork for the idea that life develops from fibers or cells. God and supernatural forces took on a new role. Variants of agnosticism emerged and were promoted by Thomas Henry Huxley (1825–95), who in 1869 founded the journal *Nature*, and induced disputes about desacralization and disenchantment of nature (Bernal, 1970; Seeland, 1988). This is reflected in the saying of Jean-

Baptiste Lamarck: "We do not need the hypothesis of God any more" (see Box 4.7).

One important contribution of biologists of the nineteenth century on environmental literacy has been their insight into the distribution of life over space and time. An early, noteworthy, expedition which focused on imperialism and biological exploration sailed under the command of James Cook (1728–79). Cook was accompanied by the wealthy botanist Sir Joseph Banks (1743–1820), who was a member of the Royal Society, Robert Brown (1773–1858), and Daniel Solander (1733–82), a scholar of Linné, to the South Pacific Ocean and Australia. Banks paid for half of the expedition and the other half was financed by the Royal family. The primary scientific task was describing new, unknown species such as *Eucalyptus* and *Mimosa*; the crew collected 4000 hitherto unknown plants (Singer, 1959). Science rewarded Banks by naming of the plant genus *Banksia*. Banks' investigations provide good examples of the impacts of classification on science. When comparing Asclepiadaceae plants from Australia with European monocotyledonous orchids, an 1831 study by Brown explored nuclei in the surface layer of cells of plants which were suspected to be closely related. As a master of microscopy, Brown saw continuously moving pollen grains of 2 μm (Ford, 1992). After its reinvention and mathematical description by Albert Einstein, the phenomenon got the name Brownian motion.

Alexander von Humboldt (1769–1859), another well-known explorer, produced little from his extensive anatomy and physiology work, but made major contributions to geology and plant geography (Rádl, 1905). Von Humboldt based his classification on the holistic impression of physiognomy. Though this approach did not provide a valuable contribution to classification, Von Humboldt was one of the first who saw the fundamental effect of the physical environment on the growth of the organism, that the biotic is fundamentally dependent on the abiotic. In his self experiments, through exposure to low oxygen gas, digester gas, and underwater high pressure in diving suits, he also noticed the highly adaptive capability of the organism (Kümmel, 1985).

Von Humboldt developed his primary research focus early when in 1789 he planned to compile a documentation of plant migration, an intensive project which he expected to last 20 years (Eisnerova, 2004). He considered himself to be botanist, geologist, physicist, and meteorologist, and had an eye for the potential of the places he visited for cultivation. Von Humboldt's documentation of plants included the recording of soil, temperature, height above sea level, and so forth. He became the founder of plant geography or "vegetable geography" (Nordenskiöld, 1928, p. 315) and the term "biology" appeared for the first time in von Humboldt's 1797 volume, in the foreword written by Gustav August Roose (Jahn, 2004a). Von Humboldt's mission was

> … to conceive the physical entities in their inner, general connectivity, the nature as an animated whole. (von Humboldt, 1845, p. VI)

Thus, von Humboldt had the capability to read the environment and to holistically identify essential configurations, textures that allow for various types of organic life. This approach has been called the thinking of a general morphologist (Meyer-Abich, 1969).

During 5 years of botanic surveying in America and Europe, von Humboldt observed that, with one exception, there has been, "not a single European plant produced spontaneously by the soil of meridional America" (quoted according to Larson, 1994, p. 116). Thus, he rejected the Linnéan approach that plants migrated from one place and took an intuitive view on physical entities such as energy flows:

> Who, according to this, is able to conceive Nature with *one* sight and knows how to abstract from local phenomena/singularities, can also see how the increase of heat from the poles to the equator causes a gradual increase of animated power and abundance/richness of life. (von Humboldt, 1889, p. 181)

We call an Alexander von Humboldt environment (or simply von Humboldt environment) the setting of all (material, bio-) physical, geographical factors – climate, soil, topology, hydrology – that characterize a location or region and affect life. This early research on geographic variation, concentrating on the comparison of far distant species, laid important groundwork for development of the fields of ecology and evolution. For instance, Charles Darwin (1809–82) made his famous voyage on the Beagle in a similar way from 1831 to 1835. Despite meager equipment – no microscope and receiving his salary through the captain of the Beagle (Singer, 1959) – he brought back records of more than 200 species, including their locations. Darwin's reflections on the voyage illustrate the significance of work by earlier scientists:

> The voyage of the *Beagle* has been by far the most important event in my life, and has determined my whole career;

[…] I owe to the voyage the first real training or education of my mind; I was led to attend closely to several branches of natural history […]. The investigation of the geology of all the places visited was far more important, as reasoning here comes into play. On first examining a new district nothing can appear more hopeless than the chaos of rocks; but by recording the stratification and nature of the rocks and fossils at many points, always reasoning and predicting what will be found elsewhere, light soon begins to dawn on the district, and the structure of the whole becomes more or less intelligible. I had brought with me the first volume of Lyell's *Principles of Geology*, […]; and the book was of the highest service to me in many ways. (Darwin, 1876; quoted in Barlow, 1958, pp. 76–7)

4.9.1 Key messages
- Alexander von Humboldt was the first to cognize how geophysical, geoclimatic, topographic conditions affected, and sometimes constrained, the living space of plants, animals and humans.

4.10 Evolution

In the nineteenth century, doubts about the capability of creationism and mathematics to describe patterns of nature continued to grow, and theory-building was grounded more and more often in evidence-based data. Assumptions and theories from this period about species distribution, geology, and extinction were building blocks for Darwin's seminal work *On the Origin of Species* in 1859. In this section, we review Darwin's two-step model of variation and selection, and also subsequent developments in evolutionary and developmental biology.

4.10.1 Evolving biological knowledge in the nineteenth century

Darwin's brilliant ideas on the common descent of species and the evolution of species revolutionized the field of biology in the mid-nineteenth century. He proposed that new species of animals and plants produce offspring that vary slightly from their parents and that a process of natural selection tends to favor the survival of those whose characteristics fit best with the environment, both in the sense of adaptation to environmental constraints but also with respect to intra-specific sexual selection.

As a new idea on the origin of humans, Darwin's theory was severely disputed. Christian cosmogony was still predominant during this period, yet more and more scientists were looking to observations of nature-driven processes and dynamics for theory-building. While the interpretation and framing of theories and empirical evidence was still strongly affected by deistic or religious conceptions, literal biblical interpretations, such as a creation day or creation center from which all plants migrated, were increasingly doubted.

On the level of theory-building, more and more scientists believed that a reductionist, mechanistic, mathematical Cartesian–Newtonian approach was insufficient for understanding bio-organismic complexity. Natural researchers also noticed that the chemical processes in abiotic matter differed from those in biotic ones. Comparative studies showed that plants and animals adapt to their environment and that there are regional differences in their variability. The cellular approach to sexual reproduction had not yet been developed, and scientists focused on disputes about epigenesis, the study of how the individual organism develops from an originally undifferentiated entity. Animaculists postulated that the male spermatozoa were the source of new life, whereas ovulists believed that new individuals grew from the female ovum (Haeckel, 1911; discussed further in Box 4.6). Further, as Jahn (1990) stresses, theory-building was increasingly based on analogies and environmental impacts. Thus, the role of radiation on plant growth was increasingly discussed.

Empirical evidence became momentous and data were more systematically sampled and intensely interpreted. Experimental results, for instance on embryo development and surveys or inventories, could be used to disprove speculative theories. For instance, in plant physiology Nicolas-Théodore de Saussure (1767–1845) proved that the plant takes up nitrogen (N) from the soil and carbon dioxide (CO_2) from the air (de Saussure, 1890/1804). Also, Justus von Liebig (1803–73) or Boussingault (1802–87) revealed the role of mineral nutrients in plants (Boussingault, 1845; Liebig, 1840). Furthermore, understanding of the transformation of carbon dioxide and organic compounds for photosynthesis and organismic growth was in the making. While the capability for measurement was advancing during this period, questions about the general pattern of life – interactions between procreation, embryogenesis, regeneration, the adaptation of fitness for purpose of organs, intellectual capacities – and the diversity of morphology and species remained unanswered (Jahn, 1990).

4.10 Evolution

Box 4.7 Driving species to adaptation: Lamarckism

Jean-Baptiste Lamarck's (1744–1829) ideas of evolution were based on those of Georges-Louis Leclerc de Buffon (1707–88). Buffon, inspired by Newton, considered motion and continuity as key concepts of the organism. Thus, based on the similarities of fossils and living organisms, Lamarck postulated a continuity between plants and animals (Singer, 1959). He speculated that life is in constant motion, and the motion is caused by "l'orgasme vital," "a state of tension … which … enables the organs to contract and expand … The cause of this tension is a fluid … absorbed by all the organs of the body, but is particularly concentrated in the nervous system. This fluid is really a peculiar variety of fire, related to heat and electricity" (Singer, 1959, p. 323). These ideas are speculative and based on known or ancient concepts: but they are an attempt to formulate a physiological model of the organism.

Like von Humboldt and others, Lamarck thought that organisms adapt and that evolution includes an "automatic intrinsic drive toward perfection" (Mayr, 1982, p. 116). Lamarck also thought that the animal kingdom is ordered between worms and infusions (primitive animals generated by spontaneous reproduction) and man, who shows the highest perfection. Also, not completely consistent with this thinking, he also postulated that degeneration is an important mechanism for producing species. Thus, the ape degenerated from man; and animals who do not utilize their eyes, such as bats which cannot see anymore, lose the functionality of certain organs. Lamarck's theory was viciously attacked in 1832, in particular by Georges Cuvier (1769–1832), who has been an accepted authority by the French Academy of Sciences (Magner, 2002).

Lamarck identified the environment as the agent causing this adaptation and formulated the law of "use and disuse" (Singer, 1959). Efforts of organisms were the main mechanism driving species to adaptation. Lamarck posed questions about how, for example, does it happen that a Great Dane, bulldog, and dachshund belong to one species. He inferred that the environment is the critical agent causing the difference. Giraffes, for instance, developed from a deer-like animal that had to eat the leaves from the trees because there was not enough herbage on the ground (see Figure 4.17).

As Mayr has pointed out, Lamarck was a strong proponent of gradual evolution as a way to describe development of new species. He was not looking for a steady-state theory, but he considered the individual's, not the single species members', activities as a source of evolution. Even though he presented an incorrect mechanism, he provided one of the first comprehensive theories of "adapted evolution" (Mayr, 1982, p. 353).

Figure 4.17 Lamarck asked how different representatives of one species could develop such a different morphology. (a) The Great Dane stands 1.07 m from floor to shoulder, a chihuahua only 19 cm (photography by Deanne Fitzmaurice). He proposed that individual permanent stretching to reach food on trees resulted in longer necks, which the giraffes passed on to their offspring (b, after an original painting by Robert Hills, 1769–1844).

4.10.2 Darwin and other evolutionists

The theory of evolution had many fathers. Historians of biology dispute whether Charles Darwin (1809–82), Alfred Russel Wallace (1823–1913) or even Herbert Spencer (1820–1903) was the first to properly put forward the essential ideas of selection (Jahn, 2004b; Mayr, 1982). In this section, we discuss the assumptions and interpretations about the relation between the organism and the environment that were introduced by Darwin and others around 1850 for developing and reasoning the theory of evolution.

Darwin started his Beagle journey from the assumption that all species have been created separately at the "world center." But he was facing overwhelming evidence in contradiction to this view. Particularly, he got insight into variation and the array of species in different parts of the world. First, he noticed that there are different species and patterns of species in separated regions of the world. Here, the finding of Wallace, who identified two dissimilar zoogeographical regions separated by a sea current, which are the Malay and Australasia archipelagos (see Figure 4.18), was significant. The other has been that the set of species on islands resembled those of the neighboring continent regions. Thus he rejected the migration hypothesis, which assumes that all plants of the world migrated from one point, ruling out the hypothesis of a single center of creation.

In addition to understanding the distribution of species over space and time, another key to understanding evolution was the understanding of "geo-logics." Stenonis (see above) suggested that the Earth's crust was developing from sea sediments. Motivated by fossil findings, Buffon (1707–88) looked at the history of the Earth as a series of epochs. And after his Beagle journey, Darwin stated that the information about nature in rock layers sheds light on the history of organisms. Here the concept of geological time and paleontological knowledge became important. Richard Owen's (1804–92) findings that some fossils represent extinct species and also closely resemble existing species called for explanations.

Another critical question was why and when certain animals went extinct. The idea that "extinction was incompatible with the omnipotence and benevolence of God was very widespread throughout the eighteenth century" (Mayr, 1982, p. 347). Darwin's autobiography describes how Thomas Robert Malthus' *Essay on the Principle of Population* (Malthus, 1798) caused a theoretical avalanche.

> … I happened to read for amusement Malthus on Population, and being well prepared to appreciate the struggle for existence which everywhere goes on from long-continued observation of the habits of animals and plants, it at once struck me that, under these circumstances, favorable variations would tend to be preserved, and unfavorable ones to be destroyed. The result of this would be the formation of new species. (Darwin, 1887; quoted according to Mayr, 1982, p. 477)

Thus, Darwin applied Malthus' idea of selection, assuming that unregulated populations grow at geometric or exponential rates but that food supply grows only arithmetically or linearly, not only to explain within-species selection but also to explain between-species competition and species extinction. However, pure selection would cause a continuous reduction of species, yet the fossil record suggested that new species emerged. Linné's

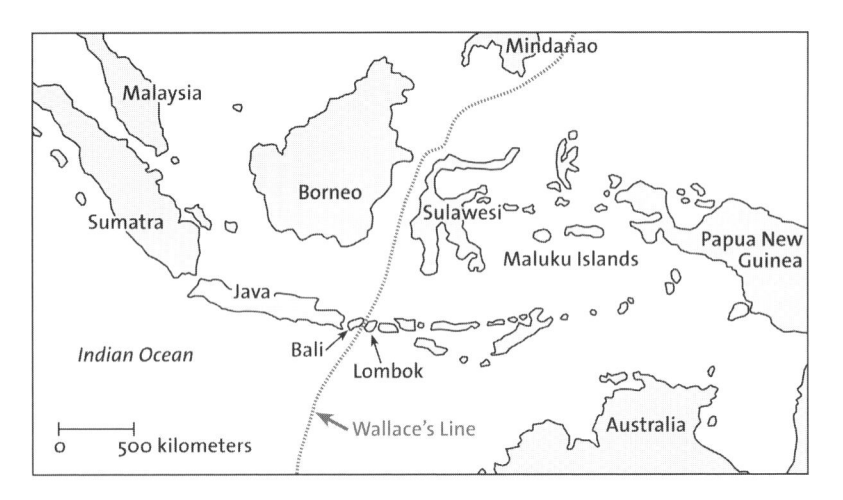

Figure 4.18 The Wallace line divides the fauna and flora of the Malay archipelago from those of Australasia and the Pacific Islands. One reason for this separation is the exceptionally strong sea current from north to south passing the strait between Bali and Lombok.

idea of crossing species for instance has been one model of generating new species. According to Darwin's biography (Barlow, 1958), two other observations were decisive. The first is that the ornithologist John Gould (1804–81) noticed that three different birds, which Darwin collected from three different Islands of the Galapagos and which were classified as different species, were actually different phenotypes of one and the same species (i.e. mocking birds). Here the idea of separation as basis for speciation was born. The second is that Darwin became interested in the human manipulation of species by breeding. He investigated how breeders of pigeons, dogs, and other animals generated diversity, an idea that also motivated Gregor Mendel's (1822–84) research (Hoppe, 2004). Later, Darwin looked at breeding as a means of experimental verification of his theory.

Another decisive conceptual component of Darwin's theory of evolution is that – contrary to the assumptions of Lamarck – diversity, variation, and adaptation occur at the species level, and at an individual level. Thus he supposed that individuals within a population are different, but acknowledging the migration behavior, in birds for example, showing that there are favorable and less favorable environments for certain species. The interaction between diversification within species and the chance to adapt to the environment is reflected in the statement:

> That the more diversified the descendants from any one species become in structure, constitution, and habits, by so much will they be better enabled to seize on many and widely diversified places in the polity of nature, and so be enabled to increase in numbers. (Darwin, 1859, p. 131)

Here, the ideas of diversity and the adaptation of species to the environment (in the sense of the von Humboldt geophysical environment) were linked and the idea of ecological niche and evolutionary ecology emerged (Pianka, 2000).

Finally, we deal with the question of how the conception of God was affected by these new theories and vice versa. Christian cosmogonists said that Darwin's strict materialist approach to the genesis of species had "dethroned God" (Mayr, 1982, p. 117). Magner (2002) points out that Lyell, who became the advocate of Darwinian thought in America, suggested that a deity might have initiated a chain of events that could proceed without or perhaps further supernatural/godlike intervention. Others take a more radical view and identified hidden goddesses of Darwinism. Sheldrake argues: Instead of God the Father, "Darwin saw in Mother Nature the source of all forms of life" with her powers of "prodigious fertility, … spontaneous variation, and … selection" (Sheldrake, 1990, p. 54). While it is difficult to decide which views dominated Darwin's thinking, we can draw the general conclusion that theories about natural dynamics became more important than conceptions of God.

Darwin and others developed a multifactor, non-volitional, natural selection-based theory, which meant that species adapt to the environment because of natural variation of offspring but not as a response to the environment. However, the generation of descending generations and "concepts of mutation, variation, population, inheritance, isolation, and species were still rather nebulous" (Mayr, 1982, p. 2). Darwin's theory is descriptive and based on few fundamental heuristics and proposals. Thus, he assumed that organisms have separate, independent corpuscular "bases," which represent intra- and inter-species variation. This idea can be considered as a solid basis for theories of inheritance. Nevertheless, the nineteenth century has provided an opaque picture on the mechanisms of diversification. Darwin did not yet differentiate between a genotype and a phenotype. And knowledge about the cell, DNA, and chromosomes were not yet developed at his time.

Darwin's theory offers important insights into how species interact with their environment. Simplified, his theory provides a two-step model. First, a species produces variation. In bisexual species, this variation can be considered as a cause of intra-specific selection, for instance in reproduction. Second, the collective of individuals of a species (S_k) are exposed to competition with other species (S_n) for environmental resources (R_l), which feeds back to the species dynamics (see Figure 4.19). This theory is naturally based on the assumption that traits are inherited.

Ernst Haeckel (1834–1919) became a prominent promoter and interpreter of evolutionary theory. Educated as a physician and biologist, he wrote many popular science books which made the theory of evolution accessible to the public. Haeckel also coined many terms such as phylogenetic (*phylon* in Greek; English "stem;" Dayrat, 2003) and metaphorically visualized the ancestry of man as an arboreal trunk with branches. Much attention has been drawn to his theory of recapitulation, suggesting that every organism evolved by the differentiation and addition of new features. Haeckel suggested that higher organisms pass (linearly) through stages of lower ones. Thus, a parallelism between embryogenesis and species history

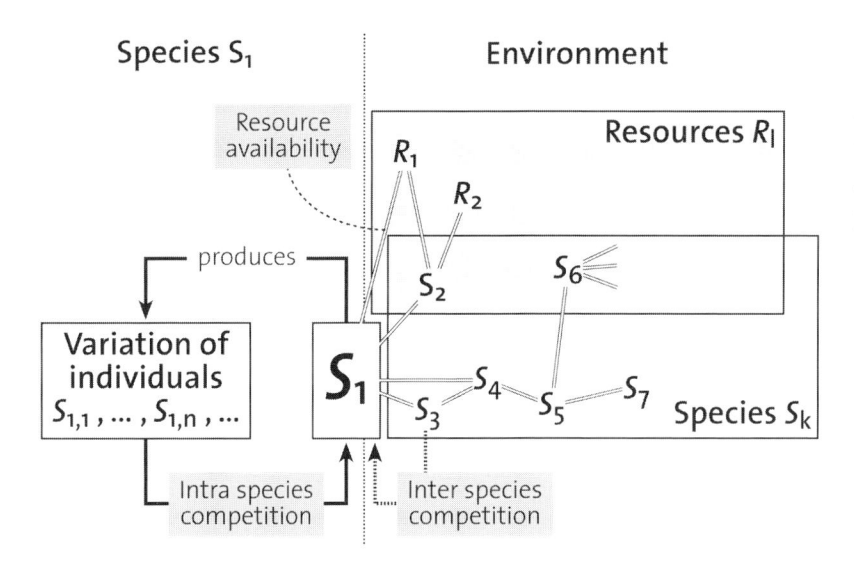

Figure 4.19 Evolution represented as species–environment dynamics for a species S_l, which produces (genetic) variation inducing intra-specific competition and causing differential inter-specific competition with species S_k and resources R_l. S_k and R_l partly overlap as species are part of the resources and the food chain.

was postulated. This has been expressed as the basic biogenetical law. Many of Haeckel's writings showed a strong anti-Christian and anti-clerical vein, evident in his pointed title *Natürliche Schöpfungsgeschichte* ("Natural History of Creation," Haeckel, 1898). An interesting question is, in what respects Haeckel's recapitulation theory is based on "fraud," biasing manipulations, idealized simplifications, or self-deception (see Box 4.8).

4.10.3 Key messages

- The theory of evolution conceptualized the interaction of species with the biotic and abiotic environments. Darwin's theory was based on assumptions leading to key concepts such as variation, speciation, and natural selection
- Evolution and the idea of common descent developed through an interplay of geological, physical-geographical, and biological knowledge and ideas from the natural economy. Physiological knowledge was irrelevant as cellular and molecular knowledge had not yet been developed.

4.11 Ecology

The interaction with nature as a general concept of the environment is inherent in the field of ecology, which emerged at the beginning of the twentieth century. Ernst Haeckel coined the term when writing:

> By Ecology we mean the comprehensive science of the relationship of the organism to the environment, which in the broadest sense includes all conditions of existence. (Haeckel, 1866, p. 286, translated by Scholz)

The complementarities between organisms and their environment that have unfolded can be taken from the etymological roots as "ecology" means the "study of the household [of nature]" (Haeckel, 1898, p. 286, translated by Scholz). The term was derived from the Greek οικοσ (household) and λογοσ (study). The first textbook on plant ecology was written by Eugene Warming (1841–1924) in 1898. The term ecology is often used in common parlance as a synonym for the natural environment and the interaction of organisms and their natural environment. When targeting the unity of organisms and the environment, Tansley (1935) proposed the term ecosystems. The Russian soil scientists Vasily V. Dokuchaev (1846–1903) and the forest ecologist Georgy F. Morozow (1867–1953) provided early contributions to ecology when focusing on the practical view of food and wood production.

4.11.1 Emergence of ecology

Mayr (1982) identified three essential lines of ecological research. One emerges from Spencer's concept of the total environment (see Chapter 1.3), another from biosphere (i.e. the interaction of organisms and their environment), and a third extended von Humboldt environments. With von Humboldt environments, the (1) geophysical characteristics of space is linked with (2) the resources, in particular the distribution of organic

4.11 Ecology

Box 4.8 Frauds, sleights of hand or simplifications in embryonic similarities? Evo-devo and abortion

The public and many scientists reject the idea that "the marvels of ontogenesis result from natural selection" (Sander, 2002, p. 532). Popular alternative theories based on creationism and intelligent design are still promoted by some biologists. For example, a 2004 creationist paper by Stephen A. Meyer (2004) passed the scientific peer-review process of Thomson Reuters ISI classified journal including "three well qualified referees" (Giles, 2004, p. 114). The paper applies information theory and argues that the complexity of living organisms cannot be explained by Darwinian evolution. Meyer argues that "Natural selection lacks foresight ..." as well as intelligent selection by "purposive or goal directed design" (Meyer, 2004, p. 225). Thus Meyer rejects Darwinism and "suggests purposive or intelligent design as a causally adequate – and perhaps the most causally adequate – explanation for the origin of ... Cambrian animals and the novel forms they represent" (Meyer, 2004, p. 236). Here we find the construct of God at work again in "evo-devo," that is in the combination of evolutionary biology and developmental biology, which some might still consider as abbreviation of "evil devil."

Ernst Haeckel (1834–1919) was a prominent promoter and interpreter of evolutionary theory and became famous for his controversial illustrations of one of his popular books (see Figures 4.20c–e). Already in 1831, embryologist Wilhelm His (1831) spoke out on Haeckel's sleight of hand and manipulations of his illustrations showing different vertebrates developing from virtually identical stages of embryonic development. In 1997, a group of English biologists reconstructed the comparative embryological drawings. The data reveal striking patterns of heterochrony, which are deviations from a supposed ideal embryological sequence in vertebrates (Richardson, 1995; Richardson et al., 1997). Given that differences are perceivable to the naked eye, Haeckel's drawings of vertebrate embryos are considered inaccurate and misleading and do not appropriately represent the differences at the early embryonic stage which, according to evo-devo, originate in an individual's DNA.

Haeckel's slip went unchecked into the pages of some top newspapers such as *The Times* and the German *Frankfurter Allgemeine Zeitung*. Christian journals such as *Creation* (Grigg, 1998) considered it as fraud and "proof" against evolutionists' arguments and the justification of abortion. Do we have the same fallacious perceptions, like those of the preformationists' miniature sperm people? Has Haeckel provided an irresponsible deception? Is the hourglass model (see Figure 4.20a) completely wrong? A second look by an experienced biologist provides a different view (Sander, 2002).

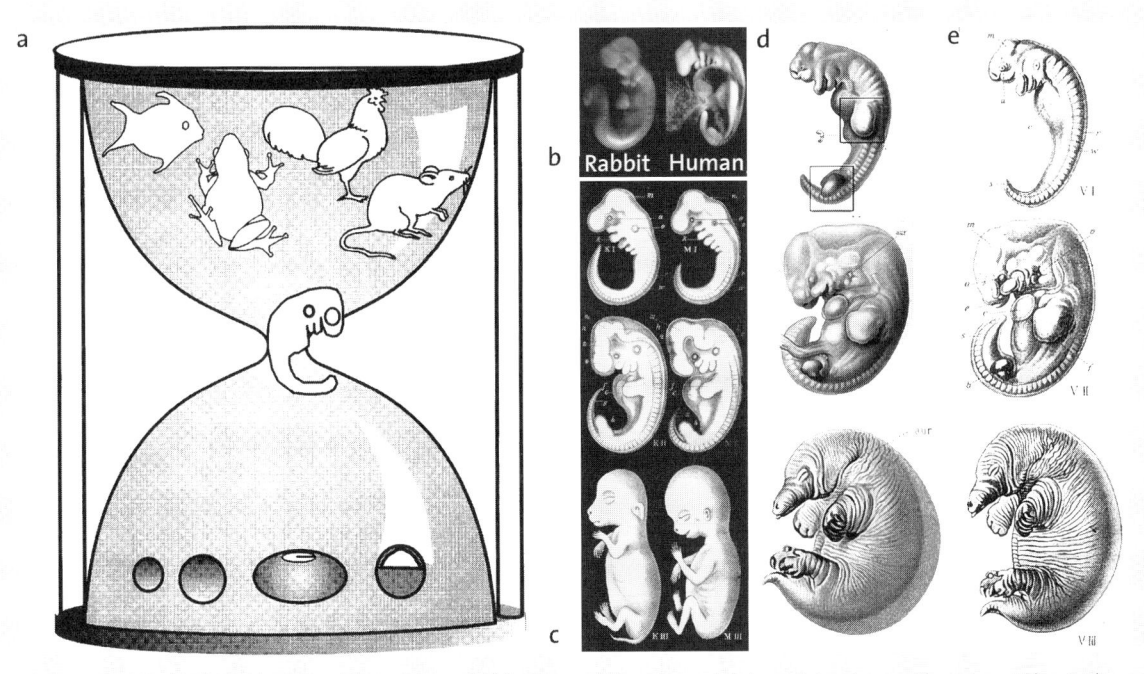

Figure 4.20 Haeckel's developmental hourglass model (a) suggested divergence at early and later stages with a conserved or similar intermediate stage (from Richardson et al., 1997, p. 92, with kind permission from Springer Science + Business Media). Images of rabbit (*Oryctolagus cuniculus*) and human embryos (b and c). Figure (b) is taken from photographs by Richardson et al. (1997) and Figure (c) from drawings by Haeckel, 1874. The deletion of the limb buds from embryo drawings by Haeckel (e) can be seen when compared with the original echidna drawings (d) from Richardson & Keuck (2001, Reprinted by permission from Macmillan Publishers Ltd). Haeckel's limb bud omission enabled him to suggest a perfect similarity between early-stage embryos from different species.

> **Box 4.8** (cont.)
>
> Haeckel's illustrations and the assumption that organisms become progressively different from the ovum are without doubt incorrect if the zygote is taken as the starting point. The question is whether this also holds true for the hourglass hypothesis. The hourglass hypothesis states that, at a certain "phylotopic" stage, the basic vertebrate body plan is established. The phylotopic stage, which for a mouse is around day eight (Irie & Sehara-Fujisawa, 2007), could be the starting point of Haeckel's illustrations. Recent evo-devo research that links comparative developmental genetics with macro-evolutionary quantitative analysis of similarities supports the hourglass hypothesis for bisexual animals and thus Haeckel's presentation (Cruickshank & Wade, 2008). There are incorrect and euphemistic elements in Haeckel's drawings with respect to size (there is a tenfold difference) and the exact developmental stage when the embryos are alike. Perhaps most irresponsible, Haeckel eliminated important details in his drawings, preventing potential further insight into evolutionary development (see Figures 4.20d–e).
>
> However, we can learn how ontogenetic and phylogenetic theories develop and can become increasingly consistent when linking data from the scale of the embryo as can be seen by the eye, a microscope view of the ovum, and genomic analysis of DNA.

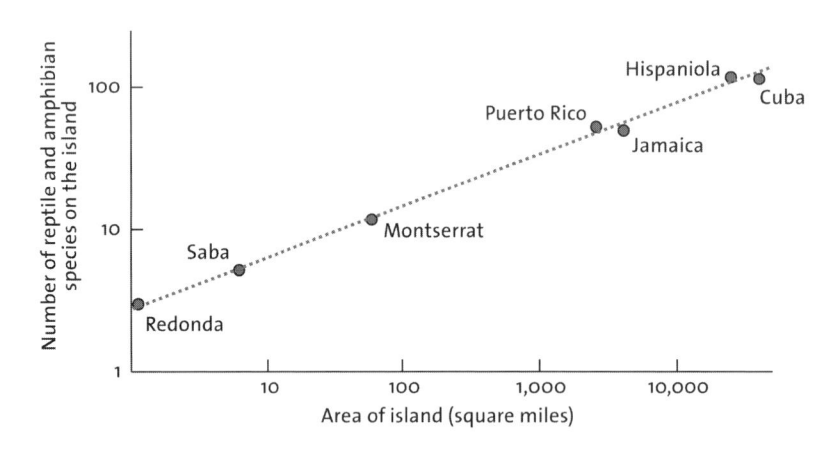

Figure 4.21 The relationship between the area of islands and the number of species living on them (adapted from Wilson & Bossert, 1971).

and inorganic matter, and (3) the genetic array of plant and animal species and their behavioral programs. This branch of research emerged from the nineteenth century period of geophysical world exploration. Studies of oysterbanks, marine ecology, or tropical ecosystems can be taken as examples. The analyses were mostly descriptive and emerging from a static steady-state system view. We can distinguish between biocenosis, which is the study of self-regulating ecological communities (Moebius, 1877), ecological niches (Elton, 1926), landscape ecology, and vegetation geography (Schmithüsen, 1942; Troll, 1939, 1971) and studies in the structure and order of natural systems (Leser, 1991; Schmithüsen, 1976).

4.11.2 Population dynamics and ecology

The second ecological research line identified by Mayr was first investigated as the Lotka–Volterra approach to logistic modeling of predator–prey relations. Here the branch of mathematical biology with single and multiple population dynamics was developed (Pielou, 1969). The main focus is the growth, decline, and cycling of interspecific competition and feedback loops (see Figure 16.11). A specific branch of this research focuses on the stability and decline of small communities (MacArthur, 1955) and the famous theory of island biogeography explaining the species richness of natural communities (MacArthur & Wilson, 1967). "Islands" can be actual, water-surrounded islands, valleys or parks and the mathematical modeling includes measures of isolation (distance to neighborhoods, time of isolation), size of the habitat, climate, the diversity and history of biodiversity, nutrient settings, ocean currents etc. It is common to model the logarithmic relationship between the area of the island and the number of species ($No_{species} = c^* \log_{10} a_{Island}$; Figure 4.21). The seminal work of Robert May (1976) demonstrated that ecological theory can be rigorously described in mathematical terms.

4.11.3 Ecosystems: energy fluxes and nutrient cycles

A third line of ecological research focuses on energy and material cycles, and considers resources or food chain analysis. One example is the paper by Raymond L. Lindeman called "The trophic-dynamic aspect of ecology," which "emphasizes the relationship of trophic or 'energy-availing' relationships with the community unit to the process of succession" (Lindeman, 1942, p. 399). Lindeman argues that the separation of zoology and botany should be overcome and asks for a bio-ecological approach:

> A more "bio-ecological" species-distributional approach would recognize both the plants and animals as co-constituents of restricted "biotic" communities, such as "plankton communities", "benthic communities", etc., in which members of the living community "co-act" with each other and "react" with the non-living environment … (Lindeman, 1942, p. 399)

Lindeman's work brought about a new view in ecology. His paper describes a careful study of the food-cycle energy flows, which was taken as a common denominator and proxy for food availability. He was referring to community economics and nutrient-cycles in water introduced by Thieneman (1926). An interesting line of thinking reads:

> [I consider] plants as producer … employing the energy obtained by photosynthesis … Animals and … plants as *consumer* organisms, feed … a considerable portion of the consumed substance to release kinetic energy for metabolism, but transform the remainder into … their own body. …. Every organism … may act as energy sources for successive categories of consumers … for … organisms … The careful study of food cycles reveals … food-cycle diagrams … (Lindeman, 1942, pp. 400–401)

As illustrated in Figure 4.24, Howard T. Odum (1957) similarly analyzed the flow of energy through four trophic levels of producers and consumers in a river ecosystem in Silver Springs, Florida. Net production is only a fraction of gross production because the organisms must expend energy to stay alive. Note that the relative difference between gross and net production is greater for animals than for the producers, which reflects their greater activity. It is also interesting to consider that much of the energy stored was lost to the surroundings by decay or by being carried downstream. Therefore, there are substantial losses in net production as energy passes from one trophic level to the next.

Lindeman's paper promoted a theoretical, modeling and mathematically oriented ecological research. However, as Eugene P. Odum stressed, the quantitative research by ecosystem ecologists (Odum, 1983) did not emerge before von Bertalanffy (1950b, 1968), Hutchinson (1944), Patten (1978), and Odum (1994; 1981) introduced system theory to ecology.

4.11.4 Succession and climax as key concepts of ecosystems

The early roots of ecology can also be seen in late nineteenth century North American biologists who investigated reasons and laws of formation for associations of fauna and flora in certain areas. Already at that time, an interesting question was whether there is a "motive force" (Clemens, 1916, p. 7) in the development of the different populations involved which results in a biotic community or organism. The focus was on vegetative development. Here, *succession* and *climax* became key concepts. Somewhat simplified, the idea of these two concepts was that vegetation starts from a mere substrate, passes through a succession of intermediate stages and ends in an ultimate, mature state or *climax*. As Trudgill (2007) states, this image of ecosystem development was derived from German philosophers in the nineteenth century. Vegetation ecosystems were believed to have a stable climax community:

> The […] climax formation is an organic entity. As an organism the formation arises, grows, matures, and dies. (Clemens, 1916, p. 3)

Tansley (1935) took a critical look at these ideas, concepts, and metaphors. He stressed that succession is often interrupted by catastrophes and that one cannot always find an upward development. The analysis should acknowledge that "the continuous effect of grazing animals … may gradually reduce forest to grassland" (p. 35). Thus, destruction is a genuine element of succession. The discussion as to whether ecosystems can be considered as organisms, quasi-organisms or even as superorganisms which develop towards stable, persistent system configuration, whether disturbance is always followed by recovery and whether the cycling of oceans is similar to the cycling of blood, goes back to the roots of biology as a discipline. Here, Tansley was skeptical and stressed that ecosystems consist "of components that are themselves more or less unstable – climate, soil, and organisms" (Tansley, 1935, p. 301). Thus, the climax and equilibrium attained are never fully perfect and the ideas of change and stability are inherent in ecosystems. In this place we want to note that the idea that the Earth is a superorganism which is capable of self-maintenance was born earlier and goes back to one of the fathers of geology, James Hutton (1726–97; 1788).

The organicism which includes the idea of ecosystems and the Earth as a "complex organism" has been disputed in ecology from its very beginnings. Ecosystems obviously include an evolutionary component which meets the idea of structure-enhancing processes and which can be conceptualized as a dynamic equilibrium. But they also include a decay, instability or destructive component. Here, "the destructive human activities" and the question "Is man part of nature or not?" were already discussed by Tansley (1935, p. 303). The elimination of species by humans builds one important aspect (see Box 4.9 and Box 4.10).

Box 4.9 Too stupid to adapt: dodos and kakapos

Explaining competition (Darwin, 1859; Spencer, 1855), survival (Wilson, 1992), invasion (Ehrlich, 1986), and extinction (Nitecki, 1984) in an often dramatically and ever-accelerating, changing biophysical environment has become a challenge of ecological and evolutionary research. Clearly, humans' lack of environmental awareness and their low appraisal of species value are factors that lead to the extinction of species. The reconstruction of the colonization of Pacific and Indian Ocean islands can serve as examples. Oblivious to the endemic, archipelago-specific niche of the dodo, settlers consumed the bird to the point of extinction.

The dodo (*Raphus cucullatus*, see Figure 4.22a) was a large, 20-kg flightless bird which became extinct in the second half of the seventeenth century. Europeans discovered the 2040-km^2 island of Mauritius in the Indian Ocean – and thus presumably the markedly noticeable bird – in 1502. Given its lack of significant predators, the dodo was entirely fearless – even of people. Its flightlessness made it easy prey for the domesticated animals that early Portuguese and Spanish explorers introduced. Dogs, cats, rats, imported small monkeys as well as pigs plundered the dodo nests. By destroying their habitat, deforestation also played a role in threatening this species (see Boxes 1.2 and 1.3). The last dodo was seen around 1690.

The story of the kakapo (*Strigops habroptilus*), a flightless, 2 kg giant parrot endemic to New Zealand, is markedly similar to that of the dodo. New Zealand was settled sometime between 700 and 2000 years ago. The island had more than ten Moas (ostrich-like birds reaching up to 230 kg) and many large mammals which were killed by the indigenous Māori hunters. On some islands, fewer than 100 kakapos managed to survive. Fortunately, "efforts to save the Kakapo began more than 100 years ago, and the Kakapo is now the subject of the most intensive single species conservation program for any bird in New Zealand and perhaps the world" (Cockrem, 2002, p. 139). The remaining birds have been given individual names and are tracked with radio transmitters.

The behavioral and genetic patterns of the kakapo, however, are special and make the conservation program difficult. They lack any escape or defense instinct and it became apparent that only zoo-like "island settings" would ensure their survival. Because of the untenable environment of the big islands for the kakapo, all breeding birds have been moved from their original habitats on the main territory to small islands, where the adults and their clutch are safe from mammalian predators. Unfortunately, some places such as Anchor Island were still too close to the big islands and rats swam the great distance to the island and spoiled the clutch (Six, 2008, see Figure 4.22).

The recovery plan was revised, and since 1997, the kakapo have only bred on Codfish Island, which is tiny at 14 km^2. There was hope that the conservation effort would maximize the chances that every fertile kakapo egg would lead to a successfully fledged chick. However, breeding of this bird, which can live to be 60 years or older, occurs naturally only every 2–5 years. Further, the proportion of females that lay eggs can vary greatly between years and seems to depend on the cycles of vegetation, in particular that of the rimu (*Dacrydium cupressinum*); research programs have barely begun to explain this phenomenon (Fidler et al., 2008).

This intense conservation work has resulted in a marked increase in successful breeding of the kakapo and its number has surpassed 100. If such a program had reached the dodos, one could assume that idioms such as "dead as a dodo" or "to go the way of the dodo" would have used the name of some other species. The case of the kakapo shows that, not only do the rates of change in the environment overburden the adaptive potential of the small kakapo population, but also shows the difficulties of finding appropriate habitats for conservation programs geared toward large wildlife, in particular flightless birds and mammals.

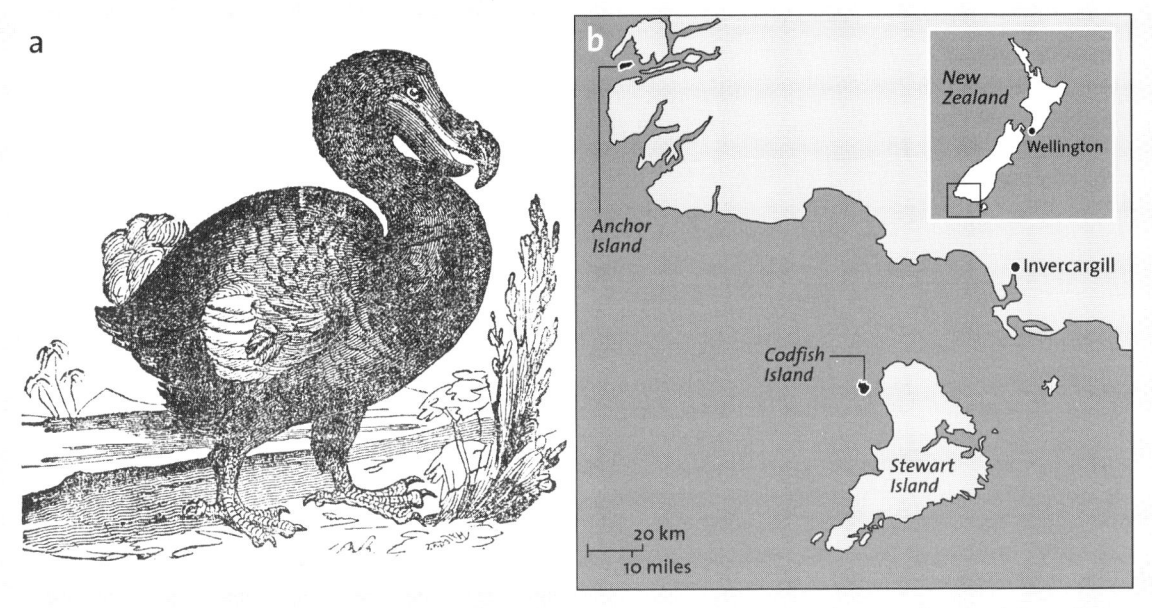

Figure 4.22 (a) The dodo (*Raphus cucullatus*) has been extinct since the late seventeenth century because of imported species (drawing by Roelant Savery, 1626). (b) Anchor Island, selected as a habitat for another large bird, the kakapo (*Strigops habroptilus*) for a New Zealand species conservation program, could be reached by rats (c. 2.5 km from the mainland); thus, the more remote and isolated Codfish Island has been chosen (c. 4 km from Stewart Island).

4.11.5 Key messages

- Ecology deals with the interrelationships among organisms and their environments.
- Ecology researchers study how energy and nutrients flow through the environment, the succession of ecosystems, the dynamics of population, the interaction between different kinds of organisms and geographic distributions. As such, ecology offers many important ideas and concepts for studying human-environment systems.

Box 4.10[1] **Climate variability: Dansgaard–Oeschger events**

The temperature increase induced by anthropogenic greenhouse gas (GHG) emissions is increasingly well understood (Intergovernmental Panel on Climate Change, 2007). This allows us to make projections about future climate change due to continuation of anthropogenic GHG emissions. However, climate researchers are far from understanding the whole climate system. Past climate variability emphasizes how natural forces can lead to abrupt, unpredictable climate changes.

Climate variability over time is measured by analyzing physical and chemical characteristics of ice core samples, such as the deuterium profile, isotope stage, grain radius, dust concentration, dielectric profile, electrical conductivity and GHG record of the entrapped air, to calculate concentrations of CO_2, CH_4, and N_2O. These measures are indicators of past temperature, volcanic activity, and atmospheric composition. Recently, analysis of an ice core sample produced a deuterium profile from the surface down to 3259.7 m, allowing an unprecedented assessment of the Antarctic surface temperature record back to about 800 000 years ago (Jouzel *et al.*, 2007). The resulting temperature profile shows 100 000 years' cycles, beginning with a short warm phase (interglacial), followed by a longer cold phase (glacial, see Figure 4.23). These results suggest that the structure of the glacial periods changed. The data show that before 430 000 years ago, the glacial periods were milder, whereas colder temperatures were measured for more recent glacial periods. The reasons for these fluctuations are still subject to

---

[1] This box has been written together with Bastien Girod.

Box 4.10 (cont.)

speculation, in part because the influence of the intensity of radiation as a consequence of the varying distance of the Earth to the Sun and changing parameters of the rotation (precession) still cannot be modeled with much precision.

Of special interest are the so-called Dansgaard–Oeschger events (Dansgaard et al., 1993), which are abrupt climate fluctuations that occurred during the glacial period. Twenty-five such events are thought to have occurred in the last glacial period (see Figure 4.23), during which temperature measurements in, for example, Greenland fluctuated between 8 and 16°C. The last time such an event occurred is about 10 000 years ago. In the northern hemisphere, these events take the form of rapid warming episodes, typically in a matter of decades, each followed by gradual cooling over a longer period. The pattern in the southern hemisphere is reversed but with lower amplitude (see Figure 4.23). This supports the thermal bipolar seesaw hypothesis, which postulates that abrupt shutdowns and restart of the Atlantic meridonal overturn circulation to produce slow warmings and coolings in the Southern Ocean/Antarctic region.

A recent, high-resolution analysis of marine sediment cores from the Iberian margin traced sea surface temperature and water mass distribution, as well as relative biomarker content (Martrat et al., 2007). This study demonstrates that the north–south coupling was pervasive for the cold phases of climate during the past 420 000 years. The authors found that cold episodes after relatively warm and largely ice-free periods occurred when the predominance of deep-water formation changed from northern to southern sources. In addition they revealed an increase in the variability of the climate: the Iberian cores register 18 oscillations during the first climate cycle, nine in the second, seven during the third, and six over the fourth.

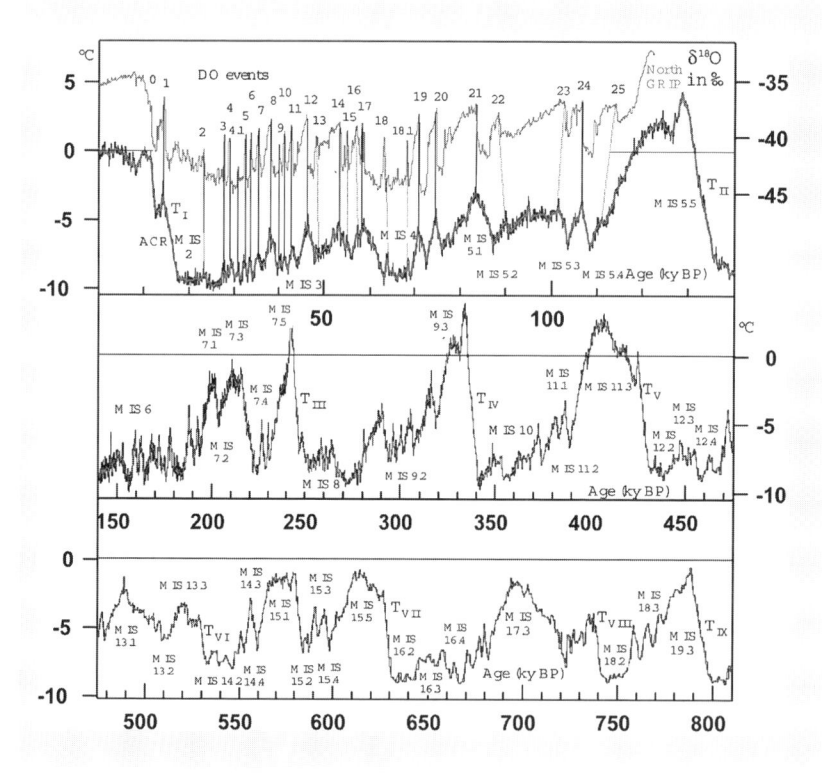

Figure 4.23 Temperature anomaly as a function of time over the past 810 thousand years (ky) from Antarctic ice cores. The upper panel, back to 140 000 years before the present (140 ky BP), shows correspondence between the Dansgaard–Oeschger (DO) events as recorded with the North Greenland Ice-core Project isotopic record (GRIP) during the last glacial period and the last deglaciation (figure from Jouzel et al., 2007, p.794, reprinted with permission from AAAS).

4.12 Biosemiotics: the first *Umwelt* research

Jakob Johann von Uexküll (1864–1944) was a zoologist who offered an important new perspective for understanding how organisms perceive the environment. He also made an important contribution to our understanding of the relation between organisms and the environment when defining a mind–body relationship for living beings. In 1923, von Uexküll

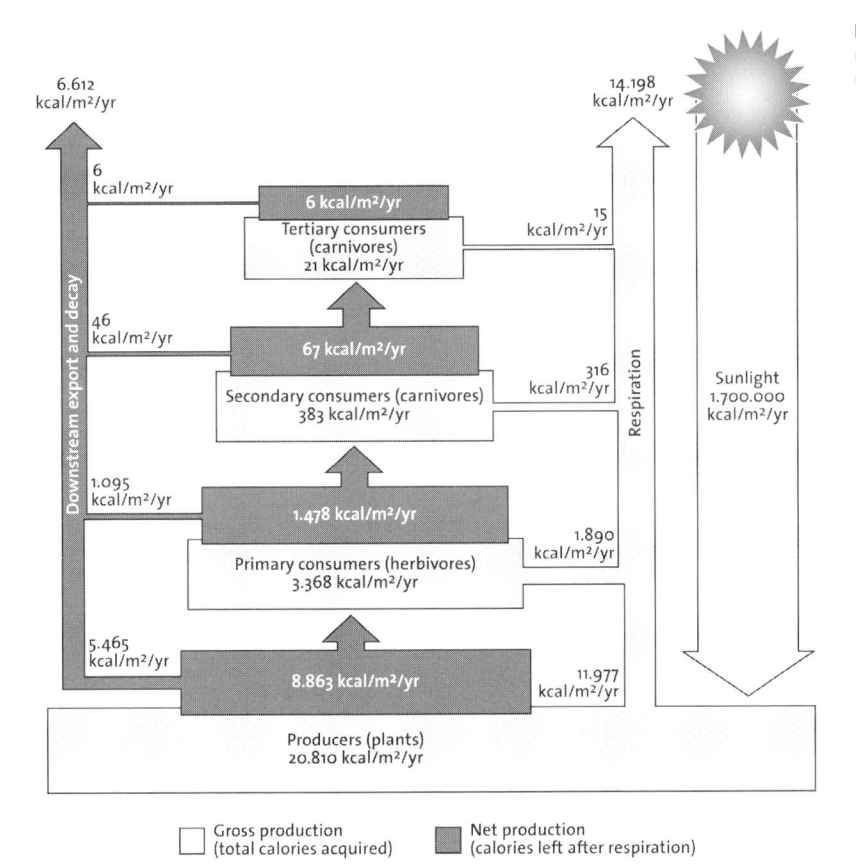

Figure 4.24 Flow of energy through a river ecosystem in Silver Springs, Florida (data collected by Odum, 1957).

founded the world's first Institut für Umweltforschung (Institute for Environmental Research) at the University of Hamburg. Von Uexküll took a two-level view on organisms (Trudgill, 2007) already including an environmental or ecological view:

> The task of biology is to expand the result of Kant's research along two lines: (1) To consider the role of our body, particularly our perceptual organs and the central nervous system and (2) to study the relationship of other subjects (animals) and their objects. (von Uexküll, 1928/1973)

The first level focused on how cells and organs interact in the body. The second level examined how individual subjects interact in families, groups, or communities. In one of his early books, *Umwelt und die Innenwelt der Tiere* ("Environment and the Inner World of Animals"; von Uexküll, 1921), he investigated how *Paramecium* (slipper animalcules), sea urchins, and dragonflies cope with their environment. He concluded that each animal has to be seen as center of its environment and is interacting with the environment as a *selbständiges* (autonomous) subject. His key message was that

> ... *signs* are of prime importance in all aspects of life processes. (von Uexküll, 1987, p. 147)

He thus established the field of *biosemiotics* (Sebeok & Umoker-Sebeok, 1992). Von Uexküll assumed that even simple animals have a mind, based on the notion that animals have a nervous system similar to that of humans. Thus, his approach provides a specific interpretation of both the mind–body and the (human–)organism–environment relationship. In his later years, von Uexküll was interested in how the meanings of signs elicit behavioral programs. This research can be seen as the cradle of ethology, which was later founded by Konrad Lorenz and Nikolaas Tinbergen.

We present J. J. von Uexküll's interpretation as it was outlined by his son, T. von Uexküll (1987, 1992). We start from his concept of *Natur*, the theory that lies behind nature, as conceived of by natural scientists. According to *Natur*, nature reveals itself through signs, which are considered the only true reality. The reader can interpret these signs as the entities represented by the "⇔" relationship in Figure 3.3*, which is

4 Emerging knowledge on biology and evolution

> **Box 4.11** Perceived environments: tick cybernetics
>
> The reflex action is an automatic neuromuscular response to certain events. This machine-like, deterministic model dominates physiological explanations of animal–environment interactions. Von Uexküll criticized this position by describing the depredating behavior of a tick to exemplify an organismic conception of perception, processing of recorded environmental information, and action. Von Uexküll postulated that "each cell … perceives and acts." The receptor cells (*Merkzellen*) and the acting cells (*Wirkzellen*) were conceived as the organism, which consists of assemblies (*Verbände*) of cells (von Uexküll & Kriszat, 1934, pp. 3–4, see Figure 4.25a). "Metaphorically speaking each animal grasps prey with two jaws" (*Gliedern eine Zange*) (von Uexküll & Kriszat, 1934, p. 6). The depredating behavior of a tick has been represented as three consecutive feedback loops: (1) the perception of "butyric acid" and the action "fall down"; (2) the perception "animal hair" and the action "run till you find temperature of bare skin"; and (3) the perception "skin temperature" and "bite/drill."
>
> This view anticipates cybernetic feedback loops or functional cycles (*Funktionskreise*, see Figure 4.25b) of perception and action, "that effectively 'couples' the ever-changing system that is the organism to the ever changing system that is the world" (Favareau, 2007, p. 32). As Barbieri noted, the functional cycle was originally designed to describe a neuromuscular cycle of marine animals and is an early formulation of the feedback principle (Barbieri, 2002, see Chapter 16, p. 3).

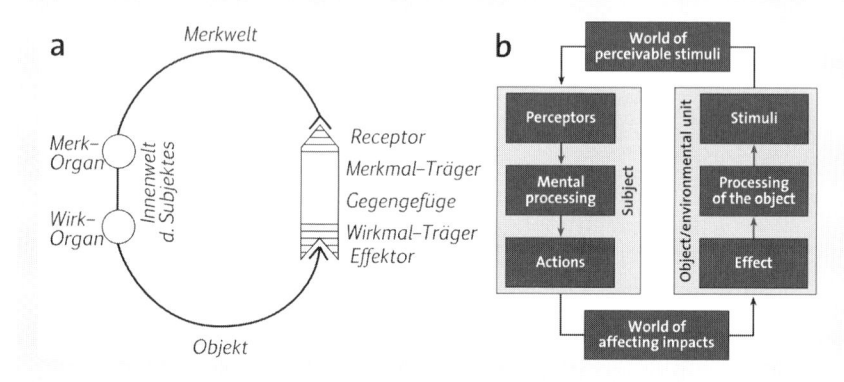

Figure 4.25 Functional cycles: (a) von Uexküll and Kriznat's (1934, p. 7, with kind permission from Springer Science + Business Media) original feedback loop and (b) a contemporary representation of the animal–environmental object feedback loop.

the relation between the material–biophysical and the social–epistemic layers of organisms. The sign processes communicate with our mind (denoted as *Gemüt*; Emmel, 1974), a concept taken from Immanuel Kant (1724–1804), which also includes the *Geist*, the spirits or the Kantian transcendental capacity, which allows us to integrate perceptions. J. J. von Uexküll considered the sign processes as the only true laws of *Natur*. The subjective "self-world bubbles" are, similar to Leibniz' monads, the core elements of reality. But, these subjective bubbles are looking for harmony with the laws of nature. The latter can be seen as a variant of the equilibration principle, which we consider a basic principle of human–environment systems (see Chapter 16, Postulate 7).

Von Uexküll also made some inferences about the relationship between the organism and the human system. He concluded that the sign relationship – self and non-self – is the most universal and refers to the code of a cell. Based on this sign, relationships differentiate between the prey or food as non-self and the hungry self which does not, for example, eat its own fingers. As he stated:

> A living cell possesses its own ego-quality (*Ichton*). (von Uexküll, 1931, p. 209)

Cells are seen as autonomous subjects, which transpose every impulse into subjective signs, which are their "ego-quality." Here, the organism and the *Umwelt* (environment) can be seen as complementary units. This approach is coherent with the definition of human systems proposed in this book (see Chapter 3.1). Von Uexküll stresses that what an organism perceives is a kind of mental bubble produced by the mind.

Thus he also dealt with the relationship between the epistemic, in the sense of what people may know, and the material–physical environment. Everything that is perceived becomes a perceived or memorized world and everything that actually takes place becomes the effect or action world. The perceived and the effect worlds become a closed unit, called the environment (von Uexküll & Kriszat, 1934, p. VIII; see Box 4.11). Clearly, the idea is that the environment is the ensemble of what is noticed, what matters, and what is relevant for the organism. Here relevance (*Bedeutung*) also includes notions of meaning and function. Thus a pebble is usually relevant as a piece of the floor. However, if we use it to banish a dog, the meaning and function of a piece of the environment is altering.

Von Uexküll considered animals as parts of functional circuits (*Funktionskreise*, Figure 4.25a). He postulated that animals show a kind of functional reconstruction of the environment, of prey, foes, and lineage, in a somewhat similar manner as humans do. Thus, von Uexküll anticipated a core idea of cybernetics (see Box 4.11), that is the functioning of a self-regulating system, where the environment can be separated in an *Umgebung*, the given surrounding world, and also an environment, which is the ensemble of perceivable things (von Uexküll, 1920, p. 58; 1921; von Uexküll & Kriszat, 1934, p. 7). The perceived parts of the environment are called perceptual cues and are based on perceptual signs, with a differentiation between distal (signs) and proximal cues. The inner or counter world (*Gegenwelt*) consists of the sensorial images of the outer world. The English translation as "self-world" (von Uexküll, 1957) is striking though rather awkward, but reflects a somewhat mechanistic conception that the environment consists of what we see. Allen *et al.* portrayed this by the heading "sustaining the Umwelt," pointing out that Uexküll assumes that the organism perceives "largely the reality that counts" (Allen *et al.*, 2003, p. 175) which is coherent with a functionalist perspective.

Clearly, biosemiotics is a general concept that has been applied in many domains (Barbieri, 2007; Sebeok & Umoker-Sebeok, 1992). Von Uexküll's conceptual framework anticipated many fundamental ideas of environmental literacy, for instance, of interpreting DNA as a sign (Kawade, 1996), the reception of therapeutic (placebo) messages in medical therapy (Moerman & Jonas, 2002), or of the immune system as a cognitive system (Cohen, 2000b).

4.12.1 Key messages

- The nature–nurture debate, particularly the discussion of what role the environment plays in the development of species, started in the midst of the nineteenth century during the origins of evolutionary theory
- Mental models of the environment and the distinction between a perceived, subjective environment and a real, affected and affecting environment had been already proposed by biologists for simple animal species
- Biosemiotics provides insight into the complementarity between a material–biophysical and social–epistemic (i.e. body and mind) layer of organisms.

Part II History of mind of biological knowledge

Chapter 5

From molecular structures to ecosystems

5.1	The human cell and its environment 95	5.6	From ecosystem functions via ecosystem services and socioecological resilience research to landscapes 120
5.2	Diseases as drivers of microbiology: from bacteria to viruses 100		
5.3	Molecular biology 103	5.7	"Landschaft" as a complex adaptive system: past or future? 129
5.4	Epigenetics: nurture in genetics 109		
5.5	Immunology 113	5.8	Homeostasis and homeorhesis on the micro and the macro levels 132

Chapter overview

Chapter 5 describes contemporary advances in biology, from cells to ecosystems, and also includes questions of securing essential ecosystem functions and services that may become the subject of transdisciplinary processes.

Chapter 5 makes the point that the cell is the basic unit of any organismic being. In Chapter 3 we defined the human being as a cellular system whose (inter-)activities of cells emerge from the fertilized human ovum (i.e. the zygote). The first parts of this chapter introduce emerging knowledge about the microstructure of organisms, the understanding of contagious diseases, and how genetics and epigenetics evolved. In some respects we thus provide insight into the rationale of the cell.

Subsequently, we offer a review of how development of microscopes and other technologies shaped progress in cell research (see Figure 14.1*) and our understanding of microstructures in organisms. In the late nineteenth century, the microscope gave researchers new opportunities to identify the cell, bacteria, amoeba, viruses, and proteins, thus equipping them to understand epidemics and pandemics. Society benefited from knowledge about the previously unexplored microscopic level. Robert Koch's (1814–77), and particularly Louis Pasteur's (1822–95), use of the microscope inspired basic research on infectious diseases.

In the next step, we reveal how physical and chemical thinking, as well as the three-dimensional representation of DNA in 1953, created progress and paradigm shifts. We highlight progress made in the understanding of biochemical and theoretical cell processes and also the reintroduction of the environment to evolutionary theory through current discussions on epigenetics. This brings biology to the molecular level. Here, we can see how the experiential level (and thus the environment) affects genetics and genetically modified organisms, which almost turns biology into an engineering science operating at the level of 100 pm (=10^{-12} m).

Following the sections on microbiology, we climb the ladder of organism–ecosystem hierarchy and exchange our microscopic lens for a macroscopic view. Here we present the *panarchy* theory of adaptive cycles, which provides insight into the stability–instability dynamics of organismic systems. We also introduce key concepts such as resilience and adaptation, which have become important concepts for the environmental sciences. Environmental literacy calls for an understanding of how the human species is embedded in environmental systems, how human action affects the environment, and what feedback loops need to be considered between these coupled systems. In the face of mounting concerns about sustainability, ecologists are adopting a socioecological perspective. Because ecosystems, unlike individual organisms, do not have built-in homeostatic responses to disturbances, humans need to step in. Now, more than ever, there is a crucial need for a transdisciplinary approach to sustainable development that involves both researchers and practitioners.

Finally, we review recent ideas in landscape ecology (German: Landschaftsforschung). Researchers in this field have been pioneers in the inclusion of human impacts on ecosystems. Today, we inhabit historically formed, built, culturally shaped, and human-impacted ecosystems and landscapes. As such, researchers and practitioners increasingly seek approaches that link and integrate the rationales of material–biophysical processes in the environment with the socioepistemic processes at work in human systems. This is the rationale behind our development of the human–environment systems (HES) framework, put forth in detail in Chapters 16–19.

5.1 The human cell and its environment

5.1.1 The microscope as a key factor in the progress of cell research

Cells are the fundamental working units of every living system. They exist in an enormous variety of structural and functional states "without losing their essence as the foundational building blocks of most life on the planet" (Hall, 2001, p. 225).

Let us briefly examine the development of biological knowledge about the cell from the progress of magnification technology perspective (Figure 5.1). The invention of the light microscope led to the discovery of the cell, a term coined in Robert Hooke's *Micrographia* (Hooke, 1665). Hooke was able to identify some pores, or box-like cell structures in cork, which resembled honeycombs using the microscope. Using the first single lens microscope, Antoni van Leeuwenhoek (1632–1723) discovered bacteria and spermatozoa in 1676 and 1677, respectively. These first microscopes were able to magnify up to 275× (James, 1994). Later, the double lens light microscope allowed a 400× magnification and visibility of a unit of about 100 nm. Through this medium, Karl Ernst von Baer (1792–1876) discovered the mammalian ovum (i.e. egg cell) in 1827, which, however, he simply called "eggs."

The light microscope also allowed for viewing the nucleus (i.e. central body) of large cells, which was first discovered by Robert Brown (1773–1858, see Chapter 4). However, discovering something and interpreting it appropriately are two different things. An important interpretation was provided by the botanist Matthias Jakob Schleiden (1804–81) and the physiologist Theodor Schwann (1810–82) when they inferred that the cell is the basic building block of plants and animals, which they based on the similarity of cell structures in both types of organisms. They also detected that the (eukaryotic) cells of plants, animals, and fungi have a nucleus with a membrane and organelles such as the Golgi apparatus or mitochondria (see Figure 5.2).

Schleiden provided a first comprehensive cell theory (Beier, 2002), including the idea that a new nucleus was formed by the crystallization of granular material within the cell contents and that the outer membrane of this new nucleus would become the new cell wall. He speculated that the nucleus originated within existing cells or crystallized within a formless fluid. Mayr emphasized that:

> … it was unthinkable for him to answer the question "How do new cells originate?" (Mayr, 1982, p. 655)

This was the case because he had no access to it because of the limits of the light microscope.

Today, we know that the fertilized egg, the zygote, contains all the necessary "instructions" for building the human body, which itself contains 100 trillion (i.e. 10^{14}) interrelated and communicating cells (Lodish et al., 2004). Creating this knowledge was only possible through new technological advances, for instance the commercial oil immersion microscope, invented by Carl Zeiss (1816–88) in 1877, which had a 2.5 times greater enlargement. The importance of theory/practice cooperation can be seen in that it was Schleiden who encouraged and coached Carl Zeiss when founding his optics firm. Using this microscope, Flemming (1879) described chromosome behavior during the division of the nucleus (Brock, 1988; see Figure 5.2). The progress and the impact of the oil immersion microscope can be taken from the number of publications between 1874 and 1878. One reviewer counted 194 papers by 86 authors (Mayr, 1982, p. 675), documenting data resulting in three main insights of cell division during this period. Core insights gained were: (1) that the division of the nucleus starts before cell division; (2) that there is a regular sequence in the changes of nuclear material; and (3) that the basic phenomena of nuclear and cell division are the same in both the plant and animal kingdoms. However, the latter had not been acknowledged in all aspects. A specific phenomenon that caused difficulty had been in defining the sexuality of plants. From Box 5.1, we understand that neither the multitude nor function of sexes was realized.

In the 1930s, the transmission electron microscope allowed scientists to study samples a few nanometers

5 From molecular structures to ecosystems

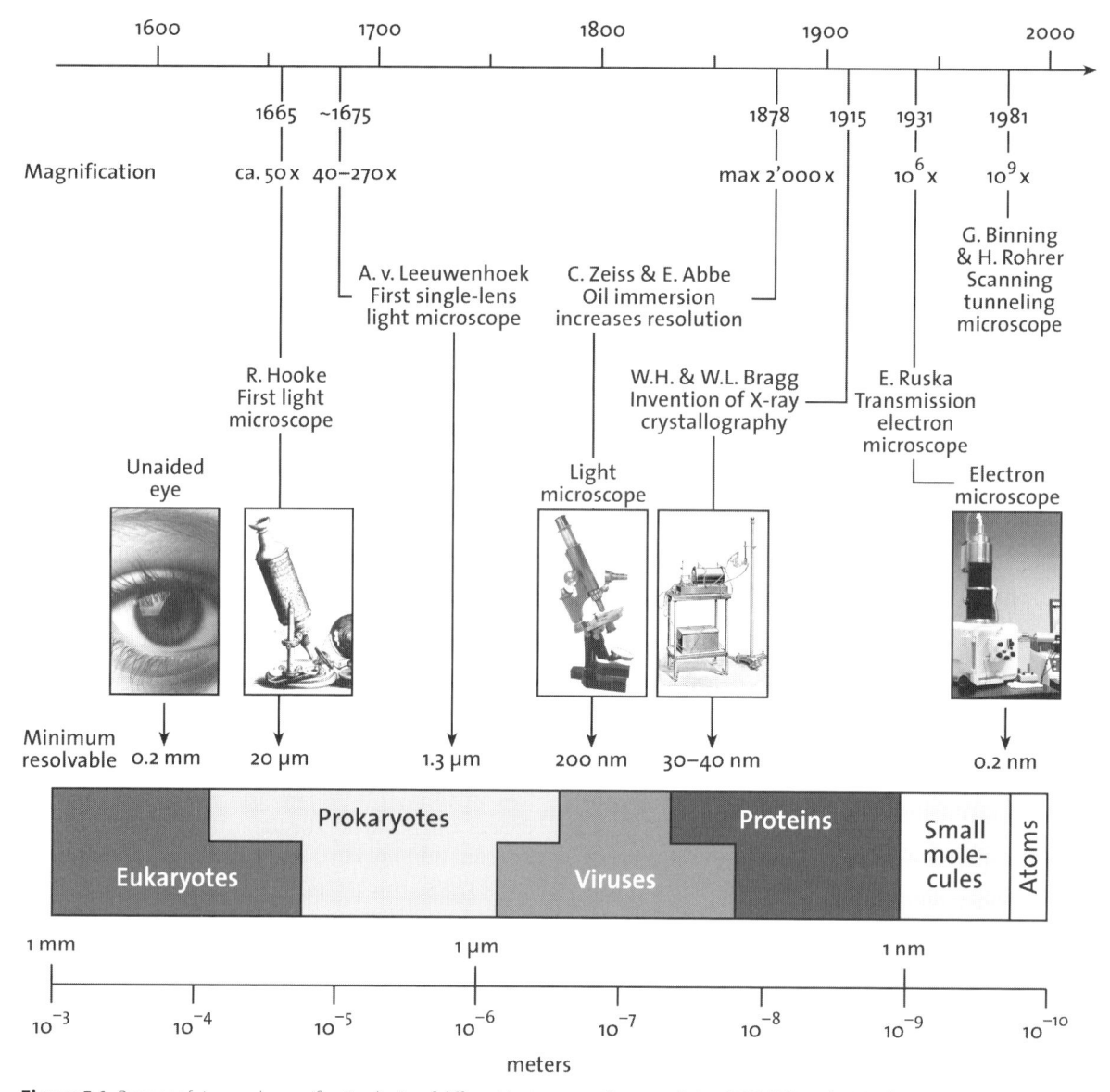

Figure 5.1 Ranges of sizes and magnification limits of different instruments (Sources: Colm, 2009; Wikipedia, 2009).

in size, thus approaching the atomic scale. Thus, with increasing magnifications, the membranes enclosing cells became visible and the similarity between the worm-shaped mitochondria and bacteria could be observed (Alberts et al., 2004). Today we know about the importance of bacteria (unicellular prokaryotes lacking a cell nucleus or other membrane-bound organelles) within the organismic environment. It has been calculated that there are about 5×10^{30} bacteria on Earth, forming a large part of the biomass in the open ocean, in soil, and in oceanic and terrestrial subsurfaces. Remarkably, well over 90% of the bacteria live below the Earth's subsurface (Rappe & Giovannoni, 2003; Whitman et al., 1998). There are ten times more bacteria in the human body than there are human cells (Berg, 1996; Sears, 2005), and some bacteria can double their colony size in 10 minutes (Eagon, 1962). The size of bacteria ranges between about 0.5 and 5 μm. The study of viruses brings us further down in scale to a range of 15–400 nm (i.e. 0.015 – 0.4 μm).

Naturally, medicine and particularly societal coping with contagious diseases have been major drivers for the development of microbiological knowledge. Today, biological research is partly carried out at

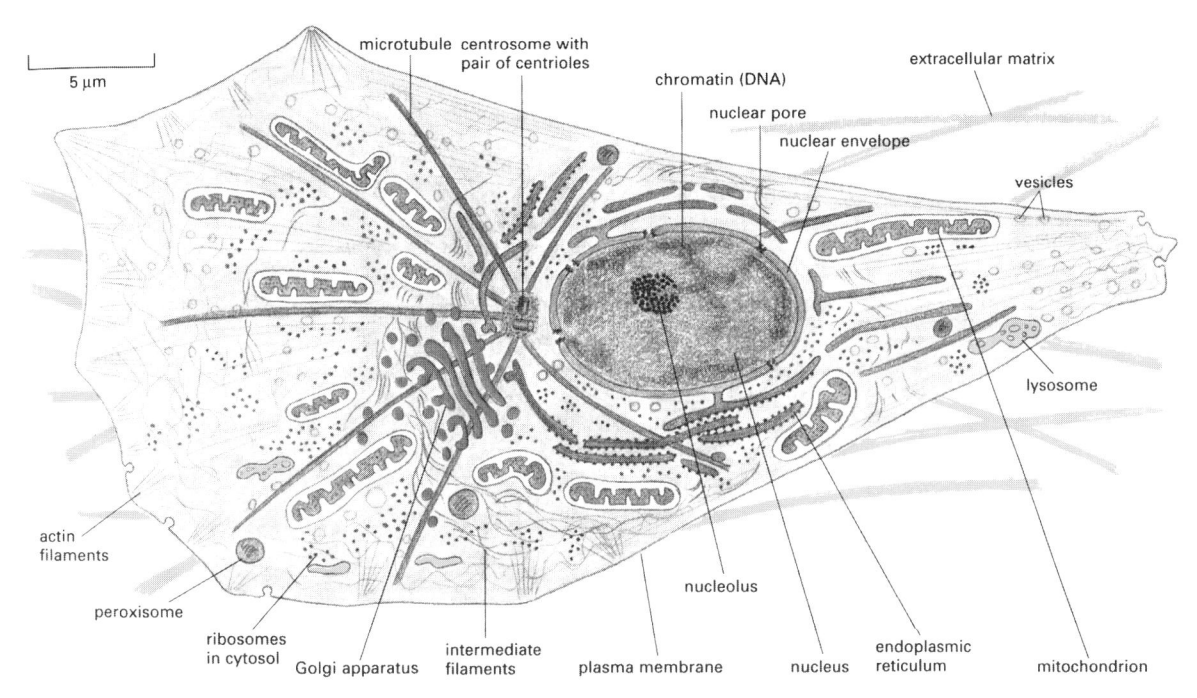

Figure 5.2 Principal features of an animal cell (Alberts *et al.*, 2004, p. 25. Reproduced by permission of Garland Science/Taylor and Francis LLC).

the atomic level and has become an engineering science in some domains (e.g. in the field of genetically modified organisms, which goes down to the level of 0.01 nm).

The transmission electron microscope allows us to see tiny cellular components such as mitochondrial membranes. To express this in the terms of J. von Uexküll (see Chapter 4), researchers utilized different signs to read the cell. Instead of beams of light bundling with glass lenses, magnetic coils were used that can read electron beams. This was invented by Louis de Broglie (1892–1987) and became practically accessible in the 1940s (Rheinberger, 2004; Williams & Carter, 1996). With access to three-dimensional representations of DNA, RNA, and proteins, the design of many intracellular structures could be understood.

From a biological perspective, the time between 1920 and 1950 was a period of intellectual immigration of physicists and chemists, leading to the emergence of biochemistry and biophysics (Morange, 1998). Physicists and engineers made the microstructure of the cell visible and provided insight into its components. However, they followed a "normal science paradigm" characterized by "puzzle solving" (Kuhn, 1996), rather than questioning the basic tenets of the discipline. X-ray data required sophisticated interpretation of three-dimensional data. By means of mathematical models such as Fourier methods, the diagrams of fuzzy X-ray signals provided insight into the three-dimensional lattices of some crystals. The path to discovering the double helix, the molecular structure of DNA, took decades and is an interdisciplinarity success story in natural science (Olby, 1974).

Before switching to molecular biology and the knowledge about the molecular level of human–environment interactions, we take a brief look at basic questions about life that relate to cell research.

5.1.2 Heritage by what?

The term *genetics* was coined in 1905 after the rediscovery of Mendel's systematic laws of hybridization (Bowler, 1989; Harper, 2005). Geneticists' key questions read: What are the carriers of genetic information? What determines the sex of an offspring? And: Are there soft factors of inheritance?

Without going into detail we review which genetic mechanisms scientists hypothesized and investigated. An important hypothesis was provided by de Hugo Marie de Vries (1848–1935; de Vries, 1889) and August Weismann (1834–1914) who assumed that a given nucleus may contain many identical replicas of a given *pangen*. Later, de Vries introduced the term *mutation*

5 From molecular structures to ecosystems

Box 5.1 A continuum of sexes in plants and animals? American Holly and Hollywood

In the history of evolution and biology, the understanding of plant sexuality has been a tough nut to crack. The ancient Egyptians knew that there were plant species with male and female individuals (i.e. dioecious plants, see Box 4.4), such as the American Holly (*Ilex opaca*), but this knowledge was later forgotten. When this phenomenon was rediscovered, a fundamental uncertainty about the universality of the sex of plants emerged. We can see two major reasons for this. First, there is a fascinating myriad of types and hybrid forms of (a) monoecious (i.e. single sex) plants, (b) hermaphrodite plants, which only have bisexual reproductive units (e.g. conifer cones), (c) synchronously monoecious plants, which have separate male and female flowers on one plant, (d) protoandrous plants that function first as males and then as females, (e) protogynous plants, which operate in the opposite way to protoandrous plants and (f) polygamodioecious plants, which have bisexual and male flowers on some plants but bisexual and female flowers on others, just to mention some of the variants (see Figure 5.3).

Moreover, even the fundamental idea that ovules need male semen to reproduce had not been proven to be universal in all plants and animals. In general, androdioecious plants can induce mutations to change the phenotype (sequential hermaphroditism). Thus, one can state that plants have gender strategy. The evolution of sexes created this pattern, which provides almost a continuum of sexes. There are even some species such as slime mold, which have more than two (i.e. 13) sexes (Anderson, 1992).

A second reason for the difficulties in understanding plant sexuality might be that "sexuality in plants continued to be widely denied well into the nineteenth century" (Mayr, 1982, p. 659). Additionally, hermaphrodites have been considered a societal taboo and are consequently not a well understood phenomenon (Fausto-Sterling, 1993). This is somewhat surprising, as intersexuality also exists in the human species and has been documented since antiquity. However, the idea of intersexuality is in conflict with biblical creationism. Depending on the definition, estimates of the prevalence of human hermaphroditism range from 1.7% to 0.018% (Sax, 2002). To retain any meaning, the term should be restricted to those conditions in which chromosomal sex is inconsistent with phenotypic sex or in which the phenotype is not classifiable as either male or female. Applying this more precise definition, the true prevalence of intersex is about 0.018%. In some cases the phenotype and the genotype mismatch, but there are many other types of cases: "For example, some people are mosaics: different cells in their body have different chromosomes. A 46,XY/46,XX mosaic is an individual in whom some cells have the male chromosomal complement (XY) and some cells have a female chromosomal complement (XX). If such an individual has both a penis and a vagina, then there is no mismatch between phenotypic sex and genotypic sex: both the phenotype and the genotype are intersexual" (Sax, 2002, p. 175). Thus, one can even state that there is a continuum between pure male/female and the pure intersexual case.

Figure 5.3 (a) The American Holly trees have either male (photo by Will Cook) or (b) female (photo by Lawrence M. Kelly) flowers. (c) Hermaphrodites (statuette of a hermaphrodite at the Louvre in Paris, photo by Marie-Lan Nguyen) are found in most, if not all, species.

Finally, also single-celled organisms can have sex. The common yeast (*Saccharomyces cerevisiae*), used to make bread and brew beer, has two mating types that are conceptually similar to the female/male gametes (eggs and sperms) of higher organisms. They can fuse or mate to produce a third cell type containing the genetic material from each cell. There is a kind of evolutionary rationality behind this phenomenon. "Such sexual life cycles allow more rapid changes in genetic inheritance than would be possible without sex, resulting in valuable adaptations while quickly eliminating detrimental mutations. That, and not just Hollywood, is probably why sex is so unique" (Lodish et al., 2004, p. 6).

to genetics as the process by which new species could originate through sudden variation in one generation. Before this, the term mutation commonly denoted any drastic change. A remarkably sophisticated, early proposition about the division of labor in the cell was provided by Haeckel:

> … the nucleus has to take care of the inheritance of the heritable characters, while the surrounding cytoplasm is concerned with accommodation or adaptation to the environment. (Haeckel, 1866, pp. 287–8)

Haeckel's statement has been based on a coarse view of the cell, which allowed differentiation between the nucleus and a few components such as mitochondria (see Figure 5.2). Insight into the physical process was given by the work of Theodor Boveri (1862–1915) and Edwin B. Wilson (1879–1964). The former stated:

> In all cells derived in the regular course of division from the fertilized egg, one half of the chromosomes are of strictly paternal origin, the other half of maternal. (Boveri, 1891, p. 410, quoted according to Mayr, 1982, p. 748)

In the course of cell division, rod-like chromosomes became visible again and again under the microscope. In 1879, Flemming described chromosomes using the advanced oil immersion microscopy, which made chromosomes of a few micrometers visible as distinct structures only when they condense in preparation for cell division (see Figure 5.2). New theories followed this observation and, in 1902, Boveri supposed chromosomes to be the origin of heredity.

Thomas H. Morgan (1866–1945) provided proof for this hypothesis using his famous "Fly Room" at Columbia University. Following Mendel's experiments, studies on *Drosophila melanogaster* elucidated the complex relationship between chromosomes, genes, and traits. A fruit fly is an ideal model case; it has a lifespan of about a month, has only four pairs of chromosomes per nucleus, and has scores of easily recognizable, heritable traits such as a "white eye mutant."

You could keep thousands in small bottles, feeding them bananas. Morgan was aware of the exceptional biotic properties of *Drosophila*, and he manipulated them so well that it was said that "God had created this Lilliputian creature especially for him" (Morgan, according to Magner, 2002, p. 403). The Fly Room created a wave of interpretable data. Because white eye pigmentation, yellow body color, and a particular wing shape segregated in male fruit flies, it could be inferred that the X chromosome carries these features (Morgan, 1910). We should note that Morgan was rather skeptical that the chromosomes could explain everything and suspected that cytoplasmic factors, which are entities within the (eukaryotic) cell besides the nucleus, played a role.

5.1.3 Inducing mutations: environment matters

Despite the progress in embryology and genetics, evolutionists were still fighting with the question: how do new species evolve? Many biologists, including Darwin, were trying to generate new species by inducing fundamental macromutations through breeding. Morgan's fruit flies were subject to radium, acids, alkali salts, sugar, and protein, and all failed to spur major mutations. The idea of providing evidence that the environment matters goes back to de Vries, who suspected that the Roentgen and other ionizing radiation might affect genetics. Finally, Hermann Joseph Muller (1890–1967), a doctoral student of Morgan, was able to identify 100 spontaneous mutations resulting from exposure to ionizing radiation. Thus, it could be incontrovertibly shown that the environment matters in evolution. Based on Muller's experimental data, scientists investigated whether cosmic radiation due to supernovae explosions had an impact on evolution (Terry & Tucker, 1968). The models, however, suggested that these would have been of minor importance.

5.1.4 Key messages

- The technical progress of microscopy and X-ray technology since the end of the nineteenth century has provided insight into the division of the cell and the nucleus, chromosomes, the process of fertilization, embryological growth, and the basic components of cells. This led to a theory of a cell-based unity of life
- The time between 1920 and 1950 was a period of intellectual immigration of physicists and chemists to the biological fields of biophysics and biochemistry. At the same time, a restricted view into the study of organisms was introduced
- The diversity of sexuality between and within species provides deep insight into the complexity and dynamics of evolution
- The understanding of the biological mechanism of heritage was a key driver of microbiology.

5.2 Diseases as drivers of microbiology: from bacteria to viruses

Contagious diseases and in particular epidemic plagues have been a main societal scourge. Epidemiologists estimated that in the late nineteenth century tuberculosis was the cause of one in seven deaths (Magner, 2002). Hays (2005) reports on 25 epidemics in the period between 1850 and 1920, all helplessly faced by a vulnerable society. The evolving but still incomplete work of Koch and Pasteur at that time contributed to the broad public hygiene movement and introduced the first convincing scientific arguments, which were major milestones in the development of environmental literacy.

Pasteur became a professor of chemistry in 1854, after which he began to study alcohol fermentation for beer and wine producers who were looking for a way to prevent product spoilage. Using a microscope he followed Schwann's (1839; see above) hypothesis that organisms participate in fermentation. Seeing small microorganisms under the microscope, he supposed that microorganisms could spread disease through worms. This approach was called the germ theory of disease. Additionally, Pasteur showed that microorganisms could be destroyed by heating them up to 55 °C (i.e. pasteurization). He further proved that milk, blood, and urine remained sterile and that no microbial growth took place if germ-transporting dust particles were prevented from invading the liquids.

Pasteur also showed that the environment matters. For example, hospitals were considered to be a hotbed for diseases. Pasteur showed that the germ content in hospitals was twice as high as that of mountain areas, thereby supporting suspicions about the physical transmission of diseases by germs. A conclusion here was that microbes associated with fermentation were brought to a substrate through the air, originally coming from contaminated places. Germ theory was able to exclude the possibility of spontaneous generation (supposing that life and diseases spring from inanimate matter; see Chapter 4.6). We should note that germ theory and bacteriology were heavily disputed. For instance, microorganisms were seen as the consequence and not as the origin of the diseases. Claude Bernard (1813–78), an early proponent of the cellular theory, focused on the contextual level and suggested that "the germ is nothing, the terrain is everything" (see Bernard, 1878, p. 113). Terminology was also shifting, as it was unclear whether bacteria were plants or animals (Köhler, 2004).

Another domain of Pasteur's work was the investigation of a silkworm disease that endangered France's silkworm nurseries. This brought him into infectious disease research, where he shed light on the riddle of diseases and gained public recognition with his insights into rabies and anthrax (see Box 5.2). Koch's challenge was tuberculosis, a disease so prevalent that it earned the sobriquet "captain among these men of death" (Daniel, 2006, p. 1862). In Berlin, the death of every third newborn and every second adult between 25 and 40 was caused by tuberculosis (Kaufmann & Schaible, 2005). Difficulties in studying the tubercle agent were that it grows exceptionally slowly, is only one-tenth the size of the anthrax bacillus, and was troublesome to stain because of its waxy cell wall, which impeded microscopic visibility. Koch eventually succeeded in discovering the etiologic agent of tuberculosis and also those of anthrax, cholera, and wound infections. A salient finding in understanding anthrax was that cultured bacteria go through a cycle in which they change into spores that can survive in soil for years. Thus, when exposed to contaminated soil, animals could become sick without being directly infected by other animals (Ullmann, 2007).

Koch and Pasteur looked at many contagious diseases and ran many experiments, mostly with animals and some with humans. Koch also helped formulate the Koch–Henle postulates to prove the existence of microbial pathogens, postulates which are still valid today

(Koch, 1882). We present them to show how cause–impact relations can be proven experimentally with respect to bacterial disease. They state that, if a living contagion from an infected organism is grown in pure, sterilized culture and is then able to infect another (inoculated) organism from which the contagion is later recovered, the contagion is proven to cause the disease.

A principal challenge was to distinguish between bacterial and viral diseases which showed the same phenomenology. Viruses are much smaller; most of them (except for the larger ones, which are the size of small bacteria) are not visible by optical microscopy. However, experimental researchers had noticed that some entities must have passed through filters specifically made to retain bacteria during potable water treatment (Loeffler & Frosch, 1898).

Transmission pathways were often demonstrated by self-experimentation; for example, purposefully allowing a mosquito which had bitten a malaria stricken person 12 days before to bite the researcher (Köhler, 2004). Experimental evidence like this and those from controlled experiments enabled researchers to construe that there must be an "organized virus which elicits the disease" (Ellernman & Bang, 1908, p. 608; translated by Scholz).

However, few viruses could be seen under the microscope; for example, vaccinia virus, later used to fight smallpox viruses. Researchers developed different theories about invisible viruses. An example is the tobacco mosaic virus, which was only detected in growing plants. Because it appeared to be invisible, some researchers postulated a diminutive, contagious living fluid (Zaitlin, 1998). Other researchers did see some small elements under the microscope, which they supposed to be virulent. However, a series of technological improvements, such as staining techniques (Herzberg, 1934), new ultrafilters, and mastery of the ultracentrifugal technique (Svedberg *et al.*, 1940) brought researchers closer to their goal. In the end it was the electron microscope that allowed reliable identification of the tobacco mosaic virus.

During World War I, bacteriologists opened a new window to understanding *bacteriophages*, a special type of virus. Bacteriophages are specialized in infecting bacteria and rapidly multiply inside the host. For instance, Frederick Twort (1877–1950) and Felix d'Hérelle (1873–1949) noticed that bacteria causing dysentery, a common disease in the nineteenth and twentieth century, were attacked by small microbes called micrococci. Max Ludwig Henning Delbrück (1906–81) investigated the bacterial genetics of the phages and considered the virus an appropriate mechanism for studying the chemical nature of hereditary material. Here, the first differentiated views of macromolecules emerged. The phage was a model system to study how the organism functions and was considered the most elementary biological system akin to how the molecule was considered the most elementary entity for physicists (Ellis & Delbrück, 1939; Morange, 1998). The phage is chemically simple, consisting only of DNA and proteins.

> Phage researchers have called bacteriophage genetics "Mendelism at the molecular level," because studying the multiplication of phage particles was essentially equivalent to studying naked genes at work. (Magner, 2002, p. 431)

5.2.1 Vaccination: engineering biological knowledge

Utilizing knowledge in practice was a key driver of the work of Koch, Pasteur, and other microbiologists, and the development of vaccines was the ultimate example of properly blending the two. According to western medical history (Stern & Markel, 2005), vaccination is attributed to the English physician Edward Jenner (1749–1823). Notwithstanding, variolation (that is, the deliberate inoculation of an uninfected person with the smallpox virus) can be traced as far back as sixteenth-century Hindu medicine or seventeenth-century Chinese medicine. Thus, the practical knowledge appeared much earlier than the scientific understanding of the underlying mechanisms.

Pasteur reinvented vaccination by chance in 1879 while working on chicken cholera. After his return from a long holiday, he injected chickens with an old culture of the bacteria. The chickens became ill but, since the culture was attenuated or weakened, they recovered. The story goes that the supply of chicken was limited and thus Pasteur experimented on the previously injected chickens with a new, aggressive culture. To his surprise, the chickens were now completely immune to the more aggressive culture. Pasteur, a gifted chemist, also noticed that vaccines generated in the laboratory had a different and reduced virulence (Plotkin & Plotkin, 2008). This observation led him to obtain attenuated cultures of anthrax with which he vaccinated some animals. After triumphant public demonstrations of the success of his anthrax vaccine in successful experiments, starting from 1881 millions of sheep and

5 From molecular structures to ecosystems

Box 5.2 From basic via applied research and public experimentation: Pasteur's quadrant

Louis Pasteur (1822–95) became a professor of chemistry at Strasbourg in 1848 and a little later the dean of the Science Faculty in Lille in 1854. His chemical research started with basic research on the polarization of light in crystals. This knowledge helped him later when he responded to a request from a beer brewery to help them prevent the beverage from going sour when separating different biomolecules in fermentation (Martinez-Palomo, 2001). Later he worked to prevent silkworm disease and became a consultant to the textile and beverage industry.

Stokes (1997) portrayed Pasteur's work as a new type of academic performance that emerged in the second industrial revolution, a time when chemistry and physics made major contributions to industry. Pasteur can be considered as a representative of use-inspired basic research (Figure 5.4a). "Pasteur was never satisfied merely to formulate the theoretical basis for a given process. Instead, he took an active interest in its industrial development and practical applications" (Ullmann, 2007, p. 384). And the mature Pasteur never conducted a study that was not applied. Thus, he contributed to both pure, fundamental research and applied, practical solution-oriented research.

In 1881, having obtained financial support, mostly from farmers, Pasteur conducted a large-scale public experiment of anthrax immunization. He prepared attenuated cultures of the bacillus after determining the conditions that led to the organism's loss of virulence. The experiment took place in Pouilly-le-Fort, south of Paris. On May 5 and 17, 1881, 25 sheep were vaccinated with an attenuated strain which causes anthrax. These and 25 other sheep, which had not been vaccinated, received an injection of a highly virulent strain of the bacteria on May 31. The experiment was a full success and was published in the daily news. Within a few days, 200 officials, politicians, farmers, and journalists gathered at the farmyard. All the control sheep were dead, whereas all the vaccinated animals survived. The success was categorical. Within a year, thousands of sheep were vaccinated and the mortality fell from 9% to below 0.6% (Martinez-Palomo, 2001; Ullmann, 2007).

Pasteur was a difficult but persistent person, and his reputation-seeking habit might have led him astray. As Robert Koch insisted and a recent study of his personal notebook revealed, in his famous anthrax experiment, Pasteur did not use the live attenuated vaccine (as he pretended) but a "dead vaccine," as suggested by the veterinarian Jean J. H. Toussaint (1847–90, Smith, 2005). Geison published this "significant and undeniable element of deception" (1995, p. 176) in the book *The Private Science of Louis Pasteur*.

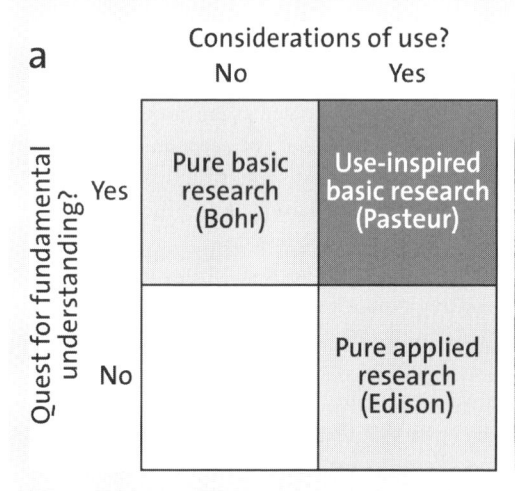

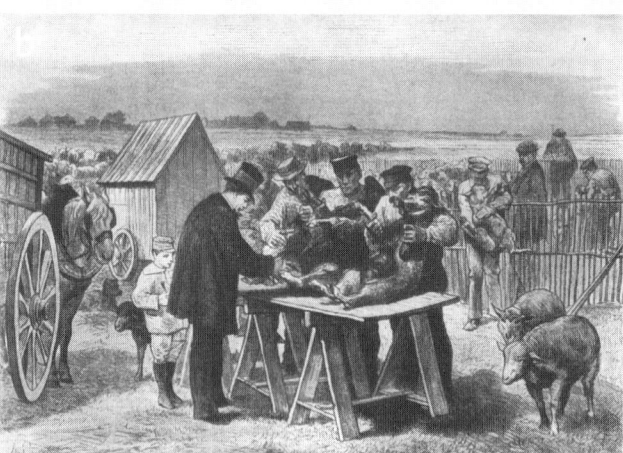

Figure 5.4 (a) Pasteur's quadrant, a classification of different modes of academic research suggested by Stokes (1997) and (b) an engraving of Pasteur's public experiment in Pouilly-le-Fort, May 31, 1881 (from de Kruif, 1927).

cattle were vaccinated (Magner, 2002). We can say that Pasteur took the perspective of an engineer and proved that vaccination worked, although he did not understand how. He proved that having had minor contact prepares the body to cope with virulent viruses.

There were, however, also erroneous conclusions made, which put research both in to the lime and the light. Additionally, there was an embittered rivalry between Koch and Pasteur, which became embroiled in the political antagonism existing between Germany and France at that time. Koch (Koch, 1882, December 28) considered Pasteur's anthrax vaccination (see Box 5.2) risky for both humans and animals, as well as unstable and impure because of incomplete standardization (Cassier, 2008). In 1890, Koch was attacked by the scientific community for incorrectly stating that tuberculin (a glycerine extract of the tubercle bacilli) inhibited the growth of tubercle bacilli in guinea pigs. Koch had supposed that it could work as a vaccine after patients injected with tuberculin, himself included, suffered skin inflammation and other symptoms. Tuberculin came into vogue as protection against tuberculosis. But soon thereafter, Clemens von Pirquet (1874–1929) discovered that the tuberculin actually only differentiated between infected and uninfected people (Daniel, 2006), and that those who had previously been injected had quicker and more severe reactions to a second injection. This hypersensitive reaction was called an allergy, the study of which opened the window to immunology.

5.2.2 Key messages

- Understanding contagious diseases, epidemics, and pandemics in plants, animals, and humans has been a major driver of biological research
- Koch, Pasteur, and others had huge success with some of these pandemics, such as anthrax, which were almost mastered by vaccinations. However, there were others, such as tuberculosis or cholera, which could not yet be mastered by vaccination
- Applying biological knowledge to disease control is in the public interest. However, as the case of Pasteur might show, it includes a delicate balance between developing proper basic knowledge, engineer-like in-vivo experimentation, commercial interests, and the researcher's drive for appreciation.

5.3 Molecular biology

We trace the genesis of molecular biology from two perspectives. We briefly introduce how data, concepts, models, and theories emerged about the functioning, development, and heredity of the human being. Secondly, we reflect on which conceptualization of the cell as a living system is linked to the molecular view. Finally, we consider whether there is a body–mind complementarity in cells.

5.3.1 The gene: an entity between chromosomes and DNA nucleoids

Historically, the early twentieth century findings that different species have different numbers of chromosomes, and that females have two X chromosomes and males an X and a Y chromosome, were milestones in genetics (Magner, 2002). At this stage of development, Wilhelm L. Johannsen (1857–1927) introduced the *gene* as a theoretical construct to explain heredity of certain traits and as a way to delineate between genotype and phenotype (1911). Johannsen considered genotype and phenotype to be abstract entities, unconfined to certain cellular spaces. But it took about 50 years from that point to determine that the 46 human chromosomes are constituted of DNA and about another 50 years to identify (in 2003) all three billion nucleotides belonging to the human genome (Gregory et al., 2006). Today, we consider chromosomes in higher organisms to be composed of double-stranded DNA molecules carrying codes for the synthesis of proteins.

An unanswered challenging question was what elements of the cell are the carriers of heritable traits. But how can one decipher genetic heritage when the science of the time can only afford images such as those in Figure 5.5? Cyril D. Darlington (1903–81; see Harman, 2005) provided straightforward assumptions for deciphering generic heritage. He (1) broke chromosomes up into genes and formulated certain laws of genetics, including that (2) the cell holds chromosomes in pairs (Darlington, 1932). In his time, Darlington was considered an outsider, and his book *Recent Advances in Cytology* was labeled as "propaganda," a "dangerous conjecture," and even "a masterpiece of mythogenesis" (Harman, 2005, p. 83).

In the 1940s the mechanisms of heredity were still unclear. In his book *What is Life?*, the renowned physicist Erwin Schrödinger (1887–1961) described chromosomes as "hereditary code script":

> In calling the structure of the chromosome fibers a code-script we mean that the all penetrating mind, once conceived by Laplace, to every causal connection lay immediately open, could tell from their structure when the egg would develop, under suitable conditions, into a black cock or into a speckled hen, into a fly or maize plant, a rhododendron, a beetle, a mouse or a woman. (Schrödinger, 1944, p. 21)

But it was new technology and other discoveries that were essential in generating today's knowledge about

5 From molecular structures to ecosystems

Today, there is evidence that each tiny human cell, which spans about 10 μm in diameter, contains approximately 2 m of DNA. This means that DNA must compact by 10 000-fold to fit into the cell's nucleus. Human DNA contains about 3.2×10^9 nucleotides, parceled out in 23 chromosome pairs. DNA is wrapped around histone proteins, which facilitate packing. We should mention one more important element, chromatin, which is the combination of DNA, RNA, and protein that makes up the chromosome. The human genome, the total DNA, consists of approximately 25 000 genes that are more or less separated by so-called "junk DNA" whose function has yet to be fully clarified (Allis *et al.*, 2007a, b). These DNA sectors have the *potential* to become genes and are thus also called protogenes (Balakirev & Ayala, 2003). In addition, this "junk DNA" is thought to have a regulating function. Genes account for only 2% of the human genome (see Figure 5.6; Alberts *et al.*, 2004), and are considered the basic physical and functional units of heredity. Molecular biology was strongly affected by ideas from physicists such as Erwin Schrödinger (1887–1961) and Linus Carl Pauling (1901–1994; Pauling, 1931), who showed that the formation of chemical bonds could be explained using quantum mechanics and could be predicted from the structure of the atoms involved (Pauling, 1931).

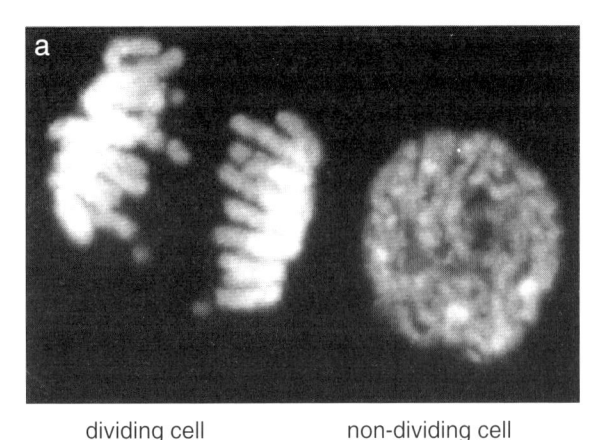

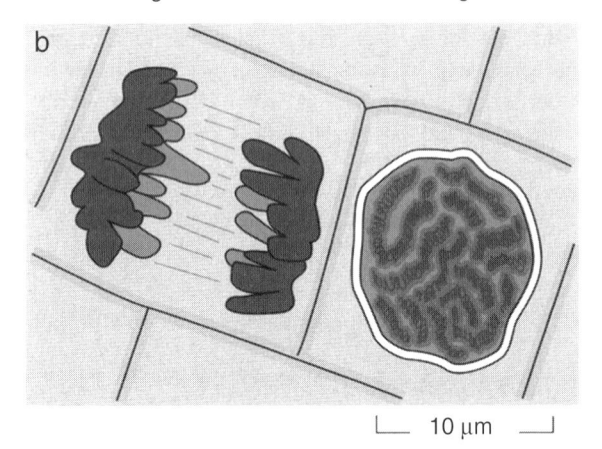

Figure 5.5 Plant cells as seen by the light microscope in normal state and directly before cell division (Figure a, courtesy of Peter Shaw, and b, after Alberts *et al.*, 2004, p. 172). Reproduced by permission of Garland Science Taylor and Francis LLC.

5.3.2 RNA, proteins, enzymes: the functional agents of the cell

Proteins, also called polypeptides, are huge molecules essential to cell function. The human cell consists of 70% water and 26% proteins. Thus, proteins constitute the main portion of the dry mass of the cell.

> Proteins … serve as building blocks for cell structures, they form the enzymes that catalyze the cell's chemical reactions, they regulate gene expressions, and they enable cells to move and to communicate with one another. (Alberts *et al.*, 2004, p. 171)

Moreover, there is a multitude of proteins – a single cell can produce about 3000 different protein types. Many

genetics. X-ray diffraction made the three-dimensional gratings of long chains of molecules observable. Around 1935 the nucleoprotein theory of a gene – up to then an abstract notion – was confirmed when X-ray crystallographers provided insight into the chemical structure of the single-stranded RNA. However, it was not until 1977 that genes could be identified with the help of subtle lysing techniques as well as DNA staining (Maxam & Gilbert, 1977; Sanger *et al.*, 1977).

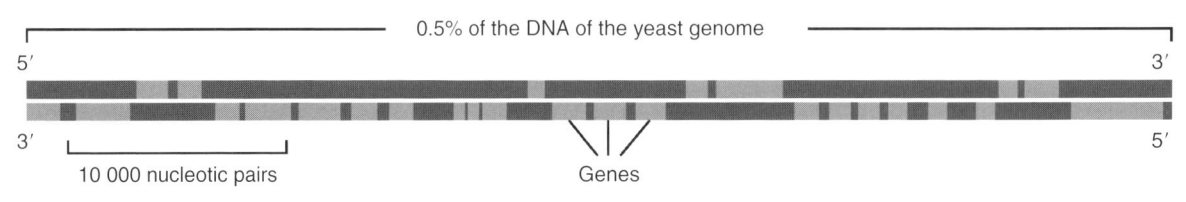

Figure 5.6 Schematic representation of a chromosome region of budding yeast containing about 6300 genes with 12 million nucleotide pairs in 16 genes (Alberts *et al.*, 2004, p. 181) Reproduced by permission of Garland Science/Taylor and Francis LLC.

Box 5.3 The central dogma of molecular biology: the protein as a sink?

With the discovery of the chemical structure of the double helix DNA, single-stranded RNA and proteins, the three basic building blocks of the cell were identified. But which of the nine possible relations among these entities are essential for cell processes (Figure 5.7a)? This was the critical question in the 1950s. Based on their experience and fragmentary, "often rather uncertain and confused experimental results" (Crick, 1970, p. 561), Crick and Watson formulated the "central dogma of molecular biology." This deals with the "residue-by-residue transfer of sequential information. It states that information cannot be transferred back from protein to either protein or nucleic acid." (ibid, p. 561)

This is presented in Figure 5.7a. The DNA transcribes information to (messenger) RNA ❶ and in a second process ❷ it is translated on ribosomes which make proteins. There is further a replication of the DNA ❼ and – detected in 1970 – a reverse transcription of RNA to DNA ❺, which is also the channel used by viruses to enter the cell.

Back translation ❹ from the more complex protein "alphabet to a structurally quite different one" of the DNA/RNA, which only includes the four nucleotide bases A (adenine), G (guanine), C (cytosine), and the structurally similar T (thymine, in case of DNA) or U (uracil, in case of RNA), was deemed improbable. Crick and Watson's belief, and here one can see the origin of the term dogma, was that the cell could not "have evolved an entirely separate set of complicated machinery for back translations" (ibid. p. 562), of which there was no trace. Thus, the dogma – "once (sequential) information has passed into protein it cannot get out again" (Crick, 1958, p. 153; Figure 5.7b) – was supposed to be the starting point.

Two points are of special interest: one is the metaphor of the cell as *machinery* and the implicit beliefs about its structure. The dogma implies "that the gene is deterministic in gene expression and therefore all gene products are fully specified by the DNA code" (Stotz, 2006b, p. 534). Once information has passed into a protein, it cannot get out again. The dogma also implies that the DNA is more or less the sole bearer of information and causality. The central dogma formulates a linear, unidirectional picture.

The second subject point deals with the use and notion of the term *information*. John Maynard Smith (2000) pointed out that technical concepts in biology such as translation, transcription, code, message, editing, proofreading, and library are concepts that completely resemble human-designed codes such as Morse code, and he pondered: "If there is 'information' in DNA, copied to RNA, how did it get there?" (Maynard Smith, 2000, p. 179).

Today, epigenetics and research in processes of gene activation has shown that the linearity of the central dogma of molecular biology is limited. New essential processes have been introduced; RNA interference by double-stranded RNA (dsRNA), for example, plays an essential role in silencing genes. But, "we are only beginning to appreciate the mechanistic complexity of this process and its biological ramifications" (Hannon, 2002, p. 250). It seems likely that the understanding of the cell as an organismic unit necessitates going beyond mere DNA-based determinism if one views "the cell as a causal *network* of genes, RNAs, proteins, and metabolites with distributed agency" (Stotz, 2006a, p. 906), that interprets different internal and environmental chemical and physical signs when facing cellular needs and environmental cues.

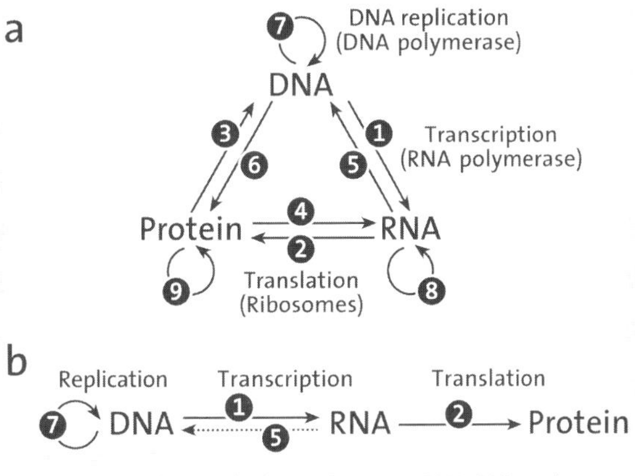

Figure 5.7 (a) Potential relations between DNA, RNA, and protein and (b) hypothesized general relations (bold arrows) and special relations (broken arrows) which are thought to exist according to the central dogma of molecular biology.

proteins are enzymes, which catalyze a vast array of biochemical reactions and processes vital for metabolism. Enzymes were originally considered to be a collection of identical, huge molecules that act as very tiny machines which freely move between the water molecules of the cell. Proteins generate movement in cells and tissue, store small molecules or ions, detect signals, and transmit them to those parts meant to respond adequately (e.g. insulin receptors), allow for communication between cells, and bind to DNA to switch genes on and off, just to mention a few of the functions (Alberts *et al.*, 2004).

This, however, does not address exactly how the cell functions and how it can reproduce. The first comprehensive answer was given by the central dogma of molecular biology (Box 5.3). The basic idea is that the gene, which is part of the double-stranded DNA, expresses itself by sending information via messenger RNA to produce the missing genes (see Figure 5.8).

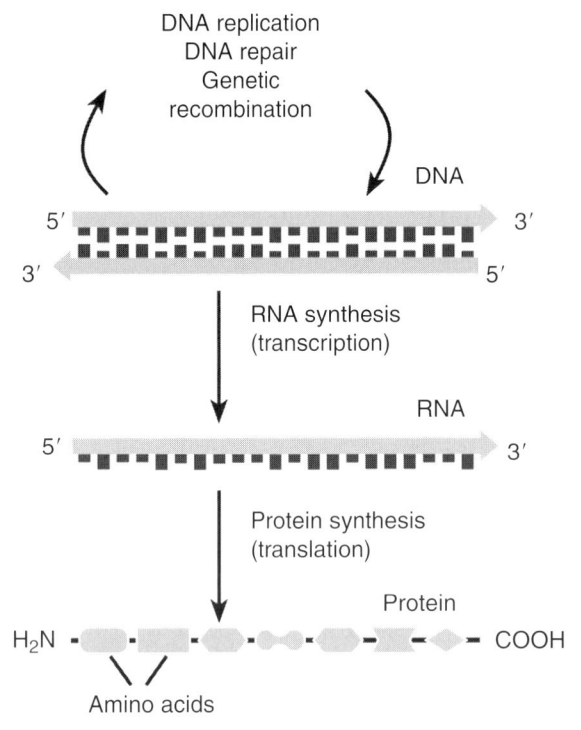

Figure 5.8 DNA replication, transcription from DNA to RNA, and translation from RNA to proteins (Alberts *et al.*, 2002, p. 331) Reproduced by permission of Garland Science/Taylor and Francis LLC.

5.3.3 The cell as a living system

> Persons who have not studied cells often cannot conceive that their degree of complexity justifies comparison of their system characteristics with those of higher-level systems. (Miller, 1978, p. 2004)

Contemporary cell biology provides a differentiated view on how the agents of a cell, in particular DNA, diverse types of RNA, proteins, and false friends such as viruses and other agents, interact. "Proteins and enzymes (which are a special type) were considered as agents of specificity" (Morange, 1998, p. 2). Thus, the cell was perceived as a chemical factory full of machinery (Reynolds, 2007). DNA acts as a manager: if a certain protein is needed, it delegates the work to the messenger RNA (mRNA), a single-stranded ribonucleic acid. This occurs when the mRNA copies a short sequence of the DNA (a gene) in a process called transcription. The transcription occurs with the help of RNA polymerases, enzymes that are similar to those that carry out DNA replication (that is, the DNA polymerases; Figures 5.7 and 5.8).

A second process is called translation. RNA, which has only four nucleotides and is simpler than proteins, which contain 20 different types of amino acids, is translated through the genetic code. As already indicated by the term translation, the RNA → protein relation is a complex "translation from one language into another language that uses quite different symbols" (Alberts *et al.*, 2004, p. 25). In essence:

> Scientists have not only cracked the code but have revealed, in atomic detail, the precise working of the machinery by which cells read this code. (Alberts *et al.*, 2004, p. 246)

5.3.4 Gene expression: switching on and off

We have looked at some of the major milestones in understanding DNA as a double helix constituted by a sugar–phosphate backbone and a vast number of directed nucleic acid building blocks, C–G and A–T (see Box 5.3). But we have also discussed that genes were considered modules of DNA separated by "junk DNA." Up to now, we have not touched on the autocatalytic processes that enable the cell to initiate reactions conducive to some of the aforementioned processes, such as DNA replication. In this regard, we only want to mention that, from an evolutionary perspective, growth is an essential property of life: if the cells' autocatalytic processes did not tend to produce more molecules like themselves, life would end (Alberts *et al.*, 2004, p. 179). This is actually a position taken by Richard Dawkins when defining a gene as a "germ-line replicator," having a selfish "teleonomy" (Dawkins, 1982, p. 81). Here teleonomy represents a special kind of functionalism (see Chapter 3) and relates to the apparent purposefulness of structure and function of living organisms and evolutionary adaptation.

However, if we consider the cell as a living system, another molecular biological topic of central interest to understand the rationales underlying human systems is the process of gene expression.

> Even the simplest single-cell bacterium can use its genes selectively – for example switching genes on and off to make enzymes needed to digest whatever food sources are available. And, in multicellular plants and animals, gene expression is under even more elaborate control. (Alberts *et al.*, 2002, p. 269)

According to a descriptive, natural science perspective, cells switch genes on and off through repressor proteins. This process is considered the cell's response pattern to signals in their environment. But, cells are supposed to have not only a script on how to react but also a "*cell memory* [which] is a prerequisite for the creation of organized tissues and for the maintenance of stably differentiated

5.3 Molecular biology

Box 5.4 Why should a mouse gene work in a fly? Eys and Seys

A critical question is whether biochemical genetic information specifies only form and function or whether cell DNA–RNA–protein interactions also allow interpretation of the meaning signs as symbols. To demonstrate the role of symbolic meaning, Maynard Smith refers to a series of genetic experiments with mice and flies conducted by Halder, Callarts, and Gehring (1995). *Drosophila*, the fly, and the mouse, despite having evolved separately for more than 500 million years, share the same master control gene for eye morphogenesis. The genes are not identical but "the proteins encoded by these genes share 94 percent sequence identity in the paired domain, and 90 percent identity in the homeodomain" (Halder *et al.*, 1995, p. 1788). For *Drosophila* there is a so-called *eyeless gene* (*ey*) that, when expressed, allows for the development of eyes. For the mouse, the homologous gene is named *small eye* (*Sey* = Pax-6). In the experiments, gene *ey* produced a protein that appeared to be the transcription factor. Mutations in gene *ey* resulted in deformed and absent eyes. An interesting finding was that, if the *Sey* gene was transferred to *Drosophila* and was activated in different parts of the developing body (such as the legs), it produced morphologically complete *Drosophila*-type (and not mouse-type) eye structures (see Figure 5.9).

A critical and controversial question is: why should a mouse gene work in a fly? (Maynard Smith, 2000). There are principally two camps on the issue. The first group, call them the purists, provides a purely genetic (genocentric), "material–biophysical" explanation when referring to genetic similarity and mutagenic-type deviations given an evolutionarily robust structure. The second group, we can call them the "symbolists," implicitly refers to Haeckel's postulate that living (sub-)systems, such as DNA–RNA–protein systems, are able to consider symbolic information. Given the physiochemical information and the received information, from the protein and other entities of the environment (at a given time and context), regulating genes respond with a specific process in the cell's translating machinery (i.e. ribosomes, tRNAs, and assignment enzymes) that results in eye development. The critical question for supporters of the latter argument is how symbolic meaning that elicits certain functions is generated (e.g. what electromagnetic or bond-affecting processes are in place?). Some suggest, for instance, introducing a genome that provides the "program that secures the formation of the mRNA and its expression in time and space" (Scherrer & Jost, 2007). Although there are still many open questions such as, for instance, which concept of information is to be taken (Winnie, 2000), or how to link the gene concept with concepts in "different levels of biological organization" (Prohaska & Stadler, 2008, p. 215, see Figure 14.1*) "the genome and its entirety has taken on a more and more flexible and dynamic configuration" (Rheinberger & Müller-Wille, 2009).

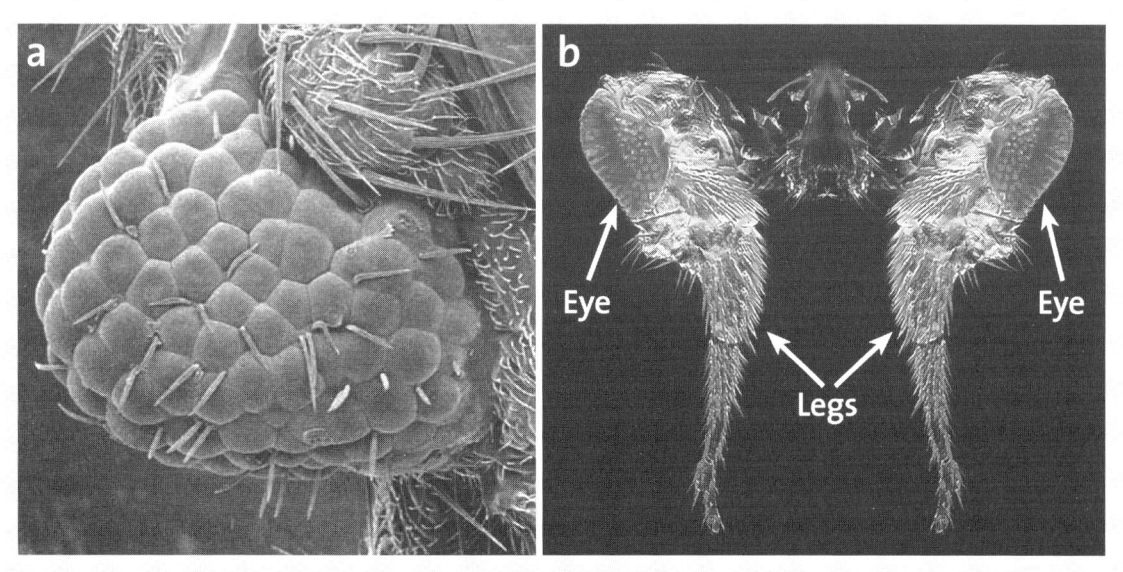

Figure 5.9 (a) The expression of a *Sey* in a *Drosophila* gene, showing a close-up of induced eye facet on a leg produced with a scanning electron micrograph (Halder *et al.*, 1995, p. 1791), reprinted with permission from AAAS, and (b) two eyes induced on the legs by targeted expression of the twin of eyeless gene (photo by Walter Gehring).

cell types" (Alberts *et al.*, 2002, p. 280). This statement raises one of the big questions of biology, "How does genetic information specify form and function?" (Maynard Smith, 2000), which is explored further in Box 5.4.

A first answer to this question, which was provided by Jacob and Monod (1961), led to the identification of mRNA as the mediator and repressor of genes and proteins (see Box 5.4). In chemical terms, Jacob and

Monod showed that, for example, the *Escherichia coli* bacteria switched on genes which started a synthesis (of the enzyme beta-galactosidase) only when lactose substrate was present. Repression of genes (or feedback inhibition) is the complementary process. In this process, enzyme activities were inhibited by the products in a synthetic process. Thus, a biochemical model to account for inducible and repressible enzyme systems was presented. This shed important light into the interactions of the trio of DNA, RNA, and proteins.

The first regulatory model explaining gene expression and gene activation was called the *Operon Model*. Two classes of genes were distinguished therein: one class consisted of the structural genes which were presumed to carry the "structural information" for the production of particular polypeptides. The other class – regulatory genes – was assumed to be involved in the regulation of the expression of structural information. As Rheinberger and Müller-Wille (2009) emphasize, a third element of DNA involved in the regulatory loop of an operon was a signal sequence linked to peptides (Cohen, 2003).

> These three elements, structural genes, regulatory genes, and signal sequences provided the framework for viewing the genotype itself as an ordered, hierarchical system, as a "genetic program" … The operon model of Jacob and Monod marked thus the precipitous end of the simple, informational concept of the molecular gene. (Rheinberger & Müller-Wille, 2009, p. 187)

5.3.5 Is there a body–mind complementarity in cells?

While humans have a limited understanding of cells and genetics, there are some data that suggest a body–mind complementarity in cells or cell systems. We review in this section how, for example, biologists describe cell processes as more than just chemical reactions. As an agent of information transfer in the cells, one could think of genes as epistemic objects. In a similar vein, some studies support the view that mechanical signals that go to genes and DNA in cells are interpreted as patterns.

Hence, "… an organism is more than a bag of specific proteins" (Maynard Smith, 2000). Cells have to be seen and understood in context. Thus, at the turn of the twenty-first century, critiques and papers were published with such titles as "Death of the gene: developmental systems fight back" (Gray, 1992).

But what do we know about what the cell is doing? Biologists construct descriptions, models, and theories based on information garnered, for instance, from the electron microscope. Moreover, they interpret this data according to their own "world view" and peer group, not unlike Linné's theories being influenced by his Swedish Protestant education and upbringing, and by his scientific peer group (see Box 4.6). But molecular biology does not perceive the processes only in a descriptive, chemical manner, particularly if an evolutionary perspective is taken. The organism is seen as a system that is in a process of mutation and selection. When describing processes, most biologists make reference to the concept of information. This is evident if one reads standard textbooks such as Alberts *et al.* (2002) or Lodish *et al.* (2004). When describing the processes, terms such as information, transcription, translation, open reading frames, codes, alphabets, etc. are ubiquitous. For instance, proteins are regarded as the messages and amino acids as the alphabet (Tauber, 1995).

We think that there are at least three levels of information that have to be distinguished, which might support Morange's (2006) suspicion that biologists' use of informational concepts is not purely a metaphorical one. First, there is a purely physical–chemical level that applies immutable natural science laws. This position, often characterized by an extreme form of reductionism, was favored because of the tremendous success of molecular biology which had shaped biological thinking for so long. This position was displaced with the rise of cell biology (Morange, 1998). Second, when talking about information, the cell is thought to have a sender–receiver framework. But, information only makes sense if there is a reference or decoding system that provides meaning to the signal (Janich, 1999). This leads to the question about the nature of the information and to how symbolic information has some semantic meaning. It is not only the physical laws that are at work. As Jakob von Uexküll's biosemiotics asserts, the cell is able to read and interpret the "signs of Nature." This is described in some detail in Chapter 4.11. Similarly, Maynard Smith argued:

> … that the inducers and repressors of gene activity are symbolic, in the sense that there is no necessary chemical connection between the nature of an inducer and its effect. (Maynard Smith, 2000, p. 188)

The third epistemic level leads us to the question of how the organism relates to the world view of the researchers, who use various metaphors such as the cell as a chemical factory in which different machines and agents are at work. Here, phrasings such as "decoding or replication systems make errors" or "viruses hide" or, when emphasizing the metabolic aspect, "loading docks" and "assembly line," can be taken as examples (Reynolds, 2007). As is the case in almost all domains of sciences, scientists use different metaphors to describe certain aspects of cell activity.

Finally, it is worth noting that as soon as we use an information metaphor to describe how the cell knows about its environment, we are calling "genes epistemic objects" (Rheinberger & Müller-Wille, 2009). We could, in principle, construct a body–mind complementarity (see Chapter 3.3) on the level of the cell that describes how its subsystems interact. It would make sense to distinguish between a material, chemophysical level system (H_m^{cell}) and an information-related level of analysis (H_s^{cell}), which has meaning beyond the pure chemophysical pattern. Such a view, however, assumes that single cells exist that – excepting random errors – are able to make decisions based on active, functional, goal-oriented processing of information. When considering the cell as a living system, we can hypothesize that the cell or parts of it are not only receiving signals mechanically but are also able "to read" complex signs emerging from the information sampled. Here, we can think about how a sign "A" is identified by a computer as such, even if the distribution of printing ink differs and one sees "A".

There seem to be proponents of this view, Maynard Smith being one of them. To substantiate his view, Maynard Smith refers to two experiments: one is the classic Jacob and Monod experiment mentioned above. If one recalls, *Escherichia coli* DNA switches on certain genes that help in metabolizing lactose when it is present. This is obviously adaptive, as it would not make sense to switch them on if there was nothing to process. Maynard Smith argued that:

> … if it was selectively advantageous for these genes to be switched on by a different sugar, say maltose, then changes in the regulatory genes that brought this about would no doubt have evolved. (Maynard Smith, 2000, p. 188)

Maynard Smith concludes that regulating genes thus utilize response schemes that induce production processes.

The second experiment he referred to is presented in Box 5.4. Here we find that certain genes from a mouse function in a fruit fly, although the genes only coincide by 90%. Thus, it seems plausible that not only the single physical elements (i.e. the nucleotides of the DNA) read piecemeal, but also that the whole pattern of information makes sense on a symbolic level.

5.3.6 Key messages

- The cell is a living system of high complexity which, in many respects, resembles higher developed organisms. A challenging question is whether cells can react to environmental impacts beyond pure biochemical processes by reacting to complex patterns (e.g. of proteins) which have a symbolic meaning. Yet a body–mind complementarity seems likely in certain cells
- From a human–environment interaction perspective, molecular biology cannot be reduced to a pure biochemical perspective or to restrictive views such as the central dogma of molecular biology.

5.4 Epigenetics: nurture in genetics

In this section we turn to a discussion of epigenetics, the study of how changes in phenotype (appearance) or gene expression are caused by mechanisms other than changes in the underlying DNA sequence. For example, factors such as diet or exposure to UV radiation have been shown to lead to changes in the suppression of tumor suppressor genes. The process of gene silencing in the slime mold *Dictyostelium discoideum*, which can switch from a single-celled to a multicellular being, is documented in Box 14.4. Given such environmental explanations for changes in gene expression, we suggest considering both a genome and an epigenome, and we conclude that such functional (survival) activities of cells suggest learning.

Epigenetics fostered a fundamental change in the view on heredity in two senses. On the one hand, one may speculate that genes have a memory (Bird, 2007) and, on the other hand, genes are considered epistemic subjects (Rheinberger & Müller-Wille, 2009). The term epigenetics was first introduced by Conrad Hal Waddington (1905–75) to denote "the causal interactions between genes and their products, which bring the phenotype into being" (Waddington, 1942a). He suspected that "there is an entire complex of developmental processes between the genotype and the phenotype" (Waddington, 1942b, p. 19) and their mutual relation. Later, Nanney (1959) used the term to explain that cells with the same genotype could have different phenotypes that persist for many generations. Holliday provided a third, slightly varied, definition which depicts epigenetics as:

> … the unfolding of the genetic program, which ultimately depends on the activation or inactivation of specific genes, or the interactions between genes and product of genes. (Holliday, 1994, p. 453)

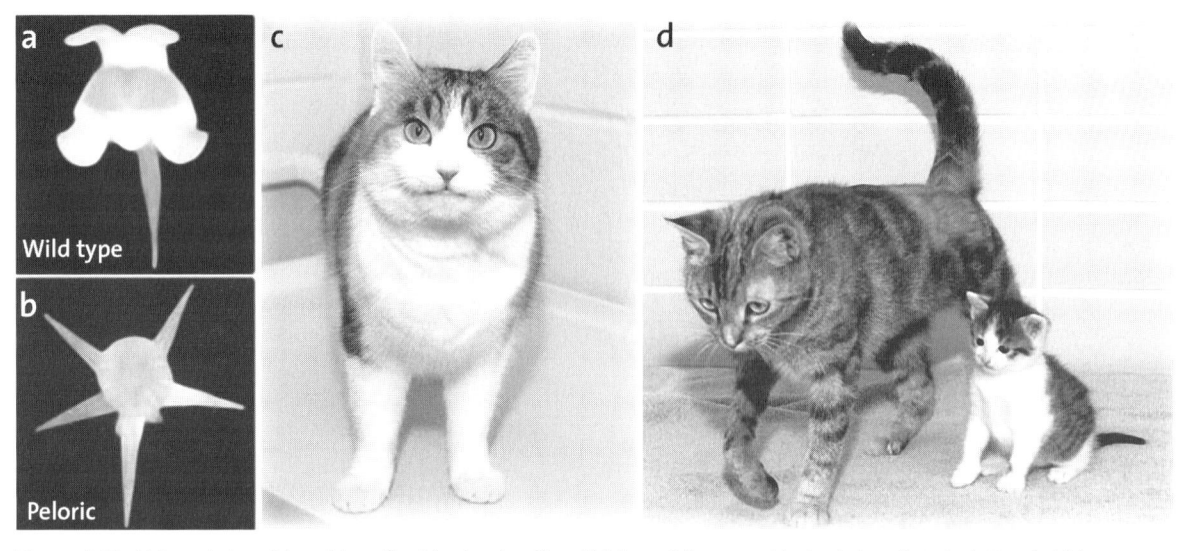

Figure 5.10 (a) Frontal view of the wild toadflax (*Linaria vulgaris*) and (b) the radially symmetrical peloric variant, the latter of which can revert in subsequent generations (Cubas *et al.*, 1999). (c) and (d) Cloned cats, despite being genetically identical, can exhibit different color patterning (Shin *et al.*, 2002, reprinted by permission from Macmillan Publishers Ltd) the nuclear donor cat is shown in (c).

The key issue is that the epigenotype of cells can be altered by external, environmental influences. This contrasts classical genetics "where the genotype is not changed by the environment" (Holliday, 1994, p. 454). Central to Holliday's definition is that "nuclear inheritance … is not based on difference in DNA sequence" (ibid. p. 454). Thus, epigenetics in the latter sense means we speak about inheritance that is nuclear (i.e. related to the cell nucleus) but not DNA-related. Epigenetic heritage has also been called paramutation or epimutation.

The difference between genetic and epigenetic can be described as follows:

> Nucleic information has the pervasiveness and static precision connoted by genetic, whereas the epinucleic information regulates the manifestation of nucleic potentialities in the dynamic, temporally responsive functioning of actual development. (Lederberg, 2001, p. 6)

The term epigenome thus denotes the overall epigenetic state of a cell. We will briefly examine three aspects of epigenetics. (1) What examples of epigenetic transfers are known or proven? (2) What mechanisms are involved in epigenetics? (3) What is the rationale of the cell's epigenetic behavior?

A classic example of proven epigenetic transfers is the existence and non-existence of flower symmetry in toadflax (*Linaria vulgaris*). It was shown that potentially stable transmission of epigenetic modifications from one generation to another is caused by heritable silencing of genes (Cubas *et al.*, 1999). This paramutation occasionally reverts in subsequent generations. A second example of non-Mendelian – that is, non-DNA-based – inheritance is the coloring in different species, such as the varying coat color phenotype of cloned cats and their surrogate mother (see Fig. 5.10). Another well explored example is the environmentally induced sex regulation in various species (see Box 5.1). *Caenorhabditis elegans*, the roundworm, for instance, uses facultative epigenetic strategies in somatic tissue to switch X chromosomes.

A final example, described in Box 5.5, comes from Donald Olding Hebb (1904–85), a pioneer in neuropsychology, who considered the flatworm *Planaria* to be simple creatures having that type of nervous system where epigenetic effects of learning and conditioning can be demonstrated (Nicolas *et al.*, 2008).

But, again, what are the mechanisms involved in epigenetics? How can we inherit something beyond the DNA sequence? What are the chemical and physical processes that induce changes of the inherited information? Clearly, a thorough answer would require more than one book. In current terms, epigenetic researchers consider DNA not as isolated in all organisms but rather as embedded in chromatin (which is a complex combination of various chromosomal proteins and DNA). There is evidence that, in eukaryotes, certain genes in the DNA are switched on/off before there are any interactions between the chromosomal proteins and the DNA. Some configurations of the chromatin are thought to establish stable "… 'epigenetic states' or

5.4 Epigenetics: nurture in genetics

Box 5.5 Eating intelligence: memory transfer?

In the early 1960s McConnell and others demonstrated through classic conditioning experiments that planaria (*Dugesia dorotocephala*), a common freshwater flatworm which can regenerate into two complete organisms when cut in half, has the capacity for learning (McConnell et al., 1959). One variant of the experiments was the classical T-maze experiment (Figure 5.11a) in which planaria were punished with electric shock when choosing to move in the wrong direction. Other carefully conducted experiments were based on studying regeneration. Conditioned planaria were cut longitudinally, transversely or into several pieces, and the learning exhibited by the subsequently regenerated planaria indicated that "part of the memory storage process was probably chemical" (McConnell & Shelby, 1970, pp. 83–4).

In a follow-up experiment, planarians trained in the T-maze experiment were fed to untrained planarians, but not to an untrained control group. The experimental data revealed that the first group learned faster in the following T-maze experiment than the control group. In 1963 the first experiment was reported in which a chemical substance extracted from conditioned animals induced recipients to learn faster. The premise that molecular mechanisms underlie memory transfer was operationalized by producing a "crude RNA mix ... from this [the] whole body preparation" and then injecting it "into untrained recipients" (McConnell & Shelby, 1970, p. 72). Although a perfect experimental design would have worked with a control group fed with untrained worms, there were further experiments with flatworms indicating that learning through eating trained organisms takes place (Westerman, 1963).

Although many doubts were voiced with respect to the specificity of the behaviors and the capacity of the animals to learn (Morange, 2006), the subject gained a new recognition when Georges Ungar published "preliminary experiments" in *Nature* showing "that an elementary form of learning, habituation to sound, can be transferred to untrained animals by injecting them with a peptide-type material extracted from the brain of habituated animals" (Ungar & Oceguera-Navarro, 1965, p. 302). The study injected a "homogenate of pooled brains from sound-habituated rats ... to albino mice" (p. 301) using a small sample of mice which included only four animals from days 7 to 14 in Figure 5.11b. Ungar viewed the active peptides as connectors in the circuits of the neurons involved in the storage of memories. Ungar, a qualified pharmacologist, was able to isolate the peptides and "after years of effort and the use of four thousand brains ..., he was able to sequence a peptide, christened 'scotophobin', that was said to be involved in the fear of darkness" (Morange, 2006, p. 324).

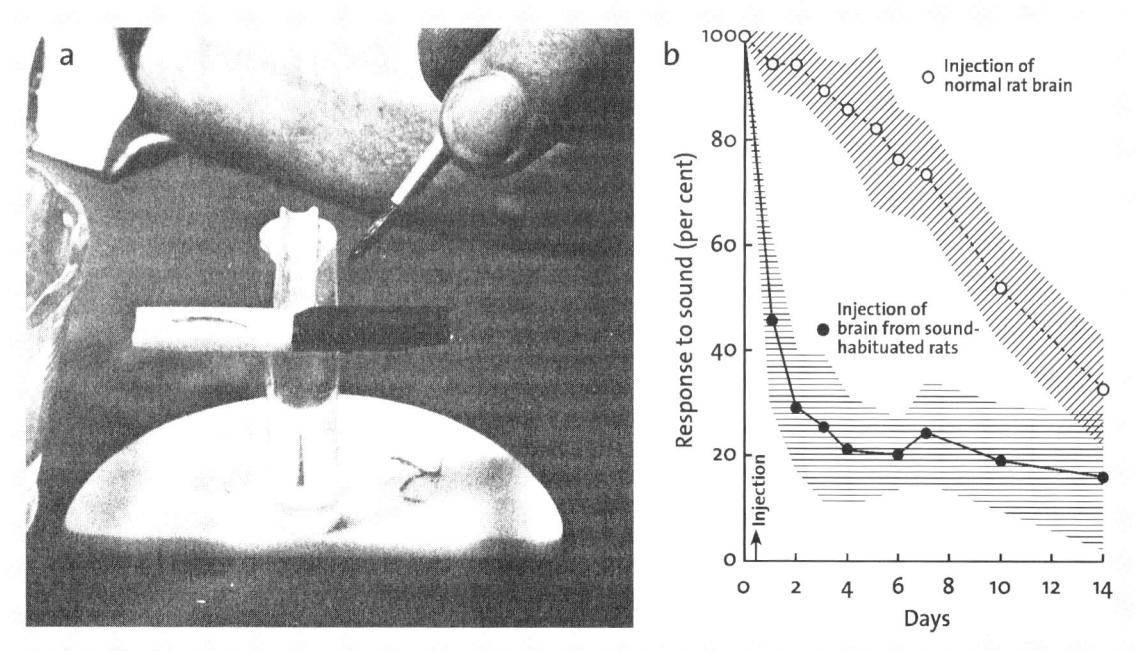

Figure 5.11 (a) T-maze used in experiments for classical conditioning of planaria (from McConnell & Shelby, 1970, p. 90, with kind permission from Springer Science + Business Media B.V.). (b) Male albino mice that received an injection of sound-habituated rat brain habituate faster than albino mice that received an injection of normal rat brain (from Ungar & Oceguera-Navarro, 1965, p. 301, reprinted by permission from Macmillan Publishers Ltd; shading indicates standard deviation).

> **Box 5.5** (cont.)
>
> One year later, *Science* published a two-and-a-half column reply article authored by 23 scientists who conducted 18 replications. None of these replications provided evidence of transfer from trained rats to untrained rats (Byrne *et al.*, 1966). The article noted that it "would be unfortunate if these negative findings were to be taken as [a] signal for abandoning … a result of enormous potential significance" (Byrne *et al.*, 1966, p. 658). Nevertheless, the hypothesis of memory transfer (that was even thought to take place between species) was deserted, and the experiments were widely forgotten, even though some interesting data had been generated.
>
> Based on today's knowledge (Allis *et al.*, 2007b; Morange, 2006), we can learn three lessons. First, given that learning changes macromolecules, which was proven by Hydén and Egyházi (Egyházi & Hydén, 1961), there is clear evidence that, in simple organisms such as planaria, or in even simpler ones such as *C. elegans* (Strome & Kelly, 2007), one can find molecule-based, non-DNA-coded intergenerational transfers of knowledge. This is the object of epigenetics. Second, there is obviously a lack of understanding of the differences with which information is processed in vertebrates and invertebrates. Third, questions such as "was it reasonable to imagine that complex behavior could be encoded in the structure of a molecule and be triggered by the simple addition of this molecule to the brain of a recipient animal?" should also have been investigated by related experimental biological and biochemical research. Rats have a completely different digestion process to simple worms and a greater capacity for learning. This has been acknowledged by behavioral researchers: "The rat, being unaware of our theoretical needs, learns a great deal more than merely to go right in the maze – it also learns that it will be handled gaily and hence habituates to a given experimenter (but not to strangers)" (McConnell & Shelby, 1970, p. 90).

means of achieving cellular memory, which remain poorly appreciated or understood" (Allis *et al.*, 2007a, p. 26). Progress has been made in the past few decades in understanding what "chemical mechanisms" underlie epigenetic control, particularly regarding the protein–enzyme families that actively modify chromatin. Important colloquial terms for describing these processes are *gene activation* and *silencing*. In technical terms the key processes are: (a) histone modifications, (b) DNA methylation, and (c) variants of chromatin remodeling (Delcuve *et al.*, 2009).

Thus, in today's modern terms, epigenetics can be molecularly (mechanistically) defined as:

> The sum of the alterations to the chromatin template that collectively establish and propagate different patterns of gene expression (transcription) and silencing from the same genome. (Allis *et al.*, 2007a, p. 29)

Finally, how can (and do) genes learn and transmit learning to future generations? Why do cells, chromosomes or genes show epimutations? While we cannot provide an answer, we can, from a systems point of view, shed some light on it based on the two complementarities introduced in Chapter 3, namely the human–environment and the body–mind (i.e. material–biophysical vs. social–epistemic) complementarities. We can state that the cell is more than the sum of its genes and that what is transferred to the cells of the offspring is more than the pure DNA sequence information. Thus, perceiving the genes as the "sole bearers of information" (Rosenberg, 2006, p. 549) seems a mistake. The structure, behavior, development activity, and heredity also depend on the environment. Thus, we should consider both a *genome* and an *epigenome*. Environmental factors such as diet (e.g. excessive alcohol) and exposure to UV radiation or contaminants (e.g. nickel, arsenic or particulate air pollution) have been found to lead to changes in DNA methylation and to repression of tumor suppressor genes (Baccarelli & Bollati, 2009; Baccarelli *et al.*, 2009; Delcuve *et al.*, 2009, p. 549; Zoghbi & Beaudet, 2007). This is a clear indicator of the high environmental sensitivity of the epigenome.

> Cells and organisms interact with each other and through these interactions they acquire epigenetic information, some of which is inherited. (Jablonka & Lamb, 2002, p. 92)

Environmental literacy in this domain is limited, and advances in research have only started to explain processes involving chromatin–proteins and DNA activation. However, understanding why and when information is transmitted is still a challenge. We argue that research has to take into account the natural science perspective and the material–biophysical system view focusing on mechanical biochemistry *and* the epistemic information-based activities of the cell as a living system when selecting, reacting, and reorganizing itself with respect to environmental information. We also argue that the concept of *information* and the *functional* aspect of cell activity have to be stressed. Conceiving the cell as the basic level of organism, a functional perspective asks for interpretation of the processes within the cell with respect to environmental information (i.e. how to acquire nutrition, how to cope with competing cells, viruses, toxins, etc.).

We can see that the understanding of epigenetics requires a thorough interdisciplinary approach and must go beyond oversimplified approaches to understand living systems. Box 5.5 documents an example of a rude scientific slip that resulted from overly simplified causal, mechanistic, and chemical conceptions of learning in worms and rats. Fifty years ago, an abstract of a paper published in *Science* read as follows:

> Rats were trained in a Skinner box to approach the food cup when a distinct click was sounded. Ribonucleic acid was extracted from the brains of these rats and injected into untrained rats. The untrained rats then manifested a significant tendency (as compared with controls) to approach the food cup when the click, unaccompanied by food, was presented. (Babich *et al.*, 1965, p. 656)

Given the complex role that eukaryotic cells play in human or animal and other species, we suspect that simplistic conceptions such as the "central dogma of molecular biology" are meaningful frameworks for better understanding material–biophysical processes. This should be complemented by a proper information-based, epistemic perspective of analysis, which is, unfortunately, largely non-existent.

5.4.1 Key messages

- There is clear evidence that some species can pass on experientially acquired traits to their offspring. Biochemical mechanisms (such as methylation) which switch genes on and off are known, but the constraints under which this is done and other mechanisms that might be involved remain a mystery.

5.5 Immunology

Immunology provides fundamental background knowledge for understanding of environmental literacy because it seeks to explain the complex processes in organisms when they adapt to toxic and pathogenic environmental conditions or to situations within the body. In this section we describe immune systems in vertebrates and review the work of some pioneers of immunology at the turn of the twentieth century. We also discuss some current perspectives from immunologists, including views on the self–non-self discrimination of the immune system and how the immune system can be viewed as a cognitive system. The latter sheds light on how the central nervous system interacts with the immune system, suggesting a body–mind complementarity. Finally, we briefly review endemics and pandemics and how societal norms of interaction affect the speed with which viral mutations spread disease.

5.5.1 The vertebrate immune system

The immune system consists of a large number of different types of cells that are distributed throughout the (human) body. Lymphocytes (i.e. a type of white blood cell) constitute the major share of these immune cells, making up 20–30% of the white blood cells in the human circulatory system. Lymphocytes represent about 1% of the body weight, totaling about 10^{12} lymphocytes per adult. Today, immunology is a rapidly developing domain at the interface between biology and medicine, and investigates how an organism protects itself from pathogenic microorganisms such as viruses, bacteria, fungi, etc. There are two types of immune system in vertebrates – natural (or innate) and acquired (or adaptive) (Alberts *et al.*, 2002). One can distinguish between active immunity, acquired through the production of antibodies within the organism, and passive immunity, established by transfer of antibodies from an actively immune individual, for instance, from the mother. Active immunity is a response to the presence of antigens acquired by, for example, infection. Thus antigens, which can include toxins, bacteria, foreign blood cells, and cells from transplanted organs, stimulate the production of antibodies. Chemically, these antibodies are glycosylated proteins or are made up of some carbohydrate substance.

We consider immunology to be central for environmental literacy, as it is a prime example of understanding the complex processes in organisms when adapting to toxic and pathogenic environmental conditions. The immune system has to: (1) *recognize* and discriminate among different molecular entities; (2) *respond*, which it does in an extraordinarily adaptive way, since it can react to previously unknown molecules, or it may decide to refrain from reacting (become tolerant); (3) *discriminate between self and non-self*, and it does this with high specificity; (4) have *memory*, a property shared with the nervous system, which biologists refer to as an anamnestic response (Coico *et al.*, 2003).

5.5.2 The cradle of immunology

At the turn of the twentieth century, immunology split from vaccinology because of contributions by Ilja I. Metchnikoff (1845–1916) and Paul Ehrlich (1854–1915), who shed light on the processes and interactions of cells involved in immune responses.

Metchnikoff's groundbreaking idea was that many diseases result from battles between different species living in the body (1905). He postulated an immune response system involving phagocytes, the principal weapon used by the host to combat the pathogenic organism actively. He suggested that phagocytes, the white blood cells that ingest foreign bacteria or noxious foreign invaders, are "… remnants in higher animals of the original primitive intracellular digestive process of lower organisms" (Silverstein, 2003b, p. 3). The conception of the human–environment thus embraces a new perspective. The human organism is believed to produce its own micro-armadas which defend the body from hostile foreign subsystems. Metchnikoff's ideas anticipated that cells have marvelous mechanisms to kill invading species. He also used the metaphor of "combat" at the cellular level.

Ehrlich's theory started from the perspective of pediatric immunology. In his efforts to find a cure for diphtheria, he became aware that newborn children are initially resistant to the illness but, some time later, they lose this resistance. In his paper "On immunity by inheritance and suckling," he argued:

> The immunity of the offspring can be caused by (1) inheritance in the ontogenetic sense; (2) the transfer of maternal antibody; and (3) the direct intrauterine influence on the fetal tissues by the immunizing agent. (P. Z. Ehrlich, 1892, p. 184; translated in P. R. Ehrlich, 1957)

To prove this, he constructed an ingenious *experimentum crucis*, which involved feeding or injecting mice with abrin, a toxin derived from plant seeds. At that time, various beliefs on genetic inheritance dominated, some of which stated that immunity could be transmitted by the father or grandfather. Ehrlich thus tested the offspring of immune fathers and non-immune mothers but noticed no protection against the toxin. On the contrary, it was the offspring of non-immune fathers and immune mothers that were almost exclusively protected against the lethal dose. He also found that the immunity lasted for several weeks but then vanished, and further experiments showed that the immunity of grandparents did not matter. Ehrlich concluded that immunity was not caused by an "altered zygote" nor paternal sperm, and inferred that the immunity of the offspring must come from the maternal antibody. The indicative experiment included the separation of the neonates from their abrin-immune mother; they were then nursed by non-immune foster mothers. He was thus able to prove that the mice acquired immunity through the mother's milk and not in utero (Silverstein, 1996).

Ehrlich also introduced the idea of side-chain receptors, which could react to infectious agents and inactivate them. Here, the metaphor of a lock and key was used to describe the binding of the receptor, which he fallaciously supposed was "… determined before its exposure to [the] antigen" (Goldsby *et al.*, 2003, p. 4). Ehrlich was also a founder of immunochemistry and he investigated how animals were capable of forming antibodies when injected with molecules coupled with carrier molecules. An interesting finding of immunology is that antibodies can react to an extraordinarily wide range of new, human-made chemical and synthesized organic compounds which do not exist in nature (Landsteiner, 1936).

5.5.3 The humoralist–cellularist combat

The French vs. German, Pasteur vs. Koch, clash reached a second stage. This time it involved a humoralist–cellularist dispute between Emil Adolf von Behring (1854–1917) and Shibasaburo Kitasato (1852–1931) from the Institute of Infectious Diseases, headed by Koch and Metchnikoff at the Pasteur Institute. Metchnikoff had not only observed that phagocytes ingested microorganisms and foreign material but also that phagocytic cells in immunized animals are more active than in non-immunized ones. Thus, the concept of cell-mediated immunity was supported.

Behring and Kitasato started from the humoral approach (see Chapter 4.4), in which all diseases are seen as based on liquid, non-cellular disturbances. Guided by Koch's work, they demonstrated that vaccination with immune serum could neutralize toxins and certain bacteria. Later on, in the 1930s, components of these serums called gamma globulins (the most significant of which are immunoglobulins) were shown to be responsible for all these immunizing processes (Goldsby *et al.*, 2003; Silverstein, 2003a).

The battle between the two camps ceased in the 1950s when lymphocytes were identified as being responsible for both cellular and humoral immunity;

it was shown that T cells (lymphocytes) are involved in cell-mediated immunity and B cells (also lymphocytes) in humoral immune response.

5.5.4 Current perspectives on adaptive and innate immunity

An important step in immunology has been the identification of diverse cell types, including subtypes of T and B cells, and their immunizing activities, as well as the identification of MHC (major histocompatibility complex) molecules. MHC molecules do not have antigen-specific receptors but rather assist T cells in the recognition process, as T cells can only recognize antigens that have previously been processed by variants of MHC. While both B and T cells originate in bone marrow, B cells mature within the bone marrow, whereas T cells migrate and mature in the thymus gland. Different types of immune system cells can degrade endogenous and exogenous antigens, the fragments of which are presented on the host cell's surface by the MHC. Figure 5.12a presents a model of how MHC processes an endogenous antigen (i.e. an antigen produced within the cell). Viral proteins are degraded within the cell into peptides, which move into the endoplasmic reticulum and are then bound to the MHC molecules before being transported through the Golgi apparatus to the cell surface.

Innate immunity is comprised of four types of relatively unspecific barriers between the host and the pathogen, including: (a) *anatomic barriers* such as the skin or membranes (involving organelles which propel microorganisms out the body, such as cilia), (b) *physiological barriers* such as increased temperature or fever, low pH (in the stomach), or chemical mediators that break down the pathogen's cell wall, (c) *phagocytic barriers* through which specialized cells kill and digest whole microorganisms, and (d) *inflammatory barriers*. The inflammatory process is an acute or chronic response to tissue injury (e.g. burn, exposure to a chemical, etc.) or infection (by microorganisms) involving the accumulation of leukocytes, plasma, proteins, and fluid. This process had already been described in Egyptian papyrus writings dated from about 1600 BC (Medzhitov & Janeway, 2000).

The receptors for innate immunity are germline-encoded, in contrast to the rearranged T and B cell receptors of the adaptive immunity in the DNA, and are expressed without rearrangement. Although these receptors have evolved with time, all vertebrates share the same type of mechanisms for innate immunity; humans, the aforementioned *Drosophila melanogaster*, and the roundworm *C. elegans* all share the genes "that encode intracellular signaling pathways" (Janeway, 2001, p. 597). The idea is that the immune system does not react to the self, because any autoreactive immune receptor would be lethal to the host (Medzhitov & Janeway, 2000, p. 185).

The adaptive immune system, on the other hand, is not germline-encoded, as the innate immune system is, but rather somatically generated, clonally rearranged, and highly specific – it can only express

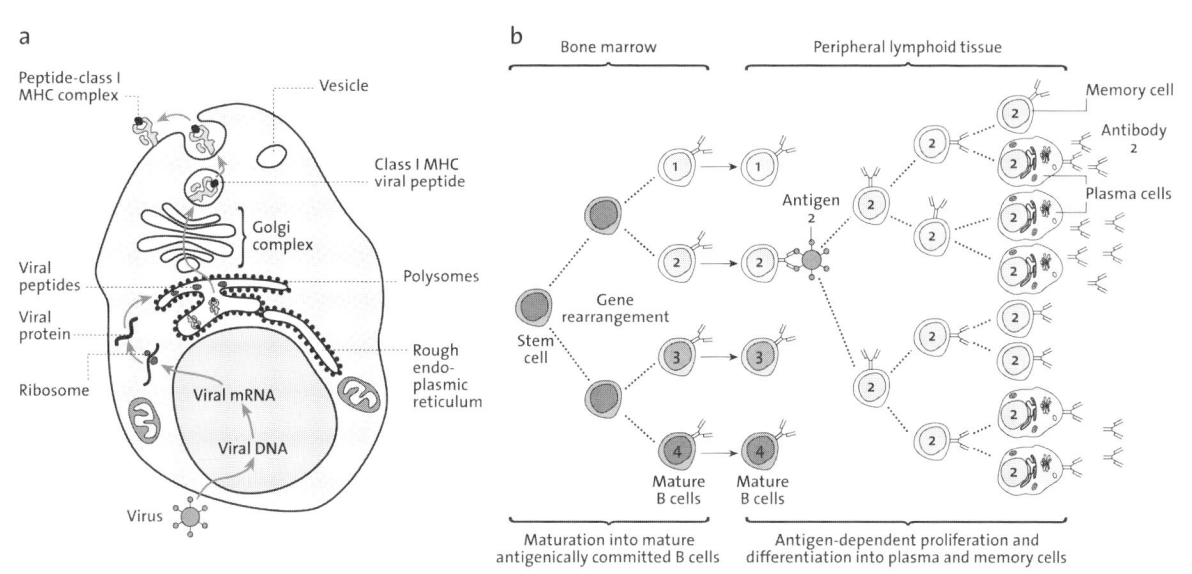

Figure 5.12 (a) MHC (major histocompatibility complex) processes an endogenous (produced within the cell) antigen. (b) Clonal selection theory of an antigen-activated B cell (number 2) leading to memory cells and effector cells (plasma cells from Goldsby *et al.*, 2003, p. 15).

a single random receptor specificity on a single cell (Medzhitov & Janeway, 2000). The adaptive immune system requires cooperation between lymphocytes and antigen-presenting cells.

A major step in the identification of immune cell processes, and how they relate to adaptation, was attained through work on the clonal selection theory by Jerne (1955) and Burnet (1957/2007). This long disputed theory states that many animals produce antigen-reactive clones of T and B cells and that the specificity of these cells is determined by the specificity of the antigen-binding receptor. The process involving B cells is presented in Figure 5.12b. The revolutionary ideas in this approach were the cellular dynamics of clonal commitment and clonal expansion (Cohn *et al.*, 2007) and the selection of the antibody receptors by the antigen. In Figure 5.12b, the stem cell produces mature B cells which have a Y-shaped protein (antibody) on their surface. Each cell has a different antigenic specificity. If a cell meets the corresponding specific antigen (see Antigen 2, Figure 5.12b), a cloning process is initiated which produces plasma cells that secrete antibodies. These antibodies react with memory cells and also specifically with Antigen 2, the antigen that initiated the antibody formation in the first place.

We should note that the specificity of antibodies is essential, such that, for example, tetanus antitoxin cannot neutralize diphtheria antitoxin.

> The inheritable "deep structure" of the immune system is now known: certain chromosomes of all vertebrate animals contain DNA segments that encode the variable regions of antibody polypeptides. Furthermore, experiments in recent years have demonstrated the generative capacity of this innate system. In proliferating B lymphocytes, these DNA segments are the target for somatic mutations, which result in the formation of antibody variable regions that differ in amino acid sequences, from those encoded by the stem cell from which these B cells have arisen. (Jerne, 1985, p. 1059)

Immune dysfunctions are allergies and asthma, autoimmune diseases, transplantation rejections, and immunodeficiency. All autoimmune diseases represent adaptive immune responses against the immune system itself (Medzhitov & Janeway, 2000). Although autoimmunity has long been viewed as dysfunctional, some researchers today conclude that there is evidence that autoimmune T cells have an important role in maintaining at least one body system – the central nervous system (Cohen, 2000a).

5.5.5 Philosophy of science views on the "self" of the immune system

We continue the journey into environmental literacy by reflecting on the conceptions of the functioning (rationale), and the notion and nature (ontology) of the immune system. Among biologists there is consensus that the immune system consists of those cells that protect the body from foreign substances, cells, and tissues by producing an immune response. This response involves other parts of the body such as the thymus, spleen, lymph nodes, lymphoid tissue, bone marrow, and particularly lymphocytes (i.e. B cells and T cells), but most importantly, antibodies. From its beginnings, two camps distinguish immunologists.

Immunochemists take a pragmatic, reductionist, molecular, pure natural science process-based view. "The molecules have no history built in beyond the fact of their formation, and no internal program built up overtime" (Silverstein, 2003b, p. 4). They try to avoid any concepts that go beyond humoral or macromolecular processes to describe immune reactivity. The role of the immune system is "body maintenance, which depends on the immune activity that we call inflammation" (Cohen, 2000a, p. 215). Thus a functional view is also inherent in this approach and is reflected in the following statement:

> The immune system does not "decide" that a stimulus is deadly or harmless and respond accordingly; the rules impose the same type of response ... solely as a function of the immunogenetic parameters ... (Silverstein & Rose, 2000, p. 177)

In contrast, immunobiologists take an organism-based, holistic approach, with roots in Metchnikoff's idea of "organismal identity as the core problem of immune reactivity" (Tauber, 2000, p. 241). With the development of the clonal selection theory in the 1970s by Jerne, the notion of "the immunological self" (Silverstein & Rose, 2000, p. 177) and the self–non-self discrimination suggested by Burnet (1957/2007; Cohn *et al.*, 2007) became key concepts of the discipline. Today, this view can be found in standard textbooks (Coico *et al.*, 2003; Janeway, 2001).

> One of the cornerstones of modern immunology is the ability of the adaptive immune system of lymphocytes to discriminate between self and nonself. (Medzhitov & Janeway, 2000, p. 185)

The conceptualization of the immunobiological approach uses decision theoretic language.

When faced with an antigen, the first decision the immune system must make is: 1) Shall I respond? If the question is yes, there are three more questions: 2) How strongly? 3) With what effector class? 4) Where? And a fifth, all-encompassing question: 5) How can I do this without destroying the tissues that I am meant to protect? (Anderson & Matzinger, 2000, p. 231)

Now, 50 years later, immunochemists and immunobiologists remain divided, and have been engaged in a lively discussion which involved 17 leading immunologists and two philosophers of science in 2000 in the journal *Seminars in Immunology* (vol. 12). In the remainder of this section, we review how self–non-self discrimination and the cognitive paradigm (Cohen, 2000b) theories of immunology, the latter of which is a follow-up model to the clonal selection theory introducing higher level information-processing mechanisms, have developed none the less. The cognitive paradigm sheds light on how the central nervous system interacts with the immune system, suggesting a body–mind complementarity.

The self–non-self discrimination is related to the dogma of immunology, the clonal selection theory of acquired immunity, and includes the assumption that the cloned cell must see the body as a bounded object.

> Any biodestructive defense mechanism must distinguish between the "bio" of self (e.g. host) and the "bio" of non-self (e.g. pathogen). (Langman & Cohn, 2000, p. 189)

Tauber criticized the self–non-self concept as anthropomorphic and stressed that "selfhood is a moral category, not an epistemological one" and that the "self-metaphor in immunology is intimately tied to notions of cognition, and … the immune system is viewed as 'recognizing', 'remembering', 'learning', and 'acting'" (Tauber, 2000, p. 242). As opposed to how information theory (and the sender–receiver metaphor) dominated cell theory (see above), immunology borrowed key concepts from cognitive science. As Cohen (2000b, p. 57) noted, the term re-cognition preserves a "sense of awareness-through action" and thus includes a certain type of knowledge component. We think that one has to reflect critically on which capabilities can be meaningfully assigned to single lymphocytes. Tauber (2000, p. 242) states that "… 'the immune self' cannot be defined as an entity, and not even as a function." Clearly, lymphocytes do not cognize, but react mechanically. At the same time, the lymphocytes of an organism react instrumentally and in an orchestrated manner as a whole, like many populations of simple organisms or insects.

A different perspective was applied to the clonal selection theory by Jerne's (1984) work on network theory. By focusing on the entirety of the immune system, the self–non-self complementarity was discarded. The key assumption here was that the immune system is an inward-directed self-seeking process which is activated if the equilibrium of maintenance is endangered or disturbed. Here, when referring to networks, the decision theoretic view is represented by an "immune dialogue" (Cohen, 2000a), a connectivity which imposes a complex dynamic pattern. The inward, self-centered view of the immune system emerged from Jerne's awareness that a human individual produces only about 10 000 different proteins but that the immune system is judged to deal potentially with 10 000 000 antigen molecules. It was thus that a different view on autoimmune disease was taken:

> I should therefore like to conclude that, in its dynamic state, our immune system is mainly self-centered, generating anti-idiotypic[1] antibodies to its own antibodies, which constitute the overwhelming majority of antigens present in the body. (Jerne, 1985, p. 1059)

Jerne's contribution can be seen as the cradle of systems biology (Tauber, 2008). He considered the cells of the immune system as an entirety which senses itself. This is clearly a cognitive view. It is interesting that Jerne justifies this approach when referring to Noam Chomsky's generative grammar, because for him it seems obvious that a basic structuring mechanism is necessary to manage the enormous complexity of the immune system.

> The immense repertoire of the immune system can be compared to a vocabulary comprised not of words but of sentences that is capable of responding to any sentence expressed by the multitude of antigens which the immune system may encounter. (Jerne, 1985, p. 1058)

5.5.6 The immune system as a cognitive system

The cognitive paradigm was proposed by Cohen (2000b) to replace the clonal selection theory. Cohen uses the term cognitive in a special way to conceptualize "meaningful interactions," which means the unique way

---

[1] Here, idiotypic refers to how antigenic determinants are specific to the region immunoglobulins that are produced by plasma cells.

in which "cognitive creatures interact with their worlds" (ibid. pp. 98–9). Meaning is conceived as the impact of information, as an outcome of interaction among the cells and the cell system and their environment.

Cohen considers the immune system as a collective of varied cells (e.g. T and B cells and macrophages) and molecules (e.g. antibodies, cytokines, enzymes, etc.), which communicate and coordinate responses in the face of antigen perception. This cognitive paradigm includes a shift from the pure material, passive object view to a subject view. Cohen's immunological homunculus (immunculus) suggests that self-reacting and self-referential receptors are kept in check by regulatory networks, and that such receptors respond depending on the context. The immune system constantly takes in new information about the external

> **Box 5.6[1]** The immune system: the unsung hero
>
> Immunologists distinguish between immunity of individuals and of species. The immune system "defines cellular individuality" (p. 5), establishes the molecular borders of a person, and is a guardian of the chemical individuality. It has friends and enemies. It "is a system that rejects foreign cells and tissues, a system that eliminates parasitic bacteria and viruses, a system that can destroy tumor cells and tissues, and a system that can destroy tumor cells arising from our own bodies" (p. 5). Thus, it has the status of individual welfare police and is considered an agent that organizes itself through experience so that it becomes capable of "organizing causality" (p. 5). But it can also be capable of friendly fire, resulting in autoimmune diseases. The molecular self is a hero because "the molecular self depends on the behavior of the immune system" (p. 6).
>
> We briefly sketch the distinguished properties of this hero. First, it is remarkable that the different agents of the system are able to detect different features of the antigenic environment. The agents form a regulatory network for feature detection (p. 149). Antigen-specific cells (T cells and B cells) and macrophages and all other cells participating in an inflammatory response have a division of labor and work in correspondence. Feedbacks are essential elements and are the basic element of self-organization. Immune systems (as cognitive systems) can make decisions, have memory storage for "images of their environment" (p. 64), and use experience to build up their internal structures in a process of self-organization. This learning at the level of the individual is called "somatic self-organization" (p. 86). Just as muscles degenerate if not used, the immune system must be activated regularly so it will not become feeble and lose memory of the surrounding environment.
>
> Cohen (2000b) argues that the immune system makes concrete, abstract, and distributed images "of the individual's internal and external environments" (p. 174), which are not made of matter but "created by processes" (p. 175). These processes, which "emerge from the ongoing adjustment of molecular patterns that reflect the ongoing needs of the body" (p. 182), allow the immune system to be viewed as a decision-maker. Here, we can again see a strong link to biosemiotics and to von Uexküll's research (see Chapter 4.12), as Cohen states: "The immune system recognizes not entities, but signs of entities" (p. 183).
>
> Cohen (2000b) sees the most essential capacity of the immune system in its ability to *arrange causality*. Here, Cohen refers to the Aristotle "four causes causality" concept and goes beyond the usual effect-related interpretation of causality, which means that environmental information or energy as input enters the system and the system transforms this input (p. 58, see Figure 5.13). Cohen (2000b) refers to a four-tiered definition, including making, form, matter, and ending of a system. In his definition, the "making" is the input–output efficiency; a "formal" cause refers to the idea that the system fulfills its functions; the "material" includes that an object exists independent of (i.e. before) an observer; and the "ending" he called "final causation." He introduces the example that the "ending" of a cake is to been eaten and thus refers to a teleological dimension of causation and thus introduces a specific notion of functionality, which appears to go beyond normal science. Thus, one can see the heroic performance of the health system as – a system between cell and tissue – working as a functional unit on a level between a single cell and a tissue.
>
> Input → [Energy, information] → SYSTEM → Output [Energy, information]
>
> **Figure 5.13** Immune system organizes causality to transform energy or information.

---

[1] This box and all unreferenced citations refer to I. R. Cohen (2000b). Irun R. Cohen was Director of the National Center of Biotechnology at Ber-Gurion University, Israel. His conceptional view is based on thorough and world-leading research, in particular about autoimmune diseases.

environment to build an image. In addition, internal, self-organizing, continuously updating processes allow the immune system to build an internal image of its self. It is able to make decisions in context, that means dependent on an "array of ancillary signals" that influence response (Cohen, 2000b, p. 151). When performing its main function of protecting the host from death by infection, the immune "system can learn to interpret context … the immune system is driven … by the antigen in context" (Cohen, 1992, p. 491). This holistic perspective "defines a system rather than an interaction that is dynamic, hierarchical, and self-organizing. The system accounts for the lack of one-to-one correspondence with antigen and receptor (degeneracy) as proposed by CST [clonal selection theory]" (Brownlee, 2007, p. 3). The view of the cognitive paradigm is nicely illustrated by the following statement, which reintroduces decision theory:

> The immune system, in short, responds simultaneously to different aspects of its target entities and to its own responses to these target features. We may call this mutual exchange of signals *co-respondence* (Cohen, 2000a, p. 218).

Cohen also takes a broader look at the immune system than that of the restricted "central dogma perspective" (see Box 5.6). He states that as the transformation of DNA information into protein information has an impact on cell activities, it generates meaning (Cohen, 2000b, p. 83).

5.5.7 The hierarchy of human systems and immunology: neuroimmunology

The immune system in humans is a network of cells and molecules that operate more cooperatively than a typical group of individual cells, but are not as connected as cells that are part of the same tissue. In this section we see how, referring to the cognitive paradigm and to studies that show that the central nervous system interacts with the immune system, one can infer a body–mind complementarity.

Given our definition of a human system, the human immune system is a human system between the cell and a tissue (Figure 5.13). It is a multicellular system of single cells located in different parts of the body which cooperate. Theoretically, it operates on a level between singular cells (protozoa) and a multicellular being. Here it is interesting to note that this boundary is not crisp, as is the case with slime mold (*Dictyostelium discoideum*, see Box 14.4), a soil-living amoeba. These amoebae live as protozoans but, given certain environmental constraints (for instance, when lacking feed), they release signal molecules into their environment (Watts & Ashworth, 1970). The protozoans create swarms and join up into a tiny multicellular slug-like coordinated creature. This creature then moves to a place where it receives regular sunlight and grows into a fruiting body that even has spores. Here, some of the amoebae sacrifice themselves to become a dead stalk, lifting the spores up into the air. The procedure does not seem to be completely predefined, as when the growing fruit is halved, the function of the cells changes.

An interesting question is whether and how the immune system interacts with higher order human systems such as the brain or a group as has been postulated in Figure 14.1* for systems from the cell upward. This brings us to psycho-neuroimmunology (Ader, 2007; Schedlowski & Tewes, 1996; Vedhara & Irwin, 2005). It is interesting that only a few of the immunology textbooks (e.g. Kuby, 1997) address this concept of the interface between immunology, neurobiology, and endocrinology (see Figure 5.14).

Recent research indicates that psycho-neuroimmunology is an expanding field. The "nervous and immune systems use a common chemical language for intra- and inter-system communication" (Blalock & Smith, 2007, p. 25). A series of studies led to the hypothesis that immune systems have special receptors and abilities to identify pathogens, tumors, and allergens that "the body cannot otherwise hear, see, smell,

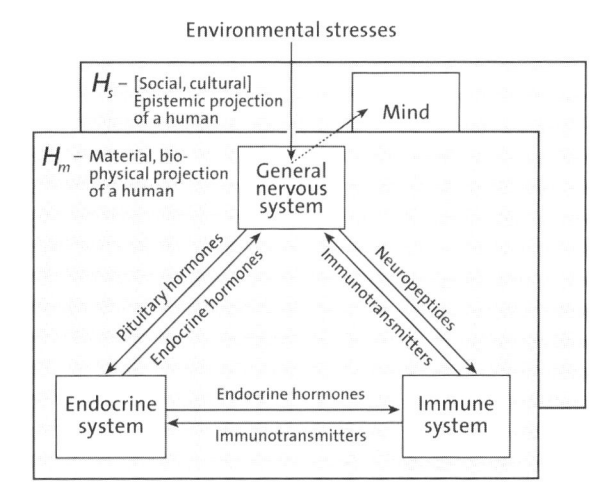

Figure 5.14 Environment–body/brain–immune system interactions. (Adapted from Khansari *et al.*, 1990, p. 173, with permission from Elsevier.)

taste or touch" (ibid. 2007, p. 25). Here, the strong metaphor of "the immune system as a sixth sense" has been coined. The understanding of the signals between different types of self and foreign cells and the idea that the cells can send danger signals to the immune system, for instance to convey the message that a virus has invaded (Matzinger, 2002) is in line with the above idea of the immune system as a cognitive system.

There is clear evidence that activity in the central nervous system, specifically the mind, affects the immune system. The two systems communicate by neurotransmitters such as acetylcholine, intestinal peptides, or histamines (Steinman, 2004). Many experiments and surveys have shown that acute time-limited stressors (such as exams or skydiving), chronic stressors (such as suffering a traumatic injury; Montoro et al., 2009; Reiche et al., 2004), or distant stressors (such as witnessing the death of a fellow soldier during combat) change, or potentially change, the activity of the natural and/or specific immune system (Segerstrom & Miller, 2004). A number of experiments have demonstrated the reverse to be true, showing that activation of the immune system can be followed by behavioral, physiological, and psychic changes. For instance, when test subjects were injected with typhoid vaccine (inducing a significant rise of interleukin-6 (IL-6) serum levels), they showed a heightened systolic blood pressure and a more negative mood resulted than their control groups (having received placebo or not performing the task; Brydon et al., 2009).

5.5.8 Epidemics and pandemics: the immune system and society

A socially relevant question regarding immune systems is how the spread of contagious diseases depends on the behavioral patterns of human groups or communities (see Figure 3.3*). If we consider a lethal droplet-transmitted disease such as the 1918 Spanish Flu, which killed between 30 and 50 million people (Johnson & Mueller, 2002), several things are important. First, the cultural contact and salutary rules such as handshaking, hugging, kissing, physical distance when chatting, as well as the way in which sick persons are nursed, all play important roles (Smieszek et al., 2009). Second, the mobility patterns of humans are important, as shown by the staggering number of deaths of indigenous North Americans during European conquests of the fifteenth and sixteenth century (see Box 8.1). The North American people were isolated from the European population and their virus pool, presumably for several tens of thousands of years. As virus mutations are accompanied by learning and thus evolution of the immune system (Boots et al., 2004; Read & Keeling, 2003), the North American populations were unprepared to combat the virus strains that European explorers brought. Third, cultural aspects, such as household and public hygiene (e.g. the use of towels), have an important effect. Chapter 18 offers illustrative discussions of epidemics and pandemics in the form of a case study.

5.5.9 Key messages

- The immune system is the health police of the human organism; it eliminates toxins, viruses, harmful bacteria, and kills tumor cells and pathogens. One of its major challenges is to find the appropriate level of activity against self and environmental "functional distractors"
- The human immune system recognizes signs of entities, has a memory, responds adaptively, and has innate components that can be found in a similar form in most higher level animals. In total, the immune system can be seen as a system representing a material–biophysical(–biochemical) and a (social–cultural–)epistemic – that is, a body–mind complementarity
- The immune system is embedded in the hierarchy of human systems and shows characteristics of a system between a cell and the human system
- The human immune system strongly interacts with higher ordered human systems, such as the brain and other organs. As epidemics and pandemics show, its activities also affect and depend on group and societal norms and interactions.

5.6 From ecosystem functions via ecosystem services and socioecological resilience research to landscapes

In this section, we shift gears from a microscopic view of the cell to a macroscopic view of large ecosystems. The question is how the human–environment relationship entered biological and ecological research.

We have sketched the emergence of ecology in Chapter 4.11. Besides Haeckel's (1898) definition of the term "ecology," Moebius' (1877) work on biocenosis was essential. The sea was seen as a microcosm, and marine ecology emerged, helping the fishing industry to better understand the reproduction of populations. Robert

E. Park (1864–1944), one of the founders of human ecology (see Chapter 8.3) who had already applied biological knowledge to the study of cities, stated:

> Probably the most important ultimate objective of ecology is an understanding of community structure from the viewpoint of its metabolism and energy relationships. (Park, 1946, p. 318)

When doing fieldwork, ecosystem researchers inevitably become aware of the impact of human systems on environmental systems (i.e. $H \rightarrow E$), for example by water pollution and the benefits that humans gain from the environment (i.e. $E \rightarrow H$), for instance, from coastal fisheries. The book *Fundamentals of Ecology* can be considered as the first comprehensive textbook (Odum, 1953) of the field. But already the subtitle of the follow-up book *Ecology – The Link Between the Natural and the Social Sciences* (Odum, 1975) reveals that ecology has been considered an interface science, although the human ecological and socioecological aspects were often relegated to the last chapters, epilogue or appendix if they were not completely ignored (Begon *et al.*, 1990). We should note that in the 1960s, E. P. Odum and others considered pollution and not lack of resources or energy as the challenge for human ecology (Odum, 1973). The social dimension was pushed into ecology by the Rio Declaration in 1992 (United Nations (UN), 1992), when the global scale of ecosystem deterioration became evident. In this section, we review how ecological approaches incorporated the human species as a major forming component of ecosystems on a macroscale. As with other chapters, we follow the history of mind of disciplinary development and focus on ideas which are important to understand contributions of biology to sustainable management of human–environment systems (HES), this time on a macroscale.

5.6.1 Disturbance ecology: from succession to adaptive cycles

In this section we review how some ecologists have questioned the idea of orderly succession, or community development, that culminates in a stable ecosystem. Ecologists have been overly sensitive about the vulnerability of human–environment relationships.

> An understanding of ecological succession provides a basis for resolving man's conflict with nature. (Odum, 1969, p. 262)

The stabilizing and destabilizing forces have been the core interest of this branch, which we call "disturbance ecology" (Jørgensen *et al.*, 2007). Naturally, ecologists are biologists and focus on the principles underlying succession dynamics and vulnerability. They thus focus what we call the "rationale of ecological systems." An important paper in this field has been Odum's (1969) presidential address to the Ecological Society of America, where he focused on a mechanistic–deterministic understanding of succession.

> Succession is characterized by an orderly, directional and predictable community development, is affected by modification of the physical environment but shaped by the internal community characteristics such as change rate, and culminates on a stabilized ecosystem. (Odum, 1969, p. 262)

To understand ecological succession, the study of energetics (including yield) and structure of community are essential. Life history, nutrient cycling, selection pressure, and overall homeostasis (including resilience and resistance to external perturbations) are considered crucial factors.

An additional important concept in ecology is the physical asymptotic stability, steady state or equilibrium, that is inherent in theoretical community ecology (Figure 5.15). However, the idea of overall homeostasis or a globally stable equilibrium state had already been questioned by Odum in his seminal paper published in 1969. Here he asked:

> The intriguing question is, do mature ecosystems age, as organisms do? In other words, after a long period of relative stability or "adulthood," do ecosystems again develop unbalanced metabolism and become more vulnerable to diseases and other perturbations?. (Odum, 1969, p. 266)

The deterministic view of succession on the one hand and the equilibrium view of ecological communities on the other have been seriously challenged by another line of research that developed in the 1970s. The former research was criticized for its strong physical, deterministic conception, which tended to conceive ecosystems as machines and neglected the uncertainty and complexity inherent in ecosystems (see Koellner, 2003). The *resilience approach* (Folke *et al.*, 2004) replaces the former views with an evolutionary, indeterministic, organic concept of ecological communities. First, global stability is replaced by local stability and a multistable state reality, in which ecological systems are considered to be prone to regime shifts towards alternative stable states or, expressed mathematically, towards different attractors (Folke *et al.*, 2004; Scheffer *et al.*, 2001). Ecosystems are described as

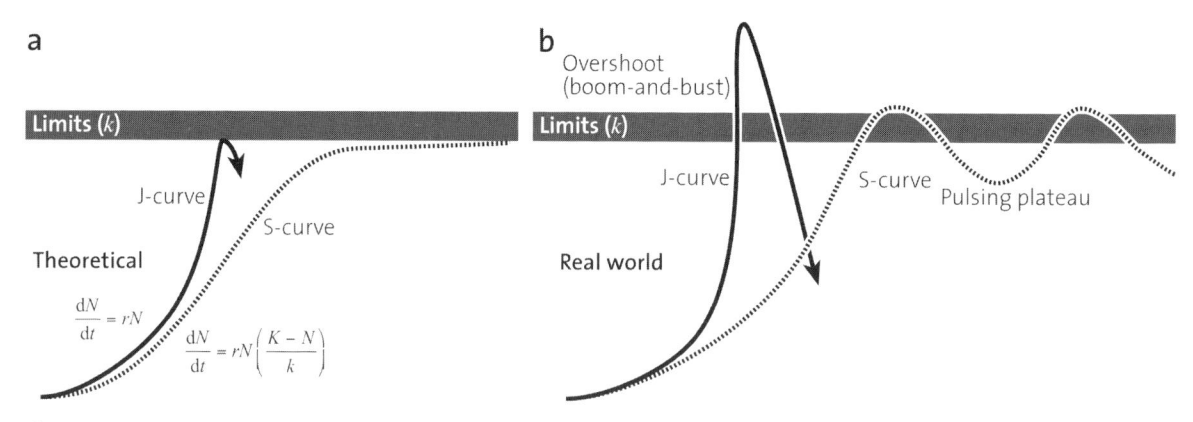

Figure 5.15 (a) Theoretical and (b) real-world growth curves of populations. N = population size; r = rate of births; K = carrying capacity (adapted from Odum, 1996).

moving targets experiencing stabilizing and destabilizing forces that undergo complex dynamics, and both internal and external disturbance may lead to decay. Second, succession does not always lead to the same climax or equilibrium, as multiple outcomes typically are possible depending on the accidents of history. Below we take a closer look at the resilience approach, which we name *disturbance ecology* (Jørgensen *et al.*, 2007).

In his panarchy theory, Holling (1973, 2001, 2004; Holling & Gunderson, 2002) states that stability, nature balance or permanent nature resilience are myths and argues for a rather evolutionary perspective dubbed "nature evolving." Panarchy represents a meta-model for ecosystem dynamics, which is comprised of the adaptive cycle, applied at first on a local scale, and the panarchy model, applied on a regional scale. The first model assumes that ecosystems follow an adaptive cycle in which the systems "grow, adapt, transform, and in the end, collapse" (Holling, 2004, p. 1). In ecosystems one can observe phases of relative stability, which are followed by sudden releases. Some of them go back to simplified species composition resulting from aging or over-maturation; others result from interactions from fast and slow processes which may occur at different scales. Development and evolution include breakdowns which open new – and in some respects unpredictable – possibilities. Here, from a system theory perspective, linking and scaling up many interacting linear processes may lead to non-linear processes. Further:

> Ecosystems do not have a single equilibrium with homeostatic controls to remain near it. (Holling & Gunderson, 2002, p. 26)

Based on these assumptions, Holling and Gunderson extend the traditional deterministic view of ecosystem succession: exploitation, e.g. rapid colonization of recently (e.g. by fires) disturbed systems by r-species; and conservation, describing slow processes of accumulation by K-species. Exploitation is related to species with a fast growth rate r, conservation to sustained maximum plateau (see Figure 5.15). Based on this, they introduce two new concepts: namely, release (Ω) and reorganization (α). When referring to the economist Schumpeter (1950), the release concept is sometimes also called the concept of *creative destruction*.

> [In ecosystems, release processes are likely if] tightly bound accumulation of biomass and nutrients becomes increasingly fragile ... by agents such as forest fires, drought, insect pests, or intense pulses of grazing. (Holling & Gunderson, 2002, p. 34)

The reorganization process includes high uncertainty and can even end with an exit of the biotic structure. From a science perspective, the processes in the reorganization phase are the most difficult ones to predict. Thus, there are four consecutive ecosystem processes of different time dimensions that can be represented as a quadruple (r, K, Ω, α). It is important to note that these phases of ecosystem development have different speeds. Ecosystems experience a rapid exploitation phase (also called structural growth), which turns slowly into a long-lasting conservation phase, which can enter a very rapid release and a fast reorganization phase. Besides potential and connectedness there is a third dimension, called (ecological) resilience, shown in Figure 5.16. The second, smaller

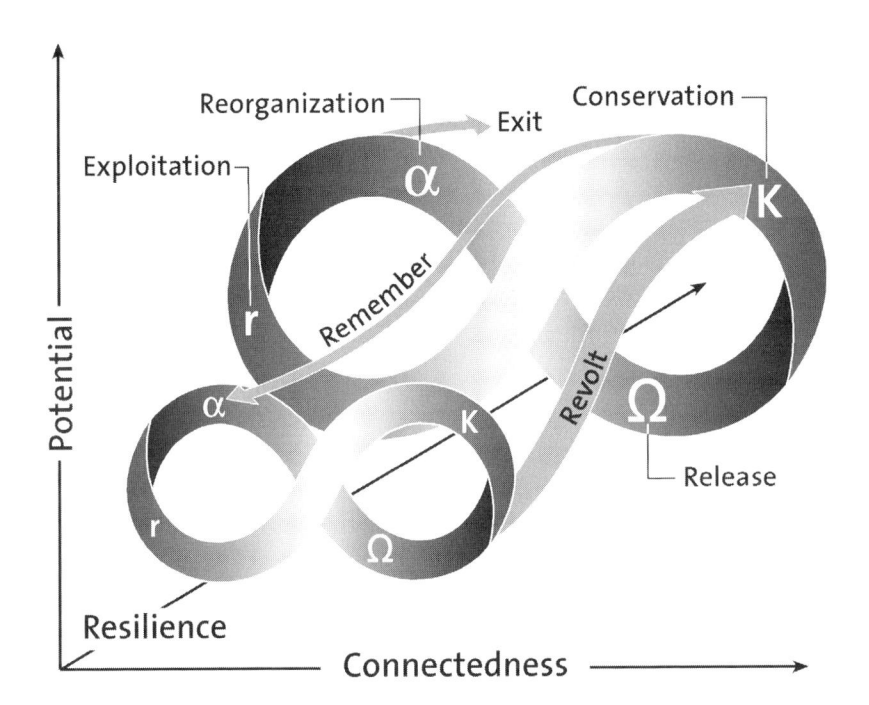

5.6 Principles, cycles, and resilience of ecosystems

Figure 5.16 The adaptive cycle. The processes (r, K, Ω, α) can have a nested structure where subsystems, for instance species, remember and affect or revolt against a certain consolidation (according to Holling & Gunderson, 2002).

adaptive cycle in Figure 5.16 represents a subprocess which can negatively affect (i.e. revolt against) the conservation phase, and also remembers weaknesses of the consolidated process. Finally, there is an exit arrow, indicating that the reorganization process may fail and a less productive or an inert system may result.

Adaptive cycles thus include the *start of succession* (depending on the genetic material available, many directions are possible owing to low connectivity of the elements), *fast growth* (structural growth; pioneer stage), *fast development* (network growth, middle succession, increasing dependencies of single species), *maturity* (information growth, captured energy needed for maintaining the system, increased vulnerability), *breakdown* (release, creative destruction, chaotic behavior possible), *reorganization* (structural and functional resources can be used to define new directions), and *reset* (a new ecosystem succession starts; see Jørgensen et al., 2007).

Proponents of the panarchy model extend the adaptive cycle metaphor from the local scale and then to the regional scale and picture "landscapes" as nested hierarchies of adaptive cycles (Holling, 2001, see Figure 5.16). We should note that the resilience approach claims validity not only for the purely ecological domain, but also for social, economic, and social–ecological systems (Allison & Hobbs, 2004; Berkes et al., 2003; Gunderson & Pritchard, 2002), which is discussed in more detail below.

5.6.2 New Ecology

The Rio Declaration and Agenda 21 persuaded some ecologists to switch to *full ecology* and to consider humans an integral part of ecology. However, this seems to be done mostly from an ecosystem management perspective when including social science knowledge. Of great importance was the formulation of 12 principles at the Fifth Meeting of the Conference of the Parties to the Convention on Biological Diversity in the year 2000 (Convention on Biological Diversity (CBD), 2007). The principles emphasized how "ecosystem managers" should "consider the effects (actual or potential) of their activities on *adjacent and other ecosystems*," "to … manage ecosystems in an economic context … within the limits of their functioning" and to "seek the balance between, and integration of, conservation and use of biological diversity." We can see that the principles, which also emphasize how an "ecosystem approach should involve all relevant sectors of society and scientific disciplines," have a strong focus on transdisciplinary activities (quoted according to Jørgensen et al., 2007, p. 2). In this vein, Jørgensen and

5 From molecular structures to ecosystems

Box 5.7 Understanding fundamentals of ecosystems: the ten principles of *New Ecology*

A major achievement of *New Ecology* is that it defines basic principles of ecosystem properties and processes. These principles are fundamental for an analysis of human environments (see Postulate 7, Table 16.1*). The principles include (Jørgensen *et al*., 2007, pp. 246–8):

(1) *All ecosystems are open systems* and receive energy–matter input from and discharge energy–matter to the environment.

(2) *Mass and energy are conserved*: thus, a thorough input–output analysis is basic. In the language of material flow analysis, this means that stocks of energy and material–matter are built.

(3) *All ecosystem processes are irreversible*: evolution proceeds in the direction of ever more complex solutions. It includes directionality and autocatalysis (i.e. higher ordered feedback loops; see Table 16.1* and Chapter 16.5). This principle is essential, as otherwise the system would return to thermodynamic equilibrium.

(4) *Carbon-based life* can only exist in relatively small domains between 250 and 350 K (−23 to 77 °C).

(5) *Carbon-based life* is based on a set of biochemical compounds of 25 elements.

(6) *All ecosystems are connected* to other organisms and to their abiotic environment.

(7) *Ecosystems are organized hierarchically* (see Chapter 16.3, Table 16.1*).

(8) *Biological processes use energy to move away from thermodynamic equilibrium* and can thereby maintain non-equilibrium states (i.e. ecosystems can grow).

(9) *Ecosystem growth and development* is based on capturing energy across boundaries and increases: (a) biomass, (b) cycling (i.e. linking ecosystem components), and (c) embodied information. The latter improves feedback control and allows the switch from *r* (reproduction-oriented) to *K* (capacity-oriented) strategies (see Figure 5.15).

(10) *Eco-exergy use* and the tendency to move the system to the furthest state from thermodynamic equilibrium are basic trends of ecosystems. Exergy is the usable energy; eco-exergy expresses the work capacity relative to the same system at thermodynamic equilibrium.

Figure 5.17 presents an example of *New Ecology* thinking. Lotka (1925) and Odum (1983) define the maximum power principle, which states that the flow of useful energy is maximized. Brown (1995) focuses the transformation of energy into work *W* and defines *fitness* as *reproductive power*, that is the rate (dW/dt) at which energy can be transformed into work to produce offspring. Thus reproductive power is a U-shaped function. This means that *useful work* and not *efficiency* is maximized. A trade-off between high rate and high efficiency is best (Jørgensen *et al*., 2007).

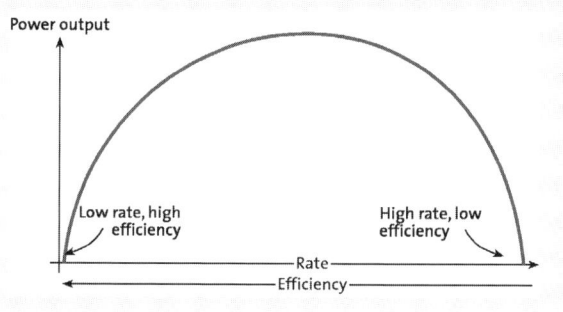

Figure 5.17 New Ecology defines reproductive power (*y* axis) as the rate at which energy can be transformed into work to produce offspring.

eight renowned ecologists wrote a blueprint for ecology, called *A New Ecology: A Systems Perspective*. The book starts with an appreciation of the Rio Declaration, but does not deal with human systems or the specific rationale and impacts of human systems after the first few pages. The index has only two terms (i.e. Human well-being [see p. 2] and Rio Declaration) that relate to the human dimension. Thus, the core of *New Ecology* has a very strong natural science accent and incorporates human systems rather from a general system theory perspective.

New Ecology theory is an ambitious venture of ecological theory building. "The theory should be considered one of the first attempts to present an (almost) complete ecosystem theory" (Jørgensen *et al*., 2007, p. 248). If we look at the principles in Box 5.7, the following organizing ideas can be identified: ecosystems are seen as open, connected, hierarchical embedded systems, exchanging energy–matter and information with other systems that allow carbon-based life processes. These life processes are irreversible, need a continuous flow of energy, can occur over a small range of temperatures, and permanently develop owing to optimizing energy flows. Not included in the principles but in the theory outline is that energy efficiency is established by management of information and complexity.

Efficient use of solar radiation, proper networking of ecosystems, and information increase (for more efficient networking) can result in an increase of biomass (primary) production. This is the practical side of ecology supporting sustainable development.

In total, the *New Ecology* approach provides an overly comprehensive view on ecosystems, perhaps with the exception that ecosystem fragmentation caused by manmade barriers (Forman, 1995) is not sufficiently represented. Although attempting to incorporate human systems, the focus remains on the biological science side, stressing the biophysical perspective. Nevertheless, *New Ecology* provides basic insights into the rationale of ecosystems in their response to severe human-caused perturbations. This is important for system limit management, a strategy for avoiding unwanted collapse of systems. System limit management is a key part of sustainability as defined in this book. However, two additional aspects of our definition of sustainability (see Chapter 2), intergenerational and intragenerational justice, are not dealt with in *New Ecology*.

5.6.3 Ecosystem functions and services

If we look at how the environment provides for and sustains humans, we are describing ecosystem services. In this section, we discuss ecosystem services and examine models that predict the direction in which an ecosystem will develop. Several models depict ecosystem functions and how they can be interpreted as systems at the biology–economics interface.

If one wants to define functions of nature or ecosystems, one is challenged to define from what perspective which functions are considered. In Chapter 3.5 we distinguished between a European tradition of functionalism, which emphasizes an evaluative feeling according to the intentionality of humans, and a pragmatic American line, which stresses learning and adaptation to given environmental constraints and information. If we look at processes and entities of ecosystem functions from the perspective of human interests, we speak about environmental services.

But is it possible to take another, an ecocentric view, for instance, if we want to describe the processes in ecosystems? From a functionalist perspective, a side change seems impossible as everything humans are doing is related to their explicit or implicit capability and intentions. This holds true if ecosystem functions (de Groot, 1992; Schulze & Mooney, 1994) are analyzed from an environmental management perspective. However, ecosystems show directionality (Jørgensen *et al.*, 2007; Odum, 1973) and, given a certain succession at a given time, future changes can be predicted to some extent. Clearly, such prediction is uncertain and the accuracy of the prediction depends on the state of the systems. Given knowledge about directionality, one can construct a model of ecosystems, including structures and relations between the elements of the system which are seen as functions. New Ecology provides the following rationale for directionality.

> Ecosystem dynamics are rooted in configurations of autocatalytic processes, which respond to random inputs in a non-random manner. Autocatalytic processes build on themselves, and in the process give rise to a centripetal pull of energy and resources into the community. Such centripetality is central to the very notion of life … (Jørgensen *et al.*, 2007, p. 78)

A critical question is whether humans are capable of speaking for nature. According to the ontological and epistemological principles of this book, anything is constructed and depends on the knowledge (which has been functionally acquired; see Chapter 3) and interests of the specific organismic or human systems. This also holds true for researchers, as elaborated by Schulze and Mooney. There is an unavoidable myriad of "tradeoffs among alternative patterns of allocation, such as growth versus defense or reproduction, roots versus shoots, competitive ability versus colonizing potential" (Schulze & Mooney, 1994, p. 502). Even the construction of key species is bound to the theoretical view of the researcher and cannot take a neutral position acknowledging all aspects of ecosystems. Ecology, as any other scientific activity, is always bound to the constructivistic, functional perspective. What is in the foreground and the background takes on the perspectives, concepts, theories, modeling tools, and, in effect, takes on the interest of the researcher.

The ecosystem service approach often follows the line of argument: "global change, loss of biodiversity, change of ecosystem functions, loss of benefits" (Beierkuhnlein & Jentsch, 2005, p. 251). Ecosystems are increasingly seen as capital assets (Turner & Daily, 2008) or critical capital (Brand, 2009) for private organizations, in particular companies, governmental organizations, and non-governmental organizations (Koellner et al., 2008). Various methods have been developed to quantify the value of ecosystem services, such as willingness to pay (see Chapter 11.4; Howarth & Farber, 2002; Luck et al., 2009; Spash, 2008). A special challenge in this context is the valuation of biodiversity, which is difficult both from a natural science perspective (one can distinguish between different measures of ecosystem, species, and genetic diversity; Koellner & Scholz, 2008; Purvis & Hector, 2000) and different measures of economic valuation.

As a framework for discussion of ecosystem services, De Groot (1992) introduced four types of ecosystem functions: (1) regulation, (2) habitat, (3) production, and (4) information function. Objects of the regulation function are the chemical compositions of the atmosphere and oceans, hydrological cycles, storage and recycling of nutrients, genetic diversity or soil fertility for biomass production. Habitat functions include refuge and reproduction space. Production functions are energy and nutrient uptake and metabolic processes. De Groot defines information functions with respect to anthropocentric functions, including "spiritual enrichment, cognitive development and aesthetic experience" (ibid. p. 395).

Clearly the functions can be constructed for the survival of different species, say human or elephant. If we follow de Groot, who defined ecosystem functions as "the capacity of natural processes and components to provide goods and services that satisfy human need, directly or indirectly" (de Groot et al., 2002, p. 394), we speak about anthropocentric ecosystem services or ecosystem services. In both cases we can start from a restricted bioecology perspective or from a full ecology perspective, and thus consider human species as part of the ecosystem. Ecosystem services depend on values and explicated needs and are those that are considered functional and valuable from a human system perspective:

> Ecosystem services are the conditions and processes through which natural ecosystems, and the species that make them up sustain and fulfill life. They maintain biodiversity and the production of ecosystem goods, such as seafood, forage, timber, biomass fuels, natural fiber, and many pharmaceuticals, industrial products, and their precursors. (Daily, 1997, p. 3)

The Millennium Ecosystem Assessment (2005) distinguishes between four types of ecosystem services. There are: (A) supporting services such as nutrient cycling, soil formation, or primary production for three anthropocentric ecosystem functions, which are (B) provisioning services like food, freshwater resources, wood and fiber, or fuel, (C) regulating functions such as climate, flood, disease, or hydrological cycles, and (D) cultural services such as aesthetics, spiritual, educational, and recreational functions. The functions A–D contribute to constituents of wellbeing: security, basic material for good life, health, good social relations, or freedom of choice and action. Here, biology enters economics as ecosystem services can go to market and become goods, for instance capital or consumer goods, which can be monetized (see Chapter 11.2) and become part of GDP.

We should note that when switching from the level of ecosystem functions to ecosystem services (or performances), different types of pricing and valuation are required which relate to the "quality of an ecosystem." The task is particularly challenging as the material world and ecosystems do not tell us which system functions are essential. What this means, for instance for biodiversity, and how the interface between biology and economics can be shaped is discussed in Chapter 18.3.

5.6.4 Socioecological resilience research

The world biosphere is facing a manmade mega-disturbance that endangers basic ecosystem services. Ecologists are concerned by the rapid global degradation of forests, marine, limnological and other natural systems, and the decline of biodiversity (see Box 5.8). Thus, ecology has been challenged to integrate humans as the most influential species in the world ecosystem. Similar to *New Ecology*, socioecological research felt pressure from the Rio Convention and follow-up programs such as the 1996 International Human Dimensions Programme on Global Environmental Change (IHDP).

An important contribution in this field has been resilience research. Originally, system resilience was conceived by Holling (1973) as a general system property:

> Resilience determines the persistence of relationships within a system and is a measure of the ability of these systems to absorb changes of state variables, driving variables, and parameters, and still persist. In this definition resilience is the property of the system and persistence or probability of extinction is the result. Stability on the other hand, is the ability of a system to return to an equilibrium state after a temporary disturbance. The more rapidly it returns, and with the least fluctuation, the more stable it is. (Holling, 1973, p. 17)

Later, Berkes and Folke defined a socioecological system as an ecosystem "that does explicitly include

5.6 Principles, cycles, and resilience of ecosystems

Box 5.8 Biodiversity and extinction: the rise and fall of species

The diversity of nature is tremendous. About 1.75 million species have been described, whereof 75% are animals (mostly insects). The estimates of today's total species diversity range between 3.6 and 111.6 million species (Heywood & Watson, 1995) and, over space and time, there is and has been a huge variety of species.

The number of individuals and species per unit area increases the closer you are to the equator. Abundant solar energy, heat, and humidity promote greater plant growth and support more organisms. Also, the stable climates found near the equator seem to support the survival and coexistence of specialist types of species (Withgott & Brennan, 2008).

Evolution is accompanied by an increase in ecosystem complexity. This is often, but not always, accompanied by an increase of diversity (see Figure 5.18a). For instance, according to paleontology, the average number of plant species found in local floras has risen steadily since the movement to land by plants 400 million years ago (Wilson, 1992b). Species extirpation and extinction occur naturally, which has advantages in that today our life might be more difficult if Earth were stuffed to the gills with dinosaurs. Paleontologists assume that more than 99% of all species that ever lived are now extinct. In the past 440 million years there have been five major extinction episodes, the most severe at the end of the Permian period 248 million years ago, in which 90% of all species were eliminated (Withgott & Brennan, 2008). Causes of extinction were sea level changes, climatic changes, oceanographic changes caused by organic carbon cycles, bolide impacts (meteorites which are supposed to have caused the dinosaur extinction 65 million years ago), and volcanism (Brenchley & Harper, 1998). *Homo sapiens'* existence is another cause. The extinction of large mammals and flightless birds coincided closely with the arrival of humans in North America, Madagascar, and New Zealand, and, less decisively, earlier in Australia (see Figure 5.18b). In Africa, where humans and animals evolved together for millions of years, the damage was less severe (Wilson, 1992b).

Today's biodiversity loss, caused primarily by land conversion, invasive species, pollution, overharvesting, and climate change, is affecting human society. The simultaneous decline of ecosystem, species, and genetic diversity may result in a critical loss of ecosystem services. As such, humans are facing a secondary feedback loop (see Figure 17.1*) that might cause a severe rebound effect which may endanger human life.

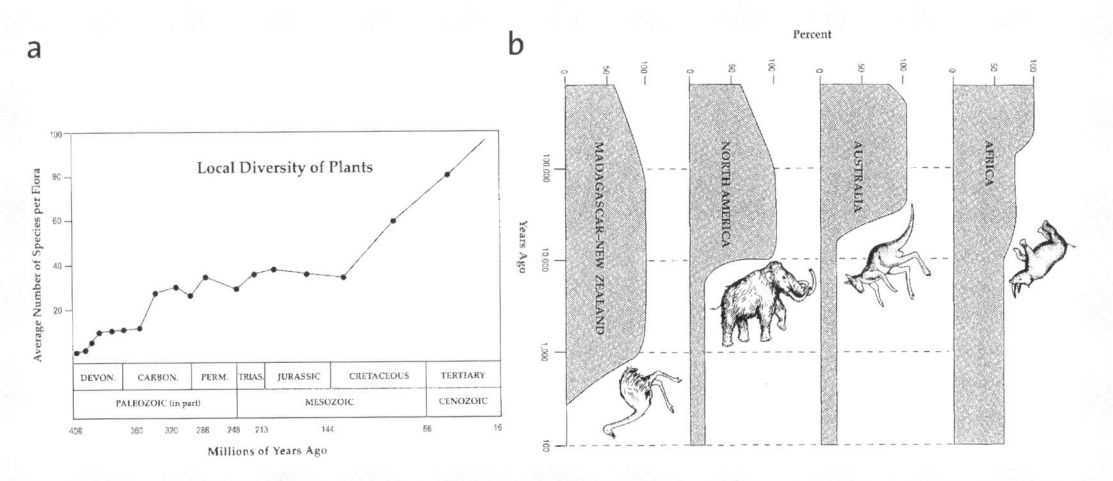

Figure 5.18 (a) Increase in the number of plant species in the past 400 million years. (b) The extinction of large mammals and flightless birds is caused by human impacts. The x axis represents percentages referring to the maximum number. Reprinted from Wilson, 1992, pp. 195, 252 by permission from The Belknap Press of Harvard University Press, © by E.O. Wilson.

humans or, more specifically, the social system" (Berkes & Folke, 1998b, p. 9). As Haberl and others noted, this "formulation, however could be misleading because it implies that society can be seen simply as a subsystem of the ecosystem" (Haberl *et al.*, 2004, p. 201). Whether a subsystem or a complementarity view is appropriate is a fundamental question. Many scholars contributing to the resilience approach think that there is a close connection between the human and the natural realm but do not define the specifics of human systems. They view the nature–culture split as arbitrary and artificial (Berkes *et al.*, 2003; Berkes & Holling, 2002), even though humans are said to have a special role in socio-ecological systems (Westley et al., 2002). It is assumed that coupled socioecological systems represent coevolutionary units of social and ecological systems and that humans have to adapt to the local natural conditions and vice versa (Folke et al., 2002).

Brand and Jax (2007) and also Folke et al. (2003) described how the original resilience concept was extended from a concept used for describing (biological) ecosystems which included, for example, the capacity to absorb disturbances, to self-organize, and to learn and adapt. It then became more or less directly a concept of environmental management, governance, or a socioecological and ecological–economic concept. Similar conception transfers can be found in sociobiology, evolutionary psychology or human ecology (see Chapter 8.3 and Box 5.9). Critical for these conceptual transfers is that biological concepts egress in a direct way to human systems and do not meet the specific rationale of different human systems. For instance, concepts of resilience research drawn from ecological systems, such as the adaptive cycle, were directly applied to economic, social, and socioecological systems (Allison & Hobbs, 2004; Kinzig et al., 2006). This makes clear that resilience researchers were looking for an integrative approach to socioecological sciences across disciplines. This search was influenced by the book by system theorist Bateson called *Mind and Nature* (Bateson, 1979) and work by others in environmental ethics (Naess, 1989), environmental history, the emerging ecological economics (see Chapter 11.5), and anthropological research about indigenous knowledge (Berkes et al., 2003; Folke, 2006). However, resilience researchers did not differentiate between the rationale of different social systems, for instance between an individual and society or between companies or communities. The characteristics of ecological resilience were rather indiscriminately applied to purely social domains as well (see Reusswig, 2007).

We agree that resilience is certainly an interesting and important concept and that a *hybrid* definition coping with both the interaction of the social side and the ecological–environmental side makes sense. Resilience can be a lens through which we analyze socioecological systems. This has been expressed when considering ecological resilience as the underlying capacity of an ecosystem to maintain desired services in the face of a fluctuating environment and human use (Carpenter et al., 2001; Folke et al., 2002).

However, instead of formulating general laws related to the material–biophysical side as *New Ecology* did, resilience research takes a "humans-in-nature" perspective (Colding et al., 2008). In Chapter 19, we elaborate on the strengths and the flaws of this approach when we compare it to the HES approach put forth in this book. The resilience approach is a coupled-systems approach emerging from biology, which has been primarily transferred to the idea of framing social learning, governance, and collective actions on the level of communities and regions. This has been also called "adaptive comanagement" (Folke et al., 2005) and provided many studies on the history and management (deficiencies) of many communities and regions. The frameworks developed in adaptive comanagement are built on few concepts, such as "ecosystem, people and technology, local knowledge, property rights" (Berkes & Folke, 1998a, p. 15). The socioecological system research of Folke and Berkes et al. can be seen as a contribution to sustainability sciences (Berkes et al., 2003).

5.6.5 Spatiotemporal dynamics and chaos

The early idea of ecosystem research has long been guided by the idea that spatiotemporal dynamics have been very simple and followed the idea that competition approaches a stable equilibrium. The study of succession showed that, on a temporal scale, adaptive cycles, including breakdowns, have to be considered. An early paper by Alex S. Watt stressed – when referring to Tansley (1935) – that the spatial patterns are important and may show partly unpredictable dynamics and "consist of mosaic patches" (Watt, 1947, p. 2) which interact with habitat factors of soil and climate – resulting in diversity manifested in spatial and temporal patterns. Watt's research on mosaic cycles in plant ecology have been largely forgotten (Stone & Ezrati, 1996). Together with Robert M. May's (May, 1976, 1981) demonstrations that irregular fluctuations and chaotic dynamics can be frequently observed, the idea that nature can be mechanistically and linearly predicted was questioned. We discuss this and the possibilities of this challenge of complex systems research in Chapters 14.1 and 14.4.

5.6.6 Key messages

- Following research on biotic factors on succession, ecology concentrated on material–energy fluxes and then moved to the study of structure

(e.g. communities) and processes (e.g. population growth). The study of succession revealed that the physical steady state model is inappropriate and disturbances become more likely in a state of conservation (e.g. by aging)
- Disturbance ecology identifies adaptive cycles with release (breakdown, start processes, creative destruction), reorganization (growth processes), exploitation (fast growth processes), and conservation (information gain) as cycling phases
- Resilience, connectedness, and potential can be considered key elements of ecosystem processes
- New ecology defines the fundamentals of ecosystems using principles that describe ecosystems as open, hierarchically organized systems with irreversible processes that optimize usable energy (exergy) which are based on carbon-based life under certain environmental constraints
- Ecosystem functions and the more anthropocentric ecosystem services are human reconstructions of processes and anthropocentric output of ecosystems. Ecosystem services such as supporting services, provision services, regulating functions, and cultural services are examples of how humans measure the worth of natural resources
- Socioecological resilience research is based on a metaphorical integration of human systems and social systems, and focuses on community and regional management
- Sociotemporal dynamics and irregular chaotic dynamics are a challenge that can be better understood by means of mathematical modeling.

5.7 "Landschaft" as a complex adaptive system: past or future?

We should not leave the topic of biology without discussing Landschaftsforschung (Engl.: landscape ecology research; Naveh & Liebermann, 1994; Schmithüsen, 1976; Scholz, 1999; Troll, 1971). First, we review the etymological roots of Landschaft, the German word for landscape, and see how it refers to something made by man and often includes nature, gestalt, form, and planning. The Landschaft view is indispensable for environmental literacy, as most of our terrestrial landscapes are comprised of physical, ecological, socioeconomic, and cultural patterns. The North American landscape research (Wiens et al., 2006), which mainly focuses on spatial ecology, will be discussed in part only. This section also reviews the biotic, abiotic, and human components of Landschaft and how landscape ecology models allow us to analyze material, energy, and information flows. Finally, we discuss how landscape ecologists were among the first to promote a transdisciplinarity approach for sustainability research.

5.7.1 Historical and cultural roots

The earliest reference in world literature to landscape, Landschaft in German, is in the book of Psalms, where landscape is related to "the beautiful overall view of Jerusalem" (Naveh & Liebermann, 1994, p. 3). Another early reference goes back to the old Germanic "Skapjan = (German) schaffen = (Engl.) making, producing, creating, managing" and the abstract concept "skapi." Here we find two notions. One is something that is made by man. The other includes nature, gestalt, form, order, and planning (Haber, 1995). The social dimension of the Landschaft concept becomes evident as, before the twelfth century, it denoted that ambit which a member of the sedentary lower socioeconomic level could travel through on foot (i.e. perambulate) within a day's journey. The term Landschaft built thus a unit of language, market, and mating (Piepmeier, 1980; see Box 5.9). However, humans developed mental representation of Landschaft – somewhat similar as the concept of environment – as a separate, valuable unit rather late in history. This is endorsed by the fact that the first pure landscape painting was drawn in 1530 (Haber, 1995) by Albrecht Altdorfer (1480–1538).

Today, landscape ecology "focuses on the reciprocal interactions between spatial patterns and ecological processes, and is well integrated in ecology" (Turner, 2005, p. 319). The North American tradition mostly focuses on the basic bioecological understanding of structure and patterns, and includes aquatic and marine ecosystems (Haber, 1995). The European tradition is broader and emerges from biogeography and geobotany, and from the concept of the *Totalcharakter einer Erdgegend*, translated roughly from German as the "total character of an earth region," put forth by Von Humboldt (see Chapter 4.9).

We follow Bürger (1935) and Troll (1939, 1971) when we define Landschaft as a spatial entity with a specific texture and character that emerges from the interaction of its internal and external spatially defined variables and properties. And we assume that the structure and process of the bioecological system are essentially

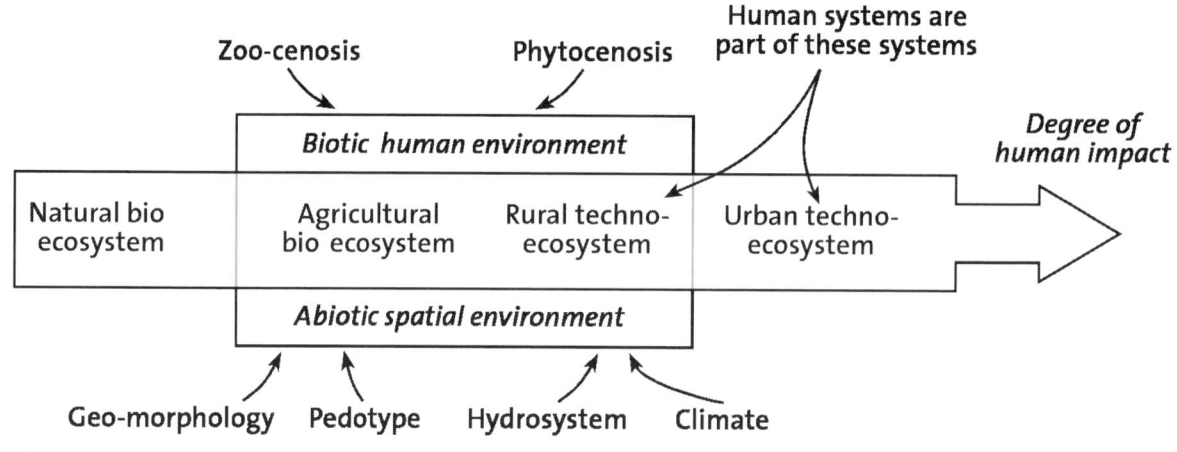

Figure 5.19 Biotic and abiotic impacts of landscape ecosystems.

formed by the activities of different human systems (see Figure 3.3*). Landschaft includes thus a (functional) constructivist purpose (see Chapter 3) and is not only a pure spatial concept but a "concept denoting a quality of consistency or texture" (Schmithüsen, 1976, p. 148, translated by Scholz). Landschaft thus also relates to humanistic, cognitive, aesthetic dimensions and to landscape architecture (Simmel, 1913). Such a view is indispensable, even from a bioecological perspective, as most of the (terrestrial) landscapes of our world are composed of "physical, ecological, socioeconomic and cultural patterns" (Wu & Hobbs, 2002, p. 361).

5.7.2 Biotic, abiotic, and human components of Landschaft

The biological subdiscipline of landscape ecology investigates how spatial patterns affect flows of materials, energy, and information and the processes that create, alter, and sometimes maintain these structures (Forman, 1995; Risser et al., 1984/2006; Wiens et al., 2006; Wu & Hobbs, 2002). The conceptual framework and the hierarchical classification of landscape ecology allows us to perform analyses on multiple scales. The classification ranges from ecotopes to microchores, mesochores, and macrochores. Ecotopes are the smallest homogeneous, holistic land units with non-excessive variations. Microchores, or land facets, are a set of ecotopes strongly related to certain properties, for example, landform (Naveh & Liebermann, 1994). Mesochores are combinations of microchores forming a reconnaissance scale, such as the North German Plains, Oregon's Willamette Valley, the Bangladesh Lowlands, or Dubai City. Macrochores are world regions such as the African savanna. Today it appears to be of utmost importance to look at the macrochores, which are the main objects of twenty-first century climate change dynamics.

Human systems, such as households, tourism, farming, trading companies, golf clubs, and armies, affect and change landscapes. To understand socioecological systems, the understanding of the biotic spatial matrix is as essential as its biotic inventory (see Figure 5.19). A critical issue for the analysis of socioecological systems such as landscapes is that conceptual and theoretical basics are missing, which makes it difficult to develop meaningful methods, models, and applications (Wu & Hobbs, 2002). Another critical question is what scientific methods and domains are most suitable. Landscape ecologists noted that general systems theory or biocybernetics, the marvel tools of the 1970s (see Chapter 14; Leser, 1991; Leser & Mosimann, 1997; Turner, 2005), are overly valuable. However, they run short as they ignore spatial heterogeneity. Landscape researchers today conceive landscapes and regional systems as partly self-organizing, humanly affected complex adaptive systems with non-linear dynamics and adaptive cycles. And the spatial patterns, as well as the actors that transform the landscape, are acknowledged (Le et al., 2008).

5.7.3 Perspectives of landscape and sustainability research

Landscape research is a genuinely interdisciplinary field in which bioecological knowledge is a key piece.

5.7 "Landschaft" as a complex adaptive system

> **Box 5.9** Sociobiological and eco-ethological adaptation: Tibet fraternal polyandry
>
> Many patterns of human behavior have evolutionary roots. Although difficult to prove, with inferences that are usually based on analog reasoning, there is much evidence that many instincts and some behavioral rules are transmitted, at least in part, phylogenetically in the course of evolution. This may explain how many behaviors that we consider to be the dark side of human behavior, such as (functional) cannibalism, inbreeding (Geist, 1978), or (military) gang-raping (Kiss Maerth, 1971; Wood, 2005), can be explained when referring to behavioral- or sociobiology. Here, the dual inheritance theory (Boyd & Richerson, 1988) should be mentioned. New findings on epigenetics (see Chapter 5.4) may shed new light, in particular on the environment of evolutionary adaptedness, which considers adaptation as a slow process. However, it is disputed whether, as some researchers suggest, there are mental modules running Darwinian algorithms optimizing fitness.
>
> However, humans have the potential to adjust their cultural and societal rules (e.g. penalties for rape). This is the object of the now-flourishing domains of eco-ecology, human behavioral ecology (Krebs & Davies, 1978), evolutionary psychology (Barrett *et al.*, 2009), psychobiology (Immelmann *et al.*, 1988), and sociobiology (Wilson, 2000). These fields of ecology measure, among other factors, differences in reproductive success between individuals in relation to behavioral strategies and their environmental–ecological setting (e.g. the size of farmland owned).
>
> "Mating is close to the heart of the evolutionary process" (Buss, 2007, p. 502), but societies' cultural rules can regulate sexual and mating behavior (see Box 2.3). To limit societal growth and societal structures in many societies, marriages are arranged by the parents. Besides the laws of marriage, inheritance law influences how some societies (see Figure 1.7*) regulate land use, eventually contributing to land fragmentation. The land belonging to a farm can be inherited by the youngest, oldest, all children in equal or different parts, or by only male or only female offspring. These land property rights form material and ecological flows and patterns. A special case is traditional Tibetan fraternal polyandry, where women usually married all brothers of a family of the same class (see Figure 5.20). This led to high percentages of unmarried females, in particular for the wealthier families who owned more land. This phenomenon is difficult to explain by sociobiological (selfish) gene arguments but its rationale is "highly materialistic … to preserve and increase the productive resource (the 'estate')" (Beall & Goldstein, 1981, p. 10). And it is obviously not targeting subsistence survival, but has to be regarded as a mechanism for reducing population growth.
>
> **Figure 5.20** A 12-year-old Tibetan bride with two of her five husbands. All of her husbands are brothers with an age between 7 and 22 years (Photo © by Thomas L. Kelly).

It serves as a center-piece of sustainability sciences as it deals with causes, processes, and consequences of land use and land cover change in a conceptually adequate manner. Earlier than most disciplines, landscape researchers gave human systems an appropriate role in sustainability research and became proponents of transdisciplinarity, as articulated by Zey Naveh, a leading pioneer of landscape research, in 1988:

> One of the major challenges for landscape ecology is … to form a unified theoretical and methodological framework for a transdisciplinary science that is oriented to both problem-inquiry and problem-solving. (Naveh, 1988, p. 23)

The HES framework put forth in this book (see Part VIII) is an answer to the call for academia and practitioners to work across disciplines to promote sustainability. Together, *transdisciplinarity* and the *embedded case study methods* (see Chapters 3, 15, and 19; Scholz *et al.*, 2006; 2002; Scholz & Tietje, 2002), are seen as our methodological center-piece. Whether the future research will be called regional science, transition management, sustainability science or receive another label is unimportant. We need scientific approaches which link and integrate the rationales of material–biophysical processes in the environment and social–cultural–epistemic processes at work in different human systems.

Given the rapid environmental changes of the twenty-first century, a special challenge for landscape ecology is certainly to take a world scale view and to cope with the entire ecosphere and biosphere (see Figure 16.3). However, the perspective derived from more than 10 large scale case studies related to landscape transitions conducted by the author of this book shows that landscape research is an invaluable research domain that can link knowledge from biology, ecology, or landscape ecology with social science research for supporting sustainable development.

> A [sustainable] natural-environmental regional and landscape research is interdisciplinary, includes the interaction of human activities with its biotic and abiotic environment, maintains complexity and holism [of landscape and regional systems], is based on historical analysis and reconstructions, organizes multiple forms of knowledge integration in a method-based form, focuses on vulnerabilities based on a holistic understanding, and organizes processes of "mutual learning" between science and society. The totality of these properties defines a new form of scientific activity, which is called *transdisciplinary research*. (Scholz, 1999, p. 36, translated by Scholz)

5.7.4 Key messages

- Landschaftsforschung (Engl.: landscape ecology research) takes a systemic approach to integrating ecological principles (e.g. of New Ecology) and human components in a historical, natural, and social science-based manner. It represents an interesting approach for research on coupled HES.

5.8 Homeostasis and homeorhesis on the micro and macro levels

> Biologists have long been impressed by the ability of living beings to maintain their own stability. The idea that disease is cured by natural powers, by a *vis medicatrix naturae*, an idea which was held by Hippocrates, implies the existence of agencies ready to operate correctively when the normal state of the organism is upset. (Cannon, 1929, p. 399)

A critical question is whether, under what constraints, and by what mechanisms biological systems maintain stability and existence. And, do the same type of regulatory mechanisms exist on both the micro and the macro levels? These questions are not new but are essential in many cases, such as when we consider the severe perturbations that our landscapes and regions (see Figure 14.1*) are facing. We will organize the discussion of the questions along the term "homeostasis," which is defined as a self-regulating process for maintaining stability of a biological system. Do large ecosystems possess the same type and organization of regulatory systems?

The ability of certain organisms to maintain a stable condition and the concept of homeostasis was first discussed by Claude Bernard (1813–1878), who considered the ability of organisms to regulate the internal setting, the "milieu intérieur," as a prerequisite of free and independent life (French: "vie libre et indépendante"; Bernard, 1878, p. 113).

For simplicity's sake, we restrict our investigation of biological processes on the micro level in the following to well researched mammals such as mice, rats, and humans. These species are also classified as regulators, as opposed to conformers (such as reptiles), which adapt, for instance, their body temperature to the environment. In regulators we can identify a large scope of homeostatic functions, including hunger, thirst, blood sugar (plasma glucose is regulated by insulin and glucagon), water poisoning (hyperhydration) of body fluids, or the ratio between naïve and memory T cell pools (Jameson, 2002; Thews *et al.*, 1999). We know that the regulation is mostly based on hormonal and nervous processes which are highly interrelated.

The regulation can be explained when we model the homeostatic control mechanism, as a control center that receives information from receptors that sample environmental information. The control center interprets these data and sends it to effectors. Depending on the signals and reactions, we discuss positive feedback if a certain activity is enhanced or negative feedback if it is depressed (see Figures 16.11 and 16.13). As research on the immune system shows, regulation is a complex process based on biochemical processes, information processing, and the interaction between control centers. Thus, individual immune system cells work as a cell system and show some similarities to the slime mold which turns from a protozoon to a multicellular animal. Also, termite mounds can be seen as governed by a kind of collective homeostatic ruling system (Turner, 2007).

5.8.1 Beyond homeostasis

But does this type of homeostasis function exist on the level of large ecosystems? Put simply, concerning this issue one can observe a dichotomy in ecology between two extreme positions: ecological organicism on the one side and ecological individualism on the other (Kirchhoff, 2007; Kirchhoff et al., 2010; McIntosh, 1995). Ecological organicism ascribes many properties to ecosystems that are typically considered to be characteristics of (single) organisms, such as unity, purposiveness, self-regulation, and spontaneity. For instance, as we have considered above, disturbance ecology views ecosystems as quite distinct unities, which self-regulate and maintain themselves by the hierarchical interactions among its components. Here, the concept of homeorhesis denotes a system's property to return to a certain trajectory of development. The view of disturbance ecology (see Gunderson & Holling, 2002) is that periods of smooth development are followed by rapid, qualitative transitions (see above). But, contrary to single organisms, which have a limited lifespan and will die, this is not the case for ecosystems.

In contrast, ecological individualism holds that there is no self-preserving system to which properties such as resilience could be ascribed, but rather that there are transient assemblages of species reacting individualistically to changes in abiotic and biotic conditions (Hengeveld & Walter, 1999; Wiens, 1984). Single species adapt to their dynamic, individually specific, more or less rapidly changing environment. However, there are no communities of species that – as a unified entity – have a fixed, homeostasis-eliciting environment to which they collectively adapt (Kirchhoff et al., 2010). There are some ecologists that champion ecological organicism. For example, Ernest and Brown use the term self-regulation to describe "the tendency of an ecosystem to maintain the approximate stability of certain properties, such as productivity, energy or nutrient flux, or biomass despite abiotic environmental perturbations or changes in biotic composition" (Ernest & Brown, 2001, p. 2119).

We can develop a feel for this if we reflect on the concept of scale. The broad-scale desertification that often follows disturbance events, such as ecosystem collapse, can be considered as a qualitative transformation that usually still maintains a biotic structure.

We have learned from different theories, such as disturbance ecology or new ecology, that, given a certain developmental level, ecosystems show a directionality towards a kind of climax state or climax community, which may exist for a certain period. Some ecologists assume that, on such a level, species maximize their intake and fitness. The ecosystem resource and energy intake will be maximized, resulting in an approximate steady state between the rates of resource supply and resource use. But even proponents of the ecosystem homeostasis comment:

> While we propose that homeostasis is indeed a general property of ecosystems, we do not claim that homeostasis should occur in all circumstances. (Ernest & Brown, 2001)

Ecologists assume that resources must be limited and that there is competitive pressure if homeostasis is at work. However, ecosystems exhibit a different type of interrelation between subsystems and have more openness than individual organisms. If environmental changes occur, new species compositions, dynamics and organizations may appear. To attempt to compare regulation in ecosystems and in individual organisms, we can assume that an ecosystem exhibits functional redundancy, coherence, networking, and coordination of control systems with many different species that are part of the ecosystem. Even then, we see that complex ecosystems draw on different regulating functions to combat perturbations than simply the regulating functions found in an individual organism.

An individual mouse or human emerges from a zygote, which then builds up a coherent, evolutionarily optimized system of different types of communicating cells, cell systems, tissues, organs, etc. None of these could live independently of the other and the genetic passport makes it difficult to become hosted in any other organism. If we consider the organs and subsystems of an organism, we notice a common organismic goal-directedness or purposeful (non-random) behavior that is teleological and is based on feedback (Rosenblueth et al., 1943). This is definitely different in ecosystems. The probability that a disturbed ecosystem recovers to a state with an identical species list as it had before a disturbance is low and depends on the number of species and thus on the complexity of the ecosystem. Thus, simple systems such as a tundra are more stable than a tropical forest (O'Neill, 2001). Here, competition among species is an essential factor. Such a competition is not a given among the organs of a normally functioning organism which builds a genuine self-regulating system. Only if the control functions derail external help, for instance by a medical doctor who administers drugs, is it necessary to correct the regulation process.

There are no regulatory functions which control, for instance, the working temperature of large ecosystems. This holds true even if we acknowledge that ecosystems are highly interdependent communities, with networks, cycling flows of energy, etc. Also, Lovelock's Gaia hypothesis (Lovelock, 1995; Lovelock & Margulis, 1974) that the living matter of Earth builds a super-organism, which modifies the planetary environment so that it can survive, has been disputed by more than one researcher (Free & Barton, 2007; Lenton & van Oijen, 2002). In addition, there is evidence for some large-scale ecosystems, such as marine systems, that the optimistic assumption of Gaia self-regulation is absent (Kirchner, 2003). The depletion of plankton, which leads to "marine deserts," can be taken as an example that ecosystems are sometimes governed by positive feedback loops which miss appropriate self-regulation.

Stability and regulation are a matter of scale. Historically, ecosystems are "intact communities of organisms that do not move as a unit, and the collection of interacting species is continuously changing" (O'Neill, 2001, p. 3278). A critical issue is that *Homo sapiens* alter the structure of the ecological earth system on a world scale in a fast and abrupt manner. From an ecological view, the human species can be considered as a keystone species operating on a global scale.

Human systems exert destructive power on ecosystems, speeding the rates of response of ecosystems disturbances, which leads to chaotic and destructive ecosystem impacts on a large scale. As such, we conclude that the regulatory functions for managing landscapes, ecosystems, regions, and the Earth ecosystem do not necessarily inherently exist, at least if the human impact is of a magnitude that it brings the world system out of the envelope of conditions that provide sufficient food and other ecosystem services. The consequence of this is straightforward: given the impacts and dynamics of anthropocenically affected ecosystems, the human mind (i.e. the socio-epistemic side of human systems; see Figure 3.3*) must reflect on what regulatory functions are missing and must be invented to keep the ecosystem within limits to sustain human life. This is a societal challenge of using environmental literacy.

5.8.2 Key messages

- Research on mammals shows that there are set point regulated processes based on hormonal and nervous input resulting in homeostatic control mechanisms
- The longstanding assumption that organismic systems or ecosystems above the level of the individual maintain homeostasis through innate healing powers of nature is not supposed to hold true in all ecosystems or organismic systems above the level of the individual organism
- Because ecosystems do not have inherent homeostatic responses to disturbances, humans need to respond to secure "ecosystem limit management" for providing vital ecosystem services. This establishes the critical need for transdisciplinary research and processes on coupled human–environment systems.

Part III

Contributions of psychology

6	Psychological approaches to human–environment interactions	137
7	Drivers of individual behavior action	190

Key questions (see key questions 1–3 of the Preamble)

Contributions from psychology to environmental literacy

- Q1 How do people perceive, conceptualize, and interpret the environment?
- Q2 What drives human and small group behavior? What are the psychological foundations of environmental awareness?
- Q3 How can models of environmental awareness and of behavioral drivers enable us to understand how human systems formulate goals, and select and evaluate strategies and actions?

What is found in this part

William James characterized psychology as the "science of mental life … its phenomena and their consequences" (for a review see James, 1890). Today, psychology is often defined as "the study of mind and behavior" (Doyle, 2000, p. 375) and the *individual* and the *small group's* activities are the primary objects. Environmental literacy raises specific interest in questions such as: what environmental aspects do humans perceive? In what way is the sensory system prepared to cognize environmental challenges properly? What mental structures are innate and which ones are (to be) learned? And what are the basic drivers of the individual and the small group?

Chapter 6 deals first with relevant landmarks of the formation of psychological theories on perception and human–environment interaction. We then look at environmental psychology. We show how cognitive and motivational structures enable humans to cope with elementary categories such as space, time, resources, or the built environment. We then present theories that explain environmental behavior and risk perception. The chapter closes with a section about group psychology. This topic takes a special place in this book and the HES framework (see key question 2 and Figure 14.1*). The small group is the next human system above the individual. We reveal that groups have their own rationale and drivers, which cannot be explained in a simple additive way from the sum of their members.

Chapter 7 introduces the key drivers of individual human behavior. We provide a conceptual map with the dimensions abstract–cognitive vs. basic–physiological and background vs. close to action processes for differentiating and relating these drivers. These processes are at work, and interact and play different roles in environmental decisions and can be used to explain environmental behavior.

Part III Contributions of psychology

Chapter 6
Psychological approaches to human–environment interactions

Section 1: Environments in psychological theories 138	6.11 Space 164
6.1 What environmental information do we perceive? 138	6.12 Time 166
6.2 Heider: the role of context and the social situation 141	6.13 Energy, waste, and resources 169
6.3 Brunswik: probabilistic functionalism in perception 144	6.14 Natural and technical disasters 170
6.4 Lewin: psychological ecology 150	6.15 Built environment 172
6.5 Hellpach: psycho-geo-ecological thought 152	6.16 Pollutants: the invisible built environment 173
6.6 Gibson: ecological theory 153	6.17 Psychological theories applied in environmental psychology 175
6.7 Barker: ecopsychology and the transactional approach 155	Section 4: Risk perception 177
Section 2: The cognitive dimension 157	6.18 Risks, hazards, and the environment 177
6.8 From gestalt to prototypes 157	6.19 Conceptions of risk 178
6.9 Piaget: genetic epistemology of environmental knowledge 160	6.20 Psychology of risk judgments 178
Section 3: Environmental psychology 162	Section 5: The rationale of a group 181
6.10 The subdiscipline environmental psychology 162	6.21 The group as a social environment 181
	6.22 Commons and public goods dilemmas 184

Chapter overview

Section 1 of this chapter focuses on how human individuals deal with environmental information and settings. We follow the history of psychology, which began investigating how the human individual perceives the environment through the senses. We progress from perceiving to storing and processing environmental information. We learn how theories in these fields are tied to fundamental assumptions about how and why environmental information is acquired. We introduce Egon Brunswik's theory of probabilistic functionalism, which is a cornerstone for the epistemology underlying this book. This theory provides basic assumptions about human–environment interactions, which are included in the HES framework (see Chapters 3, 16, and 17). Another cornerstone is Kurt Lewin's field theory.

On our journey through psychology we will repeatedly pose questions on the nature–nurture interplay (i.e. what is innate vs. what is learned), the realist–idealist dichotomy (i.e. what is real/true and what is subjectively, individually constructed), or how the human mind and behavior are adapted to the environmental setting.

We present the geopsychology, ecological theory, and ecopsychology approaches, which make different assumptions about the relationship between humans and their environment. We discuss the transactionalist concept, which is essential for the HES framework.

Section 2 introduces Jean Piaget's theory of cognitive development, in particular the concept of genetic epistemology, which relates ontogeny (i.e. the cognitive development of the individual) with phylogeny (i.e. the cognitive development of the human species).

In Section 3 we take a brief look at the emerging subdiscipline of environmental psychology. Our interest lies in understanding the psychological processes and mechanisms related to basic entities such as space, time, and resources, when interacting with the built environment or when coping with environmental disasters or pollution. The section ends by presenting acknowledged theoretical approaches of environmental psychology, such as the theory of planned behavior.

Section 4 describes how psychological research deals with risk perception and risk judgments.

Section 5 explores the rationale and drivers of a group, and how groups affect individuals. This section revisits the tragedy of the commons and relevant variables and drivers at the individual, group, and societal level for addressing this dilemma. These are issues that are also dealt with in the sociology and economics chapters and when introducing the Interference Postulate P3 (see Chapter 16.4).

Section 1: Environments in psychological theories

This section follows the history of psychology and reviews the conceptions of environment and human–environment relationships that have evolved in grand theories of perception and cognition and some predecessors of environmental psychology.

6.1 What environmental information do we perceive?

6.1.1 James and Wundt: from introspection to perception

The main camps of modern scientific psychology were built initially by the German Wilhelm Wundt (1832–1920) and the American William James (1842–1910). They shared the common conception of psychology as a natural science, and both were very skeptical with respect to the Cartesian interpretation of the matter–mind duality. Although they showed many similarities, (the early) Wundt and James differed in their theories, methodology, and conception (i.e. ontology) of the environment.

For James, the most important investigative method was introspection. James postulated that it would not make sense to distinguish between a sheet of "paper seen, and the seeing of it" (James, 1920/1895, p. 378). In other words, he contended that the relations between things are at least as real as the things themselves. This theory of knowledge, advanced by James, is known as radical empiricism. Taking it to extremes, he supposed that humans have never come across anything else except pure experience. James' position was taken up later by the first group of American environmental psychologists, the so-called transactional school (see Chapter 6.7), which assumed that the perceiver and reality are part of the same process. Physical reality is considered "the product not the cause of perception" (Ittelson *et al.*, 1974, p. 105).

However, James differentiated his extreme position by distinguishing between *percepts* and *concepts*. Perceiving is understood as a direct, unmediated, selective discovery of the real world through immediate experience. On the other hand, concepts are extracted from the immediate perceptual flow; they are "verbally fixed and coupled together" and they "let us know what is in the wind for us and get ready to react in time" (James, 1912/1976, p. 47). For him, concepts play a functional role and enable individuals to better understand the structure of the environment and the role of humans in it (Heft, 2001). One can interpret James' statement as an anticipation of the cognitive psychology view that was launched by Neisser's book *Cognitive Psychology* (Neisser, 1967) since the late 1960s. In cognitive psychology it is common to differentiate (a) perception as a process directly related to environmental stimuli or information, and (b) thinking and processing of information based on the previous perception. James differentiated percepts from concepts from a realist perspective by postulating that the truth status of concepts is ultimately dependent on how the concepts relate to the percepts. However, he argued that access to the percepts should be made by introspection.

> Introspective observation is what we have to rely on first and foremost and always. (James, 1890, p. 185)

The introspective method, which was common in the philosophy of the nineteenth century, was also a basis for his analysis of the role of the environment in the process of evolution. In 1890 James gave a lecture at the Harvard Natural History Society entitled "Great men, great thoughts, and the environment." It is interesting

6.1 Psychology and human–environment interactions

to see that he strongly attacked the pre-Darwinian conceptions of Herbert Spencer and his followers that – in his eyes – stressed the one-sided environment human direction:

> … the environment, in short, was supposed by these writers to mould the animal by a kind of direct pressure, very much as a seal presses the wax into harmony with itself. (James, 1890, p. 444)

James provided a strong plea for an interactive view – although he focused on the social environment.

> Thus social evolution is a resultant of the interaction of two wholly distinct factors, – the individual, deriving his peculiar gifts from the play of physiological and infra-social forces, but bearing all the power of initiative and origination in his hands; and, second, the social environment, with its power of adopting or rejecting both him and his gifts. Both factors are essential to change. (James, 1890, p. 448)

We should also note that James, together with Charles Sanders Peirce and John Dewey, are considered the founders of pragmatism, which is characterized by turning "towards concreteness and adequacy, towards facts, towards action and towards power" (James, 1890, p. 509). For pragmatists, the validity of a theory is proven by the practical value of its application. Finally, although James introduced experiments into American psychology, "he himself found [experiments] boring and intellectually confining" (Hunt, 1993, p. 150).

Contrary to James, Wilhelm Wundt considered experimental work to be the key method of psychological research (see Figure 6.1). This view had a tremendous impact on modern psychology. Wundt started his career in 1858 as the personal assistant of the physicist and physiologist Hermann von Helmholtz, who proved how the ear converts and transmits nerve impulses to the brain. Wundt's first experimental work followed a physical paradigm suggested by Gustav Theodor Fechner and Ernst Heinrich Weber, who suggested a law that states that the subjectively perceived intensity of a stimulus is logarithmically proportional to the physical intensity of the original stimulus (Fechner, 1860). Wundt was particularly interested in the perception of optical stimuli, and he targeted his research on psychophysiologically defined perceptual processes and brain functions. Yet, he also acknowledged that what we see is additionally dependent on psychological conditions such as arousal or evoked emotions.

As would become standard in many domains of psychological research, Wundt's experimental settings used highly reduced, artificial environments. Typical examples are stimuli from tachistoscopes, which provide a brief exposure of visual stimuli such as a series of randomly selected letters. These settings were ideal as they provided infinitely repeatable, homogeneous, reduced stimuli that could be modified in a controlled manner. The object of research was the impact of varying stimuli along one dimension (e.g. exposure time by milliseconds) on a person's response, which was

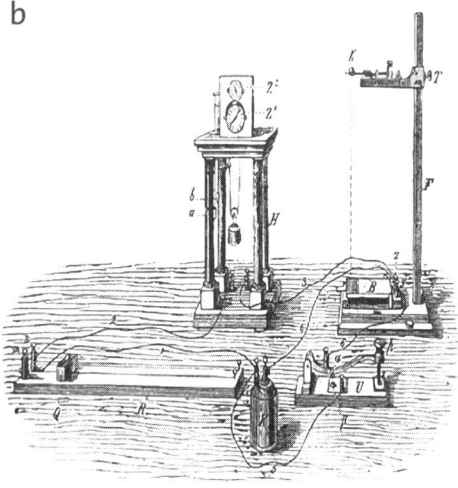

Figure 6.1 (a) The setting of psychophysiological experiments. (b) Instruments in the style of natural science served to explain the perception of environmental stimuli (Wundt, 1897/1969).

Box 6.1 Do aesthetics follow a natural law? The golden section

Some scholars have argued that, since Book VI of Euclid's (c. 325–270 BC; Euclid, 1956) *Elements*, the golden section has been supposed to be the most aesthetically pleasing visual proportion.

The golden section is a line segment divided according to the golden ratio: the total length $a + b$ is to the longer segment a as a is to the shorter segment b. Expressed algebraically: $\frac{a+b}{a} = \frac{a}{b} = \varphi$. $\varphi = 0.618$ is the negative solution of $x^2 = x + 1$ and the positive solution of this equation is the number $\phi = 1.618$.

The philosopher Adolf Zeising (1854) tried to prove empirically that the navel divides the upper and lower parts of the body according to the golden section and that all other parts relate to each other by the same ratio and applied many measurements to show that – given a certain error rate – the golden section or divine proportionality (see Figure 6.1) has been applied by many artists and architects – especially in the form of the golden rectangle, in which the ratio of the longer side to the shorter is the golden ratio. And also Zeisig believed that this proportion is aesthetically pleasing in an exceptional manner and that Leonardo da Vinci and other medieval artist should have applied this. As Schoot (2005) has revealed, there is no evidence for this assumption (see below).

Since the publication of Fechner's *On Experimental Aesthetics* (Fechner, 1971), psychologists have run countless experiments to prove whether proportions embodying the magic number $\phi = 1.618$ govern perceptual preferences. Fechner distinguished between choice (German: Wahl, for instance, pair comparison or ranking), production (German: Herstellung), and application (German: Verwendung) by examining whether objects applying the golden ratio were preferred. Some choice experiments have shown that such objects are – on average – the most preferred. Other studies, however, have failed to find any effect of the golden section on preferences. Rather, proportions such as 1:1 (e.g. the square or the equilateral triangle), 2:1, or 4:3 in landscape format, or 4:5 in portrait format are the most preferred. However, one moderating factor could be that different experimental procedures with different physical stimuli of various complexities and settings were applied. As Green, in his thorough review, concludes "I am led to the judgment that the traditional aesthetic effects of the golden section may well be real, but that if they are, they are fragile as well" (Green, 1995, p. 966). The impact of the golden section on perception – if there is one at all – is weak and cannot be compared with other psychological laws such as the Weber–Fechner law.

As van der Schoot (2005) elaborated, it was not in antiquity but rather in nineteenth-century romanticism that the golden section played a significant role. To fill the gap between nature and culture, a concept of a cognitive, basic epistemic unit such as the golden section became an attractive option. However, the structuralists' idea of a fundamental and perhaps innately cognitive element could not be empirically proven. Statements such as "We believe that the Golden Section fits the way we normally make distinctions" (Benjafield & Davis, 1978, p. 427) are oversimplified, given the complexity of human information processing.

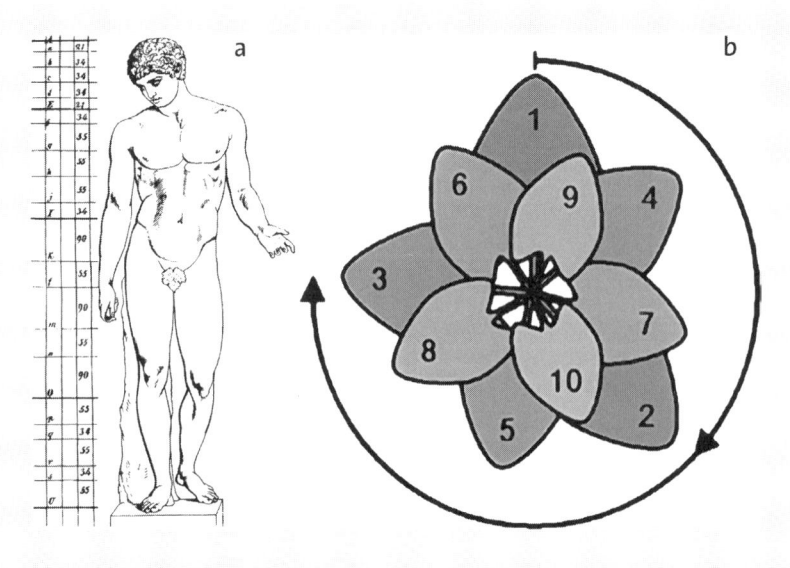

Figure 6.2 (a) The golden sections of a human being (from van der Schoot, 2005). (b) Florets and leaves of many flowers (e.g. the sunflower, *Helianthus annuus*) are arranged with an angle of 360°/Φ for optimal absorption of sunlight.

measured by simple dependent variables such as the number of letters correctly reproduced. The psychophysical approach to visual perception "is analogous to an exposure of the film in a camera, so what the brain gets is something like a sequence of snapshots" (Gibson, 1979, p. 1). Thus every physical stimulus, in a direct materialist interpretation, had its mental counterpart and vice versa.

We should note that the later works of Wundt, like James' work, also used the method of introspection as the basis of the inquiry of human mental processes. He considered the trio of sensation (German: *Empfindungen*), imagination (German: *Vorstellung*), and feeling (German: *Gefühle*) to be basic aspects of mental activity. His later work turned very general and also delved into the development of societies through cultural practices at different levels of societal development, that is, from primitive to modern society. In his research on perception and in his later writings, Wundt (similarly to James) stressed the conception of the active role of the human and introduced concepts such as volition (German: *Wille*) and self-control. This meant that individuals were supposed to actively compose elementary cognitions of complex impressions. In his theory, Wundt also avoided the separation of physical–physiological processes from mental–epistemic ones and stressed, for instance, the relation between bodily energy and mental intensity.

Nevertheless, experimental psychology, which emerged from Wundt's laboratory, followed a reductionist paradigm. Complex environments did not play a prominent role. Research focused on manipulating one or more variables of physical or environmental dimensions and investigating the individual's responses. This reductionist view was also taken for the investigation of aesthetics. Wundt's explanation of a phenomenon such as the golden section (see Box 6.1) is straightforward:

> The fact that symmetry is generally preferred for the horizontal dimensions of figures and the golden section for the vertical, is probably due to associations, especially with organic forms, such as that of the human body. (Wundt, 1897/1969, p. 166)

There are many similarities between James and Wundt. Both identified mental processes, emotions, and consciousness of a human individual as subjects of psychology. James strongly followed a functionalist, pragmatist view and stressed experience as a key factor of psychological processes. Wundt emphasized the relationship between sensory, physiological, and mental processes, and formed the discipline psychology by his experimental approach. His natural science way of experimentation became the standard in psychology, for instance in his work on perception. In this work the effects of the variation of single aspects in a *ceteris paribus* setting are investigated. However, more complex interactions of the cognitive and the environmental setting (see Box 6.2) were not the subject of empirical investigation at the beginnings of psychology.

6.1.2 Key messages

- Psychology emerged as a discipline to explain human behavior and mental processes
- James emphasized the role of experience, percepts, concepts, etc. from a functionalist perspective. He revealed the human capacity to form the material and social environment but also the strong dependence of human behavior and mental processes on the environment
- Wundt approached psychophysiological processes using controlled experiments and stressed that psychological processes have a physiological and a mental component.

6.2 Heider: the role of context and the social situation

An important step in developing a modern view on visual perception was to go beyond the analogy of the camera, which had been employed by early researchers. This occurred in the 1930s through the interaction of Fritz Heider, Kurt Lewin, Egon Brunswik, Edward C. Tolman, and Clark L. Hull. Like many psychologists of that time, Heider investigated misperception of environmental stimuli. One source of misperception was the properties of media causing perceptual noise, for instance aerosol affecting light transmission, and the interferences of the different information, for instance different beams of light, from the object (Heider, 1930a,b). Another source was sufficiency of the information. Here, Heider referred to experimental results from the Danish gestalt psychologist Edgar Rubin (1927). Rubin found that a few marked points fixed at the fringe of the inner of two rotating spinner wheels gave an illusion of circular movement. If only five (or fewer) points were visible, the impression of circular movement disappeared and was substituted by five (or fewer) pendulum images. Today there is still extensive research on

Box 6.2 The chemical and symbolic notion of odor: you stink

To understand how psychologists distinguish between perceptional and judgmental–cognitive processes (Cialdini & Trost, 1998) we consider how humans sense smell.

The human olfactory receptors are limited (Figure 6.3). Humans are certainly able to detect and to identify odors from rotting flesh, feces, vomit, or smoke, but we fail to detect other lethal chemicals, such as carbon monoxide. Whether people consider a certain smell pleasant or unpleasant differs and depends on the situational constraints. Androstenone, a hormone that is found in male urine and axillary perspiration, is described as having an unpleasant odor by some, whereas others (e.g. certain females) judge it as having a pleasant odor (sweet, floral). Furthermore, a fraction of the population cannot detect its presence at all (Araneda & Firestein, 2004). Men unconsciously avoid seats that have been sprayed with this substance (Kirk-Smith *et al.*, 1980).

However, odors are not only natural or manufactured; they can also be symbolic. Just as colors can have some meaning, certain societies, such as the Brazilian Suya, "think in smell, whereas we only react to smell" (Howes *et al.*, 2007, p. 2). Clearly, humans can distinguish body odors that depend on the other person's food consumption, life cycle stage, sex, illnesses, etc. Sense of smell differs interpersonally and is learned. But what and how much is learned depends on the environment.

Acknowledging the above facts, the Harvard anthropologist Constance Classen elaborated that the "belief that each race has a distinct odor" is "a scientifically untenable proposition" (Classen, 1992; Howes *et al.*, 2007, p. 1). But in a careful anthropological analysis, she elaborated that "you stink" is mutually attributed by races and social groups. Olfactory symbolism is used as a means of expressing and regulating cultural identity and distinction. For example, how black people think that white people have a disagreeable odor and vice versa (Dollard, 1937; Klineberg, 1935), and how indigenous tribal groups in Columbia mark their territories similarly to predatory animals can only be explained by their symbolic functions. Thus, the sense of smell has both a material, realist component and a constructivist, socio-cultural, symbolic component. The effect that believing is smelling (Cassidy, 1997) has been experimentally proven, for instance by O'Mahony (1978).

Figure 6.3 Humans' ability to sense the environment through, for example, olfaction is limited. (a) Many colorless, tasteless, odorless, and lethal gases (such as carbon monoxide) cannot be perceived. (b) Smelling is also part of communication, as the sweat ritual of the Kalahari tribes shows (Eibl-Eibesfeldt, 1989, p. 429). The capacity to detect various smells differs between persons. As the phrase "it/you stinks" indicates, perceived odor is often a result of what is subjectively constructed and thus has a socially regulatory function not only based on chemical stimuli.

similar optical illusions, mainly focusing on the constraints under which the visual system produces a certain environmental image when specific, separate information is integrated (see Adelson & Movshon, 1982; Stoner *et al.*, 1990). In their experiments, these researchers show that a few interfering entities can disaggregate the perception of geometrical information (Stoner *et al.*, 1990).

To explain these optical (mis-)perceptions, Heider (1930a, see Figure 6.4) differentiated between: (1) the real, objective, relevant stimuli, known as distal variables, D; (2) the medium, including the interfering or context stimuli, and the initial information which is directly linked to information uptake, known as proximal cues and which become subject of a mediation, V; (3) peripheral and near to stimuli information processing; and (4) the final cognitive representation, D' in the central part of the cognitive system which embeds and builds up experiences (Wolf, 2003). In original terms, Heider's theorizing is reminiscent of the functional circle of von Uexküll (see Figure 4.25a, Box 4.11).

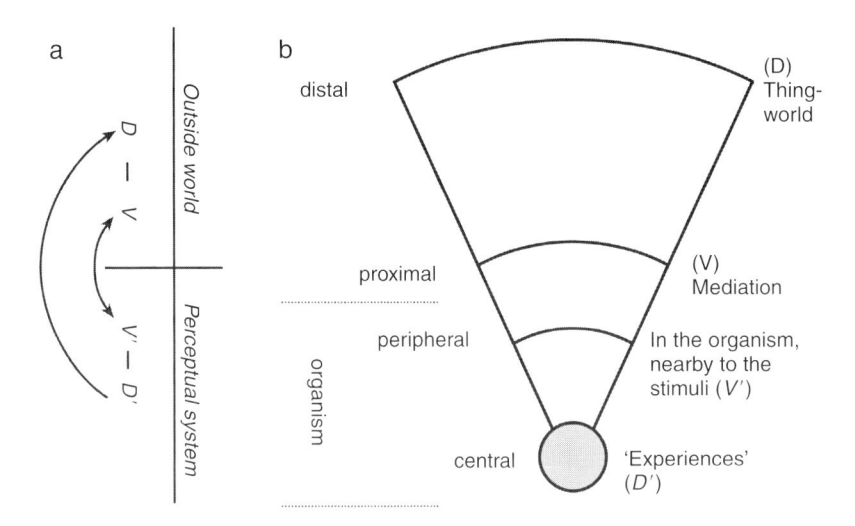

Figure 6.4 (a) Heider's model of how an organism gains access to reality (adapted from Heider, 1930a, p. 381) and (b) Brunswik's conception (adapted from Wolf, 1992, p. 23).

Heider also referred to the optical misperception presented in Figure 6.5a when stressing: "the reason for the size deception is painted around the men" (Heider, 1930a, p. 384). Thus, he concluded, the performance of the environmental system depends on the structure of the surrounding information. Heider's guiding question was: under what conditions does an individual have adequate perception and cognitive representation of the environment?

We should also note that the distinctions made in Figure 6.4 not only differentiate between processes within the organism and the outer world. It also includes the distinctions between peripheral, intuitive cognitive processes which work with visual and conceptual representations that are close to the stimuli and objects, and central, analytic processes. The latter are based on experience but also use abstract concepts. These ideas meet the intuitive–analytic thought complementarity (see Box 2.5, Scholz, 1987) or the elaboration likelihood approach which states that central processing is only at work if there is some (cognitive) need for this more expensive processing (Cacioppo et al., 1986; Petty et al., 2005).

Later on in his career, Heider was concerned with social perception. He elucidated how people explain the behavior of others and introduced the postulate of laypeople as naïve psychologists. This concept is rooted in the assumption that laypersons, similar to scientists, search for reasons (i.e. cause–impact relationships). If, for instance, a mother observes her daughter's teacher picking up a piece of paper from the sidewalk and throwing it into a garbage can, she most likely attributes the action to personal attributes (Heider, 1958) such as "this person is very orderly" instead of situational reasons such as "surely there has been an anti-littering campaign at school and so he must pick up the litter." Thus, Heider supposed a certain type of environmental illiteracy in the sense that people underestimate the power of situational constraints and overestimate personal characteristics. This tendency is called the fundamental attribution error, a pattern of causal attribution that can be found at least in individualistic cultures (Nisbett & Masuda, 2003).

6.2.1 Key messages

- In the 1950s psychologists such as Heider and Brunswik (see Chapter 6.3) realized that the medium between the distal and the proximal stimuli matters and may cause perceptual noise. Context factors can essentially affect perception
- Distinguishing between distal, proximal, peripheral, and central stimuli or processes orders human–environment relations in terms of environment (thing world), perception, and (peripheral) intuitive processes based on object-related concepts and experiences, including (central) analytic processes
- The social situation also matters. When explaining behavior, people in western societies tend to underestimate the power of social situations (i.e. what is socially expected in the situation) and to overestimate personal goals or intents. The expected social norms as well as group pressure affect judgment and behavior.

Figure 6.5 (a and b) The context is an important aspect of perception. Unnatural textures of the environment can cause misperception (see also Figure 6.12). Thus some perceptional psychologists, such as Brunswik, claim that the experimental setting should be representative for the real life settings.

6.3 Brunswik: probabilistic functionalism in perception

Egon Brunswik studied engineering sciences, mathematics, and physics. His interdisciplinary dissertation on *Strukturmonismus und Physik* (Brunswik, 1927) was supervised by the psychologist Karl Bühler and the mathematician Moritz Schlick, who was one of the founders of the Vienna Circle. Brunswik thus came in contact with logicism and probabilism, which were developed by researchers such as Rudolf Carnap, Otto Hahn, Friedrich Waismann, and Kurt Gödel, among others. Brunswik's research on perceptual and cognitive performance was strongly shaped by the principle that a thorough understanding of an object as well as the possibility of gaining information about the object is a necessary condition for psychological reasoning. (This statement includes – in principle – the essence of the Environment-first Postulate P1 of the HES framework in Chapter 16.) The message of his thesis was that the analysis of form and its isolated sensation, as well as the gestalt psychologists' pure focus on form, could not explain perceptual performance:

> Whereas one-sided sensualism was not capable of seeing the forest for the trees, structural monism cannot see the trees for the forest. (Brunswik, 1927, pp. 5–6, translated by Scholz)

Thus his theory focused on linking the parts to the whole. To understand the tremendous scope of Brunswik's theory, one should know of his collaboration with the American psychologist Edward C. Tolman, who – although being a neobehaviorist – provided fundamental ideas to cognitive psychology. For instance, he hypothesized that even rats must build cognitive maps of their environment to find their way through labyrinths (Tolman, 1948). Further, like Brunswik, Tolman was a functionalist who assumed that both rats and men show purposive behavior (Tolman, 1933).

Brunswik's theory of probabilistic functionalism was barely understood or acknowledged during his lifetime but has been extolled over the past two decades. Since the *Theory of Probabilistic Functionalism* can be considered a basic epistemological theory of the proposed framework of human–environment interactions, we present the principles of this approach in some detail. Probabilistic functionalism is shaped by a few fundamental principles:

(1) *Functionalism*: according to functionalist thinking, humans and organisms are purposive, intentional, goal-directed beings. As Goldstein and Wright (2001) point out, Brunswik's functionalism went beyond the classical evolutionary functionalism, which claims that people objectively perceive what is good for them. Brunswik acknowledges that in gaining access to object information, the individual must fight through the outer world or medium to the true distal cues of the "thing" world D (see Figure 6.4). In this process one can question: "Does our personality determine what we see?" (Hammond, 2001b, p. 123). Different notions and facets of functionalism recognized that both the distal – objective and

the motivational – subjective aspects are decisive in perception (Brunswik, 1949). Thus, Brunswik considered interest, as well as motivational and differential (i.e. personality) variables, as salient issues for the perceptional and other psychological processes.

(2) *The complementarity of human and environmental systems*: in essence, the organism and the environment are seen as separate systems, each with its own rationale (i.e. causation system). Further, Brunswik emphasized that psychological research has to deal with objective relations (Brunswik, 1937) and that the analysis of the object must precede an individual's psychological processes (Brunswik, 1936).

The following two principles emerge from the idea that perceived environmental information has a "lack of univocality" (Brunswik, 1943, p. 255).

(3) *Probabilistic information acquisition*: in many papers, Brunswik stressed the limited, arbitrary, erratic, and partly distorted process of acquiring object information. It is sometimes "not ... foolproof" owing to the "ambiguity in the causal texture of the environment" (Brunswik, 1943, p. 256). Thus the emergence of distal stimuli and the received proximal environmental information are supposed to be probabilistic in nature (see stray causes in Figure 6.6).

> Since the establishment of veridical distal environment relations is contingent upon the trustworthiness, or statistical validity, of cue-to-object relationships, and since this "ecological" validity is in turn essentially limited by the erratic nature of the environment, attainment of distal variables can never be better than probable. Environment-oriented objective functionalism is thus necessarily "probabilistic functionalism". (Brunswik, 1943, p. 58)

We should note that ecological validity in this context is defined as the extent to which the distal cues are reflected by the proximal cues.

(4) *Vicarious mediation*: the principle of vicarious mediation is "one of the most fundamental principles, if not *the* most fundamental principle, of behavior" (Brunswik, 1957, p. 22). In discussions at the groundbreaking Colorado Colloquium in 1955, Brunswik even denoted vicarious mediation as a "definiens" of psychology. Vicarious mediation is thought to be at work when one faces unreliable cue sampling. It also serves when an important cue or part of the distal object information is missing (see Box 6.3). Vicarious mediation launches substitutability to be a key aspect of robustness or resilience. This is elucidated in the following definition

> ... the environment is a causal texture (*Kausalgefüge*) in which different events are regularly dependent on each other. And because of the presence of such causal couplings (*Kausalkopplungen*), actually existing in their environments, organisms come to accept one event as a local *representative* (*Stellvertreter*) for another event. (Tolman & Brunswik, 1935, p. 17)

(5) *Evolutionary stabilization of perception*: this principle is about how human systems learn. Brunswik considers the product of the perceptual system as achievement. An achievement can be good or bad depending on the degree to which perception matches the object being observed. As can be taken from Figure 6.6, evolutionary stabilization is also probabilistic, and sometimes wrong or inadequate actions are rewarded and right or adequate ones are punished. But, in the course of a life, a stabilization takes place over time, rewarding appropriate actions.

(6) *Representative design*: Brunswik considered the dominant statistical technique of psychological research, Fisher's multivariate factorial experimental design (Brunswik, 1952, p. 30), as inappropriate. He argued that human performance cannot be measured in reduced, decontextualized laboratory environments which only manipulate single variables. Instead, he suggested systematically experimenting with representative samples of realistic situations to investigate "... one of the most important principles of functional theory ... vicarious functioning" (Brunswik, 1955b, p. 215). Thus, the researchers accompany the subjects into their natural environment rather than introducing the subjects into an artificial environment (Gigerenzer & Murray, 1987).

Thus Brunswik suggested that the natural environment, or at least the environment in which hominids and *Homo sapiens* evolved, was an environment that offered many "probable partial causes" and many "probable partial effects" (Hammond, 2001a, p. 55). Adaptation to a probabilistic world requires

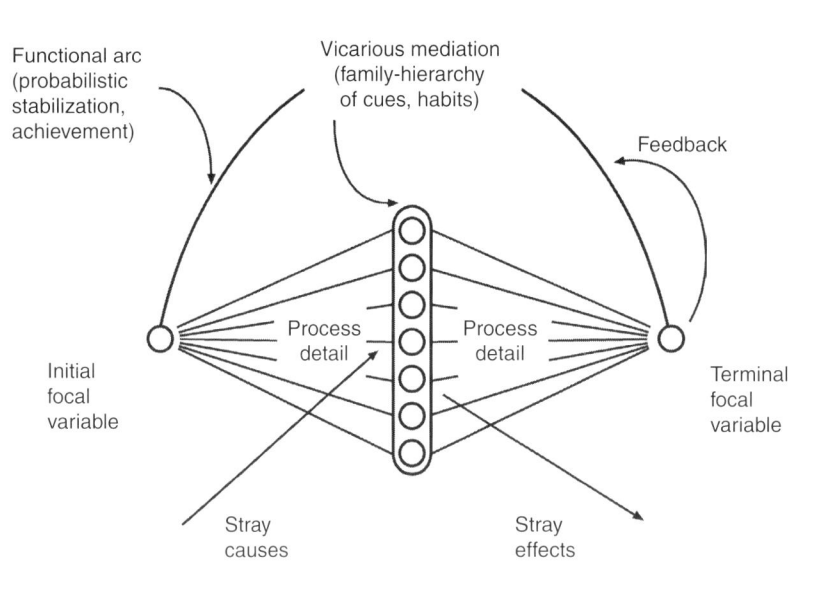

Figure 6.6 Brunswik's lens model in its original shape (Brunswik, 1952, p. 20).

Labels on figure: Functional arc (probabilistic stabilization, achievement); Vicarious mediation (family-hierarchy of cues, habits); Feedback; Initial focal variable; Process detail; Process detail; Terminal focal variable; Stray causes; Stray effects.

that the organism learns to employ probabilistic means to achieve goals and learns to utilize probabilistic, uncertain evidence about the world. Brunswik was interested in studying the conditions of successful "better than chance" achievement of an organism because of relationships between this "organism and variables in its physical environment" (Brunswik, 1943, p. 255).

> By physical environment we mean the "geographical" surroundings as well as the stages along the "historic" axis of the organism, that is to say, its past and future. (Brunswik, 1943, p. 255)

Brunswik's most encompassing and well known representation of his theory of probabilistic functionalism is the *lens model* (see Figure 6.6). On the left side is the initial focal variable, which is a part of reality (e.g. an entity of an object). For this entity an individual constructs a mental image, called a terminal focal variable, which may or may not be a "valid" representation of the initial focal variable. According to Brunswik, a valid representation is good enough for the perceiver to function successfully in its environment. The initial focal variable may be the size or quality of an environmental entity (e.g. is a dog big, does it have teeth, is it wagging its tail?), whereas the terminal focal variable may be a judgment about this entity (e.g. this dog is dangerous). The *perceptors* (i.e. the circles in the middle of the picture) represent the entities of the human system that receive and transform stimuli (e.g. the cones and rods of the retina). The arrays from the initial focal variable to the perceptors are the transmissions from the "thing" world to proximal stimuli.

The environment and the human system are complementary (see (2) above) and are linked by stray causes. Figuratively, the distal stimuli are the left tail and the proximal stimuli the perceptors. The relation between the distal and the proximal information is the stray causes since their information is probabilistic by nature (see (3) above). Uncertainty and unequivocalness also underlie the information processing on the right side of the lens, which represents the cognitive or epistemic processing between the peripheral and the central part in the organism. Vicarious mediation (see (4) above) is the capability of the perceptors to substitute missing or insufficiently represented information from another perceptor (see Box 6.3).

Finally, in Figure 6.6 the process of evolutionary stabilization (see (5) above) is represented by the functional arc.

The lens model includes the processes of decomposition and synthesis. A critical point of many applications of the Brunswikian lens model is that these two processes are mostly modeled only by means of linear regression. We think that this is certainly not the message of the principles of probabilistic functionalism, although Brunswik did not discuss interactions of the receptors. As the example in Box 6.4 shows, interactions among the receptors are an

6.3 Brunswik: probabilistic functionalism

Box 6.3 The wonder of crisp flowers: the probabilistic nystagmus

Human visual perception is organized by a system considered incomplete according to biophysical and neurological information transmission criteria. The lens of the human eye lacks full symmetry. The refraction of the light rays is suboptimally governed by the ciliary muscles, which accommodate and thus change the shape of the lens, depending on the distance of the terminal focal variable (in Brunswikian terms) from the eye to keep images in focus on the retina (see Figure 6.7a).

There are two types of photoreceptors on the optical retina – rods and cones. Cones provide high visual acuity and color vision. Rods, however, lack these qualities but allow for vision in dim light conditions. The reception on the retina is incomplete. There are, for instance, only about 6 million unevenly distributed cones and 120 million rod cells. Furthermore, the human eye has an optic disk where the optic nerve leaves the eye, which creates a blind spot (see Figure 6.7a). Altogether, given the steps described so far, Figure 6.8c could result. Finally, our eye utilizes saccadic eye movements for a visual sampling of the environment (Martinez-Conde et al., 2004). These eye movements may be interpreted in the context of the probabilistic information sampling postulated by Brunswik.

In understanding the enormous mediating and constructive work of the part of the brain responsible for the visual system, we should acknowledge that the 1.2 million nerve fibers connecting the eye to the brain produce many "stray causes" by slow, faulty, and dissipating chemoelectric transmission processes when transmitting the neural impulses to the brain.

The artist Ueli Kleeb (1997) used a computer to simulate the progress of light rays from the moment that they impinge on the cornea to reception on the retina, which functions like the film of a camera. Figure 6.8 represents a composite version with six stages. Figure 6.8a represents the object (i.e. in Brunswikian terms, the terminal focal variable), a bouquet of flowers, which is – if the reproduction is immaculate – clear, complete, and crisp. Light enters the eye from an external medium, such as air or water, and travels through the cornea into the anterior chamber (filled with the aqueous humor, a jelly-like liquid) before it passes through the lens and the vitreous humor. From an optical point of view, this process is far from perfect. Figure 6.8b presents the visual field. Figure 6.8c additionally simulates the nose in the visual field (on the left side of Figure 6.8c, the reader may experience it by closing one eye), as well as the lost crispness through the lacrimal gland and the blurring and biasing of astigmatism. In Figure 6.8, the top-down right–left retina projection is simulated (mind the flip of the picture). Figure 6.8d includes the blind spot (white arrow) and the focused crispness by the fovea centralis. Figure 6.8e illustrates the effects of the limited number of cones and rods and of the ganglions. Finally, Figure 6.8f demonstrates two effects, namely the boundary contrast amplification and the probabilistic information sampling caused by a micronystagmus that shifts the proximal stimuli about 50 times per second by five to ten photoreceptors (i.e. rods and cones).

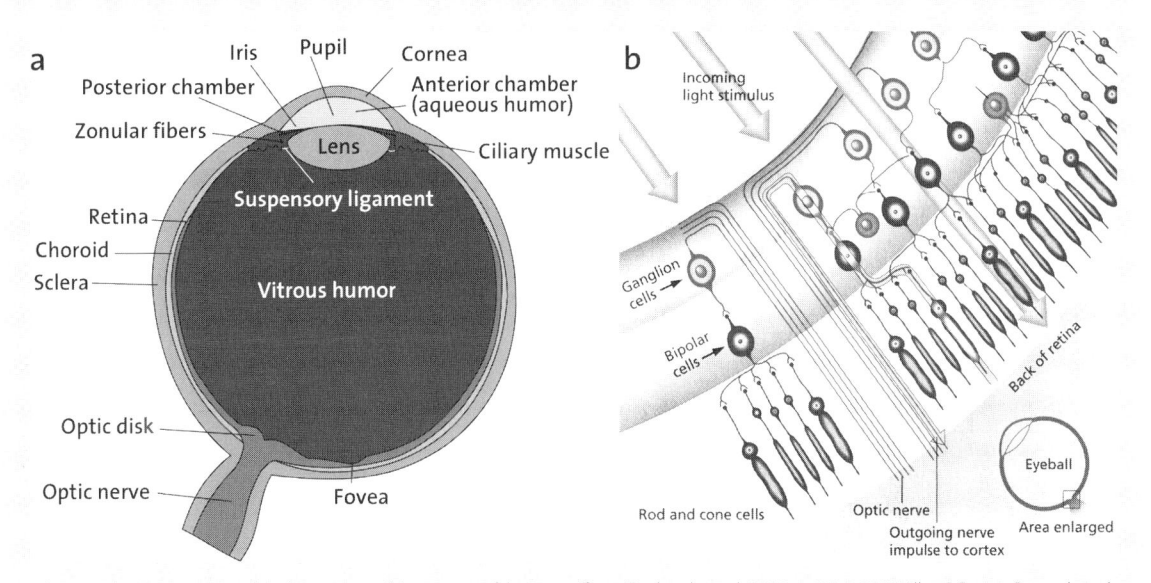

Figure 6.7 (a) Structures of the human eye; (b) anatomy of the retina (from Zimbardo et al., 2005, p. 189, © 2006 Allyn & Bacon. Reproduced by permission of Pearson Education, Inc.). Incoming light stimulates rods and cones after passing ganglion cells and bipolar cells.

Box 6.3 (cont.)

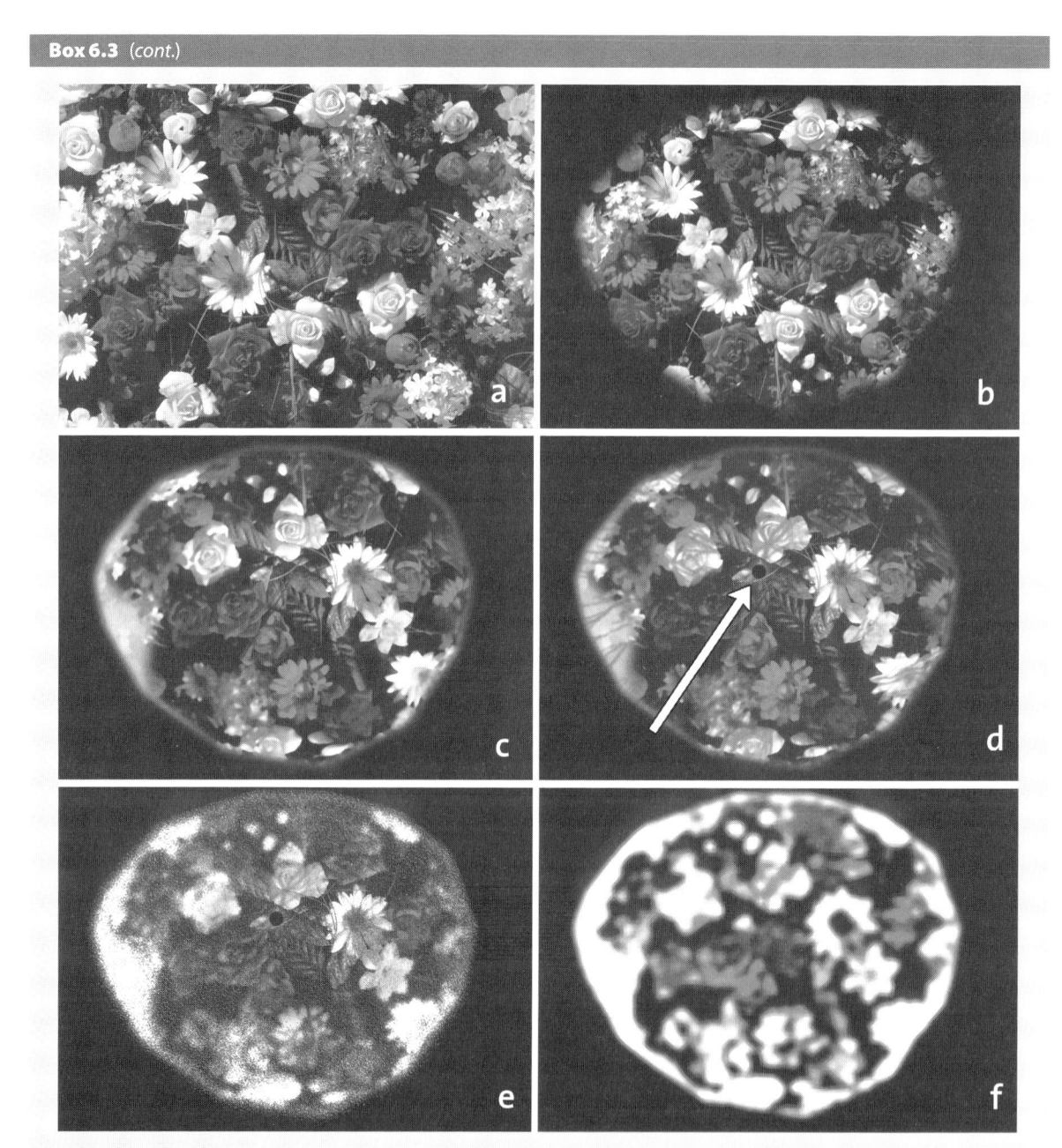

Figure 6.8 Steps of information loss in the process of visual stimuli in the human perceptional system (constructed by Kleeb, 1997).

important aspect of human information processing. As one can infer from neurophysiological studies on the interaction of photoreceptors on the retina, actual interaction on the retina goes far beyond a linear integration. Thus, vicarious mediation certainly includes complex networking.

The theory of probabilistic functionalism is by no means a theory limited to seeing or to perception but is an encompassing, epistemologically, and methodologically substantiated framework of human–environment interactions. There have been many applications of the framework to social judgment theory, evaluation

6.3 Brunswik: probabilistic functionalism

Box 6.4 Anticipating environmental information: networking "Ferrari cones"

A flash of light evokes neural brain activity with a delay of 30–100 milliseconds, a delay that, while seemingly small, is largely caused by the slow transmission of information of the visual system. The critical question is then how is it that we can still see a Ferrari speeding by at 300 km/h or a baseball at its actual place in its trajectory? It was long believed that the visual cortex constructs corrections for movement, an assumption that was later countered in a study by Berry *et al.* (1999). They showed that the process of anticipation actually begins in the retina. Working with isolated retinas of rabbits and tiger salamanders, they found that " [a] moving bar elicits a moving wave of spiking activity in the population of retinal ganglion cells" (Berry *et al.*, 1999, p. 334, see Figure 6.9), which does not lag behind the visual stimulus. In fact, the cells began firing shortly before the bar swept over them. Although such anticipating processes could also take place in the brain or in motor cells, an interaction was already shown to occur in the sensory cells. The same process can be assumed for humans. Therefore, we can assume that animals and humans anticipate the future position of an object via a networking of their photoreceptors. Anticipation by extending the perceptual field already takes place at the level of the photoreceptors.

Figure 6.9 When a moving bar with a speed of 0.44 mm/s was flashed across the receptive field center of (b) tiger salamanders or (c) rabbits, retinal ganglion cells fired in anticipation (Berry *et al.*, 1999, p. 335, reprinted by permission from Macmillan Publishers Ltd).

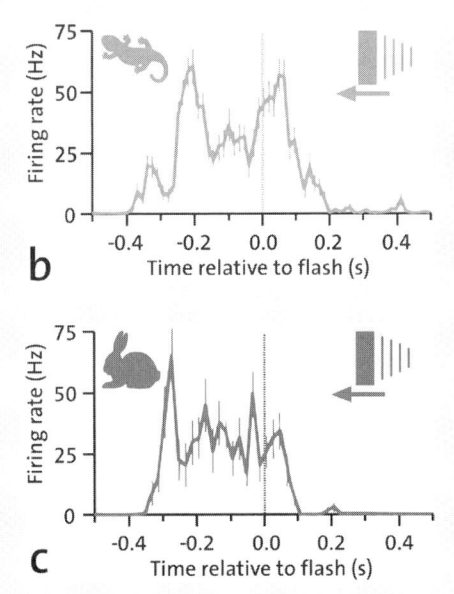

tasks, group decision-making, the construction of scenarios of complex environmental systems, and processes of knowledge integration in theory–practice cooperation (see Figueredo *et al.*, 2006; Fischer & Stadler, 1997; Hammond & Stewart, 2001; Scholz, 1999).

6.3.1 Key messages

- Perceptual and cognitive processes include probabilistic components. Brunswik's theory of probabilistic functionalism suggests that the individual copes with the uncertainties by vicarious

mediation and evolutionary probabilistic stabilization of the products of perception, which he called achievements
- The theory of probabilistic functionalism can be applied to many activities of human systems
- Brunswik's theory of probabilistic functionalism provides basic assumptions and principles such as functionalism, the human–environment complementarity, probabilistic information acquisition, substitutability (i.e. vicarious mediation) to gain a sufficient representation of the environment, learning by evolutionary stabilization both on the ontogenetic and phylogenetic levels, and representative design (i.e. measuring performance in real-world settings and not in reduced environments). These principles are essential for the theory building in HES in Part VIII of this book.

6.4 Lewin: psychological ecology

6.4.1 Field theory

Kurt Lewin (1890–1947) developed the theoretical concepts of *field theory, psychological ecology*, and the term "action theory." Field theory is the origin of social psychology. It integrates personal dispositional variables and environmental variables to explain behavior (see Box 6.5).

> In principle it is accepted everywhere that behavior (B) is a function of the person (P) and the environment (E), $B = f(P, E)$ and that P and E in this formula are interdependent variables. (Lewin, 1951, p. 25)

Lewin analyzed human behavior in relation to the field (i.e. the totality of coexisting physical and social parts of the environment at a given time). This is expressed in the above formula and has also been called psychological ecology. Statistically, the term "interdependent" from the formula means that the person and the situational variables can interact. This means, for instance, that extraordinarily high motivation can lead to high performance when coping with simple problems but can be a handicap when dealing with difficult, complex problems. Although the term "interdependent" indicates that neither a social nor a physical reductionist view should be taken, Lewin concentrated on the *social* environment and rarely referred to other environments, such as climatic conditions or the built environment.

As his wife described, Lewin aimed to provide an "applied psychology" and "… was so constantly and predominantly preoccupied with the task of advancing the conceptual presentation of the social-psychological world …" (Weiss Lewin, 1948/1997, p. 10). Nevertheless, he incorporated physical stimuli into the field concept, and even stated that "… the problem of 'time-field-units' … " is as important as that of the "… time-space-quanta for modern quantum physics," the key science of the first half of the twentieth century (Lewin, 1951, p. 79). However, with respect to the physical environment, he concentrated on what we call the immediate environment. Whatever was hidden behind the door and was not within reach of the individual's senses was not of interest (Lewin, 1943/1997).

Like Heider and Brunswik, Lewin emigrated from Germany to the USA in the 1930s. He studied anti-Semitism and the problems of minorities, such as Asian, Jewish, and Catholic people, and the poor. His observations focused on undemocratic group structures, critical food habits, and other societal deficits, such as insufficient education of lower class children in the USA. Based on his "urgent desire to make use of his theoretical insights for the building of a better world" (Weiss Lewin, 1948/1997, p. 10), Lewin (1943b) also became interested in "planned change" and "social engineering." The method of action research was thought to be a means of bringing science into society. It was described as "a spiral of steps, each of which is composed of a circle of planning, action, and fact-finding about the result of the action" (Lewin, 1946, pp. 35 and 38; see also Kemmis & McTaggert, 1988).

Lewin used the term psychological ecology when studying the person–situation–behavior relationship (Barker & Wright, 1949). This led to ecopsychological research. We note that there was a fundamental dispute between Lewin and Brunswik, which is of special interest for research on environmental literacy. Although Lewin was the founder of social psychology and related his field theory to the real-world setting, Brunswik reproached Lewin for investigating only the subjective, mental image that represents the "post-perceptual, and pre-behavioral" (Brunswik, 1943, p. 266) part of psychological processes. Lewin rejected this critique to a certain extent when acknowledging that "the 'stimulus distribution' on the retina or other receptors [was] enforced by physical processes outside the organism"

Box 6.5[1] Behavior is a function of the person and the environment: field theory

How can we use the knowledge about "personality, social relations, cognition, and motivation for the purpose of understanding, guiding, or predicting the behavior of any given individual" (Lewin, 1946/1997, p. 337)? How do needs and goals change depending on the situation? In what way does the environment change behavior, intelligence, and perception of the individual? These are key questions to understanding environmental behavior. This is expressed in the formula $B = f(P, E) = f(LSp)$, conceiving behavior (B) as a function of the life space (LSp) a composite of the person (P) and the environment (E). By conceiving the environment as a dynamic field and by making reference to physics, Lewin established this field theory approach, which focused on "the totality of coexisting facts which are conceived of as mutually interdependent" (Lewin, 1951, p. 204).

"The task of explaining behavior becomes identical with determining the function (f) which links behavior to the life space" (Lewin, 1946/1997, p. 338). Clearly, the critical questions lay in practical research on environmental behavior. So, what variables are of interest? Psychology has developed a set of theoretical variables that represent social and cognitive/epistemic factors as well as situational factors. Figure 6.10 differentiates between variables that are predominantly utilized by *psychologists* (upper part of the boxes) and by *sociologists* (lower part of the boxes). Psychologists focus on variables affecting cognitive-behavioral processes (arousal or values) and situational variables that affect behavior (built environment, density), whereas sociologists deal with variable structures and classification of sociodemographic aspects (income, social status) or sociocultural structures (cultural environment, reference groups).

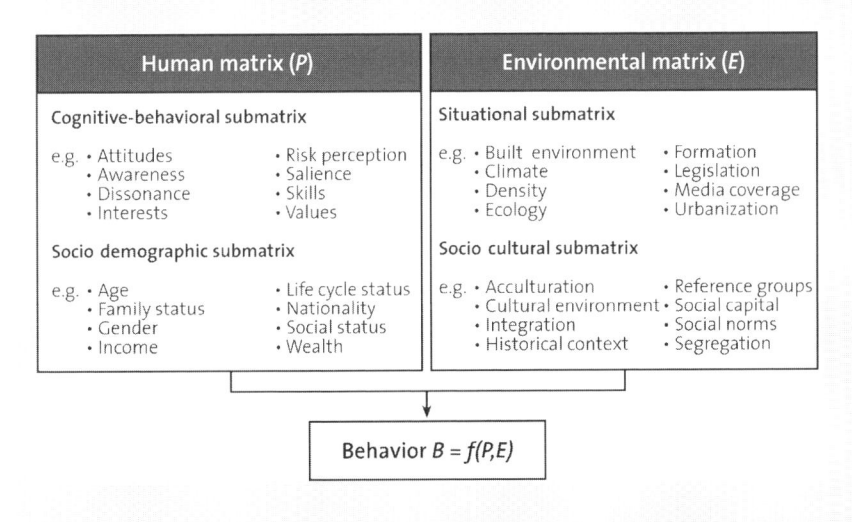

Figure 6.10 Important human and environmental variables that drive and affect environmental cognition and behavior.

Human matrix (P)

Cognitive-behavioral submatrix

e.g.
- Attitudes
- Awareness
- Dissonance
- Interests
- Risk perception
- Salience
- Skills
- Values

Socio demographic submatrix

e.g.
- Age
- Family status
- Gender
- Income
- Life cycle status
- Nationality
- Social status
- Wealth

Environmental matrix (E)

Situational submatrix

e.g.
- Built environment
- Climate
- Density
- Ecology
- Formation
- Legislation
- Media coverage
- Urbanization

Socio cultural submatrix

e.g.
- Acculturation
- Cultural environment
- Integration
- Historical context
- Reference groups
- Social capital
- Social norms
- Segregation

Behavior $B = f(P, E)$

---

[1] This box has been written together with Robert Bügl.

(Lewin, 1943a, p. 307). However, Lewin differed from Brunswik in that the uncertainties, dynamics, and the probabilities included in our outer material world were not a part of his research. This is nicely illustrated by the following quote:

> If an individual sits in a room trusting that the ceiling will not come down, should only his "subjective utility" be taken into account for predicting behavior or should we also consider the "objective probability" of the ceiling's coming down as determined by the engineers. To my mind, only the first has to be taken into account, but to my inquiry, Brunswik answered that he also meant the latter. (Lewin, 1943a, p. 308)

6.4.2 Experimental action research

Kurt Lewin was the founder of action research. He investigated social problems, such as the question of why children from certain US minorities do not benefit sufficiently from the educational programs, using case studies. Methodologically, Lewin followed a classical positivist approach. He considered the hypothesis-oriented experiment as *the* tool to empirically investigate causality in the frame of *erklären* (i.e. causal explanation; see Box 2.6), also in a complex real-life setting. Thus, his approach was called experimental action research (Cassell & Johnson, 2006).

Lewin also used the term "social engineering" as "basic social research" dealing with "different types of questions (Lewin, 1943/1997), namely the study of general laws of group life and the diagnosis of a specific situation" (Lewin, 1946, p. 36). Later, the term action research played a frequently changing, multifaceted role.

Cassell and Johnson (2006) revealed that in the past 60 years there have been completely different epistemologies and "flavors" behind these different approaches to action research. Besides Lewin's experimental action research, they also identified the following types of action research. One is "inductive action research practices," targeting *verstehen* (i.e. understanding) and thick descriptions (Geertz, 1973) of patterns of subjective meaning based on qualitative, interpretative methods. There are variants of "participatory action research," "where the researcher usually works in a consultancy role to corporate elites" (Cassell & Johnson, 2006, p. 796; see also Park, 1999) and variants of participatory action research in which researchers aspire to "an authentic dialogue with the educator/action researcher where both should 'equally know subjects'" (Cassell & Johnson, 2006, p. 799). The latter variant aspires to give voice to the disenfranchised and can establish transdisciplinary processes (see Chapter 15). Finally, there is "deconstructive action research." This approach refers to a postmodernist perspective and considers democratic consensus as an outmoded and suspect issue. It is only seen to be beneficial because it helps in understanding and deconstructing unpleasant settings. The reader may find a discussion of Kurt Lewin's action research in Chapter 15.4.

6.4.3 Key messages

- According to Lewin's field theoretical approach, individual behavior can only be described, explained, and predicted when there is understanding of the interaction between the personal variables (e.g. goals, preferences, motives) and the situational properties (e.g. built environment)
- Action research has been developed as a research methodology to investigate behavior in complex social settings and to contribute to coping with real-world problems
- Kurt Lewin's experimental action research has been the cradle of action research. Other variants are discussed in Chapter 15.

6.5 Hellpach: psycho-geo-ecological thought

Hellpach invented geopsychology and studied the effect of urban environments on humans, experimenting with factors from sunspots and lunar cycles to manmade objects and landscapes. Next, we turn to the work of James Gibson, who proposed an interactionist version of human–environment relations and studied perception as environmental conditions change. Gibson and his wife, developmental psychologist Eleanor Gibson, thought that environmental stimuli must have a pattern and structure to convey information. Roger Barker, a transactionalist (see Chapter 6.7), saw people and their environmental settings as one unit and thus thought that the behavioral setting should be natural and not contrived by an experimenter for scientific purposes. Barker employed an inductive, discovery approach to research. Urie Bronfenbrenner (1979) shared many assumptions with transactionalists, and his bioecological approach focused on interactions between the environment and the individual at the micro and higher levels.

6.5.1 From geopsychology to urban psychology

It is common knowledge that weather, climate, and landscape affect the mood and character of people. Actually, the German psychologist Hellpach was also a liberal minister of the state of Baden, and even wrote a treatise entitled *Psychologie der Umwelt* (Engl.: *Environmental Psychology*; Hellpach, 1924). As Kruse and Groffmann (1987) worked out, Hellpach followed three lines of inquiry: first, he systematically investigated the effects of the geopsychological environment. Periodic sunspot activity, the lunar cycle, and the colors and shapes of landscapes, for example, have been shown to affect behavior and wellbeing (Hellpach, 1911, 1950). Second, he investigated the effects of the social environment (i.e. the surrounding people; Hellpach, 1924). Third, Hellpach focused on all the effects of, for example, a room, furniture, a house, a hall, streets, places, roads, and vehicles, and thus turned into a "modern" environmental psychologist (Hellpach, 1939/1952). The later stream of inquiry was called "tectopsychology" and went back to Simmel's (1903/1957) notion of intensification of nervous stimulation and the heightened stimulation of the metropolitans. Thus, Hellpach reflected on the psychological effects of characteristics associated with large urban settings such as haste, vigilance, and alienation.

As Fietkau (1984) pointed out, Hellpach (1924) provided an important methodological contribution by introducing the geopsychological experiment. In such an experiment, the independent and dependent variables are linked to complex settings; for instance, investigating sleep in locations with different altitude above sea level. Thus Hellpach anticipated essential ideas of Barker's ecopsychology and Gibson's ecological theory (see below), although he was focusing on how the material environment affects the human system perspective.

Although urban development was rather the subject of anthropologists and sociologists, Hellpach was interested in the behavioral and psychological processes of urbanites. He noticed that the proportion of people living in German cities with more than 100 000 people had increased from 4.8% to 13.5% between 1871 and 1895, and that, with this urbanization, city populations decreased and crime rates increased. Provoked by Nazi statements such as "People of German blood cannot bear urban life" (German: Menschen deutschen Blutes ertragen das Leben in den Städten nicht; Hellpach, 1939/1952, p. vii), he investigated anthropological (i.e. weight, ethnicity, etc.) and geophysical factors, as well as social psychological aspects. So he wrote a treatise on the "Volatility of the will to procreate" (German: Verflüchtigung des Zeugungswillens; Hellpach, 1939/1952) and provided psychological analysis on the decline of the birth rate, demographic transitions, and affects of urban crowding. The urban crowding aspect was linked to denaturation, social alienation, and a change of the perceived "Lebensraum," a concept resembling von Uexküll's subjective environment developed together with the psychologist Stern (1938, see Chapter 4.12). Methodologically, Hellpach's work fell in line with the work of social geographers of that time. Many of the questions he dealt with are still of interest today.

6.5.2 Key messages

- Hellpach's geopsychological approach investigates how environmental factors (like weather) affect the human performance and wellbeing. The research is primarily one-directional on how the environment affects organismic and mental activities.

6.6 Gibson: ecological theory

American psychologist James Jerome Gibson proposed an interactionist version of human–environment relations by focusing on perception and cognition. He studied how people determine whether an object or the person observing the object is moving and how things are perceived as constant despite a change in the environmental conditions. In Gibson's view, perception and cognition involve the whole body.

> So we must perceive in order to move, but we must also move in order to perceive. (Gibson, 1979, p. 223)

Like Brunswik, J. J. Gibson and his wife, the developmental psychologist Eleanor Gibson, assumed that people perceive those aspects of the environment that are of value to them (see Figure 6.11). However, in contrast to Brunswik, they were not probabilists. Gibson viewed the nodes of the Brunswikian lens model as a combination of percepts and sensations, that is, activities that form the perception. Thus, Gibson was a realist who made assumptions about how information is received. His message was to switch from stimulation to stimulus information, from "physical optics" to "ecological optics" (Gibson, 1967, p. 164). Like von Uexküll (see Box 4.11), he differentiated between the physical world of the stimuli and the perceived environment in a given situation. Gibson's strong relationship to biology is reflected by the use of terms such as "animal" or "organism" to denote humans. For example, he considered organisms to be active observers using the perceived stimuli on a continuous background or layer to gain access to the visual world.

Gibson took an interactionist view of perception and action. To some extent, he denied the factoring assumption of distinguishing between perception, memory, reasoning, etc. He focused the information that is available in the environment. The Gibsons assumed that:

> … symbols, like natural objects … come in sets, not singly. And it is quite possible that the meaning of a symbol, in the mathematico-logical sense, is given by its univocality within the set. (J. J. Gibson & Gibson, 1955, pp. 449–450)

J. J. Gibson postulated reciprocity between the organism and the environment, and included ideas from gestalt psychologists. People derive "information about the environment" and the stimulus "must have pattern or structure to convey information" (Gibson,

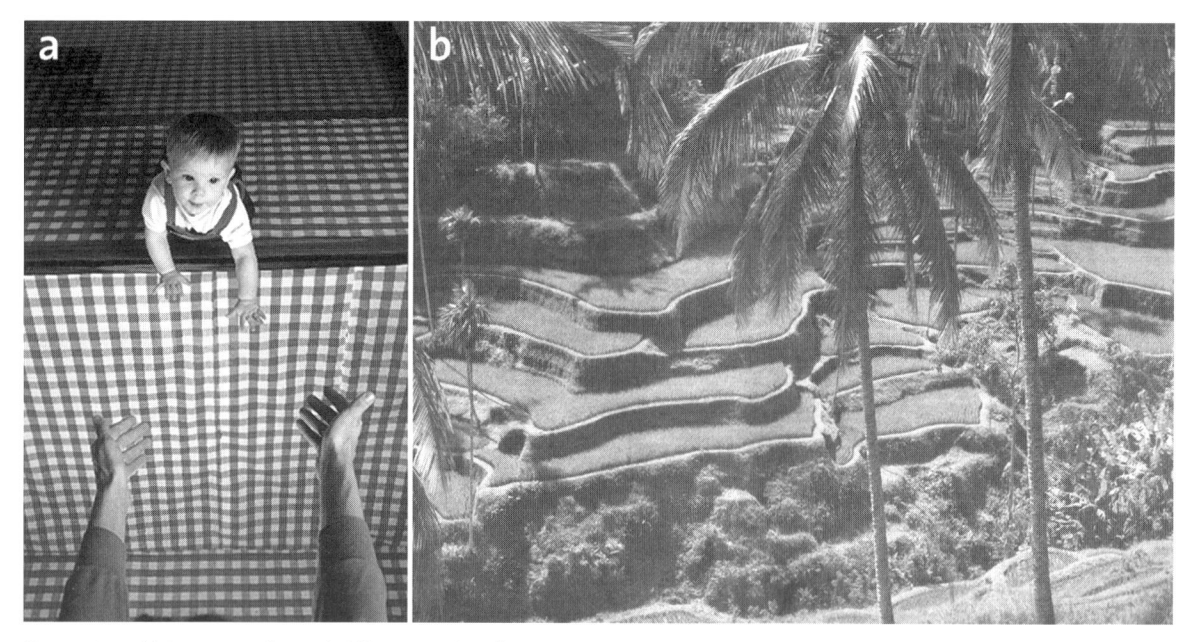

Figure 6.11 (a) Awareness of a visual cliff occurs as the infant matures biologically and gains the ability to crawl (Gibson & Walk, 1960). The environment affords utilization according to the needs of humans (e.g. a chair to sit). A piece of land affords agriculture. (b) Changing the environment, such as building the Bali terraces, to provide suitable affordances can be considered to be the core of engineering activity (photo by M. Rietze).

1967, p. 163). Thus he differentiated "physical reality" on a molecular level from the perceived "varieties of surface layout" on a molar level (Gibson, 1978, p. 416).

Gibson considered meaning or texture of the environment when asking: "how do we see what things are useful for? How do we see to do things, to thread a needle or drive an automobile?" (Gibson, 1979, p. 1). He differentiated between textures of materials from which primitive men fashioned tools (e.g. flint, clay, wood) and those of artificial environments (e.g. plywood, plaster, fiber). Gibson postulated that surface structures such as grass, sand, bushes, etc. have properties that are more easily perceivable. The same holds true for horizontal and vertical ground structures. Thus, the genesis of mental activity closely refers to the environment "in which a behaving agent is situated in the interface between organism and world" (Shanon, 2008, p. 699).

Another interesting key concept from Gibson's theory relates to what he calls affordances.

> The *affordances* of the environment are what it *offers* the animal, what it *provides* or *furnishes*, either for good or ill. The verb *afford* is found in the dictionary, but the noun of affordance is not. I have made it up. I mean something that refers to both the environment and the animal in a way that no existing term does. It implies the complementarity of the animal and the environment (Gibson, 1979, p. 127).

In other words, affordances refer to whatever the environment contributes to the interaction that occurs with humans. For instance, a niche is a set of affordances suitable for the animals that metaphorically fit. Gibson explained manmade changes in the environment as adjusting it to human needs and desired texture. In this context, he argued against the separation of nature and the artificial, as the latter are originally made from natural materials. Similarly, he argued against separating the cultural from the natural environment, and thus argued against a mind–body complementarity. Rather, he considered the affordances as properties of things that frame the interaction of humans with their material environment. Thus, an affordance is a kind of innate aptitude forming a product's appearances and its possible use (Greeno, 1994).

Affordances resemble what Chomsky and Pinker have called language instinct (Pinker, 2000). Noticing the high structural similarity of grammar in all languages of the world, Pinker suggested that the structure of language is innate and that the development of language resulted from an instinct of filling existing structures with words. The idea that people perceive things in the environment in terms of what actions they can afford has epistemological implications. Thus affordances are preconditions for activities. It could be said that the human organismic system is adjusted to the environment. People and animals have innate structures that allow perception of molar pieces of information from the environment without previous experience. From an epistemological perspective, preparedness is then implicit. This is a conclusion that can be taken from scenarios such as the visual cliff experiment (see Figure 6.11). The experimental set-up involves an apparatus that creates the illusion of a cliff's edge. The experiment shows that a child realizes the danger of a cliff as soon as it gains the ability to crawl (Gibson & Walk, 1960).

In general, we can infer from James and Eleanor Gibson's work that what we see depends on the pre-structuredness of the perceptual system that has evolutionarily adjusted to naturalist textures and to (positive and negative) affordances of the environment. This includes the drive to adjust the environment if its affordances do not meet the expectations or the needs of an individual.

6.6.1 Key messages

- Ecological theory by Gibson takes an interactionist view on HES. People shape their environment in a way that fits their affordances. Perception and information processing do not start from a *tabula rasa*. Certain environmental information is more salient, certain affordances of the environment are better perceived, and missing affordances elicit the desire to change the environment
- Ecological theory is functional, as people perceive those aspects of the environment that are of value to them.

6.7 Barker: ecopsychology and the transactional approach

We have seen that ecopsychology and the transactional approach view the environment–behavior connection as a two-way interaction. Roger Barker, a student of Lewin, introduced the term "behavior setting" (Barker, 1968, 1978). What is unique about Barker's view of a behavior setting is that the setting is an entity in itself, the setting of an organism–environment fit. Thus the individual and its environment are considered two sides of the same subject. However, Barker also refers to a fundamental idea of Brunswik when stating that the ecological environment functions "according to laws that are incommensurate with laws governing moral behavior" (Barker, 1968, p. 6), that is, the psychological drivers of an organism. He refers to Lewin's idea of psychological ecology and field theory by linking individual, psychological, and external ecological variables, but he mainly elaborates upon Lewin's later work. In line with Brunswik's idea that natural settings are essential for studying psychological facts and showing certain similarities to the power of the situation, a behavioral setting is a natural setting standing between behavioral patterns and a physical milieu. It includes the physical–spatial aspects and the social rules of our daily life.

Behavioral settings are natural and are not created by an experimenter for scientific purposes. They are physically, ontologically real, as they exist independently from anyone's perception or thinking. They hold a specifiable geographical (or topographical) location, temporal boundaries, and distinctive features. They are quasi-stable with respect to the response mechanisms elicited, exist independently of a single person's experience, and they elicit interdependent behavioral patterns (Barker, 1968). The behavioral settings form environmental perceptions and behavior. Barker thought that, to some degree, people are interchangeable. For example, when visiting a library or attending a baseball game, the way people behave is generally predictable according to the situation and setting, despite differences in individuals (see Box 6.6).

Like Brunswik and Lewin, Barker studied human behavior in real situations and established field stations where his team investigated behavioral rules, regularities, and differences among residents of two towns in the US and the UK (in Oskaloosa, Kansas for 25 years and also temporarily in Leyburn, Yorkshire). Like Heider, he concluded that "we could predict a child's behavior better from place of occurrence than from personality" (Barker, 1990, p. 511).

Barker's approach differed most from those of the other psychologists mentioned in this chapter, who were his predecessors, when it came to research methodology. His approach turned from verification research to discovery research. Whereas the former "replicates, tests, corrects, refines and elaborates previous developments and discoveries," the latter is "exploring and prospecting a new field of inquiry" (Barker, 1987, p. 1414).

Whether and to what extent the presence of a research team changes an environmental setting and in what way

> **Box 6.6** Coherent patterns of environmental settings and behavior: is this a church, Mommy?
>
> The first contribution to the transactional approach in psychological research was provided by Ames, who considered perception as a transaction between organism and environment (Ames, 1952; Ittelson, 1952). The Ames room (Figure 6.12a) provides an optical illusion in which people of similar heights appear to have tremendously varying heights. The illusion exists because the left corner of the rear wall is physically much farther away from the observer than the right one, yet at the same time much higher, thus appearing to have the same height as the right corner. Here, it seems that past experience about non-distorted rooms is the most important cue for visual perception. However, once the observers are undeceived, they can learn to recognize distortions of monocularly viewed rooms.
>
> There are many examples in life of how experience plays an important role in a subject's perception and behavior (see Figure 6.12b). A behavior setting includes a particular pattern of behaviors and certain spatial–temporal characteristics associated with these behaviors. The social environment, as well as the experience and the expectation of the subjects, matters. "When we came to Oskaloosa we found behavior settings (as we called them) to be basic facts of life for the people of the town" (Barker, 1990, p. 511). Behavior settings are programs, socioculturally shared conventions (Fuhrer, 1990) or habits as defined by the French sociologist Bourdieu (Bourdieu & Wacquant, 1992).
>
> This brings psychology into research on social milieus and lifestyles. In this context, lifestyles are those patterns of thinking and behavior that people establish (according to with whom they identify) and to which they refer when signaling social affiliation to peer groups and distinction to other groups.

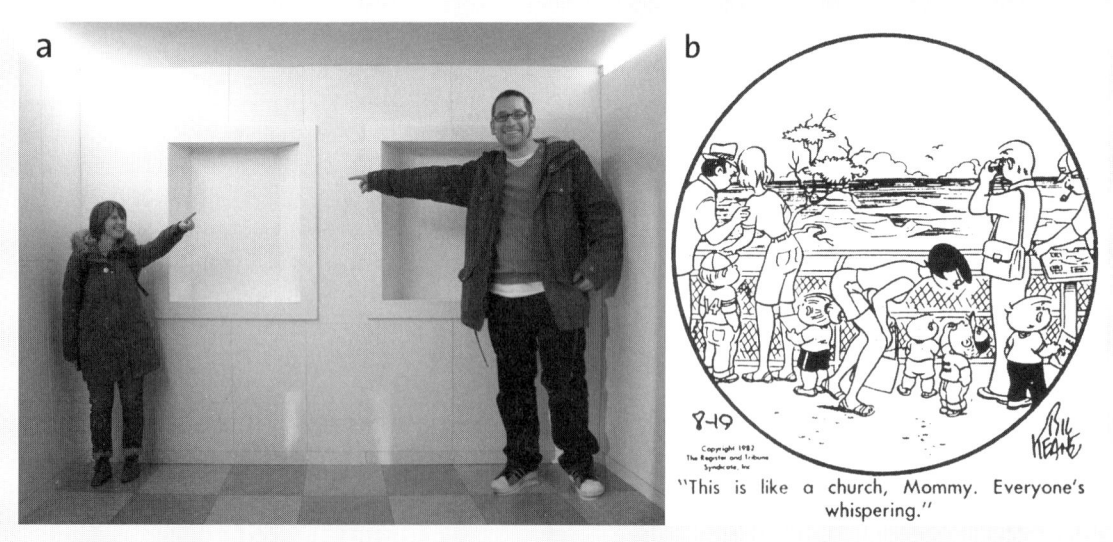

Figure 6.12 (a) The visual perception of the (trapezoidal) Ames room may violate the everyday experience (Photo by I. Stannard). (b) What we see and expect is affected by major references taken from the experience of the individual (reproduced with the kind permission of Bil Keane, Inc. and King features Syndicate).

laboratory settings are unnatural are unresolved questions. We think that in both instances the experimenter has an effect. There is clear evidence "that the location, attitudes, and behavior of the observer are aspects of the phenomenon" (Altman & Rogoff, 1987, p. 34).

Scholars of Barker stress a holistic perspective and focus on changing relations among psychological and environmental aspects of the whole. Similar to Gibson's affordances, the correlation between humans and their environment are considered "a confluence of inseparable factors that depend on one another for their definition and meaning" (Altman & Rogoff, 1987, p. 24). In contrast to experimental psychology, which defines elements of psychological variables, transactionalists do not deal with relationships between elements, components, parts, or variables but rather talk about aspects

(Werner et al., 2002). Transactionalists have a differentiated understanding of functionalism. On the one hand, they want to distinguish their approaches from the simplified organismic, teleological approaches. On the other hand, they acknowledge that human–environment systems (HES) are goal-oriented.

> Transactional approaches, in contrast with organismic orientations, de-emphasize the operation of universal regulatory principles that predetermine the course of development of a phenomenon, although they accept the idea that psychological events are purposive, intentional, and goal directed. Goals and purposes are based on short- and long term motives, social norms, emergent qualities of phenomena, and other factors. (Altman & Rogoff, 1987, p. 26)

Although the transactionalists did not deal with groups, they considered the manning of communities or organizational settings as essential. If, for instance, there are not enough people to take leadership of major action patterns or "behavior genotypes" (e.g. for education or waste management), one cannot expect that these patterns would be successfully practiced (Barker & Schoggen, 1973).

The transactional approach incorporates many concepts and ideas from psychological theories of human–environment interactions, sociology, cultural anthropology, and system theory, and it becomes difficult to order the multiplicity of interrelated meanings:

> For instance a behavior setting is not just one bunch of objects; it is many bunches of objects at the same time, because it is a different bunch for each inhabitant. (Fuhrer, 1990, p. 533)

6.7.1 Key messages

- Transactionalists such as Barker consider people and their environmental setting as one unit, which is called the behavior setting. The behavior setting includes the social rules and the biophysical aspects of the environment.

Section 2: The cognitive dimension

This section is about the cognitive dimension of the contributions of psychology to environmental literacy and begins with a discussion of gestalt psychology. Gestalt psychology deals with the idea that the whole is more than the sum of its parts. The prototype theory of Eleanor Rosch states that some members of a category are more central than others. We also discuss genetic epistemology of environmental knowledge through a review of the work of Piaget who, among other salient ideas, promotes the idea of functionalist equilibration. In the course of ontogenetic and phylogenetic development, people accommodate and construct new mental structures that are consistent with environmental structures.

6.8 From gestalt to prototypes

6.8.1 Good and bad gestalts

Gestalt psychology deals with the quantity, order, and meaning of environmental information (Koffka, 1935). The key message of gestalt psychology is that the whole is more than its parts (Ehrenfels, 1890). Gestalts are structures, configurations, or patterns of physical phenomena or information that cannot be derived by the summation of their parts (see Box 6.7). Simplified, the term gestalt means good form. The basic law of gestalt psychology can be demonstrated by many examples. A common one is that a melody remains the same independent of whether it is played in a higher key, at a faster tempo or with a violin or a trumpet (Köhler, 1971). But what is it that enables us to recognize the melody when it is played in a new key? Each element is different, yet the melody, the sum of the elements, is the same.

The founders of gestalt psychology, Kurt Koffka (1922, 1935), Max Wertheimer, and Wolfgang Köhler (1924, 1971), countered the atomist, element-based approach of Wilhelm Wundt. As was the case in early twentieth-century European psychology, the development of the theory was embedded in thorough scientific reflections that acknowledged the need to adequately acquire experimental verifications and to meet the positivists' standards. For instance, Koffka reflected on the generality of the gestalt category and asked: "… do we claim that the universe and all events in it form one big gestalt?" (Koffka, 1935, p. 22) He referred to the universal principle of causality and argued that no one would claim that any two events of the universe are supposed to be causally related. Thus his answer read:

> … to apply the Gestalt category means to find out which parts of nature belong as parts to functional wholes, to discover their positions in these wholes, their degree of relative independence, and the articulation of larger wholes into sub-wholes. (Koffka, 1935, p. 22)

Epistemologically, the question of gestalt psychology touches on the stimulus–sensation relation linked to the transformation of environmental stimuli. In what way is a human sensory or cognitive system prepared to form, normalize, restructure, or supplement environmental information? The reader may have noticed that this question is – in principle – inherent in the phenomenon of the crisp representation of environmental information (see Box 6.3), Gibson's concept of affordances, and the discussion about the golden sections (Box 6.1). Gestalt psychologists acknowledge that there are considerable interindividual differences as to what extent the individual is prepared to perceive gestalts. However, they state without doubt that such innate schemata exist (Metzger, 1963, 2006).

Gestalt psychologists formulated certain basic principles, including the principles of totality, emergence, multistability, and reification (see Box 6.7). One of the key principles of gestalt theory is the conciseness (German: *Prägnanz*) principle, which states that the mind groups similar elements together. The critical elements for conciseness are the law of closure, which means that missing elements are filled in to complete a whole, the law of proximity, which states that neighboring elements are put together into one cognitive unit (i.e. a chunk, in the language of cognitive psychology), and that symmetric images are perceived collectively (Simon, 1979).

Gestalt psychologists took the stance that gestalts appear in a natural manner. Further, perception underlies the tendency to see or to construct a *gute Gestalt* (Engl.: good or salient gestalt) everywhere. The characteristics of a *gute Gestalt* are regularity, simplicity, stability, symmetry, closedness, unambiguity, balance, scarcity, and vertical–horizontal structures. Gestalt psychologists agree that not all of these aspects have to be present and that there is a cognitive conflict among them. And as Metzger once said:

> A bad Gestalt can resemble a "gute Gestalt" but not the other way round. (Metzger, 1963, p. 66)

Kippfiguren (Engl.: reversible figures) that show multistability and visual distortions are part of the gestalt theory and show that a *gute Gestalt* can also be broken. A famous example is the twisted cord illusion, also known as the false spiral, in which the gestalt of a spiral dominates that of the closed circles (see Box 6.7; Fraser, 1908). The visual distortion of the circles is produced by misaligned parts (i.e. the differently colored strands) and the destabilizing background pattern.

Gestalt theory has been subject to multiple critiques, including some relevant to understanding environmental literacy. Heider (1930b; see Figure 6.4) criticized gestalt theory because it only dealt with proximal stimuli but not with the organization of distal environmental information. Brunswik and Gibson (Gibson, 1950), on the other hand, criticized the gestalt approach in that it only investigated the perception of artificial two-dimensional pictures. Nevertheless, Gibson – who was a colleague of Koffka's at Smith College – was strongly influenced by gestalt psychology.

Independent of the assumption of whether gestalts are innate, the gestalt perception is important for the view taken in this book, as it shows that the mind goes beyond the information given and can construct meaning.

6.8.2 Perceptual prototypes

Eleanor Rosch's theory of perceptual prototypes did not refer to gestalt theory. Rosch began working on her theory when she visited the Dugum Dani tribe in Papua New Guinea in 1969. The focus of her study was on color perception, her premise being that basic color terms are innate and generally applicable. Rosch considered colors as semantic categories which developed around natural prototypes (Rosch, 1973), and that the natural category is learned first. The Dani, for example, only had two basic color terms, one for dark, cool colors, and the other for light, warm ones (Heider-Rosch, 1971). According to Berlin and Kay (1969), this is the lowest developmental stage of color perception. Rosch supposed that the "colorly illiterate" Dani tended to consider the color category, which she called focal points, as a set of variations on the natural prototype. In her experiments, she found that the Dani could categorize objects by colors for which they had no words. She concluded:

(1) [Subjects] from a culture which did not initially possess hue or form concepts could learn names for the presumed natural prototypes faster than for the other stimuli even when the natural prototypes were not the central members of categories.

(2) Members of that culture could more easily learn hues and form concepts when the presumed natural prototypes were the central members of categories than when the categories were organized in other ways … (Rosch, 1973, p. 348).

6.8 From gestalt to prototypes

Box 6.7 How do we order the world? Gestalts, prototypes, and misgestalts

The totality principle states that experience and thinking must be considered in a molar or global manner. Gestalts are the irreducible entities – they are separated from their environment, structured but closed. The multistability principle is demonstrated with the Kopfermann illusion (Kopfermann, 1930, see Figure 6.13d). Figure d1 can be seen as either a two-dimensional star or a three-dimensional cube. We tend to see the star if the other cubes (d2 and d3) are not seen in context. However, we see only one gestalt at a time. According to Kopfermann, the stronger the two-dimensional projection is, the stronger the instability becomes (i.e. switching between alternative gestalts).

The principle of emergence (see the emerging dog in Figure 6.13b) means that a gestalt is inferred from different parts. Other principles, such as proximity of visual stimuli, can be considered to be another basic elementary principle of perception (see the black dots in Figure 6.13c).

Visual perception may be led astray, resulting in illusion, distortion, and fallacious perception. The Fraser's false spiral is a startling illusion that shows that there are interactions and conflicts in the perceptional field (see Figure 6.13e; see also Box 6.6).

The theory of prototypes has been applied to many examples of perceptual and natural prototypes and their mental representations (Scholz, 1987). For instance, a penguin is considered a less representative bird than a dove. Linguists investigated the "cuppiness" of jars and their differentiation to mugs, bowls, etc. (Labov, 1973; see Figure 6.13a). Here we find that the boundaries between cups and bowls are fuzzy. What is a prototype is socioculturally shaped. Nevertheless, also, a concept such as a cup seems to have a prototypical kernel.

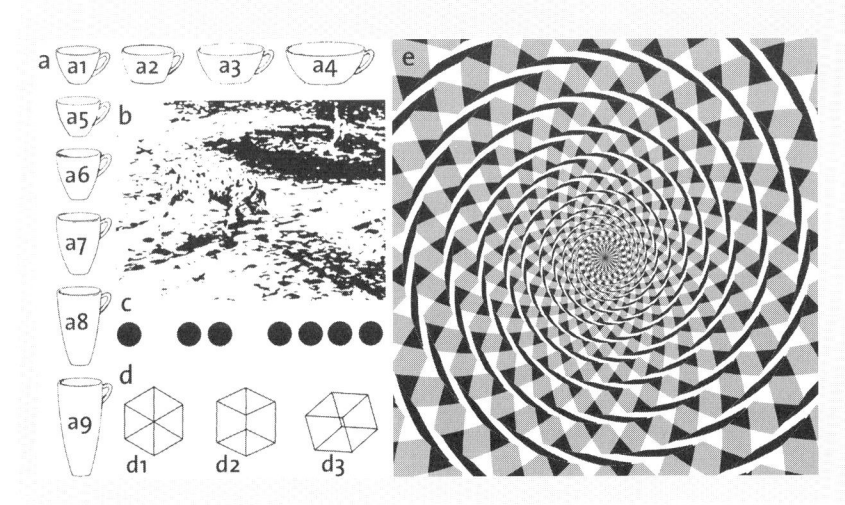

Figure 6.13 The Kopfermann illusion (d) demonstrates the multistability of an image. If the cubes d2 and d3 are covered, d1 becomes a two-dimensional star instead of a cube. (a) The cups show perceptual prototypes; (b and c) the emerging dog and the proximity grouping present basic gestalt principles (reprinted courtesy of the MIT Press from Marr, 1982, p. 101); (e) Fraser's false spiral shows us how contexts can create "misgestalts" (take a pen and prove that there is no spiral but rather closed circles).

Rosch also experimented with forms and concluded that the "categorizations that humans make for the concrete world are not arbitrary, but highly determined" (Rosch et al., 1976, p. 382). After these groundbreaking findings, Rosch (and many other psychologists) conducted several experiments on the phenomenon of prototypes of natural objects, such as fruits, musical instruments, trees, vehicles, fish, etc. (Rosch et al., 1976). The findings were verified by ethnographic methods.

A differentiated discussion on the relation of the psychological and philosophical dimension of gestalt is a key issue in the epistemics of environmental literacy. It is interesting that gestalt theory goes far beyond psychology. In his book, *A Different Universe: Reinventing Physics from the Bottom Down*, the physicist Robert Laughlin (2005) deals with the complementarity between atomist and molar macroperspectives of construction:

> A field of flowers rendered by Renoir or Monet strikes us as interesting because it is a perfect whole, while the daubs of paint from which it is constructed are randomly shaped and imperfect. The imperfection of the individual brush strokes tells us that the essence of the painting is its organization. Similarly, the ability of certain metals to expel magnetic fields exactly when they are refrigerated to ultralow temperatures strikes us as interesting because the individual atoms out of which the metal is made cannot do this. (Laughlin, 2005, p. 7)

He argues that the same holds true for physical phenomena and that gestalt-like order systems, such as emergence, are necessary in order to understand the world.

6.8.3 Key messages

- Gestalt theory shows that we perceive more than the sometimes fractionized parts and that the mind is able to construct additional information or meaning to the stimuli
- The experimental studies of Rosch on perceptual prototypes showed that the perceptual system features inherent structures that identify prototypical representatives for certain classes of environmental stimuli (material–biophysical objects).

6.9 Piaget: genetic epistemology of environmental knowledge

6.9.1 Principles of cognitive development

Jean Piaget (1896–1980) was a Swiss psychologist and epistemologist, whose initial interest and research lay in biology. At 11 years old, he published his first (albeit brief) article in a scientific journal describing his observations of an albino sparrow (Piaget, 1907). This was followed by a series of papers on shellfish, which he wrote before finishing high school. At the age of 15, he received an offer to become the mollusk custodian at the Museum of Natural History of Geneva (Ginsburg & Opper, 1988). In the 1920s Piaget investigated the emergence and development of the human cognitive system while earning his doctorate at the laboratory of Alfred Binet and Théophile Simon, the inventors of the first intelligence tests. Piaget's interest was in the interaction of the developing human with its environment, and he took a case study approach in his research. His early books were on infant language (Piaget, 1923) and reasoning (Piaget, 1924). He then focused on fundamental epistemic and scientific concepts, such as numbers (Piaget & Inhelder, 1956), space (Piaget & Inhelder, 1956), time (Piaget, 1946), probability (Inhelder & Piaget, 1958), and logical thinking (Inhelder & Piaget, 1958).

Piaget can be considered to be the founder of cognitive psychology. His theory is characterized by two basic assumptions about cognitive development that are essential to understand environmental literacy. One is that humans and human development are driven by the principle of equilibration. The other is the principle of genetic epistemology. Equilibration is illustrated in the following quote:

> The organism adapts itself by materially constructing new forms to fit them into those of the universe, whereas intelligence extends this creation by constructing mental structures, which can be applied to those of the environment. (Piaget, 1947, p. 4)

In his later work and through a structuralist perspective, he considered cognitive development to be a transformation of a mental system in relation to homeostasis and holistic functioning. Piaget strongly referred to logico-mathematical models, as he was concerned that ordinary language produced obscure, incoherent, and ambiguous psychological theorizing (Ginsburg & Opper, 1988). Piaget saw the power of logico-mathematical models in "reflecting abstraction" by performing operations among the abstracted, symbolic elements (for instance, ratio of 4:5 for a mix of four glasses of orange juice with five glasses of water). These (applied) operations are considered essential in describing cognitive operations and are supposed to mediate and give orders between the activities of the nervous system and conscious behavior. Piaget embedded the latter proposition by remarking that "this is because psychology is, in a first step, a biology" (Piaget, 1968, p. 133).

However, Piaget strongly refers to functionalism as well. For instance, he proposes that intelligent behavior means the development of appropriate adaptation. We take a quote of Brunswik to illustrate this position:

> Survival and its sub-units, which may be defined as the establishment of stable interrelationships with the environment, are possible only if the organism is able to establish compensatory balance in the face of comparative chaos within the physical environment. (Brunswik, 1943, p. 257)

Piaget did not consider equilibrium in the developing individual child or organism as something static (as in mechanics or thermodynamics). According to Piaget, equilibration in cognitive development consists of a sequence of equilibrium and disequilibrium states. When, for instance, a child is confronted with a new object, data or experience, it starts to deal with this issue by using existing (cognitive) schemata to attain equilibration, namely assimilation and accommodation. Assimilation means the modified usage

6.9 Piaget: genetic epistemology

Box 6.8 Stages of environmental literacy: the invisible man

In Jean Piaget's genetic epistemology, the pursuit of individual, societal, and scientific knowledge aims to achieve harmony with the environment. He proposed that cognitive development passes through stages, each of which is limited with respect to appropriately solving certain problems or understanding physical, mathematical, or environmental phenomena. Piaget revealed that many problems can only be understood on the highest level of abstract symbolic representation and operation. However, not all individuals and people attain the highest level. Many people maintain in lower levels, in particular in the preoperational or the concrete operational levels which link cognitive operations which can be done – more or less – with concrete physical objects. For example, Capon and Kuhn (1979) showed that less than one-third of female shoppers in a supermarket were able to employ proportional reasoning properly to decide which of two common items presented (in different packaging) was the more economical purchase, when the size ratios of the items were 2:3 vs. 3:4.

Similarly, people in the preoperational stage fail to grasp the principle of mass constancy. For example, when pouring water from one glass to another with a smaller diameter, there is a chance that water could spill over, since people in this stage only concentrate on one dimension, such as height, and forget other dimensions, such as diameter. The preoperational stage is also characterized by the phenomenon that children think that others (standing at a different place) see the same things as them. Or that they disappear when they cover their eyes with their hands. A similar voodoo-like belief has been reported in a newspaper story depicted in Figure 6.14 about the "invisible man." In both cases there is the naïve belief in operational forces – though of a different kind – to make the physical appearance invisible.

Fare Dodger Thinks He's Invisible

Lagos – In Nigeria a ticket checker discovered a stowaway on a train even though the passenger was fully convinced he was invisible. The Nigerian news agency NAN reported on Thursday that the fare dodger stated he assumed his magic amulet had rendered him invisible. But he hadn't followed instructions and had forgotten to renew his magic spell the week before, the man said: "That's why I was caught."

Like many of his countrymen, the stowaway was riding home on the roof of the train. Many people in this West African country travel this way to save on a fare or simply to be able to smoke marijuana in peace. *(SDA)*

Figure 6.14 Translation of a newspaper article about a fare dodger who claimed he was caught because a spell that made him invisible was not renewed. A refugee train in the Nubian Desert (from http://nomadsland.com), and a boy covering his eyes to "become invisible" (photo from Shutterstock images).

of existing cognitive schemata or behavioral patterns to cope with new problems or activities. This means that existing (cognitive) structures are used more efficiently. If, however, the process of assimilation is not effective or efficient, the individual can develop new schemata or patterns to deal with the new issue. This process is called accommodation. In general, adaptation is considered evolutionary because it is a change in ecological structure and function "that enables an organism to adjust better to its environment and thus enhances the organism's ability to survive and reproduce" (Wright & Nebel, 2002, p. 647).

> Intelligence is adaptation. In order to grasp its relation to life in general it is therefore necessary to state precisely the relations that exist between the organism and the environment. (Piaget, 1947, p. 3)

Note that equilibration also includes curiosity, maturation of cerebral brain structures, and experience (Inhelder & Piaget, 1958, p. 337). Thus there are five key components of cognitive development: equilibration, assimilation, accommodation, (biological) maturation, and social experience or tradition.

6.9.2 Changing structures

Piaget also developed a stage theory of cognitive development. This is a structuralist theory, as cognitive development is considered a sequence of emerging and changing relations among cognitive elements. This theory starts with the sensory–motoric stage, which ends with the development of language (i.e. the key tool of mentally representing the environment).

The second stage is the preoperational stage, in which language and symbolic play occur. In this stage, however, there are no elaborated cognitive operations. The third stage, the concrete operational stage, follows, and is characterized by mental operations on real-world objects, including classifying, ordering, and enumerating objects. Finally, the formal operational stage takes place, in which the subject is able to operate cognitively with symbols or variables. This stage forms a normative realm. Proportional reasoning, which, for example, enables one to decide which of the two fractions 5/6 or 6/7 is bigger, can only develop in the highest level of cognitive development (see Box 6.8). Each of these stages is characterized by the structure of coherent rules of inference. These structures are not presented in detail here, but Piaget's idea that the same pattern of development that one observes in a child can also be found in science and society is emphasized. This is the core idea of genetic epistemology.

6.9.3 Genetic epistemology

The purpose of genetic epistemology is to explain knowledge, and in particular scientific knowledge, on the basis of its history, its sociogenesis, and especially its psychological origins (Piaget, 1970, p. 1). Piaget hypothesized that the ontogenetic cognitive development of a human individual resembles the development of knowledge in science because scientific ideas are constructed by the human mind (Piaget, 1970). Piaget considered scientific knowledge – somewhat similarly to Thomas Kuhn – to be in perpetual evolution and to be governed by equilibration (Beth & Piaget, 1966; Kuhn, 1996).

> Piaget (along with Marx, Comte, or Spencer) believes that all development and evolution has a direction towards an ideal (orthogenesis). (Tsou, 2006, p. 205)

We make reference to Piaget's perspective on the theory of science in Chapter 16.8 and consider a non-static, evolutionary equilibration between organism and environment (Kitchener, 1986) as a fundamental challenge of sustainable development.

6.9.4 Key messages

- Jean Piaget's genetic epistemology postulates that ontogenetic cognitive development is related to the phylogenetic development of knowledge. His theory of ontogenetic development is based on five general principles which refer to the organism–environment relationship and biological factors. The principles include equilibration, assimilation and accommodation as components of adaptation, (biological) maturation, and social experience and tradition
- An essential assumption of Piaget's theory is functionalist equilibration. Cognitive development adapts and constructs new mental structures that fit to the relevant environmental structures.

Section 3: Environmental psychology

After having dealt with important *theories* on people's perception of environmental information, and how human perceptual and cognitive systems adjust and adapt to environmental stimuli, we discuss what the subdiscipline environmental psychology contributes to an understanding of environmental literacy. We take a specific view on environmental literacy and review what insights have been gained in the capability and limits of people to cope with basic environmental entities such as time, space, resources, etc.

6.10 The subdiscipline environmental psychology

6.10.1 Emergence, objectives, and institutional framing

In the first comprehensive review on environmental psychology, Craik depicted, as one option, the vision that:

> … environmental psychology will … signify a coherent framework for understanding man–environment relations from a psychological perspective. (Craik, 1973, p. 412)

There are critical voices that environmental psychology has not attained this goal (Giuliani & Scopelliti, 2009). Nevertheless, we will follow the history of this field and review some salient findings for understanding environmental literacy.

Modern environmental psychology emerged in the 1960s with the move to bounded rationality (see Box 1.4) and the environmental movement of that time. From a perspective of scientific institutions (see Figure 1.7*), three issues seem overly important. They were the 1967 decision of the Graduate Center of the City University of New York to allow environmental psychology as an area of specialization (Proshansky et al., 1970, p. vii), the foundation of the first scientific journal *Environment and Behavior* in 1969, and the editing of the first book *Environmental Psychology – Man and his Physical Setting* (Proshansky et al., 1970), in which the "sociophysical environment" was defined (Stokols & Altman, 1987, p. 1).

In the beginning, environmental psychology was conceived as "people in relation to the physical environment … or in relation to the socio-physical environment" (Sime, 1999, p. 193). In this sense, Brunswik (1933, 1934, 1935) and Gibson (1950) considered both context and the reality setting as important components of environmental psychology studies of human behavior. A more precise disciplinary definition is provided when stating that environmental psychology is "the application of psychological knowledge and the method to understand the process and implications of the human–environment transaction and applying the insight to improving the quality of the experience" (Cassidy, 1997, p. 240). However, many contributors to environmental psychology came from other disciplines, such as sociology, public policy, architecture, health sciences, and planning sciences (Bechtel & Churchman, 2002). Modern environmental psychology is an "interdisciplinary effort" (Bechtel & Churchman, 2002) dealing with "social pressure" (Proshansky et al., 1970, p. 5), and is at the least "a multidisciplinary behavioral science, both applied in orientation, whose foci are the systematic interrelationships between the physical and social environments and individual and human behavior and experience" (Veitsch & Arkkelin, 1995, p. 5), or a subdiscipline which investigates transactions between individuals and their physical settings (Gifford, 1997).

We see how the field is grounded in geography when emphasizing that environmental psychology is the "study of the molar relationship between behavior and experience and natural environments" (Bell et al., 1996, p. 6). The molar perspective is expressed by focusing on the understanding of contexts (Stokols, 1987). With contexts, psychologists understand the specific situation, including overt and covert events and processes in which individuals find themselves. Scholz and Zimmer (1997) noted that incorporating contexts asks for qualitative considerations and increases the complexity of analysis. Thus, the transactional approach views the physical environment as a "potential context for social interactions that can support, constrain, symbolize, and confer meaning upon various aspects of social relationships. … It describes physical settings in terms of dialectics, or tension between opposing influences" (Sundstrom et al., 1996, p. 491).

The environmentalist approach is clearly still visible in environmental psychology, as we read in Bechtel's *Handbook of Environmental Psychology*:

> My approach was that environmental psychology is not a simple branch of psychology but a plan for survival. (Bechtel & Churchman, 1997, p. xviii)

Thus Bechtel dedicated a chapter to "The environment will get you if you don't watch out" (Bechtel & Churchman, 1997).

Thus, with only a few exceptions, applied environmental psychology deals with factors and interventions to encourage proenvironmental behavior (Steg & Vlek, 2009). Here the research agenda consists of four steps: identifying desired behavioral change by examining the factors underlying the behavior, designing and implementing interventions, and evaluating the effects of the interventions (Gifford, 2002; Steg & Vlek, 2009).

We think that a comprehensive and thorough theory of HES is needed to understand the ongoing use and abuse of the environment (Cassidy, 1997). This requires greater reference to transactional research and a better incorporation of the analysis of environmental variables in the sense of Brunswik or the functionalistic perspective of environmental psychology. This has been, for instance, suggested by Stephen Kaplan (1972), who considered man as an adaptive system, an idea that meets ecological rationality (see Box 1.4). Such a view studies the human–environment relationship as a coupled, feedback system as suggested by the transactionalists. But here the problem is that environmental psychologists are "still lacking the analytical tools needed to deal with the transactional approaches" (Gifford, 2009, p. 388). Difficulties also arise in distinguishing between practice, applications, "basic" research, and theory (Sime, 1999). Thus, its future is somewhat open.

> Environmental psychology is institutionally unstable because it quite rightly crosses traditional educational boundaries between disciplines devoted primarily to people or environments. (Sime, 1999, p. 202)

6.10.2 Full ecology for environmental psychology

Early psychological research by Wundt, Heider, and others investigated visual perception in a static, artificial environment. The questions studied with the visual array of the environment were identification (is there something?), recognition (what is it?), differentiation (are the objects different?), and scaling (how different are they?; Veitsch & Arkkelin, 1995). This resembles a natural science approach, which contrasts with a full ecology approach (Bonnes & Bonaiuto, 2002). In the full ecology approach, the human component of the material–biophysical environment is integrated as it is required in the transactional approach. A transactional or full ecology view would also require aspects of manipulation (what can I do with the objects?) and utilization (how do I benefit from it?). This would require going beyond space, distance, visual deceptions (see Figure 6.5 and Boxes 6.5 and 6.7), and attractiveness of spatial structure to look at the environment as an anthropocentrically defined setting. We think that an enriching contribution from environmental psychology is possible although:

> … the central proponents of "sustainability science" itself have not acknowledged environmental psychology as a potential contributor … (Gifford, 2007, p. 199)

6.10.3 Key messages

- Environmental psychology aims to conceptualize the psychological processes (i.e. the rationale of the individual) in human–environment relationships
- Environmental psychology should incorporate the two-way interaction between the human and the environment, as is done in ecology. The human individual affects and is affected by the material environment. This is the perspective of the full ecology view on environmental psychology.

6.11 Space

By space we mean the three-dimensional field of physical objects and events and their order and relationships. Environmental psychology deals with space in multiple perspectives. We present insight into spaces that people want to control, mental spaces and their experiential, cultural shapes, and some theories about the perception of landscapes.

6.11.1 Personal space, territoriality, and defensible space

If we want to understand environmental awareness, personal space "with invisible boundaries into which intruders may not come" (Sommer, 1996, p. 26) and territoriality are central concepts that refer to physical spaces people want to control. The term defensible space has been developed by urban designers who created dwelling units where areas became objects of surveillance. Such design was employed to prevent crime; this can be defined as an action that is in conflict with social order and community norms (Jacobs, 1961; Newman, 1972).

Spacing and territory marking are exhibited by many animals. Hall (1959, 1966) assumed that humans have an innate distancing mechanism that helps them to regulate contact and interaction. He distinguishes four zones of personal space: the intimate (0–50 cm), personal (50–130 cm), and social space (130–230 cm), and public distance (400–800 cm), where there is no intimacy between people. Hall's work was given the name proxemics (Bechtel, 1997). According to Hall, the expressions of space can be found in the built environment (see below) and in (unconscious) behavioral rules of distancing; an issue that has been labeled as "space speaks." The boundaries differ slightly between cultures (Kaya & Weber, 2003). Thus, environmental psychologists introduced *culture* as a key variable. This aspect also introduces an anthropological basis into environmental psychology, for instance, acknowledging that there are cultures in which the individual has "little apparent privacy" (Altman & Chemers, 1980, p. 2). Here, we should note that privacy is a western concept and has different meanings in other cultures.

Psychological research on space also dealt with concepts such as population density, which is the ratio between the number of people and the size of the areas, and crowding, which denotes a subjective feeling of too many others around and a perceived lack of personal control (Gifford, 1997).

6.11.2 Mental representation of space

With the emergence of cognitive psychology research, cognitive maps are coming into use; that is, the individual, personal, subjective, autobiographic mental or mind map of spatial structures. Cognitive maps allow researchers to interpret an individual's perception of the spatial environment and, through comparison, to see how the individual's perception relates to other people and to geographic maps (Friedman & Dewinstanley,

2006; Tversky, 1991). One should note that most of this investigation is performed by researchers in the field of memory, cognition or neuropsychology, and thus has not become a core object of environmental psychology.

6.11.3 Natural and designed landscapes

The difficulty of the nature–nurture discussion can be best seen in the question of aesthetics (see Box 6.1); that is, which natural or culturally formed landscapes are liked and valued most by people. While there have been many studies that examine the question of landscape preference, the resulting theories do not offer conclusive results. On the one hand, there is evidence that people (e.g. American and European adult samples; Ulrich, 1986) like natural landscapes. When referring to Gibson's concept of affordances (see Chapter 6.6), Kaplan (1979) followed an evolutionary, functionalist view, stating that value is linked to use. This has been called "ecological aesthetics" (Gobster, 1999; Koh, 1988). However, one should consider that Gibson referred his concept of ecology to an environment that meets the needs of humans and is not relevant to the wilderness. Kaplan argued that, if aesthetics are not an underlying basis of perceiving the world, aesthetics becomes trivialized. Here she refers to the biophilia hypothesis, that humans evolved over three million years in natural environments and have an innate draw to nature. Thus, city environments, which have been around for only a tiny fraction of time, are less attractive (Gifford, 1997). Therefore, Kaplan infers that it is hardly surprising that human preference is for natural environments and, in particular, environments in which survival would be more likely. For instance, as water is essential for survival, people like rivers and lakes. She suggested that complexity and mystery are the aspects of interest that draw people to these environments. The latter refers to novelty, curiosity, and the promise of something unexpected. The other aspects are coherence and legibility (referring to the need for safety), which Kaplan called the "makes sense" aspects. Here, Stephen and Rachel Kaplan introduced the idea of a "reasonable person" model, which assumes that people should be more cooperative, helpful, satisfied, and show more responsibility when the environment supports their basic informational needs (Kaplan & Kaplan, 2009). The Kaplans hypothesize also that nature protection is caused by these mechanisms.

On the other hand, some studies show that many people (as well as certain other species) find scenic, park-like, symmetrical patterns more attractive than asymmetrical ones (Enquist & Arak, 1994; Rhodes, 2006), and indicate that structure or design of the landscape may play a role. Thus, aesthetic and environmental priorities can conflict. For example, for some people hedges might pollute aesthetics as they prefer clean, tidy woods (Parsons, 1995). Also empirical results about whether clear-cutting forests or forest harvesting is preferred provide contradictory results (Ribe, 1999; Schmithüsen, 2008). It seems that different preferences and emotions can be at work (Parsons & Daniel, 2002).

In general, the current research on landscape perception reflects the main messages of traditional research and the interactions of drivers. On the one hand, people have been shown to display care towards, attachment to, identity with, affordances, or affinities to nature, which can be explained by basic drivers such as needs, values, and motives. On the other hand, there are "aesthetic experiences from settings that have traditionally been called 'scenic beauty'" (Gobster et al., 2007, p. 959), which seem to be closely linked to social norms or positively framed and sensitized experiences. In both approaches, certain emotions can be linked to natural landscapes, depending on the personal experiences or interests, the functions that nature or landscape plays, and the role the person is taking (Vining, 1992). From a psychological perspective it seems likely that different situations, interests, and biographies elicit one or the other relation. Finally, recent experimental research revealed that Asians and Americans differ in the perception and mental recording of foreground and background variables in natural and urban environmental settings (Nisbett & Masuda, 2003).

6.11.4 Key messages

- Territoriality, when referring to the control of spaces, objects, and ideas, seems to be a basic driver of human individuals. Interpretation is open about when and under which constraints this driver can be elicited
- In general, people's (i.e. non-savants; see Box 7.5) memory of the spatial environment is very restricted, imprecise, and culturally shaped
- There are different theories, such as biophilia or socialization, and contradictory findings from experimental studies about the consistency of preference functions, the subjective value of wilderness, the natural environments, attractiveness of parks, etc.

6.12 Time

From today's environmental management perspective, time has become an overly important issue. Mitigation of climate change will show its effects decades later. High level radioactive waste has to be secured for more than 100 000 years (see Box 6.10). Also, the tilling of land for agriculture and loss of biodiversity will show negative delayed effects (see Box 16.13). Future orientation has become a central psychological factor for sustainable transitions. Compared to its significance, it is surprising that time has been a relatively neglected topic in environmental psychology. We think that this is because of the complex nature of time. Box 6.9 provides an insight into different concepts and notions of time.

We should consider that time is a construct that serves different practical and scientific purposes. It is culturally bound and shaped by religious assumptions. Quite often, especially in psychological research (Slife, 1993), the Newtonian concept is taken as a reference. General psychology distinguishes between three aspects of psychological time: first, *time period* as the persistence of events or the interval between events; second, *time chronology* as the sequential occurrence of events; and third, *time perspective* as experience or conceptions of humans about the past, presence, and future (Björkman, 1984; van der Meer, 2006). Slife (1993) describes how the dominant current western time concept includes objectivity (i.e. independence from the observer, objects, and events), continuity (i.e. having no lacunae between its parts), universality (time is the same everywhere and anytime), linearity (flowing forward at a uniform rate), and reductivity (i.e. we can consider temporal processes of certain subsystems of the universe).

In the inner field of environmental psychology, we can identify two emerging lines of research. One line relates personal time-related traits, for instance "consideration of future consequences" (Strathman *et al.*, 1994) or "time perspective" (Zimbardo & Boyd, 1999) to values or environmental behavior. Here, reference is made to Schwartz's (see Chapter 7.8, Schwartz & Bardi, 2001) approved value construct. One can find weak correlations of the universalism value with future orientation (Milfont & Gouveia, 2006). Also, the affinity to "prefer diversity and variations in bio-physical and socio-cultural living scenarios" (Corral-Verdugo *et al.*, 2009, p. 36), an issue we have dealt with in the Wilson quote of Chapter 2.2, is positively linked with incorporating long time perspectives.

The second line refers to risk perception. Here the discounting by time, the predominantly used research paradigm, is an interesting issue. If we have to invest something to avoid negative outcomes, many studies in finance have shown that the future negative values are considered less valuable and today's investment higher. Empirical studies have shown that this cannot be found in the same manner in environmental settings. This is explained by ethical and stable loss-related concerns (Bohm & Pfister, 2005). There are ambiguous studies on gender and discounting. Some indicate that men tend to discount the future more steeply than women in their behavioral choices (Daly & Wilson, 2005; Kirby & Marakovic, 1996). A study by Tonn *et al.* (2006) on cognitive representation of the future indicates that women might interpret the word "future" as being much closer to the present than men. Contrary females are able to imagine the future more clearly for all given time frames (from up to 1 year to over 1000 years).

6.12.1 Future and very long time

We briefly leave environmental psychology to reflect on whether people can anticipate future. Some insight is provided by the mental time travel theory, which is also called episodic future thinking (Atance & O'Neill, 2001). The idea of this theory is that we can simulate the future based on our episodic knowledge (that is, memories of past events and experiences), which helps us to predict, plan or avoid future events. First empirical evidence for this comparatively new approach has been given by several authors (see D'Argembeau & van der Linden, 2006; Liberman *et al.* 2002; Suddendorf & Busby, 2005). A special and new dimension of environmental literacy is how to deal with extreme time dimensions. Here, nuclear waste can serve as an example (see Box 6.10). It can be shown that the time of events occurring less than 1000 years ago can be estimated quite accurately by humans, whereas the time of events occurring more than 1000 years ago was frequently underestimated (Drottz-Sjöberg, 2006; Svenson & Karlsson, 1989). Thus, people of today have a kind of 1000-year threshold for accurate perception of past events.

6.12.2 Key messages

- Time is an overly complex concept. From a constructivist functionalist perspective, there are physical, biological, and epistemological layers suggesting a certain facet of the concept of time. Time is culturally and religiously shaped

Box 6.9 An intriguing complex concept: time

Humans have experiential access to time by days and years and by reproductive cycles, say, of domesticated animals. However, historically, it took a long time to elicit knowledge about biological reproduction cycles (see Chapter 4). This is remarkable as cause–effect relationships and learning, developmental or adaptational processes are all linked with time. The following view on time strongly refers to the work of Josef E. McGrath, a psychologist who linked the ontology, epistemology, and psychology of time (McGrath & Tschan, 2000, 2004).

The philosophical view on time is a very complex issue. The prevailing Newtonian view has dominated western modern culture. It assumes that time and space are independent. Time is an abstract absolutely scaled entity (i.e. a variable on the absolute scale having a zero point); it is singular as there is only one scale, which refers to all processes in the universe; and it is independent of persons and events. This view measures temporal processes in a linear way and – as stated – in absolute terms. Time is understood as bidirectional, allowing forward and backward measurements.

The Einsteinian view, however, proposes that space and time build an inseparable continuum called spacetime which can be thought of as a four-dimensional concept composed of a three-dimensional "Euclidian space" and a one-dimensional time space which is of a different type (Einstein, 1916; see Figure 6.15). Space and time cannot be separated, as the rate at which time passes depends on the object observed and its relation to the absolute benchmark of moving objects (the speed of light). Thus there are multiple times. The Einsteinian view can be considered real and seems to be experimentally proven by the Hafele–Keating experiment (Hafele & Keating, 1972). Atomic clocks in an aircraft circumnavigating the Earth moving in an easterly direction (the direction of the Earth's rotation) move faster than clocks that remain on the ground.

There are a number of open questions. (1) Is time *continuous* or *discrete* in the sense of quantum leaps of electrons from one quantum state to another within an atom? The problem here is whether time is discrete, whether there is a quantum of time, and, if so, what spaces are between those units. (2) Is time bidirectional, as Newton and Einstein assumed, or is time phasic and reversible in its flows as Mayan, Hinduism, and some other world views assume? (3) Time can be viewed from a realist stance (as taken in this book); it is real, showing effects and can be experienced. The complementary view here is that time can be considered an abstract mathematical entity which is relative and might be considered as "time as illusion" (McGrath & Tschan, 2004, p. 184). The classical idealist position has been depicted by Kant: "Time is therefore a purely subjective condition of our (human) intuition … and in itself, apart from subject is nothing" (Kant, 1787/1965, p. 89).

We can learn that the experience, conceptualization, and the measurement of time matter. Psychologists are interested in understanding the role of time in experience and action. As is the case with other western scientists, they mostly refer to a Newtonian construct for describing duration of processes and succession of events (in causal chains). It is of interest that the experienced time, which is considered a forward flow, is mostly not consonant with the formal measure of time. Further, there are neuropsychological indicators revealing that time is epochally perceived (Scholz, 1988; Verleger, 1988).

Time has a strong cultural and religious blend (see above). Ji *et al*. (2009), for example, demonstrated that Chinese and Canadian subjects differed with respect to their perception and cognitive representation of past information. Furthermore, time seems to be linear to the Americans but cyclic to Chinese people. Although the western world defines time in reference to celestial cycles (that is, material–physical processes), the concept itself is abstract but serves practical processes – for instance, organized social processes.

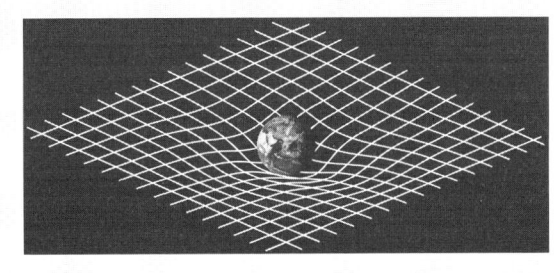

Figure 6.15 Spacetime is a concept suggested by Einstein based on a four-dimensional mathematical model in which the rate at which time passes (the length of the gridlines) is relative to the object's velocity (in relation to light speed).

- The psychological and the philosophical views on time are inextricably intertwined. The modern western world time concept primarily deals with the succession of events and duration of processes, and mostly refers to the Newtonian conception of a linear, location-independent, absolutely scaled entity. For specific questions, for instance on atomic or cosmological calculations, Einsteinian concepts or other views seem superior
- Environmental psychology has not yet thoroughly dealt with the time concept, although anticipating future processes and effects is overly important

Box 6.10[1] What means one million years? Nuclear waste

Some technical problems involve extraordinarily extensive timescales compared with the experiential frame (i.e. the lifespan) of human beings. When dealing with nuclear waste, technical experts aspire to manage a period of up to one million years for the disposal of spent fuel and for high level nuclear waste to avoid affecting ecosystems (Junker *et al.*, 2008; Nationale Genossenschaft für die Lagerung radioaktiver Abfälle (Nagra), 2002). This long-term dimension refers to hazards posed by a release of radionuclides to humans and the environment, as indicated by the extensive half-lives (e.g. ^{239}Pu = 24 110 years). One million years is about 100 times longer than the period since the end of the last glaciation (about 11 000 years before the present). Also, in view of recorded human history, this long-term dimension touches the beginnings of the human species. The timeframe of one million years poses a challenge for the risk assessment of nuclear waste management as it does not allow for (purely) quantitative probability function-based modeling. The way in which technical experts deal with this challenge is the so-called "safety case" – in essence a set of sound technogeological arguments as to why the safe storage of nuclear waste in the long run appears plausible (Nationale Genossenschaft für die Lagerung radioaktiver Abfälle (Nagra), 2002). The essence is that safety assessment must cope with critical events and dynamics on different timescales (see Figure 6.16).

However, nuclear waste is not only a technical, but a so-called sociotechnical problem (Flüeler, 2006). In the past, many initiatives to find sites for a nuclear waste repository, specifically for high-level waste (IAEA, 1994), failed because of a lack of public acceptance within the chosen host community. Here, a science-based risk assessment meets public laypeople. Hence, the decision-makers' perceptions of risk (e.g. Fischhoff, 1995; Flynn *et al.*, 1993) and trust (e.g. Siegrist *et al.* 2007; Sjöberg, 2001) became important issues.

To our knowledge, there exists only one study examining gender differences in time perception relating to nuclear energy and nuclear waste disposal. Drottz-Sjöberg (2006) found a discounting of the perceived judgment of seriousness of consequences if there were to be a leakage from a nuclear waste disposal site over five given time periods (spanning from "during the working period" to "in 1 million years") for both men and women. There was a significant difference in mean values between genders for every given time period, with men's discounting slopes being steeper than women's.

The study of Moser *et al.* (submitted) examined how such extensive timescales are cognitively understood and showed that scientists differ in their understanding of time and extensive timescales. In 18 qualitative exploratory interviews for time-related phenomena, experts from different disciplines revealed, in principle, two distinct representations of time: a more cyclic, determined one and a more open, undetermined one. Temporal representations seem to be linked to different scientific approaches: undetermined representations (which often relate to unpredictable processes) are linked to idiographic approaches, whereas determined representations are linked to nomothetic

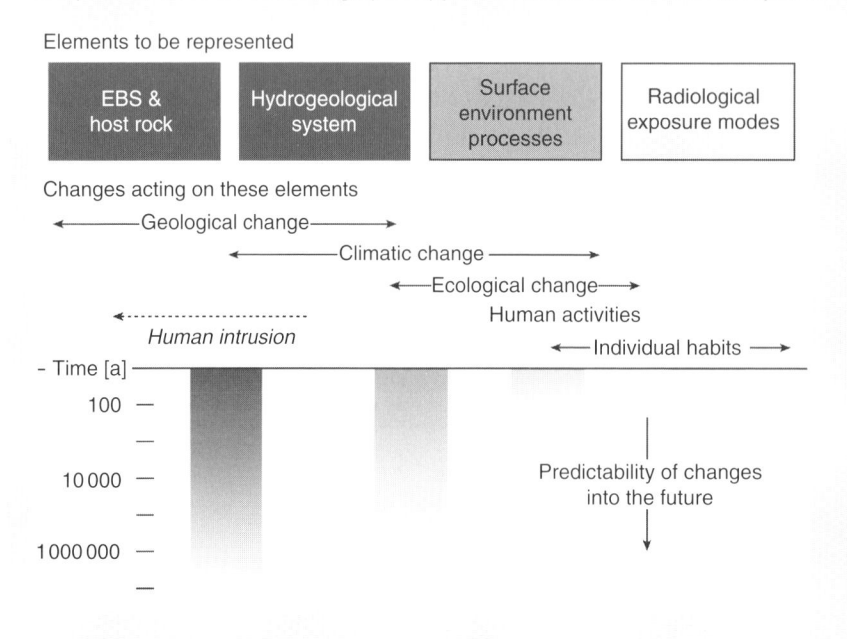

Figure 6.16 The way in which time is conceptualized in the "safety case." Schematic illustration of predictability of the different elements of a geological disposal system (adapted from Nationale Genossenschaft für die Lagerung radioaktiver Abfälle (Nagra), 2002). EBS, Engineered barrier system.

[1] This box has been written together with Corinne Moser.

approaches. Scientists from humanities and the social sciences tend to have a more open, undetermined understanding of time, whereas natural scientists tend to include both representations with a focus on a cyclic, determined understanding of time. Different understandings, however, have consequences for the understanding of the safety case approach. Hence, they can pose a challenge for communication between experts of different disciplinary domains (e.g. experts for communication and experts for risk assessment) in nuclear waste management. However, these different perspectives and understandings can broaden the focus of investigation if they are integrated successfully.

from an environmental science and sustainability view. Psychologically, time has an experiential component and seems to have cognitive representations linked to sequences of and relations between events or chronological linear, abstract mental representations.

6.13 Energy, waste, and resources

Concerns about the accessibility of essential environmental material resources are currently threatening the world. We discuss the potential scarcity of phosphorus, peak oil, and the necessity of rare materials. A definition of essential scarce resources is given in Chapter 13.3. Energy and resources can be considered to be a primary input to human life, and waste and pollutants are outputs. Both sides can cause conflicts between individual and community interests; an intriguing issue continuing under the commons dilemma label (see Box 2.3). Energy use and renewable energy has become of interest, in particular within the climate change discussion. Much research in environmental psychology relates to conservation or encouraging proenvironmental behavior.

6.13.1 Resource conservation

In 1973, public attention was drawn to high energy and resource consumption in the western world when an energy crisis hit in response to a temporary OECD (The Organisation for Economic Co-operation and Development) oil export embargo by Arab countries. Society began to reflect on resource use and psychologists began to investigate people's perceptions, knowledge, and behavior. Resource conservation efforts mostly focused on energy savings, and energy conservation became an issue of environmental psychological research. In 1982, Geller et al. (1982) reported 180 studies investigating conservation behavior (Bechtel, 1997; Gardner & Stern, 2002).

One of the early lessons learned was that changing behavior requires more than just informing people about the societal meaningfulness of resource conservation or increasing prices (Stern, 1992). Psychological studies revealed that people are resistant to behavioral change (for instance, to decreasing room temperature; Brager & de Dear, 1998), particularly if their convenience is affected. The preferred option of energy conservation is seen in technological solutions or increasing energy efficiency of heating. This holds true, in particular for people with low environmental concern (Poortinga et al., 2002, 2004).

There are also cultural differences in preferences about energy systems. Some cross-national, comparative studies on public valuation and understanding of fossil fuel energy, nuclear, and renewable energy showed that people from less wealthy European countries such as Romania and Slovakia showed less understanding of how to use renewable energy and were more strongly in favor of nuclear energy than western countries (Devine-Wright, 2003). In western countries, nuclear energy became a special issue both of societal disputes and psychological research on risk perception (see below).

There is some research on water conservation and biodiversity in environmental psychology. Psychologists' studies about water conservation focus mostly on general factors of environmental behavior, such as environmental awareness, concerns, or habits (Aitken et al., 1994; Gregory & Di Leo, 2003). However, psychology is also challenged by phenomena such as the "Tucson paradox" (Udall, 1985). In Tucson, from 1980 to 1983, 30 000 people arrived to live in the city each year. Thus, water depletion would have become a problem even if everybody did conserve water. This shows features of the dilemma of the commons (see Box 2.3) and is thus a candidate for psychological research.

6.13.2 Waste and recycling behavior

Waste, defined as unwanted or unusable material that someone wants to get rid of, is a subjective concept. What some people consider to be waste may be of value to others. Based on this idea, any waste can become a resource if we postulate that there are constraints where every material on Earth has a positive function. This is an assumption that underlies the concept of closed loop recycling management (e.g. Braungart & McDonough, 2009). However, recycling behavior is also a psychological issue, as described in some studies on recycling behavior conducted in northern developed countries. For example, according to a review by Hansmann et al. (2006), it is difficult to predict people's recycling behavior based on sociodemographic variables such as age, income, gender, household type, etc. Some studies reported significant effects for these variables, whereas many others could not find significant effects.

This is different for the following psychological variables. Recycling behavior is affected by the following factors: knowledge, the awareness of consequences of littering or believing that recycling helps the environment were associated with higher recycling rates (Bratt, 1999; Ewing, 2001); proenvironmental attitudes (Guagnano et al., 1995) or norms. Seeing recycling as a normal activity performed by neighbors and friends is associated with increased recycling rates (Thøgersen, 2008); and convenience – perceived inconvenience and awareness of the personal efforts related to recycling activities decrease the recycling rate (Ebreo & Vining, 2000).

6.13.3 Littering and different types of norms

Littering impacts on sanitation, wildlife protection, and aesthetics. Additionally, the public costs for collecting littered waste are higher than for waste placed in waste receptacles. The study of littering is of interest as it is an example of socially unwanted behavior or of a disregarded social norm. An individual's littering increases if the norm is weak. Further, people litter increasingly with the amount of litter that is already present (Finnie, 1973; Krauss et al., 1978; Reiter & Samuel, 1980). Additionally, personal costs matter. If the distance to the nearest waste receptacle is perceived to be far away or people have to inquire about the location, littering increases (Finnie, 1973; Friman & Poling, 1995; Geller, 1975).

Let us take a brief look at what norms mean psychologically (see also Chapter 7.9). Cialdini et al. (1990) and Reno et al. (1993) differentiated between descriptive and injunctive norms. A *descriptive* norm represents what is generally done in a particular setting and represents a personal norm. An *injunctive* norm is an external norm, specifying what is socially approved in a setting. The descriptive norm is linked to what "is done." The injunctive norm is what "ought to be done." In their "Focus theory of normative conduct," Cialdini et al. (1990) show that these norms are not ubiquitously at work and, in fact, elicit different patterns of behavior. The experiments were performed in clean and differently polluted places and the subjects read passages to elicit these injunctive norms. Results showed that a cognitive focus on the descriptive norm can only reduce littering in a clean place. Conversely, inducing a focus on an injunctive anti-littering norm can reduce littering behavior in both a clean and a polluted place (Cialdini et al., 1990; Reno et al., 1993).

Several field experiments by Keizer et al. (2008) indicate that people are more likely to violate a norm (e.g. show littering behavior) if they observe that others have violated it before (e.g. a graffiti is sprayed next to a "No Graffiti" sign). Further, the injunctive norm elicits different styles of appealing to the norm. Various studies show that non-commanding ambiguous information changed behavior more often than a directive method. The theory is that this challenged people to cope with the given information and they reacted positively to the challenge (Cialdini & Trost, 1998; Hansmann & Scholz, 2003).

6.13.4 Key message

- Environmental psychology focused on energy and water conservation. Psychological variables such as different types of norms, and values are usually better predictors of whether or not people practice conservation behavior than sociodemographic variables.

6.14 Natural and technical disasters

Disasters have accompanied humankind since its beginnings. Thus, natural catastrophes are rare events but also an inescapable part of life and have a substantial influence on many human cultures (Alexander, 2000; Smith, 2001). A natural disaster is caused by the destruction of a human system (e.g. a family or community) by impacts of a physical agent that is not under the control of human systems, e.g. lightning bolts, tornados, mudflows, floods, and droughts. The biggest natural disasters, such as the 1970 Bangladesh tropical cyclone, have killed between 300 000 and 500 000 people. Although this is a seemingly high number, the annual risk of being killed by influenza (1:5000 [risk of death/person/year]) or being struck by an automobile (1:20 000) in the USA is still much higher than being killed by floods in the USA or tornados in the midwest of the USA (each 1:455 000 Tobin & Montz, 1997).

Traditionally, one distinguishes between disasters caused by natural vs. human-made hazards (such as technical disasters). As we can learn from flood management, these two sources are often intertwined, for instance if insufficient levees are constructed around settlement areas. Furthermore, there are socio-economic disasters, such as wars, that may cause much higher numbers of deaths than natural disasters. Also, there can be terrorist attacks or the breakdown of financial systems – these are not discussed here.

We should note that many of the natural hazards include some ambivalence. Fires by lightning bolts, for instance, are needed for the balance of certain ecosystems and floods for watering and distribution of nutrients. Hazards and disasters are characterized by the size, proportion of human population affected, rapidity, length of future impact, and visibility. The magnitude of a disaster for a specific human system is assessed by the proportion and centrality of the human system hit.

6.14.1 Coping with disasters and catastrophes

If we identify the key capabilities of coping with disasters, we have to distinguish the capability: (1) to recognize and perhaps even to anticipate the event; (2) to define proper goals and means for managing the harms before and/or after the event; and (3) to make the right choices.

Psychological research has shown that, following a disaster, individuals often experience a post-disaster traumatic stress disorder phase (Ironson et al., 1997; Norris et al., 2002). People are stunned, numb, and experience mental health problems. Anxiety, depression, physiological distress of the gastrointestinal system and other organs, unfocused anger, and sleep disturbances are frequent phenomena. There is even indication of maternal epigenetic transmission of body reactions to traumatic experiences to offspring (Yehuda et al., 2008; see also Chapter 5.4).

Disasters disrupt communities at different scales. However, many cultures respond with social and spiritual coping strategies, and these have been effective. For example, survivors of the Asian December 26, 2004 tsunami valued these coping strategies more than formal services (Rajkumar et al., 2008). Additionally, people cope better if costs for recovery can be externalized to higher level human systems. In general, the belief and trust in community and the social network are the most important aspects of coping with disasters. Thus, coping with and managing disaster can only be adequately understood when the hierarchy of human systems is considered (Basolo et al., 2009; Norris et al., 2008; see Chapter 16.3). In general, a collective response to massive trauma can even lead to positive effects such as increased community identity and cooperation when rebuilding the damaged human environment. Here, it is interesting that disaster research uses terms such as vulnerability and resilience, which are ecological concepts (see Figure 5.16; Koenen et al., 2009).

6.14.2 Underestimation of threats

A general deficiency of human environmental literacy is the failure to take preventive action. This is a feature that can be attributed to overconfidence and also goes under the label of "disaster syndrome" (Kunreuther, 1996). It consists of two components. First, there is a systematic underestimation and high discounting of the full extent of the risk of rare disasters, which leads to subcritical interest in preparing for protection. For example, when comparing a wide range of disasters, disbelief in warnings is more common than panic (Veitsch & Arkkelin, 1995). Second, the free availability of aid from governments and private charities following a catastrophic event is problematic. Based on this worldwide and well proven phenomenon, economists infer:

> Thus, the only pragmatic solution for overcoming the "Disaster Syndrome" is mandatory natural disaster insurance. (Schwarze & Wagner, 2006, p. 5)

6.14.3 Overreaction to disasters

While the human cognitive system has a huge long-term memory (see Box 7.5), storage and retrieval of information is far from perfect. What we store and retrieve is subject to specific situational constraints, personal interests, subjective filtering, and to the limits and specifics of human information processing. Such limits are described by the term "bounded rationality" and the conception of the human as a "bounded rational being" (Box 1.4).

It has been found in studies of risk research that certain events are cognitively more salient and more available than others (Tversky & Kahneman, 1973), and that disasters show exactly these characteristics. As a result, the number of fatalities resulting from floods, tornados or accidents is overestimated (see Figure 6.17). In contrast, the mortality rates from everyday events such as heart disease, stroke, cancer, or vaccination are subjectively underestimated. Besides the cognitive

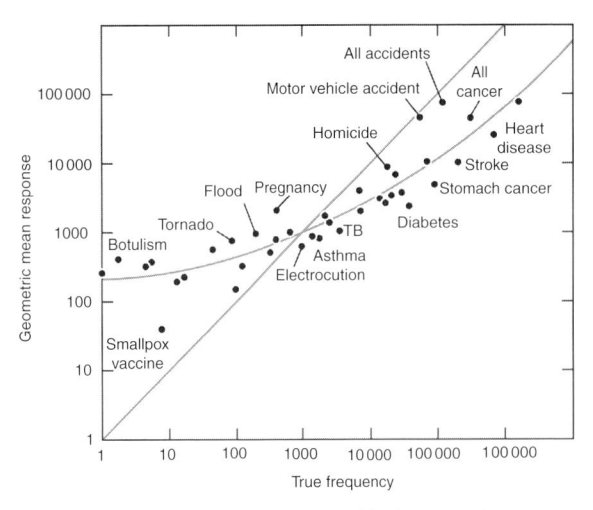

Figure 6.17 Subjects' mean estimates of fatalities (y axis) vs. true frequencies in the USA (x axis; Lichtenstein et al., 1978, p. 565).

salience, disasters are judged to be threatening, out of human control, and an involuntary risk.

Technical disasters resulting from nuclear waste, genetically modified organisms or chemical sites are also stigmatized and thus overestimated owing to moral, ethical, and political reasons (Scholz & Siegrist, 2010).

Naturally, disasters include a subjectivist, constructivist component in their assessment. This includes the hidden agendas of concerned people or those managing disasters with respect to allocating responsibility and gaining benefits. For example, some say the abysmal response to Hurricane Katrina was related to the pre-existing social and environmental injustice, the lack of preparation for disaster, and the skewed political and economic structure. Therefore, the suffering and anguish was felt most sharply in low income communities of color (Bullard & Wright, 2009).

6.14.4 Key messages

- Concerned people are not sufficiently prepared to anticipate disasters although the mortality resulting from disasters is overestimated. This seems to be caused by psychological processes resulting in an underestimation of threats and overreaction to disasters
- Coping with disasters reveals that the trust and the support of larger human systems are essential, both from a psychological point of view and with respect to the actual recovery from disaster-caused harms
- On a societal level, there seem to be backup systems, such as international relief actions, which help societies to cope with disasters.

6.15 Built environment

The built environment is an abstract term that denotes any physical alteration of the natural environment by humans. From hearths via cities and landscapes to the anthropocentically formed areas of the Earth's surface, people create the built environment and the built environment affects human behavior. The environment is a product and a producer of human behavior.

> It suggests that human groups seek to adapt their buildings to their behavioral needs or functional requirements; when the built environment ceases to accommodate behavioral requirements, people seek to correct the problem through construction, renovation, or moving to a different building. Conversely, people also change their behavior to fit the physical environment, especially when it presents limitations. (Lawrence & Low, 1990, p. 460)

This equilibration may also work without people's awareness (Proshansky & Fabian, 1986). The form of the built environment can differ and can represent functional, cultural, symbolic or other purposes. Thus, the construction of solid houses may increase suspicion and hostility. On the contrary, separated houses with spring-assisted bolt doors which automatically lock upon closure represent an individualized social structure. Society produces buildings that maintain and/or reinforce its social norms. The home is the central entity to identify the individual (Gifford, 1997), and this domicile is the focus for the formation of domestic groups. For instance, rectangular buildings are easier to extend than round ones. Thus, it is not surprising that rectangular buildings are used more frequently in sedentary societies than in nomadic ones.

6.15.1 Traffic systems and mobility

Because of their significant roles in energy consumption, material flow, and land use impacts, transportation and infrastructure are key issues for sustainable development (see Figure 6.18). In this respect knowledge from environmental psychology is valuable. We can identify different levels to which psychological knowledge could be applied, such as:

(a) Regional planning: where should settlements be planned and what type of houses best serve people's and society's needs?
(b) Traffic planning: what traffic infrastructure and technology should be implemented and what rebound effects develop with respect to energy use and CO_2 emissions?

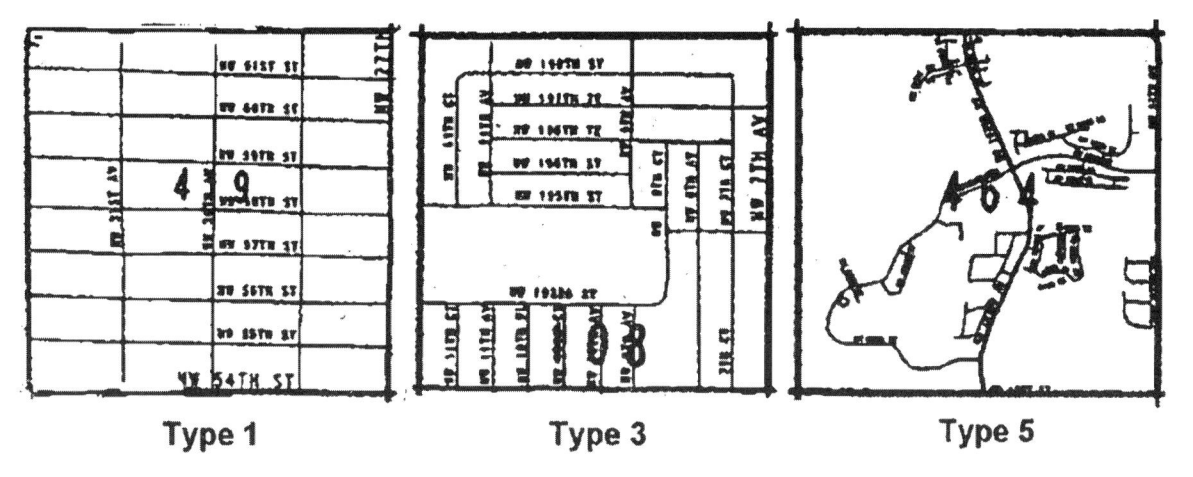

Figure 6.18 Different designs of settlement cause mobility patterns which promote different modes of traffic (e.g. individual vs. public, motorized vs. slow) and fulfill the individual's needs differently (from Ewing & Cervero, 2001, Figure 8 (a), (c), and (e), p. 102. © National Acadamy of Sciences, Washington, DC, 2001. Reproduced with permission of the Transportation Research Board).

(c) Type of transportation: what transportation means are preferred and under what conditions do people choose public or fuel-efficient options?

With respect to (a), the thinking of regional planners, decision-makers, and the public are most important. Research questions could be: what traffic and environmental impacts result from *specific* land use patterns? From the eyes of worried people, stakeholders, and experts, what are the risk aspects of greatest concern? Currently, there is work emerging the transportation sciences, but little or none from psychology (Bullard & Wright, 2009; Cervero, 2002; Ewing & Cervero, 2001; Handy, 1996; Handy *et al.*, 2005).

With respect to (b), when building new roads, railways, airports, etc., or when implementing new modes of transport, such as high speed trains, the demand balance between transport modes and rebound effects may be reshaped. For instance, a new highway may make it possible for people to travel more quickly and cover longer distances (Spielmann *et al.*, 2008). Here, some research in environmental psychology has been initiated (de Haan *et al.*, 2006, 2007).

With respect to (c), much research can be found on the reasons that people choose to use public transportation (Bamberg *et al.*, 2003, 2007) or the purchase of fuel-efficient cars or on slow-moving traffic (de Haan *et al.*, 2009; Peters *et al.*, 2008; Peters *et al.*, 2011; Train *et al.*, 1997).

6.15.2 Key messages

- The built environment is a product and a producer of human behavior

- Housing represents cultural norms, symbolic values, and social order, including the status of the owners and users
- The relationship between the planners' and decision-makers' competences, interests, and activities in the process of planning traffic systems, and the induced wellbeing, material fluxes, and environmental impacts have not yet been sufficiently investigated.

6.16 Pollutants: the invisible built environment

Edelstein called contaminants leftovers of the modern, industrial society and a "second invisible layer of the built environment" (Edelstein, 2002, p. 559). A basic psychological question is whether people can sense these impacts and how sensitive, reliable, and valid people's judgments about contamination are. The impacts of environmental contamination are broad, including changes in atmospheric conditions, effects on human performance, wellbeing (e.g. depression), and emotions (e.g. aggression). The reader can find comprehensive surveys of psychological research in textbooks on environmental psychology (Bechtel, 1997; Bell *et al.*, 1996; Veitsch & Arkkelin, 1995).

Human concern about contamination has been the cradle of environmentalism and environmental sciences (Carson, 1962). Lay people were often more sensitive than experts (see Box 8.2), although our sensory organs are blind with respect to many toxins (see Box

6 Psychological approaches to human–environment interactions

> **Box 6.11** When and why does environmental noise harm? Trains and planes
>
> Noise is ubiquitous and in some sense essential. Noise can harm and severely affect human health (Cohen *et al.*, 1977). However, measuring the impact of noise is difficult owing to three issues: defining a unit of measure, methods for measuring impacts, and the subjective and habitual components.
>
> An objective unit is the physical sound pressure in microbars. Here, we find a huge range between the perception threshold of 0.00002 microbars (equal to 0 dB) and the pain threshold of 1000 microbars (140 dB). Thus the logarithmic derivative decibel is taken as a degree of physical loudness. Psychological measures of perceived loudness are *phone* and *sone*. For instance, a 60-dB sound at 1000 Hz (which equals 60 phones) is equivalent psychologically to an 80-dB sound of 2500 Hz. A sone starts with a 1000-Hz tone at 40 dB. In general, the subjective impression depends on wavelength, symmetry, and amplitude of waves.
>
> However, phones are not linearly correlated with the impacts of noise on the human. There are many ways to measure the negative effects of noise on humans. For instance, annoyance can be measured on an 11-point scale, and high annoyance is defined as a judgment above 6. As we can see from Figure 6.19a, trains and planes cause different levels of annoyance. In Switzerland, the annoyance threshold for airplanes appears to have increased by about 10 dB between 1990 and 2000. However, annoyance is a judgment and can be "politically" biased and thus depends on the context (e.g. on who is asking the question). Therefore other physical measures (see Figures 6.19d and e) have to be assessed such as sleep disturbances. These are usually measured by apparatus-dependent psychophysical measures (Oliva, 1998). There are also almost unobtrusive, contact-free measures that assess sleep disturbances by body movements and vibrations while sleeping (Brink *et al.*, 2006). This may include different changes in the body's gravity depending on the regularity of the heartbeat (the so-called ballistocardiogram), which was invented in 1877 (Gordon, 1877). Such methods can be used to assess dose–response relationships of noise

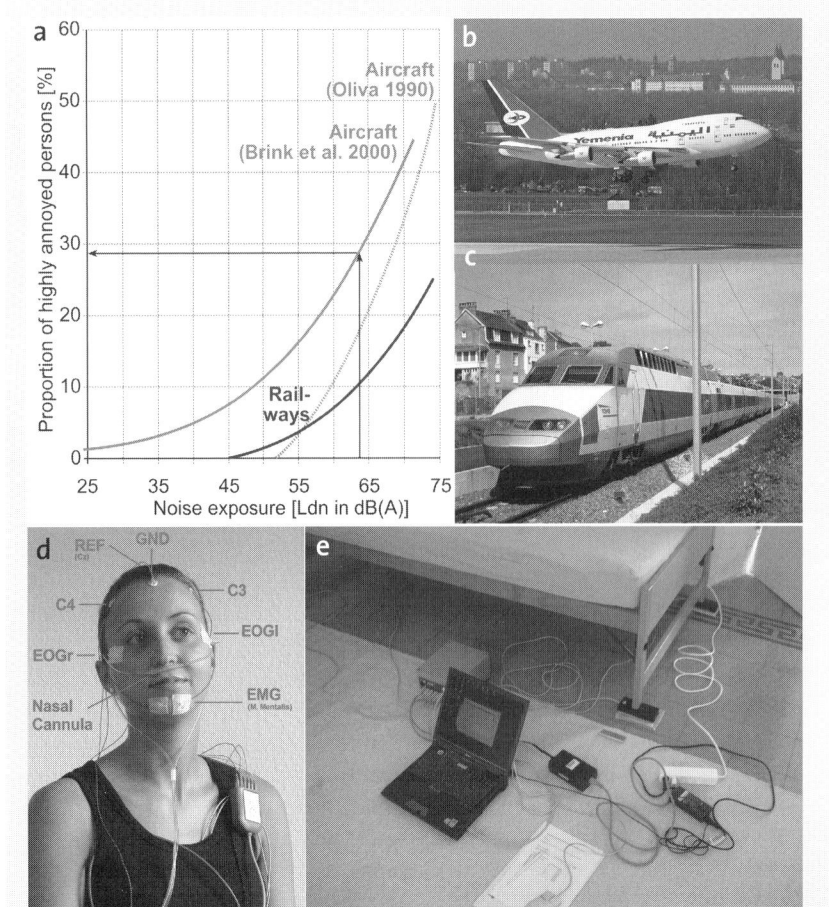

Figure 6.19 (a) The same physical sound pressure has different effects on humans. There are different measurement methods, such as inquiry, obtrusive, or (d) physiological measurements (photo by Mark Brink) and (e) unobtrusive, ballistodynamic methods. EOGl, EOGr, C3, C4, GND, and REF (Cz) represent different electrophysiological signals of brain activities and eye movement relevant for measuring sleep disturbances.

disturbances. Using this method one can show that morning flights over communities cause more harm than late evening flights (Brink *et al.*, 2006).

The health effects of noise cannot be assessed by a single measure. In current life cycle assessments, the costs of sleep and communication disturbances, as well as other effects, are estimated by medical experts (Müller-Wenk, 2004). For example, the health cost of truck noise is assessed at between 1 and 7 US cents per 1000 truck kilometers (Hofstetter & Müller-Wenk, 2005).

6.2). Feeling concerned about environmental pollutants is often linked to other aspects, such as emotional involvement (Weber *et al.*, 2001) and feeling socially suppressed (see Box 9.1).

Psychological research on environmental contamination of soil, water, and air started in the 1980s (Edelstein, 1988; May & Scholz, 1990). People who are objectively affected (high exposure group) by soil contamination judge the risk of those who live on uncontaminated ground (low exposure group) to be equal. However, to the contrary, the low exposure group, when compared to the high exposure group, judged the perceived risk for highly exposed people to be lower than the high exposure group. This is a phenomenon that can be explained by classical theories, such as the theory of illusion of (self-)control (Langer, 1975) or cognitive dissonance theory (Festinger, 1957). With regard to differences in race and gender, it was found that white males tended to differ from everyone else in their attitudes and perceptions, as they perceived risks to be much smaller and more acceptable than other people. These results suggest that sociopolitical factors such as power and status are strong determinants of perception and risk acceptance (Flynn *et al.*, 1994).

Perceived environmental contamination can cause stress, which affects mental or physical health. Psychological research suggests that this is mediated by psychosociological variables (Lima, 2004; MacGregor & Fleming, 1996). The complexity of psychological processes related to environmental stress has been well studied in the case of noise (see Box 6.11).

6.16.1 Key messages

- Feeling concerned about pollutants is often linked to variables other than being physically affected
- The illusion of self-control is at work in people exposed to pollutants.

6.17 Psychological theories applied in environmental psychology

We briefly present three theories frequently used in environmental psychology. Furthermore, the classic multi-criteria decision approach, which is often used in environmental psychology, is described. The key constructs and drivers of environmental behavior and awareness are presented in Chapter 7. All these theories and concepts were taken from other fields of psychology.

6.17.1 Theory of planned behavior

The theory of planned behavior (see Figure 6.20) is one of the most widely used theories to describe environmental behavior. It emerged from the theory of reasoned action in which attitudes towards a behavior and subjective norms are the key variables that affect the intention to act differently (Fishbein & Ajzen, 1975). As discussed in Chapter 7, we avoid the attitude concept as it is too unspecific and cannot be well differentiated from behavioral intent or planned decision. For instance, attitudes correlated with voting on a nuclear safeguard bill by 0.91 (Bowman & Fishbein, 1978). We think that attitudes in

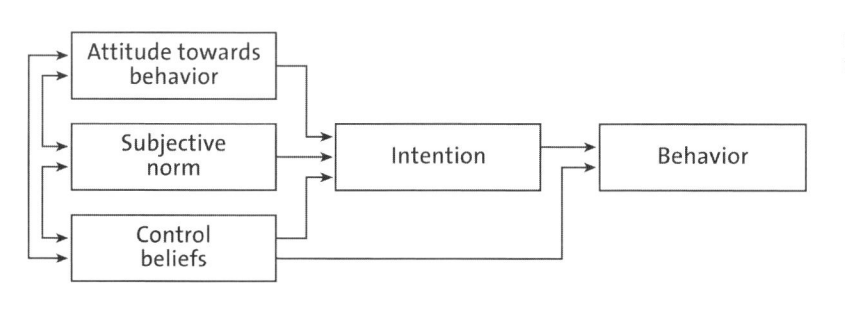

Figure 6.20 The main variables of the theory of planned behavior.

these studies often represent what people want to do. Thus the concept does not help us to become more literate about why people are showing a certain behavior.

This is a little different in Ajzen's (1991) theory of planned behavior as it supplements attitudes by normative and control beliefs (Figure 6.20). The intention to perform a behavior is related to an attitude towards an individual's own behavior, a subjective norm (i.e. the judgment of significant others' opinions of the behavior), an individual's desire to comply with or defy the subjective norm, and the individual's perceived behavioral control (Ajzen, 1991). The model can be conceived as a basic model, which can be modified. The control belief, that is the perceived behavioral control is a variable which refers to the self-perceived impact of contextual factors and represents a kind of self-efficacy. A meta-analysis of 185 studies using the theory of planned behavior has shown that it explains 27% of the variance in behavior and 39% of the variance in intention (Armitage & Conner, 2001).

6.17.2 Norm activation model

Schwartz's (1977) norm activation model – originally stemming from the study of helping behavior – identifies the influence of personal norms, which are influenced by other persons' or groups' behavior and norms. This idea extends the theory of planned behavior by incorporating external conditions and influences. According to Schwartz, an individual acts altruistically if he or she is aware of the negative consequences for others and assigns responsibility for the state of others to himself or herself. The primary external variables affecting behavior are the social norms that influence personal norms. However, to be motivated to change behavior, a person must also be aware that one's own behavior has an impact on the environment that is (potentially) harming others (see Awareness Postulate P6 in Chapter 16).

There are many similar models to the norm activation model, such as the value-belief-norm theory, which combines different types of values, belief of threats and responsibilities, and behavior-specific (situationally elicited) norms and predispositions (Stern & Fineberg, 1996; Wilson & Dowlatabadi, 2007).

6.17.3 Low-cost hypothesis

One of the pioneers of research on environmental behavior was Hans-Joachim Fietkau (Fietkau & Kessel, 1981; Kley & Fietkau, 1979), who combined environmental knowledge, value, options, and incentives of behavioral alternatives with perceived environmental impacts to derive the low-cost hypothesis. This hypothesis states that environmental concern about negative environmental impacts diminishes with increasing behavioral costs (Diekmann & Preisendörfer, 1998). To express it in other words:

> The lower the pressure of costs in a situation, the easier it is for actors to transform their attitudes into corresponding behavior. (Diekmann & Preisendörfer, 2003, p. 443)

If we express this in terms of Lewin's field theory, the low-cost hypothesis states that, if the strength of the situational factors is high, it is unlikely that what the person wants to do is realized. But, if the costs are low, there is little "inconsistency between evaluative statements, one's attitudes, and his or her self-reported and observed behavior" (Kaiser et al., 2007, p. 242). This phenomenon is often interpreted as a gap between intention and action and reasoned by the theory of dissonance. A conclusion which may be linked to the low-cost hypothesis reads that decision models are poor in predicting whether changing behavior is linked to high effort, high costs, or high involvement (Gatersleben et al., 2002).

6.17.4 Multi-criteria decision modeling

Multi-criteria decision modeling (MCDM) and multi-criteria analysis (MCA) are labels for a family of methods that model the individual's evaluation process. The family also goes under the label of multi-attribute utility theory (MAUT), which is often linked to an expected utility. The theory assumes that the choice of an individual between different alternatives can be explained by decomposing the overall attractiveness (von Winterfeldt & Edwards, 1986) of an alternative into a number of attributes. Attributes are preference-related dimensions of a decision alternative. System variables, such as CO_2 emissions, water pollution or the distribution of costs and benefits linked with the production of energy, are attributes, or utility components, u_i, of a decision alternative, A. If the importance of each of these utility components is assessed by a weight, w_i, the attractiveness or subjective utility of an alternative can be represented as the weighted sum of the utility components: $u(A) = \sum_i w_i * u_i(A)$ with $w_i \geq 0$ and $\sum_i w_i = 1$. Clearly, this term would turn into expected utility if the weights were probabilities. The MAUT is the workhorse of many domains of decision analysis, including environmental psychology.

There are many variations of this approach, assuming different cognitive processes that underlie the decomposition and composition of the decision process (Roy, 1991; Vincke, 1992; Yoon & Hwang, 1995). MAUT is not discussed in detail here, but we direct the reader to Scholz and Tietje (2002; see Chapter 16.17) for more information or similar texts (Belton & Stewart, 2002; Figueira et al., 2005).

Some attention should also be given to Saaty (1990), who developed the analytical hierarchy process as a closed theoretical framework. Here it is postulated that attributes are ordered by decision-tree-like structures.

6.17.5 Key messages

- There is a set of theories from social and cognitive-social psychology and psychological decision research that are widely applied in environmental psychology to model human behavior
- There is evidence of a trade-off between the benefit of showing proactive environmental behavior (e.g. conservation) and the costs, or losses, linked to this behavior
- Multi-criteria assessment is a general tool for decomposing and aggregating the evaluation and utility assessments in many domains of environmental psychology.

Section 4: Risk perception

6.18 Risks, hazards, and the environment

Risk is a key concept of environmental literacy. It is a basic concept that people use when reflecting on uncertainty and negative impacts resulting from human decisions. Risk is faced in situations where adverse consequences can occur with some uncertainty. All people have some understanding of what risk is. Thus, referring to Kant, risk can be called an *elementary concept* or *primitive* (Brachinger & Weber, 1997; Scholz & Tietje, 2002; Sokolowska, 2006). We distinguish risk from hazard as risk is related to decisions. In addition, we discriminate between a risk *situation* and risk *function* (see below).

The concept of risk appears to be ancient. There is evidence that there were people called "risk experts" 3000 years before Christ in Mesopotamia (Mumpower et al., 1986; Oppenheim, 1977). Natural hazards and the unreliability in shipping technology were major reasons for inventing the concept of marine insurance around 1300 in North Italy (de Roover, 1945). This was not only an economic decision but was related to psychological processes. The emotional concern emerging from the losses linked to risks is highlighted in Laplace's definition of risk. He described risk as "the probability of the events linked to the hope but also the anxiety of tomorrow" (Laplace, 1825/1999, p. 5). In this context, hope is defined as the expectation of an uncertain, but likely, benefit. Note that Laplace defined risk as the product of all harms and benefits with the probability of their occurrence, which is close to the concept of expected value. This became one facet in the concept field of risk.

Risk *perception* of individuals differs from a risk *assessment* performed by experts (see Chapter 14.5). It is interesting to see that the cradle of psychological risk research is linked again to the unreliability of technology, here specifically to that of nuclear reactors. The electrical engineer Chauncy Starr offered a "quantitative measure of benefit relative to cost for … accidental deaths arising from technological developments in public use" (Starr, 1969, p. 1232). He assumed that societies have records of accidents that happened in the past. Based on this, he claimed, it would be possible to define an optimal balance between the risks and benefits associated with a technology or activity. In the frame of cost–benefit thinking he thought that a new technology should be accepted if it has a similar risk:benefit ratio as other accepted technologies.

This suggestion was not shared by the opponents of nuclear technology, who emphasized the losses (Rammstedt, 1992). In Europe, for instance, the notion of technical risks and losses in the 1970s was almost exclusively characterized as a kind of opportunity cost to support technological progress (Scholz & Siegrist, 2010). This conception of risk, which only incorporates the negative outcomes, was later called *pure* risk and was contrasted with *speculative* risk, which incorporates both the positive and the negative outcomes related to a decision in a risk situation (Brachinger & Weber, 1997; Fishburn, 1982).

The nuclear case, and later discussions about environmental contamination, emphasized that risk was conceived differently by different people. A group of psychologists, including Paul Slovic, Baruch Fischhoff, and Sarah Lichtenstein were not in agreement with Starr's paper "How safe is safe enough," and developed an alternative approach called the psychometric paradigm (Fischhoff et al., 1978). They suggested that, from a psychological point of view, each risk has a kind of personality (see below). Judgments of riskiness rely on many aspects that go beyond quantitative data and include certain qualitative aspects such as dread or ability to control. For instance, air travel is judged to be more risky than travel by car though it is much safer on a per mile basis. The subjectivity of risk perception became evident by the judgment of fatalities caused by different events (see Figure 6.21). This is clearly due to factors such as the uncontrollability of air travel by passengers and the cognitive salience of spectacular accidents (Nisbett & Ross, 1980).

6.18.1 Key messages

- Risk is a basic, elementary concept that people use to evaluate decision alternatives with uncertain outcomes which include losses
- Risk perception of individuals usually differs from risk assessment as performed by experts.

6.19 Conceptions of risk

6.19.1 Hazard vs. risk

Many scientists distinguish *risk* from *danger* (Luhmann, 1993). If we use the term risk, it must be possible to choose between different alternatives and strategies, which may lead to different outcomes (including at least one unwanted outcome). This is the case with technologies or manmade chemical compounds, as society or industry can, in principle, decide whether they are to be produced or not. This is not the case with natural hazards, as the path of, for instance, an asteroid does not depend on human action. If we develop technological measures against asteroids, however, the danger of an asteroid is converted into a risk.

6.19.2 Risk situation and risk functions

To better define risk and understand the multiple notions of risks, we distinguish between risk situations and risk functions. In a risk situation, there is a *choice* between at least two alternatives wherein there is a loss linked to the uncertainty in the outcomes resulting from one of the alternatives. Examples are playing a lottery or not, deciding between different options of improving a levee, or choosing nuclear power or another technology as an energy source.

A risk function describes the process of risk perception. It represents *how* the alternatives in a risk situation are evaluated from the perspective of a human system, a group or society. Psychologically, the risk function represents the perceived and sensitized risk resulting from the choice of alternatives and their outcomes included in a risk situation. Practically, risk functions can often be measured by risk judgments. Technically, the risk situation a person is facing builds the *preimage* set. Following this, the perceptional, cognitive, motivational, and emotional processes elicited in a person when facing a risk situation form the *image* set of the risk function. In different risk situations, a person may have different risk functions. Also, the "values," for instance, what comprises a loss, may differ between situations or persons. A risk situation can elicit pictorial or episodic associations and related emotional processes from past experiences. But it can be also linked with certain semantic aspects, such as losing control or dreading an outcome. Some people might also have in mind numerical values related to a risk situation, for instance when stating that "the risk is low, as losing is below a 5% probability."

Thus, risk can have different notions. This refers also to the quantitative definition of risk (i.e. numerical risk functions). The risk, that is, values of the function (the riskiness evaluated for the decision alternatives) can result from the maximum loss, the expected value of the losses, or the expected value of all gains and losses linked to the outcomes from alternatives (Vlek & Stallen, 1980).

6.19.3 Key messages

- Risk is different from hazard. There are different concepts of risk, such as pure risk and speculative risks
- The risk situation describes the decision situation when risk perception is evoked
- A risk function is a mapping of the risk situation into the space of subjective risk representations. A risk function describes the evaluation of the riskiness of a decision. Risk functions or risk judgments are a component of utility and thus of decision-making.

6.20 Psychology of risk judgments

6.20.1 The psychometric paradigm

According to Slovic *et al.* (1985), the most influential factors in risk judgments are the perception of: (i) dread; and (ii) voluntariness with respect to the outcome and the conviction of; (iii) having personal control over the likelihood or the magnitude of the outcomes; and

(iv) whether the risk is known or unknown; that is, the familiarity with the risk. Familiarity is sometimes split into novelty of and knowledge about a risk situation. Naturally, the situational characteristics, such as the expected number of losses or the catastrophic potential, are decisive factors (see Renn, 1992). However, although the perceived number of fatalities correlates to the judged riskiness of a situation, the relationship is weak and generally explains less than 20% of the declared variance (Jungermann & Slovic, 1993). Finally, (v) risk is rated greater if it violates common equity principles.

The method of measuring the risk function of an individual requires people to assess, for instance, the severity of consequences (e.g. how likely is it that the consequences will be fatal) and newness (e.g. are the risks novel or familiar) for each hazard. In most studies, between 8 and 15 rating scales are used, and participants assess a wide set of hazards (Lichtenstein *et al.*, 1978; see also Figure 6.21). The above main dimensions (i)–(v) are then extracted by the method of factor analysis.

6.20.2 Modes of thought and emotions

Risk perceptions and risk judgments emerge from different psychological processes. We discriminate between *intrapersonal* factors, which are variables describing the cognitive and affective processes within individuals, and *interpersonal* variables, which are related to the psychological processes referring to the interactions with other people and human systems. The term affective here links motivational and emotional processes.

The distinction between the intuitive and analytic mode of thought or cognition is an essential one (see Scholz, 1987; Table 7.4). Some authors would argue that the intuitive process is peripheral, superficial, and cursory, whereas the analytic one is central and systematic (Cacioppo *et al.*, 1986). Here, it is important that these modes of cognition and their elicitation are linked to emotional processes. Emotions have a differential function. People switch from the intuitive to the analytic mode if they become alarmed or afraid (Kuhl, 1983; see Chapter 7.5). Risk perception is thus compiled from different psychological processes (Leikas *et al.*, 2007; Slovic *et al.*, 2004). A set of psychological factors has been provided by the psychometric paradigm. When discussing people's reactions to disasters and pollutants above, we have seen that factors such as cognitive salience, social and moral stigmatization, and feeling concerned play a central role. Coping with

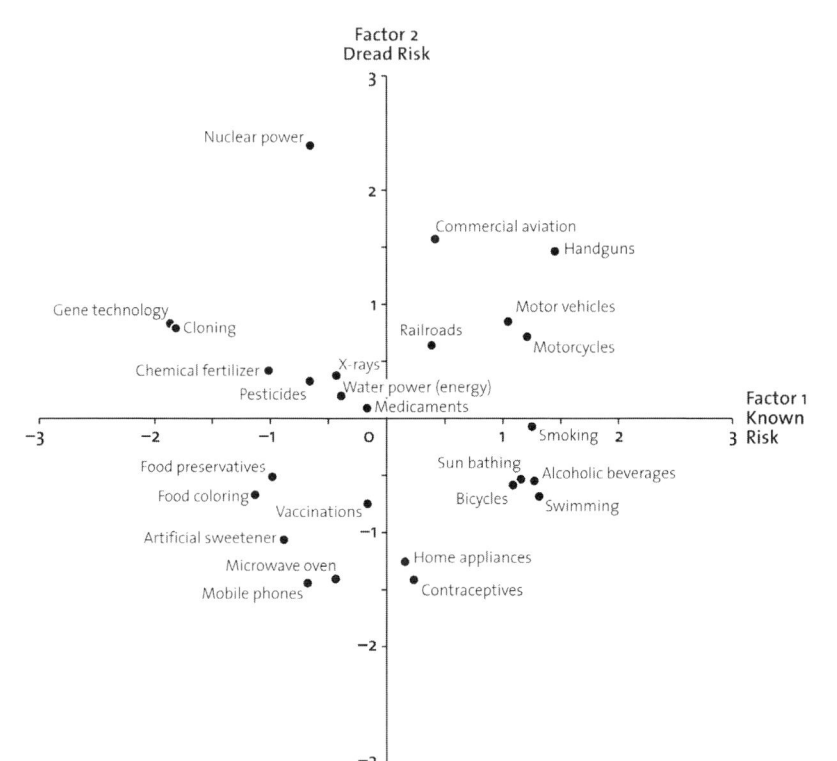

Figure 6.21 Key characteristics of risk perception of different technologies, chemicals, and activities (see Scholz & Siegrist, 2010).

risk is also accompanied by processes such as illusionary control or reducing cognitive dissonance.

A challenging finding was that risk perception is often not so much determined by the factual exposure, for instance to heavy metals, but by emotional concerns (Grasmück & Scholz, 2005). Newer approaches for explaining people's risk perception embed motivational and emotional aspects in what is called the "affect heuristic" (Finucane et al., 2000) or "implicit attitudes" (Siegrist et al., 2006). The affect heuristic assumes that people have positive and negative feelings related to a hazard. The overall affective impression associated with an object determines people's risk perception.

6.20.3 Trust and social moods

With respect to interpersonal factors we only want to make reference to *trust* and *social moods*. Trust is a key element in the initiation of cooperation. Cooperation and relying on others is an essential issue in coping with many environmental problems. People are often saturated with facts when evaluating the benefits and risks associated with technological developments. This is particularly critical if there are controversial expert opinions and/or conflicting interests or values. This is the case in fields like gene technology and nuclear power. In general, trust is most important in situations that are characterized by a high level of uncertainty, potential ignorance, and which are linked to a high degree of personal and moral importance.

In general, a person demonstrates trust in another person if he or she makes himself or herself vulnerable by passing responsibility to and supposing competence in another person. Trust, in this interpretation, becomes *social trust*. We can discriminate between trust among interrelated persons (e.g. people from one group) and trust at a distance, which is related to politicians or governmental agencies (see Siegrist et al., 2005). Siegrist et al. further distinguish between trust in persons or institutions and confidence in the performance of technologies and laws. There are many studies that show that this is a key variable in social risks.

The variable "mood" is usually used on the level of the individual in clinical psychology; however, this notion is not the interpretation here. Rather, we use mood as a societal or group characteristic; however, it still affects individual behavior, such as in group decision rules (see below). A social mood develops and changes over the years. Moods can affect behavior, as has been shown in financial markets (for a review, see Olson, 1965a). Such a social mood develops in societal discourses on the risk object in a broader context, often amplified (or attenuated) through opinion leaders and the mass media (Kasperson et al., 1988, 2003). This social mood is therefore long-lasting and will show effects at the individual level even though a certain event, having triggered its development, might have happened 10 years ago.

Figure 6.22 presents a simple model that shows different variables for predicting the acceptance of

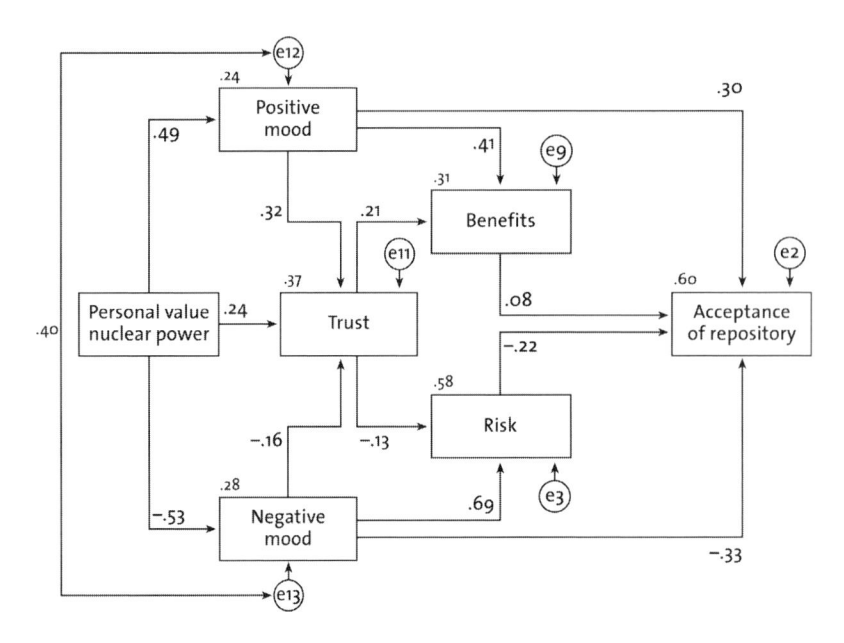

Figure 6.22 Modeling acceptance of nuclear waste repositories depending on perceived risk and benefits as affected by moods and values (Stauffacher et al., forthcoming) by means of path analysis. The numbers on the arrows represent correlation between endogenous variables that are named in the boxes and (hypothesized) exogenous variables (e.g. e12 and e13) in the circles (Arbuckle, 2006). The numbers at the upper left corners of the boxes represent the extracted variance which is extracted by a regression of the input variables (Stauffacher et al., forthcoming).

nuclear waste disposals. The model was validated based on an inquiry of 2428 Swiss respondents. Here, "risks" focused on pure risk and the "benefits" focused on the non-negative outcomes. Trust in governmental authorities, decision-makers, and nuclear industry were key variables, thus representing which values are related to nuclear power and the social (positive and/or negative) moods that affect individual psychological processes.

6.20.4 Key messages

- The psychometric approach postulates that the representation of the losses included in risk situations can be described by a set of traits such as dread, voluntariness, novelty, amount of personal control, etc.
- Risk perception is linked to different types of cognitive processes (e.g. intuitive and analytic thought), which are elicited by and linked to different emotional or affective consequences
- Risk perception is related to trust in social agents and societal moods, affecting the willingness to take or to manage risks.

Section 5: The rationale of a group

6.21 The group as a social environment

Gustave Le Bon (1896) developed the notion of a group mind as an "extra-individual" phenomenon. Le Bon's idea was to explain the influence of a crowd's behavior on the loss of individuality. However, in social psychology, groups are mostly distinguished from crowds as well as from social entities such as family or kinship associations with formal membership (Crott, 1979; Hofstätter, 1957). Also, societal units such as cultures or communities are not a primary topic in psychology.

Groups are social aggregates. Groups can be defined as "a set of individuals among whom a durable or observable set of relations exists." Group members have some prior relationship with one another and an expectation of some future relations. In general, individuals have a temporary group membership and may belong to different groups. Group formation depends on physical proximity, affective integration, and cognitive integration, including group norms (Festinger *et al.*, 1950; Tajfel, 1981). Examples of groups are friendship or residential groups (such as people living together in an apartment or housing area). Whether large groups such as environmentalists, vegetarians or animal lovers/activists are considered groups or sub-cultures depends on the perspective of investigation. Thus, the definition of group membership is rather fuzzy (McGrath, 1984).

To understand the rationale of a group, the following approach is helpful. Groups are characterized by their structure (i.e. the patterns of interpersonal relations), their process and decision rules (i.e. the activities that take place along structural paths and patterns for aggregating the input on the individual group members and the impact on group members), and products (i.e. outputs or responses following interaction, work on some task or following a goal). The action of a group depends on the individual members; for instance, their competence, goals, self-perception, as well as on the embodiment of their social environment (i.e. organization, community or social context), and the type of task, goal or problem with which a group is coping (Davis, 1969).

From an environmental literacy viewpoint, groups and the psychological research of groups are of interest from various perspectives. First, groups have their own rationale (i.e. they react and function in a different way from the individual); they do not linearly transform the knowledge and the preferences of the individual group members. Second, groups are a social environment of the individual and can transform or enhance opinions of the individual. Third, they are constituents of communities and societies. Some examples are self-help groups or support groups and even groups of neighbors that carpool.

6.21.1 Group decision schemes

Postulating a rationale for groups, however, does not mean that we consider the rules and regulatory mechanisms of groups as operating independently from the individuals' beliefs. As Bandura (2000) deduced, the behavior of the individual and the social structure of the group work interdependently. The social structures of a group are created by certain beliefs in collective efficacy.

Each group has different group decision schemes (see Box 6.12) which – dependent on the task – defines under what distribution of pros and cons the group takes a certain opinion, decision, or action. There may be unwanted trajectories underlying groupthink. The Bay of Pigs fiasco of the US government in 1961 was thought to be provoked by unrealistic group decisions (see Figure 6.23). President John F. Kennedy was

> **Box 6.12** Group decision schemes matter: the Bay of Pigs fiasco

The investigation of the role of social decision schemes as an internal regulating mechanism was first illuminated in an unpublished masters thesis by James A. F. Stoner (1961). In a questionnaire he asked the subjects to choose their preference from various risky alternatives. The experimental setting, which was frequently applied in later research, was the selection between different lotteries L_i ($i = 1,2,…, n$), which all had the same expectancy values, i.e. $E(L_i) = E(L_j)$ of a win (where $E(L_j)$ is the expectancy value of lottery L_j) but differed with respect to the probability p_i of winning a certain gain u_i with $E(L_i) = p_i*u_i$. What Stoner found out in his experiment was that groups tend to be more risky than the average individual. This is called the risky-shift phenomenon and was also labeled the group polarization effect. Later, Davis introduced a formalized theory of decision schemes (Davis, 1973; Davis et al., 1974).

The essential point of this approach is that groups have different rules for accepting a certain choice alternative. For example, as a rule, each alternative that received an absolute majority could be accepted, or the alternative that requires the lowest number of concessions, could be preferred instead. One can show, depending on the distribution of prior preferences and the applied decision rule, that a risky shift or a conservation shift (i.e. that the group decision is less risky than the average individual) can result. There have been many other explanations for this phenomenon, including that of individual shifts where group members take more or less extreme positions (see, for example, Zuber et al., 1992). However, this example demonstrates one aspect of the group rationale.

The role of critical effects of group polarization was elaborated by Irving L. Janis in his book *Victims of Groupthink: A Psychological Study of Foreign-Policy Decisions and Fiascoes* (Janis, 1982). The message of this work is that we have to be aware of psychological contagion, which can induce poor decision-making (see Figure 6.23). Group members who try to minimize conflict and groups who lack a culture of critically testing, analyzing, and evaluating the foundations of decision-making are prone to make inadequate, hasty, overconfident, and overoptimistic decisions. Groupthink is linked to structural issues, such as homogeneity of members, insulation of the group, an illusion of invulnerability on the part of the directive leadership, and a lack of quality-assuring norms and procedures.

Figure 6.23 The Excomm meetings of 29 October during the 1962 Cuba crisis (Photo: nsarchive.org). The Bay of Pigs fiasco of the US government in 1961 was thought to be provoked by decisions influenced by groupthink.

coached by a highly homogenous group of experts coming from the same academic school. Charismatic leadership and the lack of procedures for inquiring about the quality of data, arguments, and conclusions are prerequisites of unrealistic decisions. This aspect should be critically reflected, for instance in the risk assessment of nanoparticles, genetically modified organisms or any large scale implementation of new technology.

6.21.2 Group pressure

A central issue of a group's rationale is that of social or group norms.

> Norms are generally enforced only for behavior that is viewed as important by most group members. … Only the behavior that ensures group survival, facilitates task accomplishments, contributes to group morale, or expresses the group's central values are likely to be brought under normative control. Norms that reflect these group needs will develop through explicit statements of supervisors, critical events in the group's history, primacy, or carry-over behaviors from the past. (Feldmann, 1984, p. 52)

People are attracted by groups that meet their personal norms, and so group norms affect individual norms. Naturally, the group norms affect people who identify with the group more strongly (Terry et al., 1999). Asch demonstrated the power of a group on the individual opinion (see Box 6.13). It is still unresolved whether group norms (such as universality; see Schwartz, 1992) can be perceived as a kind of altruistic type of value and how they can be formed. Experimental research shows that behavior of ingroup members that deviates from

Box 6.13 Correct perceptions but incorrect judgments: group power

One of the most often reproduced studies is Solomon E. Asch's group conformity experiment (1956). In the experiment, a naïve participant was brought into a lab with seven to nine "stooges" who were accomplices of the experimenter. Usually seated last or second to last, the participant was asked to judge the length of a presented standard line (see Figure 6.24, Exhibit 1). All the others rated the length of the standard line equal to that of A or C on Exhibit 2. The naïve participant was asked for his judgment last, after all the others had provided incorrect responses. The pressure to conform was clearly shown – 37 out of 50 subjects went along with the incorrect responses of their confederates. The subjects later reasoned that they wanted to appeal to the group and not to upset the experimenter or that they believed that the judgments of the others were correct. The frequency of joining in the incorrect judgments decreased dramatically if there was one ally who provided the correct answer or if there was some degree of privacy (e.g. writing the answer behind a screen reduced conformity). Culture (i.e. whether the subjects came from an individualistic or collective culture; Neto, 1995; Perrin & Spencer, 1981) also mattered. But the experimental effect certainly also strongly depended on the acting talent of the accomplices.

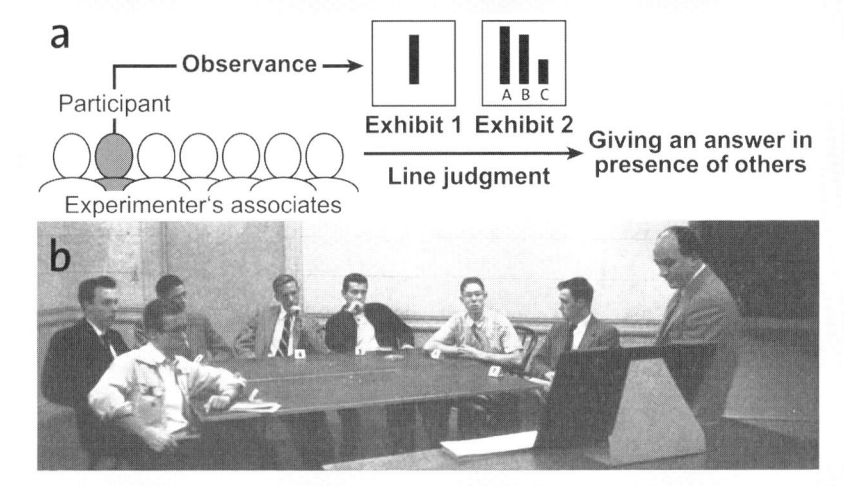

Figure 6.24 (a) The experimental design of the Asch conformity experiment and (b) a photo from the experimental situations. (Asch, 1955, p. 31, reproduced with permission © Scientific American, a division of Nature America, Inc. All rights reserved.)

the group norm is punished more harshly than deviating behavior of outgroup members. This phenomenon is known as the black sheep effect (Marques et al., 1988). We consider the formation of group norms and group pressure to induce compliance to these norms as drivers to maintain existence and functioning of groups.

6.21.3 Ingroup–outgroup relations

Ingroup–outgroup relations based on these characteristics had already been investigated by Sumner (1906) in the early twentieth century (see Tajfel, 1982). The principles of group formation include identification (see Box 6.14) with the own group and differentiation from other groups, with a tendency to compete with these groups.

Unfortunately, for a long time group research concentrated on small groups and neglected collective behavior and cultural psychological research (Graumann, 1997). However, it is evident that environmental perception, values, and behavior very much depend on cultural, religious, ethnic or political relations, and world views. A critical question from an environmental point of view is which of these grouping variables is most relevant? First, if we look at contemporary research, it is interesting to see that some studies have shown that religion (e.g. Islamic vs. Hinduism) dominates nationality (e.g. Bangladeshi vs. Indian) and language (Hagedorn & Henke, 1991; Hewstone et al., 1993). Nevertheless, the issue remains open. Religions uphold basic values, beliefs, world views, ethics, and morals, and many assume that western morals, values, and religions contradict proenvironmental eastern or indigenous concepts. However, evidence provided by Gardner and Stern suggests "that the religions/morals/values framework has only limited power in explaining why environmental problems occur" (Gardner & Stern, 2002, p. 38). Though world views such as the New Ecological Paradigm, which have a strong ecocentristic norm and religious components, exist, Gardner and Stern thought that the values had the same potential to overcome the structural and market constraints of organizing, e.g. life in the western world.

Box 6.14 Signs of environmental compliance: buttons

Religious groups, sport fans, and environmentalist groups use clothing, stickers, buttons, and symbols to mark their identity. Stickers, buttons, tags, and posters have different functions. On the one hand, they show compliance to a certain group, norm or behavior (Figure 6.25a). On the other hand, politics induce social norms through their use. Figure 6.25b shows an American propaganda poster asking for support by saving energy during the World War II effort.

Figure 6.25 (a) The buttons present group norms from the environmental and the political–environmental movement. (b) A 1943 propaganda poster from the US government urges carpooling during World War II (by Weimer Pursell).

6.21.4 Key messages

- Groups develop their own rationale and norms
- There are at least three mechanism which show that groups have drivers: these are group pressure, group decision schemes, and ingroup–outgroup relations.

6.22 Commons and public goods dilemmas

To systematize the following examples, one can distinguish two dilemma types: one referring to the usage of a resource (commons dilemma) and one referring to contributions to public good (public goods dilemma; see Dionisio & Gordo, 2006). We presented a version of the tragedy of the commons in Box 2.3. This version of a dilemma situation refers to population dynamics and how many children a person should have. In this section the issue is how many resources an individual should take from a common pool. Basically, a commons dilemma can be conceived as an N-person prisoner's dilemma game (Beckenkamp, 2006; see Box 6.15). Thus, the commons dilemma is the prototypical example of identifying interfering rationales, interests, or utilities on the individual micro and the social macro level. The problem has already been described by Thucydides (c. 460–395 BC; 1910), and Hardin (1968; 1971) referred to an essay of the British economist William Forster Lloyd (1833).

In short, a commons dilemma can be described as a specific type of social dilemma that refers to a situation in which a number of individuals share a common resource (e.g. fish, water, climate stability, or clean air) from which they can harvest. If too many members take too much from the common source, it becomes exhausted.

There are two important factors characterizing this dilemma situation (Messick & McClelland, 1983). First, one can identify a so-called social trap, which means that the gain of an action (overharvesting) feeds back to an individual, the possible losses of an overharvesting assign to all. Second, the temporal trap describes the observation that the members of a commons discount future losses. The gains belong to the present and this dominates behavior. That is, in the short term, each individual receives a higher payoff and utility for defection than for cooperation. In the long term this leads to overuse of the resource pool. However, if certain constraints such as face to face communication trust, reciprocity, etc. are given, the development of a community does not necessarily become as destructive as Hardin initially assumed (Ostrom et al., 1999). A third aspect refers to the spatial gap between those who gain the most and those who

6.22 Commons and public goods dilemmas

Box 6.15 Resource dilemmas need social solutions: Mongolian grassland

Colloquially, a commons is a good, which is equally held, enjoyed, utilized, or experienced by a number of individuals.

Historically, in Anglo-American property law, a commons has been an area of land for use by the public. In feudal England, Lords released commons, usually uncultivated land or waste, to support the commoners' livestock, for collecting firewood or for using as pasture. In the nineteenth century, the government assumed the right of approval ("Commons"; *Encyclopaedia Britannica*, 2009). There has been a dual conflict with the commons: first with the lord's right to appropriate land for his own use, provided he left enough land to support the commoners' livestock; and second, that the public not overuse the land. The famous article by Hardin (1968) considers pasture common land with open access or community management. Historically this has not been the case, because often each person had the right to graze a maximum number of cattle. However, if no appropriate regulation is given, and the resources are limited and scarce, overexploitation seems to be a general phenomenon. Historians show that the commons dilemma also shaped water, public hygiene, and fire control in medieval towns (Ewert *et al.*, 2007).

Factually, one can find many conflicts within biotic resources, such as pastoral land, fishing (St Martin, 2001), wild game (Siren, 2006), forests and fuel wood (Pandit & Thapa, 2003), deforestation (Parayil & Tong, 1998), and biodiversity (Lovett *et al.*, 2006). This is particularly enhanced by population growth and the belief of open access rights to a resource. With abiotic resources such as water, use for drinking and irrigation is under pressure in many parts of the world (Mitchell, 1976; Trawik, 2003). However, a limited common pool is emerging for oil (Lloyd, 2007). For gas, coal, and non-fuel minerals, market control as "prices increase the stocks principle," currently still dominates. This principle states that, in case of mineral scarcity, exploration, mining, substitution by other materials, and recycling will regulate the demand–supply relationship (Tilton, 2003; Wellmer & Kosinowski, 2003).

Actually, always when the tragedy of the commons appears, researchers ask for changes of behavior and effective institutions. Some answers to this are provided in Table 6.1.

Game theoretically, the situation of the tragedy of the commons can be presented by an N-person prisoner's dilemma game. Suppose there are six farmers having access to pasture that can maximally nourish six cows without deterioration from overgrazing. If each farmer puts one cow on the plot, the cow will weigh 1000 pounds. For every additional cow that is added, the weight of each animal decreases by 100 pounds. Given that the weight is the only measure of yield, the wealth of the farmer increases if he adds one cow, as he will have two cows weighing 900 pounds each. If, however, all six farmers add another cow, then there will be 12 cows that weigh 400 pounds each. Thus the farmers are trapped by the social dilemma and would do better with one cow.

Exemplarily, Mongolian grassland (Figure 6.26), its history, property rights, conflicts, and ongoing dilemmas, is a typical example of common pool resources (Li *et al.*, 2007; see also Chapter 17.3). Mongolia is one of the least densely populated countries in the world. While the average population density is just over 1 person/km^2, the population density of the southeastern Bayan-Khongor aimag region is only 0.2 person/km^2. The traditional property rights (e.g. during the Genghis Khan period) had clear hierarchical allocation to rotational rangeland and hunting ground use.

Figure 6.26 The Mongolian pastoralism has been an example for successful management of a common resource pool. Economic changes, population pressure, privatization, climate change impacts, and other factors have doubled intensity of usage (Photos by B. Lehmann).

> **Box 6.15** *(cont.)*
>
> Nobles, who were the owners, and herders took different roles and functions. A middle level of small supra-family units, of 2–20 households ("people of a valley;" Williams, 2002, p. 62), organized differential access for cows, horses, camels, and sheep herds. There was an organized collective access to commons. Property rights changed during the communist regime into a collective/state property regime, which promoted Russian urban agriculture with permanent settlements. However, there were no reports about severe rangeland degradation of pasture before the time of economic reform (China in 1980, Mongolia in 1990). "Livestock and privatization and shifting to market economy was incomparable to the several thousand years of nomadic history" (Enkh-Amgalan, 2008, p. 1). Since 1990, herder families have grown from 90 000 to 180 000. The impacts of droughts and climate change now demand innovative management schemes linking the individual herder family, groups of herders owning pasturelands, and institutional representatives such as agricultural and land officers (see Table 6.1).

suffer the most because of resource degradation (Vlek & Keren, 1992). This could be called a justice or vulnerability trap, visible for instance in the global climate issue: those who "used" the atmosphere the most by CO_2 emissions will suffer much less than those most exposed to climate change (countries like India or Southeast Asia).

The size of the community can differ from small, for instance a neighborhood, to large, such as the entire planet. A critical issue is that the commons dilemma can be seen as an appropriation situation in which different contextual and institutional framings are at work. In the basic model, the agents or appropriators are considered to have the same endowment and power. Factually, the agents can impose a variable amount of efforts and actions for use of the commons (Ostrom, 1999). For instance, in Africa many households receive a large share of their income from forests or fuel wood (Lovett *et al.*, 2006).

The dominant research method of psychologists investigating the dilemma issue is the experiment. However, as the problem is of interest to policy science, sociology, economy, and environmental sciences, psychological research on this topic has been conducted by scientists from many disciplines. Besides experiments and computer simulations (Nerb *et al.*, 1997), historical and real-world surveys and case studies have been the applied methods. Furthermore, it is interesting to see that the commons dilemma and the difficulty in finding an appropriate trade-off between the individual advantage of having more offspring and species-efficient strategies is a basic evolutionary principle. It has been observed in microbial cell interaction (MacLean & Gudelj, 2006; Pfeiffer *et al.*, 2001).

6.22.1 Collective action

Coping with limited resources requires collective action. According to Hardin (1968), the difficulty lies in human behavior. He pinpointed this by the following statement:

> We are locked in a system of "fouling our own nest," so long as we behave only as independent, rational, free enterprises. (Hardin, 1968, p. 1245)
>
> Each man is locked into a system that compels him to increase his herd without limit in a world that is limited. Ruin is the destination toward which all men rush, each pursuing his own best interest in a society that believes in the freedom of the commons. Freedom in the commons of all. (Hardin, 1968, p. 1244)

Thus, as has been expressed in Box 6.15, the commons dilemma needs collective action. Social psychologists and game theorists consider it to be an open question as to whether scarcity is enhancing or lowering cooperation.

> When a resource is being sustained at a level which does not threaten its future availability, there is essentially no conflict between individual and collective interests, so that level of group identification should have little or no effect on individual decision making. As a resource is depleted, however, individual users should alter their choices in the direction of either greater self interest or collective interest, depending upon the level of group identification as a reference point. (Kramer & Brewer, 1986, p. 210)

The situational constraints certainly matter and the threat of competitive, violent confrontations, which have been ubiquitous in prehistoric times, are inherent in many resource dilemmas. Finding an adequate solution is beyond the ability of a single agent. In the long run, competitive behavior is often less profitable than cooperation. This, however, as von Carlowitz (1713) has shown with forest management and which is as true for the current climate change policy, requires a perspective beyond the lifetime of a human being.

If we wish to summarize psychological research on the commons dilemma constructively, we have to

Table 6.1 Relevant psychological variables and drivers on the levels of the individual, "group," and society in the commons dilemma.

Relevant psychological variables and drivers	Typical situation
Level of the individual	
Problem understanding, enlightened self-interest	Understanding the situation, recognizing that the situation and not the people are malignant (Scholz & Tietje, 2002, p. 203); the individual can develop enlightened self-interest in cooperative long-term action
Cooperative motivation and norms	Normative, self-applied rules in the sense of March (1994); motivating collective action and pre-investments into a common pool (Dawes & Messick, 2000)
Empathy, trust, cooperative reputation	Empathic emotion (Batson *et al.*, 1995), open communication introduces honesty and trust (Samuelson & Messick, 1986), the reputation of keeping promises supports cooperation (Ostrom, 1998)
Learning reciprocity norm (tit-for-tat)	Tit-for-tat strategy as a behavioral strategy (Axelrod, 1984; Perugini *et al.*, 2003)
"Group" level	
Organizing collective action	Self and medium-sized organized small group interactions are effective; group formation, including working against other groups (Caporael *et al.*, 1991)
Open, face-to-face communication and dialogues	Agents have the opportunity to communicate group norms and to appeal to others (Earle & Siegrist, 2006; Kjaernes, 2006)
Sanctioning each other	Positive effects but high costs and spiraling aggression possible (Ostrom *et al.*, 1994, 1992)
Group and community identity	Introducing group norms and group identity (Bicchieri, 2002)
Reducing uncertainty	Increasing transparency in decisions of members of the group about resource size, costs, and group participation
Functional networks	Agistment networks
Group learning	Jointly developing insight into the biophysical structure of the problem; making the impact visible
Societal level	
Regulation by external authorities	An external authority controls and regulates overuse, for instance the coastwide control of lobster gangs (Acheson, 2006a, b); authorities can regulate mutual intervention between individuals (Ostrom, 1999); institutional resources for effective management needed, such as the "polluter pays" principle
Privatization	Breaking the commons down to individual owned parts, harvesting the same amount, dividing the commons – marine lobsterman (Acheson, 2006b)
Regulation and nationalizing	The government takes the leadership. Individual incentives for supporting long-term view needed. Systems perform worse (Ostrom, 1999), police-like control systems
Sustainability learning	Understanding secondary feedback loops which emerge from free-riding

notice that (social) psychological research has doubtlessly concentrated on the individual level and has considered the individual as a relatively autonomous being. Little attention has been given to decision-makers as members of a collective or a social group.

Factually, the rationale and impact of different social systems are at work. Table 6.1 summarizes the essential findings on the tragedy of the commons problem. Somewhat simplified, we distinguish between the individual, the societal, and a "group" level. With the societal level, we mean a community, state, federation, nation or similar alliance organizing collective activities and interests. In Table 6.1, the group comprises what we have defined as a (small) group from a psychological perspective. However, neighboring clusters of farmers or companies can also be included. From this perspective, the tragedy of the commons can be seen as a triple micro–macro level problem: individual vs. group, individual vs. society, and group vs. society.

From the individual's perspective, above all, it is important to recognize and to understand the nature of the conflict situation. The agent must realize the malignant structure of the situation and penetrate the dilemma each person is facing. Provided that the agents are not constrained by a situation of affliction, this would help to develop the belief that positive outcomes will result from a collaborative effort. Being aware of the chance and necessity for a collective solution and gaining insight into the necessity for a long-term strategic action (in the sense of coping with delayed environmental change and reward) are basic prerequisites of collective action. But this is in some respects a necessary and not a sufficient prerequisite of collective action.

Behaving in a non-cooperative manner, and therefore violating social norms, can foster cognitive dissonance. This dissonance between the knowledge that one should cooperate and the actual non-cooperative behavior can be an important driver toward cooperative motivation, the emerging and eliciting of social norms, and the insight for investing: "I must pay for the benefit I get" (Dawes & Messick, 2000, p. 112). There is evidence that individualistic motivation compared to cooperative motivation promotes the decline of the resource pool (Loomis *et al.*, 1995; Roch & Samuelson, 1997). However, people must believe that they can master this situation.

The initiation of cooperation is stronger if empathy and an understanding of others are developed. Empathy is an understanding of the thoughts, values, feelings, and actions of others (Cotrell & Dymond, 1949). Thus, empathy includes an emotional component (Batson *et al.*, 2002) and is often an antecedent step towards developing group identity. However, cooperative reputation also becomes a value, and trust becomes a behavioral trait. Finally, on the level of the individual, the development of strategic capacity is helpful. Here the knowledge that adequate punishment works is splendidly represented by the tit-for-tat strategy (Rapoport & Chammah, 1965). Emotional issues such as envy are also included in this topic. This has been stated by researchers from evolutionary psychology:

> ... [a] system that regulates one's level of punitive sentiment in collective action contexts is functionally specialized for removing the fitness advantage enjoyed by free riders ... (Price *et al.*, 2002, p. 231)

When considering the macro level, it is important to see the differences between the group and the societal levels. In general, collaborative action, for instance in the case of a public good, can emerge either voluntarily or be mandated from institutions such as government agencies. Clearly, privatization or police-like control and punishment are often not thinkable means at the small group level.

An essential part on the macro level is to organize collective action. "Group" and society have to cope with two issues. One is that the individual thinks that others are taking over *their* job and that their own investment is superfluous. This has been called free-riding (Olson, 1965b). The other is that the individual thinks that others are fulfilling *his/her* job and that his/her own contribution is unidentifiable. This has been termed sucking or social loafing (Latané *et al.*, 1979). The latter goes back to the motivational component of the Ringelmann effect, which states that the performance of an individual in a rope pulling group competition is proportionally reduced by an increase in the group size. In large groups, people tend to free-ride and loaf. In phenomena like this, the group size matters.

> The larger the group, the better the chances that there is another member more (less) capable than me, and the better the chances that the group could succeed without my efforts. (Kerr, 1986, p. 7)

To counter these unwanted processes, the group must organize communication and sanctioning rules. Open, face-to-face communication increases cooperation and allows the emergence and maintenance of group norms and trust. Here, trust seems to refer not only to shared social norms, but also to the concrete organization and group decision rules (Earle & Siegrist, 2006; Kjaernes, 2006). Social norms and reliable, positive feedback are more decisive with increasing resource scarcity (Dodds, 2005).

What the group does and how it sanctions must become predictable, similar to the cooperative reputation of the individual. So, with this strategy, group norms and group identity may become established. However, these measures have to take care of some unwanted phenomena of group dynamics, such as ingroup–outgroup polarization (see above). This may indicate that there is no effective cooperation between groups. If engaged group members notice that there is a large process loss by group coordination and that they would gain better access to environmental resources individually, they are likely to lose motivation to cooperate.

Also important on the group and societal level is uncertainty (Ostrom *et al.*, 2007). Uncertainty may refer to the amount of resources, the costs and effectiveness of intervention, or transparency of the interactions of others. Here, social and ecological information seem to be similarly important (Mosler & Brucks, 2003). Uncertainty can also elicit overly optimistic expectations. Finally, group learning and group efficiency are decisive outcomes based on the level of uncertainty.

The societal level provides other opportunities, such as privatization, nationalization, and regulation by external authorities, which are less a matter of psychological research. Policy scientists such as Ostrom (Ostrom *et al.*, 2007) emphasize that the governance of human–environment interactions use polycentric, integrated approaches. There is no panacea. The same holds true for policy, which must be adapted well. In general, early research recommended that structural changes to the commons are usually more effective management strategies than trying to influence individuals (Messick *et al.*, 1983; Samuelson *et al.*, 1984). Finally, we note that cultural differences in coping with the tragedy of the commons are well documented in many studies, though still disputed (Ruttan, 2006).

6.22.2 Key messages

- We can distinguish between the commons dilemma with respect to resource use or scarcity and with respect to contributing to or building up a common good
- Coping with the dilemma of the commons requires a careful interplay between individualistic and social motivation
- There are different psychological drivers on the micro level (i.e. the level of the individual), the meso level (i.e. the level of the group), and the macro level (i.e. society), which can enhance cooperation and avoid destruction in a commons dilemma situation
- The analysis, understanding, and management of commons dilemma situations include genuine psychological analysis at all levels of human systems.

Part III Contributions of psychology

Chapter 7

Drivers of individual behavior

7.1	Psychological foundations of environmental decisions 190	7.7	Motives 205
7.2	Preferences and utilities 192	7.8	Values 206
7.3	Drives 197	7.9	Norms 207
7.4	Needs 198	7.10	Motivations 209
7.5	Emotions 198	7.11	Attitudes 209
7.6	Cognitions 200	7.12	Relating and distinguishing drivers 212

Chapter overview

Chapter 6 provided insight into how individuals perceive the environment. This chapter discusses the questions: what drives individual behavior and what are the psychological foundations of environmental awareness? The idea is to gain insight into the rationale of the individual when interacting with the material and the social environment by using the concept of *drivers*. A driver is a psychological variable or construct that elicits or steers the behavior and actions of an individual. The goal of this section is to introduce key drivers and other psychological concepts that allow for an explanation of environmental literacy and behavior. The concepts and models introduced serve to explain the notions of goal formation, strategy selection or evaluation, and action (see Figure 16.12*).

As a discipline, psychology can inform us about the drivers of individual behavior and action towards the environment. This chapter begins with a discussion of the psychological foundations of environmental decisions by reviewing theoretical foundations from psychology researchers. We also present a loose taxonomy, or rough order, that allows us to interrelate a large spectrum of drivers based on the degree to which they are rooted in basic psychological processes (such as preferences, drives, needs, and emotions) and shaped by experience (such as cognitions, motives, values, and norms). Finally, we discuss drivers shaped by experiences and behaviorally relevant higher order psychological processes: motivations, norms, and attitudes.

The chapter closes with a topology that relates the discussed key drivers according to two dimensions or scales: (1) the degree to which drivers are shaped by experience vs. psychological processes; and (2) the degree to which drivers are formed by physiological vs. cognitive processes. The latter physiological–cognitive dimension shows how the body–mind complementarity is relevant from a psychological perspective.

7.1 Psychological foundations of environmental decisions

In this section, we review theories from the field of psychology that can inform us about how individuals make decisions that affect the environment. Drawing on these theories, we present a loose taxonomy or rough order, that allows us to interrelate and differentiate drivers based on the degree to which they are shaped by psychological processes or by experience.

7.1.1 Perspectives of psychology on drivers

Clearly, what is considered to be a driver depends very much on the model of rationality (see Box 1.4) and the conception of humans underlying the chosen psychological theory. The most important approaches were introduced by Zimbardo (1992). He distinguished between the following: (a) a biopsychological model focusing on biochemical, neuropsychological aspects; (b) Freud's psychodynamic approach, which explains behavioral patterns and tendencies based on the

Table 7.1 Psychological constructs serving as drivers of goal formation, strategy selection or evaluation, and action (concepts with an asterisk* are explained in the following sections; perception** is explained in Chapters 6.1–6.3).

	Psychological status of variable	Examples
A	Basic psychological processes	Arousal, curiosity, equilibration, etc.
		Preferences*, utilities*, aspiration levels, interests, etc.
		Instincts, drives*, emotions*, (basic) needs
		Perception**, cognition*
B	Experientially and socially formed lower psychological processes	Motives*, values*, dispositions, habits, etc.
C	Behaviorally relevant higher processes	Beliefs (higher needs) and motivations*, (personal and social) norms*, attitudes*, moods, etc.
D	Close to action and behavior	Decisions, intentions, judgments, valuations, beliefs, (preferences, aspiration levels, utilities), etc.

charging and releasing of drive-based tensions (see Chapter 7.3); (c) the behavioral model, which only considers basic processes, as found in category A of Table 7.1, and which regards the human as a black box with changing stimulus–response patterns; (d) the cognitive model, for which information acquisition and processing serves as the basis and which is thus predominantly a constructivist approach; or (e) the humanistic psychology approach, stressing the potential of self-actualization. The role and the impact of the environment differ completely between these approaches. At this point, we do not go into a detailed comparison of these models. However, many of the concepts that are introduced in some detail are pivotal in these approaches. For instance, instincts and drives are central concepts of the biopsychological approach (a). Drives are also considered in approach (b), whereas the concept of cognition is found in (a) and (d). In addition, needs play a role in approach (e) and preference is a key concept of (c).

The reader should note that the concepts described above are not consistently defined in the literature. Nevertheless, we suggest a first rough order or taxonomy for a large set of drivers in the literature in Table 7.1. The grouping starts from basic, less flexible, psychological drivers (A) and passes through lower processes, which are experientially and socially formed to higher, behavioral processes (C) to concepts which are close to behavior (D). This ordering helps us to understand how people cope with material–biophysical environmental problems such as air pollution, climate change, and scarce raw materials or when basic needs are not fulfilled. It is also important to know that the human mind can generate different modes of thought, specifically intuitive and analytic modes of thought. Both modes are linked to different drivers that affect goal formation and strategy selection. We deal with these phenomena in Chapter 7.6 on Cognitions.

Finally, at the end of the chapter we provide in Figure 7.13 a more differentiated two-dimensional taxonomy for the key drivers that are discussed in this chapter.

Category A of Table 7.1 includes drivers such as equilibration, curiosity, and arousal, which are shaped primarily by innate psychological processes. Body language related to emotion can be taken as a good example of a behavior resulting from type A (basic) psychological processes. Blind newborns, despite never having seen another human (and thus being unable to "learn" body language), smile to show their happiness in the same manner as those born with sight (Cohen, 2000b). The more experientially shaped and societally formed the constructs are, the closer we come to category C.

Some of the concepts in category A are elementary drivers in cognitive development. This is the case, for instance, with the concept of equilibration, which governs adaptation of the cognitive system in Piaget's genetic epistemology. However, another kind of equilibration in the sense of finding an appropriate balance of neurophysiological activation is also linked to the concept of arousal. The corresponding concepts are provided in adaptation theory (Helson, 1964) and the Yerkes–Dodson law (Yerkes & Dodson, 1908). Both approaches theorize that individuals strive to attain an efficient mid-range level of arousal or energization. States above or below this level elicit change in activities.

According to the decision theoretic and functionalist perspective taken in this book, preferences, utilities, or aspiration levels are key drivers of human behavior. The behavior of all people is supposed to be based on these drivers. In many disciplines, including psychology, the concept of utility has attracted special attention. Classical economists, for instance, apply the utility concept almost as a behavioral substitute. Therein, people are generally thought to choose the alternative with the highest utility. As we will see below, psychologists have a much more differentiated view on the concept of utility.

We use preferences, utilities and aspiration levels in two ways. On the one hand, they are basic functions inherent in *all* human systems. The underlying assumption is that, when confronted with different alternatives of what they can do, individuals, groups, and other human systems have varying levels of attraction to the alternatives or are indifferent to them. This means that, at any given moment, an individual is either able to form an opinion about his or her relative merit, preference or desirability or is indifferent with respect to the different alternatives that the environment offers. On the other hand, rational choice theories postulate that people make choices based on their preferences (Tversky, 1972), subjective utility functions (Fishburn, 1970), or aspiration levels (Scholz, 1980). In these theories the concepts often become mathematically defined entities.

Many basic key drivers of the individual, such as drives, emotions, (basic) needs (such as hunger), and instincts, are strongly rooted in biophysical processes that are also thought to exist in higher ordered animals. Tinbergen defined instinct as:

> … a hierarchical organized nervous mechanism which is susceptible to certain priming, releasing and directing impulses of internal as well as of external origin … that contribute[s] to the maintenance of the individual and the species. (Tinbergen, 1951, p. 112)

According to instinct theory, organisms are born with a set of biologically based behaviors or fixed action patterns (Zimbardo *et al.*, 2005). It is important to note that the behavioral programs linked to instincts are reflex-like and predetermined until the execution of the final action. The concept of instinct may have been defined by ethologists (Lorenz, 1937; Tinbergen, 1951) but it had already been used by, for example, William James, who considered instincts a "faculty of action" of a "general reflex type" (James, 1890, p. 384). Instincts are preprogrammed behavioral tendencies that are essential for the survival of a species. The concept has been widely abandoned by psychology but is broadly used in everyday speech to describe aggressive (e.g. hunter instinct), destructive (e.g. killer instinct) or protective behavior (e.g. maternal instinct). They are found in psychology, in particular, to explain gender differences in nurturance (Cole *et al.*, 2007).

Much of the classical work presented in this chapter is dedicated to perception and cognition. These two concepts are clearly not drivers of equilibration: the behavioral relevance of perception and cognition is, in fact, disputed. In general, they are a necessary but not sufficient prerequisite of behavior. Because of their general and elementary status, we assign these concepts to category A of Table 7.1 (see Boxes 6.1, 6.2, and 6.7).

7.1.2 Key messages

- Psychology defines a set of drivers, which are psychological constructs or concepts that underlie behavior. These concepts can be classified according to various dimensions; for instance, whether the concepts are supposed to be driven by basic psychophysiological processes or whether they are a result of higher ordered mental processes.

7.2 Preferences and utilities

Preference and utility in their basic meaning can be considered as a valuation of satisfaction resulting from the usefulness of something from the perspective of a specific human system. Naturally, this assumption does not imply *how* the valuation is done or *what* and *why* something is considered more attractive or desirable. Neither does it imply that we are referring to the mathematical definition of expected utility. However, we do postulate that curiosity and interest in something must precede the valuation and structuring of preferences.

7.2.1 Historical roots

To understand the utility concept, we first introduce the ideas of utilitarianism and hedonism from a historical perspective before touching on the mathematical, economical approach to preference and utility functions. Finally, we deal with some issues of contemporary psychological research on preference and utility.

The English philosopher, jurist, and social reformer Jeremy Bentham (1748–1832) introduced the term utilitarian to describe someone who follows the principle of utility (Hügli & Han, 2001). Bentham targeted a kind of behavioral theory starting from the following assumption:

Nature has placed mankind under the governance of two sovereign masters, pain and pleasure. … By utility is meant that property in any object, whereby it tends to produce benefit, advantage, pleasure, good, or happiness, (all this in the present case comes to the same thing) or (what comes again to the same thing) to prevent the happening of mischief, pain, evil, or unhappiness to the party whose interest is considered: if that party be the community in general, then the happiness of the community: if a particular individual, then the happiness of that individual. (Bentham, 1789/2005, pp. 1–2)

This quote embodies the hedonism principle, which assumes that pleasure (or happiness) and avoidance of pain (or suffering) are reasonable major intrinsic drivers. Bentham and utilitarianism are often linked to the simple formula that people follow the greatest happiness principle (Mill, 1863, p. 5). Hedonism is still a matter of debate among philosophers and across cultural perspectives. This is exemplarily conveyed in a quote by Nietzsche (1844–1900): "Man does not strive for happiness: only the Englishman does this" (Nietzsche, 1889/1969, p. 55, translated by Scholz).

From a history of science perspective, Bentham is the father of quantitative utility functions. The above quote includes the idea of measuring utility as a composite of both positive and negative impacts. As we will see, elements of hedonism are ubiquitous in most psychological theories of motivation, judgment, risk, and decision-making. According to the quantitative utility function, many decisions quantitatively deal with the trade-off between positive and negative aspects of alternatives.

As can be taken from the above quotes, Bentham considered the concept of utility not only applicable to the individual but also to higher, aggregated human systems.

Bentham created a hedonic calculus, which takes into account the: (i) intensity; (ii) duration; (iii) certainty or uncertainty; and (iv) remoteness of any pleasure or pain. Further, one should include the probability of: (v) the same effects being repeated; or (vi) of the same effects not being repeated; and (vii) the number of people who are affected by any pleasure or pain arising from the mentioned actions (Bentham, 1789/2005; see also Meredith, 2008).

7.2.2 Axiomatic definition

Without explicitly referring to Bentham, the mathematician John von Neumann (1903–57) and the economist Oskar Morgenstern (1902–77) defined and mathematically axiomatized preference und utility. Their aim was "to describe the fundamental concept of individual preferences by the use of a rather far reaching notion of utility" (von Neumann & Morgenstern, 1944, pp. 15–16) to find solutions that provide "a maximum satisfaction" (ibid. p. 10). They supposed that utilities are based on intuitive, basic impressions.

In the case of utility the immediate sensation of preference – of one object or aggregate of objects as against another – provides the basis. (von Neumann & Morgenstern, 1944, p. 17)

Von Neumann and Morgenstern present an axiomatic treatment of utility as a key element of their concepts of full rational behavior (see Box 1.4) and they also target the measurement of utility. The procedure they suggested assumes that people are able to show preferences between an alternative A and a lottery $L = (p, B; (1–p), C)$ that offers the occurrence of an option B with a probability p and of an option C with a probability of $1 - p$. The Neumann–Morgenstern utility functions are based on the axiom of ordering. They postulate that either: (i) one of two alternatives A and B has a higher attractiveness; or (ii) both alternatives have the same utility. Mathematically, this means if > (or <) means a greater (or smaller) utility and = stands for being indifferent that, for any alternatives A and B, $u(A) > u(B)$ or $u(A) < u(B)$ or $u(A) = u(B)$. Further axioms of this utility theory are transitivity (i.e. if A has a greater utility than B and B has a greater utility than C, then we can conclude that A has a greater utility than C) and continuity (stating that, for three alternatives A, B, and C, with increasing utility, one can define all utilities between A and B as lotteries, including A and C, and likewise for utilities between B and C). Moreover, utility theory also includes a kind of independence axiom, assuming that if C has a greater utility than A, the score of linear combinations of utility payoffs such as $p \times u(A) + (1–p) \times u(B)$ with $p \in [0,1]$ increases if A is replaced by C[1] (von Neumann & Morgenstern, 1944; Wiese, 2002).

As mentioned above, the axioms provide a utility function on an absolute scale. However, the normative approach to utility also had the unwanted precondition

---

[1] Formally the independence axiom reads: If there is a lottery L_1 and two lotteries L_2 and L_3 with $u(L_2) < u(L_3)$, then the preference is not changed if L_1 is integrated in L_2 and L_3, i.e. $u(\alpha L_1 + (1 - \alpha)L_2) < u(\alpha L_1 + (1 - \alpha) L_3)$ with $\alpha \in [0,1]$.

7 Drivers of individual behavior

> **Box 7.1** Why don't we have coherent preferences? Arrow's cyclic triads
>
> Imagine a municipality that has to choose between alternatives for supplementing future energy supply; after long discussions in parliament, three alternatives remain: gas (alternative A), nuclear power (alternative B), and a portfolio of alternative energy (alternative C). The city board decides to make the final selection based on three criteria: price (criterion c_1), impact of large accidents (criterion c_2), and a portfolio of alternative energies (criterion c_3).
>
> Unfortunately, the city board lacks an integrated overall evaluation, that is, a one-dimensional utility score that can be derived from the evaluations of the criteria. Thus, the board agrees upon the following decision rule: one alternative is preferred to another if it outperforms the other by the majority of criteria.
>
> Figure 7.1 presents expert ratings for the criteria c_1 to c_3, with "1" denoting the best and "3" the worst alternative. As one can easily figure out, gas outperforms nuclear power in the first two criteria, which can be represented by $v(A) > v(B)$. When comparing all alternatives according to the decision rule, a so-called cyclic triad is formed: $v(A) > v(B) > v(C) > v(A)$. This leads to the paradox situation that A is preferred over A, or $v(A) > v(A)$.
>
> This example illustrates Arrow's paradox from a decision theoretic perspective. Arrow (1950) showed that in general it is impossible to avoid cyclic triads if decision-makers do not decide along a one-dimensional utility score.

a

	A	B	C
	Gas	Nuclear power	Alternative energies
Criterion c_1	1	2	3
Criterion c_2	2	3	1
Criterion c_3	3	1	2

b

Figure 7.1 Comparison of three alternatives according to three criteria. (a) The numbers in the table represent ranks according to each criterion. (b) The comparison leads to cyclic triads when using the decision rule "take the alternative with the majority of better criteria."

that probabilities are to be determined without bias. Mathematical game and decision theory developed a precise definition of preference theory by means of utility functions and also defined the different types of uncertainty or subjective utility a decision-maker faces (Fishburn, 1970; Walley, 1991).

7.2.3 Limits of the axiomatized utility concept

The normative utility paradigm, which is at the core of expected utility, has been debated from many angles. The paradoxes of Arrow (see Box 7.1) and Allais (Box 7.2) played a significant role in these disputes. We present them for two reasons. First, as does the elaboration of appropriate preferences among alternatives, the paradoxes reflect fundamental difficulties in environmental literacy. Second, they have been exceedingly important for the development of psychological decision research, which describes what people prefer and why they prefer something that may seem unreasonable to others.

The Allais paradox allows for four important conclusions.

(1) *Utility is not expected utility.* The latter is a specific, useful, simple mathematical approach for integrating different options that may appear with uncertainties. Expected utility thus raises the fundamental questions not only of how these options are evaluated but also of how the uncertainties are cognitively represented and integrated (see Jungermann et al., 2005; Kahneman et al., 1982; Scholz, 1987).

(2) *Monetary amounts do not linearly correspond to utilities.* Utilities are comprehensive valuations and do not focus on money alone. Thus, the context and how the situation is framed by a subject have to be considered. By framing we conceive, for instance, whether we deal with a situation of

Box 7.2 Not following expected value: the Allais paradox

The Allais paradox is presented in Table 7.2. If Situation 1 is presented in psychological experiments about choice, people prefer Lottery 1 to Lottery 2, which means that people do not choose according to expected utilities of monetary outcomes when being confronted with lotteries. This is usually explained by the non-linearity of the utility functions, which are the marginal utility, the theory of which was first expounded by Daniel Bernoulli (1738/1954; Stigler, 1950) when anticipating the idea of the Weber–Fechner law. However, the phenomenon of marginal utility was not the focus of Allais' contribution. Rather, he wanted to show that the Neumann–Morgenstern axioms of utility functions were inadequate. Table 7.3 also shows the Allais paradox but in a slightly different way. As one can see, Situation 1 and Situation 2 differ only with respect to the outcomes A and B. However, as these outcomes are (in the language of the classical theory of utility functions) irrelevant for deciding between L_1 and L_2 in Situation 1 as well as for making a choice between L_3 and L_4 in Situation 2, the preference should not change between Situations 1 and 2. However, psychologically, the 1% chance of getting nothing in the case of L_2 is a different prospect than the (additional) 1% chance of getting nothing in L_4. The Allais paradox shows that, psychologically, the independence axiom is violated in situations with monetary outcomes. A psychological interpretation following the prospect theory (Kahneman & Tversky, 1979, 1992) is that the reference point of the subjective utility function is shifting from $1.0 M in Situation 1 to $0.0 M in Situation 2, which are the most probable outcomes in these situations.

Table 7.2 The Allais paradox. "Preferred lottery" indicates that, in experiments, a majority of people do not prefer the lottery that offers the greatest monetary reward.

Situation 1				Situation 2			
Lottery 1 (L_1)		**Lottery 2 (L_2)**		**Lottery 3 (L_3)**		**Lottery 4 (L_4)**	
p	Outcomes	p	Outcomes	p	Outcomes	p	Outcomes
1.0	A: 1.0 M $	0.89	A: 1.0 M $	0.11	A: 1.0 M $		
		0.01	B: Nothing	0.89	B: Nothing	0.90	B: Nothing
		0.10	C: 5.0 M $			0.10	C: 5.0 M $
Expected value: 1.00 M $		Expected value: 1.39 M $		Expected value: 0.11 M $		Expected value: 0.50 M $	
Preferred lottery L_1						Preferred lottery L_4	

Table 7.3 The Allais paradox (displayed in a slightly different manner from Table 7.2). Besides the potentially irrelevant shaded line, which does not discriminate between L_1 and L_2 as well as between L_3 and L_4, the lottery pairs L_1, L_2 and L_3 and L_4 respectively are identical. A, B, and C represent possible outcomes that may result with some probability.

Situation 1				Situation 2			
Lottery 1 (L_1)		**Lottery 2 (L_2)**		**Lottery 3 (L_3)**		**Lottery 4 (L_4)**	
p	Outcomes	p	Outcomes	p	Outcomes	p	Outcomes
0.89	A: 1.0 M $	0.89	A: 1.0 M $	0.89	B: Nothing	0.89	B: Nothing
0.11	A: 1.0 M $	0.01	B: Nothing	0.11	A: 1.0 M $	0.01	B: Nothing
		0.10	C: 5.0 M $			0.10	C: 5.0 M $

gain or of loss. The von Neumann–Morgenstern axioms, not having taken this into account, have been contradicted by many experiments and might thus be of limited value.

(3) There is sufficient evidence that *people do not follow the axioms of full rational behavior*. Nevertheless, human behavior should not be judged to be irrational. In many situations, intuitive heuristics can be superior to analytic processes, including elaborated utility scores (Gigerenzer et al., 1999).

(4) Utility theory is lacking in the sense that it does not include *situational properties or relations*.

Psychological and economic decision theorists, particularly Daniel Kahneman and Amos Tversky, have empirically investigated which of von Neumann

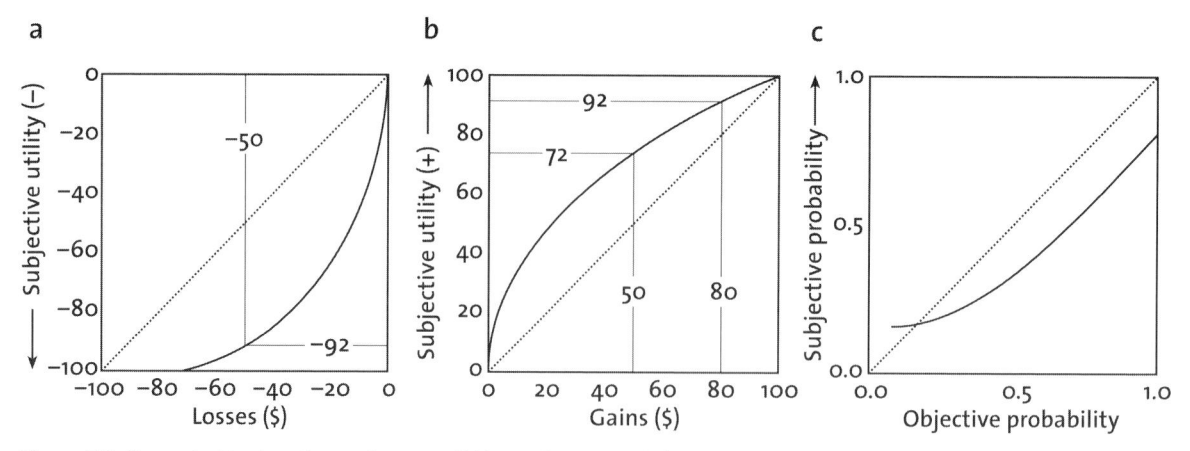

Figure 7.2 Shape of subjective utility as a function of (a) losses, (b) gains, and of (c) subjective probability as a function of objective probability.

and Morgenstern's utility axioms represent people's actual preference formation and choice behavior (for comprehensive reviews see Jungermann et al., 2005; Kahneman, 2003; Kahneman & Tversky, 2000). The empirical research originated in the notion that there is such a thing as subjective expected utility (Edwards, 1954). Most of these theories take the concept of utility in its elementary form as a basis for decision-making and a driver of human behavior. However, many decision researchers refer to expected utility in a double sense. First, they refer to an expected value according to the mathematical approach. Second, they operate with prospective (expected) choices. Certain models, such as prospect theory (Kahneman & Tversky, 1979, 1992), explicate features of subjective utilities.

Besides the abovementioned phenomena of non-linearity of both subjective utilities and subjective probabilities in the formation of expected utility (Camerer & Ho, 1991), this model specifies different shapes of the subjective utility functions in positive and negative domains. Further, it can explain contradictory decisions by pointing to different reference points. Kahneman and Tversky (1992) used the term "cancellation axiom" instead of independence axiom and stressed that (owing to the different shape of the negative and the positive part of the utility function) one and the same situation can be judged differently depending on whether there is a positive or negative reference (as was the case in the Allais paradox, where the baseline was 1 million dollars in Situation 1, whereas it was nearly zero in Situation 2; see Table 7.3).

The above arguments resulted in the prospect theory of Kahneman and Tversky (1992). This approach to subjective expected utility includes the assumption that: (i) the value of an option and, in particular, the subjective value differ on the subjective framing of the situation; (ii) the positive subjective utility function is concave for gains and negative for losses; and (iii) the curve is steeper for losses than gains. Finally, (iv) a subjective probability function is postulated (which is undefined) close to the objective probability zero and which underestimates higher probabilities (see Figure 7.2).

The situational, contextual aspects of utility have been distinguished by the different types of utilities that can be formed. Psychologically, one complementarity lies between experienced versus prospective, hypothetical utilities, while another perspective is provided by prescriptive decision utility. It is interesting to see that the current trend of looking at the genesis, foundations, and structure of utility suggests that:

> … subjective states are now a legitimate topic of study … [and that they] … strictures against a hedonic notion of utility are a relict of an earlier period in which a behavioristic philosophy of science held sway. (Kahneman, 2000, p. 760)

The difference between retrospective utility and real-time utility is also of psychological interest. Clearly, retrospective utility can be psychologically transformed, for instance, because of cognitive dissonance reduction processes, which are discussed in Chapter 7.6 on cognition.

Another interesting complementarity is that between hot utilities and cold utilities. Hot utilities are linked to emotions and affect such as joy, disgust, and sexual desire and are mostly tied to the intuitive mode (see also Table 7.4). The notion of hot utilities can be made more easily accessible by referring to the hungry shopper who buys far too much at the supermarket and who should have eaten a muffin beforehand to

Table 7.4 Features of the intuitive and analytic modes of thought (Scholz, 1987, p. 62).

Intuitive thought	Analytic thought
Preconscious	Conscious
Information acquisition	Information acquisition and selection
Processing of information	Processing of information
Understanding by feeling and instinct of empathy	Pure intellect or logical reasoning, independent of temporary moods and physiology
Sudden, synthetic, parallel processing of a global field of knowledge	Sequential, linear, step-by-step ordered cognitive activity
Treating the problem structure as a whole, "gestalt cognizing"	Separating details of information
Dependent on personal experience	Independent of personal experience
Pictorial metaphors	Conceptual or numerical patterns
Low cognitive control	High cognitive control
Emotional involvement without anxiety	Cold, emotion-free activity
Feeling of certainty toward the product of thinking	Uncertainty toward the product of thinking

keep his or her prospective utility estimates realistic. We speak about cold utilities when judgment is based on deliberate considerations, including an analytic process (Kahneman & Thaler, 2006).

7.2.4 Key messages

- Preference formation and utilities are both defined as basic, elementary processes inherent in animals and human beings as well as choice and decision-eliciting drivers, integrating other drivers
- Utility is assumed to determine behavior
- The assumption of a one-dimensional utility function is necessary to avoid inconsistencies and cyclic triads
- Concepts of expected utilities or Neumann–Morgenstern utilities incorporate uncertainties or probabilities about the outcomes and a subjective valuation of the outcomes.

7.3 Drives

The concept of drive goes back to Freud (1920), who distinguished the sexual-oriented libidinal and the object-oriented aggression drives. Both drives are considered biological, instinctual, permanently operating forces. The regulatory system behind the Freudian approach is tied to the metaphor of a steam engine. Freud, as well as the ethologist Lorenz (1963), postulated that drives accrue endogenously until they reach a certain level, at which the organism is pushed to act in such a way that the drive will be satisfied. Drives are also a key concept of Hull's (1943) behavioral theory, wherein they are described as forces that stem from physiological deviations from a homeostatic target state. For example, if one's body does not have enough water, one is thirsty.

Drive reduction and the goal of the organism to attain homeostasis – a kind of balanced condition – are key issues, core concepts used for describing drive. The regulatory mechanism can be described as a discontinuous loading–unloading dynamic process.

> The *goal* of a drive is satisfaction, once and for all ... The *object* of a drive is that upon which or through which the drive can attain its goal. It is the most variable aspect of a drive, not pristinely linked with it, but only assigned to it because of its aptitude for enabling satisfaction ... (Freud, 1915, p. 215, translated by Scholz)

Freud assumed the eros (libidinal) and thanatos (aggression) drives as universal. Intrapsychic conflicts and not environmental problems were at the root of Freud's system. Freud provided a theoretical framework in which the self, referred to as the ego, of an individual results from a trade-off between the drives of the id and the norm-shaped super-ego. The super-ego may include environmentalists' norms, as found in concepts of deep ecology. As drives are only under limited control of the individual, we also consider dynamic unconscious variables. In this context the frustration–aggression cycle of human activity (Fairbairn, 1952) is of importance, as it sheds light onto the reasons behind destructive human action. The aggression drive seems to be relevant, as aggression is object-related and secondary to environmental failure. This means that if an individual is unsuccessful at interacting with his or her

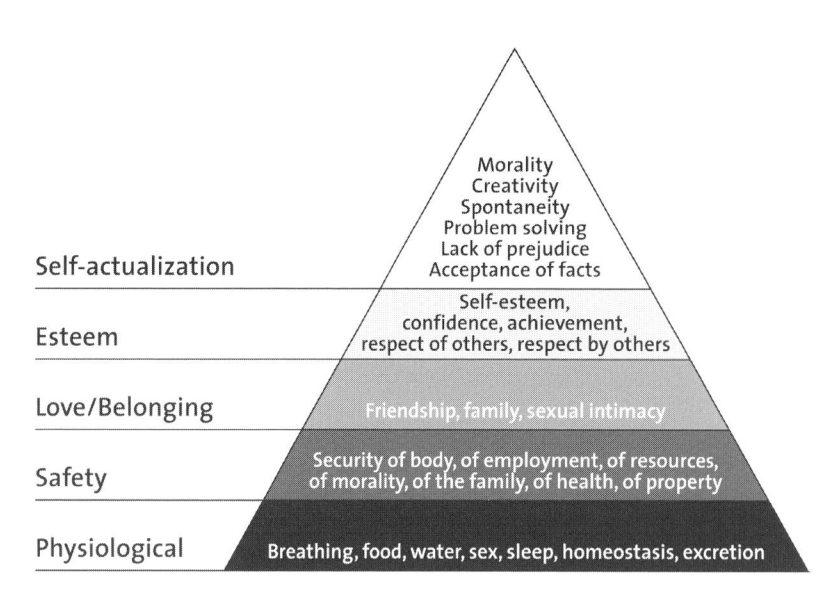

Figure 7.3 Maslow's pyramid of needs (adapted from Maslow, 1943).

A pyramid diagram with five levels from bottom to top:
- **Physiological**: Breathing, food, water, sex, sleep, homeostasis, excretion
- **Safety**: Security of body, of employment, of resources, of morality, of the family, of health, of property
- **Love/Belonging**: Friendship, family, sexual intimacy
- **Esteem**: Self-esteem, confidence, achievement, respect of others, respect by others
- **Self-actualization**: Morality, Creativity, Spontaneity, Problem solving, Lack of prejudice, Acceptance of facts

sociophysical environment, he or she is prone to show aggression, which is generally accompanied by a narrower scope of perception and thus a loss of environmental awareness in strategy formation.

> Freud was an environmentalist in the sense of suggesting that he felt the social and interpersonal environment shaped and guided the form and consequences of the person's life and death striving. (Ittelson *et al.*, 1974, p. 64)

However, form, content, and meaning of physical environments can also form drives. According to Freud, one's physical possessions (such as big, showy cars, private jets or pretentious homes) express unconscious psychosexual needs and desires or are attempts to sublimate these desires.

7.4 Needs

The humanist psychologist Maslow (1954b) proposed a hierarchy of needs in a pyramid structure consisting of the levels physiological, safety, belongingness and love, esteem, and self-actualization (see Figure 7.3). His theory is based on the prepotency principle. This principle states that needs on a lower level must be satisfied to a degree before the needs of the next level can motivate behavior. Moreover, a satisfied need ceases to motivate behavior, thus, only needs that are relatively unsatisfied are capable of motivating people.

Basic needs for food, water, oxygen, or sleep are endogenously, physiologically defined and are of a biological nature. They represent basic internal deficit conditions that must be satisfied to maintain bodily processes. Safety, the next level in the pyramid, is the need for security of the body, property, resources, or family and is considered, to a certain extent, to be a basic need causing instinct-like reactions (Maslow, 1954a). The next level, belongingness and love, includes the needs for friendship and family. Contrary to the drive approach, exogenous impacts are dominant for higher-ordered needs. Esteem includes value-like issues such as confidence, achievement, self-esteem, or respect from others. Self-esteem and confidence are considered necessary to enable a person to realize his or her potential fully. Self-actualization includes social responsiveness, openness to novelty and challenge, acceptance of facts, creativity, and morality. The relevance of this hierarchy of needs to environmental literacy becomes clear when we see how Dunlap *et al.* (1983) found that people who recycle seem to be more self-actualizing, valuing more strongly things such as aesthetics, inner harmony, and leading an exciting life; contrary social demographically similar non-recyclers were more likely to accentuate safety and security values, such as national security and world peace.

The need approach appears to have multiple environmental relevancies (see Douglas *et al.*, 1998). For example, historically, hunger and defending resources often caused territorial claims. Safety needs can be seen as a factor underlying risk aversion for new technologies such as nuclear power.

7.5 Emotions

Emotions have been contemplated by great minds, such as Aristotle, Aquinas, Spinoza, Descartes, and

Box 7.3 Emotions and cognition: anxiety-reducing snails

Emotions play a differentiated role for human activities (Russell, 2003). Current psychology postulates that there are two emotion circuits. The first is a primary, instinct-like deep brain circuitry that provides fast, automatic, unconscious, early warning reactions, for instance sweating or heavy breathing when confronted by loud noises at night or snakes or spiders (Hellmuth, 2003). These processes and physical responses are linked to the release of hormones such as serotonin or adrenaline. The other, secondary, conscious circuit is linked to cognitively anticipated action and appraisal (LeDoux, 2000), such as anticipating a first bungee jump. However, cognitions interact with the emotional dynamic, so realizing a snake to be harmless can reduce the release of hormones. Schachter (1971) assumed that emotions depend on the external situation *and* on our internal physical state. Certain emotions, such as delight, joy or happiness (see Figure 7.4), elicit the intuitive mode, while others, such as fear, activate the analytic mode (Kuhl, 1983).

Anxiety-reducing heuristics play a role in coping with environmental stress. Grasmück and Scholz investigated the relationship between emotional concern, perceived risk, and agreement on dissonance reducing heuristics of people living on a site contaminated with heavy metals. For people who agreed with statements such as "The more frequent appearance of snails shows that the situation is improving," Grasmück & Scholz (2005, p. 615) showed lower emotional concern and lower risk judgments. This suggests that dissonance reducing heuristics play an important role in emotional dynamics such as coping with anxiety, which, if left unaddressed, elicit stress.

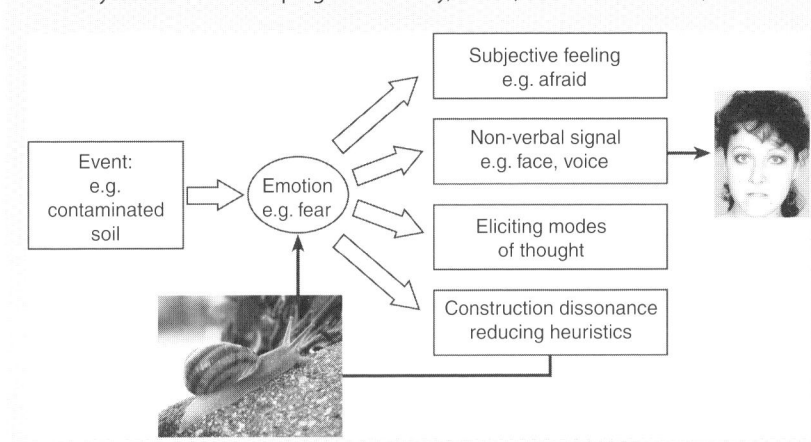

Figure 7.4 An event such as becoming aware of soil contamination can cause emotions, subjective feelings, facial expressions, and can change the mode of thought. However, certain actions, such as applying dissonance reducing heuristics, may reduce the impact of the emotional state.

Darwin. Darwin (1965/1872) proposed that individual emotional dynamics are universal to all humans and are partly derived from our evolutionary past and experiences from childhood. Emotions evolved to help organisms cope with important recurring situations (Dolan, 2002). Thus they are "specialized modes of operation shaped by evolution that increase the capacity and tendency to respond adaptively to threats and opportunities" (Dolan, 2002). They "help humans respond to survival-related problems and opportunities" (Keltner & Ekman, 2000). This position and the proposition that emotions are basic processes are also supported by the finding that fundamental emotional patterns provide the same expressions and experiential qualities in all cultures (Izard, 1971, 1977). As such, emotions are considered basic psychological processes (see Table 7.1).

Recent research identifies different emotional circuits, a primary instinctual one and a secondary one moderated and governed by mediating cognitions (see Box 7.3; Hellmuth, 2003; LeDoux, 2000). Moreover, emotions are often conceptualized in a bipolar structure, such that hate and love are complementary but contingent and ambiguous evaluations, since they are given to one and the same object *depending on the degree of satisfaction*. Russell provided a circumplex model of emotions with dimensions of pleasure/displeasure and activation/deactivation (Russell, 1980; Russell *et al.*, 1989; see Figure 7.5). Critical questions are what emotional dynamics look like and how emotions affect human behavior (see Box 7.3).

> Emotions, and feelings that follow emotions, are an integral part of the value systems necessary for laying down long-term memory and for reasoning conscious decision making involving life directing choices. (Moss & Damasio, 2001, p. 98)

It has long been known that emotions influence how we perceive and value the environment. Joy or elated

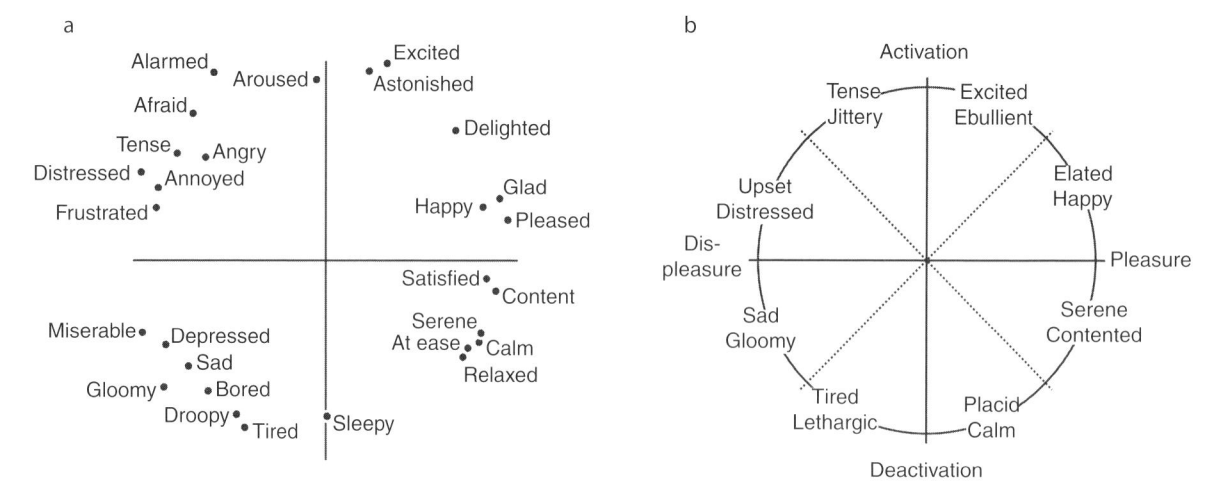

Figure 7.5 Russell's circumplex model of emotions (adapted from Russell, 1980, 2003).

happiness make it likely that the world is seen through "rose-colored glasses" (Izard, 1977). Russell and Snodgrass (1987) provided an interesting differentiation for understanding the role of emotions when interacting with the social and the material environment by distinguishing between moods and affective appraisals.

Mood on the level of the individual refers to the core emotional feeling at a given moment. Thus a mood is a temporary state of pleasure, sadness, anger or other emotion. Since, from a decision theoretic perspective, people select places, objects, or strategies with positive and pleasurable attributes, one can easily imagine that places leading to a positive mood and adequate activation level or arousal are preferred. Environmental conditions, such as level of noise, colors, and amount of sunlight, affect moods. Reciprocally, prior mood determines the perceived attractiveness of places, a phenomenon that can be called emotional priming.

An affective appraisal is an attribution of feeling or emotion to an event, place, or object. Objects may appear pleasant or disgusting. In the context of risk judgments, this phenomenon has also been called the affect heuristic (Finucane *et al.*, 2000). Finucane *et al.* experimentally influenced the affectiveness judgments of nuclear power and thus received different risk judgments. As another example, the image of "dying trees" (see Box 7.6) is thought to elicit sadness, although this has not been proven experimentally. Environmental researchers claim that emotions are important in protecting landscape (Rodewald, 1999). Naturally, fears, concerns, and hopes of decision-makers can influence their choice.

An interesting aspect is the differential role of emotions with respect to the intuitive and analytic mode of thought (see Table 7.4). Kuhl (2001) suggested that positive emotional states such as joy elicit intuitive thinking and terminate goal formation, and that negative emotions like anger or disgust induce analytic goals and evaluations. Further emotions, such as "medium levels of fear and anxiety, might induce rational evaluators and goals and a systematic search for reasons that can explain certain events" (Scholz, 1987, p. 183), and thus elicit the analytic mode of thought.

7.6 Cognitions

To cognize is related to the Latin "cognoscere," which means "to (get to) know." Since the time of Plato, the brain was postulated to be the locus of knowledge. Neisser (1976) considered cognitive processes to be based on a sensory input that is transformed, reduced, elaborated, and stored. Cognitive psychology emerged from experimental cognitive psychology and computational cognitive sciences (Eysenck & Keane, 2005), and became a huge area of research. Whereas metaphorical models reminiscent of the computer analogy were dominant during the incipient stage of cognitive psychology, today's research is rooted in neuropsychological research and cognitive neuroscience. As a subfield, environmental cognition has long been restricted to processes involved in the acquisition and representation of spatial information in real-world settings (Bell *et al.*, 1996; Gärling & Evans, 1992). In this book, we take a much broader perspective of environmental

Box 7.4 Flashbulb memory, risk perception, and fatal crashes: the impacts of September 11

The phenomenon by which we can remember details of important, emotionally packed public events has been called "flashbulb memory" (Curci & Luminet, 2006). Flashbulb memories are a special case of normal episodic memory under the constraints of high emotional involvement. Many studies conducted on the flashbulb effect in the aftermath of the events surrounding September 11, 2001 indicate that national differences in emotional involvement played an important role in memory performance (Curci & Luminet, 2006; Pezdek, 2003).

Emotional arousal also affects risk perception (Finucane et al., 2000). High dread risk events, such as the case of September 11, can serve as an example. Gigerenzer (2006) revealed that risk cognition has behavioral impacts. Americans reduced their air travel after the September 11 attack, resulting in increased highway travel for the year following the event. The number of fatal crashes subsequently increased (see Figure 7.6b), with approximately 1500 Americans dying as a consequence of avoiding the risk of flying.

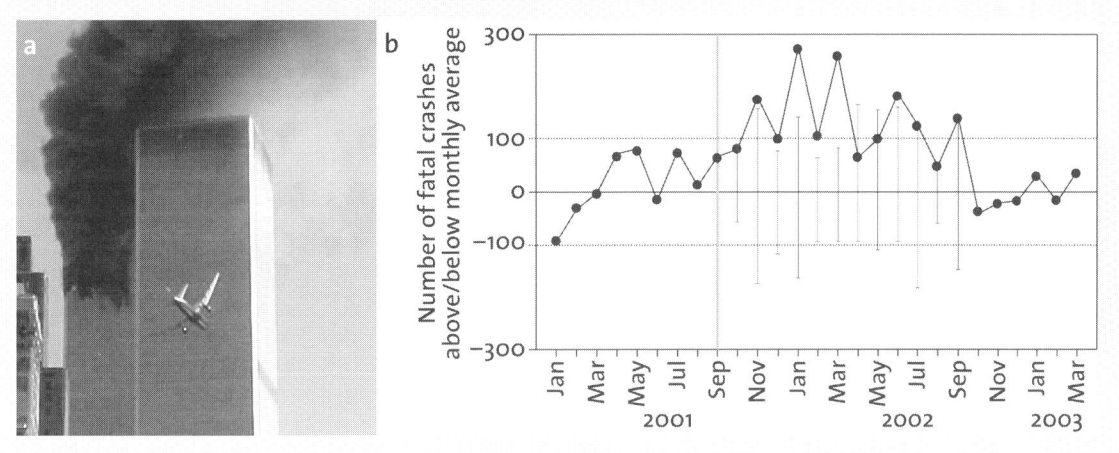

Figure 7.6 After (a) the terrorist attack of September 11, 2001 (b) the number of fatal crashes on US interstate highways increased significantly, indicating that US citizens changed their travel behavior.

cognition and consider it to be the basis for the capacity of coping with the environment in a sustainable manner, as described by Bandura:

> The capacity to exercise control over the nature and quality of one's life is the essence of humanness. (Bandura, 2001, p. 1)

As a cognitive system, the mind is an essential part of this capacity. If we want to describe and understand what knowledge individuals need if they are to interact reasonably with the environment, we should know (roughly) what people perceive, select, recode, store, retrieve, and how they process information. This is at the core of cognitive psychology. We present selected aspects from this domain to enable the reader to understand different rationales of individual reasoning and decision-making. One example of different rationales is the complementarity between the analytic and the intuitive mode of thought (see also Table 7.4). Both modes make use of different cognitive entities and processes. To understand these differences, it is necessary to understand the architecture of the mind – the different functions, abilities, and interactions of the sensory system, the working memory or the long-term memory, which has received comprehensive attention in books on this field (Hussy, 1984; Zimbardo et al., 2005).

The memory is the foundation for developing environmental literacy, and the sensory system is the immediate direct interface between the human system and the environment. The sensory system can encode and even store a large amount of information for a very short time. The short-term and the working memory capacity is limited to the magical number 7 (± 2) of cognitive units (Miller, 1956) that can be maintained for up to 20 seconds. The cognitive units of the short-term memory are called chunks (Simon, 1974). People can interact more efficiently with the environment when they learn to integrate segregated information units into more complex chunks that present more environmental information. Individual memory capacity differs considerably. Cognitive psychology distinguishes between different units of the long-term memory. One is the distinction between declarative and operative knowledge

(Anderson, 1983). Simply speaking, declarative knowledge is the data stored in the long-term memory. The capacity is unknown but seems to be almost unlimited (see Box 7.5). The operative knowledge includes the tools of operating and inferring with the elements of the declarative knowledge (e.g. knowing how to calculate 221/17). Furthermore, goals or a goal system can be considered to be part of the cognitive system.

The models of declarative memory represent different aspects of what we remember. We distinguish between episodic, pictorial, feature, and semantic network models of long-term knowledge. These different models are of interest for environmental literacy, since they represent the environment in different ways and have different behavioral relevance. Episodic memory stores personal experiences of events or episodes. It is a kind of autobiographical memory; it is directly accessible and thus rapidly and holistically available. The same holds true for pictorial memory, the "mind's eye." Certain images, such as the picture of how the airplanes crashed into the World Trade Center towers on September 11, are vividly stored by many people (see Box 7.4). The feature model of semantic long-term memory actually represents encyclopedic language-related attributes or features of words or concepts. This can be in a vector-like representation (e.g. "bird" refers to a sequence of features such as "feathers," "wings," "flying," "eggs," "two legs," …) or as semantic networks. The semantic network idea represents knowledge in a directed graph consisting of vertices (i.e. concepts such as cat, bird, …) and edges representing the relations between the concepts (a cat is a mammal, a cat eats birds).

What we have stored in our memory is differentially accessible. In principle, one can distinguish between directly accessible, simple everyday knowledge and higher ordered, indirectly accessible knowledge (Scholz, 1987). Directly accessible knowledge consists of experientially based concepts, such as chair, tree, water or wind, and different types of heuristics, such as simple everyday heuristics. Simple everyday knowledge is spontaneously accessible upon waking. Higher ordered knowledge or cognitive processes (such as Piaget's formal–operative knowledge) usually require intermediate steps of mental activation to become reliably accessible. This type of knowledge does not develop spontaneously. It is acquired in formal education (e.g. schools) or didactical situations, which allow for the learning of new concepts and algorithms (Brousseau, 1997).

Based on the concepts above, we can introduce three significant illustrations that demonstrate the role of cognition in coping with the environment. The first (A) is the intuitive and analytic modes of thought and the different achievements provided by these types of cognition. The second (B) is Leon Festinger's classical theory of cognitive dissonance that shows how the individual copes with conflicting information. Third (C), we introduce Albert Bandura's social cognitive theory of learning. Bandura's theory describes how internal, personal factors such as cognitive, affective, and biological processes interact with environmental and behavioral events.

7.6.1 Intuitive and analytic thinking (A)

Analysis vs. intuition is a well defined complementarity in cognitive science (Fischbein, 1975; Gigerenzer & Murray, 1987; Hogarth, 2001; Scholz, 1987; Springer & Deutsch, 1993). These modes of thought are based on qualitatively different cognitive processes and mental representations. Based on the theoretical concepts presented above, the intuitive mode of thought works with simple everyday knowledge and heuristics. Intuition is a less goal-driven form of cognitive activity that is not linked to strong cognitive control. Conversely, the analytic mode of thought utilizes higher ordered heuristics and knowledge, and is governed and controlled by the goal system.

Both modes of thought can be defined operationally by a list of attributes. Experienced raters may assign these attributes based on thinking aloud or written protocols that report the individual's information processing (Scholz, 1987).

Analytic thinking (see Table 7.4) is featured by conscious information acquisition, selection, and processing, pure intellect or logical reasoning (independent of temporary moods), and sequential, linear, step-by-step, ordered cognitive activity, as well as by separating details of information. Analytic thinking is at a high level of cognitive control, mostly independent of personal experience, based on conceptual or numerical patterns, can be described as "cold" and emotion-free, and is accompanied by high uncertainty towards the product of thinking.

Intuitive thinking, conversely, is characterized by preconscious information acquisition, understanding by feeling and empathy, sudden, synthetic, parallel processing of a global field of knowledge, and treating the problem structure as a whole, and thus involves "gestalt" cognition. The processing is shaped by personal experience, pictorial metaphors, and low cognitive control, and is accompanied by emotional involvement. In the end, there is a feeling of certainty

7.6 Cognitions

Box 7.5 What drives cognitive performance? Beyond savantism

Human interaction with the environment is based on an extraordinarily complex, structured, organized, multiprocess psychophysical system and the emotional system provides insight into this complexity. A snake immediately elicits fear and avoidance reactions. These reactions can also be triggered by other brain circuitries, for instance, an immediate route from the retina (see Box 6.4) via the visual thalamus to the amygdala. The amygdala is that part of the limbic system that affects physiological reactions, such as those from the endocrine system as well as arousal related to fear and reward. But there is also a slow route to the amygdala, from the visual thalamus via the visual cortex, establishing conscious formation processing that can affect its immediate processes on the amygdala (see Davidson, 2001; Zimbardo et al., 2005). Clearly, this is a very simplified model. Neuropsychologists claim that the brain does not have a single emotion center (Davidson et al., 2000) and that, instead, various, coordinated neural processes regulate emotion. Even so, emotion is only one driver of behavior (see Table 7.1).

Focusing studies on the "average" individual is one strategy for understanding drivers of environmental behavior. However, defining the average individual is no easy task, as there are strong social and cultural components that shape the way we deal with emotions, information or social norms. For instance, Asian people are taught to have emotionless facial expressions. Furthermore, there are gender differences with respect to panic disorders and depression. In general, empirical studies show a high degree of universality of differential emotion patterning but also significant cultural differences "in emotion elicitation, regulation, symbolic representation, and social sharing" (Scherer & Wallbott, 1994, p. 310). Thus, drivers have biophysical physiological communality among people, *but* also result from sociocultural differences.

Focusing on a specific individual is another strategy. This can help to identify theoretical limits of human cognitions. Neuropsychologists such as Oliver Sacks (1973) and Darold A. Treffert (1988) have taken this path when studying savants; that is, people with single extraordinary mental abilities. Some savants have an immense encyclopedic memory; some have a photographic memory (see Figure 7.7; Mottron et al., 2006). Most savants are intellectually impaired – there is evidence that certain disorders are correlated with certain dysfunctions. For instance, the ability to read both

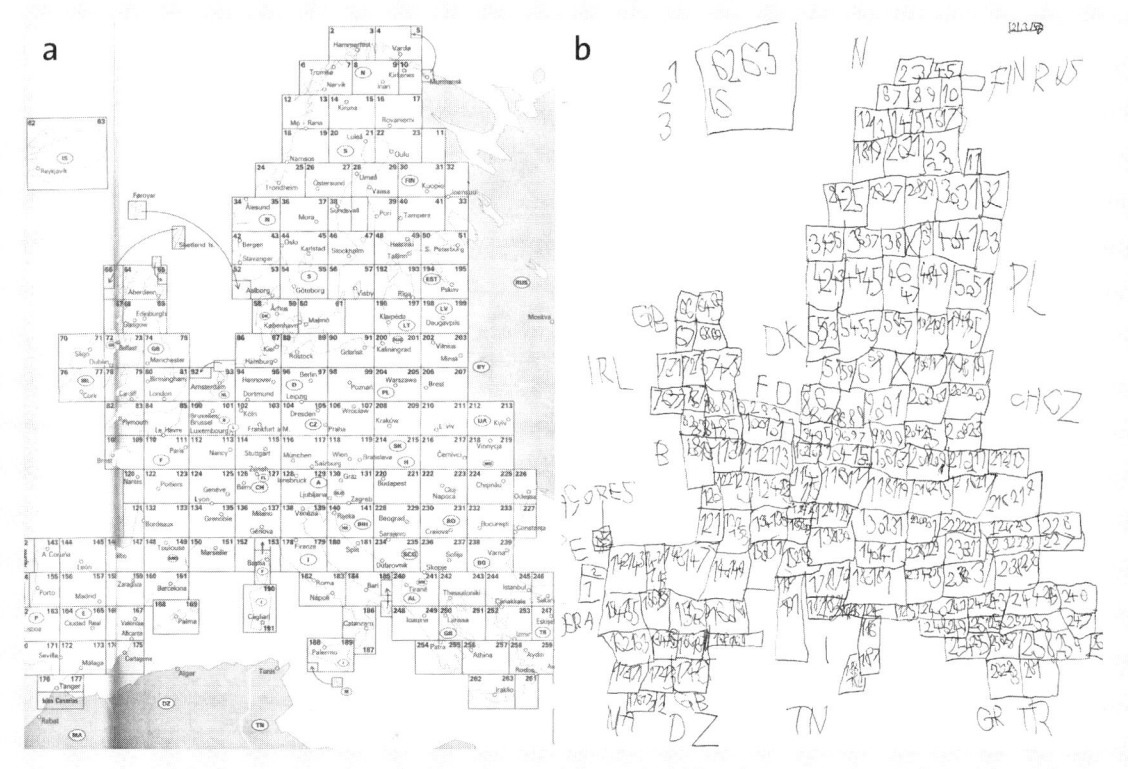

Figure 7.7 A day after flipping through an atlas (a) a seven year old boy with childhood autism was drawing the outline map from memory (b, © T. Sonne). People with this form of autism often have an eidetic memory.

> **Box 7.5** (cont.)
>
> pages of an open book simultaneously seems to be linked to the absence of the corpus callosum which connects the right and the left brain hemispheres. Islands of genius; that is, the savant syndrome (also called monosavantism), are often linked to certain identifiable disorders, impairment of mental capabilities, low general intelligence, reduced adaptivity, and are often related to autism (Heaton & Wallace, 2004). However, disorders and exceptional abilities are based on a spectrum of disorders stemming from, for example, the overfunctioning of brain regions related to primary perceptual processes (Mottron et al., 2006).
>
> With respect to environmental cognition, we can draw the following conclusions. (i) Knowing something or developing an appropriate representation of a perceived object/thing does not necessarily lead to reasonable utilization. (ii) The proper interaction of brain functions on the macro scale (i.e. neuroanatomy) and on the micro scale (see Box 6.4) is important in the sense of situational maladapted functionalism (Brunswik, 1936). A balance between lower and higher ordered processes and an adequate involvement of (hyper) performance in ecological situations is essential for good environmental cognition. (iii) Modeling the drivers of cognitive performance is the art of complexity reduction. The task of the researcher is to identify the essential aspects of the cognitive and the neuropsychological system that drives the utilization of cognitive elements and thus of behavior.
>
> Good environmental cognition (and research) results from a sufficient, ecologically valid representation (which – in the sense of Brunswikian ecological validity – means that the information sampled is adequate for representing the information objectively present) and efficient coordination of different cues and not from a single or well adjusted hyperperformance.

towards the product of thinking but skepticism with respect to the process.

The idea that the analytic mode of thought is superior to the intuitive has long dominated cognitive psychology. Yet "analyticity may lead [one] astray." The mode that provides a better result depends on the task at hand, the capabilities of the individual, and on the constraints of the decision situation. This is at the heart of ecological rationality as elaborated by Gigerenzer and colleagues (Gigerenzer & Selten, 2001; Todd & Gigerenzer, 2007). As conveyed by the book title *Simple Heuristics That Make Us Smart* (Gigerenzer et al., 1999), domain-specific, simple, everyday heuristics – called fast and frugal heuristics – are often superior, as they are better able to meet cognitive limitations as well as constraints of time and economy. Such heuristics perform adequately when they follow the "satisficing principle" (i.e. meaning satisfying and sufficing). In one such simple heuristic, the take-the-best heuristic, an individual first looks for the most relevant criterion, then checks whether it differs between the alternatives and, if so, selects the one that does best with respect to this criterion. Otherwise, the procedure is applied with the second-best criterion. As one can see, this procedure represents a different choice algorithm than the von Neumann–Morgenstern utilities discussed above.

7.6.2 The theory of cognitive dissonance (B)

Leon Festinger (1919–89), a student of Kurt Lewin, provided one of the most powerful psychological models of cognition. His theory of cognitive dissonance states that individuals who voluntarily undergo undesirable experiences or behave in contradiction to their goals, beliefs, or values, suffer an uncomfortable state of cognitive dissonance (Festinger, 1957). Cognitive dissonance is a state of tension that occurs whenever an individual holds two cognitions (ideas, cognized values, beliefs, opinions, perceived own behavior) that are incongruent. Post-decision dissonance is a widely observed phenomenon (Lawler et al., 1975) wherein shortly after making a difficult decision, people reduce dissonance by putting more space between their personal beliefs and preferences and the positive attributes of the rejected alternative. In other words, they convince themselves that the chosen alternative is more attractive than originally thought. Thus, if a person is conflicted over purchasing a diesel vehicle, which has lower fuel consumption but higher emissions, vs. purchasing a low-emission vehicle, and he or she decides on a diesel car, the negative impacts of emissions are suddenly viewed as less important.

7.6.3 The social cognitive theory of learning (C)

Albert Bandura's social cognitive theory of learning (Bandura, 1986) is based on the idea that people learn from observation and experience. In line with Lewin's perspective (Box 6.5), he proposes a linked triad between personality, environment, and behavior, and his theory gives answer to *how* we gain mastery of our life circumstances. From a humanistic psychology perspective, he took a positive world view rooted in an agentic perspective, including moral and normative references. He considers people's beliefs to be at the interface of different hierarchy levels, including neuropsychological processes, the regulatory and self-processes of the individual, and the dynamics of human–environment interactions (in the sense that people are both producers and products of social systems). This optimistic view on the forming power of cognition is expressed as follows:

> Among the mechanisms of personal agency, none is more central or pervasive than people's beliefs in their capability to exercise some measure of control over their own functioning and over environmental events. (Bandura, 2001, p. 10)

Bandura's early theory can be summarized by the following aspects. (i) Learning, in the sense of changing knowledge, preferences, or action, is achieved through personal factors and states such as affective and biological conditions (e.g. a certain degree of arousal). (ii) Human actions or beliefs affect cognitions, feelings, and neurobiology. Thus, for instance, women who live together will synchronize their menstrual cycles (Weller *et al*. 1999). (iii) Learning can be accomplished by watching others perform useful behaviors that result in achievement (success). We learn vicariously by watching others and wanting to model those behaviors that lead to mastery. (iv) By modeling cognized goals we anticipate achievement. Goals are thus important determinants of performance. The effort spent on a task correlates with the difficulty of the goal. (v) The individual's behavior affects the environment and vice versa.

Awareness makes people capable of forethought; that is, anticipation of the likely consequences of an action. However, Bandura further assumes that individuals develop a sense of personal agency, a condition in which they believe that they can make things happen that will benefit themselves and others. In an almost humanistic manner, he postulates that humans want to make a contribution to society.

Bandura included a macroperspective through the concept of proxy. He argues that people do not have direct control over the social environment; that is, the conditions and situational practices that affect their everyday lives. Under these circumstances, they seek "their well-being, security, and valued outcomes through the exercise of proxy agency" (Bandura, 2001, p. 13). For example, people build coalitions, partnerships, groups, etc. with those who have access to resources, knowledge, or expertise. "No one has the time, energy, and resources to master every realm of everyday life" (Bandura, 2001, p. 13). Thus collective structures are excellent tools for expanding one's capacity. Two arguments are of interest here. First, he believes in the impact of subjectively perceived collective efficiency; that is, the power of social systems. Second, while biological factors are one of the three basic components of his theory, he strongly opposes the biologizing and genetification of human behavior, particularly when a societal functionality is taken into account.

> Knowing how the biological machinery works tells one little about how to orchestrate that machinery for diverse purposes. … There is little on the neural level that can tell us how to develop efficacious parents, teachers, executives, or social reformers. (Bandura, 2001, p. 20)

According to the humanist perspective, the nature–nurture relationship is one of high interdependence in that "socially constructed nature has a hand in shaping human nature" (Bandura, 2001, p. 20). For Bandura it is important to look at the nature–nurture relationship in terms of a bidirectional biology–culture evolution or – in other words – in the frame of cultural systems. Biological potentialism instead of biological determinism is required.

Bandura's concept of self-efficacy refers to the belief that one is capable of performing in a certain manner or of attaining certain goals. This concept strongly affected theories that were often applied in environmental psychology, such as Ajzen's (2002) theory of planned behavior (see Chapter 6.17), as well as the idea of self-regulation through one's own ideas or norms, as suggested in Schwartz' (1977) norm activation model.

7.6.4 Key messages

- Cognitive and mental processes include intuitive and analytic processes that are based on various types of declarative and operative memories
- Cognitive processes are affected by motivational and emotional processes that are difficult to control.

7.7 Motives

Motives are personal, internal, partly experientially shaped forces pertaining to or eliciting certain behavior. They are linked with concepts such as "instinct, arousal, need, and energization" (Weiner, 2000), which are also considered to be basic psychological process variables (see Table 7.1). Motives are relatively stable personal dispositions or goals, including, for example, hostile dominant behavior, self-affirmation, saving, jealousy, and the intimacy motive (Horowitz *et al.*, 2006; Murray, 1938). The concept of motive was described by McClelland (1965), who combined behaviorist, hedonistic, and expectancy-based concepts for explaining behavior.

> A motive becomes a strong affective association, characterized by an anticipatory goal reaction and based on past association of certain cues with pleasure and pain. (McClelland, 1951, p. 466)

Originally, many psychologists thought that motives were learned or socially acquired. Given a limited ability to unlearn, motives such as the achievement motive (Atkinson, 1953) are considered as personality traits. We will thus utilize the concept of motive as a basic and intrapersonally stable concept, whereas motivations are considered situational drivers. Motives depend on societal, economic, and cultural constraints.

Theories of motives and of motivation have been strongly shaped by the concept of expected subjective utility. Gerrig and Zimbardo (2008) call this a cognitive approach to motivation. Atkinson and McClelland simply defined the motivational tendency to succeed, $E(S)$, as the product of the subjective probability of success $p(S)$ and the subjective value or utility of success, $u(S)$, that is $E(S) = p(S) \times u(S)$. According to their view on differences between individuals, they then introduced personal variables for both the motive for seeking success $m(S)$ and the motive for avoiding failure, $m(F)$.

Thus they formally defined the individual's expectancy to succeed E as the interplay or sum of positive and negative valence:

$$E = E(S) + E(M)$$
$$= m(S) \times p(S) \times u(S) + m(S) \times p(S) \times u(S)$$

We do go not into further detail here. However, we will highlight a few important assumptions found in many theories. First, the positive, beneficial aspects are differentiated from negative, risky consequences, which make sense from a psychology perspective. Second, the concept of expected utility is taken as a decision criterion. Third, there is an assumption that different individuals' motives and personal traits interact in expectancy formation.

In line with the above arguments, we suggest utilizing the motive concept for representing the different, relatively stable individual behavioral patterns such as the saving motive (Goldman, 1978; Lusardi, 1998). Further, we think that symbolic motives play an important role in explaining certain aspects of environmental behavior, such as why people purchase large, environmentally unfriendly vehicles (Peters *et al.*, 2011; Steg, 2005; see Box 7.7).

7.8 Values

Values regulate how an individual adjusts to the environment by ordering preferences and representing features of desired end states of the environment. Like needs, values represent intrapsychic constraints of what is important, desirable, meaningful, worthwhile, and good. Values differ from needs in that they have a stronger cognitive component. And, like norms, values include both a descriptive component representing what people do and a prescriptive, normative component. They are less in number, less specific but more central and fundamental than attitudes.

Values transcend specific situations and are personal and internal, rather than consensual. They represent and operate in favor of three basic requirements of human existence: these are biological needs, requisites of coordinated social interaction, and demands of group survival and functioning. Relying on Rokeach (1973), Schwartz (1977, 1992) developed a taxonomy of 10 circularly arranged values: they include a group of values relating to self-enhancement, power (wanting authority and wealth), achievement (being ambitious and successful), hedonism (targeting pleasure and joy), and stimulation (wanting an exciting life and accepting risk). Another group of values touches on self-transcendence: self-direction (aspiring to freedom and creativity), universalism (looking for social justice and being liberal), and benevolence (forgiveness and being helpful). A group of values (see Figure 7.8) being linked with conservatism includes tradition (wanting humility or devoutness), conformity (being polite and self-disciplined), and security (targeting social order).

Schwartz provided empirical evidence that the prevalence of these values is rather stable across cultures (Schwartz & Bardi, 2001). Fietkau (1984) argued that environmental values are becoming increasingly relevant, because the individual suffers a loss of control in modern society. Thus, environmental values and commitments are reactions against the rapid change of the environment.

Today, there are only a few thorough studies on how values affect environmental behavior and cognition. Environmental values, consumption values, and

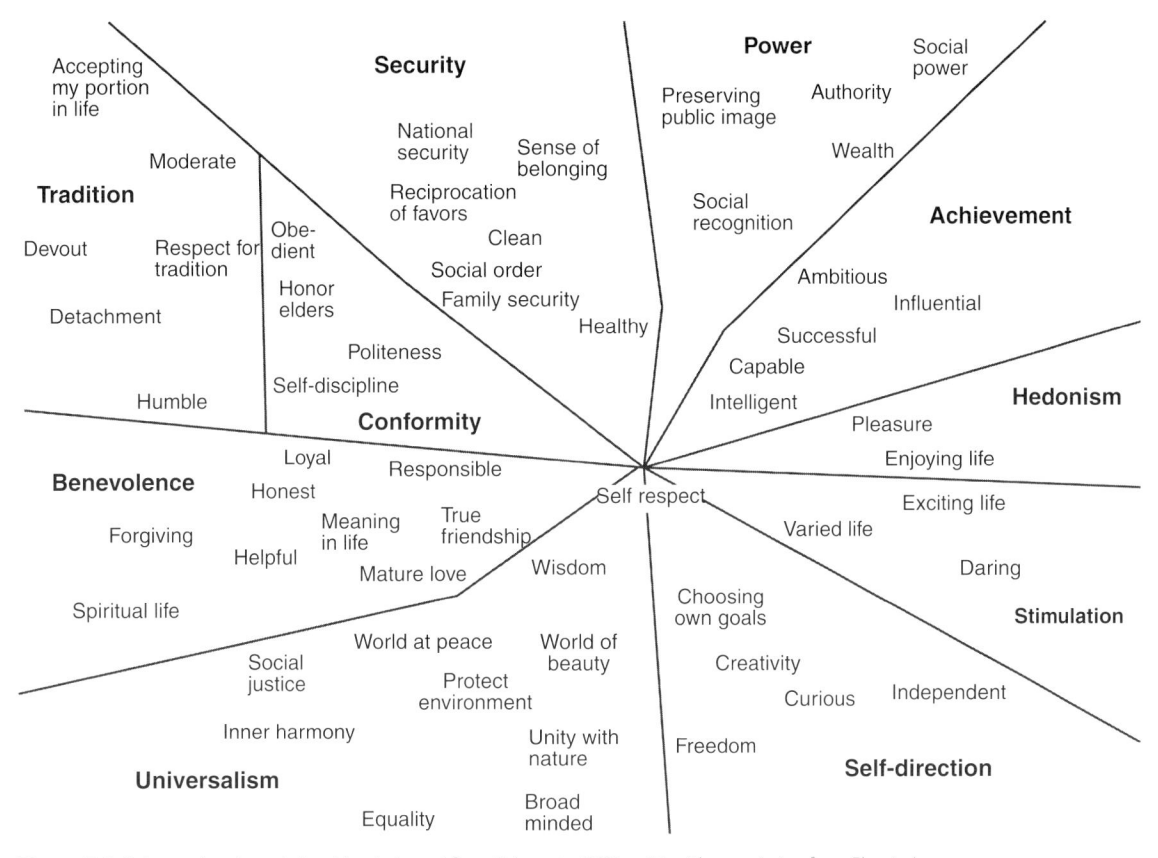

Figure 7.8 Schwartz's value relationships (adapted from Schwartz, 1992, p. 24, with permission from Elsevier).

eating habits have often proven to be predictors of behavior (Prentice, 2000). Some authors (Dietz et al., 2005; Reser & Bentrupperbäumer, 2005; Stern, 2000) postulate that values affect world views, as seen in the differences between the human exceptionalist vs. natural environmental paradigms (see Table 8.1), as well as personal beliefs of how certain interventions can change the world, and personal norms (Dietz et al., 2005). The role of values on actions, however, is not completely clear. Another open question is whether environmental values, that is, values related to the biophysical environment, actually exist in the average person. Lisbach (1999), for example, could not show an effect of environmental values on perceptual judgments (see Box 7.6). None the less, we think that specific environmental values are a reasonable construct that can have a strong impact on individual behavior.

7.9 Norms

Norms represent constructed standards of what an individual considers appropriate or typical. The use of norms can be considered a kind of cognitively represented reference point, which provides a kind of benchmarking. The concept of norms used in psychology differs from that in sociology, wherein norms are considered as stable and enduring properties of groups. In psychology personal norms, applied for describing individual behavior, differ from group norms (see Boxes 6.13 and 6.14) or from norms in ethics and economy (Dietz et al., 2005).

With respect to personal norms one has to distinguish between descriptive norms and injunctive norms. Descriptive norms are personal norms in the sense that they describe how people actually behave. Injunctive norms represent a person's belief of how others expect him or her to behave. This notion of norms shapes Schwartz's oft-cited norm activation model (Schwartz, 1977; see Chapter 6.17). According to Schwartz, altruism and environmentally friendly behavior can become a social norm and thus induce further environmentally friendly behavior.

Collectives of people form norms (Sherif, 1936). Social norms affect and constrain behavior without legislation. They develop by interaction and can be

Box 7.6 Do values affect perception: "le waldsterben"

There has been substantial concern and a huge public debate about forest decline in Europe since the early 1980s. At the time, newspapers published pictures of completely degraded forests in the vicinity of the coal mining areas in Eastern Europe, raising public awareness. In addition, forest monitoring recorded a strong decline of the silver fir in the mid-1970s, thus alerting scientists as well (Ulrich, 1990). Germans called this phenomenon *Waldsterben* (Engl. dying forest), bringing to mind the principle of *humanization of the environment* (see Box 9.3). Similarly and ironically, the French use the term "le waldsterben" when alluding to the German nationalist, romantic, infatuated portrayal of forests and when denoting a prototypical example of "formations discursives." The latter means that discourse is established by (historically shaped) concepts (Foucault, 1971; Gerbig & Buchtmann, 2003).

Traditionally, the procedure for evaluating forest health and environmental stress has been through visual assessment of the trees' crown transparency by trained foresters at ground level (see Figure 7.9; Innes, 1993). Clearly, one would guess that the foresters are environmentalists. Thus, one may ask whether their crown transparency ratings have been biased by their environmental values. Such a hypothesis goes back to the New Look paradigm or the directive state theory of perception (Allport, 1955), which emerged in the 1940s. The idea of this theory was that needs such as hunger (Bruner & Goodman, 1947; Bruner & Kretch, 1949; Bruner & Postman, 1949) affect perception and perception-based judgments. The latter has been proven by Bruner and Goodman in a ground-breaking experiment of 30 10-year-old children wherein a sample significantly overestimated the size of coins when compared to a control group presented with cardboard disks of identical size. Moreover, poor, lower class children judged the coins as significantly bigger than those of the wealthy upper class.

A study by Lisbach (1999) investigated the relationship between values and the judgments on trees' crown transparency by 48 students of various faculties. The research was motivated by a finding that, in an international comparison, French foresters judged the crown transparency of 186 trees to be 6% lower than their German and 12% lower than their British colleagues. In the Lisbach experiment, the values of the student subjects were measured, along with being given a training course in forestry. They were then given value-laden instructions, hinting that air pollution stemming from industry was a cause of forest decline. The student sample showed a 5.6% higher (and statistically significant) judgment on crown transparency than the baseline from professional raters. However, no significant correlations (i.e. no significance level below $p = 0.28$) could be found between crown transparency and four values of the Schwartz value inventory (Bilsky & Schwartz, 1994; Schwartz, 1992, 2006) as well as three environmental attitude components (Siegrist, 1996). Thus, no effect of environmental values on the perception-based crown transparency judgments could be shown in this study.

Figure 7.9 Three pictures from the training material on crown transparency from the Eidgenössische Forschungsanstalt (Müller & Stierlin, 1990), used for calibrating the judgments between professional raters. A critical question is whether professional raters are biased because of their proenvironment values. Current research has not provided evidence of this.

Today, the natural science view on the German Waldsterben discussion provides a differentiated and ambiguous view. The forest can be stressed and harmed by pollutants such as sulfur dioxide. However, some pollutants, such as acid rain, can also support forest growth for many species under certain conditions. It seems that the Waldsterben debate of the 1980s was related to a real forest decline owing to an interaction between the impacts of the traditional smokestack industry, the cold winter of 1978, drought periods (Krause *et al.*, 1986), and related insect defoliation. Although visually judging tree vitality involves the potential of value- and interest-biased misjudgments, appropriately using the procedure is still considered a valuable method, particularly for early warning purposes (Dobbertin, 2005).

explicit or implicit (Cialdini & Trost, 1998). Of special importance for environmental literacy are some specific social norms such as the reciprocity norm or the equity norm. The reciprocity norm postulates that, particularly in conflict situations such as the tragedy of the commons (see Box 6.15), what the other is doing (e.g. littering or not littering) is both affected by and affecting others. Both the reciprocity and the equity norm refer to Homans' idea of social exchange (Homans, 1955). Interaction and cooperation among individuals should be arranged such that the ratio between benefits and costs (e.g. environmental burdens between neighboring farmers) is balanced. We want to note that norms and motives are overlapping concepts.

7.9.1 Key message

- Personal and social norms such as equity or environmental conservation may become reference points of behavior.

7.10 Motivations

Motivation is a widely used concept. It is a "general term for all processes involved in starting, directing and maintaining physical and psychological activities" (Zimbardo *et al.*, 2005, p. 370). Motivation is a close-to-behavior variable that connects action and observable behavior to internal states. Motivations can be based on drives, instincts, needs, motives, values, beliefs, and other psychological variables. Motivations can be intrinsic or extrinsic, the latter stemming from incentives outside the person. In general, motivation is a summary or integral variable that describes how people are stimulated or attracted by incentives when referring to other constructs of Table 7.1.

We thus use the term motivation in a rather colloquial sense. An exception is made when using the concept to describe behavioral tendency in conflicts or social dilemmas. Decision psychologists (Liebrand & van Run, 1985; MacCrimmon, 1976) developed the

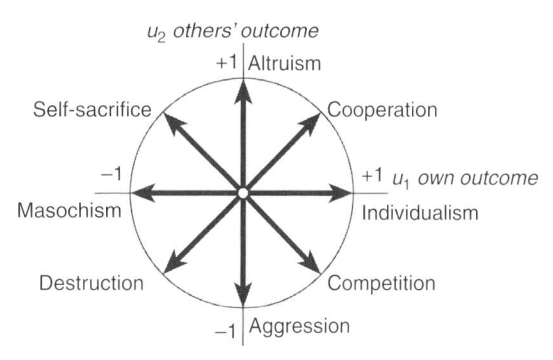

Figure 7.10 The circle of social motivations. If all points on a unit circle present possible payoffs, the arrowheads mark the most preferred points belonging to the various motivations.

circle of social motivations in conflict games (e.g. of the prisoners' dilemma game type). Let us assume that all the utilities two players can obtain are located on the unit circle (see Figure 7.10). The particular relation between u_1 and u_2 to which players aspire reveals their social motivations. If, for instance, player 1 wants to maximize $u_1 - u_2$ then his motivation is competitive. Similarly, player 1 is altruistic if he aspires to maximize u_2.

Batson *et al.* (2002) discussed altruism and collectivism in the context of community involvement. They report that empathy-related altruism is a powerful motive.

7.10.1 Key messages

- Motivation is a general concept and can be based on drives, instincts, needs, motives, values, beliefs, and other psychological variables. The circle of social motivations classifies different motivations that are at work when people interact.

7.11 Attitudes

The American engineer and quantitative psychologist Louis L. Thurstone (1887–1955) defined attitude as "the affect for or against a psychological object," that is, to denote overall degree of favorability (Thurstone, 1931, p. 261). Referring to the approach used by Atkinson

Box 7.7 Fuel efficiency in car purchase: Hummer's symbolic motives

In 2000, energy consumption from vehicles worldwide was responsible for 13% of anthropogenic CO_2 emissions[1] (and constituted 17% of oil consumption[2]). The dependent variables of such emissions are choice of travel mode (Bamberg et al., 2003), distance traveled per year, and the fuel efficiency of cars. Clearly, vehicles with hybrid or solar engines produce much fewer CO_2 emissions than regular cars. For example, a Hummer emits 470 g CO_2/km, whereas the Toyota Prius only emits 104 g CO_2/km.

Figure 7.11 Human choices such as car purchase behavior are affected by personal and social norms, what people know and believe about the impacts of fuel emissions, and other psychological variables. One variable relates to how certain types of cars are linked with symbolic motives such as power, potency, shelter, status, etc.

Given the long-term costs to society, fuel emissions should be reduced. Yet people still purchase gas-guzzlers. Why do they show this behavior? Social and environmental psychologists approach this question from the level of what an average person is doing and examine what psychological variables are involved in car purchase behavior. In general, psychological models include cognitive factors, which are sometimes linked with emotions such as worry (i.e. environmental awareness that climate protection is necessary) or the belief in the efficacy of buying hybrid cars (response efficacy), and social norms (i.e. what other people expect one to do). In line with the norm activation model (Schwartz, 1977), these factors are thought to affect the personal norm; that is, what a person considers to be an appropriate action when buying cars. The personal norm is also closely related to values (such as universalism) or social motivation (such as altruism). The latter demonstrates environmental literacy because, owing to the slow development of climate change impacts, a person's choice to make an environmentally friendly purchase benefits future generations' environmental health and not their own. In some approaches, such as the theory of planned behavior (Ajzen, 1991), control knowledge (i.e. perceived behavioral control), or the belief in agency (Bandura, 2001) are also incorporated. With a car purchase, this is the knowledge about effective action (e.g. what car – diesel, gas, solar – is the right one). Finally, depending on gender, age, and cultural and personal traits, a car can represent power, potency, shelter, status, lifestyle, etc. (see Figure 7.11). Like clothes or pets, vehicles should fit a person. Thus the symbolic motives linked with emotions play an important role (Steg, 2005). Furthermore, attitudes are incorporated as an umbrella concept of general disposition or valence towards an object (e.g. the fuel efficiency of a car).

Peters et al. (2011; see Figure 7.12) investigated whether the above variables would allow a prediction of the intention to buy a low fuel-consuming car as their next vehicle. Applying structural equation models on a

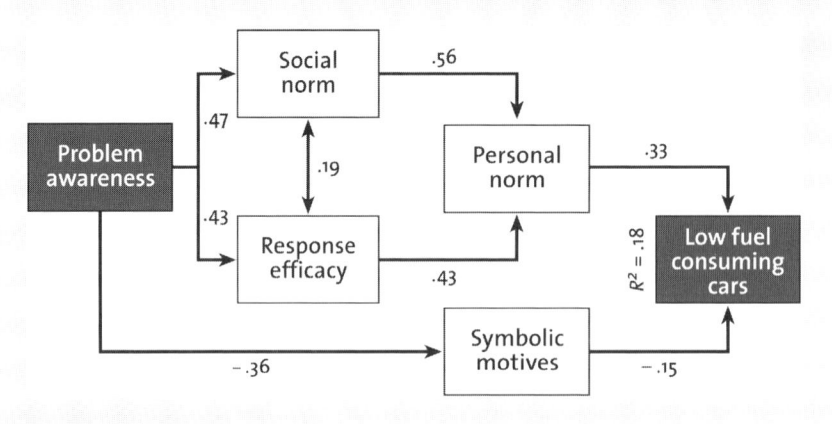

Figure 7.12 The intention to buy "low fuel consuming cars" is dependent on various psychological drivers and impact factors. The figure represents the results of a structural equation model for a sample of Swiss consumers ($n = 595$). The psychological variables (described in the text) explain 18% of the variance of importance ratings on "low fuel consumption" ($R^2 = 0.18$). The numbers along the arrows represent correlations.

---

[1] http://irps.ucsd.edu/dgvictor/publications/Victor_Article_1999_Global%20Passenger%20Travel.pdf
[2] http://www.ipcc.ch/publications_and_data/ar4/wg3/en/figure-spm-1.html

questionnaire with Swiss car owners revealed that the dependent variable low fuel consumption is highly correlated with personal norm and (negatively) with symbolic motives. The personal norm is dependent on the social norm, response efficacy, and (the cognitive variable) problem awareness, the latter of which negatively correlates with symbolic motives. The other variables mentioned above were incorporated in the model but did not show significant impacts.

and McClelland when they integrated aspects of the achievement motive from psychology with the idea of constructing expected subjective utilities from decision theory, Fishbein and Ajzen (1975) formulated the theory of reasoned action. They considered overt behavior (B) as a function of the weighted sum of measures of behavior intention (I), the attitude toward a behavior (A), and subjective norms (N). Formally this can be written as:

$$B = f(w_1 I + w_2 A + w_3 N) \text{ with } w_i > 0$$
$$\text{for all } i \text{ and } \sum_i w_i = 1.$$

As we can see, this approach shows similarities to expected utility and thus has also been called the expectancy value approach. However, in contrast to expected utility, probabilities of certain events occurring are not defined.

Attitudes are thought to be specific and object-related. People are thought to hold different attitudes simultaneously towards an object in the same context. Here, in line with dissonance theory (see Chapter 7.6), strong attitudes are expected to be relatively resistant to change when new information is provided.

Some proponents of the concept of attitudes refer to neuropsychological research that states that the evaluative categorization of attitude formation involves right and left brain hemisphere activity. However, not all researchers in the field agree:

> This evaluative meaning arises spontaneously and inevitably as we form beliefs about the object. (Ajzen, 2001, pp. 30–1)

The above statement would seem to indicate that the process is of an intuitive and not of an analytic nature (see Table 7.4).

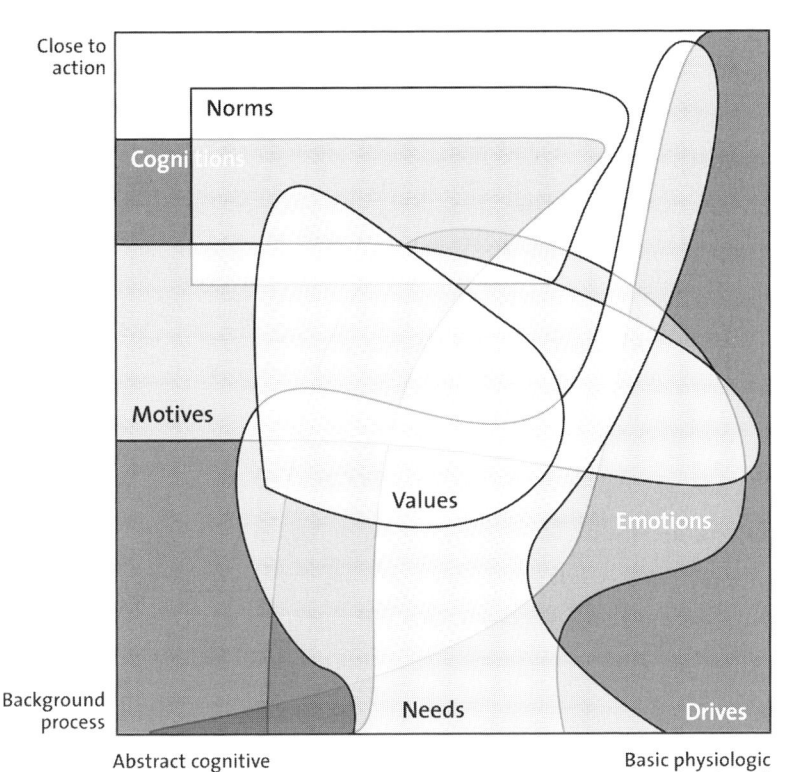

Figure 7.13 Heuristic outline of the relationships between the drivers of human behavior relevant for analyzing environmental literacy based on the dimensions of cognitive processes to physiological process and background process to close to action processes.

Just as there are questions about how uncertainties are conceptualized by the individual, ideas about how cognitive processes are involved with attitude formation, and the relation to other concepts such as values, needs, and emotions are also not very clear. Ajzen stated that "favorable valences associated with such abstract concepts as freedom and equality are known as values" (Ajzen, 2001, p. 42). Thus values or needs such as "security through order, humanistic and expressive concerns ... are related to liberal versus conservative attitudes" (Ajzen, 2001, p. 42). Symbolic aspects are also included in attitudes (see Box 7.7). Today the concept of attitudes has become very broadly defined, taking on different functions, including a "value expressive function of attitudes, the knowledge function, the ego-defensive function, the social-adjustive function, and the utilitarian function" (Ajzen, 2001, p. 41). Similarly to motivations, attitudes serve as an integrative and summary concept. We thus prefer to represent the valence of certain options, alternatives, and objects through variants of the utility and preference construct.

7.11.1 Key messages

- Attitude is a broad, mostly fuzzily defined, summary concept that is supposed to affect choice and behavior.

7.12 Relating and distinguishing drivers

We have seen that the drivers of individual behavior, which constitute appropriate environmental awareness, literacy, and behavior, are related. However, they differ in their relations to the concepts of human that underlie psychological research. Some are more closely linked to decision-making and behavior than others. Additionally, they are differentially established by basic physiological or higher ordered mental processes (see Table 7.1). This is illustrated by Figure 7.13, which is valuable for analyzing environmental literacy because it distinguishes two dimensions by which the discussed drivers can be ordered.

According to a pure behaviorist view based on basic physiologic processes, all motivations and driving forces are due to drives and emotions. These two concepts build the bottom right part of Figure 7.13. In certain situations the sexual, aggression, or thanatos drives and emotions seem to be at work. However, drives have the potential to affect action more directly than emotions.

If we look at the top left part of Figure 7.13, we find cognitions, norms, and values. This is in line with the assumption that conscious or unconscious cognitive processing is involved in almost all decisions and actions. Such a view had already been elaborated in the 1950s (Hebb, 1955). Norms, in particular social norms, are seen as relatively stable aggregates of own experiences and social experiences. They are considered to be closer to action than cognitions which, if it comes to real actions, may become more object of a "decision filter" than norms (Scholz, 1987).

Values overlap with norms and motives but are, in particular, considered personal and internal. Thus, they also touch on lower domains of internal processing. Motives are seen as a relatively central layer formed by higher ordered mental and social processes. However, motives also have strong psychophysical components and thus overlap with the domains of emotions and drives. Finally, referring to Maslow, it is not easy to categorize needs simply as basic (such as hunger) – spiritual needs have to be considered, too. Thus needs include components of drives as well as problem-solving capabilities (see Figure 7.3), which may be closer to action than norms and cognitions.

Clearly, Figure 7.13 is a subjective view (from the author) on drivers. The map depends on the definitions of the concepts, as well as the problem to be investigated and the perspective by which this is done. Nevertheless, the main message of this figure is that the psychological processes underlying individual behavior are of varying nature and must be precisely differentiated.

7.12.1 Key messages

- Drivers are psychological constructs or concepts that underlie behavior. These concepts can be classified according to the basic psychophysiological vs. higher ordered mental processes and background processes vs. close to action processes.

Part IV: Contributions of sociology

8 Traditional sociological approaches to human–environment interactions 215
9 Modern sociological approaches to human–environment interactions 231

Key questions (see key questions 1–3 of the Preamble)

Contributions from sociology to environmental literacy

Q1 How is the society–environment nexus conceptualized by different sociological theories? When and how was the material–biophysical environment incorporated into sociological theories?

Q2 According to sociological theories, what are the key drivers of societies and which ones are relevant for the interaction between human and environment systems? How do social structures affect individual actions, particularly environmental behavior, and vice versa? How are environmental processes picked up by public discourses and reconstructed as "environmental problems?"

Q3 How can sociological findings about societal decision-making and learning processes help humans cope with environmental problems? What roles should different societal actors (e.g. academia, industry, the public) play in coping with "environmental problems?" At what spatial level (e.g. regional, national, supranational) are environmental problems best addressed?

What is found in this part

Sociology is the science of human societies, which are "the primary organizational subdivisions of the human population" (Lenski et al., 1991, p. 17). With respect to environmental literacy we are interested in identifying key drivers of societies. We wish to understand in what way, why, and when the dependence and influence of society on natural resources has been incorporated into sociology. We will see that only a few sociologists have endeavored to include these aspects in their theories.

After providing a definition of society, Chapter 8 starts with the emergence of sociology and reviews how environmental issues were dealt with by the classical sociological theorists Marx, Durkheim, and Weber. We then step into human ecology and follow the emergence of environmental sociology. Finally, we go into evolutionary ecological theory. We extend this theory to a resource-based socio-technological evolutionary theory (ReSTET), which conceptualizes the evolution and adaptation of societies to environmental change (see Figure 8.3*).

Chapter 9 focuses on contemporary sociological theories, which enable a better understanding of the social rules at work on the level of society when dealing with problems from the material–biophysical environment. With respect to the components of society presented in Figure 1.7*, we thus concentrate on the sociology of the "cultural and social order system" and do not include, for instance, sociology of law, sociology of science nor sociology of education. The functioning of the economic systems that make strong reference to sociological thought, in particular the new institutional economics, are presented in the economics chapters (Chapters 10 and 11).

Part IV Contributions of sociology

Chapter 8
Traditional sociological approaches to human–environment interactions

- 8.1 What is a society? 215
- 8.2 Origins, Marx, Durkheim, and Weber 216
- 8.3 The rise of human ecology 219
- 8.4 The cradle of environmental sociology 222
- 8.5 The resource-based socio-technological-evolutionary theory (ReSTET) 225

Chapter overview

Society is a highly aggregated human system. It is considered an autonomous social system which organizes subsystems of the human population. As we will see, the material–biophysical environment has not been widely treated in sociological theories.

After discussing the role that the material–biophysical environment played in the theories of Marx, Durkheim, and Weber, we show how the Chicago School for Sociology, often called simply "the Chicago School," utilized concepts from biology in their sociological theories in the early twentieth century. This led to the foundation of human ecology, an academic discipline which led later to the formation of environmental sociology. The chapter concludes with the ecological–evolutionary theory from Lenski. We present an extended framework, namely the resource-based socio-technological evolutionary theory (ReSTET). This approach suggests that societal development is based on the interaction of sociocultural, natural resource-based (i.e. material, biophysical) and technological trajectories, and that the direction of societal development is driven by two objectives (i.e. wealth and power). The significance of each driver depends on the degree of democracy present in a society.

8.1 What is a society?

Exactly which questions belong in the realm of sociology is disputed, even among sociologists themselves. A very general view states that sociologists study the development, structure, interaction, and collective behavior in societies (Weber, 1921/1972, p. 1). In contrast to psychology, sociology has long emphasized the role of socially formed settings on groups, organizations, etc. and, in general, does not deal with internal processes at the level of the individual. Others characterize sociology simply as the study of human societies, which is defined as:

> ... a politically autonomous group of people which engages in a broad range of cooperative activities. (Lenski et al., 1991, p. 8)

From a functional perspective, sociologists say that societies organize the existence of the human species. A "society organizes a population, which is a subsector of the human species" and is "politically organized" (Parsons, 1977, p. 111).

> Societies are, in effect, the primary organizational subdivisions of the human population as a whole. (Lenski, 2005, p. 17)

According to this definition, societies include and produce various products, such as population (e.g. genetic and demographic variables), culture (e.g. language, social heritage, experience, symbols, information conveyers, customs, societal norms, moral values, corruption, retaliation systems, ideologies, consumption patterns), material products (e.g. trade products, built environment, capital goods, machines, energy systems), social organization (e.g. social order, strata, classes, roles, positions, primary groups such as family, cliques or elites), and social institutions (such as kinship/marriage, religion, and education). In this section we deal only with macro-sociological questions, which focus on social structure. Some issues of micro-sociology, such as rational choice theory, are discussed in the psychology Chapters 6.17, 7.7, and 7.11.

Some sociologists have criticized the neglect of relevance and care for contemporary problems of today's sociology. For instance, in the 2004 presidential address to the American Sociological Association, Michael Buroway stated that "professional sociology of today tests narrow hypotheses" and "that the field has [thus] fallen victim to methodological fetishism and an obsession to deal with trivial objects" (Buroway, 2005, p. 15). This harsh critique refers, on the one hand, to a communicative, public, dialogue orientation that is missing in sociology (which is addressed by transdisciplinarity; see Chapter 16.7). On the other hand, the critique suggests that the bigger questions, such as "Why does this age differ from every other age?" (Berger, 2002, p. 29), tend to get lost. From an environmental literacy perspective, this question can be translated as follows. In what way are societies dependent on natural resources and the biophysical world? Or, when do societies become aware of anthropogenic impacts? Or, as included in key question 3, how do human and environmental systems interact? And what rationales can we find in human systems – here societies?

8.1.1 Key messages

- Societies can be defined as politically autonomous, primary organizational subdivisions of the human species that secure the sustaining of human populations.

8.2 Origins, Marx, Durkheim, and Weber

The French philosopher and mathematician Auguste Comte (1778–1857) coined the term sociology in the eigtheenth century. For Comte sociology took a double role. It was a positivist science of society but also a kind of final science, which provides knowledge of the laws of the human order, merging with philosophy and integrating all other sciences. As the title of Comte's book *Treatise on Sociology Instituting the Religion of Humanity* (see Bordeau, 2008) indicates, sociology should become a kind of doctrine of the universal knowledge system. Comte's initial aspiration was nicely sketched by Park and Burgess, both of whom can be considered pioneers of environmental sociology in the twentieth century:

The earlier and more elementary sciences, particularly physics and chemistry, had given man control over external nature; … sociology, was to give man control over himself. (Park & Burgess, 1921, p. 2)

The establishment of sociology as a scientific discipline happened a century later and can be seen in the foundation of the first sociological journal by Émile Durkheim in 1896 (Poggi, 2000). From the perspective of environmental literacy, it is essential to recognize that environmental sociology has distanced itself from classical tradition of sociology, although today, at least in the USA, it has integrated to a great degree with sociology again. As Buttel (see Haberl et al., 2004, p. 199) pointed out, this has led to the loss of theoretical links to potential and seminal roots in classical sociology, where important contributions to understanding society's impact on the environment and dependence on resources have been made. In the contemporary discussion about re-establishing the roots of environmental sociology, Marx, Durkheim, and Weber (Buttel et al., 2002) have played key roles. Some of these classicists have strongly tended toward the human exceptionalist paradigm (HEP; see Table 8.1), which assumes that all (environmental) problems can be solved through human ingenuity and development of proper technologies. This holds true, for instance, for Durkheim, who considered social systems to be almost independent from natural resources (Humphrey & Buttel, 1982).

Before we present these classical sociologists we briefly look at their philosophical foundations. Schluchter (2006), for instance, emphasizes that Durkheim followed a positivist philosophical line, whereas Weber was shaped by subjectivist idealism, as formulated by Kant. In contrast, Marx's world model emerged from putting Hegel's dialectic reasoning in an objectivist, realist frame. Here we examine how Marx, Durkheim, and Weber contributed to a better understanding of human–environment systems (HES).

8.2.1 Marx's view on production means and private properties

Karl Marx's[1] (1818–83) work stressed the negative impact of private property. His historical materialism focused on technology (production means or forces

---

[1] The contributions of Karl Marx are dealt with in Chapter 10 (Economics) and Chapter 12 (Industrial Ecology). In Chapter 10, emphasis is given to the early Marx concern on the social situation of the workers and the role of

Table 8.1 Two fundamental paradigms of western sociology in coping with environmental issues (Catton & Dunlap, 1978).

Human exceptionalist paradigm (HEP)	Natural environmental paradigm (NEP)
Humans are unique among the Earth's creatures, for they have culture	Human beings are but one species among the many that are interdependently involved in the biotic communities that shape our life
Culture can vary almost infinitely and can change much more rapidly than biological traits	Intricate linkages of cause-and-effect feedback on the web of nature produce many unintended consequences from deliberate human action
Thus, many human differences are socially induced rather than innate; they can be socially altered and inconvenient differences can be eliminated	The world is finite, so there are potent physical and biological limits constraining economic growth, social progress, and other societal phenomena
Finally, cultural accumulation means that progress can continue without limit, ultimately solving all social problems	

of production; German: Produktivkräfte) and social organization, such as the division of labor, and production modes (the concept of relations of production or the ownership of the means of production; German: Produktionsverhältnisse). His theory of historical materialism aspired to explain transitions between societies (e.g. between the agrarian and the industrial society). This theory also includes two assumptions about drivers of society. According to Marx, societies strive for the development of means of production (i.e. technology) and toward a classless communist society in which the needs of the people are fulfilled. Thus his theory includes a teleological assumption. For environmental and resource sociology, this approach is interesting for two reasons. First, it has a natural materialist core. Second, both production means and modes of production are "critical of the arrangements or trends in modern society" (Buttel, 1996, p. 61). The latter is closely linked to questions of justice and equitability because access to resources is a matter of power and distributional conflicts (Schnaiberg *et al.*, 1986), and individuals depend on the material conditions to determine their production.

For Marx, use–value results from the combination of labor and natural or raw material (German: Naturstoff; Marx, 1890/1970, p. 193). Here, nature is imbued with an economic notion and value. Marx provided a series of statements that human life is dependent on eating, drinking, and housing:

> The human is *living* from nature. This means: the nature is the body and the human has to stay in a permanent process with it in order to survive. (Marx, 1844/1968, p. 515, translated by Scholz)

A central characteristic of Marxist theory is the concept of labor, which has been considered as the "Quelle des Reichtums" (Engl.: source of wealth). The importance of labor becomes evident in an early writing of Friedrich Engels, cofounder of Marxism, published in 1876 and entitled *Anteil der Arbeit an der Menschwerdung des Affen* (Engl.: The Part Played by Labor in the Hominization of Apes; Marx, 1876/1971):

> Labour is, first of all, a process between man and nature, a process by which man, through his own actions, mediates, regulates and controls the metabolism between himself and nature. He confronts the materials of nature as a force of nature. When he affects nature … he himself changes his own nature. (Marx, 1890/1970, p. 444, translated by Scholz)

Thus, Marx not only acknowledged the relationship between man and his natural environment but also recognized the secondary feedback loops involved in human–environment interactions (see Chapter 16.5). Fischer-Kowalski (1997; see socioecological system analysis in Chapters 9.3, 12.6, and 19.4) and particularly Foster (2000) describe how Marx utilized the concept of metabolism of labor as a key concept for understanding the impacts of human-induced material fluxes.

> It [the labor process] is the universal condition for the metabolic interaction [Stoffwechsel] between man and nature, the everlasting nature-imposed condition of human existence. (Marx (1887/1990), quoted according to Foster (1999, p. 380))

The concept of metabolism played a larger role in Marx's later work, stemming from interdisciplinary

---

labor and technology and natural resources. Chapter 12 focuses on metabolism (German: Stoffwechsel). The reader should read these passages to learn how Marx incorporated natural resources into his theory.

inspiration. He used the concept to deal with two central environmental problems of the mid-nineteenth century. One was the problem of decreasing soil fertility. Here, Marx referred to the work of Karl Liebig (1840) and particularly to Julius R. Mayer's (1814–78) work on *The Motion of Organisms and Their Relation to Metabolism* (Mayer, 1845/1978). Marx was concerned about "the vitality of the soil" (Marx, 1863–1865/1981, p. 949) by removing constituents from fertile soil and substituting them with scarce artificial fertilizer. The other was the relationship between the city and the countryside. Here he again looked at the missing closedness of the metabolic cycle, for instance "the excrement produced by 4½ million people … pollute the Thames … at monstrous expense" (Marx, 1863–1865/ 1981, p. 195). Today, in the twenty-first century, we find the same problem in the north–south frame. We can conclude that the material environment and natural resources availability thus played a role in Marxist theory (Gross, 2001; see Chapter 10.3).

8.2.2 Durkheim's reality of social facts

The French sociologist Émile Durkheim (1858–1917) focused on social phenomena, social and moral norms, and the sociocultural environment, and considered social facts to be real (Durkheim, 1893/1977). These are, in some respects, the basic assumptions of de Landa (2005), presented in Chapter 3.4. According to Durkheim, the power of social norms is tremendous and is a major factor of society. This is expressed by his famous statement that "social facts can only be explained by other social facts" (Durkheim, 1895/1982, p. 13). The concept of moral imperative can be taken as an example. People and societies, their morality, customs, and beliefs, which cannot be viewed through the lens of individual psychology, create rules and imperatives that govern interaction, law, rewards, and sanctions (Durkheim, 1924/1976). Durkheim – at least in his earlier writings – follows a strictly realist and positivist view. Social morals and laws are real and can be empirically measured and identified. At least in his early work, Durkheim followed Comte's idea that sociology, as the science of human societies, incorporates other sciences.

Durkheim was the first to investigate division of labor as one of the means of facilitating the operation of complex societies. Durkheim's message was that a suitable, functional organization of society is able to cope with overpopulation and resource scarcity (Durkheim, 1930/1992, p. 325). In his early work Durkheim tried to exclude cosmological, bioclimatic causes, but later he had to acknowledge that certain seasons of the year affect the activity of human life.

Durkheim considered the natural environment to be static: the physical world would remain unchanged, if one neglected those factors that have a social origin (Durkheim, 1930/1992). Nevertheless, it is interesting to see that even a pure social scientist contends with the environment when interpreting social data. In his major work, entitled *Suicide*, Durkheim (1897/1973) recognized that, in all the countries investigated, the highest suicide rate occurs in June and July, the period of highest human activity in the northern hemisphere. Durkheim is a telling example that, methodologically, it is hardly possible to separate bioclimatic from social impacts. Today, social medical research suggests that both bioclimatic and sociodemographic data have impacts and interact with each other (Chew & McCleary, 1995).

For many environmental sociologists, such as Catton and Dunlap (Catton & Dunlap, 1978; Dunlap, 2002), Durkheim was seen as a "principal intellectual source of sociology's environmental blindness" (Rosa & Richter, 2008, p. 183). However, some sociologists elaborated (see Gross, 2001; Rosa & Richter, 2008) that at least Durkheim did not rule out the laws of nature in his late work but that he focused the social facts and symbolic aspects on the level of society.

However, against the background of the material–social complementarity view taken in this book (see Chapter 3.3 and the Complementarity Postulate P1 in Chapter 16), Durkheim was a pioneer in elaborating that societal activities are formed by social–cultural rules as a self-referential system. However, the cascaded relationship between the social and the material became clearly stated in Durkheim's late work:

> The physical environment presents a relative fixity. … The organic world does not abolish the physical world and the social world has not been formed in contradiction to the organic world, but together with it. (Durkheim, 1911–1912/1983, pp. 69–70)

8.2.3 Weber's interpretative sociology

Max Weber (1864–1920) was a great influence in sociology, economics, law, and many other sciences. His focus was social organization, wherein he distinguished three types of power: political, economic, and ideological (Poggi, 2005). His major emphasis was on the political form of power, which becomes more significant when linked to legitimacy, and which transforms into domination: legal domination by state, law, institutions, and bureaucracy, traditional domination such as patriarchies and feudalism, and charismatic domination through family or religious structures, for example in the Indian caste system.

Methodologically, Weber was not a positivist – rather, he believed in the interpretative, hermeneutic method. He stated that research can never be fully inductive. With respect to human behavior, he defined four different ideal types of rationality: zweckrational (goal rationality), wertrational (value rationality), affektual (emotional rationality) and traditional (custom, unconscious habit; Weber, 1921/1972). The term ideal type (German: Idealtypus) refers to Kantian thought that, in the end, conceptual descriptions provide understanding (German: verstehen). In his theory of bureaucratization, for instance, he argued that a rational bureaucracy should shift from a traditional, value-oriented system of authority to a goal-oriented organization and action. Thus, Weber's approach also included assumptions about goals. For instance, he assumed that as a (formal) bureaucracy grows it tends to take an increasingly conservative position.

Although Weber did not deal with environmental issues explicitly, his views are of major importance, at least for environmental policy. In his principle work *Wirtschaft und Gesellschaft* (Engl.: Economy and Society; Weber, 1921/1972), he outlined the way that the law of succession and inheritance rights affect land use and what type of "oikos," or agricultural farms, follow.

Similar to Durkheim, Weber focused on the social drivers of society. Thus he considered the availability of mineral raw materials to be of only marginal importance to the rise and success of technological society in Europe. Religions were seen as major drivers of societal development. Weber argued that the Protestant ethic promotes capitalist development, as it suggests an inner asceticism but unlimited access to commodities, thus giving importance to the cultural side of the social environment. At the same time, Weber was active in several environmental conservation efforts, including the first German association of nature and cultural heritage protection, called the *Bund Heimatschutz*, and an initiative of 1500 conservative Swiss and German scientists who protested construction of a hydroelectric power plant at the Rhine cascades in Laufenburg (Lange, 2002). Still, in his writings, Weber did not deal with either the negative impacts of the Industrial Age on the natural environment nor with rebound or secondary effects of industrial activities on human health or development (Gross, 2001).

8.2.4 Key messages

- Natural resources and particularly the availability of raw materials as a foundation of labor played a role in Marx's materialistic approach. This has not been the case to a similar extent for the contributions of the classics, such as Durkheim and Weber
- A major contribution of Durkheim's work was that he elaborated that social–cultural facts and rules are key components and forces of society, which organizes and rules societal activities. We consider this to be a first outline of the rationale of society (see key question 2)
- Weber considers religions as major drivers of societal development. He describes how legal regulations (for instance, law of succession and inheritance rights) affect land use.

8.3 The rise of human ecology

8.3.1 A first start between geography and sociology

In most textbooks of sociology, the emergence of human ecology is cited as occurring in the 1920s. However, some recent papers dealing with the question of how the environment was incorporated into sociology dated the first use of the term "human ecology" to 1908 (see Gross, 2004). According to Gross, the term human ecology first emerged in a discussion about relations between geography and sociology. In a discourse between the sociologist Hayes (1868–1928) and the geographer Goode (1862–1932), the latter considered human ecology as a prerequisite of sociology as it provides "a study of the geographic conditions of human culture" (Goode, quoted in Hayes, 1908, p. 395). Here the view "physical environment affects human system" was taken. As Gross (2004) points out, Hayes later considered the relationship between the physical geographic, the inherited and acquired traits of populations, the technical, and social layer of HES. Thus, also actions of human systems on the environmental system were incorporated by Hayes' sociological theory (Hayes, 1914). Unfortunately, the use of human ecology was interrupted for some years, presumably as the topic got lost somewhere between geography and sociology for a while.

8.3.2 A second start between sociology and plant ecology

The biological and social realms were first linked in Spencer's definition of the total environment (see Chapter 1.3; Alexandrescu, 2009). A well known work by sociologists focusing on environmental issues was developed by the so-called second generation of the Chicago School (sometimes denoted as the Ecological School). This group consisted of Robert E. Park, Ernest W. Burgess, and their followers. In their seminal book *Introduction to the Science of Sociology* (Park & Burgess, 1921), the chapters on accommodation and assimilation broke new ground. Referring to Baldwin (1896), they focused on structural adjustments that are "congenital and inherited" but stressed that the

> … term accommodation, while it has a limited field of application in biology, has a wide and varied use in sociology. All the social heritages, traditions, sediments, culture, technique, are accommodations – that is, acquired adjustments that are socially and not biologically transmitted. (Park & Burgess, 1921, pp. 663–4)

Thus, the early Chicago School of Sociology held that social units, for instance, an urban community, live in a kind of biocenosis, which is similar to how plants and animals live in a biotope. For instance, homeless migrants clustering in a central business district in Chicago build "hobohemia" as the "main stem" of the city (Burgess, 1925/1967, p. 54). The city is the "natural habitat of civilized man" (Park, 1925/1967, p. 2). Besides the concept of accommodation, Park and Burgess utilized the concepts of stability, domination (subordination and superordination), invasion succession, and the struggle for existence in evolutionary processes. The strong impact of the biological model for societal systems is also reflected in the following quotes by McKenzie:

> The general effect of continuous processes of invasions and accommodations is to give to the developed community well-defined areas, each having its own peculiar selective and cultural characteristics. Such units of communal life may be termed "natural areas," or formations, to use the term of the plant. (McKenzie, 1924, p. 300)

> The plant ecologist is aware of the effect of the struggle for space, food, and light upon nature of a plant formation, but the sociologist has failed to recognize that the same process of competition and accommodation are at work determining the size and ecological organization of the human community. (McKenzie, 1926/1967a, p. 288)

The Chicago School has also been equated with human ecology (Hawley, 1944), and even its critics have acknowledged that, since the early 1920s, the Chicago School's human ecology has become "one of the definitive and influential schools in American sociology …" (Alihan, 1938, p. xi). In their practical work, their field study area and their laboratory were the city of Chicago. They sought evidence to determine whether urbanization, increased population density, and social mobility had been the causes of contemporary social problems. The sociologists believed that the community, which they regarded as a natural environment, was a major factor in shaping human behavior and that the city functioned as a microcosm. The strong similarity with biological ecology could be seen by the methods applied:

> Ecological studies consisted of making spot maps of Chicago for the place of occurrence of specific behaviors, including alcoholism, homicides, suicides, psychoses, and poverty, and then computing rates based on census data. A visual comparison of the maps could identify the concentration of certain types of behavior in some areas. Correlations of rates by areas were not made until later. (Cavan, 1983, p. 415)

Park later made distinctions between a biotic level and a cultural level of societies. On the biological level the community is a competing organism, but on the cultural level it functions as a kind of control mechanism; that is, "competition is limited by custom and culture" (Park, 1936, p. 29). Park thought that societal conflicts need to be resolved by accommodation and assimilation and thus recommended investigating how biotic and social balances and equilibria can be attained or destroyed:

> The political problem of every society is a practical one: how to secure the maximum values of competition, that is, personal freedom, initiative, and originality, and at the same time to control the energies which competition has released in the interest of community. (Park & Burgess, 1921, p. 21)

Human ecology acquired much credence through urban sociology. Burgess studied the history of urban development and concluded that cities had not expanded purely at their edges. Rather, all major

> **Box 8.1** The myth of permanent population growth: pestilence and Old World diseases
>
> The world is encountering a currently unexplainable exponential growth of population, which is highly correlated with energy consumption or CO_2 particles in the air (see Goudie, 2006).
>
> However, both Europe and the Americas have experienced dramatic decreases in population. The main reason for these decreases has been a special kind of interaction with biota (i.e. epidemics) and germ warfare. The European dynamics of epidemics (such as the Black Death, cholera, and typhoid fever) is well known. As can be seen from Figure 8.1a, the two Black Death epidemics, which occurred in Europe in 542 and around 1350, reduced the population by about one-third.
>
> The largest known decline of population on a continent, however, occurred between 1500 and 1750 in the Americas (see Figure 8.1b). The geographer Denevan recounts that, prior to initial European contact, civilizations in the New World had developed slowly for more than 15 000 years and comprised about 53.9 million inhabitants. Old World diseases spread by European exploration on the continent dramatically reversed this trend. "Indian populations declined in Hispaniola, a major island in the Caribbean, from 1 million in 1492 to few hundred 50 years later, or by 99 percent; in Peru from 9 million in 1520 to 670,000 (92 percent), and in North America from 3.8 million in 1492 to 1 million in 1800 (74 percent). An overall drop ... to 5.6 million in 1650 amounts to an 89 percent reduction. ... As replacement with Indians by Europeans and Africans was initially a slow process ... the overall hemispheric population was about 30 percent of what it may have been in 1492" (Denevan, 1992, p. 372).

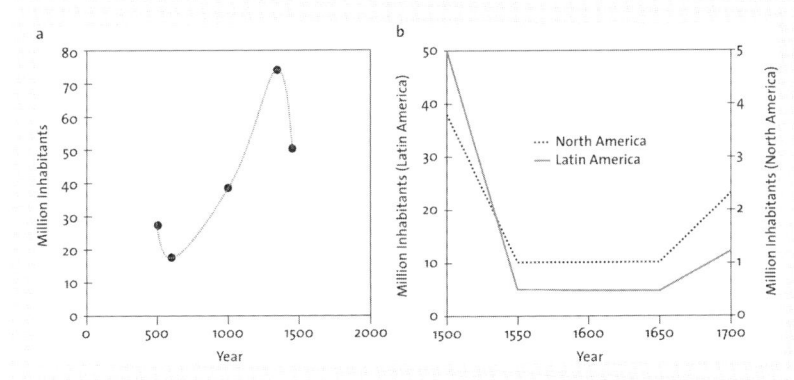

Figure 8.1 (a) Fluxes in population in medieval Europe and (b) the Americas from 1500 to 1700.

cities formed through radial expansion from the center in concentric rings which he described as zones. Zone areas include the business area in the center, the slum area (called the zone in transition) around the central area, the factory zone, the zone of working men's homes farther out, the residential area beyond this zone and then the bungalow residential section and the commuter zone on the periphery (Park & Burgess, 1921). Star-like commuter and mobility patterns, showing travel between zones, also suggest that cities form through radial expansion (McKenzie, 1926/1967b).

In general, we can conclude that the human ecology approach explains social organization of urban or regional systems by referring to theoretic principles of building structures from ecological systems, given a specific biophysical environment. Although on a small scale, applications of the organization of the world web of city centers depending on transportation and communication facilities and the concentration of intelligence (i.e. knowledge) in the centers of cities also became an object of research (McKenzie, 1927).

8.3.3 Key messages

- There have been various attempts to introduce human ecology as a hybrid and boundary scientific domain between physical geography and sociology and plant ecology and sociology. The two-way interaction between human and environmental systems have thus been recognized by early sociology
- In the early twentieth century, the Chicago School of Sociology used the term human ecology to describe the social structures and growth of cities.

Concepts such as accommodation, domination, invasion, succession, or evolutionary struggle for existence were taken from biology to describe dynamics in social systems. This approach provided an object-oriented spatial sociology.

8.4 The cradle of environmental sociology

Environmental sociology was born as a subdiscipline around 1978 as the domain of sociology began to recognize environmental variables. Conceptions of the environment ranged:

> … from the "manmade" (or "built") environment to the "natural" environment … In fact *the study of interaction between the environment and society is the core of environmental sociology.* (Catton & Dunlap, 1978, p. 44)

This section begins with a description of some of the growing environmental concerns from the 1950s, 60s, and 70s that fueled the formation of environmental sociology as a subdiscipline. We also look at how these concerns led researchers to formulate models for understanding environment–society interactions, including Catton and Dunlap's population, organization, environment, and technology (POET) framework and the IPAT formula (see Chapter 9.3) for assessing environmental impact. Finally, we discuss how the new ecological paradigm emerged in the 1970s to counter the human exemptionalist paradigm, which ignored environmental and ecological issues.

8.4.1 Environmental concerns

As early as 1285, people recognized smog as a problem (see Box 10.2). Later the problem was studied by the physicist des Vouex, who coined the term in 1911 (Bell & Davis, 2001). Still, it was not until the London Smog of 1952, which resulted in between 4000 and 12 000 deaths, that an environmental movement began.

Environmental and technological impacts on social organization had been investigated in the 1950s by sociologist William F. Cottrell who, having worked as a railroader, focused on the evolution of society through energy and technology. His book, *Energy and Society: the Relation between Energy, Social Change and Economic Development* (Cottrell, 1955), thoroughly described the role of natural resources on technology development. Further, Cottrell's paper on "Death by dieselization" (Cottrell, 1952) describes how, because of the rebound effects of technology development, workplaces and small communities that supplied water to steam locomotives closed down.

Environmental sociology should not be confused with social ecology, a philosophy suggested by the left-wing, green, anarchistic activist Bookchin in 1961 (Light, 1998) and led by the "conviction that nearly all of our present ecological problems originate in deep-seated social problems" (Bookchin, 2007, p. 19). Social ecology in the tradition of Bookchin later stood strongly juxtaposed to deep ecology, a branch of ecology that emphasizes the equal value of human and non-human life and the importance of ecosystems and natural processes.

In the 1960s, public awareness of environmental problems grew in the USA as well as in other parts of the world. Rachel Carson's (1962) book *Silent Spring* was doubtlessly a trigger. Public awareness further developed in the 1970s in some parts of American society from cases of soil contamination such as in Love Canal in 1978 (see Box 8.2), the civilian nuclear plant accident at Three Mile Island in 1979 (see Box 16.14), and from the 1973–74 energy crises. Such public concern clearly spurred the development of environmental sociology, a term originally coined in 1970 (Dunlap, 1997; Vaillancourt, 1995).

8.4.2 From environmental concern to POET and IPAT

The sociologist and human ecologist Otis D. Duncan had already used the term "ecological complex" in 1959, which influenced William R. Catton and Riley E. Dunlap when designing their "Analytic framework for environmental sociology" (Dunlap & Catton, 1979). Catton and Dunlap looked at the interactions between environment and society and studied how "human populations make considerable use of social organization and technology in adapting to their environment" (Dunlap & Catton, 1979, p. 251). As Catton and Dunlap describe, their focus was on web-like interdependencies among population, organization, environment, and technology – the POET model (Duncan, 1961). By stressing the reciprocal relationship between these elements, they developed an *ecosystemic* approach. They pointed out that their view went beyond human ecological thought, which was restricted to considering the environment as a spatial constraint "devoid of any physical substance" (ibid.). In their view, human ecology had "lacked a basis for becoming concerned with contemporary environmental problems" because they did not consider the interaction and neglected "aspects of the ecosystem that are not human or derived from human action" (ibid.).

In another effort to assess society–environment interactions, biologists (Ehrlich & Ehrlich, 1970) and chemists (Meadows et al., 1972) showed the relationship between population, resources, and the environment with the *IPAT* algorithm (Ehrlich & Holdren, 1971). The total environmental impact I can be formulated as a product of population P, affluence (which can be measured by GDP) A, and technology T ($I = P \times A \times T$). When a system exceeds its carrying capacity through, for example, uncontrolled exponential population growth rates or urbanization, trends such as pollution, potential food shortages, and jeopardized ecosystems can result. This approach has recently been elaborated further (see the STIRPAT model in Chapter 9.3). In the 1970s sociologists began to deal with problems such as those related to natural resources, albeit in a rather distant and theoretical way. For example, they began to acknowledge the limited carrying capacity of the Earth's system (see point 3 of the NEP paradigm in Table 8.1). However, the focus of application either remained at the local or regional level, in areas such as contamination of neighborhoods or acid rain in Europe ("le waldsterben", see Box 7.6), or was dedicated to theoretical questions such as environmental attitudes (Dunlap, 1997).

A second, less well received, line emerged in the 1970s under the name "treadmill of production." It takes a materialist–realist–objectivist approach (Buttel, 2004; Schnaiberg, 1975, 1980). We discuss this North American neo-Marxist approach in Chapter 9.7 when comparing it with environmental modernization.

8.4.3 HEP vs. NEP

Generally in sociology, the dominant paradigm in the 1950s and 60s was the human exceptionalist paradigm (later called the human exemptionalist paradigm, HEP; see Table 8.1). In its entirety, the HEP paradigm was anthropocentric, technologically optimistic, and ignored environmental or ecological issues. This paradigm served "to blind sociologists to the significance of environmental problems" (Dunlap, 2002, p. 334). This exclusion of environmental variables also held true in the analysis of modern industrial society, leading to what Dunlap called sociocultural determinism. The HEP approach sprang from a characterization of an abundant environmental supply and the doctrine of progress in western culture. Its focus on the boundless belief in the social environment crowded out consideration of the physical environment owing to a complete neglect of habitat (Michelsen, 1976). The term *environment* referred "entirely to society's symbolic environment (cultural system) or 'social environment' (environing social system)" (Catton & Dunlap, 1978, p. 43).

In the late 1970s, however, springing from the Rural Sociological Society, a new paradigm of environmental sociology emerged which contrasted with the ubiquitous HEP paradigm (Catton & Dunlap, 1978, p. 42; see Table 8.1). The natural environment paradigm (NEP) can clearly be seen as a change in the rationality concept (see Box 1.4) and as a first, tentative attempt to integrate environmental system properties into sociological thinking. Looking back, it is interesting that the initial label (presented above) was later changed to the new ecological paradigm, because its creators wanted to stress the ecological foundations of human societies (Dunlap, 2002, p. 336). Similarly, the term exceptionist in HEP changed to exemptionist, a term that was coined by E. O. Wilson:

> … *exemptionalism* … holds that since humankind is transcendent in intelligence and spirit, so must our species have been released from the iron laws of ecology that bind all other species. (Wilson, 1993, p. 237)

The rise of environmental sociology in the USA was strongly curbed during the conservative policy era of Ronald Reagan, from 1981 to 1987. As Dunlap outlined (Dunlap, 1997, pp. 24–5), US voters were terrified of facing the limits of growth and of Earth's capacity as well as inconvenient restrictions of access to nature, space, and resources. Further, the public was not willing to commit to anything that required a sacrifice to the national economy. In the intellectual arena the limits-to-growth concept was rejected by arguments about how human ingenuity makes people the "ultimate resource" (Simon, 1981; Simon & Kahn, 1984). Thus, environmental issues fell out of vogue, and environmental sociologists were hard pressed to secure job offers and research funds. Membership of the

> **Box 8.2** People, not experts, have brought the toxification of nature and people to the forefront: what role should experts take?
>
> Rachel Carson's (1962) contributions to identifying the effects of DDT on wildlife and the subsequent banning of pesticide can be seen as the beginning of the environmentalist movement (see Chapter 9.2).
>
> *Love Canal*, a New York neighborhood built on highly toxic soil, is another historically important case. Residents were the first to report concerns that trees, vegetables, and flowers in the area began to discolor, that the rubber soles of their children's shoes began to disintegrate at a faster rate than normal, and that dogs developed sores that did not heal (Gibbs, 1982). Reasons for these human and environmental health problems became obvious when it was discovered that the houses of the community were built on a former channel that was used as dump for more than 80 chemicals (including dioxins, benzene, and chloroform) between 1920 and 1950 (Janerich *et al.*, 1981).
>
> *Sellafield* (formerly called Windscale), a former nuclear processing and electricity generating site in England, offers a third example (Figure 8.2). In 1970 local claims were made by ordinary people in the vicinity of a nuclear reprocessing complex about the numerous cases of childhood leukemia as a possible consequence of a nuclear accident that took place in 1957 (Arnold, 1992). "These observations persisted despite official denials not only by the operators British Nuclear Fuels but by the public health authorities" (Wynne, 1996, p. 49). An expert commissioned by the local residents was not even allowed access to public health data. Only after a 1983 national television documentation did a commission from the health department look at the data related to the high prevalence of childhood leukemia (Arnold, 1992). The correlation to residential areas was confirmed, but, owing to the probabilistic nature of cancer, "it seems unlikely that any health effects will ever be observed which can be attributed to the Windscale accident" (Arnold, 1992, p. 153). As Wynne pointed out, only the authorities' findings were presented. The Sellafield case documents further examples of fallacious expert work, pointing to assessment of the impacts of the Chernobyl fallout on sheep. Initially, experts denied the potential of any relevant contamination, only to amend their statement later by pointing to a scientific study showing that the radioactive cesium isotopes had either been washed out by thunderstorms or had negligible impact after 6 weeks because of their short half-life. This statement was later amended once again, and the public lost more faith in the experts. An indefinite ban for some of the area of Sellafield was placed, because the contamination levels remained high because of effects from the 1957 accident, weapons testing fallout from the 1950s, and other emissions. The farmers' knowledge about various forms of cesium contamination in different parts of the hilly area, as well as about the behavior of the hill sheep (e.g. in field experiments), was dismissed by nuclear experts' scientific routine structures (Wynne, 1992).
>
> These examples illustrate how in some cases laypeople can demonstrate a great degree of environmental literacy and offer important data, and that expert knowledge sometimes needs an appropriate frame to cope with specific complex contexts.

Figure 8.2 Aerial view of the Sellafield nuclear reprocessing complex in 2006. (Photo: CIRIA.)

American Sociology Association Section Environment and Sociology dropped from 312 members in 1976 to 274 members in 1983.

8.4.4 Key messages

- Environmental disasters between 1950 and 1970, such as the London smog of 1952, Love Canal, and DDT impacts on nature as established in Rachel Carson's book, etc., are normally seen as the origins of the environmental movement. This movement affected US sociology relatively late, however, fueling the emergence of environmental sociology as a subdiscipline
- Conceptual and quantitative models such as POET (population, organization, environment and technology) or IPAT (impact equals the product of

population, affluence, and technology) were built to describe human impacts on the environment. These models go beyond human ecology, as they consider not only the impacts of the environment on human systems but also the human–environment interactions, including human adaptation

- Sociologists (with the exception of agricultural, rural sociologists) have not dealt with environmental and resource issues for long. The belief in the unreserved substitutability of resources falls under the human exemptionalist paradigm (HEP). With the rise of the environmental movement in the 1970s, a contrasting paradigm emerged, namely the new ecological paradigm (NEP). The HEP and the NEP view represent two world views for coping with environmental problems.

8.5 The resource-based socio-technological-evolutionary theory (ReSTET)

8.5.1 What drives how societies evolve?

For the framework of this book, a critical question is – what are the key drivers of society? To understand how sociologists describe the drivers and functioning of society, we refer to Lenski's ecological–evolutionary theory (Lenski, 2005). To emphasize the role of natural resources and the availability of technologies, we develop a slightly extended version called the resource-based socio-technological-evolutionary theory (ReSTET). Lenski argues that, although societies are imperfect systems in which one part can be in conflict with another and in which the actions of the whole are often harmful to its parts, two major goals in societies can be identified. Figure 8.3* depicts ReSTET, where the two main drivers of society are pursuing wealth by maximizing production and maintenance of power. The first is "the maximization of production and resources on which production depends" (Lenski, 1966, p. 42). The driver of wealth is based on the proposition that human societies realize that the conditions of life can continually be improved. Wealth refers to the valuable properties that characterize affluence or prosperity. Wealth, which is contrary to poverty, is related to valuable possessions and thus accentuates the economic side, but includes subjective valuation. The second goal is that societies' leading classes, parties, or cultural groups have a tendency to attempt to determine the direction of the development. This can be expressed as the minimization "of the rate of internal political change," the maintenance of the status quo, or as a drive for the maintenance of power (see Figure 8.3*). The first goal is dominant in less stratified societies and the second in societies where power is monopolized by a few. Clearly, the material environment plays an important role concerning the first goal.

The degree of democracy as a societal trajectory is important when deciding which of the two drivers is dominant. With respect to Figure 8.3*, natural resources are an important layer affecting the interaction between technological and societal trajectories. The availability of natural resources is supposed to be a necessary condition when promoting potential wealth.

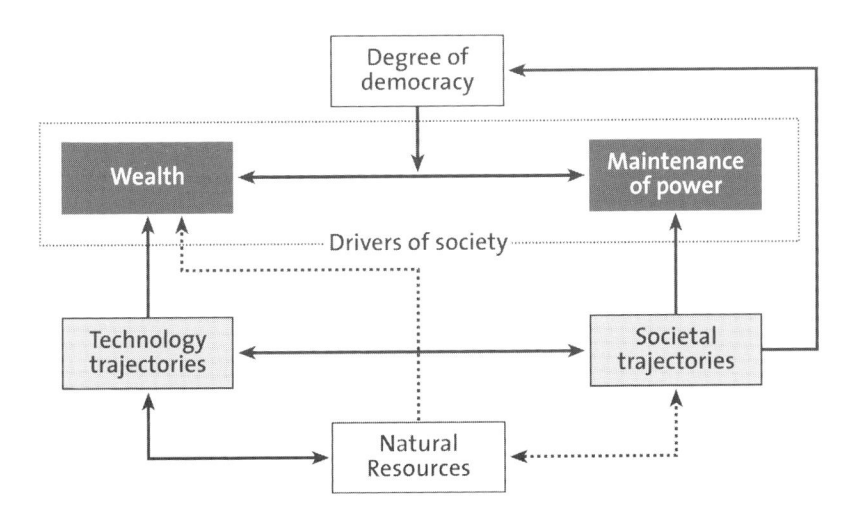

Figure 8.3* The ReSTET framework by Lenski and Scholz shows interactions between drivers of social change (wealth and maintenance of power), development trajectories (technological and societal), and key components of societal dynamics according to Lenski's evolutionary–ecological theory. The bold lines represent strong ubiquitous relationships and the dotted lines represent relationships that show strong relationships for some cases.

Here, natural resources are conceived in a broad sense and include climate (including rain and sunlight). This has already been stressed by anthropologists who state that changes in the natural environment affect the cultural setting. We should note that Lenski emphasizes the role of technology for societal development.

> Technology is a principal causal agent in human affairs. … For Lenski, technology reduces largely to knowledge or information which grows and expands as a result of human exploration, experimentation, and manipulation. (Sanderson, 2007, p. 198)

Deviating from Lenski, we consider technology not from the epistemic side, but rather as the material manifestation of ideas implemented to better utilize natural resources.

Some sociological approaches investigate the social and cultural characteristics of a society by examining its history and genetic heritage in interaction with the biophysical and social environment (Figure 8.4). This holds true in particular for evolutionary approaches (see, for example, Lenski, 2005; Sanderson, 1999; Wallerstein, 1974). Other approaches emphasize the role of energy, resources, and technology (see, for example, Humphrey et al., 2002; Rosa et al., 1988). In Figure 8.4, this is included in the upper box as a society's biophysical environments. The prior, historically rooted, social and cultural heritage of the society and also species' common heritage are seen as other components affecting society.

8.5.2 Societies, natural resources, and technologies

A key question studied by some macrosociologists is the adaptation of human societies to environmental resources. In their view, growth and decline of societies essentially depend on access to resources. This was noted as early as the eighteenth century by Thomas Malthus, who argued that population growth – if not interrupted by natural disasters or regulated by war – will always exceed the food supply.

Today's macrosociologists differentiate this view and include human capital (including the sociocultural rules and knowledge), economic capital which here stands primarily for the financial capital, and available technology as key drivers of the standard of living or economic surplus. Lenski suggests the following equation for calculating the economic surplus, with the error term ε denoting the effect of other factors:

$$\text{Economic Surplus} = \frac{\text{Resources} \times \text{Technology} \times (\text{Capital}_{economic} + \text{Capital}_{human})}{\text{Population Size}} + \varepsilon$$

Lenski proposes that to "neglect any of these variables in theory construction, or in research, is equivalent to attempting to construct a multistory building without first laying a proper foundation" (Lenski, 2005, p. 80).

According to Lenski (Lenski et al., 1991; Nolan & Lenski, 1998), the most fundamental differences among societies over the course of human history have to do with: (1) subsistence technologies, and (2) their environments. Both affect one other important entity: energy use and efficiency (Humphrey & Buttel, 1982). This is roughly presented in Figure 8.5*, which presents a "system of classification grounded in a general theory of human societies based in clearly specified covering principles" (Lenski, 2005, p. 81). This classification of societies suggests that the

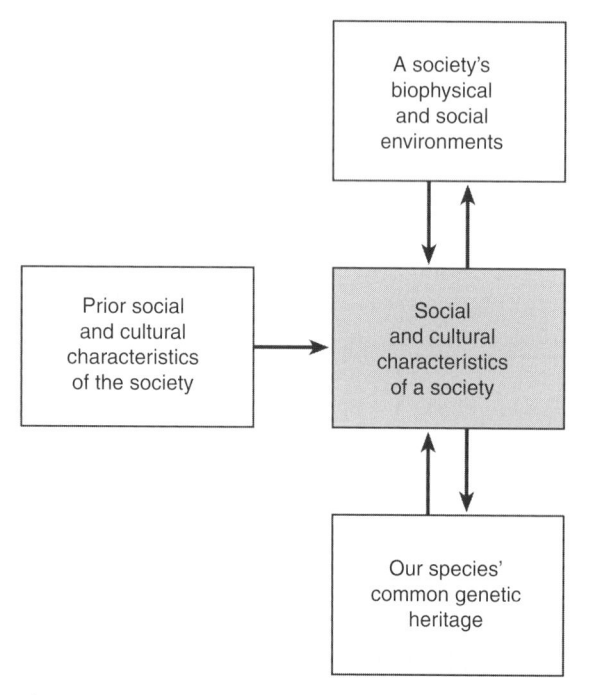

Figure 8.4 A basic model for investigating societal change from a macrosociological perspective when incorporating the interaction with the material–biophysical environment (Lenski et al., 1991, p. 18).

8.5 Resources and technology: the ReSTET theory

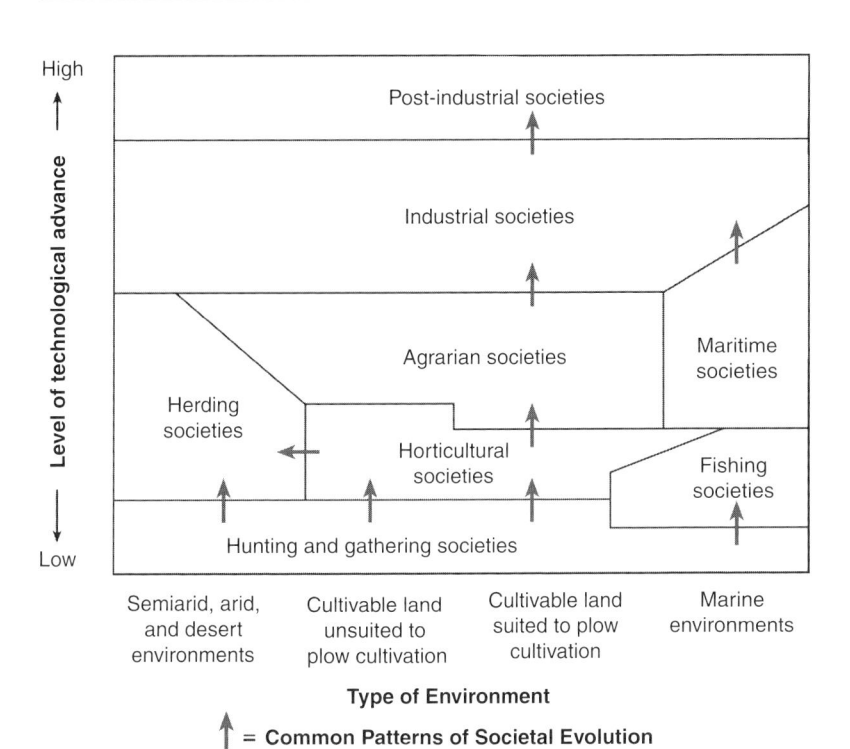

Figure 8.5* A view on the evolution of societies that relates to the level of technology and type of environment (adapted from Lenski et al., 1991, p. 73).

materials, energy, basic environmental constraints, and the available technology tell us more about a society and why it is in a specific state than any other fact. The classification also goes beyond differentiations provided by archeologists between, for example, Paleolithic, Neolithic, Chalcolithic, Bronze-, Iron- and Steel-Age societies. Such earlier distinctions, for instance, between the Paleolithic period (Old Stone Age) and the Neolithic period (New Stone Age), focused on developments in food collecting and food production based on technological knowledge that allowed societies to evolve to practices of true farming.

An illustration of common patterns of societal evolution, depending on the level of technology and type of is shown in Figure 8.5*. The vertical axis shows how differences in technology lead to different societies, and the horizontal axis shows the geographic and environmental impact settings on different societies. Thus, agrarian and maritime societies have the same (epistemic) technological level (vertical axis) but differ in their social–cultural features because of differences in environmental settings (horizontal axis). Each of the societal types reflects a distinct combination of environmental and technological characteristics, which can be called operative technologies (Lenski, 2005). Technologies induce population growth, increased societal complexity, weapon effectiveness, change of power, energy consumption, energy efficiency, and so on. For example, hunting and gathering societies depend on wild game and wild plants as resources, and on trapping, hunting, and collecting instruments as technologies. Simple hunting societies only used the spear, whereas advanced hunting and gathering societies have acquired both the spear and the bow and arrow. Horticulture societies, on the other hand, depend on the mastery of cultivation of plants in gardens and, often, the domestication of animals. Simple horticulture societies have no metal tools – digging is accomplished with wooden sticks. Advanced horticulture societies have metal tools, in particular the hoe. The change from horticulture to the agrarian society (about 3000 BC), for instance, was accompanied by the invention of the wheel, pottery manufacture, harnessing of wind power (for sailing), writing, numerical notations, and the calendar and, notably, plows drawn by animals. Agrarian societies can thus cultivate large fields.

8 Classical sociology and the environment

Box 8.3 Resource scarcity drives societal and ecological decline: West Africa's emptied sea

According to the ReSTET approach (see Figure 8.3*), environmental resources are a key basic layer that govern social and technology trajectories in the relationship between wealth and power maintenance. As we see in the current case of West African migration, different resource systems can interact to affect wealth and power dynamics. There are two critical impact chains: a social impact chain (I) and an ecological impact chain (II; see Figure 8.7).

Considering the diagram in the frame of ReSTET, we see that the fish population in West Africa's Gulf of Guinea is under pressure because of at least two societal trajectories. One stems from the need for food and proteins by the local population, and the other is over-fishing by developed countries (and is thus partly the same issue; Figure 8.6). Lafraniere (2008) reports that the African motorized pirogues cannot compete with the EU fishery fleet, which uses trawling and dredging methods along the sea floor. The catch with the pirogues does not even cover the cost of fuel.

The fishing equipment is an important technology trajectory. On the societal governmental level there is short-term interest in overselling fishing rights. Whether this is a matter of democracy (see Figure 8.3*) or of using science is an interesting research question. Local fishermen comment on the situation: "The EU has the money, so it has the power. It is easier to sacrifice the local fishermen" (cited according to Lafraniere, 2008, p. 6). The consequence is emigration because of survival and wealth incentives. "In 2007 about 31 000 Africans tried to reach Canary Island, a prime transit point to Europe ... About 6000 died or disappeared according to one estimate cited by the United Nations" (Lafraniere, 2008, p. 1). Societal decline, including an exodus of young men, is a typical phenomenon of societal transitions linked with survival and geographically uneven distribution of wealth.

Brashares *et al.* (2004) thoroughly and quantitatively investigated the ecological chain (II). The key to their analysis is based on the protein limitation hypothesis, which says that a population needs a minimum amount of protein

Figure 8.6 A Fisherman waiting for rising side to start fishing in Guinea-Bissau, Africa. Here fishermen, who – less than a decade ago – were buying more boats, now complain they are in debt and are looking at other ways to make a living (Photo 2006 by Ernst Schade/Panos Pictures).

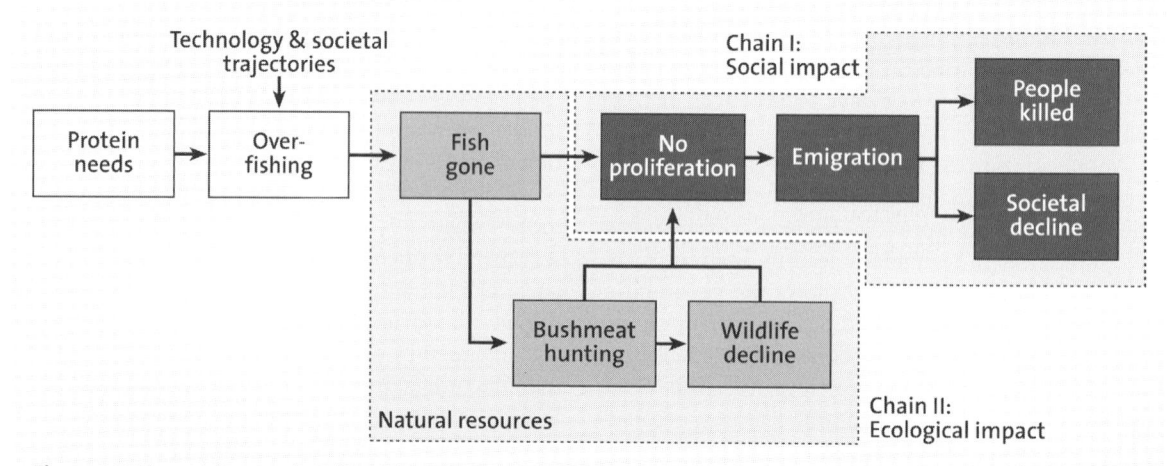

Figure 8.7 The social impact chain (I), and the ecological impact chain (II).

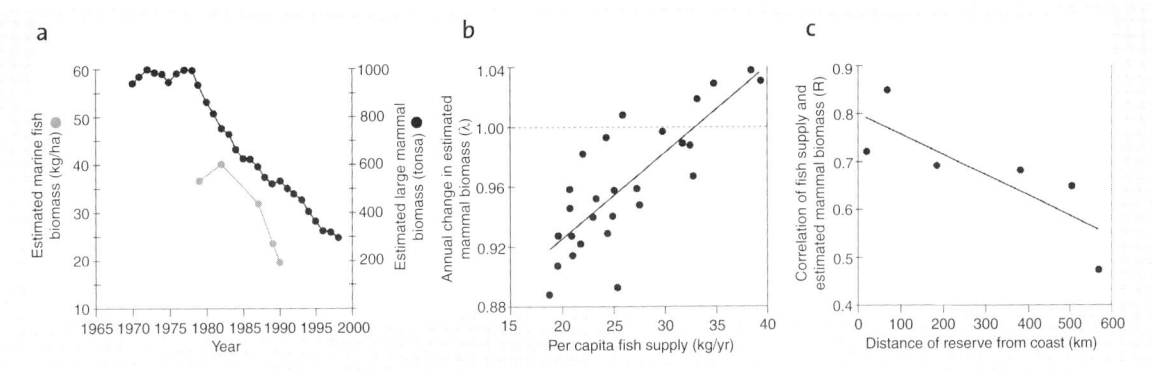

Figure 8.8 (a) Estimates of marine fish biomass in the Gulf of Guinea (gray circle) and large animal biomass in Ghana (black circle); (b) fish supply (i.e. fisheries' landings, plus imports minus exports) and estimates of biomass of large animals ($p < 0.001$); and (c) distance of reserves to the coast, which was negatively related to the correlation of fish supply and wildlife decline ($p < 0.01$). (Figures taken from Brashares et al., 2004, reprinted with permission from AAAS.)

as a basic natural resource. Starting from this, Brashares and colleagues provide a kind of causality analysis based on 30 years of data from Ghana. In a sophisticated cascaded sequence of quantitative analyses (Figure 8.8), they showed that per capita fish supply is significantly related to: (i) annual biomass of terrestrial mammals (see Figure 8.8a); (ii) availability of bushmeat; (iii) annual counts of hunters in wildlife preserves; and (iv) the annual changes in estimated mammal biomass. Data also show that the pressure on wildlife from depletion in the fish supply is stronger the closer the reservoir is to the coast.

On the other hand, industrial societies substituted machine power, such as James Watt's steam engine, for human power in production and are characterized by mass production of the factory system. We have included the post-industrial society level because today, in certain domains of developed countries, it has become evident that the main economic surplus is primarily caused by information financial transactions of derivatives which derive their value from other financial transactions. Thus, the primary value creation is transferred from production to the post-industrial, which is presumably dominated by services and the creation, manipulation, distribution, and use of information and knowledge.

From a modern perspective, we can state that the environment and the physical and material aspects of life have been at the very fringes of traditional sociology. Lenski was one of the few sociologists who emphasized environmental and technological resources and illuminated their role in societal development. It is worth noting, for example, that Lenski's ecological–evolutionary theory (Lenski, 2005), the basis for ReSTET, resembles Holling and Gunderson's theory of adaptive cycling (see Figure 5.16). Societies embody an inner dynamic that prevents them from becoming stable. They follow a process of continuity, innovation, extinction, and evolution (Sanderson, 2007). Because such a perspective is highly relevant to environmental literacy, we introduced here macrosociology from Lenski's perspective though, as some sociologists state, he might have underestimated the importance of institutional, ideological, military or taxing issues that can be just as important (Collins, 2006; Hall, 2006; Sanderson, 2006; Tainter, 1988).

Although many theorists in sociology have avoided joining the discussion on sustainability from an environmental literacy point of view, sociology has or should have a key position in discussions about sustainability.

> In recent years, a consensus seems to be under way to regard sustainability as a problem of the interaction between society and nature. (Haberl et al., 2004, p. 199)

We think that the understanding of the rationale of societies and how they interact with subsystems and supersystems is a prerequisite for a successful adaptation to the rapidly changing environmental constraints (see Figure 14.1*, as well as the Hierarchy and the Interference Postulates in Chapter 16).

8.5.3 Key messages

- Technology is a key component in the evolution of societies
- The ReSTET approach, an extension of Lenski's evolutionary–ecological theory, considers natural resources, technological trajectories, and social trajectories as basic elements of societal development. The drivers of societal development are wealth and maintenance of power. Power dominates in undemocratic arenas, whereas wealth does so in democratic societal settings.

Part IV   Contributions of sociology

Chapter 9
Modern sociological approaches to human–environment interactions[1]

9.1 Contemporary sociological thought   231
9.2 The second wave of environmental sociology   236
9.3 System analysis on society's environmental impacts   241
9.4 Risk society   242
9.5 Communication and environmental issues   244
9.6 Constructivism   245
9.7 Ecological modernization vs. the treadmill of production   249
9.8 Macrosociology can go further   252

Chapter overview

The previous chapter presented a definition of society, key ideas of the classics, a description of the emergence of environmental sociology, and a conceptualization of drivers of societies in the frame of the ReSTET theory, which links drivers of societies with the given societal and technological trajectories and natural resources.

Chapter 9 first presents ideas of the macrosociological theories of Talcott Parsons, Niklas Luhmann, and Anthony Giddens, and reflects on whether and how the material environment has been included in these theories. We then look at the second wave of US environmental sociology, which integrates environmental issues and variables in sociology. The quantitative modeling of material flows caused by societies has been the object of study for sociologists, who focus on the material consequences of human production and consumption. Here, we take a look at the sociometabolic approach and at the STIRPAT model. We also discuss Ulrich Beck's view on the material environment. His concept of the world's risk society proposes that global risks need to be addressed by social institutions above the nation-state level.

As many sociologists define society via communication, we take a look at the role and function of discourse and communication in the view of Jürgen Habermas and the concept of cooperative discourses. Because we consider what is communicated in society to be a matter of social construction, we take a look at how societies construct environmental problems and what problems arise if laypersons and experts, who each have specific roles and rationales in society, interact.

We close with a discussion of two approaches that take fundamentally diverging positions on how societies can develop sustainably. Ecological modernization is an optimistic vision developed by European environmental sociologists on the potential to solve environmental problems by societal and technical modernization. The treadmill of production view argues that the post-industrial society will show an increase in quantitative and qualitative environmental impacts.

9.1 Contemporary sociological thought

In this section we take a tour of contemporary sociological thought by examining the perspective of three sociologists whose work is tangent to environmental literacy. We begin with Talcott Parsons, whose structuralist functionalist theory of action offers a voluntaristic perspective that assumes individuals to make choices, but regards individuals as taking on a role in society, and to be governed *by* role expectations. Parsons identified three factors that affect behavior: individual personality, the social system, and the cultural system.

---

[1] A special thank you goes to Michael Stauffacher who coached me in writing this chapter.

System theorist Niklas Luhmann used communication patterns and flows as a means of understanding and describing societies and was interested in understanding boundaries between social systems and environments. Luhmann considered power as an important, communication-based regulator of society, and trust as a key for reducing complexity and regulating developed societies. Anthony Giddens' theory of structuration proposes that all human actions are at least partly predetermined based on contextual rules under which they occur and, in turn, these structures and rules are continuously reshaped by human action. Giddens differentiates between authoritative resources and material, allocative resources. Giddens thus describes a theory of how the individual interacts with his or her social and material environment.

9.1.1 Parsons: describing human behavior through structuralist functionalism

Based on the definition of a society in the last section, we know that the size and organization of societies differed historically, for example, between hunter–gatherer and industrial societies (see Figure 8.5*). Since industrial society's beginnings, the political world regimes have played a dominant role in how societies develop. In 1885, the European countries divided Africa into colonies and restricted the political autonomy of tribal societies. After World War II, the nation-state became the dominant political organization, but new organizational schemes, such as supranational societies, have also become visible (see the Hierarchy Postulate P2 in Chapter 16).

Talcott Parsons (1902–79) developed different theories about how societies function, starting with his voluntaristic theory of action (Parsons, 1937/1949). This approach contrasted with views from behavioristic psychology, as he assumed that individuals are making choices. He described behavior ("a unit act") by a mathematical formula including many terms which relate to the situation (i.e. the structure), the conditions of action, means (including knowledge), and the ends. This approach was developed into a structural functionalistic general theory of social action (Parsons & Shils, 1951/2001).

In this theory of action, Parsons distinguished between the individual personality, the social system, and the cultural system. Parsons' contributions regarding social and cultural systems were thus considered the system-theoretic turn of sociology (Schluchter, 2009). In his general theory, the cultural system was governed and limited by values and norms, which affect any individual. It is remarkable that Parsons defined the constituents of human action in a chapter that was written together with top Harvard psychologists such as Allport, Sears, and Toleman, and the anthropologist Kluckhorn (Parsons & Shils, 1951/2001).

While "The *social system* is made upon the actions of individuals" (Parsons & Shils, 1951/2001, p. 190), Parsons looked at individual choices from a macro-sociological and cultural perspective. He stressed that he was not referring to an independent individual, but to individuals who take on a role in society and who are governed by role expectations. Role expectations are defined by a "collectivity" (ibid. p. 192), which can be a group (see Chapter 6), an organization, a community, a nation-state, etc. In Parsons' view, the development of a society is accompanied by the occurrence of a more differentiated structure (i.e. of subsystems), which becomes subject to internal (within subsystem) and external (between subsystems) regulations.

A grand-scale theorist, Parsons took not only a sociological but also a political science perspective:

> Parsons sought to understand and intervene in the functioning of the fully-fledged capitalist modernity. (Chernilo, 2007, p. 78)

One of his primary concerns was the functioning and the order of society. Parsons translated the abstract ideas of society into three entities: the social system, nation-state, and modern society (Adams & Sydie, 2002).

The nation-state was seen as a politically organized spatial reference. It corresponds to an important unit organizing modernity and is often seen as one and the same as society. We should note that Parsons, who did not emphasize the historic perspective, was not yet forced to recognize that globalization and world-scale biogeochemical flows might demand management beyond the scale of the nation-state.

Historically, modern society, in Parsons' view, is the idealized western civilization or (modern) western culture that emerged from western Christianity and which is organized in territorial states. The behavior of modern societies is shaped by characteristics of this western cultural pattern. Modern society is, for instance, achievement- and self-oriented, universalistic, and affectively neutral whereas traditional society is oriented toward status (ascription) and particularistic, collective, and emotional goals (see Adams & Sydie, 2002).

What we call a cultural and social order system (see Figure 1.7*) is included in what Parsons describes as social system and modern society. These systems are key constituents of societal trajectories in the ReSTET theory (see Figure 8.3*). From a theory of science perspective, Parsons is a structuralist functionalist, as he assumed that societies build substructures and adapt to a state where an equilibrium, or order, is attained. The material–biophysical environment did not play any role in Parsons' theory.

In his later works (Parsons, 1961), Parsons puts forth the AGIL framework, which assumes that societies have to adapt, A, primarily with respect to economic conditions to attain the goals, G, which are formulated by the social system to sustain the society. This needs policy processes that can fulfill integration, I, of these goals into the legal system. This process, however, is moderated by the drive of the social system to maintain cultural and social values, called latent structure maintenance, L.

9.1.2 Luhmann: "Society is communication"

The German sociologist Niklas Luhmann (1928–98) considered Parsons' theory to be the only systematic sociological theory that provided a general theory of the action system (German: allgemeine Theorie des Handlungssystems; Luhmann, 1985). Luhmann was a system theorist as he considered system theory a general theory that can be ubiquitously applied. However, he focused on social systems and, like Parsons, did not explicitly work on the material world or environmental topics. Rather, he postulated that analysis has to separate social, epistemic, and material systems from one another (Luhmann, 1995, p. 456).

Luhmann considered the "difference" between a social system and its environment the object of sociology (Luhmann, 1985). For Luhmann, communication between systems is the basis of any society. Or, as he repeatedly stated, "Society is communication." Thus, sociology, as the science of society, is the study of how different societal systems evolve, operate, communicate, and maintain their boundaries. A key concept is the term "resonance," as systems can only communicate with other systems or the environment using their specific set of "codes." These codes are unique to each of the (sub)systems; for instance, true–false for science, legal–illegal for law, pay–not pay for economy. This of course makes communication always selective and restricted as only the parts that are in line with the system's own code resonate (Diekmann & Preisendörfer, 2001).

Luhmann's key idea was that a system aims to cope with the difference between the system itself and its environment. Making reference to Maturana and Varela (1980), Luhmann, like other system theorists, considered the cell as a system. A cell produces proteins, nucleic acids, lipids, glycosides, and so on, to maintain its inner organization and to interact with its environment. Similarly to Jakob von Uexküll, Luhmann postulates that the cell constructs an image about the environment based on the signals it receives from the environment. Usually, social systems are not independent and autarchic, but, instead, have to interact with other systems within the environment. This corresponds to Piaget's idea of equilibration and can be considered a driver of society (see Chapter 6). It is interesting to see that Luhmann is a probabilistic functionalist. Luhmann, as does Brunswik, states:

> The environment affects the [human] system in a probabilistic manner. (Luhmann, 1985, p. 65; translated by Scholz)

The probabilistic information sampling and communication urges and induces organization, and Luhmann assumes that societies are evolving toward greater functional differentiation, resulting in increasingly differentiated subsystems, such as institutions and organizations. Here he introduces the concept of *autopoesis* to explain that human systems do not act blindly on their norms, but depending on what a system has done in the past or on what other systems are doing. Systems operate in a self-referential manner, which is not arbitrary, by following an "anticipatory recursivity" (Luhmann, 1984, p. 447).

Luhmann's system theory included three further key concepts that are central in this book. These are power, trust, and risk. Luhmann considered power as an important, communication-based, regulator of society, and trust as a key for reducing complexity and regulating developed societies:

> Trust reduces complexity by going beyond available information and generalizing expectations of behaviour in that it replaces missing information with an internally guaranteed security. It thus remains dependent on other reduction mechanisms developed in parallel with it, for example those of the law, of organization and of course,

those of language, but cannot, however be reduced to them. Trust is not the sole foundation of the world; but a highly complex but nevertheless structured conception of the world could not be established without a fairly complex society, which in turn could not [function] without trust. (Luhmann, 1979, p. 94)

We discussed the concept of trust in Chapter 6.20. An extremely significant contribution from Luhmann concerning risk research, and for promoting environmental literacy, has been the differentiation between risk and danger (Luhmann, 1990; see Chapter 6). Risk is linked to having choices or making decisions, while danger is not. This is reflected in the phrasings: "I'll take the risk," which contrasts with "you are in danger."

In a time when most sociologists were completely ignorant about the emerging environmental crises at the end of the twentieth century, Luhmann wrote *Ökologische Kommunikation* (Engl.: *Ecological Communication*; Luhmann, 1986). According to his perspective, Luhmann analyzed communication around ecological risks. His conclusions were that successful coping strategies depend on the proper operation (i.e. communication) of functional systems such as policy, law, economics, science, education, and religion; that is, of what we call the principal components of society (see Figure 1.7*). The material world did not play any significant role in the theory of Luhmann, as all that matters is communication (see also section on constructivism below).

9.1.3 Giddens: human action and social structure

British sociologist Anthony Giddens provided a seminal contribution to the relationship between the society and the individual, or between social structure and action.

Anthony Giddens' theory of structuration proposes that all human actions are at least partly predetermined based on contextual rules under which they occur. The term structuration means to structurate or to build structures. In his theory social structures are built by individual action. Giddens proposes a complementarity relationship between agency and society, where humans shape the structure and rules that, in turn, affect human actions. Thus, Giddens translates the relationship between micro and macro levels, i.e. the duality of "individual and society," into the complementarity of "agency and society" (Giddens, 1984, p. 162). Similar to a transactionalist view, reciprocity and recursivity, including self-regulating feedbacks between individual action and social structures, are emphasized.

Amongst many other points, Giddens differs from Luhmann as he acknowledges a material–biophysical environment. He even focuses on the intimate relationship between production and reproduction. Giddens calls the material environment an allocative resource, including material and technological aspects, as well as goods and financial capital. These resources differ from authoritative resources that are "nonmaterial resources involved in the generation of power" (Giddens, 1984, p. 373; see Table 9.1). Resources, especially unequally distributed resources, give power and a means of domination over other humans. Here, Giddens comes close to what we call full ecology or what the resource economist Norgaard (1984) called the coevolution of social and ecological systems.

Giddens is aware that human systems affect ecosystems and the material–biophysical environment, and that these will in turn be altered and provide feedback to the human system. Referring to Giddens, Binder (2007) linked the agent–environment and agent–social structure relationships (Figure 9.1) to show a double feedback relation. The agent–social structure can be viewed as a micro–macro interaction which builds up and uses authoritative and allocative resources. The social structure emerges

Table 9.1 Allocative–material and authoritative–social resources (Giddens, 1984, p. 258).

Allocative resources	Authoritative resources
1. Material features of the environment (raw materials, material power sources)	Organization of social time–space (temporal–spatial constitution of paths and regions)
2. Means of material production/reproduction (instruments of production, technology)	Production/reproduction of the body (organization and relation of human beings in mutual association)
3. Produced goods (artifacts created by interaction of 1 and 2)	Organization of life chances (constitution of changes of self-development and self-expression)

9.1 Contemporary sociological thought

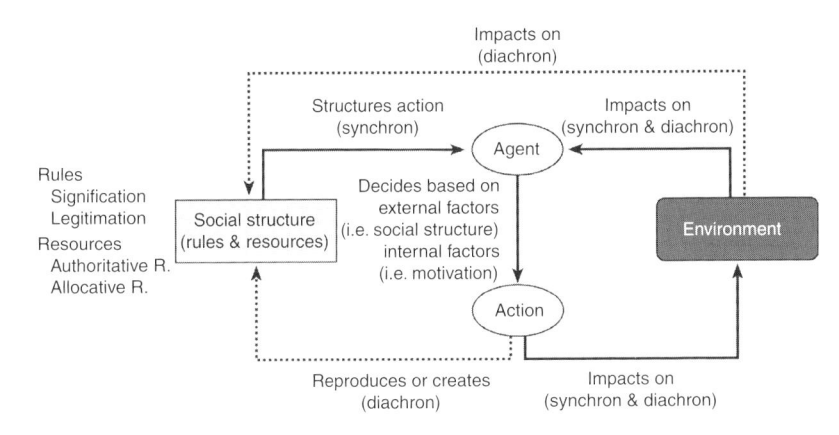

Figure 9.1 Feedback loops between the agent and his/her environment and the social structure (Binder, 2007, p. 1607). The dotted lines represent diachron impacts on the social structure.

diachronically, but only the existing social structures affect the agent. Binder stresses that this is different to the feedback between the agent and the environment. Both the impacts on and the feedback resulting from an action, via the environment, can be delayed. Thus we have different developmental dynamics on the side of the social structures and on the side of the environment. In a similar way, Benno Werlen, a Swiss geographer, adapted Giddens' structuration theory for social geography (Werlen, 1997, 2000). Yet, for Werlen, space (environment) has no direct influence on human action but its meaning is constructed by individuals in action and thereby influences future action. Woodgate and Redclift denoted the changes of the natural, biophysical environment as evolution, the development of society as structuration, and the interaction between evolution and structuration as coevolution (Woodgate & Redclift, 1998).

Thus, Giddens' view of sociology can relate social systems with their material environmental systems and allows understanding of how the actions of an individual interact with the social structure. Further, it becomes clear that we have to consider different time–space boundaries of social and material systems. Giddens describes the relationship between human and environmental systems as coupled complementary systems. The dual nature of the environment and also the way in which humans build and shape their environments affect the "mind" of human systems. Woodgate and Redclift noted that we are facing a "social construction and reconstruction on socioenvironmental spaces" (Woodgate & Redclift, 1998, p. 15).

Like Parsons, Giddens was interested in the institutional dimension of late modern societies. An interesting idea is the extension of the traditional industrial–capitalist–military triad by adding a fourth institutional dimension called surveillance. Giddens recognized that the complexity of late industrial societies poses a need for increased monitoring and processing of information about people, social and policy systems, as well as environmental systems (Giddens, 1990). From a sociological perspective, tracking is important because each of these institutional dimensions can present a risk. For instance, capitalism, because of its erratic economic patterns and unregulated growth, introduces financial risk. Ecological and environmental limits are also acknowledged as risks by Giddens.

Since 1990, environmental crises and globalization have been a central position in Giddens' thinking and therefore his work has been well received by environmental sociologists (Redclift & Woodgate, 1995, 2005) and environmental policy-makers (Mol, 2006). In the past few years, Giddens has considered two issues that are of importance for environmental literacy. The first is the challenge of a new type of risk emerging from climate change. Society is challenged to build structures that can deal with future risk when ordinarily this would be ignored in everyday life (Giddens, 2007). Political decision-makers are overburdened as the structure and constitution of the nation-state prevents making and rapidly implementing substantiated decisions (Giddens, 2009). The recent works of Giddens indicate that in a time of globalization, where justice becomes an issue and industry is working at a global scale, societal organization needs systems above the nation-state.

9.1.4 Key messages

- Parsons and other sociologists consider individual human action to be shaped by social and cultural

structures. Nation-states, social systems, and cultural variables are seen as the texture of society
- Luhmann considered the "difference" between a social system and its environment to be the object of sociology and the core of system theory. For Luhmann society is communication between social (sub)systems. For Luhmann, successful coping strategies responding to ecological risks depend on the proper operation (i.e. communication) of functional systems such as policy, law, economics, science, education, and religion
- Giddens' theory of structuration links human systems at the micro and macro levels, the individual actor, and the social structure. Individual agency is constrained and enabled by social structure and, in turn, creates and changes social structure. Further, Giddens differentiates between authoritative resources (embedded in the social structure) and material, allocative resources. Giddens thus describes a theory of how the individual interacts with its social and material environment.

9.2 The second wave of environmental sociology

In the 1980s, many of the same concerns about environmental degradation, technical disasters like the Chernobyl accident, and related human health issues that were fuelling the environmental movement also inspired the emergence of a second wave of environmental sociology research in the USA. Early studies focused on issues of overuse of natural resources and pollution and, since then, have been shifting towards investigations of multifaceted questions about sustainability. The 1990s saw a resurgence in European environmental sociology work, with an emphasis on employing sociology theory and methodology, sometimes in the context of transdisciplinary processes, to investigate environmental issues.

9.2.1 Fuelled by environmental movement – disasters

Environmental sociology in the USA was revitalized during the late 1980s, after some years of decreasing support during the Reagan administration. A primary motivator was a series of environmental disasters such as Bhopal and Chernobyl (see Box 9.1), which were intensely discussed in the media. Additionally, some regional problems, such as the contamination of the Atlantic coast and the discovery of the dramatic, steady decline of ozone in the stratosphere since 1980, also fuelled the resurgence in environmental sociology (Dunlap, 1997, p. 26). This revitalization was accompanied by a fundamental paradigm change, as pronounced by Dunlap and Catton at the end of the 2002 review on environmental sociology (Dunlap & Catton, 2002).

Environmental sociologists started to think on a larger scale about a multifaceted scope of issues. For instance, they became increasingly aware of the limits of their discipline, which were not all inherently caused by the strong impact of the human exceptionalism paradigm (HEP). Now, somewhat surprisingly, sociologists advocated interdisciplinarity:

> ... environmental sociology ... crosses boundaries not only with the conventional field of sociology, but with a range of relevant sibling disciplines in the social and the behavioral sciences as well as natural science and various applied disciplines and professions. (Dunlap & Michelson, 2002, p. 16)

If one interprets such statements literally, one could argue that environmental sociologists were pleading for a new, integrated type of environmental science that should include knowledge from both natural and social sciences. This new view, which would strongly affect the education methods of environmental scientists and sociologists, is also apparent in the following statement:

> In our view, at its best, environmental sociology includes detailed analyses both of environments and social phenomena. (Dunlap & Michelson, 2002, p. 15)

In this way, environmental sociology converged with the core message of probabilistic functionalism put forth by the psychologist Brunswik (see Chapter 3). One may add that there are different types of rationales and thus a variety of drivers in organisms (e.g. human systems) and in environmental systems, because parts of the latter are shaped by biophysical causation and natural laws.

9.2.2 US focus: from pollution to ecosystems

Catton and Dunlap, the founders and long-term leaders of US environmental sociology, focused intensely on incorporating variables such as living space, supply,

Box 9.1 Environmental injustice: landfills, Chernobyl, and the north–south problem

Figure 9.2 (a) A demonstrator at the 2006 Bhopal "Die-in" at Seattle (photo by International Campaign for Justice in Bhopal; ICJB). (b) The Chernobyl nuclear plant after the meltdown (photo from nukeworker.com). (c) The Bingham Canyon Mine at Salt Lake City is the largest open-pit copper mine in the world (picture from University of Southern California).

Following the publication of *Silent Spring* at the beginning of the 1980s, the American public became aware of an uneven distribution of environmental burdens among different parts of society. Many regional or case studies demonstrate that toxic waste disposals are built in working class communities of color (Checker, 2005). One study found that working class communities of color are most affected by hazardous waste treatment, storage, and disposal facilities. For instance, the North Carolina state officials decided in 1978 to locate a PCB (polychlorinated biphenyl) landfill in a predominantly African–American community in Warren County (McGurty, 2007). In response, the concerned people protested, lay down in the streets, and blocked trucks carrying toxic waste to the landfill. Over 500 people were arrested. This act of civil disobedience was the cradle of the Environmental Justice Movement and the first time anyone was jailed for trying to halt a toxic waste landfill (Bullard, 1990).

In a contemporary historical study, Washington (2005) showed that elite whites built certain areas of Chicago as places for "social and political lepers" in society. These outcast groups included both immigrant groups and African–Americans. These cases show that ethnic and class issues interact with the culture of a country. The Bhopal accident in 1984 was an explosion of a tank of 40 tons of methyl isocyanate (MIC) gas that overheated. About 3000 people were killed immediately, 8000 more during the first weeks, and about 20 000 on remote damages (Eckermann, 2004). Other environmental accidents, such as the 1986 Chernobyl disaster, can also be analyzed in the frame of environmental justice (Figure 9.2). Environmental justice exists if environmental risks, hazards, and burdens, as well as benefits, are equally distributed socially.

Environmental justice is often linked and interwoven with social justice. Thus, environmental concerns or protests might be able to serve as a substitute for social injustice. Thus, also old and new issues of the north–south problem, such as impacts of mineral depletion on the environment and indigenous people, surface and underground mining (Bridge, 2004), as well as hazardous waste exports (Wallbaum, 1991; Wong *et al.*, 2007), can be seen in the frame of social justice.

Table 9.2 Some typical contemporary environmental issues and questions posed by environmental sociologists.

Environmental issues	Questions from environmental sociologists
Actual and potential contamination by toxic and other hazardous wastes in neighborhoods and towns	Understanding cause and impact relationships, developing conservation practices, environmental justice (Bell, 2004; Capek, 1993)
Catastrophic technological disasters and accidents such as Bhopal and Chernobyl	How are social structures and disasters related? How do communities or regions cope with disaster impacts? How do they incur physical damages and losses to their routine functioning? How do people deal with long-term threats? Surprisingly, this has not been a dominant topic, and has only been addressed by Kreps (1984, 1985)
Built environment	"Understanding how environments get built; what objectives are chosen; how products and artifacts are distributed; and who benefits, who loses and in what way" (Dunlap & Michelson, 2002, p. 4).
Loss of the natural environment, clearing of rainforests	Sociologists assess " … the meaning of wilderness and wildlands to people …" (Dunlap & Michelson, 2002, p. 6)
Natural hazards such as hurricanes	How are these hazards encountered, experienced, and mitigated? How do race, ethnicity, income, and household level affect concern and pattern of geographical distribution (Peacock et al., 1997)?
Ozone depletion, global warming	What underlies the social construction of global environmental threats (Buttel & Taylor, 1992; Carolan, 2006; Ungar, 1992, 1995)?

and waste into sociology research. They later shifted their focus to sustainability, which, however, had a strong orientation towards efficiency:

> Moving toward a more sustainable society, for example, will require using natural resource far more efficiently in order to minimize both resource withdrawals and pollution resulting from resource, extraction, use, and disposal. (Dunlap et al., 2002, p. 11)

Thus, sociologists have noticed and reacted to the concept of oversized ecological footprints put forth by biologists and natural environmental scientists since the early 1970s (Ehrlich & Ehrlich, 1970; Meadows et al., 1972, see Box 9.2). Table 9.2 presents an exemplary list of environmental issues as they were listed in relevant reviews at the turn of the twenty-first century. However, one may critically assert that there has been a very limited understanding by environmental sociologists of the complexity of human–environment interactions. A broader perspective would go far beyond investigating resource shortages, a concept developed in some cultures as early as the eighteenth century, when sustainable forest management practices were developed.

However, upon closer inspection, this viewpoint is slowly changing in contemporary American environmental sociology. In fact, we suspect that environmental sociology is on the verge of another paradigm change. This new paradigm would regard environmental sociology as a genuinely interdisciplinary field. This is expressed by a few statements in Dunlap et al. (2002, p. 11) demanding that environmental sociologists consider physical settings, and explain how societies influence and react to environmental change.

The new paradigm also reflects a switch from environmental issues or systems to environmental functions (Dunlap & Catton, 2002). Quite remarkably, one can speculate that this switch was influenced by biology, considering that the idea of ecosystem functions has been increasingly discussed since the late 1980s (see Daily, 1997; de Groot, 1992). We should note that the issues of environmental assessment and environmental valuation have obviously played a marginal role in American environmental sociology and are usually linked, once again, to biological ecology. The consideration of ecosystem functions for human and other ecosystem units is a central issue. This approach fundamentally differs from former approaches incorporating ideas from biology to sociology.

> This ecological perspective can be distinguished from other sociological ecologies (e.g. human ecology, organizational ecology) in that it asserts the inherently bio-ecological underpinnings of society and the ongoing

9.2 The second wave of environmental sociology

> **Box 9.2** The environment as an input–output system: New York's Fresh Kills waste disposal
>
> When referring to Freese (1997), Dunlap and Catton describe the perspective of the environment in current environmental sociology as follows: "Technically, it is not 'the environment' but 'ecosystems' that serve the three functions for humans – and for all other living species" (Dunlap & Catton, 2002, p. 243). Dunlap and Catton describe issues related to the three functions: living space, supply depot, and waste repository.
>
> - *Living space*: the "home" for humans and other living beings. *Issue*: overuse of this function results in the destruction of habitats for other species
> - *Supply depot*: sources of renewable and non-renewable natural resources such as air, water, forests, fossil fuels, etc., which are necessary to sustain human society. *Issue*: avoiding "shortages" of renewable resources and "scarcity" of non-renewable resources
> - *Waste repository*: "The environment … acts as a 'sink' for absorbing the waste products of modern industrial societies" (Dunlap & Michelson, 2002, p. 10). *Issue*: exceeding the ability of ecosystems to absorb wastes creates pollution, which may harm humans and ecosystems.
>
> Sociologists acknowledge the oversized ecological footprint of the human species. This is reflected by Figure 9.3b, which shows the currently exceeded carrying capacity of the Earth with respect to the required ecosystem or Earth system functions.
>
> The overflow and waste problem can be seen at the Fresh Kills landfill in the New York City borough of Staten Island (Figure 9.4). Opened in 1948, it became the world's largest garbage dump and the highest human-made hill on the East Coast of the USA, even taller than the Statue of Liberty. Under local pressure and with the support of the US Environmental Protection Agency (EPA) the landfill was closed in 2001, and then temporarily reopened after the 2001 attacks on the World Trade Center to receive and process much of the debris from the destruction. Since its definitive closure, 13 000 tons of originally disposed municipal waste has had to be exported to other facilities (Lloyd & Mitchinson, 2006). About 900 tons, corresponding to Staten Island's daily municipal waste, were then brought 750 miles south to an Allied Waste landfill in South Carolina (Government of New York City, 2007) as the municipal authorities had not made a plan for coping with the garbage.

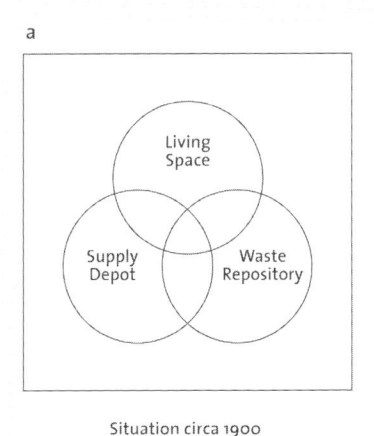

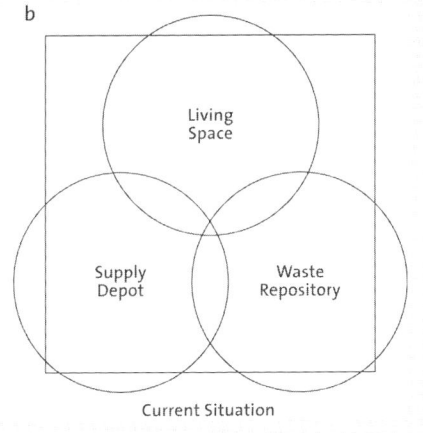

Figure 9.3 Competing functions of the environment according to Dunlap (2002, p. 245).

(a) Situation circa 1900

(b) Current Situation

enmeshment of social activities in nature. Put another way, society is seen as embedded in and deeply dependent upon natural systems. (Lutzenhiser, 2002, p. 7)

As a result of acknowledging the embedding of local ecosystems and environmental threats on global risks and environmental changes, the focus of environmental sociology research is shifting "from specific geographical sites to dynamics involving all humankind" (Dunlap et al., 2002, p. 7). We think that a further challenge for the discipline is finding a proper mix of application and theory so that environmental sociology can enter transdisciplinary discourses.

Figure 9.4 (a) The Fresh Kills landfill (photo from cryptome.org) illustrates the limited capacity of ecosystems with respect to human production (of waste). Contrary to other countries, such as Japan and Switzerland, which focus on incineration or "alternative thermal methods" of waste treatment, the US policy of municipal waste still targets (b) landfill deposits. (c) In many countries, such as the USA and UK, anxiety towards incineration has caused many environmental protests.

9.2.3 European focus: society–environment interaction

A European environmental sociology group developed in the 1990s as a special subfield of sociology. In Germany, for example, there have been impacts from the fields of biology and human ecology, and in the Netherlands "… (social) ecology is taken as the obvious frame of reference" (Spaargaren, 1987, p. 63).

Spaargaren (1987) noted that this type of social ecology concept refers to an early work of Nellisen (1970) reflecting on the Chicago School. Nellisen made the distinction between a (biotic) "community" and "society," with society standing for culture and with community standing for the biological roots of ecology and sociobiology. Becker and Jahn (2006) conceive social ecology as a scientific object that investigates the relation between societies and natural resources. This contribution is of interest because it relies on sociological theory yet is directed toward using this knowledge in transdisciplinary processes (see Chapter 15). The difference between the North American and European approaches has been nicely described by Mol:

> To put it shortly and bluntly, whereas US environmental sociologists were more worried about getting environment into sociology, European environmental sociologists were preoccupied with getting sociology into studies on the environment. (Mol, 2006, p. 11)

9.2.4 Key messages

- A series of environmental disasters in the 1980s, such as Bhobal, Chernobyl, Seveso, and Schweizerhalle, revived environmental sociology in the USA. Research has been widely atheoretical but has aspired to incorporate environmental variables such as living space, supply, and waste into sociology research
- European environmental sociologists, on the other hand, developed later but followed a more theoretical approach, focusing on a more integrative perspective of society and environment.

9.3 System analysis on society's environmental impacts

Theories about the relationships between societal activities and the state of the human-shaped environment have become a new object of research. From the human to environment perspective, there are perceptible effects on land degradation, changing biodiversity or climate change that also allow for quantitative analysis on the level of society; for example, the nation-state. Thus, the grand verbal theories can be supplemented by empirical approaches that allow for a quantitative check and some verification. A critical question, however, is, what variables are used? We briefly present two approaches that are strongly affiliated to macrosociological thought, although they also add valuable contributions to industrial ecology, ecological economics, and the field of environmental sciences.

9.3.1 Social ecology and industrial metabolism

A new framework for integrative human–environment research emerged in the 1990s in Austria from the work of the sociologist Marina Fischer-Kowalski and her co-workers, biologists Helga Weisz and Helmut Haberl. Their approach, which represents an elaborate framework of human–environment interactions, is known as social ecological system analysis, social ecology, or ecological sociology. Although Fischer-Kowalski's approach has occasionally been discussed under the label human ecology, it has no strong reference to the central human ecology assumptions (such as the concepts of competition, adaptation, symbiosis). Strong emphasis on the concept of metabolism brings social ecology into the context of industrial ecology (Ayres & Kneese, 1969; Erdman, 2007). The approach is discussed in some detail in Chapters 12.6 and 19.4 (Industrial Ecology) and we present only the basic idea here.

The group of Marina Fischer-Kowalski considers material and energy flows as a main layer of societal activities when using a double complementarity (see Figure 12.6). One is the complementarity between human society and nature. The second is that between biophysical environments and abstract culture (Fischer-Kowalski, 1998). The latter differentiation meets the basic complementarity between the material environment, including the material–biophysical part of human activities (i.e. H_m and E_m), see Figure 3.3*, and the social, cultural, epistemic layers of human and environmental systems (i.e. H_s and E_s). The research aim of this approach is to explain how human societies affect environmental change over long time periods and vice versa (Weisz et al., 2001).

The key term in this approach is industrial metabolism as the major mechanism for global environmental change. The quantitative method for representing this is material flow accounting based on the quantitative estimate of annual, national, or global extraction of biomass, fossil energy carriers, or industrial or construction minerals (Krausmann et al., 2009). The goal is to gain insight into material flows in relation to the population growth, economic development, energy use or material intensity. The latter is defined as the use of natural resources (i.e. materials) required per unit of GDP. Thus the variables are similar to those used in the ReSTET framework (see Figure 8.3*). The primarily historic analyses provide seminal insight into the dynamic human–environment systems (HES):

> So far there is no evidence that growth of global materials use is slowing down or might eventually decline and our results indicate that an increase in material productivity is a general feature of economic development. (Krausmann et al., 2009, p. 2696)

9.3.2 From IPAT to STIRPAT

The *POET* and *IPAT* models introduced in Chapter 8 provide insights into how key societal variables affect the environment. The variables are also called "driving forces" (York et al., 2003) of global environmental impacts. Recall that the *IPAT* model reads: environmental impacts (*I*) equal population size (*P*) times affluence (*A*) times technology (*T*).

$$I = P \times A \times T$$

An extension of the *IPAT* model, called the *ImPACT* model, was developed by Waggoner and Ausubel (2002), who were interested in gaining a better understanding of dematerialization. Dematerialization here means that the (environmental) intensity of using income, which can be represented by the impact per unit of consumption of GDP (T'), becomes smaller. If we define a variable consumption per unit GDP (*C*), a slight renovation of the *IPAT* model reads:

$$I = P \times A \times C \times T'$$

Already this slight differentiation provided some valuable insight into sustainable emissions. In this way, modeling in the frame of human–environment interactions becomes a powerful tool for predicting environmental impacts.

> Examples from energy and food, farming and manufacturing, and steel and water show that declining *C*, called dematerialization, can temper the sustainability challenge of growth ($P \times A$), and that innovation or efficient technology that lowers T' can counter rising consumption ($P \times A \times C$). … An annual 2–3% progress in consumption and technology over many decades and sectors provides a benchmark for sustainability. (Waggoner & Ausubel, 2002, p. 7860)

We should note that the Waggoner and Ausubel *ImPACT* model ran under the ambitious heading "A framework for sustainability science." Besides the sparse number of variables included in these models, the simple multiplicative (or additive if the logarithm is taken) structure of the relationship of the variables is critical. This provides limited insights, as it hides non-linear interactions between the variables, as they have been conceptualized, for instance, by the Kuznets curve. This has been one challenge that led to a further extension of the *IPAT* model proposed by quantitative sociologists. York *et al.* (2003) introduced the concept of ecological elasticity to investigate how, for instance, environmental impacts are dependent on the interaction between GDP and the related technology variable or technological subvariables. To allow for cross-national or cross-regional comparisons, they also introduced a stochastic error variable, *e*. One option to represent this extended STIRPAT (STochastic Impacts by Regression on Population, Affluence, and Technology) model reads:

$$I = a \times P^b \times A^c \times T^d \times e$$

In this equation, the variables *a*, *b*, *c*, *d*, and *e* are statistically estimated variables, or more complex functions that represent the interaction between these variables. We do not go into the technical details here but note that the modeling approach provides a view on the demography-based environmental impacts of societies (York, 2008; Zagheni & Billari, 2007).

9.3.3 Key messages

- Some quantitative approaches for modeling social metabolism and related environmental impacts for societies and the world are emerging from material flow analysis and the *IPAT* model. These models can provide benchmarks for sustainable development, but require a thorough sociological embedding of key variables.

9.4 Risk society

According to Ulrich Beck's world risk society perspective, humans are experiencing a second age of modernity, with supranational issues at the fore. Beck characterizes this second age of modernity as a time when people no longer see nature as independent from human systems, leaving no boundaries in time and space with respect to risk-related natural and technical disasters. He sees the traditional concepts of insurance, causality, and responsibility to be no longer relevant, particularly in cases such as the Chernobyl nuclear power plant accident, and proposes that a more proactive, anticipatory approach to compliance is needed. Beck proposes that local environmental injustices and global changes to the traditional family model are linked to inequities between nations. Recognizing the supranational nature of many current issues, he calls for a cosmopolitization of societal decision-making that involves global coordination.

9.4.1 Cosmopolitization for environmental risk management in the twenty-first century

To demonstrate the power of macrosociological analysis in the understanding of ever-changing human–environment relationships, we examine Ulrich Beck's idea of a world risk society. One should also note that the term "risk society" (German: Risikogesellschaft) emerged in the context of a series of technological disasters in the late 1980s. These events supported Beck's thesis that modernization is a major contributor to the ubiquity of risk on a global level. Clearly, nuclear technology has played a specific role in this. Humanity's use of atomic power and subsequent events, such as Hiroshima and the Chernobyl incident, have led to labeling this mastery as a technological pitfall; these were the first occasions in human history that a worldwide impact resulted from something in the microcosmic scale (Krohn & Weingart, 1987).

In Beck's view (1996), human society has now encountered a second age of modernity. Whereas the first age featured social problems, such as class struggle, the second age of modernity is shaped by relations between and beyond nation-states (Beck, 2000). Thus Beck is focusing on a supranational level. However, adequate and functional management and the establishment of a cosmopolitan democracy would involve dissolving nation-based institutionalism. Nationalism and cosmopolitanism are considered "normative antagonism" (Beck, 2004, p. 133), and the term cosmopolitan should not be confused with those of international and global. Associations such as the World Trade Organization and the World Bank have to be considered as a kind of institutionalized internationalization. Cosmopolitization would require more than institutionalized internationalization of society. According to Beck:

> … cosmopolitan relations presuppose international relations among others, but they also transform them by opening up and redrawing boundaries, by transcending or reversing the polarity of the relations between us and others, and not least by rewriting in cosmopolitan terms the relationship between state, politics and nation. (Beck, 2004, p. 143)

Beck characterizes this second age of modernity as a time when people no longer see nature as independent from human systems. In some of his writings Beck discusses the complementarity between humans and nature (and not between humans and different types of environments). His message is that there are no boundaries in time and space with respect to the risks, hazards, and natural disasters that were so widely discussed at the end of the twentieth century. He speaks of ecological interdependence, risks for food chains, epidemics, energy, pollution, etc. Thus, risk management has transcended nation-state boundaries. Because of the global ubiquity of human impact, the complementarity between nature and humans is an anachronism and is dysfunctional.

For Beck, it is impossible to talk about the destruction of nature, ecology, or the (natural) environment either on the level of a national society or from an objective perspective without first integrating human societies. Modern society is facing risks, hazards, collateral damage, and the question of insurability, all in a symbiotic human–nature setting on a global scale that requires regulatory systems different than those of the classical industrial age. Beck proposes that security mechanisms, such as private insurance, which have been developed in the past to protect individuals, families, and companies against technological or environmental risks, have become inapplicable because of the changes in magnitude of environmental impacts. For example, costs for the Chernobyl accident amounted to hundreds of billions of US dollars. Belarus, for instance, has estimated the losses over 30 years at US $235 billion (International Atomic Energy Agency (IAEA), 2006). As can be learned from current risk assessments, one can see that the traditional concepts of causality and responsibility are no longer applicable, such as the risks associated with genetically modified food (Beck, 1996, p. 131). Rather than punishment for the violations of norms, proactive and cooperative programs that foster compliance with sustainable norms for lifestyles and business are needed.

9.4.2 Global coordination for environmental justice

The microscopic, ubiquitous scale of human impacts on the environment and the altered nature of the human–environment relationship can also be demonstrated by the example of a small nanotechnology company. Beck argues that such a company could comply with all national laws and regulations but can cause significant environmental impacts. Consequently, Beck pleads for new types of transnational or supranational systems or cosmopolitization – that means coordination at the global level. In this sense, Beck recognizes not only how great the flow of natural resources for industrial use is that occurs on a global scale, but also that a world society is developing with global variability, global interconnectedness, and global intercommunication.

Cosmopolitization ... must be understood as a *multidimensional* process that has irrevocably changed the historical "nature" of the social world and the status of individual countries within those worlds. (Beck, 2004, p. 136)

Here he is speaking about global generations, suggesting that "we need a cosmopolitan outlook to understand the generational dynamics that exacerbate inter-generational tensions within nations and intra-generational affinities and conflicts between nations" (Beck & Beck-Gernsheim, 2009). This globalization on the individual level is linked to injustices between nations. Nations act antagonistically in terms of global justice and access to natural resources because "the nation-state principle functions as a legitimization for global inequalities" (Beck, 2004, p. 148). The performance principle, which is used as justification for social inequalities on the individual level, is also used to justify differences between nations. As Beck argues, the nation-state framing of societies in combination with the achievement principle (the understanding that you achieve less when your GDP is low) is used to justify the tremendous injustice between the north and the south, creating the north–south inequality.

9.4.3 Key messages

- The globalization of the world includes production, material flows, consumption, communication, and family patterns on all levels. Thus, the organization of societies on a nation-state level is anachronistic
- The nation-state allows, in combination with the achievement principle, justification for injustices between north and south.

9.5 Communication and environmental issues

In this section, we examine communicative action and discourse as they relate to environmental literacy. In his ideas about communicative rationality, the German sociologist Jürgen Habermas expresses concern about how society is increasingly administered at a level that is remote from citizen input. We examine how Habermas is a proponent of processes that embrace participation. He proposed that democratic public life only thrives where institutions enable citizens to hold thoughtful, open debates about matters of public importance. Habermas offers an optimistic suggestion about reviving meaningful, accessible public discourse through an open discovery process he calls deliberative democracy. Based on Habermas' idea of deliberated and substantiated reasoning, Renn developed the cooperative discourse process, which builds consensus around planning options by employing mediation, expert analysis, and deliberation by citizens.

9.5.1 Habermas' theory of communicative rationality

Jürgen Habermas starts from three ideas about society. Societies must reproduce and thus depend on tradition. They show communicative rationality as a kind of collective rationality. In addition, they need democracy as a basic prerequisite for efficient functioning. Habermas is a late representative of the Frankfurt School of critical theory, which derived its name from the critiques of twentieth-century western capitalism, as well as from simple interpretations of Marxism (Adams & Sydie, 2002).

Habermas is of interest for environmental literacy and the reflection on participatory processes, as well as for transdisciplinarity (see Chapter 15) since he focuses on knowledge generation by communication, which means the creation of epistemics. His hypothesis is that societies benefit from undistorted communication which can be realized if the "constraint-free force of the better argument" (Habermas, 1984, p. 24) can become real. Habermas postulates that purposive rationality is inherent in everyday communication. The model of ideal speech is realized if: (i) all people capable of speech can participate; (ii) all people have individual rights and can unrestrictedly bring in their reasons and views; and (iii) no individual can be excluded. He asserts that ideal speech enables participants to generate reasoned, rational, valid arguments that resist critique. Thus, in his view, rationality is "cognitive-instrumental" (Habermas, 1987, p. 28) and allows participants to pursue truth through theoretical discourse and move toward meaningful, morally justifiable actions (ibid. p. 45). Thus, an ideal and well functioning society aspires to reach a communicatively achieved consensus. Communication of this type can serve to establish a deliberative democracy.

Habermas acknowledges different types of interest, such as technical interests, which assist humans in gaining control over nature (cf. appropriation of nature), practical interests for getting things done (cf. pragmatism), and emancipatory interests, which is the notion of freedom of expression, a cornerstone of democracy. Habermas is critical in the sense that he recognizes that this is ideal and can be spoiled and distorted by, for example, scientism or expertocracy, or the naïve belief that all problems have technical solutions. Expertocracy makes people feel powerless and leads to depolitization and a loss of public competence.

In 1987, before the discussion about climate research, Habermas had already pointed out that "the intervention of large-scale industry into ecological balances, the growing scarcity of non-renewable natural resources" (Habermas, 1987; quoted according to Wehling, 2002, p. 153), as well as population bombs, are not tangible enough for individual comprehension. This is in contrast to the disruption of the countryside by housing developments or impairment of health effects that people can see right in front of them and then discuss at local, in-person meetings.

Habermas' ideal of deliberative democracy can be countered from many perspectives, for instance by the game theoretically reasoned idea of "deliberative lies" (Goodin, 2008). However, we think that Habermas' approach is meaningful for reflecting on what "society's socially structured relationships to nature" (Wehling, 2002, p. 160) look like. This relationship to the biotic and abiotic environment is central for societal capacity-building and is a key function of transdisciplinary processes (see Chapter 15.1–4). We should note that there are other theories of societal discourses, such as Foucault's analysis of knowledge, truth, and power in society, where the episteme is central. An episteme is a set of meanings, presuppositions that organize what counts in societal discourses (Foucault, 1971).

9.5.2 Cooperative discourse

Based on Habermas' idea of deliberated and substantiated reasoning, Renn and colleagues (Renn, 1999; Renn *et al.*, 1993) developed the cooperative discourse, a systematic procedure of consensus-building in the field of risk management. This approach is of specific value as it has been practically applied. The approach starts from the interests of stakeholder groups and assesses the specific values and criteria by which alternative technical solutions should be assessed. Then experts evaluate possible alternatives and, in the third step, a discourse on consensus building is organized. The model of cooperative discourse has been applied in environmental management projects, such as identifying a site for municipal waste (Renn & Webler, 1997). A more detailed discussion of this approach can be found in Chapter 15, where we examine transdisciplinarity.

9.5.3 Key messages

- Communication is essential for societal reproduction and is thus purposive. Habermas supposed that in an ideal speech situation, communication can become a means of both truth-finding and consensus-building
- According to Habermas, both utilizing expert knowledge as well as generating knowledge in discourses are necessary but have limits. Management of environmental problems by experts is particularly needed for abstract problems that are not (easily) accessible by the rationale of normal citizens
- There are procedures, such as cooperative discourse or transdisciplinary processes, that can link societal capacity-building, consensus-building, and the use of expert knowledge to promote and develop environmental literacy.

9.6 Constructivism

Based on our assumption that people take a constructivist approach to understanding the environment, we examine how people express concern about the environment through personal and cultural filters. Then we review reasons as to why scientists have, or should have, limited capacity to provide expert assessments for environmental problem-solving.

9.6.1 Constructivism and environmental concern

There are many notions of constructivism and we have briefly outlined that we take a functional social constructivist view from a realist stance. This means that we assume that human systems construct their reality from a social, subjective perspective based on their capabilities in an intentional or purposeful way (see Chapter 3).

From this viewpoint, macrosociology and anthropology can provide insight into the diversity of constructions underlying environmental awareness, concern, and responsibility between different cultures, genders, alternative epistemological positions or various motives for environmentalism. When referring to Douglas and Wildavsky (1983), Beck (1996) portrays culture differences in risk perceptions of technology. Whereas French families feel safe when visiting nuclear power plants (and Chernobyl has not changed this attitude very much), neighboring German families feel anxious.

In addition, gender differences regarding environmental notions reveal a kind of *circulus vitiosus*. Macrosociologists can trace gender differentiation and stratification to as far back as the hunter–gatherer societies, where gender roles resulted in divergent relationships to natural resources. Whether and in what way this differentiation continues today, or whether it is functional or dysfunctional (Beck, 1996), are matters of ongoing research and discussion.

Another interesting issue is whether events involving environmental destruction and risks can serve as indicators or symbols for social inequality or oppression. Thus, in many cases, the destruction of landscapes owing to copper mining funded by international capital may better demonstrate inequitable power structures than can statistics on poverty.

A critical question is: what exactly are people worried about when they express concern about the environment? The theoretical basis of an answer to such a query must deal with whether people's concerns are rooted in sustainability thoughts or in the perceived risk to themselves, to human health, to nature, and to wild animals. Naturally, psychologists, sociologists, and other social scientists such as anthropologists should deal with this question.

Yearley (2005) called this phenomenon the humanization of nature (Box 9.3) and stresses, as anthropologists have suggested, that concerns about nature are worries about humanity that we inscribe onto nature (Douglas & Wildavsky, 1983). The rationale behind this point of view can be exemplarily seen through analogies that cultures draw between humans and their environment. For instance, the dietary prohibitions of religions are best interpreted in terms of similarities to food that certain animals eat. Beck drew a similar conclusion: "The ecological movement is not an environmental movement but a social, inward movement which utilized 'nature' as a parameter for certain questions" (Beck, 1995). Environmentalism should thus be considered as a facet of the late-modern risk society; we worry about forces that regulate risk, safety, and resilience on our behalf. If we consider the special case of the human-affected environment, the relevant environment in the Himalayas began to disappear when Toyota Landcruisers and Caterpillar-made trails allowed human access to every inch of wilderness.

> Environmental damage matters to us because we suddenly sense that we have unwittingly taken control of a system that we thought was external to us. (Yearley, 2005, p. 201)

Likewise, symbolic interactionism (Blumer, 1969) can be utilized to explain certain environmental metaphors, such as the dying of forests [German: Waldsterben] for forest decline, and which societal groups show stronger commitments to environmentalism than others (Hannigan, 2006).

> Humans perceive nature through their cultural filters. (Pepper, 1984, p. 9)

9.6.2 The role of science experts

While psychologists investigate how expert and layperson judgments differ, sociologists reflect on what role experts and laypeople take, or should take, in environmental decision-making processes. One major reason for this research is the conflicting roles and sometimes surprisingly poor performance of experts in classical, well documented environmental accidents and disasters.

Several lines of thought have developed about why scientists have, or should have, a limited role in providing expert assessments for environmental problem-solving:

First, many environmental problems are scientifically open and contested. Thus, if a chemical such as phthalate, or a new technology such as nanoparticles, is disseminated to the environment, there are no previous cases and data that scientists can use to predict the fate and transport of these materials.

Second, conflicting recommendations from different scientists come up in many environmental decision-making processes and can be linked to how the precautionary principle is interpreted. This is partly because of different standards of acceptable risks or fatalities and sometimes because of different scientific methods (e.g. epidemiological vs. animal

9.6 Constructivism

Box 9.3 Humanization of the environment: the home alone fish

It was not until the 1960s that the social movement called environmentalism began in the western world. Rachel Carson's (1962) book *Silent Spring*, which described how insecticides and pesticides enter the food chain and affect the biological world, can be seen as the catalyst for this movement. The book had a huge impact on the ban of DDT, the emergence of environmentalism, and the founding of Greenpeace in 1971.

Yearley (2002) argues that environmental concern is often full of social constructions, which explains why people attribute preferences and values to specific environmental settings. From an environmental constructivist perspective, environmental concern emerges when we project our own worries about ourselves or other humans to animals, plants, and other environmental objects, e.g. a "poor stone." These projections from our own experience are based on subjective values and cultural norms.

A striking example is the story of the home alone fish (Figure 9.5). Yearley (2002, p. 227) reports that, in 1993, the Royal Society for Prevention of Cruelty to Animals, based in Britain, pressed charges against a man who left his tropical goldfish at home when working (*Guardian*, 13 June 1993), while simultaneously showing strong antipathy towards escaped ornamental drakes.

Animal protection practices can be biased based on symbolic attributes. Which animals are included on the protected species lists, such as wolves, is shaped, in part, by cultural values. The cultural bias can also be linked to the subjective definition of nature, resulting in, for example, people considering highly cultivated landscapes to be natural.

Figure 9.5 (a) The wellbeing of goldfish kept in (b) an English worker's home became of interest to environmentalists. As Yearley pointed out, the transference of human values or need to animal, plants, and nature is a typical constructivist approach to the environment.

experiments). Ambiguous statements can irritate the public.

Third, most experts are specialists for generalized facts in a particular subject area, e.g. wetland biology, and are not masters of a specific *context* such as redeveloping land polluted by a now-closed paper factory. Contextual knowledge can be highly developed in people from practice, who often have the best insight into the complexity and reality of in-vivo situations, and are sometimes called reflective practitioners. Thus, scientific subject expertise must be combined with the contextual knowledge of practitioners.

But what makes the experience of the expert scientific? Is it different from a layperson's judgments? Some authors (Gibbons and Nowotny, 2001; Nowotny *et al.*, 2001) argue that this is not the case and that scientific expertise, if applied to real-world problems, would have to compete in the agora, a kind of marketplace for the exchange of ideas. This position is, to some extent, taken by postmodernists. One certainly has to reflect that laypeople often lack subject knowledge and can be susceptible to prejudices, biases, and fallacies (Kahneman *et al.*, 1982; Kahneman & Tversky, 2000; Nisbett & Ross, 1980). At the same time, we should acknowledge that the same judgmental biases can also

Table 9.3 Six lay criteria for judgment of sciences (adapted from Wynne, 1992, p. 298).

Category	Criterion
A	Is scientific knowledge useful?
B	Do scientists incorporate knowledge of practitioners and other system experts from non-academia?
C	Is the form of knowledge, the epistemological status (in particular the uncertainty and time/resources invested in getting the results), as well as the content recognizable?
D	Are scientists open to criticism?
E	What are the social/institutional affiliations of experts?
F	What issue "overspill" exists in lay experience?

be inherent in science experts. A suggestion on how knowledge from scientists and practitioners is best utilized in different phases of a process has been suggested by Stauffacher *et al.* (2008).

Fourth, often experts are asked to make a judgment about the risk, environmental impact, future development, and so on, related to environmental issues for which no solid and adequate methods, studies, knowledge or data are at hand. Such a situation occurred in the early 1980s in Germany, when contaminated soil became an issue and no scientist could conceptually and quantitatively master the full range from soil contamination, bioavailability, transfer paths, or dose–response effects for different types of exposure related to a societally acceptable risk level. As neither a statistical/epidemiological nor an inductive solution was at hand, the soil researcher A. Kloke suggested a list of thresholds (Kloke, 1980). The list did not yet look at soil conditions (such as pH values), but simply included limit values on the concentration of certain heavy metals in topsoil. These limits were based on heuristics on safeguards (e.g. for vegetables consumed by humans), but were related to reference concentrations, which were considered as natural concentrations, and qualitative considerations about safety factors. The list was widely accepted, in part because Kloke was known as a lobby-free scientist. Suitable models for quantitatively assessing the risks for heavy metals in soil were developed later (Lühr *et al.*, 1991; May *et al.*, 1991; Scholz *et al.*, 1991).

Fifth, some science experts are biased because of explicit or implicit corruption (see Table 9.3). If they invented chemicals and technologies related to the issue under discussion and can often benefit from the wide consumption and application of these chemicals and technologies, they can be unrealistically overoptimistic and blind with respect to negative impacts.

Further, scientific experts often serve as (often paid) representatives of certain organizations, companies, or state agencies and thus their voice is constrained by financial incentive. As a result, in the view of the public and of non-governmental organizations (NGOs), scientists have become unreliable companions who are no longer perceived exclusively as guardians of objective truth. They claim that corporate funding has "become the driving force of research" and that "privatization of science has become official policy in all industrialized nations since the early 1980s" (Haerlin & Parr, 1999, p. 499). The skepticism of the public towards science is far reaching owing to scientists' failure to predict negative outcomes using reductionist methods and because of corruption in institutional and financial relations, such as providing value-biased science arguments on genetic superiority to justify the Holocaust (Carlson, 2006).

Sixth, further, there is the argument of imbalance of power. By their very status (education, social position, etc.) in society, scientists have more power to influence societal decision processes than, for example, the public at large. Hence, as a counterbalance it is necessary to give laypeople a voice, although this might turn out to be difficult, in particular, if extensive technical knowledge is necessary.

9.6.3 Key messages

- Constructivists such as Yearley emphasize the subjectivist perspective, recognizing that environmental problems are viewed through personal or cultural filters. These approaches explain how environmental perceptions and valuations are correlated with conflicts in other domains of human activities

- According to Wynne, science experts have lost some reputation because of a lack of appropriate case-related, contextualized knowledge, as well as unreflective application of theories, and missing degrees of freedom because of institutional constraints
- Scientists, practitioners, and laypeople each represent expertise for systems of different abstraction, and each form of expertise is at once unique and limited. For instance, many contextualized environmental (e.g. toxicological) problems remain scientifically unsolved. Laypersons and practitioners (e.g. people living or working in an environmental system) are often the "experts of contexts."

9.7 Ecological modernization vs. the treadmill of production

We close the discussion of sociological theories by contrasting ecological modernization and the treadmill of production. Ecological modernization is an approach primarily developed by European sociologists, in particular Joseph Huber, Martin Jänicke, Arthur Mol, Maarten Hajer, and Gert Spaargaren (Mol et al., 2009). As Jänicke states, the concept goes back to an argument over a 1982 initiative in the Berlin city parliament that suggested that modernization should take place "in industry, in the energy sector, in the mobility sector, and in the construction sector" (Jänicke, 2009, p. 30). Ecological modernization calls for modernization of politics and environmental re-adaptation of economic growth and industrial development. This approach is in contrast to the North American "treadmill of production" theory (Schnaiberg, 1980; Schnaiberg et al., 2002), a sideline approach of US environmental sociology that emerged in the late 1970s. The treadmill of production model suggests that advances in technology lead to a cycle of production that depends upon continuous growth in production. Proponents of this view propose that breaking away from this unsustainable cycle requires restructuring of political and economic approaches and a move away from growth dependence.

9.7.1 Theoretical foundations of ecological modernization

Industrial ecologists consider Joseph Huber as the founding father of ecological modernization. In his comprehensive book *Allgemeine Umweltsoziologie* (Engl.: *General Environmental Sociology*; Huber, 2001), he takes a macrosociological perspective and starts from the distinction between primitive, traditional, and modern societies. The modernization trajectory is considered the most important force in the developmental theory of societal innovation. In his view, innovation started in Europe in the thirteenth and fourteenth centuries. Innovation is a key component of his developmental theory (German: Entwicklungstheorie, Huber, 2001, p. 105). When referring to Schumpeter life cycles of industrial innovation (Schumpeter, 1911/1934), he infers that societies run through several phases, called latency/emergence, prime unfolding (take-off), maturation, and retention/phase-out (conservation, followed by preservation or decline/dissolution, see Figure 9.6). Huber considers that ecological problems, such as material flux or metabolic problems, emerge from industrialization.

To define society, Huber utilized the concept of noosphere (i.e. the sphere of human thought). The noosphere prolongs the chain of ontological levels reaching from the geosphere, biosphere, to the psychosphere. Huber defined society as a system of "inter-subjective and supra-personal communication and organization" (Huber, 2001, p. 24, translated by Scholz), and stressed that the natural environment of today is historically formed and shaped by different societies.

Huber postulates that, following the work of Maturana and Varela (1980), there are networks of processes in and between the biosphere, psychosphere, and noosphere that follow homeostatic cybernetics. He also postulates that dynamics that follow different time frames result in interferences. The governance process is based on a formative structure, which corresponds to the social–cultural–epistemic level (i.e. the mind) of human systems.

9.7.2 European focus on green industrial society

Huber's theoretical work supplemented ecological modernization, which, at "its beginning […], was essentially a political program" (Mol & Jänicke, 2009, p. 17). This program was thought to provide an alternative to the top-down command policy of nations, which had failed to preventatively manage the environmental problems of the 1970s and 80s.

Ecological modernization emphasized that the state and market actors should reconsider their roles.

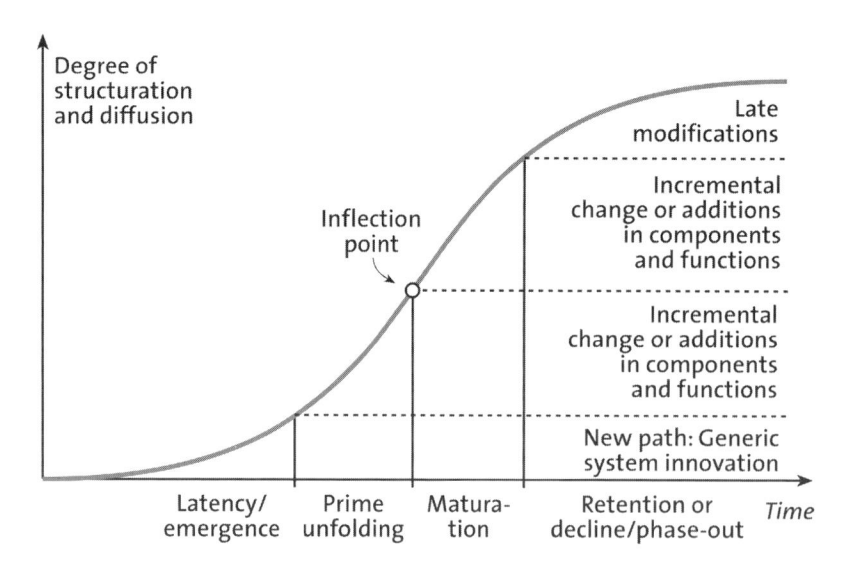

Figure 9.6 The life cycle of modernization (adapted from Huber, 2004, p. 266).

The frame of modernization was the modern welfare state and a market-oriented economy. The approach is very much in line with new institutional economics (see Chapter 11), aspiring to a self-organizing solution among the different actors.

> Ecological modernization refers to a restructuring of modern institutions to follow environmental interests, perspectives and rationalities. (Mol, 2009, p. 456)

In some respects this can be viewed as a European dream of a modern environmental state which is decentralized and facilitates networking among "credit institutions, insurance companies, the utility sector, and business associations, …" (Mol, 2009, p. 461), all following the track to political modernization. The environmental norms of civil society play an important role in bringing together industry and politics.

Ecological modernization supports technocratic and sociographic approaches to environmental management, and the example of acid rain management in the US is an example of its success (Hajer, 1995; see Chapter 13.4). The technocratic route emphasizes the role of smart technologies. A sociographic approach could also take a transdisciplinary route and asks for participation (Van Tatenhove & Leroy, 2003). Ecological modernization integrates environment and ecology in free market economics and provides a vision of a green industrial society. This vision has been denoted as an "unflappable sense of technological optimism" (Hannigan, 2006, p. 184). Its basis is the belief in an undisturbed "ecological switchover as the new Schumpeterian phase in the maturation of the industrialization process" (Mol, 2009, p. 459).

> … the dirty and ugly industrial caterpillar will transform into an ecological butterfly. (Huber, 2001 cited in Murphy & Gouldson, 2009, p. 281)

The ecological modernization approach has been criticized, for example, for underestimating and insufficiently theorizing the notion of power in policy. Furthermore, ecological modernization is intractably linked to democracy as the political system. However, as mentioned, the role of the state and the work of its institutions are going to change.

> While the environmental state remains an important institution in safeguarding environmental quality, it needs to be restructured: moving from a bureaucratic, hierarchical, reactive, command-and-control state towards a more flexible, decentralized and preventive institution that creates networks with other social actors and applies a variety of approaches and instruments to guide society into directions of sustainability. (Mol & Jänicke, 2009, p. 19)

9.7.3 North American focus on industrial production

In contrast to Europe's ecological modernization stands the treadmill of production from American environmental sociology, which was developed primarily by Allan Schnaiberg (1939–2009). After education and practical work in chemistry and engineering, Schnaiberg's focus shifted to environmental sociology via demography, an often neglected but relevant facet of environmental literacy. The contribution of Schnaiberg and his colleagues has been recognized by one of the major environmental sociologists, Fred Buttel:

> The treadmill of production is arguably the single most important sociological concept and theory to have emerged within North American environmental sociology. The treadmill of production is ... distinctly sociological – in other words, it is based on sociological reasoning and is not a biological or ecological analogy ... The treadmill of production departs from mainstream sociology in that its key or penultimate dependent variable – environmental destruction, or "additions" and "withdrawals" – is a biophysical variable. (Buttel, 2004, p. 323)

Schnaiberg conceives the environment as the dynamic process in ecosystems containing the biotic and abiotic elements, and the biosphere – the total life system of planet Earth (Schnaiberg, 1980). The societal meaning of the environment is that it builds a sustenance base for society. Here, Schnaiberg is critical with respect to environmental movements which "just" focus on "irritants in man's home: in air, water, and land" (Schnaiberg, 1980, p. 11) and do not perceive the fundamental changes and threats.

Schnaiberg and colleagues are often denoted as Neo-Marxists. Factually, Marx is only cited once in the founding work *The Environment: From Surplus to Scarcity*. However, similarities to Marxian assumptions are mentioned in some places. Presumably, one can also exchange the term treadmill of production with industrial production without confounding the core concepts therein. Thus, a fundamentally different view on the environmental impacts of industrial production is taken which contrasts with Huber's from caterpillar to butterfly vision. This is reflected in Figure 9.7. This is one basic assumption that seems justified, given the current analyses on biogeochemical cycles (see Chapter 13) and climate change. However, Schnaiberg often refers to the chemical dimension of environmental impacts.

When Schnaiberg refers to technology he has in mind technique, or "a utilized method of production" (Schnaiberg, 1980, p. 114) rather than the notion of technology as the material product of technological production, which is taken in this book, nor that of technology as knowledge (or information). His view is based on two assumptions.

(1) The "ontological structure of the treadmill of production is that there are powerful forces leading to capital-intensive-economic expansion, which in turn comes into contradiction with a biosphere that is essentially finite in nature" (Buttel, 2004, p. 323). The logic is that pursuit of market competitiveness by international companies producing, for instance fertilizers, pesticides, or genetically modified plants, is not without environmental harm.

Schnaiberg assumes that technology is not as "*deus ex machina*" but is a "*social creation*" serving the interests of society, governments, and firms. He assumes that the primary interest of these companies is to expand and to compete. Technology development is a competitive means to survive in the market, which is the primary goal. Environmental externalities are a core interest of Schnaiberg's and, in 1980, he assumed that, with progress in technology, the qualitative impacts on the environment would decrease (see Figure 9.7).

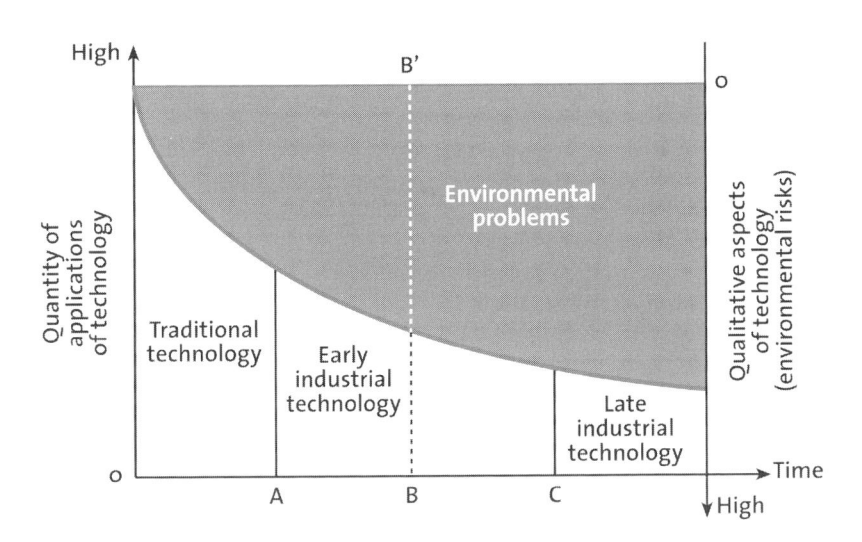

Figure 9.7 Changing relationships between quantitative aspects of technology (left y axis) and qualitative aspects (right y axis) over points in time A, B, and C (on the x axis) which may represent different boundaries between stages of societal development. The line B–B' indicates that the qualitative aspects became superior to the quantitative aspects somewhere in the middle of industrial society (adapted from Schnaiberg, 1980, p. 116, by permission of Oxford University Press).

However, he is skeptical about decreasing impacts without decreasing quantity.

> Every technology, if widely enough practiced, will generate environmental problems. (Schnaiberg, 1980, p. 115)

The environmental impacts result from the withdrawals and additions; that is, the activities of industrial metabolism, in a symbiosis of nature and technology.

(2) The treadmill of production process focuses on the production process and less on the consumption process. Schnaiberg assumes that, in the end, the "tail is not wagging the dog." However, he clearly acknowledges that consumption, and consequently lifestyles, are "a causal factor in environmental degradation" (Schnaiberg, 1980, p. 161). Here it is interesting to see that Schnaiberg already anticipated the current findings that post-industrialism does not mean "substituting new forms of service consumption for goods consumption" but rather "adding new services." The Schnaiberg group is "extremely cautious in accepting arguments about 'hypermaterialism' (superefficient technologies) as predicted by ecological modernization theorists" (Gould et al., 2004, p. 308). The belief from ecological modernization that a new industrial revolution will restructure production processes along ecological lines is doubted by them. This is a message that the reader can also find at the conclusion of late industrial ecology work (see Chapter 13).

Schnaiberg and colleagues take a *global* view on production patterns of capital-intensive production and on the relation of the American political economy to the natural environment (Gould et al., 2008). Schnaiberg provides a production-centered version of the *IPAT* model, which substitutes P by the number of producers × average work force, affluence A by the capital per worker × production per unit capital, and technology T by average withdrawals and additions per unit product (Schnaiberg, 1980, p. 231). The formula reflects that the treadmill of production "was originally an economic change theory" (Gould et al., 2004, p. 297). The theory was warning of naïve promises of unlimited energy and mineral resource availability.

The treadmill of production is a comprehensive theory that can be seen in opposition to ecological modernization. Based on a sociological analysis of production and the interests included in production, the theory has

> … the power to nullify commonly proposed and often popular nonstructural solutions to environmental problems (i.e. efficiency, recycling, appropriate technology, ecological modernization, ecotourism, population control, attitude adjustment, voluntary simplicity, etc.). (Gould et al., 2004, p. 312)

9.7.4 Key messages

- Ecological modernization and the treadmill of production take fundamentally different views on the future development of environmental impacts resulting from human production and consumption. Ecological modernization postulates a reduction of environmental impacts by post-smokestack industrial society, whereas the industrial processes-centered treadmill of production postulates no essential reduction of environmental impacts given a development of societies in the present social structures
- Ecological modernization acknowledges a new role of the nation-state in safeguarding environmental quality. At the core, this role is based on networking and action in the frame of a decentralized democratic modern welfare state at the level of nation-states and international coordination. All are supposed to develop interests and actions that do not conflict with environmental quality. On the contrary, the treadmill of production identifies diverging and conflicting interests of states, institutional actors, firms, etc. Transnational forces receive a major role but no vision on transforming society is provided.

9.8 Macrosociology can go further

Environmental sociology established itself as a subdiscipline during the past three decades (Freudenburg, 2008). In the mid 1990s it became multifaceted:

> Environmental sociology has different dimensions that have their own research foci; environmental attitudes and the environmental movement, social impact analysis, risk assessment, response to toxic siting and discovery, natural hazard research and so forth. (Laska, 1993, p. 7)

As we can see from this list, environmental sociology dealt with the microsociological perspective focusing on social action of individuals, small groups (e.g. couples and families) and their relation to society at large (see Diekmann & Franzen, 1995; Diekmann & Preisendörfer, 2001). From this a large body of theories has emerged, including action theory and role theories, symbolic interactionism, rational choice theories, and conflict theories. While not all of these approaches are genuinely sociological, many of them strongly overlap with approaches in social psychology, behavioral economics, or anthropology. For instance, environmental attitudes have been most thoroughly investigated in psychology (see Chapters 6.17 and 7.11), although this issue has some prominence in sociological journals and has garnered the attention of some great American contemporary sociologists who have constructed attitude scales along these lines (Dunlap et al., 2000; Stets & Biga, 2003).

Because the macrosociology perspective sheds light on the rationale of human systems at the level of the society, it has been the foreground of this part of the book. We have introduced the work of three sociologists "who in one way or another, address the 'big questions'" (Berger, 2002, p. 29). However, we also feel that today's sociology and environmental sociology can take this work further: under what constraints can world society establish intergenerational justice? Here, Beck's idea that social inequalities need reframing is a challenge to the basic units of world society, i.e. family and nation-states receiving new form and functions (Beck, 2007a). Another question reads: can lifestyles in developed societies that have high environmental impacts be substituted by a widely dematerialized social status (Bourdieu, 2003)? Furthermore, environmental sociologists tend to overlook or "deemphasize the political-economic dimensions of their work" (Lutzenhiser, 2002, p. 8). This means that it is unclear whether the sociological analysis is done to explore theoretical questions by deriving a descriptive, critical view, or, on the other hand, to solve a certain environmental problem with a primary focus on contributing to practice. We guess that environmental sociology would benefit from transdisciplinary discourses and vice versa.

Many of the environmental problems listed in the left column of Table 9.2 require appropriate expertise from the social sciences for environmental policy research and governance. There seem to be some promising quantitative approaches that deal with these big questions, such as the social metabolism approach from Fischer-Kowalski. Embedding these approaches in grand theories to provide a better understanding of how societies can change and using this knowledge in transdisciplinary discourses would be an attractive vision for environmental literacy.

9.8.1 Key messages

- Sociology is an important domain for understanding the impacts of social structures on individual action and vice versa
- Sociology has not been well prepared to incorporate the environmental dimension and to provide a theory on the coevolution of social, technological, and natural resource systems. Abstractly speaking, it does not define which social structures seem suitable for promoting sustainable development
- System theory seems promising if it relates material flows with activities of key societal actors causing critical environmental degradation. Here, a close link to industrial ecology seems useful
- The pragmatic equalization of nation-state and society, which emerged after World War II, seems to be losing its explanatory power. The global flow of goods and materials asks for global social structures, such as supranational systems, and perhaps for a redefinition of the concept of society.

Part V

Contributions of economics[1]

10 Origins of economic thinking and the environment 257
11 Contemporary economic theories dealing with the environment 281

Key questions (see key questions 1–3 of the Preamble)

Contributions from economics to environmental literacy

Q1 What parts of the material–biophysical environment have been incorporated into economic theories, why, and when?

Q2 Based on economic theories, what rationales can we expect/do we see from markets and market actors when natural resources are involved?

Q3 How can the various subfields of economics, such as macroeconomics, microeconomics, and institutional economics, inform environmental literacy?

What is found in this part

The economic system is considered a principal component of society (see Figure 1.7*). This book focuses on the material–biophysical environment, which includes machines and the built environment, but also discusses the socio-epistemic environment, which encompasses economic rules (see Figure 3.3*). The latter entities are not excluded from this part, although natural resources are in the foreground. Through both chapters we reflect on whether and how the value of natural resources and the negative impacts of economic action on the environment are incorporated in the theory and practice of economics.

In common terms, economics is the science studying the production, distribution, and consumption of goods and services among economic agents. Different human systems, such as individuals, organizations (e.g. companies), or societies (e.g. countries), can become economic agents. Goods can be material (e.g. commodities), both biophysical (e.g. plants) and technological (e.g. machines), and ideas can also become goods. We can learn from economics that scarcity is a prerequisite for economic transaction and for the production of goods. Chapter 10 takes a view on the preclassical era and discusses how classical and neoclassical theories conceptualize scarcity, economic actions, and what role the material environment takes in these theories.

Chapter 11 discusses rationales of markets and market actors when natural resources are involved and reflects on how common goods such as water, air, and soil, humans (e.g. slaves), what human services, or what parts of the social–cultural epistemic layer (e.g. ideas) can be traded. We reveal that the policy and legal system as well as the cultural and social order (see Figure 1.7*) can constrain what is considered a common or public good, what is considered a club good, and what receives another status. Here we will see that the potential of excluding (i.e. excludability) an economic agent from access to a wanted commodity or good and rivalry to obtain scarce goods are key issues for environmental economics.

But today there are also other areas, such as institutional economics, that cannot be subsumed to the common domains of economics. The reader will see how these perspectives of economics contribute to a better understanding of human–environment interactions.

---

[1] A special thank you goes to Bernard Lehmann who coached me in writing this chapter.

Part V Contributions of economics

Chapter 10 Origins of economic thinking and the environment

10.1 The preclassical era   257
10.2 The classical era   263
10.3 Karl Marx's socialist economic thought   268
10.4 The neoclassical era   271
10.5 Welfare theory   276
10.6 Keynes' macroeconomics and the idea of permanent economic growth   279

Chapter overview

This chapter starts with a discussion of economic thought in antiquity, including not only the use of the term economics, but also many important economic ideas that were coined at that time. We then discuss how markets and agriculture were understood and regulated in the preclassical idea of economics. From the perspective of the human–environment complementarity, economic transactions can be thought of as a transfer of ownership of parts of the material–biophysical–technological environment among human systems. There are properties or estates that are owned by individuals, groups, families, organizations (e.g. companies or cooperatives), communities, nations or other human systems. These properties serve to maintain productivity, and are of interest to other human agents.

We discuss origins of the thinking about exchange, selling, and trading. Trade emerged in societies in which not everybody is a "jack of all trades." Sometimes specialized techniques, tools or scarce materials had to be acquired or bought from outside the family, reminding us how access to certain natural resources is essential.

After dealing with economic thought in the agrarian society, in which land played a key role, the journey through economics in this book then passes through the entrance door into modern economics, which is Adam Smith's magnum opus on *An Inquiry into the Nature and Causes of the Wealth of Nations* (when the resource is fixed in absolute size but extraction costs rise with the rate of extraction; Smith, 1776/1961). We discuss Thomas R. Malthus' theory on the shortage of natural resources and Ricardo's land rent theory. We see how labor, land, and capital become fundamentals of economics and learn how Marx considered negative collaterals of the industrial society. We can then see how marginal-utility theory and modern concepts such as willingness-to-pay entered neoclassical economics and how welfare theory and the accounting of external effects and environmental damage became parts of economics.

10.1 The preclassical era

Anthropologists provide interesting views on how ancient societies might have explained scarcity as a phenomenon of "economic crises:"

> [The] "giving environment" [produced scarcity because the] "ever-providing parent" (i.e. the spirits governing the forest) may have bad temper because of not receiving proper behavior and conduct, as the offspring deviated from the behavioral rules of their ancestors. (Bird-David, 1990, p. 190)

This section does not cover prehistoric societies in the detail with which the topic was covered in Chapters 4.2 and 4.3 (biology). We just want to remind the reader that prehistoric societies had economies, with markets based on reciprocity and markets that involved the environment. We start with a description of economic thought in antiquity. Here we see how Aristotle's rejection of interest relates to today's concept of fair trade and how he differentiated between utility value and exchange value, thus establishing the principle of opportunity costs. Next, we turn to economic thought in scholasticism during the European Middle Ages, where Thomas Aquinas thought well of private property, considered trade useful but vulgar, and (also) rejected the idea of interest. During the Late Middle Ages, double entry bookkeeping emerged and there was a focus on supply and demand and on "capital," a new word used to describe impersonal, transferable assets. We then look at how economic thought in mercantilism, the principal line of economic thought between 1450 and 1750, focused on accumulation of wealth as a means for national dominance. For the first time, international flow of goods and major national infrastructure projects redistributed and degraded natural resources. In contrast, cameralism, a German–Austrian form of mercantilism, emphasized nutrition and was the first to view agriculture as an industry. Finally, we cover physiocratic economics, where soil is the sole economically productive unit, making physiocrats pioneers of ecological economics.

10.1.1 Economic thought in antiquity

The Greeks lived in *poleis*, small city-states consisting of several different villages built by enrolled citizens, each with a population of a few thousand. Poleis were political units subscribing to certain rules and laws, such as that of providing shelter to the military, and had the capacity to prescribe rules and norms within or between them. The *agora* was the place of trade, where the ruling king or council declared what could and could not be traded. Humans (slaves) could be bought and sold as silver miners, craftsmen, farmers, household servants, etc., yet other items, such as those dealing with devotional issues (e.g. indulgences), obviously could not.

In antiquity, economics was already a subject of reflection. The first writings on household management, entitled *Oikonomikos*, written by Xenophon (426–355 BC), included instructions on how to deal with sheep, courtesans, slaves, and, in particular, "husbandry of which the gods are no less lords of the agricultural chors than of war activities" (Xenophon, about 370/1815; V (19)). Here we can see reference to the divine law (Latin: ius divinum). For Xenophon, wealth was the aim of sound economic behavior. The economic agent of his considerations was a small farmer. The division of labor, for example, between cooks and farm workers was the object (Ziegler, 2008).

Aristotle's (384–322 BC) treatise on economics dealt with the subject in a systematic manner. According to Aristotle, the ultimate goal of human endeavor is to achieve happiness (eudaimonia), which refers to both a certain state of mind and a certain living standard (Kraut, 1979). Virtues, both intellectual (prudence or wisdom) and ethical, were of great importance.

When discussing economics, Aristotle differentiates between the use of material means for the improvement of life, the "oikonomike," and the purchase of these means, the "chrematistike" (Söllner, 2001).

> There are … two arts of wealth acquisition … that which takes the form of trade, and that which is the art of household. The latter deserves respect, the first, however, relies on the turnover of money and is justifiably dispraised, because it does not follow nature, but mutual exploitation. It is paired with usury business, which is abhorred by good reasons, as it generates its earning out of the money itself and not from the issues, which have been introduced. (Aristotle, quoted according to Schmölders, 1962, pp. 137–8; translated by Scholz)

Aristotle's position is of interest, as he refers economic action to what is called the law of nature (Greek: physicon dikaion or physei), that is, to the rules that are inherent in nature for reproduction. For Aristotle, the art of war was also a natural oikonomike resembling hunting, a necessary activity to the detriment of those animals "which though determined by nature to serve [man] do not voluntarily serve him" (Schmölders, 1962, p. 139, translated by Scholz).

According to Aristotle, chrematistike should always be the means to an end and not the end in itself, and he addresses the possession and expenditure of money in this context. In his opinion, money has three functions: it is a medium of exchange, a means of stocking values, and a measure of the value of goods. While Aristotle was in favor of private property, he rejects the idea of interest on loans as immoral. His scholar Theophrastus was less abstinent but suggested an interest rate of about 9%, which was less than the common rates of that time (Homer & Sylla, 1996). Regarding trading as an act of exchange, he places great value on commutative justice, which demands that all parties who sign a contract be treated equally. Moreover, he also emphasizes the importance of the dignity of the persons involved and, it follows, the meaning of personal relationships during the process of the exchange. Aristotle's idea that the price could only be fair in this way, lay an early foundation for the idea of Fair Trade, which – besides the banning of interest – has come to embody business ethics (Rabe, 1984).

Furthermore, Aristotle differentiates between utility value and exchange value, thus establishing yet another fundamental principle, namely opportunity costs, without actually calling it by this name. With regard to the use of natural resources and in relation to the "environment" as a collective good, Aristotle places great emphasis on the mediating nature of economic transactions and money. Although agriculture and thus soil were the bases for production in Greek society, these issues played a marginal role in economic thought and were considered the most "slavish" of activities.

10.1.2 Economic thought in scholasticism

Under the Roman Empire, economic thought concentrated on the state of affairs in agriculture. In scholasticism, the dominant philosophy and theology of the European Middle Ages, Thomas Aquinas (1225–74) was a key scholar (see Chapter 4.5) whose work can be considered as representative of the economic line of reasoning. His principal work, *Summa Theologica*, dealt with the subject of private property, of which – like Aristotle – he was in favor. Although he considered trade to be "useful," he rated it as "vulgar" from an ethical point of view. His deliberations on the subject of fair prices must be viewed against the background of the exchange trade relationships customary in the Middle Ages, which were governed according to the power and personal position of the people involved. In the Late Middle Ages, anonymous trade transactions were only known to a limited extent in larger towns. Aquinas stated that the price is "fair" when the value of both elements of a transaction – that which is bought and that which is sold – are "equivalent" (Söllner, 2001). He focused his attention primarily on the supplier of the goods and his costs ("own work and other costs"). In his opinion, it was essential that the supplier should be able to cover his costs so that he would not be forced out of the market as, so to speak, a victim of competition. A fair wage was viewed as a special case of a fair price. According to Aquinas' reasoning, demand as such played a subordinate role. He assumed that, in the interaction between supply and demand, there is divergence of ideas among the parties involved. Thus, since potential participants are excluded on both sides, he did not regard the market place as the sole institution for the reconciliation of interests. He proposed an authority for monitoring and regulating prices.

For Aquinas, money was a measure of value. It follows that he attached great importance to the problem of the depreciation of money without, however, suggesting any definite measures to combat it. Furthermore, he absolutely condemned as a sin the application of interest and proposed that credits be prohibited from transaction in consumption (i.e. the "Konsumptivkredit;" Sombart, 1929/1950). This targets the prevention of usury in loan contracts (mutuum), which included often cunning and unfair resale and interest rules. The rejection of interest is based on the concept that money is a commodity and not a good that may be exploited to make a profit.

In the Late Middle Ages, there was a marked increase in the striving for profits following the introduction of double entry bookkeeping and new industrial organization forms. It became more and more difficult to differentiate between "fair" and "unfair" profits. Cheating became an issue. The term "capital" emerged to describe impersonal, transferable assets. Also, the influence of religion on economic reasoning was dwindling. The "value of money," which Aquinas viewed purely from a "nominalist" point of view, was increasingly dictated by the value of the metal that was used to mint the coins ("metallism"). In this phase of the Late Middle Ages, the general opinion was that a fair price was that price which resulted from the interaction of supply and demand such that it would be accepted by all players on the market. In his work, *Treatise on Monetary Depreciations*, Nicolaus Oresmius (1325–85) discussed money as such and its value. He demanded that the state ensure the stability of the value of money (Söllner, 2001). As Schumpeter emphasized, scholasticists were focusing on the role of monetary capital, particularly "commercial and speculative activity" (Schumpeter, 1954, p. 99). They did not "work out any theory of the physical aspects of production" ("real capital;" ibid., p. 101).

10.1.3 Economic reasoning in mercantilism

Mercantilism, the principal line of economic thought between 1450 and 1750, was the era of merchants and financiers in absolutist nations. Its main characteristics were the quest for worldly prosperity and the strengthening of the national economy. Precious metals were considered a means to an end, and inflows of these facilitated trading by metal money. The English banker Richard Cantillon's (1680–1734) book *Essai sur la Nature du Commerce en Général* (Cantillon, 1755) made a major contribution to understanding of the speed with which money flows between the rural sectors, the urban sectors, and the landlords, the latter of which receive rents from farms and invest them in the urban sector.

The key entities of mercantilism were thus money, trade, and labor. The mercantilist era saw the beginnings of the global flow of goods. There were, for instance, two types of European nations. One was the colonial states of England, Holland, Spain, and Portugal, which imported and processed raw materials and which competed in the European market via custom barriers and strict governmental control (Ott, 1995). The other was the continental states of France, Prussia, and Austria. Mercantilism stimulated many infrastructure investments, and construction of the canal systems, roads, and sea ports changed traffic routes and settlement structures (see Box 10.1). These constructions ran under the label of stewardship "with special links to God … as an extension of the divine rights of a … good monarch … to manage his territory well" (Mukerji, 2007, p. 128). So, engineering for mercantilist purposes changed landscape and material flows, as the acquisition of natural resources became of great importance.

Major works on mercantilist economics include *England's Treasure by Forraign Trade* by Thomas Mun (1571–1641), which focused on economics as the accumulation of wealth:

> The ordinary means therefore to increase our wealth is by forraign trade, where wee must ever observe this rule: to sell more to strangers yearly than wee consume of theirs in value. (Mun, 1664, p. 11)

Although principally governed by monarchs or nobility with unlimited power and absolute sovereignty, the state was vital to the mercantilists. In contrast with the scholastics, mercantilists absolutely rejected state intervention in price control or criticism relating to the application of interest. Striving for profit was regarded as useful and honorable. The mercantilists recognized that economy is cyclic by nature, and the works of Cantillon, Mun, and others did indeed contribute to establishing economics as a science in its own right, for instance, by describing the agriculture and manufacturing sectors, coining the term "rent," and describing the money flows between these sectors. There was no uniform concept of economics in mercantilism. In addition, there was a great divergence of concepts, particularly in the field of the theoretical structure of value. The value theory proposed by Petty (1662), which assumes that "land" and "labor" are the determinants of value, is of special interest in the context of natural resources.

The mercantilists were preoccupied with ensuring their country exported more than it imported. The aim was building up stocks of gold which could serve as a war chest.

Mercantilism was hardly concerned with people's attitudes towards and treatment of nature and natural resources. Natural resources are mentioned as strategic resources, similar to the workforce. Land is viewed as economically valuable, which co-determines the economic process by serving as a resource and generating revenue for its owner. However, production in the agricultural domain remained a marginal issue and was only represented by the yield numbers in business records (Rabe, 1984). The ultimate goal of mercantilist economic activities was to fill the baronial and royal treasuries for building infrastructure, tilling land, tapping new silver mines, etc. (Schmölders, 1962). Forest decline was a result of these activities, and thus the idea of sustainable forest management became an issue (see Box 2.1).

10.1.4 Cameralist accounting for taxes, agriculture, and people

Cameralism was a German–Austrian form of mercantilism. It is of interest because it relates three domains of governance to economic actions: (1) the management of royal finances (translated from the Latin *camera*); (2) Polizeysachen (Engl.: police matters), which actually included an official, institutional public policy also controlling economic transactions; and (3) the (public) administrative management of goods, agriculture, and forests. Thus the various university chairs of cameralism can be regarded as predecessors of chairs in business sciences (Schäfer, 1995).

From an environmental literacy perspective, what is of interest is that agriculture was finally seen as an industry. Here, Johann Heinrich Gottlob von Justi (1717–71) was the pioneer:

> I have demonstrated in various writings that the wealth of a country, and the power and felicity of a nation depends primarily on the culture of soil and the flourishing state of the agriculture. The populace, nourishment supply, manufactures, and even commerce relies on the tuft of agriculture, on its solid and immovable ground … (Justi, 1758, p. 152, translated by Scholz).

Justi also emphasized that taxes should only be derived from the yields and not from the assets and estates (Söllner, 2001). In cameralism, the central role of agriculture in nutrition was emphasized. The emphasis

10.1 The preclassical era

Box 10.1 Mercantilism's collateral damage: changed landscapes

Promoting international trade for a nation's prosperity not only required a supply of capital and bullion (gold, silver, etc.) but also adequate transportation infrastructure and war technologies to be successful. Between the sixteenth and eighteenth century, sea ports were built. These allowed for the trading of commodities and goods such as wood for construction, corn, fuel, iron, meat, fish, exotic fruits and plants, wool, cotton, weapons, etc.

There were severe environmental impacts from these activities. The high demand for wood and the deforestation caused by iron and silver mining raised concerns about the sustainability of forests (see Box 2.1). Canals such as the 240-km long Canal du Midi and the 194-km long Canal de Garonne (see Figure 10.1a), commissioned in 1665 by Jean-Baptiste Colbert (who was the finance minister of Louis XIV), as well as the Bridgewater Canal in England, which opened in 1761 and featured an aqueduct crossing the Barton River, changed the landscape, hydrology, and settlement patterns of the surrounding environments. Canals also allowed the manufacturing industry to spread to inland areas and created profits from transportation fees. Naturally, as a secondary feedback loop, agriculture along the canals became more profitable because of improved trading options.

Another type of collateral damage came from mercantilist era trade wars. For instance, the Thirty Years' War (1618–48) was featured by trade wars such as the four Anglo–Dutch wars between 1652 and 1784. Economically speaking, the state-chartered monopolistic trading companies exemplified imperfect competition but received full royal support (Irwin, 1991).

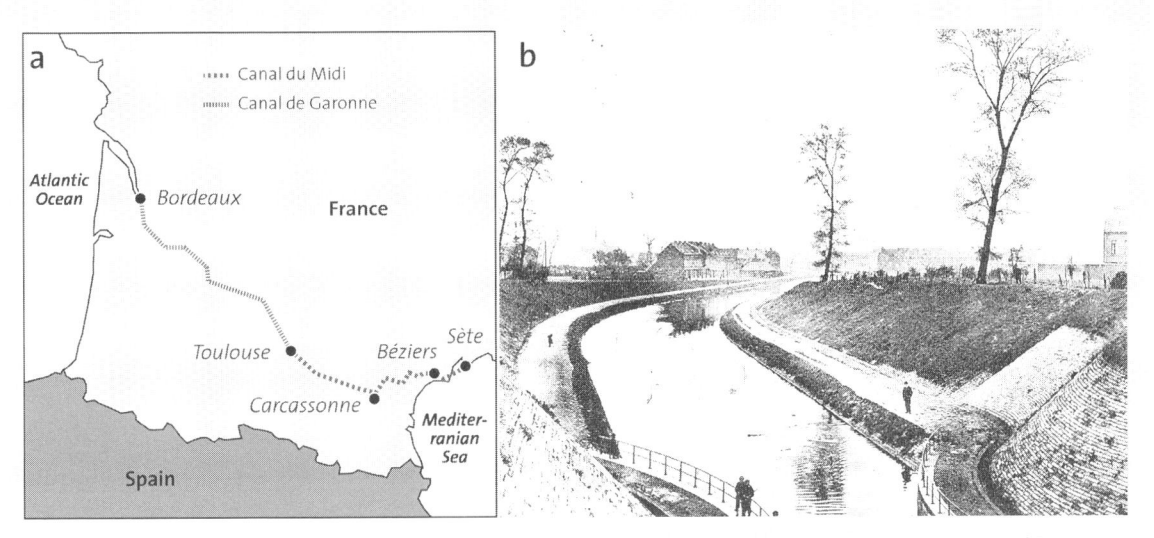

Figure 10.1 (a) Canals required huge investments that changed European trade routes and landscapes in the seventeenth century. (b) The Canal de Tourcoing near the French town Roubaix at the end of the nineteenth century.

on agriculture was also because of the consequences of destructive conflicts such as the Thirty Years' War (1618–48). Societies found themselves in a conundrum, faced simultaneously with poor agricultural production and a cultural norm of promoting population growth (seen as the strength of a nation). Linné's attempts to grow exotic fruits in Sweden (see Chapter 4.9) can be taken as an example. Here, it is interesting to see that Justi also discussed rural settlement structures as well as the trade-offs between dispersed but agriculturally efficient settlements and the clustering of farms in villages owing to lower costs of "public police control" (German: Policey Aufsicht; Justi, 1758, p. 205).

Obviously, the ordered state control took priority over attaining efficient agricultural yield.

10.1.5 Physiocratic economics based on soil and yields of nature

Physiocracy means "domination of nature." Physiocrats assume the existence of a natural order (ordre naturel); that is, of prescribed human behavioral standards based on reason. This is closely associated with the fundamental concepts of liberty, private property, and competition, which served as the basis for the slogan "laisser faire et laisser passer [English: let the market play]." Under

Colbert's (see Box 10.1) mercantilist economic policy, manufacturers in particular enjoyed support, while agriculture was protected and regulated but received no subsidy. The physiocrats had a different point of view and sought to promote primary production as a whole to obtain a "surplus." They regarded land as the only productive resource (Quesnay, 1768). For them, only soil, which was God's gift and natural order, could generate an added economic value.

The physiocrats were the first to have a comprehensive view on the whole economic cycle and to model the quantitative relationships between economic and social variables (Schmölders, 1962). They developed an input–output table in which they linked different economic sectors, for instance, "landowner, agriculture, trade." As a doctor, Quesnay (1694–1774) recognized similarities between economics and the human circulatory system. To improve understanding of that economic system, the flows of goods and money between defined actors were recorded and quantified. Consequently, the money flow linked to the national budget could be calculated by an input–output table (see Table 10.1). This kind of table was useful with regard to the general understanding of economics. Used appropriately, one can notice that profits are a typical part of the net product, which is – in the eyes of physiocrats – generated by agriculture. The social output includes the means of production (and thus the technology). The only way in which growth and accumulation can take place is through a part of the net product (Vaggi, 1985).

As added value could only be gained by deriving yields from soil, taxes were only directed at the landowners. Voltaire (1694–1778) cynically commented on this (Schneider, 2001), stating that a landlord with smallholdings that gave 40 thalers in rent had to pay an "impôt unique" (Engl.: "single tax"), whereas a retired manufacturer with an income of 400 000 thalers had to pay nothing.

In addition to the "Tableau Economique," the physiocrats also made other important contributions to economics. To start with, there is Turgot's (1767/1914) concept of time preference. He points out that a sum of money that is immediately available may well be of greater value than the same sum available at a later time. Furthermore, the money is productive if used at present. For these reasons, the levying of interest is considered justifiable. Secondly, the concept of diminishing marginal means (also Turgot, 1767/1914) must likewise be mentioned. He inferred that the effort involved in the cultivation of crops becomes less and less profitable since the fertility of the soil is exhausted sooner or later. Thirdly, mention must be made of Turgot's (1770) thoughts on the subject of price fixing. He established that the number of actors on the supply side or on the demand side played a decisive role in the fixing of prices. Asymmetry in their numbers would lead to market power.

There was no consensus among the physiocrats when it came to the theory of values. They differentiated, as was customary, between the exchange value (price) of the goods and the utility value. When fixing a price, they strongly referred to the production costs. Basically, they thwarted their intentions. On the one hand, in the era of rapidly growing cities, they sought to prevent the rural exodus (in an attempt to prevent structural change), and on the other hand they simultaneously sought to improve productivity and liberalize trade in general (laisser faire).

For physiocrats, land was one of the most important production factors. The flat rate tax on land ownership proposed by them can be viewed in the same light as a modern tax on revenues from land use. Farmers were seen as *classe productive*, landowners as *classe distributive*, and craftsmen and merchants as *classe sterile*. Given the emphasis placed on the importance of nature, the physiocrats can be regarded as pioneers of the fundamental environmental principles of economics and society (ecological economics).

10.1.6 Key messages

- In the preclassical era, economists studied three issues: the fairness of prices, the question of forbidding interest, and money as a means of payment and fortune
- Ancient Greek economic thinking emerged from household management, which was linked to both divine law and the law of nature. Profiting from money markets was considered immoral
- Regulating markets and introducing metal coins for trade and speculative business characterized scholasticism and economic thought during the European Middle Ages
- The mercantilist economy was a strictly state- and absolute governance-regulated affair featured by international trade and competition. National budgets served to build huge infrastructure facilities such as canals, levees, roads, and bridges to facilitate the transportation of raw materials and trade. Monetary affairs and usury were core issues.

Table 10.1 Quesnay's physiocratic Tableau Economique of national economy as an input–output table.

| | Input | | | |
Output	Agriculture	Craft and trade	Land ownership	Sum of input (production)
Agriculture	200	200	100	**500**
Craft and trade	100	–	100	**200**
Land ownership	200	–	–	**200**
Sum of output (goods)	**500**	**200**	**200**	**900**

Mercantilists strived for export surplus to build up gold stocks as a war chest

- Cameralism, a German–Austrian form of mercantilism, focused on state-administered accounting in agriculture and on population growth and management as pillars of a functioning society without poverty and famine
- Physiocrats considered the soil as the most important and the only economically productive unit. Thus, they took a view of the natural resources-driven economy (see Figure 8.3*). An improvement of societal wealth was assumed to depend on improving the soil quality related to crop yield. They were the first to model business cycle and monetary flows, for which they used a circulatory system model.

10.2 The classical era

The classical era began with the Industrial Revolution and the transformation from agricultural-based to industrial-based societies. Machine-based manufacturing (particularly in textiles), all-metal machine tools, the Watt steam engine, which could directly drive the rotary machinery of a mill, and also eighteenth-century discoveries in chemistry were major prerequisites of industrialization. Thus the classical era was shaped by a technological revolution made possible by engineering knowledge in mining and metallurgy. Here, England was the leading country. The transition was only achieved because of the contiguousness of science and industry (Bernal, 1970). However, because vast amounts of natural resources, like metal and coal, were used, air pollution and environmental destruction resulted (see Box 10.2).

Economic theories in the eighteenth century felt considerably influenced by utilitarian philosophers such as John Locke (1632–1704), Francis Hutcheson (1694–1746), David Hume (1711–76), and Jeremy Bentham (1748–1832). They took a critical look at human behavior, and it was suggested that behavior could not be divided into good and bad. Indeed, it was far more important, for example, that behavior should contribute to human happiness (Hume, 1752). The preclassical economists paved the way for those of the classic era in that their ideas served as a selective source of inspiration. In particular, classical scholars exhibited great interest in those concepts that gave precedence to the freedom of the actors in trade and competition and in those that restricted state intervention.

10.2.1 The liberal economic thought of Adam Smith and Jean-Baptiste Say

Adam Smith (1723–90) is considered one of the most prominent economists of the eighteenth century. He personifies the beginnings of the classical school of economics and his 1776 book, *An Inquiry into the Nature and Causes of the Wealth of Nations*, was a milestone in economic thinking. For him, a barter economy followed the natural order of things where individual interests and the public interest converge in natural harmony. In his book, Smith declares human labor to be the foundation of a nation's wealth:

> The annual labour of every nation is the fund, which originally supplies it with all the necessities and conveniences of life which it annually consumes, and which consist always either in the immediate produce of that labour, or in what is purchased with that produce from other nations. (Smith, 1776/1961, I.I.1)

This statement is reminiscent of Locke's utilitarian view that "the intrinsic natural worth of anything consists in its fitness to supply the necessities or serve the conveniencies of human life" (Locke, 1696, p. 66). Setting aside the shared utilitarian accent, the major divergences from preclassical economic thought were that, in the "classe productive," farmers were substituted by laborers and that the capital of entrepreneurs and the rent from land owning were regarded as the key components of economy. Smith acknowledged that

10 Origins of economic thinking and the environment

Box 10.2 Energy in economics: the curse of coal

Mining is an old subject. Flintstone mining can be traced back to about 40 000 years ago, copper mining to about 5000 years ago, and iron mining to 2500 years ago (Hooke, 2000). Coal use in Britain was first mentioned in 731 by the English monk Saint Bede the Venerable (c. 672–735). Coal smoke was used to drive off serpents (Freese, 2003, p. 17). Lacking environmental literacy about the nature of coal at that time, many considered coal to be living vegetation and suggested applying manure to the coal to help it to grow (Freese, 2003). Hard coal mining started around 900. Hard and soft coal, in addition to peat and charcoal, were used by blacksmiths and other craftsmen in the thirteenth century. This situation dramatically changed when cities grew larger. Wood declined around the cities and coal, which was close to the surface and therefore easily accessible, was used more often. The forest Charter of 1217 assigned ownership of everything that was on the land to the estate holder, which actually fostered urban coal use as more forests became privately owned.

People suffered from severe illnesses owing to the coal air pollution as early as this period. Queen Eleanor fled from Nottingham because she could not abide the smell of coal smoke and feared for her health. An interesting voluntary commitment to reducing the pollution was seen in the thirteenth century. In 1285, King Edward I of England initiated a process for reducing coal smoke in London, and in 1298 a group of London smiths voluntarily declared that they would not work at night because of the health impairments from coal and negative impacts on neighbors (Hughes, 1975). As these voluntary efforts were not sufficient, Edward issued a proclamation prohibiting the use of soft coal in ovens (Mumpower et al., 1986).

Thus, the London smog has a long history (see Figure 10.2a). People were especially troubled by the smell of sulfur detectable in the coal smoke because they believed that sulfur – commonly known as brimstone – characterized the atmosphere of the demonic underworld. In short, in the Middle Ages, coal had quite an image problem, associated as it was with disease, death, and the devil (Freese, 2003).

The Industrial Revolution cannot be thought of without coal. Steam engines required fuel that could not be addressed by available wood supplies, as they were exhausted, nor by hydropower from rivers, which did not produce enough power and were available only at specific locations. In 1690, there was an output of 3.0 million tons of coal; in 1790 it had risen to about 10.3 million, and by 1850 it was about 64.6 million tons (Pollard, 1980). In 1913, 3179 mines were being exploited, and about one million workers were employed in the British coal industry (Buxton, 1978). Coal mining has pervasive health impacts (considered opportunity costs), such as those caused by the smog produced from combusting coal (McIvor & Johnston, 2007).

Coal mining and transportation required the construction of canals, railways, and other infrastructure. We should note that "coal mining is still one of the most significant human earth moving endeavors, accounting for 30% of the mineral production in 1988 … and 25% today" (Hooke, 2000, p. 844). The absolute amount of earth that humans move has dramatically increased since the Industrial Revolution: from about 5 tons of earth per capita annually in Britain in 1900 to 30 tons for US citizens in 2000 (Hooke, 2000). The large amount of coal moved is illustrated in Figure 10.2b.

Figure 10.2 (a) London was densely settled, as shown by the picture of Gustave Doré from 1870, and had much air pollution, which caused environmental diseases from the thirteenth century at the latest. (b) George Walker's 1814 picture of a miner in front of huge hills of manually excavated earth and the first locomotive (i.e. the Blenkinsop) drawing loaded wagons.

the economic rent of land increases in future economic development because of the scarcity of land (Söllner, 2001). Furthermore, Smith stressed that individual interest as such should not be consciously restricted. In his opinion, conscience and empathy (i.e. the ability to put oneself in someone else's position) are all that is required to direct one's own actions.

As with all early economists, Smith started by studying the price of goods. He differentiated between the natural price and the market price and saw the natural price as a kind of cost price or the equivalent of the standard unit costs. On the other hand, he saw the market price as the sum that can be realized through the sale, respectively, of the value that is paid for the purchase of a good. He described the situations arising in the case of shortages and surpluses. He also outlined the way in which the market system always, from a medium-term point of view, swings back towards a state of equilibrium by narrowing the gap between the natural price and the market price. In the event of a shortage (i.e. when the market price is higher than the natural price), suppliers would earn more. This would generate an incentive to produce, resulting in an increase in supply, which would bring the market price back down again.

Smith also studied economic growth. In his opinion, labor productivity is the main driving force behind growth, and this can be improved by increasing the division of labor (specialization) and the accumulation (saving as a virtue) of capital. He envisaged division of labor not only on a national but also on an international level and, as such, was an advocate of free trade.

Smith viewed money as a medium of exchange. Therefore, in his opinion, money did not necessarily have to be associated with precious metals; "paper money" could also be legal tender.

As a result of his concepts, Smith favored a liberal regulatory policy combined with liberal rules on competition. The "invisible hand" of competition would ensure the convergence of individual and collective interests. For him, the dynamics of competition play a central role. He did not explicitly deal with the issue of market equilibrium. Consequently, he saw little leeway for state intervention.

Jean-Baptiste Say (1767–1832) studied the balance between supply and demand. He inferred that prices should assume a regulatory function between these two sides of a coin. Nevertheless, Say's law of market states that supply constitutes its own demand, because everybody produces goods to have money to buy the products of others. Also, supply will find its customers.

Thus, given a free market and a limited unemployment rate (Baumol, 1999), prices adjust to a level where supply and demand are in equilibrium. Say's supply-side economic view places production as a source or cause of consumption, thus setting supply over demand in the hierarchy of economics. But he also stressed the role of use value and argued that price is determined by the intrinsic value of a good and not by the labor time invested to produce it.

10.2.2 Shortage of natural resources: Thomas Robert Malthus'"The Club of Rome in the Classical Era"

Malthus realized that the growth of the population did not follow the same mathematical progression as the growth of food production and foresaw a food shortage (Malthus, 1798). This idea was based on the insights in the economic theory of marginal yield. If the population is growing (exponentially), less productive land or mines will eventually have to be used. Thus, even with intensified labor and improved technology, the production level eventually falls short of the subsistence level. To a certain extent, it is possible to recognize parallels to the conclusions reached by the Club of Rome (Meadows *et al.*, 1972), which identified the same widening gap between limited resources (renewable and exhaustible) and the consumption of resources in a growing economy.

He proposed two ways in which the gaps between these developments could be narrowed. First, there is the possibility of the "positive check." In this case, increasing poverty (e.g. low labor productivity, insufficient purchasing power for food, wars owing to shortages of land or revenue) leads to a decline in the growth of the population which would, in turn, result in an improvement of the situation. As presented in Figure 10.3a, this process would repeat itself in ever-recurring cycles. According to Malthus, the second possibility is the "preventive check." This possibility would involve precautionary control of the population growth, which would, however, demand a change in reproductive behavior.

Malthus gains further relevance in view of today's discussion on poverty and the provision of food, which has been the focus of worldwide attention since the world 2007 hunger crisis. Here the critical question is whether the human species has an infinitely increasing carrying capacity (as is held in the Cornucopian perspective) or whether it is so smart that it can increase the (food) production fast enough to balance the growth. Figure 10.3b is a model of the world population that

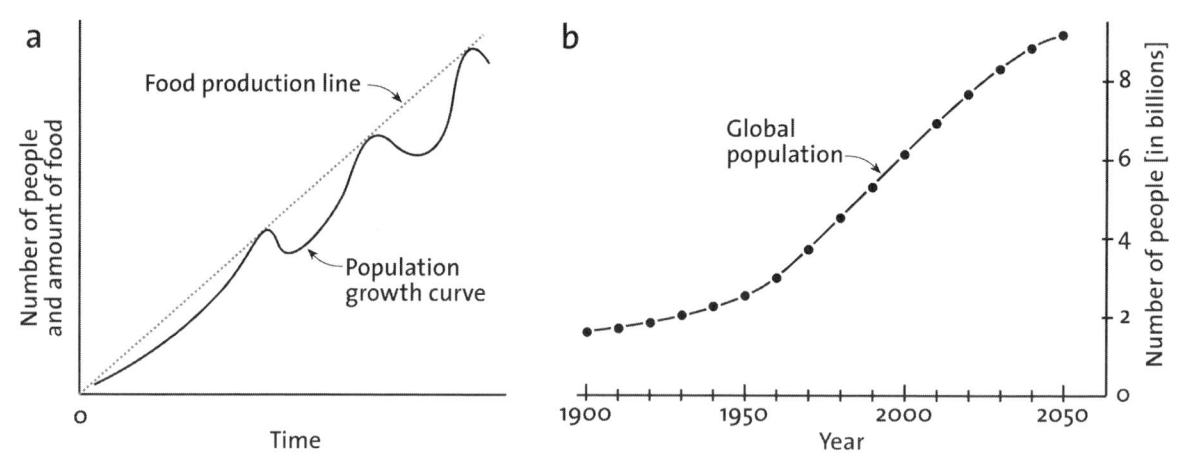

Figure 10.3 (a) Illustration of the Malthusian idea of limits of growth, assuming a linear growth of food production (straight line) and a geometric or exponential growth of population that is reduced by famine or war conflicts over limited resources. (b) Human population growth, which is currently growing steadily and resembles a sigmoid curve (adapted from United Nations, 2004).

hides the fact that different dynamics can be at work in developing and developed countries. Today, environmental economists distinguish between different types of Malthusian scarcity, such as Malthusian stock scarcity (when the resource is fixed in absolute size and extraction costs are constant) or Malthusian flow scarcity (when the resource is fixed in absolute size but extraction costs rise with the rate of extraction; Pearce & Turner, 1990).

10.2.3 Further development in the classical era: David Ricardo

The classical school reached its zenith in the year 1817 with David Ricardo's (1772–1823) *On the Principles of Political Economy and Taxation* (Ricardo, 1817). Ricardo was the first to create a self-contained theory of economics. He explored all the fields of economics that had been studied up until that time: value and price theory, distribution and growth theory, and also foreign trade theory.

With regard to the value and price theory, he adhered to the idea that prices are determined by the costs arising from the production of the goods (value of labor theory based on a subsistence wage). In his opinion, the demand for goods is merely the determinant that dictates the quantity to be produced. However, in the case of non-reproducible goods, it is possible that prices may be influenced by scarcity and demand. Any increase in the costs of labor (the most important production factor) for all other goods would be reflected directly in the price, whereby this can vary depending on the labor intensity. Therefore, price relationships may be staggered over time. An important economic concept is the so-called Ricardo effect (later introduced by Hayek, 1942), which proposes that a rise in wages will encourage capitalists to substitute machinery for labor and vice versa (Gehrke, 2003). The Ricardo effect, while not overly important from an environmental literacy view, already includes an economic rationale for when human labor is substituted by machines.

Ricardo further developed Malthus' concepts relating to the growth and distribution theory. In addition, he assumed infinite labor supply elasticity and introduced negative feedback loops in the market–population relationship. He proposed that if the supply of laborers is too big, wages would fall below the subsistence wage such that people could not afford more children (or would die earlier). The subsistence wage is viewed as the natural minimum wage (Ricardo, 1817). Even in his era, discussions were already under way as to whether the minimum wage should be stipulated by the state. But Ricardo – as Malthus had – argued against helping the poor with charity ("poor law") because this would destabilize society.

Turning his attention to the field of capital, Ricardo examined the question of land rent in detail. We have seen that land had been an important entity in various previous economic theories. But Ricardo noticed that land is a critical issue because of its scarcity. He also realized that land rent depends on the quality of the land. His analysis starts from the assumption that a free market (without subsidies) imposes the same price, e.g.

for corn, and that the price is independent of where it is produced. Thus, those producers who have the best land benefit from the favorable economic conditions and receive a land rent. With increasing population, less productive land will be used. Ricardo states:

> … that corn which is produced with the greatest quantity of labor is the regulator of the price of corn, and the rent does not and cannot enter in the least degree as a component part of its price. (Ricardo, 1817, p. 67)

Figure 10.4 explains this mechanism and shows how the three key concepts, (minimum) subsistence wages, marginal yield, and land rent, interact. Our aim is to explain the relationship between the variables in a simple way, which is understandable to non-economists. Let us consider two parcels of land, one (Figure 10.4a) with high quality Good Land, the other (Figure 10.4b) with low quality Bad Land. If we consider different intensities of labor L (i.e. hours, weeks or minutes of working on one piece of land, e.g. one hectare), then it is clear that for a given intensity, L, one can determine how much an "additional hour" of work would yield. This additional yield is called marginal yield or marginal revenue, which we denote as $MY = f(L)$. If the marginal yield is zero, land use does not provide any return when more labor is added. But people need a minimum wage for living. We call this minimum return subsistence wage w_{min}. This subsistence wage would result from the total amount of work L^* that a person provides for gaining w_{min}. We call the amount of maximum feasible working hours, L_{max}.

As can be seen from Figure 10.4a, in the case of Good Land, if a landowner employs laborers for the subsistence wage (as the marginal yield, i.e. what he would get for an additional hour of paid labor, is above the minimum wage), he would receive a rent. This would not be the case for Bad Land. Also, a landowner could not gain profit if he owned less land than needed for his reproduction if an average working hour provides less than that needed for subsistence. Ricardo's rent model assumes an unlimited and arbitrarily flexible labor market which can provide an infinite labor supply and no other land use competing with agriculture. He did not consider the effect of new technologies on the efficiency of production. He applied a similar model to growth and the allocation of money and – based on his rigid assumptions – concluded that a stagnant economy would result. He also thought that, therefore, investment (e.g. for buying land) would decline as the land available for use and purchase has lower and lower quality.

In addition to addressing these central economic issues, Ricardo also developed a well known foreign trade theory, which is of interest from an environmental perspective. He argued that, when considering two isolated states, for instance England and Portugal, it is beneficial to divide production between the two countries so that each can concentrate on that field of production which generates the lower production costs. If we consider cloth and wine, England and Portugal, respectively, have different opportunity costs for producing these two goods. From this, he derived the advantages of free trade (wine for cloth and cloth for wine). Ricardo's concept of comparative cost advantages requires that the production factors capital and labor are not mobile. Today, these premises do not apply unrestrictedly, given the high mobility of capital

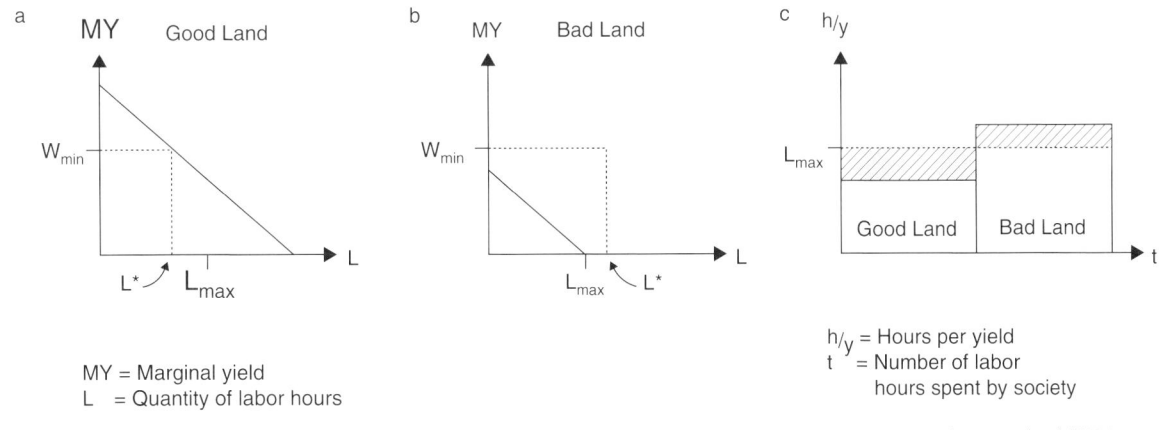

Figure 10.4 Ricardo's model of land rent given minimum wages w_{min}, a maximum labor time per person L_{max}, and marginal yield (MY; see text) for (a) Good Land and (b) Bad Land. (c) Illustration that societies first use Good Land and decrease poverty by exceeding the maximum possible labor time w_{min}.

MY = Marginal yield
L = Quantity of labor hours

h/y = Hours per yield
t = Number of labor hours spent by society

(including land ownership) and high workforce mobility. Nevertheless, international trade has the potential to promote prosperity, just as it did back then.

The natural environment changed dramatically at the beginning of the industrial age, in particular in Britain. Natural resources in primary and secondary production, in particular in agriculture, lost economic importance. This is nicely reflected in the following quote by John Stuart Mill (1806–73), which we present in three parts:

> It is only in the backward countries of the world that increased production is still an important object; in those most advanced, what is needed is a better distribution … There is room in the world, no doubt, and even in old countries, for a great increase in population, supposing the arts of life to go on improving, and capital to increase. But even if innocuous, I confess I see very little reason for desiring it. The density of population necessary to enable mankind to obtain, in the greatest degree, all of the advantages both of cooperation and of social intercourse, has in the most populous countries been attained. A population may be too crowded … (Mill, 1857/2004, Book IV)

Here, Mill identified the decreasing role of agriculture (primary production), and implicitly the advantages that countries from the north can have by setting primary production to the south. The following quote takes a close look at the change from a nature-driven to a man-driven environment (Messerli *et al.*, 2000), and – when referring to the term satisfaction – indicates that Mill is taking a utilitarian view.

> Nor there is much satisfaction in contemplating the world with nothing left to the spontaneous activity of nature: with every rood of land brought into cultivation, which is capable of growing food for human beings; every flowery waste or natural pasture ploughed up, all quadrupeds or birds which are not domesticated for man's use exterminated as his rivals for food, every hedgerow or superfluous tree rooted out, and scarcely a place left where a wild shrub or flower could grow without being eradicated as a weed in the name of improved agriculture. (Mill, 1857/2004, Book IV)

That the natural–biophysical environment may have some value is reflected in the last part of the quote. Here even the aesthetic value, and the essence of Wilson's suggestion (see Chapter 2.2) that the world would not die if population and economic growth spoiled half of the species, is somewhat anticipated.

> If the earth must lose that great portion of its pleasantness which it owes to things that the unlimited increase of wealth and population would extirpate from it, for the mere purpose of enabling it to support a larger, but not a happier or better population, I sincerely hope, that for the sake of posterity, they will be content to be stationary long before necessity compels them to it. (Mill, 1857/2004, Book IV)

Thus, Mill was "foreshadowing later developments in environmental economics" (see Chapter 11.4) and "the thinking of conservationists" (Perman *et al.*, 2003, p. 6).

10.2.4 Key messages

- Adam Smith's classical economics was shaped by the settings of the industrial revolution and manufacturing. It focuses on human labor, capital, and land resources as fundamentals of economics and promotes a free trade market. Therein, land is considered a scarce natural resource
- Economic and societal impacts of a limited carrying capacity were first systematically outlined by Robert Malthus, who stated that populations grow faster than the technological progress and efficiency of producing food
- David Ricardo focused on the relative value of land. He introduced the idea of land rents that were dependent on soil quality and concluded that population growth results in land that provides sufficient yield for landowners to survive. Ricardo also explained international trade as a consequence of efficiently using the sources of nature, for instance, for producing wine in the south and sugar beets in the north.

10.3 Karl Marx's socialist economic thought

Karl Marx (1818–83) has been and is one of the most and most controversially discussed economists, sociologists, and political reformers. From an economic theory perspective, Marx followed the classical ideas of Smith and Ricardo when considering labor and capital (e.g. ownership of land or the means of production) as the key variables of economics. Further important concepts of his approach were scarcity, as well as his relation of use and exchange values.

Marx developed his economic theories during a time when manufacturing, industrial production, mining, and burning coal in Britain and Central Europe was creating terrible societal and working conditions (see Box 10.2). These conditions drew the attention of mill owners, industrialists, and manufacturers such as Robert Owen (1771–1858) and Friedrich Engels (1820–95), who contributed to the theory of socialism, putting forth the idea that private property is the major, or even ultimate, source of societies' illnesses. For instance, at the age of 24, Friedrich Engels (1845/1972) conveyed to Marx that Manchester represented the vision of a social hell, which changed the prevalent conception of the "old hell" and consequently also the whole hierarchical cosmos of Old Europe. Child labor, poverty, famine, air pollution, and public health were critical issues of the time, as described in Owen's book, *Observation on the Effect of the Manufacturing System: with Hints for the Improvement of those Parts of it which are Most Injurious to Health and Morals*, dedicated most respectfully to the British Legislature (Owen, 1817) and the references to public health reports in Marx's major economic book *Capital* (Marx, 1887/1990; see Figure 10.5).

We have already briefly discussed Karl Marx's ideas in the sociology chapters. For Marx, labor was a key concept from multiple perspectives and he saw it as the unique source of value.

> Labor is the source of all wealth, say political economists. It is labor – besides nature – which provides the matter for changing it to wealth. But it is infinitely more than that. It is the first basic condition of any human life, but to such a degree that we have to state in a certain sense: It has created human itself. (Marx, 1876/1971, p. 444, translated by Scholz)

Labor – in combination with capital (i.e. machinery or technology which results from former labor) and with natural matter (German: Naturstoff) – generates use value. Here, Marx's utilitarian perspective originating from Bentham (1789/2005) became apparent.

Besides the abovementioned health effects of labor, the effects of human labor on the environment is of interest from an environmental literacy point of view. Here, the concept of "exchange of material matter" (German: Stoffwechsel), which later came to be known as metabolism, is important. Thus, we can see that Marx had insight into the transformations made by human labor in environmental systems and considered natural resources (material matter) as a prerequisite of economic activity. This is illuminated in the following quote.

> In the labour-process, therefore, man's activity, with the help of the instruments of labour effects an alteration, designed from the commencement, in the material worked upon. The process disappears in the product: the latter is a use value, Nature's material adapted by a change of form to the wants of man. Labour has incorporated itself with its subject: the former is materialised, the latter transformed. (Marx, 1887/1990, p. 156)

Marx considered natural resources as worthy, cost-free services. They can become scarce or resist the efforts of labor to produce surplus value. As Ricardo had already noticed, the quality of land can decline and lose fertility; mining becomes less and less productive as resources are depleted.

Marx was already aware of the exhaustion of soil in England and the necessity of worldwide material fluxes, for instance the "forced … manuring of English fields with guano" (Marx, 1887/1990, p. 348). Marx was also aware of the industrialization of agriculture and the changed relation of human to soil:

> The soil is to be a marketable commodity and the exploitation of the soil is to be carried on according to the common commercial laws. There are to be manufacturers of food as well as manufacturers of twist and cottons, but no

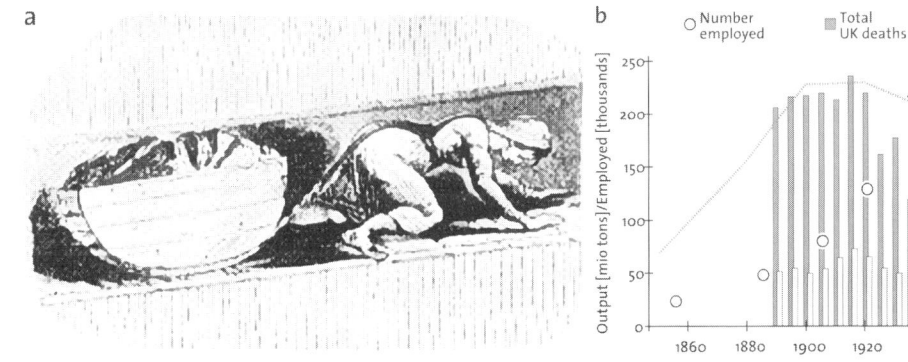

Figure 10.5 (a) Moral (from Del Col, 2002) and (b) public health impacts of industrial labor in Great Britain (data taken from McIvor & Johnston, 2007) have been drivers of socialist economic thought.

Box 10.3 Incorporating transportation costs in rents: von Thünen circles

Johann Heinrich von Thünen (1783–1850) was a nobleman, social reformer, and economist with a good knowledge of mathematics, who was a successful, practical leasing farmer in the plains of Mecklenburg. He was a pioneer in many domains. For instance, he developed a wage theory by linking the subsistence value w^* and the value of the labor $v(L)$ by assessing the wage s with the formula $s = \sqrt{w^* * v(L)}$.

An interesting idea in von Thünen's book *Isolated State* might, in some respects, reflect and idealize his farmstead, which was situated in the middle of the great East German plains. Therein, he suggested a thought experiment: "Think about a very big town located in the middle of a fertile plain without navigable rivers or canals. There is a homogeneous soil allowing for cultivation in any place. At large distance, the plain ends in an untilled wilderness separating the town from the rest of the world" (von Thünen, 1910, p. 11, translated by Scholz). Given such a setting, he postulated seven concentric circles of land use, starting with vegetable gardens, followed by forest (as the weight of the wood used for construction and fuel would require close vicinity), and then corn cultivation and finally grazing (see Figure 10.6). Mining and salt refineries are close to the city, as it is the only location of the market.

The von Thünen rent graph defines the rent of land R (e.g. per acre) as a yield function which depends on the market price of the products per unit p (e.g. on what is harvested on an acre) subtracted by the production costs per unit c and the product of the freight rate f (i.e. the transportation costs per unit of product per mile) times the distance in miles m from the central market, thus: $R = Y(p-c-fm)$. Von Thünen can be considered as a pioneer in regional sciences focusing on transportation as transaction costs.

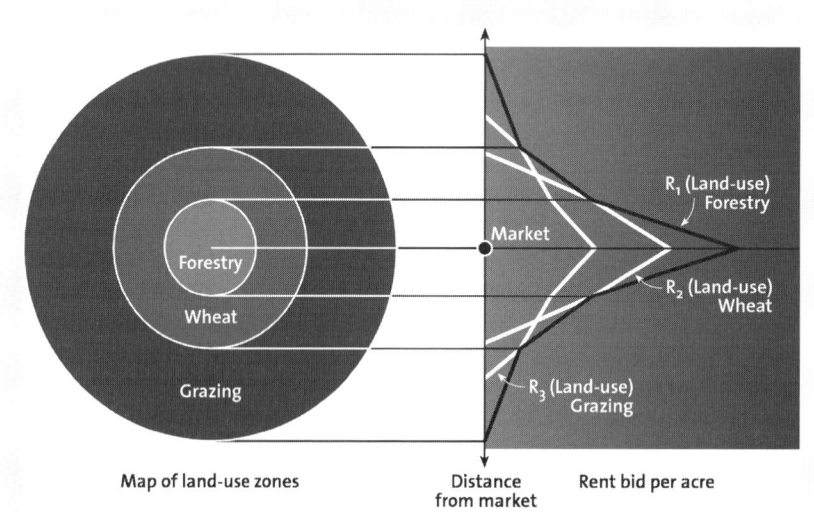

Figure 10.6 The von Thünen land use circles and the rent functions for different land uses depending on distance to market.

longer any lords of the land. (Marx & Engels, 1851–1853, p. 333)

Accordingly, nature and natural resources played an important role in Marx's economic theories, something that has not been the case with most economists. This becomes evident in the following quote:

> Man *lives* from nature, i.e. nature is his *body*, and he must maintain a continuing dialogue with it if he is not to die. To say that man's physical and mental life is linked to nature simply means that nature is linked to itself, for man is a part of nature. (Marx, 1887/1990, p. 348)

In addition to natural resources, Marx saw technology as important (see Figure 8.5*). The value of a good is defined by the labor time which – given a certain stage of technological development – is necessary to produce it. The surplus value is the ratio between the value of labor and that of constant capital. The profit of a manufacturer's land or mine and the exploitation of workers results from a rent of the labor.

Technology is seen as a major driver of societal and economic change. Because technology development results in using more constant capital and less labor, capital is needed to produce use values. By this law of declined surplus value, capital becomes more concentrated, small and medium enterprises cannot survive, and in the end only a few monopolies own land. This economic rule would lead to the end of capitalism; that is, to a private ownership of production means. Such a state would be called socialism, which would result in a class-free society, which – according to Marx's theory of historical materialism (Marx, 1859/1961) – represents a natural law of societal development. Marx's scholars discussed why this theory did not become true. For example, the German minister of finance during the Weimar Republic, Rudolf Hilferding (1877–1941) suspected a self-stabilization of capitalism by international cartels of banks (Marx, 1859/1961). Others, such as Rosa Luxemburg (1871–1919), thought that the continued existence of capitalism in the emergence of new markets in developing countries allowed extra profits because of ultra-low wages and subsistence values (Schmölders, 1962).

According to Marx, the evolution of societies as well as the drivers of societies are based on technology, which is "the *forces* of production" and of social organization such as "the *relations* of production", as a special type of societal trajectory (see Figures 8.3* and 8.5*).

10.3.1 Key messages

- Karl Marx's theory considered the private ownership of production means as capitalism. According to technology development, he assumed that the rate of variable capital (labor) and fixed capital (i.e. private ownership) becomes smaller (law of the progressive decline of the surplus rate), and that socialism would follow capitalism (law of historic materialism)
- Marx's theory was motivated by negative collaterals of industrial labor on health and justice, and recognized processes of metabolism of material matter
- Marx considered environment and natural resources as fundamental for economy.

10.4 The neoclassical era

Neoclassics laid important foundations for modern microeconomics. Let us first look at how the label "neoclassical" developed. The term "classical" was coined by Karl Marx in 1847 to brand Ricardo's formal mathematical approach. Marx contrasted the classical with "romantic economics" or with his "close to the people economics." The label neoclassics denotes a period "somewhere between 1776 and 1890" (Colander, 2000, p. 130). It was introduced by the US economist Thorstein Veblen (1857–1929; Veblen, 1900), an outsider who was critical of the upper class living habits (e.g. inconvenient way of dressing), as a negative description of Alfred Marshall's (1842–1924) approach. Veblen, along with Arthur Cecil Pigou (1877–1959), Francis Ysidro Edgeworth (1845–1926) and John Maynard Keynes (1883–1946), built the "Cambridge neoclassics" (Aspromourgos, 1986).

The new idea of the neoclassics has been called the marginal utility school, which was formed by William Stanley Jevons (1835–82), Carl Menger (1840–1921), and Léon Walras (1834–1910). Marginal utility refers to the cardinal utility of the last unit of a product that has been bought. Cardinal here means that we can measure the utility or satisfaction of consuming goods and services – like physical weight – on an absolute scale. This concept of marginal utility strongly referred to Daniel Bernoulli's (1700–82) theory of declining marginal returns but has also a strong relationship to the Weber–Fechnerian law of psychophysics (see Chapter 6.1).

Augustin Cournot (1801–77), Hermann Heinrich Gossen (1810–58), and Johann Heinrich von Thünen (1783–1850; see Box 10.3) were other leading neoclassicists. Colander nicely composes the key attributes of neoclassics and states:

> [Economic transaction starts] … with individual rationality, and the market translates that individual rationality into social rationality. (Colander, 2000, p. 134)

Neoclassics further focuses on scarce resources at a given time. It stresses utility and thus demand (thus contrasting Say's supply approach, see Chapter 10.2), and it strongly relies on calculus as a key modeling tool. The approach postulates that economic agents build rational preferences among outcomes and decide according to cardinal utility functions. Here a hedonist perspective, including cost–benefit considerations in the line with Bentham's pleasure–pain perspective is introduced. Pleasure or utility and expenditures of activities have to be traded off in a way that the "sum of beneficial use of the whole life becomes optimized" (Gossen, 1854, p. 1, translated by Scholz).

Further, the idea that market solutions can be described by the physical concept of equilibrium is a key idea of neoclassics. Using mathematical models based on calculus, neoclassicists worked with simple, idealized assumptions. One such assumption was that individuals, as well as companies, do not only maximize their utility or profit, but they do this as economic agents having full information. This can be related to the idea of optimal depletion of the environment, which became a key idea of resource economics.

10.4.1 Consumption theory

As was already the case in parts of the classical school of thought, value theory represents a central aspect of the consumption theory. According to Gossen, the value of a good depends on its subjectively perceived utility (see Chapter 7.2). Following the marginal utility concept, the exchange value of a good (price) is to be treated as equivalent to the utility of the last unit of this good which was consumed at a given income. This reveals that the price is not determined by the so-called *importance* of goods, such as basic water or foodstuffs, but rather by their *availability*. Consequently, so-called unimportant goods for subsistence, such as diamonds, are not cheap if they are considered valuable and they are scarce. We speak about the paradox of value or the water–diamond paradox.

The consumer maximizes the utility from which he benefits and the more he consumes of a good, the lower its marginal value becomes (saturation according to Gossen's First Law). Because of the scarcity of these goods, consumption is so low that the marginal value is high. A nice example of this market mechanism comes from the 1980s, when the European meat market became oversaturated, which caused image and subsequent market problems for this sector (Vonalvensleben, 1995).

For the consumer, maximum utility at a given income is reached when the utility relationship between the goods consumed corresponds to the relationships between the prices (Gossen's Second Law). It follows that an economic subject maximizes his utility/benefit when the marginal value of the money spent on all the goods consumed is the same, that is, when the marginal utility of the goods exhibits the same behavior as their prices; or, to express it in other terms, the marginal utility of all goods consumed are equal.

Marshall (1890/1920), Fisher (1896), Hicks (1939) and Samuelson (1947) expanded this concept. Marshall (1842–1924) derived individual demand functions from the utility functions. The more a consumer has already used of a good, the lower his marginal utility becomes and the less he is prepared to pay for an additional unit of the good. In the book *Principles of Economics* Marshall (1890/1920), derived the demand function in a market as an aggregation of the individual demand graphs and used price elasticity to describe demand reactions arising from changes in prices. Marshall introduced the concept of consumer rents. Here, Marshall proposes that the price of a good – which has validity to all consumers – is defined by the marginal utility of the last consumer who bought the good.

> The price which a person pays for a thing can never exceed, and seldom comes up to, that which he would be willing to pay rather than go without it. … The excess of the price which he would be willing to pay rather than go without it, over that which he actually does pay, is the economic measure of this surplus satisfaction. It has some analogies to a rent; but is perhaps best called simply consumer's surplus. (Marshall, 1890/1920, p. 191)

Though not always adequately recognized, the concept of consumer rent was a precursor to the willingness-to-pay principle (Hanemann, 1991), which became important in assessing the value of environmental services (see Chapter 11.4). Thus the price consists of a willingness to pay and a consumer rent component, which allows us to measure welfare.

Fisher (1867–1947) was the first to differentiate between *substitutive* and *complementary* goods. In the case of two substitutive goods, a surplus in consumption of the first good reduces the marginal utility of the consumption of the second good. In the case of complementary goods, the marginal utility of the second good would rise. Furthermore, it involves the classification of goods depending on demand reaction in the event of changes in income, called income elasticity. In the case of a relatively superior good, income elasticity is greater than 1. This means that, as income increases, the share of the income spent on the respective product also rises with the same share at least. The opposite applies to relatively inferior goods. In addition, he also differentiated between absolutely superior and absolutely inferior goods. In contrast to the previous concept, it is interesting to note the fluctuations in quantities resulting from changes in income. The criterion for differentiation is income elasticity and whether it is positive or negative.

A new theoretical view on the utility function came from Hicks' book *The Theory of Wages* (Hicks, 1932). Here, when substituting the cardinal, real number-based utility concept with indifference curves (Edgeworth, 1881/1932), Hicks assumed that, given a certain overall price, a consumer is

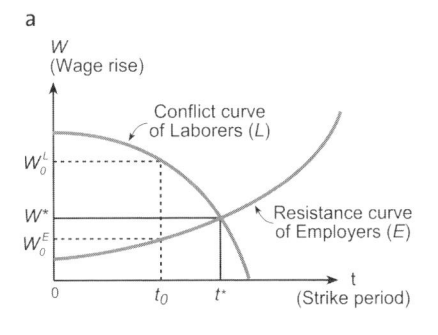

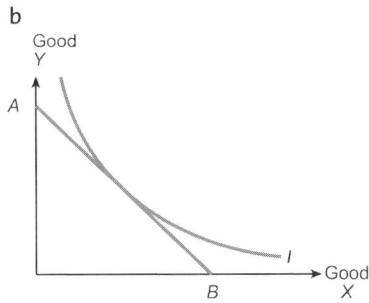

Figure 10.7 Hicks indifference curves as an example for non-cardinal utility modeling, including (a) the equilibrium thought for labor negotiations. (b) The indifference curve I of a consumer for quantities of two goods Y and X. See text for explanation of symbols.

indifferent between the different combinations of goods (see Figure 10.7b).

An example of indifference curves presented by Hicks deals with wage negotiations as part of collective action. He assumed that both parties, the laborers and the entrepreneurs, are bargaining for a wage rate w and a strike period p. Without strike (see Figure 10.7a), the employers would provide a small wage rise and the laborers would claim a very high wage rise. Hicks assumed that – given a certain strike period t_0 the laborers would be willing to accept wages above w_0^L and the employers would provide a maximum wage rise of w_0^E. The indifference curves, which represent the conflict curve of the laborers and the resistance curve of employers, include all points where the respective parties switch between accepting and rejecting if one of the two parameters changed. The point of intersection between these two curves, (w^*, t^*), represents the situation where the parties come to an agreement. Published in 1932, Hicks' book was somewhat eclipsed by the 1929 world economic crises, which introduced completely irregular economic constraints. Nevertheless, starting from indifference curves, it became possible to derive the optimum consumption or demand function based on price variations (Söllner, 2001). According to Hicks, a consumer consumes that combination of goods whereby the marginal utility divided by the price is equal.

However, since only relative prices play a role in the indifference graph analysis, the absolute price level must be determined arbitrarily when constructing the demand graphs (in which absolute prices are decisive). Taking note of Slusky's (1915) view, Hicks investigated in greater detail the distinction between substitution effects and income effects in case of relative price changes.

Consumption theory was also extended by Paul A. Samuelson (1948). His contribution focused on the preferences exhibited by economic subjects. Conclusions relating to preferences are drawn on the basis of consumption behavior. The object was to derive a consumption theory based solely on consumption behavior and independently of utility functions. Although this approach was originally regarded as a substitute for the utility concept, it finally proved to be an important addition in that implicit requirements were put forward concerning the utility concept. Modern economists do not refer to the utility concept to explain behavior because it is a normative concept used to describe rational economic subjects (utility maximizing economic subjects). Instead, concepts such as bounded rationality emerged (see Box 1.4).

The consumption theory is directly and exclusively associated with private goods which, by definition, have a price. Rivalry and excludability (see Table 11.1) are prerequisites for this. Therefore, consumption theory only dealt with goods that either were associated with rivalry or excludability. It follows that common goods, club goods or public goods, goods that are central to many environmental problems, were not an issue.

10.4.2 Production theory

Von Thünen is the pioneer of production theory, which deals with the question of the optimal conditions for production. He developed the marginal productivity theory of factor payment in his work *The Isolated State* (see Box 10.3), which was formalized by Wicksell (1851–1926). Just like consumption theory, the marginal value principle is the focal point of production theory. The central element is the production function, with declining marginal returns and increasing utilization of resources (Söllner, 2001). It is assumed that companies maximize their profits. Consequently, they use production factors and other resources so that their price corresponds to their product's marginal value. The factor payment corresponds to the company's marginal productivity.

Production theory consists of three partial concepts: the optimal *production quantity*, the optimal relationship between *production factors* (e.g. in agriculture labor, capital, and fertilizer), and the optimal *production portfolio*. The price and the marginal productivity, substitution relationships, play a central role in connection with each of these optimality requirements. One refers to decreasing, constant or increasing economies of scale depending on the way in which the output changes in relation to the input. Optimal factor utilization quantities can be derived from the concepts of production theory. As a result, cost functions can be defined and the so-called cost theory can be established (total costs, average costs, fixed costs, variable costs, marginal costs).

John B. Clark (1847–1938) also interpreted the marginal productivity theory as a distribution theory whereby he regarded the factor utilization quantities (that is, the factor payment) as the basis for the distribution of the social product. Hicks (1932, see Figure 10.7) made some important contributions to production theory. He defined substitution elasticity, which correlates the relative change in the factor utilization relationship with the relative change in the factor price relationship. The influence of factor price changes on functional income distribution can be determined from this. Hicks also investigated the influence of technical progress on optimal factor utilization and the distribution of income. Production theory also focuses on private goods and their prices. As with consumption theory, rivalry and excludability are prerequisites of production theory, so it follows that common goods, club goods or public goods were not an issue in this domain.

10.4.3 Price theory

> In the theory of a capitalist market economy, price has been one of the central problems, if not *the* problem. (Rothschild, 1947, p. 299)

Price theory is the third fundamental concept of marginal analysis. It deals with the overall supply, which, at a given demand, determines the price.

The differing environments governing competition between the actors are described by Cournot (1838) and von Stackelberg (1948). They consider markets with only one (monopoly), many small (polypoly), and few (oligopoly) suppliers. Prices result as equilibriums from the interactions among different competitive firms. The question is what amount of output should be produced in non-cooperative markets. The language and tools of analysis are differential equations. Here the modeling followed Newton's basic law of mechanics; namely that every action is opposed by an equal reaction. The goal was to determine an equilibrium price. Marshall centered his attention on price formation in the interplay between overall supply and demand from both a short- and long-term point of view. This revealed a change in the variability of the production factors.

Price theory is likewise directly and exclusively associated with private goods. Prices are formed by rivalry and excludability processes, so it follows that common goods, club goods or public goods were not an issue. Environmental resources or external costs have not yet been an issue. To better understand neoclassical thinking, we look at the economy through the eyes of Edwin G. Nourse, one of the leading economic advisors in the post World War II era. Nourse portrayed economic activity to be much like moves in a chess game (see Box 10.4), with actors following full rationality (see Box 1.4).

> While, of course, the conditioning environment imposes rigorous limitations on the price administrator's freedom of action in a capitalist society dedicated to "free enterprise," he devises and implements business plans in ways broadly similar to those of military command. A general must operate within the limitations of the terrain on which he fights and of the personnel and material at his disposal – to say nothing of meteorological conditions. But at the same time, much depends too on the strategy which he and the high command devise and the specific tactics by which he and his officers seek to carry it out. It seems appropriate, therefore, to discuss price policy in terms of business strategy and tactics. (Nourse, 1941, pp. 189–90)

10.4.4 Production factors in marginal analysis

Neoclassics deals with the production factors labor, capital, and land in marginal analysis. Households are thought of not only as demanders of goods, but also as owners and suppliers of production factors.

Labor time is primarily a reduction of the benefits of leisure time and – according to the marginal utility approach – has a declining marginal utility. The activity of work does not generate any benefits in the form of

10.4 The neoclassical era

Box 10.4 Economics as a kind of chess game: Nash equilibria

Economic agents in classical and neoclassical microtheory were fully rational actors, who could process an infinite amount of information. When making choices, they were able not only to determine what their best alternative would be reliably, but could also often postulate what (all) other market players would be doing. Many of these models postulate that market actors have complete information about the other market actors' utility functions.

Such an idea underlies the Cournot equilibrium (Cournot, 1838), from which the famous Nash equilibrium (Nash, 1951) emerged. Here, we explain the Nash equilibrium to demonstrate what type of actor behavior is postulated in game theory. In general terms, a Nash equilibrium of a game as a model of an economic situation is a tuple of the strategies of all players so that each player's strategy is an optimal response to the other players' strategies. Such a conception asks for chess-like thinking.

Take two competing market actors, which – in (economic) game theory – are called players. In the simplest case, each of them has only two strategies. Let us assume that they are each about to negotiate a contract for selling beer at a festival and that the contracts will prescribe the price per beer for the duration of the festival. Figure 10.8a presents the situation. A Cournot–Nash-like economic model would ask that both firms know in advance the outcomes (or utilities) that would result from the choice of the price strategy. The solution of this non-cooperative game structure follows the following "chess-like" strategy: the other player – like me – has all available knowledge and his target is to gain the maximum out of the market. If my Firm A chooses the price £5 (and Firm B might get to know my strategy), it would be in Firm B's best interest to do the same. In this case, the result would be a zero win and loss for both sides. If my Firm A went for £7, then Firm B goes for the lower price to get the majority of customers and I will lose. As the same argument can be made for Firm B, both would stay with the low price. Thus – given the rationale of game theory (see Chapter 14) – the unfavorable solution (0,0) results.

Figure 10.8b presents a saddle point describing the minimum of a row y that is also the maximum x of a column in the payoff matrix – it represents the shape of a saddle. In two-person zero-sum games, as illustrated in the example above, such "Nash equilibria" and "saddle points" are synonymous terms. Here x represents the strategies of Firm A, y the strategy of Firm B and z the payoff of Firm A. As we have a zero-sum game, $-z$ is the payoff of Firm B. The saddle point is the outcome that "game theoretical rational players" choose as it is the minimum of a row that is also the maximum of a column in a payoff matrix. We can illustrate saddle points using a geometric illustration (see Figure 10.8b). Independent of what Firm A is doing, Firm B chooses the strategy to bring the payoff down as far as possible. Thus the saddle point $\phi(x,y)$ at the point (X^*, Y^*) represents the point where the minimum payoff of Firm A is a maximum and vice versa for Firm B.

a

	Firm B	
	Price £5	Price £7
Firm A — Price £5	0/0	50/–10
Firm A — Price £7	–10/50	20/20

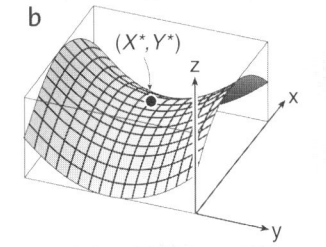

Figure 10.8 (a) A 2 × 2 matrix game and (b) visualization of the saddle point.

pleasure or anything similar. It generates money, which is transformed into utility through consumption. The utility maximization problem results from the aggregate of the benefits of leisure time and the utility of the wage in the form of goods consumed. The relationship between leisure time and work is determined on the basis of the marginal principle by the equivalence of the two marginal utilities.

Von Thünen and Marshall dealt primarily with the production factor land. Von Thünen is the founding father of location theory, which, in contrast to Ricardo's theory, gives due consideration to the distance between production locations and consumer centers (see Box 10.3). For his part, Marshall placed emphasis on balance in land use. In this case, balance is achieved when marginal productivity is identical in all possible utilizations. He used this as the reasoning behind the premise that land rent should be included in the (price determining) costs, such as wages or interest. Marshall spoke of a permanent scarcity rent drawn by landowners. Of all the production factors, land has the strongest relationship to the environment.

However, changes in the quality of this resource were not the subject of analysis in the neoclassical era. Yet awareness about land scarcity was definitely a growing concern and led to formulation of the law of diminishing returns when utilizing already used and newly cultivated land.

> … our law states that sooner or later (it being always assumed that there is meanwhile no change in the arts of cultivation) a point will be reached after which all further doses will obtain a less proportionate return than the preceding doses. (Marshall, 1890/1920, p. 153)

The capital theory provides the framework for dealing with the production factor capital. Wicksell (1898) concentrated on the optimal point in time for the sale of capital stock. In his theory of capital and interests he acknowledges that goods provide a certain natural rent or natural rate interest, which can decline after a certain point (e.g. for wine). The natural rate of interest can be contrasted with the interest on the capital market where interests are a means to govern the business cycle by the basic interest rates of national banks.

An important idea from an environmental literacy point of view is Fisher's theory of interest. His ideas refer to Eugen Ritter von Böhm-Bawerk (1851–1914), an Austrian economist who also served as minister of finance. Böhm-Bawerk used psychological arguments when supposing that economic agents underestimate the value of a good which will be available in the future. This idea is also linked to a technology-related increase in efficiency, which makes a future product cheaper, a conclusion drawn by the American economist Irving Fisher. Even today, Böhm-Bawerk's comprehensive capital theory is largely undisputed. In modern terms, the central element of the theory can be called cognitive discount rate. The Böhm-Bawerk idea is one side. It refers to the relationship between the capital used for consumption and capital staked as an investment, which pays back in the future. Fisher had already noted that this is relevant both for an individual and at a societal level.

10.4.5 Key messages

- Neoclassicists postulate that economic agents are rational, hedonistic (maximizing total life utility), cost–benefit-oriented people, and valuate goods based on Bernoulli's marginal utility principle. The subjective value (use value) and scarcity define price and consumption. The price of goods, including environmental goods, is relative and depends on what economic actors value
- The main production factors are labor, capital, and land. Market as well as surplus value theories are based on these variables. Von Thünen provided the origin of location theory when including transport costs
- Neoclassicists refer to private goods where excludability and rivalry feature on the demand side. Common goods are not an issue in neoclassics. Scarcity of material, or biophysical scarcity, is thought to be regulated by demand–supply market mechanisms
- Marshall's concept of price as a composite of a willingness-to-pay component plus a consumer surplus or consumer rent is of interest also for common goods or environmental services
- Markets function according to the Newtonian action–reaction principle between supply and demand, and are ordered by equilibrium theory, assuming competing (rivaling) and mostly completely informed subjects
- Neoclassical theory does not see value in the material–biophysical environment itself. The issue of public goods was not incorporated in economic theories of this period.

10.5 Welfare theory

Welfare economists took a societal perspective and their ideas were grounded in a normative orientation that focuses on generating high utility for a society. They started from a utilitarian perspective, as has been formulated in Gossen's laws on the individual level (see Chapter 10.4). Facing the difficulty of measuring individual utilities, they considered money as a proxy for subjectively assessing utility (Ziegler, 2008). Welfare economists aspired to maximize the utility of society as the sum of the utilities of all individual members.

Marshall approached welfare from the perspective of consumer and producer rents. He investigated the effects of a monopolization of supply on the consumer rent, as well as the influence exerted by taxes and subsidies. Marshall's scholar Pigou (1932) can be seen as the founder of old welfare economics. Pigou based his work on partial models (i.e. looking at parts of the economic system).

From an environmental literacy point of view, the Pigovian tax is very important. Pigou recognized the

10.5 Welfare theory

Box 10.5 Internalization of negative external effects: Pigovian taxes

Imagine a factory whose operation pollutes water that negatively affects a downstream laundry. The investments for cleaning the water are additional, external costs. According to Pigou, the state has to charge the polluting company. This tax should be exactly as big as the external abatement costs (see Figure 10.9a).

In general terms, Pigou means that "for every industry in which the value of the marginal social net product is less than that of the marginal private product there will be certain rates of taxes, the imposition of which by the state would increase the size of the national dividend and increase economic welfare; and one rate of tax, which would have the optimum effect in this respect" (Pigou, 1932, p. 91).

Figure 10.9b explains this in terms of the marginal utility school. The x-axis presents the pollution (by Factory A) in kilograms of pollutants emitted. The y-axis represents the costs. We can find two functions. One is the marginal abatement cost $MD(x)$, which represents the marginal cost by avoiding one unit of pollution for Factory A, for instance by using less chemicals or by producing less. The function $MC(x)$ represents the marginal external costs. An omniscient government would set the tax at t^* where the marginal abatement costs (which are a kind of demand function) meet the marginal external costs (which is a kind of supply function). Then, the tax would exactly compensate these costs at the point x^*. One can also easily see that better environmental technologies which would provide the emissions with a lower price would induce a reduction of taxes. The Pigovian tax is a state-controlled means to include the costs from negative externalities into the market.

The emission trading among companies of a state per produced unit of a product is an alternative to the Pigovian tax. The state assesses the optimal quantity of pollution according to the given external and abatement costs and market for the industry of a country. The competition among the companies can work to decrease the abatement cost function. Sulfur trade starting in the 1970s in the USA following the Clean Air Act can be taken as example (see Box 13.2).

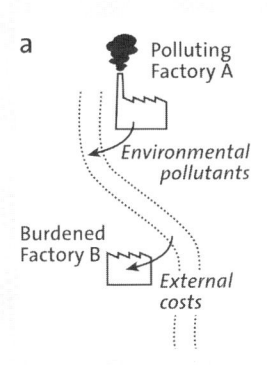

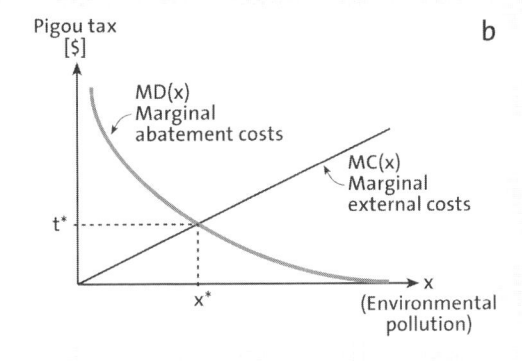

Figure 10.9 (a) The Pigovian tax rule results from the idea that the costs of company A's pollution for the public (represented by factory B) should be taxed. (b) Pigovian tax t^* as the intersection of the function of marginal abatement costs and marginal external costs as a function of environmental pollution x.

conflict between the individual industrialist and societal interests:

> In general industrialists are interested, not in the social, but only in the private, net product of their operations ..., self interest will bring about equality in the values of the marginal private net product of resources invested in different ways. But it will not tend to bring about equality in the values of the marginal social net products except when marginal private net products and marginal social net products are identical. When there is a divergence between these two sorts of marginal net products, self interest will not, therefore, tend to make the national dividend a maximum; and consequently, certain specific acts of interference with normal economic processes may be expected, not to diminish, but to increase the dividend. (Pigou, 1932, p. 172)

Thus the idea of *internalizing external costs* was born (see Box 10.5). Pigou broke away from the mainstream neoclassical microeconomists' reservations with respect to state intervention into the market. Today, among economists, "it is generally recognized that externalities are an important form of market failure" (Verhoef, 1999, p. 199). Not only do externalities occur with respect to environmental issues, but also, markets cannot fully reflect social costs or other goods where property rights are not perfectly defined. The Pigovian tax became a basic model of environmental economics (see Chapter 11.4), and many economists thought that it would be realistic to determine environmental taxes factually according to this approach. However,

10 Origins of economic thinking and the environment

> **Box 10.6** Constructing new agents for improving environmental quality: Pareto optimality arguments
>
> Following the principle of Pareto optimality, no allocation should be accepted for which there exists another allocation where one of the (economic) actors (called Players) can get more and the others do not receive less. Let us consider two Players looking at the different alternatives $A, B, \ldots, F$. Given the setting of Figure 10.10, only the alternatives A and B fulfill the condition of Pareto-optimality.
>
> Targeting such an alternative asks the actors to behave as the ideal classical microeconomic agents. This also means that the Players should focus exclusively on maximizing their own utility. According to the assumptions underlying the construction of Pareto optimality, any comparison with the utilities of others is excluded. Thus, getting more or less than the other, in the sense of a competitive motivation, is not of interest. We consider Pareto optimality as a kind of minimum collective rationality (see Box 1.4).
>
> Pareto optimality has become a "dogma of economics" and is a standard paradigm taught in economic departments. Our experience shows that the Pareto optimality rationale can be successfully utilized as a practical argument for improving the "environmental quality" of building projects. Take the example of an urban transition project in which we participated. The situation was that the landowners of a former industrial site and the City of Zurich were negotiating for a redevelopment of a former industrial area for residential and business use. Because environmental issues were not a top priority, environmental aspects were generally neglected. Here we used the Pareto optimality rationale as a door-opener.
>
> We argued that the area owner should focus on his favorite alternatives. He (Player 1) should imagine an additional environmental player (Player 2). The rules of the game were that Player 2 should allow Player 1 to get the most preferred alternative variants. However, if alternatives can be generated where the environmental player (Player 2) can improve his payoff then, according to Pareto rationality, this should be allowed. This is actually the situation shown in Figure 10.10. If C was the preferred alternative of the property owner (Player 1) before Player 2 entered the game and it was possible to construct an alternative B, Player 1 should have nothing against it. Moving over to alternative A, however, would be rejected by Player 1 because of reducing utilities.
>
> **Figure 10.10** There are different alternatives $A, B, \ldots, F$ which provide utilities u_1 and u_2 for Player 1 and Player 2. In the case of only five alternatives, only the solutions A and B are Pareto-optimal.

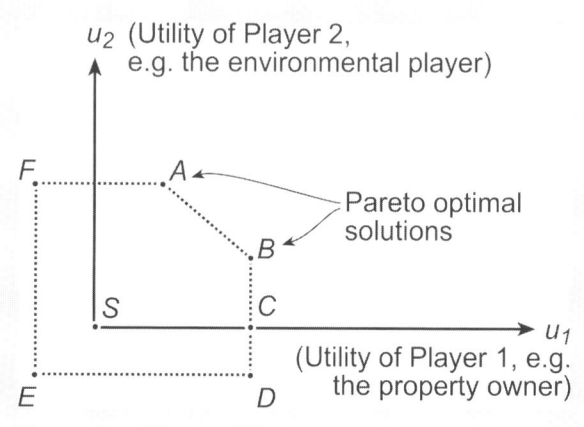

determining externalities is a "slippery business" and the factual environmental tax collections seem to be lower than the real external costs (Goodstein, 2003).

Pigou also supported the idea that income should be distributed as equally as possible, citing the declining marginal utility of rising income. Societies where total resources are allocated close to an equal distribution have a higher total sum of utility and thus a higher welfare than societies with unevenly distributed resources. This approach was criticized and elaborated on in the 1960s (Coase, 1960) and is discussed in Chapter 11.4. An exception might be seen in the contribution of Viner (1931), who differentiated between pecuniary externalities, which – though strongly affecting market relationships – do not represent market failure, and technological externalities or impacts which require the state intervention demanded by Pigou.

The contribution of Pigou was an exception. Through the beginning of the twentieth century, economists viewed resources and the material–biophysical environment as irrelevant or non-existing (Crocker, 1999; Kolb, 2004; Schneider, 2001; Söllner, 2001; Ziegler, 2008). The zeitgeist was instead shaped by the world wars, labor conflicts, economic depression, and questions of social justice.

Welfare economics investigates and models the income distribution within an economy and the allocation efficiency. As a microeconomic approach, welfare economics takes a bottom-up approach, starting from the individual economic actor (e.g. a consumer or firm; see Figure 14.1*). Welfare is measured by cardinal utility functions or money units. This has been done by the traditional approach utilizing marginal utility and interpersonal comparable utility functions, or cost–benefit analysis. New welfare economics rather works with ordinal preference or utility functions or indifference curves.

Nowadays, cardinal utility measurement has once again taken its place in the sphere of national economics. Here a top-down perspective is taken, starting from the nation (society) as an economic actor. This approach permits, for example, quantitative analysis in intertemporal resource allocation. When dealing with natural resources, the integral of the discounted welfare value of different generations is maximized. Clearly, in such an abstracted modeling, cardinal utility measurement and interpersonal comparability are essential.

New welfare theory distinguished between two ideas of social welfare. One is that all available gains or outcomes should be efficiently utilized. Economists talk about distributive efficiency. This led to the concept of Pareto efficiency or Pareto optimality. The other is the social welfare function which, however, is still a challenge to sustainable economics.

Vilfredo Pareto (1848–1923) started his theory with two important "weak" prerequisites. First he postulated no cardinal but only ordinal utility or preference measurement (Pareto, 1909). Second, he did not postulate interpersonal comparable utilities, but only that the individual maximizes utility. This individual effort is made while, at the same time, other economic agents try to optimize their utility.

Pareto supposed that the ideal situation has been reached when it is no longer possible to increase one person's utility without diminishing the utility of another individual. The question of how the utilities are distributed among the individuals is dealt with separately. We present the idea of Pareto optimality and how it can be employed for improving environmental quality in Box 10.6. However, we want to note that the economist Kenneth Arrow (1950; Arrow & Debreu, 1954) proved mathematically that, under certain idealized conditions, a system of free markets lead to a Pareto-efficient solution. In other words, the equilibrium of a competitive (unbiased) economy is Pareto-optimal. A second theorem proves – more or less vice versa – that taxes can secure Pareto-optimal solutions. The possibility of money transfers between subjects via taxes opens up new ways for allocation efficiency since there is greater freedom in the allocation sector if third parties receive compensation for any utility losses they may suffer (at least status quo).

However, Arrow and Debreu (1954) also introduced the impossibility theorem (see Box 7.1). This theorem showed that given ordinal, non-interpersonal comparable utilities and excluding dictatorship, there is no method producing stable solutions that are steadily accepted by all participants (see Box 7.2). This was later expanded by Amartya Sen with his liberty paradox which shows that, given a minimum liberty on the market, there is no way to induce Pareto optimality or to avoid unequal distributions.

Hicks and Kaldor (1908–86) looked at the subject of compensation (tax revenue, transfers). According to their criterion, an economic policy intervention has a positive effect on welfare when the winners can compensate the losers and still retain a net profit themselves (Kaldor, 1939). Indeed, it suffices that this is feasible. The compensation as such must not necessarily be realized.

Welfare economics incorporates – especially since Pigou – the environment and natural resources into the economic process. External effects were defined and demands for state intervention were put forward. From this time on, economics studies the characteristics of goods. It follows that this necessitates collective action, either to balance out market failures or to prevent them from occurring (rights of use).

10.5.1 Key messages

- (Old) welfare theory took a bottom-up view. The welfare of a society is often defined as the sum of the welfare of all the individuals in the society
- The Pigovian tax is the first economic concept of incorporating accounting external effects of environmental damage into economic means (internalization of external effects). It can be regarded as the origin of environmental economics
- Pareto efficiency becomes possible when allowing for individual maximization of utilities (without reducing own utility); this is a key concept of (old) welfare theory. Pareto efficiency or Pareto optimality can be regarded as a minimum collective rationality. If the environment is introduced as an additional actor, the identification of Pareto-optimal solutions can help to increase environmental performance without decreasing the utility of other actors.

10.6 Keynes' macroeconomics and the idea of permanent economic growth

John Maynard Keynes (1883–1946) is considered the father of modern macroeconomics. Much of his theoretical work was on critical stages of post World War I economics, such as stagflation, a situation in which economic stagnation and inflation occur simultaneously, a phenomenon that contradicted many economic theories. Keynes set forward a macroeconomic new welfare theory. Facing the high unemployment rates of the 1920s and the 1930s, his focus was the social welfare of a nation which was in some respect defined by avoiding economic breakdown (Toye, 1997). Keynes' major work was written in the red thirties, a time when orthodox Marxists interpreted the "Great Depression as confirming the validity of the view that capitalism is inherently unstable" (Minsky, 2008, p. 6). In 1936, Keynes, who was a man from the left, strongly criticized the microeconomic assumptions that each economic agent behaves fully rationally and in a reasoned manner. In his experience, businessmen did not have even a vague idea about the consequences of economic actions. Keynes doubted that the future could be predicted. His theory employed new concepts such as expectations, uncertainty, risks, or "spontaneous urge to action rather than inaction" (Keynes, 1936, p. 162). Thus, although taking a macroeconomic perspective, Keynes had a specific model of the actual attitudes of workers. As did many other scientists, he formulated fundamental psychological laws for economic behavior, such as the marginal consumer rate.

> [Men] are disposed, as rule and on the average, to increase their consumption as their income increases, but not by as much as the increase in their income. (Keynes, 1936, p. 96)

Keynes' welfare idea focused on preventing economic breakdown. Being an economist, he was more concerned about "unsaleable surpluses of food and raw materials" (Toye, 1997, p. 229) than about famine or poverty or the instance that England, the largest food supplier of that time, could have provision difficulties.

Material matter (e.g. resources) or biophysical issues (such as breeding in agriculture) were not an object of Keynes' economic reasoning. Even in his unpublished paper *Population* (Keynes, undated) and his reflections on the "Malthusian devil," the availability of physical resources did not play a role (Tarascio, 1971). Keynes believed in effects, such as that a reduction in the interest rate increased investment demand and led to a more equal distribution of wealth, which would finally decrease savings and consumption (Tarascio, 1971). Here a proactive role of the government is defined. Another effect was to promote and consolidate political consensus through the widespread implementation of welfare programs, independently of the economic cycle and the state of employment (Trigilia, 2002). This model of the state and its role in the economy became the Keynesian welfare state. Social policy has been dominated by economic policy, for instance public spending as a regulator to stimulate economy at a time of a slump.

From an environmental literacy point of view, Keynes might be of interest from different perspectives. On the one hand, Keynes did not consider any environmental variables, and even ignored land as a resource. Building upon his theory between the world wars, the idea of "man as irrational being" (see Box 1.4) might be dominant. This might be reflected in Keynes' famous ambivalent statement "Long run is a misleading guide to current affairs. In the long run we are all dead" (Keynes, 1923, p. 80). Keynes took a short-term view and did not deal with the evolution of society or economics, which may have prevented him from observing and considering environmental degradation. And, perhaps, compared to war impacts, the environmental damages resulting from economic behavior seemed secondary. After World War II, the zeitgeist changed and economic growth promoted the conception of man as rational being and Keynes moved into policy.

10.6.1 Key messages

- Environmental variables did not play a role in Keynes' macroeconomic approach. Keynes did not even consider land as a variable for theory-building.

Part V — Contributions of economics

Chapter 11
Contemporary economic theories dealing with the environment

11.1	Changing zeitgeist in economics 281	11.6	Weak vs. strong sustainability as economic principles 299
11.2	Public goods go to market 283	11.7	From institutional economics to institutional ecological economics 299
11.3	Resource economics 286		
11.4	Environmental economics 289		
11.5	Ecological economics 296		

Chapter overview

We have discussed the different concepts of rationality that scientists assume underlie human behavior (see Box 1.4). These concepts, such as man as a fully rational being or man having bounded rationality, represent different world views. The same holds true for different beliefs on how human systems or human species cope with scarcity or with limited resources. This chapter thus starts with a historical landmark, i.e. some key ideas preceding ecological economics, which is characterized by a different world view to economic thought. We then introduce two pairs of concepts: public and private goods, and the efficiency and the welfare function of markets. These are essential for understanding contemporary economic thought on how actors deal with common natural resources. After this, we present resource economics, which provides basic ideas on scarcity and substitutability inherent in the conception of weak sustainability.

Besides resource economics and ecological economics, environmental economics is a third line of economics focusing on material–environmental systems. Environmental economics overlaps with resource economics but is often related to the biotic, green environment. In particular, environmental economics emerged from the economic valuation of pollution costs in a similar way as Pigou defined them (see Chapter 10.5).

We then trace how ecological economics emerged from environmental economics. We show and critically discuss how thermodynamic ideas affected ecological economics, which stresses geophysical constraints of human life and economic activity. The chapter finishes with institutional economics and new institutional economics. This interdisciplinary field takes a broader view on the market actors, their property rights, and the contracts and enforcements of formal and informal rules that govern economic transaction. We discuss this approach and the disciplinary economics and interdisciplinary ingredients and their contributions to environmental and resource economics.

11.1 Changing zeitgeist in economics

We saw in Chapter 10 that microeconomic theories introduced economic agents and conceptualized the rules of economic transactions by various variables such as capital (e.g. money), labor, and land. According to traditional economics, the efficiency of markets is a major societal goal. Many theories did assume, and still assume, that efficiency is created by an "invisible hand" as, for instance, Samuelson (1947) assigned to Adam Smith. It is apparent that most theories considered wealth (see Figure 8.3*) the goal of economic action. The dominating dogma reads that a perfect, free market would create efficiency. Wealth was mostly linked to efficient economics in the sense of economics maximizing the value of assets owned by a country or by economic actors. However, the manner in which wealth should be distributed or allocated remained disputed. Today, economists dealing with sustainability consider both "fairness and efficiency criteria … [as] crucial to determine sustainable growth paths" (Bretschger, 1999, p. 232).

From a theory point of view, the variables capital, labor, and natural resources were rather abstract entities. Historically, some economists considered the availability of natural resources, in particular land, as a determinant of national wealth. The main material–biophysical dimension was the scarcity and limited production capacity of land. We can see that economists took a differentiated view on land. Whereas Malthus focused on absolute scarcity, others, such as Ricardo and von Thünen, considered varying land quality (see Chapters 10.2 and 10.4). The role of natural resources diminished in classical economics with the increasing importance of technology. In neoclassical approaches, however, the natural resource issue did not even arise:

> What is noticeable in early neoclassical growth models is the absence of land, or of any natural resources, from the production functions. Classical limit to growth arguments, based on a fixed land input, did not have any place in early neoclassical growth modeling. (Perman et al., 2003, p. 7)

We have learned from history that the Industrial Revolution in Britain was spurred by the availability of natural resources, in particular coal. But, as indicated by the term "the curse of natural resources" (Sachs & Warner, 2001), as will be discussed later, the availability of natural resources may have an ambivalent role.

Environmental material–biophysical resources were historically addressed via the variables land, scarcity, and negative environmental impacts. These issues had already been comprehensively discussed by Pigou by around 1930 (see Box 10.5).

As discussed in Chapter 9.2, environmental problems and particularly pollution became an issue in the 1960s in industrialized countries. This was also the period in which some economists switched from the HEP (human exceptionalist paradigm) to the NEP (new environmental paradigm; see Table 8.1). Among them was Kenneth E. Boulding (see Box 11.1), who was strongly influenced by systems theory which, with his thorough understanding of the subject, he considered the backbone of science. Boulding countered traditional economics:

> Man is finally going to have to face the fact that he is a biological system living in an ecological system, and that his survival power is going to depend on his developing symbiotic relationships of a closed-cycle character with all the other elements and populations of the world of ecological systems. (Boulding, 1965)

Such a statement coming from an American economist might be surprising. This statement and the paper "Earth as a spaceship" were written 3 years before Boulding became the president of the American Economics Society. Referring to the changing zeitgeist, as explained in Box 1.4, Boulding (1978) and others took an NEP approach which, as we will see, created a new line of economics called ecological economics (see Chapter 11.4).

Before turning to ecological economics, however, we follow the development of environmental and resource economics to discuss the question of pollution and negative externalities, as well as common goods.

Box 11.1 No more endless plains: the end of cowboy economics

Kenneth E. Boulding (1910–93) was a British Quaker. Originally educated in Oxford in classical Keynesian economics he worked on opportunity costs, capital theory, and international trade, but became one of the pioneers of general systems theory, evolutionary economics, and ecological economics (Boulding, 1978, 1991). He earned the presidency of the American Economic Association despite the fact that his thinking contrasted with the traditional economist paradigm which derived from the idea of scarce but, in principle, unlimited resources (see Box 11.5), the ubiquitous dogma of "prices increase the availability of resources," and the concept of marginal utility driving a stable steady state as "goals" of markets.

Boulding used the term "cowboy economics" to sketch the common perception that the natural environment is a virtually limitless plane with an unfixed frontier (Boulding, 1965; all following quotes in this box are taken from this paper). He started from the understanding of energy and material flows and frequently used "Spaceship Earth" as a metaphor to describe the Earth as a closed system (Figure 11.1a). Thus he wrote, "For millennia, the earth in men's minds was flat and illimitable. Today, as a result of exploration, speed, and the explosion of scientific knowledge, earth has become a tiny sphere, closed, limited, crowded, and hurtling through space to unknown destinations." He concluded, "We cannot have cowboys or Indians, for instance, in a spaceship, or even a cowboy ethic" (see Figure 11.1b). We will illuminate Boulding's meticulous critique both in traditional social science (including economics) and natural science.

The social science perspective became increasingly important. Boulding saw the need to create "institutions" which organize "cybernetic" or "homeostatic control with individual freedom and mobility" over "a price system and a market economy of a limited and controlled kind." He sought "mechanisms for preventing the overall variables of social systems for going beyond a certain range" as well as "controlling conflict processes and for preventing perverse social dynamic processes of escalation and inflation" (see Chapter 16.4). Boulding did not consider GDP as an appropriate measure of economic performance. In his economic modeling, he substituted wealth (as the goal of society) by a wide, but difficult to measure, concept of capital stocks, including human minds and bodies (i.e. health). Finally, when considering the issues discussed by welfare economics approaches, he suggested appropriate limitations of free market economics to secure distributive fairness.

Boulding's concerns focused on the fundamental change of world material and energy fluxes and the increase of entropy by industrial society considering the interconnectedness of systems, particularly the "symbiotic closed-cycle character with all the other elements and populations of the world of ecological systems." This requires starting with an understanding of the "environmental world matrix" before any economic analysis can occur (see Chapter 16, Postulate P7). Boulding was skeptical of natural scientists' HEP-like "we know everything" habit. He criticized biology for hardly moving "beyond the level of bird watching." And, long before the climate change shock of the 1990s, he argued: "We know practically nothing, for instance, about the machinery of long-run dynamics even of the physical system of the earth. We do not understand, for instance, the machinery of ice ages, the real nature of geological stability or disturbance, the incidence of volcanism and earthquakes, and we understand fantastically little about the enormously complex heat engine known as the atmosphere. We do not even know whether the activities of man are going to make the earth warm up or cool off". Clearly, this has partly changed. Even though we can now easily "update" Boulding's message with respect to climate dynamics (among other topics), we still have no definite knowledge about the interactions and feedback loops that are created therein.

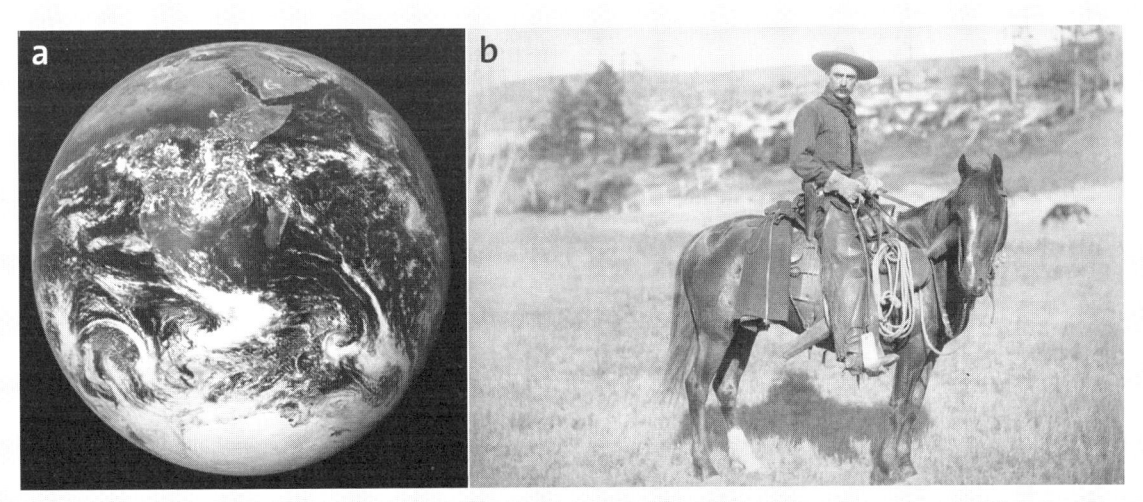

Figure 11.1 (a) Boulding often evoked the metaphor of "Spaceship Earth" (photo from NASA) and, as was also stressed by others, suggested an end to (b) cowboy economics (photo by John C. H. Grabill), typified by a free market which postulated unlimited resources (e.g. the never-ending plains).

11.1.1 Key messages

- Since the 1960s, new, alternative approaches emerged in economics, which were influenced by cybernetics, systems theory, and the changing zeitgeist. Ideas of "Spaceship Earth," describing limits to space and natural resources, and bounded rationality emerged.

11.2 Public goods go to market

Markets and economics are based on private goods. In economic terms, private goods (see Table 11.1) are rivalrous and excludable. By excludability we mean that it is possible to prevent potential consumers from gaining access to a good (e.g. if they are not willing to pay a certain price for it). Rivalry exists if the consumption of

Table 11.1 Characteristics of private and public goods.

	Excludability (marketable)	**Non-excludability (open access)**
Rivalry	**Private goods** • Food, clothing, furniture, car • Agricultural land • Fish stock in private lake (license)	**Common natural goods/common pool resources** • Water, air • Natural park (no entrance fee) • Ocean fishery (open access, outside territorial waters)
Non-rivalry	**Club good (restricted access)** • Wilderness area, natural park (entrance fee) • Cable TV	**Public goods** • Landscape, scenery • Institutional frame for food safety, animal welfare, groundwater protection, defense • Free radio programs

a good prevents or hinders its consumption by another person. Public goods are non-rivalrous and non-excludable. This is the heart of the commons dilemma (see Box 2.3).

Paul A. Samuelson is the father of the theory of public goods. He defined:

> ... collective consumption goods ... [as goods] which all enjoy in common in the sense that each individual's consumption of such a good leads to no subtraction from any other individual's consumption of that good ... (Samuelson, 1954, p. 387)

Samuelson started from a full rationality perspective and postulated that actors consistently behave according to their individual utility functions. Thus, no group decision rules or other forces that might be assumed to describe economic behavior are postulated:

> I assume no mystical collective mind that enjoys collective consumption of goods; instead I assume each individual has a consistent set of *ordinal preferences* (collective as well as private), which can be summarized by a regularly smooth and convex utility index ... (Samuelson, 1954, p. 387)

We speak about pure public goods when there is a situation of non-rivalry and non-excludability. This means that the consumption of the good by one does not affect the consumption of this good by others. In the case of private goods, the market dictates the optimal conditions. However, when non-excludability is considered, which means that the access to a good is open to everyone, the state intervention in the case of public goods (e.g. access to landscapes) may come into play.

Between private goods and purely public goods, there is a range of "mixed goods," which exhibit either the characteristic of rivalry or excludability (see Table 11.1).

11.2.1 Market failures and social welfare functions

Economic theories have "proved that under limited conditions, a perfectly competitive economic is efficient" (Samuelson & Nordhaus, 2005, p. 30). We speak about market failures if markets do not produce the most efficient outcomes. Market failures are often related to imperfect competition, positive externalities (e.g. by scientific inventions) or negative externalities (e.g. by environmental pollutions), or public goods. At the level of a society there are numerous public goods available (such as the air, seashore, etc.) that everybody – without exclusion – can use or consume. Consequently, since there is no incentive for individuals to state their preferences and manifest their willingness to pay, there is a tendency to enjoy the benefits of these goods as "free-riders" (see Chapter 6.22). In economics, non-excludability and the commons problem are denoted as a source of market failure. Besides incomplete markets, information asymmetries, and some other features, externalities (e.g. air pollution, loss of submarine fauna), as well as non-rivalry, are considered a source of market failures (Hanley *et al.*, 1997). Without state intervention, both private goods and public goods can only be supplied to a limited extent and with unsatisfying distribution. Even if the economic activities are efficient (in the sense of

Pareto optimality), the allocation and the distribution of access to the common goods can be a problem that requires regulation by taxation (Lindahl, 1918) or other public authorities.

Efficiency and fairness as an aspect of coping with scarcity and unequal distribution of wealth have been central issues of welfare economics. This brings us to the definition of a "social welfare function [as] representing a consistent set of ethical preferences among all the possible states of the system" (Samuelson, 1954, p. 387), for instance, a national or supranational world economy. We have shown (see Chapter 10.5) that social welfare has been a central element of the neoclassical era. But Samuelson again stresses that social welfare brings up a typical conflict in that we need "normative judgments concerning the relative ethical desirability of different configurations involving some individuals being on a higher level of indifference and some on a lower" (Samuelson, 1954, p. 387). Here Samuelson clearly states that it cannot be "a 'scientific' task of the economist" (ibid., p. 389) to define this function. Instead, economists need normative or sociological analyses.

We think that economics cannot get away from incorporating the distribution of the utility within societies (i.e. nation-states) and between nations in their theoretical models. If we consider sustainability as the regulating idea of the post-industrial society and conceive it as an ongoing inquiry in the frame of intergenerational and intragenerational justice, there are two key questions. The first is how a "good" welfare function should look from an economic perspective. The second is a non-economical one and refers to how a welfare function is established in a market by national, international, and supranational regulation or consensus. With respect to the first question, we can refer to the older and well established Gini coefficient. This measures the statistical dispersion of incomes within individuals or households in a population (e.g. of a town, a region, a country, or of the world). The Gini coefficient has also been called the Robin-Hood index (Kennedy et al., 1996). A similar measure is the "equity parameter ε" (Kämpke et al., 2003). There have been other suggestions for welfare functions, such as Rawls' welfare function which maximizes the worst player's function, or the Nash–Bernoulli function, which maximizes the weighted product of all market actors (Maskin, 1977).

A critical point with welfare functions is that the utility functions between the different market and society actors' utilities must be compared. This asks for generation of utilities, for instance by monetization procedures, which are accepted by all players. Further, the desired "good" welfare functions should neither positively nor negatively influence efficiency and vice versa. Also, market anomalies (such as incomplete or asymmetric information) should be considered, and one could test welfare functions in a kind of competition of computer-programmed welfare functions. Such a competition took place in game theory when different strategies in repeated prisoner dilemma games were playing against each other (Axelrod & Hamilton, 1981). Thus, one could think about how to check by computer simulation which welfare function would provide the "best" economic dynamics from a multi-attributive view, for instance when looking at the total utility, the stability (variance, collapses) of the market, etc. Here, the range of settings could be different situations in which some of the prerequisites of efficient markets (see Table 11.2) are violated.

The previous remarks on welfare functions are important as the protection of natural environmental resources becomes a welfare function.

While market economies aspire to efficiency of markets, social welfare functions target the distribution of wealth and protection of natural resources requiring normative values. We suggest that economics is challenged with deliberating which properties of welfare functions would have what impact on market anomalies.

In mainstream economics welfare functions are secondary and market efficiency dominates. In the beginning, the principle of marginal value for optimum determination, according to Walras and Pareto, formed the basis for the general equilibrium which represents an optimal allocation of resources. Technically, the optimum is attained after a number of factors are entered until their marginal sum corresponds to the price of a good. Table 11.2 presents a list of prerequisites for markets of common goods or environmental services to be efficient. The prerequisites of markets (1) from economic analysis are that there are private property rights for all goods and services and (2) that markets are perfectly informed and (3) behave perfectly competitively. Further, there should be no externalities (4), and the utility function should be "well behaved" in such a way that there are monotonic or U-shaped relations between the accessibility of a quantity of goods and the utility functions (5).

Table 11.2 Prerequisites for efficient markets (adapted from Perman et al., 2003, p. 124).

Factors for efficient markets
1 Private property rights are fully assigned in all resources and commodities (there are no public goods)
2 Market exists for all goods and services
3 Perfect information for all (trans-)actors
4 Perfect competition and maximization behavior of agents
5 No externalities
6 Utility functions are concave or "well behaved"

11.2.2 Key messages

- The distinctions between rivalry and non-rivalry and excludability and non-excludability, which differentiate private goods, club goods, common pool/natural goods, and public goods, are essential for environmental literacy. Different interests, incentives, and regulatory systems are related to these types of goods
- The way in which the efficiency of markets is attained has been a primary goal of economic theories. Social welfare functions targeting the distribution of wealth and protection of natural resources require normative values. We suggest that economics is challenged to deliberate about which properties of welfare functions would have what impact on market anomalies
- Markets, market efficiency, and the validity of economic models ask for many prerequisites which, in practice, are often not fulfilled.

11.3 Resource economics

11.3.1 Emerging awareness of resources

The traditional theoretical basis of resource economics was based on labor and human-made capital. The preclassical physiocrats stated that this would not suffice for the production of vitally essential goods and took food or mined products as examples. In addition, they emphasized that natural resources, such as mineral resources, fertile land or soil, and plants and animals, are the foundations of economic production (see Chapter 10.1).

David Ricardo and other classicists based the exchange value of a good on its scarcity and the labor required to provide it and thus inferred that the best or most fertile land or mines would be used first. However, natural resources, of which there are an (assumed) adequate supply and which are not producible, have no exchange value. Accordingly, they would rate as free goods. From this classical view of economics, there is a separation between the world of nature and the world of economics. However, as soon as human effort is required, the characteristics of goods (resources) undergo a change, and, because of this transformation process, they become marketable and have an exchange value. Scarcity and effort are the key issues. From the classical point of view, land as such was never regarded as a free good, as there have been more or less strict property rights of use. Thus, the classicists assigned an exchange value to individual natural resources, due either to scarcity or the effort involved in accessing a good.

> The idea of diminishing returns to effort expended upon the land became generally accepted in economics. (Kneese, 1989, p. 283)

Abundant resources were not included in the analysis at all. Moreover, scarcity was considered relative; even Ricardo noted the possibility of discovering new mineral deposits and developing new mining technologies (Tilton, 2003).

The neoclassicists declared that the exchange value was the use value representing the utility of a good. The concepts of scarcity and utility value drew closer together. According to neoclassicists, every useful economic good must fulfill the following conditions: (i) be purchasable; (ii) have a value and be suitable for exchange; (iii) be (or could be) manufactured industrially and be reproducible; and (iv) be available only in limited quantities. If a commodity or good is abundant then it is not an economic good but rather a common good. It follows that neoclassicists incorporate labor plus capital and partly land as production factors. Technology (e.g. the cultivation technique) was seen as a marginal factor. The natural–biophysical environment has been thought to have the characteristics of a public good.

> Nature is thought to contribute only inert, abundant, passive, indestructible building blocks to which labor and capital add whatever value the product does. (Daly, 1999, p. 641)

Environmental awareness and concern about the fundamental scarcity of resources were introduced not by economists but by public figures such as US

President Theodore Roosevelt (1858–1919), who had an educational background in natural history, and the governor of Pennsylvania, Gifford Pinchot (1865–1946), who was the first chief of the US Forest Service. These pioneers of the conservation movement demanded the "prudent use of resources, which goes far beyond the economists' concept of efficiency" (Tilton, 2003, p. 9).

> The five indispensable essential materials in our civilization are wood, water, coal, and agricultural products … We have timber for less than thirty years at the present rate of cutting. The figures indicate that our demands upon forest have increased twice as fast as our population. We have anthracite coal for but fifty years, and bituminous coal for less than two hundred. Our supplies of iron ore, mineral oil, and natural gas are being rapidly depleted, and many of the great fields are already exhausted. Mineral resources such as these when once gone are gone forever. (Pinchot, 1910, pp. 123–4)

It was during these times that the romantic idea of conservation led to the foundation of the world's first national park in 1872, Yellowstone National Park. We can learn from this that a value or welfare effect has been assigned to an ecosystem.

11.3.2 Defining optimal depletion

Neoclassical modeling of scarcity aimed at developing a theory of optimal depletion. In the twentieth century, it became apparent just how dependent human activities are on natural resources, regardless of whether they are renewable or exhaustible. Malthus' idea was taken up by the marginal utility school (see Chapter 10.4), where the focus was on key natural services, for instance, food and energy. Here, an idea from Jevons (1871/1970) imported the logic of macroeconomics to equate marginal costs and marginal benefits (see Box 11.2). Cost can result from the exhaustion of a non-renewable resource, and benefits from utilizing it.

The idea is that the access to environmental goods becomes more difficult with increasing quantity (see Figure 11.2) and that, from a certain level (e) on, it becomes futile to increase the quantity (Q). The marginal utility (MU) is countered by the marginal

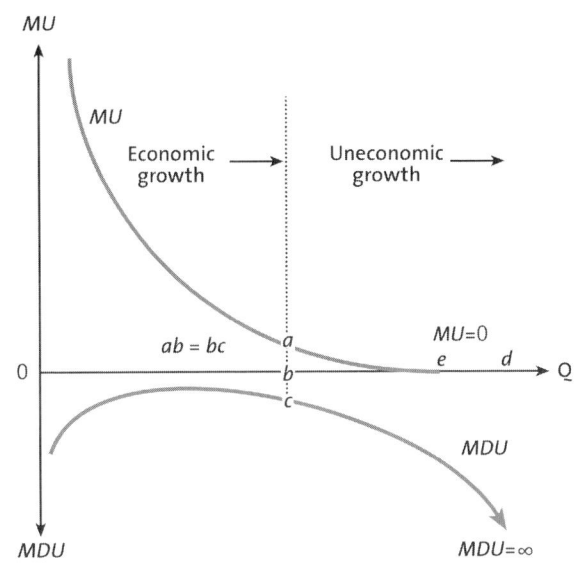

Figure 11.2 Optimal depletion of the environment (according to Daly, 1999, p. 641). Note: MU = marginal utility from consuming produced goods and services (Q). MU declines because, as rational beings, we satisfy our most pressing wants first. MDU = marginal sacrifices made necessary by growing production and consumption, for example, disutility of labor, sacrifice of leisure, depletion, pollution, environmental destruction, and congestion. Marginal sacrifice eventually increases, assuming we sacrifice least important values first. b = economic limit; e = utility limit; $MU = 0$ (consumer satiation).); d = catastrophe limit; $MDU = \infty$ (ecological disaster). At point b, the product of the economic limit (b) and MU (a) equals the product of economic limit (b) and MDU (c), which can be expressed as $ab = bc$.

Box 11.2 Mineral reserves become larger if prices rise: Earth's mineral resources economic cycle

Mineral economists consider the thesis of limits to growth which denotes the idea that mineral resources will sooner or later run out as naive (Brooks & Andrews, 1974; Tilton, 2003; Wellmer & Becker-Platen, 2002). On the other hand, however, it is "equally naive to jump from the premise that mineral resources are economically infinite to the conclusion that they could be produced in enormous volume without major social or political problems" (Brooks & Andrews, 1974, p. 13).

The rationale of the economists' reasoning is presented in Figure 11.3. Supply and demand is seen as the key regulating mechanism. If the demand for a mineral increases (e.g. because of increasing population), prices will rise. This, at least, spurs processes linked to human creativity. First, societies will initiate research activities looking for new technological solutions or increased recycling rates, substitution, etc. Second, exploration efforts are increased (and riskier ones are taken). New ores, as well as lower grade ores, are detected. Thus, larger reserves are detected, scarcity ends, prices decrease, and the Earth's mineral resource economic cycle starts again.

Box 11.2 (cont.)

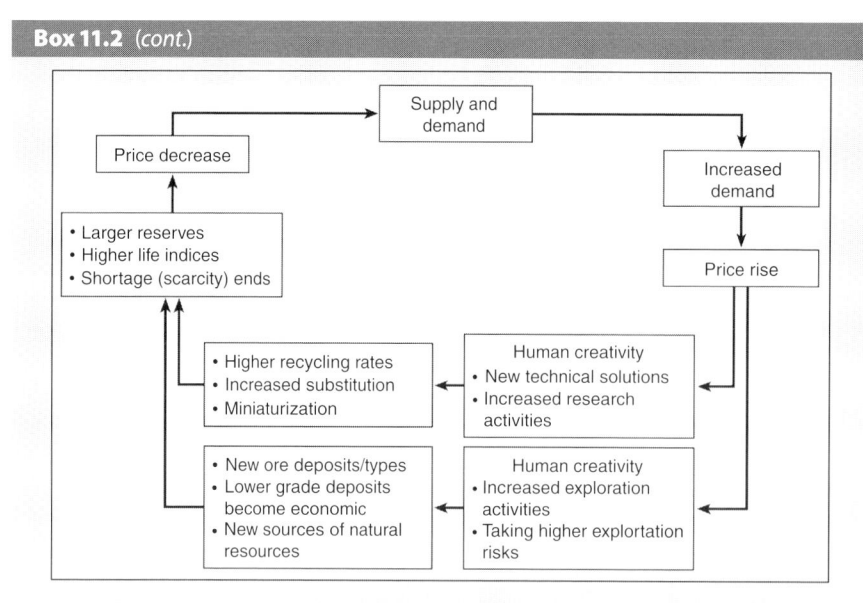

Figure 11.3 The feedback control system through which a rise in price leads to discovery of larger reserves for mineral depletion (see Wellmer & Becker-Platen, 2002).

damage utility (or marginal disutility, *MDU*) function. The somewhat *U*-shaped profile of this function is explained in that the damages decrease in some domains if more advanced technologies are applied (e.g. by substituting artisan mining by industrial mining). Here, we get a Jevonian macroeconomic view on the limits of growth. Thus, one can distinguish a domain of economic growth and of uneconomic growth. As the environmental depletion will eventually strike back at human systems, there is also a catastrophe limit *d* of disastrous depletion (Daly, 1991, p. 641).

Hotelling's rule (Hotelling, 1931) refers to a similar rationale. Harold Hotelling investigated the economic treatment of exhaustible resources such as coal or oil. His idea can be presented as the Hotelling equation, which reads:

$$P_t = P_0 e^{st} \ (s, t > 0)$$

In this equation, the price P_t of a resource in any period t is equal to the price P_0 in some initial period 0 multiplied by an exponential discount factor depending on the discount rate s (and the time span from 0). One can illustrate the increasing costs of depletion underlying Hotelling's rule by summing up the marginal utility and marginal disutility function (Figure 11.2), assuming that the *x*-axis is time and that the conditions of mining (e.g. the knowledge about the resources) do not change. The question of exhaustion is nicely stated in the following quote:

> What is the value of a mine when its contents are supposedly fully known and what is the effect of uncertainty of estimate? If a mine-owner produces too rapidly, he will depress the price, perhaps to zero. If he produces too slowly, his profits, though larger, may be postponed farther into the future than the rate of interest warrants. Where is the golden mean? (Hotelling, 1931, p. 282)

Extraction becomes more expensive the more the resource has already been depleted. Hotelling defined efficient exploitation by equating the change in the net price per unit according to the above equation with the normal discount rate. Thus, for instance, it is not profitable for suppliers such as Arab oil producers to sell all accessible oil today while expecting higher market prices in the future. The relation between the rate and the normal rent that money would gain on the market defines the speed of exploitation of a resource. Or, to express it in economic terms, if the owner wishes to maintain his assets and gain interest on its given opportunity costs (i.e. what he would gain by an alternative venture or from medium-term interest rates in the market), the price for the utilization of the resource must increase at a rate that corresponds to the rate of interest (Hotelling's rule). Clearly, this modeling of market and depletion activities starts from the assumption of perfect competition (see Table 11.2) and infinity:

> Not only is there infinite time to consider, but also the possibility that for a necessity the price might increase without limit as the supply vanishes. (Hotelling, 1931, p. 283)

The individual optimum of the owner of the resource must allow him/her to be indifferent to the choice between holding on to the resource and investing (the same amount as the mining costs and the gains resulting from mining) in an alternative venture. If the owner wishes to maintain assets and gain interest on them at opportunity costs (alternative venture), the price for the utilization of the resource must increase at a rate that corresponds to at least the average rate of interest. If the owner has an infinite planning horizon (see Box 11.2), (s)he will, from a social point of view (with reservations due to the assumptions), make the best possible use of the resource.

In the case of renewable resources, the opportunity costs of not harvesting also have to be taken into account (Hotelling, 1931). Several authors have developed theoretical time paths for price and rent/costs both for renewable and non-renewable resources (see Cleveland & Stern, 1999). Clearly, Hotelling's rule starts from overly simplified assumptions, such as that technology does not improve or that knowledge does not increase (see Box 11.2).

11.3.3 Economic dynamics of depletion

Resource economists not only deal with Pareto optimality in the use of resources but also with the intertemporality of the utilization of exhaustible natural resources. This is based on a utilitarian social welfare function (i.e. maximization of the integral of the discounted utility of present and future generations). The results obtained from this kind of concept are rated as unfair since the effects of discounting the utilization of the resource are immediately high and tend towards zero in the long term. Rawls (1972) offers an alternative by incorporating intergenerational justice in the social welfare function by demanding a constant degree of utilization over all generations (Dasgupta, 1974). In the opinion of Solow (1974), this is only possible when the natural resources can be replaced continually by capital so that consumption of the resources can still tend towards zero in the long term. This is the foundation of Hartwick's rule (Hartwick, 1977), which presupposes that the utility of the accumulated capital gradually replaces the utility of the exhaustible resource. In addition, resource economics deals with the potential and the effects of technical progress, the latter of which are integrated into the models illustrated above.

The first dynamic optimization models in the field of renewable resources are found in the sectors of forestry and fishery. From an economic point of view, the utilization of renewable resources is orientated towards the interest paid on money in the capital market. The investor's standpoint is that the resource will only be used if the interest paid over time is higher than the rate of interest for opportunity costs. First and foremost, the economic challenge is the sustainability of the utilization and the inclusion of any possible external costs such as, for instance, diminished biodiversity.

In the course of the mathematization of economics, further premises were introduced, including the complementarity between production factors and extensive substitutability. Finally, in the second half of the twentieth century, neoclassical economists rated the role of natural resources more highly.

As a result, attention was given to works that had hitherto been largely ignored, such as Faustmann (1849) on sustainable forestry, Scott Gordon (1954) on the optimal allocation of renewable production factors, and Gray (1914) or Hotelling (1931) on the economic use of exhaustible resources. Mathematical models also dealt with uncertainty and irreversibility of preservation and conflicts in resource management (Arrow & Fisher, 1974; Fischer & Krutilla, 1972).

11.3.4 Key messages

- Resource economics deals with optimal depletion, taking into account increasing efforts (declining marginal utilities) of mining and the increasing marginal environmental damages associated with the removal of natural resources
- Resource economics is based on certain fundamental rules, such as the idea that prices increase the amount of available resources or the decreasing role of natural resources for national resource economics (e.g. the "curse of natural resources").

11.4 Environmental economics

11.4.1 Origins

Whereas resource economics concentrates its attention on the inputs, environmental economics deals with degrees of outputs into the environment. Pearce (2002) and Bromley (2005) denoted the 1952 foundation of the independent research organization Resources for the Future as the beginning of environmental economics, the aim of which was to review

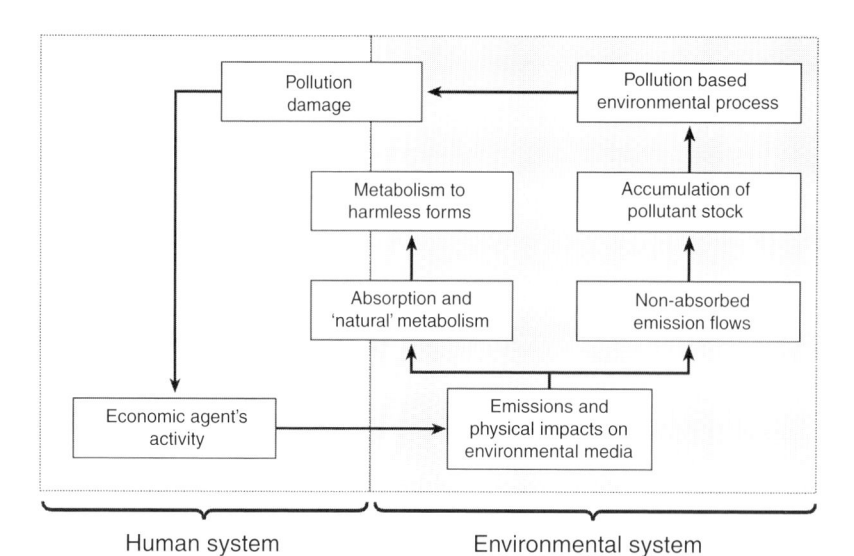

Figure 11.4 Economic activities, material flows, and environmental damages in environmental economics (based on Perman et al., 2003).

the future supply of minerals, energy, and agricultural resources during the post World War II reconstruction. Further, a paper by Emery N. Castle (1952) on uncertainties of farm management brought nature into economic models. After Rachel Carsons' warning on the costs of agrochemicals (Carson, 1962), which was in line with the NEP paradigm (see Table 8.1) and Boulding's Spaceship Earth metaphor (see Box 11.1), the former resource, mining, energy, and agricultural economics was extended by the consideration of harmful environmental impacts and by a relationship between natural and economic resources.

Environmental economics has a specific focus on the fields of nature and natural resources. It views the environment as a "collecting system" for all the refuse produced by consumption or manufacturing activities (see Figure 11.4). This environmental system, used as a reservoir for pollutants and residues, has a certain absorbing, regenerative capability. An economic problem arises when the absorption capability is no longer guaranteed and pollution damage emerges. The quality of the environment deteriorates, which is detrimental to the use and production functions of third parties and or even the polluter. Thus, economic activities face second-order feedback loops (see Feedback Postulate P4 in Chapter 16.5). Economic agents, even those responsible for this situation, may suffer from the resulting negative external effects (Perman et al., 2003). This problem is considered a byproduct of industrialization.

Environmental economics also became an important means of environmental policy evaluation. As can be seen from Box 11.3, methods such as cost–benefit analysis (CBA) or cost-efficiency analysis, including eco-efficiency, emerged early in the twentieth century.

11.4.2 Theoretical roots

Environmental economics follows the trend of classical and neoclassical concepts. It extends these concepts and enables them to cope with situations where the environment can no longer absorb emissions from economic activities. The theoretical cornerstone of environmental economics is classical economics which adopted the premise of rationality and egoism. Classical economics focused on production and distribution of profits (prosperity and wealth of nations). In contrast, neoclassical economics concentrated on the relationship between the economic subject and the material environment (goods), with emphasis on consumption (which had until then been neglected). The concept of equilibrium, optimum, and welfare was the logical result of this utility perspective (i.e. declining marginal utility). Thus, the analyses and interests of neoclassical economics centered on competition in resource allocation and the associated decision process.

While classical economists viewed personal motives as the driving force behind the economy, the neoclassicists developed a methodic assumption: the monetization of individual preferences. Until then, individual utility had been regarded as non-comparable. Individual utility was compared by means of the

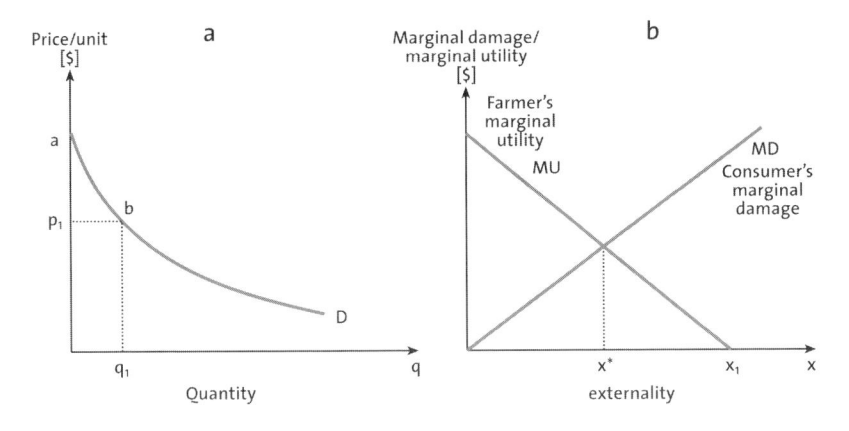

Figure 11.5 (a) The willingness-to-pay functions for the demand D of an environmental good (e.g. clean water); the y-axis p denotes the price per unit or marginal unit (e.g. per liter of clean water) that a person is willing to pay given a certain quantity q. (b) MD denotes the marginal damage costs on the environment and MU the marginal utility of a polluter (e.g. a farmer) resulting from an externality (pollutant) x.

concept of willingness-to-pay, applicable to both private goods and non-negotiable goods. The value of the goods is determined by the willingness-to-pay which, in turn, is governed by the subject's ability to pay (see Figure 11.5a). The willingness-to-pay is sometimes contrasted to the willingness-to-accept. Willingness-to-accept represents an amount to give up a good; the substitutability of market goods is decisive therein. For market goods with imperfect substitutes that cannot be easily bought back, such as personal health, the willingness-to-accept is higher (Hanemann, 1991; Knetsch & Sinden, 1984; Shogren et al., 1994). Although often applied in practice, the willingness-to-pay method has several flaws since it does not reflect the intensity of the preference and it assumes that all persons have the same marginal utility of income, i.e. the change in utility for an additional unit of the good is the same (Asafu-Adjaye, 2005).

We should note that CBA is not only used for the direct evaluation of environmental interventions but also for the evaluation of environmental monitoring projects or the costs and benefits resulting from an environmental impact assessment (EIA; see Glason et al., 2005; Scholz, 2001). There are other methods of environmental valuation, such as contingent valuation, choice experiments (Boxall et al., 1996), travel cost methodology or hedonic damage pricing (Scott et al., 1998).

Marshall investigated individual demand functions (see Chapter 10.4) based on internal and external savings. While, for example, internal savings correspond to economies of scale (i.e. the increase of efficiency by larger numbers produced), external savings result from the environment in which the company is situated (such as an industrial zone) regardless of the size of the company. This concept later became crucial in environmental economics. The concept of externality is based on Marshall's description. Externalities emerge as the economic system fails to recognize the value of certain goods when a transaction takes place.

Based on Pigou's works (see Chapter 10.5), science has developed concepts designed to aid in the comprehension of this situation and its resolution. In wider circles in the economic community, externalities became a prominent issue associated with environmental pollution. Gradually, an independent branch of economics developed – environmental economics.

The rationale of environmental economics can be described as follows: emissions from consumption and manufacturing activities have reached such a level that the environment can no longer absorb them. Since there is no market price pertaining to the environmental resources affected, negative external effects occur. By way of definition, the market process cannot achieve a Pareto optimum. An effort must be made to arrive at a new status whereby the marginal costs of an emission (pollutant), in the form of a loss of utility for the affected third parties, are equal to the marginal utility of this emission for the offender. This shows that the optimal level of pollution is not zero. The costs and benefits of environmental damage must be weighed against each other.

11.4.3 Coase theorem

Coase's theorem (Coase, 1960) states that when "externalities can be traded without transaction there will result efficient outcomes regardless of the initial allocation of property rights." Coase suggested private, negotiated settlements as a way of solving problems relating to environmental economics. Thus, his theory can be seen as a response to the Pigovian approach, which saw governmental interventions by taxes as a solution. The Coasian solution assumes that the free market system solves the problem through bargaining between the affected parties. This means, however, that property rights can have different forms but must be established. As well, the market actors face a perfect market with perfect information. This reciprocal relationship ensures that both bargain until an equilibrium point is attained. Coase suggests a quasi-symmetric situation between the polluter and the damaged party, as the damaged person contributes to the problem by his or her existence. However, given the high transaction costs owing to the often large number of parties involved, solutions are commonly offered by the government. The state has jurisdiction in fiscal matters and can therefore levy taxes on environmental damage. The criterion for this is the alignment of the private and social marginal costs of environmental damage. This internalization encourages the parties responsible for the pollution to behave as they normally would in the market for private goods.

11.4.4 Contemporary approaches

There are different possible alternatives to the free-market economy. In principle, we can distinguish between market-based instruments and command and control systems. Examples for market-based instruments are charges, subsidies, marketable permits, deposit-refund schemes, etc. (Asafu-Adjaye, 2005). Examples for command and control systems are emission standards, zoning, bans, etc. The regulation standards (pollution limits) can be set according to a Pareto optimal value. Theoretically, both approaches are equal in their effect on the environment, but they differ in their distribution effect. Both approaches demand that government authorities or market actors are in possession of a great deal of information. This is not possible as a rule and so approach procedures are opportune. For example, it may make sense to reach a politically decreed emission reduction via cost efficiency or cost-effectiveness. Taxes could be fixed and their sum could be determined and then fine-tuned (standard price approach according to Baumol & Oates, 1971). We should note that these approaches are threatened by governmental failures just as the free-market approach is threatened by market failures (see Table 11.2).

Environmental economics has developed an alternative to the standard price approach, namely, the standard quantity approach (Dales, 1968). This is applied in emission trading where emission certificates are issued by the state and then auctioned (Tietenberg, 1985). Since the extent of pollution reduction can be defined, this method is an improvement on the levying of taxes. However, the political process leading to its definition could prove difficult.

Finally, we want to present cost–benefit, cost-effectiveness, and eco-efficiency as concepts from environmental economics.

All the above concepts refer to simple input–output efficiencies. CBA (presented in Box 11.3) simply maximizes the relation (often the difference) between the monetized outputs and the monetized inputs (in dollars) of an activity, project or good. The method is limited if the preferences cannot be appropriately monetized. Cost-effectiveness analysis, on the other hand, looks at more than just the monetary terms in the evaluation of projects by, for instance, modeling preferences in terms of multi-attribute utility representations (see Box 15.1; Table 15.3, Step V; Scholz & Tietje, 2002). Eco-efficiency relates the environmental impacts to the economic costs. The guiding principle of eco-efficiency approaches is to optimize the ecological–economic ratio of desired economic output and necessary environmental inputs (Schaltegger & Burritt, 2000). From the onset, eco-efficiency has focused on the micro level: products, production processes, and companies.

There are two perspectives on eco-efficiency. The first represents the idea of creating more value with less impact, as has been formulated by the World Business Council for Sustainable Development (Schmidheiny, 1992; World Business Council for Sustainable Development (WBCSD), 2000). The idea is that economic outcome per environmental impact (e.g. measured by a life cycle instrument; Frischknecht et al., 2005) should be improved. The determination of allowable environmental impact has not yet been addressed. One can consider this approach as an ecologically defined social welfare function. A society should aspire to maximum economic benefit with minimal ecological impacts. The other option is called operational eco-efficiency (Scholz & Wiek, 2005). Here, perspectives are taken of individuals, companies, organizations, etc. which must choose between investing in environmental protection projects or different projects, A_k. Here, the maximal environmental benefit $u_{ecol}(A_k)$ should be attained for a given investment given a certain amount of utility $|u_{econ}(A_k)|$, which is the absolute value of utility as investments are usually

11.4 Environmental economics

Box 11.3 Cost–benefit analysis: a Faustian bargain?

Historically, CBA is an applied welfare economics technique. The first formal application of CBA goes back to a calculation of the net benefits of the Forth and Clyde Canal in Scotland in the late eighteenth century (Asafu-Adjaye, 2005; see Box 10.1; Paxton *et al.*, 2000). In 1808, US President Jefferson's Secretary of the Treasury Albert Gallatin stressed the need to compare the benefits and the costs of waterway improvements (Kneese, 1993). The intellectual father of general CBA, the French economist Jules Dupuit (1804–66), wrote the often-quoted study *On the Measure of the Utility of Public Works* (Dupuit, 1844/1952), which contrasted consumers' surplus with public costs. Dupuit also studied groundwater flows, flood management, and supervised the construction of the Paris sewer system. Another step in the establishment of CBA was the US Federal Reclamation Act of 1902, which introduced a rule system for irrigation to the American West requiring generation of sufficient returns to be able to recover project costs. Later, the Flood Control Act of 1936 stated that improvements for flood control could be undertaken "if the benefits to whomsoever they may accrue are in excess to the estimated costs" (US Congress, 1936). Here, we can find a practical application of the Kaldor–Hicks compensation rule between winners and losers of new technology or infrastructure facilities, which was introduced as a prerequisite for congressional authorization by the US Reclamation Project Act in 1939 (Kneese, 1993).

Clearly, applying these laws in practice required sound engineering and natural science knowledge, and thus books such as *The Economics of Project Evaluation* (Eckstein, 1950) appeared. CBA has been standardized in the US administration since the 1960s; opportunity costs were introduced in 1958 (Asafu-Adjaye, 2005). However, the way in which they were assessed and how discount rates were introduced (if at all) depended on how many projects the government wanted. During the great dam-building era, CBA was linked with multiobjective evaluation and trade-offs, e.g. including: (a) the quality of the total environment; (b) regional economic development; (c) national economic development; and (d) the well-being of people (Jaffe, 1980).

The director of Resources for the Future, Allen V. Kneese, labeled CBA a Faustian bargain (Kneese, 2006) when the technique was applied to the evaluation of nuclear energy and nuclear power plants. The concern of ecological economists against the application of CBA in this area concentrates on the timescale linked to nuclear energy (see Figure 11.6). They argued that a large-scale fission economy would require society to monitor waste disposals continuously for several hundred thousand years, which seems infinite given that *Homo sapiens* only appeared 150 000 years ago. The production of nuclear waste has been criticized as neglecting ethical norms, moral standards or sacred values which do not allow for trade-offs (Atran & Axelrod, 2008; Hanselmann & Tanner, 2008; Tetlock, 2003). Besides the timescale issue, critical ecological economists postulated that the size of nuclear accidents and possible terrorist attacks might cause non-compensationable damages.

An open question discussed in economics, philosophy, and social psychology is the question of whether there are sacred values or dimensions of utilities that are not commensurable (see Figure 11.6b). If such a dimension exists, we are facing a Faustian bargain (Tetlock, 2003).

Figure 11.6 (a) Faust, detail from the title page of the 1616 edition of *The Tragical History of Dr. Faustus* by Christopher Marlowe. (b) The two farmers who rejected any financial compensation for an exploration drill for a planned deep geological repository in Wolfenschiessen, Switzerland (from Scholz *et al.*, 2007).

negative. Thus, operational eco-efficiency defines the ratio between environmental utility and economic utility (i.e. $ee(A_k) = \dfrac{u_{ecol}(A_k)}{|u_{econ}(A_k)|}$).

A critical issue of eco-efficiency is to find appropriate methods to assess environmental damages or improvements. Besides the willingness-to-pay and monetizing of life cycle assessment (LCA;

Frischknecht, 2000), other pricing methods can be applied (Scholz & Wiek, 2005).

11.4.5 Thermodynamics meets resource economics

Physics has long affected the social sciences as the leading science of the twentieth century. Concepts such as the lever principle applied to the equality of the ratios of marginal utility, asymptotic stability or macroeconomic equilibrium can be taken as examples. Another well known example of econophysics (Giffin, 2009), where physical methods such as statistical mechanics are applied to economic problems, is in statistical finance. Mathematical axiomatization has also been common. Certain physical laws were believed to be valid for social systems. This concept is certainly linked to Adam Smith's belief in the invisible hand, and results in the macroeconomic concept of equilibrium as "a state of balance between opposing forces," (Mankiw, 1994, p. 487), "in which economic aggregated or economic agents such as markets have no incentive to change their economic behaviour" (Pearce, 1992, p. 129). Here, it sometimes looks as if economics aspires to measure with the standards of natural sciences (Turk, 2006).

> My impression is that the best and the brightest in the profession proceed as if economics is the physics of society (Solow, 1985, p. 330). [But] I suspect that the attempt to construct economics as an axiomatically based hard science is doomed to fail. (ibid., p. 328)

There has been much and fundamental critique on the limits of analogies, such as of transferring principles of energy to utility modeling, without acknowledging that there is no rationale for the conservation of energy in terms of utilities of market actors (Mirowski, 1991). As reflected in the Complementarity Postulate P1 of the HES framework (Figure 3.3*; Chapter 16.2) and in the assumption that human and environmental systems follow different rationales, we share many of the critical arguments, at least on the micro level.

The view while looking at the material–biophysical–technological processes underlying economic activities and processes is completely different than the transfer of physical laws to human market actors. Here, thermodynamics played an important role.

The first law of thermodynamics is the law of the conservation of mass and energy which states that energy and matter can neither be created nor destroyed. It can only be converted from one form to another. The fundamentals of this law are already inherent in Quesnay's Tableau Economique (see Table 10.1) and underlie Leontief's input–output, which comprises inter-industry relations of an economy and later became a basic method of industrial ecology (see Chapter 12.5; Hendrickson et al., 1998). The application in environmental economics centers on material, energy, and pollutant flows underlying and resulting from production.

> The so-called mass-balance principle states that mass inputs must equal mass outputs for every process (or process step), and that this must be true separately for each chemical element. (Ayres, 1998, p. 190)

From an environmental economics perspective, the environmental impacts and any increase in material flows are essential. And, as any product output produces a corresponding money flow and environmental burden, it provides a chance to link money and material flows (Binder et al., 2004; Hendrickson, et al., 1998). However, we should note that Einstein's mass–energy equivalence is an extension of the simple material–matter–mass conservation and lacks, at least in some domains, reversibility.

More critical and disputed has been the use of the second law of thermodynamics, the entropy law (see Box 11.4):

> If the system is isolated and closed, so that it does not exchange matter or energy with any other system, its entropy increases with every physical action or transformation that occurs inside the system or in the universe as a whole. When the isolated system reaches a state of internal equilibrium, its entropy is maximized. When two systems interact with each other, their total combined entropy also tends to increase over time. (Ayres, 1998, p. 190)

Based on the work of Georgescu-Roegen (1971), Herman Daly and others used the second law of thermodynamics to criticize traditional and particularly neoclassical economics. From a system theoretic perspective, there are two ideas inherent in many of their arguments. The first is that economics is seen as "a part of a larger finite system" (Daly, 2005, p. xxxvi). Here, the Spaceship Earth metaphor (see Box 11.1) comes to the fore. Second, the entropy law is directly applied to the idea of an isolated Earth system to argue "that energy and matter in the universe move inexorably towards a less ordered (less useful) state" (Daly, 2005, p. 28).

> In the absence of an influx of low entropy into a system the physical and chemical gradients are irreversibly diminished as physical and chemical processes occur, and the potential to do work in the future is reduced.
> (Ruth, 1999, p. 856)

We deal with this approach in some detail, since, from an environmental literacy point of view, it is important to use physical laws, models, and knowledge properly. Insight on this is provided in Boxes 11.4 and 11.5 and the following.

Before we turn to ecological economics, we should mention that at the end of the twentieth century, biology, which is now often called life sciences, became the leading science. Daly's paper *On economics as a life science* (Daly, 1968) states that the life process has been ignored in economics. Daly sees economics as "the part of ecology which studies the outside-skin life process insofar as it is dominated by commodities and their interactions" (Daly, 1968, p. 392). He aspired to combine the "traditional economic (outside skin) and traditional biological (within skin) views of the total life process ... both in their steady-state aspect and their evolutionary aspect" (ibid., p. 392). Through these ideas, we can see (see Box 11.1) the birth of ecological economics.

11.4.6 Key messages

- Environmental economics focuses on the outputs and effects (externalities) of economic activity. The idea of monetization of ecosystem functions and environmental changes is central for economic tools of environmental policy evaluation such as cost–benefit analysis, cost-effectiveness analysis, etc. Targeting input–output efficiency, environmental economics focuses on the individual market actor's willingness to pay with respect to different policies dealing with nature conservation, toxic substances, solid waste or climate protection
- Environmental economics can identify sources of market failures and the need of policy interventions
- Coase's theorem has been an answer to the Pigovian public authority and mostly tax-based incorporation of externalities in markets
- The physical idea of input–output analysis inherent in environmental economics provided the foundation for utilizing other physical laws such as thermodynamics to illustrate the weaknesses in assumptions made by traditional economics
- Environmental and resource economics have been shaped by ideas from Newtonian mechanics. When economics refers to physics (e.g. thermodynamics) or biology, sometimes proper and sometimes improper analog models can result.

Box 11.4 The second law of thermodynamics and economic processes: Ayres vs. Daly

Entropy, full world, and irreversible dissipation structure became key concepts of Daly's (1974, 1999, 2005) contribution to ecological economics. His approach, however, has been criticized by other researchers, including Robert U. Ayres, a mathematical physicist by training and one of the promoters of ecological economics, who complained that "The term *entropy* is too much used and too little understood" (Ayres, 1998, p. 190), emphasizing that it was even sometimes described as having a "mystical aura" (ibid., p. 191). Although we believe that dissipation and irreversibility are key concepts of environmental sciences, we are aware that a proper adoption of natural science laws to economics is important for environmental literacy. Thus, we introduce the dispute between Ayres and Daly on the role of the second law of thermodynamics.

The second law of thermodynamics states that entropy never decreases in an isolated and closed system. Although matter–energy is constant, the quality may change. Daly states that the "measure of quality is entropy, and basically it is a physical measure of the degree of 'used-up-ness' or randomization of the structure and capacity of matter or energy to be useful to us. Entropy increases in an isolated system. We assume the universe is an isolated system" (Daly, 2005, p. 32). Figure 11.7, based on the work of Georgescu-Roegen (1971), presents the universe hourglass, which represents the Sun–Earth relationship as an isolated system. Daly writes "that there is a continual running down of sand in the top chamber and accumulation of sand in the bottom chamber" (Daly, 1991, p. 227). Daly considers entropy as "time's arrow in a moment to moment concept" (ibid., p. 31) and stresses that the hourglass cannot be turned upside down. He then continues: "With a bit of license, we can extend the basic analogy by considering the sand in the upper chamber to be the stock of low-entropy fossil fuel on earth, depicted in the right-hand part of Figure 11.7. Fossil energy is used at a rate determined by the constricted middle of the hourglass, but unlike a normal hourglass,

Box 11.4 (cont.)

humans alter the winds … Once consumed, the sand falls to the bottom of the chamber, where it accumulates waste and interferes with terrestrial life processes. … Unlike exchange value, the flow of throughput is not circular; it is a one-way flow from low-entropy sources to high-entropy sinks. … Energy, by the entropy law, is not recyclable at all" (ibid., p. 31). The relationship to economics is drawn in the subsequent statement: "From the entropy law we know that the entropy increase of an ecosystem is greater than the entropy decrease of the economy. As the stock and its maintenance throughput grow, the increasing disorder exported to the ecosystem will at some point interfere with its ability to provide natural services" (ibid., p. 35).

Ayres pointed out flaws in this argument. He points out that the thermonuclear energy produced by the Sun "is expected to be sufficient for 7–10 billion years" (Ayres, 2004, p. 195). He stresses that plants metabolize only about 3% (a little more in the case of algae) of the solar energy, that photovoltaics can already provide 30%, and that energy is – in principle – abundant (see Box 11.5). Solar energy is neither scarce nor a near-time limiting factor and thus human life on Earth is not dependent on fossil energy, although humans are using up fossil fuel stockpiles 1000 times faster than they have been built. He concludes, "Given enough exergy, any element can be recovered from any source where it exists, no matter how dilute or diffuse, … in principle" (Ayres, 2004, p. 197). In another recent paper on the use of the life cycle metaphor he concludes: "Attempts to use ecological concepts in an economic context are often misleading and unjustified" (Ayres, 2007, p. 115).

We want to clarify that these arguments do not mean that dissipation or entropy is unimportant for economics or sustainable development. Moreover, irreversibility (e.g. of species extinction) is a major issue as well as finding proper measures for evaluating (economically) meaningful energy or material flow processes by physical measures (e.g. exergy).

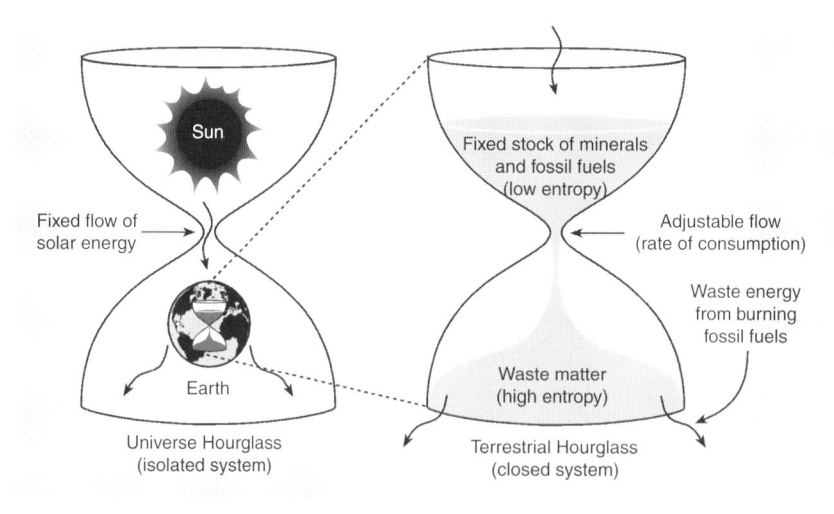

Figure 11.7 The entropy hourglass.

11.5 Ecological economics

In classical economics, besides land and Pigovian taxation of externalities, processes and quality of the material–biophysical environment did not play any role. Following the NEP paradigm (see Table 8.1), ecological economics represents an alternative form of environmental economics:

> Ecological economics addresses the relationship between ecosystems and economic systems in the broadest sense …. It implies a broad, ecological, interdisciplinary and holistic view of the problem of studying and managing our world. (Costanza, 1989, p. 1)

Table 11.3 provides a differentiated comparison between traditional economics, ecology, and ecological knowledge. Ecological economics and traditional economics differ fundamentally with respect to their cosmology: the basic assumptions on how the processes of the material–biophysical environment function (see Table 11.3). Ecological economics assumes:

> … economy as a part of a finite system. … Thus in ecological economics, optimal scale replaces growth as a goal, followed by fair distribution and efficient allocation, in that order. (Daly & Farley, 2004, p. xxvi)

Table 11.3 Comparison of basic cosmological assumption of "conventional" economics, ecology, and ecological economics (adapted from Costanza et al., 1991, p. 5)

	"Conventional" economics	"Conventional" ecology	Ecological economics
Academic stance	• Disciplinary • Monistic, relying on mathematical tools	• Disciplinary • More pluralistic but few rewards for integrative work	• Interdisciplinary and transdisciplinary • Pluralistic, problem-focused
Basic world view	• Mechanistic, static, atomistic • Individual preferences determine action • Unlimited resource base and infinite substitutability	• Evolutionary, atomistic • Evolutionary genetics is the dominant force • Limited resource base • Human species just one among others	• Dynamic, systems, evolutionary • Human preferences, understanding, technology, and organization reflect ecological constraints • Human responsibility for sustainable management
Species frame	• Humans only • Plants and animals rarely included for contributory value	• Non-humans only • Starting from the idea of "pristine" ecosystems untouched by humans	• Whole ecosystem, including humans • Acknowledges the human–environment relations
Timeframe	• Short • 50 years maximum, 1–4 years usual in practice	• Multiscale • Days to eons, but timescales often define non-communicating subdisciplines	• Multiscale • Days to eons, multiscale synthesis
Space frame	• Local to international (nation-based) • Framework invariant of scale • Actors change from individuals to firms to countries	• Local to regional • Traditionally focusing on small ecosystems • World ecosystems analysis starts	• Local to global • Hierarchy of scales of human systems and environmental systems
Primary macro goal	• Growth of national economy	• Survival of species	• Ecological economic system sustainability
Primary micro goal	• Maximum profits (firms) or maximum utility (individuals) • Individualist behavior maximizes wealth • External costs and benefits usually ignored	• Maximum reproductive success • All agents following micro goals leads to macro goal being fulfilled	• Must be adjusted to reflect system goals • Social institutions at higher hierarchic level necessary to ameliorate conflicts caused by myopic pursuit of micro goals at lower levels
Assumptions about technical progress	• Very optimistic • HEP paradigmatic	• Pessimistic or no opinion	• Prudently skeptical • NEP paradigmatic

Economy is seen as an "open subsystem of that larger 'Earthsystem'" (Daly & Farley, 2004, p. 15). Ecological economics is a genuinely interdisciplinary and problem focused approach. Economic analysis goes beyond pure (human)capital × labor analysis, reflects ecological constraints, and considers the Earth's organism layer, including all species. Thus it includes a strong ethical, normative component, for instance, when considering responsibility for sustainable management. The founders consider ecological economics the science and management of sustainability (Costanza, 1991). They take an eon (i.e. 1 million years) as the time perspective. Ecological economists suppose traditional economics to take a much too short time perspective.

> **Box 11.5** Cost, not physical availability, matters: unlimited solar energy?

Access to solar energy can help to re-establish the formation of Earth's structure, which has been changed by human activities. This has been covered by the mineral economist John Tilton (all unspecified references in Box 11.5 are to Tilton, 2003, p. 25). In principle, the availabile solar power reaching the Earth's upper atmosphere is the solar constant multiplied by the area of the Earth presented to the Sun. The rate of arrival of energy per unit area perpendicular to the Sun's rays on Earth is given by the solar constant and equals 1350 watts per square meter (Giancola, 1997). Multiplying this by the surface area provides 1.73×10^{17} watts of solar power reaching the upper atmosphere. Because only about 50% of this energy reaches the ground (Giancola, 1997), the total solar power reaching the Earth's surface amounts to 8.6×10^{16} watts. Accounting for the number of hours in 1 year (24×365) and converting this to kilowatts results in 7.9×10^{17} kilowatt-hours of solar energy reaching the Earth's surface annually.

Tilton compares this estimate with that derived from global petroleum production. In 2001, the amount of energy in a barrel of oil averaged about 5.8 million British thermal units for oil consumed in the US, i.e. 1700 kilowatt-hours. Given the average annual global crude oil production in 1997–99, which was about 2.4×10^{10} barrels, the total output from this system is 4.0×10^{13} kilowatt-hours of energy or approximately 0.005% of the solar power reaching the Earth's surface every year. As crude oil production accounts for 40% of global energy output, total energy output currently equals 0.012% of the available solar power. In effect, the available solar power is 8000 times greater than the combined total of the world's current energy production, which promotes large-scale photovoltaic energy production (see Figure 11.8).

Resource economists such as Tilton stress that, besides the difficulty in imagining the globe "completely smothered with solar panels," ... "it is costs [including the environmental costs], and not physical availability, that ultimately determine the availability of energy commodities".

Figure 11.8 Large-scale photovoltaic energy production has substituted agricultural production in areas suited for agriculture. (a) The largest US photovoltaic solar power plant at Nellis Air Force Base. (b) Sketch of possible infrastructure for a supply of power to Europe, the Middle East, and North Africa (from Desertec Foundation, 2009).

Although we agree that the temporal view on business cycles both at a nation-state and business level are short, compared to biological systems cycles, we should note that infinite time is inherent in many models, e.g. the mathematical modeling assumptions underlying Hotelling's rule (Hotelling, 1931). Ecological economics also takes a hierarchical view, as is typical in ecology (see Chapter 16.3).

The entropy law is a cornerstone for ecological economics (Georgescu-Roegen, 1971):

Burning a piece of coal changes the low-entropy natural resource into high-entropy forms capable of much less work. ... There is no way of reversing this process. (Daly & Cobb, 1989, p. 195)

Given solar input, all things could be, in principle, recycled or reconstituted. However, taking species extinction or coal combustion as examples, certain processes are irreversible. Quantity (e.g. of energy) can be restored but quality cannot. A general entropy law on the dissipation of material matter in the concept of Earth as a closed system, however, is too simple (Rennings, 1994), and ignores, for instance, the input of solar energy on organismic reproduction and order.

11.5.1 Key messages

- Ecological economics acknowledges ecosystem and geophysical constraints of economic action and human–environment relations in an encompassing way
- As ideas of classical, Newtonian physics are the basis of many economic models, principles from ecosystem biology build the basis for ecological economics
- Ecological economics includes normative ideas such as sustainability.

11.6 Weak vs. strong sustainability as economic principles

We have seen that there are two camps in resource, environmental, and ecological economics with respect to environmental risks and prudence. This is best reflected by the complementarity between weak and strong sustainability. The concept of weak sustainability is based on the contributions of neoclassical economists such as Robert Solow (1974, 1992) or John M. Hartwick (1977). Weak sustainability can be interpreted as an extension of neoclassical welfare theory (Hediger, 2000), and is based on the belief that only manmade capital (e.g. knowledge) is of consequence. In the eyes of weak sustainability believers, it does not matter whether the current generation uses up non-renewable resources or pollutes the atmosphere "as long as enough machineries, roads and ports are built in compensation" (Neumayer, 2003, p. 1). Substitutability is the panacea:

> As was pointed out …, physical exhaustion is not the issue. We will not literally run out of resources. Scarcity may push the costs of some mineral commodities sufficiently high to preclude their widespread use, but resources will remain in the ground, and so will be available at some price. Higher prices, as we have seen, increase the challenge of achieving sustainable development, but do not necessarily preclude it. (Tilton, 2003, p. 106)

Strong sustainability does not consider natural resources as arbitrarily substitutable (Ayres *et al.*, 2001; Brand, 2009). The main postulates of strong sustainability are: (1) pollutant emissions are allowed only to the level to which they can be absorbed; (2) use of renewable resources has to remain within the scope of regeneration capability (without extinction of species); (3) use of exhaustible resources has to be coupled with the creation of alternative resources; and (4) the principle of caution is a basic attitude of any action. The essence of "strong sustainability" is that capital can only substitute resources to a very limited extent and only for specific "unessential exhaustible resources".

The principles of strong sustainability were inherent in various initiatives in which environmentally committed natural scientists cooperated with visionary business leaders, such as the Natural Step Framework (Nattrass & Altomare, 2001; Robert, 2002), the World Business Council for Sustainable Development (World Business Council for Sustainable Development (WBCSD), 2000), and similar initiatives (Elkington, 1994).

11.6.1 Key messages

- The discussion about weak vs. strong sustainability entered business sciences as a subfield of economics in the 1990s based on initiatives of natural scientists. The ubiquitous assumption of substitutability of non-renewable resources, the limiting of emission to the absorption capacity of biological systems, and the use of renewable resources in the regeneration limits, were key assumptions of strong sustainability.

11.7 From institutional economics to institutional ecological economics

We close this chapter by describing institutional economics (IE), new institutional economics (NIE), and institutional ecological economics. NIE developed as an alternative to the neoclassical approach and attempts to incorporate a theory of institutions into economics (North, 1995). The approaches can be seen as endeavoring to provide alternative, more realistic, explanations to what has often been called market failure. For instance, NIE used bounded rationality instead of the *homo oeconomicus* assumption of economic actors (see Box 1.4). It aspires to show that neither the market nor the state can cope with key problems all on their own. All three approaches stress the importance of transaction costs, i.e. the costs related to exchanging a good. In addition, they emphasize property rights, which include the right to use or transfer a resource. "NIE (as well as the two other approaches) borrows liberally from various social science disciplines, but its primary language is economics" (Klein, 1998, p. 1). Thus, IE approaches go beyond the inner domain of the economic system (see Figure 1.7*). Environmental issues did not traditionally play a role in this approach. However, in recent years the management of natural resources, the dilemma of the commons, and international governance have become core areas of NIE (Paavola & Adger, 2005; Young, 2002).

11.7.1 Notions of the term institution

Historically, the term institution goes back to 1551 and denotes an "established law and practice" and the "establishment or organization for the promotion of some charity" (Harper, 2001). There is no specific economic definition of institution, and the variants of IE overlap with "sociological neo-institutionalism" (Jepperson, 2002). In this book the term institution is used in two different ways. In all chapters, institution denotes a special type of organization introduced at the level of society (i.e. founded and controlled by the government) to secure and sustain a community and society. It is conceived as a specific element of the level hierarchy of human systems (see Figure 14.1*; see Chapter 16.3). A definition may read as follows: "Institutions are the people representing public organizations, statutory corporations, or formal rules and regulations established by society and state agencies" (see Chapter 16.3). Institutions are the people and the rules establishing courts, environmental agencies, armies, police, schools, etc. This definition meets the first definition of political institution made by Herbert Spencer when referring to his *Principles of Sociology* (Spencer, 1898/2004, p. 408). Spencer conceived of society as a system which differentiates itself according to functional perspectives (Dubiel, 1976) and thus introduces different types of institutions. In the language of IE approaches this definition refers to what they consider organizations.

A second meaning of institutions makes strong reference to sociology:

> "Institutions" can be defined as the sets of working rules that are used to determine who is eligible to make decisions in some arena, what actions are allowed or constrained, what aggregation rules will be used, what procedures must be followed, what information must or must not be provided, and what payoffs will be assigned to individuals dependent on their actions. (Ostrom, 1990, p. 51)

In other words, institutions are "a set of formal and informal rules, including their enforcement arrangement" (Furubotn & Richter, 2005, p. 7). As mentioned, "organizations [can be seen] as the personal [biophysical] side of the institution" (Schmoller, 1900, p. 561).

Further, we distinguish between formal institutions, which are introduced by legal systems and legally backed policy, and informal institutions, which are simply the rules that can be found in the social and cultural system of societies and organizations. The label institutional economist is used "to indicate any economist who gives institution[s] a central role in his analysis" (Rutherford, 2004, p. 180).

11.7.2 Institutional economics

Institutionalism in economics dates back to the beginnings of the neoclassic era (see Chapter 10.4) and was formed by Thorstein Veblen (1857–1929) and John R. Commons (1862–1945). Veblen criticized the focus on pure market economics. He introduced a psychological basis of economic behavior and thus undermined the abstract methods of classical and neoclassical theory. He stressed that making money and making goods are two different things (Veblen, 1919/1990). Veblen, a social theorist, sarcastically branded practices of conspicuous leisure and consumption in the Golden Twenties. He viewed consumption preferences as highly subjective but socially framed, serving to advance status and power. Attention was paid to the "psychological economist" (Commons, 1931, p. 648) and the cognitive aspects of institutions, including interpersonal and interrelational aspects (Veblen, 1900, 1919/1990). Veblen also stated that free competition was impossible in an industrial society (economictheories.org, 2007) due to guile, dishonesty, corruption, and prestigious decadence (Meštrović, 1995).

There are a couple of major characteristics in which institutionalists differ from neoclassicists. We now outline a review by Samuels (1995), who identified the following issue: institutionalists dissent from the conception of the economy as organized and guided by the pure market. A group of American economists, including Veblen, Commons, and C. E. Ayres, definitely had "no interest in détente with marginal analysis" (Bateman, 2004, p. 198). Institutionalists think like sociologists, as they consider "individuals and culture … mutually interdependent" (Samuels, 1995, p. 572). They deny "automatic mechanisms and laws" (ibid., p. 573) and consider constructs such as the free market itself as a system of social control or collective action. Moreover, they showed no interest in "strengthening a Protestant America" following this idea (Bateman, 2004).

Institutional economics shares some key ideas with neoclassical economics, such as scarcity and competition. But institutionalists see the written and unwritten rules behind economic transaction. The institutions and power structures are at least as important as the "invisible hand:"

> [Values are formed by social norms, culture,] individual identities, goals, commodity preferences and lifestyles. …The growth of mass consumption society, for example, which parallels the growth of industrial society, has been a cultural phenomenon …. generated … by the behavior of certain individuals and groups in society. (Samuels, 1995, p. 574)

According to Veblen, human economic actors inevitably first seek pleasure and avoid pain. This assumption met Mill's hedonistic principle (see Chapter 7.2). Second, they were assumed to be driven by self-interest. The individualist motivation was supposed to contribute to the wealth and well-being of the community. Third, self-interest and competition for property, place, and power were thought to promote societal growth and wealth.

Institutional economists took a constructivist, social–psychological perspective on economic transaction:

> But the smallest unit of the institutional economists is a unit of activity – a transaction, with its participants. Transactions intervene between the labor of the classic economists and the pleasures of the hedonic economists, simply because it is society that controls access to the forces of nature, and transactions are, not the "exchange of commodities," but the alienation and acquisition, between individuals, of the rights of property and liberty created by society, which must therefore be negotiated between the parties concerned before labor can produce, or consumers can consume, or commodities be physically exchanged. (Commons, 1931, p. 652)

Institutionalists consider technology to be a key issue and democratic pluralism as an important societal constraint. Institutions and technology are conceived as dichotomous. Scarcity, as a prerequisite of economic transaction, is seen relative to technology.

> [We find] a relative scarcity or relative plentifulness of all resources … [depending on] the state of the industrial arts. (Ayres, 1957, p. 27)

Methodologically, institutional economists were empiricists and followed Durkheim's meticulous recording of data rather than neoclassical theoretical visions. Veblen, for example, recorded actions of leisure behavior (Cordes, 2009). The material–biophysical environment did not play any role in IE. However, by including psychological, subjective, motivational aspects and pointing at the impossibility of a free, undistorted market, Veblen opened the door for a new, behaviorally based institutional approach to economic thinking.

11.7.3 Basic assumptions of new institutional economics

The term "new institutional economics" (NIE) was first coined by Oliver Williamson (1975). The origin of this approach is closely linked to his theory of transactions costs, which is directly rooted in the Coase theorem. NIE argues that multiple costs of a transaction matter and cannot be ignored. Transaction costs exist because it is costly to obtain information.

> The assumption of zero transaction costs, as applied by Coase to [the] issue of externalities, effectively substituted comprehensive private ordering for legal centralism. Given zero transaction costs, private parties will costlessly bargain to an efficient result. This was the "Eureka" position to which Stigler referred. Although on reflection, it is too simplistic. (Williamson, 1995, p. 10)

However, as Douglass C. North pointed out, proper political institutions contribute to reducing transaction costs by improving the security of property rights and enforcement of contracts, which are two further important components of NIE (Menard & Shirley, 2005a).

IE provides a complex view on economic actors, including a study of "the interrelationships between the legal–governmental–political and the economic–market spheres" (Samuels, 1995, p. 581). It provides a societally shaped view on a market actor as a "cultural … not just a 'rational economic person'" (Pearce & Turner, 1990, pp. 15–16). The interdependence of the market actors is an issue, for instance, in commons dilemmas and other situations where externalities and social costs matter (Coase, 1960). Coase provided a set of examples, such as how railway companies did not have to compensate those who suffer because of fires "caused by sparks from an engine" (ibid., p. 29), or how "straying cattle which destroy crops on neighboring land" (ibid., p. 2) or "contamination of a stream" (ibid., p. 2) also went uncompensated.

Furubotn and Richter (2005) developed the basic assumptions of NIE. Following IE, the approach is built on: (1) individuals, who are goal-oriented, purposeful, self-interested; (2) bounded–rational actors; who (3) try to maximize utility within the given limits and organizational structures by; (4) showing opportunistic, sometimes not praiseworthy, dishonest behavior. This is the basis for (5) forming an economic society, including the "formal and informal rules or norms that assign sanctioned property rights to each member of society" (ibid., p. 5), resulting in; (6) governance structures. The latter result from enforcement structures of the rules defined by; (7) the institutions, which themselves include the norms that can turn into social capital, if they can solve social dilemmas. The degree to which institutions are believed to mold individuals has been nicely stated by Coase:

> … if self-interest does promote economic welfare, it is because human institutions have been devised to make it so. (Coase, 1960, p. 29)

Today, NIE has become a multidisciplinary and interdisciplinary venture. As a vivid, popular field, it has many ramifications. From an economic perspective, "the main building blocks of NIE will remain transaction-cost economics, property rights analysis, and contract theory" (Furubotn & Richter, 2005, p. xviii).

NIE strongly refers to game theory (see Box 10.4 and Chapter 14.6). For instance, based on the motivational assumptions of a self-interested human actor in NIE, the principle of Nash equilibria shines as a meaningful solution. NIE only includes non-welfare-focused motivations (Paavola & Adger, 2005), or, as has been noted from a Japanese cultural perspective: "For neo-institutional economics it is difficult to explain 'welfare state' that substitutes the role of a family …" (Sasaki, 2006, p. 189). In addition, another link to game theory is seen in that an institution can be characterized as a "self-sustaining system of shared beliefs as to a salient way in which the game is repeatedly played" (Aoki, 2001, p. 8).

11.7.4 Time scales and institutional change

North (2005) critically notes that economics is made of a static body, assuming an idealized economic system with unchanged political and social institutions and culture. As North implicitly stresses, scarcity is the main driver causing institutional change. However, societal demands and values result from how an interplay of

Table 11.4 Drivers of institutional change (based on North, 2005, p. 22)

Proposition on institutional competition and change
i. The *interplay between institutions and organizations* in the economic setting of scarcity is the main driver of institutional (societal) change
ii. An *incremental altering of institutions* is caused by changing desires and evaluation of (consumption) opportunities emerging from competition and continuous investment in, and development of, new skills and technologies
iii. *Institutional structures determine* the kind of skills and knowledge having the *highest individual and organizational payoffs*
iv. *Perception* and *desires are mental constructs* and thus of a socio-epistemic nature
v. The *institutional matrix* defines society and pathways of change

given technologies is institutionally framed and allocation is defined, thus determining who is getting what payoff (see Table 11.4 iii).

We can see that, in its incipient stage, NIE, on the level of society, did not focus at all on natural resources. The focus was on individual cognition and the societal rules:

> Mental models are the internal representations that individual cognitive systems create to interpret the environment; institutions are the external (to the mind) mechanisms individuals create to structure and order the environment. (North, 1994, p. 363)

But the institutions and societal rules change – some of the drivers of this change are presented in Table 11.4.

In their critique of classical and neoclassical economics, ecological economists argued that the timeframes were too narrow. Williamson (2000) introduced four levels of "social analysis" defined by timescales on which stability is required. The levels are seen in a hierarchical structure, where the higher levels constrain the lower levels. Compared to classical and neoclassical economics, a wider timespan is viewed by Williamson, but it still only ranges between 1 and 1000 years. Williamson presents four time scales, which are: L1: continuous; L2: 1–10 years; L3: 10–100 years; and L4: 100–1000 years.

The timescale L1 refers to resource allocation and employment. Here, prices and quantities are the subject: Williamson suggests that these entities can be best described by neoclassical economics. With L2, governance is at play: the focus here is on contractual relations supported by well functioning legal orders. L3 covers the timescale between 10 and 100 years, and refers to the establishment of polity, judiciary, and execution structures (i.e. institutions in the notion as used in this book). Finally, L4 looks at the long-term perspective of institutional rules (i.e. the social–cultural–epistemic structure) and of societies, including norms, religion, and customs. Economic activities are considered embedded in the institutional rules of this level.

11.7.5 NIE as an aid to political economics and policy

IE has often been promoted with the slogan "from government to governance," which is often related to the idea that the coercive power of the state (government and societal institutions) is not at work in (most) governance structures. We follow Paavola, who identifies that "the greatest obstacle for the further extension of the new institutional approach … [is in] its mostly implicit definition of 'governance'" (Paavola, 2007, p. 94). He argues that instead of constructing a monolithic image of a government, the state should rather be understood as the arenas and instruments of collective action. The type of self-governance which is at work, for instance, in some developing countries when resource-users govern themselves under customary institutions does not negate the fact that the state also has a role. "The involvement of the state … entails a different distribution of power than self-governance solution" (Paavola, 2007). This view on the "state" is similar to the definition of "society" in the conception of level hierarchy presented in Chapter 16.3. The concept of governance does not substitute the concept of governments (or "society"):

> … governance is what governments do. (Paavola & Adger, 2005, p. 354)

11.7.6 Institutional ecological economics

As Ostrom (2005) stresses, NIE is an interdisciplinary venture. Besides the initial framework of NIE and the critique of Coase's theorem, ecological and environmental issues do not play a significant role in NIE (see Menard & Shirley, 2005b). However, NIE has been frequently used for analyzing forest and natural resources management (Behera & Engel, 2006; Paavola & Adger, 2005).

Research in the past 15 years has shown the attractiveness of the NIE conception for understanding the management of environmental resources, particularly natural resources. We present here the key concepts that are important to understand:

(1) Interdependencies of action: how what one person is doing affects others. We have defined different interdependencies in Table 11.1. The production of private goods may cause externalities for present and future resource-users, and common pool resources involve the commons dilemma.

(2) The important role of property rights and environmental regulations as institutions (Paavola & Adger, 2005, p. 356).

(3) New ecological economics assumes positive transaction costs owing to bounded rationality (i.e. limited cognitive capacity), individualistic motivation (not disclosing information to others), the unknown and changing environmental states (which makes it difficult to assess what elements of the environment are in what state and what regulations are necessary).

(4) Game theory has allowed for better description of conflict structures, asymmetries in power and information, and thus makes certain phenomena and dilemmas more accessible. Here, one can find links to behavioral economics, which has strong roots in psychological decision research (Kahneman *et al.*, 1990).

> [This type of] new institutional economics recognizes population growth, technological change and changes in relative prices or scarcity, power structure, and preferences as factors that explain institutional change. Yet it tends to focus on collective action motivated by private interests and to ignore other sources of change. (Paavola & Adger, 2005, p. 362)

11.7.7 Key messages

- There are different notions of institutions, ranging from institutions as specific organizations established by society to secure the reproduction of society, to various conceptions of institutions as the set of formal and informal rules. The latter, sociologically shaped definition, is used by institutional economics (IE) and new institutional economics (NIE)

- IE takes a constructivist, sociopsychological perspective on transactions but considers the interaction between technology and available resources as a basis of economics
- Institutional economists are empiricists
- IE and NIE incorporate key ideas from sociology, psychology, anthropology, law sciences, and policy science, and can be considered interdisciplinary fields
- The Coase theorem and its critiques have been starting points of IE and NIE
- NIE deals with a hierarchy of formal and informal institutions that constrain market activity. NIE allows insights into changes of market structures (which are not provided by other economic theories) and permits understanding the establishment of environmental regulations.

Part VI: Contributions of industrial ecology

12 The emergence of industrial ecology   307
13 Industrial agents and global biogeochemical dynamics   320

Key questions (see key questions 1–3 of the Preamble)

Contributions from industrial ecology to environmental literacy

Q1 What type of environmental impacts does industrial ecology conceptualize? Which flows and stocks are critical for sustainability?
Q2 What environmental impacts are caused by industrial production? How does recycling affect the accessibility of natural resources?
Q3 What are the disciplinary roots of and links in industrial ecology?

What is found in this part

Industrial ecology is a rather new discipline that emerged at the end of the twentieth century. It falls under the tradition of industrial engineering, which deals with the integrated system of people, information, material, and energy. It relies on "skills in the mathematical, physical, and social sciences, together with the principles and methods of engineering analysis" (Maynard, 1971, p. 22).

Industrial ecology goes beyond manufacturing and includes operations management of all industries and services. The primary human system considered in industrial ecology is a company or firm. The consumer has recently gained importance as well. The focus of industrial ecology is on the flow characteristics of energy, material, and other resources used by industrial processes and products, and on the biophysical environment and economy.

Industrial ecologists initially aimed to reduce high environmental costs from industrial production by closing energy and material cycles. Reducing waste, increasing recycling, and establishing efficient inter-firm networks to reduce environmental impacts of production and consumption have been seen as further objectives.

Chapter 12 introduces the origins and visions of industrial ecology. We introduce the eco-industrial park, an industrial symbiosis concept that involves a cluster of companies who share resources and cooperate to maximize profits while reducing pollution and waste. We then introduce the methodological tools that industrial ecologists use to describe, evaluate, and transform industrial systems. We describe how the understanding of environmental impact of products and industrial processes has gained meaning and depth by incorporating the consumer as a decision-maker and by including rebound effects that emerge from new products or consumption patterns. The chapter closes by introducing the concept of social metabolism (see also Chapters 9.3 and 9.4).

Chapter 13 sketches a specific, new, and potentially emerging vision of how the management of the global biogeochemical cycles may become the objective of industrial ecology. We deal with material flows and biogeochemical cycles on a global scale by referring to flows resulting from agricultural activities, industrial activities, and service society activities. The chapter closes with a brief discussion on management options.

Part VI Contributions of industrial ecology

Chapter 12: The emergence of industrial ecology

12.1 Engineers and human–environment systems 307
12.2 The language of industrial ecology 309
12.3 Interdisciplinary roots of industrial ecology 310
12.4 Industrial symbiosis: the eco-industrial park idea 313
12.5 Methodological tools: the workhorses of industrial ecology 313
12.6 Industrial metabolism and the sociometabolic approach 317

Chapter overview

We start with a brief look at the history of engineering and review how the field of industrial ecology emerged to help industries respond to growing concerns about pollution and resource scarcity. We then discuss the different notions of the concept of ecology as they relate to industrial ecology. After identifying the roots and basis of industrial ecology, we present the eco-industrial park idea, an industrial symbiosis concept that involves a cluster of companies who run their business while sharing resources and cooperating to minimize pollution and waste. As an engineering science, industrial ecology relies on quantitative methods. We discuss the role of the most important methods, including material flow analysis, life cycle assessment, and system dynamics. The chapter closes with a discussion on the sociometabolic approach.

12.1 Engineers and human–environment systems

Engineers were originally designers and builders of engines. They began with military engines and then moved on to fields such as civil, mining, electrical, chemical, and sanitary engineering (Davis, 1998). Engineers are results-driven and generate and apply technical knowledge to design and implement machines, materials, systems, and devices. An engineer's primary goal is to find an optimal solution given certain constraints. Engineers focus more on practical applications and thus differ from other scientists. They "look first for the forest before trying to get species names for the trees" (McCready, 1998, p. 5). Engineers ask *what* works and *why*, whereas scientists ask *how*. Ancient master builders such as Leonardo da Vinci often received admiration for their work, as signaled by the etymology of the term engineer, which originated from the Latin *ingeniator*, meaning the ingenious one.

The First Industrial Revolution, which began in the eighteenth century, fundamentally changed human–environment interaction. Able to transform energy into useful mechanical motion, engines quickly replaced brawn and thus created a more indirect relationship between humans and many activities affecting the material-biophysical environment. Engineers promoted the development of technology that served society. With the founding of the French École Polytechnique in 1794, engineering sciences began to be considered at the same academic level as conventional sciences. As Torstendahl (1993) elaborated, there were two societal motives for founding technical universities. On the one hand, they were meant to meet the demands of industry for a skilled labor force for designing, constructing, and building machines and processes. On the other hand, technical universities were understood as providers of research competence and general knowledge to guide the development of society (Scholz et al., 2004). Here a kind of general technical and natural science-based knowledge (German: Technische Allgemeinbildung) was targeted at a group of people who would work in civil service after completing university.

The Second Industrial Revolution at the end of the nineteenth century brought with it a radical societal transition. It was industrial engineers who designed and managed mass production and distribution systems, which were made possible thanks to progress in mechanical, chemical, and electrical engineering. As was pointed out in the discussion on ReSTET theory (see Chapter 8.5), cars, airplanes, designer food, and telecommunication equipment are examples of products that change man's environment.

> The engineer typifies the twentieth century. Without his genius and the vast contribution he has made in design, engineering, and production on the material side of our existence, our contemporary life could never have reached its present standard. (Alfred Sloan, cited according to Auyang, 2004)

12.1.1 From industrial engineering to industrial ecology

In engineering sciences, we can see two predecessors of industrial ecology. The first, (modern) industrial engineering, was defined in 1955 and deals with "design, improvement, and installation of integrated systems of men, materials, and equipment" (Hammond, 1971, p. 1/5). We should note, however, that the bridge to industrial ecology was built slowly. For example, the index in *Maynard's Industrial Engineering Handbook*, published in 1971, did not include the terms environment, pollution, recycling or cleaner production. The other predecessor is environmental engineering, a branch of engineering promoted in the USA after the foundation of the Environmental Protection Agency and the passage of the Clean Air Act Amendments of 1970. These and other regulations from that time were the first to prod industry to incorporate environmental cost into product and process design (Jacko & LaBreche, 2001). Clean manufacturing consequently became an interesting topic to industrial engineers. Thus, industrial engineering and environmental engineering were important forerunners of industrial ecology:

> Industrial engineers study industrial metabolism – that is, the linkages between suppliers, manufacturers, consumers, refurbishers, and recyclers. Environmental engineers, on the other hand, study environmental metabolism – that is, the linkages between entities such as biota, land, freshwater, seawater, and atmosphere. (Stuart, 2001, p. 530)

This call for cleaner production and pollution control, along with prevention-minded ideas about how materials are extracted, processed, and used, emerged in parallel with the field of environmental economics. Thus, the rising awareness of the environmental impacts of the Industrial Age at the end of the twentieth century and the promotion of sustainable development introduces a third sort of industrial revolution requiring new types of resource management and material flows in production and consumption. Here, a new type of engineering knowledge is necessary, characterized by expertise in the analysis and shaping of the environmental impacts of production and industrial activities. Industrial ecology can be taken as an innovative and integrative field of engineering that is closely related to sustainable development (Lifset & Graedel, 2002) and which includes industry, agriculture, and consumption.

> Industrial ecology (IE) is a new ensemble concept in which the interactions between human activities and the environment are systematically analyzed. As applied to industry, IE seeks to optimize the total industry materials cycle from virgin materials to finished product to ultimate disposals of waste. (Graedel, 1994, p. 23)

Here, tasks such as adaptation and complex systems (management) appear that go beyond, and sometimes seem to contrast with, the traditional purpose and orientation of engineering.

The focus of industrial ecology is nicely represented in Figure 12.1, provided by the logo of the International Society for Industrial Ecology. On top is production, serving individual consumers (largest arrow), which is based on resource extraction. Both production and consumers are seen a second time, under secondary production and secondary consumption, which are related by the arrows labeled recycling, reuse, byproduct, fewer resources extracted, and minimal waste.

A critical question disputed among industrial engineers is the role that normative aspects should take. There are two camps: one camp aspires to remain on a purely descriptive side and to keep the normative and political side out of the business (Allenby, 1999). The other camp explicates their normative objectives:

> [The] … goal of industrial ecology is the evolution of the world's industrial activity into a sustainable and environmentally benign system. (Socolow et al., 1994)

Members of this camp stress that even the objective of not surpassing desirable carrying capacity includes normative aspects (Boons & Roome, 2000). We agree

Figure 12.1). Initially, industrial ecology did not explicitly distinguish between the human system of a company (i.e. the owners or employees, $H_S^{company}$; see Figure 3.3*) and the company's material flows ($H_E^{company}$), which are part of the material environment. This is beginning to change, and both the producing firm and the consuming individual or household is increasingly seen as a decision–maker and their interests, which are part of the socioepistemic layer, become the subject of research. The machines and products of a company are seen as part of the material-biophysical environment.

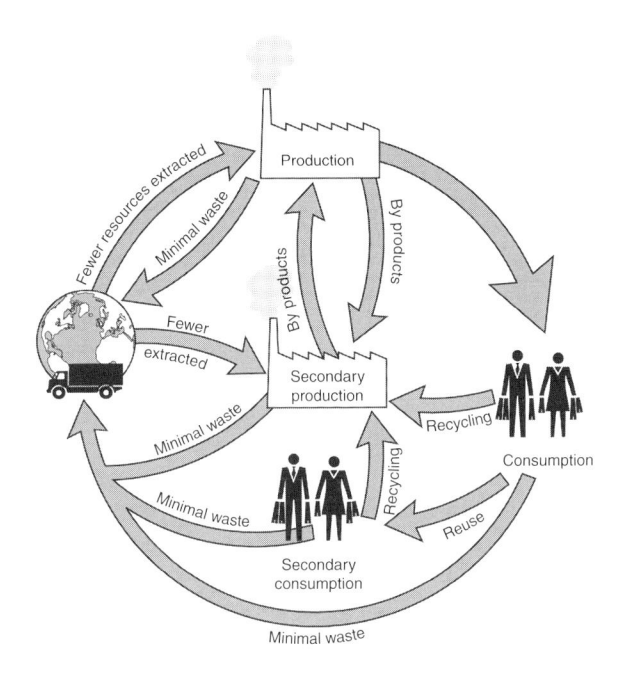

Figure 12.1 The industrial ecology process. Secondary production is a result of minimizing waste, recycling, and using fewer resources (International Society for Industrial Ecology (ISIE), 2009).

12.1.2 Key messages

- Industrial ecology is an engineering science emerging from industrial engineering, which focuses on optimizing the material cycles along the supply chain by recycling, reusing, utilizing byproducts, using fewer extracted resources, and minimizing waste
- Industries, from the firm to the farm, are at the core of industrial ecology, where efficiency and optimizing material flows build the starting point. Industrial ecologists are starting to incorporate the human systems that are part of these organizations into analyses.

12.2 The language of industrial ecology

The term industrial ecology relates to two fields: *engineering*, focusing on design, production, and material flows, and *ecosystem ecology*, which studies "the relationship of the organism to the environment" (Haeckel, 1866, p. 149). The term itself was coined by Robert A. Frosch and Nicolas Gallopoulos, both from General Motors. At a meeting of the National Academy of Engineering in 1988, they adapted Robert U. Ayres' concept of industrial metabolism for describing "pathways of some materials and their transformation through industry and through society" (Frosch, 1992, p. 800). Clearly, such a view includes cleaner production by technological innovation.

The field of industrial ecology also has a consumer side. This has been nicely expressed by Robert White, a former president of the US National Academy of Engineering:

and think that sustainability is definitely a regulating idea of the post-industrial society. For us, it includes the idea of avoiding unwanted collapses in the sense of organizing system limit management and allowing for intergenerational and intragenerational justice (Laws *et al.*, 2004), which is certainly normative. But it also allows for interpretation and challenges engineering knowledge to understand the causes of collapses and to establish means to avoid them.

From an industrial ecology perspective, as Socolow expressed it, firm or farm are considered to be change agents who have the potential to be active in policy-making: "Industry becomes a policy-maker and not a policy taker" (Socolow, 1994b, p. 13). However, the rationale, mechanisms, and political constraints for engaging industry in policy-making have yet to be deeply examined.

Given our environmental literacy agenda, we take an interest in how industrial ecology views the human–environment relationship. In our terminology, industrial ecology focuses on the material–biophysical flows caused by a company, a company network (see Figure 12.3), an industry, a nation-state (see Figure 12.4), or world population (see

Industrial ecology is the study of the flows of materials and energy in industrial and consumer activities, of the effect of these flows on the environment, of the influences of economic, political, regulatory, and social factors on the flow, use, and transformation of resources. The objective of industrial ecology is to understand better how we can integrate environmental concerns into our economic activities. (White, 1994, p. v)

In this statement, consumer activities are the key entities, as also conveyed in Figure 12.1. It would be surprising if engineers became specialists of consumer behavior; however, the industrial ecology community has clearly acknowledged that the demand side determines what is produced and to what extent recycling is viable (see Box 12.1). When the *Journal of Industrial Ecology* published the 2005 Oslo Declaration on Sustainable Consumption, sustainable consumption became part of the dialogue of industrial ecologists worldwide (Tukker *et al.*, 2006). Still, the consumer perspective is not yet fully integrated in industrial ecology. We can see that this view potentially contrasts with the optimization of profits as a primary rationale for enterprises.

12.2.1 Nature as a model

Before we discuss the roots of industrial ecology further, we briefly consider the concept of ecology and the nature term applied therein. Isenmann (2007), for example, found that the concept of nature is referenced in various ways. In a strong reference, industrial ecology takes the "natural environment as a model" (Tibbs, 1992, p. 2), since "nature is the measure of man" (Socolow, 1994b, p. 4). Here, very close analogies are assumed in the way that industrial ecologists "dream that, as it is in nature, waste equals food" (Cohen-Rosenthal, 2003, p. 14). A weaker reference is given by analogy, for which we find formulations such as how the "industrial system can be seen as a certain kind of ecosystem" (Erkman, 1997, p. 1). Both positions, the normative, ontological vs. the metaphorical, epistemological, can lead to learning in the sense that "biological systems … are instructive" (Wernick & Ausubel, 1997, p. 7). In contrast to physics, which often centers on asymptotic steady-state modeling, ecological systems show adaptive cycles.

As discussed in Chapter 16, particularly when dealing with the Complementarity Postulate P1, the nature reference and analogy has the shortcoming that the interests, values, and cultural dimension of the human population, societies, companies, and individuals such as social entrepreneurs are not appropriately acknowledged. For example, in the anthropocene, a world that is dominated by the top predator human species, more than 50% of the world elements are disturbed by human action.

12.2.2 Key messages

- Industrial ecology views human–environment interaction from the perspective of material and energy flows in industrial production. The world dialogue among industrial ecologists also incorporates the concept of sustainable consumption. The economic impacts are a key driver and the consumer activities are the triggers
- There are different definitions of ecology and nature in the industrial ecology community. Some employ nature as a metaphor for looking at industrial processes, while others consider industry as an ecological system.

12.3 Interdisciplinary roots of industrial ecology

This section discusses how industrial ecology has roots in disciplines including industrial engineering, geography, economics, agriculture, physiology, sociology, and biology as well as the work by some countries toward sustainability. As early as the First Industrial Revolution, concerns about the "momentous consequences" of human action (Marsh, 1864/2003) were formulated by, among others, geographers and economists. The focus at that time was plainly on pollution. However, as a result of the import of raw materials (e.g. wool, silk, cotton) by the European proto-industry around the same time, the quality of material and energy flows was the object of interest. Karl Marx even pondered the subject. He considered labor, man's ability to use tools and instruments (production means), both as a key aptitude of humankind and, in a second step, as a driver of material flows.

> All the phenomena of the universe, whether produced by the hand of man or indeed by the universal laws of physics, are not to be conceived of as acts of creation but solely as a reordering of matter. Composition and separation are the only elements found by the human mind whenever it analyses the notion of reproduction … [be it when] secretions

of an insect into silk, or some pieces of metal are arranged to make the mechanism of a watch. (Verri, 1771, quoted according to Marx, 1956, p. 34)

This Stoffwechsel, defined by John R. Ehrenfeld, one of the pioneers of this field, as the "physical exchange of matter" (Ehrenfeld & Chertow, 2002, p. 334), had already been of interest to eighteenth-century Italian economists. Thus the term was used in the eighteenth century despite the opinion of some authors that it had been initially used by Marx in discourses with nineteenth-century German agricultural physiologists such as Justus Liebig (Fischer-Kowalski, 2002). Today, Stoffwechsel is often translated as metabolism, based on its more literal meaning of involving change. This notion has been around since 1743 (Harper, 2001). The concept of industrial metabolism has been used in different ways. Baccini and Brunner, two civil engineers, used the term to describe material flows in regional systems (Baccini & Brunner, 1991), and others applied the concept on a societal level.

From an interdisciplinary standpoint, industrial ecology has also been influenced by, for example, ideas from the founders of ecological economics such as Boulding (see Box 11.1), Daly, and Georgescu-Roegen, as well as Kneese and the physicist Ayres, who presented the first material flow analysis (Ayres & Kneese, 1969). The primary idea of this approach was to control and avoid unfavorable entropy by monitoring the resources and material flows.

We have already introduced the *IPAT* equation in Chapter 8.4.2, developed after the work of biologist Paul R. Ehrlich and physicist John P. Holdren who raised concerns about population growth in the 1980s (Ehrlich & Holdren, 1971). The model links environmental impacts to population growth, resource utilization, and consumer affluence.

The biological roots of ecology have been presented in Chapters 4.11, 5.6, and 5.8. Industrial ecology refers to Tansley and Lindeman's energy- and nutrient-centered definition of ecosystem (see Figure 4.24) or the overshoot, adaptation, and recovery of population dynamics (see Figure 5.15).

Historically, different countries have dealt with key questions of industrial ecology. Erkman asserts that, in the early 1980s, Japan was "the only country where ideas on industrial ecology were ever taken seriously and put into practices on a large scale" (Erkman, 2002, p. 33). Although we acknowledge that Japan pioneered the field in their own way, German history also reveals how industrial ecology thinking became important with natural resource scarcity, economic stress, and environmental concern (see Box 12.1).

12.3.1 Key messages

- Industrial ecology was shaped by the emphasis on energy and nutrient cycles in the work of the ecologists Tansley, Lindeman, and H. T. Odum. Further influences include ideas from ecological economics and concerns of population growth voiced by, among others, P. R. Ehrlich
- The concept of industrial metabolism (German: Stoffwechsel) has two interpretations: one refers to describing material flows in industrial processes and is derived from the notion of the exchange of material matter that emerged in the eighteenth century. The other relates to the biological notion of the organismic exchange of nutrients and energy and is applied at the societal level.

Box 12.1 Cycling drivers for recycling: GDR's SERO

Recycling is a key concept of industrial ecology. Yet what becomes recycled by industry depends on various factors, primarily values (i.e. environmental awareness) and economic drivers (e.g. caused by shortages of natural resources), both of which can change rapidly.

Exploring changes in recycling between 1933 and 1990 in Germany is an interesting case that allows us to compare how the issue is dealt with in a society at peace and at war, and in a society under socialist planning vs. a free market economy.

Warfare is destructive. Between 1933 and 1945, the Nazi army did not recycle. Conversely, German households and companies did recycle. The National Socialist government was the first to decree separate waste collection nationwide. At the end of World War II the garbage sector was directed to "total war," and the police were authorized to control the separate waste collections in private households (Huchting, 1981). Thus, there were two institutions in one society, the army and the police, following conflicting recycling directives.

Box 12.1 (cont.)

After 1945 the Nazi waste collection structure continued to operate in the Soviet occupation zone, yet lost its compulsory nature. However, soon after the end of World War II, East Germany found itself locked in a socialist economy and running short on commodities. This prompted the German Democratic Republic (GDR) to introduce an incentive scheme in 1951 to encourage East Germans to recycle in exchange for bartered goods. Central collection stations provided five rolls of wallpaper or ten rolls of toilet paper for 2.5 kg of waste paper (Maier, 1997).

The isolated economy of the GDR faced difficult times and, in the early 1980s, the GDR broke down due to bankruptcy. This stimulated a flourishing recycling industry. The GDR even imported five million tons of waste (toxic and recyclable) from West Germany annually, 40% of its industrial waste in 1984, and thus gleaned about 500 million USD (Petschow et al., 1990). As waste became a valuable product, a fundamental shift in language came about in the GDR.

Garbage or waste (German: Müll or Abfall) began to be called secondary raw material (German: Sekundärrohstoffe or SERO). In 1949, the waste industry was called the German Trade Center for Old Matter (German: Deutsche Handelszentrale Altstoffe), and it was again renamed in 1953 as the People's Enterprise for Raw Material Reserves (German: VVB Altrohstoffe), which received scientific support from the Institute for Secondary Raw Materials (Calice, 2005; Schulz, 1986). It is worth noting that the terms byproduct (German: Abprodukt), a term taken from chemical engineering, and old material (German: Altstoffe), indicate that no waste existed that could not be recycled – they were all considered usable residues. We should note that, even during the oil crisis of the 1970s, when countries around the world struggled to deal with limited resources, the GDR supported recycling according to the moral pattern of internationality (Calice, 2005). In sum, GDR policy is an example of a formal recycling regime.

Conversely, the post-World War II West German economic engine took the end-of-pipe track, importing raw materials from the world market and exporting waste to the GDR. No drivers for recycling existed before emerging environmental awareness in the 1980s highlighted environmental values.

Today, the term informal recycling is used to denote the hundreds of thousands of people whose livelihoods depend on collecting waste (Figure 12.2). There are various types, such as door-to-door waste buyers or those that pick waste from dumps or streets, who have positive economic and critical social and health effects (Wilson et al., 2009, 2006). About 2% of the Asian and Latin American urban population earn their livelihood by waste picking (Medina, 2000).

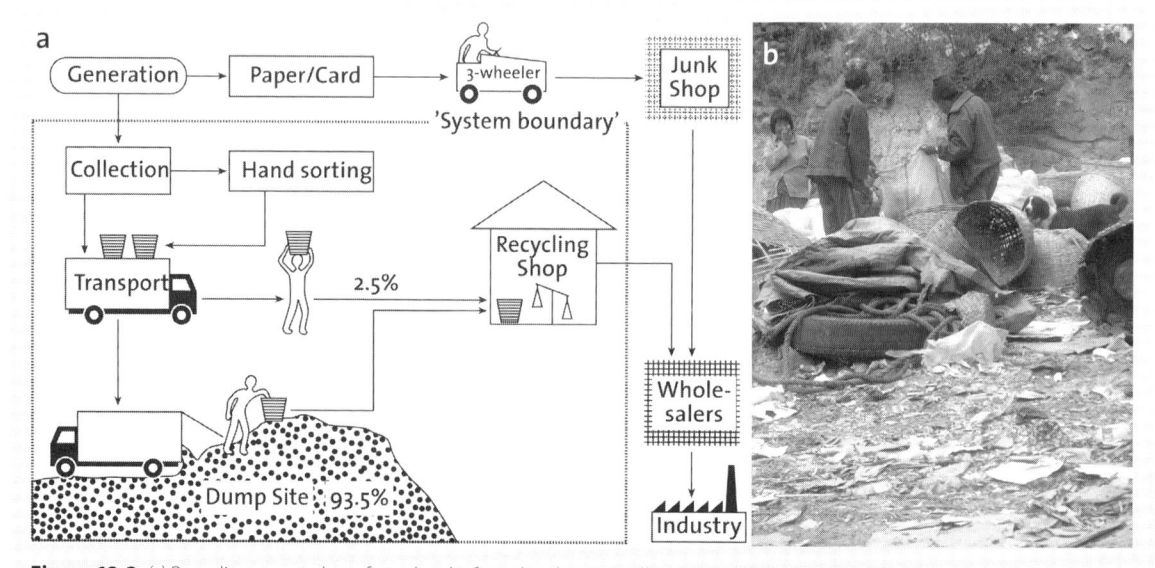

Figure 12.2 (a) Recycling streams have formal and informal pathways in all countries of the world. Material flows from Bangkok (from Wilson et al., 2006) and (b) open waste pickers in Bhutan (photo by Scholz).

12.4 Industrial symbiosis: the eco-industrial park idea

In the terminology of industrial ecology, an eco-industrial park engages companies from different industries that traditionally operate separately into a collective, cluster-based (Scholz & Stauffacher, 2007) cooperation to create competitive advantage by optimizing "the physical exchange of materials, energy, water and by-products among several organizations" (Ehrenfeld & Chertow, 2002, p. 334). Geographical proximity, location-based exchanges, and finding an optimal combination of industry types are essential here.

This is also reflected by the use of the biological term symbiosis, which denotes the "living together in more or less intimate association or even close union of two dissimilar organisms" (Merriam-Webster, 2002). The idea behind eco-industrial parks, that organizations show a better environmental performance if they operate collectively, can be interpreted as a kind of collective environmental rationality on a company level (see Box 1.4). The objective of relating different sectors by allying different companies through material flow management defines a win–win situation from an ecological and economic perspective (Jacobsen, 2006). Clearly, this idea is not completely novel and can be considered a material stream-shaped variant of Alfred Marshall's (1890/1920) industrial district or a clustering along material flows instead of geographical proximity along a production chain (Porter, 1998; Scholz & Stauffacher, 2007).

The best known example of industrial symbiosis is the Danish Kalundborg complex (see Ehrenfeld & Chertow, 2002), where a couple of enterprises, including a power station, a huge oil refinery, a pharmaceutical and biotechnology company, a wall-board producer, and a soil remediation company planned an eco-industrial park. What was exciting about the plan for this eco-industrial park was that the firms wanted to increase efficiency (in an ecological manner) by using each other's by-products. The Kalundborg eco-industrial park has developed since 1959, reducing toxic and CO_2 emissions and saving water. It is interesting that scarcity of natural resources, that is groundwater scarcity, had been the central driver. The park, for instance, has been able to reduce freshwater consumption by 25%. In 2007, almost 100 eco-industrial parks similar to Kalundborg could be identified (Posch & Perl, 2007). Figure 12.3 presents such a project at the Humber Estuary on the east coast of England (Mirata & Pearce, 2006). The issue of industrial symbioses has been subject to scientific study since 1989 (Chertow, 2007).

From a qualitative systems view, industrial ecology requires integral design of industrial systems to minimize resource consumption (Mulder, 2009). In a well designed eco-industrial park, the wastes (i.e. the output) of one company become the resources (input) of another, and cascades of energy (a concept similar to how water flows from higher to lower levels in natural waterfalls) should be followed. Cost–benefit considerations are essential. A park should include repair facilities, a decomposer industry to dismantle and discard critical products, re-manufacturing plants to bring discarded products to specification, and scavenger industries, which "gobble up" wastes just as microorganisms do in nature (Reinders, 2009). Recent studies have analyzed eco-industrial parks as agent-based, adaptive systems (Batten, 2009), and thus provide a better understanding of the incentives and dynamics of inter-firm networks. This is remarkable as an eco-industrial park challenges new business practices such as open book calculations and reduced flexibility of production.

12.4.1 Key messages

- An eco-industrial park yields cost advantages for companies from different industries if they can utilize waste and byproducts from neighboring companies.

12.5 Methodological tools: the workhorses of industrial ecology

Without doubt, the *methods* applied in the context of industrial ecology are a strong part of this young discipline. We distinguish between methods for: (i) *describing/representing* industrial ecological systems; (ii) *evaluating* "industrial systems or processes;" and (iii) *transforming* these systems. Here we introduce how key methods such as material flow analysis (MFA), system dynamics (SD), and life cycle assessment (LCA) help industrial ecologists to understand the interaction of industrial (ecological) and human systems. The reader can gain further information about these methods in Scholz and Tietje's *Embedded Case Study Methods* (Scholz & Tietje, 2002), and other books specializing in one of these methods (see Baccini & Brunner, 1991; Guinée, 2002). We also present variants of environmentally extended

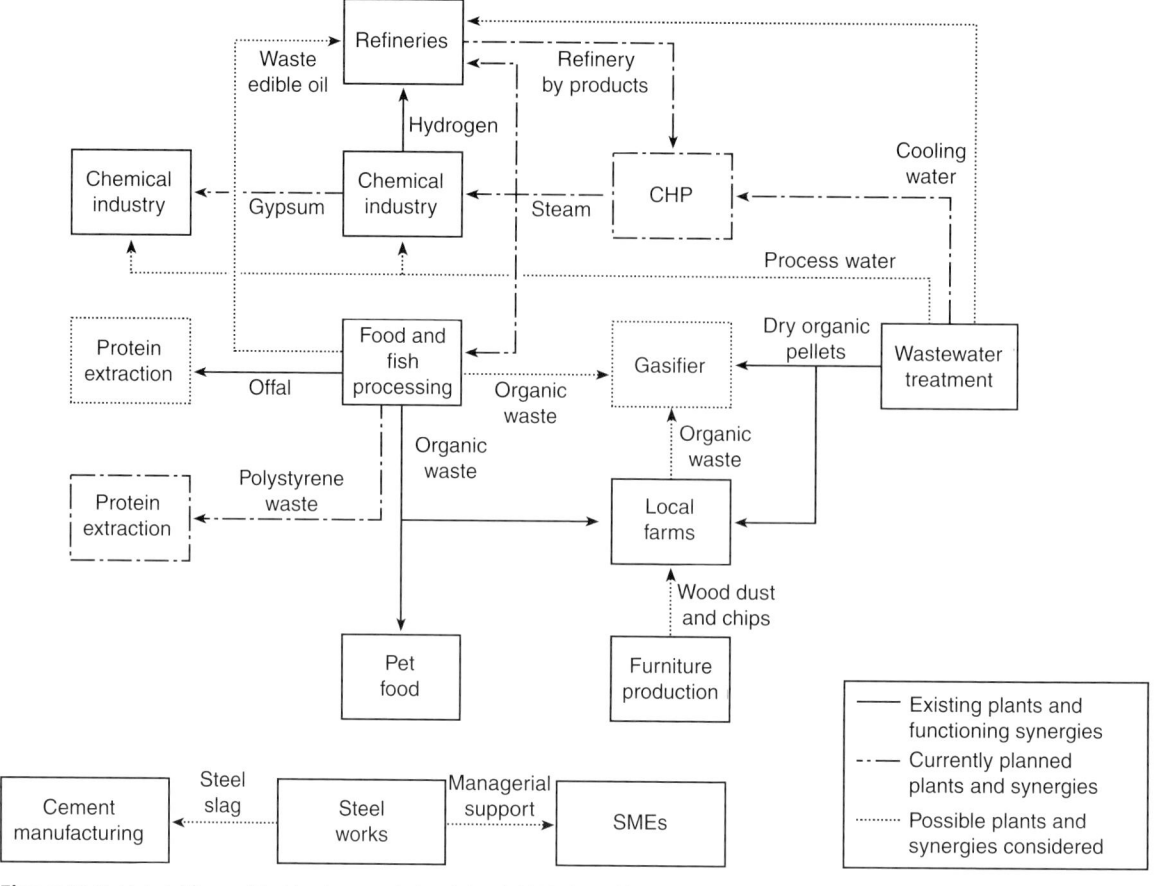

Figure 12.3 Material flows of the Humber eco-industrial park, UK (adapted from Mirata & Pearce, 2006, p. 92). CHP, combined heat and power; SMEs, small and medium enterprises.

input–output analysis (EE-IOA), which began from material and energy flows but was later linked to economic and social dimensions.

12.5.1 Material flow analysis

One of the first methods of industrial ecology, MFA, is now frequently used. MFA analyzes throughputs of process chains in manufacturing, agricultural production, chemical processes, or subprocesses such as recycling (see Figure 12.2a). It serves to answer the following questions. Which material flows are relevant and substantial for a process, product or regional system? Which flows and stocks (the key terms of MFA) are critical or can be taken as indicators for sustainability? Stocks are what accumulates in the "boxes" (see Figure 12.4) and flows are the transfers between the different boxes. MFA can be used for modeling flows in networks of companies or integrated chain management (Udo de Haes et al., 1997). From a sustainability perspective, the guiding idea is cleaner production, which involves the use of less hazardous material flows (i.e. detoxification) through more ecological design, as well as the use of fewer materials to obtain a desired service (i.e. dematerialization) by presenting an accounting of material flows.

Historically, MFA originates from input–output models (Leontief, 1966) and is based on the linear model (see Chapter 14.2). For instance, the analysis of nitrogen fluxes (i.e. flows) in a region may reveal that high nitrate concentrations in the groundwater may be influenced by agricultural practices (Baccini & Oswald, 1998). In another example, eutrophication of the North Sea was found to be caused by household food consumption (Udo de Haes & Heijungs, 2009; van der Voet et al., 1996).

MFA is used for system characterization and problem identification. It helps analysts to "put things on

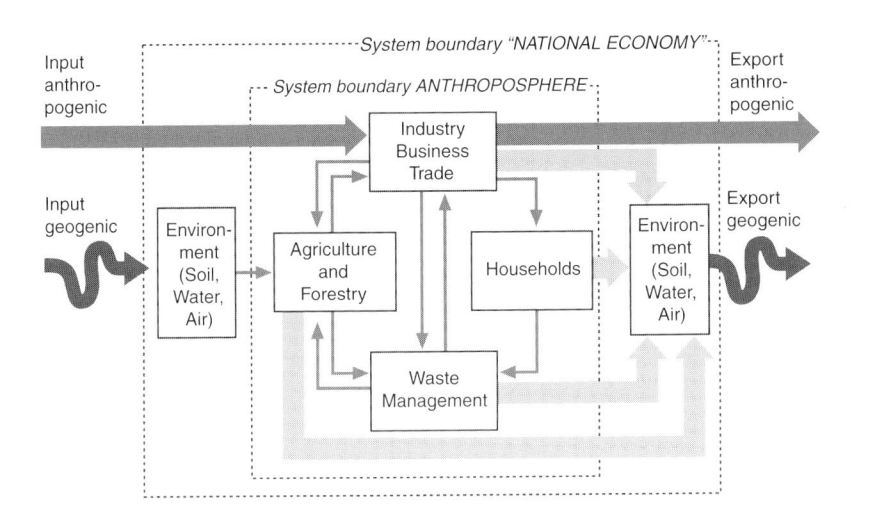

Figure 12.4 Material flows related to a national economy (adapted from Baccini & Bader, 1996, p. 21).

the radar." In principle, an MFA is applied if there is concern about relevant flows, thus finding the most relevant processes. In the frame of environmental literacy, we want to know by which processes what flows become critical. Flow for pollutants, energy, and materials can also be modeled with the System of Environmental and Economic Accounting (SEEA), a UN statistical database recording the contribution of the environment to the economy and the impact of the economy on the environment.

The question of stockbuilding is also of interest. Thus, we can ask where toxic materials are accumulating or how much of a scarce material, for instance copper, can be found in buildings or urban systems (Brunner & Rechberger, 2003). The primary challenges for conducting a meaningful MFA are the definition of the system boundaries, the specification of the activities, processes, goods, and products to be analyzed, the selection of indicative elements, and the assessment of output flows (e.g. a pollution potential).

MFA has some limitations when it comes to modeling material flows, one being that the flows aggregate over time and space, and MFA does not adequately represent specific local conditions. A second is that MFA contains no explicit valuation of the importance of relevant material flows, so assessment must follow MFA. Another practical problem is that MFAs are difficult to calculate if there are many different flows. For the Humber flow chart (see Figure 12.3), for instance, MFAs for about 10 different substances (in a very rough manner) must be combined.

Because relevant flows of materials are linked to financial flows, and in the end to GDP, nations utilize MFAs (see Figure 12.4). For example, the European Environment Agency computed the total material requirement of the European Union. Economy-wide MFAs are also called physical input–output analyses, referring to the monetary input–output analyses carried out by most national statistical offices.

MFA has strengths and critical limitations: on the positive side, it provides a thorough analysis of the material flows in the environment. If, for instance, a country decides to support bioethanol as a vehicle fuel (see Chapter 18, case 3), MFA would markedly reveal what flows will be involved on a global scale. Thus, MFA is a basic tool for environmental analysis (see the Environment First Postulate P7, Table 16.1*). Unfortunately, MFA does not offer a direct analysis of the human systems – the actors, decision-makers, and agents – involved. This must be supplemented (Binder, 2007; Binder *et al.*, 2004; Kempener *et al.*, 2009). Moreover, MFA is linear and deterministic. These constraints, however, can be overcome.

12.5.2 Industrial systems modeling and system dynamics

In principle, system dynamics is nothing more than modeling non-linear dynamics. The key research questions read as follows. What type of dynamic behavior does the (industrial) system exhibit? Which (impact) variables seem to be relevant for the system and possibly for the case? Which developments in the case may occur or are expected? System dynamics also allows for incorporating decision-makers, their changing behavior, values, and preferences (e.g. purchase behavior and agents) in a model (see Chapter 14.3).

System dynamics was initially developed for modeling mechanical problems with multiple mass points by ordinary differential equations (Dieudonné, 1985). An important further step was general systems theory, which was well received in natural and social sciences and was even denoted as a "skeleton of science" (Boulding, 1956, p. 197). Forrester's *Industrial Dynamics* became a pioneering work in the field of environmental literacy and world modeling (Forrester, 1961; see Chapter 14.3 and Box 16.7). The system dynamics model is descriptive and requires assumptions about the cause–impact effects of known or hypothesized relationships (Bossel, 1994). Knowledge about the key variables, the model structure, and functions are all essential. The reader can find a discussion on the strengths and flaws of system dynamics in Scholz and Tietje (2002; see also Chapter 14.3).

12.5.3 Life cycle assessment

LCA takes a cradle-to-grave view of the supply chain and addresses the following questions. Which of several alternative products or services that are assumed to fulfill the same functions should be selected because of their minimal environmental impacts? How and where within the life cycle of a product or service is it possible to reduce the environmental impact of a decision alternative, product or service? Thus, in some respect, LCA constructs a causal link between production processes that are designed by certain human systems, the associated environmental stresses (e.g. on ecosystem quality or human health), and the products that are demanded by certain consumers or other producers (Hertwich, 2005; Udo de Haes *et al.*, 2004).

An LCA consists of four stages. After (1) goal and scope definition there is (2) a life cycle inventory (LCI) which provides a model of the product system. This model represents the industrial, production, and economic processes involved in the life cycle and quantifies the flows between these processes and the environment (Heijungs & Hofstetter, 1996). The (3) life cycle impact assessment (LCIA) is mostly dependent on multi-attributive utility modeling to evaluate the flows of the LCI. Here, different impact categories function as attributes. In principle, the focus on impact categories in the LCIA can be input- or output-related.

Input-related categories focus on resource depletion, particularly on non-renewable resources, biodiversity, and land use. Output-related categories concentrate on impacts such as emissions affecting the climate, the stratospheric ozone layer, inducing human or ecotoxicity, eutrophication. An LCA concludes with (4) an interpretation, including an improvement analysis. Contrary to the environmental impact assessment, which focuses on local impacts, LCA takes a global perspective.

In the past two decades, a number of different LCIA methods have been developed that provide an impact assessment on different impact categories, such as human health, ecosystem quality, or resource scarcity. A series of different methods, for instance, Eco-indicator 99, is discussed in Udo de Haes and Heijungs (2009). Clearly, these methods depend on the weights allocated to these categories (Mettier & Scholz, 2008) and are thus highly value-dependent. There are thoroughly designed databases such as Ecoinvent (Frischknecht *et al.*, 2005), which model average product systems.

Recently LCA research took a more systemic perspective. *Consequential* LCA aims at assessing changes in life cycles which are indirectly induced by a decision on the level of a product system. For instance, as production capacity increases or technologies improve, more time or money is made available to the consumer for other products or services (Finnveden *et al.* 2009; Scholz & Girod, 2010). *Prospective LCA* strives to provide information on environmental impacts of future life cycles. A prospective LCA includes changed accessibility of resources and evaluation criteria in future settings and can either be consequential or attributional (Ekvall *et al.*, 2005; Spielmann *et al.*, 2005). An attributional, prospective LCA focuses on the environmental impacts of future, average product systems.

A key lesson learned in LCA is that both the inventory as well as each impact assessment are *functional* and depend on the specific perspective and interests of the modeler. Thus the results are ambiguous (Werner & Scholz, 2002). LCA is an interesting tool but involves many fundamental weaknesses. One of the most critical issues is that, when comparing two products, for example two chairs, the functional unit of each may differ. A carefully designed chair can serve as an aesthetic art object and a sturdy chair can function as a step ladder. The function an object serves depends upon its actual use.

12.5.4 Input–output analysis

We have introduced physical input–output analyses in the frame of MFA above. An environmentally extended input–output analysis (IOA), also known as an input–output LCA (IO–LCA) combines a monetary IOA with LCIA information and provides information about the environmental impacts of countries and industries. Whereas LCA presents the flow of physical products related to the life cycle, to production, use and disposal processes, IO–LCA presents the transfers among industries or other aggregated units. For instance, Hendrickson and colleagues elaborated an IO–LCA for the US economy, accounting environmental impacts of about 500 sectors (Hendrickson, 2006; Hendrickson et al., 1998). Such modeling requires huge investments in data acquisition. The method can also be applied to evaluate or compare sustainable investment funds with other funds (Koellner et al., 2007), which is possible if one succeeds in allocating companies to industries and then applies the available tools providing comparisons between industries (Duchin, 2004; Hendrickson et al., 2002; Suh, 2008).

A challenge of IO–LCA is to consider changes that are induced by technology improvements. If products become cheaper, require less resources (including space), or transportation technologies run faster, the additional money, resources, and time will be used in other domains. This is the idea of rebound effect analysis (Hertwich, 2005; Spielmann et al., 2008; Thiesen et al., 2008). From the study of human–environment systems (HES), rebound effects are of special interest, as they switch the perspective from an isolated (and often context-free) analysis of physical processes and products to the behavior and decision of consumers.

Analyzing the influence of different determinants of technological or input change on output indicators has been called structural decomposition analysis (SDA; Hoekstra & van den Bergh, 2004). SDA also allows for a better understanding of individual decision-makers and household consumption patterns.

12.5.5 Key messages

- MFA is a modeling tool that allows representation of material flows. It is a method to describe environmental dynamics and, while it can be linked with agent analysis, it does not incorporate the rationale of human agents and decision-makers
- LCA takes an important role in environmental literacy, as it provides an environmental evaluation of impacts resulting from complex industrial processes. Contrary to environmental impact assessment, which focuses on local impacts, LCA takes a global perspective
- As in many sciences, (general) systems theory and system dynamics play a major role in describing processes. This is also an important tool for industrial ecology
- The methods used in industrial ecology focus on the environmental impacts resulting from products, processes, and services. The human system and consumer perspective has, however, been neglected.

12.6 Industrial metabolism and the sociometabolic approach

The sociometabolic approach (see also Chapter 19.4) relates industrial processes to human societies. The concept thus provides a coupled systems view and goes beyond a direct transfer of the ecosystem concept to industrial systems as "a living complex organism, 'feeding' on natural resources, materials and energy, 'digesting them' into useful products and 'excreting' waste" (Johansson, 2002, p. 73). This simple transfer, which dominated the first two decades of industrial ecology, is critical because it suggests that machines behave like living cells.

Ayres (1989) was the first to use the concept of metabolism in the frame of industrial ecology. Socolow, who also stressed a strong link between industrial ecology and the human dimension and society, pointed out that "the term industrial metabolism refers to the totality of civilization" (Socolow, 1994a, p. xviii). This general perspective was also suggested by veterinarian and biologist Stephen Boyden, the director of UNESCO's Man in the Biosphere Program (Boyden, 1981). According to Boyden (1992; see Figure 12.5), nature and human society are engaged in a process of biometabolism (i.e. the input–output of materials and the throughput of energy in the living organism) and technometabolism (i.e. the internal metabolism and outputs of materials and energy performed outside the human body; Boyden, 2001). Further, artifacts are defined as non-living products of human technological activities, including buildings, roads, machines, clothing, etc. (Boyden, 2001). Artifacts are considered part

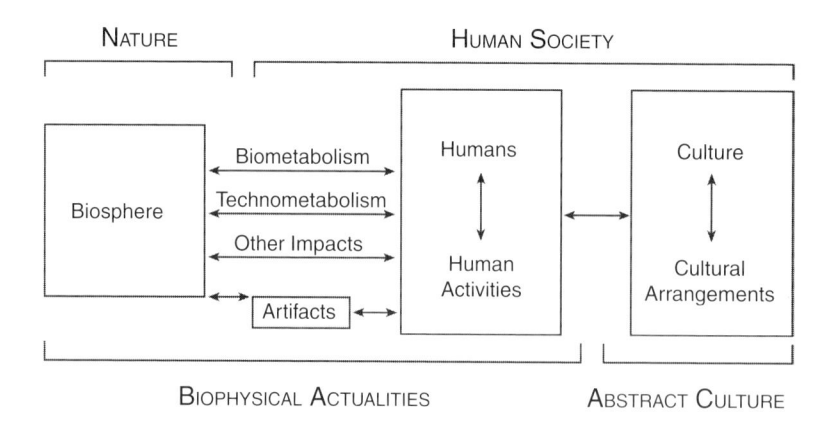

Figure 12.5 Boyden's human society and the biosphere model (see Boyden, 1992, 2001). The arrows between the boxes represent exchange of material matter or impacts; the arrows within the boxes represent relations between systems (humans, culture) and specific realizations of these systems (human activities, cultural arrangements).

of human society, which is in contrast to the definition of society in this book. Boyden also distinguished biophysical actualities, which correspond to processes in the material–biophysical–technological world, and abstract culture. Cultural abstractions, as defined by Boyden, include beliefs, assumptions, attitudes, values, and cultural arrangements, the latter of which includes the economic system, political organization, and institutional structures (Boyden, 1992). Cultural abstractions are part of the sociocultural layer of human systems (see Figure 3.3*).

Figure 12.6 is a variant of Figure 12.5 that is emerging by intersecting human society with biophysical actualities. This intersection includes the material–biophysical side of human systems, as presented in Figure 3.3* in the left circles. Culture receives a broad definition and is defined as "the total socially acquired life-style of a group of people including patterned, repetitive ways of thinking, feeling and acting" (Harris, 1991, p. 9). We should note that, in the sociometabolic conception, the concept of society includes what has been called the artifacts (e.g. the built environment; see above), the definition of which differs from that presented in this book (see Chapter 3):

> In our understanding, society comprises a cultural system, as a system of recurrent self-referential communication and material components, namely a certain human population as well as a physical infrastructure such as buildings, machines, artifacts in use, and animal livestock, which in their entirety can be denoted as "biophysical structures of society". (Haberl et al., 2004, p. 201)

Some similarities in the framework of Figure 12.6 and the complementarities presented in Figure 3.3* are linked to the concepts of program and representation. In simple terms, program relates to the mind–body relationship (our mind can plan and evoke body action). There must be a kind of "software designed to work on the human body" (Fischer-Kowalski, 1997, p. 129). From an industrial ecology perspective, this software is the technology knowledge of a human system, organizational principles, firm's philosophy, etc. Conversely, the physical stimuli we receive and the cells' activities result in a representation of the environment. In other words, at the level of the individual, this means looking from the body to the mind. In more general terms, this refers to the fundamental assumption of philosophy, psychology, or cognitive sciences that the socio-epistemic layer represents the experience gained by an individual or society. We do not, however, agree on the use of labor and nature as aspects of the causation of natural systems (see left side of Figure 12.6). In the Fischer-Kowalski framework, the terms are used in reference to Marx:

> So far, therefore, as labor is a creator of use-value, is useful labor, it is a necessary condition, independent of all forms of society, for the existence of the human race; it is an eternal nature-imposed necessity, without which there can be no material exchanges between man and Nature, and therefore no life. (Marx, 1887/1990, p. 36)

Instead of the term labor, we prefer the simpler and more general term human activities (Scholz & Binder, 2003, 2004). Any physical human activities (see Figure 17.1*) induce changes in the material environment, although, of course, an industrial company's activity shows a specific quality. Finally, in this book we avoid the use of the term nature and prefer the concept of material–biophysical environment. In contrast, the Fischer-Kowalski & Habert (2007a) refer to the French economic–anthropologist Gaudelier, who states that humans "have a history because they transform nature"

12.6 The sociometabolic approach

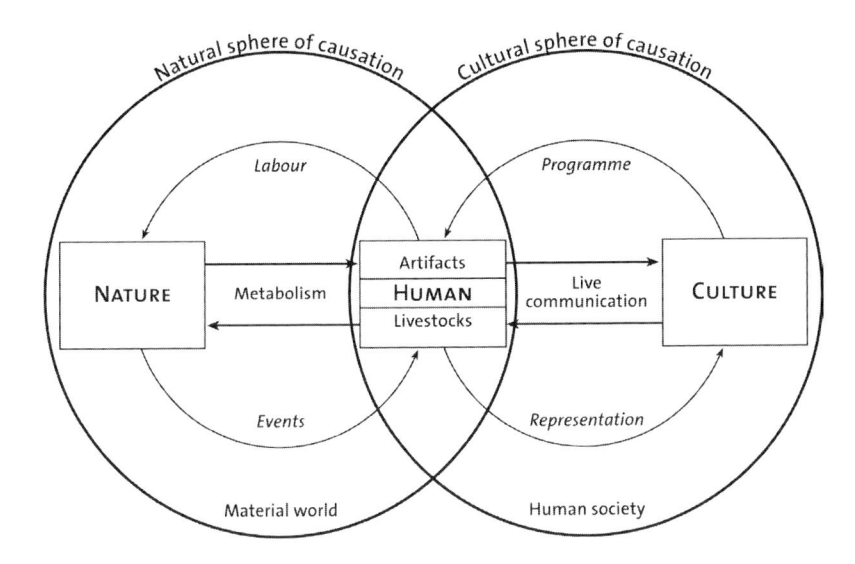

Figure 12.6 The interaction model of society and nature, with metabolism and live communication as interaction modes (modified from Fischer-Kowalski, 2003, p. 315).

(Gaudelier, 1986, p. 1). Domesticating animals is a special form of this transformation or appropriation of nature.

The term colonization, as used by Fischer-Kowalski, presents agriculture as a type of human labor that replaces natural ecosystems. Thus colonization refers to the transformation of natural processes by human manipulation, such as gene technology and land use changes, and cannot be considered a pure material flow related to the input–output relationship (Haberl, 2001). Colonization thus stresses the biophysical impacts on the environment and supplements the metabolism concept with a spatial unit, including land use and deviations from a natural reference state (Koellner & Scholz, 2007). It also emphasizes the energy efficiency and changing material intensity of society (Rosa et al., 1988).

The Fischer-Kowalski group argues that an adequate concept of human–nature/environment relationships is a prerequisite for progress towards sustainability:

Accepting the premise that sustainability is a problem of society–nature interaction means that we must observe societies, natural systems and their interaction over time and answer the following questions: (1) Which changes do socio-economic activities cause in natural systems? (2) Which socio-economic forces drive these changes, and how can they be influenced? (3) How do changes on natural systems impact on society? (4) How, if at all, can society cope with the changes it has set in motion? (Haberl et al., 2004, p. 200)

The social metabolism approach is further discussed in some detail in section 19.4. This approach constricts analysis to the society level and does not deal with the "farm and firm." How societies and enterprises potentially interfere is not conceptualized in this approach. We thus introduce a Hierarchy Postulate P2 and an Interference Postulate P3 in Chapters 16.3 and 16.4.

12.6.1 Key messages

- The sociometabolic approach introduces the relationship between human systems and technological activities (metabolism) into industrial ecology.

Part VI Contributions of industrial ecology

Chapter 13
Industrial agents and global biogeochemical dynamics

13.1 From the local to regional and global scales 320
13.2 Impacts caused by global biogeochemical cycles of primary production 322
13.3 Flows of different metals 327
13.4 Management of global pollutants 330
13.5 Changing global cycles of the post-industrial society 331
13.6 The future of material flows 333
13.7 Perspectives of industrial ecology 335

Chapter overview

This chapter takes a global- and future-oriented perspective on the impacts of industrial processes on biogeochemical cycles and material flows. With respect to human systems, we deal with the material–biophysical flows that are caused by global industry and the economy. Thus we take an international and supranational view in which the industrial activities and consumption of nations, world regions, and world industries are seen as agents. The global perspective reveals that essential biogeochemical cycles of the Earth system are dominated by human and industrial activities.

Global biogeochemical flows are of interest from two perspectives, one being the change of the material–biophysical regime, such as the change of carbon or sulfur flows. Here, industrial ecology shows a strong link to ecosystem management. A second perspective is the optimization of the material flows from an economic standpoint. Closing certain material cycles, optimizing recycling, and avoiding unnecessary dissipation of industrial systems can be beneficial from a macroeconomic perspective. This aspect sometimes goes under the label of industrial metabolism. We discuss how industrial metabolism is different from the natural metabolic processes, given that there are no "primary producers (analogous to photosynthetic organisms) in the industrial world" (Johansson, 2002, p. 73). We discuss how meaningful biophysical flux management based on solar energy input may reverse unfavorable entropic dynamics.

In addition to the previous points, the chapter focuses on the question of whether there are *essential critical* materials. We first discuss the agrarian sector and what constraints are present within primary production when taking phosphorus as an example. We proceed to the traditional industrial sector, specifically looking at metals. Here, copper serves as a case study. We then look at the service or information society and reflect on the role of rare elements, such as indium and rhenium. We close the chapter by reflecting on the global material flows with respect to their potentials and limitations in the transition to a post-industrial society.

13.1 From the local to regional and global scales

Chapter 12 covered the material and energy flows in corporations and businesses, eco-industrial parks, industries, and regional systems. The first part of this chapter deals with biogeochemical cycles and material flows resulting from production and consumption on a global scale.

Environmental knowledge about global biogeochemical cycles emerged in the last century. An important first step was Svante Arrhenius' (1859–1927) contribution showing the dependence of Earth's temperature on atmospheric CO_2 levels. It was then that interest arose in studying the atmosphere. A eureka moment occurred with the insight that very low concentrations of trace gases can cause critical effects on the atmosphere. Here, the detection of the ozone hole was an important milestone (see Conrad, 2008; Waert,

2003). We should note that the concentration of ozone (O_3) is only between 0.000001 and 0.000004 % (i.e. 0.01–0.04 ppm) at ground level. A seminal book on biogeochemical cycles entitled *Atmospheric Change: An Earth System Perspective*, written by the chemist and geophysicist Thomas E. Graedel, one of the founders of industrial ecology, and the civil engineer and meteorologist Paul J. Crutzen, is the centerpiece for understanding what we call the "rationale of environmental systems" (Graedel & Crutzen, 1993). It includes chapters entitled "Earth and its driving forces," "The atmospheric circulation: transporters of chemical constituents," "The water cycle and climate," "Ancient Earth: chemical histories," "Global change: the last few centuries," "Global futures," and "On change and sustainability." As indicated by the chapter headings, the science of global change focused on natural system knowledge and did not focus on economic development.

13.1.1 From modeling to managing the Earth systems' geochemical cycles

Natural sciences have provided data and evidence on the importance of biochemical cycles and the cycling of scarce elements in the atmosphere since the 1960s. About 99.7% of continental ground-level air is composed of nitrogen (N_2 makes up 78.1% of the volume), oxygen (20.9% O_2) and argon (0.93% Ar). Carbon dioxide (CO_2) makes up only 0.035% and sulfur dioxide (SO_2), which comes mostly from anthropogenic sources, can reach up to 0.20 ppm. Sulfur's critical role in the biosystem is to keep proteins three-dimensional (Smil, 2002). However, high levels can be toxic, as exemplified by the London Smog Disaster of 1952, when more than 10 000 people were killed by air pollution containing SO_2. Figure 13.1a presents the global sulfur cycle, including flows through the lithosphere, biosphere, hydrosphere, and atmosphere. Sea spray is depicted as the largest natural source of sulfur, but the current anthropogenic flows can outperform even this. Thus, in the early 1960s, scientists concluded:

> In terms of the biogeochemical cycles, society is no longer small in comparison to nature. (Holmberg et al., 1994)

Ten years later, when an even better database was available, Klee and Graedel (2004) reviewed the material flows of 79 of the first 92 elements in the periodic table. They revealed that for 70% of these more than half of the flows were anthropogenic (see Figure 13b; Klee & Graedel, 2004). We live in the *anthropocene* period (Allenby, 2004b; Crutzen, 2002), a time in which all material flows and physical aspects of our planet have a human signature.

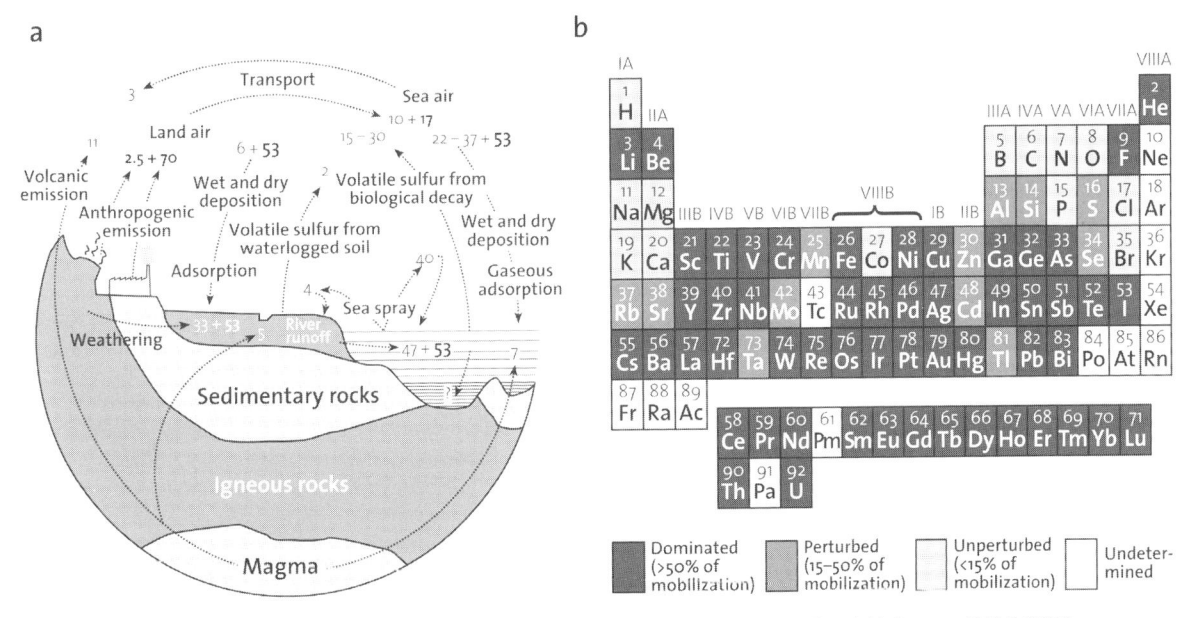

Figure 13.1 (a) Flows in the global sulfur cycle (numbers represent units in teragrams per year; Graedel & Crutzen, 1993 © 2003 by W. H. Freeman and Company. Used with permission); (b) the periodic table and shares of flows caused by human action (Reprinted with permission from Klee & Graedel, 2004, © Annual reviews).

Industrial ecology has a strong engineering component, with industrial ecologists interested in systems engineering and systems limit management. Thus, it is not surprising that some industrial ecologists go beyond the firm and farm level, and consider knowledge about global environmental geochemical cycles as "Earth systems engineering and management" (Allenby, 2007) and as a component of the developing sustainability science (Yarime, 2009; Yarime et al., 2010).

13.1.2 Flows from food, services, and industrial production

When we look at the various global geochemical flows, we ask how the emerging discipline can manage them. One option would be to structure flows according to categories of materials. Some authors, for instance, suggest grouping metals, fossil fuels, chemicals, agricultural products, and construction materials (Ayres et al., 2004). However, we question where water cycles or carbon flows resulting from fossil energy are best placed. We argue that these two cycles have to be included in industrial ecology, although this young discipline cannot substitute for hydrology or energy sciences. However, it has to deal with critical material flows from the perspective of the firm and farm. We aspire to do this from a global perspective considering certain industries, technologies, nations or world regions as subsystems (Graedel et al., 2005; Weisz & Schandl, 2008).

We classify flows and biogeochemical cycles according to both the evolution of societies (see Figure 8.5*) and the most essential vital materials and chemical elements. We first look at primary production shaping agrarian societies and identify essential flows. The critical questions are: Where do potassium (K), nitrogen (N), phosphorus (P), or other fertilizers end up? And, are there chemical elements that might become scarce with respect to food production? Secondly, we look at metals: the key ingredients of the industrial society. Third, we look at the material flows of the service or information society and their management. This last type of society can be considered nearest to the post-industrial society. Here, we analyze what are the critical cycles and material flows. A special question in this field is whether future societies will be dematerializing and what role industrial ecology will take. Finally, following the line from knowledge to decisions, we discuss what regulations and incentive schemes can be created for the critical energy flows of industry. Naturally, carbon trade and management, the challenge of the early twenty-first century, serves as an example.

13.1.3 Key messages

- The Earth's biogeochemical cycles and thus the biophysical regime of the world are seriously affected by industrial production and consumption. The elemental flow and concentrations of many global biogeochemical cycles, such as carbon, nitrogen, or sulfur, have been essentially changed. The impacts of these changes have to be assessed and can become a matter for Earth system management.

13.2 Impacts caused by global biogeochemical cycles of primary production

Primary production is the production of organic compounds that build the base for the food chain. Industrial ecology takes a different view to agricultural, forestry, and fishery sciences, despite sharing the same

---

Box 13.1[1] A critical element of sustainable agriculture: phosphorus

In the past, humans have not always managed to cope with scenarios of nutrient shortage. Montgomery (2007) explains that after a few centuries of agricultural exploitation, even the most fertile valleys lost their fertility. Because of slow erosion and inefficient nutrient cycling, people were forced to move to less propitious environments. The Romans, for example, who managed land use from the far-removed urban courts of Rome, treated the problem of soil conservation lightly. When the fertile valleys of Italy were reduced to naught, they moved their farming to conquered territories of North Africa. There again, they degraded the soil base. Newman (1997) studied phosphorus (P) flows from past agricultural systems, spanning periods from early dynastic Egypt to modern times. His results show declines of soil P over time in all the systems he studied. This suggests that dependence on outside sources of nutrients is an inherent property of sedentary agriculture.

---

[1] This box has been written together with Marion Dumas.

13.2 Global biogeochemical cycles and primary production

Yet, there are at least three important exceptions to this pattern – the centuries-old farming cultures of China, Korea, and Japan (King, 1911/2004). In these societies, all organic waste, both in the cities and the countryside, was collected and recycled. People even collected the runoff water from the mountains, which was rich in nutrients, to apply to the fields. King calculated that six times more P was wasted per million of adult population in Europe and the USA than was recovered by the Chinese through their recycling. And yet, it has been enough to maintain soil fertility for at least two or three millennia while feeding a rather dense population. What should be noted here is the high cost of maintaining nutrient balances in the absence of outside inputs, and the level of knowledge and environmental analysis needed to recognize the means and the necessity of so doing.

In the likely scenario of a declining production of P from mineral rock phosphate deposits, how should we respond? The most natural response to this scarcity would be to close the cycle, primarily by recycling P in organic matter and perhaps by returning the P dispersed in sediments to the soil. There is an important challenge inherent in this. It is clear that organic matter recycling requires a sophisticated long-term mass balance. P in soil is extracted by the yield and lost by runoff, erosion, etc. Thus, agrarian soils that are currently depleted in P can only sustain production if they receive P in the form of fertilization. Soil management options therefore have to take soil types and geographic factors into account.

Current prices of mineral fertilizer are too low to create a market incentive to invest in this transition. Indeed, the discount rate for a 100-year horizon of resource scarcity is too high to influence current market prices. However, the P that we waste today through inefficient use will be costly to recover. Furthermore, we meanwhile may lose the opportunity to use rock P to bring the poorest agricultural soils to levels of P that would allow them to maintain sustained production (see Figure 13.2), an opportunity that will be out of reach once the cost of P extraction from mineral rocks becomes too high.

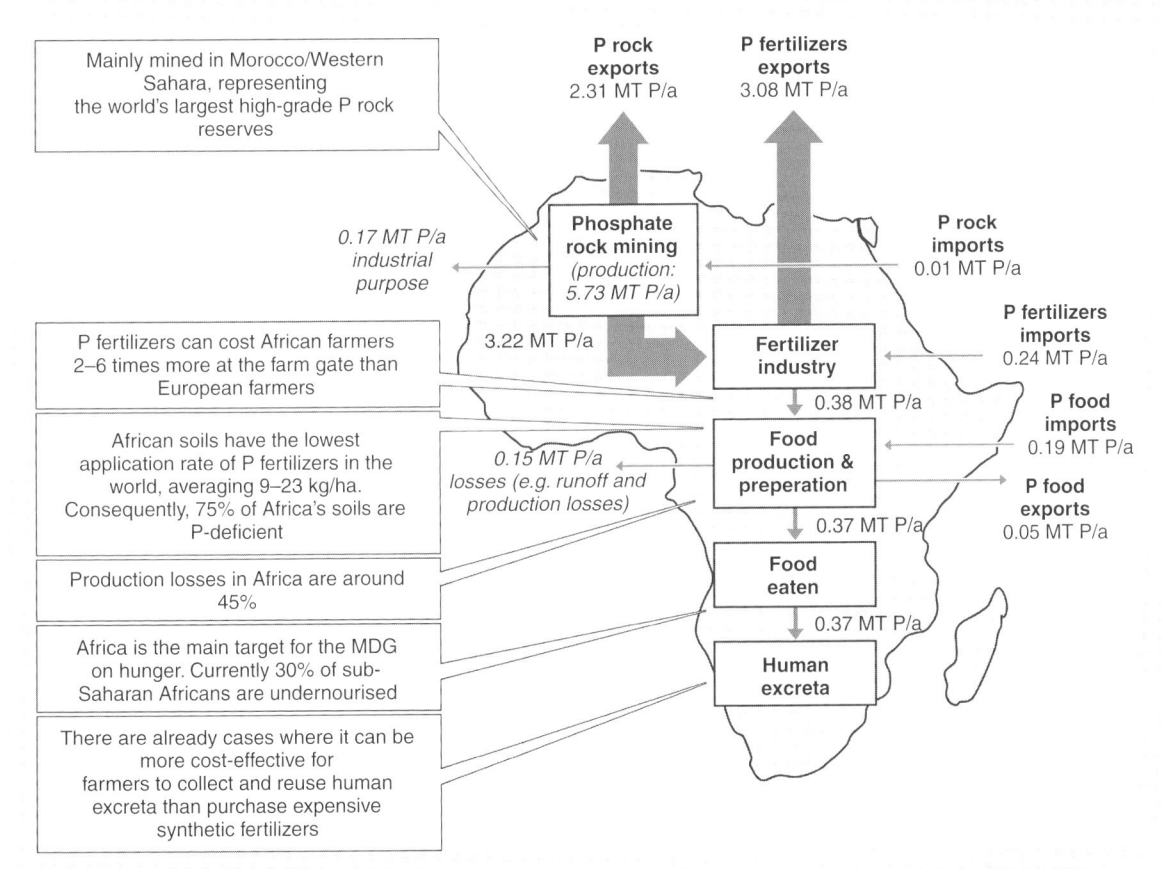

Figure 13.2 The story of African phosphorus flows reveals the role of North Africa's phosphate mining in world food production (adapted from Cordell et al., 2009, p. 296, with permission from Elsevier). MDG, Millenium Development Goals; MT P/a, megatonnes of phosphorus per year.

subject – that is, primary production. For industrial ecology, it is not efficient yield but rather the understanding and modeling of nutrient or fertilizer cycles resulting from primary production yield that are in the foreground. Recently, the energetic content of biomass, that is biofuel production, has also become an interesting topic in industrial ecology (see Chapter 18.3).

Because of its various properties, phosphorus has become one of the most critical elements in primary production. Phosphorus does not enter the atmosphere, has highly dissipative properties with respect to erosion and leaching, and is relatively scarce. It does not fall in the top 10 with respect to mass either on land or in the hydrosphere, yet it is essential for agriculture and food production (see Box 13.1). The potential of mineable rock phosphorus (e.g. apatite) as a fertilizer for plant growth was detected in the twentieth century, enabling an increase in food production. As crops (including grass) are the crucial livestock fuels, we restrict our consideration to their production. In a simplified heuristic, potassium, nitrogen, and phosphorus are the most essential elements consumed by plants in crop production. Potassium is particularly essential for cellulose growth, whereas nitrogen and phosphorus are important for every aspect of plant growth. Potassium makes up about 1.5% of the weight of the Earth's crust, is the seventh most abundant element, and is mostly mined from underground deposits. Nitrogen is also abundant; it is available in air, water, and saltpeter ores (Graedel & Crutzen, 1993). Obtaining nitrate for agriculture is simply a matter of harvesting the required energy. For instance, the Haber–Bosch process (see Box 15.4) produces ammonia fertilizer through an iron-catalyzed reaction of nitrogen and hydrogen, the two most abundant elements of the atmosphere. This is completely different with phosphorus, which is exclusively obtained via mining. Thus, we take phosphorus as an example to study critical global fluxes of agrarian production (Cordell et al., 2009).

13.2.1 Phosphorus flows and their relevance

Historically, agriculture was based on recycling manure and organic matter (see Figure 13.3). Therefore, productivity was limited by the abundance of nutrients initially present in the soil. Since the early nineteenth century, farmers began to add guano, that is, bird droppings deposited over millennia, to the soil. The great guano rush can be seen as an early illustration of industrializing agriculture. The use of rock phosphorus began in the mid nineteenth century and later ballooned together with population growth after World War II.

In the twentieth century, management of phosphorus changed completely. The discovery of mineable sedimentary phosphate rocks and the resulting production of fertilizers caused a tremendous increase of agricultural production. With the separation of livestocks and crop production on the farm level, feed was imported for animal production, and thus phosphorus and nitrogen inputs were not returned to the soil to sustain crop production. This can be seen as an important reason why fertilizer consumption in most parts of the world, such as the USA, dramatically increased. In numbers, nitrate use increased by a factor of 22 between 1940 and 1980 in the USA (Lanyon, 1995). The changed regime of agriculture altered the patterns of biogeochemical cycles. For example, phosphorus was lost by flows from terrestrial to aquatic systems through erosion and runoff. Also, modern sanitation systems transport human waste by water, thus diluting it and prioritizing efficient and sanitary disposal over nutrient recovery.

The global biogeochemical cycling of phosphorus, the quantity of phosphate rock reserves, and the potential of a phosphate rock scarcity during the twenty-first century have been well investigated in the past decade (Cordell et al., 2009; Kauwenbergh, 2010; Smil, 2000; Villalba et al., 2008). In 2005, the US Geological Survey (USGS) provided an estimate of the security of the phosphorus supply based on the reserves of phosphate rocks. The reserves are the identified resources that can be economically mined with current technology. Given the knowledge about today's consumption, some scientists suggest that the global commercial, terrestrial phosphorus reserves will be depleted within 50–120 years if we assume a fixed reserve base (see Figure 13.4). Others judge the lifetime to be much longer, e.g. 300 years (Kauwenbergh, 2010). Whether depletion occurs at the higher or lower boundary of the range depends on population growth, changing consumption patterns, technology development, and the prospective detection of new reserves. We take a critical look at this type of prediction below when discussing copper flows and consumption. Here, we simply acknowledge that phosphorus is *essential* for biomass production – and it *cannot be substituted*. Thus, food production might radically decline if rock phosphorus becomes unavailable. We consider this threat to be far more imposing than peak oil, as we can foresee the application of some substitutes for this.

Let us briefly look at the important role of phosphorus by its allocation in living systems. The concentration of phosphorus in living organisms is typically low. For example, crop plants and grasses contain on average 0.5% phosphorus, which constitutes an uptake of 15–35 kg phosphorus/ha of cultivated land in productive regions of the world. In contrast, vertebrates contain 2–3% phosphorus, mainly found in the bones, which themselves contain about 10% phosphorus. The primary producers, initiators of the food chain, draw all of their phosphorus from the soil. Phosphorus concentrations in soil vary greatly both on the local and global scale, from 0.01% up to 0.3% and more. In any case, all soils are formed by weathering of the bedrock which, on average, contains no more than 0.11% phosphorus, making phosphorus the eleventh most abundant element in the Earth's crust. The figures above indicate that phosphorus concentrations in terrestrial ecosystems are relatively low, despite the phosphorus requirements of life. This explains why phosphorus is often one of the decisive factors limiting ecosystem productivity, including forests, aquatic systems, and agricultural systems.

In general, we denote a material or chemical element to be essential when human reproduction or the existence of societies depends on its steady flow within a certain range. Thus, its accessibility must be given in the long run. We need to distinguish between *availability* and *accessibility*. According to the law of mass conservation, no element will be lost in the universe. However, it might not be sufficiently accessible because of placement, rarefaction, or other reasons, such as geopolitical policies, for instance, if a few powerful nations exclude certain countries or societies from access. Thus, some material matter might in principle be available but not accessible. As we discuss industrial ecology, it makes sense to illustrate accessibility with a zoological example. There may be enough pollen in the world to feed the furthest flung bee colony, but if the energy required to fly to the food source exceeds the energy of the pollen, it becomes rather inaccessible. Special issues of accessibility are the deep sea deposits which are currently not mineable.

Let us take a closer look at the rock phosphorus reservoirs and reserves. Phosphorus is unevenly distributed and is a resource that is geopolitically sensitive. About 85% of the production in 2007 came from seven countries only, including China and the USA. About one-third of world production currently comes from Morocco and Tunisia (Cordell *et al.*, 2009). Phosphorus exploitation is a relevant factor in Morocco's current occupation of the Western Sahara, and China has drastically reduced exports since 2005; nations' policies have a significant impact on material fluxes. The average mining grade (i.e. quality) of phosphorus rock declined from 15% in the 1960s to below 13% in 1996 (Cordell *et al.*, 2009).

The fertilizer industry is starting to recognize the emerging scarcity. Theoretically, there is the possibility of mining the ocean for inorganic phosphorus. The data are optimistic. Reportedly, there is about 93 gigatonnes (Gt) phosphorus in the oceans, but 85 Gt are located in the deep sea and only 8 Gt is near the surface (the top 300m). In addition, 20–25 Mt phosphorus/year are believed to be buried in marine sediments (Smil, 2000). We find a comparatively smaller reservoir of 35–40 Gt inorganic phosphorus and 5–10 Gt organic phosphorus in the top 50 cm of soil (Smil, 2000, p. 59). Whether humankind will mine the oceanic reservoirs depends on the assumptions made about future costs of extraction, the market prices of the resource, and the costs of increasing organic farming, among others. There we have to consider that "less than 0.2% of all oceanic P is in coastal waters" (Smil, 2000, p. 60).

13.2.2 Phosphorus life cycle characteristics

Historically, fertilizer sources were manure, guano, human excreta, and other organic material such as leaf litter in forests. Since 1950 the use of rock phosphorus for fertilizers sky-rocketed (see Figure 13.3), with only a brief depression, caused by the collapse of the Eastern Communist Block after 1989. Based on estimates for the year 2000, Smil found that 75% of the rock phosphorus mined annually was used for fertilizers (Smil, 2000, 2002). Box 13.1 shows how essential phosphorus is for food production.

However, in agriculture, about 55% of the phosphorus used is lost between "farm and fork" (Smil, 2000). The inefficiency is also reflected in the 25–30 Mt phosphorus/year lost to erosion and runoff. A similar conclusion is drawn in a recent study that states that the annually mined amount of inorganic phosphorus is equivalent to the entire amount of phosphorus transported in rivers (Cordell *et al.*, 2009). There is thus much potential for increasing the efficiency of its use in this sector.

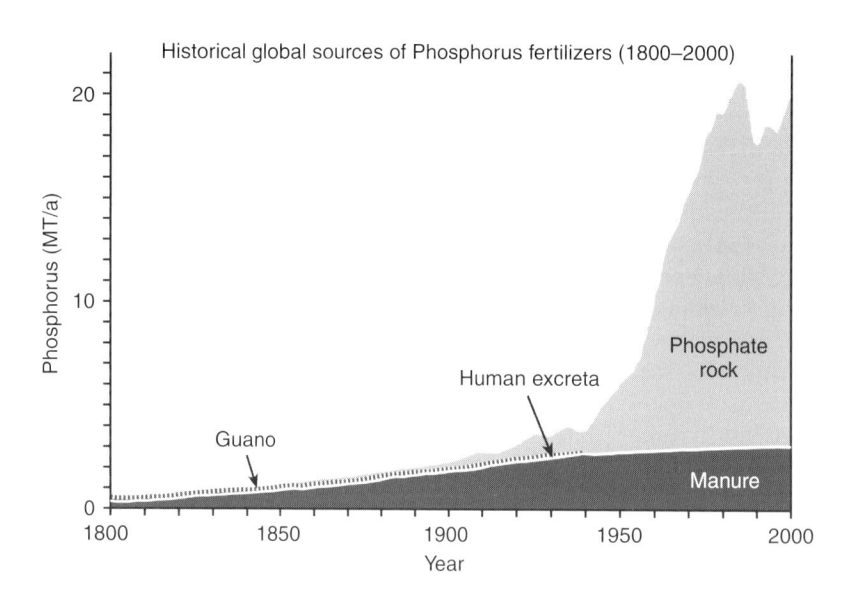

Figure 13.3 Consumption of different sources of phosphorus fertilizers (see Cordell *et al.*, 2009, p. 293).

Mining is another sector where efficiency comes into play. In 2005, 19.4 Mt phosphorus was produced from 148 Mt of phosphate rock. However, mining efficiency is low (67%), as about one-third is lost by processing and metallurgical operations. In China the efficiency is even lower. Also, phosphorus is used for a wide range of industrial purposes. For instance, in the Netherlands 34% of phosphorus in surface water is attributed to industrial consumption (Cordell *et al.*, 2009).

We have mentioned the example of critical, unsustainable flows of material from the separation of livestock and crop production that resulted in the reduced use of manure as fertilizer. Another critical flow is related to the north–south relationship. Here, difference in wealth corresponds to the flows of phosphorus. From the 5.73 Mt phosphorus mined annually from North African high-grade rocks, 5.39 Mt are exported either as rocks or as fertilizers that were produced in local plants (Cordell *et al.*, 2009). Only 0.38 Mt phosphorus are used for food production in all of Africa. Developed countries not only import inorganic phosphorus fertilizer but also large amounts of food from developing countries, which is another component of the imbalance in the global phosphorus fluxes.

13.2.3 Trade-offs around phosphorus recycling

Closing the material fluxes is an aim of industrial ecology. A challenge in this domain are the trade-offs that must be made; we briefly illuminate two, namely the use of sewage sludge and animal waste from meat production.

There is a daily consumption of 1.5 g/capita, which is above the recommended allowance (Smil, 2000). The average content in a human body is about 400–500 g phosphorus. As previously mentioned, humans and all vertebrates accumulate some phosphorus in their bones and meat, but most of the ingested phosphorus is excreted. Again, a large share of this goes to latrines or sewers, thereby barring the reclamation of the phosphorus. Moreover, in many industrialized countries, the use of sewage sludge as fertilizer has been banned. This, however, is opposed by both fertilizer associations, which can be considered practical experts, and by some non-governmental organizations. The trade-off here is between food safety, and thus human health, vs. closing nutrient cycles (see Chapter 18, case 4).

This is different in some Asian countries, where untreated city wastewater is used for vegetable production on a large scale. Moreover, one should note that only 20% of all wastewater worldwide is thought to be treated and that some farmers pay much more for wastewater than for regular irrigation water (Ensink & van der Hoek, 2007). Wastewater is also an important livelihood resource that can increase food security for poor people in urban and peri-urban domains. Also of interest in this context is the fact that two-thirds of the fish farmed globally are fertilized by wastewater (World Bank, 2005). Here, the development of proper precautionary measures is required. Wastewater treatment that reduces coliform bacteria might be seen as an option. New technological knowledge may also help. In Switzerland, for instance, mono-incinerators were developed to allow recycling of 90% of the phosphorus from the ashes (AWEL, 2008). The ash from special incinerators burning only sewage sludge shows a concentration of phosphorus of 5–6% (Franz, 2008). In Switzerland, this fraction of phosphorus almost exactly amounts to the quantity of imported phosphorus (AWEL, 2008; Lamprecht et al., submitted).

Another trade-off, which is extensively discussed in Chapter 18.4, is related to the efficient use of carcasses, in particular animal bones, in the agro-material cycle as a component of fodder or fertilizer. Here, the mad-cow disease (BSE) crisis in the late 1980s played an important role.

13.2.4 Disruption of coupled global cycles linked to phosphorus

The major biogeochemical cycles (such as the C, N, and P cycles) are coupled with each other. The rationale here is that the Redfield ratios C:P and N:P in the water, air, and soil reservoirs show specific relationships in primary production on land, coastal zone, atmosphere, etc. Models provide strong evidence that human-induced fossil fuel emissions, land use change, the application of chemical fertilizers on land, organic sewage discharge, along with the temperature rise, has triggered a "net loss of C, N, and P from phytomass and soil humus" (Lerman et al., 2004, p. 3).

13.2.5 Key messages

- Securing food and biomass production becomes an increasingly important issue in the current phase of the Industrial Age
- Phosphorus is an essential element of biomass production. Because of its dissipative properties in agriculture, and limited phosphorus reservoirs of high grade, phosphorus management should become a primary issue of environmental management
- The necessity of phosphorus and the analysis of its cycle reveal the limits of the concept of weak sustainability, as well as arbitrary growth and substitutability.

13.3 Flows of different metals

Metals are the foundation of the modern industrial society. Despite suggestions drawn from the names of previous periods in the development of human culture (e.g. Bronze Age, Copper Age, etc.), only 2.5% of the copper smelted by humans was mined before 1900 (Lifset et al., 2002). Iron, copper, chromium, and aluminum have become symbols of the Industrial Age. Additionally, metals such as vanadium, magnesium, and platinum are commonly known as key components of steel production and the machinery industry. However, technology continues to develop. In a seminal paper entitled "Dining at the periodic table: metals concentrations as they relate to recycling," Johnson, Harper, Lifset, and Graedel highlighted the new quality of material flows because of information technology (Johnson et al., 2007). They showed that the metals included on the Intel circuit board may increase from 11 elements, present in the 1960s, to over 60 in the near future. Those not acquainted with the periodic table will have to learn a new vocabulary. The spread of next-generation technologies and electrical equipment will eventually oblige us to know something about the new gold, as tantalum/coltan, rhenium, niobium, and indium are sometimes denoted. As supply disruptions can be caused by these metals we also speak about critical (raw) materials. This will be examined later in this section.

Because of globalization and the increased speed of material flows, we have to look at these metals from a global perspective. We take the well studied example of copper to discuss scarcity, accessibility, and essentiality, as we did previously for phosphorus.

13.3.1 No more copper?

In North America (USA, Canada, including Mexico), 170 kg of copper is used per capita in building and construction, infrastructure, domestic and industrial equipment, and transport; in-use copper and zinc per capita in inner Sydney are 605 kg and 420 kg, respectively (Gordon et al., 2006). Copper seems like a staple food of modern civilization. Given the high annual increase of copper use in China, India, and other developing and emerging countries, one can extrapolate these data and check whether we would have enough copper if the whole world population wanted to meet North American copper standards.

To understand availability and accessibility of minerals and metals, the McKelvey diagram (McKelvey, 1972) is useful. Two dimensions have to be considered

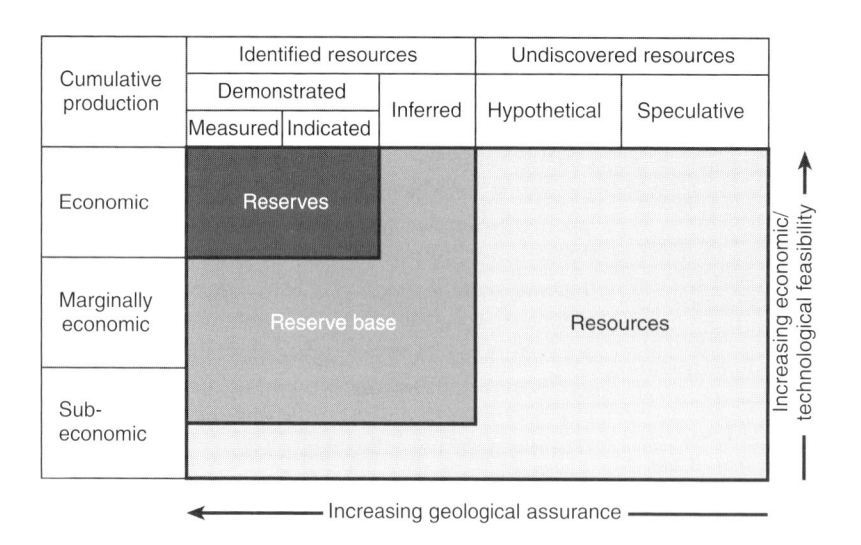

13 Production and global biogeochemical dynamics

Figure 13.4 The McKelvey diagram of resources (slightly modified from US Bureau of Mines and US Geological Survey, 1980, p. 5.)

if we want to have access to a resource: whether and where it exists, and whether it is worthwhile mining or accessing a resource from a cost–benefit perspective. These are the two axes (geological assurance and economic/technological feasibility) in Figure 13.4. We can distinguish between demonstrated and inferred identified resources and hypothetically or speculatively undiscovered resources. Furthermore, we can categorize economic, marginally economic, and sub-economic resources. The latter are available in principle but are not realistically accessible. The use concept of reserve base is currently critically discussed.

Reserves are the economically viable and identified part of the reserve base. Estimating the reserves based on historical data of newly detected copper deposits, the total copper reserve is estimated to be about 1600 Tg. Assuming a world population of 10 billion in 2050 and assuming that each world citizen wants or has the right to have access to 170 kg (the North American standard) results in a demand of 1700 Tg, thus surpassing copper reserves (Gordon et al., 2006).

However, the estimation method for copper reserves has been criticized by Tilton and Lagos (2007). These authors argue that the USGS has provided a higher estimation of 3700 Tg. The most critical objection is that the boundaries between the reserves and the resources are historically flexible and depend on the prices that people are prepared to pay. We partly agree with this argument, at least for the case of copper. But a critical question is what part of the Earth's crust copper resources, which would provide for 83 million years of consumption at current annual consumption rates, is effectively accessible (Hille, 1995)? Thus we can question whether we are facing some accessibility scarcity and whether copper is an essentially scarce material in the same way as phosphorus (see Box 13.1).

We also have to acknowledge that, according to the proposed definition, copper is *not* an essential element as is phosphorus. It has excellent properties with electric and thermal conductivity but can be substituted by other elements. Easy access may be hampered but not in a critical way. Furthermore, we have the option of urban mining, an idea spread by the urban activist Jane Jacobs (1969). Other options include better, more efficient, product design, complete recycling of newly used copper, and reuse and redesign of inefficiently used copper (Brunner, 2007). Also, the mining efficiency is for the most part less than 80% and can be improved.

As it relates to industrial production, the economic importance of scarcity is of interest for industrial ecology. It is generally argued that as commodities become increasingly scarce, they also become more expensive. This is a premise that was strongly supported by environmental researchers in the 1970s. In 1972, the copper reserves were estimated to be 380 Tg (Meadows et al., 1972). But in 2005 this estimate was increased to 480 Tg, despite the fact that 300 Tg had already been mined in the meantime (Bundesamt für Geowissenschaften und Rohstoffe (BGR), 2006; Frondel & Schmidt, 2007). Also, the annual world production has increased from about 7 to 14 Tg/year and a decrease in the prices by about 28% during that time would suggest sufficient accessibility (US Geological Survey (USGS), 2005). If we take a longer perspective and compare manufacturing wages of 1870 to modern ones, you find deflation in copper prices by a factor of 24.3 (Brown & Wolk, 2000). What holds true for copper is also valid for other metals. The majority of natural resource prices have deflated by the consumer price index in this time period (which represents the Industrial Age). In general, the share of mineral commodities in world GDP and national wealth has declined. In general, the share of mineral commodities in world GDP and national wealth has declined. This may reflect a phenomenon referred to as a "resource curse hypothesis," which says that countries with abundant natural resources, in particular minerals, are supposed to have low wealth compared to resource-poor countries, which take on more profitable economic activities (Sachs & Warner, 2001).

13.3.2 Multilevel cycling

Industrial ecology also started to monitor the life cycles of other metals from mining and refining, fabrication and manufacturing, different and changing phases of utilization by consumers, and end-of-life treatment on a global level. The methodology chosen for understanding the world scale has been called multilevel cycling. Starting from a national level, which is the common level of data recording for world nation clusters such as Europe, Africa, Middle East, or the Commonwealth of Independent States, the discard flows of different materials and elements are calculated to better understand the dynamics of the world cycle. One can identify the "personality" of the regional cycle and become literate on whether municipal waste, industrial waste, or end-of-life vehicles, just to mention a few, are critical for a closed loop cycle. We learn that discard flows are highest for industrial waste (Graedel et al., 2005) since the pure economic incentives for recycling are low. One could ask what a world supranational environmental institution could impose in sustainable waste management.

For instance, with zinc we find a discard rate of 17% in production, which necessitates environmental evaluation. Zinc is not only lost to the lithosphere but, in the course of its production or in burning coal, oil or biomass, it is also lost to the atmosphere. Here, we face dissipative structures (Graedel et al., 2005):

> The fraction of the stock of recoverable resources in the lithosphere already placed in use or in wastes from which it will probably never be recovered is currently ≈26% for copper and ≈19% for zinc. (Gordon et al., 2006, p. 1213)

The monitoring of metal flows is ongoing. Up to now we have developed flow literacy about tin (mostly used as anti-corrosion coating on steel), nickel (the stainless steel additive), platinum (often used as a catalyst in cars), lead (as an additive in gasoline), and silver (important in electronics; see Mao & Graedel, 2009). Based on the USGS, extensive knowledge about these metal flows is growing.

Historically and also prospectively, the flows of metals are not only of interest because of scarcity issues, which relate to the input side of industrial production, but also with respect to toxicity, the output. This holds true for heavy metals, such as arsenic, lead, cadmium, chromium, copper, mercury, silver, uranium/plutonium, and zinc (Ayres, 1991). Industrial ecology converges with toxicology and ecological risk assessment (Giller et al., 1998; Paustenbach, 2002; Scholz et al., 1990).

13.3.3 Rare elements

We have learned that some metals might become inaccessible and have to be substituted. But what issues are posed by the increased use of rare materials in advanced industrial production and particularly in information technology? A first definition of rare material has been suggested by Skinner, who denoted materials as "sarce" (rare) if they constitute less than 0.01% of the Earth's crust (i.e. about the abundance of copper) (Skinner, 1979). Let us take rhenium as an example. Rhenium was discovered in 1925. It is one of the rarest metals on Earth, with an average concentration of 0.000 7 ppm in the Earth's crust or about 0.000 000 07% of the crust's mass (Noddack, 1929). It does not have any (known) biological function and is not prevalent in the human body. One could say that circulation started in 1950 by advanced technology and efficient large-scale mining and smelting. Today it is of interest for technical purposes because it has one of the highest melting points of all metals. In 2006, about 45 tons of rhenium were mined in molybdenum ores and processed into nickel-based alloys for jet-turbine parts; these contain up to 6% rhenium. The toxicity of rhenium is not yet well explored.

New technologies have induced a surge in demand for previously ignored elements of the periodic table attractive for the production of electrical and electronic equipment, photovoltaics (e.g. tantalum, indium), batteries (e.g. cobalt, lithium), catalysts (e.g. platinum, palladium), and new energy systems such as wind turbines (which require dysprosium, another rare element). In addition, each motor produced in 2009 for the Toyota Prius hybrid required 1 kg neodymium (sharing 22 ppm of the Earth's crust; see Bradsher, 2005), and each battery required 15 kg lanthanum (sharing 17 ppm).

A critical issue is that the rare metals are mined as coupled products with abundant metals such as zinc. Even in the "ores" of these metals, it becomes inefficient to mine them if there is no accessibility or market for the non-rare-coupled metals. Bismuth serves as an example since it is generated mostly as a byproduct of lead smelting, and recycling strategies are unknown (Carlin, 2007). From a global perspective, the geopolitical risk linked to rare metals and earth elements is currently extraordinarily large, given that China produced 95% of these resources in 2008. None the less, this is only a matter of accessibility – the metals are available around the world but have simply not been mined yet (Bernet, 2009).

The assessment of rare elements and dealing with their surrounding issues is in its incipient stages. The analysis concentrates on variants of supply risks, for example, by physical scarcity, critically low reserves compared to annual demand, unexpected demand growth, and recycling restrictions (see European Union, 2010; US Department of Energy, 2010). The latter can be caused by a lack of price incentives, high scale of dissipative applications, or technophysical limitations of recycling (Buchert et al., 2009). We can find this dissipative application with tantalum applications in cell phones. With gallium and germanium, no recycling technologies are yet available on a commercial scale.

13.3.4 Key messages

- An increasing number of elements of the periodic table, some of which have not circulated before, have become subject to human-induced material flows
- For selected metals, such as copper, some researchers assume that half of the (economically profitable) reserve base has already been mined
- The concepts of reserves, reserve base, and non-economically accessible metals has increasingly become of interest. This holds true in particular for rare elements.

13.4 Management of global pollutants

With the presence of both scarcity and abundance of hazardous entities, industrial ecology collaborates with economics to respond in new and diverse ways to global environmental problems (see Chapter 11 for further principles of resource and environmental economics). The detection of global atmospheric ozone depletion by chlorofluorocarbons (CFCs) in the 1960s, the discussions about health impacts or *Waldsterben* (English: forest decline; see Box 7.6) by the smokestack industry in Central Europe, and the current concern about climate change are strongly related to industrial emissions. The USA was the first to supplement common command-and-control strategies with market-based instruments, which resulted from efforts to curb the occurrence of acid rain. The idea of mechanisms for an emission allowance, trading and auctioning program (see Box 13.2) was first implemented by the US EPA in the 1990 Clean Air Act Amendments, which was aimed at coal-fired power plants (Stavins, 1998). Empirical studies showed that the emissions trading program was effective with regard in particular to SO_2 emissions reduction (Schmalensee et al., 1998), and that costs were lower than with a command-and-control strategy (Carlson et al., 2000). The benefits in these cost-effectiveness assessments were primarily related to health effects, such as reduced mortality and morbidity (Burtraw et al., 1998). The model of the SO_2 trading was later applied to govern industrial CO_2 emissions (Stavins, 1998).

In 2005, the EU launched the European Union Emission Trading System, and other countries are now following. These trading schemes are also linked with the Clean Development Mechanism (CDM) under the United Nations Framework Convention on Climate Change. Within the CDM, developing countries can sell the reductions in emissions they achieve by introducing new, cleaner technologies from the developed world (Hoffmann, 2007). These mechanisms show that material flows, which have been outside the human economic system, can become objects of economic transactions.

Unfortunately, from an institutional perspective, the UN have only the status of an international organization with limited power of enforcement. Here the EU, which is a supranational organization, has some decisive advantages. Facing the disorder of the world financial systems, a stronger framing of world economics seems necessary.

13.5 Global cycles of the post-industrial society

Box 13.2 Trading and auctioning pollution: acid rain

Acid rain was a key environmental concern of the 1960s (see Box 7.6). Acid rain results when sulfur dioxide (SO_2) or nitrogen oxides (NO_x) react with water, oxygen, and other chemicals in the atmosphere to form sulfuric and nitric acids. The US coal-based electric power generation companies were the first to abide by an emissions trading mechanism.

The US Clean Air Act, beginning in 1970, issued standards for 188 toxic air pollutants. Title IV of the 1990 Amendments established a market among electric utilities for SO_2 (Environmental Protection Agency (EPA), 1995), "the primary precursor of acid rain" (Hahn et al., 2003). In Phase I, individual emission limits were allocated to 263 units of 110 coal-fired power plants east of the Mississippi River which had emissions greater than 2.5 pounds of SO_2/million British thermal units and a generating capacity of more than 100 megawatts (Winebrake et al., 1995). In the second phase almost all electric power-generating units were brought into the system. The program for NO_x worked similarly and was based on national emission allowances allocated to individual states in accordance with the emission allowances in their own laws.

A new competitive market of emission trading was established by which emitting companies could buy, sell, and bank emissions based on the idea of cost-effectiveness (see Chapter 11.4). The mechanism relies on the efficiency principle of free market economics and assumes that those companies that can reduce emissions at the lowest cost will do so, as they are assumed to have more incentives to sell their allowances to those for whom reductions would be more costly. In addition to the private bilateral trade, an annual government-sponsored auction of 3% of the emissions "was established by EPA, with revenues distributed to utilities on the basis of their original allocations" (Stavins, 1998, p. 71). Furthermore, a penalty of $2000 per ton of pollutant was introduced with a binding requirement that the original reductions should be compensated for the next year.

The discussion about this process brings us to the link between industrial ecology and economic policy. Economists estimated that the trading and auctioning program saved 25–34% in costs, as compared to a hypothetical command-and-control policy (Schmalensee et al., 1998). New York's non-participation, for instance, would have increased the state's cost by 240% (Winebrake et al., 1995).

Allowance prices were low. Thus, some critics postulated market deficiencies. Indeed, market-based instruments for pollution trading are by no means a panacea. There are many prerequisites to success, such as a sufficiently high number of "players," and there are also many uncertainties. Nevertheless, from 1970 to 1998, SO_2 emissions in the US decreased by 37%, with emissions from five of the seven major pollutants decreasing (i.e. SO_2, volatile organic compounds (VOCs), CO, lead, and PM_{10}, as well non-traded lead, which decreased by a factor of 50; see Figure 13.5). Emissions from only two pollutants slightly increased (i.e. $PM_{2.5}$ by 5% and NO_x by 17%).

The EU Emission Trading System for CO_2 reduction and the Clean Development Mechanism of the United Nations Framework Convention on Climate Change (UNFCCC) emerged from the US experience on SO_2 reduction.

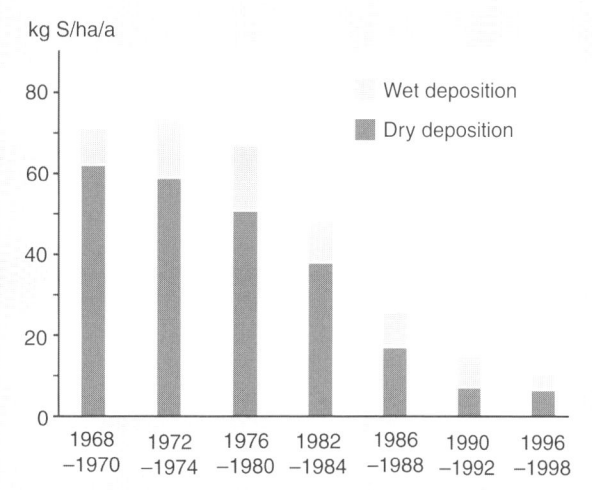

Figure 13.5 The decline of SO_2 emissions in the US following emission trading on the Chicago Stock Exchange and auctioning by companies. Kg S/ha/a, kilograms of sulfur per hectare per year.

13.4.1 Key messages

- By national and supranational laws, some global pollutants have been included in the economic system through market mechanisms. New economic mechanisms such as the Clean Development Mechanism were invented to transfer the cost of pollutants to economic goods.

13.5 Changing global cycles of the post-industrial society

We conjecture that most developed countries will be entering a post-industrial stage (see Figure 8.5*). However, taking a more integrated perspective, we see that even these countries will still be exposed to the flows of agrarian, industrial, and post-industrial societies. Critical material flows of agricultural production, including hypothesized challenges such as "peak phosphorus" (Cordell *et al.*, 2009; Dery & Anderson, 2007), as well as some of the potentially critical flows for maintaining industrial production mentioned, still play a role. We will thus briefly deal with the issue of essential flows. We subsequently deal with the carbon cycle, which is of major environmental concern owing to the imposing threat of climate change and the increasing energy demands of the industrial society.

Although we do not know what a post-industrial society looks like, some outlines of this society might already be visible. It is clear that information technology, the internet, and the generation, transfer, and use of information are inducing changes as fundamental as those that have already occurred. Before we reflect on the possible advanced structural changes of societies, we take a closer look at the trends which seem to be foreseeable.

13.5.1 Scarcity management of essential material

There are two controversial positions that divide thinking on scarcity. One is related to Boulding's (1965) "spaceship Earth" metaphor (see Box 11.1) and Daly's dissipation/entropy concept (see Box 11.4). The idea is that, from a material flux perspective, Earth is a limited system. Given the current size of the human population its growth, dissipation, and entropy will cause serious disturbances because of material shortage. This is the pessimistic, precautionary perspective, mostly linked to the principles of strong sustainability, which is not only restricted to scarcity but also to the undermining of the functioning of ecosystems that support society. The other is the optimistic weak sustainability view, closely related to the principle that reserves increase with prices (see Box 11.2), and which culminates in the belief that the scarcity or inaccessibility of all elements can be efficiently managed by technological progress, recycling, substitutability, etc.

However, there are strong arguments that substitutability, which is not restricted to the material level, is limited:

> … there is no substitute for chrome in stainless steel, cobalt in wear-resistant alloys, scandium in shock-resistant aluminum–scandium alloys, silver in printed RFID tags, indium in transparent indium–tin–oxide electrodes for displays, neodymium in strong permanent magnets

and germanium in lenses in infrared optics. (Fraunhofer Institut für System- und Innovationsforschung ISI, 2009, p. XVI)

We introduce the concept of essential materials and distinguish two types of essentiality. The first is ecosystem service-related, for which we can take phosphorus and food production as an example. We have outlined that phosphorus is an essential material for biomass and food production, and that substitutability seems absolutely unlikely, if the attempts to design genetically modified plants that provide biomass with minimal phosphorus are not successful. A second is technological essentiality linked to scarcity of materials to maintain a certain technology that is indispensable for maintaining basic societal technologies. This type of essentiality is also called criticallity (US Department of Energy, 2010).

13.5.2 Electronic information technology changes world perception and material flows

The start-up of the industrial age was characterized by heavy engineering and certain foundational technologies, such as the steam engine and rail technology. Today, in many developed countries, service industries are generating the major share of the GDP.

The evaluation of new technologies is not restricted to the evaluation of the material flows linked to their production and processes. Electronic communication and information technologies are part of infrastructure. And infrastructure changes the matrix of behavior, consumption, and thus of environmental impacts. We can take the invention of a jet plane as an example. From an industrial ecology perspective, a comprehensive environmental impact assessment should not only include the airports or the electronic information systems needed in their operation (such as radar equipment) but also the changed travel habits and the immense tourist infrastructures built in exotic travel destinations. The behavioral change and the subsequent activities, for instance, the huge tourist centers built in places such as Mallorca, Thailand, or Barbados, have had a much bigger impact. This also holds true for the internet (Allenby, 2004a). Major environmental impacts are not caused by construction and thus the direct material flows, but rather by the changes in environmental flows induced by changes in the way that humans perceive and manage the material-biophysical world, and by what they want and consume. The impact of the television as it permeates and changes traditional cultures in developing countries can be taken as an example. Here, so-called gamma services (Graedel, 1998), which are linked to economic activities based on a minimum of physical matter movement through electronic transaction of information, can implicitly change the behavior and demand of the customer.

13.5.3 Rebound effects of technology development

The structural changes induced by new technologies are of general interest. In technical terms, we are facing a variant of rebound effects or secondary feedback loops. The intended improvement of environmental performance aspired to by a new technology is confounded with unintended collaterals (Binswanger, 2001; Hertwich, 2005; Spielmann et al., 2008). We can see that the implementation of new technologies is linked to structural change, which requires careful reflection. We deal with the phenomenon of anticipating unintended and undesired secondary rebound effects in the Feedback Postulate P4 (see Chapter 16.5).

13.5.4 Key messages

- The evaluation of electronic communication and information systems does not only ask for assessment of the environmental impacts caused by their production but also particularly from the structural changes of behavior, consumer demands, and unintended secondary feedback loops
- There has been no essential scarcity in industrial societies so far. The question of whether certain rare metals become essential in the sense that their scarcity critically affects sustaining essential infrastructure in the post-industrial society is an open question and deserves investigation.

13.6 The future of material flows

We have organized this chapter according to the transitions of society (agricultural, industrial, post-industrial). Here we should note that the 192 countries that are currently members of the United Nations are completely differently developed with respect to these transitions. There are countries which are mostly shaped by agriculture, such as Rwanda or Bhutan. There are others, such as Singapore, that consist of a single urban agglomeration and which have electronic road pricing directing traffic flows.

The material flows in agrarian societies are much lower than those of industrial societies. Historically and worldwide, materials use increased eightfold during the twentieth century (Krausmann et al., 2008). Today, there is a huge disparity between the south and the north. This is well reflected by the gross domestic product (GDP) of countries. There is a factor of 427 between the gross national income of Somalia ($140 GDP/person and year) and Switzerland ($59 880 GDP/person and year). We have to acknowledge that the driver of wealth will move populous countries such as India ($950 GDP/person and year) and China ($2360 GDP/person and year) closer to the level of the USA ($46 040 GDP/person and year; all data for 2007, United Nations International Children's Emergency Fund (UNICEF), 2009). Thus, it appears quite likely that the world material flow will dramatically increase in the next decades. These include critical flows of phosphorus and carbon, and related cycles such as water owing to system changes of the processes that underlie agriculture and industry.

13.6.1 Environmental Kuznets curve

A critical question is whether consumers in highly developed countries can reduce the demand of material flows and environmental impacts, as there are some technological indicators for dematerialization and eco-efficiency. Thus, the possibility of a decline of environmental impacts in the post-industrial society exists. This is the idea of the environmental Kuznets curve, which refers to the idea that income inequality and environmental impact within societies first rises and then falls as economic development proceeds (Kuznets, 1955). The environmental Kuznets curve (EKC) suggests that:

> … at higher levels of development, structural change towards information-intensive industries and services, coupled with increased environmental awareness, enforcement of environmental regulations, better technology and higher environmental expenditures, result in leveling off and gradual decline of environmental degradation. (Panayotou, 1993, p. 1)

Stern (2004) provides a critical review on the history of this hypothesis in which he goes beyond a single element and purely descriptive approach. The latter would have provided some evidence for the Kuznets curve, for

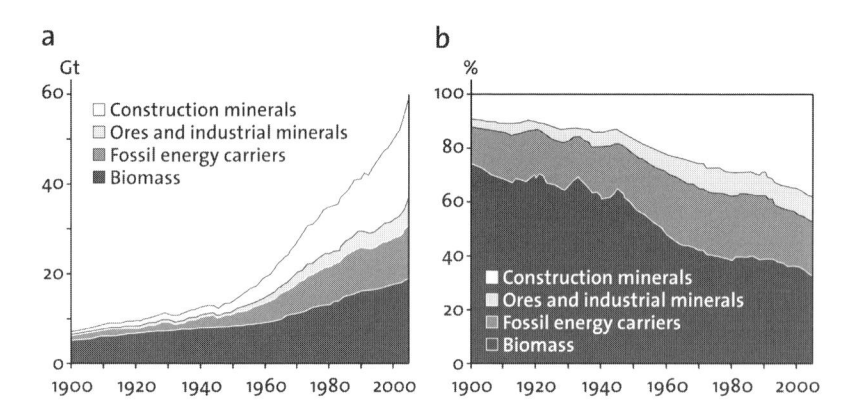

Figure 13.6 (a) Increase of different fractions of world material flows, in gigatonnes (Gt) between 1900 and 2005 and (b) the percentage change of consumption patterns (Krausmann et al., 2008, p. 2699, adapted with permission from Elsevier).

instance, when taking sulfur emissions as an example. He argues, however, "that it seems easy to develop models that generate EKCs under appropriate assumptions" (Stern, 2004, p. 1422). A comprehensive review of economic models, on the other hand, shows the integration of a broad set of variables, such as population growth, growth in GDP, scale effects (i.e. production becoming more efficient with the amount produced), changing input mixes (e.g. gas instead of coal), and various output mixes implied by different product types demanded by the consumer (Dasgupta et al., 2002). The critical issue is that many calculations referred to a single pollutant, such as sulfur, and looked at dynamics on a national level. But even for the example of sulfur emissions, the calculations for the "turning point" (i.e. the income level at which emissions are at their maximum) of the Kuznets curve differed dramatically, ranging from $3317 (Panayotou, 1993) to $101 166 (Stern & Common, 2001). Stern summarized the current state of knowledge in the following way:

> The only robust conclusions from the EKC literature appear to be that concentrations of pollutants may decline from middle income levels, while emissions tend to be monotonic in income. (Stern, 2004, p. 1426)

The essence here is that, because of GDP and population growth, further increases have to be expected. The dynamics are of interest here. A critical aspect is that, during the twentieth century, global population rose by a factor of four. And, mostly owing to technological progress, global GDP grew by a factor of 24 (Krausmann et al., 2008, p. 2701). Given this picture, one cannot expect that eco-efficiency alone can counter the increase but that major structural changes by consumers will be necessary as well. Statistical analysis indicates that material use increased at a slower pace than the global economy but faster than world population. As a consequence, material intensity (i.e. the amount of materials required per unit of GDP) declined, while materials use per capita doubled from 4.6 to 10.3 tons/capita/year. Furthermore, the shift in its structure and composition is of interest. Given the data from Figure 13.6b, we can conclude that we are transitioning to a mineral society, as the use of minerals is going to outperform biomass and fossil energy carriers.

13.6.2 Biomass/food growth as a critical domain

The changing consumption pattern seen in Fig. 13.6b deserves some further consideration. The share of crops in total biomass extraction increased from 21% to 35%. This is countered by declines of wood from 15% to 11% and of livestock fodder from 47% to 30% (Krausmann et al., 2008). Biofuel production entered as a further consumer of biomass, primarily to substitute fossil vehicle fuel. Given the hypothesized threat of potential long term phosphorus scarcity, the changing distribution of irrigation, and population growth, it is clear that current food insecurities and poverty, as well as the distributional justice between north and south, will increase through the global pressure on biomass. There are also other impacts bundled with the increasing pressures on food production, such as biodiversity loss or the greater income disparities across countries, which did not exist before the "epoch of modern economic growth" (Berger, 2007). Here, we are reminded of the paradox that the economic value of food production is increasingly falling while simultaneously fertilizer flows, which are essential for food production in the south, are transported to the north. Land use will face increasing trade-offs between necessary food production and other functions of terrestrial systems.

13.6.3 Key messages

- At the beginning of the twenty-first century, there are no indications that the increase of material flows will reverse. On the contrary, the environmental impacts generated by the agricultural, the industrial, and the post-industrial society are adding up
- Electronic and information technologies provide the potential to increase quality of life without increasing environmental impacts. However, it is not yet obvious as to whether and what extent human impact on biogeochemical cycles can be reduced
- It seems likely that biomass and thus food and primary production may become an increasingly critical issue and will increase intragenerational injustice.

13.7 Perspectives of industrial ecology

Given the current challenges discussed above and our knowledge about emerging new technologies, the young discipline of industrial ecology is already at an interesting crossroads. One option would be that industrial ecology follows the main road and focuses on the material flows of certain elements that are related to industrial processes. This would mean the following:

> As a field, industrial ecology can simply become a relatively minor and increasingly unimportant set of tools focused on material and energy flows – indeed, this is the current path, and it is far the comfortable one. (Allenby, 2009, p. 181)

But how do the other roads look? We think that some answers come out of the human–environment system (HES) view presented in this book. Some answers have been given in the paper "The industrial ecology of emerging technologies" by Braden Allenby (2009).

A *first* view would see industrial ecology going beyond mere technocratic study to assess environmental impacts of a new technology in a ceteris paribus way. A new view will be taken, if industrial ecology is also to account for the structural changes of the sociotechnological system induced by a new technology. A new transportation technology, such as the Swissmetro, a vacuum underground Maglev transportation, can serve as example. A new, extended, non-static view would mean that not only the environmental impacts per kilometer transport per person are assessed (person-kilometer). A new, systemic view would include the changes of spatial patterns (e.g. by upgrading and downgrading the attractiveness of certain regions for housing and commerce) or of traffic system use (a new technology affects the use of others such as cars, trains, and planes; slow traffic might change). This type of extension has already started (see Box 16.8; Girod *et al.*, 2011; Spielmann *et al.*, 2008), and seems to be in feasible reach of industrial ecology. It introduces a systemic view and focuses on considering feedback loops and rebound effects elicited by product change and new technologies. This is an example for a consequential LCA.

A *second* line would ask that industrial ecology includes research about qualitative changes induced in consumer behavior and social structures by new epoch-making technologies. These new technologies may shape the transition to post-industrial society. Learning from history, we have to ask which technologies are of the same quality as the invention of the plough, wheel, stem-vessel, gasoline engine, telephone or computer which elicited, characterize or presumably induce transitions of societies. A special challenge here is to anticipate what changes in HES (in the sense of Figure 3.3*) will result when a society uses these technologies on a large scale. Clearly, an intellectual challenge is to deliberate on how these new technologies affect the material-biophysical environment of the forthcoming post-industrial society and which new social–cultural–epistemic structures may be related to them. This looks like more than a Herculean task and might lead to a new type of engineering science. We suspect that this new branch of industrial ecology will reflect on the impacts that new stages of the appropriation of nature will have both on consumers and human systems. We realize that industrial ecology will become a completely different science if these questions are posed. This is much more than the task indicated by Allenby's metaphor of looking at the "Five Horsemen of technological evolution," (see below) which makes reference to four horsemen of the biblical apocalyptic story.

A *third* line might even go beyond the second to touch on the fundamental form of the socioepistemic layer that new epoch-making technologies such as the internet elicit in the post-industrial society. Here Allenby argues that we have to investigate the "policy implications of increasingly rapid technological change and the entrained cultural, social, economic, and institutional implications (indeed even the theological, …)" (Allenby, 2009, p. 178). This type of argument would go beyond mainstream industrial ecology. It is, however, compatible with the Lenski–Scholz ReSTET theory (see Figure 8.3*), as it focuses on the fundamental impact that technology has on social and societal structures. The following statement sheds light on this third line:

> For example, it may well be the case that the changes we are currently beginning to experience mark, in fact, the end of the great Enlightment project of radical democratic power. (Allenby, 2009, p. 178)

13.7.1 Anticipating structural changes in post-industrial societies' industry

The Swissmetro example has already revealed how industrial ecology can help cope with the unintended and implicit rebound effects resulting from new technologies. Allenby (2009) demonstrates what dramatic changes were historically implied with the case of the masterpiece of technology railways. The scope of impacts ranges from the impacts of railways on landscapes and settlement structures, the increased pressure for communication systems on a national scale, the change of local-times for different towns and railways companies to regional standard time in 1917, new management principles, trust-based financial markets, and the switch of the USA from an agrarian- to a industry-based society.

A critical question is whether and by what scientific methodology reliable knowledge about these future transitions can be generated. Here, we face a double dilemma. The first part is that, in common terms, a single discipline seems overburdened as knowledge about technologies, their human use, the impacts on social structures, as well as the environmental impacts have to be combined. Here, one has to take a cross-disciplinary view. We will discuss in Chapter 14.3 whether general system theory or other strategies may help. The other half of the dilemma is what track society will take; whether synthetic designer food becomes the societal standard depends on the values, societal norms, and the ethical principles forming the individual and society. We suggest that such questions should become a matter of transdisciplinary discourses which look for socially or sociotechnological robust solutions. This is discussed in Chapter 15.

13.7.2 The new quality of industrial impacts

We already know some epochal technologies that may feature in the post-industrial society. Allenby (2009) named them the Five Horsemen (NBRIC): nanotechnology, biotechnology, robotics, ICT (information and communication technology), and applied cognitive science. We suppose that this squadron of potential epoch-making technologies may gain some comrades, in particular, new efficient technologies in the energy domain. Now let us briefly consider the fundamental changes that are linked to these Horsemen from the perspective of human–environment relationships.

As we have elaborated in Chapter 5, on biology, crucial progress in understanding life was achieved by means of new measurement technologies that allowed the study of molecular structures, leading, for instance, to the identification of the sequence of DNA. Genetically modified organisms represent a new stage of appropriation of nature and allow industry to work at the core of the organismic hardware.

The production of nanotech particles brings engineering to the atomic level. Here the issue is that industry produces tiny new molecules that the world has never seen before, as far as we know, but which are spread across the compartments of the world ecosystem. Some are composed of a few elements, such as Nano-TiO_2. Others, such as fullerenes or carbon-nanotech tubes, have a highly robust structure. Many of them are very stable. Others show a high biodegradability. Some may show some toxic effects. But assessing the toxicity of nanotech particles leads to analytic frontiers with very low concentrations (Helland et al., 2007). This view conforms to a direct environmental impact assessment and thus to the conventional side of industrial ecology. More fundamental impacts of human species on the environment are expected if we look at programmable nanotech particles which may affect the lifespans or other properties of organisms and which may thus have multiple feedback loops.

We have already gained some experience with the third class of epoch-making technology – the sophisticated electromagnetic transmission and digital storage of information. There is tremendous computational and storage capacity which outperforms the human mind in many areas. This has radical impacts on society, as codified knowledge is available in a completely new manner. The wider accessibility of many types of information, historical and otherwise, for different users has dramatically changed human demands and affects social power structures. Likewise, if certain groups of society are excluded, this changes informal power and thus also changes societal structures. The building of virtual worlds, the production of certain types of artificial intelligence that can drive computers and substitute human action, and computational neurosciences may create completely new interfaces and new types of intelligent agents. This may also lead to completely new qualities of HES according to the definition proposed in this book. Human cells may interact with the environment in new ways. It remains to be seen what this means for material flows, customers' needs, and the form of human life. However, a new type of science that reflects the possible pathways that the evolution of technologies may take and on the impacts these evolutionary changes may have on HES is certainly desirable. As the pathway this evolution may take and its desirability is strongly related to societal values, ethics, etc., this new science has doubtlessly to take an interdisciplinary as well as a transdisciplinary route.

13.7.3 Key messages

- The evaluation of emerging technologies, such as nanotech, genetically modified organisms, intelligent databanks or robotics, opens new doors for industrial ecology which go beyond traditional energy–matter environmental impact analysis. This task asks for interdisciplinary and transdisciplinary work.

Part VII

Beyond disciplines and sciences

14 Integrated systems modeling of complex human–environment systems 341
15 Transdisciplinarity for environmental literacy 373

Key questions (see key questions 1–3 of the Preamble)

Contributions from a systems perspective to environmental literacy

Q1 Do we need an anthropocenic (re)definition of the environment? What does the complexity of environmental systems mean? What methods do we have to represent environmental systems and their dynamics?

Q2 How can integrated modeling and transdisciplinary processes help to better understand the rationales of causation and processes of human–environment systems?

Q3 What are transdisciplinarity and transdisciplinary processes? How does transdisciplinarity supplement disciplinarity and interdisciplinarity?

What is found in this part

This part leaves the disciplinary perspective. Instead, we take a system perspective. Chapter 14 first provides an anthropocenic redefinition of environmental systems in the range from the molecular level to the biosphere. Human–environment systems (HES) are conceived as inextricably coupled, complex systems which require special methods to understand them and special (transdisciplinary) processes to manage them.

Chapter 14 inquires from a systems perspective what integrated modeling can contribute to understanding of structure, causation, and processes, and how it allows the anticipation of the dynamics of HES. This leads us beyond disciplines as integrated systems modeling follows a system- or problem-oriented perspective.

Chapter 15 inquires whether and how coping with environmental problems can benefit from the integration of knowledge from theory and practice. This brings us beyond sciences from an actor and object perspective. We show the products, functions, benefits, limits, and risks of transdisciplinary processes. Both chapters refer strongly to our conception of sustainability as an ongoing inquiry of system limit management in the frame of intragenerational and intergenerational justice.

Part VII Beyond disciplines and sciences

Chapter 14

Integrated systems modeling of complex human–environment systems

Roland W. Scholz, Justus Gallati, Quang Bao Le, and Roman Seidl

14.1 Integrated modeling of coupled HES 341	14.5 Uncertainty, probability, stochastic processes, and risk assessment 361
14.2 The world of the linear model 351	14.6 Game and decision theory 366
14.3 Systems theory, systems dynamics, and scenario analysis 354	14.7 Modeling human–environment interactions with cellular automata 367
14.4 Self-organization, "critical transitions," and chaos 359	14.8 Modeling human–environment interactions with multiagent systems 368

Chapter overview

The definition of environmental literacy in Chapter 2 included: (i) the proper presentation and representation of the state of the environment and its dynamics; (ii) the understanding of the human impacts on the environment and environmental feedbacks; (iii) the identification of options for actions for mitigating unwanted impacts; and (iv) the generation of knowledge for successfully coping with (i) to (iii). This chapter shows how integrated systems modeling can serve to generate this knowledge. This is done by first specifying the nature of complementarity of human and environmental systems. HES are conceived as inextricably coupled complex systems.

One challenge here is to deal properly with different types of complexity and fundamental system traits such as continuity and discontinuity. Another challenge is to acknowledge that the human factor is also forming the processes of the natural environment. Humankind has become a geological factor and thus environmental systems from the micro to the macro scale require an anthropocenic redefinition.

We deal with modeling as an efficient scientific reasoning tool for gaining a better understanding of the system elements, their relations, and interactions, thus allowing for anticipating dynamics of complex HES. We summarize the epistemic requirements arising from the attempt to understand coupled HES comprehensively. From this we demonstrate the potential contributions of different modeling approaches to meet these requirements.

The essence of this chapter is to reveal what additional insights into HES can be attained by different types of modeling, which goes beyond what we can describe without making reference to formal modeling and what is essential for understanding HES.

The second part of this chapter appraises fundamental key ideas of selective modeling approaches. We make explicit how these approaches, and in particular integrated systems modeling can genuinely establish interdisciplinarity by integrating subsystems or variables that are dealt with in different disciplines. Thus, integrated modeling is considered a mode of "knowledge integration." The essence of this chapter is to reveal what additional insights into HES can be attained by different modeling approaches, (qualitative) insights that go beyond what we can describe without making reference to formal modeling techniques, but which are essential for understanding HES.

14.1 Integrated modeling of coupled HES

Given the fact that the human is a major agent in world material flows and the biosphere, environmental literacy requires the analysis and management of inextricably coupled human and environmental systems. This analysis has to take into account that these interactions

14 Integrated systems modeling of complex human–environment systems

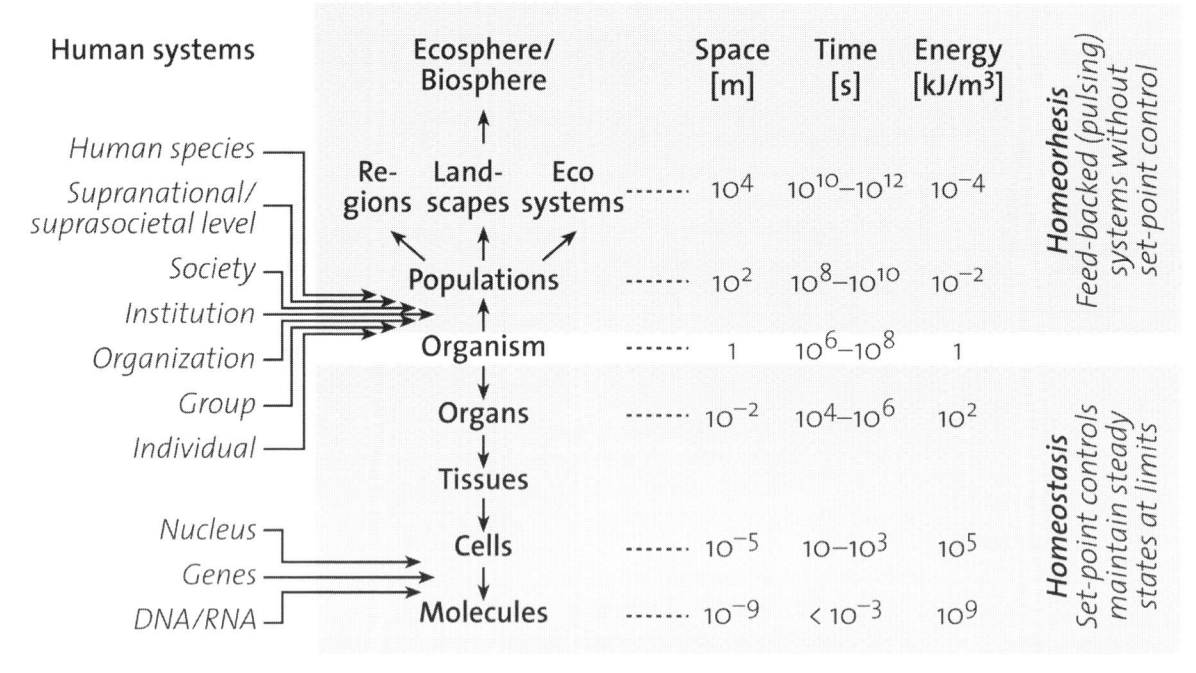

Figure 14.1* The hierarchy levels of human systems (left upper fringe scales; see Table 16.2) as part of the organism–ecosystem hierarchy (see Figure 8.2; Jørgensen et al., 2007; Odum, 1996).

take place at different hierarchical levels and that they are often intrinsically systemic, dynamic, complex, and subject to evolutionary change. The systems' dynamics embodies feedbacks and non-linearities, possibly causing – under certain conditions – "critical transitions" which can also qualitatively change the system's structure. Understanding complexity in HES is a problem deserving particular attention.

We argue that mathematical modeling can be considered a "microscope" for investigating HES, allowing for the anticipation of possible dynamic patterns of complex systems. Mathematical models that can be used to investigate HES range from linear models, systems thinking, and system dynamics to complex multiagent and adaptive systems, each providing specific contributions to these investigations.

14.1.1 Starting from an anthropocenically redefined environment

We have shown (see in particular Chapter 13) that human activities and technology have caused fundamental global change of many kinds. We are facing, for instance, innumerable new chemical elements, new technologies producing never-seen engineered nanoparticles or genetically modified species, changed water and biogeochemical cycles, fundamentally modified Earth surface structures and accelerated climate dynamics causing scarcity, pollution, and detrimental social dynamics. Biologists acknowledge that large parts of the patterns of terrestrial systems are formed by human activities (O'Neill, 2001; Turner, 1990; Vitousek et al., 1997).

Human-made changes show the same gravity as classical natural hazards, many of which are also already strongly affected by human activities. Paul Crutzen called this new period the age of the anthropocene (Crutzen, 2002a). Facing the changing world, Crutzen infers that:

> … humankind is … a noticeable geological force, as long as it is not removed by diseases, wars, or continued serious destruction of Earth's life support system … (ibid. p. 10/4)

Following this diagnosis, we speak about an anthropocenic redefinition of the environment. Considering the major impact humankind has had on the entire biosphere since the onset of agriculture and on the shape of land since the Middle Ages, the start of the anthropocene can be set even earlier than Crutzen suggests (Küster, 1999; see also Box 14.1). Today, many dynamics of lithosphere, hydrosphere, atmosphere, and biosphere are fundamentally affected by human activities. The human is a major agent in world material flows and the biosphere. This means that environmental literacy requires the analysis of inextricably coupled human and environmental systems, at least if we deal with the anthroposphere, the relevant environment of human life.

A critical issue is that the dynamics of HES are affected by processes that take place on a broad range of temporal–spatial scales. Figure 14.1* presents the scope of systems, ranging from the molecular level to the world level, represented by the ecosphere and biosphere. The two scales at the left fringe represent human systems above the scale of the individual, which builds the kernel of this book, and few levels between molecule and cells, which are essential to understand life and the interaction of the immune system with the environment. As shown in Chapter 5, systems above the individual are – in general – not assumed to be governed by homeostasis – that is, set-point controls which maintain steady states at limits – but rather a homeorhesis, that is pulsing systems returning to and following trajectories. Thus, to keep the environment within limits adequate for human living, intentional control and governance that affects the trajectories can become necessary. A simple example of what the governance of limit management with respect to inflow and outflow of material (and financial) resources looks like is presented in Box 14.2.

14.1.2 Complexity in HES

In the time of the anthropocene, the environment and thus most environmental problems have been shaped by human systems and history. It is not that humans usually directly cause flooding, but that flooding is the result of human settlement and related alterations of rivers and land, and, in particular, whether or not humans build levees. Global warming becomes a problem because of human production and fossil energy use, and soil fertility because of agriculture. All these problems are overly complex, in particular if they are investigated in real-world settings. In addition, we are facing multiple causalities. In general, what caused Chernobyl, Seveso, or the impact of hurricane Katrina does not refer to one single "cause" only. Cross-scale or cross-level interactions play a challenging role in complex HES, because reciprocal effects and (first and higher order) feedback loops are frequent and can establish tricky rebound effects. Thus, coping with multiple impacts and interactions is a special challenge.

Responses of a system's environment may be delayed, and developments can go on hidden from awareness or evolve in a manner too slow and distributed to be easily recognized (Liu *et al.*, 2007a, b).

Box 14.1[1] **Climate impacts of the agrarian society: pest impacts?**

Until 2003, researchers thought that the anthropocenic era had begun 150–200 years ago, when the industrial revolution began producing CO_2 and CH_4 at rates sufficient to alter their compositions in the atmosphere. However, in 2003, Ruddiman (2003) proposed that the first human influences on the climate occurred thousands of years ago. His hypothesis is based on three arguments. First, cyclical variations in CO_2 and CH_4 driven by Earth-orbital changes during the last 350 000 years show predictable decreases throughout most of the Holocene. However, as shown in Figure 14.2 (left), the CO_2 trend began an anomalous increase 8000 years ago, and the CH_4 trend did so 5000 years ago. Second, published explanations for these mid- to late-Holocene gas increases based on natural forcing can be rejected based on paleoclimatic evidence. Third, a wide array of archeological, cultural, historical, and geologic evidence points to viable explanations for climate change that are tied to anthropogenic changes resulting from early agriculture in Eurasia, including the start of forest clearance about 8000 years ago and of rice irrigation about 5000 years ago. In recent millennia, the estimated warming caused by these early gas emissions reached a global mean value of 0.8 °C and roughly 2 °C at high latitudes. Based on analyses from two kinds of climatic models, this rate of warming was large enough to prevent a predicted glaciation in North-eastern Canada. Ruddiman (2007) also proposes that CO_2 oscillations of ~10 ppm in the last 1000 years are too large to be explained by external (solar–volcanic) forcing, but they can be explained by outbreaks of bubonic plague that caused historically documented farm abandonment in western Eurasia (see Figure 14.2b). His studies suggest that forest regrowth on these abandoned farms sequestered enough carbon to account for the observed CO_2 decreases. Moreover, plague-driven CO_2 changes are also thought to have been a significant causal factor in temperature changes during the Little Ice Age (1300–1900 AD).

---

[1] This box has been written together with Bastien Girod.

Box 14.1 (cont.)

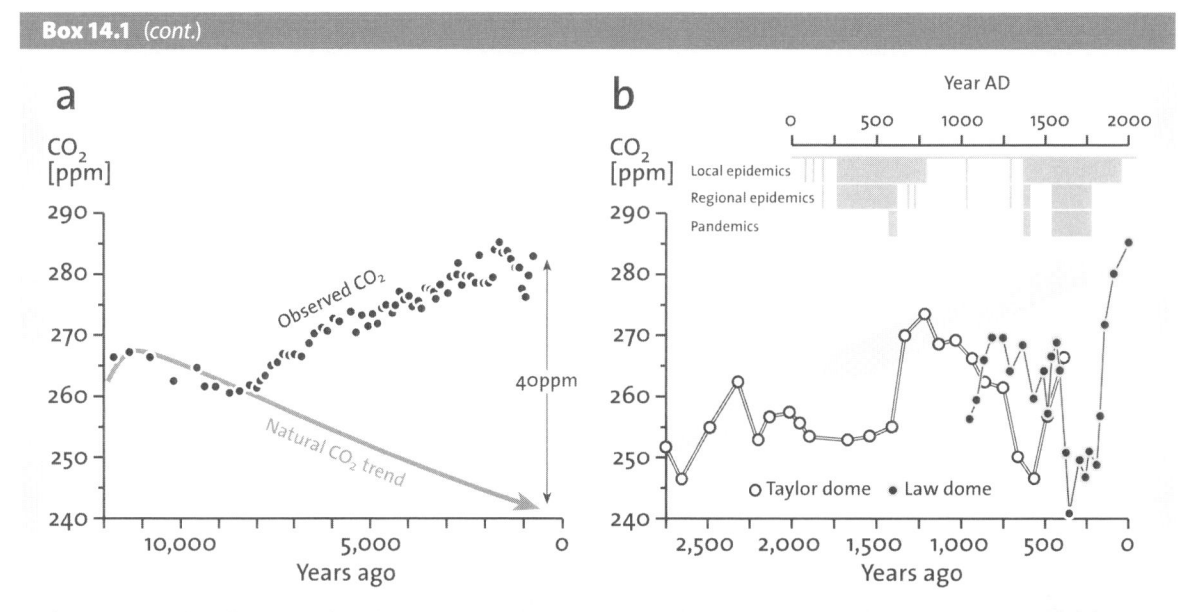

Figure 14.2 Early anthropogenic hypothesis. (a) Human activities during the late Holocene caused increases in CO_2 in contrast to the downward trends during previous interglaciation. Late Holocene greenhouse gas increases prevented much of the natural cooling that occurred in previous interglaciations (from Ruddiman, 2007, p. 2). (b) Rising CO_2 trend during the Holocene (light grey) interrupted by decreases during the past 2000 years. Several of the CO_2 drops may have been caused by reforestation, resulting from massive mortality during pandemics (from Ruddiman, 2007, p. 3).

Nearly every aspect of this early anthropogenic hypothesis has been challenged: the accuracy of the timescale and therewith the resulting anomalies in CH_4 and CO_2 emissions, the issue of use of the isotopic stage 11, the ability of human activities to account for the gas anomalies, and the impact of the pandemics. However, Ruddiman's response to this criticism (2007) reconfirms that the late Holocene gas trends are anomalous in all ice timescales; greenhouse gases decreased during the closest isotopic stage 11 insolation analog; disproportionate biomass burning and rice irrigation can explain the methane anomaly; and pandemics explain half of the CO_2 decrease over the past 1000 years. However, Ruddiman (2007, p. 1) admits that "only ~25% of the CO_2 anomaly can be explained by carbon from early deforestation." Considering the influence on the global climate of early agriculture, Kutzbach et al. (2009) conclude that a full test of the early anthropogenic hypothesis will require models that include ocean and land carbon cycles (see also Box 4.10). Calculations of the radiative forcing of climate with special resolutions support Ruddiman's thesis on the anthropogenic influence of epidemics and even population decline through warfare on the climate. Pongratz et al. (2009) point out that although some affected regions may be too small to have caused impacts on global climate, local climate may have been altered significantly.

This confronts us with legacy effects and time lags. A system may experience sudden changes, which in fact have evolved subliminally until a certain threshold is reached. These critical tipping points have to be explored to understand non-linear behavior. General characteristic properties of complex systems are circular causality, feedback loops, the issue that small change can cause large impacts, as well as emergence and unpredictability (Érdi, 2008; Matthies, Malchow & Kriez, 2001; Allen & McGlade, 1989). We can see that complexity has many origins and, as Allenby (2009) stressed, is not a unitary concept. We distinguish five types of complexity.

First, *static complexity*; this type of complexity (which others might call "complicatedness"; see Ottino, 2004) is defined by the number of system elements and their relations. In environmental sciences this complexity can often be taken from system graphs, such as Forrester's world model (see Box 16.7), or a flow chart representation of technical systems such as a nuclear power plant (see Box 16.11). Referring to the "architecture of knowledge" we conceive this type of knowledge

Box 14.2 Societal complexity, unsustainable resource flows, and indebtedness: the western Roman Empire

The rise and fall of the western Roman Empire reveals the difficulty that societies have in establishing sustainable resource flows. The great successes at the beginning of the Roman Empire (starting around 27 BC) stemmed from its expansions and the economic returns during the conquering phases. The return rates during this military and warfare-based expansion phase were initially very high. The accumulated resources of the conquered land increased the wealth of the mother country. However, the returns declined after the military process ended and was followed by an expensive administrative phase, in which the conquered country had to be organized and the defeated were no longer exploited to such a high degree. In short, the high return during expansion was only possible by ongoing conquering of new land and by taking resources from the defeated. Once this process decelerated or stopped, what was left was the huge Empire whose management caused large internal costs.

The late Roman Empire tried to solve these problems no longer by additional expansion but through its primary problem-solving institutions, i.e. the government and the armed forces which increased in complexity, size, power, and cost. "It became a coercive, omnipresent state that tabulated and amassed all resources for its own survival" (Tainter, 2000, p. 21).

What the Romans did not foresee was that their military and administrative growth was no longer complemented by growth in wealth (from external sources). Instead, beginning with Nero (in 64 AD) they tried to solve the problem by debasing their currency several times. This "was a rational solution to an immediate problem" (Tainter, 2000, p. 34) because the costly metals, especially silver, for the denarius coins became scarce. In the long term, however, the costs of self-sustaining of the empire exceeded its capabilities, because it stopped expanding while its enemies grew in number and strength.

A society must create a sustainable regime and prevent a consumption of its own substance. This is reflected by the devaluation of the money (see Figure 14.3), which represents the indebtedness of a system that has not found a sustainable flow of resources. The path the Romans were on meant "increasing the size, complexity, power, and the costliness of the primary problem-solving" systems, which in turn increased the vulnerability of the whole empire (Tainter, 2000, p. 23). Thus, the "Roman Model" of problem-solving led to unsustainable development, with increasing complexity and costs that had been subsidized by new resources through further expansion. The currency debase was a viable strategy only in the short term; but the ongoing use of the same strategy ("more of the same"; Dörner, 1989) led to bankruptcy (see Figure 14.3).

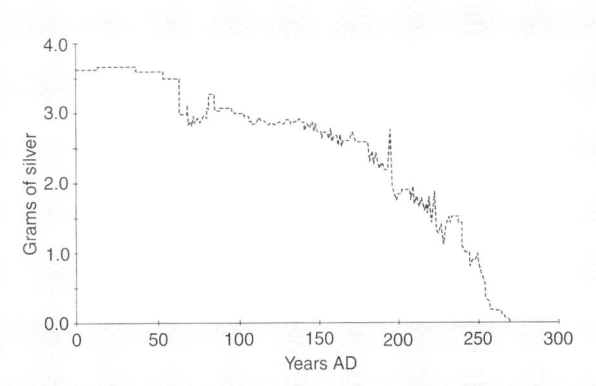

Figure 14.3 Debasement of the denarius to 269 AD (With kind permission from Springer Science + Business Media: Tainter, 2000, p. 21).

---

as "conceptual complexity," which represents the cognitive process of *begreifen* (conceptualizing; see Box 2.6). A static system model asks for definition of system boundaries and usually includes qualitative statements about cause–impact relationships, which already include a dynamic, temporal dimension and is linked to *erklären* (explaining; see Box 2.6).

Second, *dynamic complexity* arises if factors among the systems interact and induce temporal changes. Unexpected behavior emerges by the continued actions of positive feedback loops in the system. Capturing complexity of this type requires definition of quantitative relationships in the language of mathematics. We thus speak about formal models which serve to explain the complexity of systems through simplified representations of the system. The dynamic aspect of complexity is very pronounced in modern societies with accelerating technological development and information communication.

We distinguish two types of dynamics. One is predictable, gradual, and "continuous," while the other is a seemingly unpredictable, unsteady, "chaotic" process in which the fundamental system properties may change.

Third, according to Allenby (2009), *wicked complexity* emerges from the imponderabilities involved in human systems and the fuzzy goal definition when

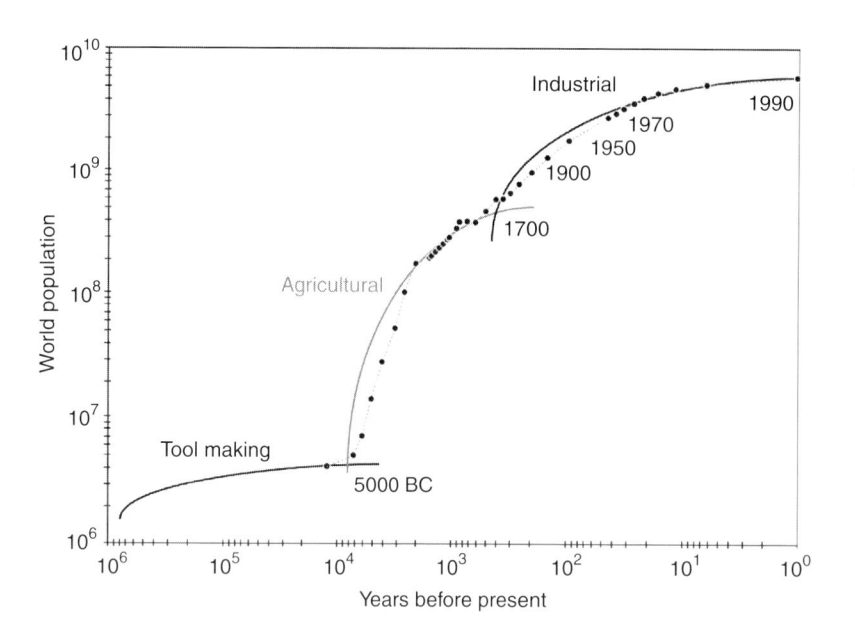

Figure 14.4 Logarithmic population growth in transitory successions of human social system. Adapted from Deevey, 1960 with permission. © 1960 Scientific American, a division of Nature America, Inc. All rights reserved. (Data from Kates, 1997; United Nations (UN), 1993).

dealing with transitions, for instance, towards sustainable development, of real-world systems. That is, the problems are ill-defined and sometimes there is no consensus what the problem *is* or which *goal* one wants to achieve (Dörner, 1989; Scholz & Tietje, 2002). This type of complexity is highly dependent on the sociopolitical structure and dynamics that constantly and inherently interacts with natural systems to general emergent behavior of coupled HES.

Fourth, *complexity of scale* emerges as we are dealing with HES on completely different spatiotemporal scales (see Table 9.1).

> [That] the Anthropocene arises because humans are impacting not just local environments and resource regimes but the global framework of physical, chemical, and biological systems is new and challenging, in that no discipline or intellectual framework enabling rational understanding at that scale yet exists (Allenby, 2009, p. 176).

The different temporal dynamics of the interacting systems can lead to interferences between subsystems and supersystems (see the Interference Postulate P3; Chapter 16.4). Further, the behavior of these systems can be qualitatively distinctive at different phases of development (see Figure 14.4). If we refer to Holling's adaptive cycle, critical issues are in particular revolting subsystems in highly connected mature supersystems in the conservation stage (see Figure 5.16). To understand this type of complexity, investigations focus on the connectivity of different systems' elements, their interactions, and their organization across various scales.

Fifth, we face *historic complexity*. This results from the cumulative, historic nature of systems, including path dependency. How a system is reacting or how and what a decision-maker is factually doing is not only dependent on the situation but also on the way they got there. Each increment of complexity is built on preceding steps, so that complexity seems to grow exponentially in time. Moreover, an increasing complexity because of technological and societal advances demands more complex societies to manage such complex systems (Tainter, 2000).

14.1.3 Non-linearities and "critical transitions"

There is no absolute linearity in real-world systems. Thus, non-linear functions or stepwise approximation by linear functions seem to be the most reasonable strategies to investigate relationships between certain variables. But in many systems, phases of relative continuity are following discontinuity. This can be exemplarily seen when modeling the population growth of human systems (see Figure 14.4). The population growth suggests that there were fundamental system transitions between the food-gatherers and hunters, the plowmen and herdsmen of the agricultural society, and the fuel-driven engine users in the industrial society. Here the invention of technologies such as the wheel, plow or steam engine can be seen as drivers of

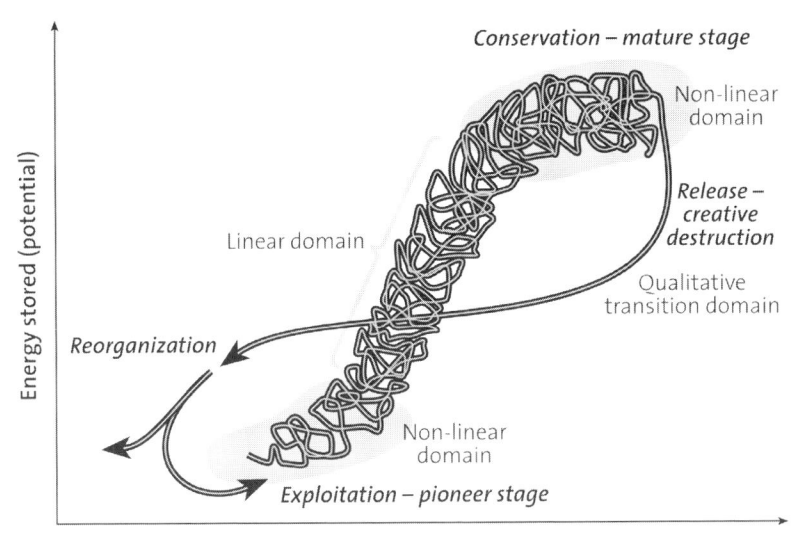

Figure 14.5 Energetic potential as a function of structural complexity in the development (succession) of ecosystems (based on Jørgensen et al., 2007, p. 158, © Elsevier).

regime shifts (see Chapter 13). The growth at each level of society seems to be limited by the environmental constraints and the technological capacity to utilize these resources efficiently, and the type of social organization (see Figure 8.5*). Assuming that the trajectories of population growth at each stage of society follow law-like regularities, one could postulate that:

> … we are in the midst of the final doubling of … the third great surge of the population. (Kates, 1997, p. 37)

Another type of dynamics has been presented in Holling's conception of adaptive cycles. Figure 14.5 illustrates the maturation process as it has been presented for ecological (see Figure 5.16) and social systems along the dimension connectedness, which means structural complexity and stored exergy (i.e. the maximum amount of usable energy), which also is denoted as potential. The graph illustrates further that there is a kind of erratic or stochastic nature. We can distinguish three types of domains: a middle domain in which the relation can be described by a mean – apparently stochastic – linear relation between "potential" and "connectedness;" and the upper and lower bounds of the linear domain where we face a non-linear mean relation. Also, there is the domain of collapse. If a certain level of saturation is attained, one can observe a "critical transition," which means here an abrupt decline of potential.

Clearly these regime shifts are most difficult to anticipate. We know many examples from the real world of such releases or collapses, such as the shift of the Sahel-Sahara from a region that had abundant vegetation during the mid-Holocene to desertification nowadays, the 1929 Wall Street stock market crash, or abrupt changes in political systems such as the fall of the Soviet Union. Some of these "critical transitions" may be described by dynamic systems theory, such as chaos (Mandelbrot, 2004), catastrophe (Thom, 1970) or the bifurcation theory. Yet, we are still at the very beginning of understanding qualitative "critical transitions" (Scheffer, 2009; Sornette, 2003), which are an important component for environmental literacy.

14.1.4 What and what not to expect from modeling

> It is commonly held that mathematics began when the perception of three apples was freed from apples and became the integer 3. (Davis & Hersh, 1981, p. 127)

Numbers, lines, variables, and functions are abstractions or mental models or constructs. A model strips away everything that is sensible and transforms a system from a concrete real-world system to an abstracted model. Mathematical modeling means to switch from experience to mathematics; for instance, from our intuitive spatial understanding to geometry (Aristotle, 350 BC). If we use the language of Jakob von Uexküll (see Chapter 4.12) in modeling we are operating on the (abstract) sign level. In the conception of this book, models are part of the socio-epistemic world.

Models serve to better conceptualize, understand, explain, and anticipate reality and they do it in

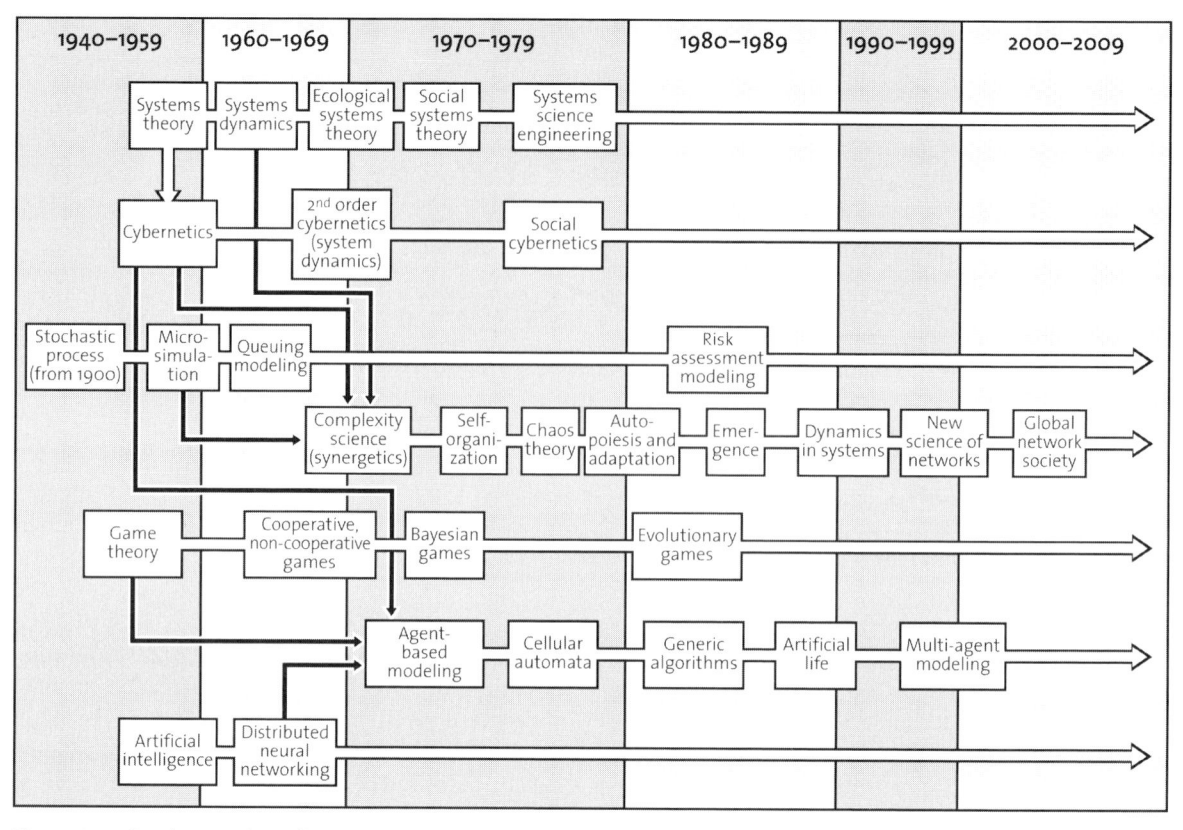

Figure 14.6 Development lines of human literacy in system complexity (modified from Wikipedia http://en.wikipedia.org/wiki/Complexity).

a systematic manner (Epstein, 2008). Figure 14.6 gives an overview of the development of different branches of *modeling* as a tool to enhance human literacy in system complexity. This holds true for a material model of a ship, which is used in a flow channel to explore flow resistance. It also holds true for conceptual or mathematical models on cause–impact relationships or dynamics of complex systems. In natural or engineering sciences, scientists are often formed by the idea that the truth is in the model. We call this the Newton and Leibniz way of thinking or knowledge integration, which comes from the belief that the laws of classical mechanics are universally valid and omnipotent models. From our perspective this is not true. We follow the understanding that each theory or model has a prototypical domain of intended applications in which the model can almost perfectly describe what is ongoing in the environment (Sneed, 1971). But for each model there are fringes where other factors and laws are at work. For instance, Hooke's law that the power of any spring is proportional to the extension of a spring is only "approximately valid" in a tiny domain of relatively small deformation of certain springs. Models are not copies of the world but constructs of our mental capability to better understand certain aspects of reality.

Mathematical parameters, variables, and functions are means of describing what is happening. They cannot explain, for instance, a causal relationship, but can only describe it. Explanations emerge from theories. Thus we conceive modeling and mathematical modeling in particular as a tool or language, which helps us to utilize an abstract representation for describing more precisely and gaining insight into the dynamics and interactions of systems. The knowledge that can be gained by modeling goes beyond verbal or graphical (maximum four-dimensional representation, including time in an animated 3D graph).

Models allow us to represent our assumptions about causalities, but do not explain them. To explain we need conceptual systems, which include terms whose role and interaction can be described, for instance, by mathematical formula. Here it makes sense to distinguish between conceptual and abstracted systems or models. Conceptual systems are built by terms from normal languages. Abstracted systems are: (i) formal, mathematical systems presented in a formal, symbolic language, that (ii) refer explicitly to natural or social science theories (Miller, 1978; Sneed, 1971).

14.1.5 Mathematical models as a "microscope"

A critical question with respect to environmental literacy is whether and in what manner models can help us to better access critical properties of HES. Can we magnify our image and penetrate what is ongoing – for example – in current or future food or energy supply on a world system level? We have seen in the history of microbiology and molecular biology, that the microscope has been the major instrument promoting progress in understanding the structure and function of microstructures of the organism (see Figure 5.1). Clearly, the data and signs from the microscope have to be interpreted. Similarly we can postulate that mathematical modeling provides new data, which can help to better describe, understand, and explain dynamism, trajectories, and regulatory mechanisms in HES.

Simulation, i.e. the imitative representation of the functioning of a real-world system by mathematical signs, can help to mimic complex systems without intervening into the real-world setting, which is often impossible for various reasons. We shed light on major modeling approaches and exemplarily reveal what core ideas they provide for an extended environmental literacy. Simulation, in particular, offers the opportunity of anticipating and investigating possible future dynamics of a complex system. As such, it is considered a tool for sustainability learning.

14.1.6 Epistemic requirements to analyze HES

The previous sections and chapters have demonstrated that a variety of epistemic requirements arise from the attempt to understand coupled HES comprehensively. These include the following epistemic operations:

- Capture complexity
- Include and quantify uncertainty
- Acknowledge complementarity of human and environment systems, including hierarchies of systems and potential interferences between them
- Take into consideration the dichotomy of agency and structure
- Understand decision-making of different actors, taking into account alternative, potentially conflicting, strategies
- Include feedbacks in human and environment systems
- Adopt a long-range view and anticipate potential future developments
- Capture non-linearities and "critical transitions," including self-organization of systems and emergent properties
- Capture heterogeneity of states and processes in space and in time, coupling multiple human and environmental agents
- Provide reliable evidence with limited availability of data and information.

These requirements pertain to properties of complex systems in general and to characteristics of HES in particular. Perhaps these requirements do not all apply at the same time but depend on the phenomenon at hand.

14.1.7 A taxonomy of models for HES

It has been argued that modeling is a promising approach to analyze HES. Here we present a taxonomy of models, referring to the epistemic requirements stated above. It will be demonstrated that the different modeling approaches show specific strengths with regard to these requirements (Table 14.1). We start with the "linear model," which is based on linear equations or transformations. These are the basic fundamentals of modeling and date back to 2000 BC (Struik, 1987). Systems thinking and system dynamics models, as the next important tool to model complex non-linear systems, go beyond linear models and are based on differential calculus, which was developed in the time of Leibniz (1646–1716) and Newton (1643–1727). Scenario analysis, and, in particular, formative scenario analysis, are options to address complex problems in case a cardinal operationalization of the variables and/or functional relationships among the variables cannot (yet) be provided.

Other than differential equation-based models, which are essentially deterministic, stochastic processes are non-deterministic in the sense that they include random processes. A system element or variable is called random if there is no way to predict its outcome exactly. This leads to uncertainty, probability

Table 14.1 Strengths of various modeling approaches with respect to epistemic requirements for understanding the dynamics of HES (indicated by x).

Epistemic requirements	Linear model	System dynamics	Probability, stochastics, risk	Game and decision theory	Cellular automata	Multiagent system
Uncertainty	x		x	x	x	
Complexity:	x	x		x	x	x
static complexity		x		x	x	x
dynamic complexity						x
wicked complexity						
complexity of scale						
historic complexity						
Complementarity of human and environment systems	x	x		x		x
Hierarchy, interference of systems		x		x		x
Agency/structure dichotomy		x				x
Conflicts, alternative strategies		x		x		x
Understanding decision-making		x	x*	x		x
Feedback		x			x	x
Anticipation, long range view		x				x
Non-linearity, thresholds		x			x	x
Self-organization and emergence		x			x	x
Heterogeneity: - in space		x		x	x	x
- in time				x	x	x
Dealing with limited availability of data and information	x	x	x	x		

* This refers to risk.

distributions, and risk assessment. As we will see, uncertainty and probability modeling is ubiquitous. Thus, as we can learn from physics and chemistry, unforeseeable dynamics of systems, qualitative change of structures, "critical transitions" or unforeseeable collapses can also result from non-stochastic system dynamics. We briefly touch on this under the label self-organization, instability, and chaos.

With game and decision theory, actors, together with their strategies, interests, and rationales, are introduced explicitly. While game theory offers the opportunity to formalize and to understand conflictive situations and social dilemmas, decision theoretical models address the mental process within individual actors prior to action. Finally, we focus on multiagent systems, which have their root in game theory, cellular automata, and artificial intelligence. Other than in system dynamics, multiagent models operate at the level of individual agents following a set of decision rules, resulting in emergent properties of the system at the level of the observed phenomena. The issue here is that this way of modeling allows for the modeling of coupled material bio physical and social systems, and thus is an interesting approach to HES.

To conclude this overview on different modeling approaches, we would like to point out that beyond these requirements there are further challenges for interdisciplinary modeling to take into consideration. These pertain in particular, first, to reconciling different disciplinary views with regard to model validation, and second, to the problem of leaving space for the unknown when addressing future developments. These challenges have to be kept in mind and investigated further when applying modeling approaches for the analysis of HES.

14.1.8 Key messages

- The anthropocenic redefinition of the material–biophysical environment systems acknowledges the strong human impact in all environmental systems, from the atomic–molecular to the global biogeochemical–cycle level. This requires incorporation of the human factor in (modeling) the material (natural) environment dynamics
- Human and environmental systems are seen as hierarchically layered, multiscaled, historically shaped, partly predictable, inextricably coupled complex systems, which may face linear, smooth or non-linear qualitative system transitions
- Modeling is considered a promising approach to analyze HES that has the potential to comply with a range of epistemic requirements that arise from the attempt to understand these systems thoroughly.

14.2 The world of the linear model

14.2.1 The linear model: the working horse for systems modeling

The key concept of the linear model is *proportionality*, which simply means that one variable x has a constant ratio to another y, which means x/y = constant = a. Historically, the linear model developed early and is the simplest approach of quantitatively relating different variables.

The linear model is still a powerful tool of many sciences, including the environmental sciences. Practical work with the *IPAT* model ($I = P \times A \times T$; see Chapter 9.3) refers to the linear model. When calculating the environmental impact per person ($P = 1$), the environmental impact I, which is often measured by CO_2, depends linearly on the affluence A, which is often represented by the GDP and the level of technology, which is taken as a constant.

For instance, life cycle analysis (LCA; Heijungs & Wegener Sleeswijk, 1999; Udo de Haes & Heijungs, 2009), material flow analysis (Baccini & Bader, 1996), and all variants of input–output analysis (Hendrickson et al., 1998; Leontief, 1966) are based on linear algebra, the mathematical field dealing with linear transformations. To understand the idea of the linear model in LCA, the reader can think about vector Y as all relevant environmental impacts that can result from one product (or process), X as all material, biophysical input resources needed to produce this product, and the cells of the variables A = $a_{i,j}$ to denote in what way the input resource x_i affects y_j. This idea also runs under the term input–output analysis (Leontief, 1966).

14.2.2 The key idea

The linear model is presumably the most simple and most frequently used approach to gain insight into the mutual relationship between two or more system variables. In the elementary case we consider two variables x and y that can be defined or measured by rational (or real) numbers and where the values of the one variable are proportional to that of another. We can represent this by the simple equation (see Figure 14.7):

$$y = ax + b \qquad (14.1)$$

In general, we can denote the linear model as a rough, simple but in many cases useful and beneficial instrument or tool of modeling the interaction of variables in HES and other systems. The wide range of use of the linear model is revealed by the terminology. Depending on the context of application, the variable y is also called the dependent, response, observed, measured, explained or outcome variable. The variable x is called the independent, manipulated, exposure, predictor or input variable.

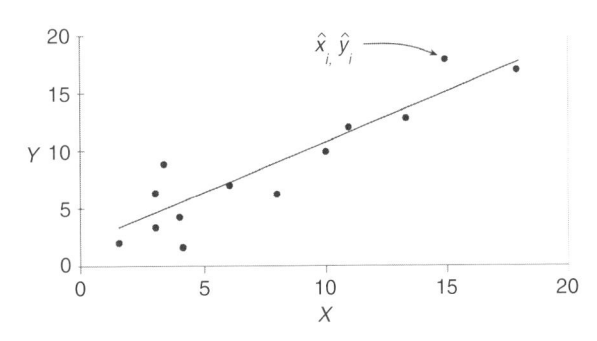

Figure 14.7 A linear regression of two variables x and y.

In the following, we want to sketch how environmental literacy is fundamentally extended by the linear model using four examples. These are: (1) the utilization of matrix algebra to represent how multiple variables (linearly) transform a vector of resources in environmental impacts. (2) We demonstrate how complexity measurements can be reduced by linear correlation and regression. This is done by the use of the linear model in statistics, which includes methods such as regression analysis, factor analysis, and analysis of variance. Here, we exemplarily illustrate how the complexity of a model and measurements can be reduced by (3) factor analysis. Finally, (4), we briefly illuminate the interaction of variables, which can be considered as the entrance of multivariate literacy before we sketch the linear programming models.

14.2.3 Representing multiple linear processes

We are often not only interested in one or more dependent variables y_j which are considered as being a function of a set of other variables x_i. The dependent variables can be different types of emissions. The reader can think about the vector $Y = (y_1,\ldots,y_j,\ldots,y_m)$ as a set of relevant environmental damages that can result from $X = (x_1,\ldots,x_i,\ldots,x_n)$, which are different material resources needed to produce a product.

If we assume that the values of the variables y_i can all be assessed by a linear combination of the vector of variables $X = (x_1,\ldots,x_i,\ldots,x_n)$ (e.g. $y_j = \sum_i a_{i,j} x_i$), we can present this complex transformation by the simple equation

$$Y = XA + B \tag{14.2}$$

where $B = (b_1,\ldots,b_j,\ldots,b_m)$. Here A is a matrix with m lines and n rows and each cell $a_{i,j}$ describes by what amount the variable x_i contributes to y_j.

We should note that the linear model is powerful and widely applied as it is often reasonable to transform non-linear relationships into linear ones by suitable functions. If variable x_i has a multiplicative impact on a dependent variable, then logarithmization or other transformations may linearize these relationships.

14.2.4 Regression analysis for data complexity reduction

An important field using the linear model is statistics, for instance (linear) correlation and regression analysis. Let us assume that we have a certain set of $k = 1,\ldots,k,\ldots,K$ observations $(\hat{x}_i, \hat{y}_i)$ of the variables x and y (see Figure 14.7). If we assume a linear "stochastic" relationship this leads to a linear correlation and related methods. The correlation coefficient tells us how strong the relation between the two variables is. If all points are on the straight line, we have a perfect correlation and the correlation coefficient is 1. If there is absolutely no linear relationship, the coefficient is 0. The term "stochastic" refers to a fundamental assumption of linear regression that the variables show factually a linear correlation, but random factors affect the values of the (measurement of the) variables. If we consider ε as a stochastic error term, the equation (14.1) turns into:

$$\hat{y} = a\hat{x} + \varepsilon \tag{14.3}$$

Clearly, we can also work with more than one dependent and independent variable, which leads to different variants of multiple regression analysis (Bortz, 2005; Draper & Smith, 1998).

14.2.5 Finding a minimum number of system variables

An important and widely applied approach is factor analysis. The issue is again complexity reduction. Let us assume that we have a large number of random variables $Y = (y_1,\ldots,y_j,\ldots,y_m)$. For each of these variables we have a set of observations or measurements of system states (i.e. $\hat{y}_{i,1},\ldots,\hat{y}_{i,m}, m \geq n$). Factor analysis is a means of reducing the number of initial variables by postulating a small number of k factor variables $F = (f_1,\ldots,f_e,\ldots,f_k)$ which incorporate all variables y_i as a linear combination (i.e. $y_i = \sum_i w_i f_e$, w_i are called factor scores).

The key idea of factor analysis is reflected in the term "principle component analysis," which is often synonymously mentioned. This procedure provides for a larger number of correlated variables a reduced number of uncorrelated variables, principle components, which can predict all variables by linear combination.

If we can find few variables f_e (i.e. factors or principle components) that can well explain the variation of the dependent variables or observed measurements $\hat{y}_{i,1}$ then we may have reduced the structural complexity.

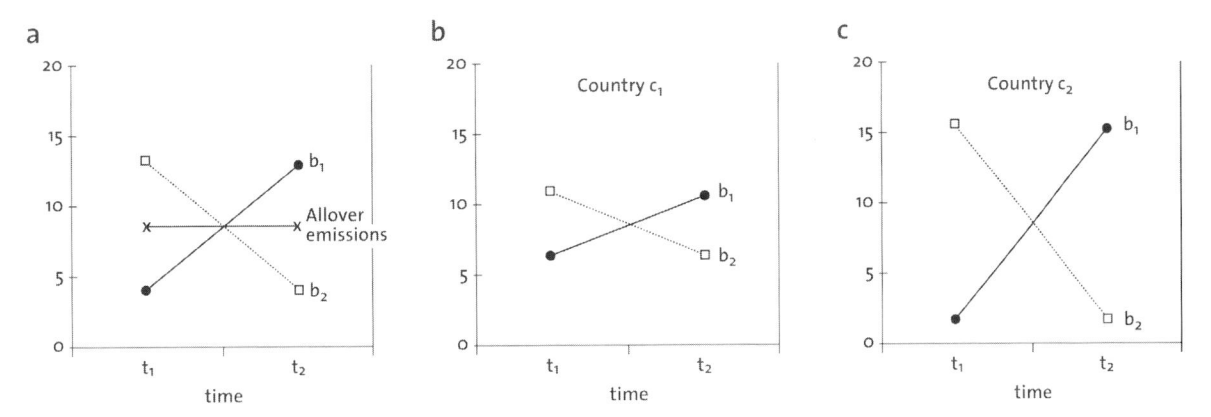

Figure 14.8 Graphical representation of first order and second order interactions of two variables by representing the means x and y over time t.

14.2.6 Understanding interactions among variables

To understand the interaction of three or more variables is an important aspect of environmental literacy. We introduce the idea through a simple example: consider the situation where one dependent variable y (e.g. environmental emissions) depends on two independent variables, which may be time t and different branches $b = b_1, b_2, \ldots$ of industry. A survey conducted by environmental scientists may be interested in the question of how the emissions y from different branches b change over time.

This is exemplarily shown in Figure 14.8a. The middle line shows that the total emissions do not differ between the two points in time t_1 and t_2. However, there is an interaction between the variable time t and the branches b. Whereas for branch b_1 the emissions are increasing between time and, the emissions are decreasing for branch b_2. This is called first order interaction among the independent variables t and b on the dependent variable y.

A slightly more sophisticated interaction is demonstrated in Figure 14.8b–c. Here we consider besides t and b a third variable c, which can represent countries, for example. In our example we consider only two countries, c_1 and c_2. As can be seen from Figure 14.8b and Figure 14.8c, in both c_1 and c_2 the emissions are increasing for b_1 and decreasing for b_2. One may diagnose (and often statistically verify) that the decrease–increase difference is (significantly) smaller under c_1 than under c_2. Here we are facing a second order interaction of three variables, t, b, and c on a dependent variable y. There are many other cases with variables of different levels of scale. The theory of interaction is a field of statistics, in particular the analysis of variance (ANOVA; see Cox et al., 1984; Hays, 1963).

14.2.7 How is interdisciplinarity achieved

In a rigid sense, interdisciplinarity is established by merging concepts or methods. Thus it is essential how and which variables are related. We can see two major forms of merging. One is by establishing causality between two variables originating in different disciplines. For example, when we face the emotional response of anxiety regarding a nuclear power plant and this response can also be measured by increased bloodflow to the amygdala, then this is a case of neuropsychological research about the perception and evaluation of new technologies. This example shows organized interdisciplinarity by integrating variables from different disciplines, which are neuropsychology and technology assessment, in a (linear) model. A second option is given by integrating variables from different disciplines in a (linear) model.

14.2.8 Key messages

- The linear model allows multiple system variables to be related, and is a quantitative representation of how a multitude of input variables affect outputs, given certain presuppositions

- The linear model can be generalized by various means, e.g. by linearizing the variables (by scale transformations) or by logarithmization of multiplicatively related variables. There is a large set of statistical models developed under the heading generalized linear models
- The linear model in statistics is a basic model for reducing complexity by interacting, reducing, and empirically testing interactions between system variables.

14.3 Systems theory, systems dynamics, and scenario analysis

14.3.1 Rationale and historical background

Systems thinking, system dynamics, and the theory of dynamic systems seek to understand and to describe the behavior of complex systems over time.

Box 14.3 Dynamic patterns in coupled HES: collective irrigation management

Water is a primary natural resource and, as such, a fundamental condition for increasing agricultural productivity in rural areas to improve food security and to generate additional income streams for farmers through the cultivation of cash crops (Molden, 2007; Norton & Food and Agriculture Organization of the United Nations, 2004). Farmer-managed, collective systems are widespread and have been found to operate often very efficiently (Vermillion & Sagardoy, 1999). In many cases, however, cooperation may fail, leading to deterioration of irrigation infrastructure, tensions between upstream and downstream users, and possibly violent conflicts. Hence, the question is, whether and under what conditions cooperation can be successful. This problem is considered an example of a coupled HES, where human actions (decisions with regard to cooperation in collective irrigation management) and the effects of these actions (condition of the irrigation system and water productivity) are integrated into a dynamic feedback system. To study possible dynamic patterns in collective irrigation management, a system dynamics model has been developed (Gallati, 2008).

This model refers to the theory of a critical mass in collective action (Granovetter, 1978; Oliver & Marwell, 2001; Schelling, 1978), and includes a number of influence factors that have been found conducive for successful common property resources management (Baland & Platteau, 1996; Lam, 1998; Ostrom, 1990; Wade, 1988). The model is based on case studies in Kyrgyzstan and Kenya. The fundamental feedback structure of the model is shown in Figure 14.9.

The model has two equilibria: the first corresponding to successful, and the second corresponding to unsuccessful cooperation. Model results are presented for the Kyrgyz case, showing a decline of cooperation, and in turn of condition of irrigation infrastructure (Figure 14.10a). This decline has been observed between the early 1990s and 2005. Variation of initial conditions or of specific functional relations, representing for example the effect of receiving sufficient water on farmers' decisions for cooperation, may shift total cooperation to the upper equilibrium.

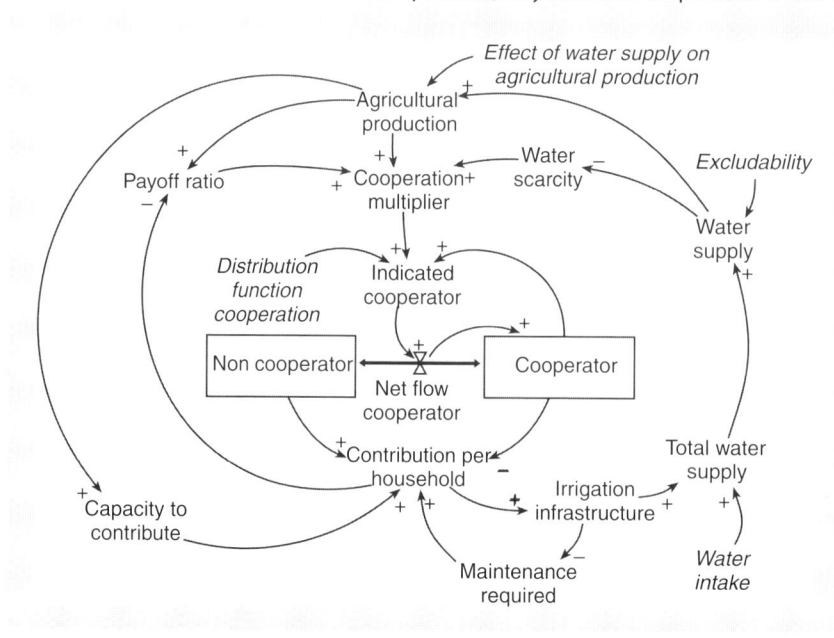

Figure 14.9 Fundamental feedback structure of the system dynamics model for collective irrigation management. The number of users intending to cooperate in future depends (i) on the number of current cooperators, (ii) on the performance of the irrigation system (water supply), and (iii) on the capacity of the users to contribute. Distribution function cooperation refers to an overall distribution of users' thresholds for participation in collective action, shaped by prior experiences as well as by individual economic resources.

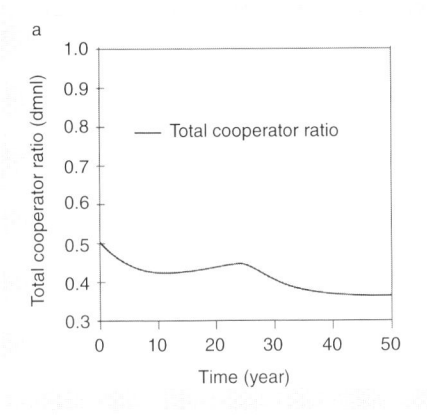

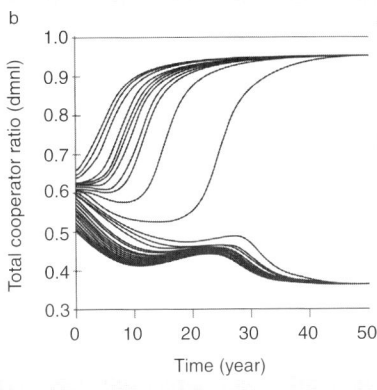

Figure 14.10 (a) Decline of cooperation and settle to the lower equilibrium of unsuccessful cooperation. (b) Slightly higher initial cooperation may result in successful cooperation (upper equilibrium). Note: x axis, time; y axis, total cooperation ratio (dmnl: dimensionless).

Historically, these approaches refer to physics and the description of the motion of physical bodies using – and further developing – the mathematical language and tools available at that time. The 17-year-old Galileo, for instance, was interested in finding a law describing the swing of a pendulum. He could see that the wider the swing, the faster the motion was. He was surprised, when using his pulse as a chronometer, that it took the same number of pulse beats for a pendulum to complete one swing no matter how far it moved (Ginoux, 2009). This led to the invention of pendulum clocks but also to differential equations. The term was first used by Leibniz in 1684 and Newton to describe the Galileo pendulum using a second order differential equation. In its simple form, a (first order) ordinary differential equation describes the change dx of a target object (or variable) x within an infinitesimally small time dt. Mathematically this reads:

$$\frac{dx}{dt} = g\left(x(t), \vartheta\right) \quad (14.4)$$

where g is a continuous function describing the change, t is time, and ϑ a vector describing location, motion, and other state variables.

Mathematics became a major tool in the search for knowledge. The primary domain in which mathematical theory developed was the investigation of properties of motion of physical bodies in idealized settings, ignoring some environmental constraints such as air resistance. The concepts used were those of idealized entities of "space, time, weight, velocity, acceleration, inertia, force, and momentum" (Kline, 1985, p. 98). One challenge was to identify the right variables, the other to develop the mathematical apparatus to describe the behavior of these entities over time. These were differential equations, which became a powerful tool to describe behavior of nature. Social scientists, in particular economists, utilized these tools to describe dynamics of economic and social systems.

However, the description and analysis of a system by means of differential equations also has its limitations, which arise on the one hand from mathematical properties of (non-linear) dynamic systems and on the other from the modeling process itself. Even elementary dynamic systems, such as the motion of three celestial gravitationally related objects, show unstable, chaotic behavior, as has been pointed out by Poincaré (1892). Moreover, it has been recognized that chaotic behavior is a property of many non-linear dynamic systems (Lorenz, 1963; Schuster & Just, 2005). Thus, the world of systems described with differential equations is not as "smooth" as one might expect. On the other hand, one has to be aware that a model of a complex problem is always only one of many possible descriptions that is valid from a particular perspective.

In the late 1940s, cybernetics evolved as a theory of (feedback) control of a system with regard to the environment (Wiener, 1948), and in the late 1950s system dynamics was developed to describe and analyze feedback mechanisms in socioeconomic systems (Forrester, 1961, 1969, 1971a, b, 2007a, b; see Box 16.7; Sterman, 2000). General system theory (von Bertalanffy, 1955) addresses characteristics of systems from a more general point of view. Finally, scenario analysis is an approach to deal with an uncertain future in a qualitative, yet meaningful, way.

14.3.2 Cybernetics

Cybernetics is a theory of control systems based on communication (transfer of information) between system and environment and within the system, and control (feedback) of the system's function in regard to the environment. (von Bertalanffy, 1968, p. 21)

In cybernetics, which concentrates on the regulatory mechanisms in systems, we find mathematicians (Wiener, 1948), organizational scientists (Beer, 1959), social scientists (Bateson, 1979), and psychiatrists (Ashby, 1956). In this book, we widely subsume cybernetics under systems theory. It is relevant in our context as one of the main targets of environmental literacy is the identification of regulatory mechanisms. However, we should mention that modern control theory is a new, mathematical domain in engineering that deals with modeling of feedback systems in engineered systems (Åström & Murray, 2008; Brogan, 1990). This approach is dealt with in some detail in Chapter 16.5 (Feedback Postulate P4).

14.3.3 Systems theory

Systems theory originated in the 1920s from biology when explaining the interrelationship between organisms and their related ecosystems by a "discovery of the principles of organization at its various levels" (von Bertalanffy, 1968, p. 12). Besides the hierarchical structure and the inner organization, the interaction with the environment (see Figures 3.3* and 17.1*) has been a key topic.

> Living systems are open systems, maintaining themselves in exchange of materials with environment, and in continuous building up and breaking down of their components. (von Bertalanffy, 1950b, p. 23)

This distinction between *open* and *closed* systems was particularly important and led to further investigation of open systems and irreversible processes in physics, chemistry, and biology. The cell and the organisms, for example, are open systems and thus can never attain true chemical equilibrium.

Systems theory is general, as we can speak about physical, biological, human, social, technical, and many other systems. Systems are described verbally or with formal, mathematical languages, and a distinction between concrete real-world systems, abstracted verbally described and abstracted mathematically described systems has been introduced. Systems theory has been formed by giants of thinking, among them biologists (Miller, 1978; von Bertalanffy, 1950a), economists (Boulding, 1956; Parsons, 1951), psychologists (Rapoport, 1996), or physicists (Capra, 1996; Nicolis & Prigogine, 1989; von Foerster, 1987), who all made important contributions to the philosophy of science.

14.3.4 System dynamics

System dynamics provides a set of concepts and tools that enable us to better understand the structure of complex systems. Designing effective policies has been the mission of the founders of system dynamics (Forrester, 1961, 1969, 1971a, b, 2007a, b; see Box 16.7; Sterman, 2000). Causality and feedbacks among variables in dynamic systems build the basic constituents.

As a modeling approach, system dynamics relies on three major constituents: first, the concept of feedback loops, including autocatalytic processes; second, computer simulation which allows us to utilize differential equations as a descriptive tool; and third, the notion of mental models which allows the participation of people concerned in utilizing these bodies of science for a better understanding of societally relevant phenomena (see Box 14.3).

The concept of feedback loops involves the collection of information about the state of the system, followed by some influencing action, which in turn changes the state of the system (Lane, 2000). The essence of the feedback concept is a circle of interactions that provides a closed loop of action and information. The patterns of behavior of any two variables in such a closed loop are linked, each influencing, and in turn responding to, the behavior of the other (Richardson, 1991b). Delays and non-linearities are essential properties of system dynamics models. Non-linearities, in particular, are necessary to cause shifts in the dominant structure of a model. This means that different parts of a system become dominant at different times (Richardson, 1991b). An extended discussion on feedback loops is provided in Chapter 16.5.

Computer simulation is the major tool used to describe the dynamic behavior of the system and to anticipate possible future states. Computer simulation is essential in particular for uncovering unanticipated side-effects (Sterman, 2000) and counterintuitive behavior. We should note that computer simulation is not a pure analytic formula-based activity but is based on many constraints, assumptions, and errors underlying the numerical, algorithmic approximation. Thus there is a second layer of modeling and simplification underlying simulation (Oliviera & Stewart, 2006).

The third element of system dynamics has to do with the involvement of practitioners and people who are facing specific problems in the modeling process. We call these people case agents and deal with the collaboration between case agents and scientists in Chapter 15, on transdisciplinarity. It has been recognized that one of the major contributions of a system dynamics modeling process is to stimulate learning processes amongst involved people (Morecroft & Sterman, 1994). It can also be said that the goal of a system dynamics intervention is to create shared inter-subjective meaning (Vennix, 1996) and to provide insight to key agents in a presumed functioning of the system, for instance for anticipating potential policy actions:

> The goal of a system dynamics policy study is understanding – understanding the interactions in a complex system that are conspiring to create a problem, and understanding the structure and dynamic implications of policy changes intended to improve the system's behavior. (Richardson, 1991b, p. 164)

This understanding is built around a dynamic hypothesis, which is the idea that a certain causal structure explains a certain dynamic behavior (Lane, 2000). This notion of a "dynamic hypothesis" is fundamental and is closely connected to the endogenous perspective adopted in system dynamics. The system dynamics model thus belongs to the category of causal, theory-like models claiming to generate the "right output behavior for the right reasons" (Barlas, 1996, p. 186). Thus, we consider a major benefit of system dynamics modeling to be that it shows the dominance of the underlying system's structure over the effect of a change of single variables to achieve a desired system change.

System dynamics makes the fundamental claim that the evolutionary behavior of a social system over time is explainable in terms of feedback loops and state variables (Richardson, 1991b). The concept of feedback thinking in system dynamics is intimately linked with the concepts of interdependence and circular causality, with a rich history in the social sciences (Richardson, 1991a). This observation points to a substantial affinity between system dynamics and the social sciences. System dynamics allows "that the environment that controls human decision making is *itself* made by human decisions" (Lane, 2000; emphasized in the original) is to be taken into account. Lane (2001), with reference to Giddens' structuration theory (see Figure 9.1), considers system dynamics an approach that has the potential to implement agency and structure integrating theories as it acknowledges the relevance of agency as well as of structure in their mutual (dynamic) relationship. Thus, by describing and analyzing such reflexive structures (see Figure 16.13), system dynamics has the potential to contribute to the social sciences.

This may illustrate that in the context of environmental literacy the roots of system dynamics are more related to social science and to the "systems movement" (Schwaninger, 2006) than to cybernetics. Richardson (1991a) argues that the origins of system dynamics are to be found in the servomechanical thread, oriented towards the combination of economics and engineering, and interested in understanding complex problems with feedbacks, rather than in adjustment processes and information transfer, as it applies to cybernetics.

14.3.5 How is complexity reduced

Reduction of complexity is achieved, first, by structuring "messy," ill-defined or "wicked" problems, second, by evaluating and exploring individual components of the models and thus assessing their influence on the behavior of the whole model, and third, by attempting to discern model structures that are as simple as possible yet which still generate the essential dynamics of the system. Such model structures are referred to as "generic models" (Lane & Smart, 1996), although a rigorous definition of generic structures is still missing.

The first point includes problem definition, which comprises the definition of the system boundaries and the problem-solving focus by a guiding question. This is an important step in structuring messy problems. This process of problem-forming, however, is shaped by perspectives and knowledge of the people involved. As such, it can be considered a first step in a mutual learning process creating a shared reality among the people involved (Checkland, 1981; Eden *et al.*, 1983; Phillips, 1984; Vennix, 1996).

The second and third points are intimately related to the primary focus of system dynamics on structure, generating a certain observed behavior. A (system dynamics) simulation model offers the opportunity to assess the influence of individual components of a model, and also of alternative model formulations, on the behavior of the model. As such it can be considered an experimental design that can be used to explore the effect of individual parts on the overall model behavior, possibly leading to a reduction in complexity. Generic models in particular seek to identify models that concentrate on the minimum structure necessary to create a particular mode of behavior (Forrester, 1968).

14.3.6 How is interdisciplinarity achieved

System dynamics is considered a valuable candidate for interdisciplinary and transdisciplinary studies for several reasons. From its origin the method is not related to any specific discipline and, as such, it qualifies for interdisciplinary studies – on the condition that the problem can be properly described in a way that is adequate to system dynamics (by means of state variables, feedbacks, and causal relations). In particular, it is possible to construct causal relations between variables that belong to different disciplinary-rooted domains. In practical modeling, however, one has to take into consideration that, owing to the high level of aggregation, these relationships may represent pure (non-linear) correlations or contingencies rather than genuine causal explanations.

With its orientation towards creating a shared and improved understanding of a problem, system dynamics contributes to interdisciplinary and transdisciplinary studies, providing both a language and a process of group interaction (Lane, 2000). In this perspective the process of building a model starts from the different perceptions of the participants and, by systematically eliciting and sharing their "mental models," aims at creating a more complete problem representation.

14.3.7 Semiquantitative system analysis by scenario analysis

When dealing with complex systems such as regional or organizational systems, we are often able to identify the key system elements or variables that we consider essential for describing the state and the systems change without being able to represent the dynamics and the interrelationship between aspects by differential equations or real number-based mathematical functions. This is the situation in which scenario analysis can be applied (Scholz & Tietje, 2002).

The method of scenario analysis was developed after World War II in strategic military and business planning. The initial idea was that scenario analysis was to describe what alternatives may emerge and how a system develops (Kahn & Wiener, 1967).

In simple terms, scenario construction is built on incorporating all essential variables that are necessary to describe the current state and the future development of a system. Here, we can distinguish between endogenous, action-based variables involved which represent the measures of the key case agents. In addition, there are exogenous variables that represent the salient changes in the environment. A critical question to be answered in scenario construction is whether the action scenarios, which describe the decision alternatives or strategies of the case agents based on endogenous variables, and the frame scenarios have to be defined separately or whether they are incorporated in an integrated model.

To describe the current state of a system and possible future states (both of action and frame scenarios), scenario analysis uses a set of variables or descriptors $D = (d_1,\ldots,d_i,\ldots,d_n)$. These descriptors are not necessarily seen as many variables, but they can also be seen as discrete variables which can have two or very few levels. These descriptors are called impact factors and should have the capability of describing the current state and the future changes of a system adequately and sufficiently.

In the case of urban development, for instance, the variables can come from the social (e.g. values, social classing), economic (e.g. income, energy prices), or environmental (e.g. emissions, impacts on ecosystems) side. If we can define different levels of each of these impact factors (e.g. d_i^1 = low income, d_i^2 = high income) we can define scenarios: a scenario is simply a "complete combination of levels of impact factors" (Scholz & Tietje, 2002, p. 105).

14.3.8 Formative scenario analysis

A formative scenario analysis (FSA) provides, on the one hand, a strictly organized method to construct scenarios that include the same specifying information on a fixed set of variables. This is already included in the above definition of scenarios. FSA also includes techniques to select a small set of scenarios that represent the possible states in a sufficient manner. This can be done by consistency analysis, i.e. by only selecting those scenarios whose combination of levels of impact variables are considered possible or not overly unlikely (Tietje, 2005). On the other hand, the term *formative* indicates that the process of defining scenarios is giving form to the case for which the scenario is built. This can be understood in an epistemic mental way, as has been described above for system dynamics. In particular, when future scenarios are jointly formed by stakeholders and scientists in a transdisciplinary process (see Chapter 15, Box 15.1), the forming of scenarios may have an impact on the case itself as all act with a desired or undesired future scenario in mind. Factually, this has been the case in many transdisciplinary case studies on sustainable transformations of regional, organizational, and political settings and processes (Scholz et al., 2006). Thus, the process of scenario construction may have a formative component.

The FSA approach includes various techniques to construct mental models of how the impact factors are related, what relative importance they have in changing the status quo scenario, and what combinations and future scenarios are considered impossible because of (logical) inconsistencies between the levels that subsets of scenarios can take (Scholz & Tietje, 2002).

14.3.9 Intergovernmental Panel on Climate Change scenarios

The scenario technique became famous by the Intergovernmental Panel on Climate Change (IPCC) scenarios. These scenarios are a combination of different assumptions about key impact variables with regard to different social, economic, and technological developments. The scenarios have changed since 1990 regarding underlying assumptions and applied methods (Girod *et al.*, 2009). Based on these scenarios, the future development of greenhouse gas emissions, temperature, and precipitation in different regions have been calculated with classical differential equation-based models.

> Future greenhouse gas (GHG) emissions are the product of very complex dynamic systems, determined by driving forces such as demographic development, socio-economic development, and technological change. Their future evolution is highly uncertain. Scenarios are alternative images of how the future might unfold and are an appropriate tool with which to analyze how driving forces may influence future emission outcomes and to assess the associated uncertainties. They assist in climate change analysis, including climate modeling and the assessment of impacts, adaptation, and mitigation. The possibility that any single emissions path will occur as described in scenarios is highly uncertain (Intergovernmental Panel on Climate Change. (IPCC), 2000, p. 3)

The last sentence of this quote stresses that there is no technique which allows for describing the likelihood of scenarios.

14.3.10 Key messages

- Systems thinking contributes to an improved, reflexive understanding of feedback structures of coupled HES by providing mental models on how systems may evolve. System dynamics supports this exploration by providing quantitative simulation models, focusing on explaining model behavior by its underlying structure
- One goal of systems thinking and system dynamics is understanding and anticipating implications of human policy and its changes
- System dynamics is a *descriptive tool*. The current practice involves and relates variables of material–biophysical systems with those of social–epistemic systems in an open way. Special attention should be paid to reflecting causalities and different rationales of human and environmental systems as well as discontinuities
- Formative scenario analysis is a coarsened system model in which only levels of impact factors (and not real value-based functions) serve for system description and for constructing possible future states.

14.4 Self-organization, "critical transitions," and chaos

In this section we come back to non-linear dynamic systems. These systems can be described by differential equations, as has been stated above, and be classified according to their equilibrium states, which may be stable fixed points or limit cycles. A third possibility is that the system might show an erratic, aperiodic behavior and settle onto a "strange attractor" (Strogatz, 1994, 2001). Under the condition that the system is also sensitive to initial conditions, chaotic behavior is generated (Schuster & Just, 2005). This sensitive dependence on the initial conditions is called the "butterfly effect" (Lorenz, 1963), which reads in its classic shape that the flapping of the wings of a butterfly in South America today could result in a typhoon in Japan next month. Thus, non-linearity is a necessary, but not a sufficient, condition for the generation of chaotic behavior. "The observed chaotic behavior is due neither to external sources of noise nor to an infinite number of degrees of freedom" (Brown, 1996, p. 53), but precisely because of this property of separating initially close trajectories. For these reasons, this behavior has been termed deterministic chaos (Schuster & Just, 2005).

Different routes to chaos have been investigated mathematically and verified experimentally (see Schuster & Just, 2005). To illustrate a particular route, we refer to the one proposed by Grossmann and Thomae (1977), Feigenbaum (1978), and Coullet and Tresser (1978). They noticed that in simple difference equations, such as the logistic difference equation (equation 14.5) used to describe populations in ecology, the equilibrium states of the system change, depending on an external parameter. At a distinct value of this parameter,

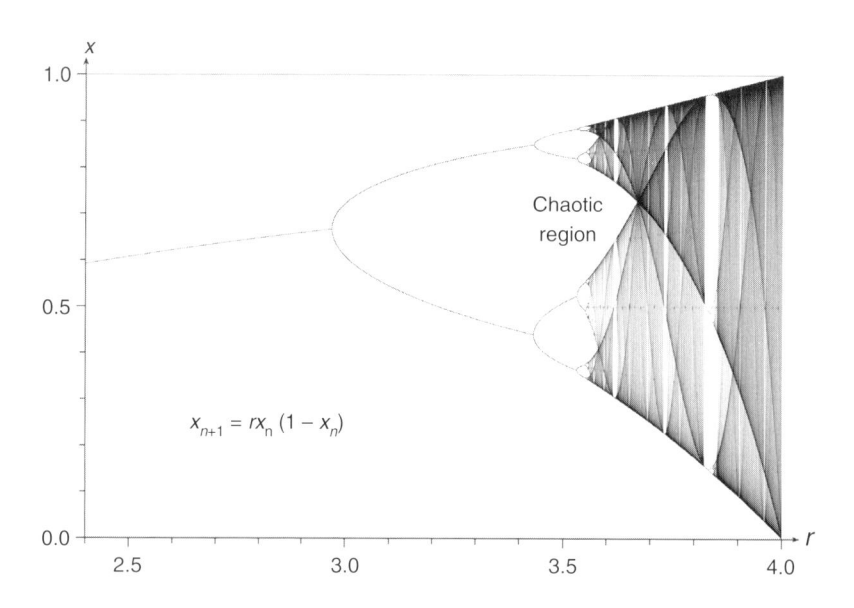

Figure 14.11 A route to chaos: bifurcation diagram of the logistic difference equation. Convergence to fixed point for $r < 3$, oscillation for $r > 3$, with x_n oscillating between four, eight and, finally, an infinite number of values, which means chaos.

the single stable fixed point becomes a limit cycle, where the population oscillates between two values. This transition is called a bifurcation. With increasing values of the external parameter, the population oscillates between four, eight and, finally, an infinite number of fixed points, where the variation in time of the population becomes irregular and chaos is attained (Schuster & Just, 2005; see also Figure 14.11, which shows a bifurcation diagram of the logistic difference equation).

The logistic difference equation reads as follows, with r as the external driving parameter:

$$x_{n+1} = rx_n(1 - x_n) \tag{14.5}$$

Though the cradle of these theories is definitely in the natural sciences, in particular physics and chemistry, there are many attempts to apply the key concepts and dynamics to social systems. The idea is that – from a complexity perspective – fundamental system principles, which govern certain material–biophysical systems dynamics, also govern the dynamics of socio-epistemic processes in human systems (Helbing, 2007; Schweitzer, 1997; Sornette, 2000; Weidlich & Haag, 1983). To illustrate this we focus here on selected aspects of the theory of non-linear systems that are relevant for the study of HES, in particular on "critical transitions" and hysteresis as well as on pattern formation, self-organization, and emergence.

The study of transitions between different dynamic states is at the heart of non-linear dynamics. If a system has alternative stable states, "critical transitions" and hysteresis may occur (Scheffer, 2009). Critical transitions take place if these alternative stable states are separated by an unstable equilibrium and no "smooth" transition is possible (see Figure 16.10). Hence, if the environmental conditions change, the system may undergo a "catastrophic" shift to an alternative stable state. It has to be noted that, owing to this configuration, the "forward" and the "backward" shift between these states do not occur at the same environmental conditions. This property, known as hysteresis, implies, first, that such critical transitions are not easy to reverse, and second, that the history of the processes matters.

An interesting issue is that we face a seemingly analog property in the perceptual processes with respect to ambiguous figure-ground configuration. If we look from the left to the right of Figure 14.12, the figure-ground configuration flips later than when starting from the right side (Haken, 1996).

It was one of the major objectives of the theory of non-linear systems to describe the formation of (temporal, spatial and spatiotemporal) patterns and to discern general principles governing self-organization of elements in systems independent of their nature (Érdi, 2008). Haken has proposed a theory of self-organization, for which he coined the term "synergetics," as "a theory of the emergence of new qualities at a macroscopic level" (Haken, 1996, p. 23). One of the central concepts in this theory is the concept of order parameters, which says that, close to the transition or instability points, the behavior of a complex system can

Figure 14.12 Hysteresis in human perception. A switch from a human face to a kneeling woman or vice versa takes place later if your eyes move from the left to right compared to if they move from right to left.

be characterized by only a small number of parameters. An example of self-organization in biology is presented in Box 14.4, with a transition of amoebae-like cells to a single, larger organism.

Self-organization and emerging structures in social systems have been widely investigated (Gilbert, 1995), and it has been recognized that in social systems certain properties on a societal level emerge from the local rules that individual actors follow. A well known example is the segregation model of Schelling (1971), showing segregation of different groups of agents because of local preferences about their neighborhood. The issue of emerging properties in complex systems will be further pursued in Chapter 14.7.

14.4.1 Key messages

- Material–biophysical and socio-epistemic systems can show instability, chaos, processes of self-organization, and pattern formation
- Even deterministic systems can show unpredictable and chaotic behavior. As a consequence, non-linear modeling tools are required
- The history of a system matters: in non-linear systems hysteresis may occur, meaning that the response of a system to a certain cause depends on the previous development of this system.

14.5 Uncertainty, probability, stochastic processes, and risk assessment

We start with a clarification of the terms used in this section. *Uncertainty* appears if we have incomplete knowledge with respect to the state of a specific object or system. *Probability* is a key concept of the language of uncertainty and will be dealt with in some depth in Chapter 14.5.2. *Stochastic processes* are sequences of events where there is real or cognized indeterminacy about future states. *Risk* is a genuine concept that links the behavioral, decision-based dimension of human systems' actions with uncertain negatively evaluated environmental impacts emerging from these actions. All four concepts can be found in all domains of sciences and life. Coping with stochasticity and properly incorporating uncertainty and risk in decisions systems understanding opens a new perspective and is an essential aspect of complexity management. Here, we only briefly pinpoint key essentials of environmental literacy.

14.5.1 Uncertainty and ignorance

Uncertainty is in knowledge or in the data. This fundamental statement departs from the socio-epistemic vs. material–biophysical complementarity (see Figure 3.3*). It leads to a similar fundamental ontological question whether randomness or random processes are something that exist in material–biophysical systems. When answering this question, we can see only two fields where uncertainty is fundamental.

The *first* is radioactive decay. The standard quantum mechanics theory is that there is a genuine stochastic process on the atomic level. It is impossible to predict when a certain atom decays, though by the concept of half-life we can predict what percentage of a pool of atoms will decay in a certain period. The discussion about the hidden variable theory goes back to 1935 and is rooted in the assumption that there are unknown deterministic theories about decay inducing processes, which allow for a prediction. However, after 40 years of experimentation "a conclusive experiment falsifying in an absolutely uncontroversial way local realism [of stochasticity] is still missing," thus there "are strong indications favoring standard quantum mechanics" (Genovese, 2005, p. 385). We can conclude that quantum mechanics leaves the question open as to whether "nature," here meaning the material–biophysical dimension of the world, is intrinsically probabilistic or whether quantum mechanical random phenomena derived from our ignorance of some theories about hidden parameters and therefore underlying deterministic

> **Box 14.4** Emerging systems: slime mold
>
> The slime mold appears to be a fungus-like organism, which can be found all over the world in soil, forest floors, mulch, and similar settings. It starts the life cycle as amoebae-like cells and spend most of its life as an amorphous setting of distinct single-cell units, each moving separately from the others. Under certain conditions, when the environment is less hospitable, the myriad of single cells coalesces into a single, larger organism. Some crawl across the floor, consuming rotting leaves and wood as they move. They are an example of equipotent cells, which may self-organize into well distinguished clusters and regions (Haken, 1983, p. 9).
>
> The cellular slime mold (*Dictyostelium discoideum*) "forms a multi-cellular organism by aggregation of single cells. During its growth phase the organism exists in the state of single amoeboid cells. … these cells aggregate, forming a polar body along which they differentiate into other spores or stalk cells; the final cell types constitute the fruiting body" (Haken, 1983, pp. 11–12). The reason for this is that the cells emit specific chemicals (called adenosine 3′5 monophosphate, i.e. cAMP molecules) as pulses into their environment. These messenger molecules are perceived as stimuli by other cells, resulting in a collective emission process which leads to a concentration gradient of the messenger chemical. Thus the single cell can measure the direction of the gradient and migrate to the center. The resulting process is displayed in Figure 14.13a. Figure 14.13b shows how a sample of slime molds ("they") has become an organism ("it"). Slime mold can be taken as an example that the boundary between a protozoal being and an organism may be fuzzy (see Chapter 5.5 on the immune system as a cognitive system).
>
> The process is an example for an environment-driven regulatory mechanism based on efficient, genetically programmed (cAMP emission-based) adaptation strategies. The process of slime molds can be mathematically programmed and visualized (Resnick, 1999).

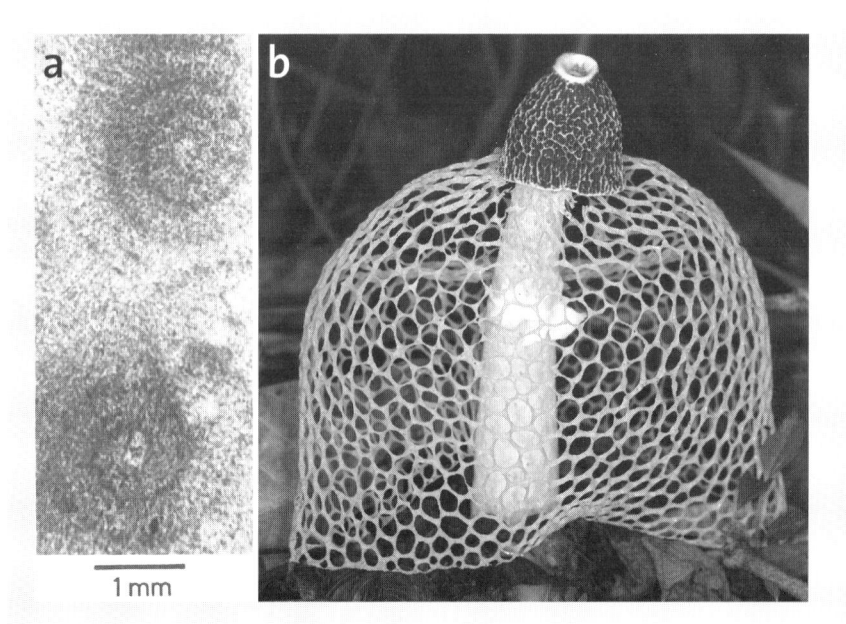

Figure 14.13 (a) Wave pattern of chemotactic activity in dense cell layers of slime mold (taken from Haken, 1983, p. 23, with kind permission from Springer Science + Business Media). (b) The tropical long net stinkhorn mushroom (*Phallus indusiatus borneo*) is up to 30-cm tall and girded with a net-like structure, the indusium (photo by David Kolba).

processes are at work. Einstein's famous quote does not give the answer:

> Quantum mechanics is certainly imposing. But an inner voice tells me that it is not yet the real thing. … I, at any rate, am convinced that *He* [God] does not throw dice. (Einstein, 1926/1971)

There is a *second* theorem of quantum mechanics, which is commonly seen as the materialist cradle of randomness that is the Heisenberg uncertainty principle. This principle states that we cannot determine both position and momentum of an electron and thus uncertainty in the knowledge is indispensable. However, this theorem does not make any statement about the intrinsic stochastic nature of the material world. The uncertainty refers to the unknowable ignorance and not to the stochasticity of the material–biophysical world.

14.5.2 Probability

As it is with many of our language terms, probability has many notions. We speak about the concept field of probability (Scholz, 1987). This field has a main dividing line between *objective* (data-oriented) and *subjective* (knowledge-oriented) probability. Both subfields have many subfacets. For instance, there is the Laplace probability, which is based on equal probability; there is the von Mises–Reichenbach frequentist probability, or area size-based probability; and there is the de Finetti definition of probability as a degree of belief. All these notions of probability have prototypical applications such as the toss of a coin or the throw of a dice, or being hit by a snowflake. All these concepts of probability also have in common that they meet the axioms of Kolmogorov (Fine, 1973; Hacking, 1975). Each notion or facet of probability is linked to an idea about the generation or mechanism of defining the likelihood of events.

At the fringes of the concept field we find quantitative conceptions of subjective probability, such as the Shafer–Dempster belief functions (Shafer, 1976), Baconian probability (Schum, 2001), or Cohen's probability theory (Cohen, 1977), which are variants of subjective probability. In accordance with Bayesian reasoning, these approaches provide conceptions on how the cognized uncertainty about the truth of a certain state can be meaningfully adjusted if new information and evidence is provided. However, the concepts at the fringes of the probability field do not follow the Kolmogorov axioms. For instance, the probabilities of all possible events may not add up to 1. This also holds true for the fuzzy set-based approach towards uncertainty (Zadeh & Kacprzyk, 1992).

An interesting variant of subjective probability is Walley's concept of imprecise probability. This approach assumes that the exact subjective probability is unknown but that the range in which it is located can be estimated if we take:

> … the lower probability to be the maximum rate at which you are prepared to bet on an event, and your upper probability to be the minimum rate at which you are prepared to bet against the event. (Walley, 1991, p. 3)

We should note that random variables usually map events to real numbers (or into a measurable space), and that probability distributions are a well established procedure to include uncertainties of such variables in modeling.

A critical question of stochastic modeling is what type of probability distribution is assumed to be adequate, for instance, for an environmental variable, say the air pollution in the vicinity of a heavy metal-emitting source of contaminants. The answer is that this depends on how the different terms causing variability are related. According to the central limit theorem, the sum of an infinite number of independent random variables (with about the same finite mean and variance) is normally distributed. In addition to such additive effects and in accordance to the central limit theorem, many small multiplicative effects lead to log-normally distributed variables. Both forms of variability stem from a variety of effects, each one acting independently.

14.5.3 Random processes

The theory of stochastic processes has been the backbone of statistical physics. Stochastic processes provide models for physical phenomena such as Brownian motion, which assumes a great number of independent and randomly moving and colliding molecules. Other examples are thermal noise, causing fluctuating voltage in electric networks or – if we refer to biology – the spatial spread and distribution of plant and animal communities or the spread of epidemics.

A critical question is whether random processes are also at work in socio-epistemic processes. Clearly, we can use Markov chains or more sophisticated operations such as the Langevin equation, which is central for Brownian motion, and apply these concepts to the study of human systems, e.g. the interaction of individuals on the stock market (Sornette, 2000, 2003). Still, the answer is no *and* yes. No means that the processes are affected by unforeseeable discrete events, which are not appropriately captured by the abovementioned concepts. Yes means that some dynamics of human system show definite similarities with material–bio physical dynamics, which as a consequence may provide a conceptual and mathematical framework to be applied to social systems.

14.5.4 Risk assessment and decisions

We focus here on a selected number of fundamental aspects of risk and risk assessment relevant for environmental literacy:

Risk is a component of decisions, because the riskiness of a decision alternative is incorporated into the evaluation of the desirability, attractiveness, and the anticipated satisfaction or utility of the outcomes related to a decision alternative.

Risk is a primitive, one of the simplest and most basic elements of a body of knowledge, and thus is a fundamental concept in the sense of Kant. Everybody knows that risk has something to do with the negative uncertain outcomes related to decisions.

We differentiate *risk* from *hazard* (Luhmann, 1990). If we talk about risk, the exposure of a human system depends on decisions and on how the human system is behaving. A hazard is given independent of an action.

Risk is a concept field. In this field we can distinguish between pure and speculative risk. Pure risk only focuses on the negative outcome. Speculative risk, however, weighs the negative and the positive outcomes, the prospective outcomes of a decision reflecting the common sense thought: if you are facing risk, there is a chance (to win). In economic terms we distinguish between benefit risk and cost risk. Which concept of risk, pure or speculative, is in the foreground, has varied historically and shifts between situations. The origins of the probability-based risk concept date back to the era of Chevalier de Méré (1607–1684), when it was the noblemen's dilemma to indulge in gambling with the objective of keeping losses to a minimum. Later, with the rise of the industrial age, it was merely the costs to gain access to the benefits tied to new technologies that were in the foreground. This led to the simple formula: risk is probability times losses. Risk is different from danger (see Chapter 6.19).

Risk can be formally defined. From a decision theory perspective, risk is a function of accessing the choice alternatives $A_i \in A$ and the probabilities $(p_{i,1},…,p_{i,Ni}) = p_i \in P$, which are linked to the possible outcomes $(E_{i,1},…,E_{i,Ni}) = E_i \in E$ resulting from choosing alternative A_i. These concepts describe a risk situation. A *risk situation* is a model that precisely describes the situation g when dealing with risk. A *risk function* r is a mapping from the set of all risk situations to the space of risk cognitions of a human system. The space of risk situations is $R = (A,P,E)$ and the image set represents what a certain decision-maker understands by risk in a specific situation. This can be quantitative risk scores, like the expected loss, the largest loss, or the probability of having a loss, to mention but three variants of the quantitative facets of the concept field of risk. But there can also be qualitative representations of risk, such as considering something dreadful and uncontrollable, as has been outlined by the psychometric paradigm (see Chapter 6.20).

Risk *assessment* is part of risk *management*. Contrary to risk *perception*, which denotes the psychological process of an individual in a risk situation, risk assessment is based on well defined methods, which define what situation (and thus what negative outcomes or hazards) we are facing. In the past few decades it has become a dominant scientific tool to protect public health and the environment.

14.5.5 Steps of risk assessment

In principle, risk assessment includes five stages (cf. Paustenbach, 2002; Scholz & Siegrist, 2010).

First, the risk situation is defined. This includes determining the specific alternatives that are looked at in a specific risk assessment. In public health this may mean that we have to define which people should be looked at in what place and at what time. The decision alternatives may result from distinguishing between being in a critical situation and being not in a critical situation.

Second, in the case of pure risk the negative effects have to be described. In the public health case this would require specifying whether we look at the general wellbeing, a specific dysfunction of an organ, or just at a critical concentration of a contaminant in a certain compartment or organism.

Third, for a given risk situation the factual exposure of a safeguard object has to be defined. This may mean that we have to define how much of the contaminant is ingested by food, breathing, or other paths, and in what circumstances.

Fourth, the dose–response relationship (i.e. the sensitivity) has to be assessed. This means clarifying how much the negative impacts depend on exposure.

Fifth, the risk has to be characterized by estimating the frequency and the meaning of a negative effect under the various conditions of exposure described and with respect to the object of interest.

A possible method of risk assessment is inductive–stochastic risk assessment. The idea behind this is that when we are facing a risk (resulting from action or of being exposed), we have to construct the impact from the source of the risk (e.g. a cadmium-contaminated ground) to a critical compartment in a safeguard object (e.g. the cortex of the kidney of a home gardener eating vegetables from this ground). Thus, we are viewing chains such as contaminated ground → contaminated plant (modeled by transfer coefficients) → contaminated food → intake of contaminated food (modeled by combining contaminants of the home garden) → adsorbing the contaminants → transporting it to the sensitive safeguard object (modeling toxicokinetics to the cortex of the kidney) → assessing the probability or

14.5 Uncertainty, probability, stochastic processes, and risk assessment

Box 14.5 The Prisoner's Dilemma: strategies and the emergence of cooperation

Robert Axelrod (1984a) developed a model based on the concept of an evolutionarily stable strategy in the context of the Prisoner's Dilemma game. This game is an embodiment of the problem of how to achieve mutual cooperation, and thus provides the basis for Axelrod's analysis. He performed multiple tournaments of such games (on a computer), with different strategies playing against each other: cooperation or defection. The payoff for T = temptation to defect is the highest, followed by mutual cooperation. The problem is, if both players want the highest return by defection, this leads to punishment (P; see Figure 14.14). These games can be played once or repeatedly. One result of these tournaments is that for the repeated games there seems to be a strategy that rules all others out in terms of reward. This strategy is called tit-for-tat, which means "cooperate as long the other cooperates and defect when or as soon as the other defects."

Axelrod believed that very sophisticated rules should not do better than the simple ones. Indeed, strategies that are too complex to be comprehended by the other "player" (or strategy) can be dangerous, leading to a worse payoff. However, Delahaye and Mathieu (1994) showed that more complex strategies can be superior to tit-for-tat. The authors added to some of their strategies the possibility of quitting the game (renunciation). This adaptation may also fit better with reality, because sometimes it is better to stop interaction with a player "that seems too weird or too aggressive" (Delahaye & Mathieu, 1994, p. 2). In their tournaments they found that renunciation in fact proved a successful extension. Only three of the first 40 strategies worked without renunciation.

	Player B C Cooperation	Player B D Defection
Player A C Cooperation	$R = 3$ Reward for mutual cooperation	$S = 0$ Sucker's payoff
Player A D Defection	$T = 5$ Temptation to defect	$P = 1$ Punishment for mutual defection

Figure 14.14 The Prisoner's Dilemma game as a matrix. The payoff to player A is illustrated by arbitrary numerical values. Per definition $T > R > P > S$ and $R > (S + T)/2$.

probability function of dysfunctions. The critical issue is that, for each "→," we have to define a probability distribution that models the uncertainty in the specific transfer, input–output, or cause–impact relationship (McKone & Bogen, 1991; Scholz et al., 1992; Wallsten & Whitfield, 1986).

14.5.6 From risk assessment via vulnerability to resilience

While risk and risk assessment may be considered a static approach, risk management points to the dynamic, prospective aspects of risk. In this perspective, *vulnerability* and *resilience* of systems, which are inherently systemic and dynamic concepts, receive particular attention. Although vulnerability research is composed of several research streams, such as research on the vulnerability of socioecological systems, the sustainable livelihoods approach, and vulnerability to climate change (Adger, 2006), these strands have in common that vulnerability is based on exposure, sensitivity, and adaptive capacity of a system with regard to certain effects (see Polsky et al., 2007). In the context of climate change, for example, vulnerability has been defined as:

… the degree to which a system is susceptible to, or unable to cope with, adverse effects of climate change, including climate variability and extremes. Vulnerability is a function of the character, magnitude, and rate of climate variation to which a system is exposed, its sensitivity, and its adaptive capacity (McCarthy, 2001, p. 995).

Adaptive capacity relates to the means upon which a human system can rely to take adaptive actions aimed at avoiding or coping with the negative effects of a risk situation (Metzger et al., 2006; Füssel, 2007). This means that when facing a risk situation, action can be taken that alters the exposure of that system to a specific hazard and/or the sensitivity with which the system is affected by this event. In other words, the decision-maker has the capability to change the risk situation. In that sense, vulnerability is a component of the wider concepts of risk governance (e.g. International Risk Governance Council (IRGC), 2009) and adaptive risk management (reference). The vulnerability concept can be defined quantitatively or qualitatively. As will be further elaborated in Chapter 14.8, on multiagent systems, the options of specific agents for adaptive actions are supported or constrained by their environment, which is made up of human components (e.g. political and institutional rules) as well as of biophysical components. This multifaceted challenge possibly leads the agents to explore new strategies qualitatively.

14.5.7 How is interdisciplinarity achieved

Risk assessment and management are highly interdisciplinary fields. Conceptualizing and defining the exposure is based on a behavioral model of the human system in relation to environmental hazards. Thus, exposure needs a coupled social and material–biophysical systems view. In addition, assessing sensitivity analysis (i.e. analyzing how much does the risk vary if some parameters vary) requires knowledge from the natural and social sciences. The system's sensitivity of an individual with respect to contaminants also depends on the mental (i.e. socioepistemic) processes and not only on the material–biophysical processes. The same holds true for risks of companies and financial markets.

14.5.8 Key messages

- Risk primarily represents the evaluation of trade-offs (i.e. negative environmental feedbacks) resulting from a decision under uncertainty
- Risk assessment provides a structured analysis of a risk situation, thus reducing the complexity of a case analysis
- Risk management and risk governance involving the concepts of vulnerability and resilience introduce a systemic and dynamic perspective. As such, risk is related to exposure, sensitivity, and adaptive capacity of a human system to a hazard.

14.6 Game and decision theory

14.6.1 Malignant and benign conflicts

When coping with environmental problems, there are often conflicts and social dilemmas among different human systems with respect to environmental resources. By social dilemma we mean situations in which a difficult choice has to be made between unfavorable alternatives between different agents or between individuals and collective interests (for an example in a contribution dilemma, see Box 14.5). Game theory can help us to conceptualize, describe, and understand these conflicts. We have dealt with risk perception in Chapter 6.18 and with the psychological foundations of decisions in Chapter 7.1. Here we only illuminate in what way game theory can provide insight into the *malignence* and *benignity* of certain situations and the different rationales of game theory. We define a situation to be malignant if, without transforming the situation, deterioration of some participants or systems or a progressive, unfavorable dynamic will result. A benign conflict is free of this peril.

Warfare situations or duels are usually represented as malignant situations, where, as a simplification, players can only win while (some of) the others are losing. Here we speak about antagonist conflicts or zero-sum games.

Decision theory can be considered a special case of game theory as there is only one player. Milnor (1954) also called decision situations as games against nature.

14.6.2 Historical background and basic concepts

Game theory originated in finding the best strategy in military conflicts and policy, and can be traced back to the Chinese art of war (e.g. Sun Tzu, 500 BC/1910) or ancient governance strategies (Thucydides, 550 BC/1910). The game of chess can be taken to illustrate the key elements of strategic games. A strategic game $\Gamma = (A,S,U)$ includes the definition of a set of players or actors $A = \{1,\ldots,n,\ldots,N\}$. Each player has a set of strategies $S_n = \{s_{1,n},\ldots,s_{k,j},\ldots\} \in S$ and of utility functions $u_n \in U$ which evaluates each outcome. Outcomes result when all players (N) have chosen a strategy and result from a strategy tuple $(S_{k1},\ldots,S_{kn},\ldots,S_{kN})$. The utility functions evaluate the attractiveness, relative satisfaction or desirability of the outcomes for each strategy tuple; that is, for all situations when all players have set up their decisions.

A strategy can also be defined verbally in the following way:

> A strategy of a player is a complete behavioral plan, which describes for each situation a player can come in, the behavior, i.e. the decision that is made in this situation. (translated from Burger, 1959, p. 10, translated by Scholz)

Chess is a finite game. Game theory shows that there is a win strategy for white, a win strategy for black, or strategies which compel a tie, as in any finite games there is a tie or one player wins (Burger, 1959). However, it is difficult to identify these win or tie strategies as, for instance, in chess a strategy includes more possible draws (i.e. decisions) than there are atoms in the universe (Mandecki, 1998). Thus, we can already see that game theory is mostly a theoretical approach.

This is not the place to go into the details of game theory. We just want to illuminate the broad range of (game) situations and their representation in game theory. We can discriminate between games in a normal form, which corresponds to the strategic games as defined above, and games in extensive form, which are presented as trees in which the nodes are the decisions the players have to make. There are 2- and N-person, zero- and non-zero-sum, and non-cooperative and cooperative games. In cooperative games, groups of players may enforce cooperation by coalition building. This aspect is of interest if we deal with supranational players, who have the power to induce cooperative behavior among loosely and voluntarily interacting nations. We make reference to this issue as the world game for climate mitigation is a non-cooperative game between the 192 nation-state players that are members of the United Nations; a solution seems easier if a supranational player is involved who represents penalties for not cooporating.

In extensive games (which can be represented by trees), there are games with *perfect* and *imperfect information* in which at least one player does not know all the actions of his opponent. We find games with *complete* or *incomplete information* in which the player does not exactly know against what player he is playing and what utilities may result. There are many excellent books about game theory and we recommend reading them to learn more on the subject (see for instance Gintis, 2009a,b; Osborne, 2004).

14.6.3 Contributions of game theory to interdisciplinarity

Game and decision theory provide an excellent language to describe the interaction between different agents or players. The language is general as it can be used to describe decisions at all levels of living systems, from the cellular level via the individual up to populations. Thus, game theory is genuinely interdisciplinary as games can include agents from different systems and the description of the different rationales of the agents – for instance, a virus that is attacking a human community – may have completely different behavioral repertoires and rationales.

We see the main benefit of game theory for interdisciplinarity in crisply describing the synergetic potential, conflict, or dilemma that is inherent in both human–environment interactions or in interactions between different levels of human systems. The latter refers to the commons dilemma problem, which is inherent in many problems of sustainable development.

14.6.4 Key messages

- Game and decision theory provide a general language to describe and classify the decision processes and conflicts involved in human–environment interactions as well as between different levels of human systems
- Game and decision theory denote human systems as decision-makers, agents, players, etc. who can choose among strategies which generate outcomes that are evaluated by preference or utility functions.

14.7 Modeling human–environment interactions with cellular automata

Cellular automata (CA) are identical cells that are arranged in a grid structure (comparable to a checkerboard). The cells can represent units of a biophysical environment or human individuals or collective actors (e.g. countries or states). Each cell is surrounded by other cells that build its environment. The cells (or CA) exist in one of a finite set of possible cell states. These states depend on the states of the surrounding cells. As time advances in discrete steps, each cell looks at its neighbor cells (whereas different kinds of neighborhood relations are possible, e.g. Moore or von Neumann neighborhood; see Gilbert & Troitzsch, 2005) and updates its own state depending on the implemented rules. The modeled system is homogeneous with respect to both (i) the set of possible states and (ii) the set of transition rules that apply to each cell. We briefly illustrate here a famous example of CA, the "Game of Life" of Conway (1970; see also Figure 14.15). The status of dark cells is "on" (or "alive"), that of light cells is "off" (or "dead"). Depending on the status of the neighbor cells, a cell in this game turns "alive" or "dead" when following certain rules. The results are often astonishing patterns. Figure 14.15a shows the initial random distribution. On the right (14.15b), the situation after some time steps is depicted. What one cannot see in these static pictures are the "movements" of living cells across the cellular grid.

The fundamental claim of CA is that temporal and spatial evolutionary behavior of systems can be

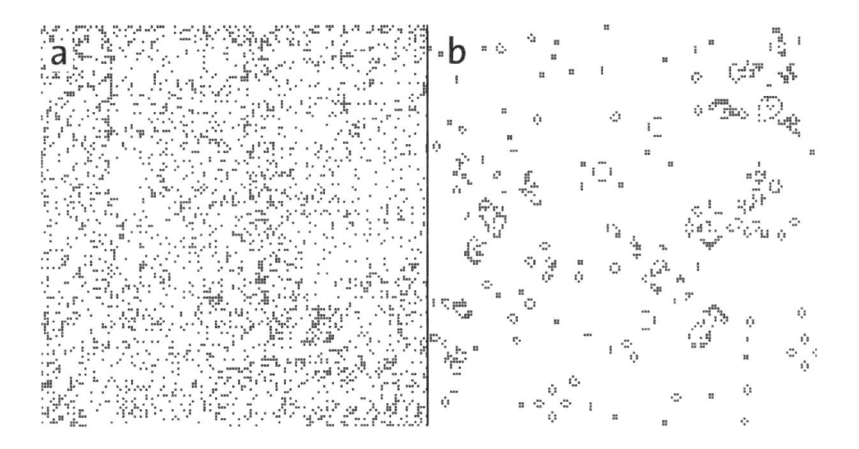

Figure 14.15 Game of Life. (a) Initial random distribution. (b) Situation after some time steps in generation 384. What one cannot see in these static pictures are the dynamics, caused by "movements" of on-cells across the grid, constantly forming new patterns (Wilensky, 1998).

explained in terms of local interactions of autonomous units. It is conceptually very simple and comprehensible compared to other approaches, but still produces emergent properties based on mimicking behavior of many constituent units of nature and society. Further, CA address the two fundamental properties that are inherent in real social and natural systems: parallel operation (synchronized self-control), and the important role of local variations in explaining the system dynamics. These features are new in comparison with the other modeling approaches described earlier in this chapter. In the context of environmental literacy, this may illustrate that new insights on regular patterns of human and environmental dynamics are rooted in diverse local actions.

The discrete nature of CA allows application of discrete mathematics, such as Markov processes, in the investigation of complex human and environmental dynamics. A Markov chain is a discrete random process where all information about the future is contained in the present state. If Markov processes are applied for CA models, cell states depend probabilistically on temporally lagged cell state values (Parker et al., 2003). In particular, CA have been applied successfully to several spatio-temporally explicit environmental dynamic models (e.g. landscape models of land use and land cover changes) that capture the macro outcomes emerging from millions of simple micro scale interactions. A classic example is the use of CA models to simulate residential discrete choices and land use conversion in urban areas. In the latter example, spatial housing patterns emerge from local decisions of single CA. These decisions depend on the neighbors' state (Hegselmann, 1998; Schelling, 1971).

In sum, CA have proven to be useful to model biophysical aspects of environmental dynamics, but they face challenges when incorporating complex human decision-making process. It is necessary to use heterogeneous, hierarchical rule-sets to differentiate between the kinds of decision-making that apply to groups of cells, such as resources tenure structure (Parker et al., 2003). Moreover, one has to be aware that pattern "movements" across the grid are not the results of autonomous agents but represent the succession of single cell states, each very restricted with regard to their decisions and spatially fixed. With respect to this, the multiagent system (MAS) approach offers additional potential.

14.8 Modeling human–environment interactions with multiagent systems

An MAS[5] is a community of agents, situated in an environment. "Agent" refers to autonomous decision-making entities within the system. "Environment" is the space that surrounds agents and supports or constrains their activities. "Community" refers to an organized collective of agents in which each agent plays specific roles and interacts with the environment (possibly including other agents). This interaction works according to protocols (rules) determined by the roles of the involved agents (Gilbert & Troitzsch, 2005; Zambonelli et al., 2003).

As an extension of CA, MAS provides all the fundamental features of CA; that is, macro phenomena that emerge from micro interactions, parallel operation, and the crucial role of local heterogeneity, and social networks. The concept of the agent originates in the so-called players of game theory.

---

[5] Sometimes MAS is also used for multiagent simulation.

14.8 Modeling human–environment interactions with multiagent systems

In addition, compared to CA the agents in MAS can be more complex with regard to their structure and behavior. While CA are uniform in terms of the set of their state, the variable set of MAS agents can be different across agent typologies. Moreover, instead of being homogeneous with CA's rules, agent's rules can be specific for every agent typology. This heterogeneity is central to multiagent systems since it acts as a trigger initiating different behavior among agents in the system, sequentially leading to emerging system complexity (cf. Axelrod, 1997; Holland, 2006).

Behavioral rules of an agent can be rather simple causal if–then rules, in which behavioral decisions directly reflect the changing states of agents or their environment. However, agent behavior can also be driven by the evaluation of equations over particular state variables, or assessment of outputs of dynamic models encoded in the agent-based systems. Regarding this aspect, MAS can be understood as a decentralized system template in which multiple mathematical formalisms (e.g. simple causal if–then rules, linear functions, differential equations, and so on) can be concurrently embodied. Several types of agent architectures, based on different behavioral theories, were proposed. Most frequently, if–then rules are mixed with other kinds of behavior formalizations such as parameterized functions, or competitive tasks, or belief–desire–intention (BDI), or evolutionary metaphor-based (e.g. genetic algorithm) architectures (Bousquet et al., 2004). This claim is different to that of earlier modeling approaches.

Environmental behavior of human agents is embedded in social interactions that are shaped by social networks. Informal institutional constraints or opportunities of environmental behavior (e.g. norms, collective attitudes, and behavior innovations), as well as their enforcement characteristics, are mediated by social networks. Social networks are important in understanding the dynamics of HES as they contribute significantly to resilience and adaptability by creating channels for information flows (Deffuant et al., 2002), for management and distribution of resources (Epstein & Axtell, 1996), and for diffusion of technological innovations. MAS modeling is particularly relevant in representing social network structures and associated interaction rules, as well as the role of inequality of the individual agent within a network (i.e. nodes). With MAS, individual agents can form wide-ranging networks of many "weak ties" that facilitate information transfer and innovation diffusion, and ultimately promote economic development. Kohler and Gumerman (2000) showed that MAS approaches were particularly useful where there is evolution of social norms or social structure, or the presence of social dilemmas and common pool resource issues. Through connecting agent characteristics and knowledge of the position of agents in social networks, van Eck et al. (2011) demonstrated the important role of the network in mediating the impact of opinion leaders on diffusion processes of new products.

The "environment" formalized in MAS largely depends on the object system that is to be modeled. In real-world applications, we often see three types of applied MAS models: (i) MAS for social systems; (ii) MAS for material–biophysical systems; and (iii) MAS for coupled HES. When MAS is used to represent a social system, the "environment" of an agent refers to other human agents, which can be similar or different in typology (Edmonds et al., 2007). The real biophysical environment is often assumed to be static, without an important role in explaining the system dynamics, thus ignored in the formalism. With MAS applied to biophysical worlds, the "environment" is truly the material-bio physical environment, which consists of other natural agents (like fish, trees, or land units; cf. Grimm et al., 2005), without human agents in it. In MAS designed for coupled HES, the "environment" of an agent (being either human or biophysical objects) becomes a coupled community of decision-making human agents and autonomous biophysical agents (Jager et al., 2000).

In MAS, due to the autonomy of the agents, interactions in the system are no longer mathematically traceable. Thus we meet what is called mathematical intractability of processes in time and space, as we can also find them in the system dynamics approach (Axelrod, 1997).

14.8.1 Multiagent systems for coupled HES

A growing number of MAS models for coupled HES have been developed in the past decade. To exemplify we refer here to a specific class of models of this type that has been formulated to represent the coupled human–landscape systems in a highly integrative manner (e.g. Le, 2005; Le et al., 2008; Matthews, 2006). Here we have the integration of different kinds of social agents (decision-making, politics, and organizations) and natural processes (e.g. regarding soil, crops, and climate). Therefore, "human agents" integrate attributes of social actors like environmental and policy knowledge and capabilities, whereas "land agents" represent the relevant parts of the environment with which the human agents interact (see Figure 14.17).

Box 14.6 Social dilemmas: contributing to a land reclamation system or not?

Using MAS as a means for representing real-world HES and understanding the complex system dynamics can be demonstrated by a case of farmers in the Odra region of Poland who are caught in a social dilemma. In the catchment, while, in principle, the existing land reclamation system (LRS) of ditches and canals can absorb the negative effects of extreme weather conditions, its proper functioning requires collective action with regard to maintenance. Thus it is important that the acquaintance and/or friendship relationships that exist amongst farmers are utilized appropriately. The agent-based model SoNARe (Social Networks of Agents' Reclamation of land, cf. Krebs et al., 2008a, b) simulates the collective decision-making (Olson, 1965; Ostrom, 1990; Dawes, 1980) of typical landowners in the catchment under fluctuating socioenvironmental boundary conditions. The model is coupled to a hydroagricultural model that simulates the hydrological dependencies that exist between farmers located along the LRS. In the model, landowners decide about investing in the maintenance of their local section of the LRS and are provided with feedback in terms of their attained crop yield under certain climatic conditions. Farmer agents keep a balanced attention to their attained crop yields and their social endorsement resulting from their opinion regarding LRS maintenance. Farmers find themselves in a twofold dilemma: (1) the working LRS is most pertinent under unpredictable extreme weather conditions because it protects crops against excess water stress; (2) the beneficial effects of the LRS may only be achieved by high degrees of mobilization of the involved farmers, and beneficial effects differ depending on the farmer's position along the channel.

The conditions encountered on individual land parcels depend highly on the amount of LRS maintenance performed on other (connected) land parcels. In wet periods, for example, LRS neglect leads to a loss of yield on neighboring land parcels upstream since the runoff of excess water is blocked, whereas LRS maintenance has the opposite, beneficial effect since it facilitates runoff. The latter effect arises even if the upstream neighbors do not themselves maintain their section of the LRS (free-riding). Maintenance of the LRS thus enables farmers to overcome environmental shocks like flooding with only reduced yields or even with no losses at all, but it requires a collective effort.

The effects of several possible combinations of the farmer agents' success can be simulated. Two extreme scenarios are described here in short: the first scenario suggests that under the assumption of selfish farmers who only consider their individual farming success, roughly one-fifth of farmers show free-riding behavior. This points in the direction of a social dilemma induced by the hydrological interdependencies of farmers' land parcels. Scenario 2 adds social influence to the decision process, which results in the emergence of a positive social lock-in. The results indicate that, under the given circumstances, the presence of an active social network and of mechanisms of social influence dampens phases of high volatility in opinion dynamics and instead leads to a coherence effect. The frequency of LRS volatility is reduced with the introduction of social influence and further social pressure leads to a positive lock-in that even prevents free-riding behavior. Figure 14.16 shows the yields of the farmer agents. Compared to scenario 1 (see Figure 14.16a), in the long run in scenario 2 (see Figure 14.16b) the average crop yields increase and the deviation between farmers decreases.

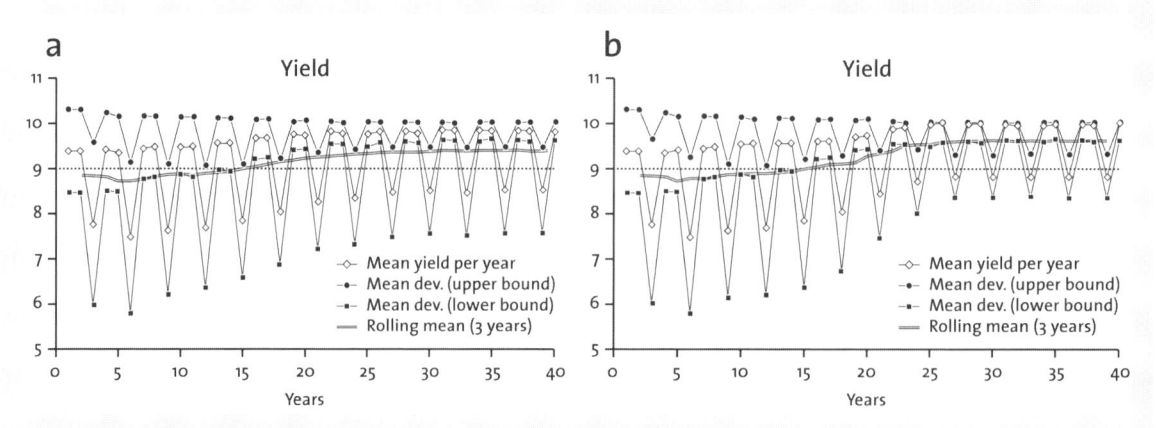

Figure 14.16 Average yields of farmers (y axis normalized, dimensionless) bounded by the mean deviation and the rolling mean over 3 consecutive years. The dotted line indicates the yield threshold. (a) Scenario 1; (b) scenario 2 (Adapted from Krebs et al., 2008b, p. 5 & 7).

14.8 Modeling human–environment interactions with multiagent systems

Box 14.7 Interactive household and landscape agents: Vietnam deforestation

Assessment of future socioecological consequences of land-use policies can support decisions about what and where to invest for the best overall environmental and developmental outcomes. However, such assessment work is highly challenging because of: (i) the inherent complexity of coupled human–landscape systems; (ii) the long-term perspective; and (iii) the multidimensional criteria required for sustainability assessment. The Land-Use DynAmic Simulator (LUDAS) is a spatially explicit multiagent model developed to simulate land-use change and the interrelated socioeconomic dynamics on the community/catchment scale (Le et al., 2008). The human community is represented by heterogeneous household agents, which integrate individual, environmental, and policy information into land-use decisions. The natural landscape was modeled as a group of landscape agents, which represent land units hosting natural processes that change in response to interventions by human agents. LUDAS embodies the material-social complementarity (see Figure 3.3*) within household agents, by for example coupling material household assets with a cognitive sub-model of land-use decisions ($H_m \Leftrightarrow H_s$), and within landscape agents, by for example representing the switch from biophysical land potential to perceived land utility ($E_m \Leftrightarrow E_s$). The model embodies the human-environmental complementarity ($H \leftrightarrow E$) by representing the impact of human systems on environmental systems ($H \rightarrow E$), for example a farmer's vegetation clearance, and the resulting benefits from the environment by crop production ($E \rightarrow H$) (Figure 14.17).

The model structure is designed by mimicking the common hierarchical levels (see Figure 14.1*) of land-use systems: household/farm – household group/landscape neighborhood – whole community/landscape. The interactions between the three layers occur through high-order feedback loops (vertical arrows in Figure 14.17). Household's land uses lead to a change in the larger human–landscape environment that reshapes the future decisions of household agents. Thus, the model represents the co-determination, and co-adaptation between human and environment systems. LUDAS illustrates many epistemological assumptions of HES described in Chapter 3.

As a virtual computational laboratory, LUDAS is able to systematically generate spatiotemporally explicit land-use change and interrelated socioeconomic dynamics resulting from land-use policy interventions. In the case of the Hong Ha catchment in Central Vietnam, comparative assessments of alternative future socioecological scenarios using LUDAS suggest that it is challenging to attain both community welfare and forest conservation by simply expanding current agricultural extension programs and subsidy schemes, without improving the qualities of these services. The results also provide new information about how policy interventions can strike a balance over the long term between 1) the need to strengthen enforcement of forest protection in critical areas of the watershed and 2) needs to simultaneously create incentives and opportunities for agricultural production in the less critical areas (Le et al., 2010).

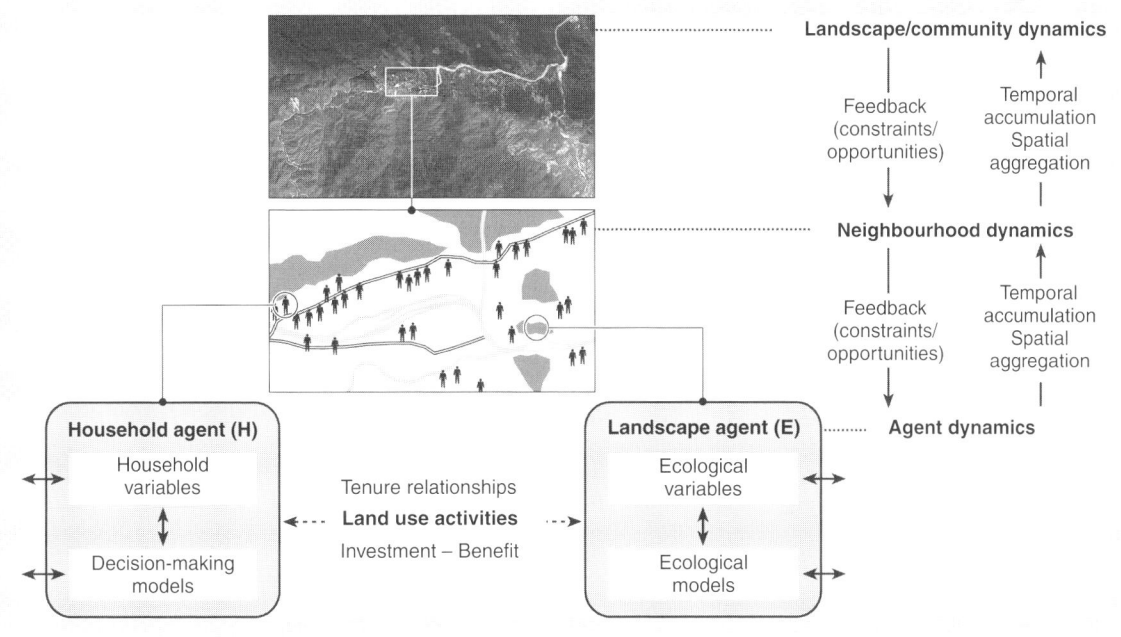

Figure 14.17 The LUDAS modeling framework mimicking natural ontologies of the coupled human–landscape system.

With MAS models for coupled HES, available models of different sociological/political/economic and ecological processes can be embodied into the structures of human and land agents. Integration of knowledge across disciplines in coupled models can be seen, for instance, in the case of MAS related to land use change (see Box 14.7 and Figure 14.17). By working as parallel subsystems that can host relevant processes possibly formulated by different analytical means, such a new class of MAS models offers a much higher degree of freedom to integrate not only knowledge from different disciplines, but also different conventional modeling methods into one integrated system model (see, for example, Box 14.6).

Within MAS for coupled HES, there are three major types of interactions: communicative interactions, physical interactions, and environment-mediated interactions. Communicative interaction means the exchange of messages among agents (e.g. concerning negotiations and exchanges of contracts, goods, or services).

Physical interactions, undertaken by agents, exert a physical action on others such as pushing or pulling. These kinds of interactions have been used in applications such as hydrology or soil physics. Environment-mediated interactions form the primary environmental feedback loop, in which human agents perceive the environment status and react to it, and the human action transforms the environment, with a retroactive effect on the decision-making process of itself and of other agents (see direct interactions between household and landscape agents in Figure 14.17). This illustrates the key advantage of MAS by mutual co-determination (Bousquet & Le Page, 2004; Krebs & Bossel, 1997) of the structure of the environment and the organization of the human agents' community.

Repeated agent interactions can lead to cumulative changes in social/economic and environmental conditions at larger scales and in the longer term. This may cause emergent changes of the whole system's performance after a delayed period (see vertical interlinks in Figure 14.17). The emergent change on the macro level has a two-path effect on the agent's behavior in the MAS: (i) it creates new opportunities or constraints for agent-based processes at the micro level; and (ii) it influences the behavior of politicians who frame policy issues that reshape interactive behavior of human and environmental agents. In this way, MAS represent secondary environmental feedback loops. By realizing these high-order environmental feedback loops, MAS models reveal the (complex) adaptive capacity of coupled HES. MAS can become complex adaptive systems (Holland, 1995, 2006); for instance, if we assume that the agents discover new rules that are more successful in dealing with an environment. This may be the case if changes in the political or climatic environment challenge household agents and/or landscape agents which have to adapt, e.g. by the mutation of rule-parts (and genetic algorithms; Holland, 1995 a,b). In this context evolutionary game theory is also relevant, as it deals with changing environments by variation of the strategy sets of the agents and subsequent adaptation (Axelrod, 1984; Smith, 1776/1961).

14.8.2 How interdisciplinarity is achieved

MAS models for coupled HES help to bridge the domains of natural and social sciences in many ways. First, like other integrated system model approaches, MAS provides feedback between variables (agents) with different ontological quality (e.g. variables of human, social, and financial assets of human individuals or organizations, and biophysical attributes of the landscape environment). Second, disciplinary knowledge in forms of available theories, models or rules/laws of different social and ecological processes can be implemented into the agents' structures (rule systems). MAS modeling also provides a kind of a meta-theoretical language in which every disciplinary statement must be reformulated. Developing MAS for coupled HES to inform and evaluate sustainability aspects (of several integrated scenarios) is a specific application of this approach. This kind of integrated modeling requires intensive discussions among the modelers from all the relevant disciplines (e.g. Barthel *et al.*, 2010; Lenz-Wiedemann *et al.*, 2010).

14.8.3 Key messages

- MAS is an interesting approach for understanding of HES as it integrates ideas from systems theory, game theory, theory of self-organization of complex systems, and allows for interdisciplinary, coupled systems modeling
- Agents' interactions within MAS allow the capture of both cross-scale (micro–macro) feedback loops and social and ecological heterogeneities. Thus, understanding self-organization and emerging structures in social and environmental systems forms a basic feature of MAS.

Part VII Beyond disciplines and sciences

Chapter 15

Transdisciplinarity for environmental literacy[1]

15.1 Goals and variants of transdisciplinarity 373
15.2 Implementing transdisciplinary processes 379
15.3 Functions of transdisciplinary processes 384
15.4 Other applied research and participatory approaches 388
15.5 Transdisciplinarity, science, and universities 394
15.6 New frontiers for science–society cooperation 399
15.7 Transdisciplinarity in action 401

Chapter overview

This chapter has three principal sections: first, we define transdisciplinarity and show how it fundamentally differs from disciplinarity and interdisciplinarity. We discriminate between transdisciplinary processes and transdisciplinary research. We then present methods to facilitate transdisciplinary processes and to nurture authentic collaboration among participants.

Second, we review how public participation, consultancy, and action research have been employed to conduct applied and participatory research and discuss how each is similar to and different from transdisciplinarity. Here we describe how community-based participatory research combines action research and participatory research and thus can become a transdisciplinary process. This section also defines different types of participation.

Third, we look at how scientists and universities can play key roles in developing environmental literacy. Thus, we discuss the power of "disciplined interdisciplinarity" when coping with current salient environmental challenges.

We discuss the potential, limitations, and risks of transdisciplinarity and discuss how it is related to the nature of democracy itself. The discussion leads us to the role and function universities and academia can play in the future of environmental literacy and sustainable development.

15.1 Goals and variants of transdisciplinarity

In this section we review how transdisciplinarity is distinguished from both interdisciplinarity and disciplinarity, and we introduce the distinction between transdisciplinarity research and transdisciplinary processes. While there have been different interpretations of transdisciplinarity, this book advocates transdisciplinary processes that involve authentic collaboration among science and society when including representatives from industry, government, administration, different stakeholder groups, and the public at large. Such collaborations, as we describe, emphasize mutual learning, joint problem definition, and knowledge integration. Transdisciplinary processes should produce relevant, socially robust knowledge that also feeds back to scientific knowledge generation and theory-building.

To implement a transdisciplinary process, facilitators can employ techniques such as embedded case study methods to structure and organize work. These methods support problem representation and modeling, problem evaluation, development, and transition of the real-world problem. After briefly introducing these methods we present a step-by-step example of a transdisciplinary process in Switzerland that has used these methods to develop a sustainable business

---

[1] Michael Stauffacher and Pius Krütli cooperated with me over many transdisciplinary case studies and theoretical discussions and helped me to compose this chapter.

future in harmony with the environment. Finally, we describe functions of transdisciplinary processes such as capacity/competence-building, consensus-building, analytic mediation, and legitimization of public policy.

15.1.1 Why transdisciplinarity?

Collaboration between science and society is often requested if uncertainty arises about substantial changes in human-environment systems (HES) such as the introduction of a new technology or new medical pharmaceuticals, diagnosis or therapy. Other problems that deserve collaborative processes between science and society are the finding of mitigation, adaptation, policy or decision strategies, for instance when facing fundamental changes of the natural or social environment such as from natural hazards, climate change, resources scarcity or changing cultural settings.

We argue that, from the perspective of society, transdisciplinarity provides an efficient use of knowledge for coping with complex, socially relevant problems. It provides societal capacity-building and bridges the growing gulf between many areas of research and the public. This equips society with a better understanding of how technologies, or the natural environment, work and how the latter interacts with human systems. Thus, in consequence, transdisciplinarity can permit us to master and cope more adequately with the new and unknown, for instance regarding inventions such as nanotech particles, both from a scientific and from a societal perspective. At the same time, it stimulates academic research by highlighting phenomena, issues, and emerging questions that require scientific reflection, and feeds experiential knowledge into the research process. Furthermore, it frees science from the cumbersome implementation problem. Instead of facing the challenge of gaining public understanding, acceptance, or appraisal of something ingenious, but that is rejected for "non-academic reasons," transdisciplinary processes put science into practice from the very beginning. Transdisciplinarity is, as we will elaborate, an efficient means to use knowledge in decision-making, at least in certain types of prodemocratic, civic societies (Almond, 2000).

15.1.2 Definition and notions of transdisciplinarity

The term transdisciplinarity is occasionally referred to as "perfected interdisciplinarity" or as the transfer of concepts or methods from one discipline to another. In the first definition of transdisciplinarity by Jantsch as the "multi-level coordination of entire education/innovation system" (Jantsch, 1972, p. 221), the "beyond" science notion of "trans"-disciplinarity is highlighted. Since there is some confusion about the distinction between interdisciplinarity and transdisciplinarity, we briefly define our understanding of these concepts.

Disciplines are characterized by objects and (core) methods by which certain problems are approached. For example, mathematics deals with relationships between symbols and numbers by the method of proof. Similarly, the purview of pharmacy is to investigate the impact of certain chemicals (called medical drugs) on diseases by the use of laboratory experiments and clinical trials.

Interdisciplinarity is established by the *fusion* of concepts and methods from different disciplines. A metaphorical example illustrating what is meant by fusion is the saxophone, which emerged from the clarinet and the trumpet. Biochemistry, which studies chemical processes in living organisms, for example, investigating reactions of proteins and other molecules, can (at least before it got established as a discipline) serve as an example of an interdisciplinary field. The experimental method is a pillar of this field. The term "industrial food web" used in the emerging domain of industrial ecology can also be taken as an integrated concept.

Transdisciplinarity is fundamentally different from interdisciplinarity. Most of today's definitions of transdisciplinarity include in their meaning that it goes beyond science in the sense that it "deals with relevant, complex societal problems and organizes processes …" that relate knowledge and values of "agents from the scientific and the non-scientific world" (Scholz *et al.*, 2000, p. 447). As there are different notions of transdisciplinarity, we provide insight into the four most salient conceptions below.

Disciplines efficiently organize the methods and systematized knowledge about the material–biophysical–technological as well as the social–cultural–epistemic world. Interdisciplinarity merges concepts and methods from different disciplines for better understanding and explanation of certain issues, phenomena, and processes that cannot be sufficiently explained from a single disciplinary perspective. Transdisciplinarity organizes processes that link scientific, theoretic, and abstract epistemics with the

15.1 Goals and variants of transdisciplinarity

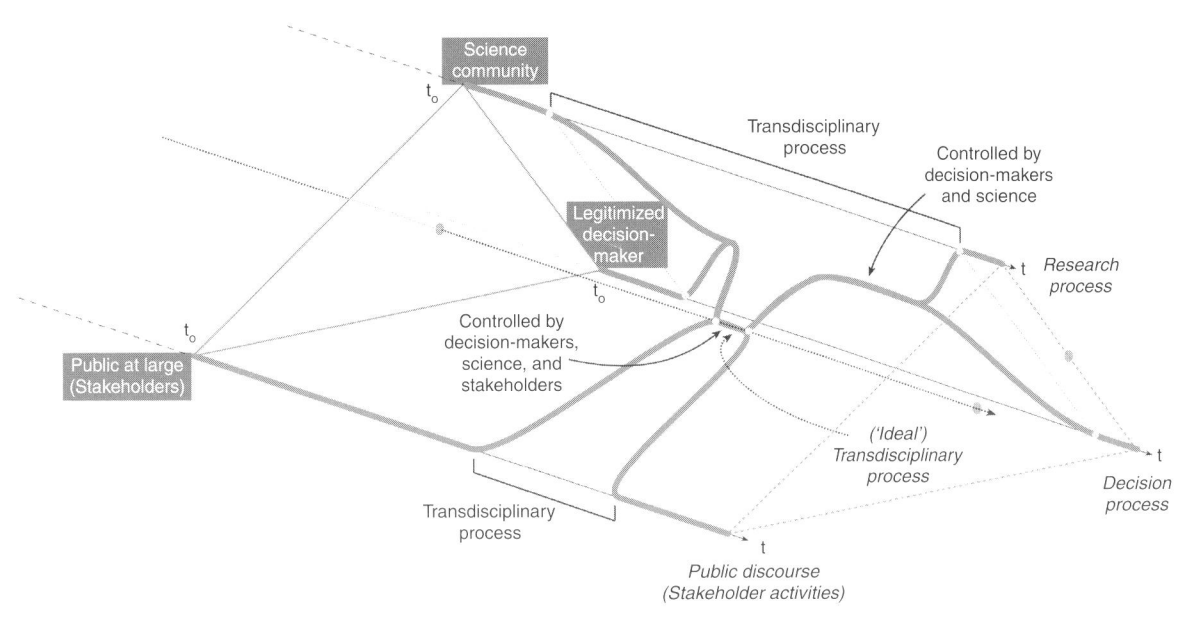

Figure 15.1 Actors from the science community, public at large, and legitimized decision-makers are involved in research processes, public discourses or stakeholder activities, and decision processes, respectively. When the actors leave their primary processes (action lines – bold gray) and join in a collaborative, power-balanced effort, we can call this a transdisciplinary process.

real-world-based experiential knowledge from outside academia. It relates human wisdom to the analytical rigor of science and academic methodology.

Distinguishing transdisciplinary *research* from transdisciplinary *processes* is important. Transdisciplinarity, according to the "Zurich 2000" definition described in the next section, organizes mutual learning among members of science and society that can generate socially robust knowledge. For the most part, this mutual learning takes place in transdisciplinary processes in which members from the science community interact with decision-makers, stakeholders or the public at large. Transdisciplinary processes differ from consultancy and contracted research with respect to power sharing and direction of involvement; that means, who participates in whose process. In consultancy and contracted research, scientists operate in the action space of the legitimized decision-maker, who lets science participate (see Figure 15.6). Members, knowledge, and results from the science community become part of the decision process. Here, the legitimized decision-maker ultimately decides how scientists' skills and knowledge are used during the process, and how outcomes and results are communicated and utilized.

Usually, a transdisciplinary process emerges if a legitimized decision-maker and members from the science community notice that they have joint interest in a complex, relevant phenomenon that can be better understood and dealt with if knowledge from practice and from science is integrated. Typical examples are sustainable transitions of regional systems caused by, for example, the overexploitation of natural resources (including pollution) or by adapting to changing environmental conditions.

The ways in which scientists, legitimized decision-makers, and the public at large can collaborate are illustrated in Figure 15.1, which shows four horizontal time axes. At bottom front, one finds the public at large, whose priority is to sustain and organize life. This public opinion-building activity, shown as public discourse, in the foreground, represents the cultural–social side of human systems. At the top, we find the time arrow for the scientific community, whose various activities (i.e. teaching, in-service training, investigating, etc.) together make up the research process. At bottom back, we find the legitimized decision-maker, whose activities make up decision processes. A legitimized decision-maker can be a national or local government, a local environmental agency that has certain responsibilities, or also a property owner (e.g. a landowner or a company directorate) who makes or plans decisions. If a legitimized decision-maker and scientists participate together in a transdisciplinary process, they leave their action spaces for a certain time and collaborate. A situation of equal control between decision-makers and scientists, a transdisciplinary process, is represented by the plane bordered by the research process and the decision process lines in Figure 15.1.

Here, scientists and decision-makers leave their primary domain (research and decision process, respectively) and establish a joint transdisciplinary process. In this process, they jointly agree on the topic or specific system to be investigated (e.g. defining the spatial and temporal system boundaries), leading to a joint problem definition.

Members of the public at large can be affected or feel concerned by the decision process and become organized into interest groups or stakeholder groups. Often these stakeholders or members from the public at large then participate in the transdisciplinary process (in a limited time period; see middle bold line). But sometimes these groups can formally (or informally) control the process. An ideal transdisciplinary process involves all three groups – researchers, decision-makers, and stakeholders – in a collaborative, power-balanced relationship. In Figure 15.1 this ideal transdisciplinary process takes place at the transdisciplinary process line that runs through the center of the triangle. We should note also that industry, as a specific stakeholder in society, may become a key player in a transdisciplinary process.

Transdisciplinary research takes place before, during or after a collaborative process, to provide preparatory research, support information or follow-up research, respectively. The transdisciplinary *process* provides important input for transdisciplinary *research*, which is controlled by scientists who can ensure that results are produced through rigorous research methods.

15.1.3 Different interpretations of transdisciplinarity

What is called trandisciplinarity and interdisciplinarity has been differently defined at different times and by different scientific domains and disciplines. We refer to the discussion in environmental and sustainability sciences. A first, moderate interpretation of going beyond science (see Table 15.1) has been suggested by Mittelstrass (1996), who "argues that transdisciplinarity is primarily a form of research addressing and reflecting on issues in the life-world" (Hirsch-Hadorn *et al.*, 2008, p. 28). Referring to a distinction of scientific activities suggested by Gibbons *et al.* (1994), one could call this Mode 1 transdisciplinarity, as such scientific work can be found in traditional disciplines.

As an example of Mode 1 transdisciplinarity, we can take Nobel Laureate Amartya Sen's work on famine and poverty. His work is characterized by classical disciplinary economics. However, many of his papers include ethical ideas (and thus are rather of an interdisciplinary type) and deal, for instance, with inequality and child survival in India, Pakistan, Bangladesh, North Africa, etc. (Sen, 1999). Though working as a classical economist, Sen's work is interdisciplinary and takes an "ultimate concern … about the lives we can or cannot lead about issues of the real world" (Sen according to Swedberg, 1990, p. 339). Mode 1 transdisciplinarity differs from isolated thinking about the environment (and applied sciences), as it includes *interperspectivity* (Giri, 2002) and *empathy*, but not necessarily collaboration with others. Sometimes pure problem-oriented research is denoted as transdisciplinary (Jaeger & Scheringer, 1998). Factually, Mode 1 or Type 1 transdisciplinarity thus does not significantly differ from applied research and is often linked to a "truth to power" theory–practice relationship (Pohl, 2008).

The second definition of transdisciplinarity (Table 15.1), which we call the Zurich 2000 definition, resulted from the participation of 500 researchers and 300 practitioners in a conference setting in Zurich in 2000. Here the definition reads:

> Transdisciplinary research takes up concrete problems of society and works out solutions through cooperation between actors and scientists. (Häberli *et al.*, 2001, p. 6)

From the perspective of scientific research, this conception of transdisciplinarity: (1) organizes processes of mutual learning among science and society; (2) integrates knowledge and values from society into research (Scholz *et al.*, 2000); and (3) provides an appropriate research paradigm that better reflects the complexity and multidimensionality of sustainable development.

Following this definition, transdisciplinarity has been declared as the appropriate methodology by which sustainable development should be investigated and promoted (Scholz & Marks, 2001). Transdisciplinary

Table 15.1 Variants of transdisciplinarity.

	Variant	Essentials
A	Mode 1 transdisciplinarity	Science becomes transdisciplinary if it reflects on real-life problems
B	The Zurich 2000 definition	There are transdisciplinary processes organizing mutual learning between the science and non-science community, and transdisciplinary research which integrates knowledge and values from practice and science. Both deal with tangible, socially relevant, real-world problems. Practice and theory (science) have different reference systems. The transdisciplinary process is featured by joint problem definition, representation, and transformation ("solving")
C	The post-normal science approach	Science becomes just one vote in an agora of arguments solving real-world problems because of the uncertainties and incompleteness of knowledge, multitude of logics, etc.
D	The Charter of transdisciplinarity conception	In addition to many aspects of B and C, a personal moral commitment is needed and a "unity of knowledge"

processes are periods of cooperation between scientists and practitioners to develop socially robust knowledge for coping with ill-defined, socially relevant problems as described above. Transdisciplinary research deals with questions emerging from these processes. For the most part this research is done without the participation of members from the non-academic world who would be normally participating in the transdisciplinary process. However, we want to remind the reader that we speak about transdisciplinary research only if it results from joint problem definition and a transdisciplinary process. This is meant by Mode 2 transdisciplinarity. Transdisciplinary research provides results that can be fed back to the body of science. Furthermore, transdisciplinary research is an element of the mutual learning process that takes place to develop robust orientations. This is the foundation for solving socially relevant problems such as groundwater management, soil protection, or living with rising sea levels in coastal countries. Thus, according to the Zurich 2000 definition, transdisciplinary research is simply research that is directed to, and done in the context of, transdisciplinary processes.

A third conception of transdisciplinarity (Table 15.1) can be called the post-normal science approach to transdisciplinarity. We exemplarily deal with the conception of Funtowicz and Ravetz (1993, 2008), which is mostly referred to as the post-normal version of transdisciplinarity. A key of the post-normal approach is the assumption that real-world issues are so complex that the scientific results become just another voice or one vote that is to be dealt with in the agora (the Greek term for forum or market place) of arguments in the real-world discourse. This is reflected in statements such as:

> The faith in the truth and objectivity of science, established by Descartes and Galileo, is overthrown. (Funtowicz & Ravetz, 2008, p. 364)

The post-normal view stresses that science has lost credibility and that "any genuine understanding of a technology must now take into account malevolence, ... also of those applying it to anti-human ends" (Funtowicz & Ravetz, 2008, p. 364). We should note that the Zurich 2000 definition and the post-normal understanding show many commonalities. Both approaches assume, for instance, that science incorporates values and world views in its investigations. However, we consider it important to distinguish between the different types of knowledge (epistemics) and constraints of producing knowledge in practice and in science. In our view, experiential knowledge and scientific knowledge differ in their foundations, epistemological status, and the roles they play in different types of real-world problems. Taking the climate change example, we see that models of sea level rise are highly uncertain. Furthermore, they include unknown and unknowable assumptions about the development of the human population, prospective CO_2 emissions, technological development, or unknown future natural systems dynamics (e.g. volcanic activities, El Niño like cycles, interactions with the biosphere), and there are huge uncertainties in many contextualized environmental research questions (van de Kerkhof & Leroy, 2000). Nevertheless, there is a community of natural scientists (Waert, 2003) developing models, theories, and predictions about climate change that differ in their genesis, status, precision, and validation strategies from statements coming from politicians, members of construction companies, etc. In addition, social scientists provide knowledge about drivers and obstacles of human behavior that are relevant to understanding adaptation processes. Thus we should also be careful when stating that "science takes place in the agora" (Gibbons & Nowotny, 2001, p. 79) of ideas, though we acknowledge that scientific statements become relative in complex contextualization. Thus, both approaches, the Zurich 2000 definition and the post-normal approach, assume that science only provides one type of epistemic, and we have to be aware that science can err.

We should note also that the Zurich 2000 definition is compatible with a constructivist perspective and acknowledges that there is a "social nature of truth" with the understanding that the science community decides what is considered as true or valid. We have, for example, shown that this even holds true for mathematics as there are many cases where no single scientist is able to verify all prerequisites included in a complex proof of a theorem (Scholz, 1998). However, in the Zurich 2000 definition, science refers to a normal science view (i.e. approaching a valid or "true" description of reality as a reference system) which is lost in the post-normal variant.

In both the Zurich 2000 and the post-normal definition, scientists can take a double role: they can contribute as facilitators or moderators to establish an appropriate process, and they can contribute as scientific experts.

A fourth definition of transdisciplinarity (Table 15.1) has been shaped by Nicolescu and the Charter of transdisciplinarity (de Freitas et al., 1994; Nicolescu, 2002). This approach shares many aspects of the Zurich 2000 and the post-normal approach of transdisciplinarity, such as "acknowledging different types of logic" (de Freitas et al., 1994, Article 2). What makes this approach unique is that it "constitutes a personal moral commitment" against the "spiritual and material self-destruction of human species" (de Freitas et al., 1994, preamble), challenging that the "dignity of the human being is of both planetary and cosmic dimensions" (de Freitas et al., 1994, Article 8). Furthermore, it addresses the unity of knowledge that targets the integration of scientific, religious, transcendent, and other forms of knowledge.

In the following sections we refer to the Zurich 2000 definition of transdisciplinarity.

15.1.4 Sustainability learning for generating socially robust knowledge

Facilitating transdisciplinary processes presents an opportunity for scientists to promote sustainability learning as a process outcome and, in turn, to use process outcomes to generate socially relevant, robust knowledge. In this section, we discuss this dual epistemical role of scientists in transdisciplinary processes.

Transdisciplinarity endeavors to use science knowledge efficiently to cope with socially relevant problems. How and whether this can be done depends on various constraints, in particular on the specific problem at hand and the given sociopolitical context. Here, sustainability, as a widely shared way of regulating ideas from many contemporary societies, is essential. Transdisciplinarity can be linked to processes where society can learn about sustainability. We call this "sustainability learning." It is related to transdisciplinarity in the following way:

> Transdisciplinarity can be said to evolve from special types of problems, i.e. real, complex, socially relevant problems, which ask for the integration of the knowledge of science and society (Burger & Kamber, 2003; Scholz et al., 2000; Thompson Klein et al., 2001). Most of these problems are strongly related to sustainable development (Blättel-Mink & Kastenholz, 2005). It can be said that planning and learning processes for sustainable development require transdisciplinarity as an approach (Meppem & Gill, 1998). This holds particularly true if the development and implementation of policies and mutual learning processes are targeted by the behavior of individuals, industries, organizations, and governments. We refer to the corresponding process as "sustainability learning". (Scholz et al., 2006, p. 231)

We argue that sustainability learning with respect to "ill-defined problems" asks for processes that go beyond traditional consultation (i.e. transfer of information from society to science) and knowledge transfer (from science to society) which are a common way of using scientific knowledge. We suggest that developing socially robust knowledge in transdisciplinary processes is a key element of societal capacity-building. Generating socially robust knowledge (Gibbons & Nowotny, 2001; Nowotny et al., 2001) involves a form of epistemics, which: (i) meets state-of-the-art scientific knowledge; (ii) has the potential to attract consensus, and thus must be understandable by all stakeholder groups; (iii) acknowledges the uncertainties and incompleteness inherent in any type of knowledge about processes of the universe; (iv) generates processes of knowledge integration of different types of epistemics (e.g. scientific and experiential knowledge, utilizing and relating disciplinary knowledge from the social, natural, and engineering sciences); and (v) considers the constraints given by the context both of generating and utilizing knowledge.

We consider "mutual learning between science and society" (Scholz, 2000) as a key characteristic of transdisciplinary processes and sustainability learning. These processes should be characterized by: (a) joint problem definition (i.e. determining what the actual problem is, what is targeted by various stakeholders, and how these sometimes diverging views can be integrated and agreed upon); (b) joint problem representation; that is, developing a language that provides a medium of representation for describing the object, content, and changes (dynamics) of the object; this step includes problem structuring (Checkland, 2000); and (c) jointly initiating a process of problem-solving (or perhaps this is more properly expressed as problem transition), which is cost-effective, socially acceptable, scientifically sound, and competitive in the market place (see Gibbons *et al.*, 1994). Mutual learning thus can be viewed as a process that can generate socially robust knowledge that can contribute to coping with challenging societal transitions.

Through the example of global warming, we take a brief look at the nature of the problems involved in these transitions and the generation of socially robust knowledge. Global warming and its environmental impacts require significant adaptations. Today, it is unclear how these adaptation processes should proceed and what they should look like. It is unclear what measures are technically feasible and socially acceptable, and how reasonable goals can be presented. Making a country more resilient and less vulnerable to climate change, for example, and to prepare a country to avoid uncontrollable damages (KfC, 2008) is a typically ill-defined problem. This holds true, for example, for countries and islands like the Netherlands, Bangladesh or the Maldives when coping with sea level rise, or adapting agriculture in semiarid regions. Nobody knows exactly what target state can and should be attained and what barriers have to be passed to reach this state. Neither is it known whether economic, environmental or social barriers are the most severe. Problem definition, for example, is a particularly difficult task, especially with respect to financial payouts that correlate to who is affected (positively or negatively) from environmental impacts. Evidently, this is all linked to the ontology of ill-defined problems.

15.1.5 Key messages

- Transdisciplinarity is essentially different from interdisciplinarity
- There are different definitions for transdisciplinarity, ranging from reflecting about society (Mode 1 transdisciplinarity) via theory–practice collaboration-based definitions referring to capacity-building and knowledge integration in transdisciplinarity processes (underlying this book) to definitions requesting a personal moral commitment
- Transdisciplinarity is considered a powerful and efficient means of using knowledge from science and society with different epistemics serving societal capacity-building under certain political cultures
- Transdisciplinarity can stimulate science in finding challenging research questions, and feeds experiential knowledge into the research process
- Transdisciplinarity organizes collaboration and mutual learning between theory and practice
- Transdisciplinary processes, which are joint-controlled processes that involve scientists, decision-makers and stakeholders, have to be distinguished from transdisciplinary research, which is controlled by researchers
- Transdisciplinarity is a means of coping with complex, ill-defined (wicked), contextualized, and socially relevant problems that are nowadays often defined in the frame of uncertainty and ambiguity. Transdisciplinary processes can organize sustainability learning and capacity-building in society. They are essential for environmental literacy
- Generating socially robust knowledge in transdisciplinary processes can be seen as a major goal of transdisciplinarity.

15.2 Implementing transdisciplinary processes

15.2.1 Sustainability learning

If we consider transdisciplinarity as a procedure for sustainability learning or establishing socially robust knowledge, a critical question is which specific methods can be used to implement transdisciplinary processes (Reeger & Bunders, 2009; Scholz et al., 2006; Scholz & Tietje, 2002). This section describes a suite of embedded case study methods for supporting and organizing transdisciplinary processes. We then present a step-by-step example of how a group in Switzerland drew on these methods to develop orientations for a sustainable business future in harmony with the environment. Acting sometimes as key facilitators during these processes, scientists draw on facilitation methods such as capacity-building, consensus-building, and analytical mediation.

15.2.2 Methods for problem representation

Common issues that community members, practitioners, and scientists jointly assess are, for example: (1) representing how an infectious disease spreads (Kruse et al., 2003); (2) identifying the causes of malnutrition; or (3) understanding the changes that have occurred in a certain system (Hellier et al., 1999). Visualization of the system or problem-structure is often organized by conceptual or iconic mapping. Flow diagrams between concepts or pictures can represent cause–impact or means–end relationships. In principle, these types of representation are mental models and are well known, simple tools from planning studies (Schnelle, 1979) or soft systems methodology (Checkland, 1981; Checkland & Scholes, 1990).

> Visualizations reveal much that is masked by verbal communication alone. (Cornwall & Jewkes, 1995, p. 1671)

Visual literacy is ubiquitous and universal and in some cases it is the only way of building a jointly shared external representation of the issue as seen by both researchers and practitioners. The concept of "rich pictures" has been used to explore conscious and unconscious perception of problems and cases (Bell & Morse, 2010a, 2010b).

Constructing a representation of a mental model includes many elements of an analytic process (see Box 2.5), as both the conceptual representation as well as the "arrows" connecting these concepts ask for abstraction. Future workshops (Jungk & Müllert, 1994), are a well-known approach that typically involves a 2-day meeting that includes community members, researchers, and administrators. These workshops result in consensus-building about the current and future state of an urban setting, an institution or other cases (Scholz & Tietje, 2002). Under certain conditions (see Chapter 15.1.2), participants can become part of a transdisciplinary process. One goal of the process is to develop a joint problem representation of complex human–environment problems that is both understandable and acknowledged by the participants and also compatible with system-theoretic scientific approaches. Facilitating identification of a joint problem representation is an art that draws on methods including verbal and system graphs from scenario analysis, system dynamics, material flow analysis (see Figure 15.2), supply chain analysis, etc. Creating a joint system representation that meets the problem at hand and that adequately deals with the different types of HES complexities (see Chapter 14.1.2) is an important part of the transdisciplinary process.

15.2.3 Methods for process management

Transdisciplinary processes require facilitation of the process of knowledge integration. Here aspects of moderation, balancing the power between the different participants, and methods of integrating knowledge are needed (Hoffmann et al., 2009; Reeger & Bunders, 2009; Scholz & Tietje, 2002). Process management includes the social processes of interaction between the participants and the content level. For the latter, certain types of integrated modeling, for instance, joint construction of scenarios (see Chapter 14.3), can also play a role in a transdisciplinary process.

15.2.4 Embedded case study methods

Embedded case study methods are a specific set of methods that have been modified, advanced, or newly developed in transdisciplinary case studies of sustainable urban development (Scholz et al., 1996; 1997; 2004, 2005), regional (Scholz et al., 1995; 2002), and organizational transitions (Mieg et al., 2001; Scholz et al., 2001; Scholz & Stauffacher, 2007). Embedded case study methods were also applied to sustainable transitions of policy processes.

In principle, embedded case methods fall into four classes A-D, as shown in Table 15.2. Transdisciplinary processes often start with case study team methods, which equip the group with strategies to work effectively together through approaches such as the experiential case encounter. Actually, we consider empathy and side change (i.e. living or working in a real-world setting) as a valuable way for researchers to gain a

Table 15.2 Embedded case study methods for transdisciplinary processes (all methods are presented and discussed in detail in Scholz & Tietje, 2002 and the types of knowledge integration are explained in Box 2.5). X indicates that this type of knowledge integration is established by the method; XX indicates that it is strongly established.

		The four types of knowledge integration			
Case study methods	Key questions	Disciplines	Systems	Modes of thought	Interests
(A) Case representation and modeling methods					
Formative scenario analysis (FSA)	What variables are crucial to the state of a system and to its change? What can be? What ought to be? What can happen?	XX	X	X	
System dynamics (SD)	Which variables are the most decisive in temporal dynamics? Which (counterintuitive) system change, outcomes, and feedback loops result from the dynamic interactions of the variables?	XX	X		
Material flux analysis (MFA)	What are critical fluxes in materials for the case? What are the sources and sinks of the system/case?	X	XX		
(B) Case evaluation methods					
Multi-attribute utility theory (MAUT)	How can different evaluation criteria be integrated? Which (mis)perceptions are inherent in an integral evaluation? What are the preferences of different stakeholder groups?	X	X		X
Integrated risk management (IRM)	How can/shall I cope with uncertainty? Which of a set of different alternatives are the least risky ones?	X	X		
Life cycle assessment (LCA)	How can the main environmental impacts (on a global level) be evaluated?	X	XX		
Bioecological potential analysis (BEPA)	How can the bioecological quality and potential of a case site be evaluated?		XX	X	
(C) Case development and transition methods					
Mediation: Area development negotiations (ADN)	What causes the conflicts among the principal-agents/key players of the case? Which (mis)perceptions do the case agents have? How can we attain Paretooptimal solutions?		X	X	XX
Future workshops	Which ideas may guide the questions: What can be? What ought to be?			XX	X
(D) Case study team methods					
Experiential case encounter	What does the case look like from the case member's perspective?			XX	X
Synthesis moderation	How can I optimize teamwork to improve the synthesis process? How can I find the right method of synthesis?			X	X

better understanding of what case agents know and understand. This step also enhances the development of mutual trust, as the practitioner notices that the researcher is willing to leave the ivory tower.

15.2.5 Implementation example: a six-step area development negotiation process

A number of processes have been developed for implementing transdisciplinarity. Here we describe how the area development negotiation (ADN) processes draw on embedded case study methods shown in Table 15.2. An example of how the ADN method was used to develop orientations for a sustainable business future for Swiss traditional industries that is in harmony with the environment is described in Box 15.1. The experience from the ADN can be applied for analytic mediation of type 3 in Table 15.4, as is explained in Box 15.2.

We see in Figure 15.3 that a crucial prerequisite for applying embedded case study methods in a transdisciplinary process is the definition of a *guiding question* (i.e. joint problem definition) in a balanced, collaborative process between science and practice on equal footing (see (i) in Figure 15.3 and 15.1). The whole transdisciplinary (research) process is directed to answer this question. The definition of the system boundaries, as well as the exclusion or inclusion of certain domains, is essential here as each alters how participants view the main drivers and rationales of the system.

"Faceting the case" (see (ii) in Figure 15.3) is the next step for a transdisciplinary approach to a complex case. The facets should sufficiently represent the most important aspects or subsystems of a case (or object of investigation) that will allow the group to answer the guiding questions. The faceting is part of the process of *begreifen* (Engl.: explaining; see Box 2.6). A simple example of faceting – not related to case studies – is the definition of subunits of the human organism by the concept of "organ," which has shaped medical sciences. In contrast to a holistic case study, during faceting the case is embedded in a conceptual grid. The grid represents the most important aspects, subsystems, and perspectives, etc. tied to a case. When analyzing – in economics or business sciences – product value chains, an effective conceptual guide is given by the nodes/stages (e.g. mining, processing, disposal) of the supply–demand life cycle approach (i.e. the facets are defined by the life cycle nodes/stages). Such a perspective allows for a focus on the critical dynamics in human systems which may be linked to critical demands that may potentially endanger the long-term supply–demand security (resilience).

A third step employs embedded case study methods from class (A), case representation and modeling methods, for joint problem representation. These methods allow participants to perform a system analysis (see (iii) in Figure 15.3) in order to identify the main impact factors, which can sufficiently describe the current state of a case (facets) and its dynamics. Problem representation can be realized by means of formative scenario analysis (FSA), systems dynamics, or material flow analysis to construct various future scenarios (see (iv) in Figure 15.3). Future scenarios can be evaluated from different stakeholders' perspectives, through a multi-criteria analysis (MCA II) (see (v) in Figure 15.3) or from a scientific data-based perspective (MCA I). There are different methods for case evaluation (type B in Table 15.2); one of them is multi-attribute utility theory which can be applied for this procedure. A characteristic of transdisciplinary processes is that the attributes that are included in multi-attribute utility assessment and the attributes with which stakeholder groups are involved become a matter of joint discussion and decision. The inclusion of the stakeholder groups in the evaluation of ADN is often a step that opens the transdisciplinary process to the public-at-large.

The selection of system impact factors, scenarios, stakeholder groups, and attributes to be included in MCA depends on the guiding question and the desired orientations at the end of the process for both the case and the scientific results. This means that collaborators must use backward planning to ensure that the early steps of the ADN process (Figure 15.3) are designed with the later steps in mind. Or, put another way, all activities have to be functional with respect to a specific question or hypothesis to be answered. Both practitioners and scientists have to be involved in all steps in a functional dynamic way (see Stauffacher *et al.*, 2008).

The interpretation of the MCA provides insight into the attractiveness of the different scenarios (orientations), the most important aspects (criteria of evaluation), and the strong and weak parts of a scenario (see (vi) in Figure 15.3). The evaluations and the comparison of the evaluations of the different stakeholder groups define dissent and consent in MCA II. Another interesting point is that the evaluations from the stakeholders are compared with those from the researchers. When

15.2 Implementing transdisciplinary processes

Box 15.1 A transdisciplinary process needs methods: sustainable future of industry

The six-step area development negotiations (ADN) method was used to develop a sustainable business future plan for Swiss traditional industries that is in harmony with the environment. This transdisciplinary process involved 101 stakeholders in a transition management process of regional clusters in textile, timber, and dairy industries in the Swiss canton of Appenzell Ausserrhoden. Scientists and regional stakeholders collaboratively planned, assessed, and discussed cooperative business strategies that would sustain a continued presence in their selected industry. It involved: (1) the construction of different regional business strategies and clustering variants through formative scenario analysis (FSA); (2) a multi-criteria evaluation of these scenarios (see Figure 15.2); and (3) a multi-stakeholder consensus process on different forms of horizontal and vertical cooperation. As a result, a regional learning process to move towards a sustainable development plan was initiated among industries, public authorities, and research institutions (Scholz & Stauffacher, 2007). Table 15.3 presents the steps and Figure 15.2 some impressions on group processes and stakeholder evaluations in the MCA IIb (see step (v); Figure 15.3).

Figure 15.2 (a) A group of stakeholders from business, unions, and junior researchers thinking about business plans of a regional textile industry (Scholz & Stauffacher, 2007) using the area development negotiations methodology.
(b) Stakeholders filling out a multi-criteria evaluation (see (v) Table 15.3 below).

Table 15.3 The steps of the transdisciplinary discourse organized according to the ADN method for a sustainable business future of Swiss traditional industries in harmony with the environment (taken from Scholz & Stauffacher, 2007, p. 2524).

Step	Reference (Scholz and Tietje, 2002)	Description
(i) Defining a guiding question	p. 84–6, 268–9	The following question was jointly defined by the transdisciplinary steering board of the project: "What are the prerequisites for a regional economy that can sustainably operate in harmony with the environment and regional socio-economic needs"
(ii) Faceting the case	p. 55–6	We identified, jointly with the stakeholders, perspectives or subsystems (facets) that allow for sufficient representation and extrapolation: textile, sawmill, and dairy industries
(iii) System analysis	p. 48–54, 87–8, 241–6	We investigated history and dynamics of each industry using document analysis and analysis of relevant data from the national statistical office. We conducted structured interviews with the owners or CEOs of around 20 companies, covering many topics and including confidential economic (e.g. annual turnover) and environmental data (e.g. energy use)
(iii–iv) System representation	p. 89–105	We selected a set of impact factors considered sufficient to describe the current state of the system and its dynamics with respect to the guiding question within a 20-year timeframe
(iv) Creating scenarios using formative scenario analysis (FSA)	p. 105–16	We defined two to three levels of development for each impact factor. A scenario, then, is defined as a complete combination of levels of all impact factors. Using consistency analysis (Scholz and Tietje, 2002, pp. 105; Tietje, 2005), those scenarios exhibiting high inconsistency scores were discarded. The final selection of scenarios was done jointly with the stakeholders

Box 15.1 (cont.)

Table 15.3 (cont.)

Step	Reference (Scholz and Tietje, 2002)	Description
(v) Multi-criteria decision (MCA)	p. 143–73, 197–224	Each of the three working groups used a small set of six to nine evaluation criteria in two different approaches of MCA: (i) calculations based on data, literature and/or expert interviews (data-based evaluation MCA I); (ii) for each industry, different stakeholder groups provided assessments (stakeholder-based evaluation, MCA II) to explore and to test differential and fallacious evaluations. Assessments were made in two steps: (MCA IIa) intuitive (holistic) overall assessment of the scenarios; and (MCA IIb), which is still intuitive, but uses the criteria from the MCA I
(vi) Developing robust orientations	p. 268–9	We discussed jointly with the stakeholders the results of the above steps in the reference groups and the case study team. Based on this transdisciplinary discourse, we developed orientations for the stakeholders

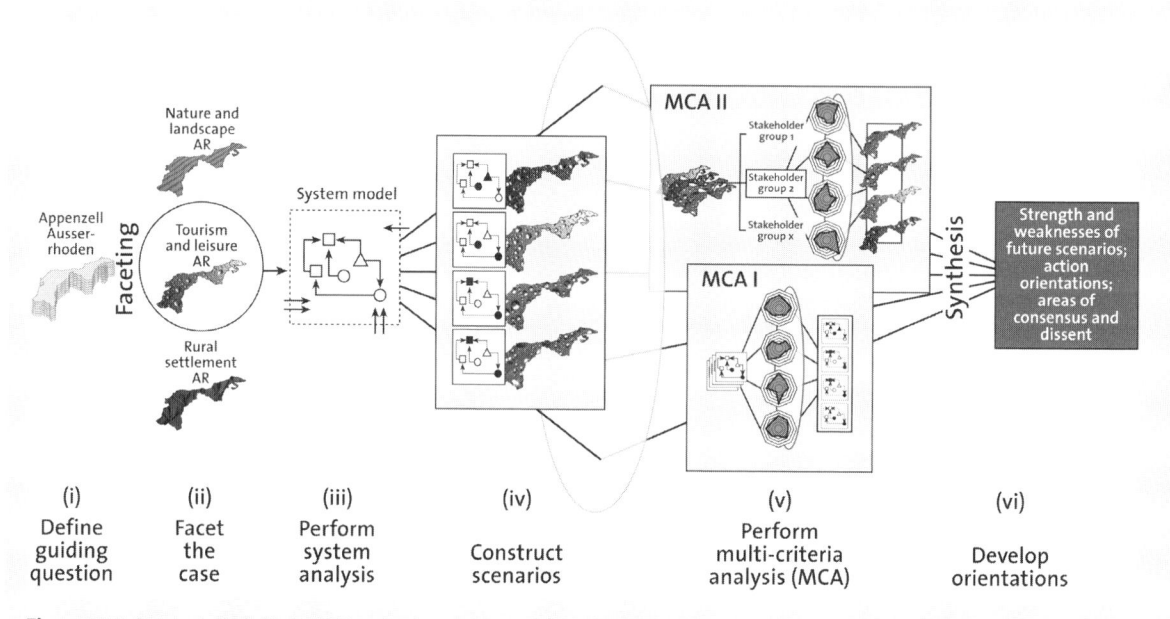

Figure 15.3 Six steps included in the ADN method as an organizing frame for elaborating robust orientation on the future development of an area.

comparing evaluation profiles from stakeholders (MCA II) with those from research (MCA I), differences in theory and practice judgments are compared. Which one is more appropriate with respect to the guiding question has to be negotiated. The methods and details, and specific methodological recommendations on how to apply these methods are described in Scholz and Tietje (2002).

The development of orientations can be the subject of a seventh step, which is the factual negotiation process among the stakeholder groups participating in MCA II. For this step, the scientist must receive a certain mandate.

15.3 Functions of transdisciplinary processes

Because scientists are knowledge system experts, they often act as facilitators during transdisciplinary collaborations and therefore have the opportunity to shape the process and, in some cases, cultivate collaboration among participants. Looking back at almost two decades of large-scale transdisciplinary projects that are in line with the Zurich 2000 definition (Scholz et al., 2006; Thompson Klein et al., 2001), we see that four general

Table 15.4 Functions of transdisciplinarity.

	Function	Essentials
(1)	Capacity-/competence-building	Collaboration in transdisciplinary processes integrates knowledge from scientists from different fields and from representatives from different stakeholder groups and decision-makers, respectively, about the concrete social, regional, and technological processes of adapting to changing environments
(2)	Consensus-building	In consensus-building experts reduce uncertainty, formulate state-of-the-art knowledge, and (a) inform the public, the media, or colleagues about what science knows and does not know or (b) organize or participate in discourses such as public consensus conferences in which answers are provided to a citizen panel's questions for paving the way to reach consensus about the issues at hand
(3)	Analytic mediation	Being concerned about environmental impacts or accessibility to environmental resources is a matter of dispute, causing conflicting interests between different stakeholders and parties. Scientists can operate as facilitators assisting in problem definition, constraining uncertainty (see 2), and creating processes that can generate acceptable solutions when applying (analytic) methods of bargaining, negotiation, and conflict resolution
(4)	Legitimization	Scientists and science knowledge can be used to legitimize political action programs. However, the impacts of transforming science into political instruments require critical reflection

facilitation approaches (i.e. functions) are most common; they are listed here and described in Table 15.4:

(1) *Societal capacity-building*, for sustainable transitions. This is generally the key function and refers to the following characteristics of socially robust knowledge: (i) using state-of-the-art science knowledge and (iv) integrating experience and science-based knowledge.

(2) *Consensus-building*, which refers to (ii) attracting consensus.

(3) *Analytic mediation* is a policy process alternative to expensive litigation or administratively focused arbitration, where scientists serve as facilitators.

(4) *Legitimizing* by informal power which reveals the involvement of science knowledge in policy processes (referring to (iii) and (v) of the characteristics of socially robust knowledge).

It is worth noting that capacity-building has been denoted as substantive integration of arguments by some experts from the case and scientific experts. Consensus-building, on the other hand, is related to the normative side as group norms or societal norms are formed. Both analytic mediation and legitimization can be seen as variants of an instrumental function (Fiorino, 1990).

15.3.1 Capacity-/competence-building

The most essential aspect of competence-building is the integration of different types of epistemics from society. On the one hand, we have scientific knowledge – disciplinary or interdisciplinary – which is formally represented in scientific articles and books, and aspires to be generally valid. In general, scientific statements refer to abstracted systems and represent general rules, principles, or laws. A statement becomes an accepted scientific assertion or theory if it meets certain standards that are defined and reviewed by the scientific community. As scientific knowledge is usually based on causation, which is linked to the rules of propositional logic, *erklären* (Engl.: explaining), it is the core rationale of scientific epistemics (Scholz & Tietje, 2002). On the other hand, we have experiential knowledge which is based on *erfahren* (Engl.: experiencing), the practice-based real-world (life) experience referring to concrete systems (see Figure 2.6). We consider experiencing as a prerequisite of understanding, which includes empathy.

In transdisciplinary processes, scientists leave the terrain of abstracted systems and approach the real world. Phenomena and problems are contextualized and embedded in complex, specific, unique, historical matrices. Scientists and practitioners can relate their epistemics on the level of *begreifen* (Engl.: conceptualizing). To produce meaningful outcomes, *begreifen* should involve joint definition of the most important aspects needed for sufficiently describing the state of the real-world system under review, as well as the system's dynamics (or for an adaptation process). These aspects are: (1) the system boundaries (spatial, temporal, content-related); (2) the goal of the transdisciplinary process; and (3) joint problem definition, as described in the paragraphs in which we defined transdisciplinarity.

15.3.2 Consensus-building

Consensus-building is a special procedure related to negotiation (Susskind *et al.*, 1999) or group decision-making. The term consensus conference emerged from public meetings held in the 1970s that were designed to include the public and their experiences in assessment of technological developments. Consensus conferences are a specific method for consensus-building. In Denmark, for example, these conferences ran under the stewardship of the Danish Board of Technology. Insofar as a consensus conference "is a dialogue between experts and citizens" (Andersen & Jæger, 1999, p. 332), it can be considered as a special form of transdisciplinarity. An example of a science-based consensus conference was the "Consensus Conference on Protection of the Environment" of the Norwegian Academy of Science and Letters (Oughton & Strand, 2002). Thus, we can distinguish two types of consensus conferences: (a) a primarily or purely academic activity that builds consensus among scientists and transfers this to the public by journal articles or media conferences (more or less following the sender–receiver metaphor); and (b) a transdisciplinary process in which experts discuss "attitudes and recommendations" (ibid., p. 332) related to questions from citizens. Whether or not processes such as consensus conferences of type (b) are factually considered as transdisciplinary processes has to be reflected upon. Again, joint problem definition and setting the agenda are critical here.

Consensus conferences are just one of a plethora of different techniques applied in such contexts (Rowe & Frewer, 2005). A consensus conference can help to contribute to a broader band of information and deliberation about social and ethical aspects of science and technology (Glasmeier, 1995, p. 68).

15.3.3 Analytic mediation of environmental disputes

Climate change has led to a worldwide dispute about the negative impacts of human activities. There are hypothetical and real winners, losers and unaffected people. Different countries and stakeholders hold disparate values, priorities, and refer to different assumptions about the future dynamics of temperature, precipitation, extreme events, etc. in different parts of the world. Here, analytic mediation as a policy process in which scientific knowledge is involved (in contrast to mediation and arbitration as administrative or adjunctive procedures; see Box 15.2) is a meaningful tool. The idea of analytic mediation emerged in the 1990s when environmental conflicts about brownfields, incinerators, soil pollution, and contaminated water became worldwide issues. Analytic mediation, where scientists act primarily as facilitators, has been a good alternative to conflict resolution through expensive and inefficient juridicial litigation.

There are multiple activities that scientists can facilitate in such processes. Scientists have often participated in problem definition, in order to guide definition of the system boundaries and the issues to be negotiated and to be excluded. This is a particularly difficult task, especially with respect to financial payouts that correlate to who is affected (positively or negatively) from negative environmental impacts. A second task is the constraining of uncertainty. The conflicting parties have specific interests, call for different levels of risk acceptance, relate to different scientific "schools," and evaluate the unknown in a completely different manner. The parties are often opinionated and mistrust the other party and sciences, especially in disputes about highly contested issues such as genetically modified organisms, nuclear energy, waste disposal, or sometimes climate change. Finally, scientists can facilitate the process of finding a solution that is considered fair and acceptable by all parties or stakeholders. As environmental disputes usually include multiple trade-offs, this is usually supported by a multi-criteria assessment (Scholz & Tietje, 2002). The discourse about the essential dimensions of different options, in terms of utility, reveals domains of consensus and dissent among stakeholders. While this can develop into negotiation or the generation of alternatives, the final decision and agreement can sometimes be ratified outside the transdisciplinary process. There are, however, few transdisciplinary processes in which the scientific partners receive the legitimized mandate for analytic mediation (Scholz *et al.*, 1996).

15.3.4 Legitimization by policy–science cooperation

The relationship between science and policy has varied historically between different countries and different time periods. The separation between scientific

Box 15.2 From parsons to Napoleon: variants of mediation

Mediators take on the role of political scientists, particularly in international conflicts or labor management disputes. Mediation became a major conflict resolution technique for environmental conflicts in the course of the 1990s land use and brownfields disputes (Susskind, 1975). Mediation has been applied to many environmental subjects, such as harmonizing environmental and economic concerns of airport management (Driessen, 1999), river estuary management (Innes & Connick, 1999), or soil cleanup negotiations (Scher, 1999).

Usually mediators are consulted if the conflict resolution process has come to a halt and the parties involved realize that breaking off negotiations is unfavorable. Mediation differs from arbitration (see Bacow & Wheeler, 1984, p. 156), which is a more formally, and often legally prescribed, procedure performed by established institutions.

Mediation procedures provide mechanisms for facilitating agreements in conflicts (Susskind et al., 1999). These procedures are a subset of alternative dispute resolution methods. Within these methods, Susskind and Madigan (1984) differentiate between unassisted negotiation, facilitated policy dialog, collaborative problem-solving, passive (or traditional) mediation, active mediation or mediated negotiation, non-binding arbitration, binding arbitration, and adjudication. A specific approach is analytic mediation, such as the ADN, in which the assessment of consensus and dissent of multi-criteria utility assessment tools are combined with multi-party negotiations (Scholz & Stauffacher, 2007; Scholz & Tietje, 2002).

Figure 15.4 shows four different types of dispute resolution and the relationship between them. ADN develops at the interplay between social psychology and policy processes. These methods can be applied to transition management in transdisciplinary processes and specifically to conflicts concerning environmental impacts or commodities at the local level (e.g. land zoning and contaminated soil management; Susskind et al., 2003). Mediation has become an important tool for balancing human aspirations with respect to limited environmental resources such as water (Dryzek & Hunter, 1987).

(1) *Mediation as a policy process* has a long history as a feature of democratic conflict resolution. Knoepfel (1995), for instance, reports on the 1481 "Helvetische Mediationsakte" (Engl.: Helvetic Mediation Act) of Stans, in which a parson took the role of a mediator. In 1803, Napoleon Bonaparte mediated a dispute between the different cantons of Switzerland. In the USA, mediation practice was established at the end of the nineteenth century. The state of Massachusetts allowed mediation in 1898, and in 1913 the Board of Mediation and Conciliation was founded. Over the past few decades, many environmental conflicts in the context of the "Not In My Backyard" (NIMBY) phenomenon, for instance, the disposal of radioactive waste (Kraft & Clary, 1991), have become matters of public concern.

(2) *Mediation and arbitration as adjudication tools*: from the very beginning, mediation was a supplementary tool of the legal system. Whether and how far it is accepted as a genuine part of the legal system differs from country to country.

(3) *Negotiation and bargaining* is a field of decision sciences; it analyzes individual, organizational, and societal decision processes.

(4) *Bargaining and negotiation* is usually employed by one party to get as much as they can. This is often referred to as distributive bargaining, a topic beyond the scope of this book.

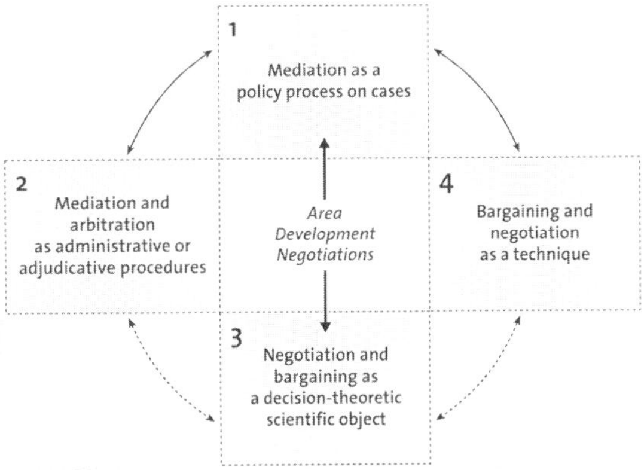

Figure 15.4 The relation between four types of activities and research in the field of mediation, arbitration, negotiation, and bargaining related to ADN. The arrows indicate which relations are more established (dotted lines indicate weak relations). Transdisciplinary processes can include (1) mediation as a policy process and (3) as the empirical decision-theoretic facilitation of the negotiation and bargaining process (see Scholz & Tietje, 2002, p. 200).

and social/political activities has been linked to methodological discussions in scientific associations (for instance, in the distinguished Verein für Socialpolitik; see Weber, 1922/1988). This is also seen in the discourse surrounding the differentiation between the social and natural sciences. For example, after World War I the participation of chemists in the production of chemical weapons led to alienation between science and the public. In contrast, particularly in non-democratic countries such as the Soviet Union or Hitler's Germany, social scientists helped politicians in justifying their regime or the inequity between peoples by race theory. In democratic countries, there have been left wing and right wing attempts to politicize the sciences. Critical voices have called the science system "a product of a bourgeois and capitalist society, unable to fulfill its inherent social mission" (Lengwiler, 2008, p. 192) and claimed that science should serve social functions (Bernal, 1939).

Transdisciplinarity is strongly rooted in the transformation of sciences into policy processes, the principle of freedom of research, the social responsibility and accountability of scientists, and the "Taylorist" principle of separation of roles and functions between scientists and decision-makers. This also holds true for scientific contributions to understanding climate change. Employing scientific research and knowledge to legitimize political actions comes into play when prioritizing research questions and directions, determining what research will be funded and also which policy measures will be implemented.

The notion of the Zurich 2000 definition of transdisciplinarity starts from the conception that the economic system, the political and legal system, the cultural and social system, and the scientific systems are the main components of society (see Figure 1.7*). The role of the university as a specific societal institution is to produce knowledge demanded by the cultural and social systems that allows for proper societal reproduction. The knowledge produced by scientists must serve this goal and, thus, should follow an inner science reference (i.e. approaching truth or validity; see above). At the same time, it should serve to answer the essential questions of society. However, if the boundaries between science and policy or between science and industry become fuzzy, the predicament of biased, politically formed or corrupted science emerges.

15.3.5 Key messages

- We can distinguish four functions of transdisciplinarity: *capacity-building* for coping with ill-defined socially relevant problems is the key function; *consensus-building* and *mediation* among stakeholders with conflicting interests in which scientists function as mediators, if they receive this mandate; *legitimization* of decisions is the final function.

15.4 Other applied research and participatory approaches

Participation is "a dazzling term" (German: "ein schillernder Begriff;" Renn, 2005). The most critical issue is that often it remains unclear *who* (what people) is participating and in *which* venture with *what* rights. In this section, we look at three other applied and participatory research processes: public participation, consultancy, and action research, and compare each with a transdisciplinarity process.

Returning to our "transdisciplinary triangle" diagram (Figure 15.1), we can distinguish modes of participation for each of these processes. When the public at large participates in the decision process this effort is commonly called public participation, as shown in Figure 15.5. When the public participates in research it is called participatory research. When the science community participates in the legitimized decision-makers' activities, this is usually called consultancy or contracted research. The latter is, for example, a common occurrence in medical and pharmaceutical research.

When the science community, which is in most countries publicly financed, moves over to practice and follows a (primary or important) goal to assist stakeholders, we call this value-driven "action research." An important aspect of participation is that the process remains under the formal control of one side, and the others usually cannot define goal or framing of the process. Participation differs from collaboration.

According to the Zurich 2000 definition, a basic feature of transdisciplinary processes is that the (collaborative) process is controlled by more than one party. This would mean that a new, transdisciplinary process is controlled by more than one partner. Graphically, this process is best represented on the area between the activity lines of the two agents. If the process is controlled by all three parties, then a new "cooperation space" is defined in the inner part of the cylinder space. This idealized view of an equally controlled cooperation space is the goal of transdisciplinary processes (see Box 15.1).

15.4.1 Public participation

The idea that the public at large (citizens) should participate in societal decisions is one of the most basic ideas of democracy. Different conceptions of democracy can underlie claims for participation. There are

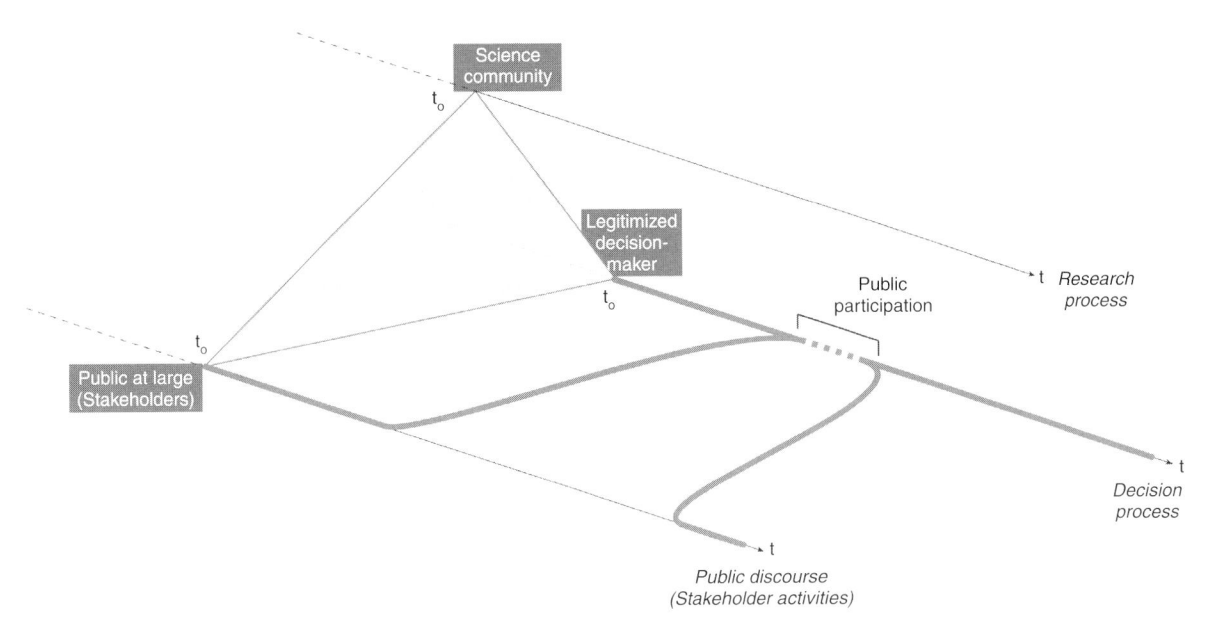

Figure 15.5 Public participation.

extreme views of democracy, such as the elitist idea that public participation begins and ends with the vote (often linked with Max Weber and Josef Schumpeter, who took an elitist perspective; Dryzek, 1993). This is contrasted by the position of an open, discourse-oriented, argument-based democracy where everyone has the same rights (Habermas, 1984, 1987), which is generally referred to as "deliberative democracy." Participation is a way of linking citizens in western societies to political life and supplements traditional forms of voting by interacting (Verba & Nie, 1987). Public participation sometimes also goes under the label of participatory appraisal, for instance in Britain where "local people at all stages, from priority setting to solution implementation" are involved in decision processes (Pain & Francis, 2003, p. 47).

In its pure form, members of the public at large, or stakeholders, participate in a decision process for a certain period, related to a specific issue (see Figure 15.5). One prototypical example (there are many) is the United Nations Economic Commission for Europe (UNECE) Convention on Access to Information, which involves over 70 international professional organizations and other governmental organizations in dealing with crucial environmental issues (see Kastens & Newig, 2005; Newig, 2005; UNECE, 1998).

Let us look at public participation in which public authorities have control. Here, a critical issue is the manner (mechanisms) of involvement, the coping with the results, and the relationship to existing traditional democratic procedures. Fiorino (1990) identified different mechanisms of public participation. Hearings and initiatives generally have a clear, legally fixed, procedure. On the contrary, public surveys (to gather the opinions of the affected but uninterested public), negotiated rule-making (to avoid escalating confrontation; Susskind *et al.*, 2000), and citizen review panels (Fiorino, 1990) follow procedures that are not defined by laws, but rather by the parties that design the survey, negotiations or panel.

An interesting circumstance is that one of the pioneers of citizen participation was the Protestant theologian Peter C. Dienel, who invented the concept of *Planungszelle* (Engl.: planning cell or citizen panel) in the late 1960s (Dienel, 1970/1991) to involve citizens in urban planning. In the US, the approach was introduced by Crosby *et al.* (1986). The model has been internationally and widely received:

> The planning cells always function according to the same model. Groups of five are randomly selected from the register of residents, to elaborate recommendations for the city, county, or state. To enable them, experts assist them. "All participants work wholeheartedly because they get paid for the suspension of earning," said Professor P. C. Dienel. (Schicke, 2006, p. 31; translated by Scholz)

Though the planning cell was initially applied on a local level, it was later also applied in Germany to energy policy (Dienel & Garbe, 1985). Renn et al. (1993) did a study with 24 citizen panels, each including about 25 participants, which, for instance, "unanimously rejected a high energy supply scenario and opted for an energy policy that emphasized energy conservation and efficient use of energy. Nuclear power was perceived as non-desirable … The panelists recommended stricter environmental regulation for fossil fuels even if this meant higher energy prices" (Renn et al., 1993, p. 202). Though this study was launched by the German Ministry of Research and Technology, a scientific advisory board, including administrators and representatives of stakeholder groups, criticized the role and function of the citizen panel. Arguments targeted the lack of accountability, the possibility of manipulation, the lack of training for citizens to understand complex scientific data and arguments, and the neglect of organized interest groups (Renn et al., 1993). We should note that some of these arguments are often also used by organized stakeholder groups such as NGOs, organizations that often reject the legitimacy of citizen panels, presumably because they think that they have the right to do the citizens' job. We suggest distinguishing between public participation of members of society *at large* if we deal with unorganized citizens and members of society (which can be done in its extremes by random selection of participants who represent themselves; Chilvers, 2004), and participation of non-organized or organized stakeholders if we deal with organized interest groups (representatives of companies, institutions, NGOs, etc.).

In principle, citizen panels follow the idea of critical theory and Habermas' idea of dominant free discourses, in which rationality and consensus are attained in a communicative process that is open to all relevant (sometimes conflicting) arguments from participants, thus honoring their equal rights and duties. This claim is an attractive vision but often does not meet the constraints of reality (Newig, 2007). One of the constraints is simply that the "competition of arguments is insufficient if those who may have problems articulating their arguments are left out" (Renn, 2006, p. 36).

15.4.2 Key messages

- Participation defines an asymmetric relationship. Thus it has to be clear who is participating in whose venture
- We speak of public participation if members of the public at large participate in the decision process of a legitimized decision-maker
- We distinguish transdisciplinarity from public participation as, in general, transdisciplinary processes are not *directly* related to a decision. Transdisciplinary processes ideally form a symmetric process in which the participants from practice, including a legitimized decision-maker, whose actions may be a subject of the process, and scientists have equal control over the goals and course of the process.

15.4.3 Consultancy

In consultancy, scientists participate under the control of the legitimized decision-maker(s) in the decision process (see Figure 15.6). We do not want to go into too much discussion about "classical" consultancy, as the latter is commonly known. We mainly want to reflect on the question of whether the university–government–industry "triple helix" of Leydesdorff and Etzkowitz (1996) should be put under the category of consultancy. The focus of this approach is knowledge-based innovation, which is necessary for solving many challenging environmental problems of the twenty-first century. The triple helix model addresses concrete and pressing problems of government, academia, and industrial policy embodied in "incubators." This approach stresses new intersections of industry, government, and academia. The idea is that:

> Academic institutions participate in various ways in capitalization of knowledge and its transmutation into factors of production. (Etzkowitz, 2001, p. 19)

In this way, the university has taken on a "third mission" of direct contributions to industry, "a change that would make participation in the progress of economic development into a core value" (Gibbons, 1999; in Stephens et al., 2008, p. 332). This plays out in some countries, such as the Netherlands, where we see "professors who are researchers of high calibre in natural sciences and engineering are also consultants for industry" (Gibbons et al., 1994, p. 141). The triple helix model might be seen as an approach countering the request posed in some countries (e.g. the Swedish Research 2000 Report; see Etzkowitz & Leydesdorff, 2000) that universities return to their medieval role as learning centers. Instead, we propose that the university should play a key role in knowledge-based society and is also a potential engine of industry and wealth creation. Clearly, this position contrasts that of the emancipatory "New production of knowledge" approach and might even be perceived as total capitulation of academia for industrial interests. However, as Shinn (2002) worked out, the triple helix approach receives much more attention in developing countries, which focus more on increasing wealth than the "New production of science" paradigm, whose audience is based dominantly in the northern hemisphere.

The key issue is that industry remains in control about when (some of the) results are published, which is in contrast to the idea of equal control in transdisciplinary processes.

When referring to policy sciences, Elzinga and Jamison (1995), Jasanoff and Wynne (1998), and Pohl (2008) reveal that consultancy can be linked to a market and a "selling-driven" science, bureaucratic policy culture (i.e. consulting policy-makers, where speaking the truth supports a power-driven culture), whereas participatory research as well as action research are linked to an open civic policy culture.

15.4.4 Shallow action research

Action research is a process that ideally brings research resources to real-world problem-solving. Figure 15.7 shows an extreme variant of action research in which members of the science community move to stakeholder activities. This variant has been called shallow action research as there is a strong focus on the real-world problem and inputs from and to science are low (Cornwall & Jewkes, 1995). In Chapter 6.4, we were dealing with action research when referring to Cassell and Johnson (2006), who describe consultancy to corporate elites by researchers as a variant of action research. This is in large part a result of the lack of differentiation between the legitimized decision-maker and the public at large.

In the following, we review the origins of action research and describe other forms of action research that involve scientific knowledge and theory-building, in particular participatory action research (PAR) and community-based participatory research (CBPR). These approaches share many features with transdisciplinarity which are important for environmental literacy.

15.4.5 The cradle of action research

At the end of World War II, the founder of social psychology, Kurt Lewin, was a professor at MIT where he established the Research Center for Group Dynamics. He wanted to understand why democratic processes fail and why ethnic minorities do not

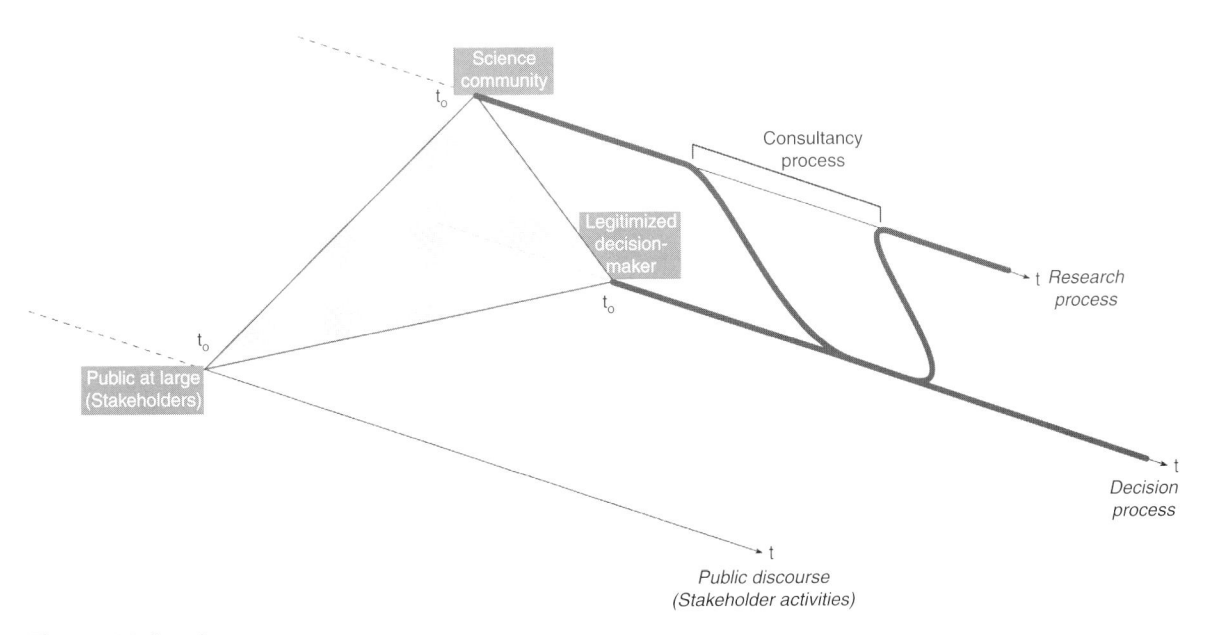

Figure 15.6 Consultancy.

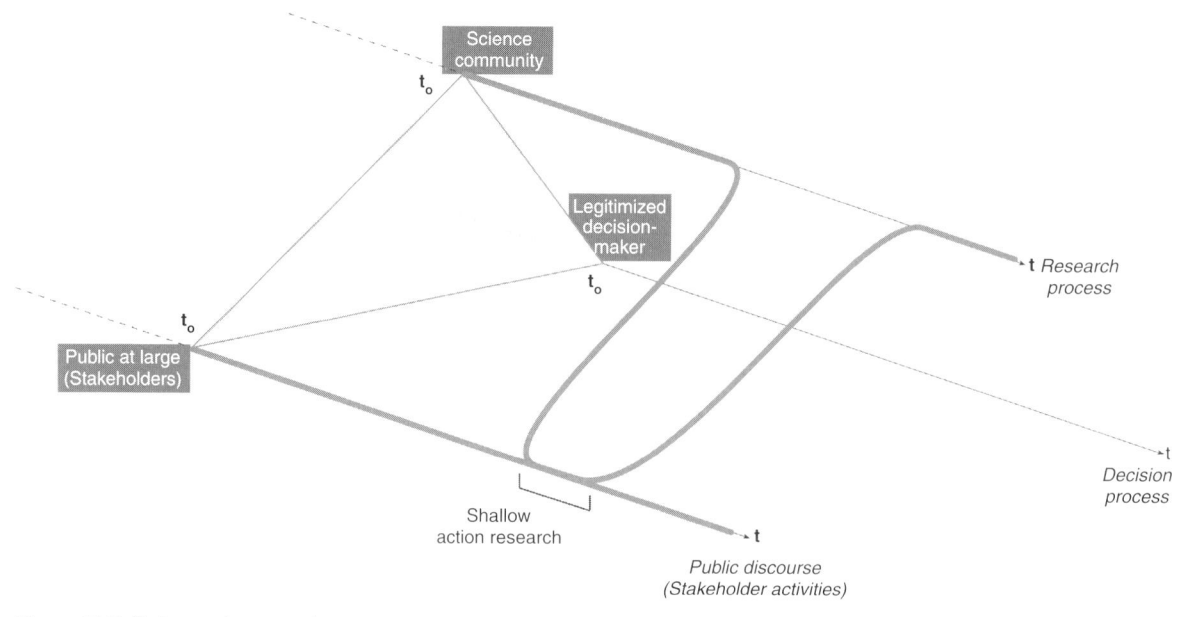

Figure 15.7 Shallow action research.

sufficiently benefit from the educational system. To explore these questions, Lewin built a praxis forum, including 41 agents and decision-makers from communities, schools, unions, and other social institutions. This forum can be seen as one of the first transdisciplinary processes.

Lewin and his research group discussed critical phenomena, but also discussed the design of research, data, and findings in order to best utilize them both for scientific theory-building and for practical purposes (Barker & Schoggen, 1973; Bradford et al., 1964; Greenwood & Levin, 1998; Lewin, 1946). Lewin's application of methods of experimental psychological research, for instance for systematically investigating the effect of different treatments (e.g. teaching styles), has been called experimental action research. Experimental action research thus can be understood as a kind of social science variant of the transfer from in vivo to in vitro research.

15.4.6 Participatory action research

PAR emerged from efforts by organizational, social, health, and developmental geography sciences to bridge the growing gulf between university research and societal practice. The great sociologist William Foot Whyte claimed that PAR – and we think also transdisciplinarity – can help science "get out of the academics morass" (Whyte, 1989a, p. 382). Science can also realize some revealing insights when investigating real-world domains:

> The complexities of working life industry are too great to permit realistic experiments in which one or only few variables are manipulated and other variables are kept constant. (Whyte, 1989b, p. 504)

This is, by the way, an idea which perfectly meets the idea of representative design proposed by Egon Brunswik (see Chapter 6.3) when suggesting a research methodology for investigating human perception and decision-making in complex environments. For some researchers, PAR is research that "involves practitioners both as subjects and co-researchers" (Argyris & Schön, 1989, p. 612).

> Participatory action research is a form of action research that involves practitioners as both subjects and core-searchers. It is based on the Lewinian proposition that causal inferences about behavior of human beings are more likely to be valid and enactable when the human beings in question participate and test them. (Argyris & Schön, 1991, p. 86)

As we will discuss below in the participatory research section, one critical question with respect to PAR is whether a normal science view or a post-normal science view should be taken. Another critical question is how the roles and relationships are defined.

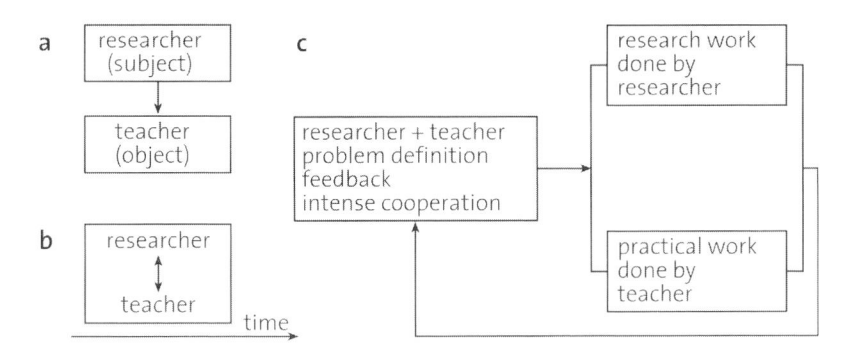

Figure 15.8 (a) The relationship between researchers and practitioners in traditional research; (b) action science research, researcher, and participant roles are undifferentiated; and (c) participatory action research (transdisciplinarity; taken from Scholz, 1978, p. 30).

Figure 15.8 exemplarily shows three different types of relationships between action researchers and teacher practitioners, as they were described in the 1970s (Scholz, 1978). In the "classical" relationship the researcher considers the practitioner (here the teacher) who participates in research as an object from whom to collect data (Figure 15.8a). In a variant of action research that was prevalent in the 1970s (Figure 15.8b), both the researcher and the practitioner are "in one boat" and there is no differentiation between the roles or the major activity domain of each. We call this variant *undifferentiated action research*. Conversely, in Figure 15.8c, the researcher and practitioner cooperate intensely. Both sides are responsible for problem definition and give mutual feedback in a jointly organized transdisciplinary process before they go back to their core domains. This type of relationship, which meets the basic ideas of theory–practice cooperation from the Zurich 2000 definition of transdisciplinarity has been called the Scandinavian, *cogenerative style of action research* (Elden & Levin, 1991; Greenwood & Levin, 1998; Rudvall, 1978; Scholz, 1978).

We should note that we often meet *shallow action research* that has not been reflected upon if "embarrassment or threat [comes] into play" (Argyris & Schön, 1991, p. 86) or when emancipating intent promotes "critical consciousness which exhibits in political as well as practical action to promote change" (Grundy, 1987, p. 154) and is too dominating. Neglecting to delineate between researchers and practitioners introduces the risk that findings emerge that do not go beyond the solution of the practical problem to inform research (Hodgkinson, 1957), or that "discoveries of action research … distort the discoveries themselves" (Argyris & Schön, 1991, p. 86).

15.4.7 Participatory research and community-based participatory research

In a strict sense, participatory research involves subjects or organizations who deliberately and consciously participate in a research project, taking on a role that goes beyond the subject or test person.

> Participatory research focuses on a process of sequential reflection and action, carried out with and by local people rather than on them. (Cornwall & Jewkes, 1995, p. 1667)

In clinical research, for instance, patients are recognized as experts who have experience and skills that complement those of researchers. This is because patients have first-hand knowledge about what it means to suffer from a particular disease, have a subjective feeling about impacts of a medicine or treatments, and they can describe, based on first-hand experience, unintended side-effects.

The extreme situation displayed in Figure 15.9 reflects a situation in which the patient has no self-interest and exclusively contributes to the research process, and the power and control of the situation are aligned exclusively to the researcher. Cornwall and Jewkes distinguish among contractual, consultative (researcher asks practitioners), collaborative (the researcher initiates and manages collaboration with local people), and collegiate forms of decision research. The latter corresponds to mutual learning as has been typical for ideal transdisciplinary processes (Cornwall & Jewkes, 1995).

Community-based participatory research (CBPR) is a special form of theory–practice cooperation that is very similar to the ideal form of transdisciplinary processes. However, it often involves a strong democratic and egalitarian approach, in particular in developmental research.

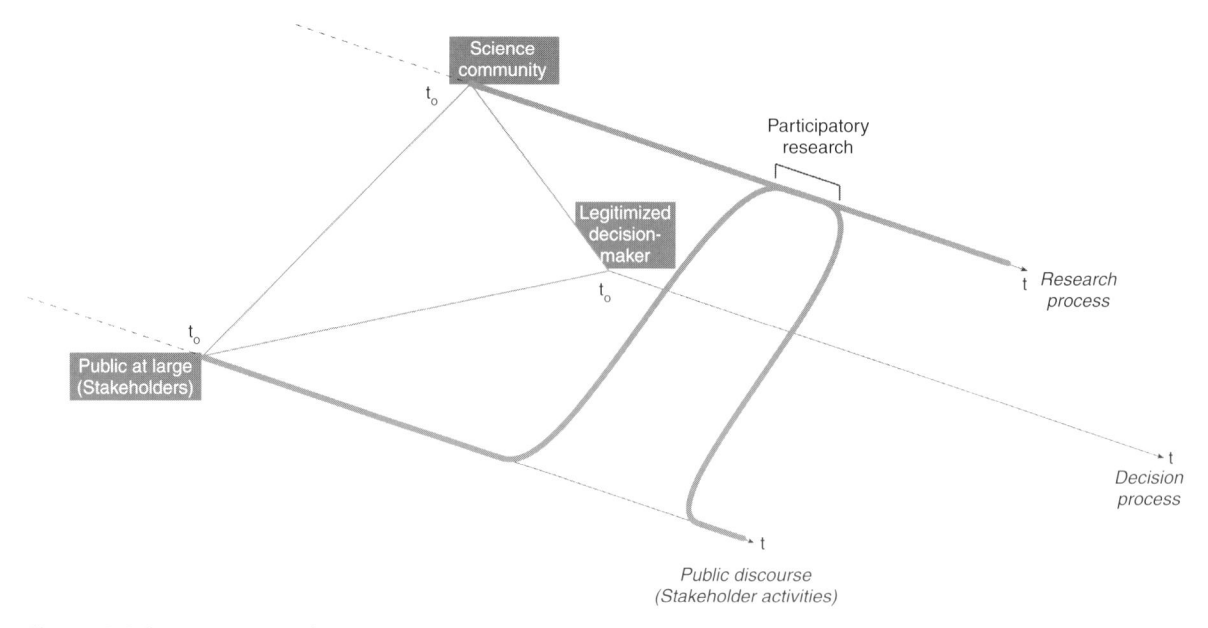

Figure 15.9 Participatory research.

CBPR traditions in the third world nations were premised on the importance of "breaking the monopoly over knowledge production by universities" not as anti-intellectualism but rather in recognition of the "different sets of interests and power relations" (Hall, 1999, p. 35) to which academic researchers and the community people they study are linked. (Leung *et al.*, 2004, p. 501)

A comprehensive sample of principles for CBPR has been described for health sciences by Israel *et al.* (1998), and includes a list of rules and characteristics such as: CBPR (1) is participatory, (2) is a cooperative process to which community members and researchers contribute equally, (3) is a co-learning process, (4) involves local capacity-building and systems development, (5) includes informal empowerment of all participants, (6) leads to action which is related to research and thus keeps a balance between action and research, (7) acknowledges the social–cultural side of human communities, and (8) requires long-term commitment. Clearly, these principles are compatible with transdisciplinarity as a process of mutual learning on equal footing for attaining socially robust solutions and capacity-building.

15.4.8 Key messages

- Consultation denotes the participation of scientists in a decision process that is almost completely controlled by a legitimized decision-maker
- There are different types of action research. During shallow action research, scientists slip into a stakeholder role and do not differentiate from practitioners
- Community-based participatory research resembles transdisciplinarity
- During participatory research, people from the public at large participate in the research process in roles that go beyond the mere proband or test person role. The researcher keeps full control over the research process.

15.5 Transdisciplinarity, science, and universities

Looking forward, we want to elaborate how scientists and universities can cultivate environmental literacy by effectively taking over key roles during transdisciplinarity processes. First, we discuss how universities enable disciplines to bring their unique skills, methodologies, and knowledge to support transdisciplinarity by serving as a place for rigorous, discipline-specific knowledge production and theory-building. We call this disciplined interdisciplinarity in transdisciplinary processes.

Next, we examine how theory–practice cooperation continues to evolve. Specifically, we review how Mode 2 research focuses on real-world problems and includes cognitive sciences, information technologies, environmental or social sciences. We assert that there is yet greater potential for science to support real-world problem-solving and propose that transdisciplinarity does become a third, complementary mode to disciplinary and interdisciplinary academic activity that can support basic (Mode 1) and various types of applied research. Here, knowledge integration from a post-normal perspective for integrating knowledge and values from practice, and a realist stance for evaluating the quality of science are important.

Finally, we discuss topics relevant for implementation of transdisciplinarity, or "transdisciplinarity in action." We first point out that some scientists, as well as decision-makers, need to overcome their fear of losing control over the transdisciplinary process, so that they can become effective facilitators for mutual trust-building and collaboration. We discuss the need to avoid "social contamination" of science, prevent misuse of transdisciplinary processes, and recognize geographic and situational limitations regarding where transdisciplinarity can effectively be practiced. We close by postulating sustainability as the engineering task of the twenty-first century, with the university as a key agent for change.

15.5.1 Role of universities for environmental literacy

The need for disciplined interdisciplinarity in transdisciplinary processes requires answering questions such as: who should have and provide what knowledge? What is the role of the university? What function and role do scientific arguments take in coping with challenging environmental problems? We show that answering these questions depends on two issues: the first is the societal, historical, institutional framing of science, in particular of the role of the university. The second is the conception of science. Here it matters whether a normal or post-normal position (see A and B Table 15.1) is taken and whether transdisciplinarity is conceived to substitute disciplinary or interdisciplinary research.

In this section, we review how the sciences first became institutionalized at universities and how some professions, such as medicine, law, and technical occupations, existed before formal education was available.

We then review how, during the second industrial revolution, industries in the USA approached universities to support research and development that would fuel economic competitiveness. Next, we examine how offering a general education about natural science (German: Naturwissenschaftliche Allgemeinbildung) and environmental literacy have fallen under the purview of universities. This suggests that the sustainability sciences have become the societal didactics of sustainability learning ("sustain–abilities", see Chapter 20.4.1). Finally, we offer a discussion of the science disciplines as a clearinghouse for transdisciplinarity.

15.5.2 Institutionalizing the sciences

The scientific and educational system has become an integral part of decision-making worldwide. To better understand how the need for collaborative problem-solving models like transdisciplinarity came about, we review the role of various sciences as they became academic disciplines and the role of university. The first university institution, which was explicitly named the University in 1088 AD, was the one of Bologna (Rosenthal & Wittrock, 1993). The role of the first western universities (see Box 15.3) was education and grading, including preserving and passing on existing knowledge ranging from ethic, ethnic, and historic knowledge to mathematics. Initially, the alpha sciences, philosophy/theology, history, languages, literature, arts, etc., and contemplation as a methodology, dominated the university landscape. Later, the beta sciences, such as the natural sciences, entered the university system and, much later, gamma sciences – that is, social sciences, evolved. At universities, each discipline developed its unique methodologies and major strategies for validation. However, knowledge and theories necessary to understand and master the material side of the environment were developed in academies and not in universities. The academies discussed and deliberated on the quality and state of the human being and the sociocultural rules of society.

Similarly, medical schools were initially not attached to universities. The first medical school was founded in Salerno (Rashdall, 1895). The practical side of medicine provided financial benefits and a "schola-collegium medicorum" was established in 1372, primarily to collect donations (Fried, 2000). Pure natural science faculties followed much later. For instance, in

15 Transdisciplinarity for environmental literacy

> **Box 15.3** Science, not only a matter of universities: building the Tower of Babel
>
> In India, China, North Africa, and Europe master builders developed skills in mathematics and physics for building design and construction long before universities were built. Without science and the knowledge of the theorem of Pythagoras, the construction of the Cheops Pyramid or the Tower of Babel would not have been possible. Such buildings do not allow for even the slightest deviation from the building plan, and this requires precise mathematics. The Tower of Babel, which was commissioned by Nebuchadnezzar in about 560 BC, is believed to have been about 100 meters high (Seipel, 2003) and must have required sophisticated building design and construction knowledge (see Figure 15.10).
>
> University-like institutions in India can be traced back to about 450 BC (Altekar, 1965), in North Africa to the ninth century (Treadgold, 1997), in China ("Shuyuan") to about the eighth century (Yang, 2005), and in Greece to about 250 BC (Harig & Kolletsch, 2004), long before the foundation of the university of Bologna. Finally, the industrial age required politechnics in the eighteenth century to cultivate the mastery of mathematical–technological knowledge to support the emergence of the steel, chemical, machinery, and electrical industries.

Figure 15.10 (a) The painting of Raphael Sanzio (1483–1520) shows the teaching of Euclid at the School of Athens (1509–10); (b) Pieter Brueghel's painting of the Tower of Babel (1563; Kunsthistorisches Museum Wien); and (c) the Krupp factory in Essen (1910, Photo: City of Essen).

Germany the first faculty of natural sciences and mathematics was founded in 1863 in Tübingen; a time when new technologies needed the generation of basic common knowledge. This was when the term "scientist" was coined:

> We need very much a name to describe a cultivator of science in general. I should incline to call him a *Scientist*. (Whewell, 1874, p. cxiii)

It is noteworthy that professions such as medicine, law, and mechanics existed in many countries before universities offered a formal education in these fields. In the USA a judgeship, for instance, was at first attained through apprenticeship. Law later became an academic discipline by codifying the knowledge in books (Langdell, 1887; Quinn, 1994). Science became a domain providing work for hundreds of thousands at the beginning of the twentieth century.

15.5.3 The university and industrial research

During the second industrial revolution, as industry and society became masters of electricity, engines, metal processing, railways, communication systems, and military means, much research (particularly in military sciences) was performed outside universities (see Box 15.4). However, universities received increasing funding from government to generate basic research, and large investors such as Rockefeller, Mellon, and Ford indicate that there was a general interest from industry. The maintenance of a country's international competitiveness became an important driver supporting universities.

Research has long been a "byproduct" of teaching and was cultivated in institutions such as the Academy of Sciences that were built starting at the turn of the eighteenth century. For instance, the "Societas Scientiarum Brandenburgica" was founded by Leibniz in 1701. It was not the material–technical dimension but rather the sociocultural dimension that was the main object of these public research academies.

> Research emerged, initially in philology and in other disciplines, from a concerted effort to revive classical learning in the eighteenth century. (Etzkowitz, 2001, p. 19)

With the foundation of universities, the distinction between basic, or fundamental, and applied research emerged. Ben David defines basic or fundamental research as "research which produces contributions to knowledge in a field cultivated by a recognized community" (Ben David, 1968, p. 17). This definition excludes research of companies, which is not communicated, as well as research resulting exclusively in the solution of practical problems (applied research). However, basic fundamental research can become applied research and vice versa.

Since the beginning of the twentieth century, much research has been done in industrial and non-university institutions (see Box 15.4). In addition, military research was a main component in the Cold War's rivalry between the superpowers after 1945. In 1961, for instance, the US government financed 64% of research on electricity, 46% on tool-building or mechanical engineering, and 89% on aviation. Most government grants for science went to industry and not to the universities, and much was invested for military purposes (Bernal, 1970). Environmental issues became a central object of university research after the first severe evidence of large-scale environmental impacts such as the London Smog in the 1950s.

15.5.4 The emergence of environmental and sustainability science

Clearly, societal needs in western countries during the twentieth century went far beyond military and purely industrial professional purposes. One societal demand that emerged was the general natural science education (German: Naturwissenschaftliche Allgemeinbildung) in the spirit of Wilhelm von Humboldt's conception of general education (German: Allgemeinbildung). One goal here was educating state officers who both understood and could oversee the practical development and the long-term effects of the processes of industrial society (Scholz et al., 2004, p. 26). This brings us back to the key question of what societal demands are put on environmental literacy, or, to express it in other terms, what are the urgent societal questions to be solved in the twenty-first century related to the material–biophysical environment and human society? And, what role should science play with regard to these questions? If we look at urgent problems, such as food, water, climate change, ozone hole, biodiversity, scarce raw materials, and energy, we can consider them all as related to ecosystem functions (Hassan et al., 2005) that contribute to human existence and wellbeing. Here, environmental science, as a new discipline, provides key knowledge for answering the above questions if the mission is defined in the following way: environmental sciences study shows how the environment works and how human systems interact with the environment.

Sustainability science, as an evolving discipline goes beyond explaining specific issues and look at critical transitions from a world perspective. This broader perspective would include, for example, a better understanding of the biogeochemical cycles, of how to manage phosphorus scarcity in agricultural biomass production, or of what ecosystem changes result from climate change. If we consider didactics to be the theory of systematic instruction and theory of learning, university research and teaching in environmental and sustainability sciences can be considered as societal didactics of sustainability learning if done in a transdisciplinary way. This is linked to capacity-building as a function of transdisciplinarity.

15.5.5 Disciplines as a clearinghouse for transdisciplinarity

> The modern connotation of disciplinarity is a product of the nineteenth century and is linked with several forces: the evolution of the modern natural sciences, the general "scientification" of knowledge, the industrial revolution, technological advancements, and agrarian agitation. (Thompson Klein, 1990, p. 21)

Disciplines became the building blocks and fabric of universities. We will briefly reflect on the role of disciplines and whether they are likely to be replaced by transdisciplinarity. There are sociocultural, epistemological, and object-related reasons for the organization of universities in a disciplinary matrix. At the end of the nineteenth century, the department was just starting to take shape in US universities, where groups of professors "with exchangeable credentials … collected in strong associations" (Abbott, 2002, p. 207) and began building disciplines. Thus the social network dimension is one constituent of disciplines.

The epistemological reason is linked to the efficient reproduction of knowledge and the coherent, dense organization of basic knowledge. Disciplines are a product of specialization and each has developed its own unique, internal structure, subdisciplines, methods, procedures, theories, concepts, paradigms, standards, and norms that allow freshmen to pass through the discipline efficiently. Disciplines also provide the template for the organizational structure of many universities through the faculty composed of chairs, which are the major units in European universities (Ben David, 1971), in which subdisciplines generally appear at the level of institutes.

The structure of disciplines has remained widely unchanged in North American universities over the past 100 years (Abbott, 2002, p. 210). According to Abbott, disciplines will have a future in top universities:

> Disciplines will survive in the elite sector. … Outside the elite level, the social structural foundations of disciplines are steadily eroding … (Abbott, 2002, p. 226)

From an institutional perspective, the disciplinary structures are mostly floating at the undergraduate major level. Interdisciplinarity enters scientific work in order to address complex questions, broad issues, or real-world problems. While interdisciplinarity requires integration, individual disciplines will continue to play a vital role as the "compartments" where

---

Box 15.4 Academic–industrial transformation of the world food production: Haber–Bosch

The industrial synthesis of ammonia, known as the Haber–Bosch procedure for nitrogen fixation (Figure 15.11), can be considered a much bigger scientific invention than nuclear energy or the internet (Smil, 2004). Artificial fertilizer made possible the growth of the world population from 1.6 billion people to 7 billion. Here, cooperation between German chemist Fritz Haber (1868–1934), who became a discovery electrochemistry scientist at the Universität Karlsruhe, and German metallurgist and mechanical engineer Robert Bosch (1861–1942) at the leading chemical company BASF, can be seen as a typical traditional engineer-like theory–practice cooperation.

The natural science foundations for this invention can be seen in Justus von Liebig's understanding of plant growth. "Carbonic acid, water and ammonia, contain the elements necessary for the support of animals and vegetables" (von Liebig according to Smil, 2004, p. 8). Further, Nicolas T. de Saussure's (1767–1845) experiment showed that plants can grow in clean sand as water can carry all nutrients, including nitrogen, to their roots. Liebig (1840) showed that plants can grow solely without inputs from inorganic compounds.

The story of this invention begins with Haber's early findings on nitric oxide formation in high-voltage arcs, which attracted the attention of Ludwigshafen BASF chemical company, which asked "if he would enter into a binding contractual research agreement with BASF" (Smil, 2004, p. 76). From 1908 on, Haber participated in what we consider a typical consultancy working (see Figure 15.6) for industry and was, among others, involved in inventing chemical warfare gases. Likewise, the engineer–deviser Bosch, who started with a fine mechanics workshop in 1886, was contracted by BASF, but remained an independent industrialist owning production sites for different technological inventions worldwide. The invention of the Haber–Bosch procedure, however, only became possible as BASF had a top research department and Alwin Mittasch (1869–1953), who was a researcher at BASF, converted the catalytic properties of metals for the Haber–Bosch procedure. This patent-oriented cooperation between university and practice can be seen as a variant of consultancy.

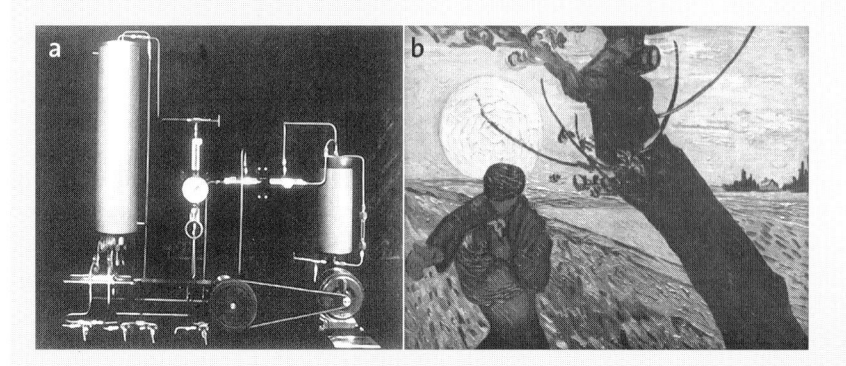

Figure 15.11 (a) The first Haber apparatus for ammonia synthesis. (b) The industrial production of fertilizers was essential for the evolution of modern agriculture; manual work became outdated ("The sawman" by van Gogh, 1888).

basic ideas, concepts, theories, and methods are neatly defined, crisply separated, and efficiently condensed, and where they continue to evolve. Famous examples of this can be found in the history of mathematics. David Hilbert's (1862–1943) program aimed at reducing mathematics as a whole "to finitist combinations" (Jahnke & Otte, 1997) based on a number of axioms. Nicolas Bourbaki, a pseudonym for a collective of (for a long time) anonymous French mathematicians, published 40 volumes between 1939 and 2007, striving for rigor and generality of set theories and axioms. These are examples for the clearinghouse function that are necessary as scientific statements lose validity if methodological rigor is missing (Scholz, 1998).

15.5.6 Key messages

- Science became an increasingly differentiated system much like "industry" with specialized disciplines. Sciences, including humanities, serve general societal and professional functions. The various main fields (faculties) of sciences, humanities, natural sciences, social sciences, and, later, engineering sciences, developed their own domains of reasoning (epistemics) with unique validation and reference systems
- Research has also always occurred, to a large extent, *outside* academic institutions. Patent-oriented collaboration of university with industry is one variant of theory–practice cooperation, which often becomes consultancy
- Universities served to secure Allgemeinbildung and Naturwissenschaftliche Allgemeinbildung. Thus, transdisciplinarity can be considered a means for capacity-building of the public at large for complex societal adaptations
- Disciplines can be considered as clearinghouses of science, supporting transdisciplinary processes with coherent, consistent statements that have to be properly inclined by transdisciplinary researchers and used for knowledge integration.

15.6 New frontiers for science–society cooperation

In this section, we discuss how theory–practice or science–society cooperation may continue to evolve. First, we review how, after nearly two centuries in their ivory towers, many scientists were switching to problem-oriented Mode 2 science, focusing on real-world problems. We discuss under what constraints ideal transdisciplinary processes can take place. Second, we elaborate, from an inner science perspective, that transdisciplinarity can become a third mode of research complementing disciplinary and interdisciplinary research. Third, we distinguish knowledge integration from a normal and from a post-normal perspective. Fourth, we discuss how a realist stance enables scientists to evaluate the quality of science assertions of causation, allowing verification of hypotheses.

15.6.1 Mode 2: bringing science to society

During most of the past two centuries the principle of division of labor by disciplinary differentiation and specialization at universities has been dominant. To fulfill the goals of science and engineering disciplines to generate consistent and cohesive theories and methods, knowledge production at universities was intended to be academic, free, curiosity-driven, socially and politically neutral, and shaped by specialty and cryptic language. This type of science has been called Mode 1 (Gibbons *et al.*, 1994). However, only a relatively small number of primarily western European and top North American universities came close to realizing this traditional "ideal." As a consequence, research came to be seen as an isolated, elitist endeavor conducted from an "ivory tower."

As Gibbons, Nowotny and proponents of the "new production of knowledge" sociology proposed (Gibbons, 1998, 1999; Gibbons *et al.*, 1994; Nowotny *et al.*, 2001), in the 1970s, the role of the university changed dramatically, taking on new roles that went far beyond a small, elitist perspective:

> … the modern university has become a hybrid institution, with multiple and sometimes incommensurable missions. (Scott, 2007, p. 214)

Universities became an important participant and driver of industry and regional development. Today, this not only holds true for top universities but even for second and third rate institutions. Already in 1968, about seven million students with 43% in the 18–21-year age group, were enrolled in institutions of higher education in the US (Ben-David, 1981/1972). Some countries endeavor to have more than half of their youth earn a university degree. For instance, in Germany, the number of universities increased from 34 in 1949 (25 in West and nine in East Germany) to 350 in 2000 (Kehm, 2004). Many former vocational training schools in remote areas became universities of applied sciences. Though traditionally running under the label "application and development," they are moving towards the label of "research," cooperating with local business and administration on regional developmental problems.

However, also the production of knowledge has changed in top universities. Scientific and technical work is increasingly performed by temporary teams, which deal with a specific, real-world problem rather than theory.

> Mode 2 science and technology include cognitive science, computers, environment studies, biotechnology, and aviation. … It is non-academic in the sense that its ties are with society and social issues. Society determines which problems are to be explored and resolved. (Shinn, 2005, p. 742–3)

According to Gibbons *et al.* (1994), Mode 2 science provides a new epistemology and asks for a rethinking of science. The Mode 2 proponents look at four principles that govern the new form of sciences research: the coevolution of science and society, contextualization, the production of socially robust knowledge, and the construction of narratives of expertise (Gibbons, 1999, p. 3). Such research brings science out of the ivory tower to work with industry, government, and laypeople to generate socially robust knowledge or socio-technical robust solutions. At the same time, scientists can use data from the "narrative of experiential expertise" that comes out of transdisciplinary processes for knowledge integration and theory-building. We suspect that this will be at work at all levels, from small regional colleges to worldwide operating universities.

Mode 2 shares commonality with the Zurich 2000 definition of transdisciplinarity, and Gibbons *et al.* state that "Mode 2 is transdisciplinary" (Gibbons *et al.*, 1994, p. 11). Scientists are required to go into open theory–practice discourses, which require them to cope with complexity and contextualization. Here, they meet the following situation:

> Collective narratives of expertise need to be constructed to deal with the complexity and the uncertainty generated by this fragmentation. … Experts must respond to issues and questions that are never merely scientific and technical and must never address audiences that only consist of other experts. (Gibbons, 1999, p. C83)

15.6.2 Transdisciplinarity as a third type of research

Based on our previous views, we suppose that transdisciplinarity becomes, or should become, a third form of academic activity. We propose that many problems ask for knowledge of the disciplinary, interdisciplinary, and transdisciplinary type.

We see at least four arguments supporting this proposal. First, we have presented the argument of disciplinary communities as a clearinghouse. Second, we have referred to social contamination of science, which is closely related to the freedom of research. Two additional reasons relate to differentiating among disciplines and distinguishing between scientific causation and non-scientific causation:

> Many antidifferentiationists refuse cognitive and social differentiation, and beyond. They deny the division between nature and culture, science and society, science and technology, and between research and enterprise. (Shinn, 2005, p. 744)

Specifically, we describe the third argument, for differentiation among different types of sciences or disciplines, and the fourth argument, for the differentiation between scientific causation and non-scientific causation. We consider it problematic not to distinguish between knowledge generated in practice and knowledge generated through scientific processes. An important issue related to the third argument is that, also, sciences include different types of rationales in reasoning and validation. There are, for instance, different reference schemes for validation; on the one side, practical efficacy, while on the other side, scientific coherence.

An example of differentiating and relating different types of knowledge is provided in the ADN, when we compare the science-based assessment (MCA I; see Figure 15.3), which relies on disciplinary data and methods, and the other, based on stakeholder judgments (MCA II). There are doubtlessly different epistemics at work in these different evaluations.

We advocate against abandoning the divisions between the modes of thought of science/theory and society/practice, as well as between different disciplines. Transdisciplinary processes require knowledge integration, both for quantitative and qualitative knowledge; for instance, intuitive vs. quantitative knowledge (Table 15.2). An interesting argument against the antidifferentiationist approach, which partly underlies the Mode 2 approach, is that the abandonment of the differentiation between and among science causation and social agent causation finally puts everything into pieces. This has been pointed out in the following statement:

The New Production of Knowledge [i.e. the variant of Mode 2 suggested by Gibbons *et al.*, 1994] posits atomistic learning and social interaction. (Shinn, 2005, p. 744)

15.6.3 Knowledge integration from a normal science perspective

A science perspective endeavoring to "integrate knowledge and values" from practice requires understanding which values are held by which particular groups in society. The other way round, as sociologists claim, is that society also acknowledges certain types of science (but not other types of science). Shinn stressed that society-based evaluation criteria for judging the propriety and worth of research findings have been historically absent.

> We think that in the course of developing environmental literacy, also society has to develop and to explicate quality criteria of what different branches of sciences should contribute to what stage of problem solving. (Shinn, 2005, p. 743)

This brings us back to the question of how knowledge integration between theory and science, between different disciplines, perspectives, modes of thought, cultures, etc. (see Box 2.5) should be organized. Antidifferentiationists postulate an open, seemingly power-equilibrated "unceasing composition, decomposition and recomposition of groups" (Shinn, 2006, p. 320), knowledge, epistemics, etc. Countering this "anything goes" postmodern position (Feyerabend, 2003), we postulate that methods of knowledge integration are needed as well as designing integration-driven social arrangements to promote efficient knowledge use by mutual learning in transdisciplinary processes.

Finally, we come back to the importance of the realist stance (see Chapter 3.4), which postulates the autonomy of the material environmental entities such that they are independent from how we think about them. This holds true for entities from the material and from the social environment. If we do not go with such a postulate, as is the case in most post-normal approaches, we can never verify whether a specific hypothesis is true or valid or not. Also, scientific statements in contextualized real-world settings can be objectively wrong or at least inadequate, whereas others show just the opposite quality.

15.6.4 Key messages

- Transdisciplinary research does not and should not substitute disciplinary and interdisciplinary research, but complements these types. It is a third mode of scientific activity that is based on transdisciplinary processes
- Transdisciplinarity should avoid antidifferentiationist approaches, at least with respect to two dimensions: first, the different rationales that are at work in intuitive, experience-based judgments of practical experts, and, second, the different types of causation and statements inherent in different sciences. Also, the specific role an individual takes matters for his or her reconstruction and evaluation
- Knowledge integration is a key issue of transdisciplinarity. Knowledge integration should acknowledge the different epistemics from various participants and is best carried out using method-driven procedures (see Table 15.2).

15.7 Transdisciplinarity in action

This section discusses topics relevant to implementation of transdisciplinarity. We first discuss how some scientists, as well as decision-makers, need to overcome their fear of losing control of the transdisciplinary process to become effective facilitators for mutual trust-building and collaboration. We then discuss the need to avoid social contamination of science, prevent misuse of transdisciplinary processes, and recognize geographic as well as situational limitations in which transdisciplinarity can effectively be practiced. Next we note how, as a relatively new approach, transdisciplinarity processes have yet to undergo many rigorous evaluations. Finally, we discuss the critical work of positioning the university as a change agent in transdisciplinary processes. Here, we envision sustainability as the engineering task of the twenty-first century and the university as an agent for change.

15.7.1 Preparing scientists for transdisciplinarity

Transdisciplinarity not only means *science for* but also *science with* society. When participating in collaborations to address today's global environmental problems, scientists leave their key domains of competence and have to cope with new research perspectives.

> The global environmental problems we face are so immense that no single company, nation, or region can deal with them effectively. Global cooperation must extend from one country to another and between institutes and universities. Researchers speaking many different tongues have to learn to work together and find common goals. This is more easily said than done. (Eliasson, 2003, p. 8)

True collaboration is a critical prerequisite for establishing "ideal-typical" transdisciplinary processes, including the public at large and various stakeholder groups. A barrier to such collaboration could come from concerns from (normal) scientists about losing control over what becomes a matter of investigation. Similarly, decision-makers sometimes fear losing control of decision-making in transdisciplinary processes. Building mutual trust and finding a suitable process for joint problem definition and ratifying the guiding question and rules of collaboration can be effective in countering decision-makers' fears. Further, it is of course essential that scientists refrain from political persuasions and stay within their functional role as providers of valid knowledge and truth.

15.7.2 Avoiding social contamination of science

> [The contribution on the New Production of Sciences] point to the end of disciplinary science, universities, laboratory-rooted research, and differentiations between scientific knowledge *per se* and society. (Shinn, 2002, p. 609)

This has been the message of books on the post-normal variant of transdisciplinarity (Gibbons *et al.*, 1994; see Table 15.1; Nowotny *et al.*, 2001). Many papers have been written about balancing science research between relative autonomy and its absorption by other interests or frames of thought (Abbott, 2002; Shinn, 2005). As we have mentioned above, the relationship between science and policy changed historically and differs between countries (Lengwiler, 2008). Today, there is no doubt that the relationship between science and society must be rethought and considered under the perspectives of the requirements of the upcoming post-industrial society.

From our perspective there are many reasons why scientists participating in transdisciplinary processes should take care not to enter the often-postulated "agora" where *all* scientific statements are haggled over as though at a medieval market, but with no safety net. In practice, there is often a huge societal pressure and need to arrive at solutions quickly. However, scientists should not be prevented from conducting basic research and theory development that expands the core of the body of scientific knowledge, work that allows scientists to bring unique value to transdisciplinary processes. Thus, scientists must be ever-vigilant, as stakeholders, government or industry collaborators can, albeit sometimes unwittingly, attempt to shape the internal development of science. In this way, "transdisciplinarity risks social contamination of scientific activities" (Scholz & Marks, 2001, p. 249). This risk is linked to the legitimization function (see above) and is shaped by a short-sighted understanding of how basic sciences can be co-opted to serve societal needs. The view that only science which provides direct societal benefit should be funded is imprudent. Here the example of number theory can help. It has been a domain of pure mathematics since ancient time in Greece, India, etc. This seemingly useless playground of the mind became overly important for coding algorithms and we cannot now imagine that advanced communication theory could exist without number theory (Yan, 2002).

Business practitioners, governments, and scientists have different societal roles and functions. Further, different stakeholders and science do not only follow different interests but also often follow different timescales.

> Reliable knowledge may have been best produced by such cohesive (and therefore restricted) scientific communities. But socially robust knowledge can only be produced by more sprawling socio/scientific constituencies with open frontiers. (Gibbons, 1999, p. C84)

We agree, but argue that it needs disciplinary academic culture to provide solid ground for interdisciplinary and transdisciplinary activities. This is one reason why the reader can find five disciplinary parts in this book. Each of them thoroughly discusses disciplinary theories and findings that can be used in different settings and transdisciplinary processes. We see transdisciplinary processes as a form of mutual learning in which practitioners and scientists temporarily cooperate to cope with relevant societal problems.

Another critical aspect, from our perspective, is the claim for "joint research" or "joint decision-making." Clearly, practice at large can reasonably participate in research processes and scientists contribute via consulting to make decisions. However, in the end, practitioners are not responsible for the quality of scientific papers, nor scientists for the final decision of a legitimized decision-maker (though the involvement in transdisciplinary processes implies accountability and responsibility). The problem with completely dissolving boundaries between science and practice is underscored in an early critique of action science of an undifferentiated type of theory–practice relationship (see Figure 15.8b):

> If the teacher becomes a researcher there is no reason why she should not be made to become a guidance counselor, nurse, musician, or psychiatrist, in the name of "professional responsibility." (Hodgkinson, 1957, p. 142)

15.7.3 Preventing misuse and abuse of transdisciplinary processes

The manner in which transdisciplinary processes are initiated can be influenced by (intellectual or financial) power. Here the question of who has the capability to facilitate an effective process is critical. The answer to this question may depend on one's conception of democracy. Critical questions here are how to prevent abuse and falsely claimed legitimacy, or how a false consensus can be prevented. A robust strategy involves a legitimized leader from practice (e.g. the president of a community or state or the CEO of a company), that the objective of problem orientation is dominating, and that a learning space is created which allows the decision-maker and the stakeholders to better cope with the problem at hand without explicitly preparing the final decision in detail. This can be accomplished if orientations are based on a type of socially robust approach for systems analysis and priority setting.

Transdisciplinary processes are influenced by the different types of involvement of stakeholders and financial resources. Here the interests, values, and potential hidden agendas may lead to science being used in an explicit or implicit way. This holds true for transdisciplinary processes but also for other research processes. Research is primarily performed in domains where money is available explicitly for research funding. Whether similar principles are present in transdisciplinary processes should be critically reviewed.

15.7.4 Constraints of transdisciplinarity

The idea of "enlightenment," where laypersons and citizens are not disenfranchised and are considered to have an intellectual competence, is a prerequisite for transdisciplinary processes (Allenby, 2009). Given this cultural requirement for transdisciplinarity, it is probably not surprising that transdisciplinary projects have developed furthest in countries such as Switzerland, Sweden, and the Netherlands (Pohl, 2008; Scholz et al., 2006). Interestingly, there are examples to be found outside the western hemisphere in places such as Bhutan (Gurung & Scholz, 2008) or the Seychelles (Stauffacher et al., 2006) where transdisciplinary processes are possible and are successfully applied. However, the author of this book experienced difficulties in establishing transdisciplinary processes in other cultures and political contexts, such as in China or Great Britain, owing to different cultural discourses and fear of process distortions by aggressive media. Whether the approach can be applied to other sociopolitical contexts will certainly be explored in the future.

15.7.5 Evaluating transdisciplinary processes

As a relatively new knowledge integration approach, transdisciplinary processes have rarely been critically questioned with respect to effectiveness and efficiency, as few evaluations have been initiated up to now (Bergmann et al., 2007; Maasen & Lieven, 2006; Walter et al., 2007; Zierhofer & Burger, 2007).

Transdisciplinary processes are usually exceedingly complex processes. They are unique and specific with respect to the constraints and contextualization of both (a) the real-world problems dealt with, and (b) the transdisciplinary process and its participants. Though it is difficult to find empirical methods to evaluate the outcomes, Walter et al. (2007) were able to show that "network-building" and the acquisition of "transformation knowledge" promoted "real" decision-making capacity in large-scale transdisciplinary processes among decision-makers and stakeholders involved. There is further evidence from the many extensive reports, that case studies are wanted by decision-makers from governmental institutions in countries such as Switzerland, Sweden, and Austria. This request can be interpreted in the way that transdisciplinary processes are at least perceived by relevant societal actors as efficient and effective means to approach complex societal problems.

15.7.6 Making the university a change agent

Taken at face value, proposing that the university become a change agent might seem to contradict our precautions about avoiding social contamination of science. Along with the privilege and opportunity of teaching and conducting research, universities have a responsibility to support society with researchers' skills, information, and discoveries. This holds true for private universities as well. In many countries, such as Switzerland, sustainability and environmental protection are part of the (federal) constitution. Transition management towards adapting to climate change, peak oil, closed fertilizer loops etc. can be considered examples of the kinds of societal tasks that universities can support. Moreover, finding new pathways toward more sustainable practices and lifestyles can be considered an issue to be tackled by transdisciplinary processes as society-wide priorities and perspectives shift (Kates et al., 2001).

Clearly, the university has a hybrid mission. The Massachusetts Institute of Technology, for example, has always been an institution with close connections to industry, a place where the private sector often goes to get advice. However, attaining socially robust solutions for sustainable transitions encompasses new directions and conceptions of research, teaching, and university activities.

We would like to end this chapter by recalling a vision. The classical vision of an engineer at the beginning of the twentieth century was that of a professional who designed engines, buildings, and infrastructure such as bridges or technologies that convert chemicals and information for human use. Thus, the construction of a technical solution was the goal. The major task of "engineer-like" scientists at the same time was to anticipate what impacts would result from new technologies and societal changes. The latter is an objective of sustainability sciences and transdisciplinary processes, as well as mutual learning. Environmental literacy is a prerequisite for finding satisfying solutions. We thus conclude that sustainability is the engineering task of the twenty-first century. If this becomes real, the university truly becomes an agent for change (Stephens et al., 2008).

15.7.7 Key messages

- Mode 2 shares communality with the Zurich 2000 definition. It implies a certain type of transdisciplinarity. A critical question is whether it reasonably should replace or complement disciplinary, inner science-driven Mode 1 activities
- Transdisciplinarity is not only *science for*, but also *science with* society
- Transdisciplinarity can be misused and risks social contamination of science
- Transdisciplinary processes involve different epistemics and rationales
- Transdisciplinary processes require that scientists, legitimized decision-makers, and the public at large (stakeholders) develop trust and put aside their concerns about losing full control over the complete transdisciplinary process
- The effectiveness and efficiency of transdisciplinary methods should be evaluated. This asks for the development of new evaluation methods
- Transdisciplinary processes on sustainable transitions can be conceived as a kind of new "engineering task of the twenty-first century"
- Transdisciplinary processes can be considered an efficient means to attain environmental literacy.

Part VIII

A framework for investigating human–environment systems (HES)

16 The HES Postulates   407
17 The HES framework   453
18 Applying the HES framework   463
19 Comparing the HES framework with alternative approaches   509

Key questions (see key questions 1–3 of the Preamble)

Contributions from transdisciplinarity to environmental literacy

Q1 How can researchers or transdisciplinary groups define, investigate, and access environmental systems that are inextricably related to human systems?

Q2 What rationales and drivers do we find in different human systems? What drivers, dynamics, and interactions are perceived in environmental systems? How can the HES framework help to conceptualize and better understand interacting human and environmental systems?

Q3 How can we draw on the unique value of various disciplines to analyze HES in an interdisciplinary way? What is the role of transdisciplinary processes here?

What is found in this part

Part VIII presents the HES framework and seven Postulates upon which it is based. The main function of the Postulates and the framework is to assist in coping with the complexity of most relationships within and between human and environmental systems.

The seven HES Postulates of Chapter 16 explicate the ontological and epistemological approach for dealing with HES drawing on the unique value of grounded disciplinarity. Research based on these Postulates allows us to describe the different *rationales* inherent in different human systems and environmental systems and their interactions. The Postulates help to conceptualize the structure and dynamics of human–environment interactions in a way that disciplined interdisciplinarity becomes possible by efficiently using the potential of natural, social, and engineering sciences.

Chapter 17 shows how and why the HES Postulates present a basic framework for investigating and understanding the structures and processes of HES. A major strength of the HES framework is its structuring potential, which supports the understanding of complex HES, particularly in an era when many processes on Earth are shaped by human activities. Here we make reference to the term *anthropocene* and refer to the anthropocenic redefinition of the environment suggested in Chapter 14, within the frame of complementary, inextricably coupled systems.

Chapter 18 proves how the HES Postulates can be used for research and sustainable transitions in various fields. Four case studies are presented. In each of the case studies all Postulates are applied to gain a comprehensive and thorough view of the human–environment interactions. Case 1 is on pandemics, case 2 deals with the future of traditional industries in periurban areas, case 3 on proper use of agro-fuel, and case 4 with the trade-off between keeping fertilizers, in particular phosphorus, in the food chain through animal feed yet preventing epidemic infections. We show how the framework can be used in transdisciplinary processes.

Finally, Chapter 19 compares the HES framework with alternative approaches to better understand the potentials, limits, and added value of the HES framework. We discuss similarities and differences seen in alternative frameworks for dealing with HES.

Part VIII — A framework for investigating human–environment systems

Chapter 16 — The HES Postulates

- 16.1 Postulates for what?   407
- 16.2 Complementarity Postulate P1   410
- 16.3 Hierarchy Postulate P2   413
- 16.4 Interference Postulate P3   427
- 16.5 Feedback Postulate P4   429
- 16.6 Decision Postulate P5   443
- 16.7 Awareness Postulate P6   445
- 16.8 Environment-first Postulate P7   448

Chapter overview

Starting from the epistemological assumptions presented in Chapter 3, Chapter 16 presents seven Postulates for structuring human–environment relationships under the perspective of improving environmental literacy.

These Postulates serve different functions and can be grouped differently. For example, the Complementarity (P1), Hierarchy (P2), Interference (P3), and Feedback (P4) Postulates enable proper modeling of coupled systems. In combination with the Hierarchy and the Interference Postulates, the Decision (P5) and the Awareness (P6) Postulates are central for understanding the drivers of social systems when making reference to key social science disciplines (see, for example, Chapters 6–11). These two Postulates allow also for a game- and decision-theoretic modeling of human systems such as individuals, groups, companies, institutions, or societies when they interact with each other and their material environment. Finally, the Environment-first (P7) Postulate stresses that each analysis of a HES should start with a thorough understanding of the functioning dynamics of the environment and its responses to human decisions. This highlights the nature of inextricably coupled human and environmental systems and acknowledges unwanted rebound effects (P4), which are ideally avoided.

16.1 Postulates for what?

Investigating environmental literacy requires dealing with multiple complexities. On the one hand, there is the *environment*; or better, a multiplicity of environments. The primary distinction among environments is between the material–biophysical components (which include the built environment) and the socio-epistemic dimensions. There are further distinctions, for instance between the abiotic and biotic to denote layers of the material–biophysical or between the visible and invisible environment to indicate what parts can be perceived by the eyes and what parts need other sense organs or technological aids to perceive them. Likewise, there is a myriad of *human systems*. The world human system now includes nearly seven billion people, acting as individuals and as parts of societies, institutions, nations, or organizations at different stages of societal development. As social systems are part of the environment, environmental literacy calls for looking at these systems over different scales, ranging from the individual to the world population.

As statements and assumptions that function as the basis for our argument, we can consider the proposed Postulates to be practical devices for exploring the relationship that humans have with their environment. Whether the Postulates are right or true cannot be proven; they have to be considered as a kind of dicta or rules ("Maxim (Regel)," Kant, 1796).

> A postulate is a practical, directly evident statement or possible principle, which regulates possible actions, for which we suppose that the way that they are performed is immediately certain. (Kant, 1800, Log. § 38 (IV 123), translated by Scholz)

The seven HES Postulates are derived from two decades of theoretical, methodological, and practical work in many fields of sustainability science and environmental science, including hundreds of operational, hands-on projects of transition towards more

Table 16.1* HES Postulates: Postulates of the human–environment systems (HES) framework.

Number	Label	Contents
P1	Complementarity	Human and environmental systems are complementary
P2	Hierarchy	Human and environmental systems both have hierarchical structures
P3	Interference	There are disruptive interactions among and within different levels of human and environmental systems, in particular between the micro and macro levels
P4	Feedback	There are different types of feedback loops within and between human and environmental systems
P5	Decision	Human systems can be conceived as decision-makers who have drivers and who act to satisfy goals
P6	Awareness	Human systems have different types of environmental awareness
P7	Environment-first	The effective analysis of inextricably coupled human and environmental systems, as well as the planning for sustainable human–environment interactions requires a thorough analysis of the material and social environment that builds the matrix of HES

sustainable systems at the level of the community, municipality, region, private, and public company or organization. Most of these projects have been performed by the Natural and Social Science Interface (NSSI) research group of ETH Zurich (Scholz et al., 2006). The HES framework and how it may serve a better understanding of complex human–environment interactions is presented in Chapter 20 (Scholz & Binder, 2003, 2004).

The HES framework is not a scientific *theory*, in the sense that it does not provide a set of definitions, principles or theorems to explain specific phenomena. The HES framework is rather a *heuristic tool* for structuring the investigation of human–environment interactions and for trying to cope with their complexity. The framework provides operative concepts for an organized exploration of a certain domain. Although this framework is not tightly enough organized to derive precise predictions, we think that it contributes to theory-building in the sense of:

> A realistic theory of society–nature relations must be applicable to a wide range of historical conditions and long-term mutual feedback between society and nature. (Fischer-Kowalski & Weisz, 1999, p. 217)

The first four Postulates represent epistemological assumptions and refer to systems theory. The Complementarity and Hierarchy Postulates P1 and P2 (see Table 16.1*) assert that HES are coupled systems with hierarchical structure. The Interference Postulate P3 states that there are interactions among and between different hierarchical levels, from the micro to the macro level (see Box 16.2). Identifying and analyzing the different levels of human systems are specific challenges for establishing dynamics of sustainability. This is because of the differentiation of the levels of human system in the course of phylogenetic and historical development (Chapple & Coon, 1953, see Box 16.3). The complexity of human societies depends on the degree of development (see Figure 8.5* and Box 16.1).

The Feedback Postulate P4 describes proactive environmental literacy, in that it describes how anticipation of secondary feedback loops is crucial for sustainability learning. This postulate supports the investigation of the question: under what constraints do human systems adapt, or fail to adapt, to unintended environmental impacts?

Two further Postulates are rooted in basic assumptions on how human–environment interactions can be conceptualized. The Decision Postulate P5 refers to functionalism (Chapter 3.5) and assumes that all human systems act purposefully and that actions are based on goals and utility functions (Chapter 7.2). The way in which systems make choices can be described with a decision- and game-theoretic approach.

The Awareness Postulate P6 considers different types and levels of sophistication in environmental awareness. It focuses on the prerequisites of knowledge and attitudes for a sustainable interaction of human systems with the environment.

Finally, the Environment-first Postulate P7 emerges from a realist stance. It states that purposeful research on human–environment interactions should be based on an analysis of the material and of the social envi-

16.1 Postulates for what?

Box 16.1 The human animal is building environments: small huts and Las Vegas

Etymologically, anthropology refers to the comparative study of humans, which researchers approach in a number of ways, including studying the development of human beings in different areas of the world in relation to physical character. This is the view of how the environment affects human life.

"The human animal is a hairless, tropical primate, and is not comfortable in a temperature of much less than 70° Fahrenheit, without clothing. He is omnivorous, in that he eats both meat and vegetable matter, but he cannot eat grass and his vegetable diet is limited to fruits, seeds, roots or tubers, and a few kinds of succulent leaves. He normally requires food every day, and has no bodily storage facilities. If he goes more than two or three days without food, the results are serious. They are even more serious if he goes without water. Although he needs water every day, the amount required depends on the temperature and humidity of his environment, and on the nature of environment, and on the nature of his activities" (Chapple & Coon, 1953, p. 86).

This portrayal makes it clear that human environments must meet basic physiological needs. But humans also adapt to and change the environment. Living in a desert environment depends on technologies that allow for utilizing the environmental opportunities and resources. Targeting the transition from tribal to modern societies, the following examples illustrate how anthropologists in the 1940s saw human–environment interactions in various climate zones.

"The Dry Lands: the opportunities offered for human occupancy by the dry lands vary … In completely waterless sections, as in the Libyan desert, this land is wholly unoccupied except by travelers following caravan routes and carrying their water with them" (Chapple & Coon, 1953, p. 87).

"The Tropical Forest Land: … the tropical forest provides no difficulty for the human organism in respect to temperature and drinking water, it supplies insufficient food, except with the use of advanced agriculture techniques" (Chapple & Coon, 1953, p. 88).

"The Mid-Latitude Mixed Forest Lands … in Europe and Asia, were looked upon by the Greek as too cold and inhospitable for human occupancy, and were avoided by the highly civilized peoples of the Near East in antiquity" (Chapple & Coon, 1953, p. 89).

Historically, natural landscapes without human alteration have sustained human life. But human technology can establish an artificial human-built environment and overcome the pristine settings (illustrated in Figure 16.1). Technology and societal trajectories have undergone a tremendous change since the survey by anthropologists Chapple and Coon. Salient examples are Las Vegas, the plantation agriculture of Indonesia, or the Dutch people's living below sea level. Humankind is building and fundamentally changing the natural environment. These systems can only be maintained by highly complex social and technological systems, an extraordinary change of the natural environment, and high consumption of natural resources (see Box 14.1). How a sustainable environment should look is one of the challenging questions of the twenty-first century.

Figure 16.1 (a) For relatively isolated tribes like the yanomami people in Brazil (© Fiona Watson/survival) the tropical and dry lands allowed only for a small population density of about 1–20 persons per km^2 (e.g. indigenous communities in South America; The Field Museum, 2003; Toledo *et al.*, 2003). (b) In the year 2009 the city of Mumbai (photo by Sebastian Jude) had more than 22 000 inhabitants per km^2 (World Gazetteer, 2010).

ronment. An effective environmental analysis depends upon a sound problem definition and understanding of the environmental problem under scrutiny.

16.1.1 Key messages

- The HES framework is a tool for analyzing, and gaining a better understanding of, interactions between human and environmental systems
- The postulates can be grouped according to how they are formulated: systems theoretical postulates (P1-P4), decision theoretical postulates (P5-P6), and one methodologically formulated postulate (P7).

16.2 Complementarity Postulate P1

Postulate 1 states that human and environmental systems are complementary. We first review how various disciplines have investigated the human–environment relationship. Next, we discuss the area of coupled systems research. Finally, three examples illustrate the practical value of viewing situations and systems as complementarities.

16.2.1 Research tracks for studying HES

Since Spencer's definition of the total environment (Spencer, 1855; see Chapter 1.3), many disciplines have included the human–environment relationship. Some disciplines, such as geography, have separate subfields, such as social and physical geography. Others, such as human ecology (Barrows, 1923; Bresler, 1966; Hawley, 1944; Sargent, 1974) or anthropology, are rooted in both social and natural sciences. Often these sciences were segregated into subdisciplines that only partly dealt with relations with the other systems.

An early modern approach dealing with the complementarity of human and environmental systems has been called "biohistory."

> Biohistory is defined as a coherent system of knowledge, or field of study, which reflects the broad sequence of happenings in the history of the biosphere and of civilization in the beginning of life to the present day. (Boyden, 1992, p. 3)

This research came out of the UNESCO's Man and Biosphere program, starting in the 1970s, while investigating one of the most congested areas of the world, Hong Kong and its hinterland. Boyden considered a double complementarity (see Figure 12.5). The first is between nature and human society. This complementarity is not well defined. For instance, which biophysical processes in human bodies belong to what side of this complementarity is an open question. Second, a complementarity was seen between abstract culture, including "beliefs, assumptions, values, world view, knowledge, understanding, etc." and the "biophysical actualities and society," and the cultural arrangements, including "economic system, legislation, social hierarchies, educational system, etc." (Boyden, 1992, p. 98).

An important step of environmental literacy was the pressure–state–response approach in the late 1970s. This approach followed the feedback loop thinking that human activities are considered as *pressure* which alters the *state* of the environment which in turn requires a *response* of behavioral adaptation to the new environmental settings. This was a framework for capturing and statistically accessing causalities and feedbacks between human and environmental systems (Organisation for Economic Co-operation and Development (OECD), 1993).

In the follow-up to the 1980s environmentalism movement, scientists concerned about the global environmental problems caused by human activity called for investigation of human–environment interactions.

> We need a second environmental science – one focused on human-environment interactions – to complement the science of environmental processes ... (Stern, 1993, p. 1897)

The key topics touched by Stern have been population growth, energy and CO_2 emissions, environmental degradation, and also poverty in the north–south frame. As a psychologist, Stern initially focused on forces that drive human activities. He proposed that research should focus on interventions to prevent environmental decline. Further, future scenarios of environmental development should be of interest. Presumably because of a basic behavioral psychological perspective, Stern (Gardner & Stern, 2002; Stern, 1993, 2000) focused on the individual and did not systematically conceptualize feedback loops. The environment of an individual human being consists of a *material* environment, a *biophysical* environment (e.g. the bacteria, plants, animals, people in our vicinity), and a *technological* environment, (e.g. the cars, elevators). Additionally, the people and their settings represent a social environment. This is nicely expressed in the following statement:

> "Environment" must be broadly construed to include things as the physical environment of urban ghettos, the

human behavioral environment, and the epidemiological environment. (Ehrlich & Holdren, 1971, p. 1212)

A different, but similarly important, contribution has been provided by the ecologists Gunderson and Holling (2002; see Chapter 5.6). For them the relationship between ecology and economic growth has been critical. Their starting points were theories of panarchy. These models come from a biological background. Holling and co-workers distinguish between ecological, economic, and social systems (Holling et al., 2002) and distinguish four cyclic stages of system development: release (or innovation, collapse), reorganization, exploitation, and conservation (see Figure 5.16). They focus on two interactions of ecosystems, one with economy, the other with ecology. The relationship is viewed from a coevolutionary perspective (Folke, 2007), whereas sometimes the understanding of a managed ecosystem emerges (Reusswig, 2007). The vocabulary of social–ecological systems (Folke, 2007; Folke et al., 2005) is also used in other domains such as hazard management (Gardner & Dekens, 2007) and the study of how cognitive processes relate to the environment (Beratan, 2007; Broderick, 2007).

A fifth track to HES is presented in this chapter. It is rooted in Brunswik's (1952) approach for understanding human systems as intentional, purposeful systems and also in the assumption that the human systems have a different rationale than the material environment. The latter aspect is further tied to our Hierarchy Postulate P2.

> Human decision making is considered a key factor in this model, as humans can regulate and control the type of interaction within HES. ... The environmental system reacts and gives feedback to human action, allowing for humans to learn and adapt their behavior. (Scholz & Binder, 2003, p. 1)

As we will see below, we consider the conceptualizing and mastering of different types of feedback loops essential for environmental literacy. For instance, we conceive sustainability learning to be a result of providing proper responses to second order feedback loops.

16.2.2 Coupled systems research

Coupled social–ecological systems may have very different dynamics when compared with the ecological component because the social domain contains the element of human intent. Currently, the complementarity of human and environmental systems, integrative human–environment research, integrative modeling, and assessment runs under the label of "coupled systems research." The focus is on reciprocal "effects and feedbacks" (Liu et al., 2007, p. 639; Scholz & Binder, 2004, p. 792).

In Chapters 3.2 and 3.3 we discussed different notions of complementarity. One can glean from psychological theories such as ecological theory (Gibson, 1979), ecopsychology (Barker, 1968), or the bioecological theory (Bronfenbrenner, 1979), that the relationships between humans and the environment go far beyond the simple set theoretical relationships: human systems properties and aptitudes fit into environmental structures and vice versa. This is also called transactionalism and can be applied both to interactions with the material and the social environment (Stokols & Shumaker, 1981). Others, such as Lovelock's Gaia hypothesis (Lovelock, 1995, p. 144) view the Earth as a single organism, in which the individual elements coexist in a symbiotic relationship.

Against this background, we can take a closer look at the human–environment complementarity and the material–biophysical vs. social complementarities. With the first, we assume that human and environmental systems both represent different subsystems of the universe. As human and environmental systems have different rationales, which we want to understand, we consider them as two separate but coupled systems. Investigating one in isolation, however, does not make sense, at least not from an environmental literacy perspective.

Both human and environmental systems include a complementarity of material and social dimensions, and these dimensions represent one and the same entity. This is in line with Heisenberg's position with respect to wave–corpuscular duality of light:

> We have two contradictory pictures of reality; separately neither of them fully explains the phenomenon of light, but together they do. (Einstein & Infeld, 1938)

The material–social complementarity, which, on the level of the individual, is also called body–mind duality, is captured well by the figure-ground phenomenon (Figure 16.2). While we can certainly see both images, one at a time, by changing our viewpoint, it is impossible to see the structure (and meaning) of both at once. At the same time, one could not exist without the other. However, this metaphor also seems limited, as the meaning of the relationship of the complementary pictures seems arbitrary, which is not so for the body–mind hypothesis or the wave–corpuscular duality.

Figure 16.2 The figure-ground phenomenon; it provides insight into one type of complementarity. You can define the common base, where two entities/meanings emerge. It is impossible to see both simultaneously, but we can switch to and fro. The picture that dominates depends on the perspective of the observer. If you focus on the middle, the vase appears. If you defocus, the faces build the starting point.

Some physicists (Capra, 2000) state that eastern physicists, contrary to their western colleagues, had fewer difficulties accepting the wave–corpuscular duality. When explaining this phenomenon, Hideki Yukawa (1907–81) expressed it using the words "… we in Japan have not been corrupted by Aristotle" (Rosenfeld, 1963, p. 47). We can gather from this quote that Aristotle metaphysics claims that "substance" is a combination of *matter* (i.e. the material stuff; also called potentiality) and *form* (i.e. the functions a substance actually takes; also called actuality; Aristotle, 350 BC, pp. 1045a–b). But also for Aristotle, both matter and form constitute a *unity* (Cohen, 2003). In contrast, Taoist philosophy is related to yin–yang opposites such as masculine–feminine, sunny–shady, heavy–light, rational–intuitive, so that particle–wave is just one of the ubiquitous complementarities of the world.

We can see from the above text that the human–environment and the material–social complementarity have fundamentally different notions. This is primarily because the first complementarity is conceived as two physically separate entities, whereas the second deals with two interpretations of one and the same entity.

16.2.3 Examples for complementarity in research and action

Through three examples, we illustrate the practical value of the Complementarity Postulate P1. The first refers to human health research, which traditionally focuses on the environment (E) to human (H), $E \rightarrow H$, direction. The second example reveals how the Complementarity Postulate underlies research in environmental evaluation, where the critical relation $H \rightarrow E$ is often employed for assessing the impacts of human activities on the environment. The third example illustrates how mitigation of and adaptation to climate change can be best conceived using the interaction of the HES (i.e. $E \leftrightarrow H$).

Example 1 refers to the interaction between microorganisms and human beings. We define the material side of an individual as all cells that originate from the totipotent stem cell; the fusion of ovum and sperm that has the capacity to form an entire organism. Based on this biological definition, the microorganisms that enter the baby's body, when leaving the amniotic sac, are guest organisms. These guests outnumber human cells by 10 to 1 and amount to about 2 kg (ScienceDaily, 2008). Some are helpful and necessary for survival, while others, called pathogens, cause diseases. A few of the pathogens, such as the plague or the malaria germ, threaten human life. Human cells, supported by "helpful" bacteria can in most cases recognize and neutralize invading germs. The immune system consists of an elaborated network of molecules, proteins, antibodies, cells, fluids, organs, and tissues, interacting in an elaborate and dynamic network (see Chapter 5.5), which adapt to bacteria. Thus the human body can become immune by learning. Likewise, pathogens adapt and change strategies. We take a closer look at this coevolution in case 1 of Chapter 18.

The significance of the complementarity between human and environmental systems can be demonstrated by the placebo effect in the case of some infectious diseases (Bausell, 2007). When a patient is treated by an authority or by acquired information, the human body (cells, tissues, organs, organ systems) can show improvement in response to the non-material, social impacts given by the cognized presence of a medical expert (doctor or nurse), treatment information or to medication containing no active substances. The cells belonging to an individual are following one program and those of the pathogens another.

Example 2 refers to the environmental burdens caused by products, from production through usage and disposal, which can be evaluated through a life cycle assessment (LCA) approach. LCA assesses a functional unit of a product, such as 1 kg of New Zealand lamb consumed in Switzerland (International Organization for Standardization/European Standard (ISO/EN), 1997). The proper definition of a functional unit is one of the challenges of environmental evaluation, because essentially the utility has a subjective component (Werner, 2005). Thus, 1 kg of New Zealand meat can be judged differently compared to similar Swiss meat with respect to environmental impacts. A second challenge for LCA is the definition of the environmental burdens. A common approach is to refer to safeguard subjects, which are entities of the environment affected by the functional unit of a product. An example is the consumption of 1 kg of lamb, whose utility is judged by its consumer (or by the constructer of the LCA), according to his or her own values. Here again the values of the consumer (i.e. elements of his or her cultural system) are related to effects in the material environment (i.e. of human health of those people not consuming the meat), ecological health, and destructive resource use (Goedkoop et al., 1998). This relationship highlights the advantages of a complementary understanding of sociocultural systems and the environment.

Example 3 refers to climate change, one of the major challenges of the twenty-first century, stemming from dramatically increasing greenhouse gas emissions caused by activities of human systems. The complementarity between human and environmental systems can help to better elucidate the essence of the concepts mitigation and adaptation. Mitigation involves the programs to reduce greenhouse gas emissions and to enhance sinks to reduce global warming. Thus, it refers to human programs affecting the environment, the $H \rightarrow E$ relationship, while adaptation refers to the other direction: $E \rightarrow H$.

Here, adaptive capacity, the potential of a human system to adapt, relates directly to the environmental hazard (or potential hazard), such as climate change. The resulting impact is defined by the parameters of exposure and sensitivity in certain geographical settings. Exposure refers to the magnitude of the influence of the environmental change on a human system, whereas sensitivity refers to the degree to which a system will respond to this influence. Adaptation analysis thus requires the definition of those environmental impacts that are critical for human systems. Because climate change refers to manmade changes, adaptation can be considered a secondary feedback loop (see Postulate 4; Chapter 16.5).

16.2.4 Key messages

- A material (–biophysical) definition of human systems ("Omne vivum ex ovo;" Figure 3.3*) provides a coherent approach for consistently conceptualizing the complementarity between human and environmental systems from the level of the cell to the human species
- The material–social (body–mind) complementarity is a second, significant complementarity referring to the human–environment relationship (see also Chapter 3), which is essentially different from the human–environment complementarity.

16.3 Hierarchy Postulate P2

The discussion of Postulate 2, how human and environmental systems can be investigated according to hierarchies, begins with a discussion of hierarchy theory. Following this is a review of four basic types of hierarchy used to describe systems. Next is a discussion of the epistemology and ontology of level hierarchies. Here a critical question is whether we utilize the hierarchy concept for organizing our own knowledge of the world or we consider the hierarchy as "a model of how living systems are self-organizing and functioning" (Pavé, 2006, p. 39). The latter includes clarifying whether hierarchies are genetically predefined or whether the hierarchies that are emerging depend on the situation. We then explore control and level hierarchy of human systems and, finally, describe a series of hierarchy levels of human systems: individuals, groups, organizations, institutions, societies, and supranational/suprasocietal systems.

16.3.1 Hierarchy theory

Similar to complementarity, the term hierarchy is applied and defined in social systems, music, language, mathematics, computation, physics, and biological systems, just to mention a few. Because the term hierarchy has been used differently by researchers in these fields, we first define a concept of hierarchy that is useful for conceptualizing HES.

The basic assumption of hierarchy theory is that living systems have to be structured if they become too

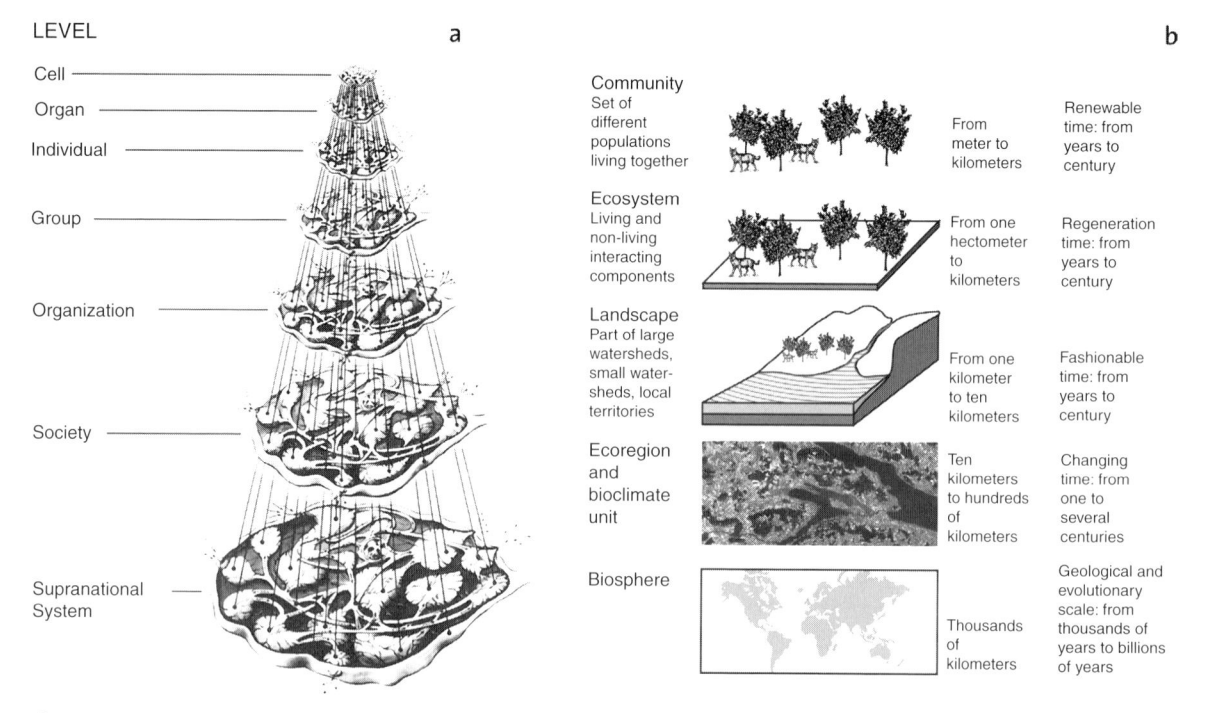

Figure 16.3 Hierarchical systems. (a) Miller's levels of generalized living systems. Miller identified 19 subsystems which process matter–energy, such as ingestor, matter–energy storage, or motor and information such as decoder or memory (that can be identified on all levels). These are connected by the lines in the figure (Miller, 1978, p. 4). (b) The ecological hierarchy.

big or complex and want to operate efficiently (Ahl & Allen, 1996; van der Pligt & de Boer, 1991). This theory emerged in the 1960s through the work of ecologists (Odum, 1996), biologists (Miller, 1978), physicists (Pattee, 1973b), developmental psychologists (Piaget, 1968), cognitive scientists (Simon, 1973), and engineers (Whyte et al., 1969), who wanted to understand how complex systems are organized and work. All these scientists were interested in the relationship between biotic and abiotic, as well as between natural and human systems.

The idea of hierarchy level has been influenced by the rise of genetics. The concept of integrative levels "recognizes as essential for the purpose of scientific analysis both the isolation of parts of a whole and their integration into the structure of the whole" (Novikoff, 1945, p. 209). The relations between the elements were considered at least as important as the nature of the element itself.

> Life is characterized by a hierarchy of structure and of functional control that spans the range from the minuteness of atomic and molecular interactions to the relatively enormous communication distances achieved by human society in ordering itself. (Grobstein, 1965, p. 38)

However, biologists acknowledged the different rationales of different levels:

> The rules governing the genetic structure of a population are, nevertheless, distinct from those governing the genetics of individuals, just as rules of sociology are distinct from physiological ones, although they are in fact merely integrated forms of the latter. (Dobzhansky, 1937, p. 11)

Figure 16.3 presents two examples of hierarchies. One is the seminal *living systems hierarchy* of Miller (Miller, 1978). The other is *ecological hierarchy* with hierarchy levels such as a biotic community that includes all populations of a designated area. It includes the ecosystem, which represents how life and Earth function together (Odum, 1996; Tansley, 1935), and also landscapes as parts of large regional units, ecoregions or ecoscapes, which denote a grassland region or ocean (Lidicker, 2008).

We take the hierarchy of human systems presented at the upper left side of Figure 14.1* as a reference. Although, from a system theory point of view, there are structural similarities between human systems on the level of the individual and above, single cells or unicellular eukaryotes, this chapter focuses on environmental literacy and human action above the level of the individual.

16.3.2 Definitions of hierarchy

A simple, formal definition of hierarchy was given by the biologist Grobstein:

> If S is a set consisting of identifiable components $A,B,...,N$ then the components make up a level of *order* when they are in a determinate association R which is the sum of relationships among the components.
>
> The set belongs to a system *which is hierarchical* when the set so defined has components which are also sets and when it itself is a component of a more encompassing set. Thus in a hierarchical system each component is a set $S = [A,B,...,N]^R$ where S itself is a component of a set and the elements A,B,C and N also have comparable sets. (Grobstein, 1973, pp. 31–2; italics not in the original text)

The nature of the relationship R is of interest. R defines the association (the neighborhood) of the system elements on a certain level (i.e. among different elements or components). We speak about levels if the ordered systems are nested (i.e. the system S is part of a higher level system, say $S^+ = [S,S_1,...,S_K]^{R+}$. A strong model describes living systems by building a nested hierarchy of units in which cells are elements of organs, organs are elements of organisms, organisms are elements of populations, etc.

Clearly, this definition depends on what the relationship R looks like. We distinguish four different types of hierarchy.

Order hierarchy: if the elements of S (e.g. persons) can be ordered according to the values of a variable (e.g. weight) assigned to them, we speak of order hierarchy. Since order hierarchy does not refer to relationships and interactions that comprise a hierarchy, not even on a certain level of order in the above definition of Grobstein, we do not include this notion.

Inclusion hierarchy: another somewhat less restricted notion of hierarchy is inclusion hierarchy, which refers to recursive operations. This concept was suggested by Herbert A. Simon (1973) in reference to the Chinese boxes, which are containers that contain other containers. The Chinese box can be considered a special case of order hierarchy, where the ordering number is simply the number of boxes that one opens before arriving at the particular box of interest. The same holds true for the Tower of Hanoi (see Figure 16.4b). The idea of recursion is that the same operation is repeatedly applied in a nested form, and such is also the case at the level of the (recursive) inclusion hierarchy.

Control hierarchy: a control hierarchy refers to a "control system in which every entity has an assigned rank, and all power is concentrated in the (usually

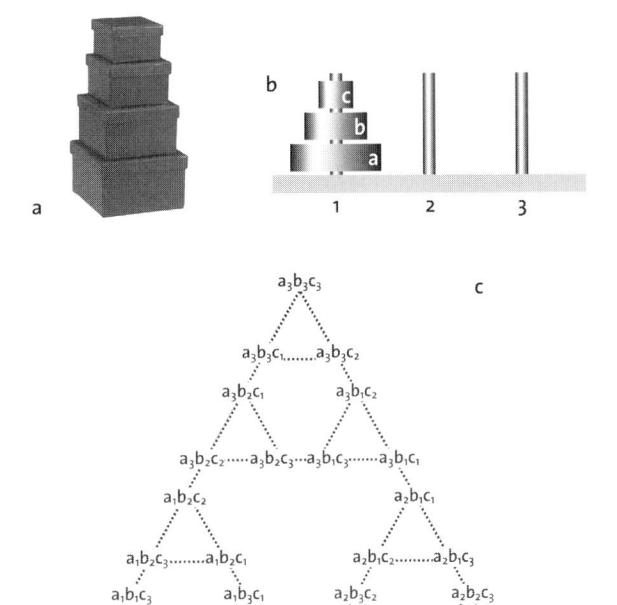

Figure 16.4 (a) The Chinese Box and recursive structures such as (b) the Tower of Hanoi are examples of the inclusion hierarchy. The Tower of Hanoi task requires one to find the fastest solution for packing a pile with N disks from peg 1 to peg 3, by only moving one disk at a time and without putting a larger disk on a smaller one. (c) The graphical solution. The lower left triangle represents the solution of $N = 1$ disk (moving disk c from 1 to 3). Tripling this operation (down left three triangles) you get the solution $N = 2$ disks. Tripling the latter, the solution for $N = 3$ results, etc.

assigned) entity with the highest rank" (Lane, 2006, p. 85). We speak of a chain of command, which can be conceived as a hierarchy of goals. The issue is that higher level goals control the settings for the subsidiary goals (Heylighen & Joslyn, 2001). Orders and commands flow downwards and requests flow upwards. These mechanisms dominate some social systems, such as churches, armies, and hierarchical companies. Technical systems can also be organized like this.

Level hierarchy: this concept refers to hierarchies in which different systems exist at different levels (see Figure 16.3). Entities and rationales on different levels are in some sense considered autonomous. There are physiochemical (e.g. elementary particles, atoms, molecules) or linguistic (e.g. letters, words, phrases, sentences, paragraphs, texts) standard examples (Polanyi, 1968; Trewavas, 2006).

There are entities of, and processes in, these systems through which they interact.

Entities at a given level may, through their interactions, construct and maintain entities at higher levels, and higher-level entities may be, at least in part, composed of lower-level entities: these are often described by the term upward causation. (Lane, 2006, p. 85)

The level hierarchies are characterized both by downward and upward causation. With downward causation, higher level entities change the lower level system. In organismic systems, we speak of upward causation when the individual members (e.g. of a group) affect the behavior of the upper level (e.g. change the group decision rules or split the group).

Some researchers, primarily those who do not distinguish between the above types of hierarchies or who consider hierarchies as outdated, propose that all living systems should be modeled as networks. Capra (1999), for instance, suggests that ecosystems should be represented as networks. The nodes should represent organisms. If each node could be blown up to a network of subsystems, then we would have a nested design. This would match nature, as there is no top or bottom. As one can see, Capra's idea is included in the above definition by Grobstein. A nested network that can be considered a system of systems is a special case of the hierarchy definition. One can also see that Capra's notion of blowing up related elements (e.g. atoms, cells) can be expressed in terms of levels.

16.3.3 Epistemology of level hierarchies

Hierarchy is a construct that can be defined from two perspectives. The objectivist view states that the laws of nature are entirely objective and determine the structural levels of matter (e.g. atoms, molecules, or cells). Matter is everything that has mass and occupies physical space. The subjectivist view assumes – even while acknowledging an objective reality – that our models of nature are structured by the level of description we choose (Pattee, 1973a). Some subjectivists hypothesize that hierarchy as a concept of "classification is itself a function of hierarchical systems of social order" or "hierarchical ideologies" (Ellen, 1979, p. 25), which may match with the scientists' world view. Thus the two views refer to: (i) whether nature is actually organized hierarchically; or (ii) whether we organize our perceptions of nature according to a hierarchy.

These two views show how basic assumptions about epistemology and ontology are intertwined. For instance, different schools of quantum physics, such as the Copenhagen, the Bohm, or the Multi-World Interpretation schools (Tegmark, 1998), make different basic assumptions on the ontology of the material universe. A critical issue in this field, which might well match the epistemic vs. ontological challenges of the hierarchy assumption, is the determinacy assumption. The questions read as follows. Is there true probability in Nature? Do we assume that Nature plays dice? Is the probability measure proportional to a measure of existence? The answers given by quantum physicists depend on basic epistemological and ontological assumptions underlying physical theories. For instance, the Multi-World Interpretation dissolves paradoxes of the Copenhagen School, such as the Schrödinger cat, by assuming that the world differs for different observers. It provides a deterministic theory for the physical universe. It is a "complete and consistent theory, which agrees with all experimental results obtained to date" (Vaidman, 2002). Multi-World Interpretation explains why a world appears to be non-deterministic for human observers.

The only possible meaning for probability is an ignorance probability, but there is no relevant information about whether an observer performing a quantum experiment is ignorant (Vaidman, 2002; see also Chapter 14.5). Yet there is also a subjectivist interpretation of quantum mechanics called *many minds interpretation*, in which the different worlds are only in the minds of sentient beings (Albert & Loewer, 1988). Objectivity would require that "the physical description of the system is objective because the definition of any later state is not dependent on measuring conditions or other observational conditions" (Faye, 2008, p. 1).

16.3.4 Ontology of level hierarchies

Hierarchies of higher aggregated human systems, at least those above the level of the organ, differ from hierarchies constituting atoms. Thus, no one would expect a consistent theory for hierarchies that applied identical constructs at all levels. However, the idea that systems of systems can be well conceived by level hierarchy (with each level having its own goals, drivers, and regulatory systems) is a basic ontological assumption made in this book.

If we want to describe the nature and structure of a hierarchical system and the relationships between system elements, we can do so assuming that a single fertilized cell can produce a fully functioning, organized complexity, with the same DNA in each cell of an organism. In other words, the information contained

16.3 Hierarchy Postulate P2

Box 16.2[1] **Level hierarchies in biological systems: controlled genes?**

Many biologists assume that life is characterized by a hierarchy of structure and of functional control (Golley, 1998; Grobstein, 1973; Pavé, 2006; Pumain, 2006; Smith, 2000; Trewavas, 2006). Fundamental questions of cell biology have been about how a complex organism develops from a single, totipotent cell and how DNA supplies information for growth, division, interaction, and control of different types of cells. Here one should note that there are around 10^{14} cells in an adult human organism. In addition, the organism knows (given normal constraints) where to stop growth development. The process of cell differentiation produces unipotent cells that can only express a certain number/pattern of genes (Alberts, 2008). Once developed, unipotent cells no longer have the ability to switch into other cell types, although they still contain all the DNA information.

The different cells of an organism are considered to use different portions of one and the same genome, contained in all of them. Thus, for instance, "the genes that make our hair grow are turned on on our head, and on in a few other places on our body, and are turned off in such places as the throat" (Bonner, 1973, p. 56). Many cell biologists explain this phenomenon through hierarchical control programs, which must have the capability of turning on the right genetic specifiers at the right moment in the right place. The material basics include multiple steps from DNA transcription to protein activation, including multiple regulation mechanisms. Thus part of the DNA is utilized for governing and coordination. These processes can be described by a model that shows level hierarchy with upregulation and downregulation.

Biologists assume that cells continuously test the (micro)environment and check external and internal signals to decide what function to take and what action is meaningful (Alberts, 2008). Bonner called this procedure "the hierarchical concept that a growing cell and developing organism is continuously performing tests of its environment." This testing ability differs between species. Some animals even have the ability to regenerate lost body parts; for example, the salamander axolotl (*Ambystoma*): when losing a limb, the cells remaining at the stump dedifferentiate into a stem cell state, similar to the state of the cells of an embryo growing limbs. The lost limb then regrows in processes similar to when an embryo grows its limbs (Gardiner et al., 2002; this is illustrated in Figure 16.5). Certain sea stars show similar capabilities and have the ability to regrow whole legs. Scientists have also been able to breed mice with scarless wound healing capabilities (cf. Masinde et al., 2005). The dedifferentiation and organization of the cells is a process not present in normal wound healing. Humans do not show such capabilities, with minor exceptions: children up to the age of 10 are reported to be able to regrow fingertips (Weintraub, 2004), and a liver can regrow from as little as 25% remaining tissue owing to the capability of self-regeneration of the hepatocytes.

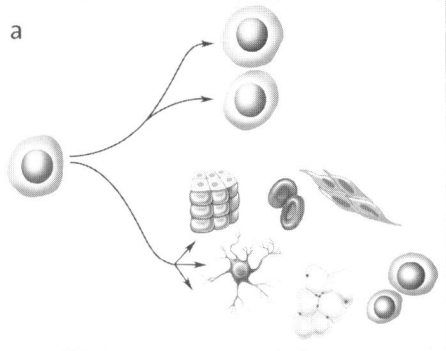

Figure 16.5 (a) The development of a human organism is a process of one totipotent cell developing into 14 trillion cells (from National Academy of Sciences, 2008). All cells have the same DNA but different functions. (b) In some animals, such as the salamander axolotl (*Ambystoma*), the cells contain programs to reproduce a whole lost leg (photo from Wikipedia).

---

[1] This box has been written together with Stefan Zemp.

within each body cell includes all information required to make the whole creature. An example is the way callus cells are produced at the root of a cutting to make a new plant. A similar view is taken by ecologists:

> The life processes take place in cells, because they have a sufficiently high specific surface to allow an exchange rate with the environment that is suitable. Cells are, therefore, the biological units that make up organs and organisms. Nature must, therefore, use a hierarchical construction …: atoms, molecules, cells, organs, organisms, populations, communities, ecosystems, and the ecosphere. (Jørgensen et al., 2007, p. 262)

16.3.5 Control and level hierarchy of human systems

An essential idea of the hierarchy concept is that hierarchy serves to cope with complexity. This idea was spun by the founder of general systems theory, Ludwig von Bertalanffy (1901–72). He stated that reality consists of a "tremendous hierarchical order of organized entities" (von Bertalanffy, 1950a, p. 164), and that the complexity of systems increases with the levels of integration. As we can learn from anthropology and sociology, this is particularly true for social systems.

There is no fixed determinist hierarchy structure for level hierarchies of social systems above the level of the individual. We think that the levels of social systems differ from those of physical and biological systems in their genesis, their rationale and interaction among the elements, and in their relations between the levels. Though we can prove that small groups have an effect on individual behavior, the social organization of human beings is rather flexible. Even mother–child (parenting, family) relationships can be substituted by collective childcare (e.g. in Nazi, Bolshevist settings). But, besides those individuals living in complete isolation (e.g. as alleged by Kaspar Hauser), people are embedded in durable and observable social networks. They belong to groups and other social agglomerates, which have rules, regulations, and imperatives that affect individual behavior (see Chapter 6.21). Thus there is a top-down influence. Likewise, a single person can affect, split, or destroy groups.

Control hierarchies and level hierarchies can be defined for different subsystems or principal components of society. When defining or dealing with hierarchies of human systems, it is not only the type of hierarchy (e.g. control vs. level hierarchy) that is important. We can define hierarchies with respect to principal components of society, such as the political, economic, or social systems (see Figure 1.7*). Hierarchies may differ socioculturally. When investigating behavior and life in the early twentieth-century European urban population, the classical social stratification of, for example, upper, middle, and lower classes, may make sense when answering certain questions. Referring to the cultural component when studying India, one certainly refers to the caste system: "… the ranking system of caste groups was validated by differential control over the productive resources of the villages" (Dumont, 1970, p. 76).

We can identify elementary level hierarchies of human systems, which follow different rationales when interacting with the environment. As we can learn from anthropology, the type of hierarchies existing developed and adapted historically. Societies of hunters and gatherers comprise(d) between 30 and 100 individuals. They did not have a state society led by kings or governments, which establish law and institutions. Thus, the specific levels of social systems that are formed depend on the level of societal evolution (see Figure 8.5*), and on societal trajectories such as culture, political order, etc. (see Figure 8.3*). Thus, a typical society of hunters and gatherers did not have organizations, if we do not consider the family as an institution (see below).

16.3.6 Hierarchy levels of human systems

Table 16.2 presents a hierarchy of social systems (J. G. Miller, 1978, 1990; J. G. Miller & Miller, 1990; J. L. Miller & Miller, 1992, 1993, 1995; Scholz & Binder, 2003, 2004). The reader may notice that the distinctions between the different levels above the individual are not always absolutely clear-cut and are subject to the evolution of societies. Thus, sometimes it is difficult to decide whether a bank is a private organization and thus a simple business company, or whether it is a public utility as it is state owned, which would bring it one step up on the ladder of human systems. One may introduce another level, such as communities, as communities have elected governments and institutions such as law or police systems. As elaborated above, what levels exist differ socioculturally.

Before defining the different levels, we want to stress that the goal system, which is an important component of the rationale of the human system, and the way in which a human system has control of its subsystems and subordinate (lower) systems, are both essential for defining the hierarchy levels.

16.3.6.1 Individuals

Individuals are any born, living people. According to the definition of the human individual (see Chapter 3.1), a person is a potentially self-reproducing system that processes matter, energy, and information. Individuals differ in their genetic heritage, meaning their chromosomes and genes (*nature*) as well as in their experiences, education, diet, exposure to environmental chemicals, radiation, etc. (*nurture*).

Human behavior is conceived as purposeful, intentional, goal-directed (which is not to say that this is necessarily bound to conscious processes), and relying on physiological and cognitive processes. The

Table 16.2 Hierarchy levels of human systems.

Hierarchy levels of human systems	Definition, properties, and function
Individual	A born, living, person, distinct from and purposefully interacting with the material and social environment. In this book, the human individual is defined as all living cells and their (inter) activities, which emerge from the fertilized egg
Group	Two or more individuals having enduring, direct, and observable relations; social groups have rules, regulations, and imperatives for organizing certain tasks
Organization	An organization is an assembly of people who are jointly planning, coordinating purposeful action; organizations are characterized by formal membership
Institution	Society establishes organizations that are under governmental control and whose task is to support the existence of society
Society	A society is a federation of individuals and/or groups, one example being an ethnic group, which is an enduring, politically autonomous association that exists as a primary organizational subdivision of the human population. A goal of societies is to survive and sustain themselves, and their structures may have principle components such as economics, law and policy, culture, and/or science
Supranational/ suprasocietal level	An organization having executive power and legislatures that can direct nation-states and determine nation-society's law while maintaining independence from the sovereign national level. Supranational systems have Kompetenz-Kompetenz (i.e. the authority to change or to extend the legal order) and are partly financially independent from the nation-states

individual's perception and mental processing is limited, functional, adaptive, shaped by different drivers (see Figure 7.12), and may show different degrees of environmental awareness (see the Awareness Postulate P6).

How people behave is affected by the groups and organizations to which they belong (see Boxes 6.12 and 6.13), and their cultural and societal setting. According to Figure 3.3*, the individual interacts with his or her material and social environment, which includes impacts on higher levels.

16.3.6.2 Groups

Groups consist of two or more individuals who, over a period of time, relate to one another directly and have an enduring, observable relationship. While groups can exhibit both upward and downward causation, a primary goal and driver of a group is to develop structures of downward causation. They develop regulations, implicit norms, explicit rules, and expectations affecting the behavior of the individual and collective action. Members of groups, such as leisure groups, friendship groups, work groups, etc. contribute to carrying group processes. Following social psychology, groups have no formal affiliation (*formal membership*) or echelons. (Parts of) Families can be conceived both as groups, if looking at them as people who live together or meet frequently for leisure purposes, and as institutions, if legal rights and responsibilities linked to families are considered.

Members of a group show identity (affiliation) with and differentiation from other groups; they have a set of shared values helping them to maintain their overall pattern of activity, and they have a specific goal or interest they wish to achieve. Group members can be partly interdependent and affect each other (social influence, see Arrow *et al.*, 2000; Hare *et al.*, 1994; McGrath, 1984; Sherif, 1936; Turner, 1984). Intergroup categorization is important to define groups (Tajfel, 1981). A critical issue of defining groups is describing their size. We use the term group when referring to small, specified groups comprising less than 100 individuals.

If one wants to analyze how group rules affect individual behavior, one has to restrict the analysis to smaller group sizes. From a social hierarchy level perspective, social categorization such as black–white or male–female are substrata of society. In some cases we also include large groups, such as the environmentalists, animal welfare activists, consumer or lifestyle groups, etc. However, for large groups it is unclear what *direct* relation the individual has to the other group members. The same holds true for virtual "internet groups." Usually, often anonymously, these groups interact from remote places and have no physical meeting place. They nevertheless develop a "we-ness" and a feeling of togetherness (McKenna & Green, 2002, p. 117). However, group norm influence on changing an individual's opinion seems lower than in normal groups (Sassenberg & Boos, 2003).

Thus, we prefer the use of the group concept in the level hierarchy analysis of human–environment interactions to those groups where direct face-to-face interaction plays a significant role. This assumption is in line with the requisite that the group should be part of the (material and social) environment so that both bottom-up and top-down processes can be at work.

16.3.6.3 Organizations

Besides formal membership, a critical difference between groups and *organizations* is that organizations have at least two echelons, as is characteristic for the control hierarchy. Organizations such as companies, clubs, associations, foundations, charitable societies, NGOs, and unions have become the dominant form of cooperation and regulation in society. The typical characteristics of organizations are: (1) free but formal membership and, in general, penalty-free disassociation; (2) free design of processes and some organizational structures by the owners or leading members of the organization; (3) free definition of organizational goals that align with the organization's function(s) (Türk & Reiss, 1995). The latter suggests that, in general, organizations are different from medieval guilds, most churches, secret societies or sects, which request that members adjust all domains of life. As another example, the Mafia could be viewed as a typical organization, as the Mafia are "a kind of organized crime being active not only in several illegal fields, but also tending to exercise sovereignty functions – normally belonging to public authorities – over a specific territory …" (Airoma, 2008). Thus, if we view the Mafia as having appropriated the authority of the state, "clans" can also be considered as institutions (see below). Governmental organizations such as the police are considered as institutions (e.g. as part of the execution). In some cases, as it is with banks, the post or churches, the boundaries between organizations and institutions are fuzzy.

We work with a somewhat restricted notion of an organization. Based on the definition of the human individual we include all activities of people conducted as memberships under the label *organization as a human system*. What owners or members of a company, for instance, do in their spare time is not part of the activities of an organization. Based on the human individual-based definition of organization, the factory buildings that are owned by a company such as General Motors are not considered as part of the organization but are of the (material) environment of this organization. As the example General Motors shows, organizations can act on the global level.

16.3.6.4 Institutions

Institutions occupy an enduring and cardinal position in society. Institutions represent, maintain, and stabilize society through regulatory systems. In our understanding, institutions are both the people representing public organizations, and statutory corporations, and the formal rules and regulations established by society and state agencies. This is in line with Durkheim's (1930/1992) assumption that institutions assist the functional organization of society. The purpose of institutions is to stabilize society by bringing it into a kind of equilibrium state. Examples of institutions are families, schools, universities, police departments, courts, and municipal or state authorities. Policy scientists claim that the genesis of institutions is because of the self-interest of actors, such as political parties, representing their economic or environmental interests. For instance, industry companies and organizations want to have stable boundary conditions and decision rules.

Institutions as *human systems* include all members (e.g. civil servants, people working for public agencies) and their actions related to their membership with that institution. The idea behind this definition is to include not only the abstract rules but also the persons representing the institution. Every institution is based on the consensus and the practice of the individuals representing the institution (Scharf, 1991).

In general, our definitions are in line with a sociological and macro-historical perspective (Capoccia & Kelemen, 2007; Pierson, 2000; Thelen, 1999). Here, institutions are conceived as public organizations and "formal political institutions, which are products of conscious design and redesign" (Pierson, 2000, p. 475). The boundary between institutions and businesses is sometimes fuzzy, as many publicly owned businesses have become privatized. This can be illustrated by many examples from the energy, water, waste, or banking sectors. If a state-run company becomes a stock corporation, which is owned by the state, the limit of liability is restricted by the capital stock. This is not so in the case of state ownership.

There are definitions of institutions other than the one above. Some disciplines, such as in institutional economics, restrict what is meant by an institution simply to the (formal) rules of the game; that is, what we call *institutional rules* (Behera & Engel, 2006; Hérétier, 2007; North, 1990; Williamson, 1979) which can be viewed as the "mind of an institution." These are the practice, customs, and behavioral patterns that organize society. This position is taken by some representatives of the new institutional economics (see Chapter 11.7).

According to the functionalist perspective, the type of institutions a society needs depends on its complexity. Primitive cultures have to assure procreation and the care and conditioning of infants. Thus the family, clan, sibship or kinship, including gender-based divisions, are the most elementary of institutions. These institutions are found at all levels of technological complexity, having a patriarch or matriarch leadership and diverse, culturally shaped forms of marriage (e.g. fraternal polyandry; see Box 5.9). Family status has legal impacts. Since family rules differ, families can be considered *cultural* institutions.

Chapple and Coon (1953) elaborated that political institutions emerge as danger, poverty, and threat demand regulation of the social structure. In tribal societies, leadership for community, justice, and warfare issues are often linked. The army is a political institution organized by a strict control hierarchy. As with most domains, the complexity of an army increases with technology. As far back as in the times of the Roman army, one could find infantry, cavalry, and artillery using ballistic war machines.

Society's basic needs are secured by institutions (Dubiel, 1976). Thus, educational and science systems emerged with increasing demands on technology. As societal complexity, departmentalization, and bureaucracy increases, institutions are needed to address concerns about public safety, welfare and health, civil service, education, environment, foreign affairs, etc. Thus institutions have a multitude of functions, such as regeneration (with institutions such as families), education (e.g. schools), distribution of goods and food (e.g. agriculture and trade agencies), and governance (government, courts, police; see Lipp *et al.*, 1995).

New institutions emerge at periods of critical societal change, which are characterized as critical junctures, turning points, crises, and unsettled times (Pierson, 2000). The US Environmental Protection Agency was founded in 1970, at a time of rising environmental concern and awareness.

Finally, as a central element of culture, there are religious institutions. Early institutions were represented by a shaman or priest leader, who had the ability to rectify one's internal emotional disturbances through ordering or intensifying rites (Chapple & Coon, 1953). Disturbances could relate to the death of family members, the planting time in spring, harvesting in autumn, or the seasonal hunting of caribou or salmon. Historically, the separation between religious and public or governmental institutions is sometimes fuzzy (Finer, 1999).

Institutions are often part of legal domination. They can control individuals or organizations in their dominion, depending on the type of family rights or political order. Institutions can severely affect behavior of individuals, organizations, and societies. They are generated and affected by societal trajectories (see Figure 8.3*) and rule the power game in and between societies. However, the "tragedy of the commons" (see Boxes 2.3 & 6.15) shows that there are situations where the effect of institutions on the micro level can be weak.

The goal of institutions is to address and satisfy core societal needs and drivers. Institutions offer formal and informal programs and rules to reach goals that are formulated by a politically or legally authorized body (e.g. government, city council, parents, etc.). Here the distinction between ministers who, in democratic countries, are politically elected persons and ministries/agencies is of interest. Ministries and their institutions execute what the ministers want and the constitution allows. Naturally, there are different types of legitimization of institutions, for instance by variants of democracy or absolute monarchy. Institutional bureaucracies can develop their own momentums and power plays, as in the case of groups and organizations (Weber, 1947).

We consider the creation and establishment of appropriate environmental institutions such as agencies for nature conservation, raw materials, food, agriculture, forestry, nuclear risks, and radiation to be most important for developing environmental awareness and literacy (Wertz-Kanounnikoff & Chomitz, 2008). Institutions are authorized with a special role with respect to property rights (Hanna *et al.*, 1996) and natural resources. In this way, institutions can be considered engines of environmental literacy. Moreover, the way these rules are introduced and represented by the members of the organization, their reliability, truthfulness, etc. is as important as the rules.

16.3.6.5 Society

We have defined *society* as a primarily organizational, politically autonomous subdivison of society (see Chapter 8.1). In general terms, human society is "the dominant feature of the human environment" (Cohen,

16 The HES Postulates

> **Box 16.3** An impact of technological and societal changes: cosmopolitical institutions?
>
> A critical question of policy science is: where do institutions come from? Anthropologists can give a clear answer on the origins of institutions. Starting from primitive societies, Chapple and Coon (1953) showed that environmental resources and technology determined the acquisitions of natural materials, husbandry, transportation and trade, division of labor, and the building of institutions. A global map of the levels of complexity in the 1940s is provided in Figure 16.6. The most elementary institution is the single-family unit, which expands to extended family, and on to non-familial (i.e. bands, hordes, etc. where a non-family member takes control outside the circle of the kindred), to political institutions, to associations, and to many complex institutions.
>
> Societal change, shifts in availability of resources, technologies, and incentives require new institutions (Krasner, 1988). Globalization presents a special challenge. A critical question is whether worldwide flows of toxic environmental pollutants, scarce energy and material resources, worldwide migration, and adaptation to climate change asks for supranational, or, as Ulrich Beck formulated (Beck, 2006), cosmopolitical institutions.

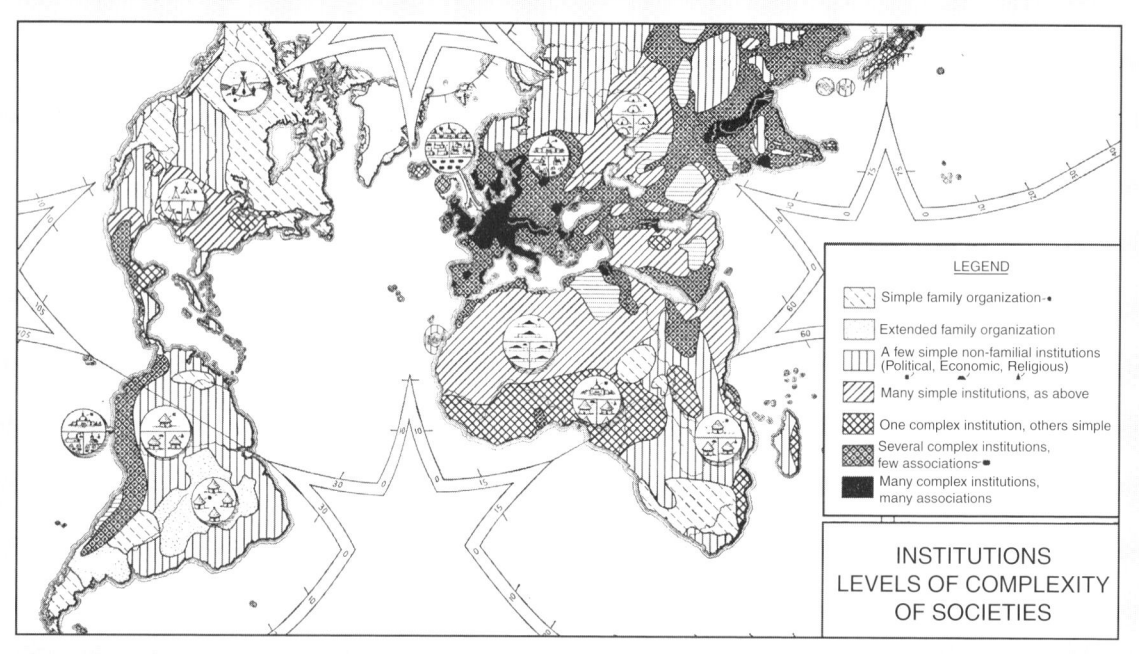

Figure 16.6 The levels of complexity of societies (adapted from Chapple & Coon, 1953, pp. 452–3; according to the conditions found in the early 1940s). This analysis includes companies (as economic "institutions" and not only public organizations controlling markets or finance such as public banks introducing fiat money) and associations (as voluntary medical or environmental associations), which we subsume under organizations.

2000b, p. 76). Society is thus "any system of interactive relationships of a plurality of individual actors … It contains within itself all the essential prerequisites for its maintenance as a self-subsisting system" (Parsons et al., 1951, p. 26). The nation-state is a modern form of society (see Box 16.4). In the "West since the seventeenth century, nation-states have claimed legitimacy in terms of largely common models" (Parsons et al., 1951, p. 113). We consider a nation-state as a purposive actor. Societies have principal components, such as economic, political and legal, cultural and social, and science and educational systems (see Figure 1.7*). Culture, while sometimes being locally shaped, is an essential element of societies, many of which can incorporate several different cultures. From a functionalist perspective, cultures create collectivities. And societies create institutions for securing sustenance.

According to the ReSTET approach (see Figure 8.3*), the drivers of society are increasing wealth and power. This holds true on an intra- and inter-societal level. Institutions are a primary means to attain this.

The nation-state has dominated the world regime in the past centuries, whereas for many it is undefined whether, for example, Palestinians or Kurds should be considered societies. Societies can be considered creatures of power and competition. This, at least, seems

16.3 Hierarchy Postulate P2

Box 16.4 Parallel societies: Sinti, Roma, and Amish people

The world regime after World War II was strongly shaped by the idea of societies as nation-states. Today, only a few, small, politically and economically independent societies of indigenous people exist in some parts of the tropical forests. However, there are ethnic and religious groups, such as the Roma, Sinti or the Amish people, that function in many ways as independent societies within nation-states.

There are about 15 million Sinti and Roma spread out in many parts of the world. The 8–10 million European Roma and Sinti have common genetic origins and speak dialects of their own Indo-Aryan language, called "rromani ćchib." Roma and Sinti consider themselves as ethnic groups. Their basic organizational units consist of extended families, who live in clusters called "njymuris," and of the "vitsa," which is headed by the elders. A vista and subunits reclaim a certain area for business activities, including sales and trade of craftsmanship. Sinti and Roma have special rules of refrainment, such as not informing others about their own language. Their societal structure and rules of refrainment work to genetically isolate these founder populations that immigrated into Europe in the late medieval period from northwestern India and South Asia. Sinti and Roma do not have a government but are requesting ethnic minority rights in certain countries. Thus the Sinti and Roma can be considered as a nested, ethnic, hierarchical subsociety.

The Amish people are a religious group of 200 000 to 250 000 people in various states of the USA, mainly concentrated in Pennsylvania. The Amish split from the Swiss Brethren in 1693 under the leadership of Jakob Ammann and emigrated from Europe from the seventeenth to the eighteenth century (Hostetler, 1993). The Amish can be considered as a religious rule-governed society. Each member has to follow the Ordnung strictly (i.e. the rules of the church) to avoid being shunned or excommunicated. Today's Old Order Amish people are not allowed to have access to power-line electricity, telecommunication, automobiles, etc. for most purposes (illustrated in Figure 16.7). Amish keeping the Amish Ordnung refuse military service and social security, visit their own schools, and avoid contact with non-Amish people. Specific rules are governed by the settlements, which aspire to live independently of other communities. The Beachy Amish, for example, have adapted to the larger society within which they live in that they now drive black cars. The Amish sell and occasionally buy products from non-Amish people according to the community rules. The community government is led by nominated preachers, bishops, and deacons. The decision against using electricity, for example, came from the Order principle stating that one should not farm more than one can manage with horsepower and manpower.

Originally, the Amish were physiocrats (Steiner, 2006; see Chapter 10.1.5). Their worldview resembled in many respects the basic idea of eighteenth-century economists with respect to the idea of the damned cities, and they believed that wealth and normal life was derived solely from the value of agricultural pursuits or land development. But population growths (e.g. of 4.5% per year, which doubled the population in 20–25 years; Amish Studies, 2010) and increasing land prices forced Amish people to serve as industrial workers in many places. They can be considered as a prototype of a parallel society.

Figure 16.7 The Amish people have a religious order preventing marriage with or visiting schools of "English" people and can be seen as a quasi-autonomous parallel society. Physiocracy (i.e. a pure agrarian-based economy) has been a key issue in their development.

to hold true for nation-states. Sometimes subunits of nation such as state or communities as societies.

16.3.6.6 Supranational level

Since the beginning of the industrial age, technological and economic development has generated huge global flows of materials, energy, and information. Many environmental issues, such as climate, seawater and air protection, preservation of endangered species, and overfishing oceans, cannot be efficiently solved on the level of the nation-state. Environmental damage recognizes no international borders, thus damage occurs internationally regardless of the location of the activity that caused it. Thus, globalization and environmental management require new forms of world governance. From a functionalist perspective, there is a need for multilateral cooperation, treaties, and institutions above the level of society and the nation-state. Traditional means for this are bilateral and multilateral international treaties and contracts.

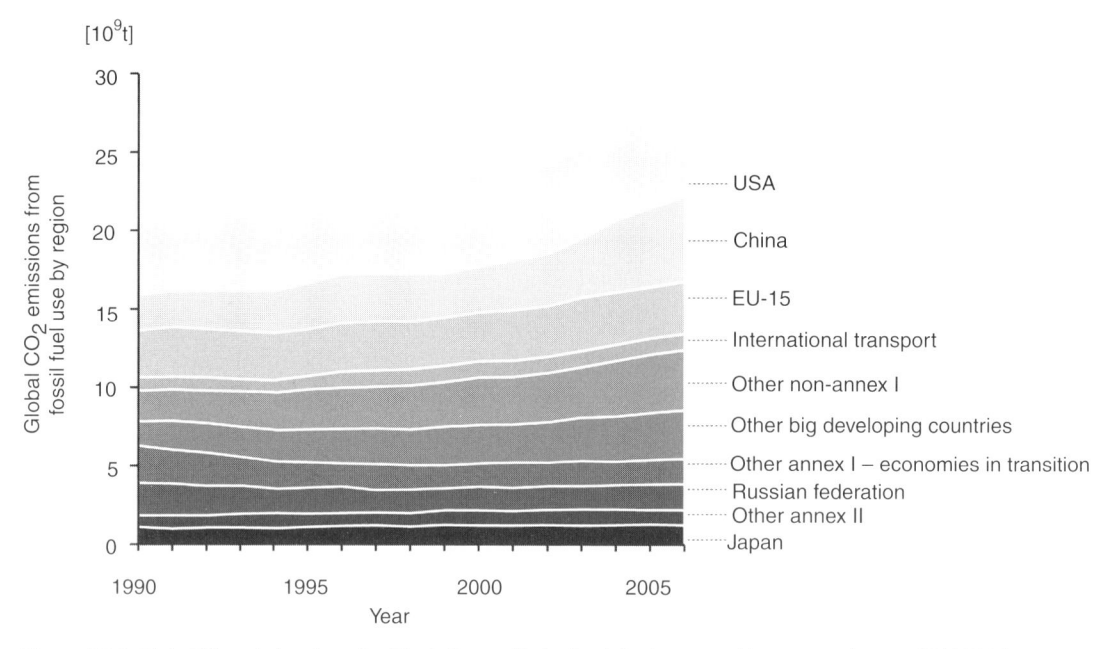

Figure 16.8 Global CO_2 emissions from fossil fuels (Source: Netherlands Environmental Assessment Agency (PBL), 2008).

Critical questions about global environmental management are: what design should institutions have above the level of the national society? What type of institutions, contracting, or institutionalization is appropriate? From an environmental literacy point of view, there is a need for sustainable governance and strategic environmental management on a world level, which we call a suprasocietal or supranational perspective. To highlight the potential role of a supranational institution, let us briefly consider the case of mitigating climate change. Figure 16.8 presents the fossil fuel-borne CO_2 emissions of different countries and country clusters of the world. The big players are USA and China, both causing about one-fifth of the fossil fuel-based CO_2 emissions.

The 1997 Kyoto protocols were not signed by the USA. Moreover, they had no impact on China, because they followed the principle of common but differentiated responsibilities and China was classified as a developing country. Similarly, most other countries that were classified as Annex II (developing country) were not challenged to reduce CO_2 emissions. One can assume, that in the current world game for wealth and power (see Figure 8.3*), there are no short-term incentives for the USA and China to reduce CO_2. According to game theory, we have a non-cooperative game. This means that no institution in today's nation-based world regime can compel the USA or China to reduce CO_2 emissions for the purposes of combating global climate change in the coming years. However, if supranational institutions existed, they could enforce compliance penalties on violating nations. From a game theoretical perspective, we face a "tragedy of the commons" dilemma structure, where it is beneficial for the USA and China to keep emitting without being penalized.

Developing human systems that go beyond the competence and the competitive habit of nation-societies is a question of international politics and law. In the following discussion, we distinguish between *inter*national and *supra*national (synonymously used as suprasocietal) levels and describe each. Societies or nations can build intergovernmental organizations by a charter, treaty, or convention. The rules to be followed for international and supranational organizations are international law, such as maritime law or international criminal law. The United Nations (UN) is considered a typical international organization. The European Union (EU) is an example of a supranational organization. The difference between them is that the EU has representatives (i.e. European Parliament electing a European Council) and legislatures (e.g. the European Court of Justice) who can dictate to the members how to apply their law. Laws of nation-states become inapplicable when conflicting with those of the supranational organization. Thus, sovereignty, providing basic competencies, and allowing for regulation of internal societal affairs are assigned to a supranational body. In this way, an autonomous legal system that can determine nation-society's law is decisive for a supranational system. For example, if an EU country does not follow EU law, it can be accused and penalized by the EU Court of Justice. This is not the case with the Universal Declaration of Human Rights (UDHR, signed in 1948). The UN can only induce moral pressure to states violating articles, but does not allow for intervention. One exception is the International Criminal Court. Thus, supranational organizations are less likely to show the failure that many international institutions show, of not being able to put a regime in place or of having a set of rules that do not become reality (Bernauer, 1995; Ostrom, 1990).

International organizations follow intergovernmental rules, whereas supranational systems have what is called Kompetenz-Kompetenz, which is the ability to extend their authority independently of the agreement of single nations. Supranational systems differ from nation-states because they do not have *original* but only *derivative* sovereign rights. Supranational institutions and structures deal with environmental problems that cannot be dealt with effectively at the national level.

From a rationalist, neofunctionalist perspective, supranational institutions are more efficient. Economic explanations refer to transaction-cost and principal-agent theory (Ross, 1993). A principal (e.g. a country afraid of negative environmental impacts from a new environmental technology) delegates certain functions to an agent (e.g. to a supranational organization) with the understanding that the agent will deal more efficiently with the problems than the principal would. Thus, for instance, the collective optimum of Pareto efficiency can be attained.

The EU offers a good model for understanding some specifics about level hierarchy. For some authors, the EU shows an ambiguous, hybrid nature between intergovernmental cooperation and federalism (Schlesinger, 2007). The EU is characterized by a multilevel polity, with regulation on two levels: the EU and the national level. EU Community institutions such as the EU Court of Justice can limit national autonomy. This type of level hierarchy is characterized by two doctrines. One is the doctrines of direct effect, which states that community law "must be regarded as the law of the land in the sphere of application of Community law" (Weiler, 1994, p. 51). The other is that of supremacy, which means that in conflicting cases the EU law "trumps" national law. The EU Community law is the higher law. However, an important issue is that the national court renders final judgments. Thus, the member states do not disobey their own courts.

Though not explicitly included in the European treaties, "… supranationalists hold that the EU is a polity in its own right" (Kelemen, 2004, p. 6). An important element is what is called constitutionalism, which means that there are well-defined rights and duties (e.g. the same right of workers to be employed in each country), which can be easily explained from a functionalist perspective.

So far, we have been dealing with supranational systems as rather impersonal systems. But it has been acknowledged that finally it is the individuals representing the institutions and their self-conception that matter. In this context, an extension of principal-agent theory suggested by Tallberg is of interest. For him, supranational systems function as supervisors, "engaged by national governments for the purpose of monitoring actual member state behavior and enforcing compliance with Commission rules" (Tallberg, 2003, p. 26). Because of the inherent commons dilemma structure, the members of the supervisory agency, in the case of the EU in particular of the EU Court of Justice, must have the aptitude to follow the interests of the supranational institutions. This clearly asks for sage management of game theoretical dilemmas (Axelrod, 1984a, 1987).

An interesting issue is how international and supranational systems in western and eastern worlds are shaped by cultural aspects. For the Islamic Ummah, for example, the part of the world system where most Muslims reside, the Cairo Declaration on Human Rights in Islam[2] (CDHRI, 1990) establishes an Islamic international organization supplementing the UN.

Thus, there is a strong, implicit layer on the socio-epistemic side (see Figure 3.3*), introducing an unwritten rule system. This might differ from the ideas of western constitutionalism created in the eighteenth and nineteenth century, which proposed that the rights of nations can be restricted by a written constitution (Cottier & Hertig, 2003). A simple advantage of supranational institutions is that they, the Court and Commission of the EU, have different time horizons than the politicians (Alter, 1998).

---

[2] Article 25 – The Islamic Shari'ah is the only source of reference for explanation to the Islamic Shari'ah (CDHRI, 1990).

A critical issue from the perspective of environmental literacy and sustainability is: what are the drivers of international and supranational institutions and regimes? And, what or who determines who are the losers and who are the winners?

One of the drivers of founding the EU after World War II was the idea of establishing the same human rights in all countries as a stabilizing transnational structure. In times of international conflict, the malign antagonistic competition between nations was substituted by the idea of a non-zero-sum relationship. Today the freedom of mobility in the frame of worldwide intercontinental competition is a driver. Many international institutions have emerged to strengthen the position of winners rather than those of the marginalized during international power-play. Clearly, the dilemma about a sound mix of competition and cooperation is ubiquitous in supranational institutions. We think that both the history of the EU and the cosmopolitical vision (Beck, 2004, 2006; Beck & Sznaider, 2006) provide insight on how to cope with this issue meaningfully.

The German sociologist Ulrich Beck (see Chapter 9.4) provided a vision of environmental literacy on the supranational level by drawing on the eco-sociological concept of world risk society (Beck, 1996). Beck proposes cosmopolitanism requiring supranational institutions. A cosmopolitan view emerges from the *awareness* of ubiquitous manmade risks that emerge from a permanent transformation, accumulation, and multiplicity of distinct worldwide risks.

The sources of these risks are not only ecological but are also biomedical, social, economic, financial, symbolic, and informational. These risks are experienced in all places in the world and are amplified by global media in building a communicative space for transnational discourses and action:

> The birth of a global risk consciousness is a trauma for humanity. (Beck, 2006, p. 341)

Risks motivate people or nation-states to seek "the meaning of life in the exchange with others and no longer in the encounter with like" (Beck, 2006, p. 331). Beck speaks about enforced cosmopolitanism resulting from "involuntary enlightenment." One side of this enlightenment is nations noticing the necessity and efficiency of worldwide cooperation for mastering manmade environmental impacts such as climate change. The other side is linked to the social side of sustainability. Accounts of most catastrophes show that *catastrophic risks are following the poor*, which can be considered a kind of global environmental racism (see Box 9.1). Salient examples were the 2005 Hurricane Katrina, the 2004 Asian Pacific Boxing Day Tsunami, 1986 Chernobyl, or the 1997 Asian financial crises.

Beck considers risk as the involuntary, unintended compulsory medium of communication in a world of conflicting differences. Worldwide concern has the potential to push people to go beyond what is around them. Publicly perceived risk compels communication between those who do not want to have anything to do with one another. Here the media play an important role.

> Global risk has the power to tear down the facades of worldwide irresponsibility. (Beck, 2006, p. 339)

Beck clarifies that he does not understand cosmopolitanism in the Kantian sense to be an obligation that one should order the world. He rather targets a cosmopolitan realism (Beck, 2004) and is looking for a "cosmopolitan realpolitik" (Beck, 2006, p. 343). Here it is interesting that the ideal of cosmopolitanism is already imbued with different notions. For instance, Diogenes is supposed to have answered "I am a citizen of the world [kosmopolitês]" (Diogenes Laertius VI 63) when being asked where he came from. One interpretation of his response is as a rejection of what was conventionally understood as living in accordance with nature and such. Others refer to moral cosmopolitanism, stating that all humans belong to a single community and that this community should be cultivated.

There is a kind of commons dilemma included in supranational institutions. Also, supranational institutions can be used in the world power game. Many western societal trajectories target human rights, global justice and struggles for a new grand radical–democratic globalization. As Beck argues (Beck, 2006), this cosmopolitical vision is used to legitimize the invasion of other countries. As the Iraqi war shows, this is often done not only through a benign idea of common wealth but by some powerful nations' interest in gaining access to natural resources.

Internationally, there are established global institutions that cope with environmental problems, such as the United Nations Environmental Programme (UNEP), the Food and Agriculture Organization (FAO), the World Health Organization (WHO), and the World Meteorological Organization (WMO). A key event was the 1992 United Nations Conference on Environment and Development (UNCED) held in Rio de Janeiro. The UNEP, founded in 1972, was intended to play a catalytic role coordinating the environmental activities of other UN agencies. However, from the beginning its role was downplayed owing to an unreliable financial base in the face of the international tasks at hand (DeSombre, 2006).

The UNEP's primary activities are what can be called system-wide Earthwatch, the aim of which is to keep a review of the global environmental situation. Besides activities on species conservation, UNESCO helped establish World Heritage Sites such as the Galapagos Islands (1978). However, the UN is far from having the status and executive power of the EU. This becomes evident when looking at the Intergovernmental Panel on Climate Change (IPCC), which was established by the WMO in 1988 to monitor climate change and to design strategies.

IPCC is an intergovernmental body. Thus, IPCC documents should undergo both peer-review by scientific experts and review by governments. In this way, IPCC practices transdisciplinarity. The strength of the IPCC has been its relative independence from single nation or stakeholder interests. The weakness is the lack of connection to a powerful supranational organization such as the European Court of Justice, which can implement mitigation strategies. Here a thorough analysis of the role of the supervisor in a principal-agent setting seems promising.

The necessity to think about Earth system management has also been introduced by policy science (Biermann, 2008). A large-scale project on Institutional Dimensions of Global Environmental Change (IDGEC) was run in the frame projects of the International Human Dimension Programme on Global Environmental Change (IHDP; Young *et al.*, 2008). Whether and what type of world environmental organization is needed that has rights beyond the UNEP is disputed (Najam, 2003). The ideas from policy science include that UNEP should be replaced and that a world organization for sustainable development is needed.

16.3.7 Key messages

- There are different definitions of hierarchies (i.e. order, inclusion, control, and level hierarchy). Level hierarchy, based on downward and upward causation, is a valuable concept for the study of HES
- Biologists dispute whether the concept of hierarchy is simply an epistemological one or whether the cell and its super- and supra-systems exist ontologically. This question is even more open when looking at human systems above the level of the national society
- Phylogenetically, hierarchies of human systems emerge with the increasing complexity of society. Currently, in most societies we can differentiate between the level of the individual, group, organization, institutions, society, and supranational systems. Intermediate systems, such as local communities, can be defined
- Supranational systems are above the nation level. The European Union can serve as an example; it has a Kompetenz-Kompetenz (i.e. the authority to change or to extend the legal order of nation-states without having an original sovereignty belonging to the national level). Supranational systems for providing incentives to avoid free-riding are missing.

16.4 Interference Postulate P3

The term interference has general and specific meanings in different disciplines. In general terms, interference denotes the disruptive effect of a certain internal *action* or *process* on another external or internal action or process. In physics, interference denotes the combination of two or more waves of the same type that reinforce or cancel each other. Looking at sound waves we can get silence, increased noise intensity, or other unwanted effects. Colloquial interference denotes the disturbing or undesired hampering of something, for instance if the growth of a virus in a host is prevented by another virus or by another superordinate process. We speak about an RNA interference (RNAi) if in certain cells the gene expression, that is, the activation or deactivation of genes, is affected by other RNAs, for instance by injected genes (Hannon, 2002). In psychology, Stroop identified that if you are asked to say the color of the letters making up a word (e.g. red letters), and they do not match what the word says (e.g. green), the response time elongates. Here, obviously visual and verbal mental processes are interfering. Stroop called it inhibiting (Stroop, 1935).

From a system-theoretic perspective, interferences between the micro and the macro level can be defined as interactions between hierarchy levels. According to the Hierarchy Postulate P2 the different hierarchy levels of human systems have different drivers. At one hierarchy level of human systems, for instance the members (minus a specific person) of a group and the rules of a group can be part of the environment of this individual person (see Figure 3.3*), the group rules may interfere with what the individual wants. The Interference Postulate P3 refers to the interaction of regulatory mechanisms between human systems on the same level and, in particular, between systems from different hierarchy levels. In contrast with feedback loops, which occur after human systems take action, interferences occur during the decision-making process and before the action. Interferences have the greatest impact if interests from different human systems conflict. A prominent interference that occurs between the micro and the macro level is seen in the case of the commons dilemma problem. This problem has been dealt with in various places of the book (see Chapter 6.22, Boxes 2.3 and 6.15). It refers to the extinction of effective rules for resources conservation on the societal level by competing, and thus mutually enforcing, (short-term) interests of resource depletion at the individual level. Such dilemmas are characterized by the divergence of collective and individual rationality.

16.4.1 Conflicts between and within human systems

Interferences within a hierarchy level occur, for example, from interaction between different societal systems. For the societal decision-making process during the dying of forests (German: Waldsterben) debate, interference between the different human systems – science, policy, media, and economy – can be observed. According to Luhmann, interference between these systems is influenced by different communication within the systems (see Chapter 9.1) and mutual dependencies among them. Interference within the same institutional system, such as a federal government, is also possible. This occurs when, for instance, a federal environmental agency wants to increase the residual flow to water bodies to preserve a certain biodiversity and, in contrast, the federal energy agency resists because this action reduces hydropower. Interferences can even occur within a person, since the same person can be part of different human systems. For instance, imagine a person who runs a business and is at the same time a member of parliament. Faced with a decision about environmental legislation that would increase overall welfare for her constituents but lead to high costs for her business, interferences between this person's different roles will result.

Recognizing and coping with interferences or trade-offs between human systems on different hierarchy levels is a key ingredient of sustainability learning. Often these interferences occur because of different rationalities on a different hierarchy level, regarding a certain action.

Box 16.5 Interfering interests between nations, banks, and individuals: green housing

Housing has many impacts on the environment. In developed countries of the north, about one-third of energy is used for housing. A central problem is urban sprawl, which often induces negative impacts on landscape, biodiversity, infrastructure costs, and transportation needs. Whereas in former times houses were built on agriculturally unproductive land, this is no longer the case owing to the current low prices of agricultural land. The result is a dramatic increase in construction at the expense of rural and wildlife areas, aesthetic amenities, and other issues relevant from an environmental point of view (Hammer et al., 2004; Robinson et al., 2005).

If one could sum up the major recommendations for sustainable land use in Europe, they would include: (i) do not build on the greens (i.e. the isolated areas); and (ii) do not build in areas without public transport. Nevertheless, most advertisers, promoters, banks, and investors ignore the recommendations of land use planners that developing green edge spaces is unsustainable (Figure 16.9).

Figure 16.9 Advertisements for Swiss homes appeal to different interests. The top figure recommends: "The construction fair: You should look before you construct." The bottom figure is a promotion from the Swedish bank SEB and reads: "Get a web loan starting at 3.49% interest; fulfill your dreams, and we will give you a free voucher for gasoline."

Land use is a typical example of interferences. Box 16.5 illustrates interfering interests and regulatory mechanisms at the level of national society, the company, and the individual. Sociologists consider the increase of welfare benefits to be one of the aims of societies. The Swiss Constitution, for instance, includes articles explicitly dedicated to the principle of sustainability (e.g. Art. 2 or Art. 73). Switzerland is facing a rapid sealing of soil by buildings and road infrastructures (Umwelt Verkehr Energie und Kommunikation (UVEK), 2000). By a linear extrapolation, the current rate would theoretically lead to a doubling of the Swiss settlement area in 93 years. Given that only 30% of the land is appropriate for human settlement and that 0.24% of the potential settlement area is built on each year, half of Switzerland would theoretically be sealed in 113 years. An effective strategy for limiting the extent of settlement will prevent development on meadows and residential sprawl while preserving agricultural land. Housing and agricultural land uses conflict. Further, landscape ecologists recommend that society should develop the landscape slowly so that a more sustainable development would result in all dimensions of sustainability. These goals and drivers on the level of society interfere with the interests of individuals who want to build on the green and cause a commons dilemma-like interference.

Many motivations are behind an *individual's* choice for living in a rural area. Although the Swiss Constitution (Art. 6) includes the prescription that "each person should support, according to their own potential, the tasks of the state and the society" (translated by Scholz), it is clear that from a hedonistic perspective individuals act to satisfy their own and their family's self-interest. In some countries the desire to build a house in a greener area is widespread (see Figure 16.9).

Investors, banks, and real estate developers represent business organizations that utilize customers' interests for their own business purposes. Clearly, their primary interest is the maximization of profits. Nevertheless, even investors are beginning to question how much their customers give priority to environmental issues or sustainability issues and to think of ways to incorporate these issues into their ventures (Pivo & McNamara, 2005).

16.4.2 Key messages

- During the decision process leading to a certain action, interferences can occur between human systems on the same level and, in particular, between systems from different hierarchy levels
- Understanding and coping with unwanted, commons dilemma-like interferences between processes on the micro and the macro level are a challenge for environmental literacy.

16.5 Feedback Postulate P4

There are different types of feedback loops within and between human and environmental systems. Feedback loops are essential to learn how human systems adapt to the environment and what the primary mechanism for this is. We begin our discussion of the Feedback Postulate P4 with a review of the history and roots of the feedback loop concept. Here we describe some early uses of the concept "feedback loop" and describe how system dynamics has been a major tool for analyzing feedback loops in HES. We also describe how various scientific disciplines, from engineering to economics, have employed feedback loops in their work. Next is a discussion about understanding and using feedback loops. Here, we describe uses for feedback loops in several systems and also cause–effect relationships observed in systems. We also articulate our assertion that sustainability learning has to incorporate system limit management to avoid unnecessary crises and collapses, and instead to promote timely and proper transitions. Then we move into discussions of primary and secondary feedback loops, feedback loops in engineering, biology, and social systems analysis. Finally, we come full circle to environmental literacy and explore in more detail the critical role feedback loops play in sustainability learning.

16.5.1 History and roots of the feedback loop concept

Feedback loops are considered to be general patterns of life. This concept was used by the physiologist Claude Bernard (1813–78) and by Walter Bradford Cannon (1871–1945) when he coined the term "homeostasis" (Cannon, 1932). Homeostasis serves to explain a multitude of self-regulation causal loops in living organisms. We have dealt with the historical background of systems theory, system dynamics, and cybernetics in Chapter 14.3. As the title of Wiener's seminal book *Cybernetics or Control and Communication in the Animal and the Machine* (Wiener, 1948) indicates, the question "Can there be a unified theory of complex adaptive systems?" (Holland,

1995b, p. 45) reflects the "grand utopia that metaphysics, logic, psychology and technology can be synthesized into a unified framework" (Érdi, 2008, p. 43).

In the modern sciences, there have been different terms for feedback, such as circular system, circular process or servomechanism. As feedback systems are often also viewed from the perspective of control and regulation, we have also suggested the term "regulatory feedback and control systems" (Scholz & Binder, 2004). The variation in these names indicates that several communities representing different branches of science contributed to the understanding of feedback loops. The core cluster of these branches includes systems theory, cybernetics, control theory, system ecology, and system technology.

The pioneers of systems research had different backgrounds. For example, the inventor of cybernetics, Norbert Wiener (1948), came from mathematics and dealt with technology mainly from a theoretical point of view. Heinz von Foerster (1982) was a physicist, originally researching linear beam vacuum tubes but then switched to biology and memory research, and emphasized concepts like self-organization, autonomy, cognition, and the role of the observer. Ludwig von Bertalanffy (1950a), the designer of general systems theory, was a biologist working from the idea that organismic systems substitute all their parts but keep their identity. Howard T. Odum started as a zoologist and switched to ecological modeling (Odum, 1983) via the analysis of energy flows in energy systems (Odum & Odum, 1981) and then to feedback loop-based general system analysis of "nature and humanity." W. Ross Ashby was a clinical psychologist investigating biophysics who provided a formal theory of regulation for the "machine and brain and society" (Ashby, 1956, p. 4). He coined the term self-organization. Jay W. Forrester (1961, 1971b) was originally an electrical engineer, who switched via microeconomic modeling to world models. The physicist Hermann Haken's goals were to find general principles of self-organizing that could be applied or seen in physics, chemistry, biology, or computer and psychological sciences (Haken, 1983, 2000, 1985). Ilya Prigogine offered contributions linked to complexity that utilized the feedback concept to describe basic biochemical processes. Feedback loops have been essential features of all these theorists. Furthermore, all these pioneers were aware that system limit management in complex systems encounters not only non-linear, but also multistable, dissipative, bifurcation, or chaotic processes that cannot be mastered by the traditional theories.

All these pioneers had roots in mathematics or natural sciences and provided extended applications to human systems. Only a few of the theorists, such as Wiener, made a commitment to functionalism and postulate a "purposeful [acting] ... directed to the attainment of a goal," a property that has also been attributed to machines such as a "torpedo with a target-seeking mechanism" (Rosenblueth *et al.*, 1943, p. 19).

> Purposeful active behavior may be subdivided into two classes: "feed-back" (or "teleological") and "non-feed-back" (or "non teleological"). The expression feed-back is used by engineers in two different senses. In a broad sense it may denote that some of the output energy of an apparatus or machine is returned as input; an example is an electrical amplifier with feed-back. The feed-back is in this case positive – the fraction of the output which reenters the object has the same sign as the original input signal. Positive feed-back adds to the input signals, it does not correct them. The term feed-back is also employed in a more restricted sense to signify that the behavior of an object is controlled by the margin of error at which the object stands at a given time with reference to a relatively specific goal. The feed-back is negative, that is, the signals from the goals are used to restrict outputs which would otherwise go beyond the goal. It is this second meaning of the term feed-back that is used here. ... All purposeful behavior may be considered to require negative feedback. (Rosenblueth *et al.*, 1943, p. 19)

In this context the term "teleology" was used synonymously with "purpose controlled by feed-back." The author acknowledges that machines and living systems are identical from a behavioral perspective but essentially different from a functional one. However, to the author's knowledge, none of the abovementioned pioneers explicitly dealt with the difference between the material-biophysical and the social-epistemic nature of human systems.

It is interesting to look at the early roots of the feedback concept. Even though engineered feedback control systems were described by thinkers as far back as the Hellenistic period (see Box 16.6), the term *feedback* only came into use quite recently. Historically, feedback loops are linked to control temperatures, pressures, liquid levels, or the speed of rotating machines, such as James Watt's steam engines (Bennett, 1996, see Box 16.11). However, modeling these control processes can also be done implicitly. For instance, the term did not appear in James Clerk Maxwell's (1831–79) set of mathematical equations describing governors of steam engines.

16.5 Feedback Postulate P4

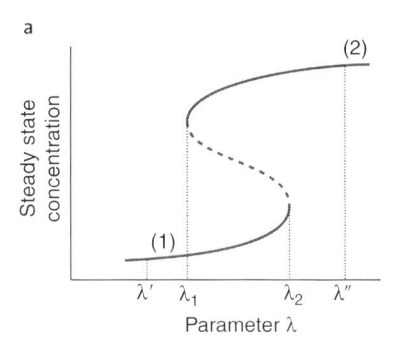

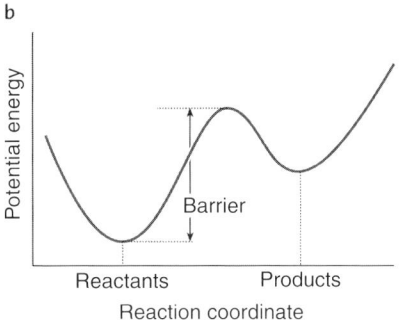

Figure 16.10 Hysteresis and bistability in dynamic processes. (a) The presented system starts on branch (1), but beyond the limit point $\lambda = \lambda_2$, it jumps to branch (2). When the system moves from λ'' to λ' it jumps from branch (2) to branch (1). The specific path of states depends on the history, which defines hysteresis. In the domain (λ_1, λ_2) the system has bistability; that is, the possibility of taking more than one stable state, depending on the history and context. (b) A chemical reaction represented in terms of motion in a double-well potential (Nicolis & Prigogine, 1989, pp. 24–5).

16.5.2 Understanding and utilizing feedback loops

Feedback thought can be considered as "a blend of intuitions and ideas from at least six intellectual traditions: engineering, economics, biology, mathematical models of biological and social systems, formal logic, and classical social science literature" (Richardson, 1991a, p. 17). Richardson identified two threads to use the feedback loop idea. One focuses on homeostatic mechanisms in biology, dynamic systems, economics, and other social or human systems. In this thread a deep understanding of the biophysical processes dominates. The other thread refers to technology, specifically engineering servomechanisms. Here, the mathematical models and differential equations serve as the basis. Servomechanisms are feedback or error-correction systems, which help to control the states of mechanical or other systems. Feedback loops link at least two unidirectional cause–effect relationships. Thus feedback loops represent circular causalities. Whereas negative feedback loops are circular causal processes that tend to counteract deviations from an equilibrium condition, positive feedback loops include an amplifying and disturbing development (Maruyama, 1963) or, in chemistry, an autocatalytic, self-stimulating process. In social systems, autocatalytic processes can be seen in arms races, market competitions, or in traffic planning where programs designed to reduce congestion induce increased traffic (Sterman, 2000).

We should note that different ideas of stabilities underlie the biological and the physical engineering track of feedback loops. Physicists often tend to focus on equilibrium in the meaning of asymptotic stability.

The same holds true for chemists, who for a long time thought "that a homogeneous, time independent state similar to equilibrium should eventually emerge from any chemical transformation" (Nicolis & Prigogine, 1989). Biologists like to refer to dynamic equilibrium; unstable processes when looking at different opposing processes. Figure 16.10 shows bistability and hysteresis, which are examples that we introduce to show that stability in natural or social systems, such as limit management, can encounter dynamic, abrupt changes. It is interesting to see that in ecology, where the concept of equilibrium has dominated ecological thought for so long, this concept has been supplemented if not dominated by the concept of panarchy (Gunderson & Holling, 2002), which postulates that periods of growth are followed by release, collapse, and reorganization. This approach implies that species, ecosystems, groups, companies, societies, etc. do not endure for eternity.

We also have to acknowledge that there are many systems where forward-looking agents are minor, such as stock markets, and lack predictability (Westley et al., 2002). We also have to be aware that presumed predictability is often unrealistic because human systems ignore the unknown (Ravetz, 1993). However, human systems have the capability to adapt. Many crises or collapses of societies seem theoretically avoidable. Human systems have forward looking capabilities and stabilizing principles such as vicarious mediation (see Chapter 6.3; Brunswik, 1952), they have multilevel rationales (see Box 1.4), and they can conceptualize and manage various types of feedback loops.

System theorists have used various languages to describe feedback loops. In general, we can distinguish between verbal, graphical, and a set of mathematical representations ranging from Boolean algebra via differential equations to multiagent models. The term *loop* is highly visual and metaphorical. Thus, though there is no difference with regard to content, system theorists prefer rope-like arrows instead of the straight arrows shown in most of the figures in this book. The difference and animosity between the biological researchers and the formal mathematical track studying feedback mechanisms is nicely reflected by von Bertalanffy:

> A *verbal model* is better than no model at all, or a model which, because it can be formulated mathematically, is forcibly imposed upon and falsifies reality. Theories of enormous influence such as psychoanalysis were unmathematical or, like the theory of selection, their impact far exceeded mathematical constructions which came only later and cover only partial aspects and a small fraction of empirical data. (von Bertalanffy, 1968, p. 24)

According to the constructivist functionalist perspective, feedback loops are, epistemologically, elements of thoughts. They are constructed to understand how human systems and environmental systems work. In a system model the *structure* is provided by the pattern of connections and by the system elements, whereas the *functions* are the flows and changes of state as energy, matter or information passes through network structures in reaction to an output function. But these elements of thought about structure and function are realized when engineers construct machines based on these ideas (see Boxes 16.6 & 16.11). Whether one can describe actual organismic processes using feedback loops is a question discussed in the next section.

16.5.3 Primary and secondary feedback loops in environmental literacy

The following analysis of feedback loops and sustainability learning starts with elementary definitions of primary and secondary feedback loops.

Figure 16.11a presents a simple case of a primary and a secondary feedback loop – that of a species' population growth being regulated by birth and death rates. The birth and death rates are assumed to be constant, and environmental stress dynamics such as population density or lack of environmental resources is not assumed in this case (see May, 1980b). If at time t_0 the birth rate is smaller than the death rate, the population declines and vice versa. Figure 16.11b presents the same story somewhat differently for the human species. Here we have goal-related decision rules, which could be the birth rate of a nation, and the environmental feedback, which could be the stress of a population. Here, Malthusian models of starvation and death rates can be applied.

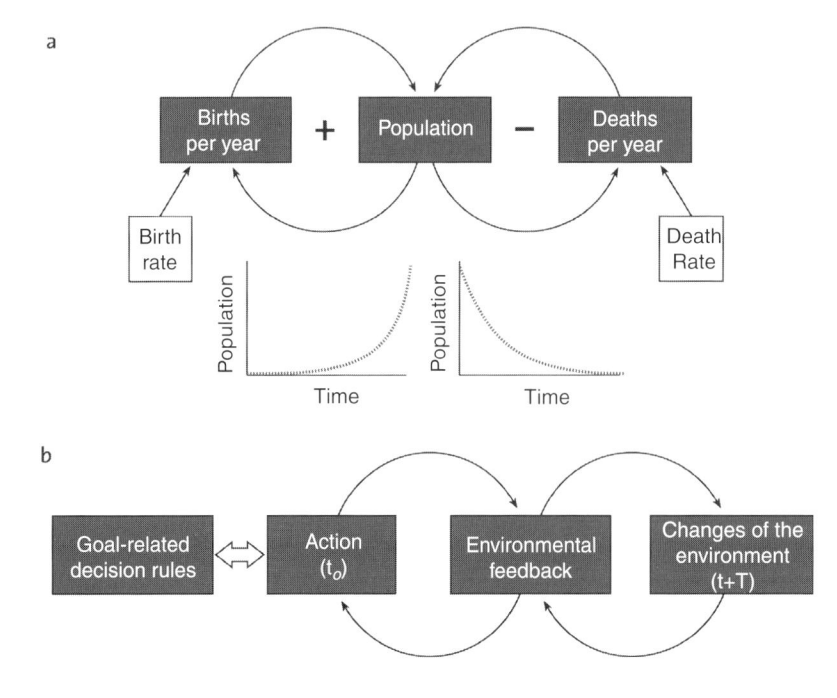

Figure 16.11 (a) Primary and secondary feedback loops in one species' population growth. (b) Human–environment interactions displayed as a simple feedback loop, including delayed secondary environmental feedback.

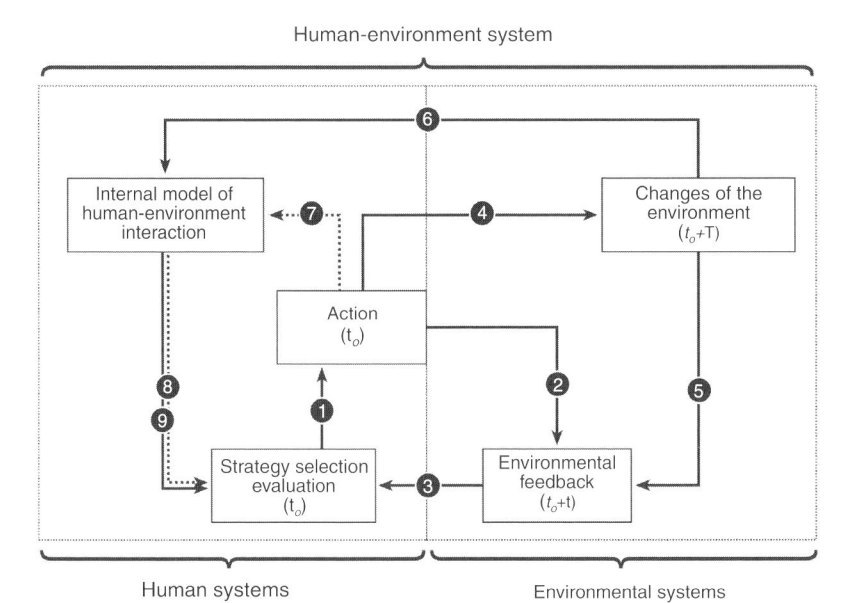

Figure 16.12* Primary and various types of secondary feedback loops in human–environment interaction. The inner cycle $F_1 = (1,2,3)$ (without the incoming and outgoing arrows) represents a primary feedback loop or single loop learning. There are three secondary feedback loops. On the right-hand side, $F_4 = (4,5,3,1)$ represents the changes of the environment, which include potentially delayed changes of the environment. The cycle $F_2 = (1,4,6,9)$ displays the information the individual receives about the changes of the environment resulting from action. The dotted cycle $F_3 = (7,8,1)$ displays objective self-reflection (e.g. when anticipating environmental feedback), which can change goals and behavioral programs.

Figure 16.11b shows a similar way of modeling environmental feedback in a secondary feedback loop for human action. The action at t_0 is assumed to be directed by a goal-related decision rule (primary feedback). However, this primary environmental feedback is coupled with a secondary environmental feedback loop that provides a *delayed* feedback at time $t_0 + T$. A similar, well known classical example of this is the Lotka–Volterra model of hare and fox population dynamics (see Figure 16.14).

To understand the different types of feedback loops that are essential for environmental literacy, we first review the most basic model of learning in *classical* and *operant* conditioning. According to a simple but powerful theory, the learner repeatedly receives one and the same stimulus under one and the same condition. In classical Pavlovian conditioning, a neutral stimulus is associated with an innate reflex. Learning here involves a simple association of the reflex-bound stimulus with a neutral response.

Another model is that of operant conditioning (i.e. giving a positive feedback to reinforce a certain behavior which is not innate). The probability that the action is shown increases as the subject receives rewards after exhibiting the desired behavior. In the opposite case, a certain action is eliminated or suppressed or the probability of its appearance is reduced if a negatively evaluated environmental feedback follows the action. Conditioning is one way in which human systems adapt to the environment (Gerrig & Zimbardo, 2008;

Rescorla, 1988). For an understanding of conditioning studies, it is important to clarify that the goal and the goal-related decision and learning rules are considered to be constant over the long term.

Figure 3.3* visualized the Complementarity Postulate P1 in the left and right parts. The human–environment complementarity is also basic for Figure 16.12*. An action of a human system is seen as part of the interface between the human system H and the environmental system E as any action is based on cellular, biophysical processes. Contrary to Figure 3.3*, Figure 16.12 does not explicate the material vs. social complementarity goal system. Actions cause intended *impacts* at time $t_0 + t$, for instance if you are harvesting vegetables. This is an example of a primary feedback loop. But this action will also affect the environmental dynamics at time $t_0 + T$ in an unintended, collateral way. Thus, arrow 6 in Figure 6.12* represents what we call a secondary feedback loop.

The above example of operant conditioning, for example, can be represented by the inner lower cycle of the sequence of actions represented by the arrows 1, 2, and 3: the feedback loop $F_1 = (1,2,3)$. This type of interaction with the environment is called primary feedback loop learning, first-order, or single loop learning (Argyris, 1977, p. 116).

A special case of this feedback loop would be that a human system might change a behavioral plan owing to anticipation of its consequences. To reflect this, arrow 7 in Figure 16.12* should be represented as a dotted line. This type of feedback loop has also been called reframing. Learning with respect to the adjustment of the decision rules depends on a static internal model of the human–environment interaction.

However, in interactions with both the social and the material environments, the state of the environment is changed by human action. Thus, learning requires not only an adaptation of the behavioral program to an unchanging environment but also the incorporation of unintended changes. This in turn can lead to redefining the goal, which can have various roots.

To understand learning as a basic component of environmental literacy, we introduce two secondary feedback loops. One is called secondary feedback loop learning: changing the internal model because of an awareness of the changing feedback characteristics of the environment. This is represented by the arrows defining the loop $F_2 = (1,4,6,9)$. A special case of this learning is when the change of the environment at time $t_0 + T$ is caused by the human system at time t_0. In reality, there are potentially negative feedback loops in many domains. An example would be that of chopping down a forest under the livelihood goal of selling the wood.

As dynamic, complex system, the environment includes non-linear changes, multiple interactions, and feedbacks among variables and subsystems. It also provides feedbacks that can be delayed or displaced. In ecology, "indirect effects may be as important as direct interaction in a food web" (Odum, 1992, p. 543). On the collapse of fish populations, ecologists report:

> In each case the pattern was similar: a period of intense exploitation during which there was a prolonged high level harvest, followed by a sudden and precipitous drop in populations. (Holling, 1973, p. 8)

Coping with slow, delayed, or non-linear changes in the environment is difficult for people, as is also shown in experimental settings (Dörner, 1996). A special characteristic of environmental secondary feedback loops is learning about spatially remote variables (Holling, 2003), which are blended from acting.

In Piaget's terminology (see Chapter 6.11), learning according to a fixed behavioral program (or internal model) which is represented by primary feedback loop learning, can be conceived as *assimilation*, whereas the change of the decision rules based on noticing environmental change can be considered as *accommodation*. The latter leads to a change of behavioral programs based on experience.

Furthermore, there is a third feedback loop to be considered, represented by the dotted line. Human systems can *reflect* on their goals and action programs. This second-order feedback loop, which has also been called a third-order feedback loop, is consciousness. It has also been called double loop learning. It means that the internal model of human–environment interaction reflects on "policies and goals as well as its own program" (Argyris, 1977, p. 116).

The changing of goals by consciousness through secondary feedback loop learning is represented by the circle $F_3 = (7,8,1)$. Here the 1 stands for an action of a human system shown at time t_0 which becomes subject to evaluation depending on the environmental feedback at time $t_0 + t$ (i.e. dependent on F_1), as well as on the environmental changes caused by the human system's action (i.e. dependent on F_2). Clearly, this loop includes an internal model of human–environment interactions. This would include a critical reflection of the "inner" purpose and adequacy of the behavior. Here an ideal/actual comparison can lead to a reformulation of the goals. This would correspond to F_3. It means that the internal model of human–environment interaction reflects on the "policies and goals as well as its own program" (Argyris, 1977, p. 116). This type of learning produces a continually changed self. If this type of reflection is paired with the secondary feedback learning, we also speak of appreciative learning (Checkland, 2000).

Duval and Wicklund (1972) introduce another, higher order, feedback loop, which is subjective self-awareness. This type of awareness is gained if one reflects on the unintended effects of one's own behavior on the environment.

The above analysis has revealed that here are at least three types of secondary feedback loops:

F_1, the secondary environmental feedback loop of unintended change; F_2, the secondary feedback loop learning about this change; and F_3, the secondary feedback loop of self-reflection. We think that all of these are essential features of environmental literacy.

16.5.4 Feedback loops in engineering servomechanisms

The first written record of the use of the feedback control loop goes back to Ktesibios' mechanical and hydrological clocks (see Box 16.6; Lepschy et al., 1992). For the study of emerging environmental literacy, it is of utmost interest to see how fundamental

16.5 Feedback Postulate P4

Box 16.6 The origins of feedback loops: Ktesibios' water clock

The feedback principle had already been elaborated by Ktesibios (285–222 BC; Mayr, 1970) when he created the clepsydra, a water clock (see Figure 16.13a) that was the first documented automatic machine applying a feedback device. These clocks were, for instance, positioned in many places in Alexandria and also served to limit the time allotted to speeches in court. When enough water trickles into the lower measuring vessel, the person indicating with the pointer is lifted by the lifting platform P.

A mathematical description of Ktesibios' clock is based on Toricelli's Law, which states that the speed of the outflow E is proportional to the square root of the height h.

Figure 16.13b presents the feedback loop structure when referring to the polarity. It provides a qualitative analysis of feedback that is utilized in systems analysis in both the natural (Thomas & D'Ari, 1989) and social sciences (Richardson, 1991a). All three feedback loops are negative, as they consist of an odd number of links, which tend to stabilize or counteract change. A description of the clock can also be shown using a simple differential equation (Lepschy et al., 1992).

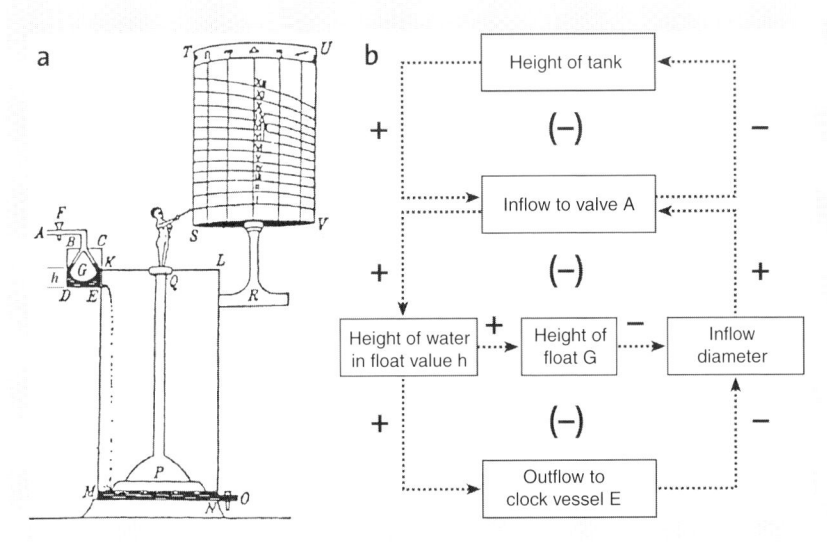

Figure 16.13 (a) The first documented feedback loop system of a float valve water clock created by Ktesibios (205 BC; adapted from Mayr, 1970, p. 12) and (b) the clock represented as a feedback loop.

ideas are implemented in technology (machines), put into mathematical notation, and subsequently used for modeling. Modern technology employs a lot of self-regulation. A special branch is automatic control theory (Bennett, 1996), which uses feedbacks to control states and outputs of dynamic systems. Feedback loops are a central means to design stability and order of systems (see Box 16.11).

Basically, secondary feedback loops in technical systems are identical to the feedback loop presented in Figure 16.11. In electrical space heating with binary on–off regulation, for example, delayed feedbacks have been employed (see also Box 16.11). The switching off or on (action at time t_0) led to slow temperature changes, which were a delayed feedback at time $t_0 + T$. Thus, the delayed feedback must be anticipated. Even in the early years of control theory, it could be shown that "the performance of a 2-position thermostat with secondary feedback was superior to the performance of continuous temperature controllers" (Roots & Walker, 1967, p. 401). Technically, the tardily reported changes of the environment (i.e. environmental feedback via F_2) have to be anticipated with a sophisticated mathematical model, which incorporates the predicted changes of the environment represented by F_3.

16.5.5 Feedback loops in biology

Regulation, homeostasis, control, and feedback systems are concepts from the field of physiology. Using these ideas, biologists have investigated the mechanisms that underlie an endothermic animal's ability to regulate its internal system to maintain a stable, constant condition (e.g. for body temperature or blood glucose levels).

A classical example illuminating feedback loops comes from early models of population dynamics. One such model was proposed independently by the

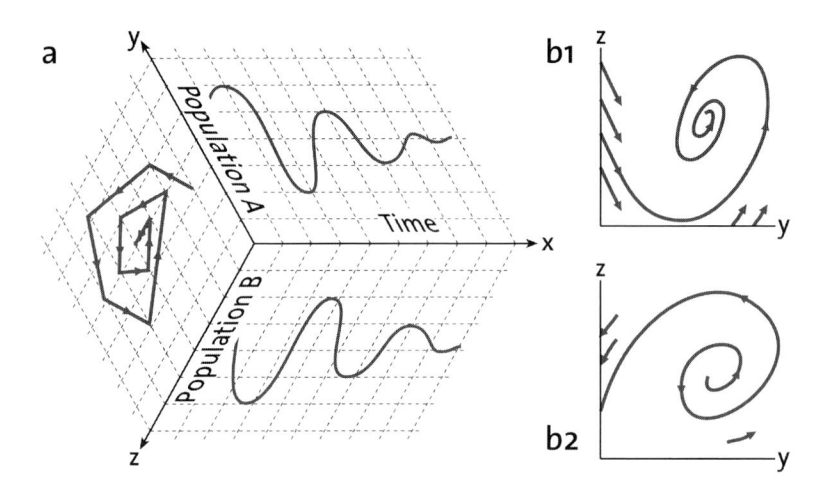

Figure 16.14 Sequential change and mutual regulation of two populations A and B over time. (a) The front graph (phase plane) on the $x-y$ plane represents stability characteristics of the interaction (adapted from Holling, 1973, p. 3). (b1) Stable dynamics; (b2) unstable dynamics (ibid, p. 4). (c) A feedback loop presentation of the two population dynamics (α, β, γ, and δ are positive real numbers).

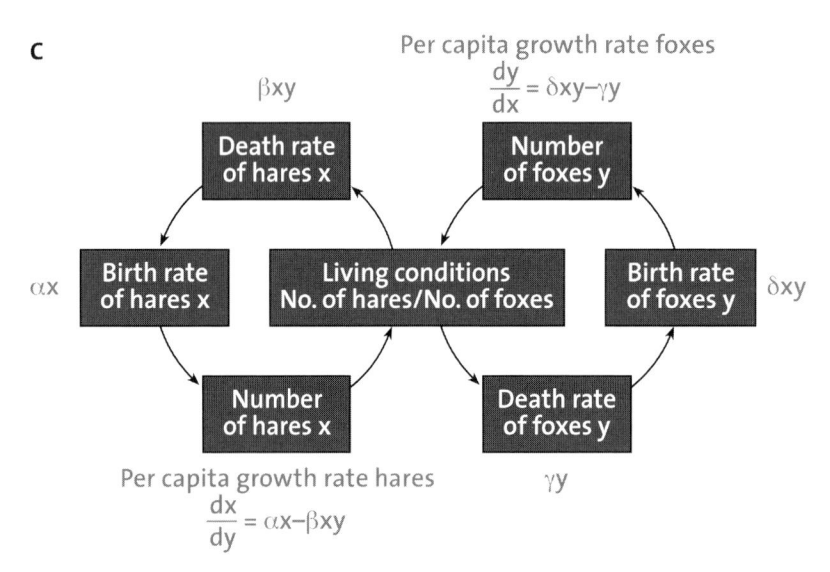

mathematical biologist Vito Volterra (1860–1940) and by Alfred Lotka (1880–1949) (Lotka, 1925), a chemist and mathematician with knowledge of ecology and demography. The basic model provides a description of two species, a prey species with unlimited food and other natural resources, and a predator species, which is exclusively dependent on the availability of that one particular prey species. Volterra's model shows how, under these simplified conditional assumptions, the environment for each species is regulated by the other species. Though it has been long stated that complex ecosystems evolve towards homeostasis, the difference to organismic body systems is that there is no fixed set-point control in ecosystems. Ecosystems return to a certain (evolutionary) trajectory in the sense of a dynamic equilibrium instead of a fixed state.

Figure 16.14c provides a feedback representation of Volterra's model. The inner species cycle can be considered a primary feedback loop, whereas the other species acts as the secondary feedback loop. The dynamics of different interacting populations can show how positive and negative feedbacks can have more than one equilibrium point or can show chaotic characteristics. The essence of the Lotka–Volterra idea is that there is a positive feedback within the populations but, in the case of hare–fox interactions, a *negative* one in the direction of foxes to hares (the more foxes, the fewer hares) and a *positive* one in the other direction (the more hares, the more foxes). We can state that under quite general conditions in non-linear systems (of first-order differential equations), a negative feedback loop is necessary for stability. The inner species dynamics can be considered as a positive primary feedback loop, whereas the feedback from the other species can be considered the response of the environment to their own actions. This can be considered secondary feedback.

Taking a closer look, the differential equations in the Lotka–Volterra model can be explained by different dynamics. For example, population equilibria (when $x(\alpha-\beta\gamma) = -y(\gamma - \delta x) = 0$ for the differential equations in Figure 16.14c) can lead to the extinction of both species or to steady states (when $y = \alpha/\beta$ and $x = \gamma/\delta$), or to oscillating dynamics, or we can meet unstable dynamics, as introduced in Figure 16.10 (May, 1980a).

Feedback loops are essential for homeostasis, and the applications range from the molecular scale (i.e. modeling of interactions on the level of DNA, RNA, and the cell) to the global scale (i.e. modeling global biogeochemical systems). We do not go into this topic in detail but should note that positive feedback loops are a necessary condition for multistationarity (Plahte *et al.*, 1995); that is, the indeterminacy of which equilibrium will be attained (Thomas & Kaufman, 2001a, b).

16.5.6 Feedback loops in social systems analysis

> It is my basic theme that the human mind is not adapted to interpreting how social systems behave. Our social systems belong to the class called multi-loop nonlinear feedback systems. In the long history of evolution it has not been necessary for man to understand these systems until very recent historical times. Evolutionary processes have not given us the mental skill needed to properly interpret the dynamic behavior of these systems, which we have now become a part of. (Forrester, 1971a, p. 54)

The tide of feedback systems research grew out of World War II engineering knowledge and was subsequently applied to business and policy models. Doubtlessly, a cornerstone for systems dynamics approach has been provided by Jay W. Forrester.

> The great challenge for the next several decades will be to advance understanding of social systems in the same way that the past century has advanced understanding of the physical world. (Forrester, 1987, p. 36)

As indicated by the progression of his book titles, such as *Industrial Dynamics* (Forrester, 1961), *Urban Dynamics* (Forrester, 1969), and *World Models* (Forrester, 1971b, see Box 16.7), Forrester initially began working from an economic viewpoint and then shifted to a comprehensive world model that included variables such as quality of life and crowding. Moreover, as he repeatedly emphasized, he was "introduced to feedback systems by Gordon Brown in the MIT Servomechanisms Laboratory in the early 1940s" (Forrester, 2007a, p. 349). A key step towards computerized and differential equation system modeling was the development of the World I software (Forrester, 1971b, 1987). The system-theoretical approach was extended to the study of business and organizational dynamics by Argyris (1977, 1978), Senge (1977), and Sterman (Senge & Sterman, 1989).

In contrast, experimental social psychologists have utilized complex computer simulations to demonstrate the exceedingly poor human capability of anticipating feedback loops, particularly in combination with time delay and non-linear functions (Dörner, 1996; Dörner *et al.*, 1994; Funke, 1999). In general, people cannot mentally simulate the simplest feedback system, a first-order positive feedback loop of exponential growth where the dynamics are usually described as $\frac{dy}{dx} = \alpha e^{\beta x}$, where x is the time, y the change of a dependent variable over time, and α and β are constants. Here we encounter bounded rationality (see Box 1.4), something we see in the case of repeated misperceptions of feedbacks in simulated market and distribution games. In this study, subjects generated costly oscillations, even where it could have been avoided with minimal effort. Moreover, the subjects ignored time delays and rebound effects inherent in commons dilemma structures. Results from these studies show pattern characteristics of many real-world commodity markets, including boom-and-bust cycles, overcapacity, price wars, and shakeouts (Paich & Sterman, 1993; Sterman, 2000).

Reflecting on how people tend not to use feedback in a way we expect, we can make several guesses as to why they do not. Perhaps people follow a single-strand causal series, or are over-focused on single aspects. One could also argue that, as is the case in many commons dilemma situations, short-term trade-offs dominate long-term trade-offs.

Clearly, these results are derived from artificial environments. Many argue that decision-makers perform better in real-world situations; however, counterexamples from accounts of training experiences using *virtual* worlds or managerial flight simulators suggest the opposite (Graham *et al.*, 1989). Learning tools that employ simulation are particularly helpful in coping with feedback structures if experiential, intuitive,

16 The HES Postulates

> **Box 16.7** Multiple feedback loops in complex systems: Forrester's world model
>
> In his book *Principles of Systems*, Jay W. Forrester argues that humans' deficit in understanding complex economic dynamics lies in the "absence of ... identifying and expressing the body of universal principles that explain the successes and failures of the system ..." (Forrester, 1968, pp. 1–2). These universal principles are system principles, open and feedback systems, and feedback loops. Based on the economic modeling of a company's order placement, he derives a couple of principles that emphasize the role of understanding feedback loops, such as:
>
> "The feedback loop is the basic structural element in these systems. Dynamic behavior is generated by feedback. The more complex systems are assemblies of interacting feedback loops" (ibid., pp. 4–5).
>
> "Every decision is made within a feedback loop. The decision controls action, which alters the system levels, which influence the decision. A decision process can be part of more than one feedback loop" (ibid., p. 4).
>
> "A feedback loop consists of two distinctly different types of variables – the levels (states) and the rates (actions). Except for constants, these two are sufficient to represent a feedback loop. Both are necessary" (ibid., pp. 4–6).
>
> System dynamics has quite a successful story. Only 9 months after Forrester's *World Dynamics* (the world model is given in Figure 16.15) came the publication of the more complex model presented in *The Limits to Growth* (Meadows et al., 1972), which strongly affected the foundation of the Club of Rome (Meadows & Meadows, 2007). The strengths of these works are that they reveal complex dynamics, exponential dynamics, declines, etc. that are not easily understood intuitively. Their flaws are, most assuredly, the difficulties in validation of and the lack of public trust in such models.

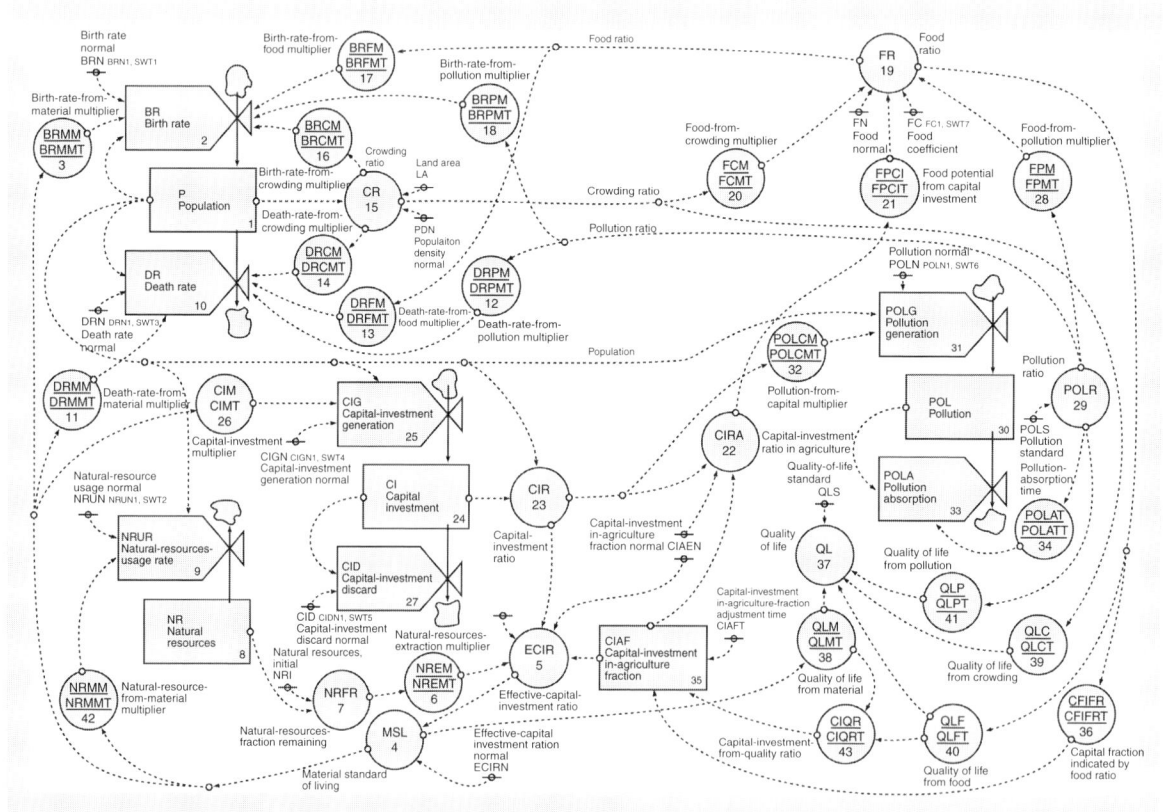

Figure 16.15 Complete diagram of the world model interrelating the five level variables – population, natural resources, capital investment, capital investment in agriculture fraction, and pollution (Forrester, 1971b, reprinted by permission of Pegasus Communications).

and analytic aspects of learning are combined and follow, for example, a "discover–invent–produce–reflect" chain. The abovementioned graphical, interactive computer laboratories can be useful to reflect on intuitively inaccessible feedback loops (Senge & Sterman, 1989). Here we find a link to hard and soft system thinking or soft operational decision-making (Checkland, 2000; Forrester, 1994; Rosenhead & Mingers, 2001). We come back to this point below when discussing the role of understanding feedback loops.

16.5.7 Feedback loops and environmental literacy

All learning depends on feedback. In many instances, learning requires more than finding the best strategy, such as an optimal response rate, in a static environment like a psychological laboratory. Many problems have to deal with constantly changing, tightly coupled actors and systems (see Complementarity Postulate P1) that display non-linear dynamics which arise from their internal structure (e.g. self-organizing characteristics) and provide counterintuitive, unexpected symptoms and developments. Many of the current problems related to resources scarcity are because of HES dynamics that are poorly understood because of multiple, spatially dislocated and temporally delayed feedbacks resulting from human systems' behavior. A critical question is how *common sense* and non-experts can access counterintuitive system dynamics. This has been widely supported by experimental psychological studies (Blech & Funke, 2006). Forrester noted:

> It has been repeatedly demonstrated that the human mind is not suited for solving high-order dynamics. (Forrester, 2007b, p. 363)

In our definition of environmental literacy (see Chapter 2), systems thinking, systems modeling, and coping with higher ordered feedback loops build the core of environmental literacy. These issues have been a key message of Forrester's systems dynamics approach (Forrester, 1969, 2007a, 2007b; Sterman, 2007).

As coping with higher ordered feedback loops is key for complex system understanding, we consider this capability as a key component of "sustainability learning" (Scholz et al., 2006).

A critical issue is that complex system graphs, such as that in Figure 16.15, are themselves an oversimplification of reality. Even worse, many researchers try to oversimplify modeling and reduce system dynamics to pure systems thinking in which only the qualitative relationships are investigated. This has been critically commented upon:

> … making systems dynamics simple [are playing] a losing game. (Forrester, 2007b, p. 361)
>
> System thinking [is] only as a sensitizer … to the existence of systems. Some people feel that they have learned a lot from the systems thinking phase. But they have got only 5 percent of the way into understanding systems. The other 95 percent lies in the system dynamics structuring of models simulations based on those models. (Forrester, 2007a, p. 355)

We think that these statements are both of importance for the development of environmental literacy and for a meaningful coping with feedback loops by systems dynamics. A critical question here is what type of system characteristics can be described by systems dynamics. Here a critical question reads:

> Why is there so little impact of system dynamics in the most important social questions? (Forrester, 2007b, p. 361)

As Forrester learnt, there is poor acceptance of the method and the findings of system dynamics by peers from other disciplines. Forrester reports on this experience when he dealt with policies for low-income housing for African Americans in the USA. His complex, quantitative modeling, which involved special variables, resulted in the counterintuitive conclusion that low-cost housing would have long-term negative effects in reviving urban settings. Thus, low-cost housing, a policy option pushed by black politicians from Harlem, should have been considered a "most damaging policy" (Forrester, 2007a, p. 349). When discussing this lesson, Forrester still optimistically claims that education and access to modeling abilities would be vital. Thus, pure modeling competence and technical knowledge (instead of just trusting in the experts and their results) are the keys to open the door to environmental literacy.

16.5.8 The qualitative and the quantitative side

Let us briefly reflect on coping with feedback dynamics in real-world settings such as flood prevention, reducing pollution, or pandemic management. As outlined in Chapter 15, we consider the division of labor, roles, and knowledge to be salient aspects of transdisciplinary processes. Here, we think that knowledge on *qualitative* system thinking is necessary for the decision-makers at every level (including common people), whereas *quantitative* system modeling knowledge is necessary for system dynamics experts.

The basic idea of transdisciplinarity is that we need the real-world experience from practitioners and the modeling experience of experts, as well as appropriate strategies to bring the two approaches together. Obviously, many naïve promoters of system dynamics seem sometimes to forget that a model is not the real world. However, a model should provide

Box 16.8 Rebound effects in introducing ecoefficient mobility systems: Swissmetro

Measures or technologies that are initially judged as environmentally positive can show unexpected and undesired environmental rebound effects. This is particularly true for transportation systems. As roads become congested with ecologically inefficient cars, public transportation is often unconditionally promoted as the preferred option, based on arguments of lower traffic volume and energy efficiency per seat-kilometer.

As the operation of vehicles is a major source of environmental impacts, high-speed Maglev underground trains appeared to be an attractive option for Switzerland. Thus, based on ideas from the 1970s, a Swiss national transportation project named Swissmetro was created (Rossel, 2001). The plan was to utilize depressurized tunnels with low air resistance, no noise emissions, and no need of surface space for rails. Potentially, the Maglev trains showed high speeds, of up to 500 km/h, and frequencies of 4–6 minutes.

A challenge is to provide a prospective environmental evaluation for this new technology. Here, a conventional way of evaluating would be simply to compare the environmental impacts E_0 of 1 kilometer traveled by a person (pkm) with an average Swiss travel mix of cars and conventional trains with the environmental impacts E_1 of 1 pkm traveled with Swissmetro (e.g. by using life cycle assessment (LCA)). We call this direct substitution, as it assumes that the pkm run by Swissmetro would directly substitute other modes of transportation. Thus we are evaluating the change of environmental impacts resulting from an action, which is an effect from a first-order feedback loop. Such an assessment includes the construction of tunnels, the utilization of the matter that was excavated, the construction vehicle technology, and the operation of infrastructure and trains, for instance in a lifetime of 100 years. An initial LCA analysis conducted in 1997, assuming 29 400 passengers per day and a capacity utilization of 61%, provided a Swissmetro energy consumption (as a proxy for E_1) of 1.40 MJ/pkm. This is almost as good as high-speed trains (1.26 MJ/pkm, given a capacity utilization of 55%) and is much better than cars (3.0 MJ/pkm, assuming a capacity utilization of 34%). Postulating that Swissmetro would substitute a normal travel mix, one could conclude that Swissmetro would cause less environmental impacts than the currently given travel modes as $E_0 > E_1$ (Baumgartner et al., 2000).

However, building Swissmetro would also bring changes to peoples' behavior and demands. These changes constitute second-order feedback loops and are based on changing travel times, costs, convenience, etc. But these changes can cause structural changes as well, such as population density increases near Swissmetro stations, reduction of conventional train services, etc. Incorporating these effects, one can assess the environmental impacts E_2, including second and higher order impact chains.

Spielmann et al. (2008) calculated a prospective LCA, assuming that people would continue travelling on average 85 minutes per day after a Swissmetro net was built (see Figure 16.16b). The calculations were done for different uses or technologies, including bicycle, car, motorbike, buses, etc. Higher ordered changes of settlement structures were not included. The results show an increase of per capita environmental impact for all considered scenarios, indicating that Swissmetro leads to higher environmental impacts than the current state (i.e. $E_0 < E_2$). If we define the environmental rebound effect $ERE = 1 - \frac{E_2 - E_0}{E_1 - E_0}$, we find that *ERE* varies between 11% and 114%, depending on mobility patterns and technological scenarios. The study allowed a better understanding of the consequences of new transport services, and facilitates the assessment of future transport technologies based on inclusion of their secondary feedback loops.

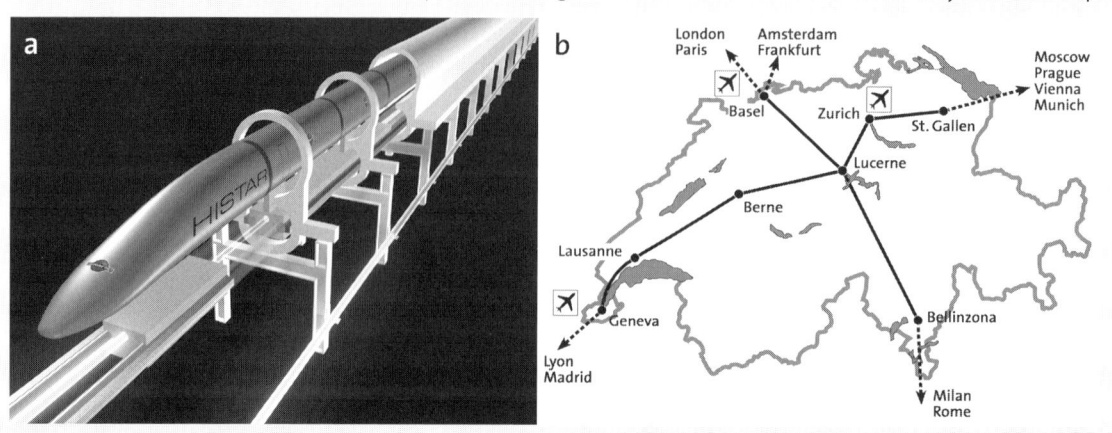

Figure 16.16 (a) Swissmetro trains are reaching a speed of up to 500 km/h in depressurized tubes. (b) The planned network of the Swissmetro, including eight stations.

16.5 Feedback Postulate P4

Box 16.9 Emerging environmental awareness: nuclear eyewatch

In his volumes *Introduction à l'Épistémologie Génétique* and other books (Piaget, 1950), Piaget outlined that the individual's cognitive development (see Chapter 16.7) follows the phylogenetic development of intelligence and scientific knowledge. Here, he was referring to Haeckel's recapitulation proposition (see Box 4.11), at least in some weak interpretation.

We can use the stages of circulation in a metaphorical way to characterize levels of environmental literacy. The stage of *primary circulation* can be attributed to very early stages of human–environment interaction. At this pre-*Homo sapiens* level, one can postulate a symbiotic, non-reflected interaction with the environment characterized by non-reflected, instinctual human interventions. This stage meets the newborn's symbiotic instinct-based interaction and ingestion by means of the sucking reflex. In the phase of *secondary circulation* children become aware of their impact on the environment (Figure 16.17). They notice that a rattle causes a pleasing noise when it is moved. This stage would portray settings of human–environment interaction where, for the first time, humans became aware of their actions causing depletion of the environment, but where the feedback of the changes in the environment were not yet noticed.

Finally, there is *tertiary circulation* (Figure 16.17). In the second year of life, the child is becoming aware of the environmental feedback resulting from his or her actions. He or she starts experimenting and testing what beneficial or negative feedback will follow what action. This stage of awareness can also be metaphorically transferred to the level of societies. It looks as though the human population is currently experimenting with the environment and exploring what feedback will result from what type of imposed stress.

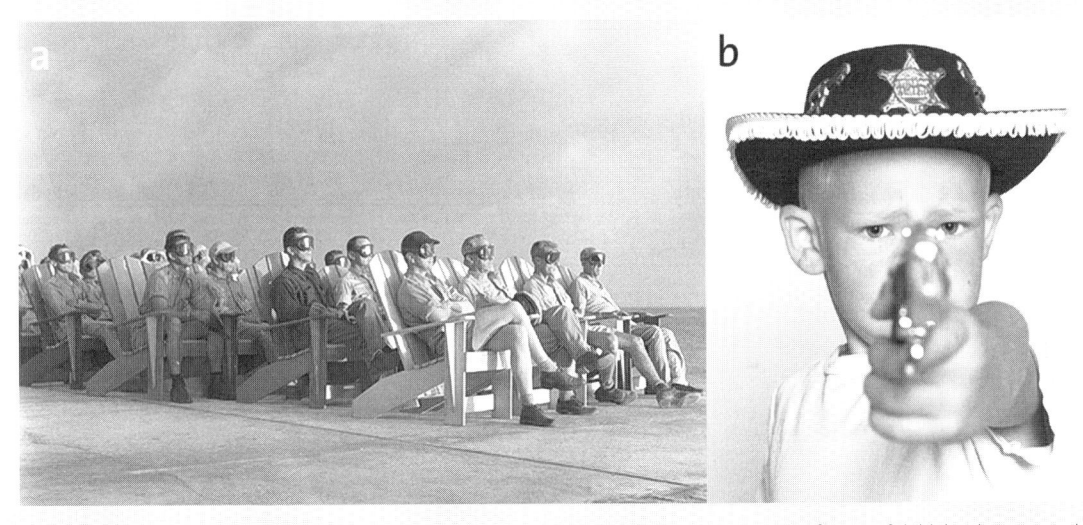

Figure 16.17 Losing the conservative characteristics and experimenting with the environment is a feature of child development in the second year of life. Piaget called this tertiary circulation. Although the child is interacting with the environment and enters a new stage of learning, he or she is not yet necessarily valuating the impact appropriately. (b) In Piaget's stage of secondary circulation, the child is only focused on the impacts without being aware of the feedbacks. A similar pattern might be seen in the "cinemas of nuclear physicists" in the 1950s. (a) VIP observers being lit up by the explosion of an Eniwetok Atoll atomic bomb in 1951 (photo from Kuran, 2007).

insight into structures and counterintuitive dynamics (see Box 16.7). This certainly requires special didactic pathways, for instance, in abandoning the opaqueness of overly complicated models (Eden & Ackermann, 2006) – a prerequisite for acceptability by policy-makers.

In summary, abstracted system thinking is not sufficient to gain insight into the dynamic structures of our material and social environments. This is also a message of Peter Checkland and colleagues' contribution to systems analysis (Checkland, 1981; Checkland & Holwell, 1998; Checkland & Scholes, 1990). Their approach emerged from operational research and moved to strategic choice and soft operational systems analysis (SODA; Rosenhead & Mingers, 2001). The soft system methodology uses less ambitious modeling tools. However, feedback loop concepts do not play such a prominent role in these approaches. Problem structuring and soft operational analysis dominate (Rosenhead, 2006), although the promoters of this approach state that "… 'soft' does not throw away 'hard'" (Checkland, 2006, p. 770).

> **Box 16.10** Secondary feedback loop management in agriculture: crop-field rotation
>
> The history of agriculture – a perfect case for understanding the course of environmental literacy – entails the development of knowledge, techniques, and measures for supplying food and materials. Agriculture involves knowing what type of soil is best for growing what food by what treatment and input. The inputs include seed breeding, irrigation, agrochemicals, fertilizers, plowing, and terracing.
>
> A most elementary secondary feedback loop is linked to fertilizing. If crops are taken from a field, the soil loses nutrients. To maintain a nutrient balance, fertilizers have to be added. Thus, in general, one question is directed to finding the conditions under which we can have sustainable farming, meaning "where we can expect the yield of food to continue undiminished for a very long time" (Newman, 1997, p. 1334). In general, there are very few areas of the world (such as the fields along the Nile) where sediment deposition has provided a balanced nutrient flow for agriculture without additives. This, in particular, holds true for phosphorus, which in modern farming comes almost completely from rock phosphate, a non-renewable resource (Lesschen *et al.*, 2007; Newman, 1997). The history of agriculture is full of examples of different fertilizing and cropping procedures, including techniques such as milpa (Figure 16.18b), swidden, that is slash-and-burn cultivation (Wenke, 1999), crop rotation, spreading manure, or parking animals overnight on arable land.
>
> Proper fertilizing requires a thorough understanding of the nutrients that plants need. This was promoted by Justus von Liebig, who brought chemistry to agriculture (Liebig, 1840). Thus the importance of nitrogen (N), phosphorus (P), potassium (K), calcium carbonate ($CaCO_3$), and magnesium (Mg) were detected.
>
> Crop rotation also seeks to balance the fertility demands of various crops to avoid excessive depletion of soil nutrients. A traditional component of crop rotation is the replenishment of nitrogen through planting legumes, which have nitrogen-fixing bacteria in their roots, or through using green manure in sequence with cereals and other crops. Crop rotation can also improve soil structure and fertility by alternating deep- and shallow-rooted plants. It is interesting to see how knowledge on how to counter the leaching of soil has developed historically. For instance, the Mayans used a farming technique called milpa which involved planting a combination of corn, beans, and squash. However, this is done in quite different ways. For example, some Mayan communities plant large milpas, of 5 hectares or more, and use the area only once before moving to a new site, whereas others apply succession techniques and "conserve forest resources in and around their fields, leave large reserves in and around their plots, and worry about sustaining biodiversity" (Nigh, 2002, p. 454).
>
> **Figures 16.18** (a) Resembling a work of modern art, green variegated crop circles cover what was once short-grass prairie in southwestern Kansas. The most common crops in this region, Finney County, are corn, wheat, and sorghum. Each of these crops was at a different point of development when the Advanced Spaceborne Thermal Emission and Reflection Radiometer (ASTER) captured this image on June 24, 2001. The image covers an area of 64 km^2 (photo by NASA, 2001). (b) Based on the ancient agricultural methods of Maya peoples and other Meso-American peoples, milpa agriculture produces maize, beans, lima beans, and squash. The milpa cycle calls for 2 years of cultivation and 8 years of letting the area lie fallow.

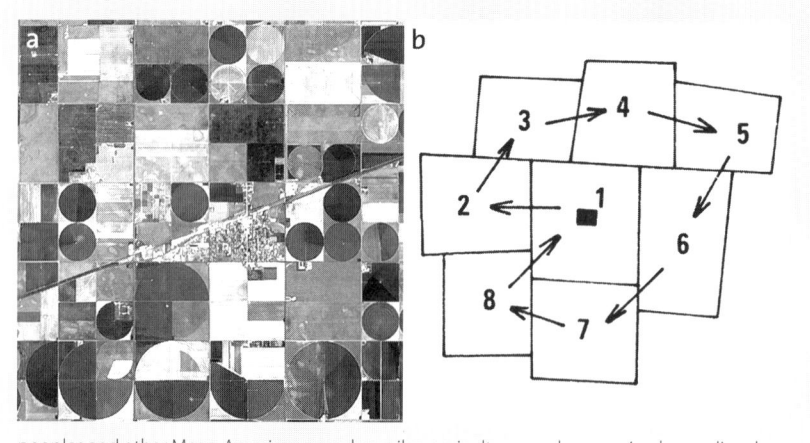

A thorough understanding of the problem has to precede a formal analysis of feedback loops. Gaining access to feedback loops in real-world settings requires merging the knowledge from real-world experts with those of system experts (Scholz, 1999; Scholz & Marks, 2001). Here, the understanding of feedback systems is crucial, if not a *conditio sine qua non* for sustainability.

16.5.9 Rebound or take back effects

The understanding of rebound or take back effects (see Chapter 14) and the capacity for anticipating unintended collaterals is considered to be a major element of environmental literacy and sustainability learning. The general issue is that technological innovation often leads to savings in energy, resources, time, expenses, etc. for each *single* good or service. This in turn may, however, cause greater *overall* consumption and environmental impacts; for instance, if the time or the money saved by a new technology is invested in additional activities or purchases whose environmental impacts surmount the reduction of environmental impacts by the new technology (Herring & Roy, 2007; Hertwich, 2005a; Hilty *et al.*, 2006; Kohler & Erdmann, 2004; see Box 16.8). Thus, from evaluating the impacts, any environmental innovation has a first-order environmental impact. This includes the increase or decrease of environmental impacts when comparing what is considered as a direct substitution. In LCA this is defined by the environmental impacts per functional unit, which is, for instance 1 km traveling by one person. The secondary feedback loops include the environmental changes that are implicitly caused by a new technology. These are called environmental rebound effects of technology innovation (Jänicke, 2008).

16.5.10 Delayed feedbacks

Not only the existence and mastering, but especially the velocity and the locating of feedback loop impacts are important dimensions of sustainable action; for example, the delayed feedback from climate change mitigation measures. Here, the desired outcomes can only be achieved if sustainability learning and practical responses by human systems are at least as fast as or faster than the creation of risks and of malignant and irreversible environmental changes.

Given that positive feedback from current action would only be realized in a couple of decades, this is difficult to understand and to believe. An even more critical issue might be that the benefits resulting from action are received by future generations. From the perspective of an individual, this might be perceived as altruistic action (see Figure 7.10), as the following generations which are benefiting from action may be considered as "others" (Hulme, 2009).

16.5.11 Key messages

- Feedback loops are general patterns of living systems (organisms), where they assure self-regulation (homeostasis)
- Feedback loops are models of circular causalities. They are a key concept in systems theory, cybernetics, control theory, system ecology, engineering sciences, and other sciences
- There are different kinds of feedback loops, such as positive, negative, primary, secondary, higher ordered, etc., which can be applied for modeling system dynamics
- Feedback loops are a major component of learning by human systems at all levels
- Human systems often lack the capacity to anticipate complex, non-linear dynamics of primary and, in particular, secondary feedback loops in environmental systems. This deficit is enhanced if primary or secondary feedbacks appear delayed or dislocated
- Learning to identify and to cope with potentially delayed, dislocated, unintended, implicit system changes (i.e. secondary feedback loops) is a major component of sustainability learning. Assessing environmental rebound effects in prospective environmental technology evaluation is a typical example of this
- There are multiple representations of feedback loops (e.g. verbal, graphical, and formal mathematical). The best representation depends on the case, on the available knowledge, and on what one wants to investigate
- Feedback loops can cause instability, chaotic behavior, bifurcation, hysteresis, and other system dynamics under certain constraints, even in systems with few interacting system components or variables
- System dynamics has become a tool to model complex systems through multiple feedback loops of human activities and interactions of human and (material) environmental systems
- Coping with the potential delayed and dislocated secondary environmental feedback loop of unintended change and the secondary feedback loop of self-reflection are essential features of environmental literacy.

16.6 Decision Postulate P5

16.6.1 Game and decision theory as a general language

Human systems can be conceived as decision-makers with goals and strategies. This Postulate roots itself in the functionalist perspective (see Chapter 3.5). We assume that all levels of human systems are active, purposeful, and intentional. This implies that human systems, and parts of them, have some control over their actions, that they have goals, and that they prefer some potential outcomes to others. As discussed in Chapter 7.2, the existence of preference functions is an essential prerequisite for this view. Referring to game and decision theory we postulate that any human system has: (i) goals and drivers; (ii) can select between different alternative strategies or actions; and (iii) selects these behavioral strategies according to preference or utility functions that meet the system's goals.

There are descriptive and normative approaches to game and decision theory. Normative approaches define optimal strategies when postulating that the different players' behavior follows certain rules. Descriptive approaches represent the action and the decision process leading to a certain behavior.

The genesis and types of goals and strategies can be different within and between human systems. The definition of goals and strategies is dealt with in respect to the disciplines and scientific fields in Chapters 4 through 11.

Decision-theoretic modeling allows for different types of rationality. Besides the von Neumann and Morgenstern (1944, p. 8) economic full rationality approach, bounded rationality or other concepts of rationality can be applied (Box 1.4). There are different goals or drivers in human systems. Different drivers of human individuals, for instance individualistic, cooperative, or competitive motivations, are presented in Figure 7.10. Different drivers of society, such as wealth or maintenance of power, are presented in Figure 8.3*.

Game and decision theory have a long history. Decisions made in the context of uncertainty and risk were dealt with by the Mesopotamians as far back as 3000 BC. When decision-makers were faced with choosing between two risky alternatives, they often called on the Ashipu, sages who were thought to be able to read positive and negative signs associated with each alternative. When taking the difference or the ratio between positive and negative signs, they determined the most favorable alternative (Covello & Mumpower, 1985; Oppenheim, 1977).

Games like chess represent simplified real-world conflicts such as wars or competing businesses. Here, the normative approach of game theory reaches its limits very quickly. Thus, for instance, game theory can prove that in chess there are optimal strategies. But they cannot be known, as chess has more moves than atoms in the universe (10^{80}). The number of possible chess games is even greater, at 10^{128} (Mandecki, 1998). Moreover, it is not possible to know whether white or black would win or whether a tie would result if the optimal strategies are applied.

16.6.2 Key assumptions of game and decision theory

Game and decision theory rely on a few concepts elaborated by John von Neumann and Oskar Morgenstern (1944). These concepts provide a general language which also allows for describing human decision behavior. Summarized, from a game-theoretic perspective, human systems are conceived as: (i) players or decision-makers, who have; (ii) strategies or decision alternatives from which; (iii) outcomes may result that become object of evaluation by; (iv) preference or utility functions.

(i) **Players or decision-makers**, who are also sometimes called agents, can be individuals or collectives (Axelrod, 1984a; Berninghaus et al., 2006; Burger, 1959; Güth, 1992; Hammerstein & Selten, 1994; Klaus, 1968; Osborne & Rubinstein, 1994; Rapoport, 1970, 1999). But theoretically, also a cell can be viewed as a decision maker. If there is only one player, we talk about decision-making (von Winterfeldt & Edwards, 1986). Decision theory can be considered as part of game theory as decisions can be conceived as games against nature (Milnor, 1954).

(ii) **The potential actions are called strategies**. In the simplest case, a player has the decision between two strategies: to act or not to act. Strategies can be represented in various ways. Game theory distinguishes between games in *extensive* form and *normal* form. Games in extensive form are represented by decision trees. The player has to decide at each node (decision situations) which consequent path is taken. Here we can speak of components of a decision or decision points. Sometimes the player has to make a decision but does not exactly know in which of two nodes he is. In such a case, we speak about extensive games with incomplete information. In a normal form game, each strategy can be represented by a natural, rational number or other sets of elements. If the player selects certain strategies with certain probability, we speak about mixed strategies represented by probability distributions. We can take from this that game theory provides a general language to describe decision situations.

(iii) **If all players have selected a strategy, an outcome results**. Formally, the outcomes can be represented by multiple strategies (or actions) of all players. What outcomes result is usually uncertain and depends on what strategies the other players take. We speak about decision-making under uncertainty, if we do not exactly know which of a known set of outcomes may result if a certain action is taken (Scholz, 1983). Note that there may be more "fundamental uncertainty," which is ignorance, if we do not know what possible outcomes may result.

(iv) **The evaluation of the outcomes can be modeled by preference or utility functions or other types of evaluative functions**. Mathematically, utility functions have an absolute scale, a zero point, and are linear (Fishburn, 1970). As these rigid assumptions are often not fulfilled, the preference function can also be described by aspiration levels or evaluation functions of lower scale.

16.6.3 Key messages

- Game and decision theory provide a *language* for formally and crisply describing what a human system is doing when interacting with the environment and other human systems
- The actions of teleological human systems that (consciously or unconsciously) pursue goals can be described by game- and decision-theoretic language, including actors, goals, strategies, and (not necessarily cardinal) utility functions
- Decision-theoretic modeling allows for assuming that different types of rationality are at work in human systems.

16.7 Awareness Postulate P6

Human systems have different types of environmental awareness.

Linguistically, concepts of awareness incorporate a series of notions, including *cognizing* something in the sense of perceiving and becoming knowledgeable about an issue. As far as we know, there is only one system "that knows with awareness: the human brain" (I. R. Cohen, 2000b, p. 57). Even if we acknowledge that various animals have self-awareness (Parker et al., 2006), humans are capable in a specific way of becoming aware of their impact on the environment by reflecting on what they notice, first-hand, about how their actions affect the environment.

The concept "environmental awareness" does not have a scientific tradition like the key concepts of the other Postulates: Complementarity (P1), Hierarchy (P2), Interference (P3), Feedback (P4), and Decision (P5). The concept of environmental awareness refers to the knowledge, mind-set, sensitivity, and the potential of becoming alert to something. In principle, environmental awareness has two components, an *epistemic-related* one and a *motivational* one. The cognitive–epistemic approach focuses on what the individual perceives, recognizes, and understands. This facet of awareness relates to the information that is gathered and cognized and which affects the behavior. The motivational approach relates to the drivers and values underlying human action. We suggest that both approaches are valuable, depending on the perspective taken on how human systems incorporate environmental issues.

16.7.1 A cognitive–epistemic view

The cognitive–epistemic approach relies on Jean Piaget's concepts of primary, secondary, and tertiary circulation (Piaget & Inhelder, 1950/1977) that characterize the development of the human child in the first 2 years of life (see Chapter 6.9). We introduce three levels of awareness: (a) ignorance or non-awareness of an environment; (b) the awareness of the impacts of the human system's action on the environment; and (c) noticing and the potential incorporation of feedback loops of human action on the environment. In many respects, these three types of awareness correspond to the levels of ontological development that the newborn child runs through in the first 2 years (Box 16.9).

Primary circulation means non-awareness of the environment (level (a) awareness). The agent's cognition is self-centered. This includes the goal formation as he or she defines "environmentally ignorant goals" (Scholz & Binder, 2003, p. 6). For instance, a newborn wants milk from the mother's breast even if she has no more milk. This means that the evaluation of outcomes disregards any environmental change induced by the action of the agent. Thus the selection of the strategy refers exclusively to the immediate utility of an action. One should note that non-awareness does not mean that the action itself must have a negative environmental impact. The idea is the exclusive focusing on one's own utility. The goals, utilities, and strategy selection are not affected by the dynamics of the environment, especially by environmental long-term impacts. This, however, does not mean that the agent does not realize the state of the environment. Environmental awareness can involve adapting one's behavior to the occurrence of both the actual available resources in the material environment and the competing demands in the social environment. When acting in a certain environment, the agent optimizes his action, but primary circulation does not include means–end analysis reflecting the impacts of behavior. In the language of feedback loops, primary circulation only includes awareness about primary feedback loops (see Figure 16.12*).

Secondary circulation means that the changes in the environment caused by the agent's action are noticed and are of interest. The states of the environment are partly integrated in the behavior. The impacts of human action on the environment are cognized and the consequences of action are evaluated. Thus the causality underlying human-made environmental impact is seen. Environmental awareness (type b) includes the capability of noticing the state of the environment and identifying human-made actions.

Tertiary circulation means that one appropriately adapts to the changing feedback that results from one's own impact on the environment. Thus, environmental awareness of type c includes reading the state of the environment and the changes made by human action, assessing the feedback loops, and adapting behavior. Thus, environmental awareness is a prerequisite for a stable human–environment interaction. The tertiary stage is a prerequisite of cognizing secondary and higher ordered feedback loops.

Though developed to describe the small child's cognitive development, the stages characterize a company's and people's behavior as well.

16.7.2 A motivational view

Environmental awareness or environmental consciousness has been widely used to describe a "proenvironmental" individual (Kuckartz, 1998) or collective behavior (Tudor et al., 2008). The concept has been primarily used in fields of environmental psychology or environmental sociology and applied journals of tourism, consumption, industrial production, farming, landscape conservation, energy management, etc. to investigate what characterizes low or high environmental pollution or energy and resources use. A typical research approach that environmental psychologists take is to investigate how people's values, attitudes, motives, knowledge, and so on, are linked to each other and to various behavioral variables (see Box 7.7; de Haan & Kuckartz, 1996; Fuhrer & Wölfing, 1997; Kals, 1996).

In the past two decades, the heightened perception of type (b) and (c) has been conceptualized with the emergence of *environmental awareness*. On the level of the individual, proenvironmental behavior can be explained by values, norms, attitudes, and by response to environmental taxes and prescriptions. However, none of these variables has been sufficient to explain proactive behavior. Some people protect the environment in one situation but not in another (Diekmann & Preisendörfer, 1996). Based on this, the concept of environmental concern has been introduced. In some studies the term (Dietz et al., 1998) is used synonymously with the terms "environmentally sensitive" or "proenvironmental behavior," and takes the role of a *dependent* variable. In others, it is taken as a variable to explain overuse and destruction of natural resources. To change this behavior, environmental concern is seen as a key *independent* variable (Bamberg, 2003; Fujii, 2006; Hansla et al., 2008; Minton & Rose, 1997; Poortinga et al., 2004).

We suggest that the term "environmental consciousness" serves as a link between the cognitive–epistemic and the motivational approach of environmental awareness. Environmental consciousness refers to being aware of *what* you are doing but also of the reasons as to *why*. Environmental awareness of type (b) encompasses all impacts of the human systems activities you are aware of at a given time. The temporal specification is made, as we explicitly want to separate the conscious from subconscious awareness. Thus, when reading this text, you are presumably not aware of your eyelid movement, breathing, or weather conditions. This example illustrates how most environmental stances are below the level of conscious awareness until something specific reminds you of them (Gerrig & Zimbardo, 2008).

16.7.3 Other views

Without detailed discussion, we just want to mention another three-level approach to understanding environmental awareness that distinguishes between anthropocentricism, biocentrism, and ecocentrism. This distinction has its roots in sociology and anthropology.

Anthropocentricism denies that non-human entities have rights. Human actions are exclusively evaluated with respect to their effects or costs and benefits for humans. Biocentrism, on the other hand, ascribes values to certain living things. Ecocentrism extends ethical consideration to whole ecosystems.

16.7 Awareness Postulate P6

Box 16.11 Loops in technical systems: Three Mile Island

Advanced technical systems need multiple controls. We demonstrate the role of primary and secondary feedback loops in technical systems. The Three Mile Island nuclear power plant accident serves as an example of challenges that may arise when human systems have incomplete information about the technical systems with which they are working and about how feedback loops can interact.

Nuclear reactors produce energy by a controlled nuclear fission reaction in the reactor core (see Figure 16.19). The heat is absorbed by water under pressure in a steam generator, which is part of the primary loop cycle. The primary loop includes relief valves and coolant pumps. The heat of the primary loop turns the water in an adjacent but physically separate secondary loop into steam.

This system requires a large set of control functions such as for the rods' activity, cooling, pressure, etc. in the primary loop. The water circulating in the steam generator interacts with the circulation in the secondary loop. Here the operation of the feed-water pumps and, in case of their defect, the functioning of auxiliary feed-water pumps is decisive.

Even this overly simplified description shows the interaction between the two loops. As we can learn from the Three Mile Island accident of March 18, 1979, erroneous manual interaction can lead to problematic situations. The incident began when the operation of the secondary loop's main feed-water pump was stopped because of a minor mechanical problem. This induced an increase in temperature in the primary loop, and the automatic safety system turned the auxiliary feedback system on. Unfortunately, maintenance workers had left two block valves closed in the auxiliary valves, and the operating crew were not aware of this because the display was hidden by a tag. Thus, another safety process began inserting graphite control rods into the core. Although the fission was stopped, the decay of fission byproducts increased and caused an afterheat, which was subject to the PPRV– positive pressure relief valve. However, in the case of Three Mile Island, this valve had not been working properly and did not reclose after venting the right amount of water. This was not reported to the operators owing to a poorly designed indicator circuit. Thus, the operators thought that the PPRV was closed and did not notice that water was pouring into the relief tank. As another technical safety measure, the ECCS emergency cooling systems injected water under pressure and the temperature dropped. However, this was overruled by the operators, because they thought that the ECCS was defective, and they feared a possible rupture of the pipes because of the high pressure injection into a closed primary loop. This resulted in a temperature increase, causing cracks in the uranium rods and the explosion of a zirconium oxide bubble.

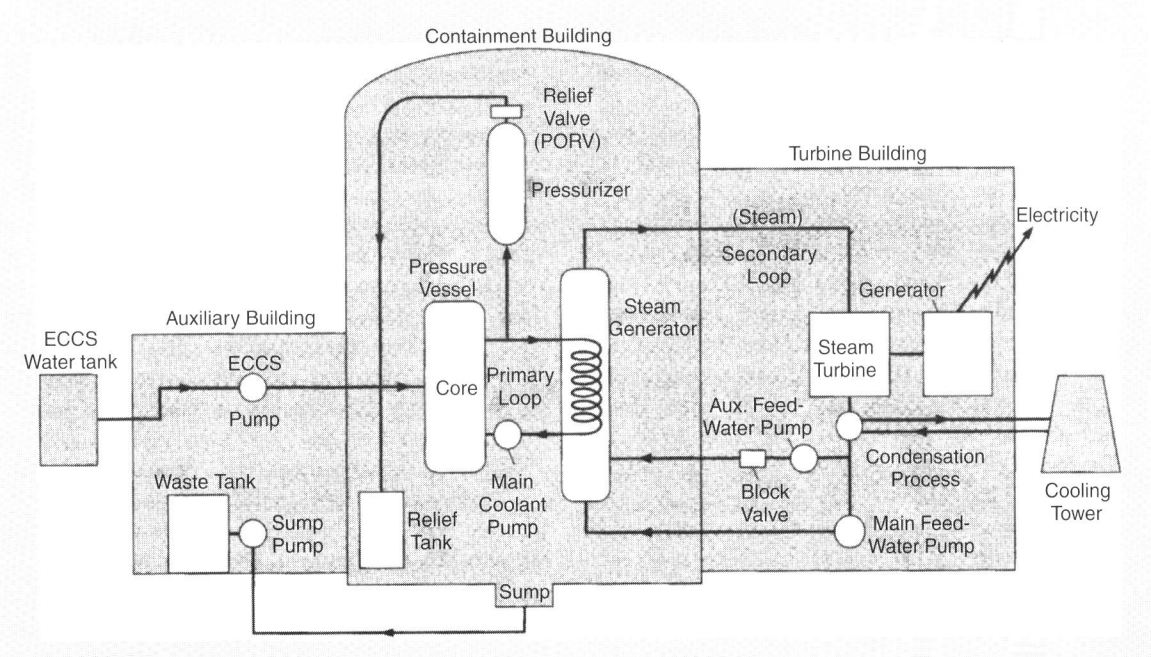

Figure 16.19 Design of a pressurized light-water type nuclear power plant, including a primary loop consisting of the core, steam generator and main coolant pump, and steam turbine and multiple feed-water pumps (from Lewis, 1980, p. 54, adapted with permission). © 1980 Scientific American, a division of Nature America, Inc. All rights reserved). PORV, pilot-operated relief valve; ECCS, emergency core cooling system.

> **Box 16.11** (cont.)
> The case revealed the following: a defect in the secondary feedback loop was the cause of the accident. The operators did not understand the state of the primary feedback loop and did not notice that the core was boiling dry. Highly complex technological systems include failure chains that can interact with human operators (Sheridan, 1980).

Many expect that ecocentrists value the wellbeing of the ecosystem over the value of the wellbeing of humans (Withgott & Brennan, 2008). Ecocentrism has different roots in ecoethology or deep ecology (Devall & Sessions, 1999; Gardner & Stern, 2002; Naess, 1989), and is somewhat similar to the motivational view. The difference is that the former is a normative approach, whereas the latter emerges from psychological research.

16.7.4 Key messages

- Environmental awareness includes an intentional, motivational and a competence-based component
- There are different conceptions for distinguishing levels of environmental awareness such as, on the one hand, focusing on oneself, becoming aware of the environmental impacts on the environment, and becoming aware of the feedbacks.

16.8 Environment-first Postulate P7

16.8.1 The material environment as a master matrix

The Environment-first Postulate P7 is a cornerstone of environmental literacy. We propose that each investigation of HES should *start* with a thorough analysis of the environment. The idea is that after an environmental problem has been defined, which emerges from the interests and needs of the human systems involved and their temporal and geographical boundaries, a comprehensive analysis of the environment is essential to identify proper decisions and actions available to cope with the problem. Problem-solving strategies may result in undesired primary and secondary feedback loops. However, this can be prevented, at least partially, with a thorough analysis of the environment and the understanding of the functioning, rationale, feedback dynamics, constraints, limits, and unknowns of environmental systems in their interactions with human systems (see, e.g., Box 16.12 & 16.13).

The Postulate is rooted in the assumption that phylogenetically, the environment is the master matrix of human life. The environment has a limited capacity. This is the ontological side of the Postulate. However, there is also a strategic, game-theoretical, and epistemological side. To define what to do properly, one must know what the other side is doing. Winning at chess requires strategies that are sensitive to what the other player is doing.

The Environment-first Postulate makes reference to equilibration, which is a basic principle of human development in the later writings of Jean Piaget (1977). Equilibration can be conceived as the basic tendency of the individual to keep a cognitive balance with the environment (see Chapter 6.9). We extend Piaget's idea from the individual to all human systems. Further, we not only talk about the cognitive balance but also about the balance between the material–biophysical dimension of systems with their environment. This means that to sustain themselves, human systems are both motivated and obliged to comprehend how the environment acts and reacts. If the cognitive or socioepistemic mechanisms are not sufficient, they have to be modified to reward or punish the actions of human systems. This is the key idea of Brunswik's theory of probabilistic functionalism (Brunswik, 1952; see Chapter 6.3). Equilibration represents a type of self-regulation and refers to the functionalist perspective taken (see Chapter 3.5).

The Environment-first Postulate prepares the human system to adapt (see Box 16.13). Following Piaget, we distinguish between two modes of adaptation. The first is *assimilation* and means that a human system is modifying existing processes or mechanisms. The other is *accommodation* and means that new processes and mechanisms are introduced. The Environment-first Postulate helps to find out what assimilation or accomodation processes or mechanisms have to be changed to cope with environmental planning.

16.8 Environment-first Postulate P7

Box 16.12 Climate changes in an uncertain world: sea level rise

Changes in climate have required humans to adapt at all times. Recent climate research revealed "that global climate could indeed shift radically and catastrophically, within a century – perhaps even within a decade" (Haug *et al.*, 2003, p. 1731).

Climate change has caused dramatic societal developments at the micro and macro scale. Consider the period of global warming between 15 000 to 8000 BC, which altered fauna and flora, raised sea levels because of melting ice, shrank habitats, and consequently changed the migration patterns of a number of large game animals and with them that of the hunters and gatherers who followed them. Human societies had to learn how to adapt to these changes. Owing to rapid climate change, substantial stores of information on the availability of plants and animals had to be altered (Jahn, 1990). The human population grew by a factor of seven, from 1.2 to 8.5 million in the period between 40 000 BC and 8000 BC.

As a consequence, the rapidly growing human population faced the situation that many parts of the Earth, such as the Sahara, were facing shrinking green areas. New technologies for food supply and forms of life had to be invented. Thus, some anthropologists and sociologists consider the emergence of horticulture, which is the domestication of wheat, barley, lentil, etc. and which developed independently in areas of Asia and America between 8000 BC and 3000 BC (Wenke, 1999; Wright, 1976) to be a successful adaptation of human societies (Nolan & Lenski, 1998).

Today, many areas at or below sea level are endangered by rising sea levels. Since the peak of the last Ice Age, about 18 000 years ago, the sea level has risen about 130 meters. However, from 3000 years ago up until the nineteenth century, the sea level rose only 1–2 mm/year (see Figure 16.20a). An acceleration of sea level rise of 1.7 mm per year was estimated between 1870 and 2004 (Figure 16.20b). Since 1993, the level has risen at a higher rate, of 3.1 mm per year. If the acceleration of sea level rise remains constant, we would have a total rise of 28–34 cm up to 2100 (White *et al.*, 2005). However, sea level rise is a product of multiple factors, including land uplift because of reduced ice shields and land uplift since the last glacial maximum, El Niño-like changes of ocean streams, marine flooding or drying of large areas, and loss of interstitial fluids (e.g. oil extraction, Houghton, 2004; Houghton *et al.*, 1995). Coping with these uncertainties and finding appropriate means against them is a challenge for countries such as the Netherlands, which has about 20% of its area below sea level.

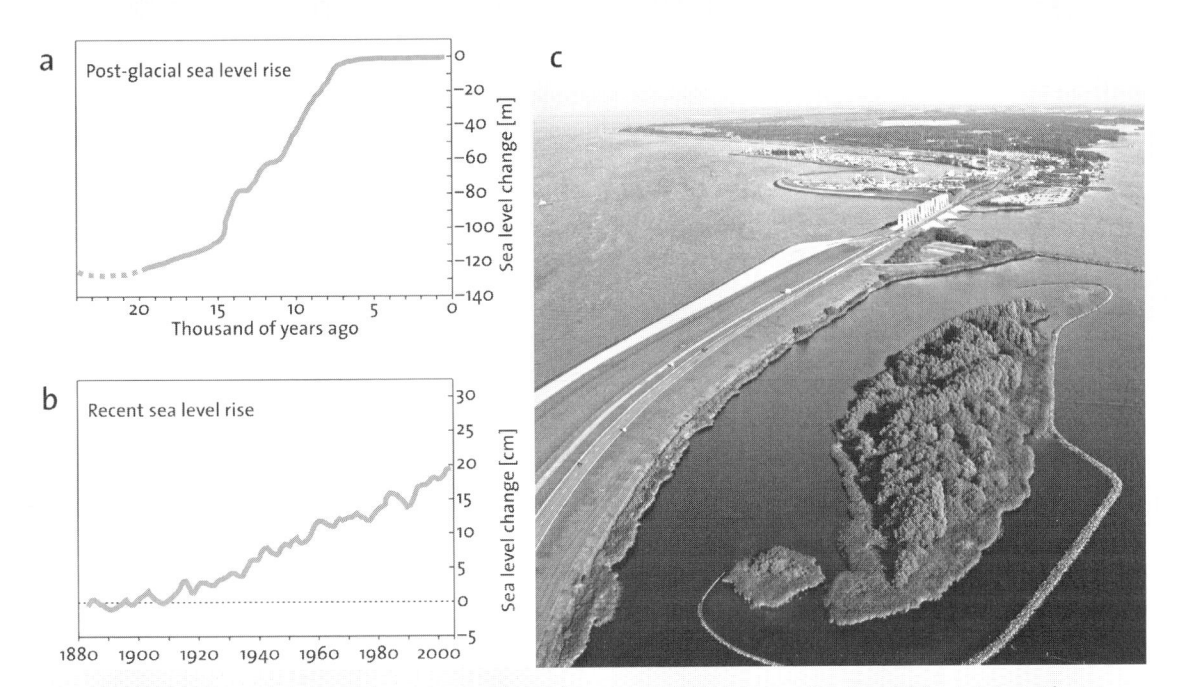

Figure 16.20 (a) The global average sea level rose smoothly and remained roughly unchanged for about 3000 years (adapted from Wikipedia, 2010), but (b) the rise has obviously been accelerating for a little more than the past 100 years (adapted from Wikipedia, 2010). However, there are many uncertainties and unknowns in the prognosis for the next century. (c) A vulnerable country such as the Netherlands, which has a relatively large amount of coastal land, has to develop appropriate adaptation strategies, which should emerge from the best scientific knowledge about what can be expected from environmental change (aerial view of Lauwersoog in the Netherlands).

Box 16.13 Understanding the environmental impacts of disasters: the Dust Bowl phenomenon

"The extension of agricultural and pastoral industry involves an enlargement of the share of man's domain ... The felling of the woods has been attended with momentous consequences to the drainage of the soil, to the external configuration of its surface, and probably, also, to local climate; and the importance of human life as a transforming power is, perhaps, more clearly demonstrable in the influence man has thus excreted upon superficial geography than in any other result of his material effort" (Marsh, 1864/2003, p. 3).

An example of the impact from humans' reshaping of the landscape can be seen in the Dust Bowl phenomenon. In the seventeenth century most of the shaded area of Figure 16.21a was a mixed deciduous forest (Williams, 2003, p. 9). The Dust Bowl is the designation for both a region in the USA and an event or a period of severe drought and wind erosion that occurred in parts of the Great Plains during the 1930s. "The drought – the longest and deepest in over a century of systematic meteorological observation – began in 1933 and continued through 1940. In 1941 rain poured down on the region, dust storms ceased, crops thrived, economic prosperity returned, and the Dust Bowl was over" (Cunfer, 2004). For 8 years, crops failed and sandy soils blew and drifted over failed croplands (Figure 16.21b). The Dust Bowl is thought to have been caused by a combination of severe drought and unusually high temperatures, accompanied by regional dust storms and normal wind erosion, leading to a collapse of the rural economy. Cunfer denotes this phenomenon as "agricultural failure, including both cropland and livestock operations" (Cunfer, 2004). However, one has to consider that other factors also contributed to the disaster, including that small farmers lacked knowledge about how to conserve their soils and the ignorance of externalities when appropriating forest land for agriculture (Hansen & Libecap, 2004).

An interesting aspect of the Dust Bowl is that it has been linked to an increase in population density (the population in the Great Plain grew by a factor of 7 between 1880 and 1930, from 0.8 up to 5.6 million). In addition, there was also widespread deforestation, changing local climate and promoting soil erosion.

Research on 1000-year tree ring data provides evidence for successive "megadroughts" of 20–40 years in the nineteenth century, presumably caused by El Niño-like changes (Herweijer et al., 2007). Against this background, it is not surprising that the recent increases in deforestation in the Amazon have been suspect in causing a Mega-Dust Bowl in the Amazonian region, although the matter is controversial (Cox et al., 2008).

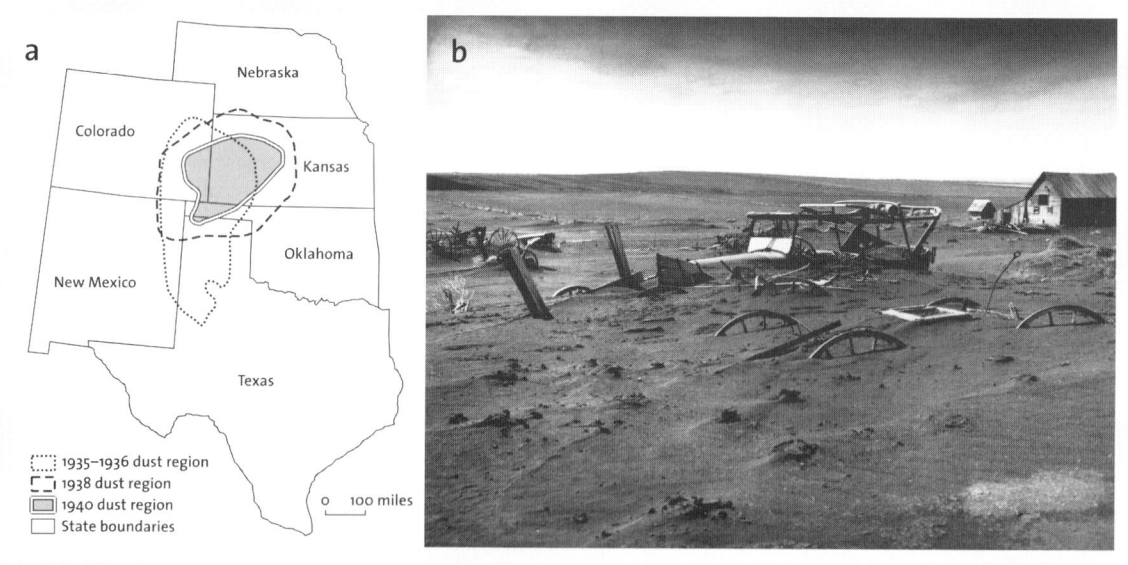

Figures 16.21 (a) Location of the Dust Bowl and (b) a picture of buried machinery in a barn lot in 1936 in Dallas, South Dakota (photo from United States Department of Agriculture).

We illustrate this by an example. Let us consider potential changes of mobility behavior with respect to climate change. Given a certain goal for reducing CO_2 emissions, for instance, limiting the increase of average global temperature to 2°C, we would seek to clarify what causes climate change and to define what an average increase of 2°C actually means before adequate actions of assimilation or adaptation can be defined.

16.8.2 Sufficient, satisficing, and state-of-the-art knowledge

If we look at major critical issues, such as energy, water, biodiversity, climate change, or public health, it is clear that a sound environmental analysis employs state-of-the-art knowledge. Furthermore, this has to be based on a thorough discipline-based interdisciplinarity. A proper analysis will consider the characteristics and dynamics of the HES under investigation to design a research plan that produces outcomes that have the desired depth, validity, and precision. The state-of-the-art analysis of the environment should follow the *sufficiency* and the *satisficing* principles. The analysis must provide enough information to establish solutions that are good enough to secure the environmental functions and services that humans appreciate. Satisfying includes both aspects, satisfying and sufficing.

A specific challenge of environmental analysis is its multilayered, overly complex, and contextualized nature. Complexity and contextualization cause uncertainties and ignorance. Thus, we need methods, concepts, and theoretical approaches to deal with the uncertainties inherent in the study of human–environment interactions. Transdisciplinary case studies (Scholz et al., 2006; Scholz & Tietje, 2002) or scientific studies that start from a real-world problem (Robson, 1993) can serve as examples. Adequately coping with the uncertainties and ignorance inherent in complex real-world settings has also been the key issue of the theory of probabilistic functionalism (Brunswik, 1952; see Chapters 3.2 and 6.3).

The complexity of environmental problems requires multiple representations and epistemics (see Box 2.6). These multiple representations often compel the researcher to cooperate with decision-makers and stakeholders to understand the goals and perspectives of all, to utilize their knowledge about a system, and to be aware of the practicable options at the disposal of decision-makers. Thus, the analysis of the environment has to be done while acknowledging the specific (functional) relation of an agent to a certain task.

The contextualization, functionality, and the multiple perspectives of an environmental setting analysis require alternative representations of uncertainty that reflect different qualities of knowledge. This means finding the right concepts and their (formal or semantic) relations in normative and cognitive modeling, which is the key to environmental literacy (Scholz & Zimmer, 1997).

Starting from the environment is not a value-free issue but is linked to what we have called functional social constructivism (see Chapter 3.6; Stauffacher et al., 2006).

Many examples from history show that humans have often made decisions without appropriately assessing and anticipating relevant environmental impacts and dynamics.

16.8.3 The material and the social environment

A special challenge of the Environment-first Postulate is that human systems are not only shaped by the material, but also by the social environment. How human systems operate depends on social constraints and on the options they have for gaining access to codified knowledge (e.g. science). Thus, all human systems are partly formed by their social environment. The evolution of language since at least 40 000 years ago can be seen as a major transition to human systems that operate in response to social dynamics. Thus, human systems are not only psychophysically, but also socially, shaped. This led social psychologists to stress the role of the social environment (Hiebsch, 1968). As elaborated in Part I of this book, we take a broader and integrative perspective and consider the action of human systems to be a function of both their own (material and social) constitution and history, as well as their (material and social) environment (Irle, 1975; Lewin, 1939; Martin, 2003).

For a better understanding of the Environment-first Postulate P7, and to show how the material and the social environment are indispensably related, we can exemplarily look at public health. A study done in New York showed that the number of hospitalizations of young patients increased fourfold between 1990 and 2006 (Lin et al., 2008). According to hospitals, the diagnosis of allergic reactions to peanuts, nuts, and seeds of all hospitalized patients increased by 41%. This phenomenon can have multiple material–biophysical causes, such as new agrochemicals, overprocessed foods, and interactions with other dietary changes. However, reasons can come from the socio-epistemic environment, such as changing cleanliness, decrease in physical activity, increased willingness to visit hospitals, and different recording of physicians. Clearly, the potential impacts of these and other reasons have to be acknowledged, understood, and conceptualized.

This example reveals the complexity of and the interrelationship between the material and the social dimensions of human actions in the environment. The Environment-first Postulate P7 embeds the study of HES in a robust knowledge of the environment. This can be perceived as a strong methodological recommendation.

16.8.4 Key messages

- The Environment-first Postulate is a cornerstone of environmental literacy. The material–biophysical environment and the social environment are the matrix of human life. Based on an understanding of the goals and needs of human systems, a thorough analysis of the (material and the social) environment(s) has to include the assessment of the potential and limits of environmental services and to identify unintended primary and secondary feedbacks that may result from the responses of the environment to human action
- The Environment-first Postulate is linked to the principle of equilibration. Human action should be based on sufficient knowledge to allow for decisions and actions enabling human systems to sustain. This meets the idea of probabilistic, evolutionary stabilization in the theory of probabilistic functionalism
- The analysis of the environment has to meet the sufficiency and satisficing principle and to match textbook state-of-the-art standards to prevent avoidable undesired rebound effects resulting from human action
- The analysis of environmental systems needs to cope with complex social, historic, and material contexts
- Uncertainty and ignorance are ubiquitous in the study of complex environments. Finding suitable concepts and approaches is a specific challenge
- Environmental analysis prepares proper adaptations, assimilation, and accommodation of thought and behavior to changing environmental constraints.

Part VIII A framework for investigating human–environment systems

Chapter 17

The HES framework

Roland W. Scholz, Claudia R. Binder, and Daniel J. Lang

17.1 A template to cope with HES 453
17.2 Preparing to utilize the HES framework 457
17.3 The HES Postulates in environmental research 459
17.4 The HES framework in action 462

Chapter overview

In the spirit of cultivating environmental literacy we present the human–environment systems (HES) framework as a practical way to cope with the complexity of human–environment relationships. Its design and our suggestions for its use assume that most environmental problems are caused largely by human activities, a view that could be called the anthropocenic redefinition of the environment.

Our unveiling of the HES framework describes how it is a structure–process template or methodological guide for research about HES. The HES framework highlights how the HES Postulates offer principles that foster a comprehensive analysis of HES.

Next we describe how the HES framework offers principles for environmental research and transdisciplinary processes. After an initial phase of formulating a guiding question and defining system boundaries, a first step is to conduct an analysis of the environmental situation (as stated by the Environment-first Postulate P7), which is a typical subject of transdisciplinary processes.

By offering an actor-oriented view, the HES framework institutionalizes the analysis of human systems as a fundamental part of environmental research. This involves conceiving human systems as decision-makers and recognizes, for example, that these human systems are related to each other from different hierarchal levels and show different degrees of environmental awareness. The HES Postulates offer important guidelines for scientific investigations, outlining how researchers can take a comprehensive approach when, for example, assessing the rationales of a human system.

Finally, to illustrate how the HES Postulates can be linked within the framework, we sketch the plight of Mongolian herders, whose herding activities are destroying the very grasslands they use for pasture. We put forth the HES framework, a template that brings together approaches used by a growing number of environmental researchers in their endeavors to encourage environmental literacy.

17.1 A template to cope with HES

17.1.1 The HES framework uncovers structure and process

Together, the seven Postulates of Chapter 16 constitute a structure–process framework for investigating human and environmental systems. We stress that the term *human* systems is used instead of *social* systems for two reasons. First, we differentiate between the material–biophysical and the sociocultural dimensions of human systems, as expressed by the material–social complementarity (see Figure 3.3*; see also Chapters 3.3 and 16.2). Second, we think that, from a systemic perspective, several aspects that hold true for the interaction between individuals and higher ordered human systems with the environment can also be applied to organs and lower ordered systems (see Figure 14.1*).

The HES framework presented here is a methodological guide or template to investigate human–environmental systems. As the term *structure* indicates, the framework identifies the essential *entities* (e.g. a goal system or action; see Figure 17.1*) and the *relationships* between entities (i.e. the structure) of human–environment interactions. It is based on the assumption that human systems can be conceived as actors, and that human and environmental systems

17 The HES framework

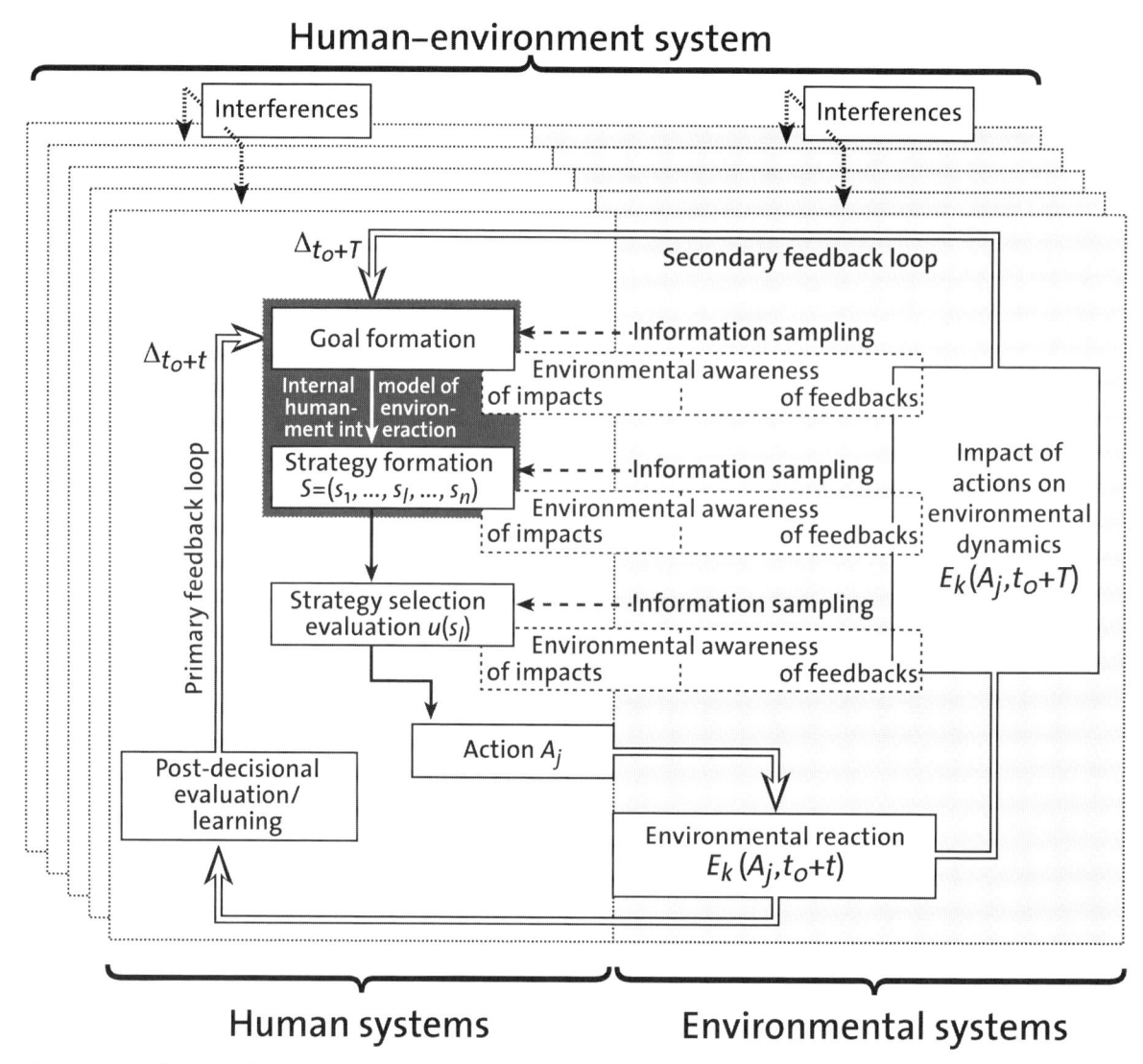

Figure 17.1* A framework for the structure and process of HES.

are inextricably coupled. The term *process* points out that the framework also treats the course of decision-making, processes of the human systems leading to action, the processes in environmental systems, and the interaction, feedbacks, adaptation, and learning of these systems.

The HES framework and the Postulates are designed to facilitate research on and coping with complex environmental questions or problems. Typically, at early stages of dealing with a complex and ill-defined environmental problem or research question, be it through (inter)disciplinary research or a transdisciplinary process, the issue appears unstructured and cannot always be clearly defined. Usually, we start from vague interests, unspecific curiosity, and fuzzy awareness about an environmental phenomenon or problem (initial curiosity). Ideally, this is followed by a more crisp definition of the matters of interest (i.e. to define a guiding question), and identifying and defining what will be looked at (i.e. constructing a system definition, including system boundaries). This requires specifying what subset of the universe we want to look at and allows for efficient initiation of a process that allows for analyzing, better understanding, and, if desired, changing the human–environment interaction.

17.1 A template to cope with HES

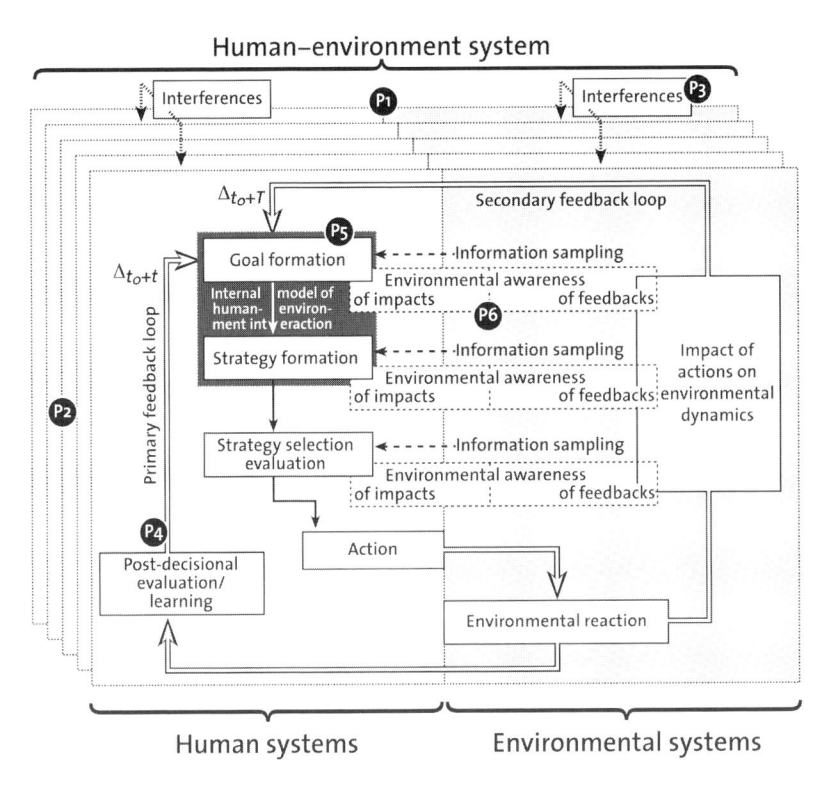

Figure 17.2 The HES postulates, highlighted by numbers in the HES framework serve as key principles for the structure and process of environmental problem-transformation.

17.1.2 A tour of the HES framework: relating the Postulates

A tour of the HES framework highlights how the HES Postulates are practical devices for guiding investigations of complex environmental questions and problems. This chapter has the purpose of *preparing* the reader for utilizing the Postulates in interdisciplinary research and for organizing transdisciplinary processes. Examples are presented in Chapter 18. The highlighted numbers in Figure 17.2 indicate where the HES Postulates fit into the HES framework.

In the following paragraphs we review how the seven HES Postulates function in the HES framework. It can be helpful to keep in mind that, theoretically, the Postulates can be classified into three categories. Postulates 1 through 4 are ontological and epistemic in nature; Postulates 5 and 6 are assumptions for a decision-theoretic conceptualization of human agents; and Postulate 7 offers a basic methodological principle for investigating human and environmental systems. Chapter 16 provides further details on this.

Complementarity Postulate P1: this Postulate represents the view that human and environmental systems are coupled, inextricably intertwined, systems. Accordingly, the HES framework illustration in Figure 17.2 shows the HES as one system (top bracket) that can be separated into two corresponding subsystems (bottom brackets). For example, while we can think of a farmer and the farmland separately, following P1 there is also the notion of a mutual adaptation process of the farmer's actions and the land.

Hierarchy Postulate P2: this Postulate proposes that human and environmental systems are both composed of hierarchical structures. The HES framework shows hierarchical levels as a deck of cards in Figure 17.2. Each layer represents a hierarchy level of human systems and environmental systems that are involved in a specific analysis. The Hierarchy Postulate applies to human systems from the individual to the supranational level (see Figure 14.1*).

Interference Postulate P3: this Postulate states that there are interactions among and within different levels of human and environmental systems, from the micro to the macro level, which cause interferences or trade-offs among these human systems. These *interferences* are visually displayed by the bracket between the card layers at top of Figure 17.2. The Interference Postulate in particular refers to commons dilemma-like problems in which we face conflicting drivers of

455

human systems on the individual (or micro) and the societal (or macro) level.

Feedback-loop Postulate P4: this states that there are different types of feedback loops within and among human and environment systems. In Figure 17.2, we simplify this idea by showing two feedback loops: a primary feedback loop and a secondary feedback loop. They represent what the human system perceives, evaluates and learns about the intended *environmental reaction* and the *impacts of actions on environmental dynamics*. We want to note that we also deal with higher ordered feedback loops, for instance, third order feedback loops for reflective learning (see Chapter 16.5.3).

The primary feedback loop is formally expressed in Figure 17.2 by the environmental reaction some time (t) after t_0. The secondary feedback loop includes possibly unintended feedbacks caused by an action, which results after a certain time T. The secondary feedback loop includes the environmental responses that are not cognized or accounted to the action in the preceding decision process. We propose rather intuitively that T represents a much longer time period than t, as secondary feedbacks are often delayed. This is the usual case. However, there are also situations that differ from these usual cases. For instance, we know that the reward of an apple (i.e. the intended environmental reaction) comes some considerable time after planting the seed. On the contrary, miners had to experience that, before a planned controlled explosion, an unintended (e.g. methane) explosion took place when lightning a match cord.

Decision Postulate P5: this Postulate articulates our assumption that all human systems are intentional systems that act to satisfy goals and drivers. As such, we assume that human systems have a goal system and that, typically, a process of goal formation takes place, shown in the left upper white box of Figure 17.2. The boxes on strategy formation and strategy selection represent the (conscious or unconscious) decisions and choices that human systems make, according to their preferences, before taking action. We suppose that goal formation and strategy formation are based on an internal model of human–environment interaction; that is, the representation of the given situation with which a human system is confronted. For strategy selection we must assume that behavior is based on preference (or utility) functions. A decision-maker, in making a single decision, may face trade-offs if he concurrently follows different goals which sometimes cannot be simultaneously fulfilled.

Awareness Postulate P6: this Postulate conceptualizes different degrees of environmental awareness. The three boxes edged with the dotted lines in Figure 17.2 represent three hypothesized, ideal types of environmental awareness. We assume that the different degrees of environmental awareness are at work during all phases of the decision process: goal formation, strategy formation, and strategy selection. First, a human system can have no environmental awareness if, for example, the human system exclusively concentrates on its own interest without reflecting on the impacts and consequences of his/her/its actions on the environment. Second, a human system can be aware of the *impacts* that may result from action, as shown in Figure 17.1* in the left part of the dotted boxes, but does not conceive what it means for itself. Third, a human system can even be aware of the *feedbacks*, shown in the right part of the dotted boxes, that may result from the (unintended) change of the environment caused by an action. We consider this highest form of accessing environmental awareness to be a key component of sustainability learning.

Environment-first Postulate P7: this Postulate states that any analysis of an HES should start with an analysis of the environmental system, shown as the right shaded side of Figure 17.2. Our experience shows that, once a guiding question and system boundaries have been developed, most effective projects begin with understanding the social and material dimensions of the environmental system(s). If we do not understand how fish and traditional fishing are being affected by overfishing by commercial fleets, how can we identify a sustainable solution to overfishing? This is illustrated in Box 8.3.

17.1.3 Mathematical descriptors for strategies, actions, and feedback

Figure 17.1* includes some mathematical notation for those readers who might be interested in modeling HES. We start from the goal formation–strategy formation box in the upper left part of the Figure. This box illustrates that, in a given situation, each human system can (consciously or unconsciously) select between different strategies. This is expressed by the strategy vector $S = (s_1,\ldots,s_l,\ldots s_n)$. Which strategy s_l is chosen is determined by the human system's preference or utility functions. The utility function (see Chapter 7.2) assigns a certain desirability, attractiveness or postulated satisfaction to the (expected) outcomes that may result if a certain strategy is chosen. The utility resulting from the choice of a strategy s_l is denoted by $u(s_l)$.

A concrete action A_j causes an intended environmental reaction $E_k(A_j, t_0+t)$, or a reaction that is directly linked to the behavioral plan (i.e. the selected strategy) after some time t. Here t_0 denotes the point in time when the action is made. The index k indicates that there is no deterministic relationship between action and environmental reaction. Thus k indicates that there is a stochastic variable in between the action and the $k = 1,...,K$ events or outcomes resulting from this action A_j. Finally, we find on the right side the term $E_k(A_j, t_0+T)$ which denotes the state of the environment at the time $T > t_0$. In principle, a specific strategy can be identical with an action. This would be called pure strategy. However, in general we consider mixed strategy in the sense that a strategy means that different concrete actions A_j $(j = 1,...,J)$ may result with certain probabilities.

The term $\triangle t_0 + t$ denotes the changes of the state of the cognitive–epistemic (i.e. by having received new information) or the biophysical–body (e.g. by having ingested food) that the human system is sensing at the point t_0+t. By comparing this changed state to the state of awareness of the human system at time t_0 we find the (intended and anticipated) consequences of a performed action A_j. Similarly, the term $\triangle t_0+T$ denotes the changes of the environment after the timespan T that are caused by an action A_j at time t_0. Based on these definitions, $E_k(A_j, t_0 + T)$ is a random variable, which represents the changed state of the environment at time $t_0 + T$. This changed state might mean that a certain action A_j, which is shown at time t_0, might cause a different environmental state at time t_0+T.

17.1.4 Key messages

- We assume that a research process on complex environmental problems emerges often from vague interests, which become specified and can be expressed by a guiding question whose answers require meaningfully defined system boundaries
- The HES framework provides a template or methodology for describing the essential processes and structures inherent in the interaction of human and environmental systems on different levels
- The Postulates can offer important *guidelines* for research projects on HES.

17.2 Preparing to utilize the HES framework

From the functionalist perspective taken in this book, it is clear that any (research) project has to start from objectives and thus a specification of the object of the research venture. This involves defining a guiding question and the system boundaries. This might appear trivial, but we know from experience that many projects can fail or experience difficulties if a guiding question and system boundaries are lacking. Clarifying these aspects is critical before any research project or transdisciplinary process is started. We begin this section by relating and distinguishing between transdisciplinary process and environmental research. Next, we discuss the first two steps in an effective investigation: identifying a guiding question and system boundaries.

17.2.1 Transdisciplinary process and interdisciplinary research with the HES framework

The HES framework was initially designed for academic endeavors that have contributed to building a theory around interactions among human and environmental systems. The primary purpose was to support environmental research when contributing to the development of an inner science perspective. However, the HES framework was designed by a team that was experienced with a large number of transdisciplinary projects. Thus, the framework was also designed to structure complex environmental problems in a way that disciplinary knowledge from the social, natural, and engineering sciences could be efficiently utilized for problem transformation.

Formulating a guiding question involves identifying the object and objectives for a project. This work can be approached from a transdisciplinary perspective based on the agendas of project stakeholders, legitimized decision-makers, or interests of the public at large, as well as on questions and interests emerging from the advancement of science. Both for practice and for research, it is necessary to reflect upon the capabilities and constraints, and in particular on the resources of the project team.

Similarly, system boundaries can be defined from a transdisciplinary, "use-inspired" perspective and also from an inner science perspective. If we take the inner science perspective, the system boundaries depend on the intentions of the researcher and the scientific necessities. Whether for instance a certain statement, say a mathematical theorem, must hold true under a special set of prerequisites or it should be valid in general depends what purpose the statement is needed for. In transdisciplinary processes, the problems emerge from real-world problems. What should be dealt with exactly depends on the purposes and environmental services in which human systems are interested and by which they are affected.

17.2.2 Formulating a guiding question

The environment is always of interest from a certain perspective. There can be subsistence, commercial, self-realization, aesthetic, spiritual or other interests. Explaining the interests and drivers of practitioners and scientists and transforming them into a well designed guiding question is a relevant step of any research project. What interests deserve attention in research projects can also be clarified or negotiated through transdisciplinary processes (see Chapter 15.1). In any case, whether from a disciplinary, interdisciplinary or transdisciplinary perspective, the guiding question should be formulated in a way that the system boundaries can be unambiguously defined.

This is by no means trivial. An effective guiding question that takes into account the relevant human and environmental systems and also the interests of scientists, stakeholders, and laypeople is an important first step in any environmental research project or transdisciplinary process. Formulating a guiding question involves understanding: (1) the goals, intentions, values, and needs that are at work in the practitioners or scientists interested in the project; (2) the system boundaries that the object and the objective of investigation requires to provide a sufficient and meaningful answer to the guiding question; and (3) capabilities, project resources, and constraints, which depend – among others – on the time, location, and available financial resources associated with the project.

Let us illustrate this important step with some examples. A typical, real-world guiding question might read: what water management measures are appropriate for the Canary Island hotel industry? Such a question could be posed, for instance, by the government of the Canary Islands, which is an autonomous state and can be considered a community or a special form of society. If an investigation is launched by the government, the guiding question should consider the interests of the whole Canary Island society. Otherwise, if the regional hotel association was heading this project, the type of strategies looked at and the system boundaries might be shaped with respect to the entrepreneurial interests of the hotel industry.

Further typical guiding questions could be: what should be done to protect the Swiss population from avian flu or another infectious disease? or what is the risk of engineered nanoparticles? These questions might be on the agenda of national health agencies (which are societal institutions, see Figure 14.1*). The characteristics of the defined HES most importantly depend, from an HES perspective, on the object, objectives, and the specific boundary conditions of a study, which are formulated in (1) to (3) above.

Doubtlessly, this book reflects an environmental sciences flavor, which is the primary disciplinary perspective taken. We think that the HES framework is not only of interest for environmental sciences but also when analyzing interactions of human systems with social environments. Thus, for instance, one could pose a guiding question about the quality of education in Switzerland: why do pupils refuse learning when confronted to certain teaching styles?

17.2.3 Defining system boundaries

Defining system boundaries means specifying what human and environmental systems should be included when investigating an environmental question or problem. Just as with formulating a guiding question, a thorough understanding of the problem has to precede a meaningful definition of the system.

As human and environmental systems are inextricably interrelated, the system boundaries have to be defined from both perspectives. Defining the system boundaries for the HES framework proceeds as follows. The guiding question(s) define(s) the primary human systems that are involved in the objects and objectives of the study. In the Canary Islands case mentioned in the preceding section, these are the hotel keepers. Then, in a second step one should define what other human systems (tourists, farmers, water supply companies, etc.) have to be included in the human system side from the perspective of understanding the water management system.

After having identified what main human systems are linked to the guiding question, we have to define what environmental systems have to be included. There is no "nostrum" that can serve to reveal how this can be done for all possible guiding questions. In principle, the system boundaries should include everything that is affected by human action. Here it is important to recognize which human systems are aware of environmental impacts via primary and secondary feedback loops.

As a project evolves, new knowledge can be generated that warrants a redefinition of system boundaries. This can occur, particularly as long-term impacts on environmental system dynamics (represented in the HES framework as the secondary feedback loops) are better understood, either through measured data or predictive modeling. Consider a study is planned on the interaction between national policies and individual behavior with respect to reducing CO_2 emissions. After the initial definition of the system boundaries, the analysis may well show that international car companies are key players, as they might have potential new technologies at their disposal that could be offered in case individuals (as market members) ask for them. Thus the analysis of secondary feedback loops and of the interaction between the micro and the macro level may fundamentally change the system boundaries. This is a key reason for the Environment(al analysis)-first Postulate. Note that this issue is dealt with in detail in the "Oil-free society – go for bioethanol" case in Chapter 18.

17.2.4 Key messages

- The definition of a guiding question and system boundaries has to precede the use of the HES framework
- The HES framework can be used for pure science purposes, in applied projects, and transdisciplinary processes
- The HES framework is suitable for facilitating transdisciplinary processes as it genuinely launches the actor's perspective. The actors, including the roles and interests of stakeholders and project initiators of the transdisciplinary process, can be defined and reconstructed according to the HES Postulates.

17.3 The HES Postulates in environmental research

In the following section we review several examples of how and when the principles and assumptions articulated as the HES Postulates can offer important guidelines for environmental researchers. This review will be twofold. First, we go through the Postulates illustrated by examples from the Canary Island case. Later (in Chapter 17.3.5) it is the other way round: a case will be looked at exemplarily through the HES framework and the Postulates.

It is important to stress that the sequence of the Postulates, if applied to a case, is not predetermined. Rather, it depends on the case at hand as to which of the Postulates are more pronounced or become specifically important. There is no specific order or sequence with respect to the process within a research project – except for the Environment-first Postulate P7. This is the reason that our tour through the Postulates starts in the next section with P7.

17.3.1 Why and when is a thorough environmental analysis necessary?

Based on the Complementarity Postulate P1 and according to the Environment-first Postulate P7 a thorough analysis of the environment should follow the definition of guiding questions and system boundaries. Again, the Canary Islands case can serve as an example. Quantitative analysis of the availability of groundwater requires, for instance, a state-of-the-art groundwater model as a prerequisite to appropriate action. Investigating secondary feedback loops, for instance a rapid depletion of small groundwater reservoirs, which cannot be anticipated without a thorough geological and socioecological analysis, might prevent shortsighted and thus unsustainable investment in groundwater supply. There are many historical examples that show that insufficient knowledge about the environment leads to inefficient action or negative rebound effects. For example, the unrealistic expectation of the amount of biofuel supply is dealt with in the bioethanol case in Chapter 18 (case 3).

Similarly, sociological and demographic research can shed important light on the social environment relevant to a particular project. Thus, the Canary Island case would require an investigation of the relevant actors, including tourists and their demands for freshwater.

17.3.2 Incompatible rationales on the micro and macro level

While the Hierarchy Postulate P2 refers to the numerous levels of human systems, from the individual to the supranational level, the Interference Postulate P3 points to the possibly conflicting or supporting interferences between these different levels. As the example of the commons dilemma shows (see Boxes 2.3 and 6.15), this problem is a result of the different goals and incentives that are at work on micro and macro levels of human systems (see Table 6.1). Referring to the concept of level hierarchy (P2) both a top-down management and regulative process and a bottom-up process of emerging patterns and structures among individuals or members of a certain set of human systems should be distinguished. A critical issue in the Canary Island case, for example, is that the different actors, for instance national and international hotel owners, hotel keepers, and the government deal with different timeframes, which induces different preferences in different types of solution. Advanced water reuse and recycling, for instance, asks for long financial recovery periods.

17.3.3 The challenge of promoting sustainability learning

Sustainability scientists can employ the HES framework to investigate the degree to which sustainability learning is occurring in a particular case. The distinction between primary and secondary feedback loops (P4) is also relevant for the Canary Island case. For example, land use changes (i.e. increasing the number of hotels) that in the short term increase monetary yield (primary feedback) may cause (accumulative) problems in the long term in several domains (water consumption, waste water, sealing of soil).

17.3.4 Conceptualizing rationales of human systems

Decision theory provides a powerful common language for describing the actions of human systems, as well as organisms. We can assume that countries, politics, companies, non-governmental organizations such as WWF (World Wildlife Fund), and human individuals, show certain preferences, based on a specific functionality, with respect to different options or alternatives. This means that decisions (Decision Postulate P5), but also the awareness of environmental problems (Awareness Postulate P6) depend on these preferences.

This is why we introduced the three boxes on the left side of Figure 17.1*; that is, goal formation, strategy formation, and strategy selection. Human systems are assumed to be capable of doing something in one manner or another and thus to have strategies. This is the essence of the Decision Postulate P5. Determining the type of decision model that is constructed on the human system side is a task of the research team. For the Canary Island case we conclude that there will be differences between actors with regard to the awareness of environmental processes (like depletion of groundwater resources). That is, not all actors will be (explicitly) aware or willing to become aware of detrimental secondary feedback loops. Additionally, even though one is aware of it, the prominent strategy could be to focus on the short-term benefit.

17.3.5 Example: a view of Mongolian herders and grassland through the HES framework

Mongolian grassland, its history, property rights, conflicts, and ongoing dilemma offer an example of the value of pooling community resources (Li et al., 2007). In this section, we examine the plight of Mongolian herders, whose herding activities are destroying the very grasslands they use for pasture, through the lens of the HES framework.

Let us imagine that we are working with a group of Mongolian herders called *khot ail* who are concerned about assuring the livelihood of their families. A first step would be to assemble a group of stakeholders, including herders, any agricultural farmers in the area (who may compete with or complement herding), purchasers of herd products, community leaders, perhaps a representative of an international agency of development, and representatives from the Mongolian government who could provide programmatic and financial support, perhaps in the form of loans. If we consider the group of Mongolian herders, we can consider this early step of the research project as part of the analysis of the (social) Environment-first Postulate.

Our work starts with defining a guiding question, such as: is enlargement of the herd an effective way to sustain the herder's livelihood? This question specifies the human system, the Mongolian herder, who operates traditionally at least at three hierarchical levels. That is, a Mongolian herder can be considered both as an individual and a head of an agricultural enterprise, who owns on average 235 animals. But he can also be looked at as a member of the khot ail, which is a group of 2–12 farmers building a herding community. We show this as P2 on Figure 17.3. As mentioned above, when planning a research project, the research team has to decide

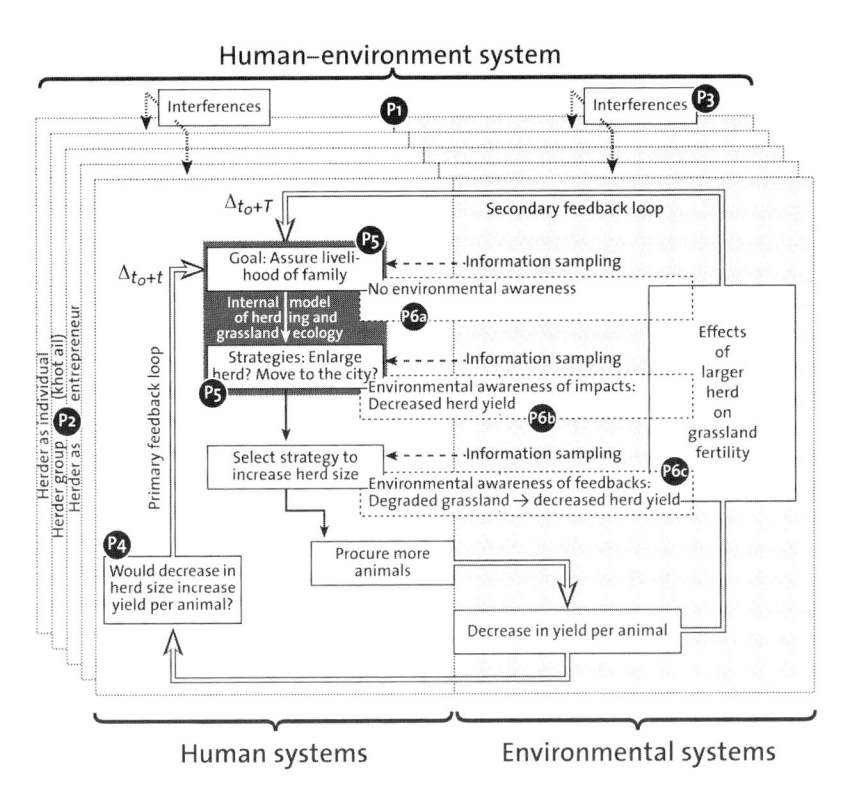

17.3 The HES Postulates in environmental research

Figure 17.3 An HES framework view of the plight of "average" Mongolian herders, whose herding activities are destroying the grassland that they use as pasture for their herds.

what system boundaries are adequate, for instance whether a single farmer or khot ail is investigated by a case study or whether a province or the whole country should be looked at. This may require the introduction of additional human systems such as the government or ministries.

Clearly, the herd is part of the environment, referring to the Complementarity Postulate P1. According to Decision Postulate P5, we can postulate that the herder's primary *goal* is to assure the livelihood of his family. Given their dire circumstances, herders may be wondering whether they should enlarge their herds or move to the city, shown as possible *strategies* on Figure 17.3, again related to the Decision Postulate P5.

Following the HES framework, before strategy formation and before taking action, we first want to understand the dynamics of the environmental system. Such an environmental investigation would involve analysis of the current ecologically sensitive arid and semiarid grasslands, as part of the material environment, as well as of the market options for herding related to agriculture, as part of the social environment. This investigation would give us an "internal model of herding and grassland ecology," shown as the Environment-first Postulate P7 on Figure 17.3.

The environmental investigation (P7) would show that Mongolia is currently facing a critical overstocking from the 42 million livestock held by 180 000 herders. The number of herders has doubled in the past 18 years as many people consider herding the only option to escape hunger and famine. The yield per livestock is decreasing and only large herds allow for a good income. An effective investigation of the environmental situation of Mongolian herders sheds light on how these conditions came about. As we use the Mongolian herder case only for illustrating the HES framework, we only sketch what an environmental analysis may include. This should be an assessment of the maximum herding intensity as – if grazing is intensified above a certain threshold,[1] in combination with low precipitation (< 350 mm/year) – the system shifts to a degraded state that could not return to its original composition even without further grazing. Much insight into the environment can be gained from historic analysis on the rotational grassland use as a basic ecological principle

---

[1] In Inner Mongolian grasslands, this threshold is assumed to be 0.49 g/u, where g = biomass in grazed area and u = biomass in ungrazed area (Christensen, 2003).

for raising nomadic livestock which provided sustenance for centuries. Further, one can study the impacts of the herding during communist times (1924–90) in which Soviet-style cooperatives became established in many places. Here we may find data about overusing the grassland. These data may be taken to reference similar current trends of increasing livestock density and declining rotational herding practice. An environmental analysis would also ask for a market model of about 180 000 nomadic herder families keeping 42 million animals. Here, a prospective agroeconomic yield model seems reasonable. This model should account for the prospective degradation caused by the doubling of the livestock in the past two decades and the fact that about 78% of the grasslands are judged to have been degraded (Enkh-Amgalan, 2009; see also Box 6.15). In addition, the impact of global climate change (e.g. increasing air temperature, decreasing precipitation) should be included.

When looking at the decision process (see P5) of the herders, whether they take action to increase herd size will depend on the degree of awareness of the environment. For example, if herders have no environmental awareness, shown as P6a on Figure 17.3, they may procure more animals and then simply wonder why it is more difficult to make a living. However, the herders may notice the decreased herd yield, P6b. Here, they would begin to wonder about whether herd enlargement is the cause of decreased yield, P4, primary feedback loop. The most environmentally aware herders would notice that increased herd size degrades grassland, which in turn decreases the yield per animal, P6c. Ideally, they would ask themselves about the effects of a larger herd on grassland fertility, P4, secondary feedback loop, and join the stakeholder group in a transdisciplinary process to review alternatives for sustaining the herds.

Underlying the entire investigation of the HES is the Complementarity Postulate P1. Based on this, we assume that the herder is making decisions, for instance, on the number of livestock in his herd. The herd is part of his environment and affects the pastureland by grazing.

17.4 The HES framework in action

We are excited to offer the HES framework, which formalizes and brings together approaches used by a growing number of environmental researchers in their endeavors to encourage environmental literacy. In the final chapters of this book we discuss how the HES framework can help us better cope with the environmental challenges of the twenty-first century. Chapter 18 describes three cases where the HES framework has been or could be applied, Chapter 19 compares the HES framework with other approaches for investigating HES questions and problems, and Chapter 20 outlines our suggestions for future research.

Part VIII   A framework for investigating human–environment systems

Chapter 18

Applying the HES framework

Roland W. Scholz, Claudia R. Binder, Daniel J. Lang, Timo Smieszek, and Michael Stauffacher

18.1 Case 1: threat of pandemics   464
18.2 Case 2: transdisciplinary case studies for sustainable regional and organizational transitions   473
18.3 Case 3: agro-fuel   482
18.4 Case 4: bones, BSE, and P   495

Chapter overview

Environmental literacy and coping with complex environmental problems can be viewed as knowledge- and experience-based artwork.

A major challenge of research on human–environment systems (HES) is to define the guiding question and system boundaries adequately, to reduce complexity by identifying the relevant actors and the relevant human and environmental system, gaining access to potentially conflicting drivers in and between human systems, and identifying with and coping feedback loops linked to decisions. This chapter shows by four exemplary cases how the HES framework can be utilized for this venture.

Case 1, the threat of pandemics, deals with the interaction of human systems with bacteria and viruses as environmental agents. Epidemics and pandemics have been major long standing threats and have disrupted societies not only by the deaths but also by the means that society has taken to cope with them. We discuss strategies that are taken by individuals, groups, and societies when coping with contagious diseases and reveal how the HES Postulates can help for better analyzing, understanding, and coping with epidemics. The HES Postulates serve to integrate the different perspectives of public health in a coupled human–environment perspective and allow insight into undesired feedback loops of different types of pandemic management.

Case 2 is based on our transdisciplinary case studies for regional and organizational transition conducted annually since 1993 at ETH Zurich. Taking the example of sustainable transitions of traditional industries in a rural area of Switzerland, we show: (i) how the HES Postulates help in structuring and mastering this complex issue at the interface of business, environmental, and social sciences; and (ii) how a mutual learning process among industry, administration, universities, and the public can be organized to sustain these industries.

Case 3 is about energy and climate, and deals with the oil-free society and the role of bioethanol, which we call agro-fuel. It starts from a societal, nation-state perspective. We analyze the pioneering decision of the Swedish government to become an oil-free society until 2020 and to push bioethanol as primary vehicle fuel. Applying the HES Postulates, in particular the Environment-first Postulate P7, reveals that environmental awareness and the identification of critical feedback loops were obviously not sufficiently acknowledged. If a macro perspective is taken, the governmental decision asks for reconsideration, in particular when acknowledging (e.g. north–south dilemma, loss of biodiversity) a portfolio view on vehicle fuel, etc. The utilizing of the Hierarchy Postulate P2 showed that the role of (missing) supranational players is essential for establishing worldwide transitions towards oil-free society.

Case 4 is called bones, BSE, and P, and investigates a possible trade-off that can arise within human–environment interactions. One and the same issue (i.e. the animal by products bones and meat) can be of interest for some perspectives but cause a threat from another perspective. The case demonstrates again the absolute priority of following a thorough investigation of the material–biophysical environment as well as the identification of the hierarchically organized actors and their interfering rationale.

18 Applying the HES framework

18.1 Case 1: threat of pandemics[1]

18.1.1 What can we learn from this case for environmental literacy?

We show how the HES Postulates and the HES framework help to gain an integral understanding of disease dynamics in society and threats of pandemics. We first detail the structure of human and environmental systems, ranging from the level of the cell, individual, institution, to the society. We then turn to a dynamic, process view that describes the different mechanisms of infection in the complementary human and environment system. Specifically, we show how the logic of a pathogen (i.e. of a causative agent) interacts with the logics of human systems and how – by means of secondary feedback loops – new system states can emerge. We discuss what this means for human societies when coping with pandemics, in particular as bottom-up and top-down processes between the hierarchy levels interact. The mechanisms discussed are general though we illustrate them for the case of influenza.

18.1.2 What is the case about?

Infectious diseases have historically been a major scourge of mankind and continue to be a primary reason for death, malaise, and economic losses (Keech & Beardsworth, 2008; Molinari *et al.*, 2007; Russell, 2004). In 2004, the third most frequent cause of death was lower respiratory infections (responsible for 4.2 million deaths worldwide, i.e. 7.1% of all deaths). Diarrheal diseases, HIV, and tuberculosis rank as the fifth, sixth, and seventh most common causes of death, respectively (World Health Organization, 2008). In high-income countries the relative burden of infectious diseases is considerably lower than in low-income countries. A much higher percentage of the population dies in developing countries because of infections. Children are at higher risk in developing countries while the elderly are more at risk in developed countries.

Guiding question
What changed pandemic threats are modern and future societies and the human population facing?

The spread of infectious diseases is thus a prime example of a complex, coupled HES.

We are facing a global threat of newly emerging diseases or recombinant strains of common viruses or bacteria for which humankind is unprepared in many respects. A special point is that this is happening in evolutionary new interaction patterns of industrialized society in which railways, passenger ships, and airplanes provide new means of transportation. Emergence and spread of pandemics are affected by strategy selection of human systems on all hierarchy levels. Here, an analysis of the causative agent and the physical properties of the inanimate environment are essential. Thus, mastering pandemics necessitates an integrative perspective (see Figure 18.1) that views disease spread as sociobiological transmission systems (Chick *et al.*, 2001; Koopman, 2005).

Content case 1

C1.1 Complementarities between humans and pathogens in pandemics are of different nature   464

C1.2 Hierarchy levels facilitate understanding pandemic spread   466

C1.3 Scientific models of influenza transmission   466

C1.4 The role of the human system in disease transmission   468

C1.1 Complementarities between humans and pathogens in pandemics are of different nature

Epidemics can be caused by bacteria (e.g. cholera, typhus, or leprosy), eukaryotes such as *Plasmodia* (causing malaria), viruses (e.g. smallpox, measles, viral hepatitis, influenza, HIV), or prions. The human–environment complementarity can be simply defined for bacteria or eukaryotes, which remain separate organisms throughout infection. This is different with viruses, which have their own DNA or RNA, but do not have an autonomous metabolism. To replicate, they enter

---

[1] Timo Smieszek is the lead author of this case; he was funded by the Swiss National Science Foundation.

Case 1: threat of pandemics

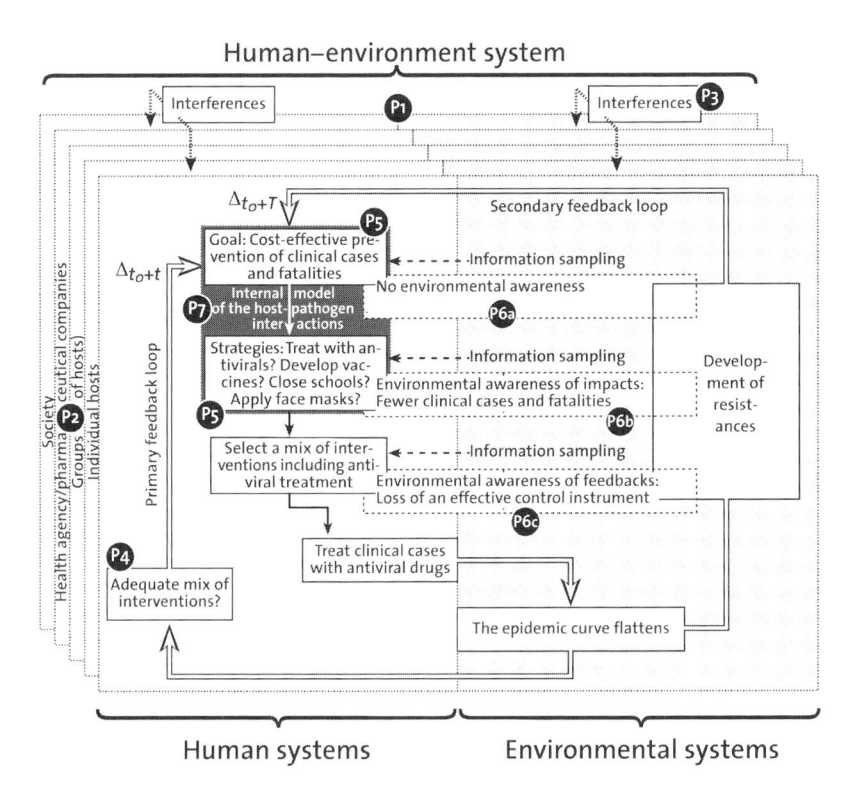

Figure 18.1 An HES framework view of the threat of pandemics. The P-circles in the framework correspond with the HES Postulates. Call-outs in the text indicate how the Postulates can be answered for this case example.

host cells, replicate, and thereby exhaust the resources of the host cell. Thus, at least at some stages of the life cycle of a virus, it is a matter of interpretation whether it is considered to be a complementary parasitic unit of the human cell or as part of the human system. In this description, we consider infecting viruses to be part of the material–biophysical environment (E_m; see Figure 3.3*). The processes underlying transmission, infection, impacts on the human cellular system, medical therapy, etc. differ fundamentally for various infectious agents.

Complementarity Postulate P1
Bacteria, eukaryotes, or viruses differ qualitatively in the nature of how they are coupled with human systems.

The interaction between human cells (H_m) and bacteria (E_m), for instance, shows many variants and cannot be classified easily. A first distinction has to be made between harmless, pathogenic, and conditionally pathogenic bacteria. This is a functionalistic distinction that is primarily motivated by a human health perspective. Large quantities of bacteria coexist with human cells, and some of them – such as the gut flora – even play a major role in the functioning of the human body as a whole (cf. Box 3.1). A human–environment problem only exists when certain bacteria start to cause harm. *Staphylococcus*, for instance, is an ubiquitous organism and a normal part of the human flora, and only causes severe diseases like sepsis under certain conditions (Fish, 2001). Other bacteria are in principle pathogenic – but their effect to the human system is not always direct. Some directly damage human cells, either by acting as an intracellular parasite (e.g. *Rickettsia* spp., the causative agents of typhus, or *Chlamydia* spp.) or by releasing toxins (e.g. *Vibrio cholerae* or *Clostridium botulinum*), while others impact on human systems by disturbing the natural flora, thereby impacting the environment of the human system.

Environment-first Postulate P7
A thorough understanding of the rationale of pathogens is a prerequisite of pandemic management.

18 Applying the HES framework

C1.2 Hierarchy levels facilitate understanding pandemic spread

At the hierarchical level of individuals, the relevant material environment includes individuals that can transmit causative agents either directly to a certain individual or by contaminating the inanimate environment with pathogenic organisms. Direct transmission is defined here as stemming from direct interaction between two individuals at a close spatio-temporal proximity such as through sexual intercourse, touching somebody, talking to somebody at a short distance apart or sneezing into somebody's face (cf. Morrow, 1980). Contaminating the inanimate environment includes releasing, for example, *Vibrio cholerae* bacteria into sources of drinking water (Snow, 1855) or *Mycobacterium tuberculosis* into the indoor air, where small droplet nuclei can stay suspended for a long time, propagated by air currents and Brownian movement (Wells, 1955). Elements of the relevant material environment also include the microclimate conditions (humidity and temperature) and the chemical and structural properties of surfaces, which determine how long a specific infectious agent can remain infectious outside a host's body (Zuk *et al.*, 2009).

Epidemics and pandemics have become a business domain. Specialized, multinational companies operate all around the world and provide vaccines and medical treatment against certain diseases.

The social environment of an individual comprises, among other aspects, the behavioral rules set with respect to a specific disease. Societal institutions such as ministries may direct curfews or school authorities the closure of schools.

Hierarchy Postulate P2
Managing pandemics asks for considering HES at least on the levels of the cell, individual, organization (including international companies and organizations), and society.

Societal norms are also part of the social environment, and these can vary among cultures. For example, in many Asian nations it is pertinent for sick people to wear face masks, whereas such a cultural norm is completely lacking in Europe. This refers to the societal level. The socio-epistemic environment includes the rules and recommendations made by WHO and other health organizations.

C1.3 Scientific models of influenza transmission

C1.3.1 How are influenza viruses transmitted?

Though (seasonal) influenza is one of the most common and ubiquitous communicable diseases in western societies, much basic knowledge needed for an adequate understanding of its dynamics is still unknown. While vast knowledge on this rather simple virus has been accumulated at the genetic level, robust knowledge about its modes of transmission is sparse. Thus (see Chapter 14.5), risk assessment can be based on thorough knowledge about the infectious agent but not about the exposure by the transmission pathways. In the case of influenza, different pathways have been identified:

(i) **Transmission by direct or indirect physical contact**: in experiments, influenza viruses were shown to remain infectious on different surfaces for more than 1 day (Thomas *et al.*, 2008; Weber & Stilianakis, 2008). Therefore, infection via shaking hands or touching contaminated surfaces seems possible.

(ii) **Droplet transmission**: various authors claim that influenza is a droplet-transmitted disease (Brankston *et al.*, 2007; Han *et al.*, 2009). Droplet transmission is defined as transmission via comparatively large droplets of infectious material that fall out quickly because of their size (Morrow, 1980) and reach other hosts within a maximum distance of 1–2 meters (Figure 18.2).

(iii) **Airborne pathways**: there is also evidence for a truly airborne pathway (Atkinson & Wein, 2008; Gregg, 1980; Musher, 2003). Truly airborne transmission or aerosol transmission means transmission by very

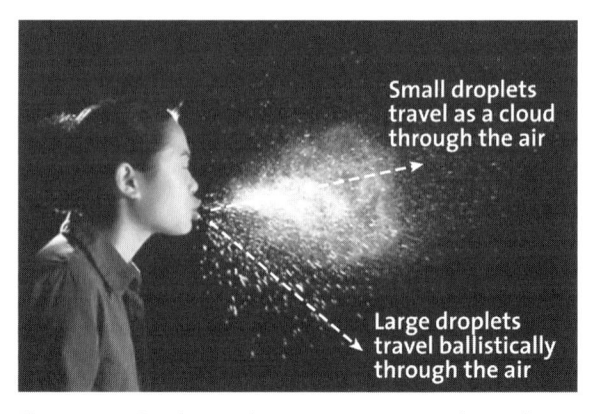

Figure 18.2 Droplets are the major transmission pathway of influenza spread (photo taken from Tang *et al.*, 2006).

small droplet nuclei, which are small enough to stay suspended in indoor air for a long time (Figure 18.2).

Environment-first Postulate P7
Even with the ubiquitous influenza, there is no unambiguous knowledge about basic processes such as the transmission pathway.

Although there is evidence that all these pathways exist, their relative importance has not been determined (Weber & Stilianakis, 2008). The issue, in fact, remains contentious, with some literature supporting the argument that truly airborne transmission is dominant, while others support the dominance of droplet transmission. In our view, there is no clear tendency from the results found in literature.

This gap in knowledge on the biophysical mechanisms for disease spread has implications for developing the required corresponding behavioral model. Both droplet and physical contact transmission are best represented by a (dynamic) one-mode network (see Figure 18.3a; Read *et al.*, 2008; Wasserman & Faust, 1994) that relates individuals to individuals: transmission can be understood to be direct from one individual to another – it is the behavior exerted upon another host that leads to transmission (e.g. to talk to somebody within a short distance or to cough on somebody).

Environment-first Postulate P7
Understanding influenza spread requires linking biomedical knowledge and knowledge about social contact networks of different sociocultural settings.

In contrast, aerosol transmission is better represented in affiliation network models (see Figure 18.3b; Wasserman & Faust, 1994) in which individuals are related to places, for instance, a bus, airplane, or classroom. A characteristic of such affiliation networks is that the precondition of temporal co-location is relaxed: the infector and the susceptible host do not have to be at a certain location at the same time, as the location may remain contaminated after the infector has left the room. The duration of contamination depends on the supply of fresh air and the time the pathogen can survive outside the human body. From the perspective of an individual, the two assumptions differ as in one case other individuals build the source of infection whereas in the other it is places where infectors have been that build the source.

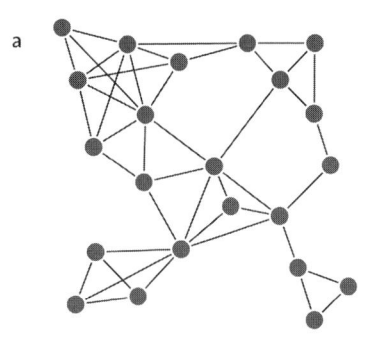

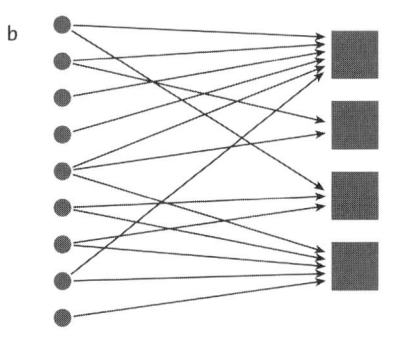

Figure 18.3 (a) A one-mode network, depicting contact and droplet transmission; (b) an affiliation network, depicting airborne transmission. Circles denote individuals, squares denote places (schools, airplanes, etc.), and lines represent connections and can be undirected or directed (arrows).

C1.3.2 Different characteristics of airborne and droplet spread

The differences between the competing theories of influenza transmission have important implications for the behavior of the corresponding transmission system models and potential mitigation measures.

(i) If truly *airborne* transmission is assumed, the infection probability in a given situation is driven by the inhaled dose: "the response to inhaled droplet nuclei contagium is quantal" (Wells, 1955, p. 123). When the risk of infection depends on the inhaled dose, the concentration of pathogens in the air and the quantity of air breathed in by a host must be determined. The contamination of (indoor) air depends on the shedding rate of pathogens by present infectors and the ventilation conditions. Such calculations have been successfully completed for specific situations (Chen *et al.*, 2006; Nicas & Gang, 2006; Riley *et al.*, 1978) and can, in principle, be expanded for population level analyses (Eubank *et al.*, 2004).

(ii) If influenza is indeed predominantly transmitted via the airborne route, this would have vast implications on the number of persons at risk, on the structure of spread models, and on the set of interventions that might be effective against a population-wide spread of the disease. Many more persons would be simultaneously at risk. Every person *currently* located or that *will* be located in the same volume of indoor air as the infector can take up small infectious droplet nuclei emitted by the infector. This can be several hundred persons, depending on the ventilation conditions (Eubank *et al.*, 2004; Gregg, 1980). From theory we know that contact structure plays a less important role when high numbers of persons are at risk (Bansal *et al.*, 2007; Read *et al.*, 2008; Smieszek *et al.*, 2009). Under such conditions, it makes no sense to target control measures on individuals (e.g. social distancing), but rather emphasis should be placed on measures to prevent contamination of places. High air change rates are advantageous (Li *et al.*, 2007), as well as measures to inactivate pathogens (e.g. subjecting the indoor air with high-energy UV light; see Terrier *et al.*, 2008).

(iii) If influenza is predominantly transmitted via large *droplets* and physical contact, the number of contact partners per day spans over two orders of magnitude (from 10^0 to 10^2). The average number of contact partners differs between countries. Germany, for instance, has a low (8.0 contact partners per day) and Italy a high average contact rate (19.8 contact partners per day; Mossong *et al.*, 2008). With such relatively low average numbers, effective social distancing interventions seem to be much more realistic than in the case of airborne transmission (Smieszek *et al.*, 2009). Another helpful intervention can be the use of surgical face masks, as recommended by several health agencies (Bundesamt für Gesundheit (BAG), 2009c; Centers for Disease Control and Prevention (CDC), 2009d; Srinivasan & Trish, 2009; World Health Organization (WHO), 2009). These masks effectively protect against transmission by large droplet particles but fail in the case of small droplet nuclei.

Finding the adequate model: This example clearly shows that there is not always a straightforward, unambiguous state-of-the-art model of the environment on which a human-environment-system analysis could base itself (Environment-first Postulate 7). We should note that the history of science is full of wrong assumptions and erroneous models (Hays, 2005). Thus, emphasizing that every scientific model must be used critically and reflected upon is trivial. Sometimes – as in the case of influenza – there are competing models. If the current empirical data and theoretical insights do not help in determining which of the competing models is most likely true, researchers trying to tackle human-environmental analysis have to think of alternatives and have to delineate the practical consequences of both models.

C1.4 The role of the human system in disease transmission

C.1.4.1 Hierarchical levels involved in disease spread

The Hierarchy Postulate P2 states that the human component of a coupled HES is divided into distinct hierarchy levels with corresponding environments. This plays an important role in understanding and managing pandemics.

Hierarchy Postulate P2
Any health program must include a comprehensive and careful analysis of the different interests and drivers inherent in the hierarchy of human systems involved.

The *individual* is the level where physical transmission of disease takes place. Infected individuals are the entities that emit pathogens. Susceptible individuals are the entities exposed to infectious material and may or may not become ill. Consequently, it is also the particular behavior of individuals – constrained or enabled by higher hierarchy levels – that decides whether they are at risk of becoming infected and also how high this risk is.

Above the individual level, there are several different kinds of *organizations*, such as pharmaceutical companies or the Red Cross and Red Crescent movements, which act and react with respect to epidemics and pandemics, and/or which suffer from such major outbreaks of disease. Pharmaceutical companies develop and provide vaccines, antibiotics or antiviral drugs. In most cases, such companies are global players and have enormous economic and political power because of their size and product monopolies. Another important kind of organization is the so-called non-governmental organizations (NGOs) such as the Red Cross/Red Crescent.

National or *regional institutions*, which are entrusted with the management and mitigation of epidemics and pandemics, are the healthcare systems, universities and research institutions, the national centers for disease control, and the national public health ministries. For instance, it is the responsibility of the bodies of the health administration to carry out the real-time surveillance of disease prevalence in society and to prepare and make decisions on interventions.

Finally, there are *international bodies* – in particular the WHO – that facilitate coordination and decision-making beyond the national level. The need for international coordination becomes more urgent as worldwide travel and migration continue to increase. Worldwide surveillance of pathogens that have a pandemic potential also helps to stop outbreaks early and to enact national interventions before the disease has actually reached a certain region.

C1.4.2 The role of disease awareness for the spread of infectious diseases

For the case of influenza, it has been shown that unambiguous, robust knowledge about the dominant pathways of transmission is missing. In practice, this has severe consequences for the effectiveness of certain interventions and thus for public health decisions.

> *Awareness Postulate P6*
> Based on different data and observations and reasoning rationales, healthcare workers develop different awareness and models about the causes of pandemics.

However, science only provides one type of knowledge. The Awareness Postulate P6 stresses that decision formation depends on the knowledge that is activated or accepted by a human system and the motivation to incorporate environmental issues. Even today, individuals, organizations, institutions, and societies can have particular perceptions of a disease and its etiology that differ fundamentally from the theories and knowledge elaborated in the scientific system. Various studies have shown that, for instance, relevant percentages of healthcare workers have models of the flu, its etiology, and the effectiveness of interventions that differ vastly from the leading theories in science. Typical causation of US health workers read:

> The shot will make me sick or give me influenza. (Steiner et al., 2002, p. 626; mentioned in 27% of all cases)
> I believe that I am not at risk of catching influenza because I am healthy and don't get sick. I have a good immune system. (ibid., p. 626; mentioned in 23% of all cases)

This contradicts standard science. Knowledge as the risk of becoming ill because of an influenza vaccination is not of any relevant order of magnitude. Nor is being "healthy" a sufficient protection against influenza. Here it is of interest that studies in England (Morrison & Yardley, 2009) and Australia (Seale *et al.*, 2009) revealed that healthcare workers have only little trust in countermeasures against influenza.

Many CDCs (e.g. that of the US Centers for Disease Control and Prevention (CDC), 2009c) assume that influenza is predominantly transmitted via large droplets and close physical contact, and not via aerosols – although there is scientific evidence for both models of transmission (see above). Concentrating on droplets is optimistic, as it allows more options for interventions. Face masks are a reasonable means of prevention, and social distancing measures make more sense than in the case of airborne transmission. Here "wishful thinking" and the "illusion of control" might be at work.

C1.4.3 Strategy formation and selection of human systems

The strategy formation and selection of a human system (P5) can hardly be separated from disease awareness, from the perception of the etiology, and the role that humans play in disease spread (P6). Different perceptions of how disease spreads may lead to completely different solution spaces. But here the internal model of how environment causes influenza is of interest. If people believe that immoral, unholy behavior is the cause of illness, they may go to church. Or they may believe in the antibiotic effect of tobacco and alcohol, and gather in pubs and tobacco stores as some Swiss men did in Switzerland during the 1918/19 influenza pandemics as the Schaffhauser Intelligenzblatt reported (Leu, 2005). This personal mode of coping strategies conflicted with local government strategies, which prohibited fairs, dance festivities, theaters, and cinemas, as well as services and festivities in churches to reduce the number of potentially contagious contacts (Leu, 2005).

Decision Postulate P5
Pandemic management includes many decision-makers following different conflicting objectives on the same and on different hierarchy levels which would ask for transdisciplinary processes.

Forbidding contacts – health authorities vs. individual interest is one example of contradictory goals on different hierarchical levels (P5) with respect to pandemic influenza. A primary goal of government strategies is to lower the total number of sick persons and fatalities (Bundesamt für Gesundheit (BAG), 2009b; Centers for Disease Control and Prevention (CDC), 2009a, b). Depending on the pathogen's infectivity and its mode of transmission, the strategy of reducing contacts by closing schools can be quite effective (Smieszek *et al.*, 2009). However, this only holds true when the avoided contacts in schools are not replaced by other eventually more intense contacts resulting from increased spare time.

Commercial vs. societal interests at the university level: a key mission of the university, as an institution, is to provide the knowledge necessary to face infectious disease and to build up the capacity in the political system to make informed decisions. This mission is quite often conflicted by a close relationship with pharmaceutical companies (hierarchy level of organizations). In many countries the current incentive schemes for scientific research reward close cooperation with the industry because of financial, public reputation, or research cooperation interest. Here the societal and individual interest may be at conflict (see Chapter 15.6).

It seems evident that transdisciplinary processes target at the capacity- and consensus-building, but which can also lead to legitimization of epidemic intervention as well as mediating between conflicting parties, are a proper means of dealing with pandemics (see Table 15.3).

C1.4.4 Pathogen evolution feedbacked with human behavioral change and vaccination

Health agencies analyze statistical health data and data from vaccination campaigns (i.e. actions) to determine whether certain interventions against a disease were effective and what kinds of behavior are associated with an elevated risk of infection. However, pandemic research is not only facing immediate feedbacks but also indirect and long-term feedbacks to human behavior.

Two very important long-term feedback loops owing to the mechanisms of evolution become evident here. One is pathogen evolution caused by changes in human contact patterns. The other is pathogen evolution caused by the use of antibiotic or antiviral drugs.

C1.4.5 Human contact patterns and pathogen evolution

There is clear theoretical (Boots & Sasaki, 1999) and empirical (Boots & Mealor, 2007) evidence that high virulence is a disadvantage in natural selection when contacts are mainly local. The reason for this is that extremely infectious and virulent pathogens tend to exploit their local resources before they have the chance to reach other neighborhoods with new, susceptible hosts successfully (Smieszek *et al.*, 2009).

Since the Middle Ages, many new means of transporting humans have been invented, resulting in mobility patterns that involve more contacts among individuals, including those in remote regions. This leads to the *small-world phenomenon*. Even when most of an individual's contacts are local, few remote links between persons connect any two individuals, even those who are separated by great geographical distances, by just a few intermediate individuals (Milgram, 1967; Watts & Strogatz, 1998). This phenomenon makes the previous disadvantage of highly virulent, aggressive strains of a pathogen disappear. Highly pathogenic viruses have a better chance of becoming evolutionarily stable, which increases the chance of disastrous pandemics.

Changes in mobility and contact frequency and intensity imply changed patterns of disease spread. In earlier times disease spreads showed wave-like patterns with a clear frontline. Such wave-like patterns were, for instance, described for the bubonic plague in fourteenth century Europe (Christakos *et al.*, 2007; Zietz & Dunkelberg, 2004), as well as for the five cholera pandemics in the nineteenth century (Hays, 2005). Nowadays, local transmission is still the major driver for many infectious diseases in wild animals, as can be seen in the example of rabies spread in Europe since the 1940s (Ou & Wu, 2006).

Outbreak maps of pandemics in the high mobility global industrialized world have already depicted characteristics of the small-world phenomenon. The influenza pandemic of 1918–19 began in the eastern coastal regions of the USA. The pandemic "diffused" from the

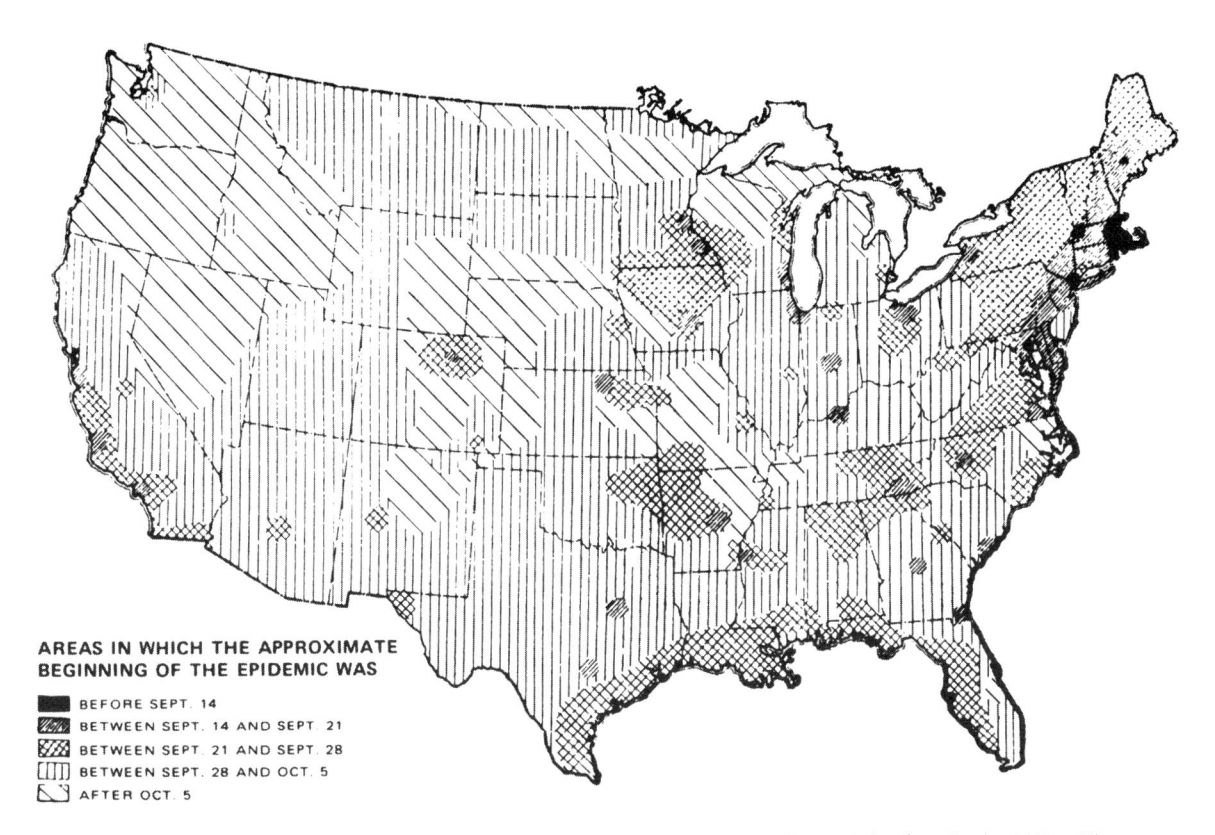

Figure 18.4 Spatial temporal patterns of the spread of the 1918/19 influenza pandemic in the USA (taken from Crosby, 2003, p. 65).

major coastal cities upcountry. However, this diffusion process, driven by local contacts, was superimposed by the fact that remote cities began facing major outbreaks quite early in the course of the pandemic (see Figure 18.4). This phenomenon was most likely driven by railway transport of persons, particularly of soldiers, who were mobilized to participate in World War I.

C1.4.6 Medical treatment and pathogen evolution

It is well known that medical treatment sooner or later leads to the development of pathogen resistance. Current strains of influenza (also the potentially pandemic strain H1N1) already show high levels of resistance against the drug oseltamivir (Chen *et al.*, 2009; Mayor, 2009).

Feedback Postulate P4
The long-term feedback loops caused by evolutionary adaptations of pathogens to changed mobility and contact patterns and large-scale drug therapy are a major peril of future pandemic management.

The development of resistance is an almost unavoidable part of natural evolution. If measures against a certain pathogen are repeatedly put into action, sooner or later some mutants will "learn" (bearing in mind that evolution has no underlying intentions) to cope with these measures. However, the dimension and the speed of resistance development are vastly determined by human behavior. The more people that take antivirals, the higher the likelihood of resistance development. As with antibiotics, it is a strategic question as to whether substances like oseltamivir and zanamivir should be restricted for severe cases and real pandemics or whether they should also be used as convenient relief for mild cases of the seasonal influenza. Thus public health authorities can influence the dimension and speed of resistance development with their strategy selection.

Other concurrent medical conditions (mainly diarrhea), overwhelming viral replication, or wrong and irregular intake of antiviral drugs can lead to inadequate viral suppression. Inadequate viral suppression is a condition favoring resistance development. Public health authorities may consider strategies aimed at improving antiviral efficacy (e.g. the use of higher doses or combinations of different antiviral drugs; de Jong et al., 2005).

However, drug resistance can evolve not only within the human patient but also indirectly via the environment. Oseltamivir is not entirely metabolized in the human body; it is excreted as the still-effective oseltamivir carboxylate (OC). OC flows through wastewater treatment plants without being decomposed and can thus contaminate natural water bodies. Avian influenza strains can then develop resistance if, for instance, waterfowl (a common host of the virus) are exposed to the OC in the surface water. Recombination of resistant avian and non-resistant human influenza genomes can become a problem for humankind (Fick et al., 2007; Sacca et al., 2009; Singer et al., 2007). Furthermore, there are initial findings that indicate that oseltamivir might be ecotoxic for *Daphnia*. Hence, this substance might also become a "classical" environmental problem for water courses.

C1.4.7 Understanding higher ordered feedback loops

In the previous sections we showed how an integral understanding of pandemics and epidemics of infectious disease can be achieved by structuring the particular human–environment problem along the HES Postulates.

Feedback Postulate P4
There are critical higher ordered feedback loops involved in large-scale pandemic prevention by drugs which deserve a better understanding to avoid critical rebound effects.

For instance, it has become clear that focusing purely on instantaneous effectiveness of intervention measures can lead to adverse long-term effects: there is good reason to consider resistance development in the societal strategy formation on the application of antibiotics and antiviral drugs. It is crucial to decide whether and which active pharmaceutical ingredients should be restricted for use only in emergency cases. However, an analysis of the actors involved makes it clear that restricting drugs conflicts with the goals of the pharmaceutical industry, which has to generate profit and finance the expensive development of new drugs. Any successful strategy for a long-term drug management has to incorporate the goals of the important actors.

Further, the release of active pharmaceutical ingredients into natural water courses is a problem to be tackled. Pharmaceuticals in lakes, rivers, and aquifers are a known problem in ecotoxicology. Moreover, we have also described an elevated risk of resistance development owing to such drugs being released into the environment. There are two strategies to deal with this problem. On one hand, pharmaceutical companies could be urged to make biodegradability one of their goals during drug development. Second, technologies like the NoMix toilet (EAWAG, 2007), which separates urine from feces, can help to avoid the loss of nutrients (in the urine) as well as the contamination of the environment by unmetabolized active agents. The urine can be separated, stored, and treated in such a way that the nutrients become available again and that pollutants can be degraded (Escher et al., 2006).

Finally, good pandemic policies try to foster research in a goal-oriented way. An adequate model of disease transmission has to understand both the biological transmission processes and their physical constraints as well as the corresponding human behavior. If there is no adequate state-of-the-art transmission model, pandemic planning has to think in alternatives and in terms of robust strategies. HES research in the field of disease spread must not only consider how effective certain intervention measures *could* be, but also how a certain intervention interacts with the goals of the hierarchical levels that have to comply. For example, it is not yet clear what kind of secondary changes in contact patterns could arise with government-imposed social distancing measures (such as school closures).

C1.4.8 Key messages

- Understanding interactions within and between individual and societal hierarchy levels sheds light on how viruses are transmitted
- The strategy formation and selection of a human system can hardly be separated from disease awareness, from the internal model of the etiology, and from the role that humans play in disease spread
- Gaps in knowledge about the biophysical mechanisms for disease spread have implications for developing the required corresponding behavioral model
- Contradictory, interfering goals exist on different hierarchical levels regarding the management of pandemic influenza.

18.2 Case 2: transdisciplinary case studies for sustainable regional and organizational transitions[2]

18.2.1 What can we learn from this case for environmental literacy?

This section links the idea of transdisciplinarity from Chapter 15 with the HES Postulates and the framework from Chapters 16 and 17 (see Table 18.1). We reconstruct one of our former case studies along the seven HES Postulates (see Figure 18.5*). We want to show: (i) how the HES Postulates may support structuring and mastering complex issues at the interface of different disciplines and serve for integrated modeling of coupled systems; and to exemplify (ii) how a mutual learning process among industry, administration, universities, and the public can be organized as a means of capacity-building. These two issues incorporate both aspects of Part VII, the going *beyond disciplines* and *beyond sciences*. It will become evident that (i) and (ii) are closely interlinked, coupled in our transdisciplinary case study (TdCS).

The TdCS is a research approach for tackling complex real-world problems in coupled HES from multiple perspectives initiating learning processes at individual, group, organizational, and societal levels. Thus the Hierarchy Postulate P2 is not only applied for analysis but also as a structuring tool for the whole research process, in particular the mutual learning between members of universities and different actors from society.

18.2.2 What is the case about?

The TdCS design evolved out of a large MSc project course (including 330 hours' student work) as a framework for learning applied research in a transdisciplinary setting. The approach is based on what we call functional sociocultural constructivism and project-based learning (Stauffacher *et al.*, 2006). Its ideas can be traced to the beginning of the twentieth century and the works of John Dewey, who advocated "learning by doing" (Dewey, 1910/1997, 1915 & 1902/2001). The TdCS includes transdisciplinary processes which support regional or organizational transition processes towards sustainable development (Scholz *et al.*, 2006).

The example we chose here is the TdCS 2002 *Future of traditional industries in the canton Appenzell Ausserrhoden* (Scholz & Stauffacher, 2007; Scholz *et al.*, 2003). The initiative for this TdCS 2002 came from the side of the canton, the president of the canton's executive council (head of state) in order to better understand options of sustainable land use in the canton Appenzell Ausserrhoden (AR), as it had been studied before by us for the German–Swiss Klettgau valley (Scholz *et al.*, 1999, 1998). The project was led jointly by the president of the canton, Dr. Altherr, and Professor Scholz, the academic project director, in a co-leadership. This represents a specific form of mutual participation; the people from AR helped with understanding the methodological and theoretical questions of sustainability, and the university team supported the canton AR in preparing for sustainable transitions. Both parties worked on equal footing, each with their own interests and goals.

> **Content Case 2**
>
> C 2.1 Guiding question and system boundaries as mutually accepted frame for the TdCS 474
>
> C 2.2 Complementarities help structuring complex human–environment interactions (P1) 476
>
> C 2.3 Environmental analysis for revealing the dynamics of the environmental system (P7) 477
>
> C 2.4 Hierarchy principle (P2) to identify different levels of the human system and interferences between and within these levels (P3) 479
>
> C 2.5 A decision-theoretic conception for understanding the rationale of human systems (P5) 480
>
> C 2.6 Improved environmental awareness for understanding potential barriers and drivers in human systems (P6) 480
>
> C 2.7 Understanding feedback loops in coupled human–environment interactions (P4) for preparing transformations 481

---

[2] Michael Stauffacher is the lead author of this case.

Table 18.1 Six steps in TdCS 2002 "Future of traditional industries in the canton Appenzell Ausserrhoden" and their link to the HES Postulates.

Steps (see also Box 15.1)	Related HES Postulates
(i) Define a guiding question	Preparing for utilizing the HES framework
(ii) Facet the case	Complementarity Postulate P1
(iii) Perform system analysis	Environment-first Postulate P7, Hierarchy Postulate P2, Interference Postulate P3; Feedback Postulate P4
(iv) Construct scenarios using formative scenario analysis (FSA)	Decision Postulate P5
(v) Perform multi-criteria analysis (MCA)	Decision Postulate P5; Awareness Postulate P6
(vi) Develop orientations	Feedback Postulate P4

C2.1 Guiding question and system boundaries as mutually accepted frame for the TdCS

After a first TdCS in the region on landscape development (Scholz et al., 2002), a joint decision was made to focus on traditional industries in the region, since the canton had been historically shaped by these. We implemented the following core elements in six steps (see Table 18.1; see also Box 15.1): (i) define a guiding question; (ii) facet the case; (iii) perform a system analysis; and (iv) a scenario development; (v) conduct a multi-criteria analysis (MCA) by referring both to science-based arguments (MCA I) and by obtaining individual preferences from different stakeholder groups (MCA II); and, finally (vi), discuss the results and develop orientations (Scholz et al., 2006; Scholz & Tietje, 2002). Overall we wanted to contribute to a

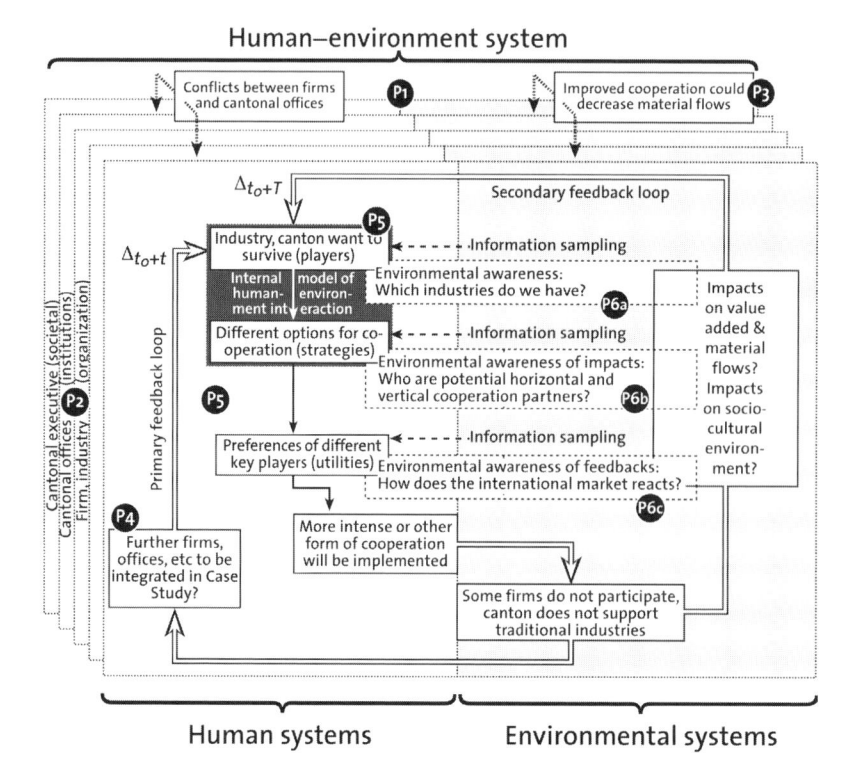

Figure 18.5 An HES framework view on a transdisciplinary process for sustainable transitions of traditional industries. The numbers P1–P7 refer to the different HES Postulates and are discussed in detail in previous chapters. Please check as well the call-out boxes for a first reading.

sustainable transition of different industries – those jointly chosen were the timber, dairy, and textile industries.

Transdisciplinarity
A transdisciplinary project between region and university is set up with co-leadership and collaboration on equal footing.

We defined the problem and the guiding question in a round of intense discussions with a "steering group." This group was strongly involved during the entire project, defining the project framework and continuously evaluating the project quality (Scholz et al., 2006). It was here that collaboration in the project was established and a substantial part of the mutual learning between science and practice took place. The steering group was composed of the following stakeholders ("key agents"; see Scholz & Tietje, 2002): head of office for the promotion of the economy, two heads of administration (agriculture/forestry, environmental protection), the cantonal historian/archivist, one farmer/mayor of a community, mayor of another community, one independent expert for tourism and landscape, and three independent experts for regional planning/development. Furthermore, the project team of ETH Zurich participated (four senior researchers). Thus, different epistemics were at play in this group and varying perspectives had to be reconciled in the process. Some preliminary media analysis, in-depth interviews with key people from the region, and first "experiential case encounters" (Scholz & Tietje, 2002, pp. 241–6) furthered our understanding of the case and helped us jointly to define the problem, which we will briefly sketch in the following paragraph.

Appenzell Ausserrhoden (AR) is a canton (i.e. a small state) of 20 communities with 53 500 inhabitants on 242 km² of land that lies between 435 and 2500 m above sea level. AR is located in the vicinity of the city of St. Gallen in the Greater Zurich Area (60–90 minutes traveling distance to downtown Zurich; see Figure 18.6). During the eighteenth century, production and sales of textiles dominated economic life (Tanner, 1982; Witschi, 2002). At the same time, 56% of the total land area of AR was agricultural, with the vast majority (98%) being utilized for dairy farming. Forests covered 29.6% of the land. AR industrialized rather early, and industries still play an important role (the proportion of people working in the secondary sector was above the Swiss average in 1910 and 2001). AR has been relatively slow in tertiary sector development (still much lower than in Switzerland) and still has an above-average proportion of agriculture. Growth of the tertiary sector has stagnated in recent years, even declining in the more isolated areas (Eisenhut & Schönholzer, 2003). Around 1880, AR was the most densely settled canton in Switzerland. From 1597 to 1794 the population even doubled, from 19 000 to 39 000 inhabitants, peaking at 57 973 in 1910 (Witschi, 2002). The population decreased to 44 500 in 1941 and slightly increased thereafter, until another decrease around 1980, which was followed by a definite growth period. This permanent struggle to maintain the number of inhabitants took place at a time of strong population growth in Switzerland. By the end of the 1990s, communities at a large distance from the national traffic infrastructure and the bigger cities suffered from decreasing population, decreasing labor opportunities, and thus decreasing wealth. Landscape was and still is the main capital of the canton, for agriculture and tourism but also for housing (Scholz & Stauffacher, 2002).

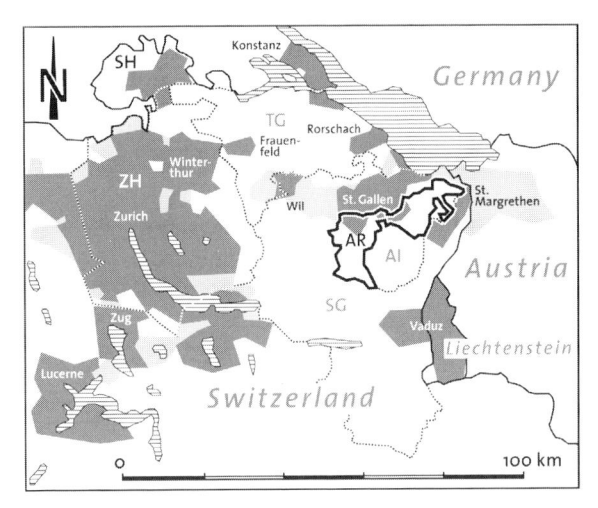

Figure 18.6 Location of Appenzell Ausserrhoden (AR) in the semi-urban and urban areas of St. Gallen, near the Greater Zurich Area. Dark and light gray areas are "agglomeration" regions according to the definition used by the Swiss Federal Statistical Office; light gray areas were designated as "agglomeration" regions between 1990 and 2000 (Scholz & Stauffacher, 2007, p. 2522).

Guiding question
What are the prerequisites for a sustainable regional economy meeting environmental and socioeconomic needs?

The jointly developed and finally agreed upon guiding question reads: "What are the prerequisites for a regional economy that can sustainably operate in harmony with the environment and regional socioeconomic needs?" More specifically, we wanted to study which options exist with respect to regional cooperation and business strategies within traditional industries and which of these options are preferred by key stakeholders. Thus, our primary interest focused on sustainable economic development, in particular maintaining employment and supporting the clusters of older industries, as these had been affected by rapid change and decline.

The system boundaries have been set at the cantonal level because industrial policy in Switzerland is still decentralized and managed at the regional level (as is also the case in some other countries, such as Austria; see Sturn, 2000). Sawmills, the dairy industry, and the textile industry were chosen because they belong to a business sector that grew in the age of agricultural manufacturing (Phelps & Ozawa, 2003), and which is under strong innovation constraints worldwide. Further, all these industries have played a decisive role in the history of AR. All three so-called traditional industries are currently experiencing a rapid process of consolidation and transition because of ubiquitous globalized product and material flows.

These traditional industries exhibit both commonalities and differences. Whereas the textile and, to some extent, the sawmill industries compete in the world market, dairy products are predominantly traded on a regional or continental level. Furthermore, these three industries allow for investigation of different contexts with respect to horizontal and vertical competition and collaboration. Companies from the textile industry cover several vertical stages of the production chain. Sawmills and dairies only cover the first transformation stage, between raw materials and industrial processing. Thus, we are dealing with horizontal competition in those cases.

System definition
A view from multiple stakeholder groups and disciplines identified traditional industries and their modes of cooperation as the key system boundary on the human system's side.

In this initial project phase, intense learning by all involved could be observed. While, for example, those from science started with the image of the canton being in a rural area, it was learned that the canton in fact extends to the fringes of a city and that industrial sites are thus sometimes in conflict with well situated housing areas. On the other hand, people from practice were first more concerned about finding alternative industries and only slowly started to realize that traditional industries might still have a future in the region.

Further, it is worth mentioning that the scientific disciplines that offered relevant knowledge for tackling the guiding question were not in the environmental (social) sciences but rather came from (economic) geography, industrial and regional economics, business and management sciences, and regional and economic development disciplines. This variety of different disciplines and the need to integrate such knowledge is not surprising given the approach followed in our case study. In fact, we turn science on its head (German: Die Wissenschaft vom Kopf auf die Füsse stellen).

Since we start (quite radically) from the case and develop together with case actors relevant questions, identifying the disciplines that are in position to tackle these questions is hence only a subsequent step. This brings some serious difficulties for science: one has to be very broad in one's knowledge and be flexible in familiarizing oneself with new disciplinary approaches. Of course, this cannot be done by one research group alone; thus, we had to integrate further expertise by contacting colleagues from other institutes (e.g. agricultural economics, regional economics, forest sciences, textile industry engineering).

To finalize this preparatory step, we then informed the greater public about the case study and its guiding question and aims with the help of the media.

C2.2 Complementarities help structuring complex human–environment interactions (P1)

Both the human–environment and the material–biophysical vs. socio-epistemic complementarities become evident when looking at traditional industries (see Table 18.2). Taking the example of the textile industry, one can distinguish between the *individual firm* and its *workers* as the human system of a company (i.e. organization; see Figure 14.1*) and the equipment and machinery as a specific company-owned material environment (E_m). The role and significance a textile company has for elected politicians and inhabitants of AR canton is part of the social environment of a company (see Figure 3.3*).

The commitment of many local firm managers to stay in the AR canton is partly due to the fact that many

Table 18.2 Basic complementarities in the TdCS 2002 "Future of traditional industries in the canton Appenzell Ausserrhoden" (see Table 3.1).

	Human systems (H)	Environmental systems (E)
Material (biophysical–technological) dimension	The living bodies (cells) and the physical activities of the owners, managers, and workers of the firms from selected industries H_m	Water, soil, air, production facilities, raw materials, products, the physical state of buyers, market competitors E_m
Social (epistemic–cultural) dimension	The knowledge and capabilities of the firm owners, managers, and workers such as business strategies, technology knowledge, affiliation to other market actors H_s	Horizontal/vertical interaction (networking) with actors of the supply chain (e.g. retailers, consumers, the image of the branch for key actors) E_s

come from families that had owned the company for more than 100 years. Further, the environment (E) of the textile industry encompasses (i) the natural systems the industry has an impact on (water, soil/land use, air (E_m)) and (ii) the horizontal/vertical cooperation within or across the production chain with other market actors (E_s). In fact, historical rights for water utilization and hydroelectric plants (E_m) still play a decisive role for some companies that require extensive consumption of natural resources, such as textile finishing. For survival, a company must also network with the unions (as a type of non-governmental organization) or the cantonal environmental agencies (i.e. institutions). This nicely illustrates the broader and specific use of the terms *human system* and *environment* followed in this book.

Complementarity Postulate P1
The activities of owners, managers, and workers of a textile firm constitute a *company* as a human system on the level of *organization*. The company owns, utilizes or affects parts of the material environment (production facilities, water, land, products, etc.). The action of the *company* (e.g. producing, buying, and selling) is embedded in a network of actors. The products produced, bought, and sold are part of a production chain which is part of the material environment.

The different relationships between human and environment systems can again be illuminated by our TdCS: traditional industries are settled in the canton of AR because of the availability of natural (i.e. material) resources such as water as the energy source for the textile industry and sawmills; water for the textile industry to wash material; forests as wood supply for sawmills; pastures offering food for cows in the dairy industry. This is the impact of the material–biophysical environment on the company as a human system ($E \rightarrow H$). But there is also a social environment. This is the existing cluster of clothing companies in the neighboring canton of St. Gallen as a potential buyer of the products of textile industry and subsistence farmers who have run out of land to cultivate acting as a labor pool for the textile industry. The latter already exemplifies a coupled human–environment system ($E \leftrightarrow H$). When a given environment is exploited, other human activities become necessary which themselves have again a new and possibly different impact on the environment. Last but not least, industries do of course have direct impacts on different environmental systems ($H \rightarrow E$). We will discuss this further in the section on the Feedback Postulate P4 below.

C2.3 Environmental analysis for revealing the dynamics of the environmental system (P7)

C2.3.1 Material and social environments

A thorough environmental analysis should follow the definition of the guiding question and system boundaries. In this regional economic study, the owners, managers, and the workers of selected companies in three industries (textile, dairy, and sawmills) build the core human system (see Table 18.2). The material environment includes the production facilities of these companies, the raw materials needed for production, products, emissions to the environment resulting from processes, retailers, and market competitors outside the canton, etc.

In the TdCS 2002, we undertook several studies that allowed a better understanding of the environmental system in this broader sense. We conducted structured interviews with the owners and CEOs of around 20 companies, covering many topics including confidential financial (e.g. annual turnover) and environmental data (e.g. energy use and related impacts).

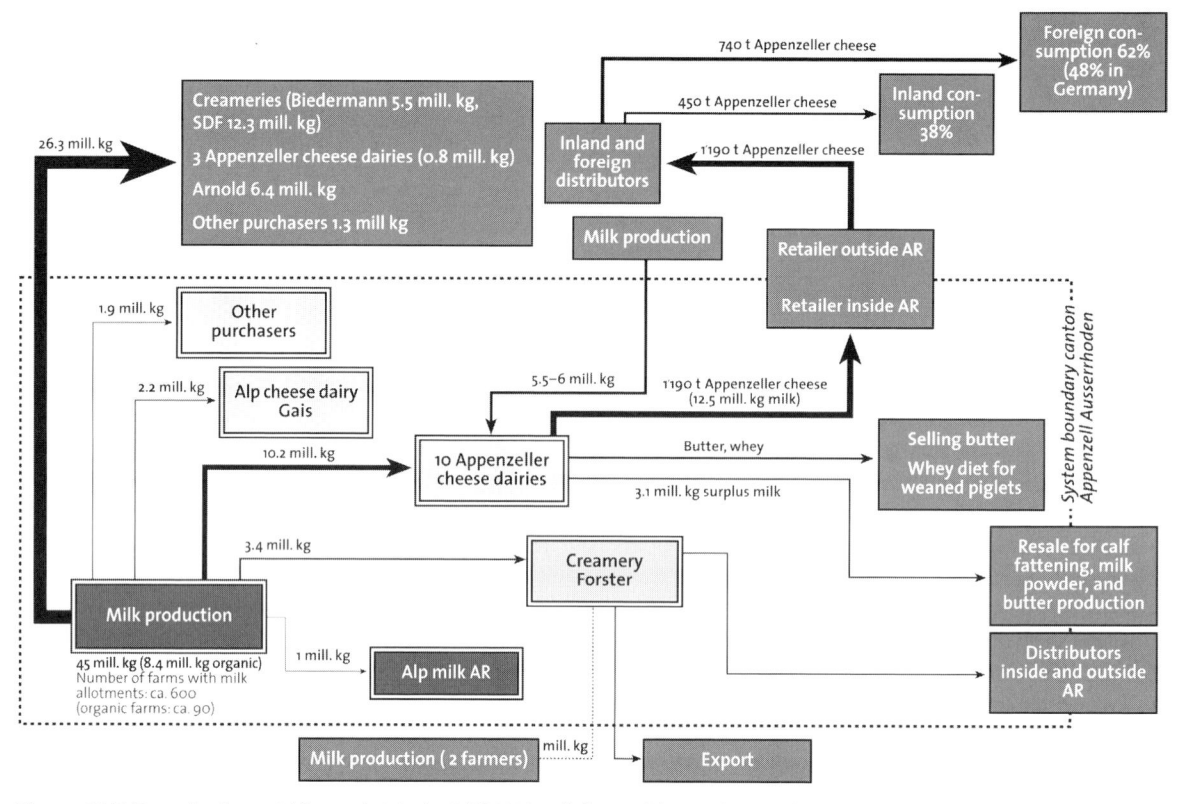

Figure 18.7 Example of material flux analysis in the TdCS 2002: milk fluxes of the production chain in the year 2001 (Scholz et al., 2003, p. 190). Of all the milk produced in AR, approximately 42% is processed in the canton. More than half of the milk goes to dairies outside the canton.

The interviews served to identify the main environmental concerns and impacts of the firms. Results of these interviews served as input for: (i) investigating the material–biophysical environment with material flow analysis (MFA, exemplarily presented for the dairy industry in Figure 18.7) and added value analyses along the production chain as well as analysis of environmental impacts (water, air, soil, energy use, i.e. E_m); and (ii) understanding the social–cultural environment by investigating the present form of collaboration within each of the three industries and between these industries and administration, politics, and science (E_s).

Environment-first Postulate P7
Material flux analysis, formative scenario analysis, and added value analyses; analysis of environmental impacts and of the social environment were essential methods to understand the case.

In addition, results from a formative scenario analysis (FSA; Scholz & Tietje, 2002; see Chapter 14.3; Wiek et al., 2007) helped us to integrate knowledge.

The system analysis part of FSA results in a semi-quantitative system model, which is composed of all impact variables required to describe the current and future states of the case and their mutual interactions sufficiently. Direct impacts between all variables were carefully assessed using an impact matrix, and the results were presented graphically (system grid, system graph see Scholz & Tietje (2002)).

The variety of disciplinary foci becomes evident again. Besides those mentioned previously (e.g. economic geography and industrial economics), industrial ecology and environmental natural science-related questions became more prominent with respect to environmental impacts. Further, knowledge of psychology and decision sciences had to be integrated to understand the social dimension.

C2.3.2 Organizing transdisciplinary processes

For each of the three industries, we established a "reference group" to discuss the project work regularly. In contrast to the "steering group," we wanted to cooperate with broader segments of the public therein. To this end, an extended range of different people were involved, such as farmers, teachers, a hotel-keeper, housewives, a medical practitioner, a bank manager, architects, planners, foresters, and a pastor. For the system analysis, these reference groups offered their detailed qualitative insights into the components and functions of the system and allowed us a better understanding of the system's functioning. Subsequent results were fed back and discussed with the reference groups.

Overall, the system analysis not only integrated knowledge from different scientific sources (beyond disciplines) but also from academia and practice (beyond science). Thereby, the understanding of the case and its problems with respect to the three industries was seriously extended, substantiated, and harmonized as well. The system analysis further helped us in detecting and discussing feedback loops (see below).

C2.4 Hierarchy principle (P2) to identify different levels of the human system and interferences between and within these levels (P3)

C2.4.1 What hierarchy levels

Hierarchy Postulate P2
Assigning all people involved in the study to their hierarchy level helped in understanding the influence they have and facilitated defining their roles.

As mentioned above, industrial policy in Switzerland is managed at the cantonal level. The canton AR is a small state, semi-autonomous with its own constitution, political system, and has administrative and budgetary independence on many issues. Thus, AR can be considered as a society (as a level of human system) which is represented by the government, which is a legitimized decision-maker (see Chapter 15). This level was included in the TdCS 2002 as the president of canton was an active co-leader of the project. Thus the orientations and outcomes elaborated in the transdisciplinary process, which should provide socially robust solutions, had a good chance of being heard. Further down the hierarchical levels, we collaborated with several institutions, like the cantonal offices of agriculture, forestry and of environmental protection and the cantonal promoter of the economy.

At the organizational level, more than 20 individual firms from textile, milk, and sawmill industries participated, each represented by their owners or CEOs.

It is on this level that most of the essential business decisions are taken (see Decision Postulate P5 below). Further, we realized collaboration with the Swiss Textile Association, labor unions, and other organizations at the national level as well. Likewise, we took into account the multilevel political system in Switzerland and the (inter)national integration of these industries since Swiss regulations and international treaties (but also national and international cooperation) are critical boundary conditions for industry in the canton AR.

At the group and individual levels, a multitude of individuals were involved. Moreover, representatives from all the above-mentioned representatives of the higher system levels are, at the same time, individuals (see the discussion on interference (P3) below). Additionally, surveys (with approximately 200 respondents) and more than 50 individual face-to-face interviews allowed the investigation of the individual perceptions of the issue and thus of the so-called "public." At the group level, the above-mentioned steering group and reference groups served as examples of where opinions were formed and group effects could be observed.

A critical issue of HES analysis is that people can take different roles. One and the same person can take the role of a worker, mother or president of the national sports association. Assigning and eliciting the specific role facilitates the analysis of the human system and conflicts among the hierarchy levels.

C2.4.2 What conflicts and interferences?

Interference Postulate P3
Potential conflicts exist *within hierarchy* (between firms) and *between hierarchy levels* (e.g. between firms and cantonal offices).

With respect to interferences, we can theoretically distinguish trade-offs and conflicts of human systems *within* hierarchy levels from conflicts *between* hierarchy levels. Further, we should note that a multilevel conflict may also occur in one and the same person. For both interferences, we would like to mention some insights gained in the TdCS 2002. A classical interference between hierarchy levels was visible from the beginning: the promoter of economic development for canton AR (institution), perceiving the canton as modern, primarily supported "modern" industries such as information technology and biotechnology, while more traditional industries (organization) were mostly neglected, leading to an interesting political lock-in

(Hassink & Shin, 2005). The value of traditional industries became apparent for this institution only after having participated in the TdCS. At the institutional level, for example, when individual CEOs negotiated with heads of the cantonal government and administration, they sought benefits for their own firms, which at times clashed with the interests of the industry as a whole. This was a conflict within a hierarchy level but was probably driven by a multilevel conflict within the same person.

A further conflict within hierarchy levels became visible between different cantonal institutions when, for example, an increased use of local wood was promoted by the office for agriculture and forestry while at the same time the office for the environment insisted on protecting natural resources. The latter office was, of course, frequently in conflict with industry (i.e. conflicts between hierarchy levels). Also interesting was the practically non-existent interference within the organizational level across different industries (see below Awareness Postulate P6).

C2.5 A decision-theoretic conception for understanding the rationale of human systems (P5)

Following a decision-theoretic conception, we have to distinguish between *players*, or agents, their *strategies* (or the combination of strategies of all players, i.e. the outcome), and *utilities*.

With respect to the guiding question on options for regional cooperation and their perception, we distinguished the following key agents in the TdCS 2002: firms at different positions in the production chain as the decision-makers and stakeholder group (e.g. sawmills vs. wood processing vs. forestry; dairy managers vs. dairy traders) and cantonal authorities, both from government (society) and administration (institution), providing the necessary framework conditions. Also, the representatives of university working on sustainability may be viewed as representatives of society (see Figure 1.7*). ETH Zurich is a federal school which is controlled and financed by the national government.

Decision Postulate P5
Individual firms (*players*) can choose between different options for collaboration (*strategies*), which are differently preferred by various key players (*utilities*).

In principle, all firms can choose between (i) more or less cooperation either (ii) horizontally or vertically along the production chain, and (iii) they can further choose between more/less cooperation outside the production chain with, for instance, cantonal administration or research institutions/universities (strategies). These different possibilities needed to be investigated more thoroughly and analytically. We used the area development negotiation (ADN) method, which includes FSA (Scholz & Tietje, 2002). The way in which this is done is described in Box 15.1, Table iii–v. Based on a thorough system (see P7 above) and consistency analysis (Scholz & Tietje, 2002, p. 105; Tietje, 2005), a few scenarios were jointly selected with all key agents. These scenarios represent different contrasting strategies. For example, for textile industry these were the following business strategies: minimal cooperation (firms remain economically independent); resource sharing (joint utilization and management of resources); AR textile network (firms remain autonomous but use a regional label following similar quality standards); full integration (all firms join a holding or merge into a single large company).

To assess utilities for all strategies, we applied an adapted stakeholder-based multi-criteria analysis (MCA) using six evaluation criteria (see Box 15.1, Table 15.3, vi; McDaniels & Trousdale, 2005; Scholz & Stauffacher, 2007). We applied two different approaches of MCA: (i) a data-based evaluation relying on literature studies and expert interviews (MCA I); and (ii) a stakeholder-based evaluation including the representatives of industry, cantonal authorities, and sustainability researchers as groups. The stakeholder-based evaluations were made in two steps, the first being overall and intuitive, and the second using the criteria from the MCA I to provide an MCA II. It was also possible to compare utilities between different players and different assessment methods (e.g. between intuitive and criteria-based evaluation, between cantonal authorities and industry, between the view of cantonal authorities vs. industry vs. the sustainability researchers). Thus, it was possible to map a spectrum of diverse views and alternatives as a component of capacity-building but at the same time to detect potentials for reaching consensus (van Asselt & Rijkens-Klomp, 2002).

C2.6 Improved environmental awareness for understanding potential barriers and drivers in human systems (P6)

Being aware of one's environment is a necessary condition for coping with it. In addition, a proper awareness has to be developed also for the social environment (see P6a–P6c in Figure 18.5). However, in the beginning this prerequisite was not met by many of the firms involved in the TdCS (non-awareness of environmental systems – both the material–technological and the social–epistemic). We often encountered a situation where other firms (which are part of the social environment E_s) were primarily or even exclusively seen as market competitors, and the potential of cooperating in clusters (see Capello & Faggian, 2005; Torre & Gilly, 2000) was not properly seen. This is not surprising if one refers to the differentiation between cooperation and learning dynamics in horizontal and vertical clusters (Malmberg & Maskell, 2002; Newlands, 2003).

Awareness Postulate P6
Other firms are often perceived only as competitors (restricted awareness of social environment). The potential market benefits were not explicitly perceived and assessed.

C2.6.1 Different benefits from cooperation and competitions in clusters

The sawmills and dairies that were investigated were involved in the first transformation stage between raw materials and industrial processing. Thus, we were dealing with horizontal clustering where, obviously, competition is much more prevalent. None the less, most of the social environment is the same for all firms (e.g. see the conflicts between hierarchy levels with the cantonal institutional and the societal level; Hierarchy Postulate P2 above) and becoming aware of this was one of the outcomes in the process. On the other hand, the clustering of the textile industry involved the vertical stages of the production chain. Hence, collaboration could be more straightforward. Yet, a lack of trust existed among the family dynasty-based companies, who have been competitors for some time, and this has prevented more intense forms of collaboration in the previous decades – just as certainly as has a lack of environmental awareness.

Awareness of impacts on the social environment caused by one's behavior has certainly grown among local participants throughout the study. This is related to changing secondary feedback (see Feedback Postulate P4). Joint dependence on natural resources (E_m), as well as the necessity for collaboration with other firms (E_s), became evident in the study and were acknowledged by all participants. As such, the TdCS offers an example of how transdisciplinarity can be a tool for capacity-building through improving environmental awareness. This could be seen by many examples. In the textile industry, the scenario of full integration was intuitively disliked by all respondents. It caused emotional resistance and contrasting traditional patterns of competition. This negative attitude was reversed in the criteria-based evaluation (MCA II). Here the full integration scenario is the most preferred. This multi-criteria-based, analytic judgment can be taken as an indication for a developing changed environmental awareness.

Missing environmental awareness often originates in cognitive lock-ins (Grabher, 1993; Hassink & Shin, 2005; Maskell & Malmberg, 1999). A thorough quantitative analysis 2 years after the TdCS, including 88 stakeholders, revealed that – besides capacity-building – cooperation and networking were the key benefits of the transdisciplinary process (Walter *et al.*, 2007b).

C2.7 Understanding feedback loops in coupled human–environment interactions (P4) for preparing transformations

Feedback loops were not a direct focus of this TdCS but were investigated in the system analysis part (see Environment-first Postulate P7 above) when impacts among different factors were assessed. FSA as a method of systems analysis includes tools to identify multiple feedback loops among the impact factors, which provides a better insight into the importance of different variables.

Feedback Postulate P4
In the TdCS, university, firms, and the canton learned how to cope with feedback loops in human and environmental systems.

Further, the whole TdCS actually serves as a tool to learn about feedback loops. Recognizing that not only individuals, but also groups, organizations, and institutions learn (Tàbara & Pahl-Wostl, 2007), a societal sustainability learning process was the focus. The very process of assessing the present situation and developing future scenarios, as well as their detailed evaluation, can stimulate learning among all participants of the TdCS (feedback loop in human system). Transdisciplinary processes empower and motivate stakeholders to contribute more actively in subsequent implementation of ideas and change future decision processes (Brown *et al.*, 2001; Sheppard & Meitner, 2005). The latter can actually lead to a feedback loop between human and environmental systems.

C2.7.1 Key messages

- The transdisciplinary case study (TdCS) at ETH Zurich was developed to support regional and organizational transition processes towards sustainable development
- The TdCS is conceptualized as a mutual learning process among science and practice. The study provided incremental deepening of insights about feedback loops when cooperating with the social environment (market competitors). This can be conceived as extending environmental awareness. The HES Postulates and the TdCS design with its six steps help in structuring transdisciplinary processes
- The TdCS starts (quite radically) from the case in developing together with case actors relevant questions; identifying the disciplines to tackle these questions is hence only a subsequent step. Thus the Environment-first Postulate P7 is essential. This asks for knowledge from multiple disciplines for analysis of the human and environmental systems involved. Thus, scientists in TdCS need a broad disciplinary background to organize disciplined interdisciplinarity in transdisciplinary processes
- The rationale of the human system can be better understood using a decision-theoretic conception distinguishing between actors (players), their strategies, and respective utilities. Various disciplines, such as psychology and economics, help to identify potential drivers and rationales at work in transition processes.

18.3 Case 3: agro-fuel[3]

18.3.1 What can we learn from this case for environmental literacy

We elaborate what (possible) benefits and unwanted negative impacts would result if the option agrofuel was taken by many countries. Using the HES framework (Figure 18.8) we reveal some unsustainable dynamics that were not satisfactorily acknowledged by science and society.

18.3.2 What is the case about?

Increased concern about peak oil, energy security, and climate change has caused many countries to explore renewable energy alternatives to fossil fuel. In some countries, biofuel has become a favorite option for substituting oil as vehicle fuel. Bioethanol is considered as renewable energy as it receives its energy from the Sun and is (fallaciously as we show) supposed to be infinitely producible. As we elaborate below, in February 2006 the government of Sweden decided to become an oil-free society until 2020 and to subsidize E85 strongly; that is, 85% ethanol fuel for private cars.

Content Case 3

C 3.1 Defining the guiding question: Sweden's decision on bioethanol, a sustainable one? 483
C 3.2 A national perspective on human systems 484
C 3.3 Why an environmental analysis should come first (P7) 485
C 3.4 Awareness of essential, critical feedback loops (P4 & P6) 491
C 3.5 Interferences from feedbacked multi-actor decisions (P3) 491
C 3.6 Conflicting variants of environmental awareness with respect to biomass' ecosystem functions (P6) 492
C 3.7 Incomplete hierarchy: the missing supranational systems (P2) 494

---

[3] Roland W. Scholz is the lead author of this case, which has been written in late 2008.

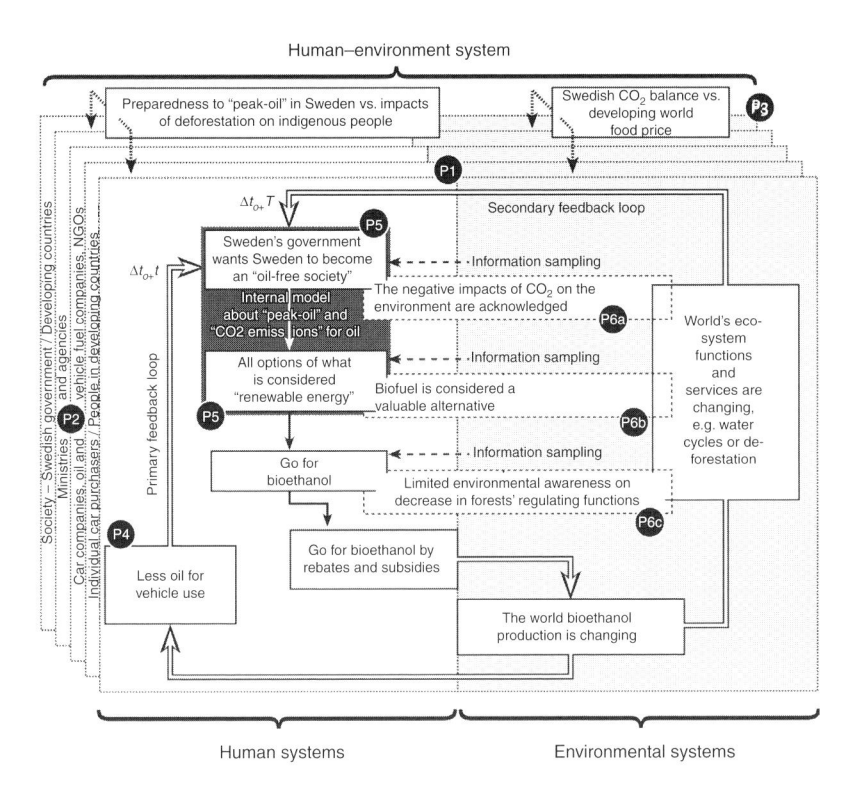

Figure 18.8 An HES framework view of the "oil-free Sweden" case.

C3.1 Defining the guiding question: Sweden's decision on bioethanol, a sustainable one?

Goals: the guiding question refers to a governmental decision about bioethanol. We first introduce some (idealized) constraints and assumptions about this decision. This is done to help to define a guiding question for the project.

Securing energy resources for the long term is a challenge of the twenty-first century. The global dilemma of peak oil and increasing awareness about the climate impacts of fossil fuel (Cavallo, 2002) made renewable energy to a top political priority. In 2000, energy consumption was estimated to contribute 61% of global manmade greenhouse gas emissions. Vehicle fuel accounts for approximately 22% of the energy consumed worldwide and 31% on the EU level and is thus often seen as a major component of mitigation (Baumert et al., 2005).

Guiding question
Is Sweden's decision to push bioethanol E85 as vehicle fuel nationally and internationally sustainable?

Sweden is known to have a high social and environmental awareness. With a low population density of 20.6 capita per km² and a big biomass potential, the government targeted biofuels for the transport sector in 2004 (Swedish Government, 2004). In February 2006, the Swedish government took a remarkable decision. Mona Sahlin, the Minister of Sustainable Development, declared:

> Our dependency on oil should be broken by 2020. (Vidal, 2006)

The decision was prepared by a commission chaired by Göran Persson, the Swedish Prime Minister. The process included a nationwide discussion, including four public hearings that were broadcast on television (Commission on Oil Independence, 2006). Science has been involved in the commission. In March 2007, the Environment Ministry introduced a rebate system of 1080 Euros per eco-car, or a car that runs on E85 (85% ethanol and 15% petrol). The crucial question (guiding question) of this analysis is whether this decision is a(n internationally) sustainable one.

When considering the rationale behind the Swedish decision, the following background about the global energy situation in 2006 should be considered.

Both science and society had recognized the problem of peak oil and the mid-term need to substitute petroleum with renewable alternatives. Biofuel was regarded as a good option:

> Biofuels may contribute to the crucial goals of enhancing energy security, energy diversification and energy access; improving health from reduced air pollution; and boosting employment and economic growth for rural communities. (United Nations (UN), 2006, p. 29)

Ethanol appears to be an attractive alternative for Sweden, at least in the long run, when the vision of a second-generation biofuel processing lignin might be economically realized. Nevertheless, we analyze whether the Swedish decision to use ethanol was a sustainable one. The case is not presented to criticize the Prime Minister's or the Swedish policy. It is rather discussed as an exemplar to illustrate the value of looking at complex problems, such as the long-term transitions of energy systems, by the means of the HES framework.

System definition
The Swedish government's decision is viewed from a "sustainable world" perspective.

Before we introduce the case, we describe some assumptions and simplifications underlying the analysis. First, we do not reconstruct the full Swedish political context but rather look at the case from an "external" world perspective. We acknowledge that, besides the bioethanol rebate decision, other strategies for reducing air pollution, such as the reduction of mileage per person and improved public transportation, were seriously discussed in Sweden. The commission preparing the oil-free society also emphasized that "no oil should be used for heating residential and commercial buildings" (Commission on Oil Independence, 2006, p. 4), a goal that is in competition with oil-free transport as society must simultaneously deal with two transformations. By means of simplification and to pinpoint critical feedback loops, the analysis rather goes face value of the decision "oil-free society" in public transport.[4] Second, we often concentrate our discussion on bioethanol as liquid vehicle biofuel as it has and can be considered a serious option for renewable energy. In 2008, 80% (expressed in MTOE – million tonnes of oil equivalent) of the global production of liquid biofuel was in the form of bioethanol (FAO, 2008). Third, we do not address the important question of whether biofuels can be more efficiently utilized in heating,[5] neither do we provide an analysis about what can be gained from using other biofuels such as biodiesel (from vegetable oils, animal fat or recycled cooking grease) or biogas from green waste. However, we do incorporate these options if it is necessary to keep the case realistic. Similarly, we do not deal with the question of whether green waste from gardens should be used for compost-based fertilizing as it is marginal, in particular for Sweden. Fourth, we aspire to look at bioethanol from a comprehensive sustainability perspective, including the north–south and thus the intragenerational justice perspective.

C3.2 A national perspective on human systems

Goals: this section presents a kind of national actor analysis by means of the Hierarchy Postulate P2 as it is usually implicitly involved in social science and policy analysis on the subject. The analysis serves to illuminate which *human systems/actors* are concerned by the decision on a national level.

C3.2.1 Level of society

The decision to become an oil-free country was made by the government, which represents the human system level of society (see the Hierarchy Postulate P2, Table 16.2, Figure 1.7*). The departments of the Swedish ministry and Swedish agencies constructing the rebate system are institutions which support the elected Prime Minister to fulfill her goals. The rebate system for biofuel affects the consumers and car purchase behavior (de Haan *et al.*, 2007), which leads us to the level of the individual.

---

[4] Factually, the formulation of the commission with respect to road transport has been more moderate: "Through more efficient use of fuel and new fuels, consumption of oil in road transport shall be reduced by 40–50%" (cited according to Aleklett, 2008, p. 97). The reader should acknowledge this lessened formulation.

[5] With respect to heating, Sweden has already reduced the oil use for heating since the mid 1970s oil crises by 70%. Thus the potential of saving oil is factually much lower in this field. Much heating switched to natural gas, as well as pellets, which in 2005 covered 7 TWh/yr with an annual growth of 19% (Commission on Oil Independence, 2006).

C3.2.2 Level of the individual

We take a brief look at the cultural and social matrix of the Swedish population. The "average Swedish citizen" is highly concerned about environmental issues. This is well documented in the Eurobarometer survey of the 27 EU countries (European Union (EU), 2008).

In the language of the HES framework, these data provide information about: (1) environmental concern on the "level of the individual citizen" (see Hierarchy Postulate P2); (2) the cultural and social order system of Swedish society (see Figure 14.1*); and (3) the awareness of the issue by the Swedish individual (Awareness Postulate P6).

C3.2.3 Level of commercial organizations

Car companies such as Saab had pioneered and others such as Volvo followed in developing flexible fuel vehicles which can run with high shares of bioethanol (see Box 18.2). Oil companies faced a new business area also, because they could sell bioethanol at their stations:

> ... oil giants operating in Sweden, such as Shell, Mobil and Statoil, told IPS [Inter Press Service] that the oil companies are quite happy with the idea of moving into biofuels – and that they don't see any threat to petroleum's future from this sector just yet. (Christodoulou, 2007)

There was a good supply infrastructure for bioethanol. There were per capita 56 times more gas stations in Sweden than in Germany and 363 times more than in Great Britain, which has not yet taken the bioethanol option. The 2006 bioethanol per capita consumption in Brazil has been 2.1 times as high as that of Sweden (EPE, 2007). Sweden's consumption grew by 17% in 2007 and Sweden was the European champion of bioethanol consumption. In 2006, there was a consumption of 209 kWh/capita, which was about five times as much as in Germany. Sweden was the country with the highest absolute consumption and production in Europe (EuroObser'ER, 2007). Diverging information has been published concerning the share of domestically produced biofuels in 2006 for road transportation. Estimates for the share of imported ethanol are in the range between 50% and 80% (Börjesson, 2009; Hansson et al., 2009). Factually, in 2008 the total renewable energy was about 4.4 TWh: of this, 2.5 TWh was ethanol-based, with more than half of this imported, mostly from Brazilian sugarcane ethanol (Swedish Energy Agency, 2009).

Also, the Swedish paper industry became active. A first research and development plant, Chemrec, was built in Piteå in the north of Sweden. The expectations of these actors were clearly pronounced.

> Chemrec could dramatically change the profitability of pulp mills and enable them to participate in the exploding alternative fuels industry. (Bernie Bulkin, VantagePoint, a US venture capital firm, quoted according to Christodoulou, 2007)

C3.2.4 Non-governmental organizations

However, some critical voices from the Green Party arose in 2007.

> We clearly cannot have the whole world using western levels of car use, running on ethanol. (END, 2007; quoted according to Christodoulou, 2007)

This first view on actors in the Swedish bioethanol case revealed that the human systems society (in particular the government and the "environmental consciousness" of Swedish culture), the Swedish energy and car companies, and a large share of the individual consumers seem happy with the decision. Only some NGOs were skeptical about transferring the Swedish model for oil independence to a world scale (Christodoulou, 2007).

C3.2.5 Supranational systems

Finally, Sweden is a member of the EU, which is a supranational system. Thus, the EU is a player on the national stage. In September 2005, the European Parliament called a binding 20% target of biofuels in total energy consumption by 2020 (European Parliament, 2005), a decision with which the EU has to comply and which does not contradict the decision analyzed. The EU also introduces other constraints for biofuels, for instance that by December 2010, 5.75% of all transport fuels are biofuels with 50% feedstock produced on EU territory (European Union (EU), 2003).

C3.3 Why an environmental analysis should come first (P7)

Enviroment–first Postulate P7
The environmental analysis constructs the constraints of and effects on material–biophysical–technological environment(s) originating from the Swedish governmental (i.e. societal) decision on "oil-free society and bioethanol for vehicle fuel." We construct the supply chain and identify the social dimensions related to it. The environmental analysis leads to a new definition of the system boundaries.

The era of "easy" energy is over. (United Nations (UN), 2006, p. 3)

Environmental impacts and external costs come to the foreground in assessing energy systems. A differentiated view for better judging the environmental impacts can be provided by a supply chain, production chain or life cycle analysis (see Schmithüsen *et al.*, forthcoming). Here, we distinguish between the: (1) production (of bioethanol and biofuel), including distribution to different countries; (2) use, including the specifics (e.g. technology efficiency); and (3) the environmental impacts.

C3.3.1 Production

Biofuel and bioethanol come from biomass. Thus critical questions are: (1) where does the biomass come from? (2) How much can be sustainably produced? Our environmental analysis is based on two assumptions (or scenarios). The first is that we take a world perspective and refer – more or less – to the current conditions of producing bioethanol fuel and importing it to Sweden (see Robert *et al.*, 2007). The second assumption checks what amount of bioethanol and/or biofuel could be produced based on Swedish feedstocks and thus takes a specific, national perspective.

The world perspective: in 2008, worldwide biomass (and waste) accounted for 10–11% of global primary energy production (FAO, 2008; Fritsche *et al.*, 2006). Of this, only 2% was used for liquid transport fuel, 18% for industrial use, and 80% for residential use (IEA, 2007). Liquid biofuels for transport were on the rise and came mainly from sugar or starchy crops (fermented to bioethanol), from oil crops (converted by esterification to biodiesel) or anerobic digestion (turning biomass to biogas). Deriving ethanol from cellulose (converted by different types of fermentation) was only possible in small pilot plants. Also in 2008 we see that Brazil and the USA accounted for almost 90% of the 80% of liquid biofuel used as bioethanol (FAO, 2008). A major exporting country, Brazil already saw biofuel as a key part of global production and trade in 2006.

How much biofuel/biomass the world can sustainably produce is the crucial point of the first question. Answering this question includes coping with the trade-offs between using food, fodder, fiber, building material, and so on, that have accompanied biomass use since the dawn of civilization. Thus, question (1) can only receive a reasonable answer if we consider also the human systems involved in these trade-offs, which leads us directly to a further question. How much liquid biofuel needs be produced to meet the needs of an oil-free Sweden and that of other countries? Here, the question is related to the fundamental prerequisites of biofuel production (i.e. land, fertilizers, water, temperature, etc.). We have to acknowledge that both the trade-offs and the prerequisites will change depending on the world demands, the technology and efficiency of biofuel production (what fermentation will be developed for, for instance, leaves, nut shells, or organic parts of municipal waste). With respect to technology, we not only speak about second-generation feedstock, but also about (genetically) improved second-generation feedstocks (FAO, 2008).

Box 18.1 provides different calculations of the world biomass and biofuel potential. Here we assume no decrease in agricultural area used to meet world food needs, which seems reasonable as 852 million (13%) people were starving earlier this decade (FAO, 2004) and the world population continues to grow. The core message is that only a small share, of the order of a very small two digit percentage, can be met by biomass, so clearly other energy sources are needed (postulating no dramatic change of lifestyles).

The limited land and trade-offs with agricultural food production provide a first critical point from a supply chain perspective. Besides land for biomass feedstock, water, seeds, nutrients/fertilizers, pesticides, energy, machines, and human labor as (a) basic resources, several other factors need to be considered during an environmental analysis. Specifically, one also needs to consider the (b) growing process of feedstocks (such as sugar cane, maize, wheat, pines, etc.), including composition, production, harvesting, transportation, and preprocessing, and (c) the technical conversion process (i.e. the processing operations for bioethanol production, including co-products such as heat, electricity, or sugar).

Let us take a look at the two major players of bioethanol production. Brazil produces biofuel primarily from sugarcane and the USA produces biofuel from corn and wheat. The conditions of feedstock production differ depending on geographic and climate conditions. In the USA, 500–4000 liters of water are needed to grow 1 liter of ethanol E100. Many sugarcane sites in Brazil do not need any irrigation (Dominguez-Faus *et al.*, 2009). However, in both locations they require a huge amount of fertilizers, in particular phosphorus (P) (Cordell *et al.*, 2009; Smil, 2000). Irrigation, fertilizers, processing, etc. demand energy input. Brazil sugarcane biofuel generates more than eight times the energy needed for its production. In the USA, for wheat the factor is a little above four and for maize less than two (FAO, 2008). This is reflected in the production prices. Whereas Brazil ethanol has been able to compete with petrol since about 2000, US wheat-based ethanol requires high subsidies. However, it is also clear that the pressure on Brazil's soils, including tilling of tropical forest, is greater than that on northern grainfields.

Environment-first Postulate P7
Producing more ethanol in Sweden will draw on agricultural production in other nations, particularly on tropical soils, making the primary biophysical environment broader than Sweden.

An environmental analysis provides a couple of conclusions that have to be taken into account. First, a world perspective must be taken. Environmental impacts have to be looked at across the whole supply chain. Second, today's biomass for ethanol is competing with biomass for food. GRAIN, an NGO supporting small farmers and social as well as biodiversity movements warned:

> … the wide-scale cultivation of agrofuels will actually make things worse in many parts of the world, notably South-east Asia and the Amazon basin where the drying of peat lands and the felling of tropical forest will release far more carbon dioxide into the atmosphere than will be saved by using agrofuels. (Christodoulou, 2007, p. 3)[6]

Third, and a point that many find frustrating, is that agricultural feedstocks are considered renewable because they gain energy from the Sun via photosynthesis. A simple view on the Feedback Postulate P4 indicates the falsity of this definition. The fact is that, for tropical soils, mineral fertilizers are required to replenish nitrogen and phosphorus. In addition, it is evident that second-generation biofuels, which are fuels from inedible plants, are only available to a limited extent (see Box 18.1).

Environment-first Postulate P7
The label renewable energy for bioethanol is misleading and excludes, for example, fertilizer shortage.

The national Swedish perspective: Sweden (like Canada) is a favored country with regard to biofuels. With 2.5 ha of forest per person, Sweden has 10–30 times more forest biomass than neighboring countries. The Commission on Oil Independence states:

> As a nation, we will utilize our enormous forest holdings, good arable land and our rich supply of fresh water. (Commission on Oil Independence, 2006, p. 11)

Scenario-based calculations are based on assumptions that only 20% of the forest area will remain as a nature reserve whereas 25% will be used for innovative plantations with forced fertilization or special biomass energy crops such as willows. The latter would increase the biomass yield by a factor of two to four. Similarly, 30% of agriculture land is assumed to be used for cultivation of biomass yield. Assumptions like this provide 177 TW h/yr biomass energy potential on the supply side.

C3.3.2 Use

On the demand side different scenarios are constructed. A thorough study by Robèrt, Hultén, and Frostell (2007) dealt with the question of whether future technology development and changes to more sustainable urban mobility behavior patterns would allow the Swedish energy demand for transportation, which amounts to 96 TWh/yr,[7] to be covered by biofuel from Sweden. These scenarios include hybrid cars with more than 70% fuel savings or ultra light solar hydrogen hypercars. This would result in an overall 55% reduction of energy per kilometer. Further, a decrease of 13% mileage per person is part of the scenarios.

The results are disillusioning. Given business as usual (according to the 2005 conditions), 100% of Sweden's biomass cannot meet the country's energy demand for the transport sector. A 65% set-aside of biomass would meet transport needs (as heating, construction, etc. also requires some biomass) only under a scenario with a large number of policy options, including "50% increased fuel price, traffic tolls, free public transport, mechanisms to decrease car ownership, technological improvements" (Robert *et al.*, 2007, p. 2095). One conclusion drawn by the authors is that other energy sources, such as solar, wind, or hydroelectric, may be better

---

[6] This aspect has not been dealt with in the Commission on Oil Independence (Commission on Oil Independence, 2006).

[7] Given this number and the statement that Sweden should produce 12–14 TWh biofuel from forest and arable land by 2020 (Commission on Oil Independence, 2006, p. 22), one might already infer that this becomes difficult.

18 Applying the HES framework

> **Box 18.1** Different views on biomass energy potential: biofuel
>
> The quest for oil-free or fossil-free energy supply poses the question of how much biofuel can be produced worldwide, theoretically and practically.
>
> According to an estimate by Kapur (2004), the total amount of energy produced by global photosynthesis is about 3150 exajoules, which is about 7.5 times as much as the current world primary energy demand of 420 exajoules. Thus, compared to the potential of solar energy (see Box 11.5), plants and (surface) biomass provide a poor yield. IEA (2006) estimates a potential of bioenergy supply up to 1100 exajoules. This would mean that more than one-third of all plants would need to be used for bioenergy, which does not appear realistic. Given technological progress in agriculture and feedstock manipulation, IEA corrected their estimation to a mid-range estimate of 399 exajoules, which would closely meet world total primary energy demand. Yet, these are energy engineer calculations.
>
> Another view starts from the land use requirements and refers to arable land, as the second-generation ethanol/biofuel is not yet available (Figure 18.9). The world's arable land used for liquid biofuel production is around 1% and could increase to 3.8% in 2030. If we take the optimistic assumption that second-generation liquid biofuels were available, this would lead the "global share of biofuels in transport demand to increase to 10%" (FAO, 2008, p. 21). This would be an agricultural scientist's estimate.
>
> We can look at two other perspectives. José Goldemberg, a physicist, figured out that "expanding the Brazilian ethanol program by a factor 10 […] would supply enough ethanol to replace 10% of the gasoline used in the world" (Goldemberg, 2007, p. 809). Simply multiplying Brazil's ethanol production by ten would show that the entire land area of Brazil could meet the current world transport biofuel needs, assuming that all areas could be used with the same efficiency, within a couple of years, as the current land for growing biomass. Fortunately or unfortunately, this does not even hold theoretically.
>
> Another calculation speculated what would happen if world supplies of wheat, rice, maize, sorghum, sugar cane, cassava, and sugar beet were used for bioethanol production. The land used for biomass production would account for 42% of the total cropland but would only replace 57% of the global petrol use. The authors come from energy and agricultural engineering but seem to take a comprehensive view, as the title of their paper suggests: "Challenge of biofuel: filling the tank without emptying the stomach" (Rajagopal et al., 2007).
>
> These calculations show that world biomass cannot cover the current biofuel demand, assuming that food production should remain part of agriculture and that the world population is not dramatically reduced. 
>
> **Figure 18.9** First-generation biofuel sources are primarily sugar cane in the South and wheat in the North. Second-generation sources might focus on rapidly growing plantage forests.

candidates or at least needed to contribute for transportation energy.

From the perspective of the technological dimension of the environment, ethanol is a favorable, historically well known, proven alternative to petroleum-based fuels (see Box 18.2). Car engines do not have to be changed too much, and modern automotive engines can use up to E25 without retrofitting. Thus, technologically, moving to bioethanol is a low-cost variant of climate mitigation (see Chapter 17.3).

C3.3.3 Environmental impacts

In 2007, von Blottnitz and Curran reviewed 47 published assessments on biofuel used for transport fuels. Of these studies, 40 were life cycle-based and included a comprehensive model of the environment. Only seven could be said to approach life cycle assessment (LCA; von Blottnitz & Curran, 2007), including not only CO_2 emissions as an environmental impact but also resource depletion, ozone depletion, acidification, eutrophication, human and ecological health, air pollution, etc. (see Chapter 12.5).

The most comprehensive seven studies "came up with divergent conclusions" (von Blottnitz & Curran, 2007, p. 607). Clearly, there were some reductions in resource use and CO_2 emissions. However, other environmental impacts, such as acidification, human toxicity, and ecological toxicity, "were more often unfavorable than favorable" (ibid., p. 606). As the von Blottnitz and Curran paper was first submitted in September 2007, it should meet the knowledge underlying the Sahlin decision.

A valid, detailed, comprehensive environmental evaluation of bioethanol is frustrated by, among other factors, the difficulty in integrating land use changes as, for instance, deforestation into an LCA (Koellner & Scholz, 2008) and also considering regional aspects. These are general limitations of (LCA) assessment.

> The ability of various bioenergy types to reduce greenhouse gas emissions varies widely, and where forests are cleared to make way for new energy crops, the emissions can be even higher than those from fossil fuels (UN-Energy, 2007, p. 5).

Environment-first Postulate P7
The technological environment that means the production technology is essential.

Further technology development progress is difficult to access. An exemplary study of Luo *et al.* (2009) revealed this issue when comparing Brazil standard technology, including co-production of sugar from sucrose and electricity from wastes (bagasse, i.e. sugarcane pulp), with a future technology. In the future technology, bagasse (a fibrous part of the plant) is mainly used for ethanol production. The difference is huge as the standard case needs 30.1 kg of sugarcane compared to 12.6 for the future technology case (a difference of a factor of 2.4). If the future changes are known, one could perform a prospective cost–benefit analysis (CBA; see Chapter 11.4). From the economic dimension the assessment of the life cycle costs is of interest. The latter has been undertaken for a Swedish plant (Larsson, 2007) and for standard and future technology-based production in Brazil (Luo *et al.*, 2009). Both studies provided positive results but included high uncertainty, in particular on how to account for external costs.

Feedback Postulate P4
Major feedback loops of the Swedish biomass decision include tropical deforestation and competition with food, which leads to the potential of famine.

The question of environmental impacts also relates to the social environment. The trade-off with food has been mentioned. One kilogram of maize gives the same environmental impacts, independent of whether it is used to move a car a few hundred

---

Box 18.2 The technological view on bioethanol production: T Fords

Ethanol has been declared the fuel of the future in the 1920s. The first Model T Ford was designed to run with bioethanol. Henry Ford (1863–1947) supported fuel from vegetation (Ford, 1925, p. 24). During the 1930s more than 2000 US service stations sold gasohol, a mixture of maize-based ethanol and gasoline. Low petroleum prices and lacking environmental awareness made the option unappealing (United Nations (UN), 2006).

Fuels derived from biomass are similar to fossil fuels, which also have their origin in biomass. Ethanol-fueled cars show better acceleration and maximum speed but have a lower cruising range as they consume approximately 30% more fuel (in terms of volume) than gasoline-driven cars. This could be partly compensated by high compression owing to higher octane in ethanol (Reschlikowski & Schmitt, 2007). In the 1980s Ford created the flex-fuel cars (i.e. engines that can run on the two sources – gasoline and bioethanol).

The 1970 Clean Air Act and its amendments since 1977 required the phasing out of lead, which was introduced in 1923 to enhance engine performance. MTBE (methyl tert-butyl ether) was introduced in 1970 to prevent engines with lead-free gasoline from "knocking." Bioethanol includes high oxygen and was allowed to substitute for MTBE. Regulations were introduced requiring that gasoline contained more than 2% oxygen during winter months, promoting bioethanol in the USA. Peak oil has been another impulse. Some car companies, such as Saab, exult: "… bioethanol – a renewable and potentially carbon-neutral fuel – is back on the agenda" (Saab, 2009); so far, on the users' side.

Bioethanol can be produced from sugar- or starch-containing plants. The different production lines are presented in Figure 18.10a. In 2006 there were around 350 ethanol plants (United Nations (UN), 2006). Currently, bioethanol is produced from major crops such as corn, barley, oat, rice, wheat, potato, sugarcane, and sorghum.

"In the agro context, it is important to distinguish between 'raw' and 'processed' bioenergy – e.g. the raw bagasse (sugarcane pulp) generated in sugar mills, which can be used to generate heat and power, versus the processed sugar that becomes a fuel in the form of ethanol" (UN-Energy, 2007, p. 12). Converting cellulosic biomass into liquid transport

Box 18.2 (cont.)

fuels requires saccharification procedures by acids or enzymes. By this method, in principle also crop residues, perennial grasses, and other cellulosic material can be converted into ethanol. Here, pretreatment of the cellulose fibers is necessary to make the cellulose more accessible to the enzymes. "However, the production cost of enzymes is still too high and requires further reduction" (Hahn-Hagerdal et al., 2006, p. 551). Costs and learning curves for reducing costs are difficult to estimate as data are only accessible on a laboratory scale and to some extent on a pilot scale.

a

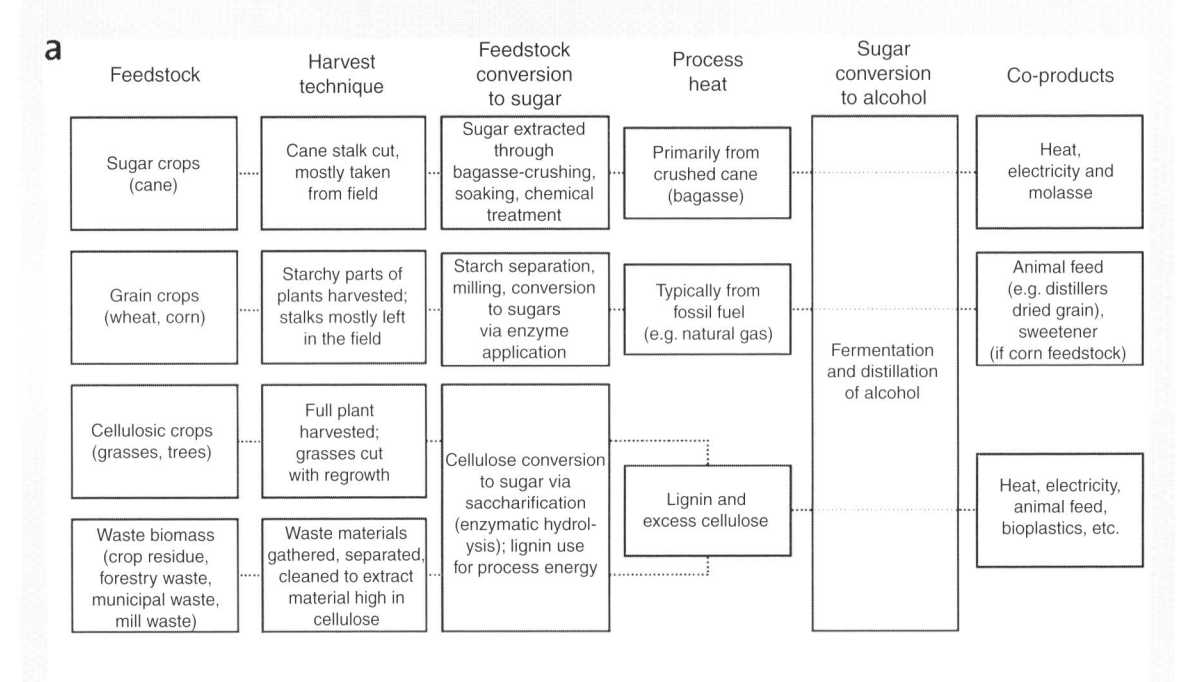

Feedstock	Harvest technique	Feedstock conversion to sugar	Process heat	Sugar conversion to alcohol	Co-products
Sugar crops (cane)	Cane stalk cut, mostly taken from field	Sugar extracted through bagasse-crushing, soaking, chemical treatment	Primarily from crushed cane (bagasse)	Fermentation and distillation of alcohol	Heat, electricity and molasse
Grain crops (wheat, corn)	Starchy parts of plants harvested; stalks mostly left in the field	Starch separation, milling, conversion to sugars via enzyme application	Typically from fossil fuel (e.g. natural gas)		Animal feed (e.g. distillers dried grain), sweetener (if corn feedstock)
Cellulosic crops (grasses, trees)	Full plant harvested; grasses cut with regrowth	Cellulose conversion to sugar via saccharification (enzymatic hydrolysis); lignin use for process energy	Lignin and excess cellulose		Heat, electricity, animal feed, bioplastics, etc.
Waste biomass (crop residue, forestry waste, municipal waste, mill waste)	Waste materials gathered, separated, cleaned to extract material high in cellulose				

Figure 18.10 (a) Illustration of ethanol production steps by feedstock and conversion technique (after IEA, 2004). (b) In 1933, at Coryell's station in Lincoln a blend of gasoline with 10% corn alcohol was sold (© Nebraska State Historical Society).

Table 18.3 Unwanted secondary feedback loops from the Swedish decision.

Global perspective	National perspective
Increasing *food* prices, more starving people	
High *water use* in northern wheat-based production	Preventing technology development (e.g. for second-generation biofuels or solar cars)
Pressure on cheap unused (fallows), natural land deforestation, impacts on biodiversity	Supporting inefficient production of feedstocks in unproductive areas

meters or to nourish a starving family. The dramatic increase in world food prices in 2007 by nearly 40% has been linked to increasing biofuel use (Lee et al., 2008). The Swedish government should have acknowledged that the bioethanol/biofuel demand could not have been fulfilled by domestic (cellulose-based) production, at least if Sweden aspired to be an oil-free country. Imports from Brazil or other countries are a possibility, with environmental impacts, such as accelerating deforestation. We can conclude from this environmental analysis that the Swedish decision is likely to have worldwide impacts on world agrobusiness. The environmental analysis thus shows that the *world* has to be considered as the environment of the Swedish society's decision (Complementarity Postulate P1).

C3.4 Awareness of essential, critical feedback loops (P4 & P6)

A brief check of the HES Postulates reveals that the Feedback Postulate P4 is overly critical for the oil-free society case. As LCA-based assessments are – in general – static, they ignore two major feedback loops. By means of simplification we split the decision under consideration and link the world perspective to the biomass to biofuel/bioethanol conversion and the Swedish perspective with the subsidies for bioethanol of the Swedish decision. Table 18.3 presents critical negative feedback loops for both of the decisions analyzed in this chapter.

We introduce another economic feedback loop. Christian Azar, a member of the Commission on oil independence, introduced an interesting economic feedback loop which focuses on the "… support systems that lead to investment in corn-based ethanol …" (Azar, 2006, p. 49). Azar's focus does not refer to the car purchase rebates but to another type of subsidy: taxation of imported bioethanol and/or subsidization of local production (such as in the USA) to promote national bioethanol production. He argues that economic and energy efficiency would be hampered by such subsidies, at least untill second-generation bioethanol production becomes real. He stresses (possibly underestimating the importance of land use change in Brazil) that much less land will be used on a world scale if the northern corn-based bioethanol production was not subsidized. Though it might be difficult to share Azar's arguments, we can see that the identification of essential secondary (or higher ordered) feedback loops is important when making or analyzing the oil-free society decision.

Which unwanted feedback loops obtain priority depends on the perspective (i.e. the human system) considered as the reference system. On the global level, biofuel production and consumption can cause increases in food prices/starvation, water consumption, and deforestation. Potential positive (primary) with negative secondary (or higher ordered) feedback loops from taxation or trade regulations have to be judged. Here the competence but also the identification of proper means to avoid these feedback loops is necessary. In the case of biofuel this leads us to the field of trade regulation, certification, World Trade Organization rules, labeling, etc. (United Nations (UN), 2006; WWF, 2006) and thus to economic actors.

We should mention that the awareness of unwanted impacts and feedback loops are the object of the *Risk Governance Guidelines for Bioenergy Policies* (International Risk Governance Council (IRGC), 2008). The motivation of the IRGC report is the same as our selection of this case:

> IRGC is concerned that the current policies may, in the longer-term, be detrimental to other fundamental challenges (International Risk Governance Council (IRGC), 2008, p. 1).

C3.5 Interferences from feedbacked multi-actor decisions (P3)

After the environmental system analysis, we take a decision-theoretic view (see Decision Postulate P5) and sketch the potential impacts of the Swedish "oil-free society" decision. As presented in Figure 18.11, the world is facing peak oil as a key problem. The Swedish society proactively followed the objective of an oil-free society. This is a natural resources-based decision of the society (for the following see also Figure 1.7*). From a societal view, two decision tracks have to be followed. One is the technology line seeking alternative energy sources, in particular for vehicle fuel. The other is providing incentives, such as the rebate system for biofuel-driven cars which affect the individual consumer (de Haan *et al.*, 2009; Peters *et al.*, 2008) when making a car purchase decision.

Environmental Awareness Postulate P6
Multi-actor decision modeling reveals unwanted feedback-loops. These ask for environmental awareness of type C.

We have already identified different market actors, such as the oil industry, which is building bioethanol plants or importing biofuel from Brazil, to run cars. We have learned that not only national actors are at work to fulfill Swedish needs. Brazil sugarcane farmers, as well as the Amazonian forest industry, can invest or preinvest into different types of forest areas. Here, we are facing a typical conflict between the market actors from the north and the south. The World Rainforest Movement, labeled "Biofuels: A threat masked in green" (World Rainforest Movement, 2006) and stated:

> ... "converting pulp mills in the North to biorefineries will drive the expansion of industrial tree plantations in the South", as the growing wood hunger leads to companies from Sweden and Finland expanding their cultivation of quick-growth forests in countries such as Brazil. (World Rainforest Movement, quoted according to Christodoulou, 2007)

The tropical wood will be cheaper than the Scandinavian wood. This will cause additional stress on tropical and other forests and land, which might turn to energy production (see Box 18.3). But future wood-based bioethanol production will also compete with pulp paper, construction, furniture, or heating wood. These are the feedback loops on the bottom of Figure 18.11.

C 3.6 Conflicting variants of environmental awareness with respect to biomass' ecosystem functions (P6)

Some of the current trade-offs regarding sources of biomass include diverging interests of human systems on the micro or macro level or the time dimension which is inherent in the Feedback Postulate P4. Here, we deal with four key trade-offs involved in biofuels. These are biomass to energy vs. biomass to: (i) fertilizers; (ii) wood; (iii) food; and (iv) other biodiversity functions.

Box 18.3 A neutral CO_2 balance means not renewable: bioethanol boomerangs

There is a lot of hope that second-generation bioethanol production will free this fuel from trade-offs with food production. That this might not be completely realistic is indicated by a statement from Jeremy Tomins from the UK National Non-Food Crop Centre (NNFCC): "Instead of just taking the grain from wheat and grinding that down to get starch and gluten, then just taking the starch, we are going to take the whole crop – absolutely everything" (http://news.bbc.co.uk/2/hi/science/nature/5353118.stm). Technological visions like this ignore the impacts on negative feedback loops of using crops for fuel.

A critical issue is the trade-off and the additional stress on tropical and other forests and land use. History is full of unfortunate cases of cellulose, pulp or fiber productions in countries such as Indonesia, India, or Brazil. The Jari Cellulose case in north-west Brazilia can be taken as an example. Starting with the purchase of 1.7 million hectares (which is 56% more than the size of Belgium or one-third of Costa Rica) by the US shipping magnate, the tropical ecosystem underwent fundamental changes. Today eucalyptus trees with a growth cycle of 5 years providing 29 m^3/ha/year dominate (Milstein *et al.*, 2004). Given the current pressure on the rest of tropical forest there are overly attractive monetary incentives for international and national market actors for utilizing ecosystems for pure energy-oriented market ventures.

Another critical judgment is labeling bioethanol as a renewable resource: this view is based on a fallacious, "short-sighted" conception of renewable resources. Agricultural feedstocks are considered renewable because they gain energy from the Sun via photosynthesis. The facts that not all minerals required for growth, in particular nitrogen and phosphorus, are returned to the land and that the tropical forest is non-renewable, are ignored.

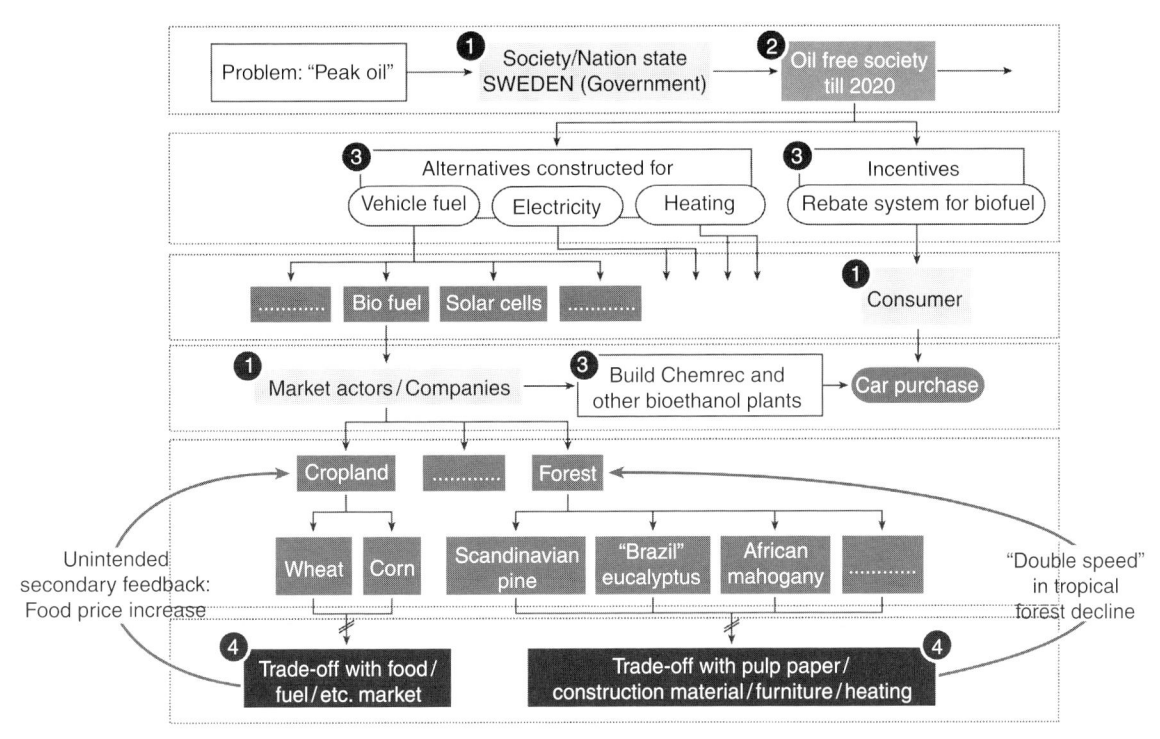

Figure 18.11 A multi-actor decision analysis of the "Sweden for bioethanol" decision. 1 = Human system/decision-maker; 2 = objectives/drivers; 3 = decisions/strategy selection; 4 = secondary feedback loops/rebound effects.

Awareness Postulate P1
Focusing on biofuel worldwide would increase tropical deforestation, rapid rock phosphorus depletion, and loss of regulative functions of forests.

C3.6.1 Fertilizers

A primary trade-off is between a company's or a country's optimization of energy production and, for instance, recycling biowaste to produce fertilizer. Here phosphorus (P) is the most critical element (see Box 13.1). It is the eleventh most abundant element on Earth, but due to its dissipative characteristics, a non-renewable resource that is depleted through most agricultural production. Prospective evaluation should acknowledge that the main phosphate stocks in Morocco, China, USA, and South Africa hold known reserves that can sustain agriculture as we know it for only 50–100 years (Cordell *et al.*, 2009, see Chapters 13.1–13.5).

Maintaining soil fertility, however, not only refers to P or nitrate but also conflicts with using the whole wheat plant and not only the fruit for second-generation biofuel. In general, using bioethanol or biodiesel would imply a severe impact on the essential geochemical cycles (see Chapter 13) and not only on the CO_2 balance. This has rarely been acknowledged as it is done in the Risk Governance Guideline of the IRGC:

> Maximise the use of waste, in particular sewage, in bioenergy generation but only deliberately use food crop residues when doing so does not lead to soil erosion of humus depletion. (International Risk Governance Council (IRGC), 2008, p. 3)

The fertilizer issue already reveals that the term biofuel could be a misleading one and should be substituted by agro-fuel.

C3.6.2 Wood

A similar trade-off can currently be seen with respect to wood. Wood fulfills many basic provisioning functions, such as construction and furniture material, pulp, or energy. Forests are declining at an overly critical speed. Approximately 40% has been lost in the past three centuries. More than 75% of the world's accessible freshwater comes from forests. Wood has been a common property in many societies. Forest decline primarily impacts the poor. About 300 million people, most of the very poor, including 60 million indigenous people, depend on forest resources (Hassan *et al.*,

2005). Humans' use of forest services for fuel wood or for road transportation is in conflict with other forest ecosystem services such as ecological (e.g. water protection), biospheric (e.g. biodiversity and climate regulation), and, as they are implicitly involved in the global departure from subterranean (fossil forest) energy use, social (e.g. cultural and recreational) functions.

C3.6.3 Food

Global food demand is expected to double by 2050 (Cassmann, 1999; Tillman et al., 2002) owing to the increase in world population, which is projected to reach between 7.3 and 10.4 billion (Lutz et al., 2001) and also to rapid increases in per capita consumption (Myers & Kent, 2003), and dietary shifts toward a meat diet. This trend may necessitate the increase of food production by expanding agriculture to formerly uncultivated areas and/or by intensifying agricultural practices. Hence, trade-offs with biofuel production will be accentuated.

C3.6.4 Other biodiversity functions

Deforestation for agriculture and energy purposes is in conflict with biodiversity. However, there are other functions, such as regulating functions, which are in conflict with biofuel production from forests. Again, in practice, many of these trade-offs neglect follow-up costs, secondary feedback loops, or other rebound effects such as flooding. A survey of 56 countries between 1990 and 2000 showed that the total cost of flooding within the decade exceeds US$1151 billion. A decrease of the natural forest of 10% provided a "model-averaged prediction of flood frequency … between 4% and 28%" among the countries modeled (Bradshaw et al., 2007, p. 2379).

C 3.7 Incomplete hierarchy: the missing supranational systems (P2)

Globalization and environmental management require new forms of world governance. From a functionalist perspective, there is thus a need for multilateral cooperation, treaties, and institutions above the level of society and the nation-state. Traditional means for this are bilateral and multilateral international treaties and contracts.

Hierarchy Postulate P2
A supranational player that could prospectively define constraints for sustainably producing biofuel is missing.

In Chapter 16.3 we distinguished between an *international* and *supranational* (synonymously used as suprasocietal) level. Societies or nations can build intergovernmental organizations through a charter, treaty or convention. The rules to be followed for international and supranational organizations are international law, such as maritime law or international criminal law. The United Nations (UN) is considered a typical international organization. The European Union (EU) is an example of a supranational organization. The difference between them is that the EU has representatives (European Parliament electing a European Council) and legislatures (e.g. the European Court of Justice) who can dictate to the members how to apply their law. If a country in the EU does not follow EU law, it can be penalized by the EU Court of Justice. This is not the case with the Universal Declaration of Human Rights (UDHR, signed in 1948), which can induce moral pressure to states violating articles, but does not allow for intervention. Thus, supranational organizations are less likely to show the failure of many institutions of not being able to put a regime in place or of having a set of rules that is not followed (Bernauer, 1995; Ostrom, 1990).

In principle, biofuel can become a meaningful component of national and the world's energy portfolios. But free market cooperation is a non-cooperative game and will not account for externalities and meaningful coping with the trade-offs mentioned above and the north–south dilemma. The case reveals that a supranational player is missing who defines appropriate constraints for sustainable development.

C3.7.1 Key messages

- The oil-free society or go for bioethanol decision by Sweden had timely intentions but, clearly, did not sufficiently follow the Environment-first Postulate P7 (including an estimate of national and global potential) and neither thoroughly reflected on trade-offs resulting from feedback loops
- An environmental analysis highlights that bioenergy can only meet a small percentage of the overall energy demand in Sweden without severe environmental impacts
- The identification of global players, that is, including human systems from the South, follows from a thorough environmental analysis

- The label renewable energy for biofuel is fallacious and misleading as it neglects – among others – the (rock P-based) fertilizers needed to replicate (sugarcane or corn) harvests in the long run. Thus, using the term agro-fuel instead of biofuel would better reflect its non-renewable nature
- The hierarchy analysis showed that most national actors had attractive economic incentives for going for bioethanol. The global level revealed conflicts and dilemmas resulting from threatening feedback loops of a North–South type. It further revealed that supranational players framing transitions to an oil-free society that can avoid negative feedback loops are missing.

18.4 Case 4: bones, BSE, and P[7]

18.4.1 What can we learn from this case for environmental literacy?

We illustrate how to apply the HES Postulates and framework presented in Chapters 16 and 17 to gain a better understanding of the history and societal learning related to bones, BSE (borine spongiform encephalopathy) and P (phosphorus) during "the BSE crisis" in Switzerland (see Figure 18.12). The specific perspective is on trade-offs in human systems and different types of feedback loops that human systems (decision-makers) are facing in strategic environmental management.

This case also focuses on the methods used when utilizing the HES Postulates. The shaded "Goal and methods" sections inform about this.

18.4.2 What is the case about?

History has taught us that P can become a valuable and scarce material. During the food crisis in the mid-eighteenth century, soils were depleted and P was sought from all sources. Thus, for instance, expensive guano was imported from Peru and desperate farmers raided "Napoleonic battlefields such as Waterloo and Austerlitz" to get hold of fertilizer (Hillel, 1991; cited in Foster, 1999). Also, in the current century it is acknowledged that P is an essential, non-substitutable resource, crucial for feeding the growing world population. In Switzerland until the year 1990, P in many waste components (i.e. organic waste, excreta, and carcasses) was put back into the agricultural system mainly in the form of compost, sewage sludge, and meat and bonemeal (MBM) or was consumed in other forms such as gelatin. In 1990, based on the first case of bovine spongiform encephalopathy (BSE) in Switzerland, stricter rules for the recycling of carcasses were established, and in the year 2001 the reuse of carcasses, including animal bones (which have a very high P content), was completely prohibited (including their use as fertilizers).

Content Case 4

- C4.1 Guiding question and system boundaries   495
- C4.2 System understanding: Postulates P1 and P7   496
- C4.3 Setting the scene: Postulates P2, P5, and P6   501
- C4.4 Understanding system dynamics: Postulates P3 and P4   504
- C4.5 Concluding remarks   507

C4.1 Guiding question and system boundaries

Guiding question and system boundaries
How can we understand the development of Swiss legislation with respect to BSE and its effect on P-flows in Switzerland?

---

[7] Claudia R. Binder, Daniel J. Lang, and Roland W. Scholz have been the lead authors of this case. We want to thank H. Lamprecht for his valuable work and input to this chapter.

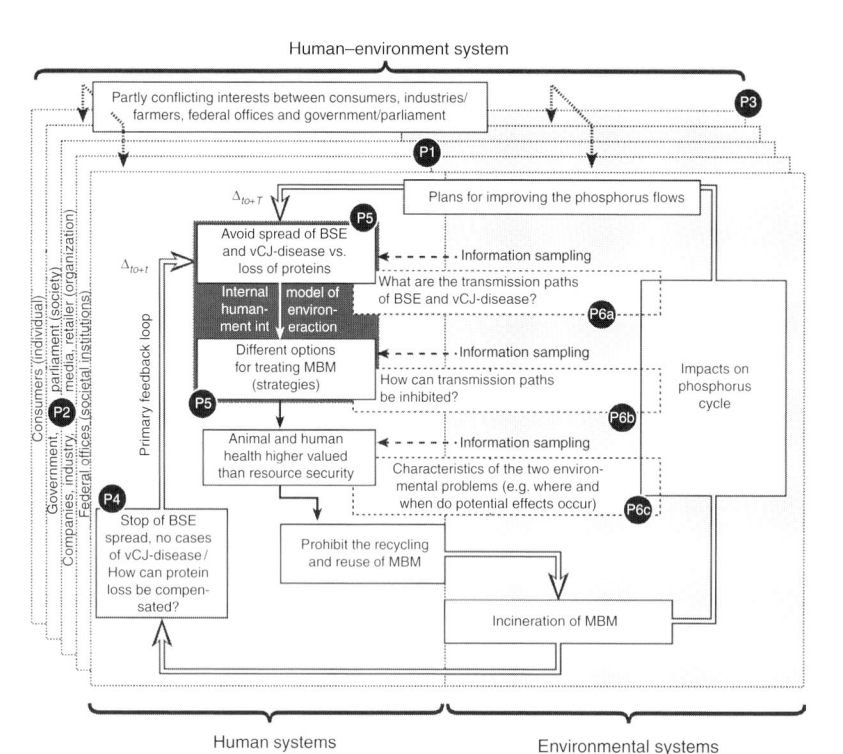

Figure 18.12 An HES framework view of the bones, BSE, and P case. MBM denotes meat and bonemeal.

The guiding question refers to how the governmental decision to prohibit the recycling of carcasses, and specifically the reuse of animal bones in agriculture, came about. It further asks what impact this decision had on the P flows in Switzerland. The aim is to better understand how this decision evolved, how different hierarchy levels of human systems interfered with each other, and which feedback loops contributed to the developments observed. To answer this question we study the time period from 1998 (before the BSE crisis in Switzerland) to 2002 (when the first "peak" of the BSE crisis was over and crucial legislations were established) within the political and geographical borders of Switzerland as our system boundaries. Though the case is analyzed from a Swiss perspective, it is relevant to keep in mind that, for both P and BSE, Switzerland has to be clearly considered as an open system, strongly interrelated with other systems, and having substantial inflows as well as outflows.

C4.2 System understanding: Postulates P1 and P7

P7: Environment-first Postulate

Goal and methods: understanding the environmental circumstances and dynamics relevant for the case on the basis of state-of-the-art textbook knowledge by means of literature review and expert interviews.

Environment-first Postulate P7
The bones, BSE, and P case included two main problems: the P/protein resource problem and the BSE health problem. These differ in timescale and geographical boundaries. The case presented shows that strategies selected to cope with a problem might lead to trade-offs.

In the case presented, which focuses on carcass and especially animal bone disposal, there is an overlap of, as well as a trade-off between, two different types of environmental problems. The first is a resource problem related to P scarcity (long-term consequence and high dependency on foreign countries) and protein availability (short-term consequence). The second is a health issue related to BSE, which is viewed as a problem where action primarily at the national level (i.e. societal level) is required.

In Chapter 13 the relevance and the fact that P is essential for life was introduced from a global perspective. Some researchers claim that there is evidence for "peak phosphorus" and that new coping strategies need to be developed with regard to the anthropogenic use of P (see, for example, Cordell *et al.*, 2009; see Chapter 13.2 and Box 13.1).

C4.2.1 The role of P in summary

We can state that: (i) P is an essential element for any organism; (ii) today's agriculture, at least in industrialized

> **Box 18.4** Endocannibalism and inbreeding: kuru, BSE, and vCJD

An ancestor of BSE is scrapie, a disease of sheep known at least since the eighteenth century. Considered incurable, the disease became a national problem for British society around 1755, a time when about one-quarter of the British population participated in wool production and when the number of sheep and the population of Britain were both around 10 million. The scrapie epidemic was supposed to be a result from inbreeding practiced to improve wool quality (Brown & Bradley, 1998). The illness was proven to be transmissible in 1936, showing a long incubation time after inoculation, from 1 to 2 years. Also, there was clear empirical evidence that the spinal cord as well as the brain were the hosts of the infectious agent.

In 1959, the American veterinarian Hadlow suspected that a phenomenological similar neurological disorder called kuru observed in a tribe of eastern highland Papua New Guinea was of the same nature. The latter disease, which had already been described two decades before, was supposed to be caused by ritual endocannibalism. This salient hypothesis could not be supported by laboratory results; however, the neuropathologist reported a striking similarity of kuru with the rare and randomly appearing Creutzfeldt–Jakob disease. In 1966, Gajdusek and co-workers "announced that chimpanzees injected with brain material from deceased kuru patients had developed a clinical syndrome akin to kuru" (Gajdusek *et al.*, 1966; cited in Lindenbaum, 2001). [Thus] "Kuru, like scrapie, appeared to be a viral disease of extraordinarily long incubation" (Lindenbaum, 2001, p. 365).

Following previous work of Alpers and others, the American neurologist Stanley B. Prusiner suggested that the "infectious agent causing certain degenerative disorders in humans and animals might consist of a protein" (Lindenbaum, 2001, p. 366). It could actually be shown that a misformation of the prion protein (PrP^C) called PrP^{SC} (sc for scrapie) was causing these diseases. Prions are unprecedented infectious pathogens that cause a group of invariably fatal neurodegenerative diseases "… [such as] Bovine spongiform encephalopathy (BSE), scrapie of sheep, and Creutzfeldt–Jakob disease (CJD) of humans …" (Prusiner, 1998, p. 13363).

Despite the various hypotheses, "the physiological function of PrP^C," as well as the concrete mechanism" by which PrP^C is converted into PrP^{SC} is still unclear" (Aguzzi & Callela, 2009, p. 1116 and p.1106).

As indicated above, for scrapie and kuru there is evidence that at least the transmission of these diseases was promoted by human practices (inbreeding and endocannibalism). There is clear evidence that the human practice of feeding animal meat and bones to cattle was a major vehicle of infection. Therefore, it seems that human–environment interactions have played a crucial role in the spread of the latter forms of transmissible spongiform encephalopathy (TSE), including BSE and new variant Creutzfeldt–Jakob disease (nvCJD) described in the text.

countries, is highly dependent on large mineral (rock) P supply; (iii) P has dissipative properties, dissolving in lakes, groundwater, and seawater; (iv) P is an essential mineral to maintain the current yield of world food production and it cannot be substituted by other elements; and (v) following precautionary thinking and principles of sustainable development, the tremendous loss of P as a non-renewable and hardly recoverable element once it is diffused should be stopped (see Box 13.1).

In this chapter we take a national perspective and focus on Switzerland as an example of a highly industrialized country (society) with a long agricultural history.

Following the Environment-first principle dealing with the guiding question and managing BSE requires knowledge of the genesis and processes underlying BSE and the variant form of Creutzfeldt–Jakob disease in humans (vCJD). The review in Box 18.4 documents both the strong impact of optimized animal husbandry and the change of natural nutrition cycles as potential causes of epidemics.

C4.2.2 The epidemic story of BSE

Although the symptoms of BSE had already been observed in a cow in the UK in 1984, it was not until 1986 that it "was first identified as a new disease" (O'Brian, 2000, p. 730). After this date the number of diagnosed BSE cases dramatically increased in the UK before a steep decline after 1992 owing to strict regulations (see Figure 18.13a).

As the numbers presented in Figure 18.13a refer to identified cases, it is likely that unreported cases exist and that these numbers may under-represent reality, especially for the first years of the epidemics. Starting in the UK, BSE spread over Europe and the first case in continental Europe was diagnosed in the Swiss Berne Jura at the end of 1990. Figure 18.13b shows that the development of BSE cases in Switzerland was similar to that in the UK, with a delay of about 2–3 years and generally lower case numbers. In total, 463 BSE cases were diagnosed in Switzerland up until 2006, and from 2007 on no additional cases accrued.

There is broad agreement that the major reason for the spread of BSE was feeding contaminated animal meat and bone meal to cattle. However, not all meat and bone residues are considered risky material. Based on this consideration, harmonized with EU regulations, Switzerland defined in 2004, for instance, three different categories of animal waste for which specific forms of disposal are allowed, respectively (Federal Veterinary Office (FVO), 2004): specified risk material (brain, eyes, bone marrow, etc. of cattle older than 6 months) and carcasses of diagnosed BSE cases fall into category 1, which is the only category where all waste has to be incinerated (with few exceptions). For the other categories different forms of recycling are allowed (among others anaerobic digestion) since 2004, but only for category 3 is feeding of processed waste to other livestock allowed (Schweizerischer Bundesrat, 2004).

In the first years after the discovery of BSE it was assumed that:

> BSE was of negligible risk to humans (c.1986–1989), because its pathology and aetiology closely resembled "scrapey," … which carries very low zoonosis [infectious disease that can be transmitted from animals to humans] threat to the general population. (Kewell & Beck, 2008, p. 136; Philips, 2001)

However, since 1996 a new variant of the Creutzfeldt–Jakob Disease (nvCJD) has been observed, especially in young patients. The subsequent identification that "BSE and vCJD encompass striking genetic similarities" (Kewell & Beck, 2008, p. 134) and that the same prion strain causes vCJD and BSE led to increasing evidence that there is a causal relationship between BSE and vCJD. Today there is broad agreement on this relationship (see Walker et al., 2008; WHO, 2002). Infectious prion agents (PrP^{SC}) related to BSE as well as vCJD are not only prevalent in the brain but also in other tissues although the infectivity of most of these tissues is lower (Walker et al., 2008). Despite the assumed causal relationship of BSE and vCJD it has to be mentioned that the latter is a rare disease and that in Switzerland no case had been diagnosed up to 2007 (FOPH, 2010). Besides the exposure to the BSE agent through ingestion there is also evidence that vCJD can be horizontally transmitted between humans, for instance via blood transfusion or surgery (Aguzzi & Glatzel, 2006; Hodges, 2005).

C4.2.3 The (Swiss) case of BSE and vCJD in summary

It can be stated that: (i) BSE has a long history and is not a new phenomenon; (ii) feeding MBM to cattle was most probably the major transmission route of BSE and thus an anthropocenic redefinition of the environment had a substantial influence; (iii) BSE can most probably be transmitted to humans via ingestion and cause vCJD; and (iv) Switzerland was among the most affected countries with respect to BSE during the first BSE crisis. Thus both from an animal health as well as from a human health perspective, substantial measures to cope with this epidemic were necessary.

C4.2.4 The Swiss P-flows

Figure 18.14 presents the main phosphorus flows in Switzerland for the year 2002 (elementary phosphorus in metric tonnes per year). The major flows were related to food production (livestock husbandry and plant production as well as meat processing), food consumption and organic (food) residues (including sewage, 7000 tP/a; waste after consumption, 3500 tP/a, and animal byproducts from meat processing, 2600 tP/a). Another aspect visible in Figure 18.14 is the large import of P into Switzerland mainly for animal feed (6500 tP/a) and plant production (i.e. fertilizers, 6400 tP/a). An interesting fact is that the highest flows occured between the husbandry and the plant production sector, with flows in the area of about 30 000 tP/a, whereas the import flows concerned about 15 000 tP/a and the flows back from the waste management sector into the agricultural sector amounted to only 4000 tP/a. With respect to the fact that P is an essential and non-renewable element, a critical aspect seems to be the substantial loss of P in landfills (around 8000 tP/a) and via runoff (2500 tP/a). Furthermore, between 2000 and 5000 tP/a contributed to stock increase in the plant production sector.

Compared to alternative materials from which to recycle P, the P content of animal bones is high. For example, in 2001 for bone coarse meal, the P concentration was 9–11% and for bone meal 7.7% (both of the fresh weight). As a comparison, the P concentration in compost is about 0.3% (Lamprecht, 2004). This high P content is one reason why bones have been in demand as a valuable commodity for a long time. For instance, as a reaction to the fundamental concern of depleting soil fertility: "[t]he value of bone imports to Britain increased from £14,000 in 1823 to £254,600 in 1837" (Foster, 1999, p. 375). In more recent years, in Switzerland and in other countries, bones served as basic materials for bone glue and, until the BSE crisis, as feeding material for animals and production of gelatin (Lamprecht, 2004). The direct utilization of P as a fertilizer played a subordinate role.

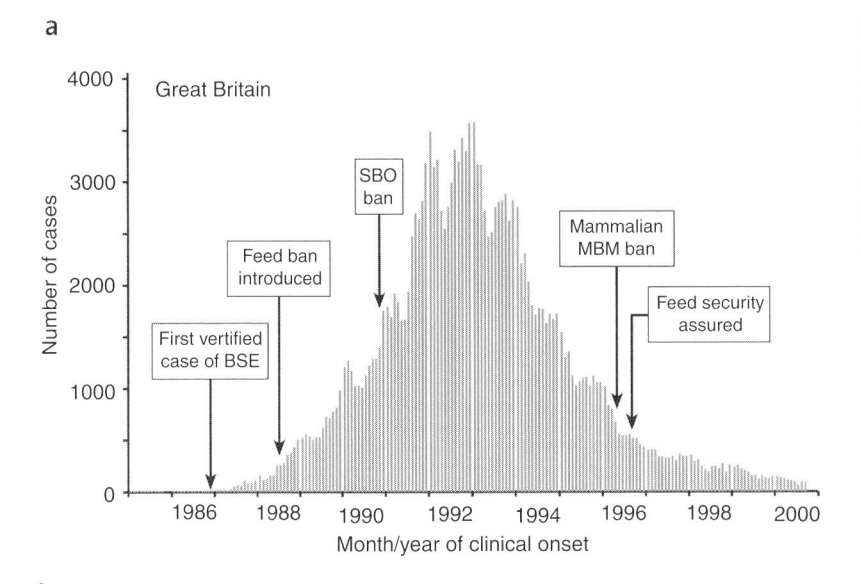

Figure 18.13 (a) Identified clinical BSE cases in Great Britain between 1985 and 2001. The figure also indicates the introduction of relevant measures to cope with the epidemic (adapted from Brown et al., 2001). (b) Yearly diagnosed cases of BSE in Switzerland between 1990 and 2008 (adapted from Federal Veterinary Office (FVO), 2009). SBO, Specified bovine offals.

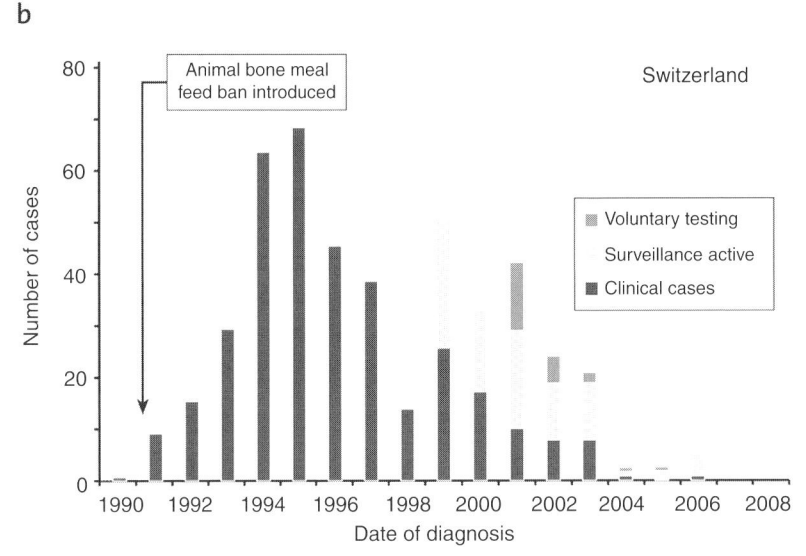

P1: Complementarity Postulate

Goal and methods: define the relevant interactions involved in the bones, BSE, and P case. Conceptualizing BSE spread by epidemic transmission networks and the P-flows by agent-oriented material flow analysis.

BSE origin and spread and vCJD: as mentioned, the interaction between activities in the environmental (animal) system and the human systems play a crucial role for the genesis and the transmission of BSE (as well as vCJD). A possible transmission appears as agriculture alters the conventional nutrition cycle, which especially includes the production of MBM from carcass and feeding these products to other animals (marked arrow between waste management and livestock husbandry in Figure 18.14). This track may keep the contagious PrP^{sc} agent unchanged, at least under certain production conditions (e.g. to low temperature and specific processing settings). There is broad agreement on the link between BSE and vCJD, thus the transmission of contaminated food between meat processing and consumption in Figure 18.14 is especially critical (see Box 18.4).

The above descriptions already show how relevant complementarities between human and environmental

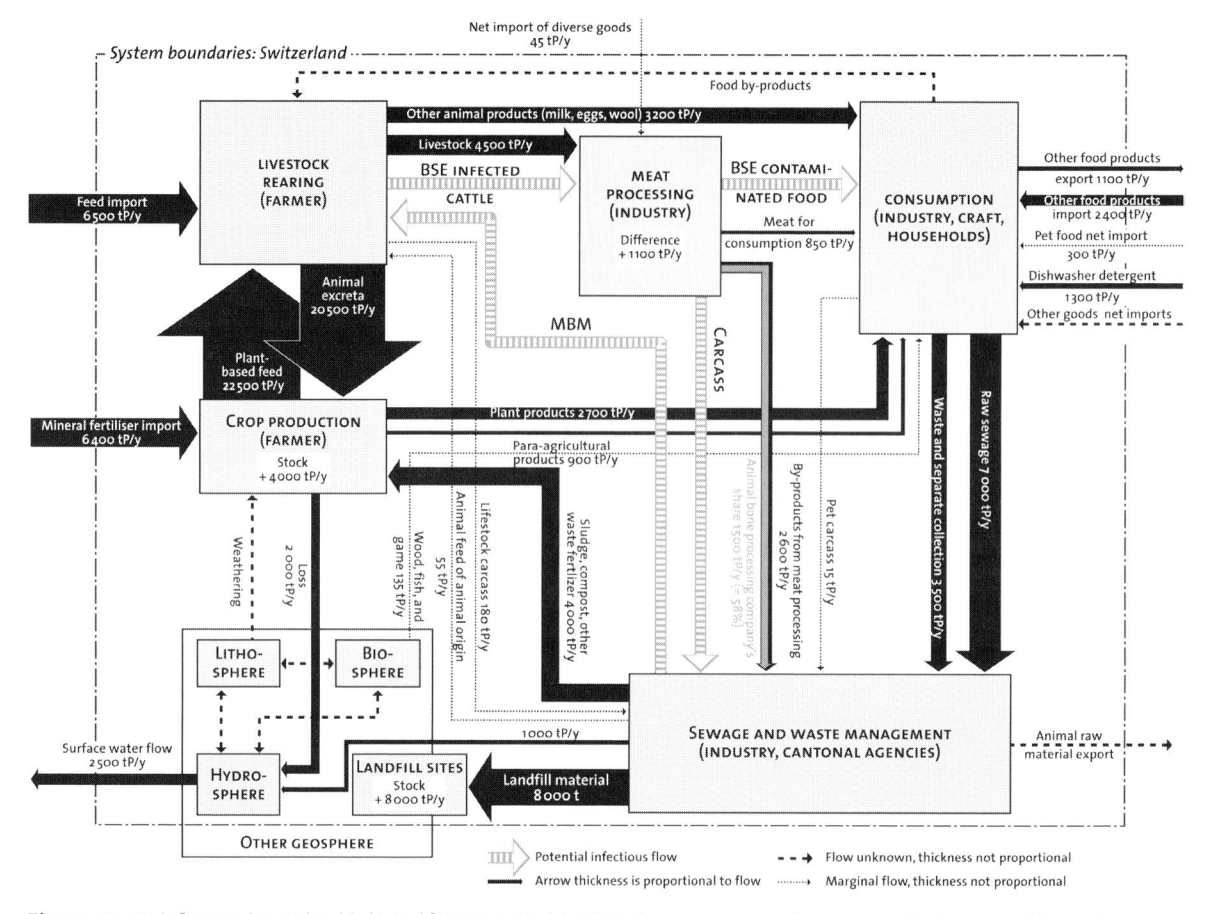

Figure 18.14 P-flows in Switzerland (adapted from Lamprecht, 2004). The complementarity between the human and the environmental system can be analyzed with an agent-oriented material flow analysis (see, for example, Binder *et al.*, 2004; Lang *et al.*, 2006, for method). This holds true for the agents directly involved in the value chain. The marked arrows indicate potential infectious flows related to BSE.

systems are in the case of BSE and vCJD transmission. This is also highlighted by Salzberger *et al.*, stating that (similar to the case of HIV):

> ... with vCJD a new dimension of infectious disease [came] into our consciousness which reveals to physicians and health-politicians how strongly our health depends on our culture, behavior or also the methods of food production. (Salzberger *et al.*, 2002, p. 707; translation by the authors)

Complementarity Postulate P1
The developments around BSE and vCJD exemplarily highlight the complementarity of human and environmental systems. With respect to physical resource management, agent-oriented MFA can serve as an interface between environmental characteristics/dynamics and the related human systems.

The management of Swiss P-flows. With respect to material flows and resource management, agent-oriented material flow analysis (MFA) can help to conceptualize the complementarity in a specific case. As shown in Figure 18.14, in this approach the processes of the system under investigation (rectangles) are defined according to the main agents directly involved in the value chain, such as meat processing companies or farmers. In so doing the material biophysical environment (physical flows and processes) and the different human systems (directly influencing the processes) are integrated in one system. Such a model can serve as a valuable interface for understanding coupled HES. It can on the one hand be extended in the direction of the environmental system by including, for instance, soil processes of P retention or eutrophication of lakes owing to P-inputs into the hydrosphere or, on the other hand, in the direction of the human systems by including agents not directly involved in but affecting the value chain (see Table 18.4) as well as other flows (e.g. information, money, regulation) between the agents (see Figure 18.16).

With respect to the P-flows in Switzerland in 2002, Figure 18.14 reveals that the main inputs into the system are introduced by farming, whose major material outputs are directly or indirectly linked to consumers. The residue disposal paths of production and consumption can generally be divided into P that is "kept" in the anthropogenic cycle and P that "leaves" the cycle, mainly because of runoff or disposal in landfills. This division takes place in waste management. The essential flows in the case of bones, BSE, and P are related to livestock husbandry, meat processing, consumption, and waste management. As mentioned above, the connections between these processes are also of utmost relevance with regard to the transmission of BSE as well as vCJD – disclosing the critical interferences between resource recovery and health protection, which are denoted by the marked arrows in Figure 18.14. Analyzing the material flows and the infectious flows in this way thus reveals first important connections between the relevant processes and the related agents.

C4.3 Setting the scene: Postulates P2, P5, and P6

Goals and method: which actors (agents) at which hierarchical levels and with which drivers contributed to critical interferences and trade-offs between health protection and resource recovery (in particular phosphorus)? How can structural agent analysis or systems modeling support the analysis.

Hierarchy Postulate P2
In the case of carcass management the individual, organizational, institutional, and societal levels were involved.

As has already been mentioned above, in the case of bones, BSE, and P, as in most complex material flow and management systems, two types of agents can be identified: the ones *directly involved* in the value chain and the ones *impacting on* the value chain. The first have already been identified in the agent-oriented MFA (see Figure 18.14). The second were identified by applying the first step of the structural agent analysis, which is "identification of relevant agents" (Binder, 2007; Binder *et al.*, 2004).

With reference to the Hierarchy Postulate P2, agents on four levels of human systems (individual, organization, institution, and society) were identified. At the individual level, the meat consumers affect the value chain directly through their demand for meat and their consumption preferences. At the organizational level, farmers, the meat processing industry, retailers, and the animal byproducts processing industry affect the value chain directly with their current activities; and media affect the chain indirectly through information distribution. The university, as an institution of society, can affect the different stakeholders by information, consulting, or in transdisciplinary processes (see Chapter 15). At the institutional level, in the present case, the interferences between different federal offices played a crucial role. At least three of them, the Federal Office for the Environment (FOEN), the Federal Office of Public Health (FOPH), and the Federal Veterinary Office (FVO), all with different societal duties, were involved in the case of carcass and BSE management.

The agents differed in their environmental awareness (see Awareness Postulate P6), as well as their goals, strategy formation, and strategy selection procedure (see Decision Postulate P5). Consequently, interferences between and among the different hierarchical levels can be identified (see Interference Postulate P3, further elaborated in the next chapter). We show exemplarily how the Hierarchy Postulate P2, the Decision Postulate P5, and the Awareness Postulate P6 can help to structure the case transparently to understand the interactions and outcomes for the case of bones, BSE, and P.

In Table 18.4 we qualitatively depict the perceived and assumed changes in environmental awareness during the years from 1989 to 2002, regarding the following issues: (i) BSE and its relatedness to carcass use for feeding animals; (ii) animal byproducts as a protein source for feeding animals; and (iii) animal bones as a relevant P source.

Awareness Postulate P6
The awareness of BSE increased over time, surpassing the awareness of resource-related concerns, such as P and proteins.

According to Lamprecht (2004), during the entire period awareness regarding BSE increased for all agents: at the individual level, for example, a study of consumers in the canton Ticino, Switzerland showed that the awareness and the fear from BSE increased concomitantly with the amount of newspaper articles

Table 18.4 Agents and changes in BSE issue awareness and related decision-making "logic" between 1989 and 2002.

Hierarchy level P2 Agent	Environmental Awareness P6 Changing awareness[1]	Decision logics P5 Goal	Strategy formation, and selection	Final strategy selected
Individual Consumer	*Individual consumers* (BSE awareness rising, Cheap meat falling, 1989–2002)	No BSE	Lower (bovine) meat consumption	Lower (bovine) meat consumption "Boycott" of processed food on the basis of animal byproducts
Organization Farmers	*Farmers* (BSE awareness rising, Protein flat, 1989–2002)	Profitable production No BSE	Reuse of waste where possible Look for new sources of protein	Higher use of other protein sources
Meat processing	*Meat processing* (BSE awareness rising, Waste m. costs rising, 1989–2002)	Profitability (sell the products) No scandals Avoid BSE Obey law	Invest to separate carcasses into three categories Incinerate carcasses	No investment Incinerate and later also export carcasses
Carcass processing	*Media* (BSE awareness rising, 1989–2002)	Profitability Reuse where possible	Define new paths for reducing risks Investment in new technologies	No implementation of new paths owing to legal insecurity and possibility of exporting carcasses

Table 18.4 (cont.)

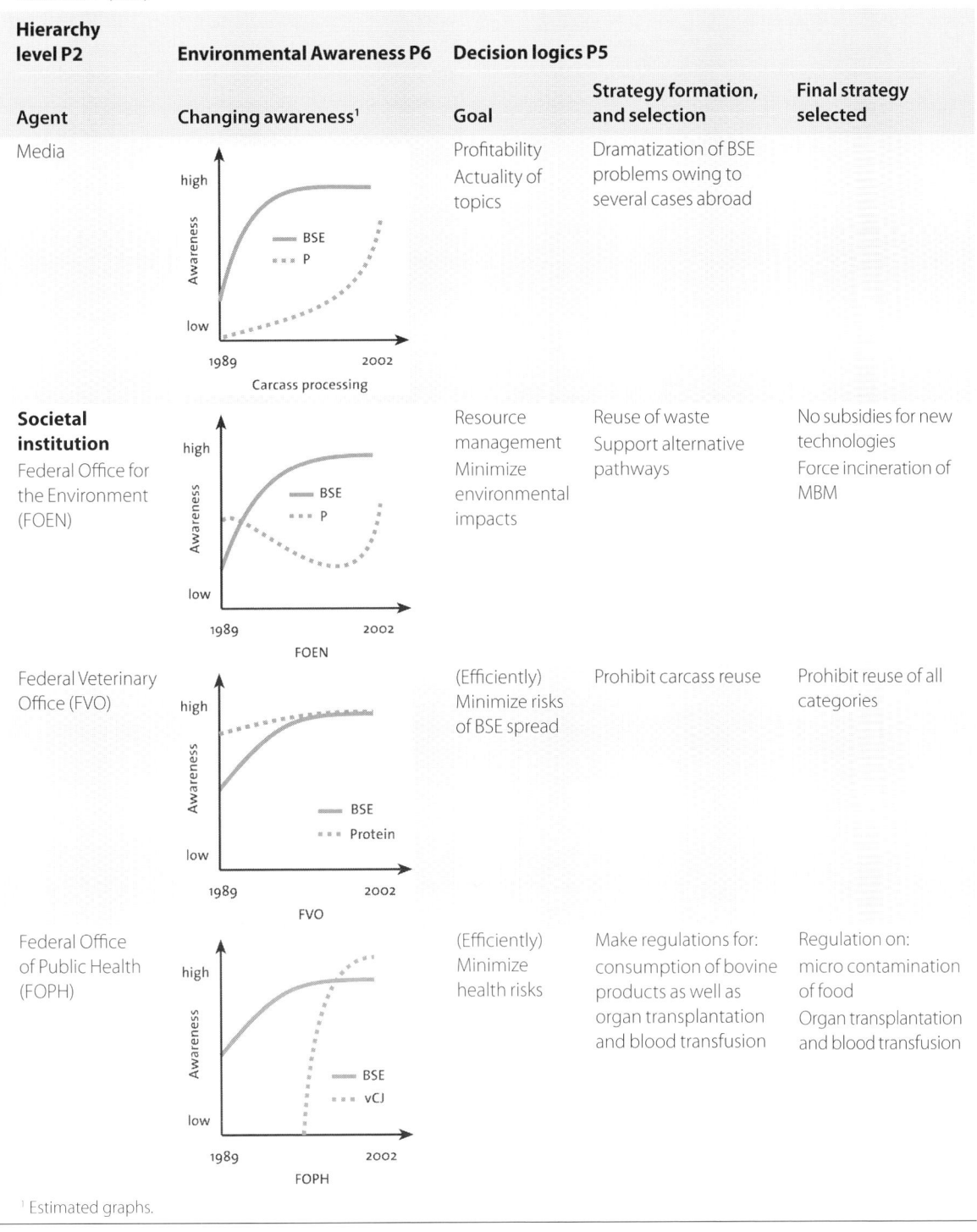

Hierarchy level P2 Agent	Environmental Awareness P6 Changing awareness[1]	Decision logics P5 Goal	Strategy formation, and selection	Final strategy selected
Media	*[graph: Awareness vs. Carcass processing, 1989–2002; BSE rises to high, P rises late]*	Profitability Actuality of topics	Dramatization of BSE problems owing to several cases abroad	
Societal institution Federal Office for the Environment (FOEN)	*[graph: Awareness vs. FOEN, 1989–2002; BSE rises to high, P dips then rises]*	Resource management Minimize environmental impacts	Reuse of waste Support alternative pathways	No subsidies for new technologies Force incineration of MBM
Federal Veterinary Office (FVO)	*[graph: Awareness vs. FVO, 1989–2002; BSE and Protein both rise to high]*	(Efficiently) Minimize risks of BSE spread	Prohibit carcass reuse	Prohibit reuse of all categories
Federal Office of Public Health (FOPH)	*[graph: Awareness vs. FOPH, 1989–2002; BSE rises, vCJ rises steeply late]*	(Efficiently) Minimize health risks	Make regulations for: consumption of bovine products as well as organ transplantation and blood transfusion	Regulation on: micro contamination of food Organ transplantation and blood transfusion

[1] Estimated graphs.

published on the issue, whereas scientific readings were not decisive information sources. An inquiry revealed a strong change of consumer behavior: only 26.9% reported no change in their meat consumption pattern at all; 36.6% reduced their consumption of bovine meat, whereof 11% did not consume any bovine meat at all; 28% also reduced their consumption of other meat types and 3.2% became vegetarian. Only one person out of the 93 respondents started to consume more meat because of decreasing prices (Volger, 2001).

For farmers as entrepreneurs, the development of environmental awareness had at least two aspects: on the one hand, the awareness of BSE increased. A fear related to the increased awareness of BSE was the potential loss of a substantial number of cattle because of culling if a BSE case was identified in the herd. Such a loss, even if economically substantially compensated, might also result in losing "valuable genetics and/ or their [the farmers] 'life's work'" (Heim & Murray, 2004, p. 190). With respect to the latter fear, the willingness of farmers to cooperate within culling programs is assumed to depend on their belief as to whether the strategy chosen was reasonable or not. On the other hand, their concern was also on how to obtain efficient protein sources (to then, a major part was provided by MBM) for feeding their animals (Lamprecht, 2004). Thus they might be confronted with direct economic drawbacks. This was of particular relevance as MBM delivers a specific mix of protein, lipids, and macrominerals, which have to be replaced by a combination of fodder and supplements (Sellier, 2003). Naturally, they were primarily concerned with the potential problem of both decreasing meat demand and meat prices.

Decision Postulate P5
In the decision-making process the goal for most of the agents was to reduce the risk of BSE and vCJD on a short and long term. The strategies selected were at each level different, whereas the policy measures were the ones with the highest long-term impact.

For the meat processing industry the increasing awareness of BSE also led to the need for innovation of the processing techniques to separate the specified risk material from the rest of the carcass. A main concern was also the potential increase in waste management costs if the waste material could not be disposed of easily and had to be incinerated in special treatment plants (Sellier, 2003).

The carcass-bones processing industry was confronted with the question of what could be gained from the waste fractions if costs for processing the wastes (see above, separation into the categories 1 to 3) grew. The awareness of P fertilizer as a potential income source increased, and they started to develop business plans for investments in innovative techniques to recover P from the bone waste fraction. They started to develop a new path for recycling, considering separation of the carcass into the three categories (Lamprecht, 2004).

At the institutional level the three offices involved had to guarantee that they fulfilled their societal tasks accordingly. In the beginning, it seemed as if only animal health was regarded as being relevant. When the first cases of vCJD emerged, the minimization of human health risk was put forward on the political agenda. As a second priority, the resource aspect, including protein and P, was brought forward by the FVO and the FOEN, respectively.

C4.4 Understanding system dynamics: Postulates P3 and P4

Goals and methods: what was the impact of the different interactions and strategy selection processes on the whole system? Which insights could be gained in this respect by systemic and feedback loop perspectives?

Interferences between and within the levels: understanding of the system dynamics in HES requires the identification of interferences and feedback loops on and between the different hierarchical levels.

Interference Postulate P3
Two types of interferences can be identified. The first, among levels (i.e. consumers at the individual level and farmers as well as meat processing industry at the organizational level), can be regarded as a *market-based interference*. The second, among the offices at the institutional level, reflects a more *societal priority setting*.

The following interferences between the levels could be identified: for consumers, human health was the main issue, and this drove them to some extent to be overcautious in their consumption decisions and to reduce their meat consumption (see Table 18.4; Volger, 2001). For the economic agents directly involved in meat and carcass processing, besides the health problem, waste management costs and the issue of investment security also played a relevant role. This situation can be viewed as a market-based interference between decreasing demand, which leads

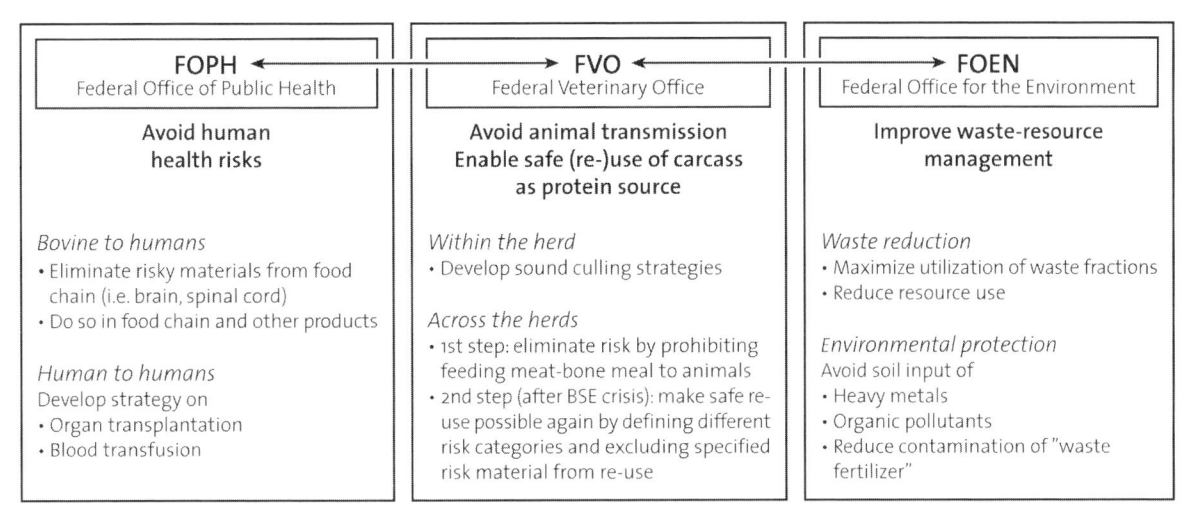

FOPH Federal Office of Public Health	FVO Federal Veterinary Office	FOEN Federal Office for the Environment
Avoid human health risks	**Avoid animal transmission** **Enable safe (re-)use of carcass as protein source**	**Improve waste-resource management**
Bovine to humans • Eliminate risky materials from food chain (i.e. brain, spinal cord) • Do so in food chain and other products *Human to humans* Develop strategy on • Organ transplantation • Blood transfusion	*Within the herd* • Develop sound culling strategies *Across the herds* • 1st step: eliminate risk by prohibiting feeding meat-bone meal to animals • 2nd step (after BSE crisis): make safe re-use possible again by defining different risk categories and excluding specified risk material from re-use	*Waste reduction* • Maximize utilization of waste fractions • Reduce resource use *Environmental protection* Avoid soil input of • Heavy metals • Organic pollutants • Reduce contamination of "waste fertilizer"

Figure 18.15 Interferences between the three main Swiss offices involved.

to reduced prices and increasing costs. This interference was alleviated to some extent by the Swiss government, who subsidized 75% of the emerging costs for 1 year as there were societal interests of maintaining farming (Lamprecht, 2004; Schweizerischer Bundesrat, 2002). However, no regulative security was created to support innovation and investment at the carcass industry level.

At the institutional level, the goals of the different offices overlapped and interfered within the level itself (Figure 18.15). For the FOPH, as for the consumers, the main goal was to minimize health risks. The main question was which materials from contaminated animals are problematic and whether a contamination from humans to humans may take place. Even though it was scientifically, but not finally, proven that vCJD is caused by BSE (Campbell & Sato, 2009; Kewell & Beck, 2008), prevention measures in the area of food and drug safety were developed. They consisted of eliminating any risky material from bovines from the food and consumer goods chain. Furthermore, research programs focusing on how the transmission of human to humans might occur were established and decrees for blood transfusion, transplantation of organs, etc. were introduced (Bundesamt für Gesundheit (BAG), 2009a).

The FOEN, and to a certain extent the FVO as well as farmers, the food processing industry, and carcass processing industries, were also concerned about the resource aspects, that is MBM as a cheap protein supply for pigs or as a source of nutrients (Sellier, 2003). This shows that there were interferences between human health issues and resource issues within the Swiss system. The health issues could have had an immediate impact on the population, and the resource issues (protein and P) were seen either as easier to solve or as long-term problems, respectively. Thus, in setting the priorities, the focus was set on the immediately perceived and relevant health problems. The resource issue has, however, been gaining importance during the past few years.

C4.4.1 Understanding of feedback loops

In this section we describe one main feedback loop that can be identified related to the case of bones, BSE, and P (see Figure 18.16). The media and public concern were the main drivers of this feedback loop. Significant cases of BSE reported from abroad and the goal of delivering first hand and excellent information led to a certain "dramatization" of the BSE problem by the media (Lamprecht, 2004). The media also put pressure on institutions, mainly the FOPH and the FVO. This was possible as human health is of concern to the whole population and securing human health is the major duty, especially of FOPH, but to a certain extend also of FVO. In the case of FVO, besides human health, also animal health and avoiding epizootic disease is of high relevance. In the latter case these aims are to be well balanced with the goal of an efficient and effective agriculture. The agencies developed several measures to affect the "infectious flows" (Figure 18.16), to ensure human as well as animal health, and to reduce any risk of further spread of BSE. The justification for these measures was partly taken from scientific results and partly based on the precautionary principle

Table 18.5 Selected P-flows in tP/a for 1989 and 2002

Flow	1989	2002	Change
Animal food originating from animal waste	2700	55	−98%
Export of animal resources	–	500	Not assessed
Fodder import	5000	6500	+30%

implemented in Swiss law (Art.1 Abs.2 USG). In 2001, a final measure prohibited the use of animal byproducts such as bone meal or bone coarse meal as fertilizer (see Sellier, 2003 for the developments in the UK and the EU). The consequences for the carcass processing industry were quite evident. As in the mean time no investments in recycling technologies for extracting P out of waste bones were made, the animal byproducts are today still either exported or burned in the cement industry (each around 50%; Binder *et al.*, 2009). This represents a significant loss in nutrients for Switzerland (about 3000 tP/a for 2006; Binder *et al.*, 2009).

Feedback-loop Postulate P4
Several feedback loops enforced themselves and led to a complete prohibition of the use of carcasses. One departed from the agent media and included the individual, institutional, and societal level as well.

The effects of the policy developments during the BSE crises can be mirrored back directly to the P-flows. Table 18.5 presents the changes in the main P-flows between 1989 and 2002. It can be noted that, except for compost and specific animal byproducts, the other

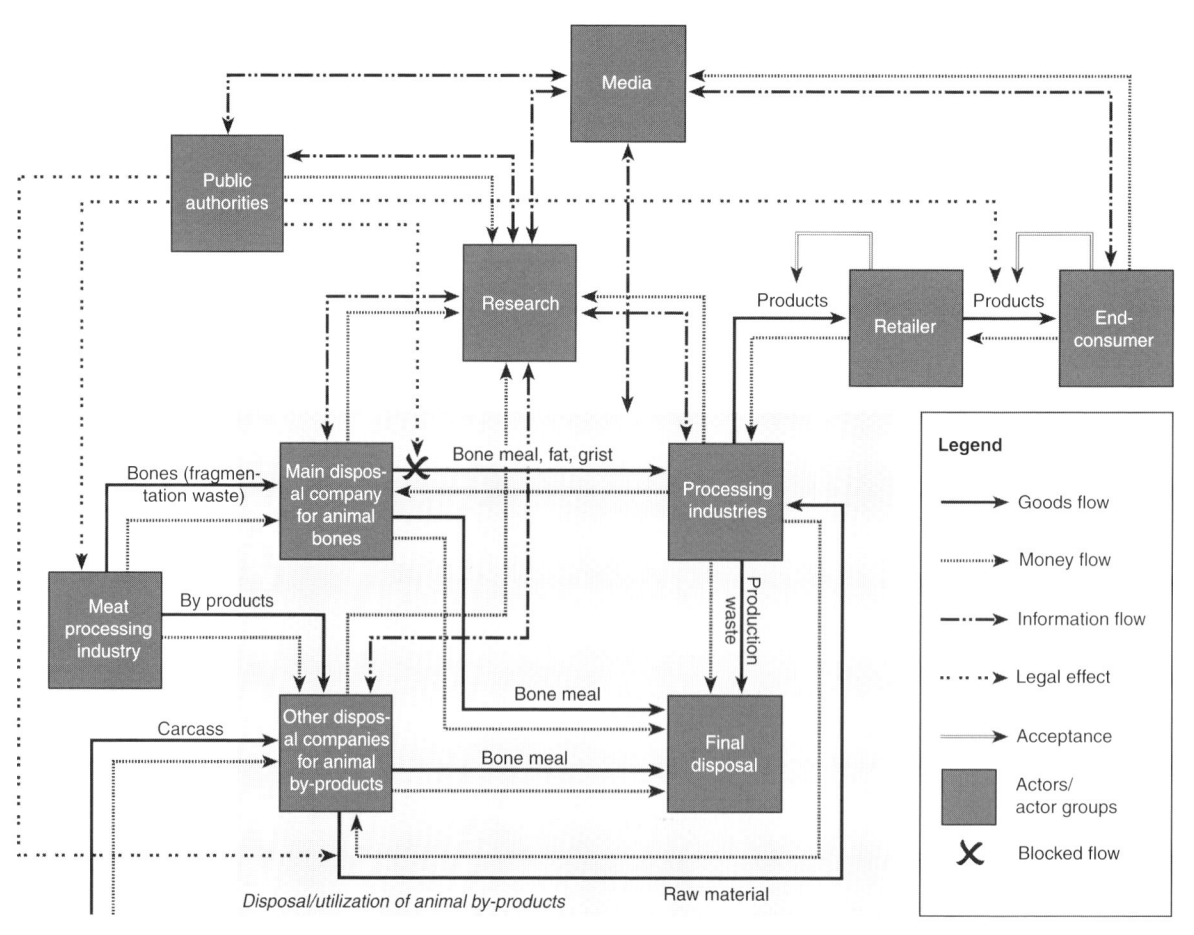

Figure 18.16 Agent scheme and feedback loops (adapted from Lamprecht, 2004).

Table 18.6 The HES Postulates, their goals, and the topics in carcass management.

	Aim	Related topic in the bones/BSE/P problematic
Analyzing the system:		
Guiding question and system boundaries	Specify the goals and define the boundaries	Understand the development of legislation with respect to BSE and its effect on P
P7 Environment-first	Problem definition	BSE spread through feeding MBM to animals, relation of BSE and vCJD
		Role of P as an essential element of life and prerequisite for modern food production
P1 Complementarity	Relate environmental problem to specific human systems (or parts thereof)	Agents directly related to the value chain
		Agents affecting the value chain
		Anthropocenic structures as relevant aspect contributing to the evolution of BSE
Setting the frame:		
P2 Hierarchy	Understand the hierarchical levels of the human system involved in the problem	Individuals
		Organizations (farmers, meat processing and carcass processing industry)
		Institutions (FOEN, FOV, FOPH)
		Society
P6 Awareness	Understand the awareness of each involved agent for the problem analyzed and the awareness to related problems	Awareness of BSE, health-related problems, and resource-related problems
P5 Decision	Identify the main goal and strategies available for each agent at each hierarchical level	Differences between the agents at the same hierarchical level (e.g. FOEN, FVO, and FOPH) or between levels (e.g. FOPH and animal byproduct industry)
System dynamics: understanding system dynamics, feedback loops, and regulatory mechanisms		
P3 Interference	Understand the interferences between the hierarchical levels	Institutions with regulative power are the strong players. Their general "problem-rating" was human health > animal health > resources (protein/P)
P4 Feedback	Understand the feedbacks between the strategies chosen and the environmental system	No BSE in Switzerland
		New protein sources required
		Export of carcass, loss of P, import of soybeans from, for instance, Brazil

flows from waste management to agriculture were reduced to almost zero, whereas the imports of fodder increased during the same time period.

In 2010, the awareness of potential phosphorus scarcity was so high that it led to changing the goals on the national level and starting the development of a concerted action on new P recycling paths (personal communication, FOEN, 2010). This is particularly interesting as before 1989 the P recycling due to the MBM use as ruminant feed was at least partly an "unintentional" one, focusing on proteins, whereas now P is the central resource. Similar activities, also with regard to the recovery of P from sewage sludge, have started already on a state level in the canton of Zurich, Switzerland (see AWEL, 2008).

C4.5 Concluding remarks

We have showed how the HES framework can be applied to structure the analysis of a complex problem in HES, i.e. the bones/BSE/P problem. The goal was to understand the dynamics behind the development of the legal framework in the case of BSE and its consequences on the P cycle.

The analysis was structured in three main parts: (i) system understanding (P7 and P1); (ii) setting the scene (P2, P5, and P6); and (iii) understanding system dynamics (P3 and P4). Table 18.6 summarizes the goals of each step and the related topics analyzed in the case of bones, BSE, and P.

After investigating the environmental system, we identified the main agents involved in the bones/BSE/P issue in Switzerland by identifying who was directly involved with and who affected the value-added chain related to P. The structuring into the different hierarchical levels and the analysis of environmental awareness, goal and strategies of the different agents allowed us to reveal interferences among the agents and put the emerging dynamics and results into a more comprehensive perspective. This provided the basis for the identification of a relevant feedback loop (Table 18.6).

Furthermore, it was shown that the HES framework facilitates integration of methods from different disciplines as well as utilizing interdisciplinary approaches. In this case we used natural science methods such as material flow analysis, which provided the link to the social science approaches, for instance, structural agent analysis. This particular characteristic, namely that the framework allows structuring of the analysis so that different parts of the system can be analyzed with methods from the different disciplines, is, from our perspective, one of the key assets and contributions of the presented framework.

C.4.5.1 Key messages

- The HES Postulates and the HES framework help to structure complex problems to utilize methods and insights from different disciplines
- Integrative methods/approaches such as agent-oriented material flow analysis help to analyze coupled human and environmental systems comprehensively
- Environmental management is facing trade-offs and dilemmas in coping with different environmental means such as avoiding risk and having access to resources
- Environmental awareness and interests in ecosystems services change, for instance, from protein to P in animal bones
- Interferences among different levels of human systems but also among different agents on the same level are crucially influencing the developments in HES.

Part VIII A framework for investigating human–environment systems

Chapter 19

Comparing the HES framework with alternative approaches

Roland W. Scholz and Fridolin Brand

19.1 Toward a unifying framework of human–environment systems (HES) 509
19.2 A comparison of the HES framework with alternative approaches 511
19.3 Natural science-accentuated: the resilience approach 512
19.4 Social science-accentuated: social metabolism 515
19.5 Action-oriented: transition management 517
19.6 Commonalities and differences among the approaches 520

Chapter overview

In the first decade of the twenty-first century, environmental sciences and sustainability research efforts have been to conceptualize human–environment systems (HES). To provide an HES framework is one of the main objectives of environmental literacy, and of the Postulates and ideas put forward in Chapters 16 and 17. In this chapter we discuss the HES framework within the context of the international research community and compare it to three alternative approaches: the resilience approach, the Vienna social metabolism, and the Dutch transition management. The objective of this section is to show the strengths and limitations of the HES framework and the alternative approaches. Furthermore, this chapter highlights the value of the HES framework for both an academic perspective, to support better analysis of HES, and from an applied project or transdisciplinary perspective, to promote inclusive communication and collaboration between academia, legitimized decision-makers, and stakeholders.

19.1 Toward a unifying framework of human–environment systems (HES)

Research on interactions within HES has been gaining momentum in the past decade. Regarding sustainability research, Kates *et al.* (2001, p. 641) state that a "new field of sustainability science is emerging that seeks to understand the fundamental character of interactions between nature and society." Further, Clark and Dickson (2003, p. 8059) state that "the multiple movements to harness science and technology for sustainability focus on the dynamic interactions between nature and society." More recently, Clark (2007, p. 1737) suggests that "a core sustainability science research program has begun to take shape that transcends the concerns of its foundational disciplines and focuses instead on understanding the complex dynamics that arise from interactions between human and environmental systems."

In a similar vein, *resilience research*, currently a highly influential area, investigates coupled social–ecological systems, as these researchers consider social and natural systems to be fundamentally linked to each other through reciprocal interactions across diverse organizational levels (Berkes *et al.* 2003; Folke *et al.*, 2002). For instance, Liu *et al.*, (2007b, p. 639) note that, although "disciplinary research continues to be important to advance disciplinary inquiries into many aspects of human and natural systems, it is not effective to study human and natural systems separately when addressing social–ecological and human–environment interactions." In addition, quite recently, Elinor Ostrom proposed a general framework for analyzing the sustainability of coupled social–ecological systems to support development of strategies to facilitate the sustainable management of natural resources (Ostrom, 2009). Thus, research on the link between human and natural systems is becoming a major frontier within both the environmental and sustainability sciences.

19 Alternative approaches for investigating HES

Table 19.1 A collection of frameworks for analyzing human–environment relationships.

Name	Primary literature	Primary disciplinary origin	Focus
HES framework	The present book; Scholz & Binder, 2003, 2004	Environmental decision research, system theory, epistemology	Relating rationales of human and environment systems in feedbacked systems, mastering complexity of human and environmental systems, preparing scientists for transdisciplinary processes
Natural science-accentuated			
Resilience approach	Gunderson & Holling, 2002; Liu et al., 2007a, b; Walker et al., 2006	Ecology, social sciences	Complex adaptive systems, resource and ecosystem services management, coupled social–ecological and economy–environment systems, coupled human and natural systems
Vulnerability analysis	Turner et al., 2003; Polsky et al., 2007	Geography	Place-based studies of coupled, sensitive, adaptive social–ecological systems
Earth system analysis	Schellnhuber et al., 2004; Schellnhuber et al., 2005	Climatology	Climate change and global sustainability
Land change science	Turner et al., 2008; Turner & Robbins, 2008	Geography	Models and assessment of land changes
Political ecology	Robbins, 2004; Turner & Robbins, 2008	Geography	Combines the concerns of ecology and a broadly defined political economy
Social science-accentuated			
Human–environment interactions	Ford, 1990; Stern, 1993; Stern, 2000	Psychology	Global environmental problems caused by human activity
Adaptive management	Ruth & Coelho, 2007; Ruth et al., 2007	Ecological economics, industrial ecology	Modeling of non-renewable and renewable resource use, industrial and infrastructure systems analysis, and environmental economics and policy
Framework for analyzing social–ecological systems	Ostrom, 2007, 2009	Political sciences	Analysis and sustainable management of resource systems
Social-ecological research	Becker, 2003; Becker & Jahn, 2006	Sociology, policy science	Theory of society–nature relationships; application to sustainability problems (e.g. water, consumption, mobility)
Social metabolism	Fischer-Kowalski & Haberl, 2007b; Haberl et al., 2004	Industrial ecology, human ecology, sociology	Energy and material flows, land use
Sustainable livelihoods	Dercon, 2004; Ellis, 2000; Robbins, 2004	Development economics	Well-being of a person or household with respect to capital assets, including natural capital
Integrated assessment approach	Pahl-Wostl et al., 2007a; Pahl-Wostl et al., 2007b	Social sciences	Collaborative water management, social learning
Action-oriented			
Great transition	Raskin et al., 2002; Raskin, 2008	Natural sciences	Analyzing alternative scenarios of development, and identifying the strategies, policies, and values for a transition

Table 19.1 (cont.)

Name	Primary literature	Primary disciplinary origin	Focus
Transition management	(Kemp *et al.*, 2007; Rotmans & Loorbach, 2009; Geels, 2004, 2005; Nelson & Winter, 1982; Rip & Kemp, 1998	Policy sciences, social sciences, sociology of technology, evolutionary economics	Conceptualizing (sustainability) transition by the use of visions, transition experiments, cycles of learning and adaptation, fulfilling specific societal functions (e.g. energy supply) based on multilevel approach (technological niche, sociotechnical regime, context/landscape)
Natural step framework	Robert *et al.*, 2002; Ny *et al.*, 2006	(Medical) System theory	Natural system limits, natural cycles

Consequently, besides the HES framework, numerous approaches have been developed worldwide to analyze and transform HES. Some of these approaches, such as the resilience approach (see Chapter 5), emerge from a natural science background. Others, such as the Vienna social metabolism (see Chapters 9 and 12), have strong roots in the social sciences. Still others, such as the natural step framework (Nattrass & Altomare, 2001; see Chapter 11) have roots in (ecological) economics, though they include genuine transdisciplinary elements and could just as well have been introduced in the transdisciplinarity chapter (Chapter 15). Table 19.1 shows that there are at least sixteen approaches advocated within the international research arena.

The approaches shown in Table 19.1 are grouped according to their disciplinary origin, indicating whether they are natural science-accentuated, social science-accentuated or action-oriented. As Ostrom (2009) argues, a common theoretical framework for analyzing HES is highly needed, as without such a framework, isolated knowledge, concepts, and models produced by numerous disciplines accumulate, thus preventing the ability to develop generalizations. Such a general framework can also be regarded as one of the main objectives of environmental literacy. Yet, as becomes obvious when looking at the diversity of approaches listed in Table 19.1, a common framework is still a far-off horizon. Thus, a first important step towards producing generalizations would be to identify the strengths and limitations of the approaches currently proposed within sustainability research and then to combine the strengths of the different approaches. To provide such a comparison is the objective of the following sections.

19.2 A comparison of the HES framework with alternative approaches

This section offers a comparison of the HES framework, as put forth in Chapters 16 and 17, with three alternative frameworks from Table 19.1, namely the resilience approach, the Vienna social metabolism, and transition management. We suggest that each alternative approach is exemplary and partly representative of the three categories specified above (these are natural science-accentuated, social science-accentuated and action-oriented), as it shows the typical characteristics, strengths, and weaknesses of the approaches within the three categories. Our analysis builds on earlier approaches to compare, for instance, the resilience approach and the vulnerability analysis (Gallopin, 2006), the social metabolism approach and transition management (Fischer-Kowalski & Rotmans, 2009), or the land change science and political ecology (Turner & Robbins, 2008). It is important to note that our comparison is focused on the specific characteristics of the HES framework to show its added value in these categories. Certainly, there might be important features of the alternative frameworks that may provide pivotal complementary elements to the HES framework.

To identify the commonalities and the differences between the HES framework, we carry out our comparison by using six features, which refer to the HES Postulates and an interdisciplinarity-related feature (see Table 19.2). Most of these assumptions are discussed in detail in Chapters 16 and 17: (i) the Complementarity Postulate P1; (ii) the different rationales of human systems compared to environmental

Table 19.2 Six specific features to compare the HES framework with alternative approaches to analyze HES.

Feature		Description
(i)	Complementarity Postulate P1	Human and environmental systems are complementary systems
(ii)	Different rationales of system types	Human systems have different rationales than the material environment; human systems on different levels have different drivers and rationales
(iii)	Hierarchy Postulate P2 of human systems	There is a hierarchy of human systems, ranging from the individual to the supranational level, with corresponding environmental systems
(iv)	Triple loop learning (see Feedback Postulate P4)	In human systems we find three types of secondary feedback loop learning
(v)	Decision Postulate P5 (decision-theoretic conceptualization of human systems)	Human systems can be better understood by decision theory
(vi)	Organizing disciplined (i.e. discipline-grounded) interdisciplinarity	A framework should allow one to integrate disciplinary knowledge and the knowledge from natural, social, and engineering sciences appropriately to better understand the nature of major environmental problems

systems (which is derived also from Complementarity Postulate P1 and Hierarchy Postulate P2); (iii) the Hierarchy Postulate P2, assuming different hierarchy levels of human systems; (iv) the triple loop learning in human systems (as illustrated in Figure 16.12* in Chapter 16; see Feedback Postulate P4); (v) the decision-theoretic conception of human systems (see Decision Postulate P5); and (vi) the ability to integrate disciplinary knowledge into the analysis of HES. Our discussion begins with the natural science-accentuated resilience approach, moves next to the social science-accentuated social metabolism approach, and finishes with the action-oriented transition management approach.

19.3 Natural science-accentuated: the resilience approach

The resilience approach, championed by the Resilience Alliance, represents a theory for investigating social–ecological systems (Walker *et al.*, 2006). While this approach is most often used to analyze managed land use systems, which are referred to as social–ecological systems or coupled human and natural systems (Liu *et al.*, 2007a, b), the theory has also been applied to other system types, such as genuinely ecological, economic or social systems (e.g. Gunderson & Holling, 2002).

If we employ the comparison criteria derived in Table 19.2, we arrive at the following picture.

(i) **Complementarity Postulate P1**: as shown in Figure 19.1, the resilience approach considers human systems and environmental systems as complementary and distinguishes, for each system type, between exogenous controls, slow controlling variables and fast variables with an additional category of human actors. The resilience approach assumes that human and environmental systems are closely connected to each other in the form of a co-evolutionary relationship (Berkes *et al.*, 2003; Liu *et al.*, 2007a, b). However, as shown in Figure 19.1, the specific interactions between human systems and environmental systems and the concrete performance and environmental impacts of the human actors are not clearly conceptualized in the resilience approach and remain rather vague.

In contrast, the HES framework clearly conceptualizes the interactions of human and environmental systems by means of the concepts of environmental awareness, action, environmental reaction, and feedback loops. Furthermore, the HES framework advocates a second complementarity between a material–biophysical aspect and a socio-epistemic (including the cultural dimension) aspect of both the human system and the environmental system, which is not taken into account by the resilience approach.

(ii) **Different rationales**: one fundamental proposition of the HES framework is that *human systems* have different rationales to *environmental systems*, and that this insight has to be integrated into models of HES. The resilience approach takes a rather vague stance on this issue. Specifically, the resilience approach acknowledges that human systems are different from environmental systems in that they exhibit the potential for abstraction, reflexivity, anticipation, and technology-building. Westley *et al.* (2002) argue that these aspects of the symbolic construction of meaning should be

19.3 Natural science-accentuated: the resilience approach

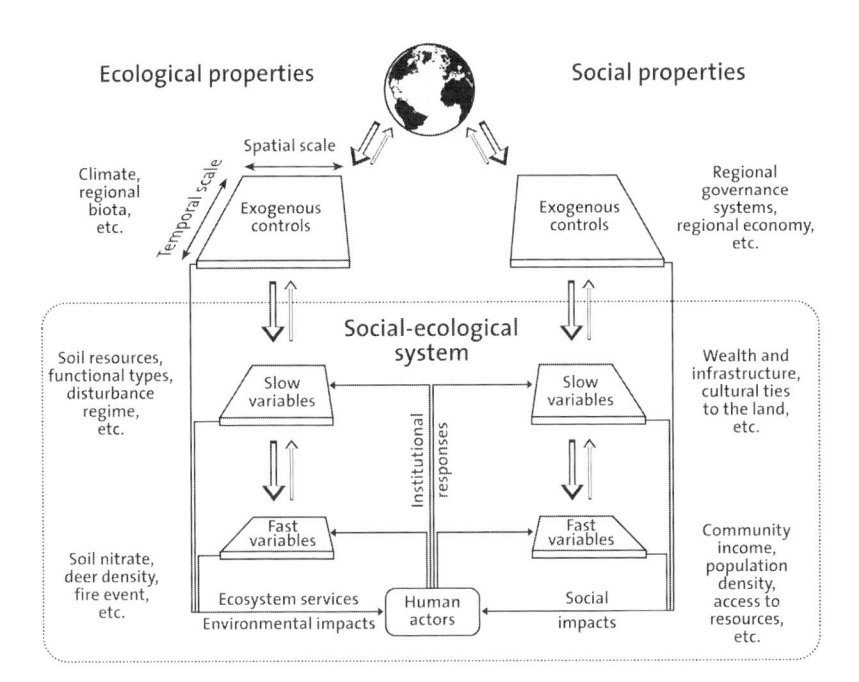

Figure 19.1 The human–environment/social–ecological system as conceptualized by the resilience approach (from Chapin et al., 2009, p. 7).

built into the models of HES to arrive at an accurate picture of reality. On the contrary, the HES approach differentiates between different levels of environmental awareness and self-reflection (see Figure 16.12*), illustrating human systems' capability to evaluate, direct, and change human action.

However, the resilience approach indiscriminately applies central characteristics of resilience theory gained in the ecological domain to social and coupled social–ecological systems. For instance, Allison & Hobbs (2004) examine the Western Australian agricultural region as an example of a large social–ecological system, by means of the adaptive cycle (outlined in Chapter 5; see Figure 5.16). The authors describe the historical changes of the region as two iterations of the adaptive cycle and propose that the Western Australian agricultural region is currently in its back-loop (release and reorganization phase). Similarly, Kinzig et al. (2006) apply the concept of ecological thresholds to social and economic systems. The authors hold that social and economic systems can also exhibit alternative stable states and are thus prone to sudden shifts between these states. In a recent key paper of resilience research, Walker et al. (2006) conclude that "it seems likely that fundamental ideas of scale, relative rate of change and thresholds apply to social and ecological systems, as well as social–ecological systems, although, of course, the specific dynamics may be infinitely varied among systems" (ibid. p. 6). By offering unspecific analyses of the status and rationales of human and environmental systems, the resilience approach leaves many open questions in its wake. In contrast, the HES framework acknowledges the specifics of human systems when utilizing the body of knowledge in the social sciences. When differentiating between different main components of society, that is, the economic, the policy/legal, the social and cultural as well as the scientific and education systems, the power of knowledge from different scientific disciplines such as economics, policy sciences, sociology, anthropology, or epistemology can be used. This allows analysis and assessment of under what constraints societies show what sustainable action and when it does not.

(iii) **Hierarchy of human systems**: in the course of the extension of the resilience approach from the original purely ecological domain (Holling, 1973) to the social and social–ecological domains in the past decade (Chapin et al., 2009; Folke, 2006; Westley et al., 2002), the existence of a hierarchy of human systems has been increasingly recognized. For instance, Chapin et al. (2009) stress the importance of analyzing social–ecological systems at multiple temporal and spatial scales as well as the cross-scale linkages, while Folke et al. (2009) explicitly refer to multilevel governance systems and the interplay between the micro and the macro levels. Such assumptions are key assumptions of the HES framework and the transition management approach (described below). The HES approach takes a system-theoretic, social–anthropological view on the formation of hierarchy levels in the evolution of societies (see Figure 8.5*). The resilience approach, however, lacks proper explication of the hierarchy of human systems that would support a meaningful, action-based analysis of human and environmental systems.

(iv) **Triple loop learning**: both the HES framework and the resilience approach use the term "triple loop learning" for similar but also slightly different phenomena. As illustrated in Figure 16.12* in Chapter 16, the HES framework distinguishes between primary feedback loop learning and two types of secondary feedback learning. With primary feedback loop learning, the human system takes into account short-term changes in the environmental system and adjusts its strategies according to these changes. With one type of secondary feedback loop learning, the human system changes the internal model of human–environment interactions because of awareness of the long-term environmental change. We call this *awareness learning* in the sense of learning to become aware of the environmental feedbacks. The other type of secondary feedback loop learning refers to the genuine reflection of human systems on their goals and action programs. We call this *reflective learning*. This type of learning refers to the reframing of the internal model of human-environment interactions, including the building of new values and norms. Thus, the term triple loop learning in the HES framework is very much referring to the specific type of environmental awareness and reflection of the human. It is important to note that awareness learning can also be delayed so that learning or adaptation takes place in following generations.

In the resilience approach, the term triple loop learning focuses on the environmental awareness in human systems but stresses the type, or depth of reflection, occurring within the human system. In this vein, Folke *et al.* (2009) differentiate between: (a) single-loop learning, which involves changing actions to meet identified management goals; (b) double-loop learning, a reflection process of evaluating underlying assumptions and models; and (c) triple-loop learning, that is the alteration of norms, institutions, and paradigms in ways that would require a fundamental change in governance. A somewhat similar notion of triple-loop learning has also been suggested recently by Pahl-Wostl (2009), pointing to changes in the actions, frames (i.e. assumptions within a value-normative framework) and the context (i.e. values, beliefs or world views). Altogether it becomes clear that the term triple-loop learning is used similarly in the HES framework and the resilience approach, but both approaches require further specification. The resilience approach does not specify which actors on what temporal and spatial scales show learning that causes changes in the framing. On the contrary, the HES framework, when making reference to different disciplines, specifies learning processes for human systems on the level of individuals and also postulates relationships and interferences of learning processes on the micro and macro levels. However, how these interactions look cannot yet satisfactorily be defined as the mechanism of how societal norms emerge, for instance, is not yet sufficiently explored.

(v) **Decision-theoretic conceptualization of human systems**: to the best of our knowledge, there are no or few references to decision theory within the resilience approach, whereas this is a fundamental part of the HES framework (see Decision Postulate P5). This is probably because, as a natural science-oriented approach, resilience theory does not incorporate decision theory. Because the HES framework strongly incorporates social science in addition to engineering and natural sciences, decision theory is a central part of the framework. This accesses decision processes and changing environmental behavior of human systems.

(vi) **Organizing disciplined (discipline-grounded) interdisciplinarity**: whether or not a common framework of HES is useful depends on, among other factors, the degree to which it allows the integration of disciplinary knowledge from the natural, social, and engineering sciences. The resilience approach has been extended in the past decade from ecological systems to social, economic, and social–ecological systems (Brand & Jax, 2007). In the course of this extension, numerous insights from the social sciences have been integrated into the framework (Folke, 2006). However, the resilience approach does not go as far as the HES framework in explicitly specifying at which points in the analysis of an HES specific knowledge is needed from the natural sciences (Postulate 7 states environmental analysis comes first) and the social and engineering sciences (Postulates 2–6).

19.3.1 Key messages

- The resilience approach and the HES framework share several features, such as the hierarchy proposition and feedback loop learning. Yet the

resilience approach lacks a clear-cut conceptualization of human systems and of human–environment interactions
- The HES framework provides added value for analyzing HES, with respect to: (i) the specification of the different rationales of human systems by a decision-theoretic approach; (ii) the differentiation of the rationales of human systems (which, for example, can show reflexive learning) and the rationales of environmental systems; and (iii) the specification of conflicts between human systems with the same and between hierarchy levels with respect to environmental resources.

19.4 Social science-accentuated: social metabolism

The social metabolism approach is advocated by the Institute for Social Ecology in Vienna, Austria, and we have delineated this approach in Chapter 9 on sociology and Chapter 12 on industrial ecology. Social metabolism is a theoretical approach for tracking flows of material and energy on a given spatial and temporal scale (Fischer-Kowalski & Haberl, 2007a; Haberl et al., 2004). This theory is based on Boyden's human society and the biosphere model, described in Chapter 12, which proposes that nature and human society are in a process of biometabolism and technometabolism. An important achievement of the Vienna approach is that it clearly shows that an environmental Kuznets curve-like improvement or decrease in material and energy flows does not take place in current industrial societies (see Chapter 13.6), but rather that we face a treadmill of production-like situations (see Chapter 9.7), which in turn requires the global management of material and energy cycles. If we again apply the specific criteria of the HES framework derived in Table 19.2, the following can be suggested:

(i) **Complementarity Postulate**: as illustrated in Figure 19.2, the Vienna approach describes two spheres of causation: (a) a natural sphere, which is comprised of natural resources and can become colonized or appropriated by human systems labor; and (b) a cultural sphere, which is conceptualized as a hybrid, comprising an autopoietic communication system (based on Luhmann's system theory) and material elements. The overlap of the two domains is represented by the biophysical structures of society, which consist of specific modes of subsistence (i.e. hunter–gatherer, agrarian or industrial society) and are characterized by typical amounts of material and energy flows as well as typical sustainability problems (Fischer-Kowalski & Rotmans, 2009; Haberl et al., 2004). In sum, the Vienna approach takes into account a complementarity of human and natural systems, as well as a complementarity of a material–biophysical and a social–epistemic dimension of human systems (Fischer-Kowalski & Haberl, 2007a, p. 11ff), but does not acknowledge the latter complementarity within environmental systems. Let us take a closer look at the nature of these complementarities.

This can be taken from Figure 19.2, which refers to the two essential complementarities in HES that were introduced in Figure 3.3*, that is, the human–environment complementarity and the material–social (body–mind) complementarity. The essential difference between the two approaches is that the Vienna

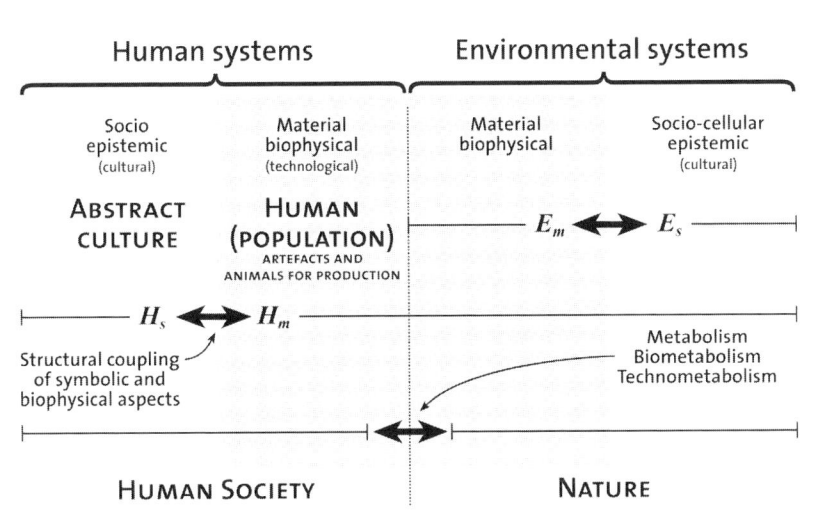

Figure 19.2 Complementarities in the HES approach and the Vienna social metabolism approach (from Haberl et al., 2004, p. 204). All terms used from Figure 12.5 of the Vienna circle are CAPITALIZED. The terms material–biophysical (E_m) and socio-epistemic (E_s) environment, as well as material–biophysical (H_m) and sociocultural (H_s) human systems are explained in Figure 3.3*.

approach makes no assumptions about the capabilities of non-human organisms to have a representation of the environment that is based on "information" (which is a genuine non-material entity). As we discussed in Chapter 4 on biology, even the immune system, that of non-human mammals, can be viewed as a cognitive system and as being able "to read the signs of nature." This has been a basic assumption of Jakob von Uexküll (von Uexküll, 1931, 1940/1970) or John Maynard Smith (Maynard Smith, 2000). What the information cells, for instance, acquire and represent about the environment depends on the communication among cells. This socio-epistemic aspect of nature is not taken into account by social metabolism theory.

Two definitions are essential here. One is the definition of human systems as all activities of human individuals, which can be assigned to this system. The other comes from a definition of human systems, from the individual up to the society, based on the cells of human individuals and their interactivities (which may construct information and symbolic meaning). A human individual is defined as all cells and their (inter) activities, which emerge from the fertilized ovum making this individual. This definition allows definition of both complementarities on all levels of human systems. On the contrary, the material–biophysical environment is managed by humans such that animals for production (see Figure 12.6) are viewed as a part of human society. Here the concept of metabolism as the exchange of material between the object of causal causation and natural causation is central. As reflected in the concept of the anthropocenic redefinition of the environment, the HES approach does not follow this differentiation. In addition, the conception of the natural sphere in the Vienna approach does not specify the biotic and ecological dimension of the environmental system. As we will point out, an essential difference is that the Vienna approach is defined from a societal perspective.

(ii) **Different rationales**: the social metabolism approach differentiates between the causation of natural systems and the causation of social systems. The overlapping of these rationales is the material side of the social system (see Figure 12.6). This material side of the social system (which, contrary to the definition of the HES approach, includes animals for production) causes on one side metabolism with the non-social material world. On the other side it interacts with culture as the abstract, symbolic layer of social systems (Fischer-Kowalski & Haberl, 2007a; Haberl et al., 2004). This sounds similar to the complementarities suggested by the HES framework but it is not. This is for two reasons. First, the hierarchy of human systems is not taken into account (see below) and the specification of the different rationales remains unstratified. The distinction here refers to the differences between rationales and drivers of the natural and the human domain, but not does distinguish between "artifacts," which include buildings, roads, farm facilities, and the natural environment since this cannot be well defined in a material–biophysical world that is increasingly appropriated by the human species (Ingold, 1986). This is why the HES approach restricts natural causation to the material–biophysical approach. The second difference is that the Vienna social metabolism approach does not differentiate between different types of causation on the level of social (i.e. human) systems.

(iii)–(v) **Hierarchy of human systems, triple-loop learning, and decision-theoretic conceptualization of human systems**: each of these characteristic features of the HES framework is not taken into account, as the social metabolism approach does not aim to propose any options for intervention (Fischer-Kowalski & Rotmans, 2009) and strongly focuses on societal material and energy flows only. For instance, social metabolism theory reduces the hierarchy of human systems (individual, group, society, etc.) to "the society." Further, as interventions of human systems play no role at all in the social metabolism approach, learning processes as well as decision theory are neither acknowledged nor needed.

(vi) **Organizing disciplined (discipline-grounded) interdisciplinarity**: the social metabolism approach strongly suggests that it is important to bridge the natural and the social sciences and to establish a link between socioeconomic variables and biophysical patterns and processes (Haberl et al., 2004). Certainly, when developing the theory of social metabolism, deep insights into human systems from the perspective of, above all, the social sciences are generated. Yet a systematic approach to integrate knowledge and methods from the natural, social, and engineering sciences is lacking.

19.4.1 Key messages

- The Vienna social metabolism approach defines two types of complementarity and different rationales of human and environmental systems. The complementarities are essentially differently defined than those in the HES framework as the built environment, animal

husbandry, etc. are seen as elements of the social system in the Vienna approach
- The Vienna approach does not aim to facilitate interventions by human systems. It does not include a concept of a hierarchy of human systems (having different rationales), a decision-theoretic conceptualization of human systems, or a concept of learning for human systems
- As the built environment, animal husbandry, etc. is part of the social system, this causes difficulties in crisply defining human–environment interactions. In addition, the Vienna approach does not include an elaborated notion of abiotic and biotic environmental systems and their rationales.

19.5 Action-oriented: transition management

19.5.1 The idea of the transition management approach

The transition management approach has mainly been developed by Dutch scientists and ministries (Loorbach & Rotmans, 2010; Rotmans & Loorbach, 2008, 2009). There are currently three interrelated research lines, focusing on societal transitions by transition experiments (Loorbach & Rotmans, 2010; Rotmans *et al.*, 2001; Rotmans & Loorbach, 2008, 2009), conceptualizing of (socio-)technical transitions (Geels, 2004, 2005; Schot & Geels, 2008), and the modeling of sociotechnical transitions (Bergman *et al.*, 2008; Haxeltine *et al.*, 2008). In this research program, a transition is understood as "a structural change in a societal (sub)system that is the result of a co-evolution of economic, cultural, technological, ecological and institutional developments at different scale levels" (Rotmans & Loorbach, 2008, p. 15), and transition management as "a new management concept that assumes complexity and uncertainty and is also known as co-evolutionary management: adjust, adapt and influence" (Rotmans & Loorbach, 2008, p. 16). Within the transition management approach, niches, which are individual technologies and actors outside or peripheral to the system, are strongly emphasized as they provide the loci for radical innovations (Bergman *et al.*, 2008; Rotmans *et al.*, 2001).

Transition management has strong roots in systems theory. Rotmans and Loorbach (2008) hold that complex adaptive systems can be defined by three key characteristics: co-evolution, emergence, and self-organization. As these complex systems exhibit a "dynamic equilibrium" (ibid., p. 17), they are prone to potentially sudden, qualitative shifts between different attractors. This notion meets the critical transition view presented in Chapter 14.4. More specifically, the transition management approach focuses on the management of structural change, which goes beyond ordering or assimilation. Here, management is defined as "influencing the process of change of a complex, adaptive system from one state to another" (Rotmans & Loorbach, 2008, p. 19).

A multi-level perspective is taken for technology transitions. Here the meso level, called regimes, is held between a macro level called landscape and a micro level called niches. The meso level is a cluster of system elements, including "actors, artifacts, institutions, rules and norms." The macro level "is described by changes in the macro economy, politics, population dynamics, natural environment, culture and worldview." The micro level variations result from innovations of "technologies and social practices" (van der Brugge & Rotmans, 2008, p. 67–8). As can be taken from this brief description, the approach has quite a metaphorical character, including a set of mixed items.

Regarding the organization of human systems, Rotmans and Loorbach (2008) follow a network approach, as opposed to a level hierarchy approach. Rotmans and Loorbach hold that "societal actors create formal and informal networks, because they have the same vested interests, and they are striving towards the same objectives which they can better achieve jointly than individually" (Rotmans & Loorbach, 2008, p. 21). The central method proposed by the transition management approach is the *cycle of transition management*, which consists of four components (e.g. Loorbach & Rotmans, 2010): (i) problem structuring, establishment of the transition arena, and envisioning; (ii) developing coalitions and transition agendas; (iii) mobilizing actors and executing projects and experiments; and (iv) evaluating, monitoring, and learning. This approach takes a rather instrumental perspective on transitions (e.g. by mobilizing actors).

A critical element of this cycle is transition experiments, which are defined as "practical experiments with a high level of risk (in terms of failure) that can make a potentially large contribution to a transition process" (Rotmans & Loorbach, 2008, p. 32). These experiments have been launched mainly by the Belgian and Dutch ministries, and are seen as necessary because of the complexity, uncertainty, and unpredictability of the outcomes. Thus, contrary to the HES framework, a thorough science-based evaluation of the environment is widely lacking. Further, the fact that solar energy is not liked by some Dutch industries, for instance, is not regarded as an obstacle for the transition of the Dutch society towards sustainable energy use, but is rather interpreted as a component of the network dynamics. These ideas of governance and social transitions are more in line with the new institutional economics, described in Chapter 11.7, and the idea and approach of "ecological modernization," critically discussed in Chapter 9.7. The latter approach believes in a societal self-organization by interactions between markets, state, and civic society.

19.5.2 The context of the transition management approach

According to the network metaphor, "transition management forms a tangled ball with no clear management structure …" (Rotmans & Kemp, 2008, p. 1007), this statement is misleading given that just three lines later one can read:

> So far all transition trajectories in the Netherlands and Belgium [which are the places where transition management takes place] operate under the flag of the government … which has adopted as an official policy line, linked through the Fourth National Environmental Plan. This means that transition policy is authorized by the Dutch Parliament and that the transition process is accountable to the Dutch Parliament (ibid., p. 1007–8).

The latter statement, which would induce a completely different interpretation based on the Hierarchy Postulate P2, is no surprise as the 3495 million Euros that the Dutch Ministry of Health, Welfare and Sports invested in 2007, was seen as money for ten transition experiments (Loorbach & Rotmans, 2010, p. 5).

It is clear that transition management is bound to the sociopolitical settings of the Netherlands and cannot really be thought of in countries such as China, Russia or Columbia. For instance, the interpretation of the transition processes has been criticized by Shove and Walker (2007, 2008) in that the transition management approach is used to "obscure their own politics, smoothing over conflict and inequality" (Shove & Walker, 2007, p. 768). Given the discussion of Chapter 15 on the role of science in transition processes, it is particularly problematic that the role of scientists in transition processes is not clearly defined in the transition management approach. It is unclear if and when scientists work as analysts, facilitators, or consultants, and what constitutes the difference. The latter appears on another model, the triple helix approach (Etzkowitz & Leydesdorff, 2000), which emphasizes that in the Dutch cultural setting the functions of top scientists are often unclear since high caliber professors rotate between science, industry, and policy positions.

19.5.3 Comparing along criteria

(i) **Complementarity Postulate P1**: because of its emphasis on developing governance strategies, the transition management approach almost exclusively considers the human system (Rotmans & Loorbach, 2008), but does not incorporate a definition of the environment. Therefore, there is no specification of complementarities either within or between human and environmental systems. Environmental issues are called structures of societal systems and are seen, besides culture and practices, as components of society.

(ii) **Different rationales**: one fundamental proposition of the HES framework is that human systems have different rationales to environmental systems, and that this insight has to be integrated into models of HES. The resilience approach takes a rather vague stance on this issue. Specifically, the resilience approach acknowledges that human systems are different from environmental systems in that they exhibit the potential for abstraction, reflexivity, anticipation, and technology-building. Westley et al. (2002) argue that these aspects of the symbolic construction of meaning should be built into the models of HES to arrive at an accurate picture of reality. On the contrary, the HES approach differentiates between different levels of environmental awareness and self-reflection (see Figure 16.12*), illustrating human systems' capability to evaluate, direct, and change human action.

However, the resilience approach indiscriminately applies central characteristics of resilience theory gained in the ecological domain to social and coupled social–ecological systems. For instance, Allison & Hobbs (2004) examine the Western Australian agricultural region as an example of a large social–ecological system, by means of the adaptive cycle (outlines in Chapter 5; see Figure 5.16). The authors describe the historical changes of the region as two iterations of the adaptive cycle and propose that the Western Australian agricultural region is currently in its back-loop (release and reorganization phase). Similarly, Kinzig *et al.* (2006) apply the concept of ecological thresholds to social and economic systems. The authors hold that social and economic systems can also exhibit alternative stable states and are thus prone to sudden shifts between these states. In a recent key paper of resilience research, Walker *et al.* (2006) conclude that "it seems likely that fundamental ideas of scale, relative rate of change and thresholds apply to social and ecological systems, as well as social–ecological systems, although, of course, the specific dynamics may be infinitely varied among systems" (ibid. p. 6). By offering unspecific analyses of the status and rationales of human and environmental systems, the resilience approach leaves many open questions in its wake. In contrast, the HES framework acknowledges the specifics of human systems when utilizing the body of knowledge in the social sciences. When differentiating between different main components of society, that is, the economic, the policy/legal, the social and cultural as well as the scientific and education systems, the power of knowledge from different scientific disciplines such as economics, policy sciences, sociology, anthropology, or epistemology can be used. This allows analysis and assessment of under what constraints societies show what sustainable action and when it does not.

(iii) **Hierarchy of human systems**: the transition management approach considers the society as "a complex network society" and the proponents state that these "networks do not have a clear hierarchical structure" (Rotmans & Loorbach, 2008, p. 21). Governments, for instance, are considered to work more and more interactively. However, a kind of level hierarchy (see Chapter 16) is introduced via spatial, temporal, and functional scales (Bergman *et al.*, 2008), and refers to a multitude of system actors at different scale levels (Rotmans *et al.*, 2001). Yet the important point here is that the transition management approach lacks a clear and analytical conception of: (i) the hierarchy of human systems; (ii) the role and drivers of different human actors; and (iii) the possibly conflicting interferences occurring between the different human systems on different levels (cf. also Shove & Walker, 2007).

(iv) **Triple-loop learning**: in transition management, system thinking is one of the key characteristics, because "unintended side effects and adverse boomerang effects can only be recognized at the system level" (Rotmans & Loorbach, 2009, p. 188). Thus, transition management acknowledges the crucial role of feedback loops. It also emphasizes the importance of learning in the process of induced change towards sustainability (Rotmans *et al.*, 2001) and differentiates between single-loop learning and double-loop learning (Argyris & Schön, 1978; van de Kerkhof & Wieczorek, 2005). We notice, however, that transition management interprets triple-loop learning as a changing of structures. This is also the case in the HES framework if triple-loop learning, which is explained in Figure 16.12*, is considered as self-reflexive learning by society. Triple-loop learning would also result in a reorganization of the values, social rules, and regulatory mechanisms of society.

(v) **Decision-theoretic conceptualization of human systems**: the transition management approach does not explicitly refer to a decision-theoretic conception with goals, drivers, and utility functions. According to the representatives, objectives are derived from an overarching, long-term vision, which functions as a frame for the whole transition process (Loorbach & Rotmans, 2006; Rotmans *et al.*, 2001). This also holds true for the transition arenas that are built on a vision of sustainable development. Clearly, the object of the transition management approach is transition processes, yet it is not the decisions of the participants that are of interest but rather the lessons learned from "transition experiments" (Rotmans & Loorbach, 2008, p. 29). These transition experiments, such as those in Dutch healthcare, were large-scale national programs. Presumably they were called transition arenas because transition managers, for instance from Ernst & Young, were coaching and evaluating the new ministerial program.

(vi) **Organizing disciplined (discipline-based) interdisciplinarity**: the "transition management approach is rooted in complexity theory …, post-normal science …, social theory …, evolutionary economics …, innovation studies …, technological transitions …, and integrated assessment" (van der Brugge & Rotmans, 2008, p. 67–8). How interdisciplinarity can be achieved and disciplinary standards be applied (for instance, for explaining decisions of actors of the transition arena) is not specified. Further, the role of scientists in transition processes is not specified. They can act as analysts, transition managers, or consultants for the initiators of transition processes (e.g. ministries).

19.5.4 Key messages

- The HES framework and the transition management approach differ in several fundamental

aspects. Transition management does not introduce a complementarity between human and environmental systems, does not refer to a hierarchy and a decision theory of human systems, but follows a network approach to human systems, and thus does not differentiate between different, potentially conflicting human systems following different rationales
- The transition management approach tends towards an anti-differentiationist approach. The role of scientists is not well defined
- Both the transition management approach and the HES approach require specific sociocultural and political constraints.

19.6 Commonalities and differences among the approaches

The HES framework and the three alternative frameworks, which are the resilience approach, the Vienna social metabolism theory, and the transition management approach, have been developed against different backgrounds. Though all approaches strongly relate to general systems theory, the approaches still differ in several dimensions.

The type of object: both the resilience approach and the HES framework explore the dynamics of complex coupled HES, including critical, qualitative system change. Yet the resilience approach concentrates on human–"nature" relations or coupled land use systems, such as managed coral reefs, lakes or savannahs (e.g. Gunderson & Pritchard, 2002). In contrast, the transition management approach focuses on radical sociotechnical or societal change in societal subsystems, while the Vienna social metabolism approach likewise investigates societal transitions, but mostly in the long run and with a specific focus on material and energy flows occurring between the society and complementary environmental systems on a macro scale, for example, regional, national, and world systems.

The type of relation between human systems and environmental systems: the resilience approach and the HES framework utilize a transactionalist or co-evolutionary complementarity perspective. The approaches originate in a biological, cell-based definition of human systems in the case of the HES framework (cf. Chapter 3) and in a holistic system comprehension in the case of the resilience approach (Kirchhoff et al., 2010). This is less clear in the Vienna social metabolism and the transition management approach. The Vienna social metabolism approach refers to a concept of society that includes the built environment, the materialized technologies, husbandry, and a complementary nature. As, besides economics, there are only a few references to social sciences and therefore ecological, resilience, and vulnerability theories build the basics, the approach is reminiscent of the beginnings of human ecology (see Chapter 8). The transition management approach almost exclusively focuses on human systems, and a clear concept of the environment is widely lacking. Thus, there is also no conceptualization of the complementarity between human and environmental systems.

The type of system model: though all models refer to systems theory, their architecture and modeling of the environment differs. Both the HES approach and the resilience approach champion a hierarchically organized coupled HES model following upward and downward causation. Yet the different rationales we find in human and environmental systems and the possibly conflicting interferences between the different levels in the hierarchy of human systems are not made as explicit in the resilience approach as in the HES framework. The resilience approach assumes a tight bond of human systems to natural systems, which is in line with the holistic system model first developed in ecological science (Kirchhoff et al., 2010). In contrast, the Vienna social metabolism uses a flow model of material and energy without specifying any hierarchy of human or environmental systems. The transition management approach, in turn, follows simulation logic, and assumes that it is possible to construct systems according to the desires of the network of human users and to the overall conception of sustainability.

The type of transition model: environmental literacy includes the knowledge of how human systems adapt. Thus, transition management or transdisciplinarity, which builds capacity for adaptation, is essential (see Chapter 15). "Transdisciplinary research takes up concrete problems of society and works out solutions through cooperation between actors and scientists" (Häberli et al., 2001, p. 6). Transdisciplinarity organizes processes that link scientific epistemics with real-world-based experiential knowledge from stakeholders outside academia. Scientists and decision-makers leave their primary domain (i.e. research vs.

decision process) and establish a joint transdisciplinary process (cf. Figure 15.1 in Chapter 15). We advocate a differentiationist perspective and sharply distinguish between experience-based epistemics of practical experts and codified knowledge generated in the scientific domain. Scientists have a clear role.

One objective in constructing the HES framework is to design and to reflect on transdisciplinary processes (see Case 2, Chapter 18). This becomes possible as it includes an actor-based approach by clearly conceptualizing the drivers of different human systems (P4) on different hierarchy levels (P2), their environmental awareness (P6), and conflicts among actors on the different hierarchy levels as well as the conflicts between the micro and macro level (P3). In contrast, an actor-based conceptualization of transition management is widely lacking in the alternative approaches. For example, in the transition management approach, the role of the scientist is not clearly defined, as they can act as analysts, transition managers, or consultants for the initiators of transition processes (e.g. ministries). Also, as a consequence of the proposed network approach in the transition management, the power game fought out between different human systems is widely ignored (cf. also Shove & Walker, 2007). The Vienna social metabolism approach does not include any recommendations or procedures for intervention and thus does not offer a transition management concept (Fischer-Kowalski & Rotmans, 2009).

The resilience approach champions a concept for managing transitions, termed adaptive co-management. Adaptive co-management is defined as a process by which institutional arrangements and ecological knowledge are tested and revised in a dynamic, ongoing, self-organized process of learning-by-doing. The aim of adaptive co-management is to examine ways of building resilience to enhance the capacity to deal with change and surprise (Berkes *et al.*, 2003; Folke *et al.*, 2002). Hereby, the tool to enhance the transition process is dubbed co-management and conceptualized vaguely as some kind of partnership between public and private actors concerned with natural resource management (Carlsson & Berkes, 2005). Again, the specification of the concrete role of scientists is rather weak and the power games occurring between the different human systems are widely ignored.

Key messages

- The HES, resilience, and Vienna metabolism approaches deal with coupled, co-evolutionary human systems. The transition management framework does not explicate this
- The transition management and the resilience approach lack an internal structure that allows for referencing to social sciences. The Vienna metabolism approach refers primarily to macro sociology, whereas the HES framework allows for systematic referral to social sciences on all scales
- Only the transition management approach and the HES framework, which are designed to be linked to and used in transdisciplinary processes, are theories that can be utilized for organizing sustainable transitions.

Part IX

Perspectives for environmental literacy

Part IX Perspectives for environmental literacy

Chapter 20

New horizons: environmental and sustainability sciences

20.1 Travel log: learning from academic disciplines   525
20.2 Provisions for an integrated approach   528
20.3 Prospects for environmental literacy and sustainability sciences   528
20.4 Two visions   533

Writing about the challenges of and possibilities for environmental literacy has been a journey of discovery and this book chronicles the learning we experienced along the way.

Our first lessons came from the academic disciplines of biology, psychology, sociology, economics, and industrial ecology. In this chapter we review what these disciplines offer and where they, as individual disciplines, fall short in supporting environmental literacy. Given that these disciplinary realms, on their own, provide an incomplete picture about what is needed to cope with human and environmental systems (HES), we explored what it means to offer meaningful insights and solutions to environmental issues and problems. A major step was that we took a different view on the environment, which we called the *anthropocenically redefined environment*.

At the very least, ecological, climate, and Earth surface processes cannot be understood without the human factor. This new view on the environment asked us to profoundly relate human and environmental systems. To meet this challenge, we turned to integrative modeling and transdisciplinarity. These new tools are integral parts of a new approach and methodology to environmental problem-solving.

This new methodology, called the HES framework, provides a guide for going beyond the disciplinary route to develop a sustainable coupling of HES. In this chapter we review how the HES framework helps us not only to model the ways that environmental systems are anthropocenically shaped, but also how they are inextricably coupled with human systems and in what ways HES may be scientifically investigated. A prerequisite for this work involves drawing from the disciplines in a "disciplined interdisciplinary" manner. We also review how transdisciplinarity is a means of integrating scientific knowledge about HES into development of strategies for sustainable action.

Environmental science has taken on the challenge of explaining how human and environmental systems interact. We propose that environmental science as well as sustainability science should develop a subdiscipline called HES. HES as a scientific field includes transdisciplinarity as a new mode of doing science. Transdisciplinarity processes are a means to support societal capacity-building for – if we take a long-term perspective – building up control systems for keeping the Earth system within limits to sustain human evolution. Sustainability science goes beyond analyzing under what constraints the Earth system may remain within limits for providing the matrix of ecosystem services and a reliable system to support human life. It also incorporates the dimension of social sustainability from an intragenerational and intergenerational perspective. This goes beyond the material–biophysical environment and opens the even more challenging dimension of social justice or social sustainability.

20.1 Travel log: learning from academic disciplines

Our journey started with a clear roadmap and travel guide. The purpose of the journey was to explore and even to experience environmental literacy. Environmental literacy is a specific art, intellectual faculty, and behavioral competence that developed in the course of evolution and history to support human adaptation to environmental settings.

One may view environmental literacy as a specific edge of human culture. We learned that the concept of environment is historically quite recent. It was invented and conceptualized by Herbert Spencer as a "total environment of organisms" and thus as the matrix of life just a little under two centuries ago. We had to realize that it is not easy to provide a general definition of the environment that can be consistently utilized for all human systems.

A general definition of environment became possible when viewing an individual human being as all living cells that emerged from the zygote and their (inter-)actions. The definition of aggregated human systems above the level of the individual, e.g. of a group, organization or society, is based on the activities of the individuals which can be assigned to this system. The environment of a human systems is a subset within the complementary set to this human system, referred to as the universe. When differentiating between the material–biophysiological and the social-epistemic, information processing aspects of the individual human's cellular activities, we were able to provide a definition of the body–mind complementarity. This definition could also be applied for aggregated systems such as groups or organizations.

Equipped with this, we started a journey through the history of mind of selected natural and social science disciplines. We made a short stop in a branch of engineering sciences called industrial ecology, as environmental literacy sometimes also requires engineered actions.

The first thing that we can report after this journey is that our simple but perhaps uncommon material–social dual definition of "environment" enabled us to work through and to structure the essential information that we could draw from all disciplines successfully. It allowed, in particular, consistent conceptualization and identification of the roles that human systems take on when interacting with their material–biophysical environment, for instance, chemical agents, pathogens, or nuclear radiation. This definition of the environment also gave us insight into the knowledge, traditions, or sociocultural rules that human systems have generated through interactions with their social environment.

20.1.1 What the disciplines offer

In the first part of the journey, we visited five disciplinary realms. The first was basic. In biology we learned how knowledge about the biotic nature of human and environmental systems proliferated, in micro and macro scales. Highlights were descriptions of how some systems are subject to homeostatic set-point controls which target steady states, while others are not. These others rather follow trajectories that depend on the environmental setting. Because ecosystems, unlike individual organisms, do not have built-in homeostatic responses to disturbances, humans need to step in to avoid critical transitions. The anthropocenically shaped biogeochemical cycles of Earth systems seem to necessitate a kind of human control system. This is particularly important as most ecosystems of the world have become human-managed. This trend of human systems to increasingly appropriate nature on all levels of scale makes it progressively impossible to exclude the human impact or factor when analyzing almost any material–biophysical environment. A significant lesson learned was that managing critical rebound effects resulting from human intrusion has become a new key task of humankind.

We also realized that change and stability are ubiquitous. Dynamic steady-state periods are often followed by "critical" system changes, which are then pursued by reorganization and restructuring of the system. We could also see that something such as a body–mind complementarity, if properly defined, can be assumed on the level of cell systems. Cells communicate by material matter, just as higher level human systems do. Cell systems, such as the immune system, seem to be cognitive systems that are able to process something that we call information or signs.

The next two stages of our journey, psychology and sociology, taught us what cognitive, epistemic, motivational, social, and cultural technological (that is, technological knowledge) factors enable individuals, groups, or societies to take the environment into account. We observed striking communalities in these realms. For instance, individuals and societies have to equilibrate by adapting. Two qualitatively different processes can achieve adaptation. One, assimilation, means that human systems utilize capabilities that are at their disposal either more efficiently or in domains in which they have not been used before. The other, accommodation, means that human systems develop new capabilities by restructuring themselves or by developing a new strategy for interacting with the environment. For instance, new technologies, such as the railways as a means of long distance traveling or freight transport or digital information processing on computers, induced sophomoric societal transitions which fundamentally changed societal settings. In these transitions, the technological trajectories and the societal trajectories interact and are essentially related to the available natural resources. In psychology and sociology we were also able to identify the main drivers of individual and societal actions. For individuals, there are sets of physiological and cognitive drivers at work which resemble what we call the hardware and software of a system. For societies, wealth and maintenance of power seem to be the principal drivers.

However, we also looked into the historic dynamics of societies. The complexity and hierarchy structure of societies and what constitutes a society is continuously changing and the hierarchy of social systems is evolving. Currently, nation-states are the prevailing form of society in the world regime. But we also met some sociologists such as Ulrich Beck, who point out the necessity of establishing suprasocietal or supranational systems that will supplement the current hierarchy of human systems.

Economics is the realm of producing, selling, buying, and consuming goods. We can see that, in the past, issues that were not considered to be in the economic realm, including public goods such as water or landscape scenery, can be given a market value. But we can also see that natural resources and assumptions about their availability play quite a different role in the theories of different economists. Here, two issues were astonishing. One is that a vast majority of economists obviously believe in unlimited natural resources, which may be substituted by other material or immaterial resources if they run short. Only a minority disagree, and underscore that there are limits with some essential materials or elements such as phosphorus, whose potential scarcity might cause critical transitions in the long run. The other issue was that macroeconomic theory almost exclusively focuses on *efficient* economic action. This is irritating, as efficiency seems to be a driver of development and evolution. Logically, efficiency is neither a sufficient nor necessary prerequisite of sustainable development. Merely focusing on efficiency ignores the limits of systems. We can easily over-demand and destroy a system by utilizing its resources by highly efficient but oversized demands.

The last stop was at a small, but rapidly evolving, branch of engineering sciences called industrial ecology. The quantitative engineering culture became visible and revealed that the majority of world material fluxes are manmade. Here, it again became evident that the human impact on the environment qualitatively differs from that of other species. Despite extraordinary technological progress and increasing eco-efficiency, material flows and adverse environmental impacts continue to increase. Humans essentially shape the world's major biogeochemical cycles. The excursion into industrial ecology opened an exciting window to the future of environmental literacy. Environmental literacy will analyze and evaluate the systemic change of new technologies and not only the environmental impacts of the new products and processes under ceteris paribus assumptions, i.e. when assuming that all things remain unchanged.

20.1.2 What the disciplines do not address

If we look back on the first part of the journey we may conclude that we could not form a sufficiently comprehensive picture about HES. In biology, we learned that humans have long described ecological processes as being separate from social processes. Moreover, unfortunately those ecological research approaches that acknowledge that human activity is a key driver of ecosystems, such as socioecological approach, do not describe the drivers of human systems. Thus, the way that biologists describe interaction between human and environmental systems is opaque.

In psychology, both the social environment (social psychology) and the material environment (environmental psychology) appeared late and the latter could be found at the fringes. In approaches where the interactions or even the reciprocal impacts between human interactions were dealt with, we could just inspect scaffold-like buildings that were missing a developed methodological repertoire and the incorporation of the material environment. The core domain of psychology provided an in-vitro impression and that complex real world-settings were missing, obviously for methodological reasons.

In sociology, a prominent researcher such as Niklas Luhmann stated that sociology has not, thus far, been prepared to incorporate environmental issues. We are missing a strong environmental sociology and likewise we lack a bigger group of approaches that systematically dealt with the adaptive processes needed in world societies owing to natural and manmade environmental change as well as the striking, seemingly increasing, disparity of wealth in world society. Besides Ulrich Beck's approach, which justified this critical disparity by the application of the achievement principle, this issue was not in the foreground in the leading approaches.

The same holds true for economic theory. We expected that issues of social justice, as well as environmental issues relevant for societal economic precautionary management or intergenerational justice, would complement the dominating efficiency discussion among economists. But this was rarely the case. Long-term external costs or irrecoverable–irreversible processes of ecosystems are not key factors of economics.

Industrial ecology brought in new views on better managing firm and farm activities. We also identified that from a world perspective, management of the global biogeochemical cycles is missing. A study of the rationale of human industrial systems is not yet sufficiently integrated in industrial ecology.

20.2 Provisions for an integrated approach

The selected disciplines provided insight into how components of human and environmental systems work. But they did not comprehensively describe how human and environmental systems *co-develop* and how human systems of different scale *interact*. Thus we supplemented our equipment, conceptually and methodologically. The key conceptual elements were the introduction of a *hierarchy* of human systems (see Figure 14.1*) and a redefinition of the environment. This redefinition acknowledges the strong impact that human systems have had on ecological systems and biogeochemical cycles, at least since agrarian societies existed. This warrants an anthropocenic redefinition of the environment as a coevolving system which is inextricably coupled with human systems activities. The ongoing appropriation of nature on an extraordinarily wide range of scales is a major driver for redefining environmental systems with an anthropocenic frame. Then, owing to the segregated disciplinary approach we looked for integrative analytic "auxiliary tools" which could help to better scientifically conceptualize the complexity and the coupling of HES. This was, in a first step, done by equipping ourselves with the tools of integrative modeling.

The second fundamental ingredient missing in the disciplines was the notion that environmental literacy in its advanced form includes a proactive component encompassing "the forming of a sustainable coupling of human environment systems" (see p. 15). This

required the ability to link scientific knowledge appropriately with the reality of human decision-making by stakeholders and decision-makers on all levels of society. The "toolbox" for this was *transdisciplinarity*, which generated socially robust solutions and serves for capacity- and consensus-building, mediation, and legitimization when utilizing disciplined interdisciplinary knowledge in society.

20.3 Prospects for environmental literacy and sustainability science

A key objective of this book was to explore what environmental knowledge societies, or "we," need to cope with the current and forthcoming environmental problems. In this final chapter, we summarize how the HES framework, linked with "disciplined" interdisciplinarity and transdisciplinarity, may offer a comprehensive approach to support the development of a sustainable coupling of HES.

20.3.1 The HES framework as a guide

The socio-epistemic layer of human systems is essential and seems to be more complex than that of natural systems. It further appears that there is less predetermination as to what drivers, rules, and programs are at work in a certain socioepistemic situation than for abiotic and basic biological processes. To understand the socioepistemic layer and how different human systems work, reference to different social sciences and humanities has to be made. They provide the bodies of knowledge that equip us to read the different rationales that are at work in various human systems and their subsystems. These disciplines are psychology as the science of the individual's behavior and mental processing, sociology as the science of society and its social institutions, and economy as the theory of production, distribution, and consumption of goods and services of and within societies. However, there are other important social sciences, such as anthropology, linguistics, political science, or law that provide relevant knowledge. In addition, the humanities, in particular history and philosophy, including aesthetics, ethics, and metaphysics, provide insight into the nature of the rationale of human systems. Here, environmental literacy requires efficient use and the ability to interrelate the knowledge of these domains. The HES framework suggests that this can be done by means of decision and game theory (see also Box 20.1).

Decision and game theory incorporate two potentials essential for the study of HES. One is a capacity to precisely describe the decision *situation*; that is, the range of potential actions at the disposal of human systems. The other is the capacity to incorporate models about the decision *process*, referring to the cognitive, mental or epistemic processes involved as well as to the social rule systems representing the situation (see Figures 17.1* & 16.12*). This allows assessment of what environmental awareness human decision-makers have about their interaction with the environment.

The latter, the assessment of the awareness that human systems have of their interaction with the environment, requires us to have a conceptual tool at hand. Here the concept of feedback loops of different type and order is essential. This allows modeling of learning processes, raising environmental awareness, anticipating rebound effects, and assessing the capability of human systems to adapt and equilibrate with environmental systems, and for keeping them within the limits required to sustain human systems. However, we see that humans, now on the cusp of the post-industrial age, often do not innately harmonize with the natural environment, but embrace resource-intensive lifestyles and societal structures that drive the material–biophysical environment to states that endanger human subsistence. Unfortunately, the evolution of human systems also has momentum as – following Thomas Aquinas – human beings embody the potential of intelligent or rational reasoning power. But whether this potential can be used to find sustainable solutions – what Kant supposed to be a categorical imperative – is an open question. Here we touch on the issue of whether "sustainability" can become a regulating idea of human systems.

Doubts about whether human systems may not have the ability or will not make it a priority to promote sustainable development are fueled by the ambiguous, evolutionary grown motivational triangle of individualistic, competitive, and cooperative motivations, motivations that are at work on the same and between hierarchy levels. We thus introduced the Interference Postulate P3. The conflicts between the macro and the micro levels are also reflected by life experiences. For example, people often believe that others (society) should take on certain responsibilities, such as recycling, yet do not themselves take on these responsibilities.

Overconfidence and the inclination of ignoring our own limited knowledge, the unknown, are two critical characteristics of human systems that prevent sustainable lifestyle adoption and decision-making. There is another worldly wisdom, which is that you do not always get what you want. This reality may be evolutionarily stimulating, but it can become a threat if a wrong expectation of what the environment can provide becomes essential. This is the reason that a thorough analysis of the material–biophysical environment must precede detailed analysis of all alternative possible decisions and actions, at least after the environmental problems have been roughly defined. The agrofuel and the animal bone cases of Chapter 18 demonstrated this.

20.3.2 Disciplined interdisciplinarity as a prerequisite

The view presented in this book is that the disciplines are clearinghouses to guarantee the consistency and validity of knowledge according to disciplinary standards. Thus scientific knowledge has to be focused according to scientific disciplines. To cope with complex environmental problems, guaranteeing textbook, state-of-the-art knowledge is a proper reference point for disciplined interdisciplinarity.

While traditional disciplines have their unique methods and knowledge about human and environmental systems, it is beyond their individual capacity to deal with the dynamics and complexity of the material–biophysical environment. In addition, they do not deal with human–environment interactions in a way in which we can expect a satisfactory answer to pressing environmental problems such as the provisioning of water, nutrients, food, energy, and biodiversity, or phenomena such as climate change or changing disease patterns in a rapidly changing world. Clearly, anthropology, geography, human ecology, and public health deal with some of these challenges. But, today, it looks as though environmental science and the newly developing field of sustainability science have the capacity and ability to bring knowledge and methods from these fields together.

However, these emerging new sciences are facing the Herculean task of coping properly with the rapidly developing knowledge of an increasing number of disciplines, which is buried in tens of thousands of journals, books, databanks, etc. This requires new strategies and standards for interdisciplinary knowledge integration. What this means and how this can be arranged is another challenge. Looking at the history of sciences,

20 New horizons: environmental and sustainability sciences

Box 20.1 The HES framework in brief: a short manual

Utilizing the HES framework can provide better understanding, representation, communication about, and investigation of HES, which by their very nature have ill-defined, complex, real-world problems. The HES framework may specifically be used to structure and support interdisciplinary collaboration, transdisciplinary processes, and strategic environmental management. While applying the HES framework should be done in a case-specific manner, the following steps should be followed (see also Figure 20.1).

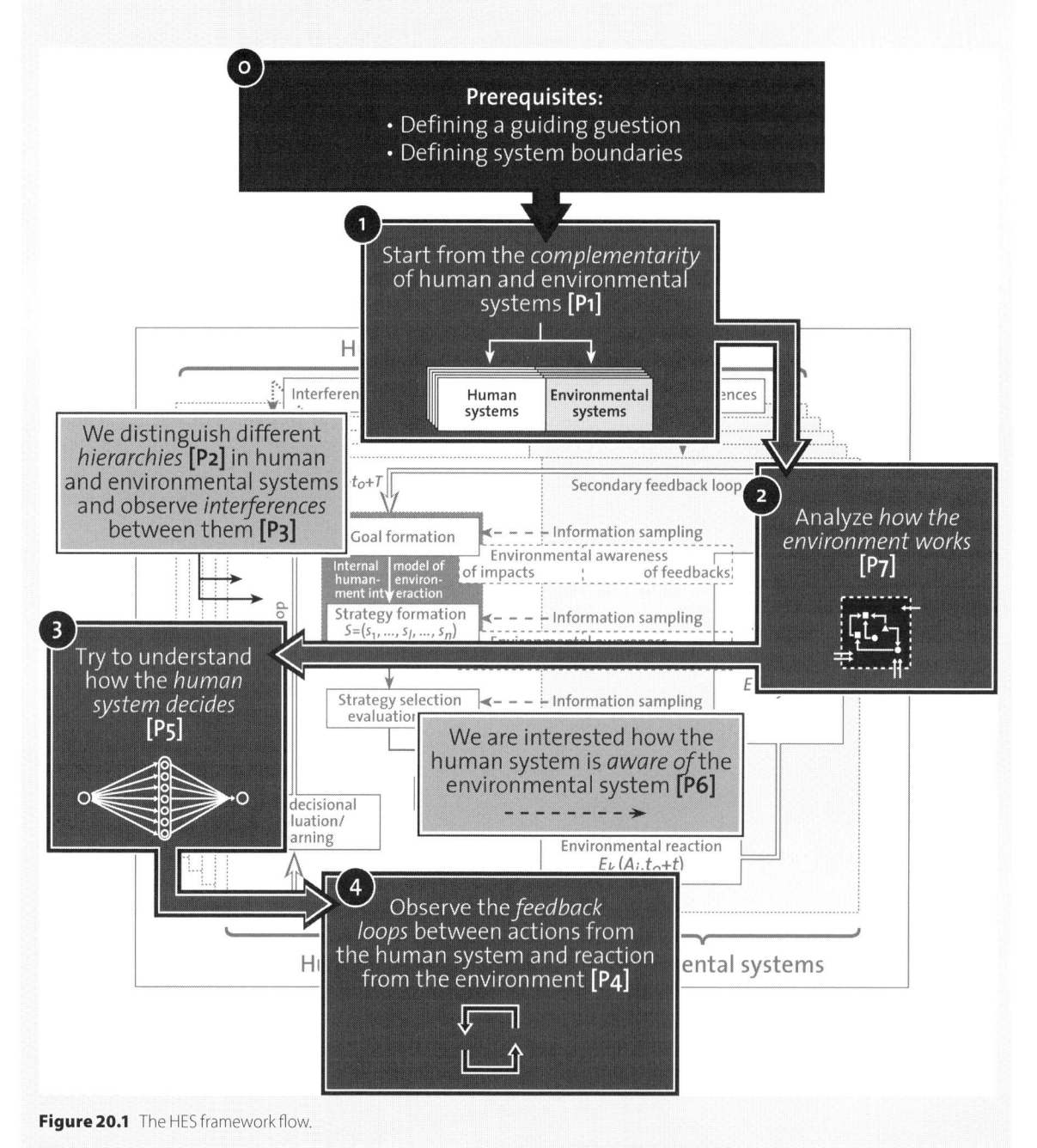

0. Prerequisites:
- Defining a guiding question
- Defining system boundaries

1. Start from the *complementarity* of human and environmental systems [P1]

2. Analyze *how the environment works* [P7]

3. Try to understand how the *human system decides* [P5]

4. Observe the *feedback loops* between actions from the human system and reaction from the environment [P4]

We distinguish different *hierarchies* [P2] in human and environmental systems and observe *interferences* between them [P3]

We are interested how the human system is *aware of* the environmental system [P6]

Figure 20.1 The HES framework flow.

> **Box 20.1** (cont.)
>
> **Prerequisites (Step 0)**: the most essential, most difficult, and often most demanding part of using the HES framework starts even before the framework is used. Defining *guiding questions* and the *system boundaries* is a fundamental first step for any investigation. Upon first becoming interested in something or facing a problem or preparing a study, one usually starts with a vague, intuitive idea, or sometimes even a half-conscious feeling.
>
> Defining a guiding question and the system boundaries brings the problem into focus and guides investigation work such as what type of data to collect. This guiding question has to explicate: the (1) perspective and thus the (primary) human system that is involved; (2) what environmental problem is of interest; and under (3) what constraints the investigation will take place. The perspective can be an individual, a company or other type of organization, a community, or a group or a subset of society. The environmental problem can be rooted in relations to natural resources, technologies (and their environmental impacts), or to human systems of a broad variety, ranging from legal systems to values represented by parts of a society. The constraints are material–technological, financial, sociopolitical, time-related, and epistemic–motivational resources available. The system boundaries depend on the perspective and the constraints of the environmental problem. As the investigation ensues, emerging knowledge sheds new light on the problem, so it is almost always the case that the guiding question and system boundary evolve during the investigation.
>
> To develop a guiding question and system boundary, consider these basic questions. (a) What material–biophysical and socio-epistemic environment must be taken into account to develop meaningful strategies for transforming the problem from the perspective of the "key" human system, which builds the focus of the study? (b) What other human (sub- or super-) systems should be considered whose action may interfere with the "key" human system? And (c) given the answers to these first two questions, what environmental systems and system boundaries can be meaningfully defined for the investigation?
>
> **Step 1**: after having defined the system boundaries, the human systems (i.e. the actors and decision-makers) and the corresponding environmental system should be crisply defined. The environmental system definition should include the material–biophysical definition, particularly the relevant parts of the universe and the timeframe which have to be included in the study.
>
> **Step 2**: according to Postulate P7, we start with an analysis of the environment. Here it is important to understand: (i) the structure of the environmental system; (ii) its productivity and performance; (iii) its dynamics, in particular and change rates; (iv) its resilience, including the buffer capacity; and (v) its dependence on and sensitivity to other systems according to the outputs and the inputs. These are the criteria of the bioecological potential analysis (BEPA) suggested by Scholz and Tietje (Scholz & Tietje, 2002). If there are social systems that are part of the environment, additional criteria such as intergenerational and intragenerational justice may be applied, which would turn a BEPA into what we call sustainability potential analysis (Lang et al., 2007). There are many methods for environmental system analysis (see Chapters 12 and 13).
>
> **Step 3**: the human systems involved in the study are examined from a decision-theoretic perspective. This analysis includes: (1) an actor analysis (what actors are essential; here the Hierarchy Postulate P2 should be applied); (2) the determination of the drivers of the human systems. This may be linked to the analysis of the motivational and epistemic (knowledge-based) components of environmental awareness (Postulate P6) of the different human systems. Based on this it may become possible to define the interferences (Postulate P3), in particular the meddling in, disturbing or hampering of other human systems, as they relate to the problem transformation of the key human system.
>
> **Step 4**: anticipating feedbacks that may result from human action is a next step. This involves understanding the intended impacts on the HES, which result in primary feedback loops, and the unintended impacts, which result in secondary feedback loops. When human systems recognize and reflect on secondary feedback loop impacts, there is the greatest potential for mature environmental literacy and sustainability learning. Methods and procedures are presented throughout the book for understanding feedback loops, particularly integrated modeling and transdisciplinary processes (Chapters 14 and 15).

we had an open-ended discussion about the unity of sciences in the 1950s promoted by Howard Odum, Paul Oppenheim, Jean Piaget, Hillary Putnam, and others. There is an ongoing discussion of what interdisciplinarity means. It is evident that each discipline uses its own language, object, method, and standard, and reference system of validation. The challenge is that the study of HES requires use of knowledge from many, if not all, of the disciplines.

The challenge of interdisciplinarity is to relate knowledge from different disciplines properly. Here, it is evident and acknowledged by the human–

environment Complementarity Postulate P1, that social sciences and natural sciences define realms with different rationales. Knowledge integration within these domains is more feasible with subdisciplines such as social psychology or biochemistry. However, there is also a third big realm, including philosophy and epistemology, that can be conceived as reflexive sciences, as it can help to identify which type of integration seems reasonable. We think that humanities will play an important role in reflecting on the potential and limits of these sciences.

20.3.3 Transdisciplinarity as a means for science working with society

Transdisciplinarity is a process for mutual learning, where scientific knowledge contributes to improving in-vivo decision-making. In turn, certain domains of sciences can benefit from incorporating knowledge from practice. Thus, transdisciplinarity and the key process of mutual learning supplements scientific consultancy, which is a traditional mode of utilizing scientific knowledge.

An intriguing question is how this learning process takes place and what type of knowledge integration can be realized. Here, we pointed out that science and practical decisions make primary reference to their own epistemics, languages, modes of thought, and interests. Thus, mutual learning can be conceived as two interrelated but distinctly embedded processes of communicating, generating, and utilizing knowledge.

Scientific knowledge from all disciplines refers to the analytic, abstracted, numeric–conceptual or formal–operative side of cognitive processes. This knowledge is also represented in codified form and is often organized in a rigid disciplinary matrix. Conversely, expertise for real-world problems is strongly rooted in experiential knowledge. This intuitive knowledge is sometimes not even consciously accessible.

Both types of knowledge have strengths and flaws. Which one is more useful depends on the guiding questions for research and the resources available to answer these questions. Scientific knowledge may provide insights for appropriately anticipating "secondary and higher order feedback loops" as key elements of strategic environmental management, something not accessible through intuitive knowledge. Conversely, scientific knowledge can often fail to acknowledge the contextualization and the multilayered nature of reality. Thus, science should consider starting from a complex real-world problem and become able to participate with stakeholders in a transdisciplinary process to jointly define research questions. Based on this, more challenges have to be met. One is appropriately acquiring and integrating the disciplinary knowledge. Another is to integrate values and knowledge from societal decision-makers and stakeholders to provide findings that can be used for generating sociotechnically robust knowledge for sustainable transitions. Transdisciplinary processes alter the mode of knowledge production: instead of "science for society" (which is the stance of applied science), transdisciplinarity means "science with society." On the side of the societal agents, the capability must be developed to interpret, value, and acknowledge scientific results properly. Developing this capability of efficient knowledge use is the main product of mutual learning as a key component of transdisciplinary processes. The 25 transdisciplinary case studies run by different European universities associated in the International Transdisciplinarity Network (ITdNet) provide models for how this can be achieved.

20.3.4 Environmental and sustainability sciences as objective

Based on the lessons from different parts of the book, we suggest that science should reorganize to become more efficiently utilized in society. The disciplinary matrix should be extended by new domains of environmental science and a properly positioned sustainability science, which may provide knowledge and methods to analyze and cope with complex, contextualized multilayered, anthropocenically shaped environmental problems.

One challenge of environmental science is to explain how human and environmental systems interact. This is a new scientific challenge. The history and dynamics of environmental systems are embedded in human systems. Similarly, the history and dynamics of human systems cannot be understood without their material–biophysical environmental matrix. This asks for the development of the new scientific field of human–environment systems as a subdomain of environmental sciences, called HES. The HES framework can be seen as a template for development of this field. HES research properly relates the body of knowledge about how different human systems perceive, anticipate, process, and evaluate environmental information, and when and why they assimilate and accommodate their behavior to changing environmental settings. HES research conceptualizes and interrelates different human and environmental systems, and the knowledge in the natural and social sciences we have about them. Environmental scientists working in the field of human–environment interactions thus have to be literate in the social and the natural (or engineering) sciences.

Traditionally, scientists have worked with great depth in one discipline. Some show a T-profile, which means that they also have the capability of building links to other disciplines. Instead, HES research needs scientists with Π-profiles, researchers whose knowledge is rooted in both in a social and a natural (or engineering) science. Such Π-profile researchers have the ability to communicate and investigate across disciplines and to develop the textbook state-of-the-art knowledge that is relevant to problems of today's anthropocenically shaped environment. This is a giant task and can – as far as we know today – presumably best be solved when HES becomes a subfield or subdiscipline of the emerging environmental science. The task is to utilize and integrate state-of-the-art knowledge about HES. This is met by disciplined interdisciplinarity in HES research.

The second challenge is related to the emergence of sustainability science and refers particularly to the social dimension of sustainable development. Sustainability has become a regulative idea of the postmodern society. Sustainable development is an ongoing inquiry targeting what we call "system limit management." This asks for proper knowledge to support the navigation of HES in a way such that unintended and unwanted critical transitions or collapses of human systems can be avoided. This requires anticipating unwanted feedback loops or rebound effects that may result from immature human action plans, for instance when utilizing new technologies. It certainly includes the prevention of destructive warfare activities. This also requires properly assessing the adaptive potentials of the social and the material environment, and the finding of proper "sustainable actions." As there is almost nothing we can get for free (and without resistance), this will ask for means to address potential conflicts with the interests of some stakeholders. We can see from these ideas that transdisciplinary processes for capacity-building and legitimation of sustainable actions are needed which go far beyond the material–biophysical environment. We think that these are issues which should be investigated by sustainability science.

Another issue is the stabilization of globalized human systems in the evolutionary triangle of competitive, individualistic, and cooperative attractors. This challenge runs under the label of "intergenerational and intragenerational justice" as competitive and individualistic motivations may cause conflict with fundamental social justice.

20.4 Two visions

After the long journey through this book, new horizons become visible. When looking forward, the question emerges of what future tasks environmental and sustainability sciences might be facing? And – when again referring to key question 3 – what knowledge and means do we need "to successfully cope with the challenging environmental problems of the twenty-first century?" At the end of this book these related questions evoke two strong visions at which the author would like to take a closer look.

One vision asks for a better understanding, defining, and generating of what may be called "sustain–abilities." These are the general abilities that are needed for establishing transitions towards sustainable HES. The other vision refers to utilizing HES-based transdisciplinary processes to develop these "sustain–abilities." This second vision links the two main streams of ideas and experience presented in this book and aims to bring them into practice. This is linking satisficing (satisfying and sufficing) analysis and understanding of complex, multilayered human–environment interactions with transdisciplinary processes which generate socially robust solutions for transitions towards sustainable HES.

When following these visions, we have to be aware that any sustainability transition is interlinked with highly value-loaded, normative aspects. Whether one wants to avoid a specific type of collapse or not, whether adaptation to global climate change or the remediation of local soil contamination is considered more important, is related to specific personal and societal values. Nevertheless, we think that there is a set of "sustain–abilities" that are valuable for any human system interested in coping successfully with relevant environmental real-world problems. These "sustain–abilities" are all related to the principles provided by the HES Postulates (see Chapter 16).

20.4.1 "Sustain–abilities" from a coupled-system perspective

Based on the seven HES Postulates (see Table 16.1*) and the examples presented in Chapter 18, we suggest that we should take a closer look at the following "sustain–abilities." We think that these abilities may be desirable for all levels of human systems, from individual to companies (business organizations) to states to supranational or suprasocietal systems, if sustainable human–environment relations are targeted.

- **Environmental potential**: reading the potential of the environment (see Postulates P7, P6 in Chapter 16 and Table 16.1*). Given certain human resources and technologies, the natural (and social) environment can only provide a limited number of benefits. A critical challenge is to know these limits and to avoid unwanted critical transitions of HES, for instance when demands change in human systems or new technologies are developed and used.
- **Rebound effects** (i.e. rebound effects; see P4, P1, P6): anticipating and coping with unintended secondary environmental feedback loops owing to human action is a key "sustain–ability."
- **Tipping points**: recognizing tipping points is a special facet of environmental awareness (see P6, P7, P4) and builds upon the first sustain–abilities (reading the potential of the environment and anticipating rebound effects). A special situation that asks for the adaptation of human systems arises if the current interactions between human systems and environmental systems are not changed before approaching or crossing critical tipping point, which may include an unwanted critical transition to a qualitatively different state.
- **Trade-offs**: human systems focus on different goals and scales of time and space. This may cause different priorities, conflicts, and dilemmas. Identifying and coping with trade-offs between different human systems (P4, P2, P3) and within specific human systems (P4, P6, P7), and supporting consensus formation and proper coping with conflicts, is a prerequisite of avoiding inefficient and destructive forms of environmental management.
- **Resilience management**: designing and realizing resilient HES to avoid collapses after exposure to negative, even unforeseen adverse impacts and threats. From a functional perspective, it describes a HES's ability to either be capable of absorbing impacts resulting from disturbances and return to its former shape, or to adapt through technical or societal innovations (i.e. structural shift in HES) to another state with sound structure and function.

These abilities should be applied in societally relevant problems. Thus the ability to identify and to tackle the most relevant problem, to define an adequate guiding question for a transdisciplinary process and related research is a general capability that precedes the above "sustain–abilities." Clearly, finding the "right" guiding question, whose answer is beneficial both for science and society, is the most difficult task of doing research in environmental and sustainability sciences. Based on our experience, the construction of a proper problem model which allows for organizing research is a prerequisite and key for meaningful and efficient research, as it serves for structuring discourses with scientists of other disciplines, stakeholders or legitimized decision-makers. A special challenge here is to properly acknowledge and to cope with the complexity of HES. The Postulates of the HES framework and the above "sustain–abilities" are considered as tools to deal with different types of complexity. These are the static and dynamic complexity, the complexity of scale involved in HES, as well as the wicked and historic complexity (which include cultural and military hostilities). Here we should remind ourselves (see Chapter 14.1.2) that the wicked complexity is due to the unreliability and to the bounded rationality of human systems, which – in general – show many inconsistencies of behaviour. We think that unreliability in goal setting is a primary cause of the wicked and historic complexity that can be mastered with the HES Postulates. Thus, we suggest another "sustain–ability:"

- **Problem definition and system representation**: the ability to define the right guiding question and to develop a proper system representation of the (HES-) problem at hand which acknowledges and copes with the different types of complexity (i.e. static, dynamic, wicked, temporal and spatial scale, and historic) in interdisciplinary and transdisciplinary discourses.

The last "sustain–ability" is a kind of master ability. Reading the potential of the environment, anticipating tipping points or rebound effects, coping with trade-offs and establishing resilient HES requires proper problem and system representation as a prerequisite. Presumably, the list is incomplete and some other "sustain–abilities" could be added. Further, the presented "sustain–abilities" do not focus on the learning process but rather present objectives of sustainability learning. The presented list shows what "sustain-abilities," which are defined from an HES perspective, may look like and what the objects of future research of HES, as a subdiscipline of environmental and sustainability science, may be.

In this place we want to highlight two abilities that we consider to be essential for sustainable action. From a sustainability science perspective, fostering and establishing **intergenerational and intragenerational justice** is a specific challenge. This challenge calls for developing proper concepts and practices dealing with the distributional and procedural aspects underlying fair and socially robust solutions. This ability refers to the way in which goals and strategies are defined (P5), an awareness of other human systems' needs and demands (P5), and the interference of different reference points of what is considered as being fair (P5).

Another sustain-ability refers to properly **utilizing the potential complementarity of different knowledge** systems for sustainable action. What different scientific disciplines and science per se can provide is – as it is the case with any other knowledge system – relative and limited (see Chapter 3.4–3.8). We have elaborated in many parts of this book that integration of knowledge of different epistemics, modes of thought, cultures, perspectives, etc. (see Chapter 3 and 15; Box 2.5) is an indispensible prerequisite to provide socially robust solutions for the challenging environmental and social problems of the twenty first century. Achieving this successfully asks for, at the very least, (transdisciplinary) processes that allow for knowledge integration as a means of capacity and consensus building, mediation (i.e. finding acceptable and fair solutions) and legitimization. But it also requires that all human systems involved in such a process are able and willing to reflect on their own knowledge, just as much as on other knowledge types, to critically recognize the strengths and limitations of each. This reflective capacity includes acknowledging the uncertainties and nescience that permeates each type of knowledge. This brings us just to the second vision, which is the ability to establish transdisciplinary processes for sustainable transitions.

20.4.2 From knowledge to decisions by HES-based transdisciplinary processes

Part VIII of this book presents the HES framework as a powerful tool to structure complex multilayered human–environment interactions. We put the focus on how coupled HES can be represented if there are a multitude of human actors receiving environmental services or critically affecting the environment. From a science perspective, the HES framework allows for systematically utilizing knowledge from different scientific disciplines (disciplined interdisciplinarity). This holds true particularly for the human system side, as the Hierarchy Postulate P2 necessitates knowledge about the rationales of different human systems by referring to psychology, sociology, business and administration science, economics, cultural anthropology, etc.

We argue that transdisciplinarity offers a new type of framing, learning, and coping with complex, tangible environmental problems. Transdisciplinary processes establish mutual learning between science and society on equal eye level. This is in particular needed for joint problem definition, representation, and transformation. Thus, transdisciplinarity relates human wisdom with analytic scientific rigor for dealing with the right problem in an adequate manner.

Coupled HES are strongly shaped by human actions (anthropocene). Thus, its *analysis* necessitates the involvement of those human systems that primarily affect it. Gaining access to their thinking, rationales, and goals is substantive for sustainable HES transitions. Involving the key agents who have caused or are concerned by environmental and social problems facilitates investigating why, how, and with what impacts they act as they do. At least from a democracy perspective, the originators and affected people need to have a say when it comes to potential action, behavioral changes or new policies.

From the perspective of the participants of a transdisciplinary process, the HES framework facilitates a reflection on what knowledge is necessary and can be provided by what scientific discipline, expert, key agent, stakeholder group or layperson. And, likewise, for understanding and transforming an environmental problem, what values from what stakeholder group or what members of the public at large should be incorporated. This includes an ongoing reflection on what sustainable future should be targeted.

Based on this, transdisciplinary processes may efficiently serve for capacity-building on both sides, i.e. science and society. Developing the above "sustain–abilities" of assessing the environmental potential, coping with trade-offs and rebound effects, identifying tipping points, or establishing resilience and reducing vulnerability of HES is seen as a key of environmental literacy. Based on experience with many transdisciplinary processes, we conclude that this asks for radically turning science upside-down, starting the process of mutual learning from a concrete, real, societally relevant, problem or case. This is necessary as otherwise the scientific discourse becomes too abstract and cannot be efficiently used for practical decisions. Contrary, when dealing with the case, one has necessarily to step out of the competitive business and politicized discourses of day-to-day business, which can be easily achieved if a longer time perspective is taken. Science can become a key agent in sustainable transitions if it develops the capacity of properly utilizing and applying disciplinary knowledge to deal with societally relevant, real-world problems. And science will be enriched as it develops novel perspectives and new types of research questions.

HES-based transdisciplinary processes are a promising tool and link scientific excellence with practical relevance in a unique manner. They serve for coping with complexity both in the analysis and in transitions of coupled HES. Transdisciplinary processes complement other forms of theory practice cooperation such as consultancy, action research or participatory research (see Chapter 15). Yet these approaches are extremely demanding and ask for rather fundamental changes both in science and society. In science, the classical conception "we provide necessary knowledge that needs just to be translated into respective policies" needs to be transformed. In society, the often-practiced conception "we selectively consult with those scientists who, and use those data and reasoning which, legitimize our decision-making" needs to be replaced by an openness to scientific results, whether or not they conflict with decision-maker interests. A much more interactive and reflective form of collaboration is necessary, guiding knowledge production and decisions in coupled HES. Such a process honors the values and roles of scientists from different disciplines, decision-makers, and other stakeholders. Though HES-based transdisciplinary processes are no panacea, they are a promising approach "to successfully cope with the challenging environmental problems of the twenty-first century."

Glossary

Terms shown in bold are listed as entries in the Glossary.

Accommodation A concept of Jean Piaget's theory. Accommodation means the creation of new cognitive schemata or behavioral patterns to deal with new problems or activities. Accommodation complements **assimilation** as a mode of **adaptation**.

Action research A form of research going back to Kurt Lewin in which scientists organize their research in such a way that it serves to improve stakeholder activities and the public discourse. Action research has to be distinguished from **transdisciplinary processes**, **consultancy**, **public participation, participatory action research**, and **participatory research**.

Adaptation A change in structure or function that enables an organism to adjust better to its environment. There are two modes of adaptation: **accommodation** and **assimilation**.

Adaptive cycle Meta-model for ecosystem dynamics that consists of four processes: exploitation, conservation, release, and reorganization (see **panarchy**).

Adaptive management A structured, iterative process of decision-making with respect to complex **socioecological systems**. The aim is to maximize multiple objectives and to accrue information needed to improve future management.

Affordance This concept, introduced by James L. Gibson within his ecological theory assumes that humans are well prepared for some environmental information and hence implies complementarity of human and environment systems.

Agent-based modeling (ABM) Modeling approach focused on actions and interactions of autonomous agents (individual or collective entities such as organizations or groups) and their effects on the system as a whole.

Analytic mediation Mediation in which the assessment of consensus and dissent is based on analytic methods such as psychometrics and multi-criteria utility assessment tools, as used in the **area development negotiation** method.

Animism The belief that souls exist not only in humans and animals, but also in plants and non-living natural phenomena of all kinds. It is the belief in an animated nature.

Anthropocene Recent period in the Earth's history, considered to start in the late eighteenth century or earlier, when the activities of humans first began to show a significant effect on ecosystems, climate, **biogeochemical cycles**, and other **Earth systems**.

Appropriation of nature Conversion of natural **resources** into objects of property by individuals and societal groups.

Area development negotiation method (ADN) A variant of **analytic mediation** in which consensus and dissent are assessed based on multi-criteria utility assessment of scenarios, followed by multi-party negotiations. It is applied to **transition management** in **transdisciplinary processes** and in many conflicts related to environmental impacts or commodities.

Aspiration level An aspiration comprises a discrete number of goals that a human system targets through actions and negotiations, and which are related to certain needs and degrees of satisfaction that are fulfilled by attaining the goals. Individuals and other human systems are supposed to have different hierarchically ordered levels.

Assimilation A concept of Jean Piaget's theory. Assimilation means the modified usage of existing cognitive schemata or behavioral patterns to solve problems. Both assimilation and **accommodation** are considered to be modes of **adaptation**.

Attribution A concept that explains why an event occurs and what cause–impact relationship underlies a certain phenomenon.

Autopoiesis This term was used by Humberto Maturana to describe the characteristics of reflexive feedback control and system dynamics in living systems.

Basic biogenetical law (see **theory of recapitulation**).

Behaviorism A dominant school of psychology in the first half of the twentieth century stressing that psychology should be entirely concerned with behavioral data and stimulus–response patterns.

Benign conflicts A benign conflict is free of malignantness (see **malignant conflict**).

Biofuel Liquid or gaseous transport fuel produced from **biomass**.

Biogenetic law This law, suggested by Ernst Haeckel, postulates that the ontogeny (see **onto-genesis**) follows the same development as the phylogeny (see **phylo-genesis**).

Biogeochemical cycle Cycle of chemicals, compounds, molecules or elements through the biosphere, lithosphere, atmosphere, and hydrosphere.

Biomass Vegetal and animal substances, as well as the biodegradable fraction of waste, that are used to produce energy. Fossil energy is excluded.

Biosemiotics A conceptualization of the mind–body interface developed by Jakob von Uexküll that provides insight into the relationship between organisms and their environment. It conceptualizes the environment cybernetically (see **cybernetics**) in terms of perceived "signs" that constitute the meaning of relevant environmental states for the individual.

Bovine spongiform encephalopathy (BSE); also called mad cow disease A fatal, neurodegenerative disease in cattle, causing degeneration in brain and spinal cord.

Cellular automata (CA) Identical cells that are arranged in a grid structure. Such cells represent units of a biophysical environment, human individuals or collective actors (e.g. countries or states). These cells are surrounded by other cells contributing to the environment.

Cleaner production Term coined in 1990 by the United Nations Environment Program (UNEP) to denote preventive environmental strategies to improve processes, products, and services to minimize environmental impact and reduce **risks** to humans and the environment.

Clonal selection theory A theory concerning the functioning of the adaptive (acquired) **immune system**. It states that the adaptive immune system works through the activities of antigen-reactive clones of T and B cells. Each B cell is specific to one antigen, as it shows only one kind of antigen-binding receptor. Once this receptor is excited by binding to its specific antigen, it multiplies by means of cloning. Some scholars consider the adaptive immune system to be able to distinguish between self and non-self (see also **cognitive paradigm**).

Club goods These goods do not rival one another. Access to these goods may be restricted, thus excluding consumers (see also **rivalry** and **excludability**).

Cognition The processes of thought and the mental capacities of an organism. Through cognition, human systems process information and use logical and symbolic abstraction and memory.

Cognitive dissonance A state of tension resulting from inconsistent or contradictory **cognitions** simultaneously.

Cognitive paradigm An approach in psychology and cognitive science which assumes that people, human or organismic systems acquire, process, and memorize information.

Cognitive psychology A discipline that investigates the mental processes related to information and thought.

Common goods/Common pool resources Resources in this category do not rival each other and are non-excludable (see also **rivalry** and **excludability**), such as an ocean fishery.

Commons dilemma A situation in which each individual receives a higher payoff and utility for defection (e.g. overuse of **resources**) than for cooperation. One example is the N-person Prisoner's Dilemma game.

Community A unified body of individuals that interact or are linked by a specific measure and often live in a common area. Communities with authorized governments can be considered to be the "lowest level of **society**".

Complementarity Postulate A postulate of the HES framework which assumes that human and environmental systems are inextricably related, show different rationales, and cannot be investigated independently on a larger scale.

Complementary goods Two goods where, given a surplus in consumption of the first good, there is an increase in the marginal **utility** of consuming the second good. The differentiation between complementary and **substitutive goods** was first introduced by Irving Fisher.

Complexity The quality or state of being complex (i.e. to be comprised of diverse interacting objects leading to dynamics that cannot be related to the single objects alone). Complexity increases with the number and diversity of the objects and their interactions.

Consciousness The mental awareness of an internal or external object, state, or mental process.

Consensus-building In consensus-building, experts help to reduce **uncertainty**, and/or formulate state-of-the-art knowledge. Activities can include informing the public, the media, or colleagues about what science knows and does not know, and organizing or participating in discourses such as public **consensus conferences**. The aim of the latter is to pave the way for reaching consensus about the issues at hand.

Consensus conference A method for assessing **technology** and objectives that involves a meeting where in both an expert panel and a lay panel of concerned citizens participate.

Consultancy A form of utilizing scientists' activities that is completely under the control of legitimized stakeholders.

Consumer's rent This rent is seen as an economic measure of the difference between a consumer's willingness to pay and the actual payment.

Cosmogony A theory which explains how the world began. It describes the origin of the perceived natural order, which is manifested in the cosmological (see **cosmology**) ideas of a people.

Cosmology The totality of notions about how the world works and is organized, and subsequently about the understanding of the world by someone. A cosmology can include a **cosmogony**.

Coupled Human and Natural Systems (CHANS) This approach employs investigation methods that integrate factors from human and natural environment interactions to better understand and predict future effects of social–ecological and human interactions and their effects.

Critical transitions Transitions between different dynamic states (non-linear dynamics), which are considered critical according to an observer's expectations with respect to system dynamics (e.g. "catastrophic" shift).

Cybernetics A theory in the field of system dynamics that focuses on the identification and analysis of **regulatory mechanisms** and **feedback loops** in complex systems.

Decision An outcome of a process leading to the selection of an option from several alternatives. Human decisions are integrated in an action cycle linking motivation, intention, and behavior.

Decision-theoretic conceptualization of human systems A decision-theoretic conception provides a general language to describe processes preceding action of different human systems, from the individual to supranational systems or nation-states.

Decision theory Approach dealing with conditions of **uncertainties** and **risk** in a given **decision**, its **rationality**, and the resulting optimal output. This theory is closely connected to **game theory**.

Dematerialization A means of **cleaner production** implementation consisting of using less material to attain a desired service by presenting an accounting of material flows. On the **macroeconomic** level, dematerialization is understood as the uncoupling of economic growth and material intensity.

Devolution A hypothesis about the relationship between humans and animals which was proposed by Plato. In contrast to the idea of humans having evolved from animals, it suggests that animals degenerated from men and, therefore, "devolved."

Disenchantment of the world The process of switching from a **cosmology** that is based on mythology, the agency of supernatural forces and belief, to a cosmology that is based on logical reasoning and scientific explanation.

Driver A psychological variable or construct that can elicit and steer cognitive activity, behavior, and action of human systems.

Earth system A more or less closed system exchanging energy and limited matter with its external environment in space. **Biogeochemical cycles** are part of the Earth system.

Earth system analysis (ESA) The development of a mathematical–logical formulation of principles applicable to **Earth systems** aiming at perceiving

Glossary

the Earth as a whole. It postulates new paradigms for the long-term coevolution of nature and civilization.

Eco-efficiency A guiding principle to optimize the ecological–economic ratio of desired output(s) and necessary input(s) (see **efficiency**).

Ecological design Any form of design that aims to minimize environmentally destructive effects through integrating such design with living processes.

Ecology The comprehensive science of the interactions between organisms and their biotic and abiotic environment.

Economic entropy In ecological economics, entropy is a semiquantitative measure of the irreversible dissipation and degradation of energy and natural materials in relation to economic activity. A decrease in entropy is supposed to limit the ecosystem's capacity to deliver natural services.

Econophysics An interdisciplinary research field focused on explaining, understanding, and/or solving economic problems by employing theories and methods that were developed originally by physicists (including stochastic and non-linear dynamics).

Ecopsychology A domain of environmental psychology that is concerned with the interplay of humans and ecosystems.

Ecosystem functions Structures and relations between elements of ecosystems that have directionality. The directionality is due to the self-organizing properties of the system which give rise to a centripetal pull of **resources** and energy into the system.

Ecosystem services Those **ecosystem functions** that support, sustain, and fulfill human life as they provide **resources**, regulate natural cycles of relevance for humans, provide cultural values, or support any of the before mentioned ecosystem functions.

Ecotoxicology A branch of toxicology that analyses effects of natural and synthetic substances on organisms and ecosystems.

Efficiency Ratio of the amount of wanted and useful output(s) from a process to the amount of input(s) to the process.

Eidos A concept used by Aristotle to describe the individual form of an animal species. This is different from its **genos**, which describes its form on a level above the individual species.

Engineering sciences Scientific studies on the design and implementation and application of machines, materials, systems, and devices.

Entropy A concept that is interpreted in at least three distinct ways: as a macroscopic, classical thermodynamics concept; as a microscopic **statistical mechanics** concept; and as an information theory concept. It has often been associated with the amount of (molecular) order or disorder in a (thermodynamic) system. Entropy is central for the **second law of thermodynamics**. Analogies have been employed in arguments in economics (**economic entropy**).

Entropy in statistical mechanics The amount of uncertainty, or disorder, about a system that remains after all observable macroscopic properties have been taken into account. Entropy is a logarithmic measurement of the density of states of a system. In equilibrium state, the disorder (entropy) of a system is greatest.

Environment A "young concept" mentioned for the first time approximately 400 years ago (see **social environment** and **material environment**).

Environmental awareness The capability to use skills based on knowledge and motivation to assess and anticipate environmental change owing to human action, and to understand the related **feedback** processes.

Environmental concern A cognitive and affective evaluation of the object of environmental protection.

Environmental engineering Branch of **engineering sciences** that involves designing and implementing technical solutions for problems affecting environmental systems.

Environmental Kuznets curve (EKC) The Kuznets curve is applied to environmental burdens. The Kuznets curve hypothesizes that economic inequalities increase before decreasing while a country is developing. It is based on the rationale that, once basic needs are fulfilled, a **society** will have the means and the will to deal with environmental issues. Thus, the use of natural **resources** and/or the emission of pollutants will first increase with the level of income and decrease beyond some turning point at higher income levels.

Environmental literacy The ability to read and utilize environmental information appropriately, to anticipate **rebound effects**, and to adapt to changes in environmental resources and systems, and their dynamics.

Epidemic An occurrence of **infectious disease** in a community or region, during a given period of time, that involves more cases of the disease than is normal.

Epigenetics An umbrella term comprising mechanisms of inheritance of ontogenetically (without mutation in the DNA, see **ontogenesis**) acquired features.

Epistemology A field of science focused on the method and foundations of knowledge, especially its limits and validity. Epistemology explains how we acquire how well we know what we know, and why.

Essential scarcity Scarcity of a material or chemical element that cannot be dealt with by substituting it with another material because of its essentiality.

Ethnobiology A discipline that investigates how indigenous people perceive, classify, and examine the organic world.

Etiology A study of causes and origins of phenomena.

Evolution A biological theory that rejects the notion of an inherently constant creation but assumes that species emerge based on the principles of variation in heredity and **selection**.

Evolutionary psychology Attempts to explain individual differences and characteristics in human mind and behavior as the product of evolutionary processes (see **evolution**).

Excludability A term meaning that it is possible to prevent potential consumers from accessing a good by (high) prices. Only excludable goods are marketable. Other types of goods are **private goods**, **club goods**, **common goods**, and **public goods**.

Feedback A term from **systems theory** that describes processes where effects, actions or outputs from a system element affect the system itself through positive (reinforcing) or negative (stabilizing) feedback.

Feedback loop The circular causal path over which output is led from an initial system element via other system elements back to the initial element (see also **primary feedback loop** and **secondary feedback loop**).

Field theory A psychological theory developed by Kurt Lewin that examines patterns of interaction between the individual and the environmental (social) field. A field is defined as the totality of coexisting facts, which are conceived as mutually interdependent.

Foraging (see **predating and foraging**).

Form–matter complementarity Aristotle's hypothesis that states that matter is a thing's reality, whereas form is a thing's potential.

Formative scenario analysis (FSA) A method describing how a **scenario analysis** can be achieved in a strictly organized (i.e. methodical and systematic) manner, while often fostering the **mutual learning** of case agents and study team (i.e. forming the construction and reflection of future visions) in a **transdisciplinary process**.

Four archetypical elements of the world regime These elements are water, earth, air, and fire. Empedocles of Acragas introduced the notion that these four are the basic underlying ontological elements of reality.

Four humors of the individual Yellow bile, black bile, phlegm, and blood. Hippocrates introduced the notion that these four humors govern the wellbeing of individuals.

Four qualities of systems According to the ancient Greek philosophy, temperaments, diseases, and other systems can show the qualities composed by hot, cold, dry, and moist.

Framework A set of ideas, conditions or assumptions that serve as a tool for structuring complexity and providing context that helps to describe how something can be approached or understood.

Full ecology perspective Perspective which integrates human activities in biological, chemical, and physical aspects with various plant and animal species of ecosystems.

Functionalist perspective This perspective assumes that processes or actions of human systems or organisms serve certain purposes.

Fundamental thermodynamic relation Based on the first law of thermodynamics, the conservation principle of energy illustrates that energy cannot be created or eliminated. Based on this principle, energy can only be transformed from one form to another form (see **second law of thermodynamics**).

Game theory Study of strategic interactions and conflicts among different players who choose between different strategies based on assumed preferences (or **utility** functions) on the outcomes of a game.

Gathering (see **hunting and gathering**).

General system theory A theory developed by Bertalanffy to address characteristics of systems from a more general point of view than **system dynamics**. This deals with and is applicable to all kinds of physical, biological, human, social, and technical systems.

Genetic epistemology An idea from Jean Piaget which postulates that the law of recapitulation (i.e. the **biogenetic law**) holds for cognitive development of a **human individual** resembles the development of knowledge in science, individual cognitive development is assumed to.

Genos (see **eidos**).

Gestalt Structures, configurations, or patterns of physical phenomena or information that cannot be derived by the summation of its parts.

Goal An end (objective) towards which effort is directed and whose achievement can be reached.

Goal formation Process of defining guiding questions and adequate system boundaries for use in both the analysis and management of complex environmental problems. It corresponds to the first step of the decision-theoretic conceptualization of human systems in the HES framework.

Great transition Approach aiming at realizing a civilization based on egalitarian social and ecological values, poverty reduction, war avoidance, and environmental protection. The goals are improved human interconnectedness and quality of life that lead to a healthy planet.

Group A level in the human system hierarchy conceived of two or more individuals having enduring, direct, observable relations. Group processes are guided by group rules, group identity, group regulations (pressure), and imperatives.

Heuristic A rapid-processing cognitive problem-solving tool (e.g. "rules of thumb," common sense) that requires relatively little storage and processing capacity and is normally acquired in the course of everyday practice.

Hierarchy A structure of levels, ranks or classes where items meaningfully treated as a whole are separated and interrelated through relational links of power, information, energy, and matter flows. Hierarchies are either determined by nature (objectivist's view) or artificially discerned to support understanding (subjectivist's view).

Horticulture A type of **appropriation of nature** and a precursor of agriculture shaped by the knowledge and art of growing plants intentionally which is linked to sedentary lifestyles.

Human–environment complementarity Describes the interrelationship and dependence between, as well as the specificities of, **human systems** and their surrounding or relevant environment.

Human–environment framework (HES) The HES framework is an interdisciplinary theoretical and methodological approach for investigating HES, developed at ETH Zurich, a university in Switzerland. It deals with structures, **regulatory mechanisms**, and decision processes on hierarchical levels, including individuals (e.g. consumer behavior), organizations (e.g. firms), **institutions** (e.g. state agencies), and **society** (e.g. socioeconomic impacts of climate change). It is a research framework for environmental, sustainability (see **sustainability science**, and other sciences for dealing with complex real-world problems or guiding **transdisciplinary case studies**.

Human exemptionalist paradigm (HEP) The HEP emphasizes the ability of humans to overcome environmental problems. Humans are seen as a uniquely superior species exempt from environmental forces also (see **new environmental paradigm**).

Human individual All human cells and their interactivities that emerge from a fertilized egg (zygote).

Human system/human system hierarchy Assumes (in current western societies) the following organizational levels: individual, **group**, organization, **institution**, society, and **supranational/suprasocietal systems**.

Hunting and gathering A type of **appropriation of nature** that is driven by socially acquired and culturally transmitted knowledge and skills. It is characterized by flexible, intentional action and includes secondary operations like storing and processing food.

Hybridization The crossing and reproduction of individuals of two different species (see **medulla–cortex theory**).

Hysteresis Phenomenon that describes how previous treatment of a system affects its subsequent response to a given force, or to information or input. It often shows up as a lagging or path-dependence of the reactions that a system shows. It reveals a kind of "memory function" of a system.

Immune system Biological structure of organisms that detects **pathogens** and protects organisms against disease. It detects a wide range of pathogens, from viruses to parasites, and other antigens.

Industrial ecology Focuses on the energy and material fluxes that relate to industrial and consumer activities. A primary question is how human management of such fluxes affects the environment and also how economic, political, regulatory, and social factors affect the fluctuation, use, and transformation of resources. The aim of industrial ecology is to develop strategies to integrate **environmental concerns** into our economic activities.

Industrial engineering A branch of **engineering science** that focuses on development, improvement, application, and evaluation of integrated systems of men, materials, and equipment.

Industrial metabolism A term put forth by Robert U. Ayres to describe the integrated group of physical processes that converts raw materials and energy, including labor, into products and wastes.

Industrial society The society that emerged from the industrial revolution; it was preceded by the agrarian society and is characterized by its dependence on finite **resources** such as fossil fuels and metals.

Industrial symbiosis Grouping of companies from different industries into eco-industrial parks with the aim of taking advantage of one another's waste and waste heat, and in doing so reducing both costs and environmental impacts.

Infectious disease A disease caused by pathogens, such as viruses, bacteria, etc., and likely to be transmitted between organisms through the environment.

Institution Institutions occupy an enduring and cardinal position in **society**. Institutions represent, maintain, and stabilize society through **regulatory mechanisms**. In our understanding, institutions are the people representing public organizations and statutory corporations as well as the formal rules and regulations established by society and state agencies.

Interdisciplinarity A field of study applied in research to describe scientific methods that use insights into different scientific disciplines or traditional fields of study crossing boundaries between disciplines, professions, and technologies.

Interference The act of, or something that results from, becoming involved, taking part, or reciprocally interacting and thereby altering, inhibiting, confusing or disturbing something. Interferences between micro and macro levels of **human systems** describe the effects of behavior at one level on the behavior of the other level.

Knowledge integration Complex real-world problems demand integration of knowledge from different fields. Here, we distinguish five types of knowledge integration: **interdisciplinarity**, integration of different systems (e.g. atmosphere, hydrosphere, pedosphere, biosphere, etc.), integration of different types of epistemologies (intuition vs. analysis), integration of different interests, and integration of different cultures.

Kuznets curve (see **environmental Kuznets curve**).

Lamarckism One class of evolutionary theory that postulates the law of use and disuse as a source of **adaptation** operating at the individual level.

Law of recapitulation (see **biogenetic law**).

Linear model A mathematical model approach based on proportionality, which means that one variable has a constant ratio to another. This model is applied in a wide range of disciplines.

Linear regression Linear regression refers in statistics to the relationship between one or more dependent variables which is/are described by a linear combination by a set of independent variables.

Macroeconomics A branch of economics investigating the performance, structure, and circulation of commodities on a national or international level.

Malignant conflicts A situation in **game theory** is seen to be malignant if, without transforming the situation, deterioration of some participants or systems or a progressive, unfavorable dynamic will result.

Marginal yield In production theory the marginal yield is the increase in income that is a result of an increase by one unit of the input of the production factors.

Markov process A random and memorylessness process (named after the Russian mathematician Andrev Markov) whose future probabilities are determined only by its most recent values.

Material environment This book subdivides the term environment into material–biophysical and social-epistemic dimensions. Material environment is defined as the interplay of life supporting physical, chemical, biotic, and materialized technological factors that act upon a **human system**.

Material flow analysis Method to model physical components, **resources**, and outputs of a system.

Material–social complementarity The dualism describes the interplay between the material–biophysical aspects and the social-epistemic aspects of organismic activities.

Mediation Mechanisms to facilitate policy processes or conflict resolution among different parties where one (or more) person take(s) the role of the mediator.

Medulla–cortex theory Linné's theory to explain the empirical phenomenon of **hybridization**. The medulla–cortex theory states that, rather than species themselves evolving in a constant manner, new species result from chance mixing of various, constant, inherited units. According to Carl Linné, plants' offspring inherit the fructification part (the medulla) from the mother and the outward appearance (the cortex) from the father. Mixing of those two units generates the offspring.

Mental representation A meaning-based internal system of semantic properties of environmental information.

Mic-Mac analysis A method to identify feedback loops between impact factors used in **formative scenario analysis**.

Microeconomics Microeconomics studies show how individuals, households, firms and other organizations, and some states, interact in markets when allocating scarce **resources**.

Mind A complex stream of **cognition** and mental capabilities in an organism. It is part of the body–mind complementarity, which reflects the material–biophysical and social-epistemic dimension of the **human system**.

Mind–body complementarity The activities of a human individual (and of other human systems) are conceptualized by material-biophysical and by immaterial and information based processes.

Mode 1 science The "traditional" knowledge production in universities, which is characterized by the hegemony of theoretical and experimental science, established disciplines, formal and formalized language, and neutral positioning, both societally and politically.

Mode 2 science A "new" approach for knowledge production that employs four principles: coevolution of science and society, contextualization of science, the production of socially robust knowledge, and the construction of narratives of expertise.

Model An intellectual construct and simplified representation of the real world, providing insight into structures and dynamics.

Modeling A scientific reasoning tool for gaining a better understanding of the system elements (agents), their relations and interactions, and thus allowing for anticipating dynamics of complex systems.

Modernity In sociology modernity refers to modern society or industrial civilization.

Multi-agent system (MAS) Community of agents, situated in an environment. Environment is the space that surrounds agents and supports or constrains their activities.

Multi-criteria analysis (MCA) Approach to structure and characterizes preferences among alternative options during a **transdisciplinary process**, where the options refer to different objectives that are determined and their attributes or indicators specified.

Mutual learning in trans disciplinary processes: Joint problem definition, problem representation, and problem transformation among practitioners and academia.

Mythic culture–nature religion A stage of cultural development in which **cosmology** is dominated by magic, supernatural conceptions, and the potency of spirits.

The Natural Step Framework (TNSF) A comprehensive model for planning in complex systems aimed at integrating sustainable development into companies' strategic planning.

Nature–nurture debate A debate about what role the environment takes in the development of an individual and to what degree it is phylogenetically (see **phylogenetics**) determined (see also **epigenetics**).

Nature religion (see **mythic culture**).

New ecology An integrative approach centered on material and energy (metabolism and thermodynamics) used for the analysis of ecosystems.

New environmental paradigm (NEP) This replaced the **human exceptionalism paradigm** (HEP) in the second half of the twentieth century. The NEP stresses the ecological basis of **societies**, points out the limitations of **resources**, and the responsibility of societies towards the environment.

New institutional economics (NIE) An interdisciplinary endeavor combining economics, law, organization theory, political science, sociology, and anthropology which relates theoretical and empirical research examining the role of **institutions** in social, political, and commercial life. NIE expands neoclassical theory, but retains some of its basic assumptions, such as scarcity and competition, which form the basis of the choice-theoretic approach in **microeconomics**. It is often used to describe transaction costs, political economy, property rights, **hierarchy** and organization, and public choice.

North–south dilemma The difference in social, economic, ecological, and political state between developed countries (mostly in the north) and developing countries (mostly in the south).

Ontogenesis The development of individuals during their lifetime.

Ontology A branch of science relating to the nature and relations of being.

Panarchy A meta-model for describing ecosystem dynamics. In contrast with climax theories of globally stable states, panarchy characterizes ecosystem dynamics as a nested hierarchy of **adaptive cycles**: exploitation, conservation, release, and reorganization. Ecosystems run through these cycles at different speeds and vary in **resilience**, connectivity, and potential.

Pandemic An **epidemic** of **infectious disease**, which becomes very widespread and affects a large region, such as a continent or even the whole world.

Participatory action research An often ambiguously defined concept in which researchers aspire to an authentic dialogue.

Participatory research Research method in which people from the public at large participate in the research process. Participants take on roles that go beyond the mere proband or test subject role. The researcher keeps full control over the research process.

Pathogen An organism such as a virus or bacteria which is the cause of a disease in the host organism.

Peak oil/phosphorus The point in time when the worldwide extraction of oil/phosphorus reaches its maximum and thereafter starts declining.

Perception The process of becoming aware of and conscious of the environment by comprehending sensory information from the **environment** that enters the **human system** through registered physical sensations.

Phylogenesis The development of a species during biological **evolution**.

Physiocracy A French concept meaning "domination of nature." Physiocrats assume the existence of a natural order, in which land is the only productive **resource**.

Pigovian tax Named after the economist Arthur Pigou which balances the negative externalities that emerge from activities related to the private costs of the activity.

Planungszelle citizen panel A variant of participation in democratic governance, in which a randomly selected number of residents elaborate recommendations for a city, country or state. The lay participants are assisted by experts.

POET Refers to a simple structural model aiming to describe the relationship between humans and their **environment**, where P = population, O = organization, E = environment, and T = technology.

Political ecology Academic discipline focused on the relation between economic, political, social, and environmental issues. In contrast to apolitical ecological studies, this approach politicizes phenomena such as degradation, marginalization, environmental conflict, conservation or control.

Positive check Malthus proposed positive checks (and **preventive checks**) to narrow the gap between limited **resources** and the consumption of resources in a growing economy: a positive check (e.g. low labor productivity, insufficient purchasing power for food, wars owing to shortages of land) leads to an increase in poverty or a decline in the growth of the population.

Post-industrial societies Societies where the main sector of economic activity has shifted from industrial/manufacturing activities to third sector, service, and information-based activities.

Postulate Statement and assumption that serves as a basis for an argument and as a practical device to conceptualize and investigate human–environment systems.

Predating and foraging A kind of **appropriation of nature** that is driven by fixed, genetically programmed behavioral rules that respond to the presence or absence of environmental objects.

Preference function A rule that defines the desirability, satisfaction, and **utility** linked to a particular alternative.

Preventive check An example of a preventive check, as opposed to a **positive check**, would involve precautionary control of the growth of the population.

Primary feedback loop (also first order feedback) Direct and intended **feedback** from the environment to the **goal**-related action of the **human system**.

Primary production This production addresses the generation of **biomass** through photosynthesis. Production of plants is seen as **primary production. Secondary production** refers to productivity of heterotrophic organisms such as animals.

Private goods These goods are both excludable and rivalrous (see also **rivalry** and **excludability**), and hence tradeable in the marketplace. Examples include food or clothes.

Probabilistic functionalism An approach by Egon Brunswik for studying and modeling human **perception** and information processing based on the following principles: (1) functionalism; (2) complementarity of human and environmental systems; (3) probabilistic information acquisition; (4) vicarious mediation; (5) evolutionary stabilization of perception; and (6) representative design.

Production factors/production function Amount of different inputs used in the process of production of various quantities of a product. Three main factors are traditionally included in the analysis of production: land, labor, and capital, with entrepreneurship sometimes included as a fourth. The production function is the maximum amount of output of an economic system, such as a company or economy, given any combination of inputs.

Public goods These goods are non-rivalrous and non-excludable (see also **rivalry** and **excludability**), and can be used by anybody without charge, such as landscape, beautiful scenery or clean air.

Public participation Processes of democratic governance in which various stakeholders participate. These processes are controlled by public (governmental) authorities and are often used in a broader, fuzzy sense for processes in which some parties have (some) control about a process and let others participate.

Random process A stochastic, indeterminate process over time.

Rationality The type of reasoning applied by human systems. There are different types of rationality, such as full rationality, bounded rationality, collective rationality or ecological rationality, each characterized by different assumptions about inferential and memory-based capacities.

Realist stance Used synonymously for ontological materialism. It refers to a philosophical point of view assuming a mind-independent existence of reality. Ontological idealism is the antagonist to this philosophical position.

Recycling A set of processes in which collecting, sorting, dismantling, and reducing harmfulness of used materials take place to provide inputs to **secondary production**.

Rebound effect Secondary and higher ordered **feedback loops** that induce an unintended effect that is most often negative. We speak about a direct rebound effect if the negative outcomes are on the same (functional) unit than those focused by the action, indirect rebounds (negative feedbacks on other units) and structural rebound effects (negative changes of the structures in systems).

Regulatory mechanism Process that regulates the dynamics and trajectories of systems, often with the goal of keeping systems relatively stable. Regulatory mechanisms can be guided and modeled by direct and indirect **feedback** loops.

Reserve (of a **resource**) Share of the **reserve base** of, for example, oil or phosphate rock which is economically accessible.

Reserve base Identified share of a **resource**.

Resilience The ability of a system to return to its original form or to provide the former services or performances after being exposed to disturbances or adverse impacts.

Resilience approach A coupled-systems approach emerging from biology, which has been primarily transferred to the idea of framing social learning, governance, and collective actions on the level of **communities** and regions. The approach aspires to involve adaptive comanagement by scientists, decision-makers, and other stakeholders.

Resource Usually considered as a physical unit that may be seen as a natural (e.g. air, water, minerals) or a human resource (e.g. workforce) that is based on properties such as quantity, availability, and utility.

Resource-based sociotechnical evolutionary theory Refers to an extended version of Lenski's "evolutionary theory," postulating wealth and maintenance of power as main drivers of society.

Ricardo effect Describes that a rise in wages encourages capitalists to substitute machinery for labor and the other way round.

Risk From a decision-theoretic perspective, risk is the evaluation of the loss potential that relates to certain decision alternatives. From a toxicological perspective it is a function of exposure (emerging from action) and sensitivity.

Risk assessment A statement on the riskiness based on a transparent, well defined procedure, including the description of the unwanted events and a characterization of uncertainties inherent in the risk-inferring process itself.

Risk function The evaluation of the loss potential of decision alternatives in **risk situations** is termed risk function. A risk function maps from the alternatives of a **risk situation** (as a pre-image set) into the space (as an image set) of risk cognition (e.g. judgments).

Risk situation A risk situation is a model of a situation in which a decision-maker perceives a risk and makes a risk judgment. In a minimal risk, a decision-maker must have the choice between two alternatives, and an uncertain loss must be linked with (at least) one of the situations. A general risk situation is represented by a set of alternatives $A = \{A_1,\ldots,A_i,\ldots,A_n\}$. Each alternative A_i is linked to various uncertain possible outcomes $E_i = \{E_{i,1},\ldots,E_{i,k},\ldots,E_{i,kl}\}$, which may occur with probabilities $p_{i,k}$. At least one event $E_{i,k}$ must be considered to be a loss.

Risk society A term suggested by Ulrich Beck which states that modern societies are producing risks which have to be coped with on a world level.

Rivalry Occurs when the consumption of a good prevents or hinders its consumption by another person. Thus, the surplus of one consumer implies a decrease in utility for another. Most goods, both durable and non-durable, are rivalrous. Non-rivalrous goods, in contrast, can be enjoyed simultaneously by an unlimited number of consumers. Rivalry and **excludability** are key concepts used to define different types of goods.

Scenario analysis Scenario analysis is a method used to construct possible future states of a system/case (called scenarios). A scenario analysis provides insights into a system and its dynamics.

Second law of thermodynamics The second law of thermodynamics states that the **entropy** of a closed system that is not in equilibrium tends to increase over time, reaching a maximum value at equilibrium. Combined with the **fundamental thermodynamic relation**, the second law of thermodynamics places a limit on a system's ability to perform useful work. Because of a transition from order to disorder, energy is transformed into lower availability levels.

Secondary (or higher ordered) **feedback loop** Feedback from the environment to goal-related action of the **human system** that was either unintended (i.e. not considered during initial decision-making) or not anticipated by the existing model of human–environment interaction.

Secondary production (see **Primary production**).

Selection A process in which the environment promotes one unit of **evolution** (e.g. individuals, populations) to the disadvantage of another unit. Natural selection, the most important concept for adaptive evolution, denotes, for example, the reduction of fertility of certain individuals in a population with the consequence that other individuals multiply more.

Self-actualization An ongoing cognitive tendency to develop, maintain, and change one's own human capacities.

Self-organization Is seen as the development of a system or a process in which an internal organization of the particular system occurs.

Service society A type of society in which the provision of services becomes the main source of revenue.

Social constructivism A perspective which states that all of our knowledge about the environment is constructed and is not a simple mechanistic representation of the material-biophysical environment.

Social–ecological system (SES) A social–ecological system is a coupled and interacting system of human systems and ecological systems.

Social environment The social environment are the information-based, non-material social, and cultural rules, and also the knowledge available in human systems, which belong to the **environment** of a specific human system.

Social facts A construct suggested by Émile Durkheim that in sociology all social phenomena (social facts) have to be explained by other social phenomena.

Social metabolism Similar to biological metabolism, social metabolism in the **social metabolism approach** assumes a continuous circulation of materials and energy for the reproduction and maintenance of socioeconomic systems.

Social metabolism approach Scientific approach and method to track flows of materials and energy between society and natural systems.

Socially robust knowledge A form of epistemics or an orientation that: (i) meets state-of-the-art scientific knowledge; (ii) has the potential to attract consensus and is understandable by all stakeholder groups; (iii) acknowledges the uncertainties and incompleteness inherent in any type of knowledge about processes of the universe; (iv) generates processes of knowledge integration of different types of epistemics, particularly (different) scientific and experiential knowledge and (v) reflects on the constraints given by the context both of generating and utilizing knowledge.

Society A level in the **human system hierarchy** describing a federation, state or similar assembly of humans that is politically autonomous and which secures subsistence, survival, and continuation. Current societies (in general nation-states) have principal components such as economics, law and policy, culture, and science.

Socioecological theory A holistic approach that couples ecological and social systems. Key concepts are **resilience**, **vulnerability**, and adaptability.

Sociology of science A domain of sociology that deals with the social conditions, social structures, and processes of scientific activity and the effects of science.

Stability The capacity to withstand stressses. Stability denotes the ability of a system to remain its form, state or performance when being exposed to a disturbance.

Statistical mechanics Approach in physics that studies macroscopic systems from a microscopic or molecular perspective.

Stimulus–response pattern The stimulus–response (S–R) model is a core concept of **behaviorism**. It describes the causal linkage of a reaction–evocation potential that results in a behavioral tendency.

Strategy formation Process of developing strategies to manage/solve complex (environmental) problems, relying on a sound analysis of the system under consideration.

Strategy selection Process of selecting a strategy from different alternatives.

Strong sustainability A sustainability conception that refers to the debate on the substitutability between the natural environment and manufactured capital. Strong sustainability aims at keeping stocks of natural capital maintained and enhanced owing to the impossibility of being replaced by manufactured capital and to avoid the emission of manmade "unnatural" chemical compounds to the environment.

Structuralism An approach that stresses the interrelation between, and the structure of, the parts of a system.

Substitutive goods Two goods where the surplus in consumption of the first reduces the marginal utility of the consumption of the second. Substitutive goods should not be confused with **complementary goods**.

Supranational/suprasocietal institution A level in the **human system hierarchy** describing an organization having executive power and legislatures that can direct nation-states and determine nation-society's law without having an original sovereignty at the national level. Supranational systems have Kompetenz-Kompetenz (i.e. the authority to penalize or to change the legal order of subsystems) and are partly financially independent from the nation-states.

Sustainability to be defined as an ongoing inquiry on system limit management (i.e. avoiding unwanted critical transitions or collapses) in the frame of intragenerational and intergenerational justice.

Sustain-ability Certain system-related cognitive and action-related abilities which enables humans to sustainably manage complex human-environment systems. Key sustainabilities are (i) reading the environmental potential of a system, (ii) anticipating and coping with **rebound effects**, (iii) recognizing tipping points, (iv) dealing with conflicting goals within and between human systems, (v) designing resilient HES.

Sustainability learning A social learning process for preventing human systems from collapsing and promoting intragenerational and intergenerational justice. Mastering **second order feedback loops** is essential for sustainability learning.

Sustainability science Scientific field that aims to provide insights into the sustainable dynamics of human–environment systems. Oriented around the concept of **sustainability** as an ongoing process of system limit management in the frame of intergenerational and intragenerational justice.

Symbolism The process of developing a capacity for symbolic representation, communication, and behavior.

System dynamics A theory for utilizing differential equation modeling that was developed by Jay J. Forrester to describe, analyze, and understand structures of **feedback** mechanisms in socioeconomic systems. System dynamics can provide useful insights for designing effective policies.

Systems theory An approach that deals with the interrelationship between organisms and their related ecosystems.

Systems thinking A way of thinking that decomposes a whole entity into component parts and their interrelations.

Technology In this book technology usually denotes materialized technical products. Technology sometimes also denotes the knowledge about technical means.

Theory of recapitulation (basic biogenetic law) A hypothesis from Ernst Haeckel that states that individuals recapitulate their species' phylogenetic (see **phylo-genesis**) history during their ontogenetic (see **ontogenesis**) development.

Toxicity The degree to which a natural or synthetic substance affects organisms (e.g. human toxicity) or ecosystems (ecotoxicity, ecological toxicity).

Trade-off A situation where gaining one benefit results in the loss of another benefit.

Transactional approach Concept used above all in economics to illustrate the restrictiveness of assumption-based research and the necessity for skepticism in this context.

Transdisciplinarity Approach of study organizing processes that link scientific, theoretic, and abstract epistemics with real-world factors that are based on experiential knowledge from outside academia. Information about real-world factors comes from relating human experiential wisdom to the analytical rigor of science and academic methodology.

Transdisciplinary case study (TdCS) A hybrid process that combines learning, research, and application for capacity-building in science and society.

Transdisciplinary processes Processes of **transdisciplinarity** that emerge if a legitimized decision-maker and members from the science community collaborate to better understand complex, societally relevant phenomena by integrating knowledge from practice and from science. **Mutual learning** is a key element of transdisciplinary process.

Transmission system A system of infectious agents, hosts, and space in which hosts move, and environments change, affecting reproduction and survival of infectious agents outside the hosts.

Triple loop learning Three types of feedback loop learning can be defined in human–environment systems: (i) primary **feedback loop** learning, i.e. the human system takes into account short-term changes in the environmental system and adjusts its strategies according to these changes; (ii) secondary **feedback loop** learning, i.e. changing the internal model of human–environment interactions owing to an awareness of the long-term environmental change; and (iii) third, the genuine reflection of **human systems** on their **goals** and action programs.

Uncertainty Uncertainty results from incomplete knowledge with respect to the state of a specific object or from the variability of or limited access to data.

Utilitarianism Is a philosophical, ethical theory suggested by Jeremy Bentham and John Stuart Mill based on judging the moral worth of an action by the degree to which it contributes to the happiness of people.

Utility A concept for defining and operationalizing **preference functions**. It is used to explain rational behavior of a Homo oeconomicus, or "economic human." Preference and utility can be considered as a valuation of satisfaction resulting from the usefulness of something in the context of a specific human system.

Vienna social metabolism approach From the social science accentuated framework that considers the complementarity of human and environmental systems and the different rationales found in human and natural systems.

Virulence The ability of a **pathogen** to cause a disease.

Vulnerability Concept that describes the degree to which a system (e.g. organism) is likely to experience harm because of exposure to physical or emotional damage, or attack. This reflects the relationship that organisms have with their environment, such as people to natural, social, institutional, and cultural forces.

Vulnerability analysis Place-based analysis of coupled, sensitive, adaptive systems based on the concept of **vulnerability**.

Weak sustainability Optimistic interpretation of sustainability that emphasizes the substitutability of **resources** by human ideas and technology knowledge (see **strong sustainability**).

World view A comprehensive conception of the environment referring to the framework of ideas and beliefs through which an individual interprets and interacts with the environment. The **new environmental paradigm** (NEP) and the **human exceptionalism paradigm** (HEP) view, for example, represent two world views for coping with environmental problems.

Zeitgeist The general intellectual, ethical, cultural, and moral state of thinking and acting during a particular era.

References

Abbott, A. (2002). The disipline and the future. In S. Brint (ed.), *The Disciplines and the City of Intellect: The Changing American Universities* (pp. 205–30). Stanford, CA: Stanford University Press.

Abel, G. A. & Glinert, L. H. (2008). Chemotherapy as language: sound symbolism in cancer medication names. *Social Science & Medicine,* **66**(8), 1863–9.

Acheson, J. M. (2006a). Institutional failure in resource management. *Annual Review of Anthropology,* **35**, 117–34.

Acheson, J. M. (2006b). Lobster and groundfish management in the Gulf of Maine: a rational choice perspective. *Human Organization,* **65**(3), 240–52.

Adams, B. N. & Sydie, R. A. (2002). *Contemporary Sociological Theory*. Thousand Oaks, CA: Pine Forge.

Adelson, E. & Movshon, J. (1982). Phenomenal coherence of moving visual-patterns. *Nature,* **300**(5892), 523–5.

Ader, R. (2007). *Psychoneuroimmunology,* 4th edn. Amsterdam; Boston, MA: Elsevier/Academic Press.

Adger, W. N. (2006). Vulnerability. *Global Environmental Change,* **16**(3), 268–81.

Agricola, G. (1565). *De re metallica libri XII* (H. C. Hoover & L. H. Hoover, Trans.). New York, NY: Dover.

Aguzzi, A. & Callela, A. M. (2009). Prions: protein aggregation and infectious disease. *Physiological Review,* **89**(4), 1105–52.

Aguzzi, A. & Glatzel, M. (2006). Prion infections, blood and transfusion. *Nature Clinical Practice Neurology,* **2**(6), 321–9.

Ahl, V. & Allen, T. F. H. (1996). *Hierarchy Theory*. New York, NY: Columbia University Press.

Airoma, D. (2008). The mafia. *Cristianita*. Retrieved March 21, 2008, from http://www.alleanzacattolica.org/idis_dpf/english/m_the_mafia.htm

Aitken, C. K., McMahon, T. A., Wearing, A. J. & Finlayson, B. L. (1994). Residential water-use – predicting and reducing consumption. *Journal of Applied Social Psychology,* **24**(2), 136–58.

Ajzen, I. (1991). The theory of planned behavior. *Organizational Behavior and Human Decision Processes,* **50**(2), 179–211.

Ajzen, I. (2001). Nature and operation of attitudes. *Annual Review of Psychology,* **52**, 27–58.

Ajzen, I. (2002). Perceived behavioral control, self-efficacy, locus of control, and the theory of planned behavior. *Journal of Applied Social Psychology,* **32**(4), 665–83.

Albert, D. & Loewer, B. (1988). Interpreting the many worlds interpretation. *Synthese,* **77**(2), 195–213.

Alberts, B., Bray, D., Hopkin, K. *et al.* (2004). *Essential Cell Biology,* 2nd edn. London: Garland Science.

Alberts, B., Johnson, A., Lewis, J. *et al.* (2002). *Molecular Biology of the Cell,* 4th edn. London: Garland Science.

Albritton, C. C. (1980). *The Abyss of Time: Changing Conceptions of the Earth's Antiquity after the Sixteenth Century*. San Fransisco, CA: Freedman, Cooper.

Aleklett, K. (2007). *Peak Oil and the Evolving Strategies of Oil Importing and Exporting Countries: Facing the Hard Truth About an Import Decline for the OECD Counties*. OECD Publishing.

Alexander, D. (2000). *Confronting Catastrophe*. Harpenden: Terra.

Alexandrescu, F. M. (2009). Not as natural as it seems: the social history of the environment in American sociology. *History of the Human Sciences,* **22**(5), 47–80.

Alihan, M. A. (1938). *Social Ecology, a Critical Analysis*. New York, NY: Columbia University Press.

Allen, P. M. & McGlade, J. M. (1989). Optimality, Adequacy and the Evolution of Complexity. In P. L. Christiansen & R. D. Parmenties (eds). *Structure, Coherence and Chaos in Dynamical Systems*. Manchester: Manchester University Press, pp. 3–21.

Allen, T. F. H., Tainter, J. A. & Hoekstra, T. (2003). *Supply-Side Sustainability*. New York, NY: Columbia University Press.

Allenby, B. R. (1999). Culture and industrial ecology. *Journal of Industrial Ecology,* **3**(1), 2–4.

Allenby, B. R. (2004a). Clean production in context: an information infrastructure perspective. *Journal of Cleaner Production,* **12**(8–10), 833–9.

Allenby, B. R. (2004b). Infrastructure in the anthropocene: example of information and

References

communication technology. *Journal of Infrastructure Systems,* **10**(3), 79–86.

Allenby, B. R. (2007). Earth systems engineering and management. A manifesto. *Environmental Science & Technology,* **41**(23), 7961–5.

Allenby, B. R. (2009). The industrial ecology of emerging technologies. *Journal of Industrial Ecology,* **13**(2), 168–83.

Allis, C. D., Jenuwein, T. & Reinberg, D. (2007a). Overview and concepts. In C. D. Allis, T. Jenuwein, D. Reinberg & M.-L. Capparros, eds. *Epigenetics.* New York, NY: Cold Spring, pp. 23–61.

Allis, C. D., Jenuwein, T., Reinberg, D. & Capparos, M.-L. (eds). (2007b). *Epigenetics.* New York, NY: Cold Spring Harbor Laboratory Press.

Allison, H. E. & Hobbs, R. J. (2004). Resilience, adaptive capacity, and the "Lock-in trap" of the Western Australian agricultural region. *Ecology and Society,* **9**(1), 185–8.

Allport, G. W. (1955). *Theories of Perception and the Concept of Structure.* New York, NY: Wiley.

Almond, G. A. (2000). The study of political culture. In C. Lockhart & L. Crothers, eds. *Culture and Politics.* New York, NY: Palgrave MacMillan, pp. 5–19.

Alpini, P. (1592). *De plantis Aegypti.* Liber. Venice: F. de Franceschi di Siena.

Alt, K. W., Jeunesse, C., Buitrago-Téllez, C. H. et al. (1997). Evidence for stone age cranial surgery. *Nature,* **387**(6631), 360.

Altekar, A. S. (1965). *Education in Ancient India,* 6th edn. Varanasi: Nand Kishore & Bros.

Alter, K. J. (1998). Who are the "masters of the treaty"? European Governments and the European Court of Justice. *International Organization,* **52**(1), 121–47.

Altman, I. & Chemers, M. M. (1980). *Culture and Environment.* Cambridge: Cambridge University Press.

Altman, I. & Rogoff, B. (1987). World views in psychology: trait, interactional, organismic and transactional perspectives. In D. Stokols & I. Altmann, eds. *Handbook of Environmental Psychology,* Vol. I. New York, NY: Wiley, pp. 7–41.

Ames, A. (1952). *The Ames Demonstrations in Perception.* New York, NY: Hafner Publishing.

Ames, P. (1994, January 24). Life's not so sweet for Europe's embattled beekeepers. *Associated Press.* Retrieved February 27, 2008, from http://delontin1.wordpress.com

Amish Studies. (2010). Population trends 2007–2008. Retrieved May 10, 2010, from http://www2.etown.edu/amishstudies/Population_Trends_2007_2008.asp

Andersen, I.-E. & Jæger, B. (1999). Scenario workshops and consensus conferences: towards more democratic decision-making. *Science and Public Policy,* **26**(5), 331–40.

Anderson, A. (1992). The evolution of sexes. *Science,* **257**(5068), 324–6.

Anderson, C. C. & Matzinger, P. (2000). Danger: the view from the bottom of the cliff. *Seminars in Immunology,* **12**(3), 231–8.

Anderson, J. R. (1983). *The Architecture of Knowledge.* Cambridge, MA: Harvard University Press.

Anderson, O. (2007). Charles Lyell, uniformitarianism, and interpretive principles. *Zygon,* **42**(2), 449–62.

Ankli, A., Heinrich, M., Bork, P. et al. (2002). Yucatec Mayan medicinal plants: evaluation based on indigenous uses. *Journal of Ethnopharmacology,* **79**(1), 43–52.

Aoki, M. (2001). *Toward a Comparative Institutional Analysis.* Cambridge, MA: The MIT Press.

Araneda, R. C. & Firestein, S. (2004). The scents of androstenone in humans. *Journal of Physiology-London,* **554**(1), 1.

Arbuckle, J. L. (2006). *Amos™ 7.0 User's Guide.* Spring House, PA and Chicago, IL: Amos Development Corporation and SPSS.

Argyris, C. (1977). Double loop learning in organizations. *Harvard Business Review,* **55**(5), 115–25.

Argyris, C. (1978). *Organizational Learning.* Reading, MA: Addison-Wesley.

Argyris, C. (1982). Organizational learning and management of information systems. *Data Base,* **12**(2–3), 3–11.

Argyris, C. & Schön, D. (1978). *Organizational Learning. A Theory of Action Perspective,* Vol. 2. Reading, MA: Addison-Wesley Publishing.

Argyris, C. & Schön, D. A. (1989). Participatory action research and action science compared. *American Behavioral Scientist*, **32**(5) 612–23.

Argyris, C. & Schön, D. A. (1991). Participatory action research and action science compared. In W. F. Whyte, (ed.). *Participatory Action Research.* Newbury Park, CA: Sage, pp. 85–96.

Aristotle. (350 BC). Metaphysics. Retrieved November 20, 2008: http://classics.mit.edu/Aristotle/metaphysics.8.viii.html

Armitage, C. J. & Conner, M. (2001). Efficacy of the theory of planned behaviour: a meta-analytic review. *British Journal of Social Psychology,* **40**(4), 471–99.

Arnold, L. (1992). *Windscale 1957: Anatomy of a Nuclear Accident.* London: MacMillan.

Arrow, H., MacGrath, J. E. & Berdahl, J. L. (2000). *Small Group as Complex Systems. Formation, Coordination, Development, and Adaptation.* Thousand Oaks, CA: Sage.

Arrow, K. J. (1950). A difficulty in the concept of social welfare. *Journal of Political Economy,* **58**(4), 328–46.

Arrow, K. J. & Debreu, G. (1954). Existence of equilibrium for a competitive economy. *Econometrica,* **22**(3), 265–90.

Arrow, K. J. & Fisher, A. C. (1974). Environmental preservation, uncertainty, and irreversibility. *Quarterly Journal of Economics,* **88**(2), 312–19.

Asafu-Adjaye, J. (2005). *Environmental Economics for Non-economists*. Singapore: World Scientific.

Asch, S. E. (1955). Opinious and Social Pressure.. *Scientific American,* **193**(5), 31–5.

Ashby, W. R. (1956). *An Introduction to Cybernetics*. London: Chapman & Hall.

Aspromourgos, T. (1986). On the origins of the term neoclassical. *Cambridge Journal of Economics,* **10**(3), 265–70.

Åström, K. J. & Murray, E. M. (2008). *Feedback Systems: An Introduction for Scientists and Engineers*. Princeton, NJ: Princeton University Press.

Atance, C. M. & O'Neill, D. K. (2001). Episodic future thinking. *Trends in Cognitive Sciences,* **5**(12), 533–9.

Atkinson, J. W. (1953). The achievement motive and recall of interrupt and completed tasks. *Journal of Experimental Psychology,* **46**(6), 381–90.

Atkinson, M. & Wein, L. (2008). Quantifying the routes of transmission for pandemic influenza. *Bulletin of Mathematical Biology,* **70**, 820–67.

Atran, S. & Axelrod, R. (2008). Reframing sacred values. *Negotiation Journal,* **24**(3), 221–46.

Auyang, S. Y. (2004). *Engineering: An Endless Frontier*. Cambridge, MA: Harvard University Press.

AWEL. (2008). *Phosphor im Klärschlamm – Informationen zur künftigen Rückgewinnung*. Zurich: Baudirektion Zürich; AWEL Amt für Abfall, Wasser Energie und Luft.

Ax, P. (1996). *Multicellular Animals: A New Approach to Phylogenetic Order in Nature*. Berlin: Springer.

Ax, P. (2003). *Multicellular Animals: Order in Nature – System Made by Man,* Vol. III. Berlin: Springer.

Axelrod, L. J. (1984b). Balancing personal needs with environmental preservation: identifying the values that guide decisions in ecological dilemmas. *Journal of Social Issues,* **50**(3), 85–104.

Axelrod, R. (1984a). *The Evolution of Cooperation*. New York, NY: Basic Books.

Axelrod, R. M. (1997). *The Complexity of Cooperation: Agent-Based Models of Competition and Collaboration*. Princeton, NJ: Princeton University Press.

Axelrod, R. M. & Hamilton, W. D. (1981). The evolution of cooperation. *Science,* **211**(4489), 1390.

Ayres, C. E. (1957). Institutional economics: discussion. *American Economic Review,* **47**(47), 26–7.

Ayres, R. U. (1989). Industrial metabolism and global change. *International Social Science Journal,* **121**(3), 163–73.

Ayres, R. U. (1991, May 20–21). *Toxic Heavy-metals: Materials Cycle Optimization*. Paper presented at the NAS colloquium on industrial ecology, Washington, DC.

Ayres, R. U. (1998). Eco-thermodynamics: economics and the second law. *Ecological Economics,* **26**(2), 189–209.

Ayres, R. U. (2004). On the life cycle metaphor: where ecology and economics diverge. *Ecological Economics,* **48**(4), 425–38.

Ayres, R. U. (2007). On the practical limits to substitution. *Ecological Economics,* **61**(1), 115–28.

Ayres, R. U., Ayres, L. W. & Warr, B. (2004). Is the U.S. economy dematerializing? Main indicators and drivers. *Economics of Industrial Ecology. Materials, Structural Change, and Spatial Scales*. Cambridge, MA: The MIT Press, pp. 57–94.

Ayres, R. U. & Kneese, A. V. (1969). *Production, Consumption, and Externalities*. Washington, DC: Resources for the Future.

Ayres, R. U., van den Bergh, J. C. J. M. & Gowdy, J. M. (2001). Strong versus weak sustainability: economics, natural sciences, and "consilience." *Environmental Ethics,* **23**(2), 155–68.

Azar, C. (2006). Separate statement of opinion to the Commission on Oil Independence. In Commission on Oil Independence, (ed.). *Making Sweden an OIL-FREE Society*. Stockholm: Swedish Government, pp. 49–51.

Babich, F. R., Jacobson, A. L., Bubash, S. & Jacobson, A. (1965). Transfer of a response to naive rats by injection of ribonucleic acid extracted from trained rats. *Science,* **149**(3684), 656–7.

Baccarelli, A. & Bollati, V. (2009). Epigenetics and environmental chemicals. *Current Opinion in Pediatrics,* **21**(2), 243–51.

Baccarelli, A., Wright, R. O., Bollati, V. *et al.* (2009). Rapid DNA methylation changes after exposure to traffic particles. *American Journal of Respiratory and Critical Care Medicine,* **179**(7), 572–8.

Baccini, P. & Bader, H.-P. (1996). *Regionaler Stoffhaushalt. Erfassung, Bewertung und Steuerung*. Heidelberg: Spektrum.

Baccini, P. & Brunner, P. H. (1991). *Metabolism of the Anthroposphere*. Berlin: Springer.

Baccini, P. & Oswald, F. (eds). (1998). *Netzstadt – Transdisziplinäre Methoden zum Umbau urbaner Systeme*. Zurich: vdf.

Bacow, L. S. & Wheeler, M. (1984). *Environmental Dispute Resolution*. (2nd printing, 1987 edn.). New York, NY: Plenum Press.

References

Balakirev, E. S. & Ayala, F. J. (2003). Pseudogenes: are they "junk" or functional DNA? *Annual Review of Genetics,* **37**, 123–51.

Baland, J. M. & Platteau, J. P. (1996). *Halting Degradation of Natural Resources. Is there a Role for Rural Communities?* Oxford: Oxford University Press.

Baldwin, S. M. (1896). A new factor in evolution. *American Naturalist,* **21**(3), 441–51.

Balick, M. J., De Gezelle, J. M. & Arvigo, R. (2008). Feeling the pulse in Maya medicine: an endangered traditional tool for diagnosis, therapy, and tracking patients' progress. *Explore – the Journal of Science and Healing,* **4**(2), 113–19.

Ballauff, T. (1954). *Die Wissenschaft vom Leben. Eine Geschichte der Biologie, Band* 1. Freiburg: Karl Alber.

Ballauff, T. (1971). Biologie. In J. Ritter, (ed.). *Historisches Wörterbuch der Philosophie,* Vol. 1. Basel: Schwabe, p. 944.

Balme, M. (1991). *Aristotle: History of Animals, books VII–X.* Cambridge, MA: Harvard University Press.

Bamberg, S. (2003). How does environmental concern influence specific environmentally related behaviors? A new answer to an old question. *Journal of Environmental Psychology,* **23**(1), 21–32.

Bamberg, S., Ajzen, I. & Schmidt, P. (2003). Choice of travel mode in the theory of planned behavior: the roles of past behavior, habit, and reasoned action. *Basic and Applied Social Psychology,* **25**(3), 175–87.

Bamberg, S., Hunecke, M. & Blobaum, A. (2007). Social context, personal norms and the use of public transportation: two field studies. *Journal of Environmental Psychology,* **27**(3), 190–203.

Bandura, A. (1986). *Social Foundation of Thought and Action: A Social Cognitive Theory*. Englewood Cliffs, NJ: Prentice Hall.

Bandura, A. (2000). Exercise of human agency through collective efficacy. *Current Directions in Psychological Science,* **9**(3), 75–8.

Bandura, A. (2001). Social cognitive theory: an agentic perspective. *Annual Review of Psychology,* **52**, 1–26.

Bansal, S., Grenfell, B. T. & Meyers, L. A. (2007). When individual behaviour matters: homogeneous and network models in epidemiology. *Journal of the Royal Society Interface,* **4**(16), 879–91.

Barbieri, M. (2002). Has biosemiotics come of age? *Semiotica,* **139**(1–4), 283–95.

Barbieri, M. (ed.). (2007). *Introduction to Biosemiotics. The New Biological Synthesis*. Dordrecht: Springer.

Barker, R. G. (1968). *Ecological Psychology: Concepts and Methods for Studying the Environment of Human Behavior*. Stanford, CA: Stanford University Press.

Barker, R. G. (1978). *Habitats, Environments, and Human Behavior: Studies in Ecological Psychology and Eco-Behavioral Science*. San Francisco, CA: Jossey-Bass.

Barker, R. G. (1987). Prospecting environmental psychology: Oskaloosa revisited. In D. Stokols & I. Altmann (eds). *Handbook of Environmental Psychology,* Vol. II. New York, NY: Wiley, pp. 1413–32.

Barker, R. G. (1990). Recollections of the midwest psychological field station. *Environment and Behavior,* **22**(4), 503–13.

Barker, R. G. & Schoggen, P. (1973). *Qualities of Community Life*. San Francisco, CA: Jossey-Bass.

Barker, R. G. & Wright, H. F. (1949). Psychological ecology and the problem of the psychosocial development. *Child Development,* **20**(3), 131–55.

Barlas, Y. (1996). Formal aspects of model validity and validation in system dynamics. *System Dynamics Review,* **12**(3), 183–210.

Barlow, N. (ed.). (1958). *The Autobiography of Charles Darwin: 1809–1882*. New York, NY: Norton.

Barrett, L., Dunbar, R. & Lycett, L. (2009). *Human Evolutionary Psychology*. Princeton, NJ: Princeton University Press.

Barrionuevo, A. (2007, April 24). Bees vanish, scientists race for reasons. *The New York Times*. Retrieved April 28, 2010, from http://www.nytimes.com/2007/04/24/science/24bees.html

Barrows, H. H. (1923). Geography as human ecology. *Annals of the Association of American Geographers,* **13**(1), 1–14.

Barthel, R., Janisch, S., Nickel, D., Trifkovic, A. & Horhan, T. (2010). Using the multiactor-approach in Glowa-Danube to simulate decisions for the water supply sector under conditions of global climate change. *Water Resources Management,* **24**(2), 239–75.

Basolo, V., Steinberg, L. J., Burby, R. J. et al. (2009). The effects of confidence in government and information on perceived and actual preparedness for disasters. *Environment and Behavior,* **41**(3), 338–64.

Bateman, B. W. (2004). Why institutional economics matters as a category of historical analysis. In W. J. Samuels (ed.). *Research in the History of Economic Thought and Methodology,* Vol. 22, Part 1. Wagon Lane: Emerald Group Publishing, Ltd, pp. 193–201.

Bateson, G. (1979). *Mind and Nature: A Necessary Unity*. New York, NY: Dutton.

Batson, C. D., Ahmad, N. & Tsang, J. A. (2002). Four motives for community involvement. *Journal of Social Issues,* **58**(3), 429–45.

Batson, C. D., Batson, J. G., Todd, R. M. et al. (1995). Empathy and the collective good – caring for one of the others in a social dilemma. *Journal of Personality and Social Psychology,* **68**(4), 619–31.

Batten, D. F. (2009). Fostering industrial symbiosis with agent-based simulation and participatory modeling. *Journal of Industrial Ecology,* **13**(2), 197–213.

Bäumer, Ä. (1991). *Geschichte der Biologie. Biologie von der Antike bis zur Renaissance,* Vol. 1. Frankfurt am Main: Peter Lang.

Bäumer, Ä. (1991). *Geschichte der Biologie: Zoologie der Renaissance – Renaissance der Zoologie,* Vol. 2. Frankfurt am Main: Peter Lang.

Bäumer, Ä. (1996). *Geschichte der Biologie: 17 und 18. Jahrhundert,* Vol. 3. Frankfurt am Main: Peter Lang.

Baumert, K. A., Herzog, T. & Pershing, J. (2005). *Navigating the Numbers. Greenhouse Gas Data and International Climate Policy*. Washington, DC: World Resources Institute.

Baumgartner, T., Tietje, O., Spielmann, M. & Bandel, R. (2000). *Ökobilanz der Swissmetro – Umweltwirkungen durch Bau und Betrieb (Teil 1) und durch induzierte Aktivitäten (Teil 2)*. Zurich: ETH Zurich.

Baumol, W. J. (1999). Retrospectives – Say's law. *Journal of Economic Perspectives,* **13**(1), 195–204.

Baumol, W. J. & Oates, W. E. (1971). The use of standards and prices for protection of the environment. *Swedish Journal of Economics,* **73**(1), 42–54.

Bausell, R. B. (2007). *Snake Oil Science: the Truth about Complementary and Alternative Medicine*. Oxford: Oxford University Press.

Beall, C. M. & Goldstein, M. C. (1981). Tibetan fraternal polyandry – a test of sociobiological theory. *American Anthropologist,* **83**(1), 5–12.

Bechtel, R. B. (1997). *Environment and Behavior: an Introduction*. Thousand Oaks, CA: Sage.

Bechtel, R. B. & Churchman, A. (2002). *Handbook of Environmental Psychology*. New York, NY: Wiley.

Beck, U. (1995). *Ecological Politics in an Age of Risk*. Cambridge: Polity Press.

Beck, U. (1996). Weltrisikogesellschaft, Weltöffentlichkeit und globale Subpolitik: Ökologische Fragen im Bezugsrahmen fabrizierter Unsicherheit. *Kölner Zeitschrift für Soziologie und Sozialpsychologie,* **36**(SI36), 119–47.

Beck, U. (1999). Democracy beyond the nation-state. *Dissent,* **Winter 1999**, 53–5.

Beck, U. (2000). The cosmopolitan perspective: sociology of the second age of modernity. *British Journal of Sociology,* **51**(1), 79–105.

Beck, U. (2004). Cosmopolitical realism: on the distinction between cosmopolitanism in philosophy and the social sciences. *Global Networks,* **4**(2), 131–56.

Beck, U. (2006). *Cosmopolitan Vision*. Cambridge: Polity Press.

Beck, U. (2007a). Beyond class and nation: reframing social inequalities in a globalizing world. *British Journal of Sociology,* **58**(4), 679–705.

Beck, U. (2007b). Living in the world risk society: a Hobhouse memorial public lecture given on Wednesday 15 February 2006 at the London School of Economics. *Economy and Society,* **35**(3), 329–45.

Beck, U. & Beck-Gernsheim, E. (2009). Global generations and the trap of methodological nationalism for a cosmopolitan turn in the sociology of youth and generation. *European Sociological Review,* **25**(1), 25–36.

Beck, U. & Sznaider, N. (2006). Unpacking cosmopolitanism for the social sciences: a research agenda. *British Journal of Sociology,* **57**(1), 1–23.

Beckenkamp, M. (2006). A game-theoretic taxonomy of social dilemmas. *Central European Journal of Operations Research,* **14**(3), 337–53.

Becker, E. (2003). Soziale Ökologie: Konturen und Konzepte einer neuen Wissenschaft. In G. Matschonat & A. Gerber (ed). *Wissenschaftstheoretische Perspektiven für die Umweltwissenschaften*. Weikersheim: Margraf Publishers, pp. 165–95.

Becker, E. & Jahn, T. (eds). (2006). *Soziale Ökologie.* Frankfurt am Main: Campus.

Becquerel, A. C. (1853). *Des Climats et de l'Influence qu'exercents les Sols Boisés et non Boisés*. Paris: Diderot.

Beer, S. (1959). *Cybernetics and Management*. London: English Universities Press.

Begon, M., Townsend, C. R. & Harper, J. L. (1990). *Ecology Individuals, Populations and Communities,* 2nd edn. Boston, MA: Blackwell.

Behera, B. & Engel, S. (2006). Institutional analysis of evolution of joint forest management in India: a new institutional economics approach. *Forest Policy and Economics,* **8**(4), 350–62.

Beier, H. M. (2002). Die Entdeckung der Eizelle der Säugetiere und des Menschen. *Reproduktionsmedizin,* **18**(6), 333–8.

Beierkuhnlein, C. & Jentsch, A. (2005). Ecological importance of species diversity. In R. J. Henry (ed). *Plant Diversity and Evolution*. Wallingford: CABI, pp. 249–85.

Bell, M. L. & Davis, D. L. (2001). Reassessment of the lethal London fog of 1952: novel indicators of acute and chronic consequences of acute exposure to air pollution. *Environmental Health Perspectives Supplements,* **109**(S3), 289–394.

Bell, M. M. (2004). *An Invitation to Environmental Sociology*. Thousand Oaks, CA: Pine Forge Press.

Bell, P. A., Fisher, J. D., Baum, A. & Greene, T. C. (1996). *Environmental Psychology*. Fort Worth, TX: Holt, Rinehart and Winston.

References

Bell, S. & Morse, S. (2010a). Rich pictures: a means to explore the 'sustainable mind'? *Sustainable Development*.

Bell, S. & Morse, S. (2010b). Triple Task (TT) method: Systemic, reflective action reseaech. Unpublished manuscript.

Belon, P. (1555). *L'Histoire de la Nature des Oiseaux*. Paris: Chez Guillaume Cauellat.

Belton, V. & Stewart, T. J. (2002). *Multiple Criteria Decision Analysis: an Integrated Approach*. Dordrecht: Kluwer.

Ben David, J. (1968). *Fundamental Research and the Universities. Some Comments on International Differences*. Paris: OECD.

Ben David, J. (1971). *The Scientist's Role in Society. A Comparative Study*. Englewood Cliffs: Prentice-Hall.

Ben David, J. (1981/1972). *Trends in American Higher Education*. Chicago, IL: The University of Chicago Press.

Benedetti, F., Mayberg, H. S., Wager, T. D., Stohler, C. S. & Zubieta, J. K. (2005). Neurobiological mechanisms of the placebo effect. *Journal of Neuroscience*, **25**(45), 10390–402.

Benevolo, L. (1980). *The History of the City*. Cambridge, MA: The MIT Press.

Benjafield, J. & Davis, C. (1978). The golden section and the structure of connotation. *Journal of Aesthetics and Art Critics*, **36**(4), 423–7.

Bennett, S. (1996). A brief history of automatic control. *IEEE Control Systems Magazine*, **16**(3), 17–25.

Bentham, J. (1789/2005). *An Introduction to the Principle of Morals and Legislation*. Whitefish, MT: Kessinger Publishing.

Beratan, K. K. (2007). A cognition-based view of decision processes in complex social–ecological systems. *Ecology and Society*, **12**(1), 27. Retrieved January 6, 2011 from http://www.ecologyandsociety.org/vol12/iss1/art27

Berg, R. D. (1996). The indigenous gastrointestinal microflora. *Trends in Microbiology*, **4**(11), 430–5.

Berger, J. (2007). Why are some countries so much richer than others? On the institutional explanation of developmental differences. *Zeitschrift für Soziologie*, **36**(1), 5–24.

Berger, P. L. (2002). Whatever happened to sociology. *First Things*, **126**(8), 27–9.

Berger, P. L. & Luckmann, T. (1966). *The Social Construction of Reality: a Treatise in the Sociology of Knowledge*. Garden City, NY: Anchor.

Bergman, N., Haxeltine, A., Whitmarsh, L. et al. (2008). Modelling socio-technical transition patterns and pathways. *Journal of Artificial Societies and Social Simulation*, **11**(3), 1–32.

Bergmann, M., Brohmann, B., Hoffmann, E. et al. (2005). *Quality Criteria of Transdisciplinary Research. A Guide for the Formative Evaluation of Research Projects*. Frankfurt am Main: ISOE.

Berkes, F. (2008). *Sacred Ecology*, 3rd edn. New York, NY: Taylor & Francis.

Berkes, F., Colding, J. & Folke, C. (2003). *Navigating Social–Ecological Systems: Building Resilience for Complexity and Change*. Cambridge: Cambridge University Press.

Berkes, F. & Folke, C. (1998a). Linking social and ecological systems for resilience and sustainability. In F. Berkes & C. Folke (eds). *Linking Social and Ecological Systems: Management Practices and Social Mechanisms for Building Resilience*. Cambridge: Cambridge University Press, pp. 1–29.

Berkes, F. & Folke, C. (eds). (1998b). *Linking Social and Ecological Systems: Management Practices and Social Mechanisms for Building Resilience*. Cambridge: Cambridge University Press.

Berkes, F. & Holling, C. S. (2002). Back to the future: ecosystem dynamics and local knowledge. In L. H. Gunderson & C. S. Holling (eds). *Panarchy: Understanding Transformation in Human and Natural Systems*. Washington, DC: Island Press, pp. 121–46.

Berlin, B. (1992). *Ethnobiological Classification. Principles of Categorization of Plants and Animals in Traditional Societies*. Princeton, NJ: Princeton University Press.

Berlin, B. (2006). The first congress of ethnozoological nomenclature. *Journal of the Royal Anthropological Institute*, **12**(SI 2006), 23–44.

Berlin, B. & Kay, P. (1969). *Basic Color Terms: their Universality and Evolution*. Berkeley, CA: University of California Press.

Berlin, E. A., Jara, V. M., Berlin, B. et al. (1993). Me'winik – discovery of the biomedical equivalence for a Maya ethnomedical syndrome. *Social Science & Medicine*, **37**(5), 671–8.

Bernal, J. D. (1939). *The Social Function of Science*. Cambridge, MA: The MIT Press.

Bernal, J. D. (1970). *Wissenschaft 1–4: Science in History – Die Entstehung der Wissenschaft, die wissenschaftliche und die industrielle Revolution, die Naturwissenschaft der Gegenwart, die Gesellschaftswissenschaften*. Reinbeck: Rowohlt.

Bernard, C. (1878). *Leçons sur les Phénomènes de la Vie Communs aux Animaux et aux Végétaux*. Paris: Baillière.

Bernard, H. R. (2006). *Research Methods in Anthropology: Qualitative and Quantitative Approaches*. London: Sage.

Bernauer, T. (1995). The effect of international environmental institutions: how we might learn more. *International Organization*, **49**(2), 351–77.

Bernet, D. P. (2009, September 13). China lässt die Muskeln spielen: Der grösste Produzent von seltenen Metallen will den Export in den Westen einschränken. *NZZ am Sonntag*, p. 39.

Berninghaus, S. K., Ehrhart, K. M. & Gueth, W. (2006). *Strategische Spiele – eine Einführung in die Spieltheorie*, 2nd edn. Berlin: Springer.

Bernoulli, D. (1738/1954). Exposition of a new theory on the measurement of risk. *Econometrica,* **22**(1), 23–36.

Bernstein, B. (1966). Elaborated and restricted codes: their social origins and some consequences. *American Anthropologist,* **66**(6), 55–69.

Bero, L. A. (2005). Tobacco industry manipulation of research. *Public Health Reports,* **120**(2), 200–8.

Berry, M. J., Brivanlou, I. H., Jordan, T. A. & Meister, M. (1999). Anticipation of moving stimuli by the retina. *Nature,* **398**(6725), 334–8.

Beth, E. W. & Piaget, J. (1966). *Mathematical Epistemology and Psychology*. Dordrecht: Reidel.

Bicchieri, C. (2002). Covenants without swords – group identity, norms, and communication in social dilemmas. *Rationality and Society,* **14**(2), 192–228.

Biermann, F. (2008). Earth system governance: a research agenda. In O. L. Young, L. A. King & H. Schroeder (eds). *Institutions and Environmental Change: Principal Findings, Applications, and Research Frontiers*. Cambridge, MA: The MIT Press, pp. 277–302.

Bilsky, W. & Schwartz, S. H. (1994). Values and personality. *European Journal of Personality,* **8**(3), 163–81.

Binder, C. R. (2007). From material flow analysis to material flow management – Part II: the role of structural agent analysis. *Journal of Cleaner Production,* **15**(17), 1605–17.

Binder, C. R., de Baan, L., Mouron, P. & Wittmer, D. (2009). *Phosphorflüsse in der Schweiz: Stand, Risiken und Handlungsoptionen*. Zurich: Soziale und Industrielle Ökologie, Geographisches Institut. Universität Zürich.

Binder, C. R., Hofer, C., Wiek, A. & Scholz, R. W. (2004). Transition towards improved regional wood flows by integrating material flux analysis and agent analysis: the case of Appenzell Ausserrhoden, Switzerland. *Ecological Economics,* **49**(1), 1–17.

Binswanger, M. (2001). Technological progress and sustainable development: what about the rebound effect? *Ecological Economics,* **36**(1), 119–32.

Bird, A. (2007). Perceptions of epigenetics. *Nature,* **447**(7143), 396–8.

Bird-David, N. (1990). The giving environment: another perspective on the economic system of gatherer-hunters. *Current Anthropology,* **31**(2), 189–96.

Björkman, M. (1984). Decision making, risk taking and psychological time: review of empirical findings and psychological theory. *Scandinavian Journal of Psychology,* **25**(1), 31–49.

Blalock, J. E. & Smith, E. M. (2007). Conceptual development of the immune system as a sixth sense. *Brain Behavior and Immunity,* **21**(1), 23–33.

Blättel-Mink, B. & Kastenholz, H. (2005). Transdisciplinarity in sustainability research: diffusion conditions of an institutional innovation. *Journal of Sustainability and World Ecology,* **12**(1), 1–12.

Blech, C. & Funke, J. (2006). Zur Reaktivität von Kausaldiagramm-Analysen bei komplexen Systemen. *Zeitschrift für Psychologie,* **214**(4), 185–95.

Bloomfield, P., Wheatley, D. J., Prescott, R. J. & Miller, H. C. (1991). Twelve-year comparison of a Bjork-Shiley mechanical heart-valve with porcine bioprostheses. *New England Journal of Medicine,* **324**(9), 573–9.

Blumenthal, A. L. (2001). Wundt, Wilhelm Maximilian. In A. E. Kazdin (ed.). *Encyclopaedia of Psychology*, Vol. 8. Oxford: Oxford University Press, pp. 289–92.

Blumer, H. (1969). *Symbolic Interactionism*. Englewood Cliffs, NJ: Prentice-Hall.

Boersema, J. J., Blowers, A. & Martin, A. (2008). The religion–environment connection. *Environmental Sciences,* **5**(4), 217–22.

Böhm, G. & Pfister, H. R. (2005). Consequences, morality, and time in environmental risk evaluation. *Journal of Risk Research,* **8**(6), 461–79.

Bohr, N. (1931). *Atomtheorie und Naturbeschreibung*. Berlin: Springer.

Bonner, J. (1973). Hierarchical control programs in biological development. In H. D. Pattee (ed.). *Hierarchy Theory: The Challenge of Complex Systems*. New York, NY: Braziller, pp. 49–70.

Bonnes, M. & Bonaiuto, M. (2002). Environmental psychology: from spatial-physical environment to sustainable development. In R. B. Bechtel & A. Churchman (eds). *Handbook of Environmental Psychology*. New York, NY: Wiley, pp. 28–54.

Bookchin, M. (2007). *Social Ecology and Communalism*. Oakland, CA: AK Press.

Boons, F. & Roome, N. (2000). Industrial ecology as a cultural phenomenon: on objectivity as a normative position. *Journal of Industrial Ecology,* **4**(2), 49–54.

Boots, M., Hudson, P. J. & Sasaki, A. (2004). Large shifts in pathogen virulence relate to host population structure. *Science,* **303**(5659), 842–4.

Boots, M. & Mealor, M. (2007). Local interactions select for lower pathogen infectivity. *Science,* **315**(5816), 1284–6.

Boots, M. & Sasaki, A. (1999). 'Small worlds' and the evolution of virulence: infection occurs locally and at a distance. *Proceedings of the Royal Society of London Series B-Biological Sciences,* **266**(1432), 1933–8.

Borchardt, J. K. (2004). Medicine of the Maya Amerindians. *Drug News & Perspectives,* **17**(5), 347–51.

Bordeau, M. (2008). Auguste Comte. *Stanford Encyclopedia of Philosophy (SEP).* Retrieved December 25, 2009, from http://plato.stanford.edu/entries/comte/#SocDouSta

Börjesson, P. (2009). Good or bad bioethanol from a greenhouse gas perspective – what determines this? *Applied Energy,* **86**(5), 589–94.

Bortz, J. (2005). *Statistik für Human- und Sozialwissenschaftler,* 6th edn. Berlin: Springer.

Bossel, H. (1994). *Modeling and Simulation.* Wellesley, MA: Peters Ltd.

Botnariuc, N. & Jahn, I. (1985). Die Biologie in der Zeit der Renaissance und des Manufakturkapitalismus. In I. Jahn, R. Löther & K. Senglaub (eds). *Geschichte der Biologie: Theorien, Methoden, Institutionen, Kurzbiografien.* Jena: Gustav Fischer, pp. 162–323.

Boulding, K. E. (1956). General systems theory: the skeleton of science. *Management Science,* **2**(3), 197–208.

Boulding, K. E. (1965). Earth as a space ship. Retrieved May 4, 2010, from University of Colorado at Boulder Libraries: http://earthmind.net/earthmind/docs/boulding-1965.pdf

Boulding, K. E. (1978). *Ecodynamics: A New Theory of Societal Evolution.* Thousand Oaks, CA: Sage.

Boulding, K. E. (1991). What is evolutionary economics? *Journal of Evolutionary Economics,* **1**(1), 9–17.

Bourdieu, P. (2003). *Die feinen Unterschiede: Kritik der gesellschaftlichen Urteilskraft.* Frankfurt am Main: Suhrkamp.

Bourdieu, P. & Wacquant, L. J. D. (1992). *An Invitation to Reflexive Sociology.* Chicago, IL: Chicago University Press.

Bousquet, F. & Le Page, C. (2004). Multi-agent simulations and ecosystem management: a review. *Ecological Modelling,* **176**(3–4), 313–32.

Boussingault, J. B. (1845). *Rural Economy, in its Relation with Chemistry, Physics, and Meteorology; an Application of the Principles of Chemistry and Physiology to the Details of Practical Farming.* London: Baillière.

Bowler, P. J. (1989). *The Mendelian Revolution: The Emergence of Ereditarian Concepts in Modern Science and Society.* Baltimore, MD: Johns Hopkins University Press.

Bowman, C. H. & Fishbein, M. (1978). Understanding public reaction to energy proposals – an application of the Fishbein model. *Journal of Applied Social Psychology,* **8**(4), 319–40.

Boxall, P. C., Adamowicz, W. L., Swait, J., Williams, M. & Louviere, J. (1996). A comparison of stated preference methods for environmental valuation. *Ecological Economics,* **18**(3), 243–53.

Boyd, R. & Richerson, P. (1988). *Culture and the Evolutionary Process.* Chicago, IL: University of Chicago Press.

Boyden, S. (1981). *Ecology of a City and its People: The Case of Hong Kong.* Canberra: Australian National University.

Boyden, S. (1992). *Biohistory: The Interplay between Human Society and the Biosphere, Past and Present.* Paris: UNESCO.

Boyden, S. (2001). Nature, society, history and social change. *Innovation,* **14**(2), 101–16.

Brachinger, H. W. & Weber, M. (1997). Risk as a primitive: a survey of measures of perceived risk. *Operations Research-Spectrum,* **19**(4), 235–94.

Bradford, L. P., Gibb, J. R. & Benn, K. D. (1964). *T-Group Theory and Laboratory Method.* New York, NY: John Wiley.

Bradshaw, C. J. A., Sodhi, N. S., Peh, K. S. H. & Brook, B. W. (2007). Global evidence that deforestation amplifies flood risk and severity in the developing world. *Global Change Biology,* **13**(11), 2379–95.

Brager, G. S. & de Dear, R. J. (1998). Thermal adaptation in the built environment: a literature review. *Energy and Buildings,* **27**(1), 83–96.

Brand, F. (2009). Critical natural capital revisited: ecological resilience and sustainable development. *Ecological Economics,* **68**(3), 605–12.

Brand, F. S. & Jax, K. (2007). Focusing the meaning(s) of resilience: resilience as a descriptive concept and a boundary object. *Ecology and Society,* **12**(1), 23.

Brankston, G., Gitterman, L., Hirji, Z., Lemieux, C. & Gardam, M. (2007). Transmission of influenza A in human beings. *Lancet Infectious Diseases,* **7**(4), 257–65.

Brashares, J. S., Arcese, P., Sam, M. K. *et al.* (2004). Bushmeat hunting, wildlife declines, and fish supply in West Africa. *Science,* **306**(5699), 1180–3.

Bratt, C. (1999). The impact of norms and assumed consequences on recycling behavior. *Environment and Behavior,* **31**(5), 630–56.

Braungart, M. & McDonough, W. (2009). *Cradle to Cradle.* London: Vintage.

Bremekamp, C. E. B. (1953). Linné's views on the hierarchy of the taxonomic groups. *Acta Botanica Neerlandica,* **2**(2), 580–93.

Brenchley, P. J. & Harper, D. A. T. (1998). *Palaeoecology: Ecosystems, Environment and Evolution.* London: Chapman & Hall.

Brentano, F. (1956). *Die Lehre vom richtigen Urteil.* Bern: Francke.

Bresler, J. B. (1966). *Human Ecology: Collected Readings.* Reading, MA: Addison-Wesley.

Bretschger, L. (1999). *Growth Theory and Sustainable Development*. Cheltenham: Edward Elgar.

Bridge, G. (2004). Contested terrain: mining and the environment. *Annual Review of Environment and Resources,* **29**, 205–59.

Brink, M., Müller, C. & Schierz, C. (2006). Contact-free measurement of heart rate, respiration rate, and body movements during sleep. *Behavior Research Methods,* **38**(3), 511–21.

Brink, M., Wirth, K. & Schierz, C. (2006). *Effects of Early Morning Aircraft Overflights on Sleep and Implications for Policy Making.* Paper presented at the Euronoise 2006: Advanced Solutions for noise control – The 6th European conference on noise control.

Brock, T. D. (1988). The bacterial nucleus – a history. *Microbiological Reviews,* **52**(4), 397–411.

Broderick, K. (2007). Getting a handle on social-ecological systems in catchments: the nature and importance of environmental perception. *Australian Geographer,* **38**(3), 297–308.

Brogan, W. L. (1990). *Modern Control Theory,* 3rd edn. Upper Saddle River, NJ: Prentice Hall.

Bromley, D. W. (2005, October 5–7). *The Emergence and Evolution of Natural Resource Economics: 1950–2000.* Paper presented at the Frontiers in Resource and Rural Economics: Rural-Urban Interplay and Nature-Human Interactions, Oregon State University, US.

Bronfenbrenner, U. (1979). *The Ecology of Human Development: Experiments by Nature and Design.* Cambridge, MA: Harvard University Press.

Brooks, D. B. & Andrews, P. W. (1974). Mineral-resources, economic, growth, and world population. *Science,* **185**(4145), 13–19.

Brosius, J. P. (1997). Endangered forest, endangered people: environmentalist representations of indigenous knowledge. *Human Ecology,* **25**(1), 47–69.

Brousseau, G. (1997). *Theory of Didactical Situations in Mathematics: Didactique des Mathématiques, 1970-1990.* Dordrecht: Kluwer.

Brown, J. H. (1995). *Macroecology.* Chicago, IL: University of Chicago Press.

Brown, K., Adger, W. N., Tompkins, E. *et al.* (2001). Trade-off analysis for marine protected area management. *Ecological Economics,* **37**(3), 417–34.

Brown, P. & Bradley, R. (1998). 1755 and all that: a historical primer of transmissible spongiform encephalopathy. *British Medical Journal,* **317**(7174), 1688–92.

Brown, P., Will, R. G., Bradley, R., Asher, D. M. & Detwiler, L. (2001). Bovine spongiform encephalopathy and variant Creutzfeldt-Jakob disease: background, evolution, and current concerns. *Emerging Infectious Diseases,* **7**(1), 6–16.

Brown, S. P. A. & Wolk, D. (2000). Natural resource scarcity and technological change. *Economic and Financial Policy Review,* **2000**(Q1), 2–13.

Brown, T. A. (1996). Measuring Chaos Using the Lyapnnov Exponent. In L. D. Kiel & E. Eliot (eds). *Chaos Theory in the Social Sciences.* Ann Arbor, MI: The University of Michigan Press, pp. 53–66.

Brownlee, J. (2007). *Cognition, Immunology, and the Cognitive Paradigm.* Melbourne: Faculty of Information and Communication Technology, Swineburne University of Technology.

Brumm, A., Aziz, F., van den Bergh, G. D. *et al.* (2006). Early stone technology on flores and its implications for *Homo floresiensis. Nature,* **441**(7093), 624–8.

Bruner, J. S. & Goodman, C. C. (1947). Values and needs as organizing factors in perception. *Journal of Abnormal Social Psychology,* **42**(1), 22–44.

Bruner, J. S. & Kretch, D. (1949). *Perception and Personality.* Durham, NC: Duke University Press.

Bruner, J. S. & Postman, L. (1949). Perception, cognition, and behavior. *Journal of Personality,* **18**(1), 14–31.

Brunner, P. H. (2007). Reshaping urban metabolism. *Journal of Industrial Ecology,* **11**(2), 11–13.

Brunner, P. H. & Rechberger, H. (2003). *Practical Handbook of Material Analysis.* Boca Raton, FL: Lewis.

Brunswik, E. (1927). *Strukturmonismus und Physik.* Vienna: Universität Wien.

Brunswik, E. (1933). Die Zugänglichkeit von Gegenständen für die Wahrnehmung und deren quantitative Bedeutung. *Archiv für die gesamte Psychologie,* **88**, 377–418.

Brunswik, E. (1934). *Wahrnehmung und Gegenstandswelt: Grundlegung einer Psychologie vom Gegenstand her.* Leipzig: Deuticke.

Brunswik, E. (1935). *Experimentelle Psychologie.* Vienna: Springer.

Brunswik, E. (1936). Psychologie vom Gegenstand her. In Congress Board of the International Congress of Philosophy (ed.). *Congrès International de Philosophie à Prague, 2–7 Septembre 1934.* Prag: Orbis, pp. 840–5.

Brunswik, E. (1937). Psychology as a science of objective relations. *Philosophy of Science,* **4**(2), 227–60.

Brunswik, E. (1943). Organismic achievement and environmental probability. *Psychological Review,* **50**(3), 255–72.

Brunswik, E. (1949). Discussion: remarks on functionalism in perception. *Journal of Environmental Personality,* **18**, 56–65.

Brunswik, E. (1952). *The Conceptual Framework of Psychology.* Chicago, IL: University of Chicago Press.

Brunswik, E. (1955a). In defense of probabilistic functionalism: a reply. *Psychological Review,* **62**(3), 236–42.

References

Brunswik, E. (1955b). Representative design and probabilistic theory in a functional psychology. *Psychological Review,* **62**(3), 193–217.

Brunswik, E. (1957). Scope and aspects of the cognitive problem. In H. Gruber, K. R. Hammond & R. Jessor (eds). *Contemporary Approaches to Cognition.* Cambridge, MA: Harvard University Press, pp. 5–31.

Brydon, L., Walker, C., Wawrzyniak, A. *et al.* (2009). Synergistic effects of psychological and immune stressors on inflammatory cytokine and sickness responses in humans. *Brain Behavior and Immunity,* **23**(2), 217–24.

Buchert, M., Schüler, D. & Bleher, D. (2009). *Critical Metals for Future Sustainable Technologies and their Recycling Potential: Executive Summary*. Freiburg: Öko-Institut e.V.

Bügl, R. & Scholz, R. W. (forthcoming). Sustainable urban lifestyles: an exploration of human–environment systems.

Bullard, R. (1990). *Dumping in Dixie: Race, Class and Environment Quality*. Boulder, CO: Westview Press.

Bullard, R. & Wright, B. (2009). *Race, place, and Environmental Justice after Hurricane Katrina: Struggles to Reclaim, Rebuild, and Revitalize New Orleans and the Gulf Coast*. Boulder, CO: Westview.

Bullough, V. L. (ed.). (2001). *Encyclopedia of Birth Control*. Santa Barbara, CA: ABC-CLIO.

Bundesamt für Gesundheit (BAG). (2009a). Creutzfeldt-Jakob-Krankheit (CJK)/Prionen. 2010. Retrieved January 6, 2011 from http://www.bag.admin.ch/themen/medizin/00682/00684/01065

Bundesamt für Gesundheit (BAG). (2009b). *Influenza-Pandemieplan Schweiz. Strategien und Massnahmen in Vorbereitung auf eine Influenza-Pandemie. Version Januar 2009*. Bern: Bundesamt für Gesundheit.

Bundesamt für Gesundheit (BAG). (2009c). Masken tragen. Retrieved January 6, 2011 from http://www.bag.admin.ch/influenza/01120/01132/10108/10115/index.html?lang=de

Bundesanstalt für Geowissenschaften und Rohstoffe (BGR), (2006). *Reserven, Resourcen und Verfügbarkeit von Energierohstoffen*. Hannover: BGR.

Burger, E. (1959). *Einführung in die Spieltheorie*. Berlin: de Gruyter.

Bürger, K. (1935). *Der Landschaftsbegriff: Ein Beitrag zur geographischen Erdraumauffassung*. Dresden: Zahn & Jaensch.

Burger, P. & Kamber, R. (2003). Cognitive integration in transdisciplinary science: knowledge as a key notion. *Issues in Integrative Studies,* **21**, 43–73.

Burgess, E. W. (1925/1967). The growth of the city: an introduction to a research project. In R. E. Park, E. W. Burgess & R. D. McKenzie (eds). *The City*. Chicago, IL: Chicago University Press, pp. 47–62.

Burnet, F. M. (1957/2007). A modification of Jerne's theory of antibody production using the concept of clonal selection. (Reprinted from *Australian Journal of Science*, Vol 20, 1957.) *Nature Immunology,* **8**(10), 1024–6.

Buroway, M. (2005). For public sociology. *American Sociological Review,* **70**, 4–28.

Burtraw, D., Krupnik, A., Mansur, E., Austin, D. & Farrell, D. (1998). Costs and benefits of reducing air pollutants related to acid rain. *Contemporary Economic Policy,* **16**(4), 379–400.

Buss, D. M. (2007). The evolution of human mating. *Acta Psychologica Sinica,* **39**(3), 502–12.

Buttel, F. H. (1996). Environmental and resource sociology: theoretical issues ond opportunities for synthesis. *Rural Sociology,* **61**(1), 56–76.

Buttel, F. H. (2004). The treadmill of production – an appreciation, assessment, and agenda for research. *Organization & Environment,* **17**(3), 323–36.

Buttel, F. H., Dickens, P., Dunlap, R. E. & Gijswijt, A. (2002). Sociological theory and the environment. In R. E. Dunlap *et al.* (eds). *Sociological Theory and the Environment. Classical Foundations, Contemporary Insights*. Lanham: Rowman & Littlefield, pp. 3–34.

Buttel, F. H. & Taylor, P. J. (1992). Environmental sociology and global environmental-change – a critical-assessment. *Society & Natural Resources,* **5**(3), 211–30.

Buxton, N. K. (1978). *The Economic Development of the British Coal Industry from Industrial Revolution to the Present Day*. London: Batsford Academic.

Byrne, W. L., Samuel, D., Bennett, E. L. *et al.* (1966). Memory transfer. *Science,* **153**(3736), 658–9.

Cacioppo, J. T., Kao, C. F., Petty, R. E. & Rodriguez, R. (1986). Central and peripheral routes to persuasion – an individual difference perspective. *Journal of Personality and Social Psychology,* **51**(5), 1032–43.

Calice, J. (2005). *"Sekundärrohstoffe – eine Quelle, die nie versiegt." Konzeption und Argumentation des Abfallverwertungssystems in der DDR aus umwelthistorischer Perspektive*. Unpublished diploma thesis. Vienna: Universität Wien.

Camerer, C. F. & Ho, T. H. (1991). Violations of the betweenness axiom and nonlinearity in probability. *Journal of Risk and Uncertainty,* **8**(2), 167–96.

Camilleri, K. (2006). Heisenberg and the wave-particle duality. *Studies in History and Philosophy of Modern Physics,* **37**(2), 298–315.

Camilleri, K. (2007). Bohr, Heisenberg and the divergent views of complementarity. *Studies in History and Philosophy of Modern Physics,* **38**(3), 514–28.

Campbell, R. & Sato, H. (2009). Policy and politics of BSE in the United States. In H. Sato (ed.). *Management of Health Risks from Environment and Food*. Dordrecht: Springer, pp. 317–38.

Cannon, W. B. (1929). Organization for physiological homeostasis. *Physiological Reviews,* **9**(3), 399–431.

Cannon, W. B. (1932). *The Wisdom of the Body*. New York, NY: Norton.

Cantillon, R. (1755). *Essai sur la Nature du Commerce en Général*. London: Fletcher Gyles.

Capek, S. M. (1993). The "Enviromental Justice" frame: a conceptual discussion and an application. *Social Problems*, **40**(1), 5–24.

Capello, R. & Faggian, A. (2005). Collective learning and relational capital in local innovation processes. *Regional Studies*, **39**(1), 75–87.

Capoccia, G. & Kelemen, R. D. (2007). The study of critical junctures – theory, narrative, and counterfactuals in historical institutionalism. *World Politics*, **59**(3), 341–59.

Capon, N. & Kuhn, D. (1979). Logical reasoning in the supermarket: adult females' use of a proportional reasoning strategy in an everyday context. *Developmental Psychology*, **15**(4), 450–2.

Caporael, L. R., Dawes, R. M., Orbell, J. M. & Van de kragt, A. J. C. (1989). Selfishness examined: cooperation in the absence of egoistic incentives. *Behavioral and Brain Sciences*, **12**(4), 683–739.

Capra, F. (1996). *The Web of Life*. New York, NY: Anchor Books.

Capra, F. (1999). *Lebensnetz. Ein neues Verständnis der lebendigen Welt*. Munich: Droemersche Verlagsanstalt.

Capra, F. (2000). *The TAO of Physics: an Exploration of the Parallels between Modern Physics and Eastern Mysticism*. Boston, MA: Shambhala.

Carlin, J. F. (2007). Bismuth. In US Geological Survey (USGS) (ed.). *2007 Minerals Yearbook*. Washington, DC: USGS, pp. 12.11–12.15.

Carlson, C., Burtraw, D., Cropper, M. & Palmer, K. L. (2000). Sulfur dioxide control by electric utilities: what are the gains from trade? *Journal of Political Economy*, **108**(6), 1292–326.

Carlson, E. A. (2006). *Times of Triumph, Times of Doubt: Science and the Battle for Public Trust*. Cold Spring, NY: Harbor Laboratory Press.

Carlsson, L. & Berkes, F. (2005). Co-management: concepts and methodological implications. *Journal of Environmental Management*, **75**(1), 65–76.

Carolan, M. S. (2006). The values and vulnerabilities of metaphors within the environmental sciences. *Society & Natural Resources*, **19**(10), 921–30.

Carpenter, S., Walker, B., Anderies, J. M. & Abel, N. (2001). From metaphor to measurement: resilience of what to what? *Ecosystems,* **4**(8), 765–81.

Carson, R. (1962). *Silent Spring*. Boston, MA: Houghton Mifflin.

Cassell, C. & Johnson, P. (2006). Action research: explaining the diversity. *Human Relations*, **59**(6), 783–814.

Cassidy, T. (1997). *Environmental Psychology: Behaviour and Experience in Context*. Hove: Psychology Press.

Cassier, M. (2008). Producing, controlling, and stabilizing Pasteur's anthrax vaccine: creating a new industry and a health market. *Science in Context,* **21**(2), 253–78.

Cassmann, K. G. (1999). Ecological intensification of cereal production systems: yield potential, soil quality, and precision agriculture. *Proceedings of the National Academy of Sciences*, **96**, 5952–9.

Castle, E. (1952). Some aspects of the crop-share lease. *Land Economics,* **28**(2), 177–9.

Catton, W. R. & Dunlap, R. E. (1978). Environmental sociology, a new paradigm. *The American Sociologist*, **13**(1), 41–9.

Cauvin, J. (2000). *The Births of the Gods and the Beginning of Agriculture*. Cambridge: Cambridge University Press.

Cavallo, A. J. (2002). Predicting the peak in world oil production. *Natural Resources Research,* **11**(3), 187–95.

Cavan, R. S. (1983). The Chicago School of Sociology, 1918–1933. *Journal of Contemporary Ethnography*, **11**(4), 407–20.

CDHRI. (1990). *Cairo Declaration on Human Rights in Islam*. Retrieved January 6, 2011 from http://www.religlaw.org/interdocs/docs/cairohrislam1990.htm

Centers for Disease Control and Prevention (CDC). (2009a). CDC guidance on helping child care and early childhood programs respond to influenza during the 2009–2010 influenza season. Retrieved November 14, 2009, from http://www.cdc.gov/h1n1flu/childcare/guidance.htm

Centers for Disease Control and Prevention (CDC). (2009b). Guidance for community settings: interim CDC guidance for public gatherings in response to human infections with novel influenza A (H1N1). Retrieved November 14, 2009, from http://www.cdc.gov/h1n1flu/guidance/public_gatherings.htm

Centers for Disease Control and Prevention (CDC). (2009c). Infection control guidance for the prevention and control of influenza in acute-care facilities. Retrieved November 15, 2009, from http://www.cdc.gov/flu/professionals/infectioncontrol/healthcarefacilities.htm

Centers for Disease Control and Prevention (CDC). (2009d). Interim recommendations for facemask and respirator use to reduce 2009 influenza A (H1N1) virus transmission. Retrieved November 14, 2009, from http://www.cdc.gov/h1n1flu/masks.htm

Cervero, R. (2002). Built environments and mode choice: toward a normative framework. *Transportation Research Part D – Transport and Environment,* 7(4), 265–84.

Chapin, F. S., Folke, C. & Kofinas, G. P. (2009). A framework for understanding change. In F. S. Chapin, G. P. Kofinas & C. Folke (eds). *Principles of Ecosystem Stewardship: Resilience-Based Natural Resource Management in a Changing World.* New York, NY: Springer, pp. 3–28.

Chapin, F. S., Kofinas, G. P. & Folke, C. (2009). *Principles of Ecosystem Stewardship: Resilience-Based Natural Resource Management in a Changing World.* New York, NY: Springer.

Chaplin, J. P. (2000). Functionalism. In A. E. Kazdin (ed.). *Encyclopedia of Psychology.* Oxford: Oxford University Press, pp. 416–9.

Chapple, E. D. & Coon, C. S. (1953). *Principles of Anthropology.* New York, NY: Henry Holt.

Checker, M. (2005). *Polluted Promises: Environmental Racism, and the Search for Justice in a Southern Town.* New York, NY: New York University Press.

Checkland, P. (1981). *Systems Thinking, Systems Practice.* Chichester: Wiley and Sons.

Checkland, P. (2000). Soft systems methodology: a thirty year retrospective. *Systems Research and Behavioral Science,* 17(S11), 11–58.

Checkland, P. (2006). Reply to Eden and Ackermann: any future for problem structuring methods? *Journal of the Operational Research Society,* 57(7), 769–71.

Checkland, P. & Holwell, S. (1998). *Information, Systems, and Information Systems.* Chichester: Wiley.

Checkland, P. & Scholes, J. (1990). *Soft Systems Methodology in Action.* Chichester: Wiley.

Chen, H., Cheung, C. L., Tai, H. et al. (2009). Oseltamivir-resistant influenza A pandemic (H1N1) 2009 virus, Hong Kong, China. *Emerging Infectious Diseases,* 15(12), 1970–2.

Chen, S. C., Chang, C. F. & Liao, C. M. (2006). Predictive models of control strategies involved in containing indoor airborne infections. *Indoor Air,* 16(6), 469–81.

Chernilo, D. (2007). *A Social Theory of the Nation State: the Political Forms of Modernity beyond Methodological Nationalism.* London: Routledge.

Chertow, M. R. (2007). "Uncovering" industrial symbiosis. *Journal of Industrial Ecology,* 11(1), 11–30.

Chew, K. S. Y. & McCleary, R. (1995). The spring peak in suicides – a cross-national analysis. *Social Science & Medicine,* 40(2), 223–30.

Chick, S. E., Koopman, J. S., Soorapanth, S. & Brown, M. E. (2001). Infection transmission system models for microbial risk assessment. *Science of the Total Environment,* 274(1–3), 197–207.

Chilvers, J. (2004, Nov). *Towards Analytic-Deliberative Forms of Risk Governance in the UK? Reflecting on Learning in Radioactive Waste.* Paper presented at the 13th Annual Meeting of the Society for Risk Analysis Europe, Paris, France.

Chomsky, N. (1975). *Reflections on Language.* New York, NY: Pantheon Books.

Christakos, G., Olea, R. A. & Yu, H. L. (2007). Recent results on the spatiotemporal modelling and comparative analysis of Black Death and bubonic plague epidemics. *Public Health,* 121(9), 700–20.

Christensen, L., Coughenour, M.B., Ellis, G.E., Chen, Z. (2003). Sustainability of inner Mongolian grasslands: application of the Savanna model. *Journal of Range Management,* 56(4), 319–27.

Christodoulou, L. (2007). Energy-Sweden: wood cellulose – alternative to Brazilian ethanol? Retrieved January 6, 2011 from http://ipsnews.net/print.asp?idnews=39416

Cialdini, R. B., Reno, R. R. & Kallgren, C. A. (1990). A focus theory of normative conduct: recycling the concept of norms to reduce littering in public places. *Journal of Personality & Social Psychology,* 58(6), 1015–26.

Cialdini, R. B. & Trost, M. R. (1998). Social influence: social norms, conformity, compliance. In D. T. Gilbert, S. T. Fiske & G. Lindzey (eds). *The Handbook of Social Psychology,* 4th edn. New York, NY: McGraw-Hill, pp. 151–91.

Clark, W. C. (2007). Sustainability science: a room of its own. *Proceedings of the National Academy of Sciences of the United States of America,* 104(6), 1737–8.

Clark, W. C. & Dickson, N. M. (2003). Sustainability science: the emerging research program. *Proceedings of the National Academy of Sciences of the United States of America,* 100(14), 8059–61.

Classen, C. (1992). The odor of the other: olfactory symbolism and cultural categories. *Ethos,* 20(2), 133–66.

Clemens, F. E. (1916). *Plant Succession. An Analysis of the Development of Vegetation.* Washington, DC: Carnegie Institution of Washington.

Cleveland, C. J. & Stern, D. I. (1999). Indicators of natural resource scarcity: a review and synthesis. In J. C. J. M. van den Bergh (ed.). *Handbook of Environmental Economics.* Cheltenham: Edward Elgar, pp. 89–108.

Coase, R. H. (1960). The problem of social cost. *Journal of Law and Economics,* 3(2), 1–44.

Cockrem, J. F. (2002). Reproductive biology and conservation of the endangered kakapo (*Strigops habroptilus*) in New Zealand. *Avian and Poultry Biology Reviews,* **13**(3), 139–44.

Cohen, A. B., Rozin, P. & Keltner, D. (2004). Different religions, different emotions. *Behavioral and Brain Sciences,* **27**(6), 734–5.

Cohen, I. R. (1992). The cognitive paradigm and the immunological homunculus. *Immunology Today,* **13**(12), 490–4.

Cohen, I. R. (2000a). Discrimination and dialogue in the immune system. *Seminars in Immunology,* **12**(3), 215–19.

Cohen, I. R. (2000b). *Tending Adam's Garden: Evolving the Cognitive Immune Self*. London: Academic Press.

Cohen, L. J. (1977). *The Probable and the Provable*. Oxford: Clarendon Press.

Cohen, S., Glass, D. C. & Phillips, S. (1977). Environment and health. In H. E. Freeman, S. Levine & L. G. Reeder (eds). *Handbook of Medical Sociology*. Englewood Cliffs; NJ: Prentice-Hall, pp. 134–49.

Cohen, S. M. (2003). Aristotle's metaphysics. *Stanford Encyclopedia of Philosophy*. Retrieved March 15, 2008, from http://www.science.uva.nl/~seop/entries/aristotle-metaphysics/#WhaSub

Cohen-Rosenthal, E. (ed.). (2003). *Eco-Industrial Strategies: Unleashing Synergy between Economic Development and the Environment*. Sheffield: Greenleaf Publishing.

Cohn, L. H., Collins, J. J., Disesa, V. J. et al. (1989). Fifteen-year experience with 1678 Hancock porcine bioprosthetic heart valve replacements. *Annals of Surgery,* **210**(4), 435–43.

Cohn, M., Mitchison, N. A., Paul, W. E. et al. (2007). Reflections on the clonal-selection theory. *Nature Reviews Immunology,* **7**(10), 823–30.

Coico, R., Sunshine, G. & Benjamin, E. (2003). *Immunology. A Short Course*. Hoboken, NJ: Wiley.

Colander, D. (2000). The death of neoclassical economics. *Journal of the History of Economic Thought,* **22**(2), 127–43.

Colding, J., Berkes, F. & Folke, C. (2008). *Linking Social and Ecological Systems: Management Practices and Social Mechanisms for Building Resilience*. Cambridge: Cambridge University Press.

Cole, E. R., Jayaratne, T. E., Cecchi, L. A., Feldbaum, M. & Petty, E. M. (2007). Vive la difference? Genetic explanations for perceived gender differences in nurturance. *Sex Roles,* **57**(3–4), 211–22.

Collins, R. (2006). Ecological–evolutionary theory: principles and applications. *Contemporary Sociology – a Journal of Reviews,* **35**(6), 553–5.

Colm, G. (2009). Virus: microbiology. Retrieved October 30, 2009, from http://en.wikipedia.org/wiki/Virus

Commission on Oil Independence. (2006). *Making Sweden an Oil-free Society*. Stockholm: Swedish Government.

Commons, J. R. (1931). Institutional economics. *American Economic Review,* **21**(4), 648–57.

Conard, N. J. (2003). Palaeolithic ivory sculptures from southwestern Germany and the origins of figurative art. *Nature,* **426**(6968), 830–2.

Conrad, J. (2008). *Von Arrhenius zum IPCC Wissenschaftliche Dynamik und disziplinäre Verankerung der Klimaforschung*. Münster: MV Wissenschaft.

Convention on Biological Diversity (CBD). (2007). Principles. Retrieved July 8, 2009, from http://www.cbd.int/ecosystem/principles.shtml

Conway, J. (1970). The game of life. *Scientific American,* **223**(4), 120–3.

Coolidge, F. L. & Wynn, T. (2005). Working memory, its executive functions, and the emergence of modern thinking. *Cambridge Archaeological Journal,* **15**(1), 5–26.

Cordell, D., Drangert, J. & White, S. (2009). The story of phosphorus: global food security and food for thought. *Global Environmental Change – Human and Policy Dimensions,* **19**(2), 292–305.

Cordes, C. (2009). The role of biology and culture in Veblenian consumption dynamics. *Journal of Economic Issues,* **43**(1), 115–41.

Cornwall, A. & Jewkes, R. (1995). What is participatory research? *Social Science & Medicine,* **41**(12), 1667–76.

Corral-Verdugo, V., Bonnes, M., Tapia-Fonllem, C. et al. (2009). Correlates of pro-sustainability orientation: the affinity towards diversity. *Journal of Environmental Psychology,* **29**(1), 34–43.

Costanza, R. (1989). What is ecological economics? *Ecological Economics,* **1**(1), 1–7.

Costanza, R. (ed.). (1991). *Ecological Economics: The Science and Management of Sustainability*. New York, NY: Columbia University Press.

Costanza, R., Daly, H. E. & Bartholomew, J. W. (1991). Goals, agenda and policy recommendation for ecological economics. In R. Costanza (ed.). *Ecological Economics: The Science and Management of Sustainability*. New York, NY: Columbia, pp. 1–20.

Cotrell, L. S. & Dymond, R. F. (1949). The empathic responses: a neglected field of our research. *Psychiatry,* **12**(11), 355–9.

Cotterill, R. (2008). *The Material World*. Cambridge: Cambridge University Press.

Cottier, T. & Hertig, M. (2003). The prospects of 21st century constitutionalism. In A. von Bogdany & R. Wolfrum (eds). *Max Planck Yearbook of United Nations Law,* Vol. 7. Leiden NL: Martinus Nijhoff Publishers, pp. 261–328.

Cottrell, W. F. (1952). Death by dieselization: a case study in the reaction to technological change. *American Sociological Review,* **16**(3), 358–65.

Cottrell, W. F. (1955). *Energy and Society: the Relation between Energy, Social Change and Economic Development.* New York, NY: McGraw Hill.

Coullet, P. & Tresser, C. (1978). Itérations d'endomorphismes et groupe de renormalisation. *Journal de Physique,* **8**(39), C5–25.

Cournot, A. A. (1838). *Recherche sur les Principes Mathématiques de la Théorie des Richesses.* Paris: Hachette.

Covello, V. T. & Mumpower, J. (1985). Risk analysis and risk management: an historical perspective. *Risk Analysis,* **5**(2), 103–20.

Cox, D. R., Atkinson, A. C., Box, G. E. P. et al. (1984). Interaction. *International Statistical Review,* **52**(1), 1–31.

Cox, P. M., Harris, P. P., Huntingford, C. et al. (2008). Increasing risk of Amazonian drought due to decreasing aerosol pollution. *Nature,* **453**(7192), 212–17.

Coyle, K. (2005). *Environmental Literacy in America: what Ten Years of NEETF/Roper Research and Related Studies say about Environmental Literacy in the US.* Washington, DC: The National Environmental & Training Foundation.

Craik, K. H. (1973). Environmental psychology. *Annual Review of Psychology,* **24**, 403–22.

Crick, F. H. C. (1958). On protein synthesis. *Symposia of the Society for Experimental Biology,* **12**, 138–63.

Crick, F. H. C. (1970). Central dogma of molecular biology. *Nature,* **227**(5258), 561–3.

Crocker, T. D. (1999). A short history of environmental resource economics. In J. C. J. M. van den Bergh (ed.). *Handbook of Environmental Economics.* Cheltenham: Edward Elgar, pp. 32–74.

Crosby, A. W. (2003). *America's Forgotten Pandemic: the Influenza of 1918,* 2nd edn. Cambridge: Cambridge University Press.

Crosby, N., Kelly, J. M. & Schaefer, P. (1986). Citizens panels – a new approach to citizen participation. *Public Administration Review,* **46**(2), 170–8.

Crott, H. W. (1979). *Soziale Interaktion und Gruppenprozesse.* Stuttgart: Kohlhammer.

Cruickshank, T. & Wade, M. J. (2008). Microevolutionary support for a developmental hourglass: gene expression patterns shape sequence variation and divergence in Drosophila. *Evolution & Development,* **10**(5), 583–90.

Crutzen, P. J. (2002a). The "anthropocene". *Journal de Physique IV France,* **12**(10), 11–15.

Crutzen, P. J. (2002b). Geology of mankind. *Nature,* **415**(6867), 23.

Cubas, P., Vincent, C. & Coen, E. (1999). An epigenetic mutation responsible for natural variation in floral symmetry. *Nature,* **401**(6749), 157–61.

Cunfer, G. (2004). The dust bowl. Retrieved May 10, 2010, from http://eh.net/encyclopedia/article/Cunfer.DustBowl

Curci, A. & Luminet, O. (2006). Follow-up of a cross-national comparison on flashbulb and event memory for the September 11th attacks. *Memory,* **14**(3), 329–44.

D'Argembeau, A. & van der Linden, M. (2006). Individual differences in the phenomenology of mental time travel: the effect of vivid visual imagery and emotion regulation strategies. *Consciousness and Cognition,* **15**(2), 342–50.

Daily, G. C. (1997). *Nature's Services: Social Dependence on Natural Ecosystems.* Washington, DC: Island Press.

Dajoz, R. (1971). *Précis d'Écologie.* Paris: Dunod.

Dales, T. (1968). *Pollution, Property, and Prices.* Toronto: University of Toronto Press.

Daly, H. E. (1968). On economics as a life science. *The Journal of Political Economy,* **76**(3), 392–406.

Daly, H. E. (1974). The economics of the steady state. *American Economic Review,* **64**(2), 15–21.

Daly, H. E. (1991). *Steady-state Economics.* Washington, DC: Island Press.

Daly, H. E. (1999). Steady-state economics: avoiding uneconomic growth. In J. C. J. M. van den Bergh (ed.). *Handbook of Environmental Economics.* Cheltenham: Edward Elgar, pp. 635–42.

Daly, H. E. (2005). Economics in a full world. *Scientific American,* **293**(3), 100–7.

Daly, H. E. & Cobb, J. B. (1989). *For the Common Good: Redirecting the Economy toward Community, the Environment, and the Sustainable Future.* Boston, MA: Beacon.

Daly, H. E. & Farley, J. (2004). *Ecological Economics: Principles and Applications.* Washington, DC: Island Press.

Daly, M. & Wilson, M. (2005). Carpe diem: adaptation and devaluing the future. *Quarterly Review of Biology,* **80**(1), 55–60.

Daniel, T. M. (2006). The history of tuberculosis. *Respiratory Medicine,* **100**(11), 1862–70.

Dansgaard, W., Johnsen, S. J., Clausen, H. B. et al. (1993). Evidence for general instability of past climate from a 250-kyr ice-core record. *Nature,* **364**(6434), 218–20.

Dantzig, G. B. (1963/1991). *Linear Programming and Extensions.* Princeton, NJ: Princeton University Press.

Darlington, C. D. (1932). *Recent Advances in Cytology.* London: Churchill.

Darwin, C. (1859). *On the Origin of Species by Means of Natural Selection, or the Preservation of Favoured Races in the Struggle for Life*. London: Murray.

Darwin, C. (1872/1965). *The Expression of the Emotions in Man and Animals*. Chicago, IL: University of Chicago Press.

Darwin, C. (1845/2007). *Die Fahrt der Beagle: Tagebuch mit Erforschung der Naturgeschichte und Geologie der Länder, die auf der Fahrt von HMS Beagle unter dem Kommando von Kapitän Robert Fitz Roy, RN, besucht wurden*. Hamburg: Mare.

Dasgupta, P. (1974). On some problems arising from Professor Rawls' conception of distributive justice. *Theory and Decision,* **4**(3–4), 325–44.

Dasgupta, S., Laplante, B., Wang, H. & Wheeler, D. (2002). Confronting the environmental Kuznets curve. *Journal of Economic Perspectives,* **16**(1), 147–68.

Davidson, R. J. (2001). Toward a biology of personality and emotion. *Annals of the New York Academy of Sciences,* **935**, 191–207.

Davidson, R. J., Putnam, K. M. & Larson, C. L. (2000). Dysfunction in the neural circuitry of emotion regulation – a possible prelude to violence. *Science,* **289**(5479), 591–4.

Davis, B. J., Maronpot, R. R. & Heindel, J. J. (1994). Di-(2-ethylhexyl) phthalate suppresses estradiol and ovulation in cycling rats. *Toxicology and Applied Pharmacology,* **128**(2), 216–23.

Davis, J. H. (1969). *Group Performance*. Reading, MA: Addison-Wesley.

Davis, J. H. (1973). Group decision and social interaction: theory of social decision schemes. *Psychological Review,* **80**(2), 97–125.

Davis, J. H., Kerr, N., Sussmann, M. & Rissman, A. K. (1974). Social decision schemes under risk. *Journal of Personality and Social Psychology,* **30**(2), 248–71.

Davis, M. (1998). *Thinking Like an Engineer*. Oxford: Oxford University Press.

Davis, P. J. & Hersh, R. (1981). *The Mathematical Experience*. Boston, MA: Houghton Mifflin.

Davis, R. & Hirji, R. (eds). (2003). *Water resources and environment-Wastewater reuse*. Technical Note, F.3. Washington, DC: The World Bank.

Davis, W. (1990). A way to stay. In W. Davis & T. Henley (eds) *Penan: Voice for the Borneo Rainforest*. Vancouver: Western Canada Wilderness Committee, pp. 635–42.

Davis, W., Mackenzie, I. & Kennedy, S. (1995). *Nomads of the Dawn. The Penan of the Borneo Rain Forest*. San Francisco, CA: Pomegranate Artbooks.

Dawes, R. M. (1980). Social dilemmas. *Annual Review of Psychology,* **31**(1), 169–93.

Dawes, R. M. & Messick, D. M. (2000). Social dilemmas. *International Journal of Psychology,* **35**(2), 111–16.

Dawkins, R. (1982). *The Extended Phenotype*. Oxford: Oxford University Press.

Dayrat, B. (2003). The roots of phylogeny: how did Haeckel build his trees? *Systematic Biology,* **52**(4), 515–27.

De Almeida, H. (2004). Romanticism and the triumph of life science: prospects for study. *Studies in Romanticism,* **43**(1), 119–34.

de Groot, R. S. (1992). *Functions of Nature. Evaluation of Nature in Environmental Planning, Management and Decision Making*. Amsterdam: Wolters-Noordhoff.

de Freitas, L. D., Morin, E. & Nicolescu, B. (1994 de November 2–6). *Charter of Transdisciplinarity*. Paper presented at the Convento da Arrábida.

de Groot, R. S., Wilson, M. A. & Boumans, R. M. J. (2002). A typology for the classification, description and valuation of ecosystem functions, goods and services. *Ecological Economics,* **41**(3), 393–408.

de Haan, G. & Kuckartz, U. (1996). *Umweltbewusstsein. Denken und Handeln in Umweltkrisen*. Opladen: Westdeutscher Verlag.

de Haan, P., Mueller, M. G. & Scholz, R. W. (2009). How much do incentives affect car purchase? Agent-based microsimulation of consumer choice of new cars – Part II: Forecasting effects of feebates based on energy-efficiency. *Energy Policy,* **37**(3), 1083–94.

de Haan, P., Müller, M. & Peters, A. (2006). Does the hybrid Toyota Prius lead to rebound effects? Analysis of size and number of cars previously owned by Swiss Prius buyers. *Ecological Economics,* **58**(3), 592–605.

de Haan, P., Peters, A. & Scholz, R. W. (2007). Reducing energy consumption in road transport through hybrid vehicles: investigation of rebound effects, and possible effects of tax rebates. *Journal of Cleaner Production,* **15**(11–12), 1076–84.

de Jong, M. D., Thanh, T. T., Khanh, T. H. *et al.* (2005). Oseltamivir resistance during treatment of influenza A (H5N1) infection. *New England Journal of Medicine,* **353**(25), 2667–72.

de Kruif, P. (1927). *Mikrobenjäger*. Zurich: Orell Füssli.

de Landa, M. (2005). *A New Philosophy of Society: Assemblage, Theory and Social Complexity*. London: Continuum.

de Roover, F. E. (1945). Early examples of marine insurance. *The Journal of Economic History,* **5**(2), 172–200.

de Saussure, N. T. (1804/1890). *Recherches Chimiques sur la Végétation*. Berlin: Königlich Preussische Akademie der Wissenschaften.

de Vries, H. (1889). *Intercelluläre Pangenesis*. Jena: Gustav Fischer.

Deal, M. (2006). An archaeological survey of Scots Bay mills and shipyards. Retrieved October 21, 2009, from http://www.ucs.mun.ca/~mdeal/MillSurvey/millsurvey.htm

Deevey, E. (1960). The human population. *Scientific American,* **203**(9), 194–204.

Deffuant, G., Huet, S., Bousset, J. P. et al. (2002). Agent-based simulation of organic farming conversion in Allier departement. In M. Janssen (ed.). *Complexity and Ecosystem Management: the Theory and Practice of Multi-Agent Systems*. Cheltenham: Edward Elgar, pp. 158–87.

Del Col, L. (2002). Testimony gathered by Ashley's Mines Commission. Retrieved October 12, 2009, from http://www.victorianweb.org/history/ashley.html

Delahaye, J. & Mathieu, P. (1994). Complex strategies in the iterated prisoner's dilemma. In A. Albert (ed.). *Chaos and Society*. Amsterdam: IOS Press, pp. 283–92.

Delcuve, G. P., Rastegar, M. & Davie, J. R. (2009). Epigenetic control. *Journal of Cellular Physiology,* **219**(2), 243–50.

Deleuze, G. & Parnat, C. (2002). *Dialogues II*. New York, NY: Columbia University Press.

Denevan, W. M. (1992). The pristine myth: the landscape of the Americas in 1492. *Annals of the Association of American Geographers,* **82**(3), 369–85.

Dercon, S. (2004). *Insurance Against Poverty*. Oxford: Oxford University Press.

Déry, P. & Anderson, B. (2007). Peak phosphorus. Retrieved January 6, 2011 from http://www.energybulletin.net/node/33164.

Desertec Foundation. (2009). Desertec EU-mena. Retrieved November 26, 2009, from http://www.desertec.org

DeSombre, E. R. (2006). *Global Environmental Institutions*. London: Routledge.

Devall, B. & Sessions, G. (1999). Deep ecology. In M. J. Smith (ed.). *Thinking through the Environment*. London: Routledge, pp. 200–7.

Devine-Wright, G. (2003). *A Cross-National, Comparative Analysis of Public Understanding of and Attitudes towards Nuclear, Renewable and Fossil Energy Sources*. Paper presented at the Third Conference of EPUK (Environmental Psychology in the UK) Network: crossing boundaries: the value of interdisciplinary research, Robert Gordon University, Aberdeeen.

Dewey, J. (1896). The reflex arc concept in psychology. *Psychological Review,* **3**(4), 357–70.

Dewey, J. (1910/1997). *How we Think*. Dover: Mineola.

Dewey, J. (1902/2001). *The School and the Society & the Child and the Curriculum*. Mineola: Dover.

Diamond, J. (1966). Zoological classification system of a primitive people. *Science,* **151**(3714), 1102–4.

Diamond, J. (1994). *Guns, Germs, and Steel: the Fates of Human Societies*. New York, NY: Norton.

Diamond, J. (2005). *Collapse: How Societies Choose to Fail or Succeed*. New York, NY: Viking.

Diekmann, A. & Franzen, A. (1995). *Kooperatives Umweltverhalten*. Zurich: Rüegger.

Diekmann, A. & Preisendörfer, P. (1996). Environmental behavior: discrepancies between aspirations and reality. *Rationality and Society,* **10**(1), 79–102.

Diekmann, A. & Preisendörfer, P. (1998). Experimental consciousness and behavior in low-cost and in high-cost situations. An empirical examination of the low-cost hypothesis. *Zeitschrift für Soziologie,* **27**(6), 438–53.

Diekmann, A. & Preisendörfer, P. (2001). *Umweltsoziologie: Eine Einführung*. Hamburg: Rowohlt.

Diekmann, A. & Preisendörfer, P. (2003). Green and greenback: the behavioral effects of environmental attitudes in low-cost and high-cost situations. *Rationality and Society,* **15**(4), 441–72.

Dienel, P. C. (1970/1991). *Die Planungszelle*. Opladen: Westdeutscher Verlag.

Dienel, P. C. & Garbe, D. (1985). *Zukünftige Energiepolitik. Ein Bürgergutachten*. Munich: FTV Edition "Technik und Sozialer Wandel."

Dietz, T., Fitzgerald, A. & Shwom, R. (2005). Environmental values. *Annual Review of Environment and Resources,* **30**, 337–72.

Dietz, T., Stern, P. C. & Guagnano, G. A. (1998). Social structural and social psychological bases of environmental concern. *Environment and Behavior,* **30**(4), 450–71.

Dieudonné, J. (1985). *History of Algebraic Geometry. An Outline of the History and Development of Algebraic Geometry*. Wadsworth Mathematics Series. Belmont, CA: Wadsworth International Group.

Dionisio, F. & Gordo, I. (2006). The tragedy of the commons, the public goods dilemma, and the meaning of rivalry and excludability in evolutionary biology. *Evolutionary Ecology Research,* **8**(2), 321–32.

Disinger, J. F. & Roth, C. E. (2003). *Environmental Literacy: its Roots, Evolution and Directions in the 1990s*. Columbus, OH: ERIC Clearinghouse for Science Mathematics and Environmental Education.

Dobbertin, M. (2005). Tree growth as indicator of tree vitality and of tree reaction to environmental stress: a review. *European Journal of Forest Research,* **124**(4), 319–33.

Dobzhansky, T. (1937). *Genetics and the Origin of Species*. New York, NY: Columbia University Press.

Dodds, W. K. (2005). The commons, game theory and aspects of human nature that may allow conservation of global resources. *Environmental Values,* **14**(4), 411–25.

Dolan, R. J. (2002). Emotion, cognition, and behavior. *Science,* **298**(5596), 1191–4.

Dollard, J. (1937). *Caste and Class in a Southern Town*. New Haven, CT: Yale University Press.

Dominguez-Faus, R., Powers, S. E., Burke, J. E. & Alvarez, P. J. (2009). The water footprint of biofuels: a drink and drive issue? *Environmental Science & Technology,* **43**(9), 3005–10.

Döpke, A., Leutert, D., Mavromati, F. & Pfeifer, T. (2007). *Phtalate – die nützlichen Weichmacher mit den unerwünschten Eigenschaften*. Dessau: Umweltbundesamt.

Dörner, D. (1989). *The Logic of Failure: Recognizing and Avoiding Failure in Complex Situations*. Cambridge, MA: Perseus Books.

Dörner, D. (1996). *Logic of Failure: Why Things Go Wrong and What We Can Do to Make Them Right*. New York, NY: Metropolitan Books.

Dörner, D., Kreuzig, H. W., Reither, F. & Stäudel, T. (1994). *Lohhausen – Vom Umgang mit Unbestimmtheit und Komplexität*. Bern: Huber.

Douglas, M. (1970). *Natural Symbols: Explorations in Cosmology*. London: Barrie & Jenkins.

Douglas, M., Gasper, D., Ney, S. & Thompson, M. (1998). Human needs and wants. In S. Raynor & E. L. Malone (eds). *Human Choice and Climate Change,* Vol. 1: The Societal Framework. Colombus, OH: Batelle Press, pp. 195–260.

Douglas, M. & Wildavsky, A. B. (1983). *Risk and Culture: An Essay on the Selection of Technological and Environmental Dangers*. Berkeley, CA: University of California Press.

Doyle, C. L. (2000). Psychology. In A. E. Kazdin (ed.). *Encyclopedia of Psychology*. Oxford: Oxford University Press, Vol 6, pp. 375–6.

Draper, N. R. & Smith, H. (1998). *Applied Regression Analysis,* 3rd edn. New York, NY: Wiley.

Dreyfus, H. L. & Dreyfus, S. E. (1986). *Mind over Machines*. New York, NY: The Free Press.

Dreyfus, H. L. & Dreyfus, S. E. (2005). Peripheral vision: expertise in real world contexts. *Organization Studies,* **26**(5), 779–92.

Driessen, P. (1999). Activating a policy network: the case of mainport Schiphol. In L. E. Susskind, S. McKearnen & J. Thomas-Larmer (eds). *The Consensus Building Handbook: a Comprehensive Guide to Reaching Agreement*. Thousand Oaks, CA: Sage.

Drottz-Sjöberg, B.-M. (2010). *Perceptions of nuclear wastes across extreme time perspectives Risk, Hazards & Crisis in Public Policy*. **1** (4), 231–53.

Dryzek, J. S. (1993). Democracy and the policy sciences: a progress report. *Policy Sciences,* **26**, 127–37.

Dryzek, J. S. & Hunter, S. (1987). Environmental mediation for international problems. *International Studies Quarterly,* **31**(1), 87–102.

Dubiel, H. (1976). Institution. In J. Ritter & K. Gründer (eds). *Historisches Wörterbuch der Philosophies,* Vol. 4. Basel: Schwabe, pp. 418–23.

Duchin, F. (2004). *Input-Output Economics and Material Flows*. Troy, NY: Rensselaer Polytechnic Institute.

Dumont, L. (1970). *Homo Hierarchicus: The Caste System and its Implications*. Chicago, IL: The University of Chicago Press.

Duncan, O. D. (1961). From social systems to ecosystems. *Sociological Inquiry,* **31**(2), 140–9.

Dunlap, R. E. (1997). The evolution of environmental sociology. A brief history and assessment of the American experience. In M. Redclift & G. Woodgate (eds). *The International Handbook of Environmental Sociology*. Cheltenham: Edward Elgar, pp. 21–40.

Dunlap, R. E. (2002). Paradigms, theories, and environmental sociology. In R. E. Dunlap, F. H. Buttel, P. Dickens & A. Gijswijt (eds). *Sociological Theory and the Environment: Classical Foundations, Contemporary Insights*. Lanham, MD: Rowman & Littlefield Publishers Inc, pp. 329–50.

Dunlap, R. E. & Catton, W. R. (1979). Environmental sociology. *Annual Review of Sociology,* **5**, 243–73.

Dunlap, R. E. & Catton, W. R. (2002). Which function(s) of the environment do we study? A comparison of environmental and natural resource sociology. *Society and Natural Resources,* **15**(3), 239–49.

Dunlap, R. E., Grieneeks, J. K. & Rockeach, M. (1983). Human values and pro-environmental behavior. In W. D. Conn (ed.). *Energy and Material Resources: Attitudes, Values, and Public Policy*. Boulder, CO: Westview, pp. 145–69.

Dunlap, R. E. & Michelson, W. (eds). (2002). *Handbook of Environmental Sociology*. Westport, CT: Greenwood Press.

Dunlap, R. E., Michelson, W. & Stalker, G. (2002). Environmental sociology: an introduction. In R. E. Dunlap & W. Michelson (eds). *Handbook of Environmental Sociology*. Westport, CT: Greenwood Press, pp. 1–32

Dunlap, R. E., Van Liere, K.-D., Mertig, A. G. & Emmert, J. (2000). Measuring endorsement of the new ecological paradigm: a revised NEP Scale. *Journal of Social Issues,* **56**(3), 425–42.

Dupuit, A. J. E. J. (1844). De la mesure de l'utilité des travaux publics. *Annales des Ponts et Chaussées*, Second series, 8. Translated by R. H. Barback as On the measurement of the utility of public works. *International Economic Papers*, 1952; 2, 83–110.

Durkheim, E. (1893/1977). *De la Division du Travail Social: Étude sur L'Organisation des Sociétés Supérieures*. Frankfurt am Main: Suhrkamp.

Durkheim, E. (1895/1982). *The Rules of Sociological Method*. London: Macmillan.

Durkheim, E. (1897/1973). *Der Selbstmord*. Neuwied: Luchterhand.

Durkheim, E. (1911–1912/1983). *Pragmatism and Society*. Cambridge: Cambridge University Press.

Durkheim, E. (1924/1976). *Soziologie und Philosophie*. Frankfurt am Main: Suhrkamp.

Durkheim, E. (1930/1992). *Über soziale Arbeitsteilung*. Frankfurt am Main: Suhrkamp.

Duval, S. & Wicklund, R. A. (1972). *A Theory of Objective Self Awareness*. New York, NY: Academic Press.

Eagon, R. G. (1962). *Pseudomonas natriegens*, a marine bacterium with a generation time of less than 10 minutes. *Journal of Bacteriology*, **83**(4), 736–7.

Earle, T. C. & Siegrist, M. (2006). Morality information, performance information, and the distinction between trust and confidence. *Journal of Applied Social Psychology*, **36**(2), 383–416.

EAWAG. (2007, 06.03.2007). Novaquatis: Nova 2 – Sanitary technology. Retrieved 15 November, 2009 from http://www.novaquatis.eawag.ch/arbeitspakete/nova2/index_EN

Ebreo, A. & Vining, J. (2000). Motives as predictors of the public's attitudes toward solid waste issues. *Environmental Management*, **25**(2), 153–68.

Eckermann, I. (2004). *The Bhopal Saga – Causes and Consequences of the World's Largest Industrial Disaster*. Hyderabad: Universities Press.

Eckstein, O. (1950). *Water Resource Development: the Economics of Project Evaluation*. Cambridge, MA: Harvard University Press.

economictheories.org. (2007). Veblen competition – supply and demand. *Economic: History of economic theory and thought*. Retrieved January 6, 2011 from http://www.economictheories.org/2008/11/veblen-competition-supply-and-demand.html

Edelstein, M. R. (1988). *Contaminated Communities*. Boulder, CO: Westview Press.

Edelstein, M. R. (2002). Contamination: the invisible built environment. In R. B. Bechtel & A. Churchman (eds). *Handbook of Environmental Psychology*. New York, NY: Wiley, pp. 559–88.

Eden, C. & Ackermann, F. (2006). Where next for problem structuring methods. *Journal of the Operational Research Society*, **57**(7), 766–8.

Eden, C., Jones, S. & Sims, D. (1983). *Messing About in Problems: an Informal Structured Approach to their Identification and Management*. Oxford: Pergamon Press.

Edgeworth, F. Y. (1881/1932). *Mathematical Physics*. London: London School of Economics.

Edmonds, B., Hernandez, C. & Troitzsch, K. G. (eds). (2007). *Social Simulation: Technologies, Advances and New Discoveries*. Hershey, NY: Information Science Reference.

Edwards, W. (1954). The theory of decision making. *Psychological Bulletin*, **51**(4), 380–417.

Egyházi, E. & Hydén, H. (1961). Experimentally induced changes in base composition of ribonucleic acids of isolated nerve cells and their oligodendroglial cells. *Journal of Biophysical and Biochemical Cytology*, **10**(3), 403–10.

Ehrenfeld, J. R. & Chertow, M. R. (2002). Industrial symbiosis: the legacy of Kalundborg. In R. U. Ayres & L. W. Ayres (eds). *A Handbook of Industrial Ecology*. Cheltenham: Edward Elgar, pp. 334–48.

Ehrenfels, C., von. (1890). Über Gestaltqualitäten. *Vierteljahresschrift für wissenschaftliche Philosophie*, **14**, 249–92.

Ehrlich, P. R. (1957). Immunology and cancer research. In E. Himmelweit (ed.). *The Collected Papers of Paul Ehrlich*, Vol. II. London: Pergamon, pp. 31–44.

Ehrlich, P. R. (1986). Which animal will invade? In H. A. Mooney & J. A. Drake (eds). *Ecology of Biological Invasion of North America and Hawaii*. New York, NY: Springer, pp. 79–95.

Ehrlich, P. R. & Ehrlich, A. H. (1970). *Population, Resources, Environment: Issues in Human Ecology*. San Francisco, CA: W. H. Freeman.

Ehrlich, P. R. & Holdren, J. P. (1971). Impact of population growth. *Science*, **171**(3977), 1212–17.

Ehrlich, P. R. & Kennedy, D. (2005). Millennium assessment of human behavior. *Science*, **309**(5734), 562–3.

Ehrlich, P. (1892). Über Immunität durch Vererbung und Säugung *Zeitschrift für Hygiene und Infektionskrankheiten*, **12**, 183–203.

Eibl-Eibesfeldt, I. (1989). *Human Ethology*. Hawthorne, NY: Aldine de Gruyter.

Einstein, A. (1916). Relativity: the special and general theory. *PublicLiterature.Org*. Retrieved November 9, 2009, from http://publicliterature.org/books/relativity/xaa.php

References

Einstein, A. (1926/1971). Letter from Einstein to Max Born, December 4th (I. Born, Trans.). In M. Born (ed.). *The Born–Einstein Letters*. New York, NY: Walker and Co, pp. 90–1.

Einstein, A. & Infeld, L. (1938). *The Evolution of Physics*. New York, NY: Simon and Schuster.

Eisenhut, P. & Schönholzer, U. (2003). *Entwicklung und Perspektiven der Ostschweizer Volkswirtschaft*. St. Gallen: Industrie- und Handelskammer St. Gallen-Appenzell.

Eisnerova, V. (2004). Konsolidierung und Neubildung von Disziplinen und Theorien im 19. Jahrhundert. In I. Jahn (ed.). *Geschichte der Biologie: Theorien, Methoden, Institutionen, Kurzbiographien*. Hamburg: Nikol Verlag, pp. 302–23.

Ekvall, T., Tillman, A. M., & Molander, S. (2005). Normative ethics and methodology for life cycle assessment. *Journal of Cleaner Production*, **13**(13–14), 1225–34.

Elden, M. & Levin, M. (1991). Cogenerative learning. In W. F. Whyte (ed.). *Participatory Action Research*. Newbury Park CA: Sage, pp. 127–42.

Eliasson, B. (2003). Introduction. In B. Eliasson (ed.). *Integrated Assessment of Sustainable Energy Systems in China*. Dordrecht: Kluwer, pp. 1–9.

Eliot, J. (1663). *The Holy Bible containing the Old Testament and the New. Translated into the Indian language*. Cambridge, MA: Samuel Green and Marmaduke Johnson.

Elkington, J. (1994). Towards the sustainable corporation: win-win-win business strategies for sustainable development. *California Management Review*, **36**(2), 90–100.

Ellen, R. F. (1979). Introductory essay. In R. F. Ellen & D. Reason (eds). *Classifications in their Social Context*. London: Academic Press, pp. 1–32.

Ellen, R. F. (1982). *Environment, Subsistence and System*. Cambridge: Cambridge University Press.

Ellen, R. F. (2006). Introduction to ethnobiology and the science of humankind. *Journal of the Royal Anthropological Institute*, **12**(SI), 1–22.

Ellernman, V. & Bang, O. (1908). Experimentelle Leukämie bei Hühnern. Centralblatt für Bakteriologie, Parasitenkunde und Infektionskrankheite. *Zentralblatt für Bakteriologie, Parasitenkunde, Infektionskrankheiten und Hygiene*, **46**, 595–609.

Ellis, E. C. & Ramankutty, N. (2008). Putting people in the map: anthropogenetic biomes of the world. *Frontiers in Ecology and the Environment*, **6**(8), 439–47.

Ellis, E. L. & Delbrück, M. (1939). The growth of bacteriophage. *The Journal of Physiology*, **22**(3), 365–84.

Ellis, F. (2000). *Rural Livelihood Diversity in Developing Countries*. Oxford: Oxford University Press.

Elton, C. (1926). *Animal Ecology*. London: Sidgewick and Jackson.

Elzinga, A. & Jamison, A. (1995). Changing policy agendas in science and technology. In S. Jasanoff, G. E. Markle, J. C. Petersen & T. J. Pinch (eds). *Handbook of Science and Technology Studies*. Thousand Oaks, CA: Sage, pp. 573–93.

Emmel, H. (1974). Gemüt. In J. Ritter & K. Gründer (eds). *Historisches Wörterbuch der Philosophie*, Vol. 3. Basel: Schwabe, pp. 258–62.

Encyclopaedia Britannica. (2009). Maya. *Encycloclopaedia Britannica Online*. Retrieved April 15, 2010, from http://search.eb.com/eb/article-9051572

Engels, F. (1845/1972). Die Lage der arbeitenden Klasse in England. In E. O'Callaghan (ed.). *K. Marx und F. Engels: Werke*, Vol. 2. Berlin: Dietz, pp. 225–506.

Enkh-Amgalan, A. (2008). *Let's Decide the Pasture Issue this Was*. Ulaanbaator: Centre for Policy Research.

Enkh-Amgalan, T. (2009). *Building up Value Added Production in Mongolia: Organization and Governance Issues*. PhD thesis, ETH Zurich, Zurich.

Enquist, M. & Arak, A. (1994). Symmetry, beauty and evolution. *Nature*, **372**(6502), 169–72.

Ensink, J. H. J. & van der Hoek, W. (2007). Editorial: new international guidelines for wastewater use in agriculture. *Tropical Medicine & International Health*, **12**(5), 575–7.

Environmental Protection Agency (EPA). (1995). Compliance results: acid rain program. Retrieved May 4, 2010, from http://nepis.epa.gov/Exe/ZyPURL.cgi?Dockey=900O0I00.txt.

EPE. (2007). *Brazilian Energy Balance*. Brasilia: Governo Federal.

Epstein, J. M. (2008). Why model? *Journal of Artificial Societies and Social Simulation*, **11**(4), 12.

Epstein, J. M. & Axtell, R. (1996). *Growing Artificial Societies: Social Science from the Bottom up*. Cambridge, MA: The MIT Press.

Érdi, P. (2008). *Complexity Explained*. Berlin: Springer.

Erdman, S. (2007). *Historischer Überblick zur Industrial Ecology*. In R. Isenmann & M. von Hauff (eds). Munich: Spektrum, pp. 41–8.

Erkman, S. (1997). Industrial ecology: an historical view. *Journal of Industrial Ecology*, **5**(1–2), 1–10.

Erkman, S. (2002). The recent history of industrial ecology. In R. U. Ayres & L. W. Ayres (eds). *A Handbook of Industrial Ecology*. Cheltenham: Edward Elgar, pp. 27–35.

References

Ernest, S. K. M. & Brown, J. H. (2001). Homeostasis and compensation: the role of species and resources in ecosystem stability. *Ecology,* **82**(8), 2118–32.

Escher, B. I., Pronk, W., Suter, M. J.-F. & Maurer, M. (2006). Monitoring the removal efficiency of pharmaceuticals and hormones in different treatment processes of source-separated urine with bioassays. *Environmental Science & Technology,* **40**(16), 5095–101.

Etkin, N. L. (ed.). (1986). *Plants in Indigenous Medicine and Diet: Biobehavioral Approaches.* London: Routledge.

Etzkowitz, H. (2001). The second academic revolution and the rise of entrepreneurial science. *IEEE Technology and Society Magazine,* **20**(2), 18–29.

Etzkowitz, H. & Leydesdorff, L. (2000). The dynamics of innovation: from National Systems and "Mode 2" to Triple Helix of university-industry-government relations. *Research Policy,* **29**, 109–23.

Eubank, S., Guclu, H., Kumar, V. S. A. *et al.* (2004). Modelling disease outbreaks in realistic urban social networks. *Nature,* **429**(6988), 180–4.

Euclid. (1956). *The Thirteen Books of Euclid's Elements.* New York, NY: Dover Publications.

EuroObser'ER. (2007). Le baromètre de biogaz. *Systèmes Solaires. Le Journal des Énergies-Renouvelables,* **179**, 51–61.

European Council for Plasticisers and Intermediates (ECPI). (2007). Plasticisers. Retrieved May 12, 2007, from http://www.ecpi.org/index.asp?page=5

European Parliament. (2005). *Report on the Share of Renewable Energy in the EU Proposals for Concrete Actions.* (2004/2153 (INI)).

European Union (EU). (2003). *Directive 2003/96/EC.* Brussels: EU.

European Union (EU). (2005). Phthalates in toys and childcare articles. Retrieved June 6, 2007, from http://eur-lex.europa.eu/LexUriServ/LexUriServ.do?uri=CELEX:32005L0084:EN:NOT

European Union (EU). (2008). *Attitudes of Europeam Citizens towards the Environment.* Brussels: EU.

European Union (EU). (2010). *Critical raw materials for the EU. Report of the Ad-hoc Working Group on defining critical raw materials:* European Commission, Enterprise and Industry.

Ewert, U. C., Roehl, M. & Uhrmacher, A. M. (2007). Hunger and market dynamics in pre-modern communities: insights into the effects of market intervention from a multi-agent model. *Historical Social Research,* **32**(4), 122–50.

Ewing, G. (2001). Altruistic, egoistic, and normative effects on curbside recycling. *Environment and Behavior,* **33**(6), 733–64.

Ewing, R. & Cervero, R. (2001). Travel and the built environment: a synthesis. *Transportation Research Record,* **1780**, 87–114.

Eysenck, M. W. & Keane, M. T. (2005). *Cognitive Psychology: A Student's Handbook.* New York, NY: Psychology Press.

Fabrega, H. & Silver, D. (1973). *Illness and Shamanistic Curing in Zinacantan: an Ethnomedical Analysis.* Stanford, CA: Stanford University Press.

Fairbairn, R. (1952). *Psychoanalytic Studies in Personality.* London: Routledge.

Faustmann, M. (1849). *On the Determination of the Value which Forest Land and Immature Stands Possess for Forestry.* English edition edited by M. Gane. Oxford: Commonwealth Forestry Institute Paper 42 (1968).

Fausto-Sterling, A. (1993). The five sexes: why male and female are not enough. *The Sciences,* March/April, 20–24.

Favareau, D. (2007). The evolutionary history of biosemiotics. In M. Barbieri (ed.). *Introduction to Biosemiotics.* Dordrecht: Springer, pp. 1–67.

Faye, J. (2008). Copenhagen Interpretation of Quantum Mechanics. *Stanford Encyclopedia of Philosophy.* Retrieved April 10, 2008, from http://plato.stanford.edu/entries/qm-copenhagen

Fechner, G. T. (1860). *Elemente der Psychophysik. Zweiter Teil.* Leipzig: Von Breitkopf und Hartel.

Fechner, G. T. (1971). *Zur Exprimentellen Aesthetik.* Leipzig: Hirzel.

Federal Veterinary Office (FVO) (2004). *Neue Wege für Schlachtabfälle.* Bern: Federal Veterinary Office (FVO).

Federal Veterinary Office (FVO) (2009). BSE-Fälle pro Jahr. Retrieved December 23, 2009, from http://www.bvet.admin.ch/gesundheit_tiere/01065/01083/01091/index.html?lang=de

Feigenbaum, M. J. (1978). Quantitative universality for a class of nonlinear transformations. *Journal of Statistical Physics,* **19**(1), 25–52.

Feldmann, D. C. (1984). The development and enforcement of group norms. *Academy of Managment Review,* **9**(1), 47–53.

Festinger, L. (1957). *A Theory of Cognitive Dissonance.* Stanford, CA: Stanford University Press.

Festinger, L., Schachter, S. & Back, K. (1950). *Social Pressures in Informal Groups: a Study of Human Factors in Housing.* New York, NY: Stanford University Press.

Feyerabend, P. K. (2003). *Wider den Methodenzwang.* Frankfurt: Suhrkamp.

Fick, J., Lindberg, R. H., Tysklind, M. *et al.* (2007). Antiviral oseltamivir is not removed or degraded in normal sewage water treatment: implications for development of resistance by influenza A virus. *Public Library of Science One*, **2**(10), e986.

Fidler, A. E., Lawrence, S. B. & McNatty, K. P. (2008). An hypothesis to explain the linkage between kakapo (*Strigops habroptilus*) breeding and the mast fruiting of their food trees. *Wildlife Research*, **35**(1), 1–7.

Fietkau, H.-J. (1984). *Bedingungen ökologischen Verhaltens*. Weinheim: Beltz.

Fietkau, H.-J. & Kessel, H. (1981). *Umweltlernen*. Königstein: Hain.

Figueira, J., Greco, S. & Ehrgott, M. (eds). (2005). *Multiple Criteria Decision Analysis: State of the Art Surveys*. New York, NY: Springer.

Figueredo, A. J., Hammond, K. R. & McKiernan, E. C. (2006). A Brunswikian evolutionary developmental theory of preparedness and plasticity. *Intelligence*, **34**(2), 211–27.

Fine, T. L. (1973). *Theories of Probability: an Examination of Foundations*. New York, NY: Academic Press.

Finer, S. E. (1999). *The History of Government from the Earliest Times: the Intermediate*. Oxford: Oxford University Press.

Finnie, W. C. (1973). Field experiments in litter control. *Environment and Behavior*, **5**(2), 123–44.

Finnveden, G., Hanschild, M. Z., Ekvall, T., Guinee, J., Heijungs, R., Hellweg, S. et al. (2009). Recent developments in Life Cycle Assessment. *Journal of Environmental Management*, **91**(1), 1–21.

Finucane, M. L., Alhakami, A., Slovic, P. & Johnson, S. M. (2000). The affect heuristic in judgments of risks and benefits. *Journal of Behavioral Decision Making*, **13**(1), 1–17.

Fiorino, D. J. (1990). Citizen participation and environmental risk. A survey of institutional mechanisms. *Science, Technology, & Human Values*, **15**(2), 226–43.

Fischbein, W. (1975). *The Intuitive Sources of Probabilistic Thinking in Children*. Dordrecht: Reidel.

Fischer, A. C. & Krutilla, J. V. (1972). The economics of environmental preservation: a theoretical and empirical analysis. *American Economic Review*, **62**(4), 605–19.

Fischer, E. F. (1999). Cultural logic and Maya identity – rethinking constructivism and essentialism. *Current Anthropology*, **40**(4), 473–99.

Fischer, K. R. & Stadler, F. (eds). (1997). *"Wahrnehmung und Gegenstandswelt": Zum Lebenswerk von Egon Brunswik (1903–1955)*. Vienna: Springer.

Fischer-Kowalski, M. (1997). Society's metabolism: on the childhood and adolescence of a rising conceptual star. In M. Redclift & G. Woodgate (eds) *The International Handbook of Environmental Sociology*. Cheltenham: Edward Elgar, pp. 119–37.

Fischer-Kowalski, M. (1998). Society's metabolism: the intellectual history of material flow analysis, Part I, 1860–1970. *Journal of Industrial Ecology*, **2**(1), 61–78.

Fischer-Kowalski, M. (2002). Exploring the history of industrial metabolism. In R. U. Ayres & L. W. Ayres (eds). *A Handbook of Industrial Ecology*. Cheltenham: Edward Elgar, pp. 16–26.

Fischer-Kowalski, M. (2003). Gesellschaftliche Kolonialisierung natürlicher Systeme. Arbeiten an einem Theorieversuch. In W. Serbser (ed.). *Humanökologie: Ursprünge – Trends – Zukünfte*. München: Oekom: LIT Verlag, Vol 1, pp. 308–25.

Fischer-Kowalski, M. & Haberl, H. (2007a). Conceptualizing, observing and comparing social ecological transitions. In M. Fischer-Kowalski & H. Haberl (eds). *Socioecological Transitions and Global Change*. Cheltenham: Edward Elgar, pp. 1–30.

Fischer-Kowalski, M. & Haberl, H. (2007b). *Socioecological Transitions and Global Change: Trajectories of Social Metabolism and Land Use*. Cheltenham: Edward Elgar.

Fischer-Kowalski, M. & Rotmans, J. (2009). Conceptualizing, observing, and influencing social–ecological transitions. *Ecology and Society*, **14**(2), 3.

Fischer-Kowalski, M. & Weisz, H. (1999). Society as hybrid between material and symbolic realms: toward a theoretical framework of society-nature interrelation. *Advances in Human Ecology*, **8**, 215–51.

Fischhoff, B. (1995). Risk perception and communication unplugged: 20 years of process. *Risk Analysis*, **15**(2), 137–45.

Fischhoff, B., Bostrom, A. & Quadrel, M. J. (1993). Risk perception and communication. *Annual Review of Public Health*, **14**, 183–203.

Fischhoff, B., Slovic, P., Lichtenstein, S., Read, S. & Combs, B. (1978). How safe is safe enough? A psychometric study of attitudes towards technological risks and benefits. *Policy Sciences*, **9**(2), 127–52.

Fish, D. N. (2001). Optimal antimicrobial therapy for sepsis. *Amercian Journal of Health-System Pharmacy*, **59**, S13–9.

Fishbein, M. & Ajzen, I. (1975). *Belief, Attitude, Intention, and Behavior*. Reading, MA: Addison-Wesley.

Fishburn, P. C. (1970). *Utility Theory for Decision Making*. New York, NY: Wiley.

Fishburn, P. C. (1982). *The Foundations of Expected Utility*. Dordrecht: Reidel.

Fisher, I. (1896). *Appreciation and Interest*. New York, NY: Macmillan.

Flamstéed, J. (1776). *Atlas Céleste*. Paris: Lamarche.

Flemming, W. (1879). Beiträge zur Kenntnis der Zelle und ihrer Lebenserscheinungen. *Archiv für Mikroskopische Anatomie*, **16**, 302–436.

References

Flüeler, T. (2006). *Decision Making for Complex Socio-Technical Systems: Robustness from Lessons Learned in Long-Term Radioactive Waste Governance.* Dordrecht: Springer.

Flynn, J., Slovic, P. & Mertz, C. K. (1993). Decidedly different – expert and public views of risks from a radioactive waste repository. *Risk Analysis,* **13**(6), 643–8.

Flynn, J., Slovic, P. & Mertz, C. K. (1994). Gender, race, and perception of environmental health risks. *Risk Analysis,* **14**(6), 1101–8.

Fodor, J. A. (1968). *Psychological Explanation: an Introduction to the Philosophy of Psychology.* New York, NY: Random House.

Folke, C. (2006). Resilience: the emergence of a perspective for social-ecological systems analyses. *Global Environmental Change-Human and Policy Dimensions,* **16**(3), 253–67.

Folke, C. (2007). Social-ecological systems and adaptive governance of the commons. *Ecological Research,* **22**(1), 14–15.

Folke, C., Carpenter, S., Elmqvist, T. *et al.* (2002). Resilience and sustainable development: building adaptive capacity in a world of transformations. *Ambio,* **31**(5), 437–40.

Folke, C., Carpenter, S., Walker, B. *et al.* (2004). Regime shifts, resilience, and biodiversity in ecosystem management. *Annual Review of Ecology Evolution and Systematics,* **35**, 557–81.

Folke, C., Chapin, F. S. & Olsson, P. (2009). Transformations in ecosystem stewardship. In F. S. Chapin, G. P. Kofinas & C. Folke (eds). *Principles of Ecosystem Stewardship: Resilience-Based Natural Resource Management in a Changing World.* New York, NY: Springer., pp. 103–25.

Folke, C., Colding, J. & Berkes, F. (2003). Synthesis: building resilience and adaptive capacity in social-ecological systems. In F. Berkes, J. Colding & C. Folke (eds). *Navigating Social-Ecological Systems. Building Resilience for Complexity and Change.* Cambridge: Cambridge University Press, pp. 352–87.

Folke, C., Hahn, T., Olsson, P. & Norberg, J. (2005). Adaptive governance of social-ecological systems. *Annual Review of Environment and Resources,* **30**, 441–73.

Food and Agriculture Organization of the United Nations (FAO) (2004). *The State of Food Insecurity in the World 2004.* Rome: Food and Agriculture Organization of the United Nations.

Food and Agriculture Organization of the United Nations (FAO) (2008). *The State of Food and Agriculture.* Rome: Food and Agriculture Organization of the United Nations.

FOPH. (2010). Press release. Retrieved February 5th, 2010, from http://www.bag.admin.ch/aktuell/00718/01220/index.html?lang=de&msg-id=4296

Ford, B. J. (1992). Brownian movement in clarkia pollen: a reprise of the first observations. *The Microscope,* **40**(4), 235–41.

Ford, H. (1925, September, 20). Ford predicts fuel from vegetation. *The New York Times*, p. 24.

Forman, R. T. T. (1995). *Land Mosaics: the Ecology of Landscapes and Regions.* Cambridge, MA: Cambridge University Press.

Forrester, J. W. (1961). *Industrial Dynamics.* Cambridge, MA: The MIT Press.

Forrester, J. W. (1968). *Principles of Systems: Text and Workbook.* Cambridge, MA: Wright-Allen-Press.

Forrester, J. W. (1968). *Principles of Systems.* Cambridge MA: The MIT Press.

Forrester, J. W. (1969). *Urban Dynamics.* Cambridge, MA: The MIT Press.

Forrester, J. W. (1971a). Counterintuitive behavior of social systems. *Technological Forecasting and Social Change,* **3**(1), 1–22.

Forrester, J. W. (1971b). *World Dynamics.* Cambridge, MA: Pegasus communications.

Forrester, J. W. (1987). Lessons from system dynamics modelling. *System Dynamics Review,* **3**(2), 136–49.

Forrester, J. W. (1994). System dynamics, systems thinking, and soft OR. *System Dynamics Review,* **10**(2–3), 245–56.

Forrester, J. W. (2007a). System dynamics – a personal view of the first fifty years. *System Dynamics Review,* **23**(2–3), 345–58.

Forrester, J. W. (2007b). System dynamics – the next fifty years. *System Dynamics Review,* **23**(2–3), 359–70.

Foster, G. M. (1988). The validating role of humoral theory in traditional Spanish-American therapeutics. *American Ethnologist,* **15**(1), 120–35.

Foster, J. B. (1999). Marx's theory of metabolic rift: classical foundations for environmental sociology. *American Journal of Sociology,* **105**(2), 366–405.

Foster, J. B. (2000). *Marx's Ecology.* New York, NY: Monthly Review Press.

Foucault, M. (1971). Orders of discourse. Inaugural lecture delivered at the College de France (R. Swyer, Trans.). *Social Science Information,* **10**(2), 7–30.

Franz, M. (2008). Phosphate fertilizer from sewage sludge ash (SSA). *Waste Management,* **28**(10), 1809–18.

Fraser, J. (1908). A new visual illusion of direction. *British Journal of Psychology,* **2**(1), 307–20.

Fraunhofer Institut für System – und Innovationsforschung ISI. (2009). *Rohstoffe für Zukunftstechnologien – Einfluss des branchenspezifischen Rohstoffbedarfs in rohstoffintensiven Zukunftstechnologien auf die zukünftige Rohstoffnachfrage*. Karlsruhe: Fraunhofer Institut für System – und Innovationsforschung ISI.

Free, A. & Barton, N. H. (2007). Do evolution and ecology need the Gaia hypothesis? *Trends in Ecology & Evolution*, **22**(11), 611–19.

Freese, B. (2003). *Coal: A Human History*. Cambridge, MA: Perseus.

Freese, L. (1997). *Environmental Connections. Supplement 1 (Part B) to Advances of Human Ecology*. Greenwich, CT: JAI Press.

Freud, S. (1915). Triebe und Triebschicksale. In S. Freud (ed.). *Gesammelte Werke*, Vol. X. Frankfurt am Main: Fischer.

Freud, S. (1920). Beyond the pleasure principle. *The Standard Edition of the Complete Psychological Works of Sigmund Freud*, Vol. XVIII. New York, NY: Norton, pp. 1–64.

Freudenburg, W. R. (2008). Thirty years of scholarship and science on environment–society relationships. *Organization & Environment*, **21**(4), 449–59.

Fried, M. (2000). 650 year anniversary of Charles University, Prague. *Obesity Surgery*, **10**(1), 70–2.

Friedman, W. J. & Dewinstanley, P. A. (2006). The mental representation of countries. *Memory*, **14**(7), 853–71.

Friman, P. C. & Poling, A. (1995). Making life easier with effort: basic findings and applied research on response effort. *Journal of Applied Behavior Analysis*, **28**(4), 538–90.

Frischknecht, R. (2000). Allocation in life cycle inventory analysis for joint production. *International Journal of Life Cycle Assessment*, **5**(2), 85–95.

Frischknecht, R., Jungbluth, N., Althaus, H. J. *et al.* (2005). The ecoinvent database: overview and methodological framework. *International Journal of Life Cycle Assessment*, **10**(1), 3–9.

Fritsche, U. R., Hünecke, K., Hermann, A., Schulze, F. & Wiegmann, K. (2006). *Sustainable Standards for Bioenergy*. Frankfurt: WWF Germany.

Frondel, M. & Schmidt, C. M. (2007). *Von der baldigen Erschöpfung der Rohstoffe und anderen Märchen*. Essen: RWI Essen.

Frosch, R. A. (1992). Industry ecology: a philosophical introduction. *Proceedings of the National Academy of Sciences of the USA*, **89**(3), 800–3.

Frossard, E., Bünemann, E. K., Jansa, J., Oberson, A. & Feller, C. (2009). Concepts and practices of nutrient management in agroecosystems: can we draw lessons from history to design future sustainable agricultural production systems? *Bodenkultur*, **60**(1–4), 43–60.

Frutiger, A. (1998). *Signs and Symbols – their Design and Meaning*. London: Ebury Press.

Fuhrer, U. (1990). Bridging the ecological-psychological gap. Behavioral setting as interfaces. *Environment and Behavior*, **22**(4), 518–37.

Fuhrer, U. & Wölfing, S. (1997). *Von den sozialen Grundlagen des Umweltbewusstseins zum verantwortlichen Umwelthandeln: Die sozialpsychologische Dimension globaler Umweltproblematik*. Bern: Hans Huber.

Fujii, S. (2006). Environmental concern, attitude toward frugality, and ease of behavior as determinants of pro-environmental behavior intentions. *Journal of Environmental Psychology*, **26**(4), 262–8.

Funke, J. (1999). Komplexes Problemlösen: Ein Blick zurück und nach vorne. *Psychologische Rundschau*, **50**(4), 194–7.

Funtowicz, S. O. & Ravetz, J. R. (1993). Science for the post-normal age. *Futures*, **7**(25), 735–55.

Funtowicz, S. O. & Ravetz, J. R. (2008). Values and uncertainties. In G. Hirsch-Hadorn, H. Hoffmann-Riem, S. Biber-Klemm *et al.* (eds). *Handbook or Transdisciplinary Research*. Heidelberg: Springer, pp. 361–8.

Furubotn, E. G. & Richter, R. (2005). *Institutions & Economic Theory*, 2nd edn. Ann Arbor, MI: The University of Michigan Press.

Füssel, H.-M. (2007). Vulnerability: a generally applicable conceptual framework for climate change research. *Global Environmental Change*, **17**(2), 155–67.

Gajdusek, D. C., Gibbs, C. J. & Alpers, M. (1966). Experimental transmission of a Kuru-like syndrome to chimpanzees. *Nature*, **209**, 794–6.

Gallati, J. (2008). *Towards an Improved Understanding of Collective Irrigation Management: a System Dynamics Approach*. Unpublished PhD thesis, University of Berne, Berne.

Gallopin, G. C. (2006). Linkages between vulnerability, resilience, and adaptive capacity. *Global Environmental Change-Human and Policy Dimensions*, **16**(3), 293–303.

Garber, D. (1985). Leibniz and the foundations of physics: the middle years. In K. Okruhlik & J. R. Brown (eds). *The Natural Philosophy of Leibniz*. Dordrecht: Reidel, pp. 27–130.

Gardiner, D. M., Endo, T. & Bryant, S. V. (2002). The molecular basis of amphibian limb regeneration: integrating the old with the new. *Seminars in Cell & Developmental Biology*, **13**(5), 345–52.

Gardner, G. T. & Stern, P. C. (2002). *Environmental Problems and Human Behavior*, 2nd edn. Boston, MA: Pearson Custom Publishing.

Gardner, J. S. & Dekens, J. (2007). Mountain hazards and the resilience of social-ecological systems: lessons learned in India and Canada. *Natural Hazards*, **41**(2), 317–36.

Gärling, T. & Evans, G. W. (1992). *Environment, Cognition, and Action. An Integrated Approach*. Oxford: Oxford University Press.

Gatersleben, B., Steg, L. & Vlek, C. (2002). Measurement and determinants of environmentally significant consumer behavior. *Environment and Behavior,* **34**(3), 335–62.

Gaudelier, M. (1986). *The Mental and the Material.* Thetford: The Thetford Press.

Geels, F. W. (2004). From sectoral systems of innovation to socio-technical systems – insights about dynamics and change from sociology and institutional theory. *Research Policy,* **33**(6–7), 897–920.

Geels, F. W. (2005). Processes and patterns in transitions and system innovations: refining the co-evolutionary multi-level perspective. *Technological Forecasting and Social Change,* **72**(6), 681–96.

Geertz, C. (1973). Thick description: toward an interpretive theory of culture. In C. Geertz (ed.). *The Interpretation of Cultures: Selected Essays.* New York, NY: Harper Collins, pp. 3–30.

Gehmacher, E. (1973). *Psychologie und Soziologie der Umweltplanung.* Freiburg: Rombach.

Gehrke, C. (2003). The Ricardo effect: its meaning and validity. *Economica,* **70**(277), 143–58.

Geison, G. L. (1995). *The Private Science of Louis Pasteur.* Princeton, NJ: Princeton University Press.

Geist, V. (1978). *Life Strategies, Human Evolution, Environmental Design: toward a Biological Theory of Health.* Berlin: Springer.

Geller, B., Winnett, R. & Everett, P. (1982). *Preserving the Environment.* Elmsford, NY: Pergamon.

Geller, E. S. (1975). Increasing desired waste disposals with instructions. *Man-Environment Systems,* **5**(2), 125–8.

Genovese, M. (2005). Research on hidden variable theories: a review of recent progresses. *Physics Reports-Review Section of Physics Letters,* **413**(6), 319–96.

Georgescu-Roegen, N. (1971). *The Entropy Law and the Economic Process.* Cambridge, MA: Harvard University Press.

Gerbig, A. & Buchtmann, P. (2003). Vom "Waldsterben" zu "Geiz ist Geil": Figurativer Sprachgebrauch im Paradigmenwechsel von der ökologischen zur ökonomischen Handlungsmotivation. *metaphorik.de, 04/2003,* 97–114. Retrieved January 6, 2011, from http://www.metaphorik.de/04/

Gerrig, R. G. & Zimbardo, P. G. (2008). *Psychology and Life,* 18th edn. Boston, MA: Pearson.

Giancoli, D. C. (1997). *Physics,* Principles with applications. Upper Saddle Point, NJ: Prentice Hall.

Gibbons, M. (1998). Changing patterns of university – industry relations. *Minerva,* **38**(3) 352–61.

Gibbons, M. (1999). Science's new social contract with society. *Nature,* **402**(6761), C81–4.

Gibbons, M., Limoges, C., Nowotny, H. *et al.* (1994). *The New Production of Knowledge.* London: Sage.

Gibbons, M. & Nowotny, H. (2001). The potential of transdisciplinarity. In J. Thompson Klein, W. Grossenbacher-Mansuy, R. Häberli *et al.* (eds). *Transdisciplinarity: Joint Problem Solving among Science, Technology, and Society. An Effective Way for Managing Complexity.* Basel: Birkhäuser, pp. 67–80.

Gibbs, L. M. (1982). *Love Canal: My Story.* New York, NY: State University of New York Press.

Gibson, E. J. & Walk, R. D. (1960). The "visual cliff." *Scientific American,* **202**(4), 64–71.

Gibson, J. J. (1950). *The Perception of the Visual World.* Boston, MA: Houghton Mifflin.

Gibson, J. J. (1967). New reasons for realism. *Synthese,* **17**(1), 162–72.

Gibson, J. J. (1978). A note on what exists at the ecological level of reality. In E. Reed & R. Jones (eds). *Reasons for Realism: Selected Essays of James J. Gibson.* Hillsdale, NJ: Lawrence Earlbaum, pp. 416–18.

Gibson, J. J. (1979). *The Ecological Approach to Visual Perception.* Boston, MA: Houghton Mifflin.

Gibson, J. J. & Gibson, E. J. (1955). Perceptual learning: differentiation or enrichment? *Psychological Review,* **62**(1), 32–41.

Giddens, A. (1984). *The Constitution of the Society.* Berkeley, CA: University of California Press.

Giddens, A. (1990). *The Consequences of Modernity.* Cambridge: Polity Press.

Giddens, A. (2007). *Europe in the Global Age.* Cambridge: Polity Press.

Giddens, A. (2009). *The Politics of Global Change.* Cambridge: Polity Press.

Giffin, A. (2009). From physics to economics: an econometric example using maximum relative entropy. *Physica A: Statistical Mechanics and its Applications,* **388**(8), 1610–20.

Gifford, R. (1997). *Environmental Psychology: Principles and Practice.* Boston, MA: Allyn and Bacon.

Gifford, R. (2002). Making a difference: some ways environmental psychology has improved the world. In R. B. Bechtel & A. Churchman (eds). *Handbook of Environmental Psychology.* New York, NY: Wiley, pp. 323–62.

Gifford, R. (2007). Environmental psychology and sustainable development: expansion, maturation, and challenges. *Journal of Social Issues,* **63**(1), 199–212.

Gifford, R. (2009). Environmental psychology: manifold visions, unity of purpose. *Journal of Environmental Psychology,* **29**(3), 387–9.

Gigerenzer, G. (2006). Out of the frying pan into the fire: behavioral reactions to terrorist attacks. *Risk Analysis,* **26**(2), 347–51.

Gigerenzer, G. (2007). *Gut Feelings: the Intelligence of the Unconscious.* New York, NY: Viking Adult.

Gigerenzer, G. & Murray, D. J. (1987). *Cognition as Intuitive Statistics.* Hillsdale, NJ: Erlbaum.

Gigerenzer, G. & Selten, R. (eds). (2001). *Bounded Rationality: the Adaptive Toolbox.* Cambridge, MA: The MIT Press.

Gigerenzer, G., Todd, P. M. & The ABC Research Group. (1999). *Simple Heuristics that Make us Smart.* New York, NY: Oxford University Press.

Gilbert, N. (1995). Emergence in social simulation. In N. Gilbert & R. Conte (eds). *Artificial Societies: the Computer Simulation of Social Life.* London: UCL Press, pp. 144–56.

Gilbert, N. & Troitzsch, K. (2005). *Simulation for the Social Scientist,* 2nd edn. Milton Keynes, UK: Open University Press.

Giles, J. (2004). Peer-reviewed paper defends theory of intelligent design. *Nature,* **431**(7005), 114.

Giller, K. E., Witter, E. & McGrath, S. P. (1998). Toxicity of heavy metals to microorganisms and microbial processes in agricultural soils: a review. *Soil Biology & Biochemistry,* **30**(10–11), 1389–1414.

Gimbutas, M. (1982). *The Goddesses and Gods of Old Europe.* Berkeley, CA: University of California Press.

Ginoux, J. M. (2009). Differential equations. In J. M. Ginoux (ed.). *Differential Equations in Differential Geometry Applied to Dynamical Systems.* Singapur: World Scientific Publishing Group, pp. 3–6.

Ginsburg, H. P. & Opper, S. (1988). *Piaget's Theory of Intellectual Development. An Introduction.* Englewood Cliffs, NJ: Prentice Hall.

Gintis, H. (2009a). *The Bounds of Reason: Game Theory and the Unification of the Behavioral Sciences.* Princeton, NY: Princeton University Press.

Gintis, H. (2009b). *Game Theory Evolving: A Problem-Centered Introduction to Modeling Strategic Behavior.* Princeton, NY: Princeton University Press.

Giri, A. K. (2011). The calling of a creative transdisciplinarity. *Futures,* **34**(1), 103–15.

Girod, B., de Haan, P. & Scholz, R. W. (2011). Consumption-as-usual instead of ceteris paribus assumption for demand integration of potential rebound effects into LCA. *The International Journal of Life Cycle Assessment,* **16**(1), 3–11.

Girod, B., Wiek, A., Mieg, H. A. & Hulme, M. (2009). The evolution of the IPCC's emissions scenarios. *Environmental Science & Policy,* **12**(2), 103–18.

Giuliani, M. V. & Scopelliti, M. (2009). Empirical research in environmental psychology: past, present, and future. *Journal of Environmental Psychology,* **29**(3), 375–86.

Gladwin, T. (1970). *East is a Big Bird: Navigation and Logic on Puluwat Atoll.* Cambridge, MA: Harvard University Press.

Glasmeier, A. (1995). Consensus conference, the media and public information in the Netherlands. In S. Joss & J. Durant (eds). *Public Participation in Science.* London: Science Museum, pp. 67–73.

Glason, J., Therival, R. & Chadwick, A. (2005). *Introduction to Environmental Impact Assessment.* London: Routledge.

Gobster, P. H. (1999). An ecological aesthetic for forest landscape management. *Landscape Journal,* **18**(1), 54–64.

Gobster, P. H., Nassauer, J. I., Daniel, T. C. & Fry, G. (2007). The shared landscape: what does aesthetics have to do with ecology? *Landscape Ecology,* **22**(7), 959–72.

Goedkoop, M., Hofstetter, P., Müller-Wenk, R. & Spriensma, R. (1998). The Eco-Indicator 98 explained. *The International Journal of Life Cycle Assessment,* **3**(6), 352–60.

Goldblatt, D. L. (2005). *Sustainable Energy Consumption and Society.* Dordrecht: Springer.

Goldemberg, J. (2007). Ethanol for a sustainable energy future. *Science,* **315**(5813), 808–10.

Goldman, S. M. (1978). Portfolio choice and flexibility – precautionary motive. *Journal of Monetary Economics,* **4**(2), 263–79.

Goldsby, R. A., Kindt, T. J., Osborne, B. A. & Kuby, J. (2003). *Immunology.* New York, NY: Freeman.

Goldstein, S., Clarke, D. R., Walsh, S. P., Black, K. S. & O'Brien, M. F. (2000). Transpecies heart valve transplant: advanced studies of a bioengineered xeno-autograft. *Annals of Thoracic Surgery,* **70**(6), 1962–9.

Goldstein, W. M. & Wright, J. H. (2001). "Perception" versus "thinking": Brunswikian thought on central responses and processes. In K. R. Hammond & T. R. Stewart (eds). *The Essential Brunswik: Beginnings, Explications, Applications.* Oxford: Oxford University Press, pp. 249–56.

Golley, F. B. (1998). *A Primer for Environmental Literacy.* New Haven, CT: Yale University Press.

Goodin, R. E. (2008). Deliberative lies. *European Political Science,* **7**(2), 194–8.

Goodstein, E. (2003). The death of the Pigovian tax? – Policy implications from the double-dividend debate. *Land Economics,* **79**(3), 402–14.

References

Gordon, J. W. (1877). On certain molar movements of the human body produced by the circulation of blood. *Journal of Anatomy & Physiology,* **11**, 533–6.

Gordon, R. B., Bertram, M. & Graedle, T. E. (2006). Metal stocks and sustainability. *Proceedings of the National Academy of Sciences,* **103**(5), 1209–14.

Gossen, H. H. (1854). *Entwicklung der Gesetze des menschlichen Verkehrs und der daraus fliessenden Regeln für menschliches Handeln.* Braunschweig: Friedrich Vieweg und Sohn.

Gottschling, D. E. (2007). Epigenetics: from phenomenon to field. In C. D. Allis, T. Jenuwein, D. Reinberg & M.-L. Capparros (eds). *Epigenetics,* Vol. 1–13. New York, NY: Cold Spring.

Goudie, A. (2006). *The Human Impact on the Natural Environment – Past, Present, and Future,* 6th edn. Malden, MA: Blackwell Publishing.

Gould, K. A., Pellow, D. N. & Schnaiberg, A. (2008). *Injustice and Unsustainability in the Global Economy.* Boulder, CO: Paradigm Publishers.

Gould, K. A., Pellow, D. N. & Schnaiberg, A. (2004). Interrogating the treadmill of production – everything you wanted to know about the treadmill but were afraid to ask. *Organization & Environment,* **17**(3), 296–316.

Goulson, D., Lye, G. C. & Darvill, B. (2008). Decline and conservation of bumble bees. *Annual Review of Entomology,* **53**, 191–208.

Government of New York City. (2007). Staten Island Railroad officially reactivated. Retrieved May 12, 2007, from http://www.nyc.gov/portal/site/nycgov/menuitem.53459261fb6a2fb3f7393cd401c789a0/index.jsp?pageID=nyc_photo&eid=12470&pc=1136&img=20839&ipos=3

Grabher, G. (ed.). (1993). *The Embedded Firm. On the Socioeconomics of Industrial Networks.* London: Routledge.

Graedel, T. E. (1994). Industrial ecology: definition and implementation. In R. Socolow, C. Andrews, F. Berkhout & V. Thomas (eds) *Industrial Ecology and Global Change.* Cambridge: Cambridge University Press, pp. 23–44.

Graedel, T. E. (1998). Life-cycle assessment in the service industries. *Journal of Industrial Ecology,* **1**(4), 57–70.

Graedel, T. E. & Crutzen, P. J. (1993). *Atmospheric Change. An Earth System Perspective.* New York, NY: W. H. Freeman.

Graedel, T. E., van Beers, D., Bertram, M. *et al.* (2005). The multilevel cycle of anthropogenic zinc. *Journal of Industrial Ecology,* **9**(3), 67–90.

Graham, A., Senge, J., Sterman, J. D. & Morecroft, J. (1989). Computer-based case studies in management education and research. In P. Milling & E. Zahn (eds). *Computer-Based Management of Complex Systems.* Berlin: Springer, pp. 317–26.

Granovetter, M. (1978). Threshold models for collective behaviour. *American Journal of Sociology,* **83**(6), 1420–43.

Grasmück, D. & Scholz, R. W. (2005). Risk perception of heavy metal soil contamination by high-exposed and low-exposed inhabitants: the role of knowledge and emotional concerns. *Risk Analysis,* **25**(3), 611–22.

Graumann, C. F. (1997). Eine historische Einführung in die Sozialpsychologie. In W. Stroebe, M. Hewstone & G. M. Stephenson (eds). *Sozialpsychologie – Eine Einführung.* Berlin: Springer, pp. 3–24.

Gray, L. C. (1914). Rent under the assumption of exhaustibility. *Quarterly Journal of Economics,* **28**, 466–89.

Gray, R. (1992). Death of the gene: developmental systems fight back. In E. Griffiths (ed.). *Trees of Life: Essays in the Philosophy of Biology.* Dordrecht: Kluwer, pp. 165–209.

Green, C. D. (1995). All that glitters: a review of psychological research on aesthetics of the golden section. *Perception,* **24**(8), 937–68.

Greene, E. L. (1909/1983). *Landmarks in Botanical History.* Stanford, CA: Stanford University Press.

Greeno, J. G. (1994). Gibson's affordances. *Psychological Review,* **101**(2), 336–42.

Greer, J. M. (2005). Collapse: how societies choose to fail or succeed. *Population and Environment,* **26**(5), 439–44.

Gregg, M. B. (1980). The epidemiology of influenza in humans. In R. B. Kundsin (ed.). *Annals of the New York Academy of Sciences: Airborne Contagion,* Vol. 353. New York, NY: New York Academy of Sciences.

Gregory, G. D. & Di Leo, M. (2003). Repeated behavior and environmental psychology: the role of personal involvement and habit formation in explaining water consumption. *Journal of Applied Social Psychology,* **33**(6), 1261–96.

Gregory, S. G., Barlow, K. F., McLay, K. E. *et al.* (2006). The DNA sequence and biological annotation of human chromosome 1. *Nature,* **441**(7091), 315–21.

Grigg, R. (1998). Fraud rediscovered. It has long been known that one of the most effective popularizers of evolution fudged some drawings, but only now has the breathtaking extent of his deceit been revealed. *Creation,* **20**(2), 49–51.

Grimm, V., Revilla, E., Berger, U. *et al.* (2005). Pattern-oriented modeling of agent-based complex systems: lessons from ecology. *Science,* **310**(5750), 987–91.

Grobstein, C. (1965). *The Strategy of Life.* San Francisco, CA: Freeman.

Grobstein, C. (1973). Hierarchical order and neogenesis. In H. D. Pattee (ed.). *Hierarchy Theory. The Complex Challenge of Complex Systems*. New York, NY: George Braziller, pp. 31–47.

Gross, M. (2001). *Die Natur der Gesellschaft. Eine Geschichte der Umweltsoziologie*. Weinheim: Juventa.

Gross, M. (2004). Human geography and ecological sociology – the unfolding of a human ecology, 1890 to 1930 – and beyond. *Social Science History*, **28**(4), 575–605.

Grossmann, S. & Thomae, S. (1977). Invariant distributions and stationary correlation functions of one-dimensional discrete processes. *Zeitschrift für Naturforschung*, **32**(A), 1353–63.

Grundy, S. (1987). *Curriculum: Product or Praxis?* Oxford: Routledge.

Guagnano, G. A., Stern, P. C. & Dietz, T. (1995). Influences on attitude-behavior relationships: a natural experiment with curbside recycling. *Environment and Behavior*, **27**(5), 699–718.

Guinée, J. B. (2002). *Handbook of Life Cycle Assessment: Operational Guide to the ISO Standards*. Dordrecht: Kluwer.

Gunderson, L. H. & Holling, C. S. (eds). (2002). *Panarchy: Understanding Transformation in Human and Natural Systems*. Washington, DC: Island Press.

Gunderson, L. H. & Pritchard, L. (2002). *Resilience and the Behavior of Large-Scale Systems*. Washington, DC: Island Press.

Gurung, D. B. & Scholz, R. W. (2008). Community-based ecotourism in Bhutan: expert evaluation of stakeholder-based scenarios. *International Journal of Sustainable Development and World Ecology*, **15**(5), 397–411.

Gustafsson, Å. (1979). Linnaeus' Peloria: the history of a monster. *Theoretical and Applied Genetics*, **54**(6), 241–8.

Güth, W. (1992). *Spieltheorie und ökonomische (Bei)Spiele*. Berlin: Springer.

Haber, W. (1995). Landschaft. In Akademie für Raumforschung u. Landesplanung (ed.). *Handwörterbuch der Raumplanung*. Hannover: ARL.

Haberl, H. (2001). *Energetische Dimensionen des gesellschaftlichen Stoffwechsels. Interdisziplinäre Analysen der physischen Wechselwirkungen zwischen menschlichen Gesellschaften und ihrer natürlichen Umwelt und ihre theoretischen und methodischen*. Grundlagen. Unpublished Habilitation thesis. Vienna: Universität Wien.

Haberl, H., Fischer-Kowalski, M., Krausmann, F., Weisz, H. & Winiwarter, V. (2004). Progress towards sustainability? What the conceptual framework of material and energy flow accounting (MEFA) can offer. *Land Use Policy*, **21**(3), 199–213.

Haberl, H., Krausmann, F., Erb, K. H. *et al.* (2002). Human appropriation of net primary production. *Science*, **296**(5575), 1968–9.

Häberli, R., Bill, A., Grossenbacher-Mansuy, W. *et al.* (2001). Synthesis. In J. Thompson Klein, W. Grossenbacher-Mansuy, R. Häberli *et al.* (eds). *Transdisciplinarity: Joint Problem Solving among Science, Technology, and Society. An Effective way for Managing Complexity*. Basel: Birkhäuser, pp. 6–22.

Habermas, J. (1984). *The Theory of Communicative Action*, Vol. 1: Reason and the Rationalization of Society. Boston, MA: Beacon Press.

Habermas, J. (1987). *The Theory of Communicative Action*, Vol. 2: Lifeworld and System: A Critique of Functionalist Reason. Boston, MA: Beacon press.

Hacking, I. (1975). *The Emergence of Probability*. Cambridge: Cambridge University Press.

Haeckel, E. (1866). *Generelle Morphologie der Organismen*, Vol. 2. Berlin: Georg Reimer.

Haeckel, E. (1874). *Anthropogenie oder Entwicklungsgeschichte des Menschen*. Leipzig: Wilhelm Engelmann.

Haeckel, E. (1898). *Natürliche Schöpfungsgeschichte. Gemeinverständliche wissenschaftliche Vorträge über die Entwicklungslehre im Allgemeinen und derjenigen von Darwin, Goethe und Lamarck im Besonderen, über die Anwendung derselben auf den Ursprung des Menschen und andere damit zusammenhängende Grundfragen der Naturwissenschaften*, 9th edn. Berlin: Verlag Georg Reimer.

Haeckel, E. (1911). *The Evolution of Man: A Popular Scientific Study* (J. McCabe, Trans.). London: Watts & Co.

Haerlin, B. & Parr, D. (1999). How to restore public trust in science. *Nature*, **400**(6744), 499.

Hafele, J. C. & Keating, R. E. (1972). Around-the-world atomic clocks – observed relativistic time gains. *Science*, **177**(4044), 166–8.

Hagedorn, D. (1994). Zum ägyptischen Kalender unter Augustus. *Zeitschrift für Papyrologie und Epigraphik*, **100**, 211–22.

Hagedorn, L. & Henke, R. (1991). The effect of multiple categorization membership on intergroup evaluation in a North-Indian context: class, caste, and religion. *British Journal of Social Psychology*, **20**, 247–60.

Hahn, R. W., Olmstead, S. M. & Stavins, R. N. (2003). Environmental regulation during the 1990s: a retrospective analysis. *Harvard Environmental Law Review*, **27**(2), 377–415.

Hahn-Hagerdal, B., Galbe, M., Gorwa-Grauslund, M. F., Liden, G. & Zacchi, G. (2006). Bio-ethanol – the fuel of tomorrow from the residues of today. *Trends in Biotechnology*, **24**(12), 549–56.

References

Hajer, M. A. (1995). *The Politics of Environmental Discourse: Ecological Modernization and the Regulation of Acid Rain*. Cambridge: Cambridge University Press.

Haken, H. (1983). *Synergetics: an Introduction. Nonequilibrium Phase Transitions and Self-Organization in Physics, Chemistry, and Biology*, 3rd edn. Vol. 1. Berlin: Springer.

Haken, H. (1996). Synergetik und Sozialwissenschaften. *Ethik und Sozialwissenschaften,* **7**(4), 587–94.

Haken, H. (2000). *Information and Self-Organization: a Macroscopic Approach to Complex Systems*, 2nd edn. Berlin: Springer.

Haken, H. (ed.). (1985). *Complex Systems: Operational Approaches in Neurobiology, Physics, and Computers. Proceedings of the International Symposium on Synergetics at Schloss Elmau, Bavaria, May 6–11, 1985.* Berlin: Springer.

Halder, G., Callaerts, P. & Gehring, W. J. (1995). Induction of ectopic eyes by targeted expression of the eyeless gene in Drosophila. *Science,* **267**(5205), 1788–92.

Hall, B. (1999). Looking back, looking forward: reflections on the international participatory research network. *Forests, Tree, and People Newsletter,* **39**, 3–26.

Hall, B. K. (2001). The gene is not dead, merely orphaned and seeking a home. *Evolution & Development,* **3**(4), 225–8.

Hall, E. T. (1959). *The Silent Language*. New York, NY: Anchor.

Hall, E. T. (1966). *The Hidden Dimension*. Garden City, NY: Doubleday.

Hall, T. D. (2006). Ecological-evolutionary theory: principles and applications. *Contemporary Sociology – A Journal of Reviews,* **35**(6), 556–8.

Halliday, A. N. (1997). Radioactivity, the discovery of time and the earliest history of the Earth. *Contemporary Physics,* **38**(2), 103–14.

Hammer, R. B., Stewart, S. I., Winkler, R. L., Radeloff, V. C. & Voss, P. R. (2004). Characterizing dynamic spatial and temporal residential density patterns from 1940–1990 across the North Central United States. *Landscape and Urban Planning,* **69**(2–3), 183–99.

Hammerstein, P. & Selten, R. (1994). Game theory and evolutionary biology. In R. J. Aumann & S. Hart (eds). *Handbook of Game Theory with Economic Applications*, Vol. 2. Amsterdam: Elsevier, pp. 929–93.

Hammond, K. R. (2001a). Organismic achievement and environmental probability [1943]. In K. R. Hammond & T. R. Stewart (eds). *The Essential Brunswik: Beginnings, Explications, Applications*. Oxford: Oxford University Press, pp. 55–7.

Hammond, K. R. (2001b). Psychology as science of objective relations. In K. R. Hammond & T. R. Stewart (eds). *The Essential Brunswik: Beginning, Explications, Applications*. Oxford: Oxford University Press, pp. 35–7.

Hammond, K. R. (2001c). Thinking about error and errors about thinking. In K. R. Hammond & T. R. Stewart (eds). *The Essential Brunswik: Beginnings, Explications, Applications*. Oxford: Oxford University Press, pp. 123–7.

Hammond, K. R., Hamm, R. M., Grassia, J. & Pearson, T. (1983). *Direct Comparison of Intuitive, Quasi-Rational, and Analytical Cognition*. Boulder, CO: University of Colorado, Institute for Cognitive Science, Center for Research on Judgment and Policy.

Hammond, K. R. & Stewart, T. R. (eds). (2001). *The Essential Brunswik*. Oxford: Oxford University Press.

Hammond, R. W. (1971). The history and development of industrial engineering. In H. B. Maynard (ed.). *Industrial Engineering Handbook*, 3rd edn. New York, NY: McGraw-Hill, pp. 1/3–1/17.

Han, K., Zhu, X. P., He, F. *et al.* (2009). Lack of airborne transmission during outbreak of pandemic (H1N1) 2009 tour group members, China, June 2009. *Emerging Infectious Diseases,* **15**(10), 1578–81.

Handy, S. (1996). Methodologies for exploring the link between urban form and travel behavior. *Transportation Research Part D-Transport and Environment,* **1**(2), 151–65.

Handy, S., Cao, X. Y. & Mokhtarian, P. (2005). Correlation or causality between the built environment and travel behavior? Evidence from Northern California. *Transportation Research Part D-Transport and Environment,* **10**(6), 427–44.

Hanemann, W. M. (1991). Willingness to pay and willingness to accept – how much can they differ? *American Economic Review,* **81**(3), 635–47.

Hanley, N., Shogren, J. F. & White, B. (1997). *Environmental Economics in Theory and Practice*. Oxford: Oxford University Press.

Hanna, S. S., Folke, C. & Mälerm, K. G. (eds). (1996). *Rights to Nature: Ecological, Economic, Cultural, and Political Principles of Institutions for the Environment*. Washington, DC: Island Press.

Hannigan, J. (2006). *Environmental Sociology*, 2nd edn. London: Routledge.

Hannon, G. J. (2002). RNA interference. *Nature,* **418**(6894), 244–51.

Hanselmann, M. & Tanner, C. (2008). Taboos and conflicts in decision making: sacred values, decision difficulty, and emotions. *Judgment and Decision Making Journal,* **3**(1), 51–63.

Hansen, Z. K. & Libecap, G. D. (2004). Small farms, externalities, and the Dust Bowl of the 1930s. *Journal of Political Economy,* **112**(3), 665–94.

Hansla, A., Gamble, A., Juliusson, A. & Gärling, T. (2008). The relationships between awareness of consequences, environmental concern, and value orientations. *Journal of Environmental Psychology,* **28**(1), 1–9.

Hansla, A., Gamble, A., Juliusson, A. & Gärling, T. (2008). Psychological determinants of attitude towards and willingness to pay for green electricity. *Energy Policy,* **36**(2), 768–74.

Hansmann, R., Bernasconi, P., Smieszek, T., Loukopoulos, P. & Scholz, R. W. (2006). Justifications and self-organization as determinants of recycling behavior: the case of used batteries. *Resources Conservation & Recycling,* **47**(2), 133–59.

Hansmann, R. & Scholz, R. W. (2003). A two-step informational strategy for reducing littering behavior in a cinema. *Environment & Behavior,* **35**(6), 752–62.

Hansson, J., Berndes, G., Johnsson, F. & Kjarstad, J. (2009). Co-firing biomass with coal for electricity generation – an assessment of the potential in EU27. *Energy Policy,* **37**(4), 1444–55.

Hardesty, D. L. (1977). *Ecological Anthropology.* New York, NY: Wiley.

Hardin, G. (1968). The tragedy of the commons. *Science,* **162**(3859), 1243–8.

Hardin, G. (1971). Collective action as an agreeable prisoner's dilemma. *Behavioral Science,* **16**(5), 472–81.

Hare, A. P., Blumberg, H. H., Davies, M. F. & Kent, V. M. (eds). (1994). *Small Group Research. A Handbook.* Norwood, NJ: Ablex.

Harig, G. (ed.). (1974). *Die Bestimmung der Intensität im medizinischen Systems Galens.* Berlin: Akademie Verlag.

Harig, G. & Kolletsch, J. (2004). Naturforschung und Naturphilosophie in der Antike. In I. Jahn (ed.). *Geschichte der Biologie: Theorien, Methoden, Institutionen, Kurzbiografien,* 3rd edn. Hamburg: Nikol, pp. 48–87.

Harman, O. S. (2005). Timeline – Cyril Dean Darlington: the man who "invented" the chromosome. *Nature Reviews Genetics,* **6**(1), 79–85.

Harper, D. (2001). Online etymology dictionary. Retrieved October 15, 2009, from http://www.etymonline.com

Harper, P. S. (2005). William Bateson, human genetics and medicine. *Human Genetics,* **118**(1), 141–51.

Harris, D. R. (2006). The interplay of ethnography and archeological knowledge in the study of past human subsistence in the tropics. *Journal of the Royal Anthropological Institute,* **12**(Special Issue 2006), 63–78.

Harris, M. (1991). *Cultural Anthropology.* New York, NY: Harper Collins.

Harrison, G. (2008). Neogeography: mapping our place in the world. *Sightlines* (2008), 60–82.

Harst, W., Kuhn, J. & Stever, H. (2006). Can electromagnetic exposure cause a change in behaviour? Studying possible non-thermal influences on honey bees – an approach within the framework of educational informatics. *Acta Systemica – IIAS International Journal,* **6**(1), 1–6.

Hartwick, J. M. (1977). Intergenerational equality and the investing of rents from exhaustible resources. *American Economic Review,* **67**(5), 972–4.

Hassan, R., Scholes, R. & Ash, N. (eds). (2005). *Ecosystem and Human Well-Being: Current State and Trends,* Vol. 1. Washington, DC: Island Press.

Hassink, R. & Shin, D.-H. (2005). Theme issue: the restructuring of old industrial areas in Europe and Asia. Guest editorial. *Environment and Planning A,* **37**, 571–80.

Haug, G. H., Gunther, D., Peterson, L. C. *et al.* (2003). Climate and the collapse of Maya civilization. *Science,* **299**(5613), 1731–5.

Hauser, R. & Calafat, A. M. (2005). Phthalates and human health. *Occupational and Environmental Medicine,* **62**(11), 806–18.

Hauser, R., Meeker, J. D., Singh, N. P. *et al.* (2007). DNA damage in human sperm is related to urinary levels of phthalate monoester and oxidative metabolites. *Human Reproduction,* **22**(3), 688–95.

Hawley, A. H. (1944). Ecology and human ecology. *Social Forces,* **22**(4), 398–405.

Haxeltine, A., Whitmarsh, L., Bergman, N. *et al.* (2008). A conceptual framework for transition modelling. *International Journal of Innovation and Sustainable Development,* **3**(1–2), 93–114.

Hayek, F. A. (1942). The Ricardo effect. *Economica,* **9**(34), 127–52.

Hayes, E. C. (1908). Sociology and psychology; sociology and geography. *American Journal of Sociology,* **14**(3), 371–407.

Hayes, E. C. (1914). Effects of geographic conditions upon social realities. *American Journal of Sociology,* **19**(6), 813–24.

Hays, J. M. (2005). *Epidemics and Pandemics. Their Impacts on Human History.* Santa Barbara, CA: ABC Clio.

Hays, W. L. (1963). *Statistics.* London: Holt, Rinehart & Winston.

Heaton, P. & Wallace, G. L. (2004). Annotation: the savant syndrome. *Journal of Child Psychology and Psychiatry,* **45**(5), 899–911.

Hebb, D. O. (1955). Drives and the CNS. *Psychological Review,* **62**(4), 243–54.

References

Hediger, W. (2000). Sustainable development and social welfare. *Ecological Economics*, **32**(3), 481–92.

Heft, H. (2001). *Ecological Psychology in Context*. Mahwah, NJ: Lawrence Erlbaum Associates.

Hegselmann, R. (1998). Modeling social dynamics by cellular automata. In W. B. G. Liebrand, A. Nowak & R. Hegselmann (eds). *Computer Modeling of Social Processes*. London: SAGE Publications, pp. 37–64.

Heider, F. (1930a). Die Leistung des Wahrnehmungssystems. *Zeitschrift für Psychologie*, **114**, 371–94.

Heider, F. (1930b). Ding und Medium. *Symposion – Philosophische Zeitschrift für Psychologie*, **114**, 371–94.

Heider, F. (1958). *The Psychology of Interpersonal Relationships*. New York, NY: Wiley.

Heider-Rosch, E. R. (1971). Focal color areas and development of color names. *Developmental Psychology*, **4**(3), 447–55.

Heijungs, R. & Hofstetter, P. (1996). Definitions of terms and symbols. In H. A. Udo de Haes (ed.). *Towards a Methodology for Life Cycle Impact Assessment*. Brussels: Society of Environmental Toxicology and Chemistry – Europe, pp. 31–7.

Heijungs, R. & Wegener Sleeswijk, A. (1999). The structure of impact assessment: mutually independent dimensions as a function of modifiers. *The International Journal of Life Cycle Assessment*, **4**(1), 2–3.

Heim, D. & Murray, N. (2004). Possibilities to manage the BSE epidemic: cohort culling versus herd culling – experiences in Switzerland. In H. F. Rabenau, J. Cinatl & H. W. Doerr (eds). *Prions, a Challenge for Science, Medicine and the Public Health System*, 2nd edn, Vol. 11. Basel: S. Karger AG, pp. 186–92.

Helbing, D. (2007). *Managing Complexity: Insights, Concepts, Applications*. Berlin: Springer.

Helland, A., Wick, P., Koehler, A., Schmid, K. & Som, C. (2007). Reviewing the environmental and human health knowledge base of carbon nanotubes. *Environmental Health Perspectives*, **115**(8), 1125–31.

Hellier, A., Newton, A. C. & Gaona, S. O. (1999). Use of indigenous knowledge for rapidly assessing trends in biodiversity: a case study from Chiapas, Mexico. *Biodiversity and Conservation*, **8**(7), 869–89.

Hellmuth, L. (2003). Fear and trembling in the amygdala. *Science*, **300**(5619), 568–9.

Hellpach, W. (1911). *Die geopsychischen Erscheinungen*. Leipzig: Johann Ambrosius Barth-Verlag.

Hellpach, W. (1924). Psychologie der Umwelt. In E. Abderhalden (ed.). *Handbuch der biologischen Arbeitsmethoden, Abteilung VI. Methoden der experimentellen Psychologie*, Teil C, Heft 3. Berlin: Urban, pp. 109–218.

Hellpach, W. (1939/1952). *Mensch und Volk in der Grossstadt*. Stuttgart: Enke.

Hellpach, W. (1950). *Geopsyche*, 6th edn. Stuttgart: Enke.

Helson, H. (1964). *Adaptation-Level Theory*. New York, NY: Harper & Row.

Hendrickson, C. T. (2006). *Environmental Life Cycle Assessment of Goods and Services: an Input-Output Approach*. Washington, DC: Resources for the Future.

Hendrickson, C. T., Horvath, A., Joshi, S. & Lave, L. (1998). Economic input-output models for environmental life-cycle assessment. *Environmental Science & Technology*, **32**(7), 184A-191A.

Hendrickson, C. T., Horvath, A., Lave, L. B. & McMichael, F. C. (2002). Industrial ecology and green design. In R. U. Ayres & L. W. Ayres (eds). *Handbook of Industrial Ecology*. Cheltenham: Edward Elgar, pp. 457–66.

Hengeveld, R. & Walter, G. H. (1999). The two coexisting ecological paradigms. *Acta Biotheoretica,* **47**(2), 141–70.

Hennig, W. (1966). *Phylogenetic Systematics*. Urbana, IL: University of Illinois Press.

Henshilwood, C. S. & Marean, C. W. (2003). The origin of modern human behavior – critique of the models and their test implications. *Current Anthropology,* **44**(5), 627–51.

Héretier, A. (2007). *Explaining Institutional Change in Europe*. Oxford: Oxford University Press.

Herring, H. & Roy, R. (2007). Technological innovation, energy efficient design and the rebound effect. *Technovation,* **27**(4), 194–203.

Hertwich, E. G. (2005a). Consumption and the rebound effect: an industrial ecology perspective. *Journal of Industrial Ecology,* **9**(1–2), 85–98.

Hertwich, E. G. (2005b). Life cycle approaches to sustainable consumption: a critical review. *Environmental Science & Technology,* **39**(13), 4673–84.

Herweijer, C., Seager, R., Cook, E. R. & Emile-Geay, J. (2007). North American droughts of the last millennium from a gridded network of tree-ring data. *Journal of Climate,* **20**(7), 1353–76.

Herzberg, K. (1934). Viktoriablau zur Färbung filtrierbarer Vira. *Klinische Wochenschrift,* **10**, 381.

Hewstone, M., Islam, M. R. & Judd, C. M. (1993). Models of crossed categorization and intergroup relations. *Journal of Personality and Social Psychology,* **64**(5), 779–93.

Heylighen, F. & Joslyn, C. (2001). Cybernetics and second order cybernetics. In R. A. Meyers (ed.). *Encyclopedia of Physical Science & Technology*, 3rd edn, Vol. 4. New York, NY: Academic Press, pp. 155–70.

Heywood, V. H. & Watson, A. J. (eds). (1995). *Global Biodiversity Assessment*. Cambridge: Cambridge University Press.

Hicks, J. R. (1932). *The Theory of Wages*. London: Macmillan.

Hicks, J. R. (1939). *Value and Capital*. Oxford: Oxford University Press.

Hiebsch, H. (1968). *Sozialpsychologische Grundlagen der Persönlichkeitsformung*. Berlin: VEB Deutscher Verlag der Wissenschaften.

Hille, J. (1995). *Sustainable Norway: Probing the Limits and Equity of Environmental Space*. Oslo: Project Alternativ Framtid/ForUM.

Hillel, D. (1991). *Out of the Earth*. Berkeley, CA and Los Angeles: University of California: Civilization and the Life of the Soil.

Hilty, L. M., Arnfalk, P., Erdmann, L. et al. (2006). The relevance of information and communication technologies for environmental sustainability – a prospective simulation study. *Environmental Modelling & Software,* **21**(11), 1618–29.

Hirsch-Hadorn, G., Biber-Klemm, S., Grossenbacher-Mansuy, W. et al. (2008). The emergence of transdisciplinarity as a form of research. In G. Hirsch-Hadorn, H. Hoffmann-Riem, S. Biber-Klemm (eds). *Handbook of Transdisciplinary Research. A Proposition by the Swiss Academies of Arts and Sciences*. Heidelberg: Springer, pp. 19–39.

Hodges, J. (2005). Editorial: developments and mysteries of BSE (Mad Cow Disease) and vCJD. *Livestock Production Science,* **97**, 27–33.

Hodgkinson, H. L. (1957). Action research – a critique. *The Journal of Educational Sociology,* **31**(4), 137–53.

Hoekstra, R. & van den Bergh, G. D. (2004). Structural decomposition analysis of iron and steel and of plastics. In G. D. van den Bergh & M. A. Janssen (eds). *Economics of Industrial Ecology*. Cambridge, MA: The MIT Press.

Hoffmann, V., Thomas, A. & Gerber, A. (eds). (2009). *Transdisziplinäre Umweltforschung. Methodenhandbuch*. Munich: Oekom.

Hoffmann, V. H. (2007). EU ETS and investment decisions: the case of the German electricity industry. *European Management Journal,* **25**(6), 464–74.

Hofstätter, P. R. (1957). *Gruppendynamik*. Hamburg: Rowohlt.

Hofstetter, P. & Müller-Wenk, R. (2005). Monetization of health damages from road noise with implications for monetizing health impacts in life cycle assessment. *Journal of Cleaner Production,* **13**(13–14), 1235–45.

Hogarth, R. M. (2001). *Educating Intuition*. Chicago, IL: University of Chicago Press.

Holland, J. (1995a). *Hidden Order: how Adaptation Builds Complexity*. New York, NY: Basic Books.

Holland, J. (2006). Studying complex adaptive systems. *Journal of Systems Science and Complexity,* **19**(1), 1–8.

Holland, J. H. (1995b). Can there be a unified theory of complex adaptive systems? In H. Morowitz & J. Singer (eds). *The Mind, the Brain, and CAS*. Reading, MA: Addison-Wesley, pp. 45–50.

Holland, W. R. & Tharp, R. G. (1964). Highland Maya psychotherapy. *American Anthropologist,* **66**(1), 41–52.

Holliday, R. (1994). Epigenetics: an overview. *Developmental Genetics,* **15**(6), 453–7.

Holling, C. S. (1973). Resilience and stability of ecological systems. *Annual Review of Ecology and Systematics,* **4**, 1–23.

Holling, C. S. (2001). Understanding the complexity of economic, ecological, and social systems. *Ecosystems,* **4**(5), 390–405.

Holling, C. S. (2003). Foreword: the backloop to sustainability. In F. Berges, J. Colding & C. Folke (eds). *Navigating Social-Ecological Systems. Building Resilience for Complexity and Change*. Cambridge: Cambridge University Press, pp. xv–xxi.

Holling, C. S. (2004). From complex regions to complex worlds. *Ecology and Society,* **9**(1), 11.

Holling, C. S., Carpener, S. R., Brock, W. A. & Gunderson, L. H. (2002). Discoveries for sustainable futures. In L. H. Gunderson & C. S. Holling (eds). *Panarchy: Understanding Transformation in Human and Natural Systems*. Washington, DC: Island Press, pp. 395–418.

Holling, C. S. & Gunderson, L. H. (2002). Resilience and adaptive cycles. In L. H. Gunderson & C. S. Holling (eds). *Panarchy: Understanding Transformation in Human and Natural Systems*. Washington, DC: Island Press, pp. 25–63.

Holling, C. S., Gunderson, L. H. & Peterson, G. D. (2002). Sustainability and panarchies. In L. H. Gunderson & C. S. Holling (eds). *Panarchy: Understanding Transformation in Human And Natural Systems*. Washington, DC: Island Press, pp. 63–102.

Holmberg, J., Robèrt, K. H. & Eriksson, K. E. (1994). *Socio-Ecological Principles for a Sustainable Society. Scientific Background and Swedish Experience*. Gothenborg: Institute of Physical Resource Theory, Chalmers University of Technology.

Holmes, J. A. (2008). Ecology – how the Sahara became dry. *Science,* **320**(5877), 752–3.

Homans, G. C. (1955). Social behavior as exchange. *The American Journal of Sociology,* **63**(6), 597–606.

Homer, S. & Sylla, R. (1996). *A History of Interest Rates*, 3rd edn. New Brunswick, NY: Rutgers University Press.

Hooke, R. (1665). *Micrographia; or, some Physiological Descriptions of Minute Bodies made by Magnifying Glasses*. London: J. Martyn and J. Allestry.

Hooke, R. L. (2000). On the history of humans as geomorphic agents. *Geology,* **28**(9), 843–6.

Hoppe, B. (2004). Das Aufkommen der Vererbungsforschung unter dem Einfluss neuer methodischer und theoretischer Ansätze im 19. Jahrhundert. In I. Jahn (ed.). *Geschichte der Biologie: Theorien, Methoden, Institutionen, Kurzbiografien*, 3rd edn. Hamburg: Nikol, pp. 386–418.

Horowitz, L. M., Wilson, K. R., Turan, B. *et al.* (2006). How interpersonal motives clarify the meaning of interpersonal behavior: a revised circumplex model. *Personality and Social Psychology Review,* **10**(1), 67–86.

Hostetler, J. A. (1993). *Amish Society*, 4th edn. Baltimore, MD: The Johns Hopkins University Press.

Hotelling, H. (1931). The economics of exhaustible resources. *Journal of Political Economy,* **39**(2), 137–75.

Houghton, J. T. (2004). *Global Warming. The Complete Briefing*. Cambridge: Cambridge University Press.

Houghton, J. T., Filho, L. G. M., Callender, B. A. *et al.* (1995). *Climate Change 1995: the Science of Climate Change. Contribution of Working Group I to the Second Assessment Report of the Intergovernmental Panel on Climate Change*. Cambridge: Cambridge University Press.

Hovi, J., Schram Stokke, O. & Ulfstein, G. (2005). Introduction and main findings. In O. Schram Stokke, J. Hovi & G. Ulfstein (eds). *Implementing the Climate Regime*. London: Earthscan, pp. 1–17.

Howarth, R. B. & Farber, S. (2002). Accounting for the value of ecosystem services. *Ecological Economics,* **41**(3), 421–9.

Howes, D., Synnott, A. & Classen, C. (2007). Anthropology of odor. Retrieved January 07, 2007, from http://alcor.concordia.ca/~senses/Consert-Odor.htm

Hubbell, S. P. (2001). *The Unified Theory of Biodiversity and Biography*. Princeton, NY: Princeton University Press.

Huber, J. (2004). *New Technologies and Environmental Innovation*. Cheltenham: Edward Elgar.

Huchting, F. (1981). Abfallgeschichte im dritten Reich. *Technikgeschichte,* **48**(3), 252–73.

Hughes, J. (1975). *Ecology in Ancient Civilizations*. Albuquerque, NM: University of New Mexico Press.

Hügli, A. & Han, B. C. (2001). Utilitarismus. In J. Ritter, K. Gründer & G. Gabriel (eds). *Historisches Wörterbuch der Philosophie*, Vol. 11. Basel: Schwabe, pp. 503–10.

Hull, C. L. (1943). *Principles of Behavior: an Introduction to Behavior Theory*. New York, NY: Appleton-Century-Crofts.

Hulme, M. (2009). *Why we Disagree about Climate Change – Understanding, Controversy, Inaction and Opportunity*. Cambridge: Cambridge University Press.

Hume, D. (1752). *Political Discourses*. Edinburgh: A. Kincaid & A. Donaldson.

Humphrey, C. R. & Buttel, F. H. (1982). *Environment, Energy, and Society*. Belmont, CA: Wadsworth.

Humphrey, C. R., Lewis, T. L. & Buttel, F. H. (2002). *Environment, Energy, and Society: a New Synthesis*, 2nd edn. Belmont, CA: Wadsworth.

Humphrey, N. (1984). *Consciousness Regained*. Oxford: Oxford University Press.

Hunn, E. (2007). Ethnobiology in four phases. *Journal of Ethnobiology,* **27**(1), 1–10.

Hunt, M. (1993). *The Story of Psychology*. London: DoubleDay.

Hussy, W. (1984). *Denkpsychologie*, Vol. 1. Stuttgart: Urban.

Hutchinson, G. E. (1944). Circular causal systems in ecology. *Annals of the New York Academy of Sciences,* **50**, 221–46.

Hutson, S. R. & Stanton, T. W. (2007). Cultural logic and practical reason: the structure of discard in ancient Maya houselots. *Cambridge Archaeological Journal,* **17**(2), 123–44.

Hutton, J. (1788). *Theory of the Earth; or an Investigation of the Laws Observable in the Composition, Dissolution, and Restoration of Land upon the Globe*, Vol. 1. Edinburgh: Transactions of the Royal Society of Edinburgh.

IEA. (2006). *World Energy Outlook*. Paris: IEA.

IEA. (2007). *World Energy Outlook*. Paris: IEA.

Imhoff, M. L., Bounou, L., Ricketts, T. *et al.* (2004). Global patterns in human consumption of net primary production. *Nature,* **429**(6994), 870–3.

Immelmann, K., Scherer, K. R., Vogel, C. & Schmoock, P. (1988). *Psychobiologie. Grundlagen des Verhaltens*. Stuttgart: Gustav Fischer.

Ingold, T. (1986). *The Appropriation of Nature*. Manchester: Manchester University Press.

Inhelder, B. & Piaget, J. (1958). *The Growth of Logical Thinking: from Childhood to Adolescence*. London: Routledge & Kegan Paul.

Innes, J. E. & Connick, S. (1999). San Francisco estuary project. In L. E. Susskind, S. McKearnen & J. Thomas-Larmer (eds). *The Consensus Building Handbook: a Comprehensive Guide to Reaching Agreement*. Thousand Oaks, CA: Sage, pp. 801–28.

Innes, J. L. (1993). Methods to estimate forest health. *Silva Fennica,* **27**(2), 145–57.

Intergovernmental Panel of Climate Change (IPCC). (2000). *Emission Scenarios. Summary for Policy Makers. A Special Report of IPCC Working Group III*. Geneva: IPCC.

Intergovernmental Panel of Climate Change (IPCC). (2007). *Climate Change 2007: Synthesis Report*. Cambridge: Cambridge University Press.

Intergovernmental Panel of Climate Change (IPCC). (2008). *Risk Governance Guidelines for Bioenergy Policies*. Geneva: International Risk Governance Council.

International Atomic Energy Agency (IAEA). (1994). *Classification of Radioactive Waste: a Safety Guide*. Vienna: IAEA.

International Atomic Energy Agency (IAEA). (2006). *Chernobyl's Legacy: Health, Environmental and Socio-Economic Impacts – The Chernobyl Forum: 2003–2005*. Vienna: IAEA.

International Organization for Standardization/European Standard (ISO/EN). (1997). *Environmental Management: Life Cycle Assessment – Principles and Framework*. Brussels: European Committee for Standardization.

International Risk Governance Council (IRGC). (2008). *Risk Governance Guidelines For Bioenergy Policies. Executive Summary*. Geneva: IRGC.

International Risk Governance Council (IRGC). (2009). *Risk Governance Deficits: an Analysis and Illustration of the most Common Deficits in Risk Governance*. Geneva: IRGC.

International Society for Industrial Ecology (ISIE). (2009). International society for industrial ecology. Retrieved October 15, 2009, from http://www.is4ie.org/

Ioannidis, J. P. A. & Trikalinos, T. A. (2007). An exploratory test for an excess of significant findings. *Clinical Trials*, **4**(3), 245–53.

Irie, N. & Sehara-Fujisawa, A. (2007). The vertebrate phylotypic stage and an early bilaterian-related stage in mouse embryogenesis defined by genomic information. *BMC Biology*, **5**(1), 1–8.

Irle, M. (1975). *Lehrbuch der Sozialpsychologie*. Gottingen: Hogrefe.

Ironson, G., Wynings, C., Schneiderman, N. *et al.* (1997). Posttraumatic stress symptoms, intrusive thoughts, loss, and immune function after Hurricane Andrew. *Psychosomatic Medicine*, **59**(2), 128–41.

Irwin, D. A. (1991). Mercantilism as strategic trade-policy – the Anglo-Dutch rivalry for the East India trade. *Journal of Political Economy*, **99**(6), 1296–1314.

Isenmann, R. (2007). Industrial ecology: Mit Ökologie zukunftsorientiert wirtschaften. In R. Isenmann & M. von Hauff (eds). *Industrial ecology: Mit Ökologie zukunftsorientiert wirtschaften*. Heidelberg: Elsevier, pp. 61–4.

Israel, B. A., Schulz, A. J., Parker, E. A. & Becker, A. B. (1998). Review of community-based research: assessing partnership approaches to improve public health. *Annual Review of Public Health*, **19**, 173–202.

Ittelson, W. H. (1952). *The Ames Demonstrations in Perception. A Guide to their Construction and Use*. Princeton, NJ: Princeton University Press.

Ittelson, W. H., Proshansky, H. M., Rivlin, L. & Winkel, G. (1974). *An Introduction to Environmental Psychology*. New York, NY: Holt, Rinehart & Winston.

Izard, C. E. (1971). *The Face of Emotions*. New York, NY: Appleton-Century-Crofts.

Izard, C. E. (1977). *Human Emotions*. New York, NY: Plenum Press.

Jablonka, E. & Lamb, M. J. (2002). The changing concept of epigenetics. *Annals of the New York Academy of Sciences*, **981**, 82–96.

Jacko, R. B. & LaBreche, T. M. C. (2001). Environmental engineering: regulation and compliance. In G. Salvendy (ed.). *Handbook of Industrial Engineering: Technology and Operations Management*. New York, NY: Wiley, pp. 589–600.

Jacob, F. & Monod, J. (1961). On regulation of gene activity. *Cold Spring Harbor Symposia on Quantitative Biology*, **26**, 193–211.

Jacobs, J. (1961). *The Death and Life of Great American Cities*. New York, NY: Vintage.

Jacobs, J. (1969). *The Economy of Cities*. New York, NY: Random House.

Jacobsen, N. B. (2006). Industrial symbiosis in Kalundborg, Denmark – a quantitative assessment of economic and environmental aspects. *Journal of Industrial Ecology*, **10**(1–2), 239–55.

Jaeger, J. & Scheringer, M. (1998). Transdisciplinarität: Problemorientierung ohne Methodenzwang. *GAIA*, 7(1), 10–25.

Jaffe, A. B. (1980). Benefit-cost analysis and multi-objective evaluation of federal water projects. *Harvard Environmental Law Review*, **4**(1), 58–85.

Jager, W., Janssen, M. A., De Vries, H. J. M., De Greef, J. & Vlek, C. A. J. (2000). Behaviour in commons dilemmas: Homo economicus and Homo psychologicus in an ecological–economic model. *Ecological Economics*, **35**, 357–79.

Jahn, I. (1990). *Grundzüge der Biologiegeschichte*. Jena: Gustav Fischer.

Jahn, I. (2004a). Konsolidierung und Neubildung von Disziplinen und Theorien im 19. Jahrhundert. In I. Jahn (ed.). *Geschichte der Biologie: Theorien, Methoden, Institutionen, Kurzbiographien*, 3rd edn. Hamburg: Nikol,Verlag, pp. 196–230.

Jahn, I. (2004b). Naturphilosophie und Empirie in der Frühaufklärung. In I. Jahn (ed.). *Geschichte der Biologie: Theorien, Methoden, Institutionen, Kurzbiografien*, 3rd edn. Hamburg: Nikol Verlag, pp. 274–301.

References

Jahnke, H. N. & Otte, M. (1997). Science as a language. *Epistemological and Social Problems of the Sciences in the Early Nineteenth Century*. Dordrecht: Reidel, pp. 75–90.

James, J. (1994). Van-Leeuwenhoeks discoveries of 1677–1678 – a look too far. *Micron*, **25**(1), 1–4.

James, V. (2001). *Spirit and Art: Pictures of the Transformation of Consciousness*. Herndon, VA: Steiner Books.

James, V. (2004). Spirit and art: paleo-shamanic iconography. *Quest Magazine*, **6**(1). Retrieved January 6, 2011 from http://www.theosophical.org/publications/quest-magazine/1238

James, W. (1890). *Principles of Psychology*. New York, NY: Henry Holt.

James, W. (1912/1976). *Essays in Radical Empiricism*. Cambridge, MA: Harvard University Press.

James, W. (1920/1895). The knowing of things together. In W. James (ed.). *Collected Essays and Reviews*. New York, NY: Longmans, Green and Co, pp. 371–400.

Jameson, S. C. (2002). Maintaining the norm: T-CELL homeostasis. *Nature Reviews Immunology*, **2**(8), 547–56.

Janerich, D. T., Burnett, W. S., Feck, G. et al. (1981). Cancer incidence in the Love Canal area. *Science*, **212**(4501), 1404–7.

Janeway, C. A. (2001). Evolution of the immune system: past present, and future. In C. A. Janeway, P. Travers, M. Walport & M. J. Shlomchik (eds). *Immmunobiology*, 5th edn. New York, NY: Garland, pp. 597–611.

Janich, P. (1999). Kritik der Informationsbegriffs in der Genetik. *Theory in Bioscience*, **118**(1), 66–84.

Janick, J. (2002). Lecture 3: neolithic revolution and the discovery of agriculture. *History of Horticulture*. Retrieved October 30, 2009, from http://www.hort.purdue.edu/newcrop/history/lecture03/lec03.html

Jänicke, M. (2008). Ecological modernisation: new perspectives. *Journal of Cleaner Production*, **16**(5), 557–65.

Jänicke, M. (2009). On ecological and political modernisation. In A. P. J. Mol, D. A. Sonnenfeld & G. Spaargaren (eds). *The Ecological Modernization Reader: Environmental Reform in Theory and Practice*. London: Routledge, pp. 28–41.

Janis, I. L. (1982). *Groupthink: Psychological Studies of Policy Decisions and Fiascos*, 2nd edn. Boston, MA: Houghton Mifflin.

Jantsch, E. (1972). *Towards Interdisciplinarity and Transdisciplinarity in Education and Innovation*. Paris: Organisation for Economic Cooperation and Development.

Jasanoff, S. & Wynne, B. (1998). Science and decision making. In S. Rayner & E. L. Malone (eds). *Human Choice and Climate Change*, Vol. 1. Ohio: Batelle Press, pp. 1–87.

Jasinski, S. M. (2009). Phosphate rock. In US Geological Survey (ed.). *Mineral Commodity Summaries*. St. Louis, MO: USGS, pp. 120–1.

Jepperson, R. L. (2002). The development and application of sociological neoinstitutionalism. In J. Berger & M. J. Zelditsch (eds). *New Direction in Contemporary Sociological Theory*. Oxford: Rowman & Littlefield, pp. 229–66.

Jerne, N. K. (1955). The natural-selection theory of antibody formation. *Proceedings of the National Academy of Sciences of the United States of America*, **41**(11), 849–56.

Jerne, N. K. (1984). Towards a network theory of immune system. *Immunological Review*, **79**, 5–24.

Jerne, N. K. (1985). The generative grammar of the immune-system. *Science*, **229**(4718), 1057–9.

Jevons, W. S. (1871/1970). *The Theory of Political Economy*. Harmondsworth: Penguin.

Ji, L.-J., Guo, T., Zhang, Z. & Messervey, D. (2009). Looking into the past: cultural differences in perception and representation of past information. *Journal of Personality and Social Psychology*, **96**(4), 761–9.

Jobling, M. A., Hurles, M. E. & Tyler-Smith, C. (2004). *Human Evolutionary Genetics*. New York, NY: Garland.

Johannsen, W. (1911). The genotype conception of heredity. *Amercian Naturalist*, **45**(531), 129–59.

Johansson, A. (2002). Industrial ecology and industrial metabolism: use and misuse of metaphors. In R. U. Ayres & L. W. Ayres (eds). *A Handbook of Industrial Ecology*. Cheltenham: Edward Elgar, pp. 70–9.

Johnson, J., Harper, E. M., Lifset, R. & Graedel, T. E. (2007). Dining at the periodic table: metals concentrations as they relate to recycling. *Environmental Science & Technology*, **41**(5), 1759–65.

Johnson, N. & Mueller, J. (2002). Updating the accounts: global mortality of the 1918–1920 "Spanish" influenza pandemic. *Bulletin of the History of Medicine*, **76**(1), 105–15.

Jørgensen, S. E., Fath, B. D., Bastianoni, S. et al. (2007). *A New Ecology: Systems Perspective*. Amsterdam: Elsevier.

Jouzel, J., Masson-Delmotte, V., Cattani, O. et al. (2007). Orbital and millennial Antarctic climate variability over the past 800,000 years. *Science*, **317**(5839), 793–6.

Jungermann, H. (1986). The two camps of rationality. In H. Arkes & K. R. Hammond (eds). *Judgment and Decision Making: an Interdisciplinary Reader*. New York, NY: Cambridge University Press, pp. 627–41.

Jungermann, H., Pfister, H.-R. & Fischer, K. (2005). *Die Psychologie der Entscheidung: Eine Einführung*, 2nd edn. Heidelberg: Spektrum.

Jungermann, H. & Slovic, P. (1993). Charakteristika individueller Risikowahrnehmung. In Bayerische Rück (ed.). *Risiko ist ein Konstrukt*. Munich: Knesbeck, pp. 89–107.

Jungk, R. & Müllert, N. R. (1994). *Zukunftswerkstätten: mit Phantasie gegen Routine und Resignation*, Vol. 73. Munich: Heyne.

Junker, B., Flüeler, T., Stauffacher, M. & Scholz, R. W. (2008). *Description of the Safety Case for Long-Term Disposal of Radioactive Waste – the Iterative Safety Analysis Approach as Utilized in Switzerland* (Technical paper as part of the project: "Long-term dimension of radioactive waste disposal: the role of the time dimension for risk perception.") Zurich: ETH Zurich.

Justi, J. H. G., von. (1758). *Staatswirtschaft oder systematische Abhandlung aller oeconomischen und Cameral-Wissenschaften, die zur Regierung eines Landes erfordert werden*, 2nd edn. Vol. 1. Leipzig: Verlag Bernhard Christoph Breitkopf.

Kahn, H. & Wiener, A. J. (1967). *The Year 2000*. London: Macmillan.

Kahneman, D. (2000). New challenges to the rationality assumption. In D. Kahneman & A. Tversky (eds). *Choices, Values, and Frames*. Cambridge: Cambridge University Press, pp. 758–74.

Kahneman, D. (2003). Maps of bounded rationality: psychology for behavioral economics. *American Economic Review*, **93**(5), 1449–75.

Kahneman, D., Knetsch, J. L. & Thaler, R. H. (1990). Experimental tests of the endowment effect and the Coase theorem. *Journal of Political Economy*, **98**(6), 1325–48.

Kahneman, D., Slovic, P. & Tversky, A. (eds). (1982). *Judgement under Uncertainty: Heuristics and Biases*. Cambridge: Cambridge University Press.

Kahneman, D. & Thaler, R. H. (2006). Anomalies – utility maximization and experienced utility. *Journal of Economic Perspectives*, **20**(1), 221–34.

Kahneman, D. & Tversky, A. (1979). Prospect theory: an analysis of decision under risk. *Econometrica*, **47**(2), 263–91.

Kahneman, D. & Tversky, A. (1992). Advances in prospect theory: cumulative representation of uncertainty. *Journal of Risk and Uncertainty*, **5**(4), 297–323.

Kahneman, D. & Tversky, A. (2000). *Choices, Values, and Frames*. Cambridge: Cambridge University Press.

Kaiser, F. G., Oerke, B. & Bogner, F. X. (2007). Behavior-based environmental attitude: development of an instrument for adolescents. *Journal of Environmental Psychology*, **27**(3), 242–51.

Kaldor, N. (1939). Welfare propositions in economics. *Economic Journal*, **49**(195), 549–52.

Kals, E. (1996). *Verantwortliches Umweltverhalten. Umweltschützende Entscheidungen erklären und fördern*. Weinheim: Belz.

Kämpke, T., Pestel, R. & Radermacher, F. J. (2003). A computational concept for normative equity. *European Journal of Law and Economics*, **15**(2), 129–63.

Kant, I. (1787/1965). *Critique of Pure Reason* (W. S. Pluhar, Trans.). Indianapolis, IN: Hackett Publishing Company.

Kant, I. (1796). Verkündung des nahen Abschlusses eines Tractates zum ewigen Frieden in der Philosophie. *Berlinische Monatschrift*, **2**, 485–504.

Kant, I. (1800). *Logik: ein Handbuch zu Vorlesungen*. Königsberg: Friedrich Nicolovius.

Kaplan, S. (1972). Challenge of environmental psychology – a proposal for a new functionalism. *American Psychologist*, **27**(2), 140–3.

Kaplan, S. (1979, April 23–25). *Perception and Landscape: Conceptions and Misconceptions*. Paper presented at the National Conference on Applied Techniques for Analysis and Management of the Visual Resource, Incline Village, NV.

Kaplan, S. & Kaplan, R. (2009). Creating a larger role for environmental psychology: the Reasonable Person Model as an integrative framework. *Journal of Environmental Psychology*, **29**(3), 329–39.

Kaptchuk, T. J. (2002). The placebo effect in alternative medicine: can the performance of a healing ritual have clinical significance? *Annals of Internal Medicine*, **136**(11), 817–25.

Kapur, J. C. (2004). Available energy resources and environmental imperatives. *World Affairs*, **V10**, N1.

Kasperson, J. X., Kasperson, R. E., Pidgeon, N. & Slovic, P. (2003). The social amplification of risk: assessing fifteen years of research and theory. In N. Pidgeon, R. E. Kasperson & P. Slovic (eds). *The Social Amplification of Risk*. Cambridge: Cambridge University Press, pp. 13–46.

Kasperson, R. E., Renn, O., Slovic, P. *et al.* (1988). The social amplification of risk: a conceptual framework. *Risk Analysis*, **8**(2), 177–87.

Kastens, G. B. & Newig, J. (2005). The current implementation of the EC water framework directive – consequences and prospects for agriculture taking Lower Saxony as an example. *Berichte über Landwirtschaft*, **83**(3), 463–82.

Kates, R. W. (1997). Population, technology, and the human environment: a tread through time. In J. H. Ausubel & H. D. Langford (eds). *Technological Trajectories and the Human Environment*. Washington, DC: National Academy Press, pp. 33–55.

References

Kates, R. W., Clark, W. C., Corell, R. et al. (2001). Environment and development – sustainability science. *Science,* **292**(5517), 641–2.

Kaufmann, S. H. E. & Schaible, U. E. (2005). 100th anniversary of Robert Koch's Nobel Prize for the discovery of the tubercle bacillus. *Trends in Microbiology,* **13**(10), 469–75.

Kauwenbergh van, S. J. (2010). *World Phosphate Rock Reserves and Resources.* Muscle Shoals, AL: IFDC International Fertilizer Development Center.

Kawade, Y. (1996). Molecular biosemiotics: molecules carry out semiosis in living systems. *Semiotica,* **111**(3/4), 195–215.

Kaya, N. & Weber, M. J. (2003). Cross-cultural differences in the perception of crowding and privacy regulation: American and Turkish students. *Journal of Environmental Psychology,* **23**(3), 301–9.

Keech, M. & Beardsworth, P. (2008). The impact of influenza on working days lost. A review of the literature. *Pharmacoeconomics,* **26**(11), 911–24.

Kehm, B. M. (2004). Hochschulen in Deutschland: Entwicklung, Probleme und Perspektiven. *Aus Politik und Zeitgeschichte,* **25**, 6–17.

Keizer, K., Lindenberg, S. & Steg, L. (2008). The spreading of disorder. *Science,* **322**(5908), 1681–5.

Kelemen, R. D. (2004). *The Rules of Federalism. Institutions and Regulatory Politics in the EU and Beyond.* Cambridge, MA: Harvard University Press.

Keltner, D. & Ekman, P. (2000). Emotion: an overview. In A. E. Kazdin (ed.). *Encyclopedia of Psychology,* Vol. 3. Oxford: Oxford University Press, pp. 162–7.

Kemmis, S. & McTaggert, R. (1988). *The Action Research Planner.* Geelong: Deakin University Press.

Kemp, R., Loorbach, D. & Rotmans, J. (2007). Transition management as a model for managing processes of co-evolution towards sustainable development. *International Journal of Sustainable Development and World Ecology,* **14**(1), 78–91.

Kempener, R., Beck, J. & Petrie, J. (2009). Design and analysis of bioenergy networks. *Journal of Industrial Ecology,* **13**(2), 284–305.

Kennedy, B. P., Kawachi, I. & Prothrow-Stith, D. (1996). Income distribution and mortality: cross sectional ecological study of the Robin Hood index in the United States. *British Medical Journal,* **312**(7037), 1004–7.

Kerr, N. L. (1986). Motivational choices in task groups: a paradigm for social dilemma research. In H. A. M. Wilke, D. M. Messick & C. G. Rutte (eds). *Experimental Social Dilemmas.* Frankfurt am Main: Lang, pp. 4–28.

Kewell, B. & Beck, M. (2008). The shifting sands of uncertainty: risk construction and BSE/vCJD. *Health, Risk & Society,* **10**(2), 133–48.

Keynes, J. M. (1923). *A Tract on Monetary Reform.* London: Macmillan.

Keynes, J. M. (1936). *The General Theory of Employment, Interest and Money.* London: Macmillan.

Keynes, J. M. (undated). Population, manuscript, King's College Cambridge, quoted according to Trascio, 1971.

KfC. (2008). *Excecutive Summary of the Dutch National Research Program.* Utrecht: Knowledge for Climate.

Khansari, D. N., Murgo, A. J. & Faith, R. E. (1990). Effects of stress on the immune-system. *Immunology Today,* **11**(5), 170–5.

Kim, K. C. & Byrne, L. B. (2006). Biodiversity loss and the taxonomic bottleneck: emerging biodiversity science. *Ecological Research,* **21**(6), 794–810.

King, F. H. (1911/2004). *Farmers of Forty Centuries: Organic Farming in China, Korea and Japan.* Mineola, NY: Dover Publications.

Kinzig, A. P., Ryan, P., Etienne, M. et al. (2006). Resilience and regime shifts: assessing cascading effects. *Ecology and Society,* **11**(1), 20 [online] URL: http://www.ecologyandsociety.org/vol11/iss21/art20.

Kirby, K. N. & Marakovic, N. N. (1996). Delay-discounting probabilistic rewards: rates decrease as amounts increase. *Psychonomic Bulletin & Review,* **3**(1), 100–4.

Kirchhoff, T. (2007). *Systemauffassungen und biologische Theorien: zur Herkunft von Individualitätskonzeptionen und ihrer Bedeutung für die Theorie ökologischer Einheiten.* Freising: Lehrstuhl für Landschaftsökologie, Technische Universität München.

Kirchhoff, T., Brand, F., Hoheisel, D. & Grimm, V. (2010). The one-sidedness and cultural bias of the resilience approach. *Gaia,* **19**(1), 25–32.

Kirchner, J. W. (2003). The Gaia hypothesis: conjectures and refutations. *Climatic Change,* **58**(1–2), 21–45.

Kirk-Smith, M., Booth, D. A., Carroll, D. & Davies, P. (1980). Effects of androstenone on choice of location in other's presence. In H. van der Starre (ed.). *Olfaction and Taste,* Vol. VII. London: IRL Press, pp. 397–400.

Kiss Maerth, O. (1971). *Der Anfang war das Ende. Der Mensch entstand durch Kannibalismus – Intelligenz ist essbar.* Düsseldorf: Econ.

Kitchener, R. F. (1986). *Piaget's Theory of Knowledge: Genetic Epistemology and Scientific Reason.* New Haven, CT: Yale University Press.

Kjaernes, U. (2006). Trust and distrust: cognitive decisions or social relations? *Journal of Risk Research,* **9**(8), 911–32.

Klaus, G. (1968). *Spieltheorie in philosophischer Sicht*. Leipzig: VEB Deutscher Verlag der Wissenschaften.

Klee, R. J. & Graedel, T. E. (2004). Elemental cycles: a status report on human or natural dominance. *Annual Review of Environment and Resources,* **29**, 69–107.

Kleeb, U. (1997). *Blumenwiese zwischen Wahrnehmung und Wirklichkeit*. Zurich: Museum für Gestaltung.

Klein, P. G. (1998). New institutional economics. *Working Paper Series*. Columbia, MO: University of Missouri at Columbia.

Kley, J. & Fietkau, H.-J. (1979). Verhaltenswirksame Variablen des Umweltbewusstseins. *Psychologie und Praxis,* **23**(1), 13–22.

Kline, M. (1985). *Mathematics and the Search for Knowledge*. Oxford: Oxford University Press.

Klineberg, O. (1935). *Race Differences*. New York, NY: Harper.

Klix, F. (1980). *Erwachen des Denkens. Eine Entwicklungsgeschichte der menschlichen Intelligenz*. Berlin: VEB Deutscher Verlag der Wissenschaften.

Kloke, A. (1980). *Richtwerte '80 – Orientierungsdaten für tolerierbare Gesamtgehalte einiger Elemente in Kulturböden* (no. 1/3). Darmstadt: VDLUFA.

Klüver, J. (2002). *An Essay Concerning Sociocultural Evolution*. Dordrecht: Kluwer.

Kneese, A. V. (1989). The economics of natural resources. In A. V. Kneese (ed.). *Natural Resource Economics*. Cheltenham: Edward Elgar, pp. 281–309.

Kneese, A. V. (1993). Economics on water resources. In M. Reuss (ed.). *Water Resources Administration in the United States: Policy and Emerging Issues*. East Lansing, MI: Michigan State University Press, pp. 23–35.

Kneese, A. V. (2006). The Faustian bargain: risks, ethics, and nuclear energy. In W. E. Oates (ed.). *The RFF Reader in Environmental and Resource Policy*. Washington, DC: Resources for the Future, pp. 43–50.

Knetsch, J. L. & Sinden, J. A. (1984). Willingness to pay and compensation demanded – experimental-evidence of an unexpected disparity in measures of value. *Quarterly Journal of Economics,* **99**(3), 507–21.

Knoepfel, P. (ed.). (1995). *Lösung von Umweltkonflikten durch Verhandlung – Beispiele aus dem In- und Ausland*, Vol. 10. Basel: Helbing & Lichtenhahn.

Knorr-Cetina, K. (1981). *The Manufacture of Knowledge: an Essay on the Constructivist and Contextual Nature of Science*. Oxford: Pergamon.

Koch, R. (1882). Die Aetiologie der Tuberculose. *Berliner klinische Wochenschrift,* **19**(15), 221–30.

Koch, R. (1882, December 28). Réponse de M. Koch à M. Pasteur. *Le Semaine Médicale*. Dole: Musée Pasteur.

Koellner, T. (2003). *Land Use in Product Life Cycles and Ecosystem Quality*. Bern: Peter Lang.

Koellner, T. & Scholz, R. W. (2007). Assessment of land use impacts on the natural environment – Part 1: an analytical framework for pure land occupation and land use change. *International Journal of Life Cycle Assessment,* **12**(1), 16–23.

Koellner, T. & Scholz, R. W. (2008). Assessment of land use impacts on the natural environment – Part 2: generic characterization factors for local species diversity in Central Europe. *International Journal of Life Cycle Assessment,* **13**(1), 32–48.

Koellner, T., Sell, J., Gähwiler, M. & Scholz, R. W. (2008). Assessment of the management of organizations supplying ecosystem services from tropical forests. *Global Environmental Change-Human and Policy Dimensions,* **18**(4), 746–57.

Koellner, T., Suh, S., Weber, O., Moser, C. & Scholz, R. W. (2007). Environmental impacts of conventional and sustainable investment funds compared using input-output life-cycle assessment. *Journal of Industrial Ecology,* **11**(3), 41–60.

Koenen, K. C., Amstadter, A. B. & Nugent, N. R. (2009). Gene-environment interaction in posttraumatic stress disorder: an update. *Journal of Traumatic Stress,* **22**(5), 416–26.

Koerner, L. (1999). *Linnaeus: Nature and Nation*. Cambridge, MA: Harvard University Press.

Koffka, K. (1922). Perception: an introduction to the Gestalt-Theorie. *Psychological Bulletin,* **19**(10), 531–85.

Koffka, K. (1935). *Principles of Gestalt Psychology*. London: Routledge & Kegan.

Koh, J. (1988). An ecological aesthetic. *Landscape Journal,* **7**(2), 177–91.

Köhler, A. & Erdmann, L. (2004). Expected environmental impacts of pervasive computing. *Human and Ecological Risk Assessment,* **10**(5), 831–52.

Kohler, T. A. & Gumerman, G. J. (2000). *Dynamics in Human and Primate Societies: Agent-Based Modelling of Social and Spatial Processes*. Oxford: Oxford University Press.

Köhler, W. (1924). Gestaltprobleme und Anfänge einer Gestaltpsychologie. *Jahresberichte für die gesamte Physiologie und experimentelle Pharmakologie für das Jahr 1922,* **3**(1), 512–39.

Köhler, W. (1971). *Die Aufgabe der Gestaltpsychologie*. Berlin: de Gruyter.

Köhler, W. (2004). Entwicklung der Mikrobiologie mit besonderer Berücksichtigung der medizinischen Aspekte. In I. Jahn (ed.). *Geschichte der Biologie – Theorien, Methoden, Institutionen, Kurzbiographien*. Hamburg: Nikol Verlag, pp. 620–41.

Kolb, D. A. (1984). *Experiential Learning. Experience as the Source of Learning and Development.* Upper Saddle River, NJ: Prentice Hall.

Kolb, G. (2004). *Geschichte der Volkswirtschaftslehre: Dogmenhistorische Positionen des ökonomischen Denkens.* Munich: Franz Vahlen.

Koopman, J. S. (2005). Infection transmission science and models. *Japanese Journal of Infectious Disease,* **58**, S3–S8.

Kopfermann, H. (1930). Psychologische Untersuchung über die Wirkung zweidimensionaler Darstellungen körperlicher Gebilde. *Psychologische Forschung,* **13**, 293–364.

Kraft, M. E. & Clary, B. B. (1991). Citizen participation and the nimby syndrome: public response to radioactive waste disposal. *The Western Political Quarterly* **44**(2), 299–328.

Kramer, R. M. & Brewer, M. B. (1986). Social group identity and the emergence of cooperation in resource dilemma. In H. A. M. Wilke, D. M. Messick & C. G. Rutte (eds). *Experimental Social Dilemmas.* Franfurt am Main: Lang, pp. 205–34.

Krasner, S. D. (1988). Sovereignty – an institutional perspective. *Comparative Political Studies,* **21**(1), 66–94.

Krause, G. H. M., Arndt, U., Brandt, C. J. *et al.* (1986). Forest decline in Europe – development and possible causes. *Water Air and Soil Pollution,* **31**(3–4), 647–68.

Krausmann, F., Fischer-Kowalski, M., Schandl, H. & Eisenmenger, N. (2008). The global sociometabolic transition. *Journal of Industrial Ecology,* **12**(5–6), 637–56.

Krausmann, F., Gingrich, S., Eisenmenger, N. *et al.* (2009). Growth in global materials use, GDP and population during the 20th century. *Ecological Economics,* **68**(10), 2696–705.

Krauss, R. M., Freedman, J. L. & Whitcup, M. (1978). Field and laboratory studies of littering. *Journal of Experimental Social Psychology,* **14**(1), 109–22.

Kraut, R. (1979). Two conceptions of happiness. *Philosophical Review,* **88**(2), 167–97.

Krebs, F. & Bossel, H. (1997). Emergent value orientation in self-organization of an animal. *Ecological Modelling,* **96**(1–3), 143–64.

Krebs, F., Elbers, M. & Ernst, A. (2008a, September 1–5). *Comparing the Impact of Different Compensation Policies on Collective Land Reclamation.* Paper presented at the 5th Conference of the European Social Simulation Association, Brescia, Italy.

Krebs, F., Elbers, M. & Ernst, A. (2008b, January 11–12). *Modelling the Social and Economic Dimensions of Farmer Decision Making under Conditions of Water Stress.* Paper presented at the 1st ICC Workshop on Complexity in Social Systems, Lisbon.

Krebs, J. R. & Davies, N. B. (1978). *Behavioral Ecology – an Evolutionary Approach.* Oxford: Blackwell.

Kreps, G. A. (1984). Sociological inquiry and disaster research. *Annual Reviews,* **10**, 309–30.

Kreps, G. A. (1985). Disaster and the social order. *Sociological Theory,* **3**, 49–65.

Krohn, W. & Weingart, P. (1987). Nuclear-power as a social experiment – European political fall out from the Chernobyl meltdown. *Science Technology & Human Values,* **12**(2), 52–8.

Kruse, L. & Groffmann, C. F. (1987). Environmental psychology in Germany. In D. Stokols & I. Altmann (eds). *Handbook of Environmental Psychology*, Vol. I. New York, NY: Wiley, pp. 7–41.

Kruse, N., Behets, F., Vaovola, G. *et al.* (2003). Participatory mapping of sex trade and enumeration of sex workers using capture–recapture methodology in Diego-Suarez, Madagascar. *Sexually Transmitted Diseases,* **30**(8), 664–70.

Kuby, J. (1997). *Immunology*, 3rd edn. New York, NY: Freeman.

Kuckartz, U. (1998). *Umweltbewusstsein und Umweltverhalten.* Berlin: Springer.

Kuhl, J. (1983). Emotion, Kognition und Motivation: II. Die funktionale Bedeutung der Emotionen für das problemlösende Denken und für das konkrete Handeln. *Sprache & Kognition,* **4**, 228–53.

Kuhl, J. (2001). A functional approach to motivation: the role of goal enactment and self regulaton in current research on approach and avoidance. In A. Efklides, K. J. & R. M. Sorrentino (eds). *Trends and Prospects in Motivation Research.* Dordrecht: Kluwer, pp. 239–68.

Kuhn, T. S. (1985). *The Copernican Revolution: Planetary Astronomy in the Development of Western Thought.* Cambridge, MA: Harvard University Press.

Kuhn, T. S. (1996). *The Structure of Scientific Revolutions*, 3rd edn. Chicago, IL: University of Chicago Press.

Kümmel, W. F. (1985). Alexander von Humboldt und die Medizin. In W.-H. Hein (ed.). *Alexander von Humboldt.* Frankfurt: Weisbecker, pp. 195–210.

Kunreuther, H. (1996). Mitigating disaster losses through insurance. *Journal of Risk and Uncertainty,* **12**(2–3), 171–87.

Kuran, P. (2007). *How to Photograph an Atomic Bomb.* Vanconver:vce.com

Küster, H. (1999). *Geschichte der Landschaft in Mitteleuropa: Von der Eiszeit bis zur Gegenwart.* Munich: Beck.

Kutzbach, J., Ruddiman, W. F., Vavrus, S. & Philippon, G. (2009). Climate model simulation of anthropogenic influence on greenhouse-induced climate change (early agriculture to modern): the role of ocean feedbacks. *Climatic Change,* **99** (3–4), 351–81.

Kuznets, S. (1955). Economic growth and income inequality. *The American Economic Review,* **45**(1), 1–28.

Labov, W. (1973). The boundaries of words and their meanings. In C. J. N. Bailey & R. W. Shuy (eds). *New Ways of Analyzing Variation in English.* Washington, DC: Georgetown University Press, pp. 340–73.

Lafraniere, S. (2008, January 12–13). Europe takes Africa's fish, and migrants follow. *International Herald Tribune,* pp. 1, 6.

Lahr, M. M. (1994). The multiregional model of modern human origins – a reassessment of its morphological basis. *Journal of Human Evolution,* **26**(1), 23–56.

Lam, W. F. (1998). *Governing Irrigation Systems in Nepal: Institutions, Infrastructure, and Collective Action.* Oakland, CA: ICS Press.

Lambek, J. & Scott, P. J. (1986). *Introduction to Higher Order Categorial Logic.* Cambridge: Cambridge University Press.

Lamprecht, H. (2004). *Beitrag von Knochen am Phosphorhaushalt der Schweiz und Phosphorrecycling unter dem Einfluss der BSE Problematik.* Unpublished diploma thesis, ETH Zurich, Zurich.

Lamprecht, H., Lang, D. J. & Binder, C. R., Scholz, R. W. (submitted). Animal bone disposal during the BSE crisis in Switzerland – an example of a "disposal dilemma."

Landsteiner, K. (1936). *The Specificity of Serological Reactions.* Springfield, IL: Charles C. Thomas.

Lane, D. (2000). Should system dynamics be described as a "hard" or "deterministic" approach? *Systems Research and Behavioral Science,* **17**(1), 3–22.

Lane, D. (2001). Rerum cognoscere causas: Part II — opportunities generated by the agency/structure debate and suggestions for clarifying the social theoretic position of system dynamics. *System Dynamics Review,* **17**(4), 293–309.

Lane, D. (2006). Hierarchy, complexity, society. In D. Pumain (ed.). *Hierarchy in Natural and Social Sciences.* Dordrecht: Springer, pp. 81–119.

Lane, D. & Smart, C. (1996). Reinterpreting "generic structure": evolution, application and limitations of a concept. *System Dynamics Review,* **12**(2), 87–120.

Lang, D. J., Binder, C. R., Scholz, R. W., Schleiss, K. & Stäubli, B. (2006). Impact factors and regulatory mechanisms for material flow management: integrating stakeholder and scientific perspectives – the case of biowaste delivery. *Resources, Conservation and Recycling,* **47**(2), 101–32.

Langdell, C. C. (1887). Address to the Harvard Law School Association, *Harward Law Review,* 1(3), 111–31.

Lange, J. (2002). *Die Geschichte des Gewässerschutzes am Ober- und Hochrhein. Eine Fallstudie zur Umwelt- und Biologiegeschichte.* Unpublished PhD thesis, Albert-Ludwigs-Universität Freiburg, Freiburg im Breisgau.

Langer, E. (1975). The illusion of control. *Journal of Personality and Social Psychology,* **32**(2), 311–28.

Langman, R. E. & Cohn, M. (2000). A minimal model for the self-nonself discrimination: a return to the basics. *Seminars in Immunology,* **12**(3), 189–95.

Lanyon, L. E. (1995). Does nitrogen-cycle – changes in the spatial dynamics of nitrogen with industrial nitrogen-fixation. *Journal of Production Agriculture,* **8**(1), 70–8.

Laplace, P.-S. (1825/1999). *Philosophical Essay on Probabilities (translated from 5th French edn, 1825).* New York, NY: Springer.

Larson, J. L. (1967). Linnaeus and the natural method. *Isis,* **58**(3), 304–20.

Larson, J. L. (1968). The species concept of Linnaeus. *Isis,* **59**(3), 291–9.

Larson, J. L. (1994). *Interpreting Nature. The Science of Living Form from Linnaeus to Kant.* Baltimore, MD: Johns Hopkins University Press.

Larsson, J. (2007). *A Cost Benefit Analysis of Bioethanol Production from Cereals in Sweden – a Case Study Approach.* Uppsala: Swedish University of Agricultural Sciences.

Laska, S. B. (1993). Environmental sociology and the state of the discipline. *Social Forces,* **72**(1), 1–17.

Latané, B., Williams, K. & Harkins, S. (1979). Social loafing. *Psychology Today,* **13**(5), 104–10.

Latsch, G. (2007, March 19). Gentechnik: Aids im Bienenstock. *Der Spiegel,* 12/2007.

Laughlin, R. (2005). *A Different Universe – Reinventing Physics from the Bottom Down.* New York, NY: Basic Books.

Lawler, E. E., Kuleck, W. J., Rhode, J. G. & Sorensen, J. E. (1975). Job choice and post decision dissonance. *Organizational Behavior and Human Performance,* **13**(1), 133–45.

Lawrence, D. L. & Low, S. M. (1990). The built environment and spatial form. *Annual Review of Anthropology,* **19**, 453–505.

Laws, D., Scholz, R. W., Shiroyama, H. *et al.* (2004). Expert views on sustainability and technology implementation. *International Journal of Sustainable Development and World Ecology,* **11**(3), 247–61.

Le, Q. B. (2005). Multi-agent system for simulation of land-use and land-cover change: a theoretical framework and its first implementation for an upland watershed in the Central Coast of Vietnam. *Ecology and Development Series 29.* Göttingen: Cuvillier.

References

Le, Q. B., Park, S. J. & Vlek, P. L. G. (2010). Land use dynamic simulator (LUDAS): a multi-agent system model for simulating spatio-temporal dynamics of coupled human-landscape system. 2. Scenario-based applications for impact assessment of land-use policies. *Ecological Informatics* 5(3), 203–21.

Le, Q. B., Park, S. J., Vlek, P. L. G. & Cremers, A. B. (2008). Land use dynamic simulator (LUDAS): a multi-agent system model for simulating spatio-temporal dynamics of coupled human-landscape system. 1. Structure and theoretical specification. *Ecological Informatics,* 3(2), 135–53.

Le Bon, G. (1896). *The Crowd.* London: Fischer Unwin.

Lean, G. & Shawcross, H. (2007, April 15). Are mobile phones wiping out our bees? Scientists claim radiation from handsets are to blame for mysterious 'colony collapse' of bees. *The Independent.* Retrieved April 29, 2010 from http://www.independent.co.uk/environment/nature/are-mobile-phones-wiping-out-our-bees-444768.html

Lederberg, J. (2001). Commentary. *The Scientist,* 15(18), 6.

LeDoux, J. E. (2000). Emotion circuits in the brain. *Annual Review of Neuroscience,* 23, 155–84.

Lee, H., Clark, W. C. & Devereaux, C. (2008). *Biofuels and Sustainable Development: Report of an Executive Session on the Grand Challenges of a Sustainable Transition.* Cambridge MA: Sustainability Program, Center for International Development, Harvard University.

Lee, J. & Feng, W. (1999). Malthusian models and Chinese realities: the Chinese demographic system 1700–2000. *Population and Development Review,* 25(1), 33–65.

Leikas, S., Lindeman, M., Roininen, K. & Lahteenmaki, L. (2007). Food risk perceptions, gender, and individual differences in avoidance and approach motivation, intuitive and analytic thinking styles, and anxiety. *Appetite,* 48(2), 232–40.

Lengwiler, M. (2008). Participatory approaches in science and technology – historical origins and current practices in critical perspective. *Science Technology & Human Values,* 33(2), 186–200.

Lenski, G. (1966). *Power and Privilege: a Theory of Social Stratification.* New York, NY: McGraw-Hill.

Lenski, G. (2005). *Ecological–Evolutionary Theory: Principles and Application.* Boulder, CO: Paradigm Publisher.

Lenski, G., Lenski, J. & Nolan, P. (1991). *Human Societies. An Introduction to Macrosociology.* New York, NY: McGraw-Hill.

Lenton, T. M. & van Oijen, M. (2002). Gaia as a complex adaptive system. *Philosophical Transactions of the Royal Society of London Series B: Biological Sciences,* 357(1421), 683–95.

Lenz-Wiedemann, V. I. S., Klar, C. W. & Schneider, K. (2010). Development and test of a crop growth model for application within a global change decision support system. *Ecological Modelling,* 221(2), 314–29.

Leontief, W. (1966). *Input-Output Economics.* New York, NY: Oxford University Press.

Leontjew, A. N. (1964). *Probleme der Entwicklung des Psychischen.* Berlin: Volk und Wissen.

Leps, G. (2004). Ökologie und Ökosystemforschung. In I. Jahn (ed.). *Geschichte der Biologie – Theorien, Methoden, Institutionen, Kurzbiographien.* Hamburg: Nikol Verlag, pp. 601–19.

Lepschy, A. M., Mian, G. A. & Viaro, U. (1992). Feedback-control in ancient water and mechanical clocks. *IEEE Transactions on Education,* 35(1), 3–10.

Lerman, A., Mackenzie, F. T. & Ver, L. M. (2004). Coupling of the perturbed C-N-P cycles in industrial time. *Aquatic Geochemistry,* 10(1–2), 3–32.

Leser, H. (1991). *Landschaftsökologie.* Stuttgart: Ulmer.

Leser, H. & Mosimann, T. (1997). *Landschaftsökologie. Ansatz, Modelle, Methodik, Anwendung,* 4th edn. Stuttgart: Ulmer.

Lesschen, J. P., Stoorvogel, J. J., Smaling, E. M. A., Heuvelink, G. B. M. & Veldkamp, A. (2007). A spatially explicit methodology to quantify soil nutrient balances and their uncertainties at the national level. *Nutrient Cycling in Agroecosystems,* 78(2), 111–31.

Leu, U. (2005, November 3). Der Besuch von Grippeerkrankten ist untersagt. *Schaffhauser Nachrichten,* p. 18.

Leung, M. W., Yen, I. H. & Minkler, M. (2004). Community based participatory research: a promising approach for increasing epidemiology's relevance in the 21st century. *International Journal of Epidemiology,* 33(3), 499–506.

Lévi-Strauss, C. (1963). The bear and the barber. *The Journal of the Royal Anthropological Institute of Great Britain and Ireland,* 93(1), 1–11.

Levin, S. A. (1999). *Fragile Dominion.* Cambridge, MA: Perseus Books.

Levy-Brühl, L. (1921). *Das Denken der Naturvölker.* Leipzig: Braumüller.

Lewin, K. (1939). Field theory and experiment in social psychology. *American Journal of Sociology,* 44(6), 868–97.

Lewin, K. (1943a). Defining the "field at a given time". *Psychological Review,* 50(3), 292–310.

Lewin, K. (1943b). Forces behind food habits and methods of change. *Bulletin of the National Research Council,* XVIII, 35–65.

Lewin, K. (1943/1997). Defining the "field at a given time". In K. Lewin (ed.). *Resolving Social Conflicts & Field*

Theory on Social Sciences. Washington, DC: American Psychological Association, pp. 200–11.

Lewin, K. (1946). Action research and minority problems. *Journal of Social Issues,* **2**(4), 34–46.

Lewin, K. (1946/1997). Behavior and development as a function of the total situation. In K. Lewin (ed.). *Resolving Social Conflicts & Field Theory in Social Science*. Washington, DC: American Psychological Association, pp. 337–81.

Lewin, K. (1951). *Field Theory in Social Science*. New York, NY: Harper.

Lewis, H. (1980). The safety of fission reactors. *Scientific American,* **242**(3), 53–65.

Lewis-Williams, D. & Pearce, D. (2005). *Inside the Neolithic Mind: Consciousness, Cosmos and the Realm of Gods*. London: Thames & Hudson.

Leydesdorff, L. & Etzkowitz, H. (1996). Emergence of a triple helix of university-industry-government relations. *Science and Public Policy,* **23**(5), 279–86.

Li, W. J., Ali, S. H. & Zhang, Q. (2007). Property rights and grassland degradation: a study of the Xilingol Pasture, Inner Mongolia, China. *Journal of Environmental Management,* **85**(2), 461–70.

Li, Y., Leung, G. M., Tang, J. W. *et al*. (2007). Role of ventilation in airborne transmission of infectious agents in the built environment – a multidisciplinary systematic review. *Indoor Air,* **17**(1), 2–18.

Libby, W. (1961). Radiocarbon dating: the method is of increasing use to the archeologist, the geologist, the meteorologist, and the oceanographer. *Science,* **133**(3453), 621–9.

Libecap, G. D. (2005). State regulations of open-access, common-pool resources. In C. Menard & M. M. Shirley (eds). *Handbook of New Institutional Economics*. Dordrecht: Springer, pp. 545–72.

Libecap, G. D. (2009). The tragedy of the commons: property rights and markets as solutions to resource and environmental problems. *Australian Journal of Agricultural and Resource Economics,* **53**(1), 129–44.

Liberman, N., Sagristano, M. D. & Trope, Y. (2002). The effect of temporal distance on level of mental construal. *Journal of Experimental Social Psychology,* **38**(6), 523–34.

Lichtenstein, S., Slovic, P., Fischhoff, B. & Combs, B. (1978). Judged frequency of lethal events. *Journal of Experimental Psychology: Human Learning and Memory,* **4**(6), 551–78.

Lidicker, W. Z. (2008). Levels of organization in biology: on the nature and nomenclature of ecology's fourth level. *Biological Reviews,* **83**(1), 71–8.

Liebig, J., von. (1840). *Die organische Chemie in ihrer Anwendung auf Agricultur und Physiologie*. Braunschweig: Friedrich von Vieweg.

Liebrand, W. B. G. & van Run, G. J. (1985). The effects of social motives on behavior in social dilemmas in 2 cultures. *Journal of Experimental Social Psychology,* **21**(1), 86–102.

Lifset, R. & Graedel, T. E. (2002). Industrial ecology: goals and definitions. In R. U. Ayres & L. W. Ayres (eds). *A Handbook of Industrial Ecology*. Cheltenham: Edward Elgar, pp. 3–15.

Lifset, R. J., Gordon, R. B., Graedel, T. E., Spatari, S. & Bertram, M. (2002). Where has all the copper gone: the stocks and flows project, part 1. *JOM-Journal of the Minerals Metals & Materials Society,* **54**(10), 21–6.

Light, A. (ed.). (1998). *Social Ecology after Bookchin*. New York, NY: Guilford Press.

Lillie, M. C. (1998). Cranial surgery dates back to mesolithic. *Nature,* **391**(6670), 854.

Lima, M. L. (2004). On the influence of risk perception on mental health: living near an incinerator. *Journal of Environmental Psychology,* **24**(1), 71–84.

Lin, R. Y., Anderson, A. S., Shah, S. N. & Nurruzzaman, F. (2008). Increasing anaphylaxis hospitalizations in the first 2 decades of life: New York State, 1990–2006. *Annals of Allergy Asthma & Immunology,* **101**(4), 387–93.

Lindahl, E. (1918). *Die Gerechtigkeit der Besteuerung*. Lund: Gleerupska Universitets-Bokhandeln.

Lindeman, R. L. (1942). The trophic-dynamic aspect of ecology. *Ecology,* **23**(4), 399–418.

Lindenbaum, S. (2001). Kuru, prions, and human affairs. *Annual Review of Anthropology,* **30**, 363–85.

Linné, C. (1735/1788). *Systema Naturae*. Leipzig: Beer.

Linné, C. (1743). Oratio de telluris habitabilis incremento. In C. Linné (ed.). *Amoenitates Academicæ*, Vol. II. Stockholm: Laurentius Salvius. pp. 430–72.

Linné, C. (1751). *Philosophica Botanica*. Stockholm: G. Kiesewetter.

Linné, C. (1751/1907). Bref och skrifvelser af och till Carl von Linné, afdelningen 1, del 4 quoted according to J. L. Larson, 1968. Stockholm: Akademiska Bokhandeln.

Linné, C. (1764). *Genera Plantarum: Eorumque Characteres Naturales, Secundum Numerum, Figuram, Situm, & Proportionem Omnium Fructificationis Partium*, 6th edn. Stockholm: Impensis Laurentii Salvii.

Linné, C. (1947/1952). *Herbationes Upsalienses: Protokoll över Linnés excursioner i Uppsalatrakten*, quoted according to L. Koerner, 1999, p. 92. Uppsala: Almquist & Wiksell.

Lipp, W., Hofmann, H. & Hubig, C. (1995). Institution. In Görres-Gesellschaft (ed.). *Staatslexikon*, Vol. 3. Freiburg: Herder, pp. 99–109.

Lisbach, B. (1999). *Werte und Umweltwahrnehmung. Der Einfluss von Werten auf die Wahrnehmung der natürlichen Umwelt*. Marburg: Tectum Verlag.

Liu, J. G., Dietz, T., Carpenter, S. R. *et al.* (2007a). Complexity of coupled human and natural systems. *Science,* **317**(5844), 1513–16.

Liu, J. G., Dietz, T., Carpenter, S. R. *et al.* (2007b). Coupled human and natural systems. *Ambio,* **36**(8), 639–49.

Lloyd, B. (2007). The commons revisited: the tragedy continues. *Energy Policy,* **35**(11), 5806–18.

Lloyd, J. & Mitchinson, J. (2006). *QI Annual: The Book of General Ignorance*. London: Faber and Faber.

Locke, J. (1696). *Some Considerations of the Consequences of the Lowering of Interest and Raising the Value of Money*. London: Awsham and John Churchill.

Lodish, H., Berk, A., Matsudaira, P. *et al.* (2004). *Molecular Cell Biology*, 5th edn. New York, NY: Freeman.

Loeffler, F. & Frosch, P. (1898). Berichte der Kommission zur Erforschung der Lehre von Bacterien. *Zentralblatt für Bakteriologie, Parasitenkunde, Infektionskrankheiten und Hygiene,* **23**, 371–91.

Loomis, D. K., Samuelson, C. D. & Sell, J. A. (1995). Effects of information and motivational orientation on harvest of a declining renewable resource. *Society & Natural Resources,* **8**(1), 1–18.

Loorbach, D. & Rotmans, J. (2006). Managing transitions for sustainable development. In X. Olsthoorn & A. Wieczorek (eds). *Understanding Industrial Transformations: Views from Different Disciplines*. Dordrecht: Springer, pp. 187–206.

Loorbach, D. & Rotmans, J. (2010). The practice of transition management: examples and lessons from four distinct cases. *Futures,* **42**(3), 237–46.

Lorenz, E. N. (1963). Deterministic nonperiodic flow. *Journal of the Atmospheric Sciences,* **20**(2), 130–41.

Lorenz, K. (1937). Über die Bildung des Instinktbegriffs. *Naturwissenschaften,* **25**(19), 289–300.

Lorenz, K. (1963). *Das sogenannte Böse: Zur Naturgeschichte der Agression*. Vienna: Borotha-Schoeler.

Lorenz, K. (1981). *The Foundation of Ethology*. New York, NY: Springer.

Löther, R. (2004). Kenntnisse und Vorstellungen über Lebewesen und Lebensprozesse in frühen Kulturen. In I. Jahn (ed.). *Geschichte der Biologie: Theorien, Methoden, Institutionen, Kurzbiografien*, 3rd edn. Hamburg: Nikol, Verlag, pp. 27–47.

Lotka, A. J. (1925). *Elements of Physical Biology*. Baltimore, MD: Williams and Wilkins.

Lotka, A. J. (1956). *Elements of Mathematical Biology*. New York, NY: Dover.

Lovelock, J. E. (1995). *Gaia*. Oxford: Oxford University Press.

Lovelock, J. E. & Margulis, L. (1974). Atmospheric homeostasis by and for the biosphere: the Gaia hypothesis. *Tellus,* **26**(1/2), 2–10.

Lovett, J. C., Quinn, C. H., Ockwell, D. G. & Gregorowski, R. (2006). Two cultures and tragedy of the commons. *African Journal of Ecology,* **44**(1), 1–5.

Luck, G. W., Harrington, R., Harrison, P. A. *et al.* (2009). Quantifying the contribution of organisms to the provision of ecosystem services. *Bioscience,* **59**(3), 223–35.

Luhmann, N. (1979). *Trust and Power: Two Works*. Chichester, NY: Wiley.

Luhmann, N. (1984). *Social Systems*. Stanford, CA: Stanford University Press.

Luhmann, N. (1985). *Soziale Systeme: Grundriss einer allgemeinen Theorie*, 2nd edn. Frankfurt am Main: Suhrkamp.

Luhmann, N. (1986). *Ökologische Kommunikation: Kann die moderne Gesellschaft sich auf ökologische Gefährdungen einstellen?* Opladen: Westdeutscher Verlag.

Luhmann, N. (1990). *Risiko und Gefahr*. Opladen: Westdeutscher Verlag.

Luhmann, N. (1993). Risiko und Gefahr. In W. Krohn & G. Krücken (eds). *Riskante Technologien: Reflexion und Regulation*. Frankfurt am Main: Suhrkamp, pp. 138–85.

Luhmann, N. (1995). *Soziologische Aufklärung*, Vol. 6. Opladen: Westdeutscher Verlag.

Lühr, H.-P., Hefer, B. & Scholz, R. W. (1991). Das Donator-Akzeptor-Modell (DAM). In E. Bütow (ed.). *Ableitung von Sanierungswerten für kontaminierte Böden*. Berlin: Schmidt, pp. 17–53.

Luo, L., van der Voet, E. & Huppes, G. (2009). Life cycle assessment and life cycle costing of bioethanol from sugarcane in Brazil. *Renewable & Sustainable Energy Reviews,* **13**(6–7), 1613–19.

Lusardi, A. (1998). On the importance of the precautionary saving motive. *American Economic Review,* **88**(2), 449–53.

Lutz, W., Sanderson, W. & Scherbov, S. (2001). The end of world population growth. *Nature,* **412**(6846), 543–5.

Lutzenhiser, L. (1992). A cultural model of household-energy consumption. *Energy,* **17**(1), 47–60.

Lutzenhiser, L. (2002). Environmental sociology: the very idea. *Organization & Environment,* **15**(1), 5–9.

Lux, A.-C. (2005). *Die Dreckapotheke des Christian Franz Paullini (1643–1712)*. Unpublished Master of Arts, Mainz: Johannes Gutenberg-Universität at Mainz.

Lyell, C. (1830–1833). *Principles of Geology*. London: John Murray.

Maasen, S. & Lieven, O. (2006). Socially robust knowledge. Transdisciplinarity: a new mode of governing science? *Science and Public Policy,* **33**(6), 399–410.

MacArthur, R. H. (1955). Fluctuations of animal population, and a measure of community stability. *Ecology,* **36**(3), 533–6.

MacArthur, R. H. & Wilson, E. O. (1967). *The Theory of Island Biography*. Princeton, NJ: Princeton University Press.

MacCrimmon, K. R. (1976). Framework for social motives. *Behavioral Science,* **21**(2), 86–100.

MacGregor, D. G. & Fleming, R. (1996). Risk perception and symptom reporting. *Risk Analysis,* **16**(6), 773–83.

Machlis, G. E., Force, J. E. & Burch, W. R. (1997). The human ecosystem. Part I: The human ecosystem as an organizing concept in ecosystem management. *Society & Natural Resources,* **10**(4), 347–67.

MacLaury, R. E. (1991). Prototypes revisited. *Annual Review of Anthropology,* **20**, 55–74.

MacLaury, R. E. (1997). Ethnographic evidence of unique hues and elemental colors. *Behavioral and Brain Sciences,* **20**(2), 202.

MacLean, R. C. & Gudelj, I. (2006). Resource competition and social conflict in experimental populations of yeast. *Nature,* **441**(7092), 498–501.

Mägdefrau, K. (1992). *Geschichte der Botanik*. Jena: Gustav Fischer.

Magner, L. N. (2002). *A History of the Life Sciences, Revisted and Expanded*, 3rd edn. New York, NY: Marcel Dekker.

Mahon, C. (2001). *Collective Rationality and Collective Reasoning*. Cambridge: Cambridge University Press.

Maier, D. (1997). *Zur Erfassung, Aufbereitung und Verwertung von Sekundärrohstoffen in der DDR: Kreislaufwirtschaft zwischen ökonomischen Sachzwängen und Umweltschutz*. Unpublished diploma thesis, TU Berlin.

Malmberg, A. & Maskell, P. (2002). The elusive concept of localization economies: towards a knowledge-based theory of spatial clustering. *Environment and Planning A,* **34**, pp. 429–49.

Malthus, T. R. (1798). *An Essay on the Principle of Population*. London: J. Johnson.

Mandecki, W. (1998). The game of chess and searches in protein sequence space. *Trends in Biotechnology,* **16**(5), 200–2.

Mandelbrot, B. (2004). *Fractals and Chaos*. Berlin: Springer.

Mankiv, N. G. (1994). *Macroeconomics*, 2nd edn. New York, NY: Worth.

Mannino, M. A. & Thomas, K. D. (2002). Depletion of resources? The impact of prehistoric human foraging in intertidal mollusc communities and its significance for human settlement, mobilty and dispersal. *World Archeaology,* **33**(3), 452–74.

Mannino, M. A. & Thomas, K. D. (2008). A research agenda for the archaeomalacological study of prehistoric human ecology in the coastal zone of NW Sicily. *Archaeofauna,* **17**, 35–45.

Mao, J. S. & Graedel, T. E. (2009). Lead in-use stock. *Journal of Industrial Ecology,* **13**(1), 112–26.

March, J. G. (1994). *A Primer on Decision Making*. New York, NY: Free Press.

Marques, J. M., Yzerbyt, V. Y. & Leyens, J. P. (1988). The black sheep effect – extremity of judgments towards ingroup members as a function of group identification. *European Journal of Social Psychology,* **18**(1), 1–16.

Marr, D. (1982). *Vision*. New York, NY: W. H. Freeman.

Marsh, G. P. (1864/2003). *Man and Nature; or Physical Geography as Modified by Human Action*. New York, NY: Charles Scribner & Co.

Marshall, A. (1890/1920). *Principles of Economics*, 8th edn. London: Macmillan.

Martin, J. L. (2003). What is field theory? *American Journal of Sociology,* **109**(1), 1–49.

Martinez-Conde, S., Macknik, S. L. & Hubel, D. H. (2004). The role of fixational eye movements in visual perception. *Nature Reviews Neuroscience,* **5**(3), 229–40.

Martinez-Palomo, A. (2001). The science of Louis Pasteur: a reconsideration. *Quarterly Review of Biology,* **76**(1), 37–45.

Martrat, B., Grimalt, J. O., Shackleton, N. J. et al. (2007). Four climate cycles of recurring deep and surface water destabilizations on the Iberian margin. *Science,* **317**(5837), 502–7.

Maruyama, M. (1963). Second cybernetics – deviation-amplifying mutual causal processes. *American Scientist,* **51**(2), 164–79.

Marx, K. (1844/1968). Ökonomisch-philosophische Manuskripte aus dem Jahr 1844. In E. O'Callaghan (ed.). *K. Marx und F. Engels: Werke, supplementary*, Vol. 1. Berlin: Dietz, pp. 471–588.

Marx, K. (1876/1971). Anteil der Arbeit an der Menschwerdung des Affen. In K. Marx & F. Engels (eds). *Anti Dühring: Dialektik der Natur*. Berlin: Dietz, pp. 444–71.

Marx, K. (1887/1990). *Capital: a Critical Analysis of Capitalist Production. London 1887* (S. Moore & E. Aveling, Trans.). Berlin: Dietz.

Marx, K. (1890/1970). *Das Kapital: Kritik der politischen Ökonomie*, Vol. 1, Book 1. Berlin: Dietz Verlag.

Marx, K. (1956). *Theorien über den Mehrwert*. Berlin: Dietz. (Reprinted from Kautsky, 1905–1910; unpublished manuscript by the author 1861/1863).

Marx, K. & Engels, F. (eds). (1851–1853). *Collected Works in Fifty Volumes*, Vol. 11. London: Lawrence & Wishart.

References

Masinde, G., Li, X., Baylink, D. J., Nguyen, B. & Mohan, S. (2005). Isolation of wound healing/regeneration genes using restrictive fragment differential display-PCR in MRL/MPJ and C57BL/6 mice. *Biochemical and Biophysical Research Communications,* **330**(1), 117–22.

Maskell, P. & Malmberg, A. (1999). Localised learning and industrial competitiveness. *Cambridge Journal of Economics,* **23**, 167–85.

Maskin, E. (1977, June). *Nash Equilibrium and Welfare Optimality*. Paper presented at the Summer Workshop of the Econometric Society, Paris.

Maslow, A. H. (1943). A theory of human motivation. *Psychological Review,* **50**(4), 370–96.

Maslow, A. H. (1954a). The instinctoid nature of basic needs. *Journal of Personality,* **22**(3), 326–47.

Maslow, A. H. (1954b). *Motivation and Personality*. New York, NY: Harper & Row.

Matthews, R. B. (2006). The People and Landscape Model (PALM): towards full integration of human decision making and biophysical simulation models. *Ecological Modelling,* **194**(4), 329–43.

Matthies, M., Malchow, H., & Krie, J. (2001). *Integrative Systems Approaches to Natural and Social Dynamics*. Heidelberg: Springer.

Maturana, H. R. & Varela, F. J. (1980). *Autopoiesis and Cognition: the Realization of the Living*. Dordrecht: Kluwer.

Matzinger, P. (2002). The danger model: a renewed sense of self. *Science,* **296**(5566), 301–5.

Maxam, A. M. & Gilbert, W. (1977). New method for sequencing DNA. *Proceedings of the National Academy of Sciences of the United States of America,* **74**(2), 560–4.

May, R. M. (1976). Simple mathematical models with very complicated dynamics. *Nature,* **261**(5560), 459–67.

May, R. M. (1980a). Modelle für zwei interagierende Populationen. In R. M. May (ed.). *Theoretische Biologie*. Basel: Verlag Chemie, pp. 48–68.

May, R. M. (1980b). Populationsmodelle für eine Art. In R. M. May (ed.). *Theoretische Biologie*. Basel: Verlag Chemie, pp. 5–22.

May, R. M. (1981). *Theoretical Ecology Principles and Applications,* 2nd edn. Oxford: Blackwell.

May, R. S. & Scholz, R. W. (1990, September 23–27). *Betroffenheit und Wahrscheinlichkeitsurteile bei umweltbedingten Gesundheitsrisiken*. Paper presented at the 37. Kongress der Deutschen Gesellschaft für Psychologie, Kiel, Germany.

May, T., Scholz, R. W. & Nothbaum, N. (1991). Die Anwendung des Donator-Akzeptor-Modells am Beispiel Cadmium für den Pfad "Boden-Weizen-Mensch". In E. Bütow (ed.). *Ableitung von Sanierungswerten für kontaminierte Böden*. Berlin: Schmidt, pp. 99–182.

Mayer, J. R. (1845/1978). Die organische Bewegung in ihrem Zusammenhang mit dem Stoffwechsel. In J. R. Mayer (ed.). *Mechanik der Wärme: Sämtliche Schriften*. Stuttgart: Verlag der J. G. Cotta'schen Buchhandlung, pp. 41–154.

Maynard, H. B. (1971). *Industrial Engineering Handbook,* 3rd edn. New York, NY: McGraw-Hill.

Maynard Smith, J. (2000). The concept of information in biology. *Philosophy of Science,* **67**(2), 177–94.

Mayor, S. (2009). Resistance of flu virus to oseltamivir is growing, US study shows. [News Item]. *British Medical Journal,* **338**, 2.

Mayr, E. (1963). *Animal Species and Evolution*. Cambridge, MA: Harvard University Press.

Mayr, E. (1982). *The Growth of Biological Thought. Diversity, Evolution, and Inheritance*. Cambridge, MA: Beknap.

Mayr, O. (1970). *The Origin of Feedback Control*. Cambridge, MA: The MIT Press.

McCarthy, J. J. (2001). *Climate Change 2001: Impacts, Adaptation, Vulnerability*. Cambridge, UK: IPCC.

McClelland, D. C. (1951). *Personality*. New York, NY: Holt, Reinhart and Winston.

McClelland, D. C. (1965). Towards a theory of motive acquisition. *American Psychologist,* **20**(5), 321–33.

McConnell, J. V., Jacobson, A. L. & Kimble, D. P. (1959). The effects of regeneration upon retention of a conditioned response in the planarian. *Journal of Comparative and Physiological Psychology,* **52**(1), 1–5.

McConnell, J. V. & Shelby, J. M. (1970). Memory transfer experiments in invertebrates. In G. Ungar (ed.). *Molecular Mechanisms in Memory and Learning*. New York, NY: Plenum Press, pp. 71–101.

McCready, M. J. (1998). *Defining Engineers: How Engineers Think about the World*. Notre Dame, IN: Department of Chemical Engineering, University of Notre Dame.

McDaniels, T. L. & Trousdale, W. (2005). Resource compensation and negotiation support in an aboriginal context: using community-based multi-attribute analysis to evaluate non-market losses. *Ecological Economics,* **55**, 173–86.

McDermott, J., Weiss, Y. & Adelson, E. H. (2001). Beyond junctions: nonlocal form constraints on motion interpretation. *Perception,* **30**, 905–23.

McDougall, I., Brown, F. H. & Fleagle, J. G. (2005). Stratigraphic placement and age of modern humans from Kibish, Ethiopia. *Nature,* **433**(7027), 733.

McGill, B. J. (2003). A test of unified neutral theory of biodiversity. *Nature,* **422**(6934), 881–5.

McGrath, J. E. (1984). *Groups: Interaction and Performance*. Englewood Cliffs, NJ: Prentice-Hall.

McGrath, J. E. & Tschan, F. (2000). *Small Groups as Complex Systems. Formation, Coordination, Development, and Adaptation*. Thousand Oaks, CA: Sage.

McGrath, J. E. & Tschan, F. (2004). *Temporal Matters in Social Psychology. Examining the Role of Time in the Lives of Groups and Individuals*. Washington, DC: American Psychological Association.

McGrew, W. C. (1992). *Chimpanzee Material Culture: Implications for Human Evolution*. Cambridge: Cambridge University Press.

McGurty, E. M. (2007). *Transforming Environmentalism. Warren County, PCBs, and the Origins of Environmental Justice*. Piscataway, NY: Rutgers University Press.

McIntosh, R. P. (1995). H. A. Gleason's "individualistic concept" and theory of animal communities: a continuing controversy. *Biological Reviews*, **60**(2), 317–57.

McIvor, A. & Johnston, R. (2007). *Miners' Lung: a History of Dust Disease in British Coal Mining*. Aldershot: Ashgate.

McKelvey, V. E. (1972). Mineral resource estimates and public policy. *American Scientist*, **60**(11), 32–40.

McKenna, K. Y. A. & Green, A. S. (2002). Virtual group dynamics. *Group Dynamics: Theory Research and Practice*, **6**(1), 116–27.

McKenzie, R. D. (1924). The ecological approach to the study of the human community. *The American Journal of Sociology*, **80**(3), 287–301.

McKenzie, R. D. (1926/1967a). The ecological approach to the study of the human community. In R. E. Park, E. W. Burgess & R. D. McKenzie (eds). *The City*. Chicago, IL: University of Chicago Press, pp. 63–79.

McKenzie, R. D. (1926/1967b). The scope of human ecology. In E. W. Burgess (ed.). *The Urban Community: Selected Papers from the Proceedings of the American Sociological Society 1925*. New York, NY: Greenwood Press, pp. 167–82.

McKenzie, R. D. (1927). The concept of dominance and world-organization. *The American Journal of Sociology*, **33**(33), 1.

McKone, T. E. & Bogen, K. T. (1991). Predicting the uncertainties in risk assessment: a California groundwater case-study. *Environmental Science & Technology*, **25**(10), 1674–81.

McNeill, J. R. (2007). Woods and warfare in world history. *Environmental History*, **9**(3), 1–19.

Meadows, D. H. & Meadows, D. (2007). The history and conclusions of The Limits to Growth. *System Dynamics Review*, **23**(2–3), 191–7.

Meadows, D. H., Meadows, D. L., Randers, J. & Behrens, W. W. (1972). *The Limits to Growth*. New York, NY: Universe Books.

Meadows, D. H., Randers, J. & Meadows, D. (2004). *Limits to Growth: the 30 Years Update*. London: Chelsea Green Publishing Company.

Medina, M. (2000). Scavenger cooperatives in Asia and Latin America. *Resources Conservation and Recycling*, **31**(1), 51–69.

Medzhitov, R. & Janeway, C. A. (2000). How does the immune system distinguish self from nonself? *Seminars in Immunology*, **12**(3), 185–8.

Menard, C. & Shirley, M. M. (2005a). Introduction. In C. Menard & M. M. Shirley (eds). *Handbook of New Institutional Economics*. Dordrecht: Springer, pp. 1–18.

Menard, C. & Shirley, M. M. (eds). (2005b). *Handbook of New Institutional Economics*. Dordrecht: Springer.

Meppem, T. & Gill, R. (1998). Planning for sustainability as a learning concept. *Ecological Economics*, **26**(2), 121–37.

Meredith, J. E. (2008). Jeremy Bentham's hedonic calculus. Retrieved January 1, 2008, from http://www.iejs.com/Criminology/jeremy_benthams_hedonic_calculus.htm

Merriam-Webster's. (2002). *Merriam-Webster's Third New International Dictionary, Unabridged*. Retrieved 2008, December 2, from Merriam-Webster: http://www.merriam-webster.com/premium/index.html#unabridged

Merton, R. K. (1938). Science, technology and society in the seventeenth century England. *Osiris*, **4**, 360–632.

Messerli, B., Grosjean, M., Hofer, T., Nuñez, L. & Pfister, C. (2000). From nature-dominated to human-dominated environmental changes. *Quaternary Science Reviews*, **19**(1), 459–79.

Messick, D. M. & McClelland, C. L. (1983a). Social traps and temporal traps. *Personality and Social Psychology Bulletin*, **9**(1), 105–10.

Messick, D. M., Wilke, H., Brewer, M. B. *et al.* (1983). Individual adaptations and structural change as solutions to social dilemmas. *Journal of Personality and Social Psychology*, **44**, 294–309.

Meštrović, S. G. (1995). Introduction to the transaction edition. In D. Riesman (ed.). *Thorstein Veblen*. New Brunswick, NJ: Transaction Publishers, pp. ix–xxx.

Metchnikoff, E. (1905). *Immunity in Infective Diseases*. Cambridge: Cambridge University Press.

Mettier, T. & Scholz, R. W. (2008). Measuring preferences on environmental damages in LCIA. Part 2: Choice and allocation questions in panel methods. *International Journal of Life Cycle Assessment*, **13**(6), 468–76.

Metzger, M. J., Rounsevell, M. D. A., Acosta-Michlik, L., Leemans, R. & Schröter, D. (2006). The vulnerability of ecosystem services to land use change. *Agriculture, Ecosystems & Environment*, **114**(1), 69–85.

Metzger, W. (1963). *Psychologie: Die Entwicklung ihrer Grundannahmen seit der Einführung des Experiments*. Darmstadt: Dietrich Steinkopff.

References

Metzger, W. (2006). *Laws of Seeing*. Cambridge, MA: The MIT Press.

Meyer, S. C. (2004). The origin of biological information and the higher taxonomic categories. *Proceedings of the Biological Society of Washington,* **117**(2), 213–39.

Meyer-Abich, A. (1969). Alexander von Humboldt as a biologist. In H. Pfeiffer (ed.). *Alexander von Humboldt: Werk und Weltgeltung*. Munich: Piper, pp. 179–96.

Michaels, S. & O'Connor, M. C. (1990). *Literacy as Reasoning within Multiple Discourses: Implication for Policy and Educational Reform*. Newton, MA: Education Development Corporation.

Michelsen, W. H. (ed.). (1976). *Man and his Urban Environment*, 2nd edn. Reading, MA: Addison-Wesley.

Mieg, H. A., Hübner, P., Stauffacher, M., Balmer, M. & Bösch, S. (eds). (2001). *Zukunft Schiene Schweiz 2 – Ökologisches Potenzial des SBB-Bahngüterverkehrs für die Region Zugersee. ETH-UNS Fallstudie 2000*. Zurich: Rüegger.

Milfont, T. L. & Gouveia, V. V. (2006). Time perspective and values: an exploratory study of their relations to environmental attitudes. *Journal of Environmental Psychology,* **26**(1), 72–82.

Milgram, S. (1967). The small world problem. *Psychology Today,* **1**(1), 60–7.

Mill, J. S. (1857/2004). *Principles of Political Economy*. New York, NY: Prometheus Books.

Mill, J. S. (1863). *Utilitarism*. London: Parker, Son, and Bourn.

Millenium Ecosystem Assessment. (2005). *Ecosystem and Human Well-being: Synthesis*. Washington, DC: Island Press.

Miller, F. G. & Kaptchuk, T. J. (2008). The power of context: reconceptualizing the placebo effect. *Journal of the Royal Society of Medicine,* **101**(5), 222–5.

Miller, G. A. (1956). The magic number seven plus minus two: some limits on our capacity for processing information. *Psychological Review,* **63**(2), 81–97.

Miller, J. G. (1978). *Living Systems*. New York, NY: McGraw-Hill.

Miller, J. G. (1990). The strategy of living systems research. *Simulation,* **54**(3), 170–2.

Miller, J. G. & Miller, J. L. (1990). The nature of living systems – introduction. *Behavioral Science,* **35**(3), 157–63.

Miller, J. L. & Miller, J. G. (1992). Greater than the sum of its parts 1. Subsystems which process both matter-energy and information. *Behavioral Science,* **37**(1), 1–9.

Miller, J. L. & Miller, J. G. (1993). Greater than the sum of its parts 2. Matter-energy processing subsystems. *Behavioral Science,* **38**(1), 1–9.

Miller, J. L. & Miller, J. G. (1995). Greater than the sum of its parts 3. Information-processing subsystems. *Behavioral Science,* **40**(3), 171–269.

Milnor, J. (1954). Games against nature. In R. M. Thrall, C. H. Coombs & R. L. Davis (eds). *Decision Processes*. New York, NY: Wiley.

Milstein, M., Hart, S. & Sadinha, B. (2004). *Jari Celilose SA*. St. Gallen: Oikos.

Minsky, H. P. (2008). *John Maynard Keynes*. New York, NY: McGraw-Hill.

Minton, A. P. & Rose, R. L. (1997). The effects of environmental concern on environmentally friendly consumer behavior: an exploratory study. *Journal of Business Research,* **40**(1), 37–48.

Mirata, M. & Pearce, R. (2006). Industrial symbiosis in the UK. In K. Green & S. Randles (eds). *Industrial Ecology and Spaces of Innovation*. Cheltenham: Edward Elgar, pp. 77–105.

Mirowski, P. (1991). *More Heat than Light: Economics as Social Physics, Physics as Nature's Economics*. Cambridge: Cambridge University Press.

Mitchell, W. P. (1976). Irrigation and community in Central Peruvian Highlands. *American Anthropologist,* **78**(1), 25–44.

Mithen, S. (1996). *The Prehistory of the Mind. A Search for the Origins of Art, Religion and Science*. London: Phoenix House.

Mithen, S. (2006). Ethnobiology and the evolution of the human mind. *Journal of the Royal Anthropological Institute,* **12**(Special Issue 2006), 45–62.

Mithen, S. (2007). *Singing Neanderthals: the Origin of Music, Language, Mind and Body*. Cambridge, MA: Harvard University Press.

Mittelstrass, J. (1996). Transdisziplinarität. In J. Mittelstrass (ed.). *Enzyklopädie Philosophie und Wissenschaftstheorie*, Vol 4. Stuttgart: Metzler.

Moebius, K. (1877). *Die Auster und die Austernwirtschaft*. Berlin: Wiegundt, Hempel and Parey.

Moerman, D. E. & Jonas, W. B. (2002). Deconstructing the placebo effect and finding the meaning response. *Annals of Internal Medicine,* **136**(6), 471–6.

Mol, A. P. J. (2006). From environmental sociologies to environmental sociology. *Organization & Environment,* **19**(1), 5–27.

Mol, A. P. J. (2009). Environment and modernity in transitional China: frontiers of ecological modernization. In A. P. J. Mol, D. A. Sonnenfeld & G. Spaargaren (eds). *The Ecological Modernization Reader: Environmental Reform in Theory and Practice*. London: Routledge, pp. 456–97.

Mol, A. P. J. & Jänicke, M. (2009). The origins and theoretical foundations of ecological modernisation theory. In A. P. J. Mol, D. A. Sonnenfeld & G. Spaargaren (eds). *The Ecological Modernization Reader: Environmental Reform in Theory and Practice*. London: Routledge, pp. 18–27.

Mol, A. P. J., Spaargaren, G. & Sonnenfeld, D. A. (2009). Ecological modernisation: three decades of policy, practice and theroretical reflection. In A. P. J. Mol, D. A. Sonnenfeld & G. Spaargaren (eds). *The Ecological Modernization Reader: Environmental Reform in Theory and Practice*. London: Routledge, pp. 3–14.

Molden, D., International Water Management Institute, & Comprehensive Assessment of Water Management in Agriculture (Program). (2007). *Water for Food, Water for Life: a Comprehensive Assessment of Water Management in Agriculture*. London; Sterling, VA: Earthscan.

Molinari, N. A. M., Ortega-Sanchez, I. R., Messonnier, M. L. et al. (2007). The annual impact of seasonal influenza in the US: measuring disease burden and costs. *Vaccine,* **25**(27), 5086–96.

Molleson, T. (1994). The eloquent bones of Abu-Hureyra. *Scientific American,* **271**(2), 70–5.

Montgomery, D. R. (2007). *Dirt: The Erosion of Civilizations*. Berkeley, CA: University of California Press.

Montoro, J., Mullol, J., Jauregui, I. et al. (2009). Stress and allergy. *Journal of Investigational Allergology and Clinical Immunology,* **19**(Suppl. 1), 40–7.

Moore, A. M. T., Hillman, G. C. & Legge, A. J. (2000). *Village on the Euphrates: from Foraging to Farming*. Oxford: Oxford University Press.

Morange, M. (1998). *A History of Molecular Biology*. Cambridge, MA: Harvard University Press.

Morange, M. (2006). What history tells us – VII. Twenty-five years ago: the production of mouse embryonic stem cells. *Journal of Biosciences,* **31**(5), 537–41.

Morecroft, J. D. W. & Sterman, J. (1994). *Modeling for Learning Organizations*. Portland, OH: Productivity Press.

Morgan, T. H. (1910). Sex limited inheritance in Drosophila. *Science,* **32**(812), 120–2.

Morrison, L. G. & Yardley, L. (2009). What infection control measures will people carry out to reduce transmission of pandemic influenza? A focus group study. *BMC Public Health,* **9**, http://www.biomedcentral.com/1471-2458/9/258

Morrow, P. E. (1980). Physics of airborne particles and their deposition in the lung. In R. B. Kundsin (ed.). *Annals of the New York Academy of Sciences: Airborne Contagion*. New York, NY (NY): New York Academy of Sciences, pp. 71–80.

Morwood, M. J., Brown, P., Jatmiko, S. T. et al. (2005). Further evidence for small-bodied hominids from the Late Pleistocene of Flores, Indonesia. *Nature,* **437**(7061), 1012–17.

Moser, C., Stauffacher, M., Krütli, P. & Scholz, R. W. (submitted). Interdisciplinary collaboration in nuclear waste management. *The crucial role of nomothetic and idiographic conceptions of time*.

Mosler, H. J. & Brucks, W. M. (2003). Integrating commons dilemma findings in a general dynamic model of cooperative behavior in resource crises. *European Journal of Social Psychology,* **33**(1), 119–33.

Moss, H. & Damasio, A. R. (2001). Emotion, cognition, and the human brain. *Annals of the New York Academy of Sciences,* **935**, 98–100.

Mossong, J., Hens, N., Jit, M. et al. (2008). Social contacts and mixing patterns relevant to the spread of infectious diseases. *PLoS Medicine,* **5**(3), 381–91.

Mottron, L., Dawson, M., Soulieres, I., Hubert, B. & Burack, J. (2006). Enhanced perceptual functioning in autism: an update, and eight principles of autistic perception. *Journal of Autism and Developmental Disorders,* **36**(1), 27–43.

Mukerji, C. (2007). Stewardship politics and the control of wild weather: levees, seawalls, and state building in 17th-century France. *Social Studies of Science,* **37**(1), 127–33.

Mulder, K. (2009). Technology for environmental problems. In J. Boersma & L. Reinders (eds). *Principles of Environmental Sciences*. Heidelberg: Springer, pp. 305–34.

Müller, E. & Stierlin, H. R. (1990). *Sanasilva-Kronenbilder mit Nadel- und Blattverlustprozenten*. Birmensdorf: Eidgenössische Forschungsanstalt für Wald, Schnee und Landschaft WSL.

Müller, F. M. (1888). *Natural Religion: the Gifford Lectures delivered before the University of Glasgow in 1888*. London: Longmans, Green.

Müller, G. H. (2001). Umwelt. In J. Ritter, K. Gründer & G. Gabriel (eds). *Historisches Wörterbuch der Philosophie*, Vol. 11. Basel: Schwabe, pp. 99–105.

Müller-Wenk, R. (2004). A method to include in LCA road traffic noise and its health effects. *International Journal of Life Cycle Assessment,* **9**(2), 76–85.

Mumpower, J., Menkes, J. & Covello, T. (eds). (1986). *Risk Evaluation and Management*. New York, NY: Plenum.

Mun, T. (1664). *England's Treasure by Forraign Trade*. London: Clark.

Murdock, G. P. (1967). Ethnographic atlas: a summary. *Ethnology,* **6**(2), 109–236.

Murphy, J. & Gouldson, A. (2009). Environmental policy and industrial innovation: integrating environmental and economy through ecological modernization. In

References

A. P. J. Mol, D. A. Sonnenfeld & G. Spaargaren (eds). *The Ecological Modernization Reader: Environmental Reform in Theory and Practice*. London: Routledge, pp. 275–94.

Murray, H. A. (1938). *Explorations in Personality*. New York, NY: Oxford University Press.

Musher, D. (2003). How contagious are common respiratory tract infections? *New England Journal of Medicine,* **348**, 1256–66.

Myers, N. & Kent, J. (2003). New consumers: the influence of affluence on the environment. *Proceedings of the National Academy of Sciences of the United States of America,* **100**(8), 4963–8.

Nabielek, R. (2004). Biologische Kenntnisse und Überlieferungen im Mittelalter (4–15. Jh). In I. Jahn (ed.). *Geschichte der Biologie: Theorien, Methoden, Institutionen, Kurzbiografien*, 3rd edn. Hamburg: Nikol, Verlag, pp. 88–160.

Naess, A. (1989). *Ecology, Community, and Lifestyle*. Cambridge: Cambridge University Press.

Najam, A. (2003). The case against a new international environmental organization. *Global Governance,* **9**(3), 367–84.

Nanney, D. I. (1959). Epigenetic factors affecting mating type expressions in certain ciliates. *Cold Spring Harbor Symposia on Quantitative Biology,* **23**, 327–35.

Nash, J. F. (1951). Non-cooperative games. *Annals of Mathematics,* **54**, 286–95.

National Academy of Sciences. (2008). Stem cells. Retrieved April 20, 2008, from http://dels.nas.edu/bls/stemcells/about.shtml

National Aeronautics and Space Administration (NASA). (2001). Crop circles in Kansas. Retrieved October 5, 2008, from http://earthobservatory.nasa.gov/Newsroom/NewImages/images.php3?img_id=17006

National Science Teachers Association (NSTA). (2007). *Resources for Environmental Literacy*. Arlington, VA: NSTA.

Nationale Genossenschaft für die Lagerung radioaktiver Abfälle (Nagra). (2002). *Project Opalinus Clay. Safety Report. Demonstration of Disposal Feasibility of Spent Fuel, Vitrified High-Level Waste and Long-Lived Intermediate-Level Waste. NTB 02–05*. Wettingen: Nagra.

Nattrass, B. & Altomare, M. (2001). *The Natural Step for Business. Wealth, Ecology and the Evolutionary Corporation*, 2nd edn. Gabriola Islands: New Society Publishers.

Naveh, Z. (1988). Biocybernetic perspectives of landscape ecology and management. In M. R. Moss (ed.). *Landscape Ecology and Management*. Montreal: Polyscience, pp. 23–34.

Naveh, Z. & Liebermann, A. S. (1994). *Landscape Ecology: Theory and Application*, 2nd edn. New York, NY: Springer.

Neisser, U. (1967). *Cognitive Psychology*. New York, NY: Appleton-Century Crofts.

Nellisen, N. J. M. (1970). *Sociale Ekologie*. Nijmegen: Catholic University of Nijmegen.

Nelson, R. R. & Winter, S. G. (1982). *An Evolutionary Theory of Economic Change*. Cambridge, MA: Harvard University Press.

Nerb, J., Spada, H. & Ernst, A. (1997). A cognitive model of agents in a commons dilemma. In M. Shafto & P. Langley (eds). *Proceedings of the Nineteenth Annual Conference of the Cognitive Science Society (1997)*. Mahwah, NJ: Lawrence Erlbaum Associates, pp. 560–5.

Netherlands Environmental Assessment Agency (PBL). (2008). China now no. 1 in CO_2 emissions; USA in second position. Retrieved November 11, 2008, from http://www.mnp.nl/en/dossiers/Climatechange/moreinfo/Chinanowno1inCO2emissionsUSAinsecondposition.html

Neto, F. (1995). Conformity and independence revisited. *Social Behavior and Personality,* **23**(3), 217–22.

Neumayer, E. (2003). *Weak versus Strong Sustainability*. Cheltenham: Edward Elgar.

Newig, J. (2005). Erleichtert Öffentlichkeitsbeteiligung die Umsetzung (umwelt-)politischer Massnahmen? Ein Modellansatz zur Erklärung der Implementationsfähigkeit. In P. H. Feindt & J. Newig (eds). *Partizipation. Öffentlichkeitsbeteiligung, Nachhaltigkeit*. Marburg: Metropolis-Verlag, pp. 89–116.

Newig, J. (2007). Symbolic environmental legislation and societal self-deception. *Environmental Politics,* **16**(2), 276–96.

Newlands, D. (2003). Competition and cooperation in industrial clusters: the implications for public policy. *European Planning Studies,* **11**, 521–32.

Newman, E. I. (1997). Phosphorus balance of contrasting farming systems, past and present. Can food production be sustainable? *Journal of Applied Ecology,* **34**(6), 1334–47.

Newman, O. (1972). *Defensible Space: People and Design in the Violent City*. New York, NY: Macmillan.

Nicas, M. & Gang, S. (2006). An integrated model of infection risk in a health-care environment. *Risk Analysis,* **26**(4), 1085–96.

Nicolas, C. L., Abramson, C. I. & Levin, M. (2008). Analysis of behavior in the planarian model. In R. B. Raffa & S. M. Rawls (eds). *Planaria: a Model for Drug Action and Abuse*. Austin, TX: Landes Bioscience, pp. 83–94.

Nicolescu, B. (2002). *Manifest of Transdisciplinarity*. Albany: State University of New York.

Nicolis, G. & Prigogine, I. (1989). *Exploring Complexity: an Introduction*. New York, NY: Freeman.

Nietzsche, F. (ed.). (1889/1969). *Götzen-Dämmerung, oder wie man mit dem Hammer philosophiert*. Vol. 3. Berlin: de Gruyter.

Nigh, R. (2002). On Maya medicine and the biomedical gaze. *Current Anthropology,* **43**(5), 451–77.

Nisbet, E. G. & Sleep, N. H. (2001). The habitat and nature of early life. *Nature,* **409**(6823), 1083.

Nisbett, R. E. & Masuda, T. (2003). Culture and point of view. *Proceedings of the National Academy of Sciences of the United States of America,* **100**(19), 11163–70.

Nisbett, R. E. & Ross, L. (1980). *Human Inference: Strategies and Shortcomings of Social Judgement*. Englewood Cliffs, NJ: Prentice-Hall.

Nitecki, M. H. (1984). *Extinctions*. Chicago, IL: University of Chicago Press.

Noddack, W. (1929). Die Herstellung von einem Gramm Rhenium. *Zeitschrift für anorganische und allgemeine Chemie,* **183**(1), 353–75.

Nolan, P. & Lenski, G. (1998). *Human Societies: An Introduction to Macrosociology*, 8th edn. New York, NY: McGraw-Hill.

Nordenskiöld, E. (1928). *The History of Biology*. New York, NY: Tudor.

Norgaard, R. B. (1984). Coevolutionary agricultural development. *Economic Development and Cultural Change,* **32**(3), 525.

Norman, A. P. (1991). Telling it like it was – historical narratives on their own terms. *History and Theory,* **30**(2), 119–35.

Norris, F. H., Friedman, M. J., Watson, P. J. et al. (2002). 60,000 disaster victims speak: part I. An empirical review of the empirical literature, 1981–2001. *Psychiatry-Interpersonal and Biological Processes,* **65**(3), 207–39.

Norris, F. H., Stevens, S. P., Pfefferbaum, B., Wyche, K. F. & Pfefferbaum, R. L. (2008). Community resilience as a metaphor, theory, set of capacities, and strategy for disaster readiness. *American Journal of Community Psychology,* **41**(1–2), 127–50.

North, D. C. (1990). *Institutions, Institutional Change and Economic Performance*. Cambridge: Cambridge University Press.

North, D. C. (1994). Economic-performance through time. *American Economic Review,* **84**(3), 359–68.

North, D. C. (1995). The new institutional economics and their development. In J. Hariss, J. Hunter & V. L. Lewis (eds). *The New Institutional Economics And Their Development*. London: Routledge, pp. 17–26.

North, D. C. (2005). Economic sociology and new institutional economics. In C. Menard & M. M. Shirley (eds). *Handbook of New Institutional Economics*. Dordrecht: Springer, pp. 21–31.

Norton, R. D. & Food and Agriculture Organization of the United Nations. (2004). *Agricultural Development Policy: Concepts and Experiences*. Chichester: John Wiley & Sons.

Nourse, E. G. (1941). The meaning of "price policy". *The Quarterly Journal of Economics,* **55**(2), 175–209.

Novikoff, A. B. (1945). The concept of integrative levels and biology. *Science,* **101**(2618), 209–15.

Nowotny, H., Scott, P. & Gibbons, M. (2001). *Rethinking Science – Knowledge and the Public in an Age of Uncertainty*. London: Polity Press.

Nunn, J. F. (1996). *Ancient Egyptian Medicine*. Norman, OK: University of Oklahoma Press.

Ny, H., MacDonald, J. P., Broman, G., Yamamoto, R. & Robert, K. H. (2006). Sustainability constraints as system boundaries – an approach to making life-cycle management strategic. *Journal of Industrial Ecology,* **10**(1–2), 61–77.

O'Brian, M. (2000). Have lessons been learned from the UK bovine spongiform encephalopathy (BSE) epidemic? *International Journal of Epidemiology,* **29**(4), 730–3.

O'Mahony, M. (1978). Smell illusions and suggestions: reports of smells contingent to tones played on television and radio. *Chemical Senses and Flavor,* **3**(2), 183–9.

O'Neill, R. V. (2001). Is it time to bury the ecosystem concept? (With full military honors of course!). *Ecology,* **82**(12), 3275–84.

Odum, E. P. (1953). *Fundamentals of Ecology*. Philadelphia, PA: Saunders.

Odum, E. P. (1969). Strategy of ecosystem development. *Science,* **164**(3877), 262–70.

Odum, E. P. (1973). *Fundamentals of Ecology*, 3rd edn. Philadelphia, PA: Saunders.

Odum, E. P. (1983). *Basic Ecology*. Philadelphia, PA: Saunders.

Odum, E. P. (1992). Great ideas in ecology for the 1990s. *Bioscience,* **42**(7), 542–5.

Odum, E. P. (1996). *Ecology: a Bridge Between Science and Society*. Sunderlang, MA: Sinauer Associates.

Odum, H. T. (1957). Trophic structure and productivity of Silver Springs, Florida. *Ecological Monographs,* **27**(1), 55–112.

Odum, H. T. (1983). *Systems Ecology*. New York, NY: Wiley.

Odum, H. T. (1994). *Ecological and General Systems. An Introduction to Systems Ecology*. Niwot, CO: University Press of Colorado.

References

Odum, H. T. & Odum, E. C. (1981). *Energy Basis for Man and Nature*, 2nd edn. New York, NY: McGraw Hill.

Olby, R. (1974). *The Path to the Double Helix*. London: Macmillan.

Oliva, C. (1998). *Belastungen der Bevölkerung durch Flug- und Straßenlärm*. Berlin: Duncker & Humblot.

Oliver, P. E. & Marwell, G. (2001). Whatever happened to Critical Mass Theory? A retrospective and assessment. *Sociological Theory,* **19**(3), 292–311.

Oliviera, S. & Stewart, D. E. (2006). *Writing Scientific Software: a Guide to Good Style*. Cambridge: Cambridge University Press.

Olson, M. (1965a). *The Logic of Collective Action*. Cambridge, MA: Harvard University Press.

Olson, M. (1965b). *The Theory of Collective Action: Public Goods and the Theory of Groups*. Cambridge, MA: Harvard University Press.

Oppenheim, L. (1977). *Ancient Mesopotamia*. Chicago, IL: University of Chicago Press.

Organisation for Economic Co-operation and Development (OECD). (1993). *OECD Core Set of Indicators for Environmental Performance Reviews*. Paris: OECD.

Orians, G. H. (1980). Habitat selection. In J. S. Lockhard (ed.). *The Evolution of the Human Social Behavior*. New York, NY: Elsevier Science Ltd, pp. 49–66.

Osborne, M. J. (2004). *An Introduction to Game Theory*. Oxford: Oxford University Press.

Osborne, M. J. & Rubinstein, A. (1994). *A Course in Game Theory*. Cambridge, MA: MIT Press.

Ostrom, E. (1990). *Governing the Commons: the Evolution of Institutions for Collective Action*. Cambridge: Cambridge University Press.

Ostrom, E. (1998). A behavioral approach to the rational choice theory of collective action. Presidential address, American Political Association, 1997. *American Political Science Review,* **92**(1), 1–22.

Ostrom, E. (1999). Coping with tragedies of the commons. *Annual Review of Political Science,* **2**, 493–535.

Ostrom, E. (2005). Doing institutional analysis: digging deeper than markets and hierachies. In C. Menard & M. M. Shirley (eds). *Handbook of New Institutional Economics*. Dordrecht: Springer, pp. 819–47.

Ostrom, E. (2007). A diagnostic approach for going beyond panaceas. *Proceedings of the National Academy of Sciences of the United States of America,* **104**(39), 15181–7.

Ostrom, E. (2009). A general framework for analyzing sustainability of social-ecological systems. *Science,* **325**(5939), 419–22.

Ostrom, E., Burger, J., Field, C. B., Norgaard, R. B. & Policansky, D. (1999). Revisiting the commons: local lessons, global challenges. *Science,* **284**(5412), 278–82.

Ostrom, E., Gardner, R. & Walker, J. M. (1994). *Rules, Games, and Common-Pool Resources*. Ann Arbor, MI: University of Michigan Press.

Ostrom, E., Janssen, M. A. & Anderies, J. M. (2007). Going beyond panaceas. *Proceedings of the National Academy of Sciences of the United States of America,* **104**(39), 15176–8.

Ostrom, E., Walker, J. & Gardner, R. (1992). Covenants with and without a sword – self-governance is possible. *American Political Science Review,* **86**(2), 404–17.

Ott, H. (1995). Merkantilismus. In Görres-Gesellschaft (ed.). *Staatslexikon*, Vol. 3. Freiburg: Herder, pp. 1121–3.

Ottino, J. M. (2004). Engineering complex systems. *Nature,* **427**(6973), 399.

Ou, C. H. & Wu, J. H. (2006). Spatial spread of rabies revisited: influence of age-dependent diffusion on nonlinear dynamics. *Siam Journal on Applied Mathematics,* **67**(1), 138–63.

Oughton, D. H. & Strand, P. (2004). The Oslo Consensus Conference on Protection of the Environment. *Journal of Environmental Radioactivity,* **74**(1–3), 7–17.

Owen, G. (2007, February 9). Who invented the environment? *The National Post*, p. A15.

Owen, R. (1817). *Observation on the Effect of the Manufacturing System: with Hints for the Improvement of those Parts of it which are most Injurious to Health and Morals*, 2nd edn. London: Longman.

Paavola, J. (2007). Institutions and environmental governance: a reconceptualization. *Ecological Economics,* **63**(1), 93–103.

Paavola, J. & Adger, W. N. (2005). Institutional ecological economics. *Ecological Economics,* **53**(3), 353–68.

Pahl-Wostl, C. (2009). A conceptual framework for analysing adaptive capacity and multi-level learning processes in resource governance regimes. *Global Environmental Change: Human and Policy Dimensions,* **19**(3), 354–65.

Pahl-Wostl, C., Craps, M., Dewulf, A. et al. (2007). Social learning and water resources management. *Ecology and Society,* **12**(2), 4. http://www.ecologyandsociety.org/vol12/iss2/art4/

Pahl-Wostl, C., Sendzimir, J., Jeffrey, P. et al. (2007). Managing change toward adaptive water management through social learning. *Ecology and Society,* **12**(2), 30.

Paich, M. & Sterman, J. D. (1993). Boom, bust, and failures to learn in experimental markets. *Management Science,* **39**(12), 1439–58.

Pain, R. & Francis, P. (2003). Reflections on participatory research. *Area*, **35**(1), 46–54.

Panayotou, T. (1993). Empirical tests and policy analysis of environmental degradation at different stages of economic development. *Working Paper WP 238.* Geneva: International Labour Office.

Pandit, B. H. & Thapa, G. B. (2003). A tragedy of non-timber forest resources in the mountain commons of Nepal. *Environmental Conservation*, **30**(3), 283–92.

Parayil, G. & Tong, F. (1998). Pasture-led to logging-led deforestation in the Brazilian Amazon: the dynamics of socio-environmental change. *Global Environmental Change: Human and Policy Dimensions*, **8**(1), 63–79.

Pareto, V. (1909). *Manuel d'Economie Politique*. Paris: Giard et Brière.

Park, P. (1999). People, knowledge, and change in participatory research. *Management Learning*, **30**(2), 141–57.

Park, R. E. (1925/1967). The city: suggestions for the investigation of human behavior in the urban environment. In R. E. Park, E. W. Burgess & R. D. McKenzie (eds). *The City*. Chicago, IL: University of Chicago Press, pp. 1–46.

Park, R. E. (1936). Human ecology. *American Journal of Sociology*, **42**(1), 1–15.

Park, R. E. & Burgess, E. W. (1921). *Introduction to Science of the Sociology*. Chicago, IL: University of Chicago Press.

Park, T. (1946). Some observations on the history and scope of population ecology. *Ecological Monographs*, **16**(4), 313–20.

Parker, D., Manson, S., Janssen, M., Hoffmann, M. & Deadman, P. (2003). Multi-agent systems for the simulation of land-use and land-cover change: a review. *Annals of the Association of American Geographers*, **93**(2), 314–37.

Parker, S. T., Mitchell, R. W. & Boccia, M. L. (2006). *Self-Awareness in Animals and Humans: Developmental Perspectives*. Cambridge: Cambridge University Press.

Parsons, R. (1995). Conflict between ecological sustainability and environmental aesthetics – conundrum, canard or curiosity. *Landscape and Urban Planning*, **32**(3), 227–44.

Parsons, R. & Daniel, T. C. (2002). Good looking: in defense of scenic landscape aesthetics. *Landscape and Urban Planning*, **60**(1), 43–56.

Parsons, T. (1937/1949). *The Structure of Social Action: a Study in Social Thought with Special Reference to a Group of recent European Writers*. Glencoe, IL: The Free Press.

Parsons, T. (1951). *The Social System*. New York, NY: The Free Press.

Parsons, T. (1961). General theory of sociology. In R. K. Merton, L. Broom & L. S. Cottrell (eds). *Sociology Today: Problems and Prospects*. New York, NY: Basic Books, pp. 3–38.

Parsons, T. (1977). *Social Systems and the Evolution of Action Theory*. New York, NY: The Free Press.

Parsons, T. & Shils, E. A. (1951/2001). *Toward a General Theory of Action: Theoretical Foundations of the Social Sciences*. New Brunswick, NJ: Transaction.

Parsons, T., Shils, E. A., Allport, G. W. *et al.* (1951). Some fundamental categories of the theory of action: a general statement. In T. Parsons & E. A. Shils (eds). *Toward a General Theory of Action: Theoretical Foundations for the Social Sciences*. Cambridge, MA: Harvard University Press, pp. 3–29.

Pattee, H. H. (1973a). The physical basis and origin of hierarchical control. In H. D. Pattee (ed.). *Hierarchy Theory: the Challenge of Complex Systems*. New York, NY: George Braziller, pp. 71–108.

Pattee, H. H. (1973b). Postscript. Unsolved problems and potential applications of hierarchy theory. In H. D. Pattee (ed.). *Hierarchy Theory: the Challenge of Complex Systems*. New York, NY: George Braziller, pp. 131–56.

Patten, B. C. (1978). Systems ecology: a course sequence in mathematical ecology. *Bioscience*, **16**(9), 593–8.

Pauli, W. (1994). *General Principles of Quantum Mechanics*. Berlin: Springer.

Pauling, L. (1931). The nature of the chemical bond. Application of results obtained from the quantum mechanics and from a theory of paramagnetic susceptibility to the structure of molecules. *Journal of the American Chemical Society*, **53**(4), 1367–400.

Paustenbach, D. J. (ed.). (1989). *The Risk Assessment of Environmental and Human Health Hazards: a Textbook of Case Studies*. New York, NY: John Wiley & Sons, Inc.

Paustenbach, D. J. (ed.). (2002). *Human and Ecological Risk Assessment. Theory and Practice* (Foreword). New York, NY: John Wiley & Sons, Inc.

Pavé, A. (2006). Biological and ecological systems hierarchical organisation. In D. Pumain (ed.). *Hierarchy in Natural and Social Sciences*. Dordrecht: Springer, pp. 39–70.

Paxton, R. A., Stirling, J. M. & Fleming, G. (2000). Regeneration of the Forth & Clyde and Union canals, Scotland. *Proceedings of the Institution of Civil Engineers-Civil Engineering*, **138**(2), 61–72.

Peacock, W. G., Morrow, B. H. & Gladwin, H. (eds). (1997). *Hurricane Andrew: Ethnicity, Gender and the Sociology of Disasters*. New York, NY: Routledge.

Pearce, D. W. (1992). *The MIT Dictionary of Modern Economics*, 4th edn. Cambridge, MA: The MIT Press.

Pearce, D. W. (2002). An intellectual history of environmental economics. *Annual Review of Energy and the Environment*, **27**, 57–81.

Pearce, D. W. & Turner, K. R. (1990). *Economics of Natural Resources and the Environment*. Hertfordshire: Harvester Wheatsheaf.

Pellet, F. (1938). *History of American Beekeeping*. Ames, IA: Collegiate Press.

Pepper, D. (1984). *The Roots of Modern Environmentalism*. London: Croom Helm.

Perman, R., Ma, Y., McGivray, J. & Common, M. (2003). *Natural Resource and Environmental Economics*. Harlow: Pearson.

Permanent Service for Mean Sea Level (PSMSL). (2009). Data. Retrieved September 20, 2009, from http://www.psmsl.org/data/

Perrin, S. & Spencer, C. (1981). Independence or conformity in the Asch paradigm as a reflection of cultural or situational factors. *British Journal of Social Psychology*, **20**, 205–10.

Perugini, M., Callucci, M., Presaghi, F. & Ercolani, A. P. (2003). The personal norm of reciprocity. *European Journal of Personality*, **17**(4), 251–83.

Peters, A., Mueller, M. G., de Haan, P. & Scholz, R. W. (2008). Feebates promoting energy-efficient cars: design options to address more consumers and possible counteracting effects. *Energy Policy*, **36**(4), 1355–65.

Peters, A., Gutscher, H. & Scholz, R. W. (2011). Psychological determinants of fuel consumption of purchased new cars. *Transportation Research Part F*, **14**, 229–39.

Peterson, C. R. & Beach, L. R. (1967). Man as an intuitive statistician. *Psychological Bulletin*, **68**(1), 29–46.

Petschow, U., Meyerhoff, J. & Thomasberger, C. (1990). *Umweltreport DDR*. Frankfurt am Main: Fischer.

Petty, E. M., Cacioppo, J. T., Strathmann, A. J. & Priester, J. R. (2005). To think is not to think: exploring two routes to persuasion. In T. D. Brock & M. C. Green (eds). *Psychological Insights and Perspectives*, 2nd edn. Thousand Oaks, CA: Sage.

Petty, W. (1662). *A Treatise of Taxes & Contributions. Shewing the Nature and Measures of Crown-Lands. Assessements. Customs. Pollmoneys. Lotteries. Benevolence*. London: printed for N. Brooke.

Pezdek, K. (2003). Event memory and autobiographical memory for the events of September 11, 2001. *Applied Cognitive Psychology*, **17**(9), 1033–45.

Pfeiffer, T., Schuster, S. & Bonhoeffer, S. (2001). Cooperation and competition in the evolution of ATP-producing pathways. *Science*, **292**(5516), 504–7.

Phelps, N. A. & Ozawa, T. (2003). Contrasts in agglomeration: proto-industrial, industrial and post-industrial forms compared. *Progress in Human Geography*, **27**, 583–604.

Philips, L. (2001). Lessons from the BSE inquiry. *FST Journal*, **17**, 3–5.

Phillips, L. D. (1984). A theory of requisite decision models. *Acta Psychologica*, **56**(1–3), 29–48.

Piaget, J. (1907). Le rameau de sapin. *Organe du Club Jurassien – Neuchâtel*, **41**, 36.

Piaget, J. (1923). *Le Langage et la Pensée chez l'Enfant*. Paris: Delachaux & Niestlé.

Piaget, J. (1924). *Le Jugement et le Raisonnement chez l'Enfant*. Paris: Delachaux & Niestlé.

Piaget, J. (1946). *Le Développement de la Notion de Temps chez l'Enfant*. Paris: Presses Universitaires de France.

Piaget, J. (1947). *La Psychologie de l'Intelligence*. Paris: Colin.

Piaget, J. (1950). *Introduction à l'Épistémologie Génétique, tomes I–III*. Paris: Presses Universitaires de France.

Piaget, J. (1968). *Le Structuralisme*. Paris: Presses Universitaires de France.

Piaget, J. (1970). *Genetic Epistemology*. New York, NY: Columbia University Press.

Piaget, J. (1973). *Die Entwicklung des Erkennens III*. Stuttgart: Ernst Klett.

Piaget, J. (1977). *The Development of Thought*. New York, NY: Viking.

Piaget, J. & Inhelder, B. (1950/1977). *Die Psychologie des Kindes*. Frankfurt am Main: Fischer.

Piaget, J. & Inhelder, B. (1956). *The Child's Conception of Space*. London: Routledge and Kegan.

Pianka, E. R. (2000). *Evolutionary Ecology*. San Francisco, CA: Benjamin Cummings.

Pickering, A. (ed.). (1992). *Science as Practice and Culture*. Chicago, IL: University of Chicago Press.

Pickett, S. T. A. & Cadenasso, M. L. (2008). Linking ecological and built components of urban mosaics: an open cycle of ecological design. *Journal of Ecology*, **96**(1), 8–12.

Pickett, S. T. A., Cadenasso, M. L., Grove, J. M. *et al.* (2001). Urban ecological systems: linking terrestrial ecological, physical, and socioeconomic components of metropolitan areas. *Annual Review of Ecology and Systematics*, **32**, 127–57.

Piek, J., Lidke, G. & Terberger, T. (2008). Ancient trephinations in neolithic people – evidence for stone age neurosurgery? Nature Proceedings. Retrieved January 8, 2009, from http://hdl.handle.net/10101/npre.2008.1615.1

Pielou, E. C. (1969). *An Introduction to Mathematical Ecology*. New York, NY: John Wiley & Sons.

Piepmeier, R. (1980). Landschaft. In J. Ritter & K. Gründer (eds). *Historisches Wörterbuch der Philosophie*, Vol. 5. Basel: Schwabe, pp. 11–27.

Pierson, P. (2000). The limits of design: explaining institutional origins and change. *Governance – an International Journal of Policy and Administration*, **13**(4), 475–99.

Pigou, A. C. (1932). *The Economics of Welfare*, 3rd edn. London: Macmillan.

Pinchot, G. (1910). *The Fight for Conservation*. New York, NY: Doubleday.

Pinker, S. (2000). *The Language Instinct: how the Mind Creates Language*. New York, NY: Harper Collins.

Pitman, A. J., de Noblet-Ducoudre, N., Cruz, F. T. *et al.* (2009). Uncertainties in climate responses to past land cover change: first results from the LUCID intercomparison study. *Geophysical Research Letters*, **36**(L14814).

Pivo, G. & McNamara, P. (2005). Responsible property investing. *International Real Estate Review*, **8**(1), 128–43.

Plotkin, S. L. & Plotkin, S. A. (2008). A short history of vaccination. In S. A. Plotkin, P. A. Offit & W. A. Orenstein (eds). *Vaccines*, 5th edn. Amsterdam: Elsevier, pp. 1–16.

Poggi, G. (2000). *Durkheim*. Oxford: Oxford University Press.

Poggi, G. (2005). Classical social theory, III: Max Weber and Georg Simmel. In A. Harrington (ed.). *Modern Social Theory: an Introduction*. Oxford: Oxford University Press, pp. 63–86.

Pohl, C. (2008). From science to policy through transdisciplinary research. *Environmental Science & Policy*, **11**(1), 46–53.

Poincaré, H. (1892). *Les Méthodes Nouvelles de la Mécanique Céleste*, Vol. I. Paris: Gauthier-Villars.

Polanyi, M. (1968). Life's irreducible structure. *Science*, **160**(3834), 1308–11.

Pollard, S. (1980). A new estimate of British coal production, 1750–1850. *Economic History Review*, **33**(2), 212–35.

Polsky, C., Neff, R. & Yarnal, B. (2007). Building comparable global change vulnerability assessments: the vulnerability scoping diagram. *Global Environmental Change*, **17**(3–4), 472–85.

Pongratz, J. T., Raddatz, C. H., Reick, M., Esch, M. & Claussen, M. (2009). Radiative forcing from anthropogenic land cover change since A.D. 800. *Geophysical Research Letters*, **36**(2), (L02709).

Poortinga, W., Steg, L. & Vlek, C. (2002). Environmental risk concern and preferences for energy-saving measures. *Environment and Behavior*, **34**(4), 455–78.

Poortinga, W., Steg, L. & Vlek, C. (2004). Values, environmental concern, and environmental behavior – a study into household energy use. *Environment and Behavior*, **36**(1), 70–93.

Popper, K. R. (1935/2005). *Logik der Forschung*. Tübingen: Mohr Siebeck.

Popper, K. R. (1959). *The Logic of Scientific Discovery*. London: Hutchinson.

Popper, K. R. & Eccles, J. C. (1977). *The Self and its Brain*. Heidelberg: Springer.

Porter, M. E. (1998). Clusters and the new economics of competition. *Harvard Business Review*, **76**(6), 77–90.

Posch, A. & Perl, E. (2007). Regionale Verwertungsnetze und industrielle Symbiosen. In R. Isenmann & M. von Hauff (eds). *Industrial Ecology: Mit Ökologie zukunftsorientiert wirtschaften*. Heidelberg: Elsevier, pp. 265–78.

Prentice, D. A. (2000). Values. In A. E. Kazdin (ed.). *Encyclopedia of Psychology*, Vol. 8. New York, NY: Oxford University Press, pp. 153–6.

Price, M. E., Cosmides, L. & Tooby, J. (2002). Punitive sentiment as an anti-free rider psychological device. *Evolution and Human Behavior*, **23**(3), 203–31.

Prohaska, S. J. & Stadler, P. F. (2008). Genes. *Theory in Biosciences*, **127**(3), 215–21.

Proshansky, H. M. & Fabian, A. (1986). Psychological aspects of the quality of urban life. In D. Frick (ed.). *The Quality of Urban Life*. Berlin: de Gruyter, pp. 19–29.

Proshansky, H. M., Ittelson, W. H., Rivlin, L. & Winkel, G. (eds). (1970). *An Introduction to Environmental Psychology: Man and his Physical Setting*. New York, NY: Holt, Rinehart and Winston.

Prusiner, S. B. (1998). Prions. *Proceedings of the National Academy of Sciences*, **95**, 13363–83.

Pumain, D. (ed.). (2006). *Hierarchy in Natural and Social Sciences*, Vol. 3. Dordrecht: Springer.

Purvis, A. & Hector, A. (2000). Getting the measure of biodiversity. *Nature*, **405**(6783), 212–19.

Quesnay, F. (1768). *La Physiocratie*, Vols 1 & 2. Paris: DuPont.

Quinn, G. (1994). *Legal Theory and the Casebook Method of Instructions in the United States*. Munich: Hampp.

Rabe, H. (1984). Ökonomie. In J. Ritter & K. Gründer (eds). *Historisches Wörterbuch der Philosophie*, Vol. 6. Basel: Schwabe, pp. 1149–53.

Rádl, E. (1905). *Geschichte der biologischen Theorie seit dem Ende des siebzehnten Jahrhunderts*. Leipzig: Engelmann.

Radtke, P. (1998). Zeit und Raum aus der Sicht behinderter Menschen. *Behinderte in Familie, Schule und Gesellschaft*, 2/1998, 11–19.

References

Rajagopal, D., Sexton, S. E., Roland-Host, D. & Zilberman, D. (2007). Challenge of biofuel: filling the tank without emptying the stomach. *Environmental Research Letters*, **2**, 1–9.

Rajkumar, A. P., Premkumar, T. S. & Tharyan, P. (2008). Coping with the Asian tsunami: perspectives from Tamil Nadu, India on the determinants of resilience in the face of adversity. *Social Science & Medicine*, **67**(5), 844–53.

Rammstedt, O. (1992). Risiko. In J. Ritter & K. Gründer (eds). *Historisches Wörterbuch der Philosophie*, Vol. 8. Basel: Schwabe, pp. 1045–50.

Rapoport, A. (1970). *N-Person Game Theory: Concepts and Applications*. Ann Arbor, MI: University of Michigan Press.

Rapoport, A. (1996). Der systemische Ansatz der Umweltsoziologie. *Kölner Zeitschrift für Soziologie und Sozialpsychologie*, **36**(special issue), 61–88.

Rapoport, A. (1999). *Two-Person Game Theory*. Chelmsford, MA: Courier Dover Publications.

Rapoport, A. & Chammah, M. A. (1965). *A Study in Conflict and Cooperation*. Ann Arbor, MI: The University of Michigan Press.

Rappe, M. S. & Giovannoni, S. J. (2003). The uncultured microbial majority. *Annual Review of Microbiology*, **57**, 369–94.

Rashdall, H. (1895). *The Universities of Europe in the Middle Ages: Salerno-Bologna-Paris*. Oxford: Clarendon Press.

Raskin, P., Banuri, T., Gallopin, G. et al. (2002). *Great Transition: The Promise and the Lure of the Times Ahead: a Report of the Global Scenario Group*. Boston, MA: SEI PoleStar Series Report no. 10.

Raskin, P. D. (2008). World lines: a framework for exploring global pathways. *Ecological Economics*, **65**(3), 461–70.

Rauch, J. N. & Pacyna, J. M. (2009). Earth's global Ag, Al, Cr, Cu, Fe, Ni, Pb, and Zn cycles. *Global Biogeochemical Cycles*, **23**(GB2001). Doi: 10.1029/2008GB003376.

Raup, D. M. (1991). *Extinction: Bad Genes or Bad Luck*. New York, NY: Norton.

Ravetz, J. R. (1993). The sin of science – ignorance of ignorance. *Knowledge*, **15**(2), 157–65.

Read, J. M., Eames, K. T. D. & Edmunds, W. J. (2008). Dynamic social networks and the implications for the spread of infectious disease. *Journal of the Royal Society Interface*, **5**, 1001–7.

Read, J. M. & Keeling, M. J. (2003). Disease evolution on networks: the role of contact structure. *Proceedings of the Royal Society of London Series B – Biological Sciences*, **270**(1516), 699–708.

Redclift, M. & Woodgate, G. (eds). (1995). *The Sociology of the Environment*, Vol. III. Aldershot: Edward Elgar.

Redclift, M. & Woodgate, G. (eds). (2005). *New Developments in Environmental Sociology*. Cheltenham: Edward Elgar.

Reeger, B. J. & Bunders, J. G. G. (2009). *Knowledge Co-Creation: Interaction between Science and Society. A Transdisciplinary Approach to Complex Societal Issues*. Den Haag: RMNO.

Reiche, E. M. V., Nunes, S. O. V. & Morimoto, H. K. (2004). Stress, depression, the immune system, and cancer. *The Lancet Oncology*, **5**(10), 617–25.

Reichel-Dolmatoff, G. (1996). *Das schamanische Universum. Schamanismus, Bewusstsein und Ökologie in Südamerika*. Munich: Diederichs.

Reinders, L. (2009). Specific problems. In J. Boersema & L. Reinders (eds). *Principles of Environmental Sciences*. Heidelberg: Springer, pp. 85–96.

Reiter, S. M. & Samuel, W. (1980). Littering as a function of prior litter and the presence or absence of prohibitive signs. *Journal of Applied Psychology*, **10**(1), 45–55.

Renn, O. (1992). Concepts of risk: a classification. In S. Krimsky & D. Golding (eds). *Social Theories of Risk*. Westport, CT: Praeger, pp. 179–97.

Renn, O. (1999). A model for an analytic-deliberative process in risk management. *Environmental Science & Technology*, **33**(18), 3049–55.

Renn, O. (2005). Participation – a dazzling concept. *Gaia – Ecological Perspectives for Science and Society*, **14**(3), 227–8.

Renn, O. (2006). Participatory processes for designing environmental policies. *Land Use Policy*, **23**(1), 34–43.

Renn, O. & Webler, T. (1997). Steuerung durch kooperativen Diskurs. In S. Köberle, F. Gloede & L. Hennen (eds). *Diskursive Verständigung? Mediation und Partizipation in Technikkontroversen*. Baden-Baden: Nomos, pp. 64–100.

Renn, O., Webler, T., Rakel, H., Dienel, P. & Johnson, B. (1993). Public participation in decision-making: a 3-step procedure. *Policy Sciences*, **26**(3), 189–214.

Rennings, K. (1994). *Indikatoren für eine dauerhaft-umweltgerechte Entwicklung*. Stuttgart: Metzger-Poeschel.

Reno, R. R., Cialdini, R. B. & Kallgren, C. A. (1993). The transsituational influence of social norms. *Journal of Personality & Social Psychology*, **64**(1), 104–12.

Reschlikowski, W. & Schmitt, M. (2007). Bioethanol im Fokus der nachhaltigen Energie- und Chemiewirtschaft. *Wissenschaftliche Zeitschrift der Technischen Universität Dresden*, **56**(3–4), 73–8.

Rescorla, R. A. (1988). Pavlovian conditioning – it's not what you think it is. *American Psychologist*, **43**(3), 151–60.

Reser, J. P. & Bentrupperbäumer, J. M. (2005). What and where are environmental values? Assessing the impact of current diversity of "environmental" and "World

Heritage" values. *Journal of Environmental Psychology,* **25**(2), 126–46.

Resnick, M. (1999). *Turtles, Termites, and Traffic Jams: Explorations in Massively Parallel Microworlds.* Cambridge: Cambridge University Press.

Reusswig, F. (2007). Exploring resilience in social-ecological systems. Comparative studies and theory development. *Regional Environmental Change,* **7**(2), 121–2.

Reynolds, A. (2007). The cell's journey: from metaphorical to literal factory. *Endeavour,* **31**(2), 65–70.

Rheinberger, H. J. (2004). Die Geschichte der Molekularbiologie. In I. Jahn (ed.). *Geschichte der Biologie: Theorien, Methoden, Institutionen, Kurzbiographien.* Hamburg: Nikol Verlag, pp. 642–63.

Rheinberger, H. J. & Müller-Wille, S. (2009, March 10). Gene. *Stanford Encyclopedia of Philosophy.* Retrieved April 15, 2010, from http://plato.stanford.edu/entries/gene/

Rhodes, G. (2006). The evolutionary psychology of facial beauty. *Annual Review of Psychology,* **57**, 199–226.

Ribe, R. (1999). Regeneration harvests versus clearcuts: public views of the acceptability and aesthetics of northwest forest plan harvests. *Northwest Science,* **73**(SI), 102–17.

Ricardo, D. (1817). *On the Principles of Economy and Taxation.* London: Murray.

Richardson, G. P. (1991a). *Feedback Thought in Social Science and Systems Theory.* Philadelphia, PA: University of Pennsylvania Press.

Richardson, G. P. (1991b). System dynamics: simulation for policy analysis from a feedback perspective. In P. A. Fishwick & P. A. Luker (eds). *Qualitative Simulation Modelling and Analysis.* New York, NY: Springer, pp. 144–69.

Richardson, M. K. (1995). Heterochrony and the phylotypic period. *Developmental Biology,* **172**(2), 412–21.

Richardson, M. K., Hanken, J., Gooneratne, M. L. *et al.* (1997). There is no highly conserved embryonic stage in the vertebrates: implications for current theories of evolution and development. *Anatomy and Embryology,* **196**(2), 91–106.

Richardson, M. K. & Keuck, G. (2001). A question of intent: when is a "schematic" illustration a fraud? *Nature,* **410**(6825), 144.

Richey, M. W. (1974). Indigenous navigation in the Pacific: review. *The Geographical Journal,* **140**(1), 114–17.

Riley, E. C., Murphy, G. & Riley, R. L. (1978). Airborne spread of measles in a suburban elementary-school. *American Journal of Epidemiology,* **107**(5), 421–32.

Rip, A. & Kemp, R. (1998). Technological change. In S. Rayner & E. L. Malone (eds). *Human Choice and Climate Change,* Vol. 2. Columbus, OH: Battelle Press.

Risser, P. G., Karr, J. R. & Forman, R. T. T. (1984/2006). Landscape ecology: Directions and approaches. In J. A. Wiens, M. R. Moss, M. G. Turner & D. J. Mladenoff (eds). *Foundation Papers in Landscape Ecology.* New York, NY: Columbia University Press, pp. 255–64.

Robbins, P. (2004). *Political Ecology: A Critical Introduction.* Malden: Blackwell.

Robert, K.-H. (2002). *The Natural Step Story: Seeding a Quiet Revolution.* Gabriola Island: New Society Publishers.

Robert, K.-H., Schmidt-Bleek, B., de Larderel, J. A. *et al.* (2002). Strategic sustainable development – selection, design and synergies of applied tools. *Journal of Cleaner Production,* **10**(3), 197–214.

Robert, M., Hulten, P. & Frostell, B. (2007). Biofuels in the energy transition beyond peak oil. A macroscopic study of energy demand in the Stockholm transport system 2030. *Energy,* **32**(11), 2089–98.

Roberts, N. (1993). *The Holocene: an Environmental History.* Oxford: Blackwell Publishing.

Robinson, L., Newell, J. P. & Marzluff, J. A. (2005). Twenty-five years of sprawl in the Seattle region: growth management responses and implications for conservation. *Landscape and Urban Planning,* **71**(1), 51–72.

Robson, C. (1993). *Real World Research. A Resource for Social Scientists and Practitioner-Researchers.* Oxford: Blackwell.

Roch, S. G. & Samuelson, C. D. (1997). Effects of environmental uncertainty and social value orientation in resource dilemmas. *Organizational Behavior and Human Decision Processes,* **70**(3), 221–35.

Rodewald, R. (1999). *Sehnsucht Landschaft.* Zurich: Chronos.

Rokeach, M. (1973). *The Nature of Human Values.* New York, NY: Free Press.

Roots, W. & Walker, F. (1967). Secondary feedback in ON-OFF electric heating process. *IEEE Transactions on Industry and Applications,* **3**(5), 401–20.

Rosa, E. A., Machlis, G. E. & Keating, K. M. (1988). Energy and society. *Annual Review of Sociology,* **14**, 149–72.

Rosa, E. A. & Richter, L. (2008). Durkheim on the environment – ex libris or ex cathedra? Introduction to inaugural lecture to a course in social science, 1887–1888. *Organization & Environment,* **21**(2), 182–7.

Rosch, E. H. (1973). Natural categories. *Cognitive Psychology,* **4**(3), 328–50.

Rosch, E. H., Mervis, C. B., Gray, W. D., Johnson, D. M. & Boyesbraem, P. (1976). Basic objects in natural categories. *Cognitive Psychology,* **8**(3), 382–439.

Rose, K. D. (2006). *The Beginning of the Age of Mammals.* Baltimore, MD: Johns Hopkins University Press.

Rosenberg, A. (2006). Is epigenetic inheritance a counterexample to the central dogma? *History and Philosophy of the Life Sciences,* **28**(4), 549–65.

Rosenblueth, A., Wiener, N. & Bigelow, J. (1943). Behavior, purpose and teleology. *Philosophy of Science,* **10**(1), 18–24.

Rosenfeld, L. (1963). Niels Bohr's contribution to epistemology. *Physics Today,* **16**(10), 47–54.

Rosenhead, J. (2006). Past, present and future of problem structuring methods. *Journal of the Operational Research Society,* **57**(7), 759–65.

Rosenhead, J. & Mingers, L. (eds). (2001). *Rational Analysis for a Problematic World Revisited.* Chichester: John Wiley & Sons.

Rosenthal, R. & Wittrock, B. (1993). *The European and American Universities since 1800: Historical and Sociological Essays.* Cambridge: Cambridge University Press.

Ross, S. (1993). The economic theory of agency: the principal's problem. *American Economic Review,* **63**(2), 134–9.

Rossel, P. (2001). Swissmetro engineers' fancy or major project. In P. Goujon & B. Hériard Dubreil (eds). *Technology and Ethics: a European Quest for Responsible Engineering.* Leuven: Peeters, pp. 271–84.

Rosser, J. B. (2009, November, 3–5). Complex Ecological-Economic Dynamics and Sustainability: Post-Keynesian Perspectives. Paper presented at the International Conference "From GDP to Wellbeing", Ancona.

Roth, C. E. (1968). On the road to conservation. *Massachusetts Audubon,* June, pp. 38–41.

Roth, C. E. (1992). *Environmental Literacy: its Roots, Evolution and Directions in the 1990s.* Columbus, OH: ERIC Clearinghouse for Science Mathematics and Environmental Education.

Rothman, H. (2005). Collapse: how societies choose to fail or succeed. *Population and Development Review,* **31**(3), 573–80.

Rothschild, K. W. (1947). Price theory and oligopoly. *The Economic Journal,* **57**(227), 299–320.

Rotmans, J. & Kemp, R. (2008). Detour ahead: a response to Shove and Walker about the perilous road of transition management. *Environment and Planning A,* **40**(4), 1006–12.

Rotmans, J., Kemp, R. & van Asselt, M. B. A. (2001). More evolution then revolution: transition management in public policy. *Foresight – The Journal of Foresight Studies, Strategic Thinking and Policy,* **3**(1), 15–32.

Rotmans, J. & Loorbach, D. (2008). Transition management: reflexive governance of societal complexity through searching, learning and experimenting. In J. van den Bergh & F. Bruinsma (eds). *Managing the Transitions to Renewable Energy.* Cheltenham: Edward Elgar.

Rotmans, J. & Loorbach, D. (2009). Complexity and transition management. *Journal of Industrial Ecology,* **13**(2), 184–96.

Rowe, D. (2002). Environmental literacy and sustainability as core requirements: success stories and models. In L. Filho (ed.). *Teaching Sustainability at University Level.* Frankfurt: Peter Lang, pp. 79–103.

Rowe, G. & Frewer, L. J. (2005). A typology of public engagement mechanisms. *Science Technology & Human Values,* **30**(2), 251–90.

Roy, B. (1991). The outranking approach and the foundations of ELECTRE methods. *Theory and Decision,* **31**(1), 49–73.

Rubin, E. (1927). Visuell wahrgenommene wirkliche Bewegungen. *Zeitschrift für Psychologie,* **103**, 384–92.

Ruddiman, W. F. (2003). The anthropogenic greenhouse era began thousands of years ago. *Climatic Change,* **61**(3), 261–93.

Ruddiman, W. F. (2007). The early anthropogenic hypothesis: challenges and responses. *Reviews of Geophysics,* **45**(RG4001), 1–37.

Rudvall, G. (1978). *Ten Years Research and Development in the Field in Malmö. Curriculum Research and Development.* Kiel: IPN.

Ruff, C. B., Trinkaus, E. & Holliday, T. W. (1997). Body mass and encephalization in Pleistocene Homo. *Nature,* **387**(6629), 173–6.

Russell, J. A. (1980). A circumplex model of affect. *Journal of Personality and Social Psychology,* **39**(6), 1161–78.

Russell, J. A. (2003). Core affect and the psychological construction of emotion. *Psychological Review,* **110**(1), 145–72.

Russell, J. A., Lewicka, M. & Niit, T. (1989). A cross-cultural-study of a circumplex model of affect. *Journal of Personality and Social Psychology,* **57**(5), 848–56.

Russell, J. A. & Snodgrass, J. (1987). Emotion and the environment. In D. Stokols & I. Altmann (eds). *Handbook of Environmental Psychology,* Vol. I. New York, NY: John Wiley & Sons, pp. 245–80.

Russell, S. (2004). The economic burden of illness for households in developing countries: a review of studies focusing on malaria, tuberculosis, and human immunodeficiency virus/acquired immunodeficiency syndrome. *American Journal of Tropical Medicine and Hygiene,* **71**(2), 147–55.

Ruth, M. (1999). Physical principles and environmental economics. In J. C. J. M. van den Bergh (ed.). *Handbook of Environmental Economics.* Cheltenham: Edward Elgar, pp. 197–214.

Ruth, M., Bernier, C., Jollands, N. & Golubiewski, N. (2007). Adaptation of urban water supply infrastructure to impacts from climate and socioeconomic changes: the case of Hamilton, New Zealand. *Water Resources Management,* **21**(6), 1031–45.

Ruth, M. & Coelho, D. (2007). Understanding and managing the complexity of urban systems under climate change. *Climate Policy,* **7**(4), 317–36.

Rutherford, M. (2004). Institutional economics: the term and its meanings. In W. J. Samuels (ed.). *Research in the History of Economic Thought and Methodology,* Vol. 22A, part 1. Bingley: Emerald Group Publishing Ltd, pp. 179–84.

Ruttan, L. M. (2006). Sociocultural heterogeneity and the commons. *Current Anthropology,* **47**(5), 843–53.

Saab (Producer). (2009) Sweden leads European bioethanol market. Retrieved January 6, 2011 from http://www.saabnet.com/tsn/press/060509B.html

Saaty, T. L. (1990). *The Analytic Hierarchy Process,* 2nd edn. Pittsburgh, PA: RWS Publications.

Sacca, M. L., Accinelli, C., Fick, J., Lindberg, R. & Olsen, B. (2009). Environmental fate of the antiviral drug Tamiflu in two aquatic ecosystems. *Chemosphere,* **75**(1), 28–33.

Sachs, J. D. & Warner, A. M. (2001). Natural resources and economic development: the curse of natural resources. *European Economic Review,* **45**(4–6), 827–38.

Sacks, O. (1973). *Zeit des Erwachens*. London: Picador.

Sahlins, M. (1976). *Culture and Practical Reason*. Chicago, IL: University of Chicago Press.

Salzberger, B., Diehl, V. & Kurth, R. (2002). BSE und vCJD (in German). *Internist,* **43**, 707–8.

Samuels, W. J. (1995). The present state of institutional economics. *Cambridge Journal of Economics,* **19**(4), 569–90.

Samuelson, C. D. & Messick, D. M. (1986). Alternative structural solutions to resource dilemmas. *Organizational Behavior and Human Decision Processes,* **37**(1), 139–55.

Samuelson, C. D., Messick, D. M., Rutte, C. G. & Wilke, H. (1984). Individual and structural solutions to resource dilemmas in two cultures. *Journal of Personality and Social Psychology,* **47**(1), 94–104.

Samuelson, P. A. (1947). *Foundations of Economic Analysis*. Cambridge, MA: Harvard University Press.

Samuelson, P. A. (1948). Consumption theory in terms of revealed preference. *Economica,* **15**(60), 243–53.

Samuelson, P. A. (1954). The pure theory of public expenditure. *The Review of Economics and Statistics,* **36**(4), 387–9.

Samuelson, P. A. & Nordhaus, W. D. (2005). *Economics,* 18th edn. New York, NY: McGraw-Hill.

Sanctorius, S. (1624/1711). *De Statica Medicina*. Leyden: Cornelium Boutesteyn.

Sander, K. (2002). Ernst Haeckel's ontogenetic recapitulation: irritation and incentive from 1866 to our time. *Annals of Anatomy – Anatomischer Anzeiger,* **184**(6), 523–33.

Sanderson, S. K. (1999). *Social Transformations: a General Theory of Historical Development*. Lanham, MD: Rowman & Littlefield.

Sanderson, S. K. (2006). Ecological–evolutionary theory: principles and applications. *Contemporary Sociology – a Journal of Reviews,* **35**(6), 551–3.

Sanderson, S. K. (2007). *Evolutism and its Critics. Deconstructing and Reconstructing an Evolutionary Interpretation of Human Society*. Boulder, CO: Paradigm Publisher.

Sanger, F., Nicklen, S. & Coulson, A. R. (1977). DNA sequencing with chain-terminating inhibitors. *Proceedings of the National Academy of Sciences of the United States of America,* **74**(12), 5463–7.

Sargent, F. (1974). *Human Ecology*. Amsterdam: North-Holland.

Sasaki, N. (2006). A reexamination of welfare states from an institutional perspective. *Journal of Economic Studies,* **33**(3), 189–205.

Sassenberg, K. & Boos, M. (2003). Attitude change in computer-mediated communication: effects of anonymity and category norms. *Group Processes & Intergroup Relations,* **6**(4), 405–22.

Sax, L. (2002). How common is intersex? A response to Anne Fausto-Sterling. *Journal of Sex Research,* **39**(3), 174–8.

Schachter, S. (1971). *Emotion, Obesity, and Crime*. New York, NY: Academic Press.

Schäfer, H. (1995). Kameralismus. In Görres-Gesellschaft (ed.). *Staatslexikon,* Vol. 3. Freiburg: Herder, pp. 276–7.

Schaltegger, S. & Burritt, R. (2000). *Contemporary Environmental Accounting: Issues, Concepts and Practice*. Sheffield: Greenleaf.

Scharf, F. W. (1991). Political Institutions, decision styles, and policy choices. In R. M. Czada & A. Windhoff-Héritier (eds). *Political Choice. Institutions, Rules, and the Limits of Rationality*. Frankfurt am Main: Campus, pp. 53–86.

Schedlowski, M. & Tewes, W. (1996). *Psychoneuroimmunology*. Heidelberg: Spektrum.

Scheffer, M. (2009). *Critical Transitions in Nature and Society*. Princeton, NJ: Princeton University Press.

Scheffer, M., Carpenter, S., Foley, J. A., Folke, C. & Walker, B. (2001). Catastrophic shifts in ecosystems. *Nature,* **413**(6856), 591–6.

Scheible, J. (ed.). (1847). *Dreck-Apotheke von C. F. Paullini,* Vol. 3. Stuttgart: J. Scheible Verlag.

References

Schelling, T. C. (1971). Dynamic models of segregation. *Journal of Mathematical Sociology,* **1**(2), 143–86.

Schelling, T. C. (1978). *Micromotives and Macrobehaviour.* New York, NY: Norton.

Schellnhuber, H. J., Crutzen, P. J., Clark, W. C., Claussen, M. & Held, H. (2004). *Earth System Analysis for Sustainability.* Cambridge, MA: The MIT Press.

Schellnhuber, H. J., Crutzen, P. J., Clark, W. C. & Hunt, J. (2005). Earth system analysis for sustainability. *Environment,* **47**(8), 10–25.

Scheper-Hughes, N. & Lock, M. M. (1987). The mindful body: a prolegomenon to future work in medical anthropology. *Medical Anthropology Quarterly,* **1**(1), 6–41.

Scher, E. (1999). Negotiating superfund cleanup at the Massachusetts military reservation. In L. E. Susskind, S. McKearnen & J. Thomas-Larmer (eds). *The Consensus Building Handbook: a Comprehensive Guide to Reaching Agreement.* Thousand Oaks, CA: Sage, pp. 859–78.

Scherer, K. R. & Wallbott, H. G. (1994). Evidence for universality and cultural variation of differential emotion response patterning. *Journal of Personality and Social Psychology,* **66**(2), 310–28.

Scherrer, K. & Jost, J. (2007). The gene and the genon concept: a functional and information-theoretic analysis. *Molecular Systems Biology,* **3**, 87.

Schicke, S. (2006, Feburary 23). *Bürgergutachten: Peter C. Dienel in Europa und Asien als Experte.* Nordwest-Zeitung, p.31.

Schiffer, M. B. (1976). *Behavioral Archaeology.* New York, NY: Academic Press.

Schlesinger, P. (2007). A cosmopolitan temptation. *European Journal of Communication,* **22**(4), 413–26.

Schluchter, W. (2006). *Grundlegungen der Soziologie,* Vol. I. Tübingen: Mohr Siebeck.

Schluchter, W. (2009). *Grundlegungen der Soziologie,* Vol. II. Tübingen: Mohr Siebeck.

Schmalensee, R., Joskow, P. L., Ellerman, A. D., Montero, J. P. & Bailey, E. M. (1998). An interim evaluation of sulfur dioxide emission trading. *Journal of Economic Perspectives,* **12**(3), 53–68.

Schmidheiny, S. (1992). *Changing Course: a Global Business Perspective for Development and Environment.* Cambridge, MA: The MIT Press.

Schmidt, D. & Hoerstrup, S. P. (2006). Tissue engineered heart valves based on human cells. *Swiss Medical Weekly,* **136**(39–40), 618–23.

Schmithüsen, J. (1942). Vegetationsforschung und ökologische Standortlehre in ihrer Bedeutung für die Geographie der Kulturlandschaft. *Zeitschrift der Gesellschaft für Erdkunde zu Berlin,* H. 3/4, 113–57.

Schmithüsen, J. (1976). *Allgemeine Geosynergetik: Grundlagen der Landschaftskunde.* Berlin: de Gruyter.

Schmithüsen, F. (2008). Common Goals for Sustainable Forest Management – Divergence and Reconvergence of American and European Forestry. In V. A. Sample & S. Anderson (Eds.), *European Forests – Heritage of the Past and Options for the Future* (pp. 216–248). Durham, NC: Forest History Society

Schmithüsen, F., Kaiser, B., Schmidhauser, A., Mellinghoff, S., & Kammerhofer, A. W. (forthcoming). *Entrepreneurship in forestry and the wood industry – Principles of business eco-nomics and management processes.* London Routledge

Schmitt, M. (2008). Carl Linnaeus, the order of nature, and binomial names. *Deutsche Entymologische Zeitschrift,* **55**(1), 13–17.

Schmölders, G. (1962). *Geschichte der Volkswirtschaftslehre: Überblick und Leseproben.* Reinbeck: Rowohlt.

Schmoller, G. v. (1900). *Grundriss der allgemeinen Volkswirtschaftslehre im deutschen Reich.* Leipzig: Duncker und Humblot.

Schnaiberg, A. (1975). Social syntheses of the societal–environmental dialectic: the role of distributional impacts. *Social Science Quarterly,* **56**, 5–20.

Schnaiberg, A. (1980). *The Environment: from Surplus to Scarcity.* Oxford: Oxford University Press.

Schnaiberg, A., Watts, N. & Zimmermann, K. (eds). (1986). *Distributional Conflicts in Environmental-Resource Policy.* Aldershot: Gower.

Schnaiberg, A., Weinberg, A. S. & Pellow, D. N. (2002). The treadmill of production and the environmental state. In A. P. J. Mol & F. H. Buttel (eds) *The Environmental State under Pressure.* London: Elsevier, pp. 15–32.

Schneider, D. (2001). *Betriebswirtschaftslehre,* Vol. 4. Munich: Oldenburg.

Schnelle, E. (1979). *The Metaplan-Method. Communication Tools for Planning and Learning Groups,* Vol. 7. Quickborn: Metaplan.

Scholz, R. W. (1978). What research has found out on the cooperation of teachers and the effect of team teaching. *Educational Resources Information Center (ERIC Document Reproduction Service No. ED 173154).*

Scholz, R. W. (1980). *Dyadische Verhandlungen: Eine theoretische und experimentelle Untersuchung von Vorhersagemodellen unter der besonderen Berücksichtigung von Anspruchsniveau-Vorhersagemodellen.* Königstein im Taunus: Hain.

Scholz, R. W. (ed.). (1983). *Decision Making under Uncertainty.* Amsterdam: Elsevier.

Scholz, R. W. (1987). *Cognitive Strategies in Stochastic Thinking.* Dordrecht: Reidel.

Scholz, R. W. (1988). Variants of expectancy and subjective probability in P300 research. *Behavioral and Brain Sciences,* **11**(3), 396–7.

Scholz, R. W. (1991). Psychological research in probabilistic understanding. In R. Kapadia & M. Borovcnik (eds). *Chance Encounters: Probability in Education*. Dordrecht: Reidel, pp. 213–54.

Scholz, R. W. (1998). Umweltforschung zwischen Formalwissenschaft und Verständnis: Muss man den Formalismus beherrschen, um die Formalisten zu schlagen? In A. Daschkeit & W. Schröder (eds). *Umweltforschung quergedacht: Perspektiven integrativer Umweltforschung und -lehre*. Berlin: Springer, pp. 309–28.

Scholz, R. W. (1999a). "Mutual Learning" und Probabilistischer Funktionalismus – Was Hochschule und Gesellschaft voneinander und von Egon Brunswik lernen können, *UNS Working Paper 21*. Zurich: ETH Zurich.

Scholz, R. W. (1999b). Region und Landschaft zwischen wissenschaftlicher Analyse und Verständnis. In R. W. Scholz, S. Bösch, L. Carlucci & J. Oswald (eds). *Nachhaltige Regionalentwicklung: Chancen der Region Klettgau*. Zurich: Rüegger, pp. 25–37.

Scholz, R. W. (2000). Mutual learning as a basic principle of transdisciplinarity. In R. W. Scholz, R. Häberli, A. Bill & M. Welti (eds). *Transdisciplinarity: Joint Problem-Solving among Science, Technology and Society. Workbook II: Mutual Learning Sessions*. Zurich: Haffmans Sachbuch, pp. 13–17.

Scholz, R. W. (ed.). (2001). *Erfolgskontrolle von Umweltmassnahmen*. Heidelberg: Springer.

Scholz, R. W. & Binder, C. R. (2003). The paradigm of human-environment systems. *Natural and Social Science Interface*. Zurich: ETH Zurich.

Scholz, R. W. & Binder, C. R. (2004). Principles of human-environment systems (HES) research. In C. Pahl-Wostl, S. Schmidt, A. E. Rizzoli & A. J. Jakeman (eds). *Complexity and Integrated Resources Management, Transactions of the 2nd Biennial Meeting of the International Environmental Modelling and Software Society*, Vol. 2. Osnabruck: Zentrum für Umweltkommunikation (ZUK), pp. 791–6.

Scholz, R. W., Bösch, S., Carlucci, L. & Oswald, J. (eds). (1999). *Nachhaltige Regionalentwicklung: Chancen der Region Klettgau*. Zurich: Rüegger.

Scholz, R. W., Bösch, S., Koller, T., Mieg, H. A. & Stünzi, J. (eds). (1996). *Industrieareal Sulzer-Escher Wyss: Umwelt und Bauen – Wertschöpfung durch Umnutzung*. Zurich: vdf.

Scholz, R. W., Bösch, S., Mieg, H. A. & Stünzi, J. (eds). (1997). *Zentrum Zürich Nord: Stadt im Aufbruch: Bausteine für eine nachhaltige Stadtentwicklung*. Zurich: vdf.

Scholz, R. W., Bösch, S., Mieg, H. A. & Stünzi, J. (eds). (1998). *Region Klettgau: Verantwortungsvoller Umgang mit Boden*. Zurich: Rüegger.

Scholz, R. W., Bösch, S., Stauffacher, M. & Oswald, J. (eds). (2001). *Zukunft Schiene Schweiz 1: Ökoeffizientes Handeln der SBB*. Zurich: Rüegger.

Scholz, R. W. & Girod, B. (2010, November 9–12). Coping with resound effects of biofuel production in LCA. Paper presented at the EcoBalance 2010, The 9th International Conference on EcoBalance, Tokyo.

Scholz, R. W., Koller, T., Mieg, H. A. & Schmidlin, C. (1995). *Perspektive "Grosses Moos" – Wege zu einer nachhaltigen Landwirtschaft*. Zurich: vdf.

Scholz, R. W., Lang, D. J., Wiek, A., Walter, A. I. & Stauffacher, M. (2006). Transdisciplinary case studies as a means of sustainability learning: historical framework and theory. *International Journal of Sustainability in Higher Education,* **7**(3), 226–51.

Scholz, R. W. & Marks, D. (2001). Learning about transdisciplinarity: Where are we? Where have we been? Where should we go? In J. T. Klein, W. Grossenbacher-Mansuy, R. Häberli, A. Bill, R. W. Scholz & M. Welti (eds). *Transdisciplinarity: Joint Problem Solving among Science, Technology, and Society*. Basel: Birkhäuser Verlag AG, pp. 236–52.

Scholz, R. W., May, T. W., Hefer, B. & Nothbaum, N. (1991). Die Anwendung des Donator-Akzeptor-Modells am Beispiel Blei für den Pfad "Boden-Mensch". In E. Bütow (ed.). *Ableitung von Sanierungswerten für kontaminierte Böden*. Berlin: Schmidt, pp. 55–98.

Scholz, R. W., May, T. W., Nothbaum, N., Hefer, I. & Lühr, H. P. (1992). Induktivstochastische Risikoabschätzung mit dem Donator-Akzeptor-Modell am Beispiel der Gesundheitsbelastung durch cadmiumbelastete Weizenackerböden. In W. van Eimeren, K. Ueberla & K. Ulm (eds). *Gesundheit und Umwelt*. Berlin: Springer, pp. 57–61.

Scholz, R. W., Mieg, H. A. & Oswald, J. (2000). Transdisciplinarity in groundwater management: towards mutual learning of science and society. *Water, Air, and Soil Pollution,* **123**(1–4), 477–87.

Scholz, R. W., Schmitt, H.-J., Vollmer, W., Vogel, A. & Neisel, F. (1990). Zur Abschätzung des gesundheitlichen Risikos kadmiumbelasteter Hausgärten. *Das öffentliche Gesundheitswesen,* **52**, 161–7.

Scholz, R. W. & Siegrist, M. (2010). Low risk and high public concern? The cases of persistent organic pollutants (POPs), heavy metals, and nanotech particle. *Human and Ecological Risk Assessment,* **16**(1), 1–14.

Scholz, R. W. & Stauffacher, M. (2002). Unsere Landschaft ist unser Kapital: Überblick zur ETH-UNS Fallstudie "Landschaftsnutzung für die Zukunft: der Fall Appenzell Ausserrhoden". In R. W. Scholz, M. Stauffacher, S. Bösch

& A. Wiek (eds). *Landschaftsnutzung für die Zukunft: der Fall Appenzell Ausserrhoden. ETH-UNS Fallstudie 2001.* Zurich: Rüegger und Pabst, pp. 13–47.

Scholz, R. W. & Stauffacher, M. (2007). Managing transition in clusters: area development negotiations as a tool for sustaining traditional industries in a Swiss prealpine region. *Environment and Planning A,* **39**(10), 2518–39.

Scholz, R. W., Stauffacher, M., Bösch, S. & Krütli, P. (2004). *Mobilität und zukunftsfähige Stadtentwicklung: Freizeit in der Stadt Basel.* Zurich: Rüegger und Pabst.

Scholz, R. W., Stauffacher, M., Bösch, S. & Krütli, P. (2005). *Nachhaltige Bahnhofs- und Stadtentwicklung in der trinationalen Agglomeration: Bahnhöfe in der Stadt Basel.* Zurich: Rüegger.

Scholz, R. W., Stauffacher, M., Bösch, S. & Krütli, P. (eds). (2003). *Appenzell Ausserrhoden Umwelt Wirtschaft Region.* Zurich: Rüegger und Pabst.

Scholz, R. W., Stauffacher, M., Bösch, S., Krütli, P. & Wiek, A. (eds). (2007). *Entscheidungsprozesse Wellenberg – Lagerung radioaktiver Abfälle in der Schweiz.* Zurich: Rüegger.

Scholz, R. W., Stauffacher, M., Bösch, S. & Wiek, A. (eds). (2002). *Landschaftsnutzung für die Zukunft: der Fall Appenzell Ausserrhoden.* Zurich: Rüegger und Pabst.

Scholz, R. W., Steiner, R. & Hansmann, R. (2004). Role of internship in higher education in environmental sciences. *Journal of Research in Science Teaching,* **41**(1), 24–46.

Scholz, R. W. & Tietje, O. (2002). *Embedded Case Study Methods: Integrating Quantitative and Qualitative Knowledge.* Thousand Oaks, CA: Sage.

Scholz, R. W. & Wiek, A. (2005). Operational eco-efficiency – comparing firms' environmental investments in different domains of operation. *Journal of Industrial Ecology,* **9**(4), 155–70.

Scholz, R. W. & Zimmer, A. C. (1997). Qualitative aspects of decision making: a challenge for decision research. In R. W. Scholz & A. C. Zimmer (eds). *Qualitative Aspects of Decision Research.* Lengerich: Pabst Publishers, pp. 7–13.

Schot, J. & Geels, F. W. (2008). Strategic niche management and sustainable innovation journeys: theory, findings, research agenda, and policy. *Technology Analysis & Strategic Management,* **20**(5), 537–54.

Schrödinger, E. (1944). *What is Life? The Physical Aspect of the Living Cell.* Cambridge: Cambridge University Press.

Schulz, H.-J. (1986). *Sekundärrohstoffwirtschaft.* Berlin: Staatsverlag der DDR.

Schulze, E. D. & Mooney, H. A. (1994). Ecosystem function of biodiversity: a summary. In E. D. Schulze & H. A. Mooney (eds). *Biodiversity and Ecosystem Function.* Berlin: Springer, p. 497.

Schum, D. A. (2001). *The Evidential Foundations of Probabilistic Reasoning.* New York, NY: Wiley.

Schumpeter, J. A. (1911/1934). *Theorie der wirtschaftlichen Entwicklung.* Berlin: Duncker & Humboldt.

Schumpeter, J. A. (1950). The process of creative destruction. In J. A. Schumpeter (ed.). *Capitalism, Socialism and Democracy,* 3rd edn. London: Allen and Unwin, pp. 81–6.

Schumpeter, J. A. (1954). *History of Economic Analysis.* London: Routledge.

Schumpeter, J. A. (1994). *History of Economic Analysis,* with new introduction by Mark Perlman. London: Routledge.

Schuster, H. G. & Just, W. (2005). *Deterministic Chaos: an Introduction,* 4th edn. Weinheim: Wiley InterScience.

Schwaninger, M. (2006). System dynamics and the evolution of the systems movement. *Systems Research and Behavioral Science,* **23**(5), 583–94.

Schwann, T. (1839). *Mikroskopische Untersuchungen über die Übereinstimmung in der Structur und dem Wachstum der Thiere und Pflanzen.* Berlin: Sander.

Schwartz, S. H. (1977). Normative influence on altruism. In L. Berkowitz (ed.). *Advances in Experimental Social Psychology,* Vol. 10. New York, NY: Academic Press, pp. 221–79.

Schwartz, S. H. (1992). Universals in the content and structure of values – theoretical advances and empirical tests in 20 countries. *Advances in Experimental Social Psychology,* **25**, 1–65.

Schwartz, S. H. (2006). Les valeurs de base de la personne: théorie, mesures et applications. *Revue française de Sociologie,* **42**(4), 249–88.

Schwartz, S. H. & Bardi, A. (2001). Value hierarchies across culture: taking a similarities approach. *Journal of Cross-Cultural Psychology,* **32**(3), 268–90.

Schwarze, R. & Wagner, G. (2006). *The Political Economy of Natural Disaster Insurance: Lessons from the Failure of a Proposed Compulsory Insurance Scheme in Germany.* Berlin: DIW.

Schweitzer, F. (1997). *Concepts and Models of a Quantitative Sociology. From Individual to Collective.* Boca Raton FL: CRC Press.

Schweizerischer Bundesrat. (2002). Verordnung vom 20. November 2002 über die Ausrichtung von Beiträgen an die Kosten der Entsorgung von tierischen Abfällen im Jahre 2003, *SR 916.406.*

Schweizerischer Bundesrat. (2004). Verordnung vom 23. Juni 2004 über die Entsorgung von tierischen Nebenprodukten, *SR 916.441.22.*

ScienceDaily. (2008, June 5). Humans have ten times more bacteria than human cells: how do microbial communities affect human health? *Science News*. Retrieved May 10, 2010 from http://www.sciencedaily.com/releases/2008/06/080603085914.htm

Scott, M. J., Bilyard, G. R., Link, S. O. et al. (1998). Valuation of ecological resources and functions. *Environmental Management*, **22**(1), 49–68.

Scott, P. (2007). From professor to "knowledge worker": profiles of the academic profession. *Minerva*, **45**(2), 205–15.

Scott Gordon, H. (1954). The economic theory of a common-property resource: the fishery. *Journal of Political Economy*, **62**, 124.

Seale, H., Leask, J., Po, K. & MacIntyre, C. R. (2009). "Will they just pack up and leave?" – attitudes and intended behaviour of hospital health care workers during an influenza pandemic. *BMC Health Services Research*, **9**, 30.

Sears, C. L. (2005). A dynamic partnership: celebrating our gut flora. *Anaerobe*, **11**(5), 247–51.

Sebeok, T. A. & Umoker-Sebeok, J. (eds). (1992). *Biosemiotic: The Semiotic Web 1991*. Berlin: de Gruyter.

Seeland, K. (1988). Ecology and the comeback of religiousness. In C. Das (ed.). *Comparative Religion and the Challenges of the Present*. Cuttack: Institute of Oriental and Orissan Studies, pp. 1–12.

Segerstrom, S. C. & Miller, G. E. (2004). Psychological stress and the human immune system: a meta-analytic study of 30 years of inquiry. *Psychological Bulletin*, **130**(4), 601–30.

Seipel, W. (ed.). (2003). *Der Turmbau zu Babel: Ursprung und Vielfalt von Sprache und Schrift – Eine Ausstellung des Kunsthistorischen Museums Wien für die europäische Kulturhauptstadt Graz*, Vol. 1: Der babylonische Turm in der historischen Überlieferung, der Archäologie und der Kunst. Vienna: Kunsthistorisches Museum Wien.

Sellier, P. (2003). Protein nutrition for ruminants in European countries, in the light of animal feeding regulations linked to bovine spongiform encephalopathy. *Revue scientifique et technique (office international des Epizooties)*, **22**(1), 259–69.

Sen, A. (1999). *Development as Freedom*. New York, NY: Knopf.

Senge, P. M. (1977). Statistical estimation of feedback models. *Simulation*, **28**(6), 177–84.

Senge, P. M. & Sterman, J. D. (1989). Systems thinking and organizational learning: acting locally and thinking globally in the organization of the future. In T. A. Kochan & M. Useem (eds). *Transforming Organizations*. Oxford: Oxford University Press, pp. 354–71.

Shafer, G. A. (1976). *A Mathematical Theory of Evidence*. Princeton, NJ: Princeton University Press.

Shanon, B. (2008). Mind-body, body-mind: two distinct problems. *Philosophical Psychology*, **21**(5), 697–701.

Sharpe, E. J. (1983). *Understanding Religion*. London: Duckworth.

Sheldrake, R. (1990). *The Rebirth of Nature: the Greening of Science and God*. London: Century.

Sheppard, S. R. J. & Meitner, M. (2005). Using multi-criteria analysis and visualisation for sustainable forest management planning with stakeholder groups. *Forest Ecology and Management*, **207**(1–2), 171–87.

Sheridan, T. B. (1980). Human error in nuclear power plants. *Technological Review*, **82**(4), 22–33.

Sherif, M. (1936). *The Psychology of Social Norms*. New York, NY: Harper.

Shin, T., Kraemer, D., Pryor, J. et al. (2002). A cat cloned by nuclear transplantation. *Nature*, **415**(6874), 859.

Shinn, T. (2002). The triple helix and new production of knowledge: prepackaged thinking on science and technology. *Social Studies of Science*, **32**(4), 599–614.

Shinn, T. (2005). New sources of radical innovation: research-technologies, transversality and distributed learning in a post-industrial order. *Social Science Information*, **44**(4), 731–64.

Shinn, T. (2006). New sources of radical innovation: research-technologies, transversality and distributed learning in an post-industrial order. In J. Hage & M. Meeus (eds). *Innovation, Science, and Institutional Change*. Oxford: Oxford University Press, pp. 313–33.

Shogren, J. F., Shin, S. Y., Hayes, D. J. & Kliebenstein, J. B. (1994). Resolving differences in willingness-to-pay and willingness to accept. *American Economic Review*, **84**(1), 255–70.

Shove, E. & Walker, G. (2007). Caution! Transitions ahead: politics, practice, and sustainable transition management. *Environment and Planning A*, **39**(4), 763–70.

Shove, E. & Walker, G. (2008). Transition Management (TM) and the politics of shape shifting. *Environment and Planning A*, **40**(4), 1012–14.

Shukla, J., Nobre, C. & Sellers, P. (1990). Amazon deforestation and climate change. *Science*, **247**(4948), 1322–5.

Shulman, L. S. & Carey, N. B. (1984). Psychology and the limitations of individual rationality. Implications for the study of reasoning and civility. *Review of Educational Research*, **54**(4), 501–24.

Siegrist, M. (1996). Fragebogen zur Erfassung der ökozentrischen und anthropozentrischen Umwelteinstellung. *Zeitschrift für Sozialpsychologie*, **27**(4), 290–4.

Siegrist, M., Gutscher, H. & Earle, T. C. (2005). Perception of risk: the influence of general trust, and general confidence. *Journal of Risk Research*, **8**(2), 145–56.

Siegrist, M., Earle, T. C. & Gutscher, H. (eds). (2007). *Trust in Cooperative Risk Management: Uncertainty and Scepticism in the Public Mind*. London: Earthscan.

Siegrist, M., Keller, C. & Cousin, M.-E. (2006). Implicit attitudes toward nuclear power and mobile phone base stations: support for the affect heuristic. *Risk Analysis,* **26**(4), 1021–9.

Silverstein, A. M. (1996). History of immunology – Paul Ehrlich: the founding of pediatric immunology. *Cellular Immunology,* **174**(1), 1–6.

Silverstein, A. M. (2003a). Cellular versus humoral immunology: a century-long dispute. *Nature Immunology,* **4**(5), 425–8.

Silverstein, A. M. (2003b). Darwinism and immunology: from Metchnikoff to Burnet. *Nature Immunology,* **4**(1), 3–6.

Silverstein, A. M. & Rose, N. R. (2000). There is only one immune system! The view from immunopathology. *Seminars in Immunology,* **12**(3), 173–8.

Sime, J. D. (1999). What is environmental psychology? Texts, contexts and content. *Journal of Environmental Psychology,* **19**(2), 191–206.

Simmel, G. (1903/1957). Die Grossstädte und das Geistesleben. In G. Simmel (ed.). *Brücke und Tür. Essays des Philosophischen zur Geschichte, Religion, Kunst und Gesellschaft*. Stuttgart: Koehler, pp. 227–42.

Simmel, G. (1913). Philosophie der Landschaft. *Die Güldenkammer,* **3**(II), 635–44.

Simon, H. A. (1956). A behavioral model of rational choice. *The Quarterly Journal of Economics,* **69**(1), 99–118.

Simon, H. A. (1957). *Models of Man: Social and Rational*. New York, NY: Wiley.

Simon, H. A. (1973). The organization of complex systems. In H. D. Pattee (ed.). *Hierarchy Theory: the Complex Challenge of Complex Systems*. New York, NY: George Braziller, pp. 3–27.

Simon, H. A. (1974). How big is a chunk. *Science,* **183**(4124), 482–8.

Simon, H. A. (1979). *Models of Thought*. New Haven, CT: Yale University Press.

Simon, H. A. (1982). *Models of Bounded Rationality*. Cambridge, MA: The MIT Press.

Simon, J. L. (1981). *The Ultimate Resource*. Champaign, IL: University of Illinois Press.

Simon, J. L. & Kahn, H. (eds). (1984). *The Resourceful Earth*. New York, NY: Basil Blackwell.

Singer, A. C., Nunn, M. A., Gould, E. A. & Johnson, A. C. (2007). Potential risks associated with the proposed widespread use of Tamiflu. *Environmental Health Perspectives,* **115**(1), 102–6.

Singer, C. (1959). *History of Biology to about the Year 1900: a General Introduction to the Study of Living Things*, 3rd edn. London: Abelard-Schuman.

Singer, C. (1921). Biology. In R. W. Livingstone (ed.). *The Legacy of the Greeks*. Oxford: Clarendon Press, pp. 163–200.

Siren, A. H. (2006). Natural resources in indigenous peoples' land in Amazonia: a tragedy of the commons? *International Journal of Sustainable Development and World Ecology,* **13**(5), 363–74.

Sismondo, S. (1993). Some social constructions. *Social Studies of Science,* **23**(3), 515–53.

Six, A. (2008). Zu dumm für diese Welt: Der neuseeländische Kakapo kann nicht fliegen und hat kein Glück in der Liebe. *NZZ am Sonntag*, pp. 75–6.

Sjöberg, L. (2001). Limits of knowledge and the limited importance of trust. *Risk Analysis,* **21**(1), 189–98.

Skinner, B. J. (1979). Earth resources. *Proceedings of the National Academy of Sciences of the United States,* **76**(9), 4212–17.

Slater, J. (1952). George Ripley and Thomas Carlyle. *Publications of the Modern Language Association,* **67**(4), 341–9.

Slife, B. D. (1993). *Time and Psychological Explanation*. Albany, NY: State University of New York Press.

Slovic, P., Finucane, M. L., Peters, E. & MacGregor, D. G. (2004). Risk as analysis and risk as feelings: some thoughts about affect, reason, risk, and rationality. *Risk Analysis,* **24**(2), 311–22.

Slovic, P., Fischhoff, B. & Lichtenstein, S. (1985). Characterizing perceived risk. In R. W. Kates, C. Hohenemser & J. X. Kasperson (eds). *Perilous Progress: Managing the Hazards of Technology*. Boulder, CO: Westview, pp. 91–125.

Slusky, E. (1915). Sulla teoria del bilancio del consumatore. *Giornale Degli Economisti,* **51**, 1–126.

Smieszek, T., Fiebig, L. & Scholz, R. W. (2009). Models of epidemics: when contact repetition and clustering should be included. *Theoretical Biology and Medical Modelling,* **6**, 11.

Smil, V. (2000). Phosphorus in the environment: natural flows and human interferences. *Annual Review of Energy and the Environment,* **25**, 53–88.

Smil, V. (2002). Global biogeochemical cycles. In R. U. Ayres & L. W. Ayres (eds). *A Handbook of Industrial Ecology*. Cheltenham: Edward Elgar, pp. 249–59.

Smil, V. (2004). *Enriching the Earth: Fritz Haber, Carl Bosch, and the Transformation of World Food Production*. Cambridge, MA: The MIT Press.

Smith, A. (1776/1961). *An Inquiry into the Nature and Causes of the Wealth of Nations*. London: Methuen.

Smith, D. R. (1991). African bees in the Americas: insights from biogeography and genetics. *Trends in Ecology & Evolution,* **6**(1), 17–21.

Smith, K. (2001). *Environmental Hazards. Assessing Risk and Reducing Disaster,* 3rd edn. New York, NY: Routledge.

Smith, K. A. (2005). Wanted, an anthrax vaccine: Dead or alive? *Medical Immunology,* **4**(5), 1–6.

Sneed, J. D. (1971). *The Logical Structure of Mathematical Physics.* Dordrecht: Reidel.

Snow, J. (1855). *On the Mode of Communication of Cholera.* London: John Churchill.

Socolow, R. (1994a). Preface. In R. Socolow, C. Andrews, F. Berkhout & V. Thomas (eds). *Industrial Ecology: Definition and Implementation.* Cambridge: Cambridge University Press, pp. xv–xx.

Socolow, R. (1994b). Six perspectives from industrial ecology. In R. Socolow, C. Andrews, F. Berkhout & V. Thomas (eds). *Industrial Ecology: Definition and Implementation.* Cambridge: Cambridge University Press, pp. 3–18.

Socolow, R., Andrews, C., Berkhout, F. & Thomas, V. (eds). (1994). *Industrial Ecology: Definition and Implementation.* Cambridge: Cambridge University Press.

Sokolowska, J. (2006). Risk perception and acceptance – one process or two? The impact of aspirations on perceived risk and preferences. *Experimental Psychology,* **53**(4), 247–59.

Söllner, F. (2001). *Geschichte des ökonomischen Denkens.* Heidelberg: Springer.

Solow, R. M. (1974). Intergenerational equity and exhaustible resources. *Review of Economic Studies,* **41**, 29–45.

Solow, R. M. (1985). Economic history and economics. *The American Economic Review,* **42**(2), 328–31.

Solow, R. M. (1992). *An Almost Practical Step towards Sustainability.* Washington, DC: Resources for the Future.

Sombart, W. (1929/1950). *Die drei Nationalökonomien: Geschichte und System der Lehre von der Wirtschaft.* Munich: Von Duncker & Humblot.

Sommer, R. (1996). *Personal Space: the Behavioral Basis.* Englewood Cliffs, NJ: Prentice Hall.

Sornette, D. (2000). *Critical Phenomena in Natural Sciences: Chaos, Fractals, Selforganization.* Heidelberg: Springer.

Sornette, D. (2003). *Why Stock Markets Crash.* Princeton, NJ: Princeton University Press.

Spaargaren, G. (1987). Environment and sociology. Environmental sociology in the Netherlands. *The Netherlands' Journal of Sociology,* **23**(1), 54–72.

Spash, C. L. (2008). Deliberative monetary valuation (DMV) and the evidence for a new value theory. *Land Economics,* **84**(3), 469–88.

Specht, R. (1995). Soziologie. In J. Ritter & K. Gründer (eds). *Historisches Wörterbuch der Philosophie,* Vol. 5. Basel: Schwabe, pp. 186–202.

Spencer, H. (1855). *The Principles of Psychology.* London: Longman, Brown, Green, and Longmans.

Spencer, H. (1898/2004). Revision of biology and first principles (October 1895–1900). In D. Duncan (ed.). *The Life and Letters of Herbert Spencer.* Whitefish, MT: Kessinger Publishing, pp. 395–424.

Spergel, D. N., Verde, L., Peiris, H. V. et al. (2003). First-year Wilkinson Microwave Anisotropy Probe (WMAP) observations: determination of cosmological parameters. *Astrophysical Journal Supplement Series,* **148**(1), 175–94.

Spielmann, M., de Haan, P. & Scholz, R. W. (2008). Environmental rebound effects of high-speed transport technologies: a case study of climate change rebound effects of a future underground maglev train system. *Journal of Cleaner Production,* **16**(13), 1388–98.

Spitzer, L. (1942). Milieu and ambiance: an essay in historical semantics. *Philosophy and Phenomenological Research,* **3**(2), 169–218.

Spix, J. B. & von Martius, C. F. (1823–1831). *Reise in Brasilien auf Befehl Sr. Majestät Maximilian Joseph I. König von Baiern in den Jahren 1817–1820 gemacht und beschrieben. 3 Bände und 1 Atlas.* Munich: M. Lindaue.

Springer, S. P. & Deutsch, G. (1993). *Left Brain, Right Brain.* New York, NY: Freeman.

Sprockhoff, A. (1880). *Grundzüge der Botanik: Ein Hülfsbuch für den Schulgebrauch und zum Selbstunterricht. Anordnungen der Pflanzen, Bau, Gestalt und Leben, Systematik, Charakteristik der Familien, Beschreibung der Gattungen und Arten,* 9th edn. Leipzig: Carl Meyer.

Srinivasan, A. & Trish, M. P. (2009). Respiratory protection against influenza. *Journal of the American Medical Association,* **302**(17), 1903–4.

St Martin, K. (2001). Making space for community resource management in fisheries. *Annals of the Association of American Geographers,* **91**(1), 122–42.

Stanford Encyclopedia of Philosophy (SEP). (2009). Anaxorgoras. Retrieved 2009, January 12, from University of Stanford: http://plato.stanford.edu/entries/alcmaeon/

Starr, C. (1969). Social benefit versus technological risk. *Science,* **165**(3899), 1232–8.

Stauffacher, M., Flüeler, T., Krütli, P. & Scholz, R. W. (2008). Analytic and dynamic approach to collaborative landscape planning: a transdisciplinary case study in

Stauffacher, M., Günther, S., Krütli, P. *et al.* (2006). Transdisciplinary case study as a tool for collaborative planning of sustainable tourism development in the Seychelles. In M. Stauffacher (ed.). *Beyond Neocorporatism: New Practices of Collective Decision Making. Transdisciplinary Case Studies as a Means for Collaborative Learning Processes in Sustainable Development.* Thesis for Doctor of Philosophy. Zurich: University of Zurich, Faculty of Arts, pp. 60–88.

Stauffacher, M., Krütli, P., Moser, C. & Scholz, R. W. (forthcoming). The societal context drives individual risk perception: an empirical investigation of the acceptance of disposal projects for nuclear waste in Switzerland.

Stauffacher, M., Walter, A. I., Lang, D. J., Wiek, A. & Scholz, R. W. (2006). Learning to research environmental problems from a functional socio-cultural constructivism perspective: the transdisciplinary case study approach. *International Journal of Sustainability in Higher Education,* **7**(3), 252–75.

Stavins, R. N. (1998). What can we learn from the grand policy experiment? Lessons from SO_2 allowance trading. *Journal of Economic Perspectives,* **12**(3), 69–88.

Stearn, W. T. (1959). The background of Linnaeus's contributions to the nomenclature and methods of systematic biology. *Systematic Zoology,* **8**(1), 4–22.

Steg, L. (2005). Car use: lust and must. Instrumental, symbolic and affective motives for car use. *Transportation Research Part A – Policy and Practice,* **39**(2–3), 147–62.

Steg, L. & Vlek, C. (2009). Encouraging pro-environmental behaviour: an integrative review and research agenda. *Journal of Environmental Psychology,* **29**(3), 309–17.

Steiner, M., Vermeulen, L. C., Mullahy, J. & Hayney, M. S. (2002). Factors influencing decisions regarding influenza vaccination and treatment: a survey of healthcare workers. *Infection Control and Hospital Epidemiology,* **23**(10), 625–7.

Steiner, P. (2006). Physiocracy and French pre-classical political economy. In J. E. Biddle, J. B. Davis & W. J. Samuels (eds). *A Companion to the History of Economic Thought.* Malden, MA: Blackwell Publishing, pp. 61–76.

Steinman, L. (2004). Elaborate interactions between the immune and nervous systems. *Nature Immunology,* **5**(6), 575–81.

Stenonis, N. (1669). *De Solido intra Solidum Naturaliter contento Dissertationis Prodromus.* Florence: ex Typographia sub signo Stellae.

Stephens, J. C., Hernandez, M. E., Roman, M., Graham, A. C. & Scholz, R. W. (2008). Higher education as a change agent for sustainability in different cultures and contexts. *International Journal of Sustainability in Higher Education,* **9**(3), 317–38.

Sterman, J. D. (2000). *Business Dynamics: Systems Thinking and the Modelling for a Complex World.* Boston, MA: McGraw-Hill.

Sterman, J. D. (2001). System dynamics modeling: tools for learning in a complex world. *California Management Review,* **43**(4), 64–88.

Sterman, J. D. (2007). Exploring the next great frontier: system dynamics at fifty. *System Dynamics Review,* **23**(2–3), 89–93.

Stern, A. M. & Markel, H. (2005). The history of vaccines and immunization: familiar patterns, new challenges. *Health Affairs,* **24**(3), 611–21.

Stern, D. I. (2004). The rise and fall of the environmental Kuznets curve. *World Development,* **32**(8), 1419–39.

Stern, D. I. & Common, M. S. (2001). Is there an environmental Kuznets curve for sulfur? *Journal of Environmental Economics and Management,* **41**(2), 162–78.

Stern, P. C. (1992). What psychology knows about energy-conservation. *American Psychologist,* **47**(10), 1224–32.

Stern, P. C. (1993). A second environmental science: human-environment interactions. *Science,* **260**(5116), 1897–9.

Stern, P. C. (2000). Toward a coherent theory of environmentally significant behavior. *Journal of Social Issues,* **56**(3), 407–24.

Stern, P. C. & Fineberg, H. V. (eds). (1996). *Understanding Risk: Informing Decisions in a Democratic Society.* Washington: National Academies Press.

Stern, W. (1938). *General Psychology from a Personalistic Standpoint.* New York, NY: Macmillan.

Stets, J. E. & Biga, C. F. (2003). Bringing identity into environmental sociology. *Sociological Theory,* **21**(4), 398–423.

Stigler, G. J. (1950). The development of utility theory II. *The Journal of Political Economy,* **LVII**(5), 373–96.

Stokes, D. E. (1997). *Pasteur's Quadrant. Basic Science and Technological Innovation.* Washington, DC: Brookings Institution Press.

Stokols, D. (1987). Conceptual strategies of environmental psychology. In D. Stokols & I. Altmann (eds). *Handbook of Environmental Psychology,* Vol. I. New York, NY: John Wiley & Sons, pp. 41–71.

Stokols, D. & Altman, I. (1987). Introduction. In D. Stockols & I. Altman (eds). *Handbook of Environmental Psychology,* Vol. I. New York, NY: Wiley, pp. 1–4.

Stokols, D. & Shumaker, S. A. (1981). People in places: a transactional view of settings. In J. Harvey (ed.).

Cognition, Social Behavior and the Environment. Hillsdale, NJ: Erlbaum, pp. 441–88.

Stone, L. & Ezrati, S. (1996). Chaos, cycles and spatiotemporal dynamics in plant ecology. *Journal of Ecology*, **84**(2), 279–91.

Stoner, G. R., Albright, T. D. & Ramachandran, V. S. (1990). Transparency and coherence in human motion perception. *Nature*, **344**(6262), 153–5.

Stoner, J. A. F. (1961). A comparison of individual and group decision involving risk. Unpublished Master's Thesis, School of Industrial Management, MIT, Cambridge, MA.

Stormsurf. (2007). Kandui. Retrieved May 26, 2007, from http://www.stormsurf.com/page2/tutorials/wavebasics.shtml

Stotz, K. (2006a). With "genes" like that, who needs an environment? Postgenomics's argument for the "Ontogeny of Information". *Philosophy of Science*, **73**(5), 905–17.

Stotz, K. (2006b). Molecular epigenesis: distributed specificity as a break in the central dogma. *History and Philosophy of the Life Sciences*, **26**(4), 527–44.

Strathman, A., Gleicher, F., Boninger, D. S. & Edwards, C. S. (1994). The consideration of future consequences: weighing immediate and distant outcomes of behavior. *Journal of Personality and Social Psychology*, **66**(4), 742–52.

Strogatz, S. H. (1994). *Nonlinear Dynamics and Chaos: with Applications in Physics, Biology, Chemistry, and Engineering*. Reading, MA: Addison-Wesley.

Strogatz, S. H. (2001). Exploring complex networks. *Nature*, **410**(6825), 268–76.

Strome, S. & Kelly, W. G. (2007). Epigenetics regulation of the X chromosomes in C. elegans. In C. D. Allis, T. Jenuwein, D. Reinberg & M.-L. Capparos (eds). Epigenetics. New York, NY: Cold Spring Laboratory Press. pp. 291–305.

Stroop, J. R. (1935). Studies of interference in serial verbal reactions. *Journal of Experimental Psychology*, **18**(6), 643–62.

Struik, D. J. (1987). *A Concise History of Mathematics*, 4th edn. New York, NY: Dover Publishing.

Stuart, J. A. (2001). Clean manufacturing. In G. Salvendy (ed.). *Handbook of Industrial Engineering: Technology and Operations Management*. New York, NY: John Wiley, & Sons, pp. 530–43.

Sturn, D. (2000). Decentralized industrial policies in practice: the case of Austria and Styria. *European Planning Studies*, **8**(2), 169–82.

Suddendorf, T. & Busby, J. (2005). Making decisions with the future in mind: developmental and comparative identification of mental time travel. *Learning and Motivation*, **36**(2), 110–25.

Suh, S. (ed.). (2008). *Handbook on Input-Output Analysis in Industrial Ecology*. Cheltenham: Edward Elgar.

Sumner, W. G. (1906). *Folkways*. New York, NY: Ginn.

Sun Tzu. (500 BC/1910). *The Art of War* (L. Giles, Trans.). London: Luzac.

Sundstrom, E., Bell, P. A., Busby, P. L. & Asmus, C. (1996). Environmental psychology 1989–1994. *Annual Review of Psychology*, **47**, 485–512.

Susskind, L. E. (1975). *The Land Use Controversy in Massachusetts – Case Studies and Policy Options*. Cambridge, MA: The MIT Press.

Susskind, L. E., Levy, P. F. & Thomas-Larmer, J. (2000). *Negotiating Environmental Agreements: How to Avoid Escalating Confrontation, Needless Costs, and Unnecessary Litigation*. Washington, DC: Island Press.

Susskind, L. E. & Madigan, D. (1984). New approaches to resolving disputes in the public sector. *The Justice System Journal*, **9**(2), 179–203.

Susskind, L. E., McKearnen, S. & Thomas-Larmer, J. (eds). (1999). *The Consensus Building Handbook: a Comprehensive Guide to Reaching Agreement*. Thousand Oaks, CA: Sage.

Susskind, L. E., van der Wansem, M. & Ciccareli, A. (2003). Mediating land use disputes in the United States: pros and cons. *Environments – A Journal of Interdisciplinary Studies*, **31**(2), 59–68.

Svedberg, T., Pedersen, K. O. & Bauer, J. H. (eds). (1940). *The Ultracentrifuge*. Oxford: The Clarendon Press.

Svenson, O. & Karlsson, G. (1989). Decision-making, time horizons, and risk in the very long-term perspective. *Risk Analysis*, **9**(3), 385–99.

Swedberg, R. (1990). *Economics and Sociology: Redefining Their Boundaries: Conversations with Economists and Sociologists*. Princeton, NJ: Princeton University Press.

Swedish Energy Agency. (2009). *Energy in Sweden 2008*. Eskilstuna: Swedish Energy Agency.

Swedish Government. (2004). Introduktion av förnybara fordonsbräslen SOU2004:133. Retrieved from http://www.sweden.gov.se/content/1/c6/03/67/01/bb423867.pdf.

Tàbara, J. D. & Pahl-Wostl, C. (2007). Sustainability learning in natural resource use and management. *Ecology and Society*, **12**(2).

Tainter, J. A. (1988). *The Collapse of Complex Societies*. Cambridge: Cambridge University Press.

Tainter, J. A. (2000). Problem solving: complexity, history, sustainability. *Population & Environment*, **22**(1), 3–41.

Tajfel, H. (1981). *Human Groups and Social Categories*. Cambridge: Cambridge University Press.

References

Tajfel, H. (1982). Social psychology of intergroup relations. *Annual Review of Psychology*, **33**, 1–39.

Takács-Sánta, A. (2004). The major transitions in the history of human transformation of the biosphere. *Human Ecological Review*, **11**(1), 51–66.

Tallberg, J. (2002). Delegation to supranational institutions: why, how, and with what consequences. *West European Politics*, **25**(1), 23–46.

Tallberg, J. (2003). *European Governance and Supranational Institutions*. London: Routledge.

Tang, J., Li, Y., Eames, I., Chan, P. & Ridgway, G. (2006). Factors involved in the aerosol transmission of infection and control of ventilation in healthcare premises. *Journal of Hospital Infection*, **64**, 100–14.

Tanner, A. (1982). *Spulen-Weben-Sticken, Die Industrialisierung in Appenzell Ausserrhoden*. Zurich: Juris Druck.

Tansley, A. G. (1935). The use and abuse of vegetation concepts and terms. *Ecology*, **16**(3), 248–307.

Tarascio, V. J. (1971). Keynes on sources of economic growth. *Journal of Economic History*, **31**(2), 429–44.

Tauber, A. I. (1995, May). *Historical and Philosophical Perspectives Concerning Immune Cognition*. Santa Maria Imbar, Italy.

Tauber, A. I. (2000). Moving beyond the immune self? *Seminars in Immunology*, **12**(3), 241–8.

Tauber, A. I. (2008). The immune system and its ecology. *Philosophy of Science*, **75**(2), 224–45.

Tedlock, B. (1992). *Time and the Highland Maya*. Revised edition. Albuquerque, NM: University of New Mexico Press.

Tegmark, M. (1998). The interpretation of quantum mechanics: many worlds or many words? *Fortschritte der Physik*, **46**, 855–62.

Terrier, O., Essere, B., Yver, M. et al. (2008, March 12–14). *Cold Oxygen Plasma Technology Efficiency against Different Airborne Respiratory Viruses*. Saariselska, Finland.

Terry, D. J., Hogg, M. A. & White, K. M. (1999). The theory of planned behaviour: self-identity, social identity and group norms. *British Journal of Social Psychology*, **38**(3), 225–44.

Terry, K. D. & Tucker, W. H. (1968). Biologic effects of supernovae. *Science*, **159**(3813), 421–3.

Tetlock, P. E. (2003). Thinking the unthinkable: sacred values and taboo cognitions. *Trends in Cognitive Sciences*, **7**(7), 320–4.

The Field Museum. (2003). Peru – report at a glance. *Rapid Biological Inventories*. Retrieved March 18, 2008, from http://fm2.fieldmuseum.org/rbi/results_per12.asp

Thelen, K. (1999). Historical institutionalism in comparative politics. *Annual Review of Political Science*, **2**, 369–404.

Thews, G., Mutschler, E. & Vaupel, P. (1999). *Anatomie, Physiologie, Pathophysiologie des Menschen*, 5th edn. Stuttgart: Wissenschaftliche Verlagsgesellschaft.

Thieneman, A. (1926). Der Nahrungskreislauf im Wasser. *Verhandlungen der Deutschen Zoologischen Gesellschaft*, **31**, 29–79.

Thiesen, J., Cluistensen, T. S., Kristensen, T. G. et al. (2008). Rebound effects of price differences. *International Journal of Life Cycle Assessment*, **13**(2), 104–14.

Thøgersen, J. (2008). Social norms and cooperation in real-life social dilemmas. *Journal of Economic Psychology*, **29**(4), 458–72.

Thompson Klein, J. & Newell, W. H. (1997). Advancing interdisciplinary studies. In J. Gaff, J. Radcliff & Associates (eds). *Handbook of Undergraduate Curriculum*. San Francisco, CA: Jossey-Bass, pp. 393–413.

Thom, R. (1970). Topological models in biology. In C. H. Waddington (ed.). *Towards a Theoretical Biology*, Vol. 3. Edinburgh: Edinburgh University Press, pp. 89–116.

Thomas, R. & D'Ari, R. (1989). *Biological Feedback*. Boca Raton, FL: CRC Press.

Thomas, R. & Kaufman, M. (2001a). Multistationarity, the basis of cell differentiation and memory. I. Structural conditions of multistationarity and other nontrivial behavior. *Chaos*, **11**(1), 170–9.

Thomas, R. & Kaufman, M. (2001b). Multistationarity, the basis of cell differentiation and memory. II. Logical analysis of regulatory networks in terms of feedback circuits. *Chaos*, **11**(1), 180–95.

Thomas, Y., Vogel, G., Wunderli, W. et al. (2008). Survival of influenza virus on banknotes. *Applied and Environmental Microbiology*, **74**(10), 3002–7.

Thompson Klein, J. (2004). Prospects for transdisciplinarity. *Futures*, **36**, (4) 515–26.

Thompson Klein, J. (ed.). (1990). *Interdisciplinarity: History, Theory, and Practice*. Detroit: Wayne State University Press.

Thompson Klein, J., Grossenbacher-Mansuy, W., Häberli, R. et al. (eds). (2001). *Transdisciplinarity: Joint Problem Solving among Science, Technology, and Society. An Effective Way for Managing Complexity*. Basel: Birkhäuser.

Thucydides. (550 BC/1910). *History of the Peloponnesian War*, book 1, section 141 (R. Crawley, Trans.). New York, NY: Dutton.

Thurstone, L. L. (1931). The measurement of social attitudes. *Journal of Abnormal and Social Psychology*, **26**(3), 249–69.

Tibbs, H. B. C. (1992). Industrial ecology: an environmental agenda for industry. *Whole Earth Review,* **4**, 4–19.

Tietenberg, J. E. (1985). *Emission Trading.* Oxford: Blackwell.

Tietje, O. (2005). Identification of a small reliable and efficient set of consistent scenarios. *European Journal of Operational Research,* **162**(2), 418–32.

Tillman, D., Cassmann, K. G., Matson, P., Naylor, R. & Polansky, S. (2002). Agricultural sustainability and intense production practices. *Nature,* **418**(6898), 671–7.

Tilton, J. E. (2003). *On Borrowed Time? Assessing the Threat of Mineral Depletion.* Washington, DC: Resources for the Future.

Tilton, J. E. & Lagos, G. (2007). Assessing the long-run availability of copper. *Resources Policy,* **32**, 19–23.

Tinbergen, N. (1951). *The Study of Instinct.* Oxford: Oxford University Press.

Tobin, G. A. & Montz, B. E. (1997). *Natural Hazards. Explanation and Integration.* New York, NY: Guilford Press.

Todd, P. M. & Gigerenzer, G. (2007). Environments that make us smart: ecological rationality. *Current Directions in Psychological Science,* **16**(3), 167–71.

Tolasch, T. (2008). Sex pheromone of Ectinus aterrimus (Linné, 1761) (Coleoptera: Elateridae). *Chemoecology,* **18**(3), 177–80.

Toledo, V. M., Ortiz-Espejel, B., Cortés, L., et al. (2003). The multiple use of tropical forests by indigenous peoples in Mexico: a case of adaptive management. *Ecology and Society,* **7**(3).

Tolman, E. C. (1933). *Purposive Behavior in Animals and Men.* New York, NY: Century.

Tolman, E. C. (1948). Cognitive maps in rats and men. *Psychological Review,* **55**, 189–208.

Tolman, E. C. & Brunswik, E. (1935). The organism and the causal texture of the environment. *Psychological Review,* **42**(1), 43–77.

Tomonari, Y., Kurata, Y., Kawasuso, T. et al. (2003). Testicular toxicity study of di(2-ethylhexyl)phthalate (DEHP) in juvenile common marmoset. *Toxicological Sciences,* **72**, 385.

Tonn, B. E., Hemrick, A. & Conrad, F. (2006). Cognitive representations of the future: survey results. *Futures,* **38**(7), 810–29.

Torre, A. & Gilly, J.-P. (2000). On the analytical dimension of proximity dynamics. *Regional Studies,* **34**(2), 169–80.

Torstendahl, R. (1993). The transformation of professional education in the nineteenth century. In S. Rosenblatt & B. Wittrock (eds). *The European and American University since 1800: Historical and Sociological Essays.* Cambridge: Cambridge University Press, pp. 109–41.

Toulmin, S. E. (1975). The discovery of time. In C. C. Albritton (ed.). *Philosophy of Geohistory.* Stroudsbourg: Dowden, Hutchinson & Ross, pp. 11–23.

Townsend, C. R. Begon, M., & Harper, J. L. (eds). (2003) *Essentials of Ecology,* 2nd edn. Malden, MA: Blackwell Science Ltd.

Toye, J. (1997). Keynes on population and economic growth. *Cambridge Journal of Economics,* **21**(1), 1–26.

Train, K. E., Davis, W. B. & Levine, M. D. (1997). Fees and rebates on new vehicles: impacts on fuel efficiency, carbon dioxide emissions, and consumer surplus. *Transportation Research Part E – Logistics and Transportation Review,* **33**(1), 1–13.

Trawik, P. B. (2003). *The Struggle for Water in Peru. Comedy and Tragedy in the Andean Commons.* Palo Alto, CA: Stanford University Press.

Treadgold, W. (1997). *A History of the Byzantine State and Society.* Palo Alto, CA: Stanford University Press.

Treffert, D. A. (1988). The idiot savant: a review of the syndrome. *American Journal of Psychiatry,* **145**(5), 563–72.

Trewavas, A. (2006). A brief history of systems biology – "Every object that biology studies is a system of systems." Francois Jacob (1974). *Plant Cell,* **18**(10), 2420–30.

Trigilia, C. (2002). *Economic Sociology: State, Market, and Society in Modern Capitalism.* Oxford: Blackwell.

Troll, C. (1939). Luftbildplan und ökologische Bodenforschung. *Zeitschrift der Gesellschaft für Erdkunde zu Berlin,* **74**(7/8), 241–98.

Troll, C. (1971). Landscape ecology (geo-ecology) and biocenology: a terminological study. *Geoforum,* **2**(4), 43–6.

Trudgill, S. (2007). Tansley, A.G. 1935: the use and abuse of vegetational concepts and terms. Ecology, 16, 284–307. *Progress in Physical Geography,* **31**(5), 517–22.

Tsou, J. Y. (2006). Genetic epistemology and Piaget's philosophy of science – Piaget vs. Kuhn on scientific progress. *Theory & Psychology,* **16**(2), 203–24.

Tudor, T. L., Barr, S. W. & Gilg, A. W. (2008). A novel conceptual framework for examining environmental behavior in large organizations: a case study of the Cornwall National Health Service (NHS) in the United Kingdom. *Environment and Behavior,* **40**(3), 426–50.

Tukker, A., Cohen, M. J., de Zoysa, U. et al. (2006). The Oslo Declaration on sustainable consumption. *Journal of Industrial Ecology,* **10**(1–2), 9–14.

Turgot, A. R. J. (1767/1914). Sur la Mémoire de Saint-Péravy. In G. Schelle (ed.). *Œuvres de Turgot,* Vol. II. Paris: Alcan, pp. 641–58.

Turgot, A. R. J. (1770). *Réflexions sur la Formation et les Distributions des Richesses.* Paris: Lacombe.

References

Turgot, A. R. J. (1970). Mémoire sur les Prêts de l'Argent. In G. Schelle (ed.). *Œuvres de Turgot*. Paris: Alcan, pp. 154–202.

Türk, K. & Reiss, M. (1995). Organisation. In Görres-Gesellschaft (ed.). *Staatslexikon*, Vol. 7. Freiburg: Herder, pp. 198–206.

Turk, M. H. (2006). The fault line of axiomatization: Walras' linkage of physics with economics. *European Journal of the History of Economic Thought*, **13**(2), 195–212.

Turner, B. L., Kasperson, R. E., Matson, P. A. et al. (2003). A framework for vulnerability analysis in sustainability science. *Proceedings of the National Academy of Sciences of the United States of America*, **100**(14), 8074–9.

Turner, B. L., Clark, W. C., Kates, R. W., Richards, J. F., Mathews, J. T., & Meyer, W. B. (Eds.). (1990). *The earth as transformed by human action – global and reginal changes in the biosphere over the past 300 years*. Cambridge: Cambridge University Press.

Turner, B. L., Lambin, E. F. & Reenberg, A. (2008). Land Change Science Special Feature: the emergence of land change science for global environmental change and sustainability. *Proceedings of the National Academy of Sciences of the United States of America*, 104(52), 20666–71. [Correction] **105**(7), 2751.

Turner, B. L. & Robbins, P. (2008). Land-change science and political ecology: similarities, differences, and implications for sustainability science. *Annual Review of Environment and Resources*, **33**, 295–316.

Turner, J. C. (1984). Social identification and psychological group formation. In H. Tajfel (ed.). *The Social Dimension*, Vol. 1: European Developments in Social Psychology. Cambridge: Cambridge University Press, pp. 518–40.

Turner, J. T. (2007). *The Tinkerer's Accomplice: how Design Emerges from Life Itself*. Cambridge, MA: Harvard University Press.

Turner, M. G. (2005). Landscape ecology: what is the state of the science? *Annual Review of Ecology Evolution and Systematics*, **36**, 319–44.

Turner, R. K. & Daily, G. C. (2008). The ecosystem services framework and natural capital conservation. *Environmental & Resource Economics*, **39**(1), 25–35.

Turner, W. (1908). Giordano Bruno. Retrieved 2009, October 20, from The Catholic Encyclopedia: http://www.newadvent.org/cathen/03016a.htm

Tversky, A. (1972). Elimination by aspects – theory of choice. *Psychological Review*, **79**(4), 281–99.

Tversky, A. & Kahneman, D. (1973). Availability: a heuristic for judging frequency and probability. *Cognitive Psychology*, **5**(2), 207–32.

Tversky, B. (1991). Spatial mental models. *Psychology of Learning and Motivation – Advances in Research and Theory*, **27**, 109–45.

Udall, J. R. (1985). The Tucson Paradox. *Audubon*, **87**(1), 98–9.

Udo de Haes, H. A. & Heijungs, R. (2009). Analysis of physical interactions between the economy and the environment. In J. Boersema & L. Reinders (eds). *Principles of Environmental Sciences*. Heidelberg: Springer, pp. 207–37.

Udo de Haes, H. A., Heijungs, R., Suh, S. & Huppes, G. (2004). Three strategies to overcome the limitations of life-cycle assessment. *Journal of Industrial Ecology*, **8**(3), 19–32.

Udo de Haes, H. A. & Klijn, F. (1994). Environmental policy and ecosystem classification. In F. Klijn (ed.). *Ecosystem Classification for Environmental Management*. Dordrecht: Kluwer, pp. 1–21.

Udo de Haes, H. A., Van der Voet, E. & Kleijn, R. (1997). Substance flow analysis (SFA), an analytical tool for integrated chain management. In S. Bringezu, M. Fischer-Kowalski, R. Kleijn & P. Viveka (eds). *Regional and National Material Flow Accounting: from Paradigm to Practice of Sustainability*. Proceedings of the ConAccount workshop 21–23 January, 1997 in Leiden, The Netherlands. Wuppertal: Wuppertal Institute for Climate, Environment and Energy.

Ullmann, A. (2007). Pasteur-Koch: distinctive ways of thinking about infectious diseases. *Microbe*, **2**(8), 383–7.

Ulrich, B. (1990). Waldsterben – forest decline in West-Germany. *Environmental Science & Technology*, **24**(4), 436–41.

Ulrich, R. S. (1986). Human responses to vegetation and landscapes. *Landscape and Urban Planning*, **13**(1), 29–44.

UVEK. (2000). *Die Siedlungsfläche der Schweiz wächst weiter*. Press release *January 1, 2000*. Bern: Bundesamt für Umwelt Verkehr, Energie und kommunikation.

UN-Energy. (2007). *Sustainable Bioenergy: a Framework for Decision Makers*. New York, NY: United Nations.

Ungar, G. & Oceguera-Navarro, C. (1965). Transfer of habituation by material extracted from brain. *Nature*, **207**(4994), 301–2.

Ungar, S. (1992). The rise and (relative) decline of global warming as a social problem. *Sociological Quarterly*, **33**(4), 483–501.

Ungar, S. (1995). Social scares and global warming – beyond the Rio Convention. *Society & Natural Resources*, **8**(5), 443–56.

United Nations (UN). (1987, December 11). Report of the World Commission on Environment and Development. General assembly resolution 42/187. Retrieved June 13, 2007, from http://www.un.org/documents/ga/res/42/ares42-187.htm

United Nations (UN). (1992). *Agenda 21. Report of the United Nations Conference on Environment and Development annex I and II*. New York, NY: United Nations.

United Nations (UN). (1993). *World Population Prospects: The 1992 Revision*. New York, NY: United Nations.

United Nations (UN). (2004). *World Population to 2300*. New York: UN, Department of Economic and Social Affairs.

United Nations (UN). (2006). *United Nations Conference on Trade and Development: The Emerging Biofuel Market. Regulatory, Trade and Development Implications*. New York, NY and Geneva: United Nations.

United Nations Economic Commission for Europe (UNECE) (1998). *Convention on Access to Information, Public Participation in Decision-Making and Access to Justice in Environmental Matters*. Aarhus: United Nations.

United Nations International Children's Emergency Fund (UNICEF). (2009). Information by country and programme. Retrieved October 2, 2009, from http://www.unicef.org/index.php

US Burean of Mines & US Geological Survey. (1980). Principles of a Resource/Reserve Classification for Minerals. US Geological Survey Circular, 831.

US Census Bureau. (2008). Las Vegas (city), Nevada. State & County QuickFacts, Retrieved June 5, 2010 from http://quickfacts.census.gov/qfd/states/32/3240000.html

US Congress. (1936). *Flood Control Act*. Washington, DC: US Congress.

US Department of Energy. (2010). Critical materials strategy. Retrieved December 28, 2010, from http://www.energy.gov

US Geological Survey (USGS). (2005). *Mineral Commodity Summaries 2005*. Washington DC: USGS.

Vaggi, G. (1985). The role of profits in physiocratic economics. *History of Political Economy*, **17**(3), 367–84.

Vaidman, L. (2002). Many-worlds interpretation of quantum mechanics. *Stanford Encyclopedia of Philosophy*. Retrieved May 10, 2010, from http://plato.stanford.edu/entries/qm-manyworlds

Vaillancourt, J.-G. (1995). Sociology of the environment: from human ecology to ecosociology. In M. D. Mehta & É. Wuellet (eds). *Environmental Sociology: Theory and Practice*. New York, NY: Captus Press, pp. 3–32.

van Asselt, M. B. A. & Rijkens-Klomp, N. (2002). A look in the mirror: reflection on participation in Integrated Assessment from a methodological perspective. *Global Environmental Change*, **12**, 167–84.

van de Kerkhof, M. & Leroy, P. (2000). Recent environmental research in the Netherlands: towards post-normal science? *Futures*, **32**(9–10), 899–911.

van de Kerkhof, M. & Wieczorek, A. (2005). Learning and stakeholder participation in transition processes towards sustainability: methodological considerations. *Technological Forecasting and Social Change*, **72**(6), 733–47.

van der Meer, E. (2006). Zeit. In J. Funke & P. Frensch (eds). *Handwörterbuch Psychologie, Band Allgemeine Psychologie: Kognition*. Gottingen: Hogrefe, pp. 75–85.

van der Pligt, J. & de Boer, J. (1991). Contaminated soil: public reactions, policy decisions, and risk communication. In R. E. Kasperson & P. J. M. Stallen (eds). *Communicating Risk to the Public*. Dordrecht: Kluwer, pp. 127–44.

van der Schoot, A. (2005). *Die Geschichte des Goldenen Schnitts. Aufstieg und Fall der göttlichen Proportion*. Stuttgart: Frommann-Holzboog.

van der Voet, E., Kleijn, R. & Udo de Haes, H. A. (1996). Nitrogen pollution in the European union – origins and proposed solutions. *Environmental Conservation*, **23**(2), 120–32.

Van Eck, P. S., Jager, W. & Leeflang, P. S. H. (2011). Opinion leaders' role in innovation diffusion: a simulation study. *Journal of Product Innovation and Management*, **28**(2), 187–203.

Van Tatenhove, J. P. M. & Leroy, P. (2003). Environment and participation in a context of political modernisation. *Environmental Values*, **12**(2), 155–74.

Vanengelsdorp, D., Hayes, J., Underwood, R. M. & Pettis, J. (2008). A survey of honey bee colony losses in the US, fall 2007 to spring 2008. *Public Library of Science One*, **3**(12), e4071.

Veblen, T. (1900). The preconceptions of economic science: III. *Quarterly Journal of Economics*, **14**(2), 240–69.

Veblen, T. (1919/1990). *The Place of Science in Modern Civilisation*. New York, NY: Viking Press.

Vedhara, K. & Irwin, M. (2005). *Human Psychoneuroimmunology*. Oxford: Oxford University Press.

Veitsch, R. & Arkkelin, D. (1995). *Environmental Psychology: an Interdisciplinary Perspective*. New York, NY: Upper Saddle River.

Vennix, J. A. M. (1996). *Group Model Building: Facilitating Team Learning Using System Dynamics*. Chichester; John Wiley & Sons.

Verba, S. & Nie, N. H. (1987). *Participation in America. Political Democracy and Social Equality*. Chicago, IL: University of Chicago Press.

Verhoef, E. T. (1999). Externalities. In J. C. J. M. van den Bergh (ed.). *Handbook of Environmental Economics*. Cheltenham: Edward Elgar, pp. 197–214.

Verleger, R. (1988). Event-related potentials and memory – a critique of the context updating hypothesis and an alternative interpretation of P3. *Behavioral and Brain Sciences*, **11**(3), 343–56.

Vermillion, D. L. & Sagardoy, J. A. (1999). *Transfer of Irrigation Management Services: Guidelines* (No. 9251043086). Rome: FAO.

Vidal, J. (2006, February 8). Sweden plans to be world's first oil-free economy. *The Guardian*. Retrieved October, 5, 2009, from http://www.guardian.co.uk/environment/2006/feb/08/frontpagenews.oilandpetrol/print

Villalba, G., Liu, Y., Schroder, H. & Ayres, R. U. (2008). Global phosphorus flows in the industrial economy from a production perspective. *Journal of Industrial Ecology*, **12**(4), 557–69.

Vincke, P. (1992). *Multicriteria Decision-Aid*. Chichester: John Wiley & Sons.

Viner, J. (1931). Cost curves and supply curves. *Zeitschrift für Nationalökonomie*, **3**, 23–46.

Vining, J. A. (1992). Environmental emotions and decisions – a comparison of the responses and expectations of forest managers, an environmental group, and the public. *Environment and Behavior*, **24**(1), 3–34.

Virchow, R. (1958). *Disease, Life, and Man*. Stanford, CA: Stanford University Press.

Vitousek, P. M., Aber, J., Howarth, R. W. *et al.* (1997). Human alteration of the global nitrogen cycle: sources and consequences. *Ecological Applications*, **7**(3), 737–50.

Vlek, C. A. J. & Keren, G. (1992). Behavioral decision theory and environmental risk management: assessment and resolution of four "survival" dilemmas. *Acta Psychologica*, **80**, 249–78.

Vlek, C. A. J. & Stallen, P. J. (1980). Rational personal aspects of risk. *Acta Psychologica*, **45**, 273–300.

Volger, G. (2001). *Risikoeinschätzung und Medien: Beispiel Rinderwahnsinn*. Unpublished Semester Thesis, ETH Zurich, Zurich.

Volkov, I., Banavar, J. R., Hubbell, S. P. & Maritan, A. (2003). Neutral theory und relative species abundance in ecology. *Nature*, **424**(6952), 1035–7.

von Bertalanffy, L. (1950a). An outline of general system theory. *The British Journal for the Philosophy of Science*, **1**(2), 134–65.

von Bertalanffy, L. (1950b). The theory of open systems in physics and biology. *Science*, **111**(2872), 23–9.

von Bertalanffy, L. (1955). General system theory. *Main Currents in Modern Thought*, **11**(4), 75–83.

von Bertalanffy, L. (1968). *General System Theory: Foundations, Development, and Applications*. New York, NY: Braziller.

von Blottnitz, H. & Curran, M. A. (2007). A review of assessments conducted on bio-ethanol as a transportation fuel from a net energy, greenhouse gas, and environmental life cycle perspective. *Journal of Cleaner Production*, **15**(7), 607–19.

von Carlowitz, H. C. (1713). *Sylvicultura oeconomica, oder haußwirthliche Nachricht und naturmäßige Anweisung zur wilden Baum-Zucht*. Leipzig: Johann Friedrich Braun.

von Foerster, H. (1982). *Observing Systems*. Seaside, CA: Intersystems Publications.

von Foerster, H. (1987). *Cybernetics. Encyclopaedia for Artificial Intelligence*. New York: John Wiley & Sons.

von Glasersfeld, E. (1996). *Radikaler Konstruktivismus*. Frankfurt am Main: Suhrkamp.

von Humboldt, A. (1845). *Kosmos: Entwurf einer physikalischen Weltbeschreibung*, Vol. 1. Stuttgart und Tübingen: Cotta'scher Verlag.

von Humboldt, A. (1889). *Gesammelte Werke: 12 Bände*. Stuttgart: Cotta'sche Buchhandlung.

von Neumann, J. & Morgenstern, O. (1944). *Theory of Games and Economic Behavior*. Princeton, NJ: Princeton University Press.

von Stackelberg, H. (1948). *Grundlagen der theoretischen Volkswirtschaftslehre*. Stuttgart: Francke.

von Thünen, J. H. (1910). *Der isolierte Staat in Beziehung auf Landwirtschaft und Nationaloekonomie*. Jena: Gustav Fischer.

von Uexküll, J. J. (1920). *Biologische Briefe an eine Dame*. Berlin: Gebrüder Paetel.

von Uexküll, J. J. (1921). *Umwelt und Innenwelt der Tiere*, 2nd edn. Berlin: Springer.

von Uexküll, J. J. (1928/1973). *Theoretische Biologie*. Frankfurt am Main: Suhrkamp.

von Uexküll, J. J. (1931). Der Organismus und die Umwelt. In H. Driesch & H. Woltereck (eds). *Das Lebensproblem im Lichte der modernen Forschung*. Leipzig: Verlag Quelle & Meyer, pp. 189–224.

von Uexküll, J. J. (1940/1970). *Die Bedeutungslehre*. Frankfurt am Main: Fischer.

von Uexküll, J. J. (1957). A stroll through the worlds of animal and men: a picture book of invisible worlds. In H. C. Schiller (ed.). *Instinctive Behavior: the Development of a Modern Concept*. New York, NY: International University Press, pp. 5–80.

von Uexküll, J. J. & Kriszat, G. (1934). *Streifzüge durch die Umwelten von Tieren und Menschen*. Berlin: Springer.

von Uexküll, T. (1987). The sign theory of Jakob von Uexküll. In M. Krampen, K. Oehler, R. Posner, T. A. Sebeok & T. von Uexküll (eds). *Classics of Semiotics*. New York, NY: Plenum Press, pp. 147–79.

von Uexküll, T. (1992). Introduction: the sign theory of Jakob von Uexküll. *Semiotica*, **89**(4), 279–315.

von Winterfeldt, D. & Edwards, W. (1986). *Decision Analysis and Behavioral Research*. Cambridge: Cambridge University Press.

Von Alvensleben, R. V. (1995). Image problems with meat – causes and consequences. *Berichte über Landwirtschaft*, **73**(1), 65–82.

Waddington, C. H. (1942a). Canalization of development and the inheritance of acquired character. *Nature*, **150**(3811), 563–5.

Waddington, C. H. (1942b). The epigenotype. *Endeavour*, **1**, 18–20.

Wade, R. (1988). *Village Republics: Economic Conditions for Collective Action in South India*. Cambridge: Cambridge University Press.

Waert, S. (2003). *The Discovery of Global Warming*. Cambridge, MA: Harvard University Press.

Waggoner, P. E. & Ausubel, J. H. (2002). A framework for sustainability science: a renovated IPAT identity. *Proceedings of the National Academy of Sciences of the United States of America*, **99**(12), 7860–5.

Waldstein, A. & Adams, C. (2006). The interface between medical anthropology and medical ethnobiology. *Journal of the Royal Anthropological Institute*, **12** (SI 2006), 95–118.

Walker, B., Gunderson, L. H., Kinzig, A. et al. (2006). A handful of heuristics and some propositions for understanding resilience in social-ecological systems. *Ecology and Society*, **11**(1), 15.

Walker, J. T., Dickinson, J., Sutton, J. M., Marsh, P. D. & Raven, N. D. H. (2008). Implications for Creutzfeldt-Jakob Disease (CJD) in dentistry: a review of current knowledge. *Journal of Dental Research*, **87**(6), 511–19.

Wallbaum, T. A. (1991). Americas lethal export – the growing trade in hazardous waste. *University of Illinois Law Review*, (3), 889–924.

Wallerstein, I. M. (1974). *The Modern World-Society. Capitalist Agriculture and the Origins of the European World-Economy in the Sixteenth Century*. New York, NY: Academic Press.

Walley, P. (1991). *Statistical Reasoning with Imprecise Probabilities*. London: Chapman and Hall.

Wallsten, T. S. & Whitfield, R. G. (1986). *Assessing the Risks to Young Children of Three Effects Associated with Elevated Blood-Lead Levels*. Argonne, IL: Argonne National Laboratory.

Walter, A. I., Helgenberger, S., Wiek, A. & Scholz, R. W. (2007). Measuring societal effects of transdisciplinary research projects: design and application of an evaluation method. *Evaluation and Program Planning*, **30**, 325–38.

Walter, A. I. & Scholz, R. W. (2007). Critical success conditions of collaborative methods: a comparative evaluation of transport planning projects. *Transportation*, **34**(2), 195–212.

Washington, S. H. (2005). *Packing Them in: an Archeology of Environmental Racism in Chicago, 1865–1954*. Lanham, MD: Lexington Books.

Wasserman, S. & Faust, K. (1994). *Social Network Analysis: Methods and Applications*. Cambridge: Cambridge University Press.

Watkins, T. (2001). The beginning of religion at the beginning of the neolithic. Retrieved April 30, 2010, from British Association of Near Eastern Archeology: http://www.arcl.ed.ac.uk/arch/watkins/banea_2001_mk3.pdf

Watt, A. S. (1947). Pattern and process in the plant community. *Journal of Ecology*, **35**(1–2), 1–22.

Watts, D. J. & Ashworth, J. M. (1970). Growth of myxamoebae of cellular slime mould *Dictyostelium discoideum* in axenic culture. *Biochemical Journal*, **119**(2), 171–4.

Watts, D. J. & Strogatz, S. H. (1998). Collective dynamics of "small-world" networks. *Nature*, **393**(6684), 440–2.

Watts, S. (1999). *Epidemics and History. Disease, Power and Imperialism*. New Haven, CT: Yale University Press.

Weber, M. (1919/1984). *Wissenschaft als Beruf*. Tübingen: Duncker & Humblot.

Weber, M. (1921/1972). *Wirtschaft und Gesellschaft. Grundrisse der verstehenden Soziologie*. Tübingen: Mohr.

Weber, M. (1922/1988). Die "Objektivität" sozialwissenschaftlicher und sozialpolitischer Erkenntnis. In M. Weber (ed.). *Gesammelte Aufsätze zur Wissenschaftslehre*. Tübingen: Mohr.

Weber, M. (1947). *The Theory of Social and Economic Organizations*. New York, NY: The Free Press.

Weber, O., Scholz, R. W., Bühlmann, R. & Grasmück, D. (2001). Risk perception of heavy metal soil contamination and attitudes toward decontamination strategies. *Risk Analysis*, **21**(5), 967–77.

Weber, T. P. & Stilianakis, N. I. (2008). Inactivation of influenza A viruses in the environment and modes of transmission: a critical review. *Journal of Infection*, **57**(5), 361–73.

Week, A. (1997). *Paracelsus: Speculative Theory and the Crisis of the Early Reformation*. Albany, NY: State University of New York Press.

Wehling, P. (2002). Dynamic constellation of the individual, society, and nature: critical theory and environmental sociology. In R. E. Dunlap, F. H. Buttel, P. Dickens & A. Gijswijt (eds). *Sociological Theory and the Environment: Classical Foundations, Contemporary Insights*. Lanham, MD: Rowman & Littlefield Publishers, pp. 144–66.

Weidlich, W. & Haag, G. (1983). *Concepts and Models of a Quantitative Sociology: the Dynamics of Interacting Populations*. New York, NY: Springer.

References

Weiler, J. H. H. (1994). A quiet revolution – The European Court of Justice and its interlocutors. *Comparative Political Studies,* **26**(4), 510–34.

Weiner, B. (2000). Motivation: an introduction. In A. E. Kazdin (ed.). *Encyclopedia of Psychology.* Oxford: Oxford University Press, pp. 314–17.

Weintraub, A. (2004). The geniuses of regeneration. Retrieved April 2008, from http://www.businessweek.com/magazine/content/04_21/b3884008_mz001.htm

Weiss, K. M., Ferrell, R. E., Hanis, C. L. & Styne, P. N. (1984). Genetics and epidemiology of gallbladder-disease in new-world native peoples. *American Journal of Human Genetics,* **36**(6), 1259–78.

Weiss Lewin, G. (1948/1997). Preface to the 1948 edition. In K. Lewin (ed.). *Resolving Social Conflicts & Field Theory on Social Sciences.* Washington, DC: American Psychological Association, pp. 10–12.

Weissert, H. & Stössel, I. (2009). *Der Ozean im Gebirge. Eine geologische Zeitreise durch die Schweiz.* Zurich: vdf.

Weisz, H., Fischer-Kowalski, M., Grünbühl, C. M., Haberl, H., Krausmann, F. & Winiwarter, V. (2001). Global environmental change and historical transitions. *Innovation,* **14**(2), 117–42.

Weisz, H. & Schandl, H. (2008). Materials use across world regions. *Journal of Industrial Ecology,* **12**(5–6), 629–36.

Wellek, A. (1953). Verstehen, Begreifen, Erklären. In V. E. von Gebsattel (ed.). *Jahrbuch für Psychologie und Psychotherapie.* Würzburg: Echter, pp. 393–409.

Weller, L., Weller, A. & Roizman, S. (1999). Human menstrual synchrony in families and among close friends: examining the importance of mutual exposure. *Journal of Comparative Psychology,* **113**(3), 261–8.

Wellmer, F.-W. & Becker-Platen, J. D. (2002). Sustainable development and the exploitation of mineral and energy resources. A review. *International Journal of Earth Sciences,* **91**(5), 723–45.

Wellmer, F.-W. & Kosinowski, M. (2003). Sustainable development and the use of nonrenewable resources. *Geotimes,* **48**(12), 14–17.

Wells, W. F. (1955). *Airborne Contagion and Air Hygiene. An Ecological Study of Droplet Infections.* Cambridge, MA: Harvard University Press.

Wenke, R. J. (1999). *Patterns in Prehistory: Humankind's First Three Million Years.* New York, NY: Oxford University Press.

Werlen, B. (1997). *Gesellschaft, Handlung und Raum: Grundlagen handlungstheoretischer Sozialgeographie.* Stuttgart: Franz Steiner Verlag.

Werlen, B. (2000). *Sozialgeographie: Eine Einführung.* Bern: UTB Haupt.

Werner, C. M., Brown, B. B. & Altman, I. (2002). Transactional oriented research: examples and strategies. In R. Bechtel (ed.). *Handbook of Environmental Psychology.* New York, NY: John Wiley & Sons, pp. 203–21.

Werner, F. (2005). *Ambiguities in Decision-Oriented Life Cycle Analysis. The Role of Mental Models and Values.* Dordrecht: Springer.

Werner, F. & Scholz, R. W. (2002). Ambiguities in decision-oriented life cycle inventories. The role of mental models. *International Journal of Life Cycle Asessment,* **7**(6), 330–8.

Wernick, I. K. & Ausubel, J. H. (1997). *Industrial Ecology: Some Directions for Research.* Livermore, CA: Office of Energy and Environmental Systems, Lawrence Livermore National Laboratory.

Wertz-Kanounnikoff, S. & Chomitz, K. M. (2008). *The Effects of Local Environmental Institutions on Perceptions of Smoke and Fire Problems in Brazil* (No. 4522). Washington, DC: The World Bank.

Westerman, R. A. (1963). Somatic inheritance of habituation of responses to light in planarians. *Science,* **140**(3567), 676–7.

Westley, F., Carpenter, S. R., Brock, W. A., Holling, C. S. & Gunderson, L. H. (2002). Why systems of people and nature are not just social and ecological systems. In L. H. Gunderson & C. S. Holling (eds). *Panarchy: Understanding Transformation in Human and Natural Systems.* Washington, DC: Island Press, pp. 103–21.

Whewell, W. (1847). *The Philosophy of the Inductive Sciences: Founded upon their History,* 2nd edn. London: John W. Parker.

White, L. (1967). The historical roots of our ecologic crisis. *Science,* **155**(3767), 1203–7.

White, N. J., Church, J. A. & Gregory, J. M. (2005). Coastal and global averaged sea level rise for 1950 to 2000. *Geophysical Research Letters,* **32**(1), (L01601).

White, R. (1994). Preface. In B. R. Allenby & D. Richards (eds). *The Greening of Industrial Ecosystems.* Washington, DC: National Academy Press.

White, T. D., Asfaw, B., DeGusta, D. *et al.* (2003). Pleistocene *Homo sapiens* from Middle Awash, Ethiopia. *Nature,* **423**(6941), 742–7.

Whitman, W. B., Coleman, D. C. & Wiebe, W. J. (1998). Prokaryotes: the unseen majority. *Proceedings of the National Academy of Sciences of the United States of America,* **95**(12), 6578–83.

Whitney, C. (1989). Francis Bacon's Instauratio: Dominion of and over Humanity. *Journal of the History of Ideas,* **50**(3), 371–90.

WHO. (2002). *Fact Sheet No. 180: Variant Creutzfeldt-Jakob Disease.* Geneva: World Health Organization.

Whyte, L. L., Wilson, A. G. & Wilson, D. (1969). *Hierarchical Structures*. New York, NY: American Elsevier.

Whyte, W. F. (1989a). Advancing scientific knowledge through participatory action research. *Sociological Forum*, **4**(3), 367–82.

Whyte, W. F. (1989b). Introduction. *American Behavioral Scientist*, **32**(5), 502–12.

Wicksell, K. (1898). *Geldzins und Güterpreise*. Jena: Gustav Fischer.

Wiek, A., Zemp, S., Siegrist, M. & Walter, A. I. (2007). Sustainable governance of emerging technologies – critical constellations in the agent network of nanotechnology. *Technology in Society*, **29** (4), 388–406.

Wiener, J. B. (2005). Collapse: how societies choose to fail or succeed. *Journal of Policy Analysis and Management*, **24**(4), 885–90.

Wiener, N. (1948). *Cybernetics or Control and Communication in the Animal and the Machine*. Cambridge, MA: The MIT Press.

Wiens, J. A. (1984). Resource systems, populations, and communities. In P. W. Price, C. N. Slobodchikoff & W. S. Gaud (eds). *A New Ecology. Novel Approaches to Interactive Systems*. New York, NY: John Wiley & Sons, pp. 397–436.

Wiens, J. A., Moss, M. R., Turner, M. G. & Mladenoff, D. J. (eds). (2006). *Foundation Papers in Landscape Ecology*. New York, NY: Columbia University Press.

Wiese, H. (2002). *Spiel- und Entscheidungstheorie*. Berlin: Springer.

Wiggins, C. L., Espey, D. K., Wingo, P. A. et al. (2008). Cancer among American Indians and Alaska Natives in the United States, 1999–2004. *Cancer*, **113**(5), 1142–52.

Wijewardene, R. & Waidyanatha, P. (1984). *Conservation Farming for Small Farmers in the Humid Tropics: Systems, Techniques & Tools*. Eschborn: GTZ.

Wikipedia. (2009). Prokaryote: structure. Retrieved October 6, 2009, from http://en.wikipedia.org/wiki/Prokaryote

Wikipedia. (2010). Sea level. Retrieved May 10, 2010, from http://en.wikipedia.org/wiki/Sea_level

Wilensky, U. (1998). NetLogo Life model. http://ccl.northwestern.edu/netlogo/models/Life. Center for Connected Learning and Computer-Based Modeling. Evanston, IL: Northwestern University.

Williams, D. B. & Carter, C. B. (1996). *Transmission Electron Microscopy: a Textbook for Materials Science*. Dordrecht: Kluwer.

Williams, D. M. (2002). *Beyond Great Walls: Environment, Identity and Development on the Chinese Grassland of Inner Mongolia*. Palo Alto, CA: Stanford University Press.

Williams, M. (2000). Dark ages and dark areas: global deforestation in the deep past. *Journal of Historical Geography*, **26**(1), 28–46.

Williams, M. (2003). *Deforesting the Earth*. Chicago, IL: University of Chicago Press.

Williamson, O. E. (1975). *Markets and Hierarchies: Analysis and Antitrust Implications*. New York, NY: Free Press.

Williamson, O. E. (1979). Transaction-cost economics: the governance of contractual relations. *Journal for Law and Economics*, **22**(2), 233–61.

Williamson, O. E. (1995). Some uneasiness with the Coase theorem – comment. *Japan and the World Economy*, **7**(1), 9–11.

Williamson, O. E. (2000). The new institutional economics: taking stock, looking ahead. *Journal of Economic Literature*, **38**(3), 593–613.

Wilson, C. & Dowlatabadi, H. (2007). Models of decision making and residential energy use. *Annual Review of Environment and Resources*, **32**, 169–203.

Wilson, C. & McDaniels, T. (2007). Structured decision-making to link climate change and sustainable development. *Climate Policy*, **7**(4), 353–70.

Wilson, D. C., Araba, A. O., Chinwah, K. & Cheeseman, C. R. (2009). Building recycling rates through the informal sector. *Waste Management*, **29**(2), 629–35.

Wilson, D. C., Velis, C. & Cheeseman, C. (2006). Role of informal sector recycling in waste management in developing countries. *Habitat International*, **30**(4), 797–808.

Wilson, E. O. (1992a). *Der Wert der Vielfalt. Die Bedrohung des Artenreichtums und das Überleben des Menschen*. Munich: Piper.

Wilson, E. O. (1992b). *The Diversity of Life*. Cambridge, MA: Harvard University Press.

Wilson, E. O. (1993, June 13). A suicidal impulse. *The Toronto Star*.

Wilson, E. O. (2000). *Sociobiology: the New Synthesis*, 25th edn. Cambridge, MA: Belknap Press of Harvard University Press.

Wilson, E. O. (2002). *The Future of Life*. New York, NY: Random House.

Wilson, E. O. & Bossert, W. H. (1971). *A Primer of Population Biology*. Sunderland, MA: Sinauer Associates.

Winebrake, J. J., Farrell, A. E. & Bernstein, M. A. (1995). The Clean Air Act's sulfur dioxide emissions market: estimating the costs of regulatory and legislative intervention. *Resource and Energy Economics*, **17**(2), 239–60.

Winnie, J. A. (2000). Information and structure in molecular biology. Comments on Maynard Smith. *Philosophy of Science*, **67**(3), 517–26.

References

Withgott, J. & Brennan, S. (2008). *Environment: The Science behind the Stories*, 3rd edn. San Francisco, CA: Pearson.

Witschi, P. (2002). Historisches Lexikon der Schweiz – Appenzell Ausserrhoden. Retrieved January 6, 2011 from http://www.hls-dhs-dss.ch/textes/d/D7476-2-1.php

Wolf, B. (1992). Die Theorie Egon Brunswiks und ihre Bedeutung für die ökologische Perspektive in der Psychologie. Unpublished Habilitationsschrift. Landoa Universität.

Wolf, B. (2003, November 6–7). *Fritz Heider and Egon Brunswik – Their Lens Models: Origins, Similarities, Discrepancies*. Paper presented at the 19th Meeting of the Brunswik Society, Vancouver, Canada.

Wolfe, V. L. (2001). A survey of the environmental education of students in non-environmental majors at four-year institutions in the USA. *International Journal of Sustainability in Higher Education,* 2(4), 301–15.

Wong, M. H., Wu, S. C., Deng, W. J. *et al.* (2007). Export of toxic chemicals – a review of the case of uncontrolled electronic-waste recycling. *Environmental Pollution,* **149**(2), 131–40.

Wood, K. (2005). Contextualizing group rape in post-apartheid South Africa. *Culture Health & Sexuality,* 7(4), 303–17.

Woodgate, G. & Redclift, M. (1998). From a "sociology of nature" to environmental sociology: beyond social construction. *Environmental Values,* 7(1), 3–24.

World Business Council for Sustainable Development (WBCSD). (2000). *Eco-efficiency. Creating More Value with Less Impacts*. Geneva: WBCSD.

World Gazetteer. (2010). World: largest cities and towns and statistics of their population. Retrieved May 10, 2010, from http://world-gazetteer.com

World Health Organization. (2008). *The Global Burden of Disease: 2004 Update*. Geneva: WHO.

World Health Organization (WHO). (2009, May 3). Advice on the use of masks in the community setting in Influenza A (H1N1) outbreaks. Retrieved November 14, 2009, from http://www.bag.admin.ch/pandemie/massnahmen/05955/index.html?lang=de&download=M3wBPgDB/8ull6Du36WenojQ1NTTjaXZnqWfVpzLhmfhnapmmc7Zi6rZnqCkkId6fHp/bKbXrZ6lhuDZz8mMps2gpKfo

World Rainforest Movement (WRM). (2006). WRM Bulletin, 112(November). http://www.wrm.org.uy/bulletin/112/viewpoint.html

Wright, H. E. (1976). The environmental setting for plant domestication in the Near East. *Science,* **194**(4263), 385–9.

Wright, R. T. & Nebel, B. J. (2002). *Environmental Science: toward a Sustainable Future*. Upper Saddle River, NJ: Pearson Education.

Wright, R. T. & Nebel, B. J. (2005). *Environmental Science: toward a Sustainable Future*, 9th edn. Upper Saddle River, NJ: Pearson Prentice Hall.

Wu, J. G. & Hobbs, R. (2002). Key issues and research priorities in landscape ecology: an idiosyncratic synthesis. *Landscape Ecology,* **17**(4), 355–65.

Wundt, W. (1897/1969). *Outlines of Psychology* (C. H. Judd, Trans.). St. Clair Shores, MI: Scholarly Press.

WWF. (2006). Bioenergy assurance schemes and WTO rules. Retrieved August 29, 2009, from http://cgse.epfl.ch/webdav/site/cgse/shared/Biofuels/Further%20Reading/WWF-Bioenergy%20Assurance%20Schemes%20and%20WTO%20Rules.pdf

Wynne, B. (1992). Misunderstood misunderstanding: social identities and the public uptake of science. *Public Understanding of Science,* **1**(3), 281–304.

Wynne, B. (1996). May the sheep safely graze? A reflexive view of the expert-lay knowledge divide. In S. Lasch, B. Szerszynski & B. Wynne (eds). *Risk, Environment & Modernity: towards a New Ecology*. London: Sage, pp. 44–83.

Xenophon. (about 370/1815). Oikonomikos. In J. G. Schneider (ed.). *Xenophontis: Oeconomicus, Convivium, Hiero, Agesilaus*. Leipzig: Caspar Fritsch.

Yan, S. Y. (2002). *Number Theory for Computing*. Berlin: Springer.

Yang, R. (2005, June 8–10). *Higher Education in the People's Republic of China: Historical Traditions, Recent Developments and Major Issues*. Paper presented at the "Challenges and Expectations of the University": Experiences and Dilemmas of the Reformation, Universitidad Autónoma de Tampico, Tamaulipas, Mexico.

Yarime, M. (2009). Public coordination for escaping from technological lock-in: Its possibilities and limits in replacing diesel vehicles with compressed natural gas vehicles in Tokyo. *Journal of Cleaner Production,* **17** (14), 1281–1288.

Yarime, M., Yoshiyuki, T., & Yuya K. (2010). Towards institutional analysis of sustainability science: A quantitative examination of the patterns of research collaboration. Sustainability Science, 5 (1), 115–125

Yearley, S. (1991). *The Green Case: a Sociology of Environmental Issues, Arguments and Politics*. New York, NY: Harper Collins.

Yearley, S. (2002). The social construction of environmental problems: a theoretical review and some not-very-Herculean labors. In R. E. Dunlap, F. H. Buttel, P. Dickens & A. Gijswijt (eds.). *Sociological Theory and the Environment: Classical Foundations, Contemporary Insights*. Lanham, MD: Rowman & Littlefield Publishers, Inc., pp. 275–85.

Yearley, S. (2005). The "end" or the "humanization" of nature? *Organization & Environment,* **18**(2), 198–201.

Yehuda, R., Bell, A., Bierer, L. M. & Schmeidler, J. (2008). Maternal, not paternal, PTSD is related to increased risk for PTSD in offspring of Holocaust survivors. *Journal of Psychiatric Research,* **42**(13), 1104–11.

Yerkes, R. M. & Dodson, J. D. (1908). The relation of strength of stimulus to rapidity of habit information. *Journal of Comparative Neurology and Psychology,* **18**, 459–82.

Yoon, K. P. & Hwang, C. L. (1995). *Multiple Attribute Decision Making.* Thousand Oaks, CA: Sage.

York, R. (2008). De-carbonization in former Soviet republics, 1992–2000: the ecological consequences of de-modernization. *Social Problems,* **55**(3), 370–90.

York, R., Rosa, E. A. & Dietz, T. (2003). STIRPAT, IPAT and ImPACT: analytic tools for unpacking the driving forces of environmental impacts. *Ecological Economics,* **46**(3), 351–65.

Yorque, R., Walker, B., Holling, C. S. *et al.* (2002). Towards an integrative synthesis. In L. H. Gunderson & C. S. Holling (eds). *Panarchy: Understanding Transformation in Human and Natural Systems.* Washington, DC: Island Press, pp. 419–38.

Young, O. L., King, L. A. & Schroeder, H. (eds). (2008). *Institutions and Environmental Change: Principal Findings, Applications, and Research Frontiers.* Cambridge, MA: The MIT Press.

Young, O. R. (2002). *The Institutional Dimensions of Environmental Change: Fit, Interplay and Change.* Cambridge, MA: The MIT Press.

Zadeh, L. A. & Kacprzyk, J. (eds). (1992). *Fuzzy Logic for the Management of Uncertainty.* New York, NY: John Wiley & Sons.

Zagheni, E. & Billari, F. C. (2007). A cost valuation model based on a stochastic representation of the IPAT equation. *Population and Environment,* **29**(2), 68–82.

Zaitlin, M. (1998). The discovery of the causal agent of the tobacco mosaic disease. *Discoveries in Plant Biology.* Hong Kong: World Publishing, pp. 105–10.

Zambonelli, F., Jennings, N. & Wooldridge, M. (2003). Developing multiagent systems: the Gaia methodology. *ACM Transactions on Software Engineering and Methodology (TOSEM),* **12**(3), 317–70.

Zeising, A. (1854). *Neue Lehre von den Proportionen des menschlichen Körpers.* Leipzig: Rudolph Weigel.

Zheng, X. Y. & Eltahir, E. A. B. (1997). The response to deforestation and desertification in a model of West African monsoons. *Geophysical Research Letters,* **24**(2), 155–8.

Zhu, W. X., Lu, L. & Hesketh, T. (2009). China's excess males, sex selective abortion, and one child policy: analysis of data from 2005 national intercensus survey. *British Medical Journal,* 338(b1211).

Ziegler, B. (2008). *Geschichte des ökonomischen Denkens,* 2nd edn. Munich: Oldenbourg.

Zierhofer, W. & Burger, P. (2007). Transdisciplinary research – a distinct mode of knowledge production? Problem-orientation, knowledge integration and participation in transdisciplinary research projects. *Gaia,* **16**(1), 29–34.

Zietz, B. P. & Dunkelberg, H. (2004). The history of the plague and the research on the causative agent *Yersinia pestis. International Journal of Hygiene and Environmental Health,* **207**(2), 165–78.

Zimbardo, P. G. (1992). *Psychologie,* 5th edn. Berlin: Springer.

Zimbardo, P. G. & Boyd, J. N. (1999). Putting time in perspective: a valid, reliable individual-differences metric. *Journal of Personality and Social Psychology,* **77**(6), 1271–88.

Zimbardo, P. G., Johnson, R. L. & Weber, A. L. (2005). *Psychology: Core Concepts,* 5th edn. Boston, MA: Pearson.

Zoghbi, H. Y. & Beaudet, A. L. (2007). Epigenetics and human disease. In C. D. Allis, T. Jenuwein, D. Reinberg & M.-L. Capparos (eds). *Epigenetics.* New York, NY: Cold Spring Laboratory Press, p. 435.

Zuber, J., Crott, H. W. & Werner, J. (1992). Choice shift and group polarisation: an analysis of the status of arguments and social decision schemes. *Journal of Personality and Social Psychology,* **62**(1), 50–61.

Zuk, T., Rakowski, F. & Radomski, J. P. (2009). Probabilistic model of influenza virus transmissibility at various temperature and humidity conditions. *Computational Biology and Chemistry,* **33**(4), 339–43.

Index

Entries in bold refer to a table or a figure. Entries in italic refer to a box.

Acid rain 331
Achievement principle 244
Action Research 391
   roots 391
   participatory 392
Adaptation 448, 526
   assimilation and accomodation 434, 448
   behavior adaption 411
      mitigation and adaptation 412, 413
   and epistemology 160, 191
   and evolution *21, 72*, 93, 99, 106, *131*
   and Lamarkism *92*
   and functionalism 47, 50
   and HES co-adaptation 372
   and immunity 115, 471
   and sustainability learning 379
   as part of Transition Management 513, 514, 520
   as part of The Resilience approach
Agro-fuel 482
   production of 486, *489*
   use of 485, 487
   potential of *488*
   environmental impact of 488
   vs. fertilizer production 493
   vs. food production *488*, 494
Ajzen, Icek 176, 205
Allenby, Braden 335, 348, 349
Ames, Paul *19*
Anthropocene 8, 31, 346, 498, 525, 528
   and material flow 321, 327
Appropriation of nature 14, 319, 336
Aquinas, Thomas 258, 259
Aristotle 34, 59, 257, 412
Arrhenius, Svanta 320
Asch, Salomon 182
Assumptions
   epistemological 29, 408
   ontological 36
Awareness Postulate
   and the cognitive-epistemic approach 445
   and the motivational approach 446
   other approaches 446

   and environmental consciousness 446
   and disease spread 469
   and BSE 501
Axelrod, Robert 369
Ayres, Robert *295*, 309, 311

Bacon, Francis 70
Bandura, Albert 204, 205
Barker, Roger 155
Beck, Ulrich 242, 246, 426
Bentham, Jeremy 192, 271
Berlin, Brent 53, 58
Bernard, Claude 429
Biogeochemical cycle
   and evolution of societies 322, 333
   and information technology 332
   of sulfur 321
   of copper 327
   of phosphorus *322*, 493, 495, 496
      role in primary production 322
      scarcity of 324
      life cycle characteristics 325
      recycling of 326, 507
      management and flow in Switzerland 498, 500
   availability vs. accessibility 325, 327
   and essentiality 332
   and multi-level cycling 329
   and scarcity 328, 332
   of rare elements 329
Biosemiotics 90
Body–mind dualism 29, 33, 34
   illustrated **39**
   similarity to material-social complementarity 31
   in biosemiotics 90
   in epigenetics 112
   in cells 108
   in immunology 119
Bone disposal 498
Botany
   emergence of 67, 72
Boulding, Kenneth 282, 311, 332
Boundaries *30*, 32
"fuzzy" 32
   Systems 315, 357, 382, 458, 463, *530*

Boussingault, Jean-Baptiste 4
   deforestation 5
Boyden, Stephen 317, 410, 515
Brown, Robert 79, 95
Brunswick, Egon 31, 37, 144, 411
Burgess, Ernest 216
Burnet, Frank Macfarlane 116

Capra, Fritjof 416
Case studies
   Pandemics, epidemics 464
   Biofuels, agrofuels 482
   Industry in Appenzell, Switzerland 473
   Nutrient recycling 495
Carson, Rachel 222
Carrying Capacity **122**, 223, 265, 308
   see also Spaceship Earth
Catton, William 222
Checkland, Peter 441
Clark, John 274
Classification 53, 73
   Linne's natural method 74
   and probabilism 77
Climate Change 4, 347, 413, *449*
Clean Air Act 308, *331*
Clean Development Mechanism 330
Coase, Ronald 291, 301
Cognitive Development 160
   stages of *161*
   and genetic epistemology 162
   and accomodation 160
   and equilibration 160
Cohen, Irun 117
Colbert, Jean-Baptiste *261*
Commons dilemma 20, 184, 185, 426, 428
   and game theory *185*
   and economics 300
   and collective action 186, 354
   free-rider problem 188
   variables and drivers **187**
Community-based Participatory Research 393
Complementarity
   material–social 33, 411, 526
   human–environment 31, 411
   wave–corpuscular 33, 411

and logic 32
relationships between **39**
and hierarchical levels **35**, 39
vs. Cartesian perspectives 34
in Linnean classification 74, *76*
program and representation 318
in industrial ecology 309
in social ecology 241
and biohistory 410
and pressure–state-response 410
and coupled-systems research 411
and panarchy 411
in traditional industry 476
examples in research 412
see also body–mind dualism
Comte, Auguste 216, 218
Constructivism 245
and humanization of nature 246, *247*
Consumer behavior 491
and BSE 501
Copernicus, Nicolaus 67
Correcting negative human impact 5
Cosmogony
and evolution 80, 83, *85*
and geology *71*, 72
and Linnean classification 74, 76
of Ancient Greece 61
Cosmology
defined 45
of Ancient Greece 59, **60**
versus cosmogony 46
of economics vs. ecology 296
Cournot, Antoine 274
Crutzen, Paul 321, 342
Cybernetics 355, 429

Daly, Herman 295, 311, 332
Darwin, Charles 11, 80
Da Vinci, Leonardo 67
Decision Postulate 443, 460, 480, 502, 516, 519
see also Game Theory
Deforestation 5, *261*
de Landa, Manuel 218
Descartes, Rene 34, 69
Dewey, John 139, 473
Diamond, Jared *21*
Disease
Creutzfeld-Jakob disease *497*, 498
conceptualization in prehistoric societies 54, *56*, *57*
and pathogen resistance 471
and the placebo effect 56, 412
see also Microbiology or Health, human
Disturbance ecology 121
and ecological organicism 133
and panarchy 122

and adaptive cycles of ecosystems 123, 346
Drivers
defined 190
and their degree of behavioral relevance 190
and instinct theory 192
and equilibration 191, 233
of institutional change 302
and societal development 206, 217, 219, 225, 233
preference and utility 192
hedonism principle 193
and utilitariansim 192
axiomatic definition 193
paradoxes *194*, *195*
Prospect theory 196
hot vs. cold utilities 196
drives 197
needs 198
emotions 198
and mood vs. affect 200
cognition 200
and memory 201, 202
and cognitive dissonance 204
and social cognitive theory of learning 204
motives 205, *210*
values 206
Schwartz Value Inventory 206, **207**
norms 207
motivations 209
attitudes 209
and reasoned action 211
and their relation/differentiation 212
foundations in psychology 190
of prehistoric societies 58
Douglas, Mary 34
Dunlap, Riley 222, 223
Durkheim, David Émile 36, 37, 216, 218, 219, 420
Dust Bowl *450*

Eco-industrial Park 313
Ecology 12, 84
defined 84
and human systems 120
and population dynamics 86, **122**
Ecological economics 296
Ecological evolutionary theory 225
and ReSTET framework 225
and degree of democracy 225
and economic surplus 226
and patterns of societal evolution 226
and resource scarcity *228*
Economics
pre-classical era 257
in Ancient Greece 258
and scholasticism 259

and mercantilsm 259
and cameralism 260
and physiocracy 261
and opportunity cost 259
classical 263
and supply-side perspective 265
and land rents 263, 266
and marginal yield 265
and Malthusian scarcity 266
and the Club of Rome 265
and theory of public goods 284
rivalry and excludability 283
and market failure 284
and social welfare function 285
and market efficiency 285
Ecosystems
energy and nutrient cycles 87
succession and climax 87, 121
equilibration principle 92
and functionalism 125
Ecosystems Services 125
and ecosystem functions 125
and the Millennium Ecosystem Assessment 126
Edgeworth, Francis Ysidro 271
Ehrenfeld, John 311
Einstein, Albert 17
Maximum Sustainable Yield *20*
Ehrlich, Paul R. 311
Ehrlich, Paul 114
Embedded Case Study Method 380
Emissions trading 322, *331*
Engels, Friedrich 217, 269
Engineering
foundations of 307
Environment
defined 526
material dimension 12
invention of 10
Environment First Postulate 448, 478, 485
and equilibration 448
and complementarity 451
sufficiency vs. satisficing 451
Environmental literacy
Defined 7, 15
of religion **23**, 32, 52, 183, 216
hunter/gatherer vs. predating/foraging societies 51
concept development 13
elements of 15
in the Dark Ages 66
in the early Renaissance 67
role of science 395
Environmental Awareness 10
and environmental psychology 162
and constructivism 246
public awareness 222
see also Awareness Postulate

Environmental Justice *237*
  and the achievement principle 243
Environmental Economics 289
  roots 290
  and externality 291
  and willingness-to-pay principle 291
  Coase theorem 291, 302
  cost–benefit vs. eco-efficiency analyses 292, *293*
  and incorporating thermodynamics 294, *295*
Environmental Engineering 308
Environmental Kuznets Curve 333
Environmental Psychology 162
  and space 164
  and territoriality 164
  and landscape preference 165
  and time 166
  and episodic future thinking 166
  and conservation 169
  and recycling behavior 169
  and norms 170
  and disaster management 170
  and the built environment 172
  and mobility patterns 172
  and pollutants 173
  theory of planned behavior 175
  Norm Activation Model 176
  Low-cost hypothesis 176
  and field theory 176
Environmental Sociology 222, 236, 252
  foundations of 222
  vs. social ecology 222
  and the POET model 222
  and the IPAT algorithm 223
    extensions of 242, 252
  Human Exemptionalist vs. New Ecological Paradigms *217*, 223
  treatment of environmental issues 236, 238
  European vs. American approaches 236, 240
  ecological modernization 249
  treadmill of production 249, 250
    and view of technology 251
Epistemics 24, *25*
  epistemiology, defined 31
  and transdisciplinarity 377
Epigenetics 109
  and complementarity 112
  examples 110, *111*
Equilibration 92, 160, 191, 233, 448, 526
Ethnobiology 53
  and formation of language 53
Evolution 80
  and geology 72

Fechner, Gustav *140*
Feedback loop 21, 429, *442*, *447*, 481, 491, 505
  example *21*, *92*
  conceptual roots 429, *435*
  and equilibrium 431
  epistemology 432
  and systems theory 431
  primary vs. secondary 432
  in biology 435
  Lotka–Volterra model 436
  and systems dynamics 437
  and sustainability learning 442
  and transdisciplinarity 439
  and rebound effects *440*
  delayed feedbacks 443
  and servomechanism 431, 434
  and pathogen evolution 470
Festinger, Leon 202, 204
Fietkau, Hans-Joachim 176
Fischer-Kowalski, Marina 241, 318
Fischhoff, Baruch 177, 178
Fisher, Irving 272, 276
Forrester, Jay Wright 316, 344, 430
Formative scenario analysis 358, 382, 478, 480
Freud, Sigmund 197
Functionalism 16, 37
  probabilistic 144
  Functional social constructivism 38
  proponents 160, 232, 430
  and cognitive development 160
  and epigenetics 112
  and perception 144

Galen 62
Galilei, Galileo 69, 355
Game theory 366, 443
  Aspiration level 191
  and cooperation 365
  malignant vs. benign conflicts 366, game situations 367
  strategy, defined 366
  Nash equilibria 275
  and pandemic management 367
Gestalt Psychology 157
Genetics 97, 103
  mutations 97
    and the environment 99
  Drosophilia melanogaster 99, *107*
Gibbons, Marta 399
Gibson, James 31, 152, 163
Gibson, Eleanor 152
Giddens, Anthony 232, 234, 357
Gossen, Hermann Heinrich 272
Gottlob von Justi, Johann Heinrich 260
Graedel, Thomas 321

Grobstein, C. 415, 416
Gunderson, Lance 229, 411
Group
  defined 181
  norms 182
  conformity *183*
  rationale 181
  groupthink *182*
  decision rules 181
  ingroup vs. outgroup 183
  voting culture 39

Haeckel, Ernst 83, 84, *85*, 99
Haber-Bosch process *398*
Haberl, Helmut 241
Habermas, Jürgen 244
Haken, Hermann 360
Hardin, Garrett 20, 184, *185*, 186
Hartwick, John 299
Hayes, Edward Cary 219
Health, human
  and bovine spongiform encephalopathy (BSE) 495, *497*
  and scrapie *497*
  and pandemics 120, 464
    and the small world phenomenon
  and pathogens 464
  and influenza transmission pathways 466, 467
Heider, Fritz 141
Hellpach, Willy 152
HES Framework 453, **454**, 528, 530, 532
  mathematical notation 456
  vs. alternative approaches **510**, 512, 520
  applied 462
  and setting system boundary 458, 476, 484, 495
  and defining the guiding question 458, 475, 496
  use in case study **465**, **474**, 484, **496**, 507
HES Postulates
  overview
  categories of 455
  relationship between 455
  in environmental research 459
  and transdisciplinary process **474**
  Awareness Postulate 445, 481, 501, **502**
  Complementarity Postulate 410, 499, 512, 515, 518
  Decision Postulate 443, 460, 516, 519
  Environment First Postulate 448, 459
  Feedback Postulate 429, 514, 516, 519
  Hierarchy Postulate 413, 459, 513, 516, 519
  Interference Postulate 427, 459

Index

Hicks, John Richard 274, 279
Hierarchy
  of human systems 38, 418
    individual 418
    organization 420
    institution 420, *422*
    society 421
    supranational level 423, 494
  defined 415
  types of 415
  theory 413
  living systems hierarchy **414**
  control hierarchy 415, 418
  level hierarchy 415, 418
    ontology 416
    epsitemics 416
  and interference 427
  and disease spread 466, 468
  in traditional industry 479
  in national energy plans 484
  and public health 501, **502**
  dimensions of **31**
    material–biophys–tech.
    social–cultural–epist.
    see also HES postulates
Holling, Crawford Stanley 229, 346, 347, 411
Hooke, Robert 95
Hotelling, Harold 288
Huber, Joseph 249
Hull, Clark* 197
Human Individual 29, *30*
Human systems
  defined 29
  dimensions of
    material.biophysical 32
    socio-epistemic 34, 36

Immunology 113
  contributions to environmental literacy 114
  humoralist/ cellularist debate 115,
  roots 114
  and clonal selection theory 116
  and the "self" of the immune system 116
  and the cognitive paradigm 117
  neuroimmunology 119
  and network theory 117
  innate vs. adaptive immunity 115
Industrial Ecology
  roots of 308, 310
  and role of consumers 310
  view on nature 310
  focus of 308
  Eco-industrial park 313
  and the IPAT algorithm 311
  Social metabolism 317, 511, 515, 520
  and view on labor 318,
  and colonization 319
  and recycling *311*
  and emerging technologies 335
  and feedback loops 335
Industrial Engineering 308
Industrial development 475
Input–Output Analysis 314, 317
Institution 420
  types of 421
Institutional Economics 299
  and time scale 302
  and governance 303
  and transaction costs 299, 301
  and game theory 302, 303
  and natural resource management 303
Interdisciplinarity 23, 27
  and linear modeling 353
  and muli-agent systems 372
  and systems dynamics 357
  and risk management 366
  and game theory 367
Interference 427
  in traditional industry 480
  in biofuel use 491
  in carcass management 504
Intergovernmental Panel on Climate Change (IPCC) 427

James, William 138
Jantsch, E. 374
Jerne, Nils Kaj 116
Jevons, William Stanley 288,

Kahnemann, Daniel 195
Kant, Immanuel 92, 177, 216, 219
Kaplan, Robert 165
Kepler, Friedrich Johannes 70
Keynes, John Maynard 271, 280
  Keynesian economics or New Welfare Theory 280
Kneese, Allen 311
Knowledge Integration 22, *23*, 401
Koch, Robert 100, 103
Koffka, Kurt 157
Köhler, Wolfgang 157
Kyoto Protocol 424

Lamarck, Jean-Baptiste 11, *81*
Laplace, Pierre-Simon 177
Landscape ecology 129
  roots 129
  components 130
LeBon, Gustav 181
Leibniz, Gottfried Wilhelm 70
Lenski, Gerhard 37
Lewin, Kurt 150, 391
Lichtenstein, Sarah 177, 178, 179

Lindeman, Raymond 87
Life Cycle Analysis 316, 413, 488
Linear Modeling 351, 352
  Regression Analysis 352
  Factor analysis 352
  Analysis of Variance 353
  Matrix algebra 352
Locke, John 263
London Smog Disaster 222, 321, 397
Loorbach, D. 517
Lotka, Alfred *124*, 433
Luhmann, Niklas 232, 233, 234

Malthus, Thomas 82, 226, 265, 287
Marshall, Alfred 272, 274, 275
Marx, Karl 216, 268, 310, 318
Maslow, Abraham 198, 212
Material flow analysis 314, 382, 478, 500
Maynard Smith, John *105*, 108
McKelvey, Vincent Ellis 327, **328**
Metchnikoff, Ilja 114, 116
Medicine
  development in Ancient Greece 61, 62
    humoral medicine 61
  Medical ethnobiology and anthropology 54
Mediation 386
Mendel, Gregor 97,
Metabolism
  and historical materialism 217
  and social ecology 241
  and socialist economics 268, 269
  industrial 308, 309, 311, 317
Microbiology 100
  germ theory of disease 100
  and epidemics 100
  and development of vaccination 101, 102
Mill, John Stuart 268
Miller, Joan **414**
Molecular biology 95
  cell, conceptual development 95
  and genetics 103
  and the central dogma 105
  and gene expression 106,
  and the Operon Model 108
Modeling Approaches
  purpose of 347–49,
  and simulation 349, 356
  cross-comparison of 349
  use of cellular automata 367
  and Markov processes 368
  see also Systems Dynamics, Linear Modeling, Multi-Agent Systems and Scenario Analysis

Modes of thought **197**
   and risk judgment 179
   and emotions 200
   and cognition 202
Morgan, Thomas 99
Morgenstern, Oskar 193, 444
Multi-agent Systems 368
   and social networks 369
   and couples HES systems 369
   and feedback loops 372
   examples of *371*
   and types of interactions 372
Multi-criteria Decision Analysis 176, 382, 480

Nature vs. nurture 43, 165, 205
Neo-classical Economics 271
   and factors of production 274
   and marginal utility 271, 272
   and utility function 272
   consumption theory 272
   production theory 273
   price theory 274
   rivalry and excludability 273
   capital theory 276
   Cournot equilibrium 274, *275*
New Ecology 123, *124*
   and the Rio Convention 123, 126

Odum, Howard 430
Oil free Society 482, **483**
Ostrom, Elinor 189, 509
Owen, Robert

Pangenisis *65*
Paracelsus 68
Park, Thomas 216, 220
Pareto, Vilfriedo 279
Parsons, Talcott 231, 232
Pasteur, Louis 100, *102*
Peirce, Charles Sanders 139
Perception
   and biosemiotics 92
   and probabilistic functionalism 144
   of environmental information 138
   see also Cognitive Psychology
Peak Oil 482, 492
Piaget, Jean 160, 434, 445, 448
Pigou, Arthur Cecil 271, 276
   Pigovian Tax 276, 277
Prehistoric societies 46
   and archeology 47
   and comparative linguistics 47
   and cultural–epistemic settings 48
   natural history 47
   and stages of prehistoric human development 50
Principal-Agent Theory 425
Probabilism
   in classification *77*
   in contemporary sociology 233
Probability 363
Psychological Approaches
   introspection vs. perception 138
   precepts vs. concepts 138
   pragmatism 139
   psychophysical approach 141
   and fundamental attribution error 143
   and misperception of environmental stimuli 141
   distal vs. proximal stimuli 142
   and Probabilistic Functionalism 144, *147*
   Brunswik's lens model 146
   Field Theory 150
   variables driving environmental cognition **151**
   Action Research 151
   Geopsychology 152
   Ecological theory 153
      and affordance 154
   Ecopsychology 155
      and behavior setting 155
   Transactional approach 156
   perceptual prototypes 158

Rationality 7, 444
   Bounded *7*, 171, 437
   in interpretive sociology 219
   and economic full rationality 444
Realism 38, 426
Reproductive mechanisms
   early thought *65*
   plant sexuality *98*
   and Linnean classification 74
Resilience Approach 126, 512, 520
Resource Economics 286, 509
   theory of optimal depletion 287
   and Hottelling's Rule 288
   and dynamic optimization 289
Ricardo, David 266, **267**, 275, 286
Risk
   concept of 177
   hazard vs. risk 178
   pure versus speculative 364
   situation 364
   in contemporary sociology 234
Risk Assessment 363
Risk Judgment 178
   and the psychometric paradigm 178
   and trust 180
   and social mood 180
   and emotion 179
   and modes of thought 179
Risk Perception 177
   and time preference 166
   and pollutants 175
   examples *201*
Risk Management 365
Risk Society 242, 426
   and cosmopolitization 242, 426
Rosch, Eleanor 158
Rotmans, Jan 517
Russell, James 199

Say, Jean-Baptiste 265
Scenario Analysis 354, 487
   scenario, defined 355
   IPCC scenarios 359
   and impact factors 358
Schelling, Thomas 361
Schleiden, Matthias 95
Schnaiberg, Allen 250
Schumpeter, Joseph 249
Schwartz, Shalom 176, 205, 206, **207**
Science
   development of scientific method 69
   and academia 395
   basic vs. applied research 397
   and industry 396, *398*, 470
   Mode 1 vs. Mode 2 research 399
   role of transdisciplinarity 400
   and social contamination 402
   and the microscope 70, 95
Sen, Amartya 376
Signs 233
Simon, Herbert 415
Slovic, Paul 178
Smith, Adam 263, 294
Social ecology 241
Socialist economics 268
Society
   defined 215, 249, 317, 421
   rationale of 218
   principal components 13, 418
   order of 232
   and the role of communication 233
   equilibration as a driver of society 233
   spread of epidemics 120
   role of academia 27, 388, 404
Societal development
   Drivers of 217, 219
Sociology, classical
   roots 216
   historical materialism 216
   and role of metabolism 217
   and social facts as reality 218
   and conceptualization of the environment 218
   interpretive sociology 218
   power and social organization 218
Sociology, contemporary
   structuralist functionalism 232
   theory of action 232